World Sustainability Series

Series Editor

Walter Leal Filho, European School of Sustainability Science and Research, Research and Transfer Centre "Sustainable Development and Climate Change Management", Hamburg University of Applied Sciences, Hamburg, Germany

Due to its scope and nature, sustainable development is a matter which is very interdisciplinary, and draws from knowledge and inputs from the social sciences and environmental sciences on the one hand, but also from physical sciences and arts on the other. As such, there is a perceived need to foster integrative approaches, whereby the combination of inputs from various fields may contribute to a better understanding of what sustainability is, and means to people. But despite the need for and the relevance of integrative approaches towards sustainable development, there is a paucity of literature which address matters related to sustainability in an integrated way.

More information about this series at http://www.springer.com/series/13384

Walter Leal Filho · Paulo R. Borges de Brito ·
Fernanda Frankenberger
Editors

International Business, Trade and Institutional Sustainability

 Springer

Editors
Walter Leal Filho
European School of Sustainability
Science and Research
HAW Hamburg
Hamburg, Germany

Paulo R. Borges de Brito
College of Business
Colorado State University
Fort Collins, CO, USA

Fernanda Frankenberger ⓘ
Universidade Positivo
Curitiba, Paraná, Brazil

Pontifical Catholic University
of Paraná (PUCPR)
Curitiba, Paraná, Brazil

ISSN 2199-7373   ISSN 2199-7381   (electronic)
World Sustainability Series
ISBN 978-3-030-26758-2   ISBN 978-3-030-26759-9   (eBook)
https://doi.org/10.1007/978-3-030-26759-9

This Springer imprint is published by the registered company Springer Nature Switzerland AG
The registered company address is: Gewerbestrasse 11, 6330 Cham, Switzerland

Preface

Global, social, and environmental trends represent pressing concerns for the future of the planet and human populations. The environmental Kuznets curve suggests that economic development initially leads to environmental deterioration, but after a certain level of economic growth, a society begins to improve its relationship with the environment and environmental degradation levels reduce.

International trade tends to increase as a result of economic development in developing countries. As countries begin to industrialize, they experience environmental degradation as a result of economic activity moving from subsistence farming in rural areas to industrial activities in urban factories.

However, as incomes continue to grow, citizens become increasingly willing to pay for cleaner water and air, and environmental quality begins to improve as cleaner technologies are adopted.

Thus, it has been shown that as international trade boosts income, the result is not a so-called race to the bottom, but instead a race to the top as wealthier countries are increasingly willing to pay for environmental quality. Similarly, it has been shown that international business in a free trade environment allows a free flow of cleaner technologies to spread across borders. More recent research suggests that international business has not only facilitated access to clean technologies but also led to more rapid adoption of these technologies by poor countries.

At the company level, the concepts of sustainability and corporate social responsibility (CSR) have been spread worldwide and are considered to be one of the most important themes to emerge in recent decades as pressing concerns for the future of the planet and human populations increase. CSR and sustainability have come to represent an important dimension of corporate strategy for business firms across the globe, with an increasing number of companies—and institutions—trying to determine, monitor, and improve the social and environmental impacts of their operations. Despite such an explosion of interest, effective incorporation of sustainability into business practices and management faces serious obstacles, raising the need for more research and implementation. One of the major obstacles is not that clean technologies are not available, but that they often come at a much higher cost than traditional technologies.

Based on the perceived need to address this gap of information on innovative technologies and adoption, the book *International Business, Trade and Institutional Sustainability* has been prepared. This timely publication showcases research, feasibility studies, case studies, and projects which aim to demonstrate a variety of ways to implement environmental sustainability in companies and higher education institutions, as well as their best practices in international management, adoption of cleaner technologies, global supply chain, greenhouse gas emission reduction, and transportation.

The book is structured as follows:

Part One: Global Sustainable Management Practices
Part Two: Global Sustainable Waste Management Practices
Part Three: Global Sustainable Food Systems and Agricultural Markets
Part Four: Global Responsible Mining and Energy
Part Five: Global Sustainable Transportation, Construction, and Infrastructure
Part Six: Global Conservation and Sustainability Innovations, Investments, and Policies
Part Seven: Sustainability Reporting

We thank all the authors for their contributions and hope this publication may be a positive driver towards greater environmental sustainability in both companies and higher education institutions, making it a vital information source to those already engaged, or willing to engage, on institutional sustainability practices.

Hamburg, Germany	Walter Leal Filho
Fort Collins, USA	Paulo R. Borges de Brito
Curitiba, Brazil	Fernanda Frankenberger
Autumn 2019	

Contents

Global Sustainable Management Practices

**Corporate Social Responsibility and Sustainability Initiatives
of Multinational Hotel Corporations**   3
Artie W. Ng and Pimtong Tavitiyaman

**Sustainable Initiatives and Practices of the Most Sustainable
Organizations in the World**   17
Waqas Nawaz, Patrick Linke and Muammer Koç

**Plastic Bag Ban in the Context of Corporate Social Responsibility:
Consumption and Trade vis-a'-vis Environmental Sustainability
Concerns**   43
Irina Safitri Zen

The Role of Public Administration in Sustainable Development   69
Fernanda Caroline Caldatto, Sandro Cesar Bortoluzzi
and Edson Pinheiro de Lima

**Dynamic Capabilities and Business Model in the Transition to
Sustainability: The Case of Bosch/Curitiba-Brazil**   81
Cristina M. S. Ferigotti, Sieglinde Kindl da Cunha
and Jonatas Soares dos Santos

**The Use of Digital Transformation as a Sustainable Mechanism:
An Automotive Industry Case**   97
Pablo Carpejani, Bárbara Luzia Santor Bonfim Catapan,
Luiz Felipe Pierin Ramos, Izabelle Cristine Hannemann de Freitas,
Camila Mantovani Rodrigues, Edson Pinheiro de Lima,
Sergio Eduardo Gouvea da Costa, Eduardo de Freitas Rocha Loures,
Fernando Deschamps, José Marcelo A. P. Cestari and Eduardo Andrade

**Comparative Study of LCIA, MFCA, and EPIP Tools
for the Environmental Performance Evaluation in Industrial
Processes**   107
Marcell Mariano Corrêa Maceno, Urivald Pawlowsky
and Thaísa Lana Pilz

Global Sustainable Waste Management Practices

**Principles of Sustainability and Circular Economy: Application
and Case Analysis of Historical Evidence and Real
Internationally-Oriented Food Processing Company in Latvia**   133
Maira Lescevica

Current Approaches to Waste Management in Belarus   151
Nikolai Gorbatchev and Siarhei Zenchanka

**Environmentally Friendly Concept of Phosphogypsum Recycling
on the Basis of the Biotechnological Approach**   167
Yelizaveta Chernysh and Leonid Plyatsuk

**Can Circular Economy Tools Improve the Sustainable Management
of Industrial Waste?**   183
Mariana de Souza Silva Rodrigues and Alvim Borges

**Decision Model for Selecting Advanced Technologies for Municipal
Solid Waste Management**   201
Douglas Alcindo da Roza, Guilherme Teixeira Aguiar,
Edson Pinheiro de Lima, Sergio Eduardo Gouvea da Costa
and Gilson Oliveira Adamczuk

**Diagnostic Model in Sustainable and Innovative Operations
for Municipal Solid Waste Management**   221
Douglas Alcindo da Roza, Edson Pinheiro de Lima
and Sergio Eduardo Gouvea da Costa

**The Development Role of Customers in the Reverse Logistics
of Industrial Waste**   245
Tamiris de Oliveira and Alvim Borges

**Challenges in Reducing Construction and Demolition Waste
Generation in Construction Sites in Curitiba**   255
Leilane Kusunoki, Eduardo Felga Gobbi and Patricia Charvet

**A Methodology for Sewage Network Maintenance Toward
the Fulfillment of Sustainable Development Goals**   271
Luciano Rodrigues Penido, Karen Juliana do Amaral,
Regina Maria Matos Jorge, Jörg Wolfgang Metzger,
Jefferson Skroch and Rafael Cabral Gonçalves

Global Sustainable Food Systems and Agricultural Markets

Internet of Things: The Potentialities for Sustainable Agriculture 291
Tehmina Khan

**Geographic Information Systems as a Tool to Display Agribusiness
and Human Development Synergy** 303
Rodrigo Martins Moreira

**The Effects of Climatic Variations on Agriculture: An Analysis
of Brazilian Food Exports** 321
Gisele Mazon, Beatrice Maria Zanellato Fonseca Mayer,
João Marcelo Pereira Ribeiro, Sthefanie Aguiar da Silva,
Wellyngton Silva de Amorim, Larissa Pereira Cipoli Ribeiro,
Nicole Roussenq Brognoli, Ricardo Luis Barcelos,
Gabriel Cremona Parma, Jameson Henry McQueen, Issa Ibrahim Berchin
and José Baltazar Salgueirinho Osório de Andrade Guerra

**Public Policy on Sustainable Food and Agricultural Markets:
Legal Perspective from Nigeria** 349
Fatimah M. Opebiyi

**Some Ways of Environmentally Sustainable Agriculture Production
in the Context of Global Market and Natural Barriers** 369
Eugeny V. Krasnov, Galina M. Barinova, Dara V. Gaeva
and Timur V. Gaev

**Indicators for Assessing Sustainability Performance of Small Rural
Properties** 385
Aleriane Zanetti Vian, Dalmarino Setti and Edson Pinheiro de Lima

**Analysis of Sustainability Indicators in Irrigated Rice Production
in the South of Santa Catarina, Brazil** 403
Tiago Comin Colombo and Melissa Watanabe

**Study of the Antioxidant and Nutraceutical Properties of Strawberry
Fragaria x ananassa Organically Grown as an Option of Sustainable
Agriculture** 415
Juan Carlos Baltazar-Vera, Ma del Rosario Abraham-Juárez,
Iovanna Consuelo Torres-Arteaga, Nancy Karina Vargas-Ramos,
Gilberto Carreño-Aguilera, Cesar Ozuna-Lopez,
Ana Isabel Mireles-Arriaga and María Elena Sosa-Morales

**Indicators for Assessing Sustainable Operations in a Poultry
Slaughterhouse, Considering Industry 4.0 Perspective** 425
Débora de Souza Soares, Marcelo Gonçalves Trentin
and Edson Pinheiro de Lima

**The Use of the Seed Germination Test to Evaluate Phytotoxicity
in Small-Scale Organic Compounds: A Study on Scientific
Production and Its Contributions to Goals 2 and 12
of the UN 2030 Agenda** 461
Isael Colonna Ribeiro, Walessa Nunes Barcellos,
Isabella Maria de Castro Filogônio, Poliana Daré Zampirolli Pires,
Jacqueline Rogéria Bringhenti, Sheila Souza da Silva Ribeiro
and Adriana Marcia Nicolau Korres

**Innovations in Agriculture: The Important Role of Agroforestry
in Achieving SDG 13** 475
Marcia Fajardo Cavalcanti de Albuquerque

**Organic Foods in Brazil: A Bibliometric Study of Academic
Researches Related to Organic Production Before
and After the Conceptual Law** 485
Juliana Fatima de Moraes Hernandes and Viviane Beraldo Rosolen

Global Responsible Mining and Energy

**Integrating Corporate Social Responsibility and Sustainability
for Deep Seabed Mining** 499
Yao Zhou

**Integrated Bio-cycle System for Rehabilitation of Open-Pit Coal
Mining Areas in Tropical Ecosystems** 515
Cahyono Agus, Enggal Primananda and Malihatun Nufus

**Best Practice for Responsible Small Scale Aggregates Mining
in Developing Countries** 529
P. Schneider, K.-D. Oswald, W. Riedel, A. Le Hung, A. Meyer, I. Nolivos
and L. Dominguez-Granda

**Assessing Sustainability in Mining Industry: Social License
to Operate and Other Economic and Social Indicators
in Canaã dos Carajás (Pará, Brazil)** 555
Thiago Leite Cruz, Valente José Matlaba, José Aroudo Mota,
Celso de Oliveira Júnior, Jorge Filipe dos Santos, Leon Nazaré da Cruz
and Eduardo Nicolau Demétrio Neto

**A Review on Multi-criteria Decision Analysis in the Life Cycle
Assessment of Electricity Generation Systems** 575
José Guilherme de Paula do Rosário, Rodrigo Salvador,
Murillo Vetroni Barros, Cassiano Moro Piekarski, Leila Mendes da Luz
and Antonio Carlos de Francisco

**Towards More Sustainable Extractive Industries: Study
of Simulation of Efficient Ventilation Systems in the Emission
Reduction of Gases for the Development of Mine Works**   591
Pablo Vizguerra-Morales, Rosa Isela Lopez-Mejia,
Juan Carlos Baltazar-Vera, Joel Everardo Valtierra-Olivares,
Roberto Ontiveros-Ibarra, Carolina de Jesús Rodríguez-Rodríguez,
Juan Esteban García-Dobarganes Bueno, Gilberto Carreño-Aguilera
and Alberto Florentino Aguilera-Alvarado

**Affordable and Clean Energy: A Study on the Advantages
and Disadvantages of the Main Modalities**   615
Pablo Carpejani, Érica Tessaro de Jesus, Bárbara Luzia Santor Bonfim
Catapan, Sergio Eduardo Gouvea da Costa, Edson Pinheiro de Lima,
Ubiratã Tortato, Carla Gonçalves Machado and Bernardo Keller Richter

**Is Energy Planning Moving Towards Sustainable Development?
A Review of Energy Systems Modeling and Their Focus on
Sustainability**   629
Pedro Gerber Machado, Dominique Mouette, Régis Rathmann,
Edmilson dos Santos and Drielli Peyerl

**Sustainable Development Goals as a Tool to Evaluate
Multidimensional Clean Energy Initiatives**   645
Karen L. Mascarenhas, Drielli Peyerl, Nathália Weber,
Dominique Mouette, Walter Oscar Serrate Cuellar, Julio R. Meneghini
and Evandro M. Moretto

**Appraising Services to the Ecosystem: An Analysis of Itaipu Power
Plant's Water Supply in Energy Generation**   659
Fabrício Baron Mussi, Ubiratã Tortato and Aline Alvares Melo

**Using GIS to Map Priority Areas for Conservation Versus Mineral
Exploration: Territorial Sea of Espírito Santo State, Brazil, Study
Case**   677
Viviane K. Bisch, Valeria S. Quaresma, João B. Teixeira
and Alex C. Bastos

**Challenges and Opportunities Due to Energy Access in Traditional
Populations: The Quilombo Ivaporunduva Case, Eldorado—SP**   691
Rodolfo Pereira Medeiros and Célio Bermann

Global Sustainable Transportation, Construction and Infrastructure

Collaborative Outsourcing for Sustainable Transport Management . . .   709
Marzenna Cichosz, Katarzyna Nowicka and Barbara Ocicka

**Public Attitude Toward Investment in Sustainable Cities
in Taiwan**   725
Meng-Fen Yen and Yuh-Yuh Li

Circularity in the Built Environment: A Focus on India   739
Usha Iyer-Raniga, Priyanka Erasmus, Pekka Huovila and Soumen Maity

Modern City in the Perception of Students–Architects   757
Olga Melnikova

**Sustainable Logistics: A Case Study of Vehicle Routing
with Environmental Considerations**   765
Aline Scaburi, Júlio César Ferreira and Maria Teresinha Arns Steiner

**Green Supply Chain Management and the Contribution
to Product Development Process**   781
Alda Yoshi Uemura Reche, Osiris Canciglieri Jr.,
Carla Cristina Amodio Estorilio and Marcelo Rudek

**Global Conservation and Sustainability Innovations, Investments,
and Policies**

Sharing Economy—Another Approach to Value Creation   797
Pawel Dec and Piotr Masiukiewicz

**International Business, Trade and the Nagoya Protocol:
Best Practices and Challenges for Sustainability in Access
and Benefit-Sharing**   813
Natalia Escobar-Pemberthy and Maria Alejandra Calle Saldarriaga

**Global New Economy: Structure and Perspectives
in Kaliningrad Region**   831
Yulia Aleynikova

**Foreign Direct Investment, Domestic Investment and Green Growth
in Nigeria: Any Spillovers?**   839
Akintoye V. Adejumo and Simplice A. Asongu

**Openness and Greenness: Pay-Offs or Trade-Offs for the Nigerian
Economy**   863
Oluwabunmi O. Adejumo

**Ecotechnology as Mechanism of Development in Disadvantaged
Regions of Mexico**   887
Lorena del Carmen Álvarez-Castañón and Daniel Tagle-Zamora

**Mapping the Industrial Water Demand from Metropolitan Region
of Curitiba (Brazil) for Supporting the Effluent Reuse
from Wastewater Treatment Plants** 899
Carlos Henrique Machado, Patrícia Bilotta and Karen Juliana do Amaral

**Selection of Best Practices for Climate Change Adaptation
with Focus on Rainwater Management** 915
Jessica Andrade Michel, Giovana Reginatto, Janaina Mazutti,
Luciana Londero Brandli and Rosa Maria Locatelli Kalil

**Implementing the SDG 2, 6 and 7 Nexus in Kenya—A Case Study
of Solar Powered Water Pumping for Human Consumption
and Irrigation** 933
Izael Da Silva, Geoffrey Ronoh, Ignatius Maranga, Mathew Odhiambo
and Raymond Kiyegga

Sustainability Reporting

**Sustainability Reporting in Australian Universities: Case Study
of Campus Sustainability Employing Institutional Analysis** 945
Gavin Melles

**Sustainability Reporting in Higher Education Institutions:
What, Why, and How** 975
Naif Alghamdi

**"Reaching for the STARS": A Collaborative Approach
to Transparent Sustainability Reporting in Higher Education,
the Experience of a European University in Achieving
STARS Gold** 991
Maria J. Kirrane, Chris Pelton, Pat Mehigan, Mark Poland,
Ger Mullally and John O'Halloran

**Comfortable Environment: The Formation of Students-Architects'
Professional Consciousness in the Paradigm of Sustainable
Development** 1009
Olga Melnikova

**Students and University Teachers Facing the Curricular Change
for Sustainability. Reporting in Sustainability Literacy and Teaching
Methodologies at UNED** 1021
A. Coronado-Marín, M. J. Bautista-Cerro and M. A. Murga-Menoyo

**Unfolding the Complexities of the Sustainability Reporting Process
in Higher Education: A Case Study in The University of British
Columbia** 1043
Kim Ceulemans, Carol Scarff Seatter, Ingrid Molderez,
Luc Van Liedekerke and Rodrigo Lozano

**Measuring the Use of Sustainable Modes of Transport
at a University** 1071
Jan Silberer, Thomas Bäumer, Patrick Müller, Payam Dehdari
and Stephanie Huber

Sustainability and Institutions: Achieving Synergies 1087
Walter Leal Filho

Global Sustainable Management Practices

Corporate Social Responsibility and Sustainability Initiatives of Multinational Hotel Corporations

Artie W. Ng and Pimtong Tavitiyaman

Abstract Multinational hotel corporations participating actively in international business developments have become aware of the significance of disclosures on Corporate Social Responsibilities (CSR) and Sustainability issues. Such development is driven by the external stakeholders and endorsed by the internal ones. Financial regulators around the world have appeared to embrace the international trend of disclosures about CSR and Sustainability by publicly listed corporations. Through case studies of major multinational hotel corporations operating in Asia and the West, this study examines the emerging emphases of disclosures on CSR and Sustainability matters as a global trend—including public relation management (PRMs), enterprise risk management (ERMs), and integrated innovation management (IIMs). The results showed different aspects of CSR and sustainability practices among multinational hotel corporations worldwide. Western hotel corporations indicate an emphasis on IIMs with global measurement and standard such as GRI standard and ISO14001 certification. In the meantime, Asian hotel corporations have diverse CSR and sustainability initiatives in terms of PRMs (e.g., staff development, staff remuneration and welfare, and good stewards of the environment) and ERMs (e.g., operational risks management, use of solar energy, and new energy conservation) approaches, but low implementation on IIMs. Various aspects of CSR and sustainability initiatives can be reinforced to enhance multinational hotel corporations' sustainability.

Keywords Corporate social responsibility · Sustainability · Hotel industry · Multinationals · Disclosures

---

Note: An earlier version of this article was presented as a working paper under the Working Paper Series organized by the School of Professional Education and Executive Development at The Hong Kong Polytechnic University.

A. W. Ng (✉) · P. Tavitiyaman
School of Professional Education and Executive Development, College of Professional and Continuing Education, The Hong Kong Polytechnic University, Kowloon, Hong Kong
e-mail: spartie@speed-polyu.edu.hk

P. Tavitiyaman
e-mail: spimtong@speed-polyu.edu.hk

1 Introduction

The process of corporate social responsibility (CSR) has been explored and strategically implemented by many hospitality organizations, regardless of their size and scale in operations (Raub and Blunschi 2013). Proposing the CSR practices from the company's perspectives reflects a lot of positive outcomes. The benefits of CSR are to increase financial performance, reduce operating costs, improve staff commitment and involvement, enhance the capacity of innovation, promote brand reputation and image, gain new product opportunities, and stimulate customer's awareness (Raub and Blunschi 2013; Vogel 2005).

In fact, many stakeholders would draw much their attention to a company's CSR practices and implementation. CSR disclosure can be one of the effective marketing tools in expressing the company responses on those issues. CSR disclosure can inform the public concern and express the actions taken on social and environmental impacts (Cuganesan et al. 2010). However, the presentation of CSR information is likely to vary across different industries. The content of CSR disclosure information, contained in both corporate annual report and standalone document would be interpreted and presented in different formats as driven by how the society may perceive (Cuganesan et al. 2010; Deegan et al. 2002). For instance, a company with a higher CSR profile could report contents in a way to change public perception while deflecting attention in comparison with a company with lower CSR profile (Cuganesan et al. 2010). CSR disclosure reporting has also been investigated with respect to the reporting characteristics, the producing organization, and how it is used by external stakeholders (Bebbington et al. 2008).

The hotel industry is one of the global industries that have applied CSR concept in its management and operations. With the uniqueness of industry's nature, the hotel industry is an industry that faces issues of energy overconsumption and unhealthy working environment (Jones et al. 2006). The company culture, hoteliers' leadership, hotel characteristics also influence CSR practices and implementations within hotels (Mackenzie and Peters 2014). For instance, if a hotel would prefer to monitor employee actual behavior of performance, the hoteliers are encouraged to set its corporate policy to mandate employees' actions (Mackenzie and Peters 2014). In contrast, a more flexible workplace environment could improve environmental awareness in both employees' and managers' perspectives compared to a more control working environment. Many hotels have introduced various CSR practices and implications to inform the public of their responsibilities such as scholarships, training courses, recycling and reducing waste at the Housekeeping and Food and Beverage Departments, and volunteer work by hotel employees to assist the community (Mackenzie and Peters 2014).

However, when it comes to disclosing CSR reports and practices, few studies have investigated the complexity of CSR disclosure reporting practices (Bebbington et al. 2008). Some top hotels and hotel executives have reported mostly on their charitable donations and human resources but seemed to lack environmental issues, vision and values (Hocomb et al. 2007). As a result, these hotels are criticized by the community on their operational practices, which in turn could cause social and environmental concerns. This can be questioned to the point that what sort of information is relevant and appropriate for disclosure to the public to reflect a hotel's performance in CSR.

In light of these issues, the purpose of this study is to: (1) explore the variations in CSR disclosures from the perspectives of global chain hotels with reference to the proposed framework, and (2) review the range of practices for CSR and sustainability among selected international hotel groups with noticeable operations in Asian and Western regions. Hospitality and tourism businesses are expanding around the world, it has become a challenge for hoteliers to create an effective strategy in aligns with CSR on a global scale. CSR disclosure can nevertheless be explored in three perspectives: public relations, enterprise risk management and an integrated management approach.

2 Literature Review

2.1 CSR and Sustainability

CSR has traditionally been considered as a model for corporations to take actions to be a good corporate citizen covering legal and ethical standards for their broad stakeholders. In more recent years, the concept of sustainability has gained further recognition to complement with the scope of CSR in explaining what corporations should further their responsibility for the environment under a globalized economy. With reference to Lele (1991), Sustainability is about "Ensuring that development meets the needs of the present without compromising the ability of future generations to meet their own needs". Gray and Bebbington (2001) characterized Sustainability as "Treating the world as if we intended to stay". More recently, Sustainability is considered as "Actions and approaches adopted by organizations compatible with, and contributing to, sustainable development (Hopwood et al. 2010)." Hopwood et al. (2010) also pointed out three main types of sustainability, namely economic, environmental and social sustainability.

Further, the international accounting authority IFAC (2011) noted, "This definition also requires organizations to take into account the wider and longer-term consequences of decisions. This is the route to achieving long-term sustainable value for investors and stakeholders, and involves considering the impact of economic activities—things bought, investments made, waste and pollution generated—on the natural and human resources on which they depend, to avoid irreparable damage to the productive capacity of these resources".

Due to the social trend of stakeholders (e.g., groups or individuals who can impact or being influenced by the success of organizational objectives), CSR becomes a central remark of the corporate strategy by many organizations. However, SCR reporting and implementation are still largely unregulated and unilateral; hence, the organization develops CSR initiatives via voluntary disclosure (Medrado and Jackson 2016).

The hospitality and tourism industry would be motivated to implement the CSR policy because it can enhancing corporate reputation, increasing customer and employee loyalty, attracting new investors, and developing new market share and productivity (Medrado and Jackson 2016).

2.2   Legitimacy Theory and CSR Disclosure

The legitimacy concept describes the notion of how the society and companies react and respond to the CSR practices legitimately. Legitimacy theory is commonly used in CSR accounting and reporting literature, which is to describe the agreement of social practices by organizations (Hopwood 2009; Mahadeo et al. 2011; Monfardini et al. 2013). In fact, there is apparently a social contract between a company and society, the society has expectations on how a company should operate its business (Cuganesan et al. 2010). The judgment of legitimacy theory is to disclose information in light of the society's expectation; however, managers could have a different perception in interpreting the concept of society's expectation (Lanis and Richardson 2013). As a consequence, CSR disclosures are like to be inconsistent across companies and industries.

In prior studies, CSR disclosure is described as the process of providing information designed to discharge social accountability and the contents might include information in the annual report, special publications or even socially oriented advertising (Gray et al. 1987). More recently, Lanis and Richardson (2013) suggested that companies should present their CSR disclosure reporting to enhance their reputation, especially in the aspects of the environmental, social and ethical issues. Moreover, some other prior studies harnessing legitimacy theory have found a positive relationship between community concern and CSR disclosure on particular social and environmental issues (Deegan et al. 2002; Lanis and Richardson 2013). As a growing trend of observing internationally acceptable practice, companies may consider adopting international reporting standard, such as Global Reporting Initiative (GRI), as a means to be perceived by stakeholders as "legitimate".

Examining the international hotel industry, Chung and Parker (2010) investigated the developments of integrating hotel environmental strategies through management control. Chung and Parker (2010) further revealed the emergence of "Triple-bottom Line Framework" of reporting in the case of the hospitality industry in Singapore. The triple bottom line includes financial, social, and environmental outcomes and impacts of its operations.

2.3   Focus on Public Relations

The CSR reporting has been criticized as being used as a tool to enhance public relations without substantial contributions (Saha and Darnton 2005; Moneva et al.

2006). Many large hotels use CSR as a marketing tool while embedding CSR concepts into company mission statements (Mackenzie and Peters 2014). For instance, according to Jones et al. (2006), the pub operators in the UK use CSR concepts in the promotion of the balance of responsibilities among producers, retailers, government regulators and consumers. These initiatives to demonstrate good relations with the public includes communicating to the stakeholders about the responsible alcohol consumption, discouraging excessive drinking, and training staff on how to sell alcohol responsibly.

In addition, other management initiatives to improve public relations consist of the provision areas of smoking and disabled customers and showing committed customer services to prospect customers. The recent concept of green marketing, for instance, would also give incentives for vendors to focus on their "green features" in their marketing and promotion.

2.4 Enterprise Risk Management

CSR disclosure can be considered a part of the process in relation to reputation risk management (Bebbington et al. 2008). The effectiveness of CSR will link to attain sustainable development of the business and to enhance risk management in three key major pillars (economic, environmental, and social), company reputation, and retain new stakeholders (MacLean and Rebernak 2007; Medrado and Jackson 2016). There have been corporations looking into enhancement of enterprises risk management as there are concerns over external environmental changes and overall business sustainability (Dovers and Handmer 1992). The notion of risk management would be pertinent to dealing with a corporation's reputational risk and legitimacy while minimizing potential adverse impacts such as the environmental impacts (Vogel 2005). In other studies, it is further noted the linkage between sustainability and risk for a society's long-term survival and climate change (Yanitsky 2000).

In order to minimize risks, there are proposed strategies in relation to control social and environmental impact in the hotel industry. Those are environmental training programs, green purchase policies, energy and water saving actions and recycling, solid waste generation, reduction in use of chemicals, and utilities monitoring (Gil et al. 2001; Meade and del Monaco 1999). These practices can benefit a hotel's long-term success because of reduced energy emissions and generating more business revenues.

However, limited disclosures about sustainability for the public are explored in a prior study (Dumay et al. 2010). CSR disclosure is found to be lacking relevance and utility and not contributing to sustainability. Ng and Nathwani (2012) argued that despite the fact that there are risks associated with climate change, such unsustainability and credit ratings of major energy corporations are not fully examined and disclosed.

2.5 *Integrated Innovation Management*

Implementing an integrated management approach attempts to deal with CSR and sustainability in an integrated manner by adopting the triple-bottom-line approach as a competitive advantage (Chung and Parker 2010). Such CSR reporting explains the effective management approaches in environmental responsiveness, company plans in reducing energy consumption and compliance with the environmental legislation in an effective manner. In addition, these hospitality companies would encourage an awareness of environmental issues to all employees, partners, and suppliers. All these practices would create positive working environments and management development (Jones et al. 2006).

Such an integrated management approach can be explained via the reporting of corporate financial reporting (CFR), corporate governance (CG), corporate social responsibility (CSR), shareholder value creation (SVC), and sustainability (Bhimani and Soonawalla 2005; Perrini and Tencati 2006; Schaltegger and Wagner 2006). In addition, focus on performance, practice and systems within an organization would enhance sustainability performance as well as strategic development in a longer run (Obrien and Parker 1999; Adams and Larrinaga-Gonzalez 2007; Inoue and Lee 2011; Dai et al. 2013).

3 Framework Development

With reference to the literature review, a trilogy of disclosure strategy is constructed as the framework for this study. It is argued that hotel groups would develop their particular CSR and sustainability disclosure approaches driven by three different concerns, namely public relations management, enterprise risk management and an integrated innovation management approach for sustainability. These three main concerns are summarized as follows:

(i) Public Relations Management (PRM). It focuses on local relationship building with one's primary stakeholders, namely customers and shareholders (existing and potential ones). It attempts to disclose social sustainability, e.g. community services and donation. It works to enhance relationship with primary stakeholders while providing a direct means to economic sustainability through recurring revenues and positive perception by the consumers and capital markets.

(ii) Enterprise Risk Management (ERM). It attempts to protect one's reputational risk and legitimacy. It would minimize potential adverse impacts, e.g. environmental impact. It aims to enhance economic sustainability through reducing and mitigating any adverse issues with CSR and sustainability. It also adopts international standards for performance measurements, demonstrating global citizenship while taking care of business risks. In disclosures, it would provide proactive reporting on CSR and sustainability matters following international standards, such as GRI.

Table 1 The six selected cases

Case	Descriptions
A1	Hong Kong listed, prestigious brand with operations in prime cities in Asia, China and overseas
A2	Dual listing in Hong Kong and Singapore with significant growth in China and expanding into North America
A3	Dual listing in Hong Kong and Shanghai with significant presence in China and expanding overseas through M&As
W1	London listed multiple-brand hotel chains with global operations
W2	Euronext listed multiple-brand hotel chains with global operations
W3	New York listed multiple-brand hotel chains with global operations

(iii) Integrated Innovation Management (IIM). It attempts to deal with CSR and sustainability in n integrated manner while adopting the Triple-Bottom-line reporting approach. It requires development of internal management systems to deal with CSR and Sustainability matters on a continuous basis. It may develop median to long-term commitments as competitive advantage.

4 Methodology

As an exploratory study, multiple-case studies were adopted to examine the issues involved (Yin 1994). Six international hotel groups are selected for this study. Three of these hotel groups are headquartered in the Asian Pacific/China region whereas the remaining three are based in North America or Europe (see Table 1). All of them are listed on major stock exchanges, such as London, New York and Hong Kong. These hotel groups have access to capital markets among Hong Kong, Shanghai, Singapore, London and New York. As all of them are publicly listed corporations, they are required to provide financial reporting and related disclosures to the public. Through such public disclosures, relevant information was collected in relation to CSR and sustainability practices. The content analysis is adopted to perform comparative analysis of these qualitative data with reference to the proposed framework.

5 Findings

Through content analysis of disclosures of the six selected cases, it is demonstrated that there are noticeable variations in their emphases on CSR and Sustainability. In particular, four of them (A1, A2, W1, W2) have disclosed their adoption of GRI standard. Two of Asian hotel groups emphasize the association of CSR and sustainability with risk management and relationship with stakeholders in standalone sustainability reports. Two of Western hotel groups adopt an integrated approach and emphasize on their competitive advantage. They provide disclosures about of programs and forward looking initiatives plans for CSR and sustainability beyond providing certain performance measurement or indicators. A summary of their key disclosures on CSR and sustainability is provided in Table 2.

Table 2 Summary of key disclosures on CSR and sustainability

Case	Capital market base	Main disclosure emphases
A1	Hong Kong	• Adopt a sustainability vision: "aspires to manage its businesses and operations to high ethical and social responsibility standards as a leading hotel and property group. We are committed to integrating sustainable practices and principles across our operations in a balanced way whilst providing an exceptional level of service to our customers" • Policy statement as the guiding principle: "We will consult with our key stakeholders and share information with them about our sustainability performance, achievements and challenges while seeking to balance their needs and goals with the Company's objectives" • Embark on developing a Group Risk Register to further strengthen how we assess and manage key strategic and operational risks that the Group faces, including sustainability, health and safety, and supply chain risks • Integration into business process through investing into staff development • Released a stand-alone sustainability report with details year-to-year measurement of results using GRI standard and specific targets to reduce energy consumption, etc.
A2	Both Hong Kong and Singapore	• Released a stand-alone sustainability report • Participate in Carbon Disclosure Project • Make use of international standards; e.g. hotels certified under ISO 14001 Environmental Management Systems, GRI indicators with year-to-year performance comparison • Promote rainwater harvesting, use of solar energy and the incorporation of composting and herb gardens to manage food waste • Develop CSR Projects: 10 to 15-year partnership with a chosen beneficiary working on children's health or education programs • "We are committed to serving as good stewards of the environment. We ensure that every property respects local traditions and culture, restores natural habitats, conserves biodiversity and manages waste, water and energy" • Target set with reference to business growth • "stakeholders recommend that Shangri-La the Hotel should take early action to integrate environmental issues into its comprehensive risk management systems"

(continued)

Table 2 (continued)

Case	Capital market base	Main disclosure emphases
A3	Both Hong Kong and Shanghai	• Obtained Platinum award, Corporate and Employee Contribution Programme, Community Chart, Hong Kong • "The Group has always focused on the improvement of staff remuneration and welfare. During the Reporting Period, the Group continued to work on improving staff remuneration and raising the fixed salary for all staff of our wholly-owned subsidiaries. The Group has also continued to optimise the supplementary medical insurance program and the supplementary commercial medical insurance coverage for retired personnel for our staff. In addition, the Group also timely provides funding assistance for those employees in need" • Continue to promote modification projects in relation to its new energy conservation technology system, completing system modification in its hotels in China involving primarily air source and residual heat-recovery, and the effect of energy conservation was considered remarkable
W1	London ('BBB' long-term corporate credit ratings)	• "Corporate responsibility (CR) is central to the way we do business and is a key part of our responsible business practices. We treat it as a strategic business issue, believing CR only makes sense if it aligns to our vision of becoming one of the world's great companies by creating Great Hotels Guests Love" • Adopted GRI standard for regular reporting and performance measurement • Issue a Corporate Responsibility Committee Report • Innovation through integration: "We acknowledge there is a tension between tourism and the environment but we believe this can be a creative one, providing an opportunity to find innovative solutions to the environmental, social and economic effects of our business" • Develop specific performance measures: kWh consumed for per available room
W2	Euronext ('BBB-' long-term and 'A-3' short-term)	• A comprehensive program with commitments to sustainability • Seven pillars, namely Health, Nature, Carbon, Innovation, Local Communities, Employment and Dialogue (with partners) • Integrated into design of hotels and daily operations • Twenty one commitments to sustainability • Specific targets are made for Health and Nature pillars; e.g. reducing use of water and responsible eating • "Reinventing hotels for a sustainable future" as a decisive competitive advantage • Adopted ISO14001 certification, UN Global Compact Indicators and GRI standard

(continued)

Table 2 (continued)

Case	Capital market base	Main disclosure emphases
W3	New York (BBB-)	• Disclosure of group policy for environmental and community initiatives • Charity donation programs in association with non-profit, charitable organizations; foundation is formed to manage funding for donation • Target set to reduce consumption of energy and water • Carbon Disclosure Project reporting is initiated as part of a global CDP movement to provide corporate climate change information to a global investor group • Other initiatives include human rights policy and global citizenship scheme • "Committed to integrating leading environmental practices and sustainability principles into our core business strategy"

6 Discussion and Implications

As illustrated in Fig. 1, there are variations among the six cases in adopting international reporting standards and their emphases on management initiatives for furthering future developments on CSR and Sustainability. All of six cases emphasize on general CSR and sustainability reporting. This result supports previous finding of Medrado and Jackson (2016). A3 appears to be a PRM that focuses on local community relationship building and enhancing their public relations, which is similar

- "PRMs" focus on local community relationship building
- "ERMs" adopt international standards for performance measurements, demonstrating global citizenship while taking care of business risks
- "IIMs" measure performance with international standards while creating competitive advantage through implementation of unique management programs and plans

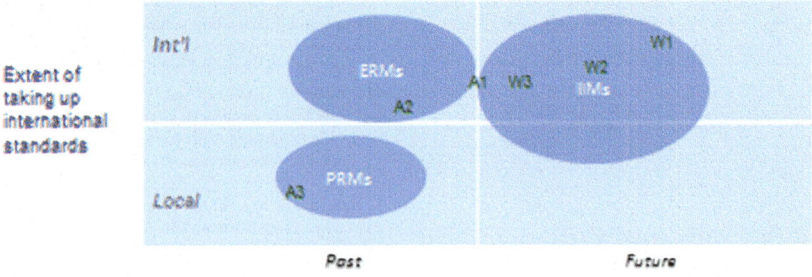

Fig. 1 Variations in adopting international standards and future emphasis

as the study of Medrado and Jackson (2016). A3 provides funding assistance for the employees in need and improve staff remuneration and welfare. A2 has the features of an ERM that adopts international standards for performance measurements, demonstrating global citizenship while taking care of business risks. The existing practices include integrating environmental issues into its comprehensive risk management systems, the use of international standard of ISO14001 Environmental Management Systems and comparison of GRI 10-year indicators. In contrast, A1 is between ERMs and IIMs. A1's CSR and Sustainability policy is set to adopt the international standards with rather minimal efforts to demonstrate their strategic developments for the future. The practices are presenting a stand-alone sustainability report by using GRI standard and integrating business process through staff development.

Nevertheless, the rest of the cases (W1, W2, and W3) carry the characteristics of an IIM that measures their performance with reference to international standards but also attempt to create competitive advantage through implementation of unique management programs and plans that would further uphold their CSR and Sustainability initiatives. Adopting GRI standard, developing specific performance measures, reinventing hotels for a sustainable future are examples of IIMs approach. These western hotels are international worldwide; the CSR disclosure would have a major influence on the overall brand image of their hotels. Creating the unique CSR programs would lead the companies in the prospect positioning.

The theoretical and managerial implications are presented as following aspects. The legitimacy theory is applicable for multinational hotel corporations in disclosing CSR initiatives and sustainability practices in the areas of public relation management, enterprise risk management, and integrated innovation management. It helps positioning the hotel corporations in the mind of stakeholders and society. Legitimacy theory can be used as a tool to maintain the reputation of the multinational hotel corporations (Aureli et al. 2017).

From the managerial implications' viewpoint, multinational hotel corporations in Asia region aim the focus of CSR practices on performance measurement of business risks and building local community relationships. Asia multinational hotels can extend the areas of CSR practice and sustainability initiatives in terms of international standard and innovative management. This might be due to the fact that these Asia hotel corporations follow good CSR practices from the major leaders in the hotel industry, which are literally the Western Multinational hotel corporations. Meanwhile, multinational hotel corporations in Western region develop advance practices of CSR, so they tend to regain the strong reputation of innovative management of CSR and sustainability implications.

7 Limitations and Future Research

This study had limitations needed to be addressed. While this is a pilot study that explores the phenomena, future studies would aim to collect additional data through in-depth interviews of the stakeholders as well as survey of a larger group of pertinent corporations in the industry so as to refine the framework proposed in this study. Despite globalization, there are growingly various local regulatory and compliance issues in different countries that would influence how these listed enterprises would present their disclosures on CSR and Sustainability. Future research studies need to take into consideration these evolving regulatory concerns and their implications for the practice and reporting of CSR and Sustainability. Further, sample size of this study was only six corporate hotel organizations. Future study can extend the sample size for both Asian and Western hotel corporations.

References

Adams C, Larrinaga-Gonzalez A (2007) Engaging with organizations in pursuit of improved sustainability accounting and performance. Acc Audit Account J 20(3):333–355

Aureli S, Medei R, Supino E, Travaglini C (2017) Sustainability disclosure and a legitimacy crisis: insights from two major cruise companies. Eur J Tour Res 17:149–163

Bebbington J, Larringaga C, Moneva JM (2008) Corporate social reporting and reputation risk management. Acc Audit Account J 21(3):337–361

Bhimani A, Soonawalla K (2005) From conformance to performance: the corporate responsibilities continuum. J Account Public Policy 24(3):165–174

Chung LH, Parker L (2010) Managing social and environmental action and accountability in the hospitability industry: a Singapore perspective. Acc Forum 34(1):46–53

Cuganesan S, Guthrie J, Ward L (2010) Examining CSR disclosure strategies within the Australian food and beverage industry. Acc Forum 34:169–183

Dai NT, Ng AW, Tang G (2013) Corporate social responsibility and innovation in management accounting. CIMA Research 9(1):1–10

Deegan C, Rankin M, Tobin J (2002) An examination of the corporate social and environmental disclosures of BHP from 1983–1997. Acc Audit Account J 15(3):312–343

Dovers SR, Handmer JW (1992) Uncertainty, sustainability and change. Glob Environ Change 2(3):262–276

Dumay J, Guthrie J, Farneti F (2010) GRI Sustainability reporting guidelines for public and third sector. Public Manag Rev 12(4):531–548

Gil M, Jimenez J, Lorente J (2001) An analysis of environmental management, organizational context and performance of Spanish Hotel. Omega 29:457–471

Gray R, Owen D, Lavers S (1987) Corporate social reporting: accounting and accountability. Prentice Hall, London

Gray R, Bebbington J (2001) Accounting for the environment. Sage, London

Hocomb JL, Upchurch RS, Okumus F (2007) Corporate social responsibility: what are top-hotel companies reporting? Int J Contemp Hosp Manag 19(6):461–475

Hopwood AG (2009) Accounting and the environment. Acc Organ Soc 34(3):433–439

Hopwood A, Unerman J, Fries J (2010) Introduction to the accounting for sustainability case studies. In: Hopwood A, Unerman J, Fries J (eds) Accounting for sustainability. Earthscan, London/Washington DC, pp 1–26

IFAC (2011) Sustainability framework 2.0: professional accountants as integrators

Inoue Y, Lee S (2011) Effects of difference dimensions of corporate social responsibility on corporate financial performance in tourism-related industries. Tour Manag 32(4):790–804

Jones P, Comfort D, Hillier D (2006) Reporting and reflecting on corporate social responsibility in the hospitality industry: a case study of pub operators in the UK. Int J Contemp Hosp Manag 18:329–340

Lanis R, Richardson G (2013) Corporate social responsibility and tax aggressiveness: a test of legitimacy theory. Acc Audit Account J 26(1):75–100

Lele SM (1991) Sustainable development: a critical review. World Dev 19(6):607–621

Mackenzie M, Peters M (2014) Hospitality managers' perception of corporate social responsibility: an explorative study. Asia Pacific J Tour Res 9(3):257–272

MacLean R, Rebernak K (2007) Closing the creditability gap: the challenges of corporate responsibility reporting. Environ Qual Manage 16(4):1–6

Mahadeo JD, Oogarah-Hanuman V, Soobaroyen T (2011) Changes in social and environmental reporting practices in an emerging economy (2004–2007): exploring the relevance of stakeholder and legitimacy theories. Acc Forum 35(3):158–175

Meade B, del Monaco A (1999) Environmental management: the key to successful operation. In: First Pan-American Conference, Latin American Tourism in Next Millennium: Education, investment and sustainability Panama City, Panama

Medrado L, Jackson LA (2016) Corporate nonfinancial disclosures: an illuminating look at the corporate social responsibility and sustainability reporting practices of hospitality and tourism firm. Tour Hosp Res 16(2):116–132

Moneva JM, Archel P, Correa C (2006) GRI and the camouflaging of corporate unsustainability. Acc Forum 30(2):121–137

Monfardini P, Barretta AD, Ruggiero P (2013) Seeking legitimacy: social reporting in the healthcare sector. Acc Forum 37(1):54–66

Ng AW, Nathwani J (2012) Sustainability performance disclosures: The case of independent power producers. Renew Sustain Energy Rev 16(4):1940–1948

Obrien PW, Parker L (1999) Environmental strategies and the international hotel industry. J Manag Organ 5(1):12–25

Perrini F, Tencati A (2006) Sustainability and stakeholder management: the need for new corporate performance evaluation and reporting systems. Bus Strat Environ 15(5):296–308

Raub S, Blunschi S (2013) The power of meaningful work: how awareness of CSR initiatives fosters task significance and positive work outcomes in service employees. Cornell Hosp Q 5(1):10–18

Saha M, Darnton G (2005) Green companies or green companies: are company really green, or are they pretending to be? Bus Soc Rev 110(2):117–157

Schaltegger S, Wagner M (2006) Integrative management of sustainability performance, measurement and reporting. Int J Acc Audit Perform Eval 3(1):1–19

Vogel D (2005) The market for virtue: the potential and limits of corporate social responsibility. Brookings Institution Press, Washington DC

Yanitsky ON (2000) Sustainability and risk: the case of Russia, innovation. Eur J Soc Sci 3:265–277

Yin RY (1994) Case study research: design and methods. Sage, Thousand Oaks, CA

Dr. Artie Ng is currently Deputy with PolyU SPEED. Dr. Ng obtained his Ph.D. degree from the Adam Smith School of Business at the University of Glasgow, Scotland. He is a qualified accountant as well as a fellow member with both HKICPA and CIMA (UK). He has research and scholarly interests in knowledge-intensive organizations and their sustainable performance.

Dr. Pimtong Tavitiyaman is a Senior Lecturer and Head of Cluster in Hotel and Tourism Management. Dr. Tavitiyaman received her PhD from School of Hotel and Restaurant Administration, Oklahoma State University, USA. Her research interests are in hospitality and tourism management and marketing.

Sustainable Initiatives and Practices of the Most Sustainable Organizations in the World

Waqas Nawaz, Patrick Linke and Muammer Koç

Abstract Sustainability management in organizations lacks a holistic operational framework; thus, learning from the best practices of selected sustainable firms may allow researchers to extend and transform the theoretical sustainability management models to practicable frameworks and standards. The objective of this study is to review the sustainability best practices from sustainability/annual reports of the systematically identified most sustainable organizations. The starting set of 100 most sustainable organizations was obtained from a sustainability ranking—Corporate Knights, Global 100. A review-selection-criteria was developed and applied to objectively screen the organizational reports between 2013 and 2017. As a result, 28 sustainability/annual reports, of six organizations, were selected for the review of best sustainability practices. The review highlights several best practices under eight sustainability themes, ranging from governance to operational level. Managers can use the findings of this study to roll-out similar initiatives in their firms, while researchers may use these practices to instill a practical approach in the theoretical sustainability management models.

Keywords Organizational sustainability benchmarks · Sustainable initiatives and practices · Sustainability management · Corporate responsibility · Corporate knights

W. Nawaz (✉) · M. Koç
Division of Sustainable Development, Hamad Bin Khalifa University (HBKU), Qatar Foundation (QF), Doha, Qatar
e-mail: waqnawaz@mail.hbku.edu.qa

M. Koç
e-mail: mkoc@hbku.edu.qa

P. Linke
Department of Chemical Engineering, Texas A&M University at Qatar, PO Box 23874, Doha, Qatar
e-mail: patrick.linke@qatar.tamu.edu

© Springer Nature Switzerland AG 2020
W. Leal Filho et al. (eds.), *International Business, Trade and Institutional Sustainability*, World Sustainability Series,
https://doi.org/10.1007/978-3-030-26759-9_2

1 Introduction

A major contribution to unsustainable growth was industrialization, which emanated from the expansion of domestic and multinational organizations throughout the twentieth century. Although sustainability management in organizations is gaining importance, as it offers competitive advantages and value creation (Bonini and Görner 2011), it remains under-addressed, if not unaddressed in many organizations (Hong et al. 2012). There are various reasons why sustainability management remains under-utilized. While some researchers believe that the barrier to operationalizing sustainability goes beyond the organizational boundaries, especially when it comes to value creation in the entire value chain (Stewart et al. 2016), others argue that the lack of practicable operational models holds back the organizational actors from integrating sustainability in strategic and operational management (Nawaz and Koç 2018). The latter is particularly important since most of the existing sustainability management models are based on conceptual and theoretical design (Nawaz and Koç 2018; Sealy et al. 2010), and therefore, the structure of these models lack a fundamental practicable approach.

In order to transform the existing sustainability management models to practicable frameworks, it is important to integrate the best practices of the most sustainable firms within the structure of these models (Nawaz and Koç 2018). Reflecting the best practices in the existing models may not only integrate a practical perspective, it will also significantly enhance the potential of their application, as it will become easier for the organizational actors to understand 'how to implement'.

In the past, many studies have been carried out to analyze sustainability in firms, but these are mostly based on quantitative measures—indicator-based assessments (Knoepfel 2001; Rahdari and Rostamy 2015). Although quantifying the performance facilitates the investors in differentiating between the organizations (Székely and Knirsch 2005), the indicator-based assessment does not allow the evaluators to broadly assess the management potential of the firms (Windolph 2011). In fact, many researchers advocate that the indicator or index-based performance assessment does not really reflect on how well a company manages sustainability-related challenges (López et al. 2007; Schneider and Meins 2012); the quantitative comparison may give 'some feel' of the efficiency and effectiveness of sustainability management in firms, but it cannot provide a holistic view of firms' sustainable capabilities.

For example, in 2015, Marks and Spencer saw an increase in its CO_2 emissions compared to the previous year. This may have been one of the many factors which demoted the firm on index-based rankings, but in reality, the change in CO_2 emissions was associated with the change in the carbon conversion factor used to calculate emissions from the United Kingdom's electricity grid. Similarly, Universal Music Group (UMG), a subsidiary of Vivendi, saw a 40% decrease in the electricity consumption between 2014 and 2015. However, the actual reason for the decrease in electricity consumption was the closure of nine UMG sites internationally. Therefore, understanding the context of performance is more important than purely relying on the quantitative indicators.

Based on the above, this study attempts to identify the sustainability best practices of the most sustainable organizations. Six organizations are systematically selected for the review and their relevant sustainability reports, and in some cases—annual reports, are studied between 2013 to 2017 to capture the sustainability themes and the corresponding best practices.

2 An Overview of Literature

Adams and Larrinaga-González (2007) emphasized on considering the organizational sustainability practices in academic research. The authors argued that in order to identify how accounting and management systems may address the organizational sustainability challenges, it is imperative to learn from the practical success of sustainable organizations. The authors called for a scholarly engagement with organizations to learn from their internal processes, especially at the micro level. Following suit, Rondinelli (2006) analyzed the principles and practices of the 25 transnational corporations (TNCs). The review revealed that many TNCs are still at the elementary or engagement level—far from being innovative, integrative or transformative. This conclusion is in line with the findings of Livesey and Kearins (2002) who argued that it is the lack of 'scientific basis' which holds back the multinational corporations (MNCs) from practicing sustainability above and beyond the optics of philanthropy.

Collins et al. (2010) analyzed the adoption of sustainable business practices in New Zealand. The authors identified common environmental and social practices (Fig. 1) opted by the member firms of the Sustainable Business Network[1] (SBN).

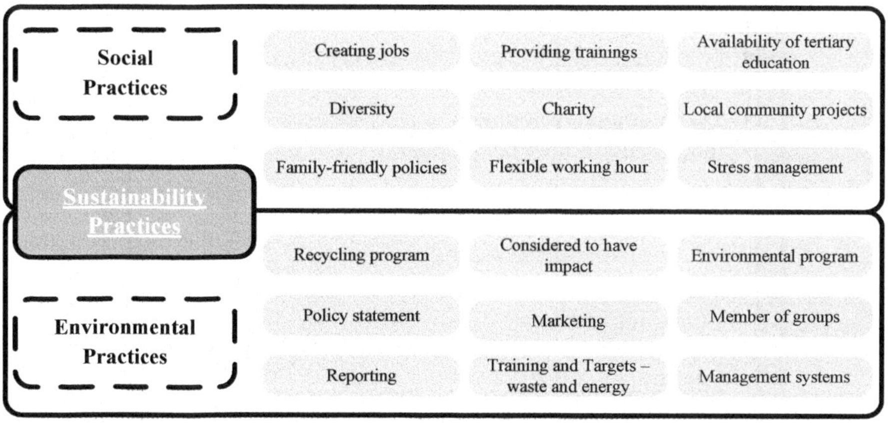

Fig. 1 Sustainability practices of the firms associated with SBN in New Zealand, adopted from Collins et al. (2010)

---

[1]SBN is a network of businesses interested in sustainability practices.

Similar to Livesey and Kearins (2002), Collins et al. (2010) found that businesses are engaged in social activities to a great extent than environmental activities.

Eweje (2011) also studied the sustainability practices of 15 large firms operating in New Zealand. The author's interview with managers revealed that the organizations are actively taking steps to roll-out sustainability initiatives to address economic, environmental and social impacts (Fig. 2). Also, the managers seemed committed to these initiatives; they believed that robust implementation benefits the corporations eventually.

Lozano (2012) thoroughly examined the literature to seek the most cited voluntary sustainability management tools. The author critically analyzed 16 of the resulting 22 tools shown in Fig. 3. Lozano (2012) highlighted that while each sustainability initiative offers benefits, the synergy between different initiatives and integration of initiatives throughout the value chain remain as challenges. The author concluded that understanding the context of business and initiating (context-specific) rational programs can orchestrate sustainability in corporations.

Williams (2015) studied the evolution of sustainable development in nine firms (with over a billion dollar revenue a year) to thoroughly examine the factors that

Environmental Initiatives

- Recycle and re-use of materials
- Industrial use of unusable materials
- Targeting zero carbon footprint
- Minimizing construction waste
- Formal environmental impact assessment
- Waste management
- Extending share of renewable energy
- Green procurement
- Technology development
- Indoor environmental quality

Social Initiatives

- Community partnership / stakeholder engagement
- Grants for research
- Fundraising for charities
- Educating employees
- Partnership with conservation groups

Fig. 2 Environmental and social initiatives of 15 large organizations in New Zealand, adopted from Eweje (2011)

Fig. 3 Most cited voluntary sustainability management tools, adopted from Lozano (2012)

helped in the transformation of business of these Green Giants, shown in Table 1. As a first step, the author has stressed on the importance of changing the mindset, followed by innovation and integration of sustainability within the business strategy. Interestingly, while the author has emphasized the importance of responsible procurement and supply chains, she has also underlined branding and marketing for

Table 1 Six key factors responsible for the sustainability success of nine billion-dollar firms

Factors of green giants	Description	
The iconoclastic leaders	*Competencies (4Cs):* – Conviction – Courage – Commitment – Contrarian	
Disruptive innovation	*Principles:* – Make it better, not just greener – Embrace the counterintuitive – Bet on yourself – Engage the problem solvers – Cultivate pervasive innovation	*Steps:* – Formalization – Inspiration – Generation – Evaluation – Realization
A higher purpose	A business purpose that goes beyond profit making; however, that purpose will eventually lead to profit	
Built-in, not bolted-in	*Structures:* – Corporate strategy – Organizational structure – Governance structure – Cost structure – Incentive structure – Reporting structure	
Mainstream appeal	Sustainability and branding and marketing	
A new behavioral contract	– Ownership and visibility of operations (supply chain) – Collaboration	

Leadership	Strategy
• Show commitment from the top • Scan business environment for potential risks and opportunities • Lead a cultural transformation	• Develop a mission statement • Consider global and local regulations, as well as voluntary standards • Consider the impact of social investors
Structure	**Systems**
• Integrated throughout organization • Effective use of human resources • Manager access to top leadership • Aligned with strategy	• Costing and capital investment • Risk management • Performance evaluation and reward • Measurement • Feedback • Reporting and verification

Fig. 4 Organizational sustainability benchmarks, adopted from Epstein (2008)

the sustainability success of organizations. Williams (2015) has clearly established that the pursuit of sustainability in an organization eventually leads to financial profitability as well.

Similarly, Epstein (2008) described some of the organizational sustainability benchmarks, as shown in Fig. 4. The author pointed out that while it is not a comprehensive list, these organizational features are key to sustainability success in organizations. However, the specific programs to be rolled out largely depend on the context and impact of the firm's activities; these may range from capital investments in new technologies or products[2] to programs promoting ethical sourcing and diversity.

Small (2017) conducted an empirical study to explore the sustainability practices which can promote profitability in the petroleum industry. The results of the survey carried out by the author revealed six types of sustainability practices which can directly affect the sustainability and financial performance of petroleum firms at the same time:

1. **Environment**: air quality, community, and general environmental protection issues;
2. **Fuels**: reduction in the use of fossil fuels, use of alternative fuels (natural gas, biofuels), and projects that reduce GHG emissions;
3. **Human Resources**: training, engagement, employment, and safety;
4. **Recycling**: waste management, water contamination, and chemical treatment;
5. **Mitigation**: flaring, greenhouse gases, and oil and gas-leaks and emissions; and
6. **Water**: sourcing, storage, transportation, recycling, treatment, and disposal.

While the aforementioned studies provide useful insights into the sustainability practices of organizations, these studies primarily lack a systematic approach for

---

[2]Hereafter, 'product' refers to 'product and service'.

the selection of the sample organizations. Furthermore, most of these studies are restricted to the organizational practices at the strategic-level and little has been done to explore the micro-level sustainability initiatives and practices. This work aims to overcome these shortcomings in the literature by selecting the sample organizations systematically and exploring the sustainability practices at the micro, as well as at the strategic level.

3 Materials and Methods

In order to be transparent and systematic in the selection of the sample, the starting universe of 100 organizations was obtained from Global 100 by Corporate Knights (2017). The other available option was the Dow Jones Sustainability Index, but Global 100 was preferred due to the ease of access to its methodology and the corresponding organizational data. Next, a review-selection-criteria was developed and applied to identify the organizations whose performance objectively aligns with the goals of this study (Fig. 5). Only those organizations which were in the Global 100 (of 2017) for at least three consecutive years passed the primary screening. This restriction alone eliminated 48 organizations from the starting pool. Secondly, only those organizations passed the screening which fall under one of the following categories: (i) best average ranking (top 5); (ii) continuously improving ranking—every year; and (iii) most consistent in the ranking (top 5 in least standard deviation in ranking).

A secondary screening-criteria was also applied to ensure the robustness of the research design in the multi-sectoral study (Fig. 5). Hoepner et al. (2008) suggested that the unconditional consideration of sectors may offset the context of the study due to the heterogeneity across industries, and therefore, sector-specific restrictions were applied based on the findings of Vermorken et al. (2008). Another reason for sector-specific elimination was to limit the reports to be reviewed to a manageable number.

The results of primary and secondary screening are tabulated in Table 2. Four organizations were selected as a result of the screening while two organizations were added to the list due to special interest: Daimler and Centrica. These final two were added to the list since both fall within the category of 15 most consistent organizations between 2013 and 2017. Hence, six organizations were finalized for the review and the total number of reports to be reviewed was 28 (more than 4000 pages).

The reports reviewed to capture the sustainability practices included the following:

- **BMW**: Sustainable Value Reports 2012–16
- **Daimler**: Sustainability Reports 2012–16
- **Schneider Electric**: Annual Reports and Strategy and Sustainability Highlights 2012–16
- **Centrica**: Annual Reports and Corporate Responsibility Reviews 2012–16
- **Enbridge**: Corporate Social Responsibility Reports 2012–16

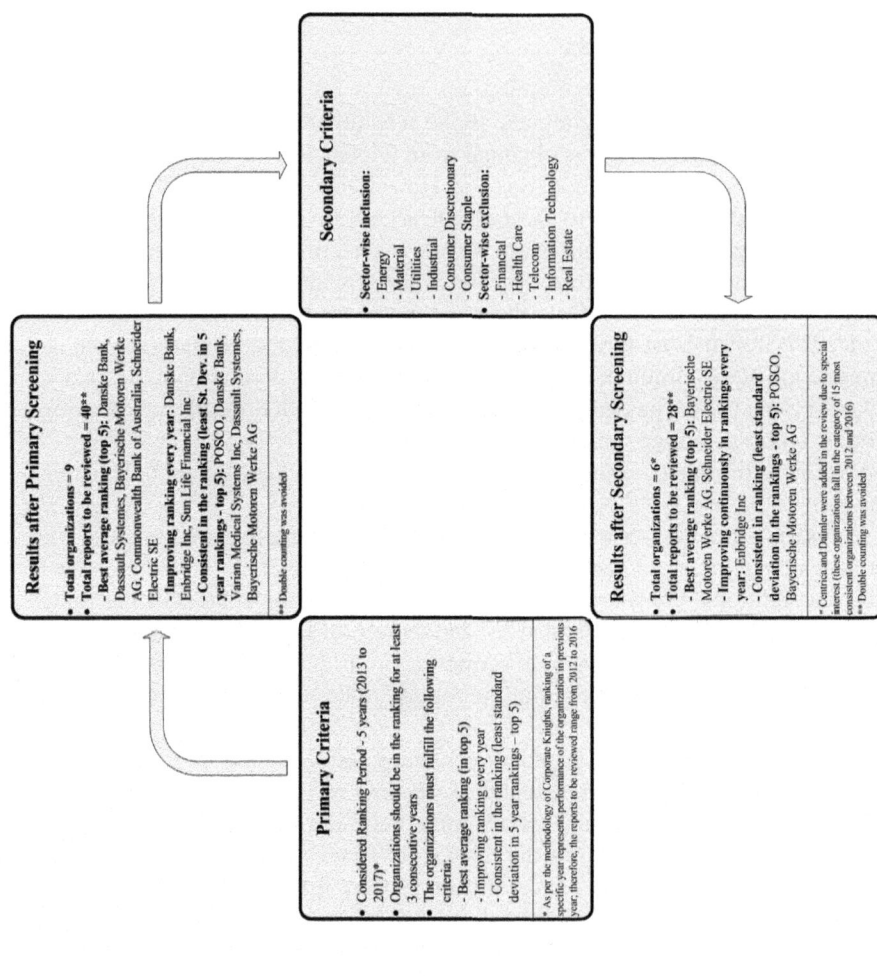

Primary Criteria

- Considered Ranking Period - 5 years (2013 to 2017)*
- Organizations should be in the ranking for at least 3 consecutive years
- The organizations must fulfill the following criteria:
  - Best average ranking (in top 5)
  - Improving ranking every year
  - Consistent in the ranking (least standard deviation in 5 year rankings – top 5)

*As per the methodology of Corporate Knights, ranking of a specific year represents performance of the organization in previous year; therefore, the reports to be reviewed range from 2012 to 2016

Results after Primary Screening

- **Total organizations = 9**
- **Total reports to be reviewed = 40****
  - **Best average ranking (top 5):** Danske Bank, Dassault Systemes, Bayerische Motoren Werke AG, Commonwealth Bank of Australia, Schneider Electric SE
  - **Improving ranking every year:** Danske Bank, Enbridge Inc, Sun Life Financial Inc
  - **Consistent in the ranking (least St. Dev. in 5 year rankings - top 5):** POSCO, Danske Bank, Varian Medical Systems Inc, Dassault Systemes, Bayerische Motoren Werke AG

** Double counting was avoided

Secondary Criteria

- Sector-wise inclusion:
  - Energy
  - Material
  - Utilities
  - Industrial
  - Consumer Discretionary
  - Consumer Staple
- Sector-wise exclusion:
  - Financial
  - Health Care
  - Telecom
  - Information Technology
  - Real Estate

Results after Secondary Screening

- **Total organizations = 6***
- **Total reports to be reviewed = 28****
  - **Best average ranking (top 5):** Bayerische Motoren Werke AG, Schneider Electric SE
  - **Improving continuously in rankings every year:** Enbridge Inc
  - **Consistent in ranking (least standard deviation in the rankings - top 5):** POSCO, Bayerische Motoren Werke AG

* Centrica and Daimler were added in the review due to special interest (these organizations fall in the category of 15 most consistent organizations between 2012 and 2016)
** Double counting was avoided

Fig. 5 A systematic screening criteria for the selection of the most sustainable organizations

Table 2 Results of primary and secondary screening

Organization	Type of business (sector)	Rankings					Average ranking	Continuous improvement in ranking	Consistency in ranking (standard dev.)	Status
		2017	2016	2015	2014	2013				
Danske Bank	Financial	4	7	10	–	–	7.00*	•	3.00**	Eliminated
Dassault Systemes	Information Technology	11	2	17	5	9	8.80*		5.76**	Eliminated
Bayerische Motoren Werke (BMW) AG	Consumer Discretionary	16	1	6	13	16	10.40*		6.66**	Selected
Commonwealth Bank of Australia	Financial	6	4	21	25	–	14.00*		10.55	Eliminated
Schneider Electric SE	Industrials	27	12	9	10	13	14.20*		7.33	Selected
Enbridge Inc	Energy	39	46	64	75	79	60.60	•	17.59	Selected
Sun Life Financial Inc	Financials	45	66	67	79	85	68.40	•	15.36	Eliminated
POSCO	Materials	35	40	36	–	–	37.00		2.65**	Selected
Varian Medical Systems Inc	Health Care	88	90	83	–	–	87.00		3.61**	Eliminated
Daimler AG	Consumer Discretionary	74	48	60	60	44	57.20		11.80	Added
Centrica PLC	Utilities	13	11	8	26	32	18.00		10.42	Added

*Belongs to top 5 average ranking
**Belongs to top 5 most consistent in rankings (least standard deviation in the rankings—top 5)
Note the rankings represent the performance of organizations in the previous year, i.e., the sustainability ranking of 2017 correspond to the report of 2016

- **POSCO**: Integrated Reports 2014–16.

Although this is a qualitative study, the objectivity in reviewing the sustainability practices was maintained by keeping in view the principles of benchmarking proposed by Schalock et al. (2016). With the help of an extensive literature review, the authors proposed qualitative principles for identifying best sustainability practices (Fig. 6). Schalock et al. (2016) also stressed that the best practices are those which have strong evidence of conformance with the principles of benchmarking.

Based on the proposal of Schalock et al. (2016), all potential sustainability best practices were initially collected in separate files, one for each organization. These practices were individually rated by the authors on the principles of benchmarking,[3] shown in Fig. 6. The practices which conformed with most of the principles were considered for the review in this chapter.

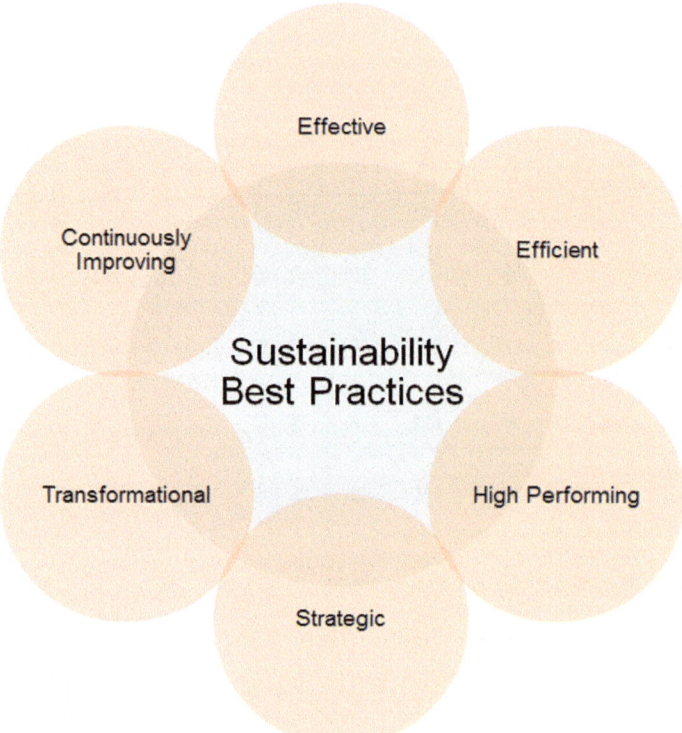

Fig. 6 Principles of benchmarking, adopted from Schalock et al. (2016)

---

[3] All six principles were weighed equally.

4 Findings and Results

The sustainability best-practices identified through the review can be divided into eight main themes, shown in Fig. 7. It is important to note that these themes were not predetermined, rather these evolved with the review process, i.e., the themes were added, renamed, and merged over the complete review process until the final eight sustainability themes were determined.

Each theme can be divided into its functional areas and the corresponding sustainability best practices. For example, 'resource optimization and minimization of waste and emissions' covers sustainability practices in the following functional areas: Energy, Products and Services, and Water. The functional areas also evolved along with the review process and were not predetermined. The sustainability themes and the respective functional areas and their corresponding sustainability practices will be discussed in the following sub-sections.

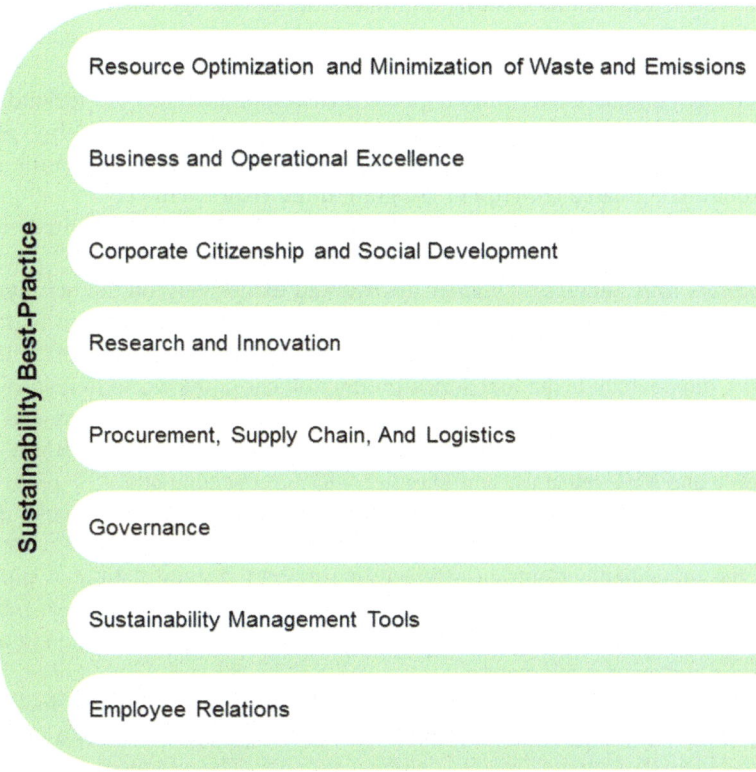

Fig. 7 Main themes representing the sustainability best practices

4.1 Resource Optimization and Minimization of Waste and Emissions

Energy: In order to be energy efficient in heat and power segment, BMW and Daimler both utilized combined heat and power systems, which resulted in an annual saving for BMW of over €1 million, in addition to the various environmental advantages. Similarly, while replacement of lights with LEDs sounds trivial, BMW reduced its energy consumption by 6,848 MWh (2800t $CO_2(e)$) in 2016 by replacing 115,000 conventional lights with LEDs.

Investment to increase the share of renewable energy is a practice that has been observed in all organizations without exception. Enbridge invested around $5bn in renewable energy projects which correspond to 2,800 MW (gross) of green power capacity—enough to meet the electricity requirements of one million homes. Likewise, Centrica keeps increasing its share of power produced from low carbon sources every year—not only for its own operations, but also for its commercial offerings. This, however, does not necessarily eliminate energy wastage. Therefore, Schneider ensures the presence of an energy manager and an energy reduction action plan at its 100 highest-energy consuming sites.

BMW, in a unique move, shifted its German computer centers to Iceland to take advantage of the colder climate and abundant resources of renewable energy (geothermal and hydro). After this move, the centers are now completely footprint neutral. Schneider also initiated a Green IT program in its Asia Pacific computation facilities. The firm customized the power settings of 15,000 PCs which resulted in a 20% reduction in power consumption and saved around 850 MW annually.

Products and Services: Daimler has worked extensively on the aerodynamics of its vehicles. The improvements resulted in a lower drag coefficient and subsequently, low fuel consumption and CO_2 emissions. Daimler has also introduced the S-Class Coupé which is the first series-production car in the world to feature a CO_2 air conditioning system. Daimler and BMW also have state of the art end-of-life vehicle recovery system. The credit goes to the design of the vehicles; BMW—95% recyclable and 85% reusable, and Daimler—85% recyclable and 95% recoverable. In addition, Daimler has started using 3D-printed spare parts for its trucks in order to be resource efficient.

On the other hand, Centrica offers its customers free insulation of pipes and infrastructure to reduce waste energy from homes. Besides the service of free insulation, Centrica offers £50 to customers after the installation of insulation in order to encourage customer participation.

Water: Besides treating its sewage water in the sewage treatment facility, and reusing it for cleaning purposes, POSCO also receives 80,000 tons of treated water every day from the Pohand district—where it operates. The treatment and subsequent use of the treated water ensures a minimum water load on the community caused by the operations of POSCO. In addition, cooling water, sprinkling water, and rainwater are also treated and reused twice before final disposal. Similarly, the technology and operations center at Enbridge, located in Ontario Canada, also captures the rainwater from its rooftop to reuse it for landscape irrigation and in cooling towers.

4.2 Business and Operational Excellence

Business Structuring: Centrica's synergy of residential energy and services activities in North America not only led to the operational simplifications, but also the firm realized more effective and efficient sales channels with reduced headcount. While Centrica decided to dispose of its entire gas and wind farm portfolio (£147-million) along with two non-core businesses (£22-million) in 2015, Enbridge acquired 24.9% shares in 400 MW Rampion offshore wind project and 100% of shares in 103 MW Creek Wind Project in the same year. These investments allowed Enbridge to enhance its Green Energy Portfolio, but in 2016 Enbridge decided to merge with Spectra Energy to be the largest energy infrastructure company in North America ($126-billion). Such business restructuring decisions help the most sustainable organizations to fully realize their potential.

Risk Management: The above-mentioned acquisitions, divestments, and mergers are all part of business excellence: to ensure that 'business keeps making profit'. In order to be vigilant in such restructuring, POSCO works on a five-scenario model (S1 to S5)—from the best scenario (S1) to the worst scenario (S5). Before deciding to revamp the business structure, the firm analyzes all scenarios and develops rigorous plans to minimize the risk of S3 to S5.

Operations: Enbridge optimizes the energy consumption in its oil pumping operations through additives and increased oil temperature which helps in reducing friction and improving flow. Furthermore, pressure cycling is avoided to reduce stress on the pipeline system. In 2014, these two techniques together resulted in the saving of 59,900t $CO_2(e)$. Similarly, to start the drivers of its gas compressors, Enbridge makes use of compressed air instead of natural gas, which eliminates a source of additional emissions. On the other hand, POSCO pioneered the Smart Factory System which integrates traditional steel manufacturing methods with modern Information and Communications Technology. Besides connecting the manufacturing machinery through the Internet of Things and artificial intelligence, highly-tuned wearables are also connected to the Smart Factory System. This helps in operations management, quality control, and employee safety—all at one time.

4.3 Corporate Citizenship and Social Development

Community: In order to reduce the energy expenditure of socially disadvantaged homes, POSCO provided solar generators in the Gwangyang area. Similarly, Schneider provided micro-off grid solar powered facilities for villages in Senegal, which do not have access to electricity. The main purpose of these solar facilities is to provide uninterrupted electricity to schools, community centers, and health centers.

Empowerment: Centrica's Ignite program offers impact investment for entrepreneurs with innovative energy ideas which can benefit industry and society. POSCO also offers various assistance schemes for supporting local SMEs. For example, besides providing the low-interest loans, POSCO covers most of the R&D expenses of SMEs. In addition, POSCO provides free support to boost the technological competitiveness of SMEs. It also shares its patents with SMEs and provides full financial support for POSCO-SME co-applicant patents. Similarly, Schneider Electric Energy Access program aims to go beyond charity by empowering the ones who are at the base of the pyramid. The purpose of this program is to help create energy businesses, especially in the rural areas, and support entrepreneurs in innovative energy solutions. Schneider also offers loans for energy-related income-generating activities in the disadvantaged communities of Ethiopia and Tanzania. On the training side, BMW runs the 'Young Engineer Dream Project' to provide one year of coaching to disadvantaged students in vocational schools. The graduating students from this program can be employed at BMW or may look for other opportunities as well.

Donation and Funding: Schneider's BipBop Mutual Fund is a channel where employees can contribute to the social activities of the company, and Schneider matches the amount raised by its employees. Similarly, Enbridge offers funding to law enforcement agencies, firefighters, and other emergency related services under its Safe Community Program to help in purchasing new safety-related equipment, obtaining professional training, or delivering educational programs which can benefit society. POSCO also supports educational and training activities through its scholarship projects which provide funding for students from high school to Ph.D. program. In addition, POSCO 1% Sharing Foundation financially supports charitable activities through the donations made by its employees (1% salary every month).

Environment: Enbridge has been successfully fulfilling ambitious environmental commitments in order to ensure neutral footprints from its activities. These commitments include:

- planting a tree for every removed tree;
- preserving the equivalent area of natural habitat for every permanently altered acre; and
- matching the kilowatt hours of renewable energy with the kilowatt hours produced by conventional electricity (since 2008).

In order to achieve these optimistic commitments, Enbridge relies on the internal carbon pricing and carbon shadow-pricing methodology for corporate planning and in the investment decision-making processes. Schneider, on the other hand, uses eco-branding criteria, Green Premium, as a business strategy to demonstrate the environmental efficiency of its products. The French multinational uses social media channels as branding tools to highlight the achievements of the firm and maintain a green reputation among its stakeholders.

Volunteering: Enbridge encourages its employees to take a paid day (or eight hours) off per year to volunteer at any non-profit organization. Schneider, on the other hand, utilizes the expertise of its retired employees as professional trainers for the non-profit organizations. The volunteers receive appropriate recognition from Schneider for their voluntary services.

4.4   Research and Innovation

Technology: Research and innovation related to autonomous driving are treasured at BMW and Daimler. The German carmakers have integrated autonomous driving features in the upcoming commercial models, e.g., i3 and i8 of BMW. Daimler has also tested the possibility of truck platooning by digitally connecting a convoy of trucks. A telematics platform communicates between the infrastructure, the convoy, and other vehicles on the road to ensure extended safety, less space on highways, and less fuel consumption.

Similarly, Centrica offers hi-tech innovative products to compete in the utilities sector, e.g., the Hive products and My Energy Live, in collaboration with Amazon's Alexa Voice Services, allow the customers to remotely monitor (through smartphones) and control (heating, lighting and other devices simply by speaking) their household energy consumption in real time.

Services and Solutions: BMW and Daimler offer various innovative services, such as car2go and DriveNow, which allow customers to rent a car without queuing in rental offices; the customers can rent a car from their nearest locations and leave it in the Home Area (numerous parking locations in a city)—even for a one-way trip. Likewise, Centrica hosts an online platform, Local Heroes, for on-demand home services. Through Local Heroes, customers can book repair and other services online with certified local traders (with a service guarantee from Centrica).

Business Innovation: Innovation can go beyond the technological advancement of products. For example, POSCO's Solution-based Platform Business, a novel business model, allows the firm to commercialize its original technologies. On the other hand, Schneider offers Climate Bonds to enhance the spectrum of technology-oriented innovation programs which can enable its customers to achieve superior CO_2 savings. The bond, launched in 2015, raised a sum of 200-million euros to finance low-carbon innovation programs.

Support Programs: Centrica Innovations, a venture of Centrica group, aims to identify and align the group activities with promising R&D and innovation opportunities. The center also serves as an incubator for the external projects which require further support to reach an acceptable maturity level. Likewise, POSCO's Idea Marketplace Program supports venture companies from idea creation to business model planning, investment, and growth management. The two-way collaboration also helps POSCO in developing seed technologies at low cost.

4.5 Procurement, Supply Chain and Logistics

Sourcing: Procurement and supply chain are the weak links in organizational sustainability since the firms have only limited control over the activities of their suppliers. However, to ensure sustainable procurement, Schneider runs a Conflict Minerals Compliance Program which allows the French firm to participate in international programs, such as the Conflict-Free Smelter Initiative. Similarly, BMW has memberships of various international associations and commissions which assist the carmaker in sustainable extraction of steel, aluminum, and cobalt, and farming of natural fibers (kenaf).

Supplier Performance: BMW and Enbridge have established clear sustainability requirements for the suppliers to participate in the tendering process. At BMW, for example, it is a requirement that the potential suppliers with more than 100 employees should be certified with ISO14001, while the firms with more than 500 employees should regularly publish a 'sustainable value report' to communicate their sustainability understanding and performance. BMW also assigns scientific sustainability goals to its top 100 suppliers every year.

Similarly, Schneider has integrated sustainability goals in the supplier evaluation and performance assessment procedures, which are evaluated through third-party audits. Although third-party audits are time-consuming and costly for the contractors, Enbridge has found a promising solution to encourage its contractors to participate in the third-party audits. Enbridge is part of a pipeline firms' consortium, Facility Audit Network (FAN), which sets out common guidelines for the contractors' audit. The audit results are shared among the firms and the successful contractors qualify for the contracts in all five firms. FAN reduces the time and the cost associated with audit activity and encourages the contractors to perform better in order to be more competitive than the others.

Freights: As far as the logistics are concerned, all firms prefer sea freight over air freight since the former is more economical and environmentally friendly. In fact, Schneider has implemented a manager's approval workflow in certain regions which challenges the manager's decision of using air freight. The manager has to provide a detailed explanation of 'why air freight is preferred over sea freight'.

Vehicles and Commuting: The gas division of Enbridge converted 648 of 853 fleet vehicles to run on natural gas, which eventually resulted in an annual saving of 500t $CO_2(e)$. The firm further saved 14t $CO_2(e)$ in 2015 by simply encouraging teleconferences instead of business trips. Likewise, Daimler's employees receive discounted yearly passes from the firm for the public transit system in order to reduce the footprint from their day-to-day commuting.

Capacity Utilization: Daimler developed an IT-based container management system with the help of 17 plants and 5 logistic centers in order to optimize the transport of reusable shipping containers. The container management system uses reverse logistics to move the reusable containers in the same shipment which moved the goods initially. This helps Daimler in avoiding about 2,200t CO_2 emissions annually. BMW also came up with an innovative solution to the capacity utilization problem.

The German carmaker uses an on-site screw compactor to compress the recyclable packaging. This helps in increasing the shipping volume to the recycling facility and saves the firm roughly 24,000 truck kms.

4.6 Governance

Guidelines: Daimler follows the established German Corporate Governance guidelines to conduct the governing activities. The use of approved guidelines for governance helps the firm to comply with the legal requirements on one hand, and to integrate sustainability within the governance structure on the other. There are various other guidelines which the organizations under discussion follow, including OECD guidelines for multinational corporations, the universal declaration of human rights, UN guiding principles on business and human rights, and many others.

Policies and Programs: In general, the organizations under discussion form policies and run programs to deal with governance-related malpractices, including ethical misconduct, discrimination, harassment, human rights violation, corruption, fraud, bribery, embezzlement, conflict of interest, power abuse, and privacy breach. In addition, to ensure compliance with these policies, the firms usually establish means to support alerts from employees; the hotline and SMS based systems, provided by independent companies, allow the employees to report any problem with complete anonymity (Enbridge and Schneider). The reported incidents are assessed by relevant departments and a working plan is approved during the board meeting to prevent reoccurrence (POSCO).

Performance Incentives: At Schneider, all members of the Executive Committee are assigned sustainability-related goals for performance-based incentives. In fact, up to 20% of the variable salary of executives is calculated based on sustainability-related performance, i.e., reduction of emissions, safety in operations, and sustainability in global supply chain, among others.

Diversity: In order to promote workplace diversity and inclusivity, Centrica runs various employee networks, including Women, Carers, Parents and Lesbian, Gay, Bi-Sexual and Transgender (LGBT) networks. In fact, Business in the Community (BITC), a business-community outreach charity, recognized Centrica's efforts for gender and race inclusion with a Bronze award. On the other hand, Daimler runs an International Assignment Program to promote diversity in the firm; employees from one location are assigned responsibilities to another, which results in a two-way flow of diverse sustainability practices.

Development of Culture: In 2016, Daimler entered into the Leadership 2020 Program, in which 144 employees from 24 countries were selected to develop proposals for new management culture at Daimler. The group of eight teams was not given any specific instructions for this task in order for them to be fearless in their thinking. The teams came up with structured processes and procedures in eight thematic areas which they believed could transform the management culture at Daimler (Fig. 8). The German carmaker started to roll out the suggested changes at the end of 2016.

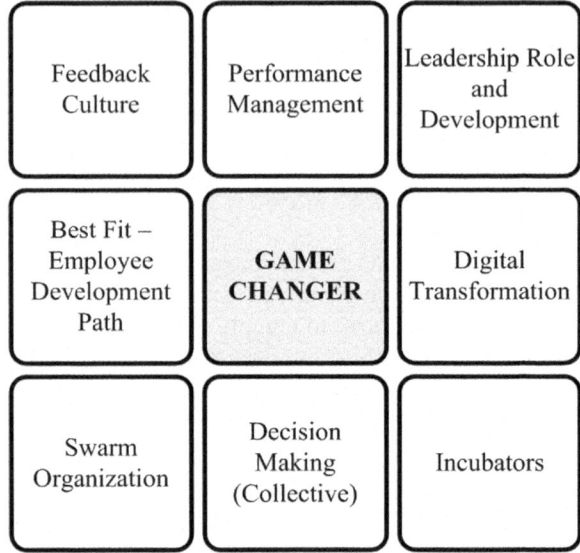

Fig. 8 Eight themes of Daimler's leadership 2020—game changer initiative

For Schneider, the commitment for the development of sustainability culture goes beyond organizational boundaries. Schneider started an Energy Leaders' Education Initiative to develop sustainability culture among societies. Under this program, Schneider aims to: (i) educate school children about energy conservation; (ii) host competition focused on energy management in homes; and (iii) inspire Green innovation.

Fairness in Disclosure: Besides being an ethical obligation, fairness in disclosure highlights a firm's confidence concerning its development roadmap. In this attempt, POSCO voluntarily shares the real-time air and water pollution data from two of its sites with the local government through a Tele Monitoring System. In addition, POSCO dedicates a part of its sustainability report to discuss media coverage (both positive and negative) related to its operations. While this unprecedented practice allows the firm to directly address the negative media coverage, it also offers an opportunity to reassess the impact of operations.

Political Donations and Dialogues: Daimler and BMW both favor political donations, dialogue, and lobbying, as long as these activities are carried out and registered in an ethically correct fashion. In 2016, Daimler supported various democratic parties during the German elections with a total of €320,000 in donations. This allowed the firm to instigate a political debate and negotiate trade agreements and emission regulations with the (to be) government.

Certified Tools	**Non-certified Tools**	**Other Management Programs**
• Quality MS (ISO 9001) • Environmental MS (ISO 14001) • Safety MS (OHSAS 18001) • Social Responsibility (ISO 26000) • Energy MS (ISO 50001) • Information Security MS (ISO 270001) • Life Cycle Assessment (ISO 14040/44) • Guidelines for incorporating eco-design (ISO 14006) • EU Eco-Management and Audit Scheme (EMAS) • Green Building (LEED) • Sector Specific Standardized Guidelines (e.g., ISO/TS 16949:2009, ISO/TC 17, ISO/TR 14062:2002, EN50581 etc.)	• Policy MS • Idea MS • Enterprise Content MS • Performance MS • Data MS • Emergency and Security MS • Risk MS • Transport MS • Freight MS • Carbon MS • Water MS • GHG Target MS • Six Sigma • Lean Manufacturing	• Stakeholder Management • Enterprise Risk Management • Operational Risk Management • Project Management • Crisis Management • Supply Chain Management • Supplier Relationship Management • Customer Relationship Management • Industrial Hygiene Management • Stress Management **Disclosure** • GRI • CDP

Fig. 9 List of essential management tools employed by the firms under discussion

4.7 Sustainability Management Tools

While governance decisions provide broad guidelines at the strategic level, the operation managers require appropriate tools to execute the strategies and monitor and report the related activities. To list a comprehensive set of sustainability management tools is not possible in a limited chapter; however, a representative list is shown in Fig. 9.

It is important to note that the application of these tools does not necessarily extend to the group-wide operations; not all sites are certified on these standards or comply with these programs. Nevertheless, in most cases, not only the organizations use these tools to manage sustainability in their operations, suppliers and contractors are also encouraged to improve their performance through these techniques. For example, along with various sites owned by Schneider, 150 suppliers (7.8% of all) also comply with the requirements of ISO50001.

4.8 Employee Relations

Similar to the governance-related initiatives, there are numerous employee-related initiatives which organizations roll-out to ensure good working relations with and between the employees, including training and development, right to bargain, grievance mechanisms, alternative work arrangements, flexible working hours, health recovery programs, wages and benefits, leave of absence and many others.

Engagement: Schneider engages with its employees via an internal social media network, Spice. The platform offers Schneider an opportunity to establish a continuous two-way communication with its employees, and also serves as a channel to develop internal bonds among employees. Schneider hosts various polls on Spice to get instant feedback from employees on the policies and programs of the company.

Similarly, Schneider's Cool Site program enhances employee engagement in the workplace by developing and transforming the office space into an attractive, inspiring, distinctive, innovative and energizing workplace which can boost creativity and collaboration among employees. After completion, the Cool Sites undergo an internal certification audit, and annual recertification audit—after the first certification, to ensure that the site conforms with the mission of the firm. Schneider believes that this initiative has made a huge difference to the work life of employees.

Development and Training: Enbridge allows its employees to develop skills and expertise based on their own career choices and interests. Therefore, every employee, in consultation with their assigned managers, creates an Individual Development Plan for personalized competence development based on their career choice and interests.

Besides the material used for training, the method of delivery is also important for the effectiveness of the trainings. Daimler has opted for a unique method of training delivery. The carmaker has developed an online game, Monster Mission, to train its employees on issues pertaining to work-integrity. Players (employees) are confronted with real-life situations and they have to make correct decisions in order to sustain the success of their virtual company. Daimler permits employees to play this game even during working hours. The company believes that the use of this game element has significantly improved the acceptance of the training material.

POSCO also designed interactive safety training by using state-of-the-art 4D technology. Disaster cases are simulated in a 4D experience and employees are trained on how to manage these situations in real time.

Satisfaction: BMW and Daimler take various steps to ensure a high level of employee satisfaction, including flexible working hours, mobile working (depending on the nature of the job), a daycare center for employees with children, task sharing (two employees working on the same task to meet the deadline), and others. Nevertheless, Daimler's 'mail on holiday' initiative is worth noting in this regard. Employees on holiday do not receive emails sent to them, instead, the correspondents are notified about the alternative point-of-contact, while the originally sent messages are permanently deleted. The initiative has been greatly welcomed by the employees, and perhaps it is for this reason that in the same year that 'mail on holiday' was launched, Daimler was ranked among the top 33% companies in the world with respect to employee satisfaction (Employee Commitment Index).

On a slightly different level, POSCO maintains an employee welfare fund from the company's profit to help employees in times of need. Interest-free loans are offered to employees for housing and living, tuition fees for children, family events, and to support handicapped family members.

Retention: Daimler's "Safeguarding the Future of Daimler" agreement with the employee-associations protects the job security of employees in Germany; it safeguards all of the employees of Daimler in Germany from being laid off until the end of 2020. Enbridge, on the other hand, recognizes the importance of talent retention through a Talent Review and Succession Planning process. The firm manages the risk of losing critical talent by offering variable compensation, rewards, and targeted career growth to certain talented employees.

5 Discussion

This research has explored organizational sustainability themes and their corresponding functional areas as well as the best practices of the systematically identified six most sustainable firms. The summary of the eight organizational sustainability themes is presented in Fig. 10. However, it is important to mention that there are at least four other sustainability themes which could not be discussed in this work due to space limitations, including (i) health, wellness, safety and security; (ii) stakeholder engagement; (iii) collaboration and partnership; and (iv) organizational reputation.

The findings of this work can be used to address the needs of integrating practical sustainability approaches in the theoretical sustainability management models, as highlighted by Nawaz and Koç (2018). The sustainability management models available in the literature largely lack in the practical dimension of sustainability, which is why these models are not considered sound from an applied perspective. Integration of sustainability practices with the theoretical models will benefit the latter in two ways. First, this will lead to a better and broader understanding of the practical limits of organizations in managing sustainability, which will subsequently enable researchers to address these issues through the design of their models. Secondly, this will bridge the gap between theory and practice, and hence increase the acceptability of the sustainability models among practitioners.

Organizational managers, corporate change agents, and policy makers can also benefit from the findings of this work to evaluate the sustainability potential of their organizations. The sustainability best-practices noted in this study can be replicated to excel other organizations. However, this may require an operationalization tool. In the regard, Nawaz and Koç (2018) developed an operational tool based on an extensive review of sustainability management and assessment literature, and the international standardized guidelines. The authors proposed a six-element cyclic structure for organizational sustainability management, namely the Sustainability Management Framework (SMF). The structure of SMF compliments the contents proposed in this work. The integration of the two (SMF and the content-framework) will result in a holistic tool for sustainability management in organizations.

6 Conclusion

While organizational sustainability has gained importance among the researchers over the years, the actual organizational sustainability best practices have largely been ignored in the literature. There are only a few studies which explore the sustainability themes in organizational setting, but these too have methodological restrictions. In order to overcome these shortcomings, we have conducted this study to review the organizational reports of the systematically identified six most sustainable firms. The goal of this study was to collect the sustainability themes and the corresponding best practices of the organizations in an objective way.

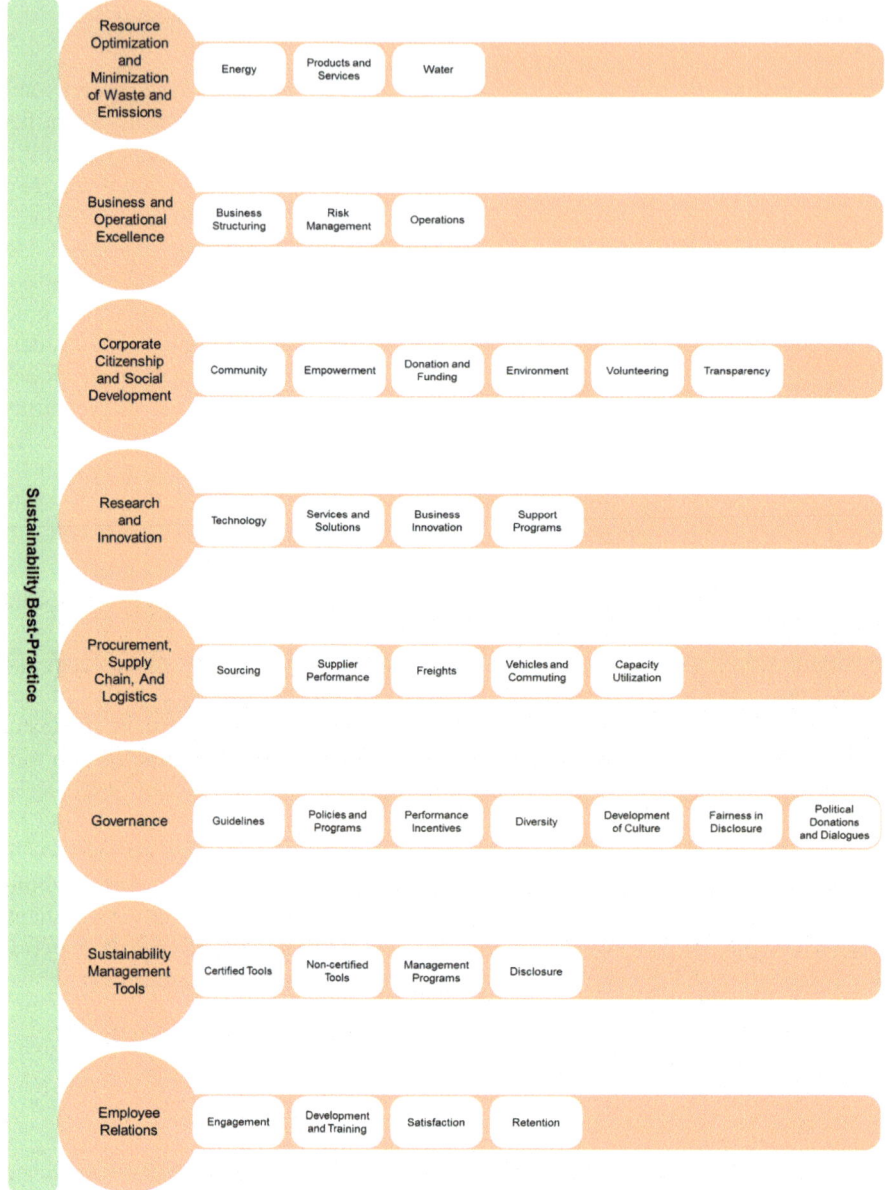

Fig. 10 Summary of organizational sustainability themes and the corresponding functional areas

Based on the review of 28 organizational reports, which comprise of more than 4000 pages, we have identified eight organizational sustainability themes and 36 corresponding functional areas which enable the most sustainable firms to achieve sustainability excellence (Fig. 10). However, there are at least four other themes which could not be discussed in this study due to space limitations, as highlighted in the Discussion section.

The present work is unique from previous studies in three ways: (1) the present work is an attempt to capture the potential reasons behind the performance of the most sustainable firms, which was limited to indicator-based comparisons in the previous works; (2) the study reflects on the evolution of the firms by reviewing best practices over a five-year period, which was mostly limited to a single year in the past, and (3) practices at micro-level are considered instead of restricting the review to strategic initiatives.

As far as the limitations of this study are concerned, the selection criteria devised to screen the sample (organizations and the respective organizational reports) may have eliminated some of the organizations from the review which, if included, would have benefitted this study. Furthermore, the practices highlighted in this study are not absolutely complete; these are representative of the many other sustainability practices of these organizations. Moreover, the sustainability practices noted in this article with reference to one organization may not be limited to that organization only; other organizations may also be undertaking the same, or similar activities, but these have not been referred to in this article due to space limitations.

The authors do not imply at any point in this study that the discussed sustainability practices are a certain cause of the organizational performance or its position in the ranking. In contrast, this work only identifies the practices which could 'potentially' be linked with the sustainability performance of organizations. In order to be certain in associating particular practices or initiatives to the organizational performance, a comprehensive causality analysis will be required, which was not included in the scope of this study. Also, while the authors ensured objectivity in each phase of the review process through systemization (selection of organizations, identification of reports to be reviewed, and collection of sustainability practices), the possibility of human error cannot be completely ruled out. Therefore, it is enormously important to validate the findings of this study before using the sustainability themes for decision making and in sustainability research.

Acknowledgement Many thanks are directed to Mr. Owen Connor (of AWARE center at Hamad bin Khalifa University) for his help in improving the writing style, language, grammar, spellings and punctuations of this paper. This research did not receive any specific grants from funding agencies in the public, commercial, or not-for-profit sectors.

References

Adams CA, Larrinaga-González C (2007) Engaging with organisations in pursuit of improved sustainability accounting and performance. Acc Audit Account J 20:333–355. https://doi.org/10.1108/09513570710748535

Bonini S, Görner S (2011) The business of sustainability: putting it into practice. McKinsey & Company

Collins E, Roper J, Lawrence S (2010) Sustainability practices: trends in New Zealand businesses. Bus Strateg Environ 19:479–494. https://doi.org/10.1002/bse.653

Corporate Knights (2017) Global 100—most sustainable corporations in the world

Epstein MJ (2008) Making sustainability work: best practices in managing and measuring corporate social, environmental and economic impacts, 1st edn. Routledge, Abingdon

Eweje G (2011) A shift in corporate practice? Facil Sustain Strateg Co 136:125–136

Hoepner AGF, Yu P-S, Ferguson J (2008) Corporate social responsibility across industries: when can who do well by doing good? Soc Sci Res Netw 38 http://dx.doi.org/10.2139/ssrn.1284703

Hong P, Roh JJ, Rawski G (2012) Benchmarking sustainability practices: evidence from manufacturing firms. Benchmarking Int J 19:634–648. https://doi.org/10.1108/MRR-09-2015-0216

Knoepfel I (2001) Dow jones sustainability group index: a global benchmark for corporate sustainability. Corp Environ Strat 8:6–15. https://doi.org/10.1016/S1066-7938(00)00089-0

Livesey SM, Kearins K (2002) Transparent and caring corporations? Organ Environ 15:233–258. https://doi.org/10.1177/1086026602153001

López MV, Garcia A, Rodriguez L (2007) Sustainable development and corporate performance: a study based on the Dow Jones sustainability index. J Bus Ethics 75:285–300. https://doi.org/10.1007/s10551-006-9253-8

Lozano R (2012) Towards better embedding sustainability into companies' systems: an analysis of voluntary corporate initiatives. J Clean Prod 25:14–26. https://doi.org/10.1016/j.jclepro.2011.11.060

Nawaz W, Koç M (2018) Development of a systematic framework for sustainability management of organizations. J Clean Prod 171:1255–1274

Rahdari AH, Rostamy AAA (2015) Designing a general set of sustainability indicators at the corporate level. J Clean Prod 108:757–771. https://doi.org/10.1016/j.jclepro.2015.05.108

Rondinelli DA (2006) Globalization of sustainable development: principles and practices in transnational corporations (No. 023–07/08). Working Paper Series 2007–2008

Schalock RL, Verdugo M, Lee T (2016) A systematic approach to an organization's sustainability. Eval Prog Plann 56:56–63. https://doi.org/10.1016/j.evalprogplan.2016.03.005

Schneider A, Meins E (2012) Two dimensions of corporate sustainability assessment: towards a comprehensive framework. Bus Strateg Environ 21:211–222. https://doi.org/10.1002/bse.726

Sealy I, Wehrmeyer W, France C, Leach M (2010) Sustainable development management systems in global business organizations. Manag Res Rev 33:1083–1096. https://doi.org/10.1108/01409171011085912

Small LB (2017) Sustainability practices that influence profitability in the petroleum industry. Walden University, Minneapolis. https://doi.org/10.1080/13562517.2013.860105

Stewart R, Bey N, Boks C (2016) Exploration of the barriers to implementing different types of sustainability approaches. In: 23rd CIRP Conference on Life Cycle Engineering. Elsevier B.V, pp 22–27. https://doi.org/10.1016/j.procir.2016.04.063

Székely F, Knirsch M (2005) Responsible leadership and corporate social responsibility: metrics for sustainable performance. Eur Manag J 23:628–647. https://doi.org/10.1016/j.emj.2005.10.009

Vermorken M, Szafarz A, Pirotte H (2008) Sector classification through non-Gaussian similarity (No. 08/032)

Williams FE (2015) Green giants: how smart companies turn sustainability into billion-dollar businesses. AMACOM
Windolph SE (2011) Assessing corporate sustainability through ratings: challenges and their causes. J Environ Sustain 1:1–22. https://doi.org/10.14448/jes.01.0005

Plastic Bag Ban in the Context of Corporate Social Responsibility: Consumption and Trade vis-a'-vis Environmental Sustainability Concerns

Irina Safitri Zen

Abstract A corporate social responsibility (CSR) initiative by supermarkets have made significant contributions to the elimination of plastic bags consumption, hence establishing a new norm in the society. This means that the various initiatives conducted in promoting a sustainable shopping lifestyle have challenged the 'status quo' of providing free single-use plastic bags for consumers' convenience. Moreover, pressure from global environmental challenges for plastic pollution and the multifaceted global trade environment affect the corporate sector's operation. This paper analyses the multifaceted involvement of the CSR initiatives of supermarkets in a global movement on the anti-plastic bag ban by examining Malaysia's No Plastic Bag Campaign (NPBC) as a case study. The studies have deployed the Carroll's global CSR pyramid framework to defragment the complexities of the supermarkets' CSR initiatives by using content analysis. Furthermore, the result reinforced by the descriptive analysis of the consumer's survey and revealed several competitive advantages of the corporate sector's involvement in anti-plastic bag initiatives. These advantages include elaborating on the concept of supermarket CSR in the context of the plastic bag ban, detangling the multifaceted discussion around supermarket CSR, provide platform to nudge sustainable shopping behaviour lifestyle practices and clarifying the crucial role of supermarket CSR in the governance of the plastic bag ban from the business sector's perspective. Finally, by deploying the global CSR pyramid framework, the study outlined the CSR initiatives of supermarkets in terms of the anti-plastic bag campaign from four key responsibilities; economic, ethical, legislative and philanthropy. Hence, the involvement of the CSR supermarket initiative's contribution to a global movement of anti-plastic bag ban is crucial.

Keywords Corporate social responsibility, CSR · Plastic bag · Supermarket · Environment · Consumption

I. S. Zen (✉)
International Islamic University Malaysia (IIUM), Jalan Gombak, 53100 Kuala Lumpur, Federal Territory of Kuala Lumpur, Malaysia
e-mail: irinazen@iium.edu.my

© Springer Nature Switzerland AG 2020
W. Leal Filho et al. (eds.), *International Business, Trade and Institutional Sustainability*, World Sustainability Series,
https://doi.org/10.1007/978-3-030-26759-9_3

1 Introduction

Plastic bag pollution is a global environmental challenge (Derraik 2002; Jambeck et al. 2015) due to its detrimental effects on the environment (Adane and Muleta 2011) as marine plastic pollution is transboundary (Vince and Hardesty 2017). Thus, it brings rise to complexities in governing the post product consumption (Dauvergne 2010) where solutions for plastic-related pollutions are complex and require many actors at various scales (Loorbach 2002; Loorbach 2010; Zen et al. 2017). Moreover, plastic bag pollution is concerned with the issue-based approach of nature sustainability science (Kates 2017) which needs to be solved in an inter-disciplinary approach (Vince and Hardesty 2018). Among the issues championed by this approach is creating a sense of responsibility to the polluter either by using 'carrot and stick' incentives or 'charges or tax' put forward by the environmental economics discipline (Pearce and Turner 1992), campaigning for behavioural changes and social value from behavioural discipline (Rotmans et al. 2004), and most recently, supporting the emerging concept of the citizen-consumer where consumers become an agent of change in the modern sustainable consumption context (Spaargaren and Oosterveer 2010). Governing the new modern society is no longer the same. Despite governance challenges about the plastic bag ban because of the lack of a global treaty to ban plastic bags (Clapp and Swanston 2009), other mechanisms to accelerate the consumer's behavioral changes is necessary. Hence, new stewardship functions are manifested in the CSR initiatives of companies to facilitate consumer behavioral changes have the potential to be explored.

The emergence of the anti-plastic shopping bag and associated regulatory policies represent global support of the new environmental norm (Clapp and Swanston 2009; Sharp et al. 2010), where various mechanisms by actors are needed. This norm is supported by emerging studies related to the eradication of plastic bags. The crucial role of supermarkets in shaping consumer's behavior in using single-use plastic shopping bag in Australia (Sharp et al. 2010), Malaysia (Richards and Zen 2016) and Argentina (Jakovcevic et al. 2014). The implementation of plastic bag tax influenced the decreased use of plastic bag waste in Ireland (Dikgang and Visser 2010; Convery et al. 2007), Israel (Ayalon et al. 2009). The non regulatory approach such as imposing plastic fee in City of New York (Akulian et al. 2007) and plastic bag charges in several states in Malaysia (Zen et al. 2013) and the regulatory approach thorugh the creation of plastic bag legislation in South Africa (Hasson et al. 2007).

Imposing plastic bag charges has been proven to be able to reduce the use of plastic bags by consumers (Sharp et al. 2010; Dikgang et al. 2012; Poortinga et al. 2013), and it is a part of a crucial role of government action against plastic bag consumption (Miller et al. 2012). Most of the initiatives help to enhance the establishment of a new environmental norm at a societal level. Nevertheless, it establishes and strengthens the legal aspect of the anti-plastic shopping bag initiative, which sends the right signals to the stakeholders involved, including supermarkets, in establishing the new social norm. However, the involvement of supermarkets in this effort, especially from the perspective of the CSR, is less studied.

From the corporate organization's perspective, the establishment of laws and regulations on the anti-plastic bag initiative reflects the new codified ethics from the society at large. It articulates fundamental notions of fair business practices as established by lawmakers at the federal, state and local levels (Carroll 2016). In this case, various acts, regulations or bills at the country, state or city level and the establishment of anti-plastic bag initiatives relates with the legal aspect of a company's CSR. Hence, it matches with the CSR pyramid framework developed by Carroll (1991, 2004, 2016) which outlines the corporate involvement into four main dimensions of responsibilities: economic, ethical, legal and philanthropic dimensions. Furthermore, this legal aspect is closely related to ethical responsibility. For organizations to embrace initiatives that contribute to norms, standards, and practices that are not codified into law, the CSR of companies becomes crucial.

At the bottom line, it is fundamental for businesses to exercise their economic responsibility to society. Society views business organizations as institutions that will produce and sell the goods and services it needs and desires. Although it may seem unusual to think about an economic expectation as social responsibility (Carroll 2016), the interdependence between corporations and society due to their direct and indirect impact to the society cannot be denied (Porter and Kramer 2006). For example, McDonald's replaced its clamshell packaging with waxed paper because of the increased consumer concern relating to polystyrene production and Ozone depletion (Gifford 1991; Hume 1991). Xerox introduced a "high quality" recycled photocopier paper in an attempt to satisfy the demands of firms for less environmentally harmful products. Companies that sell harmful products such as plastic bags may in the long-term, affect the benefit of the company due to its indirect impact on the consumers and its direct and detrimental impact on the environment. More so in the case of plastic bags which have been claimed as a global environmental challenge (Derraik 2002; Adane and Muleta 2011; Jambeck et al. 2015).

The establishment of a new social norm on the anti-plastic bag effort and the announcement of plastic bags being a global environmental challenge creates an unpleasant situation for the company. It may affect their long-term profitability due to the global movement on the plastic bag ban which will directly affect their corporate image. Due to that, it is crucial to study the extent of the involvement of supermarket CSR in facilitating the reduction of plastic bag pollution. How the CSR of supermarkets creates long-term benefits from their involvement in educating the public about Bring Your Own Bag, BYOB campaign, is not captured in many studies. From the corporate strategic perspective, supermarkets' CSR is concerned with the environment and being involved in the campaign establishes a new social norm and a sustainable lifestyle. This is achieved through their involvement in campaigns like the No Plastic Bag Campaign (Sharp et al. 2010; Zen et al. 2013; Jakovcevic et al. 2014) which may increase their corporates competitiveness. These points will be discussed further in this study.

With the current global challenges, the issues of image and reputation become increasingly important in a corporation's success. By performing the CSR programme, corporations can increase stakeholders' awareness about corporate ethical, social and environmental behavior, direct stakeholder and investors pressure, and peer pressure. CSR also creates an increased sense of social responsibility (Ernst and Young 2002), creates, develops, and sustains differentiated brand names as well as increases recognition in new and emerging forms of governance (Girod and Bryane 2003; Commission of the European Communities 2002), increases their competitiveness and creates innovative and creative solution (Commission of the European Communities 2002; Ernst and Young 2002; Girod and Bryane 2003; Porter and Kramer 2006).

Taking into consideration the extensive aspects of the CSR activities, the key issues of CSR are often clustered into four main groups: marketplace, workplace, environment, and community (Whooley 2004). It is argued that the marketplace presents a risky platform where CSR is performed using their reputation through the products and services they offer. Under workplace, the issues are related with the worker's wellbeing, while under the environment, it is where the external activity exists beyond the company territory and furthermore, the CSR activity is close to the community. According to the Commission of the European Communities (2001, 2002), the issues can be categorized based on the internal and external dimensions. On the internal side, more focus is given to socially responsible practices within the company while the latter extends beyond the company into the local community and the world beyond and involves a wide range of external stakeholders. The CSR aspects reflect the versatile yet integrative nature of CSR activities in the business environment.

In summary, the positive effect of CSR will cut operating costs by reducing ecological inefficiencies, enhancing corporate legitimacy and social responsibility, creating a competitive advantage among "green" or environmentally conscious consumers, enhancing the corporate image through environmental responsiveness, reducing the production risks associated with resource depletion, reducing energy costs, improving pollution management practices, mitigating health risks to the local community, and positioning the organization ahead of the regulatory curve (Shrivastava 1995; Bansal and Roth 2000).

This study explores the various dimensions of the CSR of supermarket programmes in the context of NPBC in Malaysia to identify to what extent it reshapes the green market and creates an enabling environment for the anti-plastic bag consumption movement.

2 Taking Stock of the CSR Initiatives of Supermarkets in the Context of Plastic Bag Ban

Earlier concepts of CSR defined by Bowen (1953) is limited to a company or corporate obligation in giving back to society as the consequences of their business activities. More complex definitions include several dimensions of CSR activities such as the legal requirements, technical aspects and economic aspects of a firm (Davis 1973). Hence, a more comprehensive CSR definition refers to "consideration of, and response to, issues beyond the narrow economic, technical, and legal requirements of the firm to accomplish social benefits along with the traditional economic gains which the firm seeks" (Davis 1973; Carroll 1999), where the CSR firm should "strive to make a profit, obey the law, be ethical, and be a good corporate citizen" (Carroll 1991: 43), "where there is a need for organizations to consider the good of the wider communities, local and global, within which they exist in terms of the economic, legal, ethical and philanthropic impact of their way of conducting business and the activities they undertake" (Whooley 2004, Commission of the European Communities (2001, 2002). The CSR concept in the context of business for social responsibility is defined as "achieving commercial success in ways that honor ethical values and respect people, communities, and the natural environment" (www.bsr.org). The CSR concept evolves from a simple effort of giving back to the society into more complex activities that need consideration of economics, business ethics, legal requirements, corporate citizenship and corporate accountability in the CSR activity.

Moreover, six major types of CSR initiatives identified by Kotler and Lee (2005) expand the corporate CSR implementations covering the: (i) marketing of CSR activities, (ii) types of approach on CSR initiatives, (iii) selection of the initiative that will do the most good for the social issue as well as the corporation, (iv) development of the CSR programmes, (v) implementation of the successful programme plans, and (vi) evaluation of the programme efforts. The multidimensional aspects of CSR relate to its dynamic interactions across major stakeholders which include consumers, employees, owners, the community, the government, competitors, and the natural environment. Hence, in the stakeholder's theory where supermarket CSR needs to consider the interest of all parties, this includes the consumer who could affect their actions and initiatives (Lantos 2001). The ability to manage a good relationship among various stakeholders and consumers, and meet their needs and demands, influence companies' long-term success (Branco and Rodrigues 2007). The implication of the supermarket's involvement in plastic bag reduction becomes more crucial due to the nature of their business. This provides more potential for corporations to play significant roles, especially in shaping the consumers' behavior.

The adoption of pyramid CSR framework developed by Carroll (1991, 2004, 2016) is used to analyze the involvement of the CSR of supermarkets in the anti-plastic bag effort and the consumer's readiness to accept the supermarket's CSR involvement in the eradication of plastic bags. The analysis will be concerned with the extent to which the plastic bag initiative is executed according to the four aspects of the CSR pyramid which are the Economic, Legal, Ethical and Philanthropic Responsibilities. **First**, the Economic Responsibilities is defined as being profitable as the foundation upon which all others rest. It is how the supermarket's CSR involvement in the no plastic bag campaign covers its economic responsibilities. **Second**, Legal Responsibilities include obeying the law in which the society's codification of right and wrong is executed lawfully. **Third**, Ethical Responsibilities is to be ethical, where there is an obligation to do what is right, just and fair, as well as to avoid harm, and **Fourth**, Philanthropic Responsibilities is to be a good corporate citizen. Contributing resources to the community to improve the quality of life.

Several studies use the CSR pyramid framework to investigate consumers' readiness to support socially responsible organizations (Maignan 2001; Ramasamy and Yeung 2009) and the consumers' perception of CSR (Ramasamy and Yeung 2009). Research on consumer perception and response toward CSR suggests that there are positive relationships between a company's CSR activity and the consumers' reactions to that company and its products (Beckman 2007). Consumers are willing to pay a higher price for that firm's products (Creyer and Ross 1997; Handelman and Arnold 1999) if they sense that the company is genuinely committed to the campaign. However, how consumers perceive supermarket CSR in the "no plastic" bag campaign and its relation to a bigger society, has not been studied.

Although most of the CSR activities serve as the external drivers to the corporations, they have the potential to be integrated into the corporation's strategic initiative for long-term benefits. Internalizing environmental sustainability into the organization's performance reporting systems, such as Balanced Scorecards, will create a measurable objective for various dimensions of environmental performance (Stringer 2009). CSR helps in identifying relevant stakeholders and guiding organizations in incorporating relevant environmental performance measures and strategies (Epstein and Roy 2001), as well as to create commonly shared values (Porter and Kramer 2006). However, how CSR plays its roles needs to be carefully identified (Epstein and Roy 2003). The CSR of supermarkets must disclose the right amount of information to the public (Girod and Bryane (2003) as it can potentially increase the stakeholder's confidence (Fernández and Souto 2009; Epstein 2018). How far the consumers perception on the CSR of supermarkets initiative on the "no plastic bag" and what their role is, needs to be analyzed further.

CSR's concerns about related environmental issues in Malaysia have begun in early 2000 due to the concern raised by non-governmental organizations (NGOs). Among the environmental issues that had initiated action were environmental pollution, health-risking products, product safety, discrimination against the handicapped and drug abuse (Abdul Rashid and Ibrahim 2002). The NGOs' [Sahabat Alam Malaysia, SAM, Federation of Malaysian Consumers Association, FOMCA, and Consumers Association of Penang (CAP)] involvement, have shown how these pressure groups influence the legislative framework related to the environmental issues. Their approach provides a classic example of CSR programmes which are not part of the corporate operations. Hence, this study will explore how supermarkets' involvement in the "no plastic bag" campaign will affect their operation.

At the organizational level, the CSR studies conducted in Malaysia involves perception, roles, and the potential expansion of CSR. They also involve perception of corporate social involvement, social reporting and social performance (Teoh and Gregory Thong 1986), managers' attitude towards CSR (Abdul Rashid and Abdullah 1991), and factors determining executives and managers' attitudes towards CSR (Abdul Rashid and Ibrahim 2002) and socially responsible activities that can provide a favorable public image (Abdul Rashid and Ibrahim 2002).

The study is an early effort to provide insight into the consumer regarding the business sector's involvement in the NPBC in Malaysia, viewed from the CSR perspective. Afroz et al. (2016) studied the people's perception of NPBC in Selangor about the factors that motivate plastic recycling. It was found out that the factor of "helping to reduce landfill use" was the most important factor while the factor of "raising money for charity" was the least important factor that motivated households to participate in recycling. The study suggested some strategies that could help the government's efforts to boost consumers' behavioral changes in the plastic bag ban but not from the corporate strategic initiative perspective. To avoid the public stigmatism of the supermarkets' involvement by charging for the plastic bags given to consumers, the campaign was announced as part of their CSR.

The formal announcement of the Nationwide plastic bag campaign in 2011 was voluntarily signed between the Ministry and supermarket chains using a letter of agreement, LA (Zen et al. 2013). The voluntary plastic bag ban such as, 'No Free Plastic Bag' started on 6th July 2009, initiated by the Penang State and involved the supermarket chains as the implementers. The NPBC by Miri City Council in Sarawak was announced and initiated on 6th September 2009 while the Selangor State of NPBD started in December 2010. They had a formal voluntary signing agreement with the supermarkets and was a part of the supermarkets' CSR. However, the extent of the public's perception about this activity as a part of the supermarkets' CSR programme and how the CSR initiative of supermarket plays an important role in multiple dimensions within a complex sustainable consumption have yet to be deeply studied.

The involvement of supermarket CSR in educating consumers to eliminate the use of plastic bags is not only found in Malaysia but also other countries. In Brunei, Friday to Sunday is slated as the "No Plastic Weekend." This initiative was supported by the government by distributing 10,000 reusable bags to the public and by the involvement of civil servants under Department of Environment, Parks, and Recreation (JASTRE), Ministry of Development (Brunei Time 2012). A Memorandum of Understanding (MoU) between the Ministries and eighteen (18) major retail stores in 2011 grew to 62 retail stores in 2012.

The creative approach of the supermarkets' CSR initiative involves stimulating consumers' sensitivity and altruism. For example, plastic bag reduction programmes and activities conducted in Singapore combines the elimination of plastic bag usage and encourages cans or polyethylene terephthalate, PET bottles to be recycled through a FairPrice's "Love Nature" scheme. Under that scheme, for approximately $10,6 million plastic bags were saved, and the donations of up to $30,000 were given to the Straits Times School Pocket Money Fund and the Singapore Disability Sports Council by the Fair Trade Foundation. About more than one million plastic bags were handed out each day in Singapore (Kaur 2002). Therefore, creating a stimulating programme that changes the consumers' behavior in reducing the plastic bag on site where the CSR supermarkets are involved is crucial to determine the success of the government's initiative to ban the plastic bag.

The global emergence of the new norm on plastic bag provision, as well as the global sustainability governance on plastic bags may create competitive advantages for the corporate involvement in supermarket CSR concerning the plastic bag ban. Globally, the social responsibility is well structured in the four components of the global CSR pyramid; philanthropic, ethical, legal and economic (Carroll 1991, 2004,

2016) where the four are empirically interrelated, but conceptually regarded as independent components of corporate social responsibility. Henceforth, assessing the supermarket involvement in the global CSR pyramid strategically strengthens their contribution to global environmental challenge of plastic pollution.

3 Methodology

In general, the study was constructed in two parts. The first part was the adoption and utilization of the Framework Analysis as the overall framework to analyze the complexities of the issues around the supermarket CSR involvement in the anti-plastic bag initiative in Malaysia. In the second part, the consumers' perceptions towards CSR and supermarkets' involvement in the "no plastic bag" campaign initiative from the survey results were discussed. Content analysis was performed to analyze the framework analysis, reaffirmed by the descriptive analysis from the consumer's survey using the Statistical Package for Social Science, SPSS version 13.0.

3.1 Framework Analysis

The framework analysis adopted was intended to familiarize and contextualize the issues related to the "no plastic bag" campaign in the context of the supermarket CSR. It was adapted to research that has specific questions, a limited time frame, and a pre-designed sample, i.e. professional participants and priori issues, i.e. organizational and integrational issues that need to be dealt with (Srivastava and Thomson 2009). The tool aims to describe and interpret what is happening in a particular setting for qualitative data in a policy research study (Ritchie and Spencer 1994). Here, the framework analysis is deployed to address the complex interrelated and multifaceted issues of NPBC, the supermarket CSR, consumers, and public policy. The overall study adopts the following steps:

Step 1: Familiarization and Contextualisation

This stage involves the content analysis of various issues related to the "no plastic bag" campaign initiatives.

Step 2: Review and Identification

A Thematic Framework which suits the CSR of the supermarket is reviewed and identified. Therefore, the global CSR pyramid framework developed by Carroll (1991, 1999, 2004) was employed as a basis of analysis. Four dimensions of responsibilities: the economic, legal, ethical and philanthropic dimensions are identified to suit the context of the "no plastic bag" campaign which involved the supermarkets. Besides, this stage also reviewed several definitions of CSR and the four dimensions of responsibilities previously mentioned. For this part, data will be collected from secondary sources such as literature review and newspapers on the issues involving NPBC.

Step 3: Development of a Definition

The definition for supermarket CSR in the context of the "no plastic bag" campaign is developed. The consumer's perception of the general CSR and CSR in the context of the "no plastic bag" campaign were also assessed from the survey's results.

Step 4: Charting

Establishing the scenario of the CSR of supermarkets' and their involvement in the "no plastic bag" campaign.

Step 5: Mapping and Interpretation

The general concept of CSR and supermarket CSR will be assessed to further extend the concept of CSR for more profound strategic policy recommendations in an anti-single-use plastic bag initiative.

3.2   Consumers Survey

The general concept of CSR which included the quality of products and services, fair prices, caring for the society and environment, and ethics (van Marrewijk 2003; Matten et al. 2003) were covered in the first part of the questionnaire. The questions were developed based on the four dimensions of responsibilities of the CSR pyramid framework that was developed by Carroll (1991, 2004, 2016), namely the economic, legal, ethical and philanthropic dimensions. These dimensions were manifested into several perceptions and statements in relation to the definition of CSR in the context of supermarkets' involvement in the "no plastic bag" campaign. This study also considered the external dimensions of CSR addressed by the Commission of the European Communities (2001) mostly due to the consumers' involvement with the supermarkets' CSR initiative in campaigning the plastic bag ban. Hence, this affects their perceptions.

The questions addressed the plastic bag charges as a mechanism to take care of the environment, fair treatment to consumers, ethical campaign, to change consumers' behavior towards the plastic bag and to educate consumers on a sustainable shopping lifestyle. Furthermore, the key issues of CSR which were clustered under four main groups; marketplace, workplace, environment, and community (Whooley 2004), were translated into the consumer's perception in the context of CSR supermarket. It reflected the complexities of CSR which were operationalized using the four dimensions basic theories; instrumental, political, integrative and ethical (Garriga and Melé 2004), the comparison of CSR and corporate sustainability (van Marrewijk 2003) and corporate citizenship (Matten et al. 2003).

Hence, several questions related to the supermarkets' involvement in NPBC were asked: (i) What is the consumers' general overview of the CSR? (ii) How is the CSR of supermarkets developed to facilitate the demand in educating the society on the prohibition of plastic bags? (iii) Why is the plastic bag use dangerous to the environment? (v) How does the prohibition of plastic bags affect the consumers' shopping lifestyle and what is the consumers' perception of the supermarkets' CSR as part of the no plastic bag campaign? (vi) What is the consumers' perception towards various dimensions of CSR, e.g., offering quality products and services, fair prices, caring for the society and environment, helping the people, and practicing ethical, legal and fair behavior? All the questions addressed will be utilized in constructing the questionnaire for the consumer's survey conducting in supermarket with the initiative 'No Plastic Bag Campaign'.

The face-to-face consumer survey was conducted using standard questionnaires with the respondents' husband/father, wife/mother or adults above 18 years. The survey was performed in selected major hypermarket chains in Selangor state in 2013. The sampling framework developed by the Statistics Department of Malaysia revealed that from the 4,500,100 population in the Selangor state (Department of Statistic 2006), a total 401 samples were selected to represent the household population. However, after the answered questionnaire were reviewed only 253 questionnaires were valid for further analysis. The descriptive analysis was conducted for the various perceptions of the "no plastic bag" campaign, NPBC and supermarket's CSR.

4   Result and Discussion

In general, the results were divided into two main parts. First, the results of the framework analysis of the CSR of the supermarket was defragmented according to the CSR pyramid framework. Then, the results were described in the scenario of the supermarkets' involvement in the "no plastic bag" campaign in Malaysia. Second, the consumer's perception towards the general definition of CSR in general and the CSR of supermarkets in the context of the no plastic bag campaign was identified.

4.1   The CSR of a Supermarket in the Context of the CSR Pyramid Framework

The CSR of the supermarkets in the anti-plastic bag movement in Malaysia and any other parts of the world contributes towards enhancing and strengthening the supermarkets' corporate roles in a multifaceted, complex interaction of consumption, trade, and environmental sustainability. Based on the framework analysis, to familiarize and contextualize the plastic bag ban in Malaysia in the context of the CSR pyramid framework, several points have been outlined:

i. *Economic Responsibilities*

One of the supermarkets' basic operation principles is to maintain the loyalty of consumers and to keep providing convenient conditions for shopping. These include providing plastic bags. The prohibition of plastic bag usage created an unpleasant situation for consumers (Warner 2009; Zen et al. 2013) but has the potential to reduce the supermarkets' operational budget and business activity. Plastic charges or plastic bag taxing; the disapproval of plastic bag provision for consumers; and the initiative to reduce the use of plastic bag as a part of the supermarket CSR have affected their operations and business activities. Therefore, their involvement functions as an example of how corporations can show an ethical manner of operating their economic activities by not providing harmful products and preserving the environment. The supermarkets' initiative to sell alternative shopping bags such as biodegradable garbage bin liners, jute bags or other types of shopping bags, facilitates the behavioral changes in establishing a new social norm in society. Several studies have shown that reusable bags (plastic or cotton) require less energy and cause lower greenhouse gas emissions, whereas craft paper bags have the greatest environmental impact due to the energy used, greenhouse gas emissions and water consumption (Greene 2011; Environment Australia 2002; Business for Social Responsibility 2010).

Observation from this study has identified several initiatives imposed independently by some supermarkets in educating the consumers and facilitating the NPBC in Malaysia. The store in Bangsar South shopping center has been a plastic bag-free supermarket since it opened in 2009 and provided only carton boxes for its customers. The Tropicana City Mall outlet has a no-plastic-bag day on Mondays, while at Bandar Tun Hussein Onn, 8 out of 15 checkout lanes give priority to customers with their own shopping bags and Carrefour aims to achieve zero distribution of plastic bags by 2020.

The charge of the plastic bag was imposed to reduce the plastic bag consumption. Standard charges of MYR 20 cents per plastic bag were imposed on the consumers at the check-out counters in supermarkets and cashiers in retail markets. It is a part of the nationwide 'No Plastic Bag Campaign Day' (NPBC) in Malaysia which started in 2011. It was mentioned that the activity is the part of the supermarket CSR which supermarkets are bound to voluntarily in the Letter of Agreement (LA). The money collected from the plastic bag charges does not contribute to the profit of the company but rather on the CSR of the company (Zen et al. 2013). However, the initiatives do not have the monitoring and control mechanism; there are hardly even clear guidelines on how the plastic bag charges were administered.

Supermarkets' involvement in educating the public to change their practices towards plastic bag has the potential to nudge consumers. In the context of a nudge as an approach to persuade consumers' behavioral changes, the CSR of supermarkets provides a place to nudge the consumer. It is where "nudged" by choice architects influences one to make the "right" choice by providing a wide range of alternatives for shopping bags offered by the supermarket. The study found that the involvement of supermarket CSR in the anti-plastic bag campaign is also one example where the nudge behavioral-based policy can be influenced. Its cheapness and effectiveness can be promoted in everyday choices as claimed by Richard Thaler and law scholar Cass R Sunstein (Thaler and Sunstein 2008). In their book, *"Nudge—Improving Decisions about Health, Wealth and Happiness,"* Thaler and Sunstein suggested that public policy-makers and other choice architects, in this case, supermarkets, arrange decision-making contexts in ways to promote behavior that is in their own, as well as society's general interests. However, most economists still favor the traditional top-down policy economic instrument such as plastic bag levy (Miller et al. 2012).

It is argued that "nudging" is the cheapest way to influence citizens' behavior without restricting their freedom of choice, or the need of imposing new taxations, or tax-reliefs (Thaler and Sunstein 2008). Overlook the role of the CSR of supermarkets, respecting the freedom of choice of consumers or the libertarian paternalism approach, is sometimes overlooked. Its function is to provide several choices which help in designing the public policy approach on anti-plastic bag usage. Instead of using plastic bags, consumers are required to opt for other choices of shopping behavior to perform a sustainable shopping lifestyle (Chan et al. 2008). In this case, supermarket acts as the architects who are instructed by policymakers, the government, to influence the consumers' shopping lifestyle and create a new social norm. Hence, this study will disclose the gaps in the complexity of governing the plastic bag ban from the perspective of the CSR of supermarkets.

ii. *Legal Responsibilities*

The need to legalize voluntary effort of the plastic bag ban may contribute to a supermarket's competitive advantage. This is to sustain the supermarket CSR involvement with various stakeholders involved. Specific regulations such as Plastic Bag Levy or tax or charges helps in gaining more legitimate action. It is part of the management and administration of the money or incentives being channelled from the consumers to the retailer. The legislation helps in managing the complex interaction between several contradicting players of the plastic bag campaign such as the consumers, retailers and corporate sectors (Fig. 1).

In the context of the no plastic bag campaign, the regulatory intervention helps in reshaping the consumers' perception and practices. The government's intervention produced skepticism and a lack of consumer confidence towards the NPBC. According to Gupta (2011), most regulatory policies emerged in 1990 to control the use of plastic bags around the world. While most of the Asian countries are still in the voluntary phase of plastic bag ban due to several concerns, some countries such as Taiwan and China have introduced regulations on plastic bags that require retailers to explicitly state the price of the plastic bags. Hence, searching for other mechanisms to accelerate consumer behavioral changes towards single-use of the plastic bag is needed.

Many multinational companies or MNCs of supermarkets involved in the plastic bag campaign realize that they are members of the wider community and therefore behave in an environmentally responsible fashion is required as a part of the global CSR stakeholder requirement (Carroll 2004, 2016). This norm is translated into

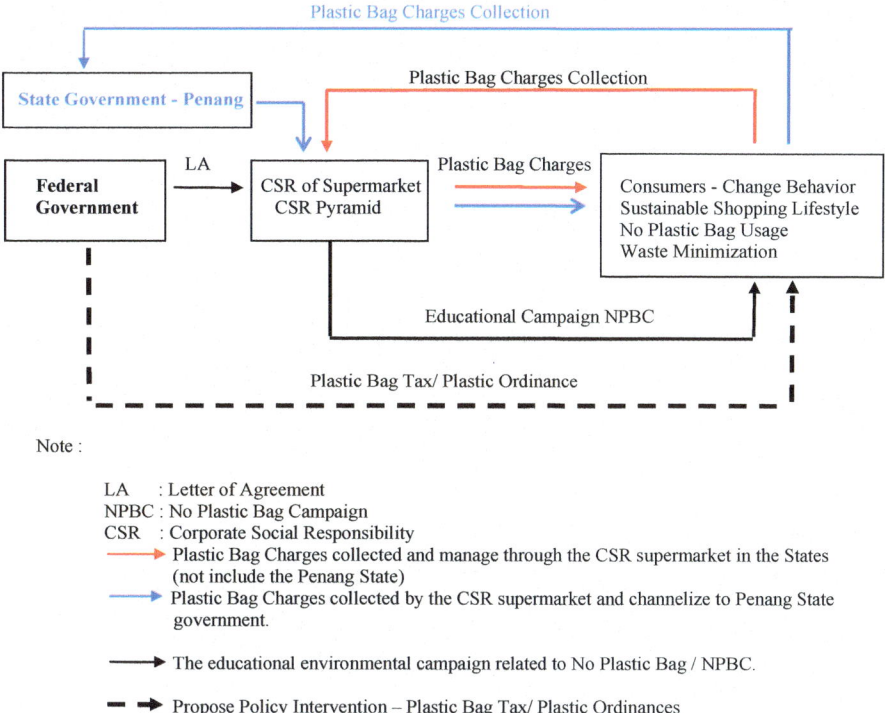

Note :

LA   : Letter of Agreement
NPBC : No Plastic Bag Campaign
CSR   : Corporate Social Responsibility
——▶ Plastic Bag Charges collected and manage through the CSR supermarket in the States
       (not include the Penang State)
——▶ Plastic Bag Charges collected by the CSR supermarket and channelize to Penang State
       government.

——▶ The educational environmental campaign related to No Plastic Bag / NPBC.

- -▶ Propose Policy Intervention – Plastic Bag Tax/ Plastic Ordinances

Fig. 1 The scenario of the no plastic bag campaign in Malaysia and the stakeholder involvement

corporations believing they must achieve environmental objectives as well as profit related objectives. In this case, the comprehensive supermarket CSR programmes related to this plastic bag reduction campaign which is initiated by the parent company policy have a strong influence in redesigning the new CSR of supermarket practices.

The CSR of supermarkets in the plastic bag ban, at the same time, provides a chance for companies to penetrate and create the niche market for consumers with a green concern which creates new competitive advantages and branding (Woerle 2011). Such supermarkets include Tesco Hypermarket from the United Kingdom, Jusco-AEON from Japan and IKEA from Sweden (Zen et al. 2013; Richards and Zen 2016). This CSR is part of the green marketing strategy and an attempt to reshape the consumer's values (Suki 2013; Piercy and Lane 2009). In general, it is also part of a company's motivation that involves competitiveness, legitimation, and ecological responsibility (Bansal and Roth 2000; Bansal 2005).

iii. *Ethical Responsibilities*

The CSR of supermarket compliance to the anti-plastic bag campaign initiative is expected to be good, not only to the level of acceptable behavior and consumers' ethical development and sustainable shopping lifestyle practice but also to obey the country's specific laws on plastic bag charges to support the decrease use of plastic bags (Carroll 1999, 2004). Voluntary supermarket CSR may support the libertarian paternalism where regulations may not exist (Dunn et al. 2014). This is not the case for the anti-plastic bag initiative where specific laws are needed to regulate the complex inter-related issue in banning plastic bags. By having a Plastic Bag Tax or Regulations or ordinance, it will legitimately strengthen the consumer compliance beyond the ethical dimension but towards legal compliance. This is related to the legal responsibility's aspect of CSR of supermarkets in the no plastic bag campaign.

Disclosure the charges collected from the plastic charges is another ethical responsibility from the supermarket CSR involving the no plastic bag campaign. Even though there is no control mechanism imposed or regulated under any certain law, the information disclosure principle is one of the crucial ethical components in the CSR (Carroll 1991, 1999). The Carrefour hypermarket in Malaysia has saved 26 million plastic bags in three years (2009–2011) of the NPBC while the AEON Co (M) Sdn. Bhd., the Japanese multinational company that runs the Jusco Supermarket chain, saved up to 10 million plastic bags since the launch of its no plastic bag campaign in January 2008 (New Straits Time 2011). Nevertheless, this shows a significant supermarket contribution to the environment by reducing millions of plastics and preventing them from entering the waste stream.

The plastic bag charges imposed to the consumers when they purchase things in a supermarket function as a basis for further regulatory intervention of Plastic Ordinances or Plastic Tax. This is implemented in China and Taiwan (Gupta 2011). In our case, the charges collected will be used for the CSR activities of the supermarket or channeled to the state government to organize the educational campaign for the various environmental programs (Fig. 1). The diagram shows no connection between the federal government and the state government on this matter. In the case of the Penang state government, it shows a closed loop interaction through the plastic bag charges collection from the CSR supermarket as a source of income for the Green Fund to support various environmental programs at the state level (in the blue line) The complex interaction is explained in Fig. 1.

iv. *Philanthropic Responsibilities*

For the applied policy on a plastic bag, it shows that the CSR of the supermarkets is seen in their involvement in the No Plastic Bag Campaign, NPBC in Malaysia which involves various stakeholders. These stakeholders include the government, industry (Malaysia Plastic Manufacturing Association (MPMA)) and NGOs, such as consumer associations [Consumer Association of Penang (CAP), Federation of Malaysia Consumer Association (FOMCA)]. Hence, supermarket needs to consider their interests in their CSR initiatives to sustain and manage their desirable relationships (Lantos 2001; Branco and Rodrigues 2007). Nevertheless, the supermarket CSR directly involves consumers.

The involvement of the CSR of supermarkets in the plastic bag ban is not common. The initiative by the Malaysian government to engage supermarkets to ban the plastic bag consumption as part of their CSR in educating the consumers matches with the needs of the new governance to tackle the plastic bag issue in a public-private partnership mode (Loorbach 2002; Loorbach 2010). This is by providing reusable shopping bags, educating consumers, and establishing a new norm of a sustainable shopping lifestyle (Zen 2018) as well as a platform for societal learning (Richards and Zen 2016). Norms refer to ideas and beliefs about what behavior is deemed appropriate (Bernstein 2001) in this case, is meant to support the anti-plastic bag initiative.

Taking on this CSR role, supermarkets create new responsibility, i.e., educating consumers on a new shopping lifestyle to help the government's initiative to reduce the plastic bag for environmental sustainability. This creates a unique CSR programme and platform to facilitate society's learning phase for the establishment of a normative behavior/new norms and values that brings changes into practice. Studies on the consumer's perception of the CSR supermarket on the NPBC in Selangor found that about 35% of households are willing to participate in this initiative (Afroz et al. 2016).

In general, the four points describe the complexity of supermarket CSR involved in the anti-plastic bag consumption initiative as shown in Fig. 1. Supermarkets interface the government, consumers and environmental responsibility vis-à-vis a CSR pyramid framework which fits into the Philanthropic Responsibility, Ethical Responsibility, Legal Responsibility and Economic Responsibility aspect of CSR pyramid framework adopted from Carroll (1999). As mentioned earlier, the federal government engages the supermarket chain in conducting the no plastic bag campaign through the Letter of Agreement, LA developed by the Ministry of Domestic Affairs and Consumerism of Malaysia (Zen et al. 2013). It is stated that the plastic bag charges collected will become part of the supermarket's CSR that is used to educate the consumers about eliminating the use of plastic bag. The complexities are also addressed in a study by Afroz et al. (2016).

The results of the study found that there were different terms of the plastic bag ban plastic bag charges at the state level, especially for the Penang State government. The Penang state government asks that the money collected from the campaign is to be channeled into the special Environmental Fund and to the poor to fund the environmental related activity and charity for the poor. From the in-depth interview with the Penang State government, this study revealed that with the 20-cent price point, the reduction of plastic bags in the Penang State alone is up to 700.000 MYR from the year of 2009–2011. It is equivalent to 3.5 million plastic bags for the price of 20 cents each plastic bag imposed by the respective supermarkets. The specific fund, 'Partner Against Poverty' has been set up by the Penang State government.

The supermarket's involvement is crucial and has gained a momentum where the creation of a conducive consumer shopping lifestyle to break the old pattern of shopping behavior helps in establishing new norms and values. Despite the prohibition of the plastic bag or imposing a fee for plastic bags, most supermarkets provide several alternatives to facilitate the behavioral changes. Among the alternatives are selling the reusable shopping bags known as green bags, providing the substitute of plastic bags with boxes during the campaign days, and providing biodegradable waste bin liners (Zen et al. 2013). Consumers' sustainable lifestyle or consumption is the ultimate objective of these initiatives. Hence, more institutionalized efforts need to be developed to accommodate the campaign by the supermarkets.

4.2 Consumer's Perception Towards CSR of Supermarket

The socio-economic profile on the consumer is described in Table 1. In general, the respondents are dominated by more females than males, with more than half of them being married (58.5%). 54.9% of the respondents were below 30 years old. 81.4% of the respondents are Malay, and 38.7% are graduate holders.

The consumers' awareness about CSR studies was assessed by asking them, 'Do you know about the corporate social responsibility, CSR?' In general, more than half of the respondents (57.3%) do not know about CSR while 42.7% are aware of CSR. However, when the CSR is linked with the NPBC, and the respondents were asked '*Do you know that the No Plastic Bag Campaign in Malaysia is part of the Supermarket CSR?*', the results show that 66% of the respondents know that the No Plastic Bag Campaign in Malaysia is part of the Supermarket CSR. Only the remaining 34% are not aware of it. It is concluded that the CSR of NPBC helps in promoting the CSR supermarket.

In detail, the consumers perceive that supermarket CSR covers '*Caring about society*' by 24.2%. This is followed by '*Offering quality products and services*' by 20.4%, and '*Fair Prices*' by 19.0% (Table 2). It reflects the complexities of CSR which are operational by using the four dimensions of basic theories: instrumental, political, integrative and ethical (Garriga and Melé 2004), the comparison of CSR

Table 1 Socio-economic profile of consumer

	Details socio economic profile	Percentage (%)
1	Sex	59.7% (151 respondents) female 40.3% (102 respondents) male
2	Age composition (years)	29.6% (75 respondents) 21–25 years old 25.3% (64 respondents) 26–30 years old 18.2% (46 respondents) 31–35 years old 9.5% (24 respondents) 36–40 years old 6.7% (17 respondents) 41–45 years old 4.7% (12 respondent) 46–50 years old 5.9% (15 respondents) 51 years old and above
3	Status	Married 58.5% (148 respondents) Bachelor 39.5% (100 respondents) Widow 1.6% (4 respondents) Widower 0.4% (1 respondent)
4	Race	81.4% (206 respondents) Malay 10.7% (27 respondents) Indian 4.7% (12 respondents) Chinese 3.2% (8 respondents) were other races
5	Education	24.9% (63 respondents) degree 13.8% (35 respondents) post graduate 37.2% (94 respondents) diploma 22.1% (56 respondents) secondary school 2% primary school

Table 2 Consumers' perception towards' general CSR definition

Consumers' perception on general definition of CSR	Sample (*n*)	Percentage (%)
Offering quality products and services	115	20.4
Fair prices	107	19.0
Caring for the society and environment	136	24.2
Help the people	56	9.9
Ethic, legal and fair behaviour	50	8.9
Kind attitudes towards the employee	49	8.7
Equal opportunities and training	50	8.9
Total	563	100

and corporate sustainability (van Marrewijk 2003) and corporate citizenship (Matten et al. 2003).

The consumers' perceptions of the supermarkets' CSR effort are exhibited in Table 3. There are ten statements measuring the consumers' perception of the activities of the supermarkets CSR involved in the anti-plastic bag effort. The study employs a 5-point Likert Scale that includes: strongly agree (1), agree (2), no option (3), disagree (4) and strongly disagree (5).

The result shows that most of the consumers support the statement, "*Supermarket caring about their consumers*" reflected in 43.5 and 9.9% of them agreeing and strongly agreeing respectively (Table 3). About 66% of the respondents understand that the *No Plastic Bag Campaign* in Malaysia is part of the supermarket's CSR program. The proposed scenario of plastic bag charges as part of the environmental fund shows that 62.1% of the respondents are skeptical about channeling the charges to fund environmental programs.

About 64.0% of the respondents supported the statement '*Supermarkets impose charges on plastic bags to help the environment*' with the 47.8% voting *agree* and 16.2% *strongly agree*. The results showcase the level of the consumers' understanding of CSR activities related to plastic bag prohibition if the society does not have environmental awareness about the impact of plastic bags. Society might view this effort as ethically correct or incorrect as different societies have different sets of ethical rules (Carroll 1979, 1991; Atan and Abdul Halim 2011). Though ordinary CSR philanthropic activities could be good in improving the society's quality of life (Carroll 1979, 1991, 1999), it needs a supermarket to specially design the activities related to the prohibition of plastic bag use. The various initiatives for plastic bags are tailored in tandem with the top philanthropic responsibilities found in the four dimensions of the CSR Carroll spectrum. Furthermore, the contribution of each dimension is explained in detail.

About 46.7% of the respondents support the statement '*Supermarket caring for the community*' with 34.8% *agree* and 11.9% *strongly agree* (Table 3). Only 9.9 and 4% of them disagree and strongly disagree with this statement respectively, and 34.8%

Table 3 Consumer's perception towards supermarket CSR and no plastic bag campaign (NPBC)

List of consumers perception on the CSR of supermarket	Strongly disagree (1) (%)	Disagree (2) (%)	No option (3) (%)	Agree (4) (%)	Strongly agree (5) (%)
1. Supermarket caring to their *consumers*	17 (6.7)	31 (12.3)	70 (27.7)	110 (43.5)	25 (9.9)
2. Supermarket uses plastic bag charges to help the *environment*	14 (5.5)	26 (10.3)	51 (20.2)	121 (47.8)	41 (16.2)
3. Supermarket cares for the *community*	10 (4.0)	25 (9.9)	88 (34.8)	100 (34.8)	30 (11.9)
4. Supermarket *cares for the consumers* by offering cheaper shopping bag	18 (7.1)	42 (16.6)	65 (25.7)	97 (38.3)	31 (12.3)
5. Supermarkets' *fair treatment of the consumers*	14 (5.5)	39 (15.4)	82 (32.4)	89 (35.2)	29 (11.5)
6. Supermarkets conduct *ethical campaign*	10 (4.0)	28 (11.1)	62 (24.5)	119 (47.0)	34 (13.4)
7. Supermarkets *educate the consumer* on a sustainable shopping lifestyle through the 'bring on shopping bag' campaign	9 (3.6)	25 (9.9)	44 (17.4)	114 (45.1)	61 (24.1)
8. Supermarkets play an important role in taking care of the *environment*	6 (2.4)	28 (11.1)	44 (17.4)	114 (45.1)	61 (24.1)
9. Supermarkets are offering **environmentally** plastic bag products such as a bio-degradable plastic bag	8 (3.2)	25 (9.9)	60 (23.7)	121 (47.8)	39 (15.4)
10. Supermarkets have *transparency* and accountability to manage plastic bag charges	6 (2.4)	1 (0.4)	16 (6.3)	82 (32.4)	148 (58.5)

of the respondents are impartial about this statement. This statement contributes to the component of philanthropic responsibilities in the pyramid CSR framework.

The statement *'Supermarket cares for their consumers by offering cheaper shopping bags'* gained a dominantly positive response from 80.6% of the respondents (Table 3). About 38.3% of them agree, while 12.3% of them strongly agree with the statement. Supermarkets offer cheaper reusable shopping bags to nudge consumers towards a sustainable shopping lifestyle (Zen 2018). The result shows high support for supermarket CSR in the anti-plastic bag campaign while extending the definition of supermarket CSR and their contribution to the component of Philanthropic Responsibilities of the pyramid CSR framework.

About 46.7% of the positive response on *'Supermarket fair treatment of the consumers'* are found with 35.2% saying that they agree and 11.5% of them strongly agree (Table 3). For this statement, only 15.4 and 5.5% of them vote to *Disagree* and *Strongly Disagree* respectively, and 32.4% of the respondents decide not to give their opinion on this statement. A higher positive response is gathered from the statement *'Supermarket conduct an ethical campaign'*. For this statement, 47.0 and 13.4% of them agree and strongly agree, respectively. Only 11.1% disagree while 4% strongly disagree with this statement while 24.5% of the respondents decide not to give their opinion on this statement. These two statements contribute to the Ethical Responsibilities of the pyramid of the CSR framework.

Most CSR company programs experience a lack of consumer's confidence (Mohr et al. 2001) and gained negative perceptions in a study conducted in Indonesia (Ali and Rizwan 2013). In contrast, the CSR supermarket in the form of the *No Plastic Bag Campaign* has its own characteristics which contribute positively to the overall image of the CSR of the supermarket as an evolving concept. Most supermarket CSR involved in the No plastic bag campaign in Malaysia are specifically involved in providing alternative shopping bags to replace the plastic bag consumption which at the same time facilitate the installation of sustainable shopping lifestyle for consumers shopping. They are actively involved in the campaign while offering various attractive packages to stimulate the consumer's behavioral changes such as by collecting points for recycling. TESCO hypermarket is an example, whereby they sell attractive alternative shopping bags with lower prices. All of these initiatives helps in nudging the consumers sustainable shopping lifestyle and contribute to sustainable lifestyle (Lehner et al. 2016)

Consumers show high confidence and a positive perception towards the supermarket's CSR that has a real impact on their shopping practice. About 69.2% of respondents support the statement *'Supermarket educates the consumer about sustainable shopping lifestyle through the 'bring on shopping bag' campaign'*. For this statement, 45.1% agree, and 24.1% strongly agree that the supermarket educates them about sustainable lifestyle through the mentioned campaign. 13.5% of the respondents disagree and strongly disagree with this statement while 17.4% of the respondents who do not give their opinion about it. Supermarkets in Johor are representative of facilitating the consumers' behavioral changes in the no plastic bag campaign (Zen et al. 2013) or known as Bring Your Own Bag, BYOB campaign, contributes to the overall societal learning process (Richards and Zen 2016) for a new environmental norm towards a sustainable shopping lifestyle (Lehner et al. 2016; Zen 2018). Most of the initiatives are part of the CSR of supermarkets which have been similarly implemented in several countries in Asia such as China, Australia, Singapore, and Malaysia.

About 69.2% of the respondents have a positive attitude towards the statement '*Supermarkets play an important role in taking care of the environment*' (Table 3). 45.1% of the respondents agree with this while 24.1% strongly agree. This attitude is supported by the statistics of 26 million plastic bags being saved due to the No Plastic Bag Campaign by Carrefour in 2009–2011 and 10 million by AEON since 2008 (New Straits Time 2011). The next statement, '*Supermarkets are offering environmentally plastic bag products such as a bio-degradable plastic bag*' has 47.8% of the respondents agreeing and 15.4% strongly agreeing which adds up to 63.2%. Only 9.9 and 3.2% of the respondents disagree and strongly disagree with this statement, respectively. 23.7% of the respondents do not give their opinion about it. Overall, these statements show positive results to the Philanthropic Responsibilities component of the pyramid of the CSR framework.

The statement '*Supermarket has transparency and accountability in managing the plastic bag charges*' has 32.4% of the respondents agreeing and 58.5% of them strongly agreeing (Table 3). In comparison, only 0.4 and 2.4% of the respondents disagree and strongly disagree with this statement respectively, while 6.3% of the respondents remain impartial to this statement. The survey shows positive results from consumers about their confidence towards the CSR efforts of the supermarkets involved in the no plastic bag campaign, which also embodies the ethical responsibilities in Carroll's pyramid of CSR framework. Furthermore, the results are supported by several public aspects of financial transparency of several multinational companies (MNCs) involved in various forms of the No Plastic Bag Campaign in Malaysia. Japanese companies such as AEON gives between 80 and 90 million reusable shopping bags to customers every year. About MYR 200,000 (USD48.300) is collected from the sale of plastic bags in the year 2010 (New Straits Times 2011). In Malaysia, the French hypermarket chain, Carrefour, has managed to channel MYR 180,000 (USD43.500) to the Malaysian Nature Society, a local environmental NGO from the sale of plastic bags during the same year (New Straits Time 2011). Hence, supermarket CSR creates an innovative approach to reporting their financial contribution from the plastic bag ban initiative. It is more unusual than the common report of the corporate CSR in financial terms (Porter and Kramer 2006). Contrary to that, the amount of plastic bag implicates to a huge number of plastic bags used from the environmental perspective, from the economic perspective of the polluter, the MYR 0.20 is not sufficient to prevent the consumers from using plastic bags and bringing their own shopping bag (Pearce and Turner 1992). Perhaps, plastic bag charges need to be increased to discourage plastic bag consumption and encourage other alternatives to change consumer behavior (Table 4).

Table 4 Extended definition of the CSR supermarket in the plastic bag ban

General definition of CSR	Extended definition of the CSR supermarket in the context of NPBC	CSRT supermarket in the context of the CSR pyramid framework
1. Offering quality products and services	1. Provision of a biodegradable plastic bag (63.2%)	The efforts of supermarket CSR in promoting the NPBC force them to provide green products as part of the consumers' education to support sustainable shopping lifestyle *Ethical Responsibilities*
2. Fair prices	2. Supermarket provides cheaper shopping bag (50.6%)	To stimulate the behavioral changes and educate the consumers about the sustainable shopping lifestyle, a cheaper shopping bag is provided. It functions to trade of the plastic shopping bag charges *Economic Responsibilities*
3. Caring for the society and environment Help people	3. Supermarket care about their consumers (53.4%), the community (46.7%), educate consumers on sustainable shopping lifestyle (69.2%) and contribute to take care of the environment (69.2%)	The CSR of a supermarket in the NPBC provides various mode on how the organization contributes to the society and environment which beyond the classic philanthropic mode *Ethical and Philanthropic Responsibilities*
4. Ethic, legal and fair behavior	4. Supermarket fair treatment to the consumers (46.7%) Supermarket conducts ethical campaign (60.4%) Supermarket manages transparency and accountability in manage the plastic bag charges (90.5%)	The CSR of a supermarket in the NPBC provides an extension mode on how the supermarket tries to be fair to the consumers despite the consumers' issues on unfair treatment due to the plastic bag charges by the consumers group association. The supermarket's efforts in mediating the plastic bag charges enforced by the government create inconvenience shopping treatment. This situation needs to be contra with any other initiatives in order to support the real objective as to save the environment *Ethical & Legal Responsibilities*

(continued)

Table 4 (continued)

General definition of CSR	Extended definition of the CSR supermarket in the context of NPBC	CSRT supermarket in the context of the CSR pyramid framework
5. Kind attitudes towards the employee	Not applicable	The role of supermarket CSR in NPBC is more towards the company external effort. Though the company has internalized the side effect in term of cost saving on operational cost from plastic bags, it rationalizes that the kind attitude the company inherits through this initiative. Observation and interview with the AEON Company found a way in training the cashier on educating the consumers on the plastic bag charges and the option to sale the sustainable shopping bag
6. Equal opportunities and training	Not applicable	This aspect is not explored in this study However, various training sessions for the cashier to persuade the consumer to pay for the plastic bags and to promote shopping bags for Bring Your Own Bag. BYOB campaign was performed by the supermarket. It is part of the supermarket's CSR initiative in promoting sustainable shopping lifestyle

5 Conclusion

Overall, the study decompartmentalizes the multifaceted complexities of the involvement of supermarket CSR in the anti-plastic bag movement. The dynamic interaction between consumers, supermarket, the corporate sector and the government in anti-plastic bag consumption initiatives in Malaysia shows the vital role of the supermarket CSR described in the Carroll's CSR pyramid framework. The supermarket CSR create an enabling ecosystem for societal learning by providing a platform to nudge consumers towards a sustainable shopping lifestyle as a new environmental norm. It is also provides a chance for green marketing and has a potential mechanism in internalizing the "no plastic bag" campaign cost as part of the corporate strategy to reduce the operational costs and create a new competitive advantage. Finally, the supermarket CSR in the context of NPBC has a potential to facilitate the creation of a new corporate mechanism in reshaping the green market for anti-plastic bag movement to support the global plastic environmental challenges.

The study extends the definition of the supermarket CSR's involvement in the plastic bag ban and outlined the economic, legal, ethical and philanthropic responsibilities of Carroll's CSR pyramid framework. Based on the economic responsibilities, supermarket's CSR creates a conducive environment for long-term profitability, be it directly or indirectly gained from the green marketing and green image of the organization involved in continuous participation in the anti-plastic bag campaign. The definition of the legal responsibilities of CSR in the anti-plastic bag effort creates a conducive situation for policy intervention for further possible legal development such as the Plastic Ordinance or Plastic Tax. This is to legitimize the Plastic Bag Charges described in Fig. 1. The Ethical responsibilities implemented by conducting the anti-plastic bag campaign and promoting a sustainable shopping lifestyle are manifested in the form of a continuous educational campaign. Moreover, the Philanthropic responsibilities of the anti-plastic bag campaign contribute to an improvement of the consumers' quality of life through promoting a new shopping lifestyle.

Finally, the results from the survey distributed to 253 consumers in the Selangor state revealed that the positive influence of the anti-plastic bag initiative in enhancing the supermarket CSR legitimacy and social responsibility towards green activity.

Acknowledgements The study was conducted as a part of the Research University Grant from Ministry of Education, Malaysia (Vot 08J98)

References

Abdul Rashid MZ, Abdullah I (1991) Managers attitudes towards corporate social responsibility. In: Proceeding of Pan Pacific conference, Kuala Lumpur, pp 218–220

Abdul Rashid MZ, Ibrahim S (2002) Executive and management attitudes towards corporate social responsibility in Malaysia. Corp Gov 2(4):10–16

Adane L, Muleta D (2011) Survey on the usage of plastic bags, their disposal and adverse impacts on environment: a case study in Jimma City, Southwestern Ethiopia. J Toxicol Environ Health Sci 3(8):234–248

Afroz R, Rahman A, Masud MM, Akhtar R (2016) The knowledge, awareness, attitude and motivational analysis of plastic waste and household perspective in Malaysia. Environ Sci Pollut Res, pp. 1–12

Akulian A, Karp C, Austin K, Durbin D (2007) Plastic bag externalities and policy in Rhode Island. Environ Res Econ 38:1–11

Ali W, Rizwan M (2013) Factors influencing corporate social and environmental disclosure (CSED) practices in the developing countries: an institutional theoretical perspective. Int J Asian Soc Sci 3(3):590–609

Atan R, Abdul Halim NA (2011) Corporate social responsibility: the perception of Muslim consumers

Ayalon O, Goldrath T, Rosenthal G, Grossman M (2009) Reduction of plastic carrier bag use: an analysis of alternatives in Israel. Waste Manag 29(7):2025–2032

Bansal P (2005) Evolving sustainably: a longitudinal study of corporate sustainable development. Strateg Manag J 26:197–218

Bansal P, Roth K (2000) Why companies go green: a model of ecological responsiveness. Acad Manag J 43:717–736

Beckman S (2007) Consumers and corporate social responsibility. Australas Mark J 15(1):27–36

Bernstein S (2001) The compromise of liberal environmentalism. New York, Columbia University Press

Bowen HR (1953) Social responsibility of the businessmen. Harper Row, New York

Branco MC, Rodrigues LL (2007) Positioning stakeholder theory within the debate on corporate social responsibility. Electr J Bus Ethics Organ Stud 12(1):5–15

Business for Social Responsibility (2010) Building long-term solutions: retail shopping bag impacts and options

Carroll AB (1979) A three dimensional conceptual model of corporate performance. Acad Manag Rev 4(4):497–505

Carroll AB (1991) The pyramid of corporate social responsibility: towards the moral management of organizational stakeholders. Business Horizons, July-August: 42. In: An alternative to the pyramid is a Venn-diagram model presented in MS Schwartz, AB Carroll (2003) Corporate social responsibility: a three-domain approach. Bus Ethics Q 13(4): 503–530

Carroll AB (1999) Corporate social responsibility. Bus Soc 38(3):268–295

Carroll AB (2004) Managing ethically with global stakeholders: a present and future challenge. Acad Manage Perspect 18(2):114–120

Carroll AB (2016) Carroll's pyramid of CSR: taking another look. Int J Corp Soc Responsib 1(1):3

Chan RY, Wong YH, Leung TK (2008) Applying ethical concepts to the study of "green" consumer behaviour: an analysis of Chinese consumers' intentions to bring their own shopping bags. J Bus Ethics 79(4):469

Clapp J, Swanston L (2009) Doing away with plastic shopping bags: international patterns of norm emergence and policy implementation. Environ Polit 18(3):315–332

Commission of the European Communities (2001) Promoting a European framework for corporate social responsibility. Available at http://europa.eu.int/eur-lex/en/comgpr/2001/com2001_0366en01.pdf

Commission of the European Communities (2002) Communication from the commission concerning corporate social responsibility: a business contribution to sustainable development. Available at http://europa.eu.int/comm?employment_social/soc-dial/csr/csr2002_en.pdf

Convery F, McDonnell S, Ferreira S (2007) The most popular tax in Europe? Lessons from the Irish plastic bags levy. Environ Res Econ 18(4):367–371

Creyer EH, Ross WT Jr (1997) The influence of firm behaviour on purchase intention: do consumers really care about business ethics? J Consum Market 14(6):419–432

Dauvergne P (2010) The problem of consumption. Global Environ Polit 10(2):1–10

Davis K (1973) The case for and against business assumptions of social responsibility. Acad Manag J 16(2):312–322

Department of Statistic (2006) Section methodology and research. Department of Statistics Malaysia

Derraik JG (2002) The pollution of the marine environment by plastic debris: a review. Marine Pollut Bull 44(9):842–852

Dikgang J, Visser M (2010) Behavioral response to plastic bag legislation in Botswana discussion. Papers dp-10-13-efd. Resources for the future

Dikgang J, Leiman A, Visser M (2012) Elasticity of demand, price and time: lessons from South Africa's plastic-bag levy. Appl Econ 44(26):3339–3342

Dunn KE, Rakes GC, Rakes TA (2014) Influence of academic self-regulation, critical thinking, and age on online graduate students' academic help-seeking. Distance Educ 35(1):75–89

Environment Australia (2002) Plastic shopping bags—analysis of levies and environmental impacts. Victoria

Epstein MJ (2018) Making sustainability work: best practices in managing and measuring corporate social, environmental and economic impacts, Routledge

Epstein MJ, Roy MJ (2001) Sustainability in action: identifying and measuring the key performance drivers. Long Range Plann 34:585–604

Epstein MJ, Roy MJ (2003) Making the business case for sustainability. J Corp Citizensh 9(1):79–96

Ernst & Young (2002) Corporate social responsibility. Available at www.ey.nl/download/publicatie/doemload/corporate_social_responsibility.pdf

Fernández B, Souto F (2009) Crisis and corporate social responsibility: threat or opportunity? Int J Econ Sci Appl Res 2(1):36–50

Garriga E, Melé D (2004) Corporate social responsibility theories: mapping the territory. J Bus Ethics 53:51–71

Gifford B (1991) The greening of the golden arches-McDonald's teams with environmental group to cut waste. The San Diego Union, August 19, C1, C4

Girod S, Bryane M (2003) Branding in European retailing: a corporate social responsibility perspective, European Retail Dig 38(2):1–6

Greene J (2011) Life cycle assessment of reusable and single-use plastic bags in California. California

Gupta K (2011) Consumer responses to incentives to reduce plastic bag use: evidence from a field experiment in urban India. SAMDEE

Handelman JM, Arnold SJ (1999) The role of marketing actions with a social dimension: appeals to the institutional environment. J Market 63(3):33–48

Hasson R, Leiman A, Visser M (2007) The economics of plastic bag legislation in South Africa. South Afr J Econ 75:66–83

Hume S (1991) Consumer doubletalk makes companies wary. Advert Age 62(46):62–64

Jakovcevic A, Steg L, Mazzeo N, Caballero R, Franco P, Putrino N, Favara J (2014) Charges for plastic bags: motivational and behaviorial effects. J Environ Psychol 40:372–380

Jambeck JR, Geyer R, Wilcox C, Siegler TR, Perryman M, Andrady A, Law KL (2015) Plastic waste inputs from land into the ocean. Science 347(6223):768–771

Kates RW (2017) Sustainability science. In: Richardson D, Castree N, Goodchild MF, Kobayashi AL, Liu W, Marston RA (eds) The international encyclopedia of geography, Wiley, Hoboken. https://doi.org/10.1002/9781118786352.wbieg0279

Kaur S (2002) 1 million plastic bags change hands each day. The Straits Times, 3rd April 2002

Kotler P, Lee N (2005) Corporate social responsibility: doing the most good for your company and your cause. Wiley, New Jersey, EUA

Lantos GP (2001) The boundaries of strategic corporate social responsibility. J Consum Market 18(7):595–632

Lehner M, Mont O, Heiskanen E (2016) Nudging—a promising tool for sustainable consumption behaviour? J Clean Prod 134:166–177

Loorbach D (2002) Transition management: governance for sustainability, paper presented at the international conference on Governance and sustainability, Berlin

Loorbach D (2010) Transition management for sustainable development: a prescriptive, complexity-based governance framework. Int J Policy Adm Inst 23(1):161–183

Maignan I (2001) Consumers' perceptions of corporate social responsibilities: a cross-cultural comparison. J Bus Ethics 30(1):57

Matten D, Crane A, Chapple W (2003) Behind de mask: revealing the true face of corporate citizenship. J Bus Ethics 45(1–2):109–120

Miller TL, Grimes MG, McMullen JS, Vogus TJ (2012) Venturing for others with heart and head: how compassion encourages social entrepreneurship. Acad Manage Rev 37(4):616–640

Mohr LA, Webb DJ, Harris KE (2001) Do consumers expect companies to be socially responsible? The impact of corporate social responsibility on buying behaviour. J Consum Aff 35(I):45

Patten DM (2002) Give or take on the internet: an examination of the disclosure practices of insurance firm web innovators. J Bus Ethics 36(3):247–259

Pearce D, Turner RK (1992) Packaging waste and the polluter pays principle: alleviation solution. J Environ Plann Manage 35(1):5–15

Piercy NF, Lane N (2009) Corporate social responsibility: impacts on strategic marketing and customer value. Market Rev 9(4):335–360

Poortinga W, Whitmarsh L, Suffolk C (2013) The introduction of a single-use carrier bag charge in Wales: attitude change and behavioural spillover effects. J Environ Psychol 36:240–247

Porter ME, Kramer MR (2006) The link between competitive advantage and corporate social responsibility. Harvard Bus Rev 84(12):78–92

Ramasamy B, Yeung M (2009) Chinese consumers' perception of corporate social responsibility (CSR). J Bus Ethics 88(1):119–132

Richards C, Zen IS (2016) From surface to deep corporate social responsibility. J Global Responsib 7(2):275–287

Ritchie J, Spencer L (1994) In Bryman A, Burgess RG (eds) Analyzing qualitativedata, pp 173–194

Rotmans Jan, Grin John, Schot Johan, Smits Ruud (2004) Multi- inter- and transdisciplinary research program into transitions and system innovations. Maastricht University, Maastricht

Sharp A, Hoj S, Wheeler M (2010) Proscription and its impact on anti-consumption behaviour and attitude: the case of plastic bags. J Consum Behav 9:4

Shrivastava P (1995) The role of corporations in achieving ecological sustainability. Acad Manag Rev 20:936–960

Spaargaren G, Oosterveer P (2010) Citizen-consumers as agents of change in globalizing modernity: the case of sustainable consumption. Sustainability 2: 1887–1908. https://doi.org/10.3390/su2071887. www.mdpi.com/journal/sustainability

Srivastava A, Thomson SB (2009) Framework analysis: a qualitative methodology for applied policy research. Research Note JOAAG 4:2

Stringer L (2009) The green workplace. Palgrave Macmillan, New York, NY

Suki NM (2013) Green awareness effects on consumers' purchasing decision: some insights from Malaysia. Int J Asia-Pacific Stud 9(2)

Teoh Hai-Yap, Gregory Thong TS (1986) Another look at corporate social responsibility and reporting: an empirical study in a developing country. Malays Manag Rev 21(3):36–51

Thaler R, Sunstein CR (2008) Nudge—improving decisions about health, wealth and happiness. New Haven, CT, Yale University Press

Van Marrewijk M (2003) Concept and definitions of CSR and corporate sustainability: between agency and communion. J Bus Ethics 44:95–105

Vince J, Hardesty BD (2017) Plastic pollution challenges in marine and coastal environments: from local to global governance. Restor Ecol 25(1):123–128

Vince J, Hardesty BD (2018) Governance solutions to the tragedy of the commons that marine plastics have become. Front Mar Sci 5:214

Warner BM (2009) Sacking the culture of convenience: regulating plastic shopping bags to prevent further environmental harm. U Mem L Rev 40:645

Whooley N (2004) Social responsibility in Europe. Available at: www.pwc.com/extweb/newcolth.nsf/0/503508DDA107A61885256F35005C1E35?

Woerle A (2011) New branding strategy meets corporate social responsibility. Reg Bus Stud 3(1):825–831

Zen IS (2018) Nudge to promotes sustainable shopping lifestyle. In: Proceedings 2(22): 1394. https://doi.org/10.3390/proceedings2221394

Zen IS, Ahamad R, Omar W (2013) No plastic bag campaign day in Malaysia and the policy implication. Environ Dev Sustain 1(11):1259

Zen IS, Kaur CR, Arisman (2017) The potential for sustainability science approach for multi-level marine science governance: a case study of Wakatobi national park. Policy Brief CSEAS (Center for Southeast Asia Study). Series 1, Indonesia. February

The Role of Public Administration in Sustainable Development

Fernanda Caroline Caldatto, Sandro Cesar Bortoluzzi
and Edson Pinheiro de Lima

Abstract Sustainability has been seen as the key to reversing the planet's plight and ensuring better living conditions in the future. Since the discussion on sustainable development has gained momentum, actions at the global level have been implemented so that, through an integrated change process, the long-awaited sustainable development is achieved. There are many inherent challenges to sustainability, however, through strategies and proper execution, it is possible to achieve good performance in all dimensions (environmental, economic and social). Thus, the objective of this study is to identify, through a literature review using the Knowledge Development Process—Constructivist (Proknow-C) intervention instrument, the role of public entities and local authorities in the implementation and execution of sustainable practices, as well as the indicators used to assess their performance. Considering that the decision maker has a fundamental role in choosing the criteria for choosing one action instead of another, an evaluation model was built supported by the Multi-Criteria for Decision Support—Constructivist methodology—MCDA-C, through which it was possible to realize that the focus of attention of the municipal administration studied is the social dimension. The limitation of the research is the construction of the model based on only one municipality, so that the methodology can be replicated in other municipalities of similar size in order to confirm the existence of a standard of vision of public decision makers.

F. C. Caldatto (✉) · S. C. Bortoluzzi · E. P. de Lima
Industrial and Systems Engineering, Federal University of Technology—Parana (UTFPR), Via do Conhecimento, Km 1, Pato Branco, Parana, Brazil
e-mail: fercaldatto@gmail.com

S. C. Bortoluzzi
e-mail: sandro@utfpr.edu.br

E. P. de Lima
e-mail: pinheiro@utfpr.edu.br

E. P. de Lima
Pontifical Catholic University of Parana—PUCPR, R. Imaculada Conceição, 1155 Curitiba, Parana, Brazil

© Springer Nature Switzerland AG 2020
W. Leal Filho et al. (eds.), *International Business, Trade and Institutional Sustainability*, World Sustainability Series,
https://doi.org/10.1007/978-3-030-26759-9_4

Keywords Sustainable development · Public administration · Performance evaluation

1 Introduction

Developing countries have experienced rapid urban population growth since the 1990s combined with major technological changes (Cohen 2006). Specifically, in Brazil, in the last population survey carried out, the urbanization rate already reached 84% (IBGE 2018). This high population concentration leads to serious environmental, social and economic problems arising from production (which is concentrated mainly in the urban area), excessive consumption and generation of waste.

Given this, it is necessary that there is a change in the way government, companies and citizens act, so that the problems already perceived are not aggravated. One way of achieving economic growth without too much environmental and social impact is through sustainable development, which enables present generations to meet their needs without jeopardizing future generations (WCED 1987).

With the growth of a town, all decisions and even management become more complex (Cohen 2006). While public administration is engaged in sustainable development, there is greater motivation for sustainable goals to be achieved (Hoppe and Coenen 2011), which is essential, considering that cities play an important role in achieving these goals (Grimm et al. 2008; Shen et al. 2011).

However, it is not an easy task. Some cities have already realized the urgency of the situation and started to seek three-dimensional sustainability (covering the environmental, economic and social spheres), but they have encountered difficulties (Kusakabe 2013). Local managers experience constant dilemmas, since environmental goals compete with political goals (Hoppe and Coenen 2011) and most of the initiatives are focused on the environmental dimension (Domingues et al. 2015; Braulio-Gonzaloet al. 2015).

The solution to urban problems has mainly focused at the local level, so the better managed cities are, the better the ability to cope with the challenges (Cohen 2006). Thus, it is essential to define, at the local level, the objectives and strategies for achieving sustainable development (Shen et al. 2011).

Faced with these dilemmas and the need for a better governmental position to achieve a sustainable urban environment, the objective of this study is to identify how cities and countries have dealt with urban sustainable development and what practices have been implemented. Therefore, the work consisted of: (1) Literature review using the Proknow-C methodology; (2) survey of the main variables for assessing urban sustainability; and (3) structuring a public sector performance evaluation model in relation to urban sustainability.

2   The Role of Public Administration

Decision makers are considered facilitators, since through the adoption of government policies they can encourage the promotion of sustainable consumption (Shao et al. 2017), as well as improving sustainability in all dimensions, by promoting behaviors of excellence in private entities (Figueira et al. 2018). Implementing policies related to sustainable issues will not only matter to local government, but also to the national government (Peng and Liu 2016) and to the population at large.

In this search for sustainability, public participation becomes very important (Kusakabe 2013) and is considered indispensable for achieving sustainable urbanization (Shen and Zhou 2014). In this sense, Chapman and Shigetomi (2018) perceived the existence of different motivations, according to the age group, since young people tend to be more individualistic and seek convenience, while older people tend to prioritize actions that reflect positively on issues related to the environment, resource management, economics and equity.

Particularly in the social arena, citizens are seen as the main influencers, since they can ensure better access to housing and education if they have active participation in the community (Lewis and Donald 2010). Angeloni and Borgonovi (2016) present some possibilities for social sustainability, from the engagement of the elderly in voluntary activities, which, in addition to generating benefits for the community involved in providing services, will improve their quality of life. It is important to note that the role of citizens is not restricted to participation as consumers of resources but should also be perceived in the monitoring of public administration performance and actions (Shen et al. 2011).

In addition to the citizens, companies have the means to collaborate for sustainable development, especially those industries that need cleaner production, which is aimed at reducing waste, energy, resource use and pollution. In this way, public administration has the function of guiding the behavior of companies (Peng and Liu 2016; Junior et al. 2018). However, the Chinese experience indicates that there needs to be a real threat for companies to adopt the clean production stance and start using it as a competitive advantage, ruling out the high-consumption, pollution and low-cost competition (Peng and Liu 2016). In addition, the implementation of mandatory clean production audits for highly polluting industries shows significant results (Bai et al. 2015).

Another highly effective business practice is the management of the supply chain for sustainable products, especially when it comes to focal companies, which may require suppliers to meet environmental and social standards. The main incentive identified in the context of the supply chain refers to legal demand, since companies always seek to reduce the risks inherent in the activity (Seuring and Müller 2008).

Identifying ways to stimulate industries towards sustainable production should be one of the priorities of the public administration, since this activity derives large amount of emissions and consumption of resources. An example of a strongly used resource is water, considered one of the most important (Chapman and Shigetomi

2018), since it is vital for human existence and the amount of water per capita in the world is very low (Tan et al. 2017).

Some solutions to water scarcity can be: investments in renewal and maintenance of networks, allowing the identification of leaks; promotion of efficient technologies; public awareness campaigns; and municipal regulations (Stavenhagen et al. 2018). Identifying existing problems facilitates population awareness (Chapman and Shigetomi 2018) and the search for solutions. Public agencies need to be aware of these issues and find ways to involve the community in the cause.

Likewise, understanding the constraints and drivers for the progress of sustainability is paramount (Hîncu 2011; Lowe et al. 2015) to evaluate progress towards the goal. To address emerging issues, policy makers can use indicators, which will guide policy identification (Shen and Zhou 2014).

In the public sector the measurement of sustainability performance is usually only partially adopted (Hoppe and Coenen 2011), one of the reasons being a lack of obligatoriness and the fact that the obtaining of resources is independent of the results of the performance evaluation (Adams et al. 2014). The motivation for performance evaluation should be internal, which demands greater knowledge about the reflex of the actions (Lundberg et al. 2009).

The selection and identification of the best sustainability indicators allows monitoring and information generation for decision-makers to plan new actions and formulate policies (Shen et al. 2011). In order to be more effective in incorporating the processes, the indicators can be adapted to the context or region and should guarantee easy interpretation (Lowe et al. 2015; Braulio-Gonzalo et al. 2015). Indicator-based approaches often vary by the selection, nature and number of indicators (Tanguay et al. 2010), data used and method for performance calculation (Rajaonson and Tanguay 2017).

The support in decision-making and policy processes are the main motivators for the creation of sustainable development indicators, which allow the monitoring of progress and the instruments used, as well as the transparency of public actions through reports (Mascarenhas et al. 2010). The challenge is to use the information generated by the indicators in decision-making processes (Lowe et al. 2015; Shen et al. 2011).

As performance measurement should be adjusted according to local interests and needs, the presentation style of these results should be adapted to the public (Lowe et al. 2015). Providing clear information is one way to inform the population of the progress towards sustainable development and to enable them to be more engaged and empowered to participate in decision-making.

Among the cities that usually evaluate their performance, small cities excel in social indicators, with overall performance being greatest in large cities (Junior et al. 2018). Another difference can be seen in the availability of data, which is not always available. When there is information limitation, some variables can be substituted (Rajaonson and Tanguay 2017), which makes it important to identify the available variables (Liu et al. 2011).

3 Methodology and Data

In order to identify the sustainable practices and policies that have been implemented around the world, a literature review was initially carried out using the intervention tool known as Proknow-C (Knowledge Development Process—Constructivist). This methodology is composed of four steps: the search process for papers for the formation of a bibliographic portfolio; bibliometric analysis; systemic analysis; and definition of research problems (Ensslin et al. 2010). In this work, only the first stage of Proknow-C was carried out in order to construct a bibliographic portfolio about the studied subject.

The search process began with the definition of the research axes (evaluation of organizational performance, sustainability and urban public management), followed by the definition of search terms and combinations of terms (using Boolean expressions). The databases were then defined (Scopus, Science Direct and Emerald) and the research was carried out. The management of bibliographical references found was done through Mendeley. After the adhesion test and the exclusion of duplicate papers, the filtering process (title alignment, scientific recognition, abstract alignment and reading in full) was performed, as shown in Fig. 1.

After the selection of papers aligned with the object of study and with scientific recognition, the second stage of the research was conducted, which consisted of identifying the most relevant items for evaluating the performance of urban sustainability. For this, among the papers in the portfolio, the categories with the highest frequency were identified, which were listed in Table 1, with their respective author. As the focus of the research was not the identification of indicators, some papers had another bias, and only part of the authors of the portfolio are listed in the Table 1.

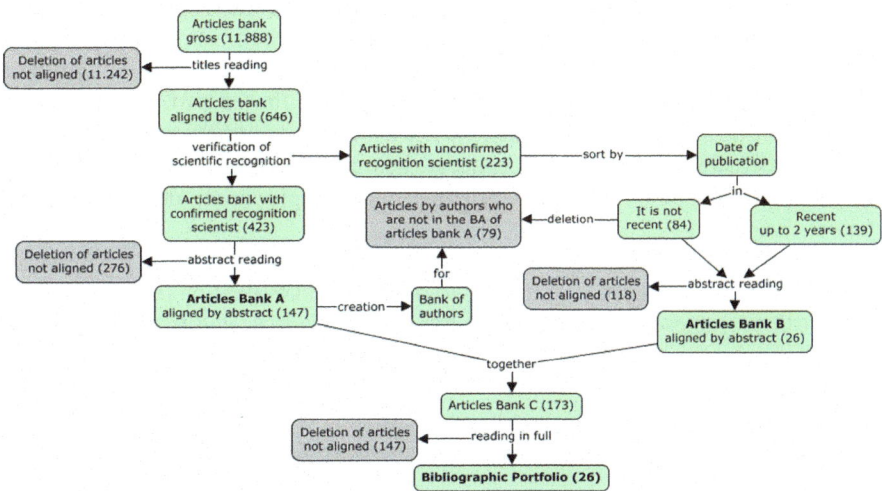

Fig. 1 Summary of the Proknow-C filtering process. *Source* Adapted from Ensslin et al. (2010)

Table 1 Categories of indicators more frequent in the literature

Indicator Category	Dimension	Adams, Muir e Hoque, 2014	Braulio-Gonzalo, Bovea, Ruá, 2015	Domingues et al., 2015	Hiroki, 2011	Kusakabe, 2013	Liu et al., 2011	Lowe et al., 2015	Mascarenhas et al., 2010	Mohammed, Alshuwaikhat, Adenle, 2016	Rajaonson e Tanguay, 2017	Shao, Taisch, Mier, 2017	Shen et al., 2011	Shen e Zhou, 2014	Tan et al., 2017	Tanguay et al., 2010
Green / protected areas	A		X	X	X		X	X	X		X	X		X	X	X
Noise	A		X	X	X	X	X		X				X	X	X	
Biodiversity	A		X	X		X			X		X		X	X		
Waste collection / treatment	A		X	X	X	X						X		X	X	X
Energy consumption	E		X	X				X	X	X			X	X	X	
Culture	S			X			X	X	X		X		X	X	X	X
Population density	A			X		X			X		X		X	X		X
Education	S	X	X			X	X	X	X	X	X	X	X	X	X	X
Job	S	X	X	X	X		X	X	X		X	X	X	X		X
Renewable energy	A		X	X		X						X	X	X	X	
Participation of citizens	S		X			X			X		X	X	X	X		X
Poverty	S		X		X			X	X	X			X	X	X	X
Water quality	A	X	X			X			X	X	X	X	X	X		X
Air quality	A	X	X	X		X	X	X	X		X	X	X	X		X
Per capita income	E			X			X	X	X				X	X		X
Restoration of urban constructions	E		X			X				X			X		X	
Reuse/recycling resource	A		X	X		X				X	X	X	X			X
Sanitation	S				X	X						X		X	X	
Population satisfaction	S	X	X	X	X							X		X		
Health	S	X	X			X	X	X	X	X		X	X	X	X	X
Safety	S	X	X			X		X		X	X	X	X	X		X
Public/alternative transportation	A		X			X	X		X	X	X		X	X		
Tourism	E		X			X	X			X			X			

A - Ambiental; E - Economic; S - Social.

Source Elaborated by the authors

Identifying the main practices implemented and concerns of several countries from the point of view of performance evaluation, the work went on to structuring the evaluation model. One caution that was taken in the research was the identification of indicators that were related to the political objectives and the existing responsibilities (Lowe et al. 2015). Considering that sustainability has a multi-criteria approach (Neves et al. 2018), the MCDA-C methodology was chosen as the basis for the construction of the evaluation model, which is based on in the view of the actors, people involved with the decision process, being the problem delineated by their beliefs and convictions, and from that in the most appropriate decisions for the context (Ensslin et al. 2000).

The MCDA-C is divided into eight stages (de Lacerda et al. 2011). However, considering that the objective is limited to obtaining an evaluation instrument, only the first three stages of the methodology were applied, related to the structuring phase. This phase is composed of: approach "soft" for structuring (through which the contextualization and identification of the problem is done); family points of view (the primary elements of evaluation being identified, as well as where one wishes to arrive and what one wishes to avoid); and construction of descriptors.

4   Results and Discussion

The model was developed for the evaluation of sustainability performance in the municipal management of a small municipality (with a little more than 20,000 inhabitants), located in the southern region of Brazil. In order to demonstrate the findings about the concerns inherent in the three dimensions of sustainability, the cognitive maps constructed from the decision maker's view are presented. In Fig. 2 the cognitive map concerning the social dimension is presented.

It can be seen that the main concerns reported by the decision-maker in the social sphere were education, health and quality of life. This evidence supports the findings of Junior et al. (2018), that the best performance in small cities is found in areas such as education, housing and human development, that is, issues related to daily life (Tan et al. 2017).

One of the issues raised about the quality of life of the population concerns employment. Countries such as Sweden show that the concern for sustainable development can be a motivator for the attraction of new investors and, consequently, generation of new jobs (Eckerberg and Forsberg 1998).

In the environmental dimension, as shown in the cognitive map presented in Fig. 3, the decision maker reported as main concerns: pollution, green areas and garbage collection. It is noteworthy that garbage collection is outsourced, and currently those who do the separation and disposal are former waste pickers, which does not exempt the municipal administration from the role of advisor. Outsourcing was a means of generating income for this part of the population (through the sale of recyclable products) and, at the same time, reducing the costs that public administration had to the activity. The interviewee commented that most of the companies' control over the pollution generated by production is carried out by environmental agencies, and local

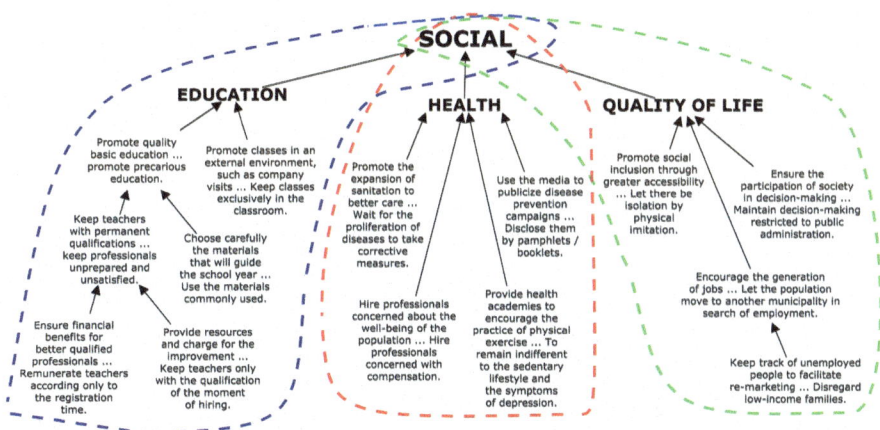

Fig. 2 Cognitive map of evaluation of sustainable practices in the social dimension. *Source* Elaborated by the authors

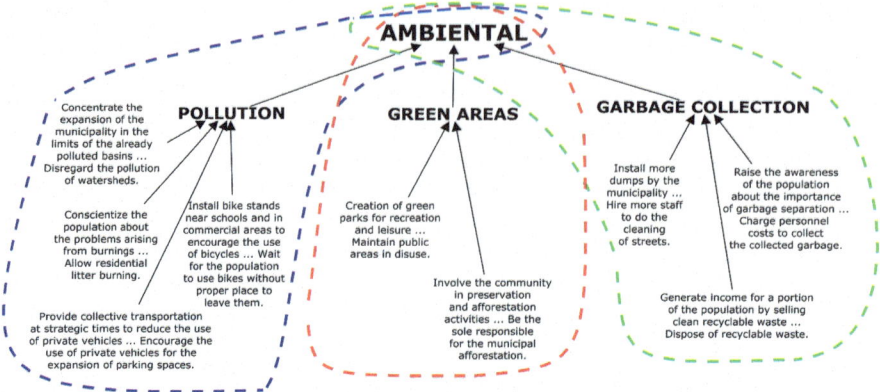

Fig. 3 Cognitive map of evaluation of sustainable practices in the ambiental dimension. *Source* Elaborated by the authors

administration is responsible for activities that are more focused on the population level.

In addition, especially in this dimension, public administration depends on the collaboration of the other organs and the population in general for the achievement of goals. In this way, the importance of awareness campaigns is evidenced by the decision maker in order to provide information necessary for a better targeting of routine practices. It happens that sometimes it is necessary to go beyond and make the care obligatory and subject to sanctions for a better reach of the objective. In this sense, a collaborative regulation for greater commitment and citizen participation is essential (Kusakabe 2013).

Many of the initiatives in other areas of concern depend on investment, as there are a number of improvements to be implemented and technologies to adhere to. However, municipal initiatives tend to be small in scope, due to the great investment demand and insufficient resources (Eckerberg and Forsberg 1998). Regardless of this, sustainable growth cannot be considered a second option. It must be considered during urban planning and seek ways to progress. Thus, Fig. 4 presents the cognitive map of the evaluation of sustainable practices in the economic dimension.

In this way, it can be seen that among the most urgent investments evidenced by the decision maker, they are related to environmental issues, such as the installation of solar panels for energy production (which can be used in public lighting, having as a bonus the reduction of costs with energy), installation of cisterns for the use of rainwater in activities that do not require the use of drinking water such as discharges or cleaning sidewalks, which besides being financially economical, is a way of preserving drinking water.

The decision maker considered the importance of each area of concern within the dimensions, which can be seen in Fig. 5.

Once again, we see the preference of the social dimension in relation to the economic and environmental, which is in agreement with the results obtained by

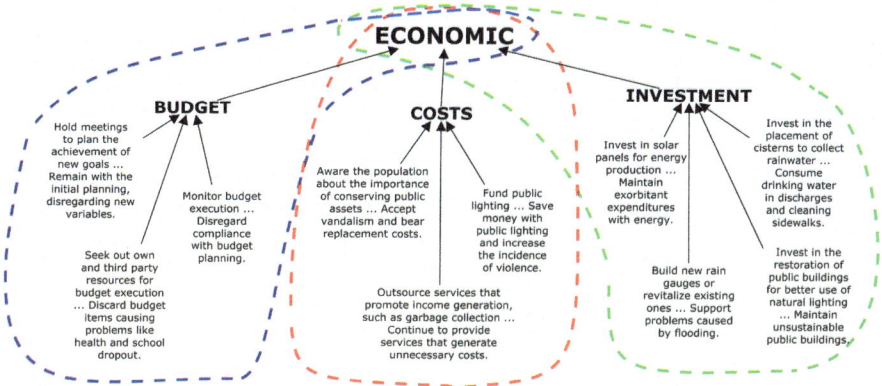

Fig. 4 Cognitive map of evaluation of sustainable practices in the economic dimension. *Source* Elaborated by the authors

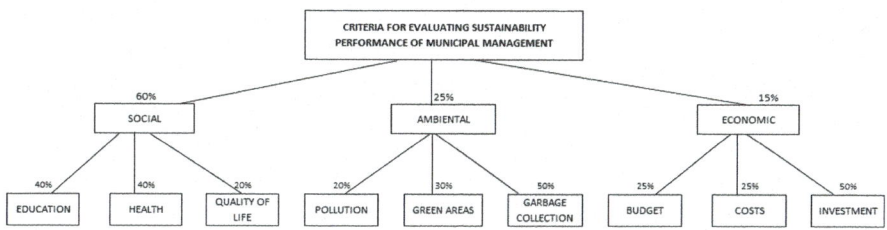

Fig. 5 Criteria considered for sustainability assessment. *Source* Elaborated by the authors

Mascarenhas et al. (2010), which indicate that the predominance of evaluation occurs in the environmental dimension. This demonstrates that the performance evaluation model reflects the values and concerns of the formulator and that the importance of each indicator may vary according to perceived needs.

5 Conclusion

The research work consisted of identifying the role of public administration as a generator and driving sustainable practices. To do so, a bibliographic portfolio of papers dealing with actions carried out in several countries and the results obtained through them were obtained through a literature review, guided by Proknow-C and supported by the Mendeley bibliographic reference manager.

The literature review, in addition to contributing to the survey providing information on the main urban sustainability practices carried out around the world, largely demonstrated the most used indicators, having extracted the categories of indicators of higher frequency in the literature (mentioned in Table 1). In order to prove the importance of such categories in the most varied contexts, a performance evaluation model was built, guided by the structuring phase of the MCDA-C methodology, according to the needs of a small municipality located in the south of Brazil.

The comparison of the categories of indicators obtained in the literature review with the criteria chosen by the interviewed decision-maker demonstrated that, although there is a pattern, it is important to make some adaptations according to the size of the municipality and the areas of greatest lack of attention. In addition, by identifying the needs and key actions to be carried out, it is important to track progress toward objectives in order to correct possible implementation failures.

Another conclusion obtained is that the local administration of the studied context understands that its role in relation to the sustainable development is more of execution than orientation or regulation, since a low number of actions mentioned as relevant was related to the population's awareness, and that regulation and policymaking that resulted in a real threat to unsustainable actions was never cited as a measure.

The limitation of research consists in the construction of only one model, which is constructed from the perspectives of a decision maker. In this way, it is suggested that other constructivist models be created in order to identify how local administrators see sustainable development, to what extent they are engaged with it, and whether there is a pattern. Moreover, given the identification that there is no interest in creating regulations, or even penalties for unsustainable actions, it is suggested to evaluate the causes of regulatory resistance and possible reactions of the population and companies if sustainable actions become mandatory.

References

Adams CA, Muir S, Hoque Z (2014) Measurement of sustainability performance in the public sector. Sustain Acc Manag Policy J 5(1):46–67

Angeloni S, Bomrgonovi E (2016) An ageing world and the challenges for a model of sustainable social change. J Manag Dev 35(4):464–485

Bai Y, Yin J, Yuan Y, Guo Y, Song D (2015) An innovative system for promoting cleaner production: mandatory cleaner production audits in China. J Clean Prod 108:883–890

Braulio-Gonzalo M, Bovea MD, Ruá MJ (2015) Sustainability on the urban scale: proposal of a structure of indicators for the Spanish context. Environ Impact Assess Rev 53:16–30

Chapman A, Shigetomi Y (2018) Developing national frameworks for inclusive sustainable development incorporating lifestyle factor importance. J Clean Prod 200:39–47

Cohen B (2006) Urbanization in developing countries: Current trends, future projections, and key challenges for sustainability. Technol Soc 28(1–2):63–80

Domingues AR, Pires SM, Caeiro S, Ramos TB (2015) Defining criteria and indicators for a sustainability label of local public services. Ecol Ind 57:452–464

de Lacerda RTO, Ensslin L, Ensslin SR (2011) A performance measurement framework in portfolio management: a constructivist case. Manag Decis 49(4):648–668

Eckerberg K, Forsberg B (1998) Implementing Agenda 21 in local government: the Swedish experience. Local Environ 3(3):333–347

Ensslin L, Dutra A, Ensslin SR (2000) MCDA: a constructivist approach to the management of human resources at a governmental agency. Int Trans Oper Res 7(1):79–100

Ensslin L, Ensslin SR, Lacerda RTO, Tasca JE (2010) ProKnow-C, knowledge development process—constructivist. Processo técnico com patente de registro pendente junto ao INPI, Florianpólis, Santa Catarina

Figueira I, Domingues AR, Caeiro S, Painho M, Antunes P, Santos R, Videira N, Walker RM, Huisingh D, Ramos TB (2018) Sustainability policies and practices in public sector organisations: the case of the Portuguese Central Public Administration. J Clean Prod 202:616–630

Grimm NB, Faeth SH, Golubiewski NE, Redman CL, Wu J, Bai X, Briggs JM (2008) Global change and the ecology of cities. Science 319(5864):756–760

Hîncu D (2011) Modelling the urban sustainable development by using fuzzy sets. Theor Emp Res Urban Manag 6(2):88–103

Hoppe T, Coenen F (2011) Creating an analytical framework for local sustainability performance: a Dutch case study. Local Environ 16(3):229–250

IBGE (2018) Instituto Brasileiro de Geografia e Estatística. https://biblioteca.ibge.gov.br/visualizacao/livros/liv64529_cap6.pdf. Accessed 25 Nov 2018

Junior CM, Ribeiro DMNM, da Silva Pereira R, Bazanini R (2018) Do Brazilian cities want to become smart or sustainable? J Clean Prod 199:214–221

Kusakabe E (2013) Advancing sustainable development at the local level: the case of machizukuri in Japanese cities. Progr Plann 80:1–65

Lewis NM, Donald B (2010) A new rubric for 'creative city' potential in Canada's smaller cities. Urban Stud 47(1):29–54

Liu Y, Yao C, Wang G, Bao S (2011) An integrated sustainable development approach to modeling the eco-environmental effects from urbanization. Ecol Ind 11(6):1599–1608

Lowe M, Whitzman C, Badland H, Davern M, Aye L, Hes D, Giles-Corti B (2015) Planning healthy, liveable and sustainable cities: how can indicators inform policy? Urban Policy Res 33(2):131–144

Lundberg K, Balfors B, Folkeson L (2009) Framework for environmental performance measurement in a Swedish public sector organization. J Clean Prod 17(11):1017–1024

Mascarenhas A, Coelho P, Subtil E, Ramos TB (2010) The role of common local indicators in regional sustainability assessment. Ecol Ind 10(3):646–656

Mohammed I, Alshuwaikhat HM, Adenle YA (2016) An approach to assess the effectiveness of smart growth in achieving sustainable development. Sustainability 8(4):397

Neves D, Baptista P, Simões M, Silva CA, Figueira JR (2018) Designing a municipal sustainable energy strategy using multi-criteria decision analysis. J Clean Prod 176:251–260

Peng H, Liu Y (2016) A comprehensive analysis of cleaner production policies in China. J Clean Prod 135:1138–1149

Rajaonson J, Tanguay GA (2017) A sensitivity analysis to methodological variation in indicator-based urban sustainability assessment: a Quebec case study. Ecol Ind 83:122–131

Seuring S, Müller M (2008) From a literature review to a conceptual framework for sustainable supply chain management. J Clean Prod 16(15):1699–1710

Shao J, Taisch M, Mier MO (2017) Influencing factors to facilitate sustainable consumption: from the experts' viewpoints. J Clean Prod 142:203–216

Shen LY, Ochoa JJ, Shah MN, Zhang X (2011) The application of urban sustainability indicators—a comparison between various practices. Habitat Int 35(1):17–29

Shen L, Zhou J (2014) Examining the effectiveness of indicators for guiding sustainable urbanization in China. Habitat Int 44:111–120

Stavenhagen M, Buurman J, Tortajada C (2018) Saving water in cities: assessing policies for residential water demand management in four cities in Europe. Cities 79:187–195

Tan Y, Xu H, Jiao L, Ochoa JJ, Shen L (2017) A study of best practices in promoting sustainable urbanization in China. J Environ Manag 193:8–18

Tanguay GA, Rajaonson J, Lefebvre JF, Lanoie P (2010) Measuring the sustainability of cities: an analysis of the use of local indicators. Ecol Ind 10(2):407–418

WCED World Commission on Environment and Development (1987) Our Common Future. Oxford University Press, Oxford

Dynamic Capabilities and Business Model in the Transition to Sustainability: The Case of Bosch/Curitiba-Brazil

Cristina M. S. Ferigotti, Sieglinde Kindl da Cunha
and Jonatas Soares dos Santos

Abstract This article focuses on innovation and sustainability from the perspective of dynamic capabilities at BOSCH Curitiba/PR. The conceptual study model articulates the sociotechnical approach of transition towards sustainability, dynamic capabilities and sustainable innovations. The general objective of the study was to analyze the development of dynamic capabilities, considering the sociotechnical approach and, the transition to sustainability at BOSCH, in order to understand the relationship between innovative actors and alternative urban mobility systems, with a view to reducing energy consumption, as well as public transitional policies for sustainability, in the Brazilian automotive industry. The study design was an exploratory-descriptive, qualitative and cross-sectional approach, with a single case study as the method. Interviews were conducted with managers and coordinators, as well as non-participant observation and secondary source research. The company under study implemented innovation in a business model for mobility, articulating cooperation among universities, research institutes, and other subsidiaries of the group, encouraging research, corporate startups, and incubators for sustainable innovation projects. Evidence suggests that the company is developing dynamic

C. M. S. Ferigotti (✉)
Getulio Vargas Foundation (FGV), Rio de Janeiro, Brazil
e-mail: cmferigotti@uol.com.br

Positivo University (UP), Curitiba, Brazil

Social Studies of Paraná Foundation (FESPPR), Curitiba, Brazil

C. M. S. Ferigotti · S. K. da Cunha
Innovation and Sustainability Research Group—Master and Doctorate Program, Positivo
University (PMDA/UP), Curitiba, Brazil

S. K. da Cunha
Institute of Economics (UNICAMP), Campinas, Brazil
e-mail: skcunha21@gmail.com

J. S. dos Santos
Robert Bosch, Curitiba, PR, Brazil
e-mail: soares.jnts@gmail.com

© Springer Nature Switzerland AG 2020
W. Leal Filho et al. (eds.), *International Business, Trade and Institutional
Sustainability*, World Sustainability Series,
https://doi.org/10.1007/978-3-030-26759-9_5

capabilities in new business sources in order to achieve superior performance in the transition to sustainability.

Keywords Dynamic capabilities · Business model · Innovation · Sustainability · Sociotechnical transitions

1 Introduction

This article focuses on a business model (BM), articulating concepts of dynamic capabilities (DC), sociotechnical transition, innovation, and sustainability at Bosch Curitiba-PR/Brazil, a member of Powertrain Solutions Latin America exclusively focused on the electromobility.

The objective was investigated how DC contributes to the development of BM for sustainable innovation in a sociotechnical transition environment at Bosch Curitiba. For this, the study describes how a company creates value by combining external and internal resources in its set of activities (Zott et al. 2011; Santos et al. 2009). It includes sustainability integrated into systems from idea to product marketing, as a key issue for value creation (Charter and Clark 2007).

The research on the transition to sustainability (Kemp and Rotmans 2008; Geels et al. 2011) is concerned with sustainable development (Brundtland 1987) and pre-supposes the emergence of BM studies in transitional organizations (Tulder et al. 2014). The sociotechnical regime (Geels et al. 2011) for the analysis of BM oriented to sustainable innovation is related to the innovation technology systems (ITS) of Bergek et al. (2015) and Hekkert et al. (2011), since they refer to infrastructures, regulations, actors. As well as the proposition of new values that integrates it and the value network of the ecosystem (Adams et al. 2016).

Rather, DC enables the company to coordinate various tangible and intangible resources to carry out activities such as product development and the design of new business models. Also, contributing to the alignment of production plans with the needs of customers (Teece 2007). Furthermore, a multinational company needs to adapt BM to the local conditions where its subsidiaries operate, value delivery relates to how activities are carried out by whom and how they are coordinated (Tallman 2014), although there is still a shortage of studies focusing on the automotive industry and transition to sustainability. This study intends to contribute to filling gaps in sustainability research articulating concepts of DC and BM oriented to innovation and sustainability, in a sociotechnical transition environment, considering "System Project" as a unit of analysis.

Thus, based on the association of the mentioned ideas and the application of a single case study with a transversal cut, this work aims to understand: How DC contribute to developing BM for sustainable innovation in a sociotechnical transition environment in Bosch Curitiba/PR? For this, the article is structured as follows: first, the Introduction, second the Theoretical Foundation with concepts of innovation and transition to sustainability, BM and DC. Third, the Conceptual Model of the study and the analytical structures, then, the Methodology followed by Context of Study with the analysis and the considerations on the empirical evidence, finally the Conclusions.

2 Theoretical Foundation

Innovation in this article is a process that adds value to ideas with practical achievements Tidd et al. (2015). However, the innovation process is increasingly fragmented and dispersed the innovative company has the collaboration of several partners such as suppliers, users, startups, specialized consultants, universities and research institutes (Simonsen and Figueiredo 2018). Thus, system innovation is a crucial strategy for implementing sustainability in sociotechnical systems (Charter et al. 2008; Geels 2005; Johnson and Suskewicz 2009).

In addition, the integrated approach to innovation (OECD 2005) presupposes inter-organizational relationships and institutional networks for interactions and diffuse new technologies (Freeman and Soete 2008). There is an emergency for studies on technological transition (Geels et al. 2011; Schot and Geels 2008) focusing on innovation and sustainability criteria in the automotive industry due to the increasing requirement of improvement of energy efficiency Borroni-Bird (2006), at the sociotechnical and niche level perspective. It is important to note that in subsystems that make up the sociotechnical regime of the automotive industry (Geels et al. 2011) is dependent on existing routines and operated by companies incorporating these routines (Schot and Geels 2008). Besides a proposal of integrated value in terms of context and functionality provided the success of a technology, in order with consumer demand (Orsato et al. 2012). Also, sustainable products thus considered those intended to protect or improve the natural environment, conserving energy and or resources and reducing or eliminating the use of toxic agents, pollution and waste (Ottman et al. 2006). From another perspective, in the condition of the uncertain environment and constant transformation, DC makes it possible (i) to create; (ii) configure; (iii) reconfigure resources (Teece et al. 1997). DC enables the company to coordinate various tangible and intangible resources to carry out activities such as product development and new BM design, as well as contributing to the alignment of production plans with customer needs (Teece 2010). What it means to destroy competencies, while creating others, so comprise must know-how to deal with innovations that strengthen the existing business, while also seeking new market opportunities. Organizational and managerial competencies are needed to: (i) perceive opportunities, (ii) seize opportunities, and (iii) manage threats by combining and reconfiguring assets within and outside the organization's borders (Teece 2007). Thus, there is a close relationship between the concept of DC and BM, the literature of BM articulates the logic, the data and other evidence that supports a value proposition for the customer, and the viable structure of revenues and costs for the enterprising delivering that value Teece and Linden (2017).

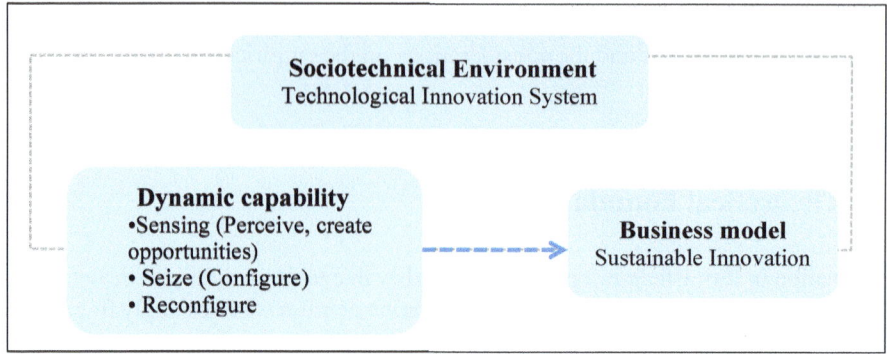

Fig. 1 Conceptual model. *Source* Own elaboration based on theoretical framework

For Teece (2010, p. 172), BM "describes the design or architecture of value creation, delivery, and capture mechanisms that are employed by it." BM is a system-level, activity-centered concept (Zott et al. 2011). Chesbrough and Rosenbloom (2002) show how the Xerox Corporation grew in part by employing an effective business model to commercialize a technology rejected by other leading companies, Calia et al. (2007) how technological innovation can trigger changes in the company's operational and commercial activities, and hence in the BM, Johnson and Suskewicz (2009) argue that in large infrastructural change (such as the transition from a fossil fuel economy to a cleantech economy), the key is to shift the focus from developing individual technologies to creating whole new systems. BM innovation in complementary markets is a consequence of the reconfiguration of downstream activities and capabilities (Gambardella and McGahan 2010). Therefore conceived the BM can be a vehicle for innovation as well as a subject to innovation (Zott et al. 2011). BM requires a systematic perspective, but always from the viewpoint of how the firm can connect to, or build up, that system while delivering a certain value proposition (Santos et al. 2009).

The ideas presented on the theoretical foundation illustrated the integration of DC with BM for innovation and sustainability in a sociotechnical transition environment defined the conceptual model illustrate in Fig. 1.

3 Conceptual Model of the Study

Figure 1 illustrates the articulation between sociotechnical transition, DC and BM for sustainable innovation. The proposed model analyzes the process of developing DC that contributes to BM in sustainable innovation, from the following research question:

How DC contribute to the development of BM for sustainable innovation in a sociotechnical transition environment at Bosch Curitiba/Brazil?

Two analytical structures (Tables 1 and 2), and Fig. 1 were used to analyze the development of DC to BM.

The analytical matrix of innovation and sustainability activities of Adams et al. (2016) in illustrate in Table 1.

Table 1 examines innovation and sustainability activities in the columns: (i) incremental efficiency in the operational context, (ii) organizational transformation, (iii) system construction when innovation affects social, organizational and technological factors, according to the sociotechnical regime. Figure 2 complements Table 1 to understand how the organization evolves from unsustainable and technical limited activities to progressively sustainable activities in line with the sociotechnical system.

Figure 2 presents the integration of sustainability criteria from strictly technical use in end of pipe solutions to systems Carrillo-Hermosilla et al. (2009), adapting to the sociotechnical system of the industry.

The analytical matrix of system functions of technological innovation was used in order to examine the contribution of DC to the development of the BM according Hekkert et al. (2011) and illustrate in Table 2.

Table 2 presents in the columns (i) the functions of the innovation system and (ii) the indicators for the evaluation of the functioning of the innovation system.

4 Methodology

The research was qualitative, descriptive and exploratory cross-sectional (Miles and Huberman 1994). The case study Yin (2015) was used to respond to "How DC contributes to developing BM for sustainable innovation in a sociotechnical transition environment at Bosch Curitiba/Brazil?".

The unit of analysis refers to the "System Project" and the choice was (i) being a piece of evidence to substantiate the study question and interpret it, (ii) enable the evaluation of the development of the BM, (iii) evidence about the adoption of sustainability of product designs. For the definition of the sample the method applied was snowball sampling, "where the researcher makes initial contact with a small group of people who are relevant to the research questions, and these sampled participants proposed other participants who have had the experience or characteristics relevant to the research" (Bryman 2012, p. 424).

The first contact with the phenomenon under study was a meeting with the innovation team of Bosch Curitiba/PR, responsible for coordinating the work done by project teams in intra-startups. From this first meeting, two responsible for the implantation of BM contributed to organize and indicate the participants of the research, setting up a sample for convenience (Hair et al. 2005), like in Table 3.

As in Table 3 the sample comprises newly graduated young people participating in innovative startup projects and specialist employees with years of experience at Bosch Curitiba.

Table 1 Sustainability-oriented analytical matrix

Dimension	Operational optimization	Organizational transformations	Systems building
1. Strategy: organizational and management process to delivery sustainability	Comply with regulations or pursue efficiency gains	Embed sustainability as a cultural and strategic norm in a shaping logic that goes beyond greening	Logic of wide collaborations and investing in systems solutions to derive new, co-created value propositions
2. Innovation process: the organizational of the innovation process to deliver sustainability, from searching new ideas to converting them into products and services and capturing value from them	Focus on internal and incremental innovation facilitated by use of tools and methods	Adopt new values and platforms (e.g. reverse innovation) and new ideation practices	Adopt new collaborative process platforms with diverse stakeholders
3. Learning: recognize the value of new knowledge, assimilating and applying it to support sustainability	Exploit existing knowledge management capabilities to identify and access relevant knowledge	Engage with key stakeholders of the firm—internal and external	Developed ambidextrous competencies enabling shadow tracking and learning from experimentation with multiple new approaches
4. Linkages: internal and external linkages crafted as opportunities for learning and influencing around sustainability	Recruit external domain experts for new knowledge	Shift focus from intra-firm linkages to collaborations with immediate stakeholders	Get the whole system in the room to diagnose problems, understand system complexity, build trust and identify levers for change
5. Innovative organization: Organization that create the conditions within which SOI can take place	Exploit existing innovation capabilities	Embed SOI culture Through the organization	Adopt new business paradigms (e.g. B-Corps)

Source Adapted from Adams et al. (2016)
SOI—sustainability-oriented innovation

Table 2 Analytical matrix of the technological innovation system

Functions	Indicators
1. Entrepreneurial activities	Most relevant actors present in industry from a structural analysis
2. Knowledge development	Amount of patents and publications
3. Knowledge exchange	Type and amount of networks
4. Guidance of the search	Regulations, visions, expectations of government and key actors
5. Market formation	Project installed
6. Resource mobilization	Physical resources (infrastructure, material, etc.), human resources (skilled labor), financial resources (investments, venture capital, subsidies, etc.)
7. Counteract resistance to change/legitimacy creation	Length of projects from application to installation to production

Source Adapted from Hekkert et al. (2011)

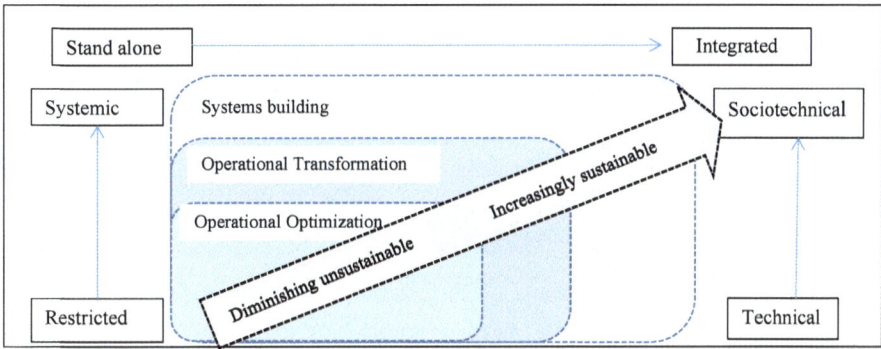

Fig. 2 Organizational dynamics towards integration with the sociotechnical system. *Source* Adams et al. (2016)

Interviews were carried out with individuals who were part of startup projects for the mobility segment, but also had the participation of two members of startup to develop a new market in Agribusiness. The observation was carried out by local visits in Bosch at where intra-startups were worked. Also, internal documents and external publications were consulted. Hence, the triangulation of sources of evidence occurred to verify the validity and reliability of the constructs (Yin 2015). Subsequently, the validation of the data occurred with a key informant of the organization and with a consultation to the legal department of the company. To analyze the empirical evidences content analysis Bardin (2011) was the techniques of treatment used. Table 4 lists the main conceptual definitions adopted in this article.

Table 3 Sample

Function	Formation	Specialization	Experience (years)	Activities
1. Scrum project	Mechanical engineer	–	5	Innovation projects
2. Team member	Public relations	Production engineering	15	Production supervision
3. Team member	Production engineering	–	1	The system project
4. Facilitator BM innovation	Economist	Executive education program: business model innovation; MBA business administration, management and marketing	11	Controlling, logistics, project management and lean manufacturing, new business developer
5. Mentor	Automation control engineering	Innovation management	9	Organizational development
6. Team member	Production engineering	–	1	The system project
7. Facilitador BM innovation	Production engineering	Product master	5	New business developer
8. Team member	Mechanical engineer	Specification in 6S quality management and master in thermodynamics	12	Senior software engineer
9. Team member	Agricultural engineer	–	2	"Sugar Project"
10. Team member	Mechatronic engineer	–	2	"Sugar Project"

Source Evidence of research

5 Context of the Study

The Bosch Group's operations comprise 440 subsidiaries, regional companies in 60 countries. In 2004, the MNE became a signatory of the Global Compact adhering to the United Nations Sustainable Development Goals (SDGs), (UN 2015).

The initiative involves commitment and development with product innovations and new BM based on sustainability (Bosch 2017). Consequently, the management and organization of the Bosch Group are geared towards sustainability with an Office, for internal and external consultations and a Board to discuss issues related

Table 4 Synthesis of the main conceptual definitions

Theoretical construct	Definition	Theoretical guideline
Dynamic capabilities (DC)	• Firm's ability to perceive, apprehend manage and reconfigure opportunities	Teece et al. (1997)
Sustainable innovation	• Sustainability is integrated into systems from idea to product marketing, as a key issue for value creation • Innovation and sustainability activities	Charter and Clark (2007), Adams et al. (2016)
Business model (BM)	• "Describes the design or architecture of value creation, delivery, and capture mechanisms that are employed by it." BM is a system-level, activity-centered concept	Teece (2010, p. 172), Zott et al. (2011)
Innovation technology systems (STI)	• Infrastructures, regulations and actors	Bergek et al. (2015), Hekkert et al. (2011)
Sociotechnical regime	• It can be understood as a series of practices, rules and shares, which dominate systems and actors	Geels (2005)

Source Based on theoretical foundation

to corporate social responsibility. The organization comprises the Robert Bosch Startup GmbH located in Stuttgart for the development and evaluation of new ideas in the face of the competitive pressures of the automotive industry such as sustainability (Bosch 2017). Concerning the principles 7, 8 and 9 for sustainable development linked to the Sustainable Development Goals (SDGs), (UN 2015): reduction of CO_2 in 32.8% compared to 2007, 54% of R&D expenditure for sustainable products; development and distribution of environmentally friendly technologies with ISO 14001 and 50001, reduced waste volume and water consumption at Bosch locations (Bosch 2017, 2018).

In 2016, Bosch Curitiba became part of Powertrain Solutions in Latin America, a division for mobility of the Group. The subsidiary operates and creates value at the country multinational nexus development the ability to understand idiosyncratic technical opportunities, research institutes, on supply and distribution chains on regulatory agencies, on labor markets, and on other local institution (Nelson 1993).

In the study three questions were important to relate the development of activities of sustainable innovation and the system of technological innovation in sociotechnical transition with DC.

1. **The Competence to Perceive and Create Opportunities: Strategic Change at Bosch Curitiba (2015–2016)**

From 2015, the Brazilian company was stimulated to change rapidly, and the strategic planning identified a need for BM to innovation and sustainability as a cultural and strategic norm (Bosch 2017). Stakeholders were consulted for structural analyzes in the automotive industries and agribusiness (Hekkert et al. 2011). In order to reconfigure the existing knowledge base at the plant in Curitiba, learning and create opportunities for new business (Adams et al. 2016; Teece and Linden 2017).

There was a strategic change in Bosch Curitiba, which operates in a sociotechnical environment dependent on existing routines (Schot and Geels 2008). Obliging it to research new regulations and government expectations, establish a shared vision with key actors for value creation that an MNC subsidiary needs to adapt its BM to local conditions (Tallman 2014). The culture of innovation for sustainability has led to the adoption of new business paradigms Adams et al. (2016).

As a result, everyone in the company could suggest ideas for Innovation and the App Bosch Curitiba, from the Bosch Curitiba Innovation community at intranet system (Bosch Connect) has been used to propose projects. Such (i) System Project, (ii) Generators, (iii) Fuel, (iv) Load (v) Sugar Project, (vi) Productivity.

The company intensified learning processes to build innovative capacity (Bell and Figueiredo 2012), mobilizing physical, human and financial resources to create and seize opportunities through an acquisition of internal and external knowledge, (Teece 2010; Teece and Linden 2017), with training and qualification in technology (Hekkert et al. 2011), as prototyping, lean startup, agile project, executive education program (business model innovation), design thinking (consumer-centered model, user experience). The training had been offered to provide subsidies for decision making in evaluation pillars namely: *Technology, Business,* and *User*, to adapt the product and processes development to the technological innovation system (Adams et al. 2016). Since BM can be a vehicle for innovation as well as a subject to innovation (Zott et al. 2011).

2. **Reconfiguration and Implementation of Changes: BM for Intrastartups (2017–2019)**

The company changed its innovation process, for the implementation of startups projects with intrapreneurship with the collaboration of clients and making agreements, teaching and development institutions (Simonsen and Figueiredo 2018). The practice suggests systems building in the innovation process and learning (Adams et al. 2016). There was a restructuring in the physical environment of the company, Bosch Curitiba created an area to innovation and development of startups and their projects related to mobility and the agribusiness market, which was not yet developing, characterizing knowledge development (Hekkert et al. 2011).

In period development of the projects of startups occurred with entrepreneurial assets and leadership orchestration. Committees with diverse manager's expertise in Brazil and in Germany decided on priorities considering value for the client, technical and economic viability, that supports a proposition for the customer and the viable structure of revenues and costs for the enterprising delivering that value (Teece and Linden 2017).

Strategic partners cooperated in the reconfiguration of competencies and implementation of the changes at Bosch Curitiba, being the Bosch Group responsible for project evaluation.

3. The System Project: Learn to Reconfigure Competencies

In 2018 a strategic decision to continue the System Project with steps of acceleration in 100 days, being evaluated in Brazil and Germany.

Sustainable technology is one of the pillars of the BM for motor drive and battery management, through sensors and software, will turn off the engine when the vehicle is idling and fulfilling the specific requirements of the battery charge. The technology will promote the reduction of fuel consumption, promoting energy efficiency and reducing CO_2 emissions in adherence to principle 7 of the SDGs. Although the technology already exists for touring vehicles, the opportunity was to develop BM "Local to Local" Ottman et al. (2006) and Gambardella and McGahan (2010) according principles 8 and 9 of SDGs (UN 2015).

The payback must reach a certain deadline so that the client feels stimulated to acquire the technology. The innovation technical in "Systems Project" were evaluated through more than 100 interviews focusing on a delivery truck and urban buses in several regions in Brazil to adapting it to the Brazilian market. Additionally, the illustrative "System Project" case up to the time of this research was in the discovery phase characterizing learning capability (Bell and Figueiredo 2012).

The research investigates the factors that contributed to developing DC in BM with innovation and sustainability in a sociotechnical transition environment. Table 5 presents the synthesis of DC development and empirical evidences.

Table 5 makes it possible to observe the relation of the DC with the empirical evidence of the development of the BM for the startup "System Project" through a series of interactions (Teece 2010; Teece and Linden 2017), according the principles for SDGs (UN 2015).

6 Conclusions

This article presents how DC contributes to developing BM for sustainable innovation in a sociotechnical transition environment, with a single case study at Bosch Curitiba.

The company created and seized the opportunity to develop dynamic capabilities to innovate in organizational structure and in the innovation process. The empirical evidence is in line with recent studies on BM and dynamic capabilities (Teece 2010; Teece and Linden 2017), including BM as a subject to innovation (Zott et al. 2011; Santos et al. 2009).

The article presented evidence of project creation for startups focused on innovation, adapting the guidelines of the parent company and the SDGs principles (Bosch

Table 5 Synthesis of DC development and empirical evidence

Dynamic capabilities	Evidence
1. Perceive, create opportunities (sensing)	• Strategic planning cycles to perceive and create opportunities for BM • Research with stakeholders to adapt BM to local conditions • Perception of innovation culture and sustainability "local to local" • Suggestion of projects for startups focused on innovation and sustainability, according principle 7 (SDGs) • Learning processes: technology, user and business
2. Configure (seizing)	• Changes in the innovation process, according principle 8 and 9 (SDGs) • Collaboration and promotion with government agencies • Creation of a physical area to intra-startups • Search of value proposition for the client/user from internal and external resources • Development of learning capacity to evaluate BM and projects
3. Reconfigure	• Human resources: entrepreneurs and leaderships allocated for the development of BM in innovation and sustainability • Adaptation of the "System Project" to the local market from existing assets in the subsidiary company

Source Elaboration of the author based on evidence and literature

2017), integrating sustainability criteria into systems from idea to product commercialization (Charter and Clark 2007), as a key issue for the creation of value.

The actions of the company are leading to the central and constitutive argument of DC as a multidimensional construct. The construction of systems suiting the sociotechnical environment by adjusting them to regulations, environmental pressures and consumer value, because system innovation is a crucial strategy for implementing sustainability in wider sociotechnical systems (Charter et al. 2008; Geels 2005; Johnson and Suskewicz 2009).

The contribution of the study is to present a management vision for DC associating with BM for sustainable innovation in the automotive industry. Though research on DC is related to processes, phenomena that occur in movement and in the evolution of time (Teece 2012), the article bring forward evidences to management practices in changes and adaptation to sustainability and the integration of SDGs into BM.

7 Limitations and Recommendations of Future Studies

The research period from 2015 to 2019 was a limitation but suggesting the evidence of DC. However longitudinal studies are necessary evaluating the adoption to the Sustainable Development Goals (SDGs), along with the other actors of the sociotechnical regime of the automotive industry.

Acknowledgements This article is the result of the Innovation in Business Model for Mobility project carried out under the Araucaria & Bosch Foundation Scholarship Program, subsidized by the Secretary of Science, Technology and Higher Education of the state of Paraná/Brazil. The authors are grateful to BOSCH Curitiba/PR and its collaborators, who kindly expended efforts and time to contribute to the study.

References

Adams R, Jeanrenaud S, Bessant J, Denyer D, Overy P (2016) Sustainability-oriented innovation: a systematic review. Int J Manag Rev 8(2):180–205
Bardin L (2011) Análise de conteúdo. São Paulo, Edições 70
Bergek A, Jacobsson S, Carlsson, B Lindmark, S Rickne, A (2008) Analyzing the functional dynamics of technological innovation systems: a scheme of analysis. Res Policy 37(3):407–429
Bergek A, Hekkert M, Carlsson B, Jacobsson S, Markard J, Sanden B, Truffer B (2015) Technological innovation systems in contexts: conceptualizing contextual structures and interaction dynamics. Environ Innov Soc Trans 16:51–64
Bell M, Figueiredo P (2012) Building innovative capabilities in latecomer emerging market firms: some key issues. In (Ed.), Innovative firms in emerging market countries, Oxford University Press. Retrieved 15 Apr 2019, from http://www.oxfordscholarship.com/view/10.1093/acprof:oso/9780199646005.001.0001/acprof-9780199646005
Brundtland GH (1987) Our common future: report of the world commission on environment and development. Geneva, UN-Dokument A/42/427. Retrieved 15 Apr 2019. http://www.un-documents.net/ocf-ov.htm
Bryman A (2012) Social research methods, 4th edn. Oxford University Press
Bosch (2018) Bosch annual report. Retrieved 14 Apr 2019, from https://www.bosch.com/our-company/sustainability/report
Bosch (2017) Connect for more sustainability. Retrieved from https://www.bosch.com/our-company/sustainability/
Borroni-Bird C (2006) The reinvention of the automobile. In: Nieuwenhuis P, Vergragt P, Wells P (eds) The business of sustainable mobility: from vision to reality. Greenleaf Publishing, Sheffield, England, pp 209–222
Calia RC, Guerrini FM, Moura GL (2007) Innovation networks: from technological development to business model reconfiguration. Technovation 27(8):426–432. https://doi.org/10.1016/j.technovation.2006.08.003
Carrillo-Hermosilla J, Del Río P, Könnola T (2009) Eco-innovation. When sustainability and competitiveness shake hands, Palgrave, London
Charter M, Clark T (2007) Sustainable innovation. Key conclusions from sustainable innovation conferences 2003–2006. The Centre for Sustainable Design, SEEDA, University College for the Creative Arts
Charter M, Gray C, Clark T, Woolman T (2008) Review: the role of business in realising sustainable consumption and production. In: Tukker A, Charter M, Vezzoli C, Stø E, Andersen MM (eds)

Perspectives on radical changes to sustainable consumption and production 1: system innovation for sustainability. Greenleaf, Sheffield, pp 46–69

Chesbrough HV, Rosenbloom RS (2002) The role of the business model in capturing value from innovation: evidence from xerox corporation's technology spinoff companies. Ind Corp Change 11(3):533–534

Freeman C, Soete L (2008) A economia da inovação industrial. Unicamp, Campinas

Gambardella AM, MCGahan A (2010) Business model innovation: general purpose technologies and their implications for industry structure. Long Range Plann 43:262–271

Geels FW (2005) Technological transitions and system innovations: a co-evolutionary and socio-technical analysis. Edward Elgar Publishing Ltd, Cheltenham, UK. ISBN 1845420098

Geels FW (2011) The multi-level perspective on sustainability transitions: responses to seven criticisms. Environ Innov Soc Trans 1(1):24–40

Geels F, Kemp R, Dudley G, Lyons G (2011) Automobility in transition? Sócio-technical development analysis of sustainable transport. Routledge Studies in Sustainability Transitions, Routledge, New York, pp 335–373

Hair JF, Babin B, Money A, Samuel P (2005) Fundamentos e métodos de pesquisa em administração. Porto Alegre, Bookman

Hekkert M, Negro S, Heimeriks G, Harmsen R (2011) Technological innovation. System analysis. A manual for analysts. Utrecht University, Report for Joint Research Center, Energy Institute

Johnson MW, Suskewicz J (2009) How to jump-start the clean tech economy. Harvard Bus Rev 87(11):52–60

Kemp R, Rotmans J (2008) Transitioning policy: co-production of a new strategic framework for energy innovation policy in the Netherlands. Submitted to Policy Sciences (2nd round of review)

Miles MB, Huberman AM (1994) Qualitative data analysis: an expanded sourcebook, 2nd edn. Sage, London

Nelson RR (1993) National innovation systems: a comparative analysis University of Illinois at Urbana-Champaign's academy for entrepreneurial leadership historical research reference in entrepreneurship. Available at SSRN. https://ssrn.com/abstract=1496195

OECD (2005) Organization for economic co-operation and development. Retrieved 14 Apr 2019, from http://www.oeDC.org/science/inno/2367580.pdf

Orsato RJ, Disk M, Kemp PR, Yarime M (2012) The electrification of automobility. In: Geels F, Kemp R, Dudley G, Lyons G (eds) Automobility in transition? A socio-technical analysis of sustainable transport. Routledge, 394 p

Ottman JA, Stafford ER, Hartman CL (2006) Green marketing myopia. Environment 5(48):22–36

Santos J, Spector B, Van der Heyden L (2009) Towards a theory of business model innovation within incumbent firms. Working Paper. Global Strategic Management. Emerald Group Publishing Limited, vol 16, pp 115–138

Schot JW, Geels FW (2008) Strategic niche management and sustainable innovation journeys: theory, findings, research agenda and policy. Technol Anal Strateg Manag 20:537–554

Simonsen LCI, Figueiredo PN (2018) Inovação e tecnologia no Brasil: desafios e insumos para o desenvolvimento de políticas públicas. Technological Learning and Industrial Innovation Working Paper Series, Rio de Janeiro, n. 1, pp 1–32. Retrieved 13 Apr 2019 from http://bibliotecadigital.fgv.br/ojs/index.php/tlii-wps/article/view/77828. http://dx.doi.org/10.12660/tlii-wps.77828

Tallman S (2014) Business models and the multinational firm. In: Boddewyn J (eds) Multidisciplinary insights from new AIB fellows (Research)

Teece DJ (2007) Explicating dynamic capabilities: the nature and microfoundations of (sustainable) enterprise performance. Strateg Manag J 28(13):1319–1350

Teece DJ (2010) Business models, business strategy and innovation. Long Range Plann 43(2–3):172–194

Teece DJ (2012) Dynamic capabilities: routines versus entrepreneurial action. J Manag Stud 49(8):1395–1401

Teece DJ, Linden G (2017) Business models, value capture, and the digital enterprise. J Organ Design 6(1)

Teece DJ, Pisano G, Shuen A (1997) Dynamic capabilities and strategic management. Strateg Manag J 18(7):509–533

Tidd J, Bessant J, Pavitt K (2015) Gestão da Inovação, 5ª Edição. Porto Alegre: Bookman

Tulder RV, Tilburg RV, Francken M, Da Rosa A (2014) Managing the transitions to a sustainable enterprise-lesson from frontrunner companies, 1st edn, Routledge

UN (United Nations) (2015). The power of principles: sustainable begins with a principle based to doing business. Retrieved 13 Apr 2019 from https://www.unglobalcompact.org/what-is-gc/mission/principles

Yin RK (2015) Estudo de caso: planejamento e métodos. 5ª Edição. Porto Alegre, Bookman

Zott C, Amit R, Massa L (2011) The business model: recent developments and future research. J Manag 37(4):1019–1042. Retrieved 13 Apr 2019 from http://www.sagepub.com/journalsPermissions.nav

The Use of Digital Transformation as a Sustainable Mechanism: An Automotive Industry Case

Pablo Carpejani, Bárbara Luzia Santor Bonfim Catapan,
Luiz Felipe Pierin Ramos, Izabelle Cristine Hannemann de Freitas,
Camila Mantovani Rodrigues, Edson Pinheiro de Lima,
Sergio Eduardo Gouvea da Costa, Eduardo de Freitas Rocha Loures,
Fernando Deschamps, José Marcelo A. P. Cestari and Eduardo Andrade

Abstract The adoption of a sustainability approach within emerged technologies is increasing across the years. It is becoming impossible to disassociate sustainable development from the use of information technology tools. The use of digital technologies has gone from optional to necessary, and including sustainability in this spectrum, is thinking toward and for future generations. The dynamism of the corporate market demands solutions with agile development. Introducing sustainability principles into agile and effective software projects ensures a rapid proliferation of the triple bottom line culture. The aim of this analysis is to present a simple but efficient solution that balances the sustainability pillars (economic, social, and environmental) within the human resources management of an automotive organization, specifically regarding employees' registration of time (time card). This article explores an online portal for clocking employees' time that is already connected with the pertinent internal systems that avoids the use of thermal paper and reduces the company's carbon footprint. Configured to engage employees in a "home office" strategy, this solution reduces CO_2 displacement.

Keywords Digital transformation · Sustainability · Human resources ·
Automotive industry

P. Carpejani (✉) · B. L. S. Bonfim Catapan · L. F. P. Ramos · I. C. H. de Freitas ·
C. M. Rodrigues · E. Pinheiro de Lima · S. E. Gouvea da Costa · E. de F. R. Loures ·
F. Deschamps · J. M. A. P. Cestari · E. Andrade
Department of Industrial and Systems Engineering, Pontifical Catholic University of Parana
(PUCPR), 1155 Imaculada Conceição—Prado Velho, Curitiba, Brazil
e-mail: pablo.carpejani@pucpr.edu.br

E. Pinheiro de Lima · S. E. Gouvea da Costa
Department of Industrial and Systems Engineering, Federal University of Technology - Parana
(UTFPR), Via do Conhecimento—Fraron, Pato Branco, Brazil

F. Deschamps
Department of Mechanical Engineering, Federal University of Paraná (UFPR), Curitiba, Brazil

© Springer Nature Switzerland AG 2020
W. Leal Filho et al. (eds.), *International Business, Trade and Institutional
Sustainability*, World Sustainability Series,
https://doi.org/10.1007/978-3-030-26759-9_6

1   Introduction: Sustainability in Industry

Organizations are adopting sustainability practices in their operations (Mangla et al. 2018; Govindan et al. 2016; Luthra et al. 2017; Sarkis and Zhu 2017), as industries adapt to the rapid and constant changes required to ensure the sustainable evolution of business from social, environmental, and economic perspectives (Stock and Seliger 2016; Elkington 2001). According to Elkington (2001), the adoption of those three perspectives, called the Triple Bottom Line (TBL), supports the organization in achieving greater sustainable development for society. More recent research considers other pillars for categorizing and measuring sustainable actions. For instance, Lemos (2010) proposes the following additional perspectives: (i) cultural, (ii) political-institutional, and (iii) spatial or territorial.

Innovation transforms the ways in which organizations manage technology in their flows and processes; the great demand of Industry has arisen from this transformation. Saltiél and Nunes (2017) explain that the industry scenario changed, because its main characteristic is the use of intelligent systems with a high degree of automation and connection. Nowadays, companies use modern information and communication technologies in the form of industrial automation, data networks, and new manufacturing technologies, such as intelligent production, human-computer interaction, 3D printing, and remote operations (Basl 2017; Khan et al. 2017; Duarte and Cruz-Machado 2017).

According to Luthra and Mangla (2017), the sustainability-oriented concept of industry encourages industrial managers not only to introduce environmental protections and control initiatives, but also to consider and promote safer workplace practices, such as resource efficiency; employee and community welfare; and, smarter, more flexible processes in their supply chains. The most significant potential for sustainability in industry—currently—is the capacity for identifying the most flexible production and the associated possibility of making more use of renewable energy sources, transferring production to periods of time when a lot of energy is fed into the network (Beier et al. 2015).

Directed through the application of intelligent devices and intelligent production systems, the current industry has the capacity to reduce production residues, overproduction, the movement of goods, and energy consumption (Kamble et al. 2018). Although the themes of sustainability and industry play an increasingly important role in the scientific environment, the scholarly literature on these themes is scarce. It is necessary to check how technological scenarios can contribute to sustainability in the industries (Stock et al. 2018).

2   Digital Transformation

The digital transformation of our modern era has seen the creation of a plethora of technologies that significantly impact both society and corporate environments. In companies, as efficiency has been improved, especially in manufacturing and logistics processes, it has increased the need to monitor specific data generated during the production lifecycle (Berman and Bell 2011; Bernhart et al. 2012). The automotive sector industries have been outstanding in the acquisition of technological solutions due to the complexity of their internal processes, both in the composition of their products and in the management of their services and projects, facilitating the interaction between the sectors where there are relevant communication gaps (KPMG 2014).

However, digital transformation has been broad tackling other areas that were limited to traditional business models. (Burkhart et al. 2011). For example, consider the use of technological tools to improve the quality of teaching provided in universities, high schools, and technical schools. This range can also be perceived through the way people's behavior has transformed in relation to urban mobility applications, such as Uber and Cabify. Digitization has become ubiquitous in many people's lives, and it is expected to expand further in the coming years as the cost of devices lowers, making them increasingly accessible to most of the population. Therefore, the need to improve the quality of services offered by businesses in order to better serve customers is also increasing. This last point has been well explored by startups worldwide, which, every year, develop solutions that have attracted the attention of distinct audiences and gained scalability in several countries.

The digital transformation is expected to bring great benefits to the economies of countries that are strategically planning for it (Ridgway et al. 2013; Siemieniuch et al. 2015). For instance, since 2011 the United States government has initiated a series of discussions, actions, and recommendations, at different levels, entitled "Advanced Manufacturing Partnership AMP" to ensure that the US is properly prepared to lead the next technological revolution in manufacturing (Rafael et al. 2014). In 2012, the German government approved the action plan "High-Tech Strategy 2020", which sets out billions of euros annually for the development of state-of-the-art technologies. As one of the ten future projects that make up this plan, "Industrie 4.0" represents the German ambitions for their manufacturing sector (Kagermann et al. 2013). The French government initiated a similar strategic review in 2013, called "La Nouvelle France Industrielle", in which 34 sectoral initiatives are defined as priorities for French industrial policy (Conseil National de l'industrie 2013). In the first half of 2017 the Ministry of Industry, Foreign Trade and Services (MDIC) in Brazil created a working group ("Task force") to make a plan to implement Industry 4.0 in the country. The Working Group has more than 50 representative institutions (including from government, companies, organized civil society, and more) charged with carrying out plans for Brazil's digital development over the next years. According to the Brazilian government (Ministério da Indústria, Comércio e Serviços 2018), the gains generated by the country will be significant. Specifically, they highlight a reduction in industrial costs around R$73 billion/year, an increase in efficiency gains of R$34 billion/year, a reduction in machinery maintenance costs of R$31 billion/year, and, finally, an energy savings of R$7 billion/year.

In Brazil, there is still an apparent difficulty in distinguishing between the terms "Industry 4.0" (Liao et al. 2017) and "digital transformation" (Bharadwaj et al. 2013; Fitzgerald et al. 2013). Some companies in the automotive sector have been piloting innumerable innovation initiatives, some at experiment and proof-of-concept (PoC) levels, which allows the concept to spread in the organization more quickly. The interaction between IT (information technology) and AT (automation technology), while still conflicting, is critical for successful project execution. There are several technical elements in which IT needs to be directly involved—for example, highlighting criteria regarding the security of information stored or sent to different types of platforms. The absence of IT involvement has generated significant delays in the execution of the projects, often leading to their cancellation (CFO Innovation Asia 2014). Therefore, the concept of Bimodal IT (interaction between traditional IT and digital IT) has been introduced and practiced recently in more structurally sound corporations because they realize that support for a product series lifecycle still needs to be maintained, but they also need to implement the necessary digital innovations that cannot be overlooked (Horlach et al. 2016).

Considering all the presented context, the objective of this work is to verify how environmental, social, and economic concepts can be used in the application of digital transformation.

3 Method

The application case intends to understand how digital transformation can be used as a mechanism of sustainability for organizations. In a way, the application cases activities have some similar issues related to a case study. Because of this, it is important to expose some fundamental concepts regarding case studies to, afterward, adopt and extends these concepts to the application cases of this research.

Although this research does not use the case study method, which aims to describe an event and to analyze the development of practice that grew from the event, thereby deepening the knowledge on the subject (Yin 2018). It follows some basic principles, such as dealing with single or multiple cases, qualitative and quantitative evidence, sources of evidence, and related to a prior development of theoretical propositions (Cestari et al. 2018).

A test case is a research strategy that uses an empirical inquiry to investigate a phenomenon within its real-world context, usually when the frontiers between the phenomenon and the context are not clearly defined (Yin 2018). It may be very useful in situations such as when the questions to be answered are based on "how" or "why", when the researcher has little control over the events and when the phenomena are complex and contemporary (dealing the day-today context). The application cases can follow the same aspects as proposed in Yin (2018) (e.g., explanatory, exploratory, descriptive).

The country of application is Brazil. Although the country has several vehicle factories and operations, only one was selected to compose this study.

4 Development

This section presents the development stage of the paper, which is composed of the context, the proposed solution, and the product vision related to the case.

4.1 Context

A company in the automotive sector found that they needed to move their administrative headquarters to a new workspace in order to reduce company costs, as well as to provide a harmonious workspace environment for their employees. Since one of the objectives of the company was to reduce costs, the company moved their headquarters' infrastructure to a smaller head office. However, with this transition, a new issue came up: the reduced infrastructure size caused overcrowding in the workspace. To solve this problem, the company established a relay system in which ninety employees needed to work from home (for, at least, several days of the week). This system had the added benefit of providing comfort and satisfaction for the staff, limiting fuel usage, traffic, and other issues.

Nevertheless, many barriers emerged regarding the need to register employees working hours, since the hardware they used to register time at the office was not available at their homes. Additionally, the company had to consider some specific legal aspects regarding the Brazilian law.

4.2 Proposed Solution

To solve the working hours registration problem for employees working from home, the company conducted a study to find the best option considering both costs and benefits. A benchmarking analysis was carried out involving three suppliers that provide some type of solution. The organization's Information Technology (IT) area was also involved in this study.

After undertaking a general analysis of the benefits that each solution could provide for the problem, the company determined which option was best for their needs. Table 1 presents some advantages and disadvantages of the solution offered by the IT of the company in comparison to the suppliers.

Aiming to solve the working hours registration issue for employees working from home, the company realized that there was no need to obtain a system that offered time record management mechanisms with reporting interfaces, edits, justifications, comments, and additional features. Therefore, the company chose to obtain the product offered by its own IT department, as it can perform the working hours registration and offers the best cost-benefit solution for the company, adopting a digital-transformation technology and low-code platform.

Table 1 Advantages and disadvantages of proposed solutions

	Advantages	Disadvantages
IT (internal solution)	– Working hours registration—lower cost solution – Direct connection to the human resources system – No need for human interaction or "report generation"	– Does not provide mechanisms to manage (edit) working time records
External suppliers	– Working hours registration—provides mechanisms to manage working time records	– Complex systems – No connection to internal systems – Provide only an interface for registering the time, but the data still needs to be inputted into the human resources systems – Higher cost solution (around 8 times higher)

Source The Authors (2018)

4.3 Product Vision

To solve the problem of registering working hours, the company's IT department developed an application called "Virtual Office". This system allows users (pre-authorized) to register the entry/exit timestamp of their workdays. The system operates similarly to the operation of a "clock card machine", but as an online portal.

Unlike the old solution, which required hardware to register the hours with the use of a card, the online portal allows the users, through a VPN/Intranet connection, to register their entry/exit time in a place other than the headquarters' office.

4.4 Product Premises and Restrictions

The system developed by the IT department has some premises and restrictions, which are presented in Table 2.

There is a person responsible for registering the users who can login and use the Virtual Office portal. To access the portal from an external Internet connection, the user needs to have the company's VPN. That way, risks related to cybersecurity will be reduced.

Table 2 System premises and restrictions

Premises	Restrictions
– Only previously authorized users can use the system – Users must be logged in within the organization's intranet to gain access to the portal	– The portal does not need to be "responsive" – The application is not available for external Internet access – It is not possible to manage/edit the time registered – The date-time is established by the company's internal server

Source The Authors (2018)

4.5 Sustainable Advantages

By bringing economic benefits to the company, the proposed solution also affects the environmental pillar of sustainability. The traditional registration system emits a paper that is 3.3 cm wide and 5.6 cm long each time the employee records their working hours—at entrance, exit, and at lunchtime. Approximately four papers are printed for each employee daily. The solution relies on the relay of ninety employees who will work a few days at home and other days at the company's headquarters. The new proposal of the online registering system offers an annual reduction of approximately 243 m^2 of thermal paper.

Among other benefits, it is worth mentioning that thermal paper contains a chemical compound called Bisphenol-A (BPA), which is highly toxic, making its recycling complex, as there is a risk of contaminating other materials. According to Otero and Carvalho (2014), BPA behaves like the estrogen hormone within the human body, because it is an endocrine disruptor capable of causing detrimental health consequences in children, young people, and adults. Previous studies have associated BPA with influencing the following health problems: prostate, breast, and ovarian cancer, as well as obesity and infertility (Otero and Carvalho 2014).

Another advantage of the company's new relay system is that it reduces their carbon footprint. Carbon footprint is an index that measures the impact of man's activities on nature based on the amount of carbon dioxide (CO_2) they emit. On average, Brazilians emit 17 kg of CO_2 per day. One of the main aspects considered in the calculation of this index is transportation. By implementing this solution, approximately ninety people will be reducing their displacement to travel to their work place. Consequently, the carbon footprint of these employees will decrease, which directly impacts the environment.

5 Discussion

The product developed by the company's IT department provided innovation gains while contributing to social, economic, and environmental sustainability. Despite being a simple product, the solution is complete. No other system reviewed by the company proposed to integrate the company's SAP. By performing the "check-in" on the system, the employee—from anywhere in the world with Internet access—is automatically integrated with the company's HR database. That is, unlike the products from external companies, the internally developed product integrates seamlessly with the company's current system. The online check-in action is the same input as the physical medium (through a card or biometrics).

The company's economic gains from using this product occur at the first moment in the cost of the product. The company saved financial resources by opting for internal development. In addition, the system enables the company to track its employees' working hours even in a home office environment, eliminating the high costs of renting or buying a new building. The environmental gains include the reduction in CO_2 emissions generated by fewer workers driving to work each day. In large corporations, like the one in this case study, this measure can contribute significantly to a reduction in pollutant emissions. Also, the online system lowers the use of BPA paper because the confirmation of the arrival time and exit time of the employee is displayed in the online platform (retiring the need for paper proof). The main social benefits are experienced by the workers. The city where the company operates has a lot of traffic, causing a displacement of—approximately—1 h for every 5 km traveled. The time saved to perform work activities can be directed to leisure activities or time with family. Also, for women with young children, working from home can fit their lifestyle better and, consequently, improve their working conditions.

6 Conclusion

This paper presents one way in which sustainability can be incorporated into the technological environment in the context of digital transformation. Activities considered simple, such as the control of employees' working hours (time card), can contribute to the sustainability of both the company and society. The case study can encourage organizations to integrate sustainability into their flows and processes, since there are economic, social, and environmental gains. Consequently, the digital transformation undertaken by these companies affects large markets.

As a limitation, this study verified examples of sustainability in only one company in a specific sector (automotive). For future research, it is recommended to expand the sample to verifying other sustainability actions that digital transformation favors and possible implementation barriers.

Acknowledgements This study was financed in part by the Coordenação de Aperfeiçoamento de Pessoal de Nível Superior—Brasil (CAPES)—Finance Code 001.

References

Basl J (2017) Pilot study of readiness of Czechcompanies to implement the principles of Industry 4.0. Manag Prod Eng Rev 8(2):3–8. https://doi.org/10.1515/mper-2017-0012

Beier G, Niehoff S, Maas A (2015) Nachhaltigkeitsaspekte von Industrie 4.0. Ökologisches Wirtschaften-Fachzeitschrift 30:8. https://doi.org/10.14512/oew300408

Berman SJ, Bell R (2011) Digital transformation: creating new business models where digital meets physical. IBM Inst Bus Value 1–17. https://doi.org/10.1108/10878571211209314

Bernhart W, Schlick T, Escobar JS (2012) The connected vehicle ecosystem: the race for the future profit pools is on. Automot Insights Roland Berg Strat Consult 14–19

Bharadwaj A, El Sawy OA, Pavlou PA, Venkatraman N (2013) Digital business strategy: towards a next generation of insights. MIS Q 37(2):471–482. https://doi.org/10.25300/MISQ/2013/37:2.3

Burkhart T, Krumeich J, Werth D, Loos P (2011) Analyzing the business model concept—a comprehensive classification of literature. In: ICIS Proceedings, Shanghai

Cestari J, Loures E, Santos E (2018) A method to diagnose public administration interoperability capability levels based on multi-criteria decision-making. Int J Inf Technol Decis Mak 17(01):209–245. https://doi.org/10.1142/S0219622017500365

CFO Innovation Asia (2014) Non-IT Departments Now Control Technology Budgets, Study Shows, http://m.cfoinnovation.com/story/8438/non-it-departments-now-control-technologybudgets-study-shows. Last accessed 11/28/2018

Conseil national de l'industrie (2013) The new face of industry in France. French National Industry Council, Paris

Duarte S, Cruz-Machado V (2017) Exploring linkages between lean and green supply chain and the industry 4.0. In: International conference on management science and engineering management. Springer, Cham, pp 1242–1252. https://doi.org/10.1007/978-3-319-59280-0_103

Elkington J (2001) Cannibals with forks. Makron Books, São Paulo

Fitzgerald M, Kruschwitz N, Bonnet D, Welch M (2013) Embracing digital technology. MIT Sloan Manag Rev 1–12

Govindan K, Seuring S, Zhu Q, Azevedo SG (2016) Accelerating the transition towards sustainability dynamics into supply chain relationship management and governance structures. J Clean Prod 112:1813–1823. https://doi.org/10.1016/j.jclepro.2015.11.084

Horlach B, Drews P, Schirmer I (2016) Bimodal IT: business-IT alignment in the age of digital transformation. MKWI Strat IT Manag 1417–1428

Kagermann H, Wolfgang W, Johannes H (2013) Recommendations for implementing the strategic initiative INDUSTRIE 4.0. Industrie 4.0 Working Group of Acatech, Berlin

Kamble S, Gunasekaran A, Gawankar SA (2018) Sustainable Industry 4.0 framework: a systematic literature review identifying the current trends and future perspectives. Process Saf Environ Prot 117:408–425

Khan M, Wu X, Xu X, Dou W (2017) Big data challenges and opportunities in the hype of Industry 4.0. In: 2017 IEEE international conference on communications (ICC), May. IEEE, pp 1–6. https://doi.org/10.1109/icc.2017.7996801

KPMG (2014) Kpmg's Global Automotive Executive Survey. 130811. KPMG International

Lemos IS (2010) Proposta de metodologia para classificação de destinos turísticos típicos segundo os princípios de sustentabilidade por meio de análise multicritério. Tese de doutorado em administração. Pontifícia Universidade Católica do Paraná

Liao Y, Deschamps F, Loures E, Ramos L (2017) Past, present and future of Industry 4.0—a systematic literature review and research agenda proposal. Int J Prod Res. https://doi.org/10.1080/00207543.2017.1308576

Luthra S, Govindan K, Kannan D, Mangla SK, Garg CP (2017) An integrated framework for sustainable supplier selection and evaluation in supply chains. J Clean Prod 140:1686–1698. https://doi.org/10.1016/j.jclepro.2016.09.078

Mangla SK, Luthra S, Jakhar S, Tyagi M, Narkhede B (2018) Benchmarking the logistics management implementation using Delphi and fuzzy DEMATEL. Benchmarking: Int J. 25(6). https://doi.org/10.1108/bij-01-2017-0006

Ministério da Indústria, Comércio e Serviços (2018) Agenda Brasileira para a Indústria 4.0. http://www.industria40.gov.br/. Last accessed 11/28/2018

Otero C, Carvalho M (2014) Bisphenol-A and the effects on human development: the breach of the rights of personality hidden in transparency and strength of plastic. http://www.publicadireito.com.br/artigos/?cod=eadd2c9c45ec261d. Last accessed 11/22/2018

Rafael R, Shirley AJ, Liveris A (2014) Report to the president accelerating U.S. advanced manufacturing. The President's Council of Advisors on Science and Technology, Washington, DC

Ridgway K, Clegg CW, Williams DJ (2013) The factory of the future. Future of Manufacturing Project: Evidence Paper 29. Government Office for Science, London

Saltiél R, Nunes F (2017) A Industria 4.0 e o Sistema Hyundai de produção: suas interações e diferenças". Anais do V Simpósio de Engenharia de Produção-SIMEP

Sarkis J, Zhu Q (2017) Environmental sustainability and production: taking the road less travelled. Int J Prod Res 1–17. https://doi.org/10.1080/00207543.2017.1365182

Siemieniuch CE, Sinclair MA, Henshaw MJ (2015) Global drivers, sustainable manufacturing and systems ergonomics. Appl Ergon 51:104–119. https://doi.org/10.1016/j.apergo.2015.04.018

Stock T, Seliger G (2016) Opportunities of sustainable manufacturing in industry 4.0. Procedia CIRP 40:536–541. https://doi.org/10.1016/j.procir.2016.01.129

Stock T, Obenaus M, Kunz S, Kohl H (2018) Industry 4.0 as enabler for a sustainable development: a qualitative assessment of its ecological and social potential. Process Saf Environ Prot 118:254–267. https://doi.org/10.1016/j.psep.2018.06.026

Yin RK (2018) Case study research and applications: design and methods, 6th edn. Sage, Thousand Oaks, 319p

Comparative Study of LCIA, MFCA, and EPIP Tools for the Environmental Performance Evaluation in Industrial Processes

Marcell Mariano Corrêa Maceno, Urivald Pawlowsky and Thaísa Lana Pilz

Abstract This study aimed to compare the environmental performance measurement in industrial processes by three different tools, LCIA, MFCA, and EPIP. Thus, a case study of a planned furniture industrial process was carried out. Visits were made in the industrial process for observation and data collection. The functional unit defined was 1 month of production, and the boundary considered was the industrial process area. Data collection addressed the identification of Environmental Aspects (EA) and Cost Centers (CC), as well as quantitative information such as masses, energy, operating times, machine power, employee numbers, and costs. Subsequently, the Spearman's Correlation Coefficient was used to analyze the sensitivity of the three tools. Thus, the most critical EAs were Mixed powder, and VOCs of drying for both EPIP, and LCIA. For the MFCA, the most critical CCs were Cyclone and Rolling mill. Through the Spearman's Rank Correlation Coefficient, it was verified that EPIP showed influence of a larger number of input parameters. This was because the EPIP has a greater number of analysis variables, presenting greater comprehensiveness in the parameters of environmental performance. Then, the EPIP showed more robustness to evaluate environmental performance compared to other tools.

Keywords Environmental performance evaluation · Industrial processes · Environmental accounting · Environmental impacts

M. M. C. Maceno (✉)
Department of Production Engineering, UFPR, Francisco Heráclito dos Santos Street 100, Jardim das Américas, Curitiba, Paraná, Brazil
e-mail: marcell.maceno@gmail.com

U. Pawlowsky
Department of Hydraulics and Sanitation, UFPR, Francisco Heráclito dos Santos Street 100, Jardim das Américas, Curitiba, Paraná, Brazil
e-mail: urpawl@gmail.com

T. L. Pilz
Production Engineering Post-Graduate Program, UFPR, Francisco Heráclito dos Santos Street 100, Jardim das Américas, Curitiba, Paraná, Brazil
e-mail: thaisa.pilz@gmail.com

© Springer Nature Switzerland AG 2020   107
W. Leal Filho et al. (eds.), *International Business, Trade and Institutional Sustainability*, World Sustainability Series,
https://doi.org/10.1007/978-3-030-26759-9_7

1 Introduction

Industrial activity contributes significantly to the levels of global pollution in the environment, due to resource consumption, generation and release of pollutants, and energy inefficiency (Maceno et al. 2018). Manufacturing activities are estimated to have consumed 37% of the world's energy produced in 2012 and released 18% of greenhouse gases into the atmosphere in 2010 (Unep 2012). In addition, the environmental quality is still degraded by the discharges of solid and liquid wastes from these industrial operations (Kumar and Pal 2013; Pizer et al. 2011).

In contrast to the generation of pollution, in the last few years, scientists, governments, and society have begun to press the industry for actions to reduce and maintain their Environmental Impacts (EI). Thus, the industrial activity began to take into account, as a marketing strategy and as a way of securing the business survival, the requirements of consumers, policy makers, environmentally conscious partner companies (Carvalho et al. 2014), international standards, more restrictive legislation, and marketing competition. To answer these demands, companies have explored environmental management strategies based on voluntary administrative tools (Herva and Roca 2013). These tools can be divided into process tools and analytical tools. The first, aims to structure the procedures for environmental improvement actions, such as the Environmental Management System (EMS) and the Environmental Performance Evaluation (EPE). The second, assists the decision making through the measurement or calculation of the environmental performance, such as the Life Cycle Impact Assessment (LCIA) methods and the Environmental Performance Indexes.

In order to facilitate the understanding and evaluation of economic, technological, and environmental systems, many scientists and marketing professionals recommend the use of the analytical tools in decision making processes (Höjer et al. 2008). These analytical tools can be used within the process tools to aid decision making by taking actions to reduce environmental impacts and improve environmental performance (EP) (Höjer et al. 2008).

In the literature, several analytical tools are available to evaluate the environmental impacts of industrial processes (Carvalho et al. 2014), and to evaluate the environmental performance of these activities (da Silva and Amaral 2009), but with a wide variety of approaches.

Among the existing analytical tools, those that are based on Environmental Impact Assessment are the most used, being cited in several scientific studies, as in Blanco Morón et al. (2009), Deng et al. (2014), da Silva and Amaral (2009), Herva and Roca (2013), and Rigamonti et al. (2014).

This environmental performance is directly related to economic performance when it comes to assessing the environmental performance of companies (da Silva and Amaral 2009). Therefore, some authors have cited analytical tools that use environmental accounting to evaluate the environmental performance in companies, or even the aggregation of accounting and environmental impact assessment tools in this evaluation (Ahlroth 2014; Finnveden and Moberg 2005; Höjer et al. 2008).

In this context, this study aimed to compare three different analytical tools in relation to its results in the measurement and decision support in the assessment of environmental performance: Material Flow Cost Accounting (MFCA) (ISO 2011), Life Cycle Impact Assessment (LCIA) composed by IMPACT 2002+ vQ2.22 (Humbert et al. 2015; Jolliet et al. 2003), EDIP 2003 (Danish Environmental Protection Agency 2005), ReCiPe (Goedkoop et al. 2013), and Water Scarcity (Pfister et al. 2009), and Environmental Performance in Industrial Processes (EPIP) (Maceno et al. 2018). In addition, it was used a case study of the planned furniture industrial process.

2 Methodology

To enrich this study, an industrial process was selected, and data collections were performed to include the three tools analyzed.

After defining the focus of the work, which it was to compare the environmental performance by MFCA, LCIA, and EPIP tools, a survey was carried out regarding the industrial processes in Curitiba and its Metropolitan Region with possibility and availability to the application of these tools.

This survey took into consideration the following criteria to be met:

- Existing contact in the company, by personal knowledge or indication;
- Predisposition and interest of the companies contacted;
- Existing data to input into the tools;
- Ease of obtaining and making available data for the study;
- The possibility of visits to the industrial process to analyze the field conditions and to confirm the consistency of the obtained data in its records. Data should be collected regarding the model and power of the machines, segregation forms and mixing of waste, and the volumes of recycled wastewater, among other factors.

Therefore, considering the fulfillment of the presented criteria, this study selected a planned furniture industrial process to apply the tools.

Then, a trial was carried out at its premises to define the system boundary (limited data) and the functional unit (1 month of production).

Subsequently, the information on the production process was collected. At this stage, the process flowchart was elaborated, and the environmental aspects referring to the raw materials, inputs, and waste in the production process were identified.

The quantification of the following items was also performed in order to fill in the data collection forms: raw materials, inputs consumed, waste generated, material unit costs used in the process, electrical energy costs, waste action costs, power and duration of machine use in the production process, and waste destinations. In addition, the Cost Centers (CCs) required for the MFCA were defined, using the physical quantities and the costs raised for allocation in these centers. For the MFCA, it was necessary to determine the labor costs.

In sequence, the environmental impact categories due to materials consumption and emission, and energy consumption, were identified, followed by the identification of the compounds and types of energy for each impact category.

The data entered on the forms were transferred to Microsoft Excel® for the completion of the tests and analysis of the results. The data collected in the field were registered in three files of Microsoft Excel® to LCIA, EPIP, and MFCA tools.

Lastly, after recording all the data into the tools, the results were obtained with the classification of the environmental aspects, being that the highest values to the environmental aspects represented the worst environmental performance. It was possible to compare the results of the tools and analyze the different responses using the Spearman's Rank Correlation Coefficient. These results made it possible to draw conclusions about the performance and amplitude of the studied tools.

3 Results and Discussion

The industrial process studied has been a local manufacturing company of planned furniture for home environments and offices, including wardrobes, kitchen furniture, and desks, among others. It is a small company, located in the City of Pinhais in Paraná, Brazil. The details of the production process, the data collection, and the results obtained by applying the three environmental performance tools are showed in the following items.

Data collection in this industrial process was carried out between the months of April and June of 2015. In the months of April and May, biweekly visits were held to understand and map the process. In June, data collection and filling of the data forms were executed. Due to the work with customized projects, this company has partners in the fields of architecture and in civil construction, with stable demands throughout the months. This fact reduces the seasonality of the production and facilitates the application of the tools for data gathered in one production month.

3.1 Planned Furniture Industrial Process

The industrial process has a wide variety of manufacturing products, using MDF (Medium Density Fiberboard) as its main raw material. The main processes mapped were cutting, finishing, treatment, and painting of the wood. Its flowchart is presented according to Fig. 1.

The beginning of the production process is given by the collection of customer orders. Its layout is cellular, and each employee is responsible for the processes of sliding a table saw as well as grinding, sanding, sealing, and varnishing, for the same group of furniture. Therefore, production starts by requesting stock materials to be sent to the production areas.

In the first process, called sliding a table saw (Fig. 2), the joiner receives MDF

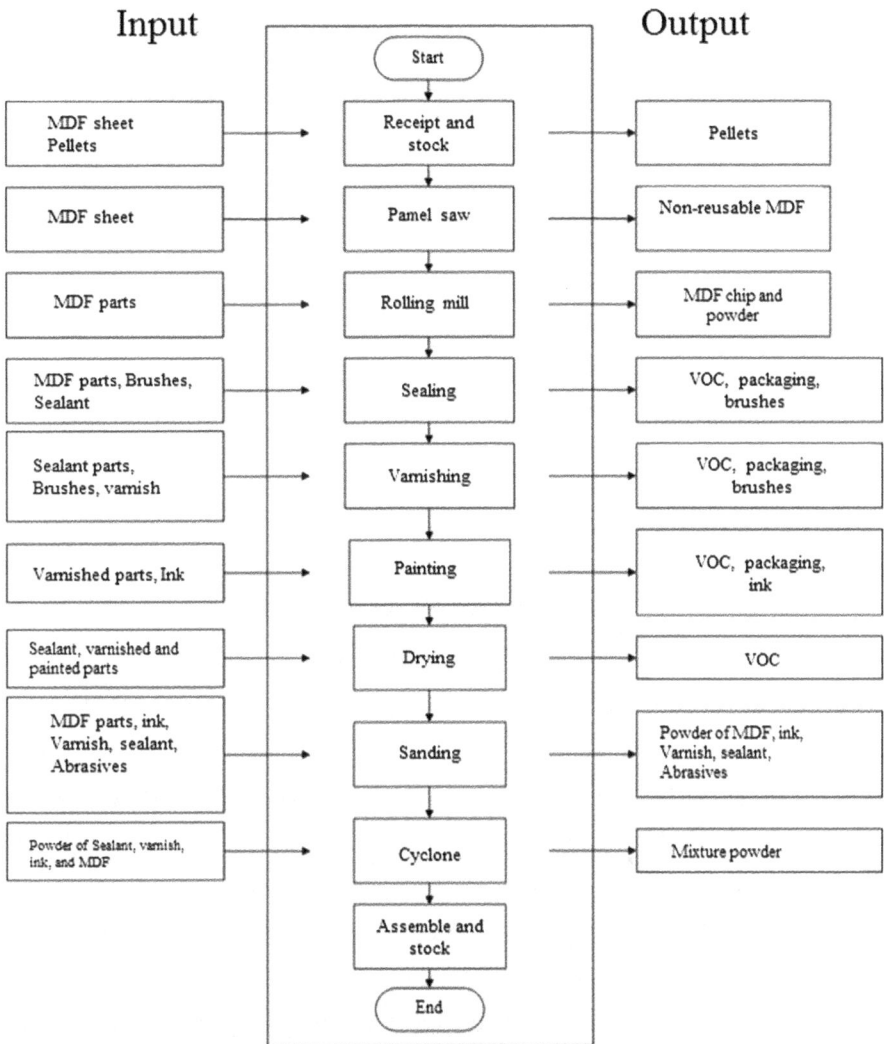

Fig. 1 Flowchart of the planned furniture industrial process

sheets with a mean dimension of 2750 × 1850 × 15 mm and performs pre-defined design cuts according to the furniture to be produced. This process is performed to obtain the larger parts of this furniture, such as doors, tables, and cabinet base, among others. After cutting the sheets into large parts, the joiner uses woodworking routers to make cuts to produce smaller parts. Subsequently, the finishing is made, including the rounding of edges, holes, and friezes. Then, the wood is sanded to smooth the furniture and to remove imperfections, followed by the pre-assembly of the furniture produced to verify possible imperfections during the cutting and finishing steps. If the furniture is not approved in this assembly test, it must return

Fig. 2 Picture of the precision sliding table saw used in a joiner workstation

to the area of the woodworking routers for adjustment. However, if approved, the furniture is disassembled to go into the visual finishing process.

In the visual finishing process, the joiner inspects the furniture, and begins applying base chemicals to the wood.

The first chemical applied is the sealer, which has the function of sealing the wood from the entrance of humidity and possibly degrading agents, such as termites. The sealing is done with a synthetic enamel applied with a brush. After, the furniture is oven dried for a time of up to 8 h. The furniture then returns to the joinery area and is sanded to remove excess sealant.

In sequence, a varnish coat is applied over furniture, and the drying and sanding processes are repeated. These varnishing, drying, and sanding processes repeat twice more. In the third varnishing coat, drying can take up to 24 h before the parts are painted.

The varnished furniture goes to the painting cabinet (Fig. 3). In this process, its parts are painted with the use of a paint gun and are sent to the drying area for a time of 8 h. Again, after drying, the parts are sanded. This process is repeated once more, and the parts can remain up to 24 h in the drying oven.

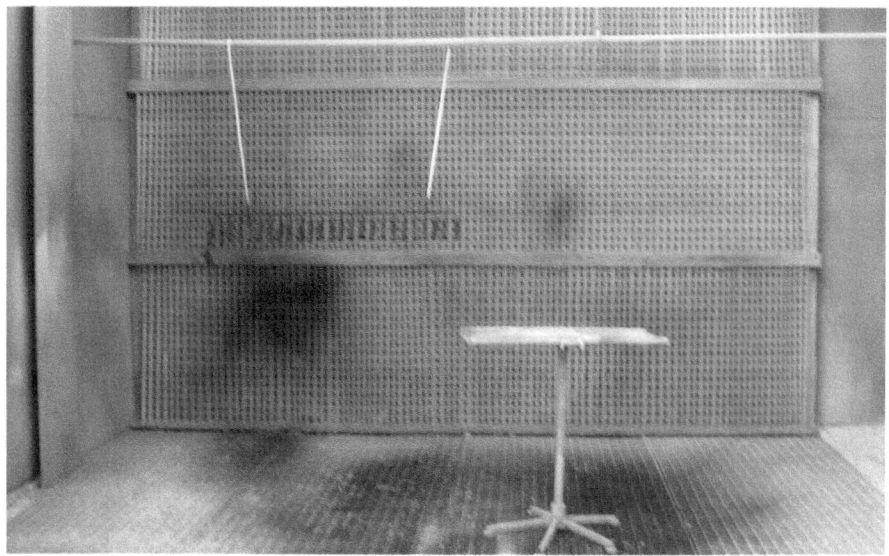

Fig. 3 Picture of painting area

3.2 Data Collection of the Planned Furniture Industrial Process

The industrial process studied is composed of five joinery areas, containing a sliding table saw and a woodworking router in each of them. Each joinery area has 1 joiner and 1 auxiliary joiner. In addition, there are two painting cabinets composed of 2 painters that receive the finished parts of the woodwork. The industrial process operates from Monday to Friday, from 8:00 a.m. to 12:00 p.m., and from 2:00 p.m. to 6:00 p.m., with a staff of 15 employees, being 1 designer, 1 administrator, 1 secretary, and 12 employees of the production. The Functional Unit (FU) used for the 3 environmental performance analytical tools was 1 month, referring to June of 2015.

Therefore, Table 3 (Appendix 1) presents the processes list, environmental aspects, and material inputs and outputs in the form of product or waste. Table 4 (Appendix 1) presents impact categories for each environmental aspect. Tables 3 and 4 have served as a support for the registration of the EPIP and LCIA tools.

Subsequently, the data already collected were organized in the form of Cost Centers to enable the application of the MFCA tool, according to Table 5 (Appendix 1). For the calculation of the MFCA results, the labor costs by activity, and the total time for each activity, were raised.

Finally, for the calculation of the results of the LCIA, the integration of the LCIA methods proposed in the EPIP was defined as a representative method.

3.3 Results of the Planned Furniture Industrial Process for the EPIP, MFCA, and LCIA Tools

The 46 environmental aspects and the 8 Cost Centers of the industrial process to be improved in their environmental performance were analyzed by the environmental performance calculations. The difference of responses to the rank among the tools was also analyzed. The results obtained for each one of these 3 tools are presented in Fig. 4.

In this Fig. 4 was verified that the A07 (Mixed powder) was the one that obtained the best ranking by the EPIP and LCIA tools, i.e. the worst environmental performance. Moreover, for the MFCA tool, the CC 3 (Cyclone) was the one with the worst environmental performance, and in this activity the A07 is found.

Furthermore, the first five environmental aspects ranked in the EPIP and LCIA tools and the five first CCs ranked for the MFCA were organized in Table 1, in order to compare the results between the tools. These environmental aspects and CCs were used because corresponded to values greater than 60% of the total environmental performance calculated for the industrial process studied.

According to the results of these first five environmental aspects in the EPIP tool, it was verified that the A07 was influenced by costs with material loss, energy consumption, and the material destination (Fig. 5), and mainly by the environmental impact due to the presence of hazardous waste and the consumption of the electric energy resource (Fig. 6). This environmental aspect accounted for 87.9% of the calculation of the total environmental performance of this industrial process and it obtained the first position in the EPIP ranking. Subsequently, the A28 (VOC of drying) was influenced for the material loss costs (Fig. 5), and mainly on the VOC potential impact in human health (Fig. 6). For this environmental aspect, the value obtained by LCIA tool moved this EA from the third position to the second position of the rank. For the A22 (Contaminated filter), ranked in the third position by the EPIP, the influence of the material loss cost and mainly the lower value of environmental impacts in relation to A28 was verified. Finally, for the A03 (MDF pieces non-reusable) and A20 (MDF powder—with varnish), ranked in the fourth and fifth positions, respectively, the material loss costs were the ones that influenced the position of these environmental aspects in the EPIP. Nonetheless, the lower environmental impact of the A03, regarded as a non-hazardous waste, made it moves down from the second position in the cost analysis rank to the fourth position in the EPIP tool.

For the LCIA result (Fig. 6), the A07 (Mixed powder) was ranked in the first position, as well as in the EPIP tool. This environmental aspect accounted for 47.7% of the total environmental impact of this industrial process according to the LCIA tool. This percentage was lower than the percentage result of the total environmental

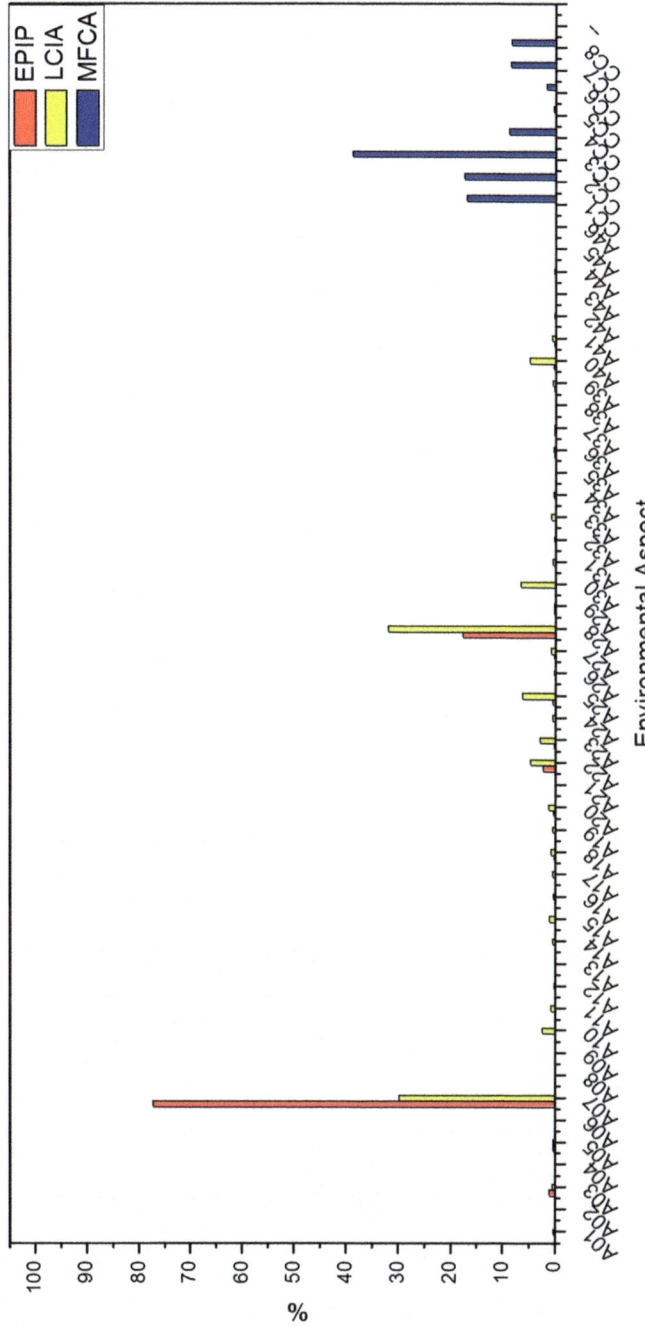

Fig. 4 Environmental performance results of the environmental aspects lifted for the planned furniture industrial process by the EPIP and LCIA tools, and the environmental performance of the CCs by the MFCA tool

Table 1 Results of the environmental aspects ranking obtained through the application of the EPIP, LCIA, and MFCA tools, for the planned furniture industrial process

Ranking	EPIP	LCIA	MFCA
1	A07	A07	CC3
2	A28	A28	CC2
3	A22	A22	CC1
4	A03	A25	CC4
5	A20	A30	CC7

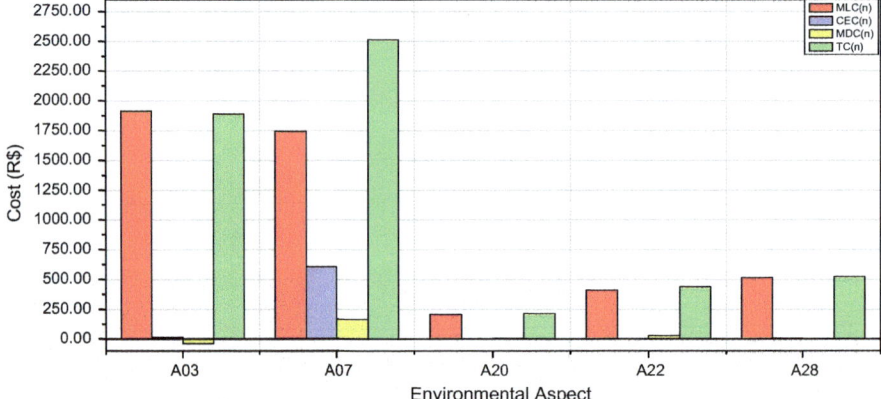

Fig. 5 Results obtained from material loss cost (MLC), consumed energy cost (CEC), material destination cost (MDC), and total cost (TC), for the five environmental aspects with the worst environmental performances in the planned furniture industrial process by EPIP tool

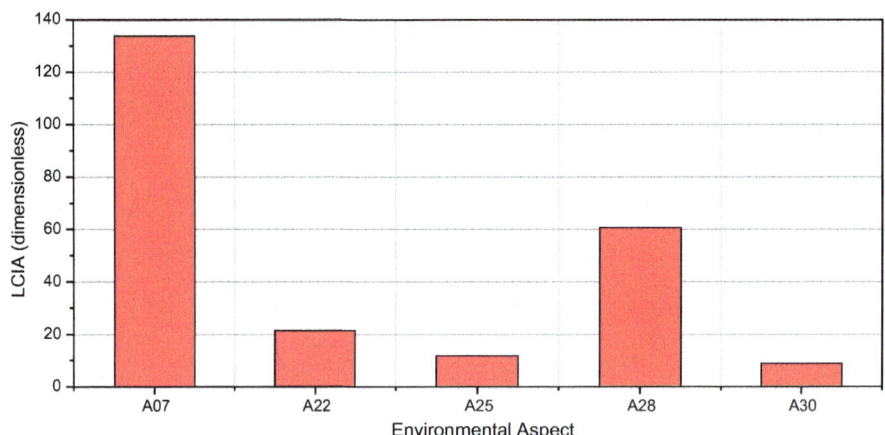

Fig. 6 Environmental performance results for the five environmental aspects with the worst environmental performances in the planned furniture industrial process by LCIA tool

performance obtained for this aspect by EPIP. That is, the values of material loss costs, energy consumption, and material destination boosted up the result of A07 in EPIP in relation to LCIA. Subsequently, A28 (VOCs of drying) and A25 (VOCs of painting), ranked in the second and fourth positions to the LCIA, had a greater influence of the impacts related to human health, although their masses were lower than A22 (Contaminated filter). Finally, A22 and A30 (Packaging waste) were influenced by the mass quantities and by the impacts related to the generation of hazardous waste, ranking them in the third and fifth positions, respectively.

Finally, the results of the CCs by the MFCA for the material loss costs, energy costs, system costs (labor), and waste management costs, are shown in Fig. 7. After analyzing these results, it was seen that CC 03 (Cyclone) achieved the worst environmental performance result among the Cost Centers of this industrial process. The influence of the costs with the consumed energy and with the material losses were predominant for the positioning of this CC in the ranking of the MFCA. Subsequently, the CC 02 (Rolling mill), CC 01 (Receipt, stock, and panel saw machine), CC 04 (Manual Sanding), and CC 07 (Painting), were ranked from second to the fifth position, and the material loss cost was the variable that most influenced the rank of these CCs.

As can be verified, each one of the tools presented specific influences for the rank of environmental aspects and Cost Centers in relation to the environmental performance calculated. Thus, the sensitivity analysis was performed to compare the responses of the studied tools. Table 2 was elaborated with the quantified values of the Spearman's Rank Correlation Coefficient.

The results presented on Table 2 show that the environmental performance quantified by the EPIP tool for the industrial process studied was influenced by five of the six analyzed variables: Mass, MLC, CEC, TC and EI. Despite this, strong correla-

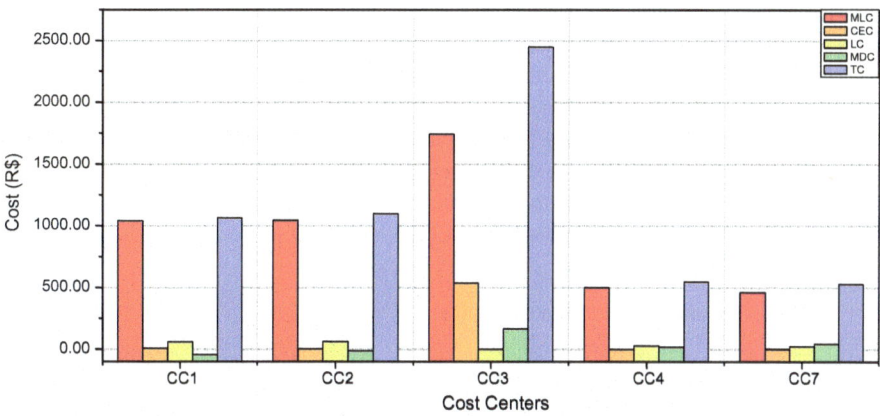

Fig. 7 Results of environmental performance of the cost centers obtained for waste: MLC (material loss cost), CEC (consumed energy cost), LC (labor cost), MDC (material destination cost), and TC (total cost) for the five CCs with the worst environmental performances for the planned furniture industrial process by MFCA tool

Table 2 Spearman's rank correlation coefficient result obtained between the input parameters considered essential for a robust environmental performance evaluation, and the three tools analyzed for the planned furniture industrial process. Trust level $\alpha \leq 0{,}05$

	EPIP	LCIA	MFCA
Mass (n)	0.395 (V)	0.086 (F)	0.659 (V)
MLC (n)	0.531 (V)	0.282 (F)	0.385 (V)
CEC (n)	−0.504 (V)	−0.098 (F)	−0.168 (F)
MDC (n)	0.190 (F)	0.223 (F)	0.028 (F)
TC (n)	0.708 (V)	0.180 (F)	0.950 (V)
EI (n)	0.720 (V)	1.000 (V)	0.180 (F)

MLC: material loss cost; CEC: consumed energy cost; MDC: material destination cost; TC: total cost (MLC + CEC + MDC); EI: environmental impact; (V): true hypothesis test; (F): false hypothesis test. $p = 0.05$

tions occurred only for the TC and EI parameters. The LCIA tool was influenced only by the EI parameter. Finally, the MFCA presented true correlations for the variables Mass, MLC and TC, and the only strong correlation obtained was with the variable TC.

4   Conclusion

Simulations performed with the EPIP, LCIA, and MFCA tools, showed differences in the results of the classifications obtained for each tool. This ranking serves as a guiding element to assist decision makers regarding items to be improved in environmental performance, in order to make the industrial process more efficient and with less environmental impact.

In a general analysis of the sensitivity analysis results using the Spearman's Rank Correlation Coefficient for the three tools, it was verified that the EPIP showed influence of a greater number of input parameters in the case study. This may be to the fact that EPIP has a greater number of analysis variables, presenting greater comprehensiveness in the parameters of environmental performance evaluation in an industrial process. In this context, the EPIP presented more robustness to evaluate the environmental performance compared to the other tools.

Appendix 1: Data Collection of Case Study

See Tables A, B and C.

Table A Process, environmental aspects outputs, costs, energy and destination data for case study

			MLC (n)		CEC (n)			MDC (n)		
Id (n)	Process	Environmental aspect	Outputs (kg)	Material cost (R$/kg)	CE (kWh)	EUC (R$/kWh)	PE (%)	Destination type (d)	Percentage destination (%)	Destination cost (R$/kg)
A01	Receipt and stock	Pallets of wood for disposal	60.0	1.32	0.0	0.6621	0.02	4	100	−0.400
A02	Panel saw machine	MDF sheets	2524.5	0.00	484.0	0.6621	0.96	0^a	0	0.000
A03	Panel saw machine	MDF pieces non-reusable	100.5	19.05	484.0	0.6621	0.04	5	100	−0.400
A04	Rolling mill	MDF parts	2469.7	0.00	193.6	0.6621	0.98	0^a	100	0.000
A05	Rolling mill	MDF chips	51.0	19.05	193.6	0.6621	0.02	5	100	−0.400
A06	Rolling mill	MDF clean powder (rolling mill)	3.8	19.05	193.6	0.6621	0.00	9	100	1.820
A07	Cyclone	Mixed powder	91.5	19.05	814.0	0.6621	1.00	9	100	1.820
A08	Manual sanding (clean parts)	Abrasives used	1.0	105.00	0.0	0.6621	0.00	9	100	1.800
A09	Manual sanding (clean parts)	MDF clean powder (sanding)	4.8	19.04	0.0	0.6621	0.00	9	100	0.820

(continued)

Table A (continued)

Id (n)	Process	Environmental aspect	MLC (n) Outputs (kg)	Material cost (R$/kg)	CEC (n) CE (kWh)	EUC (R$/kWh)	PE (%)	MDC (n) Destination type (d)	Percentage destination (%)	Destination cost (R$/kg)
A10	Sealing	VOCs of sealing	0.7	10.50	0.0	0.6621	0.00	0[a]	0	0.000
A11	Sealing	Sealant packaging	2.1	0.00	0.0	0.6621	0.00	9	100	1.800
A12	Sealing	Brushes contaminated with sealer	0.2	38.90	0.0	0.6621	0.00	9	100	1.800
A13	Manual sanding (sealing parts)	Abrasives contaminated with sealer	0.0	105.00	0.0	0.6621	0.00	9	100	1.800
A14	Manual sanding (sealing parts)	Sealant powder	1.2	10.50	0.0	0.6621	0.00	9	100	1.800
A15	Varnishing	VOCs of varnishing	0.3	59.50	0.0	0.6621	0.00	0[a]	0	0.000

(continued)

Table A (continued)

Id (n)	Process	Environmental aspect	MLC (n) Outputs (kg)	Material cost (R$/kg)	CEC (n) CE (kWh)	EUC (R$/kWh)	PE (%)	MDC (n) Destination type (d)	Percentage destination (%)	Destination cost (R$/kg)
A16	Varnishing	Brushes contaminated with varnish	0.7	38.90	0.0	0.6621	0.00	9	100	1.800
A17	Varnishing	Varnish packaging	1.3	0.00	0.0	0.6621	0.00	9	100	1.800
A18	Varnishing	Cloths contaminated with varnish	2.3	6.30 / 59.50	0.0	0.6621	0.00	9	100	1.800
A19	Manual sanding (varnished parts)	Abrasives contaminated with varnish	0.1	105.00	0.0	0.6621	0.00	9	100	1.800
A20	Manual sanding (varnished parts)	MDF powder (with varnish)	3.5	59.50	0.0	0.6621	0.00	9	100	1.800
A21	Painting	Painted parts	2620.4	0.00	66.0	0.6621	0.99	0[a]	0	0.000

(continued)

Table A (continued)

Id (n)	Process	Environmental aspect	MLC (n)		CEC (n)			MDC (n)		
			Outputs (kg)	Material cost (R$/kg)	CE (kWh)	EUC (R$/kWh)	PE (%)	Destination type (d)	Percentage destination (%)	Destination cost (R$/kg)
A22	Painting	Contaminated filter	14.7	188.60	66.0	0.6621	0.01	9	100	1.800
A23	Painting	Ink packaging	8.8	26.10	66.0	0.6621	0.00	9	100	1.800
A24	Painting	Contaminated cloths with paint	1.6	32.40	66.0	0.6621	0.00	9	100	1.800
A25	Painting	VOCs of painting	2.5	26.10	66.0	0.6621	0.00	0[a]	0	0.000
A26	Manual sanding (painted parts)	Abrasives contaminated with ink	0.2	105.00	0.0	0.6621	0.00	9	100	1.800
A27	Manual sanding (painted parts)	MDF powder (paint)	2.2	26.10	0.0	0.6621	0.00	9	100	1.800
A28	Drying	VOCs of drying	12.8	40.20	1600.0	0.6621	0.00	0[a]	100	0.000

(continued)

Table A (continued)

Id (n)	Process	MLC (n)			CEC (n)			MDC (n)		
		Environmental aspect	Outputs (kg)	Material cost (R$/kg)	CE (kWh)	EUC (R$/kWh)	PE (%)	Destination type (d)	Percentage destination (%)	Destination cost (R$/kg)
A29	Drying	Varnished and painted parts	2587.6	0.00	1600.0	0.6621	1.00	0[a]	0	0.000
A30	Packaging, stock, and shipping	Packaging waste	20.4	0.00	0.0	0.6621	0.01	5	100	−0.300
A31	Machine maintenance	Used oil	1.5	15.80	0.0	0.6621	1.00	5	100	−0.300
A32	General services	Scavenging waste	22.5	0.00	0.0	0.6621	0.00	9	100	0.820
A33	General services	Contaminated cloths	2.1	2.20	0.0	0.6621	0.00	9	100	1.800
A34	General services	Cleaning product packaging	0.9	0.00	0.0	0.6621	0.00	9	100	1.800
A35	General services	Wastewater of washing	572.5	0.00542	0.0	0.6621	1.00	8	100	0.004
A36	General activities	Wastewater	15440.0	0.00542	0.0	0.6621	1.00	8	100	0.004

(continued)

Table A (continued)

Id (n)	Process	Environmental aspect	MLC (n) Outputs (kg)	Material cost (R$/kg)	CEC (n) CE (kWh)	EUC (R$/kWh)	PE (%)	MDC (n) Destination type (d)	Percentage destination (%)	Destination cost (R$/kg)
A37	General activities	Used toilet paper	22.5	10.80	0.0	0.6621	0.00	9	100	0.820
A38	General activities	Used plastic cups	0.5	10.90	0.0	0.6621	0.00	9	100	0.820
A39	General activities	Used lamps	1.0	25.90	528.0	0.6621	1.00	8	100	11.900
A40	General activities	IPE (individual protection equipment)	15.0	1.00	0.0	0.6621	1.00	9	100	1.800
A41	Kitchen	Organic waste	105.0	0.00	0.0	0.6621	0.00	9	100	1.800
A42	Kitchen	Plastic packaging	30.0	0.00	0.0	0.6621	0.00	9	100	0.820
A43	Office	Paper	2.1	5.65	20.0	0.6621	1.00	9	100	0.820
A44	Office	Printer cartridge	0.2	0.00	20.0	0.6621	0.11	9	100	0.820
A45	Office	Plastic packaging	0.0	0.00	20.0	0.6621	0.01	5	100	−3.000
A46	Office	Computer use	0.0	0.00	207.7	0.6621	1.00	0[a]	0	0.000

[a]Zero destination because the environmental aspect in the output is a product and not a waste

Table B Impact category data to each environmental aspect n

ID (n)	W[a]	WU[a]	NRE[a]	HT[a]	RIn[a]	PO[a]
A01	Non-hazardous waste	NA	Electric energy, BR	NA	NA	NA
A02	NA	NA	Electric energy, BR	NA	NA	NA
A03	Non-hazardous waste	NA	Electric energy, BR	NA	NA	NA
A04	NA	NA	Electric energy, BR	NA	NA	NA
A05	Non-hazardous waste	NA	Electric energy, BR	NA	NA	NA
A06	Non-hazardous waste	NA	Electric energy, BR	NA	Total particulate matter	NA
A07	Hazardous waste	NA	Electric energy, BR	NA	Total particulate matter	NA
A08	Non-hazardous waste	NA	Electric energy, BR	NA	NA	NA
A09	Non-hazardous waste	NA	Electric energy, BR	NA	Total particulate matter	NA
A10	NA	NA	Electric energy, BR	VOC	NA	VOC
A11	Hazardous waste	NA	Electric energy, BR	NA	NA	NA
A12	Hazardous waste	NA	Electric energy, BR	NA	NA	NA
A13	Hazardous waste	NA	Electric energy, BR	NA	NA	NA
A14	Hazardous waste	NA	Electric energy, BR	NA	Total particulate matter	NA
A15	NA	NA	Electric energy, BR	VOC	NA	VOC
A16	Hazardous waste	NA	Electric energy, BR	NA	NA	NA
A17	Hazardous waste	NA	Electric energy, BR	NA	NA	NA
A18	Hazardous waste	NA	Electric energy, BR	NA	NA	NA
A19	Hazardous waste	NA	Electric energy, BR	VOC	NA	VOC

(continued)

Table B (continued)

ID (n)	W[a]	WU[a]	NRE[a]	HT[a]	RIn[a]	PO[a]
A20	Hazardous waste	NA	Electric energy, BR	NA	Total particulate matter	NA
A21	NA	NA	Electric energy, BR	NA	NA	NA
A22	Hazardous waste	NA	Electric energy, BR	NA	Total particulate matter	NA
A23	Hazardous waste	NA	Electric energy, BR	NA	NA	NA
A24	Hazardous waste	NA	Electric energy, BR	NA	NA	NA
A25	NA	NA	Electric energy, BR	COV	NA	VOC
A26	Hazardous waste	NA	Electric energy, BR	NA	NA	NA
A27	Hazardous waste	NA	Electric energy, BR	NA	Total particulate matter	NA
A28	NA	NA	Electric energy, BR	VOC	NA	VOC
A29	NA	NA	Electric energy, BR	NA	NA	NA
A30	Hazardous waste	NA	Electric energy, BR	NA	NA	NA
A31	Hazardous waste	NA	Electric energy, BR	NA	NA	NA
A32	Non-hazardous waste	NA	Electric energy, BR	NA	NA	NA
A33	Hazardous waste	NA	Electric energy, BR	NA	NA	NA
A34	Hazardous waste	NA	Electric energy, BR	NA	NA	NA
A35	NA	Used water	Electric energy, BR	NA	NA	NA
A36	NA	Used water	Electric energy, BR	NA	NA	NA
A37	Non-hazardous waste	NA	Electric energy, BR	NA	NA	NA
A38	Non-hazardous waste	NA	Electric energy, BR	NA	NA	NA

(continued)

Table B (continued)

ID (n)	Wᵃ	WUᵃ	NREᵃ	HTᵃ	RInᵃ	POᵃ
A39	Hazardous waste	NA	Electric energy, BR	NA	NA	NA
A40	Hazardous waste	NA	Electric energy, BR	NA	NA	NA
A41	Non-hazardous waste	NA	Electric energy, BR	NA	NA	NA
A42	Non-hazardous waste	NA	Electric energy, BR	NA	NA	NA
A43	Non-hazardous waste	NA	Electric energy, BR	NA	NA	NA
A44	Hazardous waste	NA	Electric energy, BR	NA	NA	NA
A45	Non-hazardous waste	NA	Electric energy, BR	NA	NA	NA
A46	NA	NA	Electric energy, BR	NA	NA	NA

ᵃW (waste); WU (water use); ENR (non-renewable energy); HT (human toxicity); RIn (respiratory inorganics); PO (photochemical oxidation). NA = not applied

Table C MFCA data for the Case study

CCs	Name	Materials	Input (kg)	Output—product (kg)	Output—waste (kg)	Material cost (R$/kg)	Destination cost (R$/kg)	Energy (kW h)	Labor N°	Labor Time (h)	Labor Cost (R$ h^{-1})
CC 1	Receipt, stock, and panel saw machine	MDF	2625.00	2524.50	100.50	19.05	-0.40	160.50	10	40	7.22
		Pallets	60.00	0	60.00	1.40	-0.40				
CC 2	Rolling mill	MDF	2524.50	2469.70	0	19.05	0	193.60	10	40	7.22
		Chip	0	0	51.00	19.05	-0.40				
		Powder	0	0	3.80	19.05	1.80				
CC 3	Cyclone	Powder	91.50	0	91.50	19.05	1.80	814.00	0		
CC 4	Manual sanding	MDF	2469.70	2464.95	4.75	19.05	1.80	0.00	10	40	7.22
		Abrasive	1.26	0	1.26	105.00	1.80				
		Sealer	28.59	27.39	1.20	10.50	1.80				
		Varnish	11.41	7.91	3.50	59.50	1.80				
		Ink	99.99	97.79	2.20	26.10	1.80				
CC 5	Sealing	Parts	2469.70	2469.7	0	19.05	0	0.00	10	20	7.22
		Sealer	30.60	29.93	0	10.50	0				
		Brushes	0.18	0.0	0.18	38.90	1.80				
		Packaging	2.06	0	2.06	0	1.80				
		VOC	0	0	0.67	10.50	0				
CC 6	Varnishing	Parts	2469.70	2469.7	0	19.05	0	0.00	10	40	
		Varnish	12.98	12.68	0	59.50	0				

(continued)

Table C (continued)

CCs	Name	Materials	Input (kg)	Output—product (kg)	Output—waste (kg)	Material cost (R$/kg)	Destination cost (R$/kg)	Energy (kW h)	Labor	
		Brushes	0.72	0	0.72	38.90	1.80		Cost (R$ h^{-1})	7.22
		Packaging	1.30	0	1.30	0	1.80			
		Cloths	1.57	0	1.57	6.30	1.80			
		VOC	0	0	0.30	59.50	0			
CC 7	Painting	Parts	2469.70	2469.7	0	19.05	0	66.00	N°	2
		Ink	128.50	110.8	15.20	26.10	1.80		Time (h)	180
		VOC	0	0	2.50	26.10	0		Cost (R$ h^{-1})	7.22
CC 8	Drying	Parts	2469.70	2469.70	0	19.05	0	1600.00	N°	0
		Sealer	29.93	28.59	0	10.50	0			
		Varnish	12.68	11.41	0	59.50	0			
		Ink	110.80	99.90	0	26.10	0			
		VOC	0	0	12.84	40.20	0			

References

Ahlroth S (2014) The use of valuation and weighting sets in environmental impact assessment. Resour Conserv Recycl 85:34–41. https://doi.org/10.1016/j.resconrec.2013.11.012

Blanco Morón A, Delgado Calvo-Flores M, Martín Ramos JM, Polo Almohano MP (2009) AIEIA: software for fuzzy environmental impact assessment. Expert Syst Appl 36:9135–9149. https://doi.org/10.1016/j.eswa.2008.12.055

Carvalho A, Mimoso AF, Mendes AN, Matos HA (2014) From a literature review to a framework for environmental process impact assessment index. J Clean Prod 64:36–62. https://doi.org/10.1016/j.jclepro.2013.08.010

da Silva PRS, Amaral FG (2009) An integrated methodology for environmental impacts and costs evaluation in industrial processes. J Clean Prod 17:1339–1350. https://doi.org/10.1016/j.jclepro.2009.04.010

Danish Environmental Protection Agency (2005) Spatial differentiation in Life Cycle impact assessment—The EDIP2003

Deng X, Hu Y, Deng Y, Mahadevan S (2014) Environmental impact assessment based on D numbers. Expert Syst Appl 41:635–643. https://doi.org/10.1016/j.eswa.2013.07.088

Finnveden G, Moberg Å (2005) Environmental systems analysis tools—an overview. J Clean Prod 13:1165–1173. https://doi.org/10.1016/j.jclepro.2004.06.004

Goedkoop M, Heijungs R, Huijbregts M, De Schryver A, Struijs J, van Zelm R (2013) A life cycle impact assessment method which comprises harmonised category indicators at the midpoints and endpoint level. ReCiPe 2008

Herva M, Roca E (2013) Review of combined approaches and multi-criteria analysis for corporate environmental evaluation. J Clean Prod 39:355–371. https://doi.org/10.1016/j.jclepro.2012.07.058

Höjer M, Ahlroth S, Dreborg KH, Ekvall T, Finnveden GG, Hjelm O, Hochschorner E, Nilsson MM, Palm V (2008) Scenarios in selected tools for environmental systems analysis. J Clean Prod 16:1958–1970. https://doi.org/10.1016/j.jclepro.2008.01.008

Humbert S, Margni M, Jolliet O et al (2015) IMPACT 2002+: user guide. Work 21:36

ISO (2011) ISO 14051—environmental management—material flow cost accounting—general framework

Jolliet O, Margni M, Charles R, Humbert S, Payet J, Rebitzer G, Robenbaum RK (2003) IMPACT 2002+: a new life cycle impact assessment methodology. Int J Life Cycle Assess 8:324–330. https://doi.org/10.1007/BF02978505

Kumar R, Pal P (2013) Turning hazardous waste into value-added products: production and characterization of struvite from ammoniacal waste with new approaches. J Clean Prod 43:59–70. https://doi.org/10.1016/j.jclepro.2013.01.001

Maceno MMC, Pawlowsky U, Machado KS, Seleme R (2018) Environmental performance evaluation—a proposed analytical tool for an industrial process application. J Clean Prod 172:1452–1464. https://doi.org/10.1016/j.jclepro.2017.10.289

Pfister S, Koehler A, Hellweg S (2009) Assessing the environmental impacts of freshwater consumption in LCA. Environ Sci Technol 43:4098–4104. https://doi.org/10.1021/es802423e

Pizer WA, Morgenstern R, Shih JS (2011) The performance of industrial sector voluntary climate programs: climate wise and 1605(b). Energy Policy 39:7907–7916. https://doi.org/10.1016/j.enpol.2011.09.040

Rigamonti L, Grosso M, Møller J, Martinez Sanchez V, Magnani S, Christensen TH (2014) Environmental evaluation of plastic waste management scenarios. Resour Conserv Recycl 85:42–53. https://doi.org/10.1016/j.resconrec.2013.12.012

Unep (2012) The emissions gap report 2012. A United Nations Environment Programme (UNEP) Synthesis Report. ISBN: 978-92-9253-062-4

Global Sustainable Waste Management Practices

Principles of Sustainability and Circular Economy: Application and Case Analysis of Historical Evidence and Real Internationally-Oriented Food Processing Company in Latvia

Maira Lescevica

Abstract Nowadays the society observes changes in the surroundings and the environment more and more often, probably caused by traditional ways of production and service practised for several decades—several generations have been taking, using and wasting too much resources. Now companies, especially small and medium enterprises, have to change their operations—life cycle management to match requirements of the Circular Economy. The aim of this chapter is to introduce the concept and requirements of the Circular Economy as a logical continuation and evolution of Environmental Sustainability; to find learning outcomes from observation of historical evidence and identify gaps in the life cycle of existing products in a real food processing company, and to elaborate recommendation for company improvements to be applied as fundamentals of the Circular Economy. The research consists of three sections—theoretical overview of the main definitions and concepts, practical analysis—historical observations, processing of statistical data about waste management in Latvia; and practical findings from case analyses of a food processing company. The research methods used were qualitative research methods—interviews, statistical data analysis, and company's documentation analysis for Life Cycle Assessment, and also a non-structured direct interview was carried out with the management of the company. Besides Triangulation problems, also other losses were identified after analysis of company "XXX" and turned into recommendations for the company management, such as, to attract European funds support programs, acquisition of triangular equipment and improvement of efficiency; to consider the idea of creating a glass container recycling plant as an additional niche for the company; to pay more and more attention to the sorting and management procedures for residue; to organize feedstock feed in such a way as to minimize the long-term storage of unprocessed products. The theory section includes an overview of the main concepts and definitions and differences between the Linear and the Circular Economy. Reflections of final/modern concepts and definitions of Sustainable business are also included—not

M. Lescevica (✉)
Entrepreneurship and Business Administration, Institute of Social, Economic and Humanities Research (HESPI), Vidzeme University of Applied Sciences (ViA HESPI), Cēsu iela 4, Valmiera 4200, Latvia
e-mail: maira.lescevica@va.lv
URL: http://va.lv/en

© Springer Nature Switzerland AG 2020
W. Leal Filho et al. (eds.), *International Business, Trade and Institutional Sustainability*, World Sustainability Series,
https://doi.org/10.1007/978-3-030-26759-9_8

only as a role, but also as a stage of development. The historical and today's development in the EU and Latvia is also observed and described. Some statistics about today's situation in waste and waste management has been presented and analysed. The practical research consists of socio-economic process and case analysis. One part of the practical research is devoted to the Soviet occupation period in the Baltic countries when many aspects of the Circular Economy evidence were observed, not because of luxury life, but because of poor and modest living conditions. The second section—case includes analyses of the life cycle of a food processing company in Latvia to identify ways to improve the company performance according to the requirements of the Circular Economy. This will give a strong base for the future development of guidelines for food processing companies in order to support their self-evaluation and transition towards requirements of the Circular Economy. Framework for the Circular Economy is tested as flexible and easy to use for explanation and for creation of common understanding among company management and workers. A handbook or guidelines for introduction and implementation of fundamentals of the Circular Economy could be developed for future research.

Keywords Linear economy · Circular economy · Sustainable development ·
Life-cycle analysis · Value added chain analysis

1   Introduction

Economics as a science is a study of how to most efficiently meet the unlimited needs of people with limited resources. The main objective of the modern economy is sustainable development, not only at the social and political level, but also at the ecological level. So the economy needs to be analysed at the consumer, business, municipal, state and also global levels, to cover all ongoing processes and events. The current problem today is related to the definition of economy—the preservation of limited resources. As the unlimited consumption desire of people is growing along with the advancement of technical progress and the level of globalization, on the contrary, the extraction, conservation and renewal of resources are becoming an ever-increasing challenge for businesses and for the society as a whole. The irrational use of resources leads to another even so urgent problem of the modern world—creation of excessive waste and increase of environmental pollution. The economy could be described as a loop in which the planet and the environment play an important role in the provision of natural resources and waste prevention. Such a model will remain so long as the planet is able to absorb and is not too small. Most modern manufacturers produce and consumers use the product according to the principle: used-discarded or damaged-discarded, from which it can be concluded that it is more convenient to dispose of than to repair or reuse.

The classic economic model in the industry is based on the linear cycle method, but nowadays it is difficult to call linear an efficient and growing operating model. The needs of a globalized world are changing—consumer demands are growing, besides, the various resources remain more limited. As a result, the linear economic model is no longer appropriate in today's developed world, because the take-make-waste approach does not allow for future development. In order to ensure a sustainable growth, the use of resources must be smarter and more Sustainable. The use of existing resources in the best possible way is also in the economic interest of companies (European Commission 2018).

This and other problems should be solved by the Circular Economy, it envisages the resources that have been used in the Linear Economy and have lost their value are not wasted. Instead, the Circular Economy ensures that resources are restored, maintained and used for as long as possible.

2 Sustainability, Linear and Circular Economy: Concepts and Application

During the last decade, sustainability and sustainable development are the most discussed topics at the national level, the most comprehensive and oldest definition of that can be found in the 'Burtland's report': "development which meets the needs of current generations without compromising the ability of future generations to meet their own needs" (The United Nations Report of the World Commission on Environment and Development: Our Common Future 1987).

In order to explain how Sustainability applies to entrepreneurship an explanation of triple bottom line (TBL) is used where a company focuses on the environmental, social and economic value they either add or damage. The definition from Peter Zollinger's report reveals that the term Sustainability "captures the whole set of values, issues and processes that companies must address in order to minimize any harm resulting from their activities and to create economic, social and environmental value" (Zollinger 2001).

The following is a summary of the theoretical information about the current economic model—the Linear Economy and its main differences from the economic model towards which the Latvian state and society have to move—the Circular Economy.

2.1 Linear Economy

In the development of contemporary society, people have lived up to now according to the linear economic model. Such a model is borrowed from the western countries with a relatively advanced economy. Its development was driven by abundant natural resources, cheap labour and low waste disposal costs (Benton et al. 2015, p. 27).

The Linear Economy is defined as the conversion of resources into waste as a result of the production process. The value is created by maximizing the product and selling it in quantities (Skene and Murray 2015, 241 p).

The Linear Economic model is depicted as a line with the beginning and the end—from extraction to waste disposal, based on a simple linear process: obtaining, producing, consuming and discarding (see Fig. 1). This means that the manufacturer acquires resources, manufactures and markets the products purchased by the consumer. The consumer of the product eventually throws it out of the product or it expires and returns in the landfill without paying attention to the generation of contamination in each of the stages of the model (Scott and Vare 2018, 196 p).

In each of the stages of the Linear Economic model shown in the picture, waste and pollution are generated that do not always come to their intended place, but on the ground, in the air or in the water, causing environmental contamination (Scott and Vare 2018, 196 p). Waste generation causes environmental pollution Diversion: the removal of natural resources (capital) from the environment (from the extraction of resources or unsustainable harvesting) and the reduction of the value of natural capital by contaminating the environment with waste (Skene and Murray 2015, p. 242).

Reuse of the product in the Linear Economic model is low as part of the product's components are used for low quality purposes, which reduces the value of the materials and makes the process of giving the product a new life complicated (Weetman 2017, 46 p).

The Linear Economic model was topical at a time when resources and raw materials were cheap due to their abundant quantities. The resources were low in demand and the products produced did not create high environmental pollution. Today, raw material and resource prices are increasing as the demand is growing. With the devel-

Fig. 1 Linear Economy model (developed by the author based on origin Sauve et al. 2016, 52 p)

opment of the economy and the changing trends, manufacturers' interest in various rare elements and resources is increasing, for example, more and more technological devices (television screens, mobile phones) use elements such as tertiary and yttrium that can be obtained from polymetallic and rare earth metal ores. The demand for the extraction of these rare elements and technological equipment is increasing, seeing its usefulness and high value (Scott and Vare 2018, 196 p).

Looking at the essence of the Linear Economy, that this model ignores the environmental impact of resource consumption and waste generation, resulting in excessive extraction of raw resources (pollution and waste). As these factors are not insignificant in the development of the modern economy and the resolution of environmental problems, companies and countries are thinking about the Linear Economy and how it can change.

2.2 Circular Economy

The antonymic model of the Linear Economic model is the Circular Economy, the exact time of occurrence of this idea and the author is unknown, however, its practical application in the modern economy and industrial processes has been successful since the end of the 1970s (Ellen MacArthur Foundation 2017).

The ideas are based on a variety of "schools", for example, the great role played by the German chemist Michael Braungart and the American architect Vill McDonough, "Cradle to Cradle," concept that everything that is produced is designed to be restored or modified. The Swiss architect Walter Stahel presented the idea of an "eyeglass" economy, recycling resources for reuse instead of losing them. Even the Circular Economic model has been affected by the term "industrial ecology", which applies to the entire economy—different industries utilize each other's waste (also called "industrial symbiosis") (Benton et al. 2015, p. 33).

Japan has also participated in the introduction of the economy in the form of "producer liability" legislation: product manufacturers have a duty to ensure that the product is recovered and recycled after its use. Europe has gradually moved to a mentality that aims to regain value from depleted resources while at the same time reducing environmental impact rather than dealing with the negative effects of waste. The move towards a moving economy means moving towards a model that seeks to maximize the value of the materials and products already in the economy (Benton et al. 2015, p. 35).

The Circular Economy refers to the production and consumption model, which differs significantly from the Linear Economic model that has been dominating in the society. The Circular Economy is a model in which the value of products and materials is sustained for as long as possible, waste generation and resource utilization are reduced, and when the product reaches the end of its life cycle, resources remain in the economy where they are used over and over again to create added value (Lewis 2018, p. 59).

The model is depicted not as a line, but as closed loops, in which resources are circulating in the production and consumption system (see Fig. 2). The purpose of the Circular Economy is to optimize the use of raw resources and reduce pollution and waste generation in each of the model phases, as well as replace the traditional linear economic model—fast, cheap production and cheap waste disposal, with durable products that can be recycled. The Circular Economy model contributes to resource flexibility. A production-driven economy can prolong the useful life of the product, that is, delay its final consumption and give preference to re-use before the product is actually depreciated (Sauve et al. 2016, p. 52).

The Circular Economy is not only an opportunity to reduce waste and protect the environment but also to fundamentally reorganize the entire economy by creating a new model of production, labour and consumption, creating new opportunities and jobs (Lewis 2018, p. 59). The idea of this model eliminates the main problem of the Linear Economy—products are manufactured in such a complicated way that the retrieval and reuse of their valuable material is impossible and the only option at the end of the product life cycle is to throw it away (Scott and Vare 2018, 197 p). The purpose of the Circular Economic model is to resemble processes occurring naturally, where waste is little or, if it occurs, it is used by other entities (e.g. plants). In nature there is a competition and mutual cooperation among plant species, thus preserving the effectiveness of the natural ecosystem, ensuring flexibility and adaptability. Assimilating such a metaphor to the economic system helps to ensure a healthy competition and efficient and maximum use of available resources (Sauve et al. 2016, p. 53).

Fig. 2 Circular Economy model (developed by the author based on origin Sauve et al. 2016, 52 p)

3 Introduction of the Circular Economy: EU, Latvia

The Circular Economy creates the link between the economy and the environment. The modern, traditional or linear economy is a simple model that does not always provide a solution to a serious economic and environmental problem not only locally but also globally. The existing problems can be addressed through a variety of tools such as taxes, fees, permits, licenses, patents, labelling schemes, deposit systems, and other tools that could have an impact on the particular illegal activity.

Municipal waste (household and production, excluding hazardous waste) is waste created during retail, manufacturing, and service providing or household process (LR Central Statistical Bureau 2017, ER).

The common trends show that as the number of population in the country and the level of people's income increase, the amount of waste generated by households in the country will also increase. According to available data from the World Bank, the population of Latvia is rapidly decreasing, comparing year 2012 (2,034,319) with year 2016 (1,960,424). The population at this stage has decreased by almost 74 thousand, which in a short period of time can create major changes. This situation could be explained by the poor demographic situation in Latvia, the aging of the population in the country as a whole, as well as the fact that the traveling population is growing significantly, increasing the level of migration in the whole country (World Bank 2017, ER).

Reduction in the population should also reduce the amount of waste they generate, but the statistics show that this does not happen. Figure 3 shows that the amount of the generated waste in the period from 2012 to 2015 is increasing, while in 2016 a small decrease (by 48 thousand tons) of the amount of waste generated is observed. Comparing the amount of waste generated in 2012 and 2016 in the country, this amount has increased by 181 thousand tonnes. It can be concluded that the decrease in the number of inhabitants in Latvia does not reduce the amount of the generated waste, but, on the contrary, increases which could be explained with the fact that

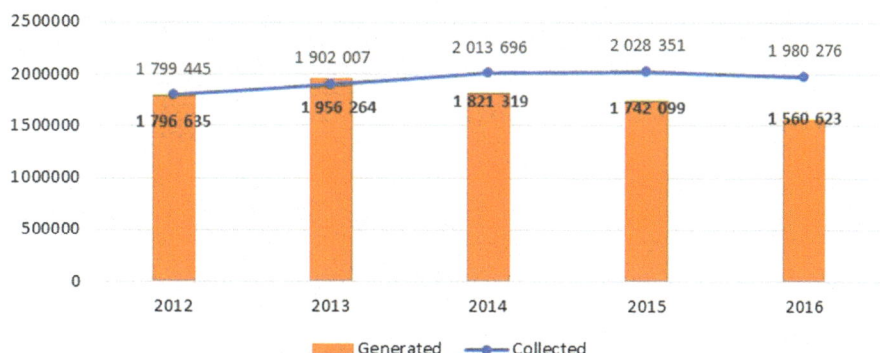

Fig. 3 Generated and collected municipal waste amount (tons) in Latvia, 2012–2016 (developed by the author based on source: LR Central Statistical Bureau of Latvia 2017)

the level of prosperity has increased, the population less often is cultivating the products themselves, buying more products in packages, and as a result, the volume of packaging has increased. Citizens have changed their mind set about consumption: they "take, make, waste", but are repairing, using them for a longer time, reusing, recycling things less often.

In turn, the amount of waste collected in the period from 2012 to 2016 is rapidly decreasing, despite the fact that the amount of waste collected in 2013 exceeds the amount of waste generated (by 54 thousand tons). Comparing 2013, when the largest amount of waste (1,956,264 tons) was collected, with 2016 (1,560,623), the amount of waste collected decreased by 395.6 thousand tons in 3 years. Such a situation may indicate a negative tendency for the resulting waste to be inadequate and could be related to the problems of a common inventory system (see Fig. 3).

Looking at the amount of waste generated per capita (see Fig. 4) and comparing these indicators with the neighbouring countries Lithuania and Estonia, as well as with the European Union (EU-28), it can be observed that the waste generated in Latvia per capita is lower than in Lithuania and the EU on average, but it should be taken into account that Lithuania is larger in terms of territory and population than Latvia and Estonia. An important feature of this Figure is the amount of waste deposited in the country, which is almost twice as large (131 kg) per capita in Lithuania (132 kg), which has a larger population, indicating a negative tendency in Latvia which might be caused by irrational use of waste use, lack of interest in sorting waste. Looking at the EU average for landfill dumping (114 kg), Latvia's indicator is more than twice the EU average.

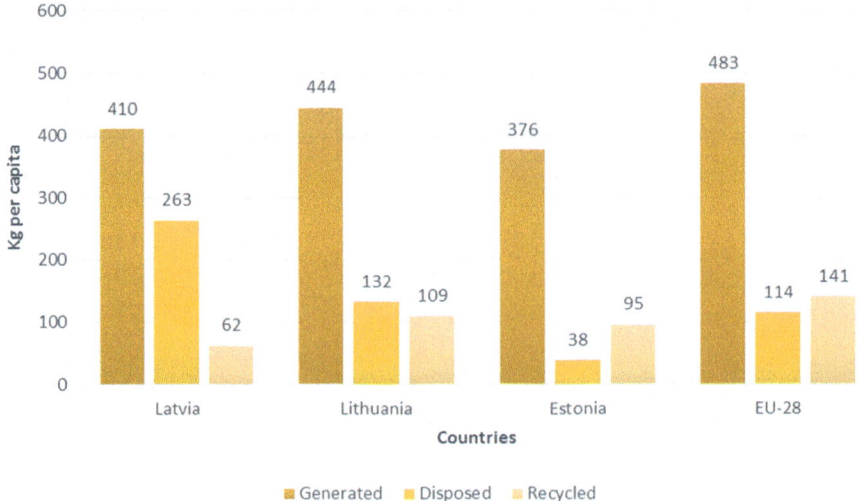

Fig. 4 Generated, disposed and recycled municipal waste in kilograms per capita in the Baltic States and the EU on average in 2016 (developed by the author based on source: Eurostat 2017, ER)

Data on the amount of recycled municipal waste in the country are indicative of a negative trend in waste generation and irrational use of resources. In Latvia only 62 kg of recycled waste per capita is processed, which is a poor indicator compared to Lithuania (109 kg), Estonia (95 kg) and the EU average (141 kg), indicating that improvements are required in the waste management system to collect as much quality material as possible for recycling or to introduce a system that ensures the efficient collection of used materials.

Waste disposal at landfills intended for them is an environmentally damaging process, therefore it is necessary to sort and recycle the waste, thus saving resources without contaminating the environment, reducing the threat to living organisms and human health.

Since 2015, the European Commission has drawn up the Circular Economy package in Brussels or rules on facilitating the transition to a more complete model of Circular Economy. The Circular Economy package contains a task, a measure, a package of proposals and a long-term plan that requires that the amount of waste disposed of in landfills is reduced and that the amount of recycling is increased—from the producer to the secondary raw materials market. The main objective of this package is to focus on the issues and challenges posed by plastics, food waste, biomass and other materials in order to bring about change at the European Union level.

In 2015, Europe lost around 600 million tonnes of waste that could be recycled. Waste conversion into resources is an important process, as this would increase the efficiency of resource utilization and ensure the transition to the economic model more fully. In order to make this transition even more effective, the European Commission has set targets for waste recycling and conservation of resources:

- In the European Union, by year 2030, 65% of waste should be recycled;
- In the European Union, by year 2030, 75% of the quantity of packaging waste should be recycled;
- Until 2030, no more than 10% of all waste should be disposed of in landfills (European Commission 2018).

The Circular Economy system means—"Reduce, Reuse and Recycle"—minimizing material yields while minimizing resources production, the products themselves are made of reusable parts and materials, and once the product is discarded, materials and components are recycled again.

Since Latvia is a part of the EU, all guidelines of introduction of the Circular Economy should be also followed and implemented.

Therefore, during this research, requirements of the Circular Economy were tested and implemented on a real internationally-oriented food processing company in Latvia for this purpose.

4 Fundamentals of Circular Economy: History and Introduction in Company

The concept of circularity has a profound historical and philosophical origin. The idea of feedback, the cycle of real-world systems is old, and comes from different philosophy schools. After World War II, industrial countries experienced a revival, when the computer-based studies on the non-linear system clearly showed how complicated, interconnected and therefore unpredictable the world in which we live is, more like a metabolic process than a robotic machine. With the current development, digital technology is capable of supporting the transition to the Circular Economy, radically increasing virtualization, material destruction, transparency and reciprocal intelligence. (Ellen MacArthur Foundation 2017).

During the Soviet occupation (till 1990) there were many existing episodes in Latvia, which were similar to the Circular Economy fundamentals (see Fig. 5). Many state companies had similar features, but not because of their urgent need for development, but because of poverty. For example,

- Glass jars and bottles had long life, they were reused and reused after sale and consumption of the content;
- There were less plastic materials, packaging, more paper—especially recycled paper;
- Buyers used special elastic bags instead of plastic bags;

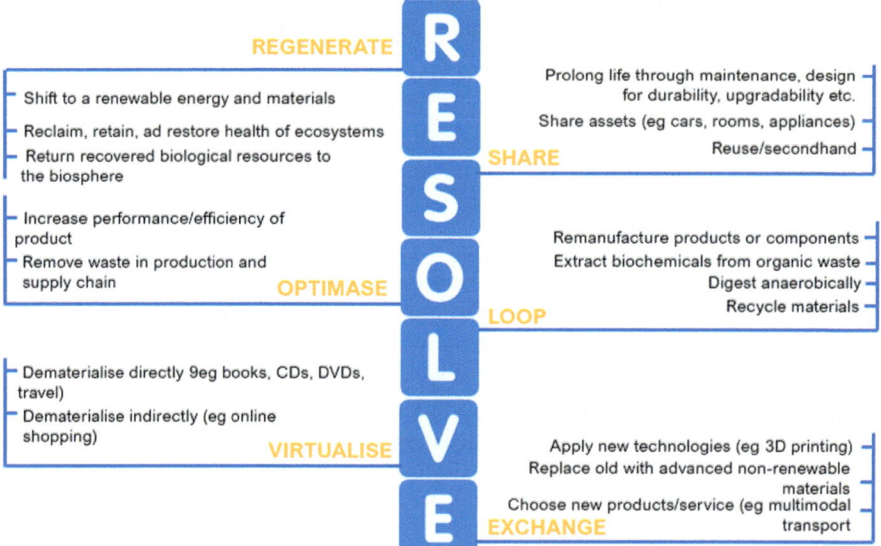

Fig. 5 Circular economy six action steps (developed by the author based on source: McKinsey 2016)

- Biological and food waste was carefully collected and delivered to huge pig farms for feeding.

All these processes were deeply rooted in people's minds, therefore easily followed and passed from one to the next generation.

McKinsey's Action Model, Fig. 5, was developed following the principles of circulation. A 6-step system of action has been developed to improve efficiency.

The basis of the Circular Economy is the efficient and prudent use of resources, where the production process also brings value after use of the product. Efficient use of resources is based on the use of as little as possible related components (materials, energy resources, human resources, financial resources, natural resources, etc.) with the aim of maximizing the benefits. The maximum benefit and utility is related to increasing the value of the already existing resources, for example, nowadays a large proportion of food and non-food products is packed in durable material packaging (glass, wood, metal, plastic, textiles) that can then be used afterwards according to the type and content of the product, thus investing less, gaining more, as well as proving the need for two or more times, using time, resources and energy to produce equivalent effects.

The framework created by Weetman in Fig. 6 was used for analysis of a food processing company. During interviews losses and waste that could be minimized or eliminated were identified.

The transition to the Circular Economy is closely linked to science and new, non-transactional solutions. EU funds have set innovative projects, including studies that promote the prudent use of natural resources, as the priority. The comprehensive benefits implementing fundamentals of the Circular Economy:

- Significantly reduced exposure to the environment in Europe and beyond.
- The possibility of reducing the continent's high and increasing dependence on imports. As this type of addiction grows, justified vulnerability may arise, as

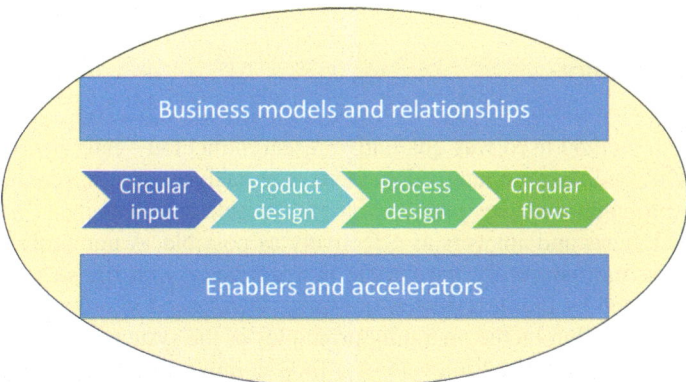

Fig. 6 Circular economy framework (developed by the author based on source: Weetman 2017)

global competition for natural resources has contributed to a significant rise in price levels and volatility.

- Energy-saving strategies could generate significant savings by increasing the competitiveness of European industry while ensuring the net benefits of job opportunities. (European Environment Agency 2016).
- Protect nature and strengthen ecological stability. EU natural capital is currently not sufficiently protected, preserved and improved.
- Stimulate the development of sustainable, resource-efficient, low-carbon technologies. The European economy is growing faster than using its raw materials, indicating the need for more efficient use of resources.
- Effectively address the environmental health and well-being risks. During the last decade, air and water contaminants have significantly reduced. However, a source of concern about air quality and noise pollution in urban areas, as well as the exposure of the chronic population to complex chemical mixtures in products remains. (European Environment Agency 2017).

The introduction of the Circular Economy is, to a large extent, a long-term process, as introduction of innovation requires energy resources and substantial financial capital. As free funds are largely limited, the acquisition of new technologies by itself would cause too much loss in the company's budget. However, various European support funds are being organized, which, according to the circumstances and the desired idea, can provide a certain amount of funding. The Circular Economy is a completely new operating method that helps to process large volumes of waste, thus creating more free space, releasing the environment from pollution, saving resources, reducing residue and improving air quality, reducing health risks. Despite the fact that the results and contributions of the activities carried out will only be possible after several years, the Circular Economy model is still called the basis of a long life which can change the current situation even on a global scale (European Environment Agency 2017).

5   Methodology

The research method used was qualitative research method—interviews, statistical data analysis, and the company's documentation analysis. It is important to create simple and thought-out questions in order to be able to comprehend and analyse the acquired views and answers as effectively as possible. A non-structured direct interview was carried out during which, in accordance with the situation and the answers provided by the interviewer, questions were asked in order to find out the necessary information on the operating principles of the system.

During the research the life-cycle of a food company was analysed in order to develop recommendations for adherence to the basic principles or fundamentals of the Circular Economy.

Life Cycle Assessment is a scientific method that allows a comprehensive assessment of the product's environmental, natural resources and human health impacts. A full life cycle assessment examines these "cradle-to-grave" effects—from the production of all the necessary raw materials, up to what and how it happens with the product when it has ended up serving and turned into waste.

A quality life cycle assessment provides information that allows:

- to deliberately reduce the environmental impact of new products even in their design phase;
- to identify the existing bottlenecks;
- to avoid problem transfer when solving one problem, creating or reinforcing another;
- to compare the impact of similar products on the environment, the consumption of natural resources and human health.

The Life Cycle Assessment method provides systematic and comprehensive data that are used for quality and informed decision making. Life Cycle Assessments are the basis for developing special requirements for products.

For that purpose, an internationally-oriented food processing company in Latvia was chosen the name of which will remain confidential because it was a requirement of the company owners. It is one of the leading manufacturers of canned food and food additives, tomato sauces, mayonnaise, preserves and other food products in Latvia. The company was founded during the 1980s. In 2008 corporate changes were introduced in the company and significant investments were made to improve and develop infrastructure, equipment and brands. The product range consists of more than 130 products designed for a wide range of customers and adapted to different market segments. More than 20% of the company's turnover is made up of exports. The company has been awarded LVS EN ISO 22000: 2006 certificate, certifying conformity of production processes to the highest quality standards not only in Latvia, but also in the international market. For the convenience, the company's name is made up—company "XXX" due to the early mentioned confidentiality reason.

Also, a non-structured direct interview was carried out with the management of the company, during which it was discovered how the Circular Economy requirements could be implemented/applied. The supply chain analysis/value chain analysis was also performed in order to find out the most appropriate parts of processes for introduction of solutions and suggestions.

6   Results

During the research the life-cycle of food company "XXX" was analysed in order to develop recommendations for adherence to the fundamentals of the Circular Economy.

The production in company "XXX" is carried out in several stages, using 56 possible production units, which are interconnected and a separate operation of a technical production unit does not reach the final product. In each technical production unit, individual components of the finished product are processed and prepared, which are transported through specialized pumps and pipes from one plant or production boiler to the next, where the next processing process takes place.

During the last stage of the production process products are packed in various types of packaging (glass jars, polyethylene bottles, polymer buckets, polymer bottles, aluminium polymer stacked packs, soft polymer packs), packed in a number of polyethylene packs, stacked on wooden pallets with a flexible transparent food packaging film and then delivered to the finished product storage warehouse until further processes. That is the stage that creates most of the losses.

The following are main findings during this research:

- Technical losses:

  - Glass jars—tearing, clinging in the production process, breaking of the jars more often (closing problems), resulting in unused glass and metal scraps;
  - Polythene-Aluminium Stacked and Soft Polymer Package Materials—in cases where it is necessary to stop the filling unit of a stacked or soft mayonnaise bag (for example, if regulation is required), a loss of several meters of packaging material occurs;
  - Washing waters from mayonnaise—production of mayonnaise occurs periodically, in large volumes, and as soon as the production process is completed, there is a special equipment washing procedure, during which a large amount of fatty mayonnaise is produced, and it is forbidden by the municipal laws to dispose of it in the public sewerage;
  - Cardboard, wicker bags, paper—supply of raw materials, packaging materials for glass jars, sugar packages, separating sections, shipping of bursting materials;
  - Unclassified waste—in the meantime, in the food industry in particular numerous volumes of different household waste occur.

- Biological losses:

  - Biological waste—food waste from the production of basic dishes;
  - Crop production—in exceptional cases, damage of production occurs, which is mainly due to external environmental factors, such as damage caused during storage due to temperature changes, physical damage to the packaging, etc. If broken products are found in the production area, the product is disposed of according to its composition, that is, the contents are delivered to the compartment for utilization of organic waste, while the packaging is separate, depending on the type (glass, polyethylene, and polymer).

- Loss of energy resources:

  - Air pollution—the heat production process produces hot saturated steam at a temperature of 165 °C with a pressure of 6.0 bars and, consequently, a large amount of flue gas emissions is produced in the air, which is released into the air at 2400 °C, resulting in tax costs for the company, because emissions are chemicals and are environmentally unfriendly;

– Coal Mica—the combustion of fossil fuels generates about 40% of slag residues from its original state. Storage of discharges is a chronic problem for the company, and requires financial, technical and human resources.

Nowadays, when the food industry is one of the dominant ones and its production is carried out at levels that are almost unavailable, it results in incalculable volumes and a valid reason to seek solutions. One of the possible solutions to the problems is the current state of affairs—the Circular Economy, which, with certain basic conditions, helps companies to adapt their existing activities as quickly and efficiently as possible, in order to minimize the use of resources, as well as sources of different resources and expenses.

7 Conclusions

The traditional up-to-date principle of the Linear Economy—"take, make, throw away"—has reached its limit based on the huge amount of readily available cheap energy and materials. The Circular Economy provides the world with an opportunity to rethink and transform how different things, goods and materials are made.

Introduction of a circular cycle in the daily production process is the beginning of sustainability and future potential.

The Circular Economy seeks to rebuild capital, whether it is financial, manufacturing, human resources, social or natural capital, in order to provide greater flows of goods and services. The life cycle of a product is the period from the introduction of the product to the termination of its production and sale, the life of the product is on the market for a limited period of time. Life Cycle Assessment helps to identify the company's residual amount of resources, find optimal and economically viable options for balancing if needed.

In 2014, the European Commission adopted a Circular economy package in Latvia that proposes guidelines for generation of energy from waste (the European Commission in co-operation with the European Investment Bank) to Industrial symbiosis, i.e., enterprise collaboration in technological processes that ensures the involvement of the product, by-products and waste in the next life cycle. If one product becomes redundant for one company, it can be useful to others as a raw material. There is an example of recommendation for one company's loss.

Air pollution—Use of flue gas emissions in triangulation (to produce cold from low-energy sources). Triangulation is a combination of electricity, heat and cold production. Triangulation is a good solution because it allows you to use heat efficiently not only in winter—for heating purposes, but also in summer—for air conditioning in large areas (Gaso 2018). Triangulation is an economically viable recovery of the auxiliary production processes. Facility functions are rebuffed in large volumes and in the long run, requiring huge initial investment. The approximate cost of the trenching plant is starting from 170,400 EUR.

The other losses after analysis are turned into recommendations for company management:

- to attract European funds support programs, acquisition of triangular equipment and improvement of efficiency;
- to consider the idea of creating a glass container recycling plant as an additional niche for the company;
- to pay more and more attention to the sorting and management procedures for residue;
- to reduce the amount of pollution in the environment from the production of thermal energy;
- to organize feedstock feed in such a way as to minimize the long-term storage of unprocessed products.

Framework for Circular Economy is flexible and easy to use for explanation and for creation of common understanding of company workers.

For the future research a handbook or guidelines for introduction and implementation of fundamentals of the Circular Economy must be created.

References

Benton D, Hazell J, Hill J (2015) Book: the guide to the circular economy: capturing value and managing material risk. Routledge, London, 27, 33, 35 p

Ellen MacArthur Foundation (2017) Circular economy system diagram. Retrieved 24.08.2018 from https://www.ellenmacarthurfoundation.org/circular-economy/interactive-diagram

European Commission (2018) Circular economy package: questions and answers. Press release database. Retrieved 24.08.2018 from http://europa.eu/rapid/press-release_MEMO-15-6204_en.htm

European Environment Agency (2016) Circular economy to have considerable benefits, but challenges remain. Retrieved 24.08.2018 from https://www.eea.europa.eu/highlights/circular-economy-to-have-considerable

European Environment Agency (2017) Critical gaps remain in Europe's environmental performance despite improvements. Retrieved 24.08.2018 from https://www.eea.europa.eu/highlights/critical-gaps-remain-in-europes

Eurostat (2017) Waste statistics. Retrieved 24.08.2018 from http://ec.europa.eu/eurostat/statistics-explained/index.php?title=Waste_statistics

Lewis E (2018) Circular economy. In: Sustain speak: a guide to sustainable design terms. Taylor & Francis Group/Routledge, London, C59 p

LR Central Statistical Bureau (2017) Household and hazardous waste, their collection and processing. Retrieved 24.08.2018 from http://data.csb.gov.lv/pxweb/lv/vide/vide__ikgad__vide/VI0040.px/?rxid=3d45ed82-37cb-4028-9e9c-9e6bde2ba4ba

McKinsey (2016) The ReSOLVE framework for a circular economy. Retrieved 24.08.2018 from https://makewealthhistory.org/2016/09/12/the-resolve-framework-for-a-circular-economy/

Sauve S, Beatrnard S, Solan P (2016) Environmental sciences, sustainable development and circular economy: alternative concepts for trans-disciplinary research. Environ Dev 17:52–53p. Retrieved 24.08.2018 from https://ac.els-cdn.com/S2211464515300099/1-s2.0-S2211464515300099-main.pdf?_tid=d7188a88-c493-439a-a9c9-459e6515e03a&acdnat=1523718061_f77699eacecd8f2243bf43888c2aa9c1

Scott W, Vare P (2018) The world we'll leave behind: grasping the sustainability challenge. Routledge, London, 196, 197 p

Skene K, Murray A (2015) Sustainable economics: context, challenges and opportunities for the 21st-century practitioner. Routledge, London, 241, 242 p

The United Nations Report of the World Commission on Environment and Development: Our Common Future in 1987 ('Burtland report')

Triģenerācija. Koģenerācija (2018) Retrieved 24.08.2018 from https://www.gaso.lv/kogeneracija

Weetman C (2017) A circular economy handbook for business and supply chains: repair, remake, redesign, rethink. Kogan, London, 46, 48 p

World Bank Group (2017) Population. Retrieved 24.08.2018 from https://data.worldbank.org/indicator/SP.POP.TOTL?locations=LV

Zollinger P (2001) The power of change: mobilising board leadership to deliver sustainable value to markets and society. The international business leaders forum. Retrieved 24.08.2018 from http://sustainability.com/our-work/reports/the-power-to-change/#.V0pxxeSK2kV

Current Approaches to Waste Management in Belarus

Nikolai Gorbatchev and Siarhei Zenchanka

Abstract Sustainable waste management in the Republic of Belarus is an urgent problem. This is mostly due to the lack of understanding of the need to sort waste, the lack of pledge containers' turnover, of sorting opportunities in rural areas. About 1500 types of waste with a wide range of morphological and chemical properties are produced in Belarus. Most waste is generated at Belaruskali JSC. That is why the improvement of waste management in Belarus, as in all developed countries of the world, is recognized as a priority and a specific problem to solve for environmental protection. The Republic of Belarus has been taking some steps in the field of waste management, within the framework of the goals and objectives of the Strategy of sustainable social and economic development and the Strategy of environmental protection. Notwithstanding a certain progress in the field of the industrial waste management, there is still much to be done in the field of municipal waste management. One of the promising solutions this respect is the public-private partnership, which means the involvement of private businesses on a contractual basis for a more efficient and high-quality performance of tasks related to the public sector of the economy in the terms of cost compensation, risk sharing, obligations, competencies. The present research was aimed at studying the public-private partnership in waste management. It involved the analysis of the national and international legislation in the field of waste management; as well as of the experience of such partnership. The article discusses the experience of partnership of state bodies of the Republic of Belarus with a leading global company. An example of such a partnership is the agreement concluded by Remondis JLLC with the administration of one of the regions of Belarus—the city of Soligorsk. The agreement will allow developing and implementing a new concept of creating a closed cycle of waste management.

N. Gorbatchev (✉)
Faculty of Retraining and Advanced Studies, School of Business of Belarus State University, 4, Nezavisimosti Avenue, 220030 Minsk, Belarus
e-mail: ngorbachev@sbmt.by

S. Zenchanka
Minsk Branch of PRUE, 40, Radialnaya Str., 220070 Minsk, Belarus
e-mail: zench@tut.by

© Springer Nature Switzerland AG 2020
W. Leal Filho et al. (eds.), *International Business, Trade and Institutional Sustainability*, World Sustainability Series,
https://doi.org/10.1007/978-3-030-26759-9_9

Keywords Sustainable development · Waste management · Green economy · Circular economy · Partnership

1 Introduction

The final documents of the Rio conference in 1992 identified the 21st century as the century of sustainable development. Agenda 21 (1992) contains a programme for sustainable development over the world. Developing the ideas of sustainable development, the Conference Rio + 20 adopted document "The Future we want" (2012) in which "green economy" is considered *"in the context of sustainable development and poverty eradication as one of the important tools available for achieving sustainable development and that it could provide options for policy making but should not be a rigid set of rules"*.

The most common "green economy" principles identified from a review of eight published sets of principles or characteristics were proposed by UNDESA (2012).

To estimate progress on the pathway to "green economy", "green growth indicators" were developed. OECD (2017) pointed, that *"several countries are at the forefront of the transition towards green growth, but no country leads in all areas. In fact, countries often advance in one dimension of green growth while remaining stagnant on other fronts. Too often, progress has been insufficient to protect the natural asset base"*.

Different green economy initiatives suggest decreasing of greenhouse gas emissions, decreasing of production and consumption, resource economy, etc. Nevertheless, the waste problem is still to be solved. To tackle it the concept of circular economy was put forward in the EU (Action Plan 2015).

The EU directives provide the framework for waste management in EU countries, formulating the premises for the development and implementation of standards at the national level. From the point of view of meeting the conditions of circular economy, they ensure the minimization of the direct disposal of reactive waste at landfills. The basic principle of waste management is recycling and recovery of materials, as well as the production renewable energy sources during waste disposal. In EU countries, a significant proportion of the primary energy demand is derived from waste. About 50% of the energy reserves of municipal solid waste (MSW) in most EU countries are of biogenic origin, and MSW should be considered as regenerative fuels to the same extent (Vehlow et al. 2007).

An important step in the development of a circular economy in the EU is the widespread use of reusable packaging. This process is regulated by the relevant EU documents (European Commission 2004). Accordance to Rigamonti et al. (2019), the EU has a significant range of recycled packaging materials, which allows delivering products from the moment the raw materials are procured to the consumer. Belarus is making just the first steps toward implementing this important practice.

Another important mechanism implemented in EU countries in the field of waste management is the participation of the private sector in competitive waste management through tenders and public-private waste recycling partnerships (Leal et al. 2014).

Belarus is taking steps in these areas to improve waste management, applying the best practices of EU countries. The article will analyze the successes and failures of the Republic of Belarus in the implementation of the European experience of private-public partnerships in the field of waste management.

2 Circular Economy as a New Paradigm of Development

One of the definitions of "circular economy" was presented by Ellen MacArthur Foundation (2012):

A circular economy is an industrial system that is restorative or regenerative by intention and design. It replaces the "end-of-life" concept with restoration, shifts towards the use of renewable energy, eliminates the use of toxic chemicals, which impair reuse, and aims for the elimination of waste through the superior design of materials, products, systems, and, within this, business models.

It is clear that "circular economy" develops the principles of "green economy" in waste management.

In December 2015 the European Action plan "Closing the loop" was adopted which suppose *"the transition to a more circular economy, where the value of products, materials and resources is maintained in the economy for as long as possible, and the generation of waste minimized, is an essential contribution to the EU's efforts to develop a sustainable, low carbon, resource efficient and competitive economy"* (Action Plan 2015). This action plan will contribute to the achievement of the Sustainable Development Goals (SDG) by 2030, in particular Goal 12 to ensure sustainable consumption and production patterns.

The Action plan considers that waste management plays a central role in circular economy: it determines how the EU waste hierarchy (Directive 2008)—i.e. prevention; preparing for re-use; recycling; other recovery, e.g. energy recovery; and disposal—is put into practice. The Action plan also defines priority areas of its implementation as plastic; food waste; critical raw materials; construction and demolition; biomass and bio-based products.

The European Commission is putting a lot of effort into the implementation of the concept of a circular economy. Thus, an ambitious package of regulatory documents was adopted for implementation, which requires the revision of many legislative acts on waste management (Traven et al. 2018). Objectives-2030 implies achieving up to 65% level of waste processing and requires a reduction in the level of waste disposal at landfills to less than 10%.

At the pan-European level, considerable efforts are being made to synchronize the indicators, to allow classification and analysis of the volumes of waste produced and processed. At the same time, some authors (Bräutigam et al. 2014) point out the difference in volumes, in particular, of food waste, depending on the selected data sources and the assumptions made.

Kalmykova et al. (2018) presented a review of theories and practices of circular economy and combined different circular economy approaches in two: Circular Strategies Database and Circular Economy Implementation Data Base which enable searching for examples of suitable strategies for CE implementation by the part of the value chain, by scope in interest and is useful for looking up the application-ready strategies (Pilot Scale and/or Market Ready).

The analysis of the legislative base and literature sources show the importance of developing a circular economy. It is expected to stimulate economic growth by creating new businesses and job opportunities, saving materials' cost, dampening price volatility, improving security of supply, at the same time reducing environmental pressures and impacts.

3   Methodology

This investigation has several tasks of:

- To study the national and international waste management legislation;
- To study the state of art in the field of waste management in Belarus;
- To study the possibility of public-private cooperation in the field of waste management.

The investigation is based on the analysis of the Belarus legislation, its comparison with international provisions, as well as on the study is of scientific reports and articles, practices of waste management in the EU and Belarus.

4   Belarusian Waste Management Legislation

Legislation on waste management is based on the Constitution of the Republic of Belarus and consists of Presidential Acts, the Law on Environmental Protection (Law 1992), the Law on Waste Management (Law 2007), other legislative acts on waste management, as well as international treaties in the field of waste management, signed by the Republic of Belarus.

The Law on Waste Management (Law 2007) provides the framework for responsible waste management. The Law requires:

- to ensure the collection of waste and its separation by type, except for cases when mixing of different types of waste is allowed in accordance with legal acts technical regulation;
- to ensure the disposal and use of waste or its transportation to waste disposal facilities, as well as their storage in authorized waste storage sites or dumping in authorized waste disposal sites;

- to provide training for workers in the field of waste management, as well as instruction, examination and improvement of their skills;
- to keep records of wastes and carry out their inventory in the manner prescribed by this Law and other legislative acts on waste management, as well as provide statistical data in the field of waste management;
- to inform citizens and local authorities about waste management;
- to develop and take measures to reduce waste volumes;
- to carry out production control over the state of the environment and prevent the harmful effects of waste, products of their interaction with the environment, the health of citizens, property, and, in the event of such an impact, take measures to eliminate or reduce the consequences of this impact.

The main body exercising control over the implementation of the state policy in the field of waste management is the Ministry of Natural Resources and Environmental Protection of the Republic of Belarus. Together with its local bodies the Ministry implements Governmental control over waste management, prevents the disposal of secondary material resources, identifies unauthorized waste disposal sites, and analyzes the availability of community waste management schemes approved in the established manner; and optimizes district schemes of waste management. In 2017 the volume of waste produced in the Republic of Belarus amounted to 55.5 million tons (Ministry 2017). The Ministry is also responsible for the collection and analysis of statistical information on produced waste, it processes and coordinates the implementation of the provisions of the Stockholm Convention on Persistent Organic Pollutants (POP) (Convention 2001) and the Basel Convention on the Control of Transboundary Movements of Hazardous Wastes and their Disposal (Convention 1989).

The municipal waste in Belarus is handled by the Ministry of Housing and Communal Services of the Republic of Belarus. The ministry deals with:

- the development of a unified state policy, development and implementation of state programs in the field of waste management, plans and activities;
- approving the list of waste related to municipal waste;
- carrying out coordination of regional programs in the field of waste management;
- establishing, in coordination with the Ministry of Natural Resources and Environmental Protection of the Republic of Belarus and the Ministry of Health of the Republic of Belarus, the composition and procedure for the development, coordination and approval of municipal waste management schemes;
- together with the Ministry of Natural Resources and Environmental Protection of the Republic of Belarus, approving the activity in the sphere of handling of secondary material resources by creating a state non-commercial, specially authorized organization—the operator in the sphere of handling secondary material resources.

The turnover of secondary material resources is coordinated by the "Operator of secondary material resources" public institution subordinated to housing and communal services (https://vtoroperator.by/). This institution is a non-profit organization aimed at coordinating activities in the field of handling secondary material resources.

This organization, reporting to the Ministry of Housing and Communal Services of the Republic of Belarus, analyzes the effectiveness of handling secondary material resources and keeps a register of enterprises engaged in the collection, sorting and recycling of waste.

The National Strategy for the Management of Municipal Solid Waste (MSW) and Secondary Material Resources (SMR) in the Republic of Belarus for the period up to 2035 provides important indicators for the prevention of waste generation (Strategy 2017). The purpose of this Strategy is to identify the major ways to minimize the harmful impact of MSW on human health, the environment, and the rational use of natural resources by preventing waste generation and maximizing the extraction of components contained in the waste, their use in the economy as additional sources of raw materials, materials and semi-finished products. This strategy establishes a set of waste management system indicators to be achieved by 2035. For example, the "Level of Use in Total Volume of Waste Generation" should be increased from 15.6% in 2015 to 50.2% in 2035. Besides, the Strategy points out that "The hierarchy of priorities in waste management has not been fully established at the legislative level. The prevention and/or minimization of waste generation in general and MSW in particular, including the introduction of deposit-return systems, resource-saving technologies and energy use, are not defined by law as the main priorities of the waste management system".

Though a system of collecting statistics on waste exists in the Republic of Belarus, there is a problem of compatibility of the national statistical reporting and the European system (Beroc 2018). The quality of waste statistics largely depends on the timeliness, reliability and international compatibility of data, and differences between National waste classification (Classifier 2007) and the EU one do exist (Notice 2018).

The institutions below are responsible for the collection of statistical reports in the Republic of Belarus:

- Ministry of Housing and Communal Services of the Republic of Belarus—for secondary raw materials and for municipal waste and secondary raw materials from municipal waste;
- Ministry of Natural Resources and Environmental Protection of the Republic of Belarus—on production waste;
- National Statistical Committee—on scrap and wastes containing precious metals on the cost of protecting the environment from pollution by industrial waste;
- Ministry of Industry of the Republic of Belarus—for scrap and waste of ferrous and non-ferrous metals.

As part of the collection, processing and analysis of primary statistical data the Republican Research Unitary Enterprise "Bel NIC Ecology" conducts the state waste inventory.

In accordance with the classification of waste in Belarus (Law 2007), all waste is classified (Fig. 1).

Production wastes (substances or objects) are formed in the process of economic activities by legal entities and individual entrepreneurs (production of products,

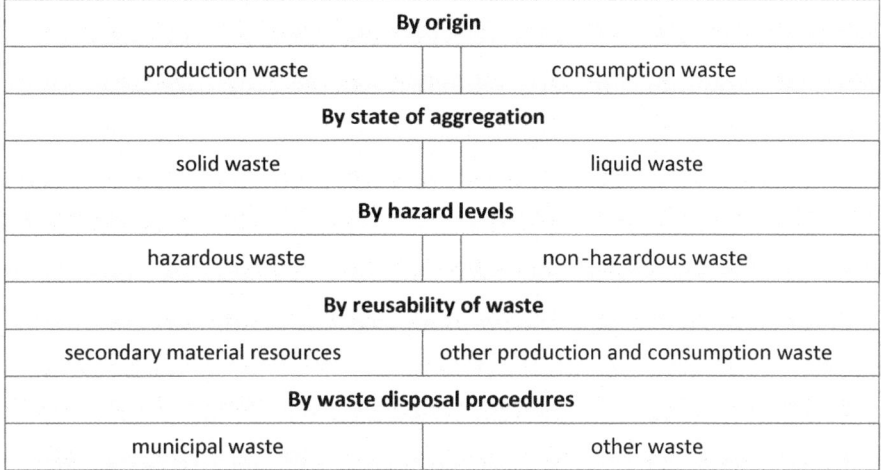

By origin		
production waste		consumption waste
By state of aggregation		
solid waste		liquid waste
By hazard levels		
hazardous waste		non-hazardous waste
By reusability of waste		
secondary material resources		other production and consumption waste
By waste disposal procedures		
municipal waste		other waste

Fig. 1 Classification of waste types in Republic of Belarus (according to the 2007 Law provisions)

energy, performance of works, rendering services), by-products and associated products of mining and mineral processing. The economic activity of a person entails the formation of industrial wastes, therefore such wastes are singled out as a separate type.

Consumption waste is waste generated in the course of human activity not related to economic activity: waste generated in garage cooperatives, gardening partnerships and other consumer cooperatives, as well as street and yard wastes generated in the common areas of populated areas.

Municipal waste is a type of waste that, depending on its origin, is divided into municipal consumption waste and municipal production waste.

The hazard class of hazardous industrial waste affects the amount of the environmental tax for the disposal of industrial waste. Indeed, depending on the hazard classes of hazardous production wastes, environmental tax rates are set for the disposal (storage) of production wastes.

Secondary material resources are wastes, which after their collection can be involved in civil circulation as secondary raw materials and for the use of which there are facilities in the Republic of Belarus.

Secondary raw materials are waste that are prepared for use for the production of products, electrical and (or) thermal energy (hereinafter referred to as energy), performance of work, and provision of services in accordance with the requirements established by technical regulatory legal acts.

Waste disposal is the isolation of waste at special sites in order to prevent their harmful effects, products of their interaction and (or) decomposition in the environment, on the health of citizens, state-owned property, property of legal entities and individuals, not providing the possibility of their further use. In turn, waste disposal facilities—landfills and other facilities designed for the disposal of waste.

Some researchers have underscored the problems with the compatibility of waste management statistics between established practices in the Republic of Belarus and the EU (Beroc 2018). Improving the quality, international comparability and availability of waste statistics will improve the information environment in this area. This, in turn, will contribute to the formation of a market for secondary raw materials, the development of processing, attraction of investments, and finally, the transition to a green economy, including a circular economy in Belarus.

5   The Experience of Belarus in Waste Management

Kovalchik et al (2011) singles out three main stages in waste management in Belarus. The first one is the establishment of control over the process of collecting and disposing of waste from the largest waste producers, the second is connected with the organization of waste processing, as well as streamlining their collection and disposal, and the third one refers to prevention of waste production and minimization of waste flow.

At the same time waste management has a great potential, which makes it possible to consider it as one of the promising areas of development of the industrial complex. In the years to come, the main focus should be on ensuring compliance of the waste management sector with environmental safety requirements.

Belarus now is on the third stage of waste management. Recycling of household waste is just beginning to develop in Belarus. The amount of waste generated per person in Belarus is about 2.5 tons per year. This is quite high, and is primarily associated p with the structure of the industrial complex. In Belarus, 24–28 million tons of industrial waste and about 3 million tons of household waste are generated annually. 3.5 thousand hectares of land are occupied by waste. The main method of utilization of municipal waste in Minsk is MSW landfills (municipal solid waste), which leads to the constant removal of land from production an increase in the degree of environmental pollution, although some of this waste can be used as secondary raw materials. There are four landfills for MSW in the capital, two of which have practically exhausted their resources, one is not operative. MSW contains up to 60% of secondary material resources like paper and cardboard, glass, plastic, metals, textiles, rubber, etc., which are potential for the industry of raw materials. However, now with sorting municipal waste, no more than 10–15% of secondary resources are extracted (Strategy 2017).

Unfortunately, the volume of MSW is increasing every year; it has already reached 885 thousand tons. In total, there are about 200 such landfills in the republic, covering an area of more than 890 ha. 60% of this area is already occupied by waste (Strategy 2017).

Waste is one of the most intensive sources of environmental pollution. This is due to the diversity of chemical (including toxic) substances in the waste, their high concentration, and to the incompatibility of most landfill sites with the regulatory requirements for their location, arrangement and operating conditions.

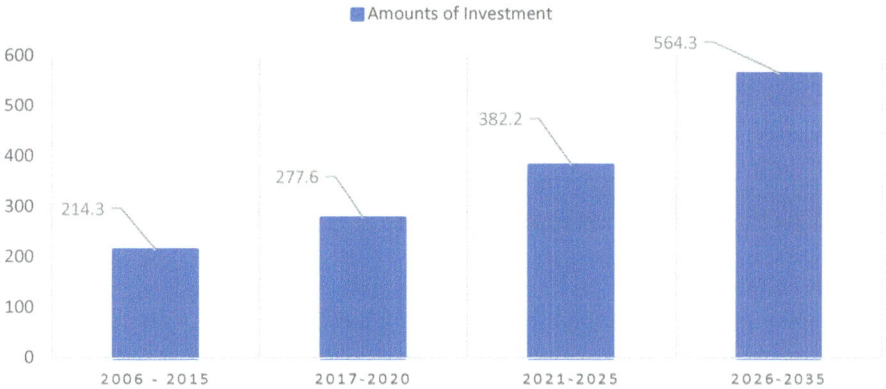

Fig. 2 Estimated amounts of investments according to national strategy for the management of solid municipal waste and secondary material resources in the Republic of Belarus for the period up to 2035, million Euro (According to Strategy 2017 data)

A business plan has been developed for the financial support of the activities outlined in the National Strategy. This strategy envisages significant increase in both public and private investments towards improving the waste management system. From 2006 to 2015 the investments amounted to 214.3 million Euros, the plan is to reach 382.2 million Euros in 2017–2020, and 564.3 million Euros in 2026 (Fig. 2) (Strategy 2017).

The business plan envisages the following stages:

- In 2017–2020. The introduction of basic technological solutions begins, as well as the improvement of legislation, to provide the legal basis for the achievement of National Strategy indicators.
- 2021–2025. This period is planned for major investments into the solid municipal waste management system primarily through investments in the construction of new landfills and transshipment stations, as well as an incinerator in Minsk. Provided there is no experience in the design, construction and operation of incinerators in the Republic of Belarus, the development of framework conditions and the announcement of an international investment competition are chosen as optimal. This approach will optimize the operational and investment costs of the project, create starting conditions and the opportunity to attract direct foreign investment.
- 2026–2035. The introduction of a disposable consumer packaging management system and the construction of an incineration plant in Minsk require the reconstruction and replacement of equipment estimated at 30 million euros for disposable consumer packaging and 100 million euros for incineration plant. In addition, it is planned to introduce aerobic composting technology in large landfills (20 cities), which will increase the level of utilization of MSW by at least 10%.

The implementation of the National Strategy involves the expansion of public-private partnerships and the attraction of private investment in the amount of 420.5

Table 1 Major international investments in waste management in Belarus

Project description	Private partner	Country of origin
Construction of a waste recycling plant in Brest	STRABAG	Germany
Introduction of a cogeneration plant at the boiler house Oktyabrsky	TEDOM	Czech Republic
Construction of a complex for the collection and recycling of biogas in Novopolotsk and Orsha landfills	Vireo Energy	Sweden
Construction of a cogeneration plant at the boiler house in Chausy	Elteco	Slovenia
Construction of a cogeneration plant at the boiler house in Chausy	TDF Ecotech AG	Austria
Construction of a plant operating on biogas obtained from municipal waste (TKO Trostenets landfill, Minsk), etc.	TDF Ecotech AG	Austria

million Euros. Let us give examples of successful private projects in the field of waste management operating in the Republic of Belarus.

6 Public-Private Partnerships for Waste Management in Belarus

To implement the Strategy, Belarus expects to attract a significant amount of foreign investments. These investments are necessary for the implementation of projects in public-private partnerships to improve the waste management system.

Recycling of MSW at landfills is the area where investors are active. As a part of these projects, electricity is generated from waste that is undergoing a degassing process.

According to the statistical data of the Ministry of Housing and Communal Services of Belarus, there will be enough work for many years as there are about 170 garbage dumps in Belarus. Belarusian, Austrian, Germany, Swedish, etc. companies are already involved in the process. The state and local authorities are investing into the construction of modern waste processing plants (Investments 2018) (Table 1).

Belarus is counting on attracting a significant amount of foreign investment to implement the Strategy. These investments are planned for the implementation of projects in public-private partnerships to improve the waste management system.

One of the important problems for investors in the recycling of household waste is the stability of supply. In accordance with the statistics of the Ministry of Housing and Communal Services, the following secondary resources prevail in the overall waste structure (Fig. 3).

Case 1. Remondis—Germany—over the world—Belarus
Remondis JLLC is a German company leading in the world in the field of waste and

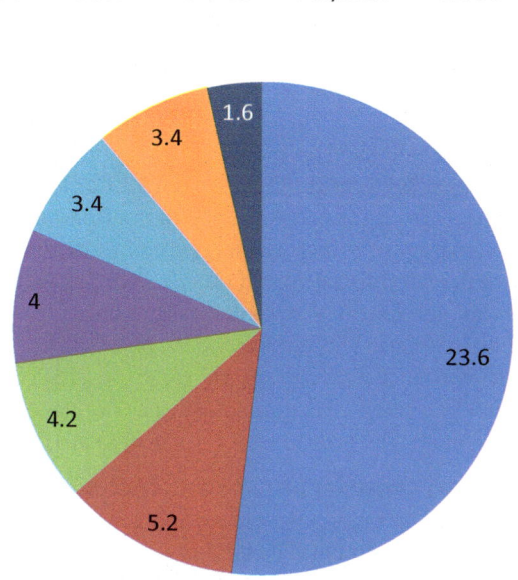

Fig. 3 General morphological composition of municipal solid waste in the Republic of Belarus (Investments 2018)

water management. Remondis has a network of over 750 offices and enterprises in 35 countries. The staff is over 30 thousand people. In 2010, the Minsk City Executive Committee and Remondis signed a framework agreement on the establishment of a Belarusian-German enterprise in waste management, Remondis Minsk, which was the first enterprise in the municipal sector in Belarus formed on the basis of public-private partnership.

Almost 80% of approximately ten million inhabitants of Belarus live in towns and cities. The country's urban hub is beyond doubt its capital city Minsk, home to practically two million people. According to Dr. Ervin Kurtbedinov (2018), managing director of Remondis Belarus, Remondis operates in three of the nine districts of Minsk, collecting waste and recovering recyclables from around one million local residents. The Company took over these operations at the beginning of the decade as part of a joint venture in which Remondis owns 51% of the shares and the City of Minsk has 49%.

The first step of newly established joint venture was upgrading the equipment for collecting the domestic waste. Several million euros were invested into purchasing new vehicles and waste containers. As a result, over 6000 old metal containers were replaced with environmentally friendly plastic bins. The company now has modern rear end loaders making waste collection more efficient. This, in turn, meant that the original fleet could be reduced by one third to just 60 vehicles that are serviced regularly in the company's own well-equipped workshop. Newly purchased vehicles are dispatched more effectively with modern GPS-controlled software and the ongoing improvements of the logistics.

In 2018, more than 800,000 Minsk residents, i.e. over 40% of the city's population were served by Remondis. The volume of waste collected has increased by 15% since the creation of the joint venture. This growth is due to two factors: an increase in the number of serviced households and an increase in the number of services provided to commercial consumers. As a result, there are 80% of orders from the municipal sector, and 20% of orders from the private sector in the Remondis portfolio.

Not limited to the territory of Minsk, the largest city of Belarus, Remondis develops partnerships with the administrations of other regions as well.

In October 2018, an agreement was signed on the creation of the first regional waste processing plant in Belarus (Soligorsk) and the possibility of the German participation in a project to introduce a pledge system for disposable consumer packaging in Belarus. It works on the money—for glass\plastic packaging return through special automatic receiver, taromat. The project cost is estimated at 100 million Euros.

Case 2. TDF Ecotech AG (Austria) http://www.tdf-ecotech.ch
TDF Ecotech AG is a leading project developer and general engineering contractor operating in the field of environmental technology, waste treatment and alternative "green" energy with offices and the focus of its activities in the Western and the Eastern Europe.

Inside the landfills, due to biological reactions, the formation of gas occurs, which causes environmental damage through methane emissions and unpleasant odors. Biogas is one of the end products of biological processing of organic waste (ADOS process). In addition, the company is engaged in the development of technology for the extraction of biogas from slurry, plant waste.

Biogas processing is one of the main activities at the existing landfills in Belarus, at waste recycling plants, in biogas plants. TDF Ecotech conducts exploration, it prepares, builds and operates networks and processing plants, provides their technical maintenance, deals with the conclusion of agreements on the supply of electricity to public networks.

Due to the fact that a significant amount of waste in Belarus is stored at landfills, there is a need for their degassing. Thanks to investments, electricity generation started at the Minsk landfill of municipal solid waste "Trostenets". The installation for active degassing of the landfill is mounted here. The 1 MW plant reduces uncontrolled greenhouse gas emissions in the amount of about 20 thousand tons of CO_2 equivalent per year, which the dump .emits into the atmosphere. Three plants reduce emissions by about 60 thousand tons per year. The unit operates on an internal combustion engine, instead of using liquid fuel it uses methane gas extracted from waste that has accumulated at the site during 50 years of its exploitation. 37 gas drainage wells with a total depth of about 1000 m are combined with a piping system. Through them, methane is supplied to the compressor part of the installation, where the generator processes it into energy.

In addition to this project, an investment agreement was signed between the Belarusian side and the Swiss company on the "Design, construction and operation of biogas complexes project". The Ministry of Agriculture and Food and the Minsk Regional Executive Committee are coordinating the implementation of the contract and assisting TDF Ecotech AG in implementing the project. According to the project, three joint ventures were created (on the basis of SPK Agrokombinat Snov, SPK Lan-Nesvizh). The construction of these complexes had a positive environmental effect in the Minsk region. The commissioning of the complexes made it possible to generate electricity from biogas produced during the fermentation of organic waste. Besides, high-quality fertilizers were produced.

Case 3. UNDP programme "Facilitating the transition of the Republic of Belarus to a green economy" (2015–2017)
The project was aimed at the implementation of the ideas of "green" economic growth and environmentally sustainable patterns of production and consumption through the support of local "green" initiatives and an information campaign. The project assisted the Republic of Belarus in economic growth based on "green" principles, including environmentally sustainable and economically viable use of natural resources, promotion of environmentally sustainable production and consumption, creation of green jobs, as well as through changing behavior and functioning patterns of target groups towards greater environmental sustainability.

The project was designed for three years; it was funded by the European Union and implemented by the UN Development Program. The Ministry of Natural Resources and Environmental Protection of the Republic of Belarus was the national executing agency.

Among other priorities, the project implementation included the creation of effective partnership mechanisms and joint activities of NGOs, local administrations and business organizations for the implementation of economically viable projects based on the ideas of "green" economic growth in the areas of waste management, ecotourism, biodiversity conservation, expanding the use of renewable sources of energy.

However, this project may be considered as a negative example of partnership between the state and private bodies. A part of this project on integrated utilization of wastes of dairy enterprises of Vitebsk region failed to attract sufficient financing from a private partner.

7   Conclusion

There are similar processes of increasing waste generation with economic growth in the Republic of Belarus and the European Union,. At the same time, waste generated in Belarus poses a significant threat of environmental pollution, since it contains a high proportion of hazardous waste. This is due to the structure of industrial production in the USSR times and the presence of large deposits of potash salt, the extraction of which causes water and soil salinization.

Over the past two decades waste management in Belarus has been dominated by processing and streamlining of waste management. In many cases, Belarus applies the positive experience of EU countries. This will improve the existing approaches to waste management in the country through focusing on preventing waste generation and minimizing waste flows. In the EU, there is a hierarchical two-tier system of waste management, supranational and national levels. The latter develops rules and actions for waste management common for all countries, which are included in national strategies. This approach is effective as it allows, first, to concentrate more powerful intellectual resources on solving the problem under consideration, and second, to use the positive experience that each country has in this area.

As waste management in the Republic of Belarus has taken shape quite recently, its contribution to the development of the national economy can described as rather modest. At the same time, it has significant development potential that allows treating it as one of the promising areas of industrial development. In the next 3–5 years the focus should be done on assuring compliance with waste management environmental safety requirements.

For waste management in Belarus, it is advisable to involve the general strategies developed in the European Union, including the quantitative indicators they provide. This will allow a more constructive approach to the waste management, and bring the national waste management system in line with the European one.

References

Action Plan (2015) Communication from the Commission to the European Parliament and the Committee of the regions. Closing the loop—an EU action plan for the Circular Economy. European Commission. Brussels, 2.12.2015. COM (2015) 614 final

Agenda 21 (1992) United Nations Conference on Environment & Development, Rio de Janerio, Brazil, 3 to 14 June 1992. Available at: https://sustainabledevelopment.un.org/outcomedocuments/agenda21. Accessible 15 Nov 2018

Beroc (2018) BEROC Green Economy Policy Paper Series, PP GE no. 2. Available at: http://www.beroc.by/webroot/delivery/files/GE_2.pdf. Accessible 15 Dec 2018

Bräutigam K-R, Jörissen J, Priefer K (2014) The extent of food waste generation across EU-27: different calculation methods and the reliability of their results. Waste Manag Res 32(8):683–694

Classifier (2007) Waste classifier of the Republic of Belarus. Available at: https://otxody.by/klassifikator-otxodov-dlya-belarusi-skachat/. Accessible 15 Nov 2018

Convention (1989) Basel convention on the control of transboundary movements of hazardous wastes and their disposal. Available at http://www.basel.int/TheConvention/Overview/TextoftheConvention/tabid/1275/Default.aspx. Accessible 15 Dec 2018

Convention (2001) Stockholm convention on persistent organic pollutants (POP). Available at: https://www.wipo.int/edocs/trtdocs/en/unep-pop/trt_unep_pop_2.pdf. Accessible 15 Nov 2018

Directive 2008/98/EC of the European parliament and of the council of 19 November 2008 on waste and repealing certain Directives. Official Journal of the European Union. L 312/3. 22.11.2008

Ellen MacArthur Foundation (2012) Towards the circular economy: economic and business rationale for an accelerated transition. Available at: https://www.ellenmacarthurfoundation.org/assets/downloads/publications/Ellen-MacArthur-Foundation-Towards-the-Circular-Economy-vol.1.pdf. Accessible 15 Nov 2018

European Commission (2004) Directive 2004/12/EC of the European Parliament and of the Council of 11 Feb 2004 amending Directive 94/62/EC on packaging and packaging waste. Off J Eur Union 18.2.2004, L 47

Investments (2018) Investment project in energy efficiency and environment. Available at: http://www.mjkx.gov.by/uploaded/investic/presentrus.pdf. Accessible 15 Nov 2018 (in Russian)

Kalmykova Y, Sadagopanb M, Rosadoc L (2018) Circular economy—from review of theories and practices to development of implementation tools. Resour Conserv Recycl 135:190–201. Available at: https://www.sciencedirect.com/science/article/pii/S0921344917303701?via%3Dihub. Accessed 25 June 2018

Kovalchik N, Struk V, Homich V (2011) Comparative waste management assessment in Belarus and European Union. Bulletin of BSU. Ser. 2. 2011. № 1 с. 91–94 (in Russian)

Kurtbedinov (2018). Interview at official web-site of Remondis. Available at: https://www.remondis-aktuell.de/en/022017/latest-news/focusing-on-sustainable-goals/. Accessible 15 Nov 2018

Law (1992) Law of the Republic of Belarus on 26 Nov 1992 № 1982-XII "On Environment Protection". Available at: http://pravo.by/document/?guid=3871&p0=v19201982. Accessible 15 Nov 2018

Law (2007) Law of the Republic of Belarus on 20 July 2007 № 271-3 "On Waste Management". Available at: http://kodeksy-by.com/zakon_rb_ob_obrawenii_s_othodami.htm. Accessible 15 Dec 2018

Leal WF, Moora H, Stenmarck A, Kruopienė J (2014) An overview of approaches towards sustainable waste management in Baltic Sea Region Countries. Res J Environ Earth Sci 6(3):134–142

Ministry (2017) Waste management. Available at: http://minpriroda.gov.by/ru/otxody-ru/. Accessible 15 Dec 2018 (in Russian)

Notice (2018) Commission notice on technical guidance on the classification of waste (2018/C 124/01). Official Journal of the European Union. 9.4.2018

OECD (2017) Green growth indicators 2017, Highlights, OECD Green Growth Studies, OECD Publishing, Paris. Available at: http://dx.doi.org/10.1787/9789264268586-en. Accessible 15 Nov 2018

Rigamonti L, Biganzoli L, Grosso M (2019) Packaging re-use: a starting point for its quantification. J Mater Cycles Waste Manag 21(1):35–43

Strategy (2017) National strategy for the management of solid municipal waste and secondary material resources in the Republic of Belarus for the period up to 2035. Available at http://vtoroperator.by/sites/default/files/doc/belarus_national_strategy_for_the_management_of_msw_and_smr_2035.pdf. Accessible 15 Dec 2018

The Future We Want (2012) Outcome document of the United Nations Conference on Sustainable Development. Rio de Janeiro, Brazil, 20–22 June 2012. Available at: https://sustainabledevelopment.un.org/content/documents/733FutureWeWant.pdf. Accessible 15 Dec 2018

Traven L, Kegal I, Šebelja I (2018) Management of municipal solid waste in Croatia: analysis of current practices with performance benchmarking against other European Union member states. Waste Manag Res 36(8):663–669

UNDESA (2012) A guidebook to the green economy. Issue 2: exploring green economy principles. UNDESA. Available at: https://sustainabledevelopment.un.org/content/documents/743GE%20Issue%20nr%202.pdf. Accessible 15 Dec 2018

Vehlow J, Bergfeldt B, Visser R et al (2007) European Union waste management strategy and the importance of biogenic waste. J Mater Cycles Waste Manag 9(2):130–139

Environmentally Friendly Concept of Phosphogypsum Recycling on the Basis of the Biotechnological Approach

Yelizaveta Chernysh and Leonid Plyatsuk

Abstract From the viewpoint of environmental security it is important to review the life cycle of phosphorite raw material transformation, from extraction to its use in the chemical industry with the forming of both useful products and secondary raw materials/waste, primarily upon the production of phosphate fertilizers with the forming of large-tonnage waste—phosphogypsum. The annually worldwide production of phosphogypsum is possibly up to 100 million tons. This paper focuses on the environmental analysis of chemical wastes as resource of technogenic genesis. To achieve the aim, the following tasks are set: the environmental impact analysis of the extraction and treatment process of phosphorous raw materials; alternatives in the field of phosphogypsum recycling within the framework of the concept of environmentally safe waste treatment development; the expediency of phosphogypsum use in environmental protection technologies. Thereupon, analytical study considers not only the accumulation of production waste in the environment, but also the products derived from this production waste when using it as a secondary raw material. Also the study includes analyses of features factors and practical recommendations for the implementation of a business strategy for different products from phosphogypsum.

Keywords Chemical industry · Phosphogypsum waste · Environment impact · Recycling · Business strategy

1 Introduction

From the viewpoint of environmental security it is important to review the life cycle of phosphorite raw material transformation, from extraction to its use in the chemical industry with the forming of both useful products and secondary raw materials/waste,

Y. Chernysh (✉) · L. Plyatsuk
Department of Applied Ecology, Sumy State University, Rymskoho-Korsakova Str., 2, Sumy 40007, Ukraine
e-mail: e.chernish@ssu.edu.ua

L. Plyatsuk
e-mail: l.plyacuk@ecolog.sumdu.edu.ua

© Springer Nature Switzerland AG 2020   167
W. Leal Filho et al. (eds.), *International Business, Trade and Institutional Sustainability*, World Sustainability Series,
https://doi.org/10.1007/978-3-030-26759-9_10

primarily upon the production of phosphate fertilizers with the forming of large-tonnage waste such as phosphogypsum.

Phosphogypsum is a chemical waste from the processing of phosphate rock by the "wet acid method" of fertilizer production, which currently accounts for over 90% of phosphoric acid production. The precipitate consists mainly of calcium sulphate dihydrate ($CaSO_4 \cdot 2H_2O$) and also contains impurities of phosphate, which is not decomposed, silicates and it can be used in the anaerobic treatment system, which is one of the promising way for phosphorus recovery. Because of trace impurities, this material has not been used to replace natural gypsum in the housing industry. phosphogypsum also has been used as agricultural fertilizers or soil stabilisation amendments. But only 10–15% of phosphogypsum is used today.

The annually worldwide production of phosphogypsum is possibly up to 100 million tons (Tayibi et al. 2009). 5.6–7.0 billion tons of phosphogypsum produced in lifetime of phosphoric acid industry to date (Aliedeh and Jarrah 2012).

There are known cases of soil, natural water and plant products contamination by interactions with phosphogypsum in different countries: Brazil, Greece, Jordan, Spain, Kazakhstan, the USA, Turkey, India, South Korea and Japan.

Currently, 15 mineral fertilizer production facilities in the Baltic Sea Area (Poland, Lithuania, Russia, Sweden, Finland) following 8 industrial sites are producing phosphate mineral fertilizers or possessing phosphogypsum waste handling facilities. A Report on implementation of HELCOM Recommendation 17/6 "Reduction of Pollution from Discharges into Water, Emissions into the Atmosphere and Phosphogypsum out of the Production of Fertilizers" (HELCOM HOD 41/2013, Document 3/17, 10.6.2013) was produced. The report recommended to continue regular monitoring and seek further solutions for possible re-use of phosphogypsum.

Currently, over 50–65 million tons of phosphogypsum are accumulated in Ukraine. Over 14 million tons of phosphogypsum are accumulated in Sumy region of Ukraine.

Environmental pressures as well as increased land costs associated with stockpiling of phosphogypsum forces the researchers to look for better utilization of this material.

Thereupon, analytical study should consider not only the accumulation of production waste in the environment, but also the products derived from this production waste when using it as a secondary raw material.

This paper focuses on the environmental analysis of chemical wastes as resource of technogenic genesis.

To achieve the aim, the following tasks were set:

- the environmental impact analysis of the extraction and treatment process of phosphorous raw materials;
- alternatives in the field of phosphogypsum recycling within the framework of the concept of environmentally safe waste treatment development;
- the expediency of phosphogypsum use in environmental protection technologies and the model forming for the implementation of a business strategy.

2 The Environmental Impact Analysis of the Extraction and Treatment Process of Phosphorous Raw Materials: Methodological Approach

The works of Gelbmann and Klampfl-Pernold (2010), Guinee et al. (2011), Brentrup and Palliere (2008), Brentrup and Lammel (2011) apply the life cycle assessment (LCA) method to the manufacturing of products, with the input flows related to categories of influence such as abiotic depletion of natural resources and the use of soil, and the output flows related to the category of global warming, acidification and eutrophication, that comply with the international standardization in the field of environmental security (ISO 14040, ISO 14044).

It should be noted that not all researchers consider all these categories during life cycle assessment, often narrowing them to the carbon footprint or energy balance analysis, as in Hsien et al. (2006), Thoms and Green (2008), Brentrup and Palliere (2008), Kneifel (2010), Lin (2010), Korre et al. (2010), ISO 14040:2004.

On the other hand, some publications expand the list of categories to include soil quality or biodiversity (Moreno et al. 2010; Nemecek et al. 2011).

Figure 1 generalizes input and output flows for the process of extraction and treatment of phosphorite raw materials on the basis of LCA.

Materials that are treated at extraction are usually on average 1/3 of phosphate, 1/3 of sand and 1/4 of clay (Energy and Environmental Profile of the U.S. 2002).

The consumption of abiotic resources upon the production of phosphate fertilizers mainly includes the consumption of such exhaustible resources as fuels and phosphate raw materials. Resources should be differentiated based on their functions, that is, those used as energy sources (coal, natural gas, oil) and those that are procured as raw materials for the production of fertilizers (phosphorites, apatites, etc.).

During the processing of phosphate mineral raw materials extracted, the yield of the main product is about 33% by weight of the raw material and 67% is waste, based on the data from (Energy and Environmental Profile of the U.S. 2002). The need for using and obtaining wet-process phosphoric acid and phosphorus-containing fertilizers causes adverse impact of production facilities on the environment.

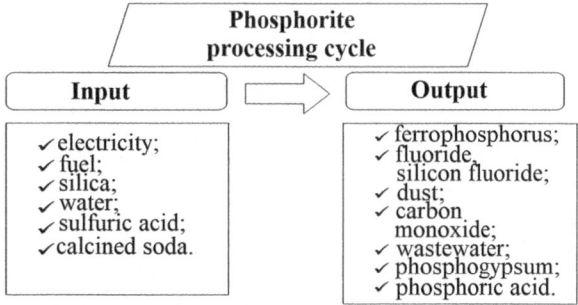

Fig. 1 Block diagram of extraction and processing of minerals based on phosphate rock

The authors of most works consider fluorine compounds and sulfur to be primary pollutants in industrial emissions generated by the production of wet-process phosphoric acid and phosphate fertilizers. The information about the development of technogenic anomalies of heavy metal accumulation in areas adjacent to facilities that produce phosphate fertilizers is ambiguous (Nemecek et al. 2011; Muravyov and Belyuchenko 2007, 2010; Savoyskaya 2017).

Possible heavy metal pollution in areas of phosphorus-containing fertilizer production is caused by the presence of metals in phosphorite raw materials in the form of ballast elements, which is described in a number of works (Degirmenci et al. 2007; Villa et al. 2009).

The main sources of wastewater in the production of phosphate fertilizers are wet purification systems used for off-gas treatment. Pollutants may include filtered solids, phosphorus, ammonia, fluorides, heavy metals (e.g. Cd, Hg, Pb). Wastewater is characterized by high chemical oxygen demand (Guidelines for environmental, health and labour protection, 2007). Condensed acid fumes may contain fluorine and small amounts of phosphoric acid.

Drainage waters from material storages may contain heavy metals (e.g. cadmium, mercury, lead), fluorides and phosphoric acid. Phosphorus and fluorine compounds, suspended solids, heavy metals and radionuclides may be dumped into the water when the thermal process is used for the production of phosphoric acid.

The sulfuric method of apatite concentrate treatment results in the generation of 4.3–5.8 tons of phosphogypsum for 1 ton of H_3PO_4 depending on the raw materials and technology used. In general, under the dihydrate mode, for 1 ton of P_2O_5 in phosphoric acid the amount of phosphogypsum generated ranges from 2 to 6 wt% that correlates with the calcium content in the phosphate raw material (Table 1).

Chemical pollutants contained in phosphogypsum in the form of water-soluble and volatile compounds pose the greatest risk to the environment.

Table 1 The basic chemical composition of phosphogypsum, depending on the type of raw material (in %)

Component of phosphogypsum	Phosphogypsum from the different type of phosphorite raw material			
	Aktyubinskij phosphoric	Apatite phosphoric	Apatite	Karatauskij phosphoric
CaO	24.4	30.0	32.4	31.2
SO_3	34.8	39.8	46.2	40.2
$P_2O_{5(total)}$	1.9	6.2	1.3	1.6
$P_2O_{5(soluble)}$	1.1	3.3	0.7	0.8
Fe_2O_3	0.9	0.7	0.1	0.1
Al_2O_3	0.8	0.8	0.3	0.4
F	0.1	0.3	0.3	0.3
Insoluble residue	21.7	4.3	0.7	8.1
Crystalline water	15.6	17.4	18.6	18.0

The content of radioactive and rare-earth elements, cadmium and other heavy metals is in direct correlation to their content in phosphate raw materials.

Thus, phosphates of magmatic origin (Kola and South African apatites) contain more rare earth elements and less cadmium than phosphates of sedimentary origin (Morocco, Florida, Senegal, etc.).

Kola apatite concentrate contains almost 10 times less radioactive elements than the above-mentioned phosphates of sedimentary origin (Tovazhnyanskij et al. 2009).

As a result, impurities contained in the phosphorite ore are distributed among the produced phosphoric acid and calcium sulfate (gypsum); mercury, lead and radioactive components, if present, mainly remain in phosphogypsum, while arsenic and other heavy metals (in particular cadmium) remain mainly in the acid and later on may move to become part of a fertilizer (Production of Phosphoric Acid 2000; Papastefanou et al. 2006).

These by-products show potential for valorisation, but transport costs, contamination with impurities and the competition with, e.g. natural resources, restrict the successful marketing. Hence, excess volumes require disposal. Many of the important considerations in the design and construction of gypsum disposal areas are based on the necessity to keep both the gypsum and the acidic stack effluent within a closed system. To avoid pollution of the subsoil and groundwater by acidic and contaminated phosphogypsum leachate and run-off (process water and rainwater), stringent preventive measures are necessary, such as seepage collection ditches, intercept wells, natural barriers and lining systems. Furthermore, to prevent or minimise pollution of the surrounding area and water systems, it is necessary to make provisions for any effluent overflow. The effluent requires appropriate treatment, such as immobilisation of soluble P_2O_5 and trace elements by neutralisation, before it can be released from the system. Besides applying control during the build-up of a gypsum stack, the run-off from gypsum stacks will require treatment for many years after the acid plant has ceased production (HELCOM HOD 41/2013, Document 3/17, 10.6.2013).

3 Phosphogypsum Recycling: Results and Discuses

3.1 Alternatives in the Field of Phosphogypsum Recycling Within the Framework of the Concept of Environmentally Safe Waste Treatment Development

Analysis of the use of phosphogypsum in world practice helped to identify an innovative approach to the use of phosphogypsum of a number of companies. With traditional building materials degrading the landscape and adding significantly to CO_2 emissions, building from environmentally friendly Rapidwall has become even more attractive. India produces significant amounts of fertilizer for worldwide use but in doing so creates phosphogypsum as a by-product in the order of millions of tonnes annually. Presently there is 31 million tonnes of excess phosphogypsum stockpiled and this is added to annually by 2.5 million tonnes. By utilising Rapid Building Systems Rapidflow calcination plant the phosphogypsum can be turned into plaster and subsequently into Rapidwall, thereby cleaning up the environment. This stockpiled gypsum is enough to build 5 million 30 m^2 Rapidwall homes (https://www.rapidwall.com.au/). A similar phosphogypsum processing project was created in Kochi RCF-Building Products Ltd. Prayon, one of the world's leading producers of phosphoric acid, for many years of research has achieved the use of phosphogypsum by 90%, focusing on the supply of phosphogypsum to consumers in the construction industry (http://www.frbl.co.in/). Thus, it's possible to produce building materials from phosphogypsum, since the product consists of 90% gypsum. But the global annual demand for natural gypsum is estimated at 84 million tons (Savoyskaya 2017).

Numerous studies have investigated the possibility of using phosphogypsum for the remediation of contaminated soils (after oil contamination, for remediation of soils contaminated with fuel oil) (Wolicka and Kowalski 2006; Wolicka 2008; Kalinina 2011; Yakovlev et al. 2013; Shaydullina et al. 2015; Ablieieva and Plyatsuk 2016).

There is a method for the remediation of oil-contaminated soils or soils located in areas of oil spill cleanup, which lies in the fact that contaminated soil is not disposed of, but instead is introduced with ameliorants based on a mixture of phoshogypsum, sand, humus and mineral fertilizers made of nitrogen, phosphorus and potassium, and is later plowed and seeded with cultivated crops. The application rate of ameliorants is determined by taking into account the oil spilled, the agronomic assessment of the soil before the spill, the duration of oil presence on the surface of the soil.

Maintaining an acidic reaction will lead to further breakdown of the organic oil component. Moreover, it will also result in the balancing of the oxidation-reduction potential and the restoration of the soil respiration response. As a result, the absorption and poisoning of the soil by CO_2 will be reduced (Shaydullina et al. 2015).

In domestic agriculture, neutralized phosphogypsum is used as a phosphorus-calcium-sulfur fertilizer and ameliorant of sodic soils. In Russia, a technology has been developed to apply this secondary product, taking into account the soil and climatic features of the region. It is noteworthy that the use of phosphogypsum up to 30% will save costs for the use of fertilizers. In the country, sodic soils cover an area of 31.4 million ha, and their restoration will require about 430 million tons of gypsum-containing materials (Savoyskaya 2017).

It is possible to use phosphogypsum in remediating drill cuttings that are collected and stored directly on the territory of the drilling site. Drill cuttings are characterized by unfavorable chemical properties—they bloat and become viscous and sticky when wet, when dry, they are consolidated and solid due to the presence of Na in the absorbing complex. Drill cuttings are also characterized by high alkalinity, their pH equals 8.68–9.10 and is detrimental to plant growth. Coagulators, with phosphogypsum being one of the most prospective ones, are to be used to displace the absorbed sodium with calcium in order to improve physical and chemical properties of drill cuttings. This is due to the fact that phosphogypsum, being a waste, is much cheaper than gypsum and has higher solubility, and the presence of water-soluble phosphorus in it enhances the ameliorative effect (Kalinina 2011).

The possibility of using biotechnological methods of drilling wastewater purification based on biodegradation of organic pollutants is reasonable. Purification efficiency largely depends both on the activity of microbial degraders and the presence of macro and trace elements in the environment. Lack of some of the most important elements—phosphorus and calcium—is a limiting factor for the process of drilling wastewater bioremediation, so the use of phosphogypsum as a cheap source of these elements is quite promising. The work of Barakhnina et al. (2011) establishes that within three days of cultivation, experimental application of 1.0% wt of phosphogypsum resulted in a 99.8% biodegradation of carboxymethyl cellulose and polyacrylamide in drilling wastewater, which is 66.8% higher than in the control sample. However, this method requires further development and improvement as its industrial-scale implementation requires the activation of the biological component of the process while taking into account all the physical and chemical properties of biodegrading materials.

Phosphogypsum can also be used in biotechnological processes of sewage sludge neutralization. In works Plyatsuk and Chernish (2014), Chernysh et al. (2014) were described a method of treating sewage sludge in anaerobic fermentation systems with heavy metal precipitation into a complex sulfide fraction using biogenic hydrogen sulfide, which is a by-product of sulfate reducers, where phosphogypsum is used as a mineral supplement to enhance bacterial culture growth (Fig. 2).

At that, in order to intensify the growth of a bacterial culture, it would be appropriate to apply the process of its immobilization in mineral media to reduce the removal of biomass from the bioreactor and to put the treatment process in a continuous mode of fermentation, which would increase bioreactor productivity.

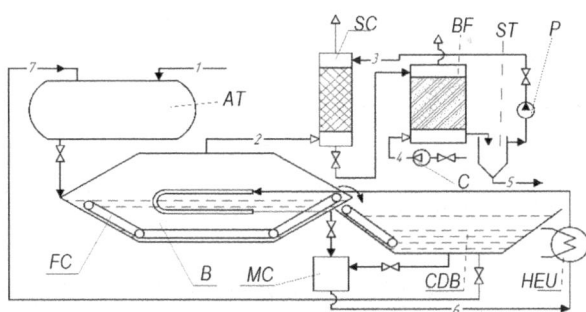

1 – sewage sludge and phosphogypsum to be treated; 2 – gaseous phase outlet; 3 – chemical agent inlet; 4 – air input; 5 – biosulfur outlet; 6 – liquid circulation; 7 – introduction of a part of the solid digested sludge fraction as an inoculum
AT – accumulation tank; B – bioreactor; FC – flight conveyor; SC – scrubber; BF – biofilter with granular phosphogypsum; C – compressor; ST – settling tank; P – pump; MC – mixing container for fluid flows; HEU – heat exchange unit; SDB – sludge drying bed

Fig. 2 Process flow diagram of sewage sludge and phosphogypsum treatment under bio-sulfidogenic conditions

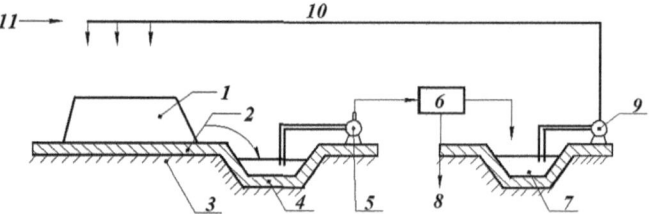

1 – phosphogypsum dump, 2 – impervious screen, 3 – soil surface, 4 – a special pond for collecting metal-rich runoff, 5, 9 – pumps, 6 – processing unit for bioleaching according to known technology, 7 – a pond for the spent solution, 8 – product withdrawal (REE, phosphates), 10 – irrigation system, 11 – bacterial suspension.

Fig. 3 Technological scheme of biological leaching of rare-earth metals from phosphogypsum dumps

Light and middle REE are concentrated into phosphogypsum. In total 21 Mt of rare earth elements (REE) has been locked into phosphogypsum over the past decades. Phosphogypsum is therefore a potential secondary source for sustainable REE production (Recovery of rare earth elements from phosphogypsum (REE–PG) 2018).

Figure 3 shows the flow chart of bacterial leaching of REE from phosphogypsum. It can be applied in practice, as it is easy to operate, has no waste after the treatment process and is an environmentally friendly installation.

The biochemical direction has several environmental advantages: reducing the amount of chemical reagents in the processing of generated phosphogypsum and wastewater, environmentally safe recycling of treatment products.

Extensive test work programs on REE recovery from phosphogypsum in Finland and in South Africa have shown that the phosphogypsum is highly variable, both in terms of physical properties and REE recovery, which prevents the industry from successfully treating the phosphogypsum dumps (Lutskiy et al. 2018).

According to Hussien et al. (2018) the bioleaching approaches offer several advantages over chemical leaching methods, where the roles of media ions (Mg^{2+} and NH_4^+ ions) and phenomena occurring at mineral- and bacteria-water interfaces can be appropriately considered. The effects of essential biological parameters (e.g. pH, ionic strength and temperature) on REE recovery can result from the competition between ions, changes in the activity of REEs and/or microbial functional groups, or changes to the electrical double layer at the microbe-water interface (Zhuang et al. 2015).

Against the background of a favorable market situation, with the demand for REE, as well as with the growing need for high-strength building materials and thanks to government support in different countries, over the past few years, the creation of phosphogypsum processing plants has increased. Incentives in the area of handling highly efficient processing product—phosphogypsum, which consist in changing the regulatory framework, providing state support, in the possibility of attracting financial resources. A wide range of use of products extracted from phosphogypsum, contributed not only to the emergence, but also the implementation of new effective business strategies.

3.2 The Expediency of Phosphogypsum Use in Environmental Protection Technologies

Fresh dihydrate phosphogypsum is a partially clumped grey-white finely dispersed powder that contains 35–40 wt% of total moisture. Phosphogypsum tends to clump, and consolidates in case of long-term storage. It exhibits thixotropic properties, that is, it can thin out under mechanical stimuli (vibration, stirring, shaking).

Dihydrate phosphogypsum is essentially a disintegrated rock treated with sulfuric acid and introduced with a lime mortar after P_2O_5 extraction; it is characterized by a high content of insoluble compounds (CaO, SO_3, Al_2O_3, Fe_2O_3, SiO_2, MgO). Since insoluble components of phosphogypsum are poorly hydrated, their mutual coagulation does not occur and pure structural aggregates do not form. The aggregate composition of phosphogypsum is mainly comprised of particles sized 0.15–0.20 mm (65–70%) (Wolicka 2008; Shuisky et al. 2016; Golova and Davtyan 2017).

Phosphogypsum can be defined as a mineral dispersed system. Relatively high dispersion of phosphogypsum is determined by the specificity of its physical state in the form of a mixture of finely ground particles of colloidal substance distributed in a homogeneous environment. Its colloidal matter is therefore characterized by a slow diffusion rate, inability to penetrate through the fine-pored membranes of living organisms' cellular structures, as well as by nonequilibrium solubility.

Phoshogypsum has a rather large specific surface area that averages 3950 m^2/g (Tovazhnyanskij et al. 2009). Phosphogypsum is characterized by a certain density that depends on its moisture content, compaction, etc. After crushing and sifting, the density of phosphogypsum equalled 2.2–2.4 g/m^3 as determined by the pycnometer method.

Phosphogypsum is not conventionally viewed as a colloidal system, however it contains a lot of ultra-fine particles of calcium sulfate, silicofluoride, sodium and potassium, sesquioxide phosphates, sulfur compounds, various aggregates absorbed on the surfaces of its particles that exhibit coagulation properties when combining this waste with organic substrates—manure (humus) (Belyuchenko 2015; Yurkova and Dokuchaeva 2016), poultry manure (Shalashova and Ivanova 2016), sewage sludge (Wolicka and Kowalski 2006; Plyatsuk and Chernish 2014) and soil (Muravyov 2010; Gukalov and Slavgorodskaya 2011; Yurkova and Dokuchaeva 2016).

The chemical composition of average phosphogypsum samples (Table 1) and the list of elements required for bacterial cell life indicate that substrate phosphogypsum can be used for nourishing microorganisms and can stimulate the metabolic processes of the bacterial cell (Fig. 4).

Composition of phosphogypsum

P_2O_5

- Phosphorus is included in nucleic acids, phospholipids, numerous enzymes, ATP.
- Phosphates play a special role in energy metabolism, carbohydrate breakdown and membrane transport.
- Enzymatic synthesis of a biopolymer array can only start after the formation of phosphoric esters of parent compounds.

SO_3

- Sulfur is included in some amino acids (cysteine, methionine), vitamins (biotin, thiamine), peptides (glutathione); is involved in synthetic processes in the restored state, that is in the form of R-Sh groups with high reactive capacity that can easily dehydrate into S-R groups. Such reactions are instrumental in the regulation of the oxidation-reduction potential in the bacterial cytoplasm.

CaO

- Calcium is a constituent of the cell wall of Gram-positive bacteria, is involved in the functioning of bacterial cytoskeleton myofibrils.

Fe_2O_3

- Iron is found in respiratory enzymes.

MgO

- Mg and Fe ions are used as enzyme activators.
- Magnesium is part of magnesium ribonucleate, which is localized on the surface of Gram-positive bacteria.
- Mg and K ions activate ribosomes during protein synthesis.

Na_2O

- Sodium is involved in maintaining osmotic pressure in the cell.

SiO_2

- For some specific soil bacteria, it is characteristic that the absorption of silicon. Silicon can enters the cell walls of these bacteria in the form of a silicate ion or oxides. The energy released by the splitting of high-energy phosphates is used directly to bind silicon.
- When the concentration of silicon esters of sugars reaches a certain level, they enter the intracellular fluid, where they are associated with sugar and the previously inorganic part of silicon. Here, the formation of various silicon organic compounds occurs, including those containing Si-C bonds, as well as insoluble silica polymer.

Zn, Cu, Mo, Co, Cr, Mn

- Trace elements are involved in the synthesis and activation of certain enzymes.

Fig. 4 The role of phosphogypsum components in the microorganism's metabolism

It should be noted that it is important to study the effect of phosphogypsum on the association of microorganisms that belong to different ecological and trophic groups, with some research existing about soil microbiocenosis.

The process study of phosphogypsum granules production and use it in bio-methanogenesis technologies was carried out to Chernysh et al. (2014). According to Plyatsuk et al. (2018) the total cost of production of 1 ton of modified phospho-gypsum granules is 348.43 Euro, the payback period is 1.8 years (with a production capacity of 1000 tons of product per year).

It should be noted that the average price of granulated phosphogypsum (Fig. 5a) is 0.41 Euro per kg, and the most common carrier in purification systems such as zeolite (Fig. 5b) approximately 1.4 Euro per kg.

Summarizing, we can thus define the main physical and chemical properties of phosphogypsum that play an important role for the expansion of its scope of application as follows:

– its content in various proportions of components (CaO, Fe_2O_3, FeO, MgO, MnO_2, Cr_2O_3, CuO, SiO_2 etc.) which depends on the content of these compounds in the source phosphorite raw material, stocks of these compounds in soils have decreased due to their removal with agricultural crops, weathering and leaching;
– pH value in the range from 3.5 to 5.0 depending on the duration of storage in disposal areas;
– solubility in the alternate humidification-drying mode equals 0.33–0.36% under laboratory conditions; all-year solubility in the soil approximates 10.5–12.6%.

It should be noted that the influence of phosphogypsum on soil mesofauna and the microbial biome of the environment subjected to elimination of contaminants (complex hydrocarbons, hydrogen sulfide, etc.) is of great importance (Wolicka and Kowalski 2006; Wolicka 2008; Plyatsuk and Chernish 2014; Belyuchenko 2015; Shalashova and Ivanova 2016; Chernish and Plyatsuk 2016). Thus, the following biochemical properties of phosphogypsum have been revealed:

Fig. 5 Carriers for biotechnological systems: **a** granulated phosphogypsum (4–5 mm); **b** zeolite (3–5 mm)

- in a microbial living environment phosphogypsum can act as an additional nutrient source and can stimulate metabolic processes of producer cells that belong to different trophic groups;
- earthworm populations increase after the application of activated phosphogypsum-laced compost;
- the interaction of phosphogypsum with organic waste results in the generation of macroaggregates, which, when dissolved in water, lead to slight acidification of the environment which has a negative impact on the development of worm eggs and leads to their death.

Thus, a number of patterns has been revealed:

- phosphogypsum-containing organic and mineral complexes form in the soil through binding of labile organic matter in stable aggregated formations with micro-particles of phosphogypsum colloids;
- solubility of phosphogypsum in the soil increases with the decrease in its particle size;
- phosphogypsum acts as a source of macro and trace elements for the development of different ecological and trophic microbial groups;
- the acidic reaction of phosphogypsum creates favorable conditions for the break-down of such organic compounds as surfactants, hydrocarbons and other substances, allowing phosphogypsum to be composted with waste that contains such substances as sewage sludge, sawdust, manure, bird droppings;
- composting different kinds of organic waste with phosphogypsum significantly improves the sanitary and epidemiological situation.

Figure 6 shows the model of the practical recommendations for the implementation of a business strategy for different products from phosphogypsum under international cooperation level.

Therefore, any scientific idea on the use of phosphogypsum should be implemented on the basis of financial considerations. Preliminary preparation of a feasibility study on the use of specific types of phosphogypsum in certain markets, the key to a successful business. This task needs to be implemented by specific manufacturers, research institutes and regulatory authorities before taking a priority decision.

In addition, it is extremely important to choose a financial accounting model for assessing cost factors and analyzing the cost of project implementation on phosphogypsum processing. In this project, the estimated costs for the treatment of phosphogypsum will include potential sources of income, since any sale of by-products implies additional cash.

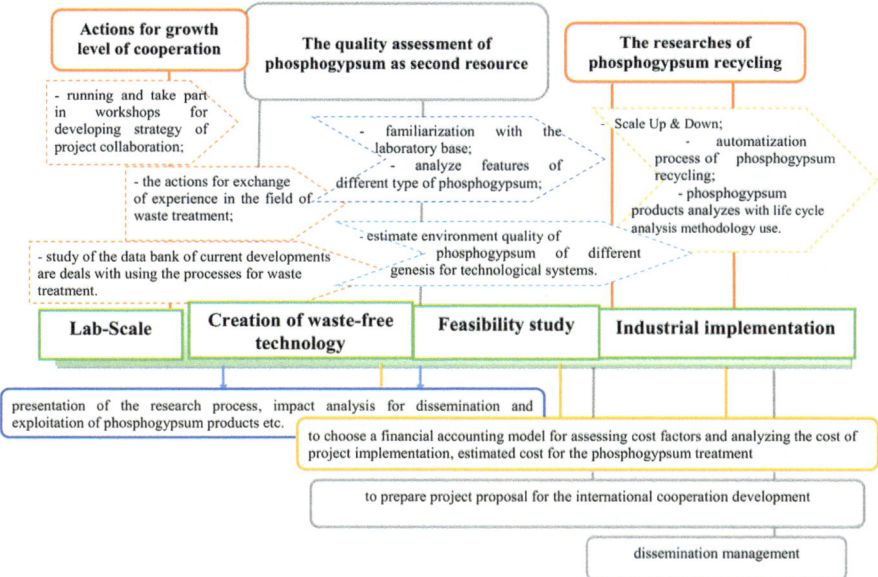

Fig. 6 Model of the implementation of a business strategy for different products from phosphogypsum

4 Conclusion

The constant search for unique technologies that prevent the formation of phosphogypsum by introducing it into various sectors of the economy gradually builds phosphogypsum into a closed-loop economy. The search for ways of reducing the environmental impact of industrial waste and using them as valuable secondary material resources remains relevant. The development of an environmentally friendly concept of phosphogypsum recycling on the basis of the biotechnological approach is prospective in this regard.

Thus, using phosphogypsum as a mineral raw material to manufacture mineral support medium for the development of useful ecological and trophic microbial groups will allow to expand the scope of phosphogypsum application in environmental protection technologies. For example, the use of product processing as a source of rare elements, as component of fertilizer or support media for bacteria growth under biotechnological solution of waste treatment.

The analysis showed that by studying the technological and economic features of the phosphogypsum waste recycling turned into a useful resource.

References

Ablieieva I, Plyatsuk L (2016) The immobilization of heavy metals during drilling sludge utilization. Environ Technol Innov 6:123–131

Aliedeh MA, Jarrah NA (2012) Application of full factorial design to optimize phosphogypsum beneficiation process (P_2O_5 Reduction) by using sulfuric and nitric acid solutions. In: Sixth Jordanian International Chemical Engineering Conference, Amman, Jordan, Mar 2012. Available at: http://www.jeaconf.org/UploadedFiles/Document/31365fa2-c30a-442c-ae60-bea341617140.pdf

Application of GFRG Panel (2018) Homepage, http://www.frbl.co.in/RAPIDWALL_FOR_HOUSING.pdf

Barakhnina VB, Hafizova AA, Kireev IR (2011) Исследование возможности использования фосфогипса при биоочистке буровых сточных вод [Investigation of the possibility of using phosphogypsum in the biological purification of drilling wastewater]. Bashkir Chem J 18(2):90–92

Belyuchenko IS (2015) Влияние дозы фосфогипса на состав и агрономические свойства сложного компоста [The effect of phosphogypsum dose on the composition and agronomic properties of complex compost]. North Cauc Ecol Her 11(1):84–92

Brentrup F, Lammel J (2011) LCA to assess the environmental impact of different fertilisers and agricultural systems. In: Proceedings of International Fertiliser Society, York, UK. http://fertiliser-society.org/proceedings/uk/Prc687.HTM

Brentrup F, Palliere C (2008) GHG emissions and energy efficiency in European nitrogen fertiliser production and use. In: Proceedings of international fertiliser society, York, UK, 24 p

Building with Rapidwall in India (2018) Homepage, https://www.rapidwall.com.au/

Chernish Y, Plyatsuk L (2016) Opportunity of biochemical process for phosphogypsum utilization. J Solid Waste Technol Manag 42(2):108–111

Chernysh YY, Plyatsuk LD, Dorda VA (2014) Ecotechnology for hydrogen sulfide removal and production of elemental sulfur. Int J Energy Clean Environ 15(2–4):189–202

Degirmenci N, Okucu A, Turabi A (2007) Application of phosphogypsum in soil stabilization. Build Environ 42(9):3393–3398

Energy and Environmental Profile of the U.S. Mining Industry (2002) https://energy.gov/eere/amo/downloads/itp-mining-energy-and-environmental-profile-us-mining-industry-december-2002

Gelbmann U, Klampfl-Pernold H (2010) Applying life cycle-oriented tools for analysing the sustainability of a regional waste management system. Reg Dev Dialogue 31(2)

Golova TA, Davtyan AR (2017) Исследование фосфогипса как эффективного вяжущего для строительных композитов [The study of phosphogypsum as an effective binder for building composites]. In: Collection of articles of the winners of the VIII international scientific and practical conference: world science: problems and innovations, Penza, 30 Mar 2017, pp 63–65

Guidelines for Environmental, Health and Labour Protection (2007) General guidelines: environmental protection. Phosphate fertilizer production, 28 p. http://www.ifc.org/ifcext/sustainability.nsf/Content/EnvironmentalGuidelines

Guinee JB, Heijungs R, Huppes G, Zamagni A, Masoni P, Buonamici R, Ekvall T, Rydberg T (2011) Life cycle assessment: past, present, and future. Environ Sci Technol 45(1)

Gukalov VV, Slavgorodskaya DA (2011) Влияние фосфогипса на водно-физические свойства чернозема [The effect of phosphogypsum on the water-physical properties of chernozem]. North Cauc Ecol Her 7(1):83–85

Hsien H Khoo, Reginald B, Tan H (2006) Life cycle evaluation of CO_2 recovery and mineral sequestration alternatives. Environ Prog 25(3):208–217

Hussien S, Patra P, Somasundaran P et al (2018) Assessment of 'bacterial (acidic)-leaching' of rare earth elements from a phosphate ore. Adv Environ Stud 2(2):91–97

Kalinina OV (2011) Использование фосфогипса для рекультивации загрязненных мазутом почв [The use of phosphogypsum for the remediation of soils contaminated with fuel oil]. North Cauc Ecol Her 7(1):86–88

Kneifel J (2010) Life-cycle carbon and cost analysis of energy efficiency measures in new commercial buildings. Energy Build 42(3):333–340

Korre A, Nie Z, Durucan S (2010) Life cycle modelling of fossil fuel power generation with post-combustion CO_2 capture. Int J Greenh Gas Control 4(2):289–300

Lin C-F (2010) Method for the sequestration of carbon dioxide. From German Offen, DE 102008054395 A1 20100617

Lutskiy DS, Litvinova TE, Ignatovich AS, Fialkovskiy I (2018) Complex processing of phosphogypsum—a way of recycling dumps with reception of commodity production of wide application. J Ecol Eng 19(2):221–225. https://doi.org/10.12911/22998993/83562

Moreno MA, Ortega JF, Córcoles JI, Martínez A, Tarjuelo JM (2010) Energy analysis of irrigation delivery systems: monitoring and evaluation of proposed measures for improving energy efficiency. Irrig Sci 28(5):445–460

Muravyov YI (2010) Перспективы использования фосфогипса в сельском хозяйстве [Prospects for the use of phosphogypsum in agriculture]. North Cauc Ecol Her 6(4):85–89

Muravyov YI, Belyuchenko IS (2007) Влияние отходов химического производства на загрязнение окружающих ландшафтов [Impact of chemical production wastes on pollution of surrounding landscapes]. North Cauc Ecol Her 3(4):77–86

Nemecek T, Elie OH, Dubois D et al (2011) Life cycle assessment of Swiss farming systems: II. Extensive and intensive production. Agric Syst 104:233–245

Papastefanou C, Stoulos S, Ioannidou A, Manolopoulou M (2006) The application of phosphogypsum in agriculture and the radiological impact. J Environ Radioact 89(2):188–198

Plyatsuk L, Chernish Y (2014) Intensification of the anaerobic microbiological degradation of sewage sludge and gypsum waste under bio-sulfidogenic conditions. J Solid Waste Technol Manag 40(1):10–23

Plyatsuk LD, Chernysh YY, Yakhnenko OM (2018) 4343 Methodical instructions for conducting practical classes, recommendations for performing calculations of cleaning installations in technological systems of environmental protection for the discipline "Technoecology", SSU, Sumy, 15 p

Production of Phosphoric Acid (2000) Booklet 4 of 8. General Product Information on Phosphoric Acid. European Fertilized Manufacturers Association. Fisherprint Ltd., Peterborough, 44 p

Recovery of Rare Earth Elements from Phosphogypsum (REE–PG) (2018) Homepage, https://www.aka.fi/globalassets/32akatemiaohjelmat/misu/hankekuvaukset/misu_sainio_ree-pg.pdf. Last accessed 2018/11/12

Report "Reduction of Pollution from Discharges into Water, Emissions into the Atmosphere and Phosphogypsum out of the Production of Fertilizers" (2013) HELCOM HOD 41/2013, Document 3/17, 10.6.2013. 41st Meeting. Helsinki, Finland, 17–18 June 2013, 40 p

Savoyskaya EV (2017) Prospect for the development and economic efficiency of raw material resources. Bull Russ Acad Sci 2:122–127

Shalashova OYu, Ivanova NA (2016) Изменение физико-химических свойств чернозема обыкновенного среднесолонцеватого под влиянием удобрительно-мелиорирующих смесей [Changes in the physicochemical properties of the chernozem of the ordinary medium-colic under the influence of fertilizer-meliorating mixtures]. Bull Omsk State Agrar Univ 2(22):72–80

Shaydullina IA, Yapparov AH, Degtyareva IA, Latypova VZ, Gadieva SH (2015) Рекультивация нефтезагрязненных почв на примере выщелоченных черноземов Татарстана [Reclamation of oil-contaminated soils on the example of leached chernozem of Tatarstan]. Oil Ind 3:102–105

Shuisky AI, Novozhilov AA, Torlina EA (2016) Методы и способы переработки фосфогипса в кондиционное сырье с учетом экологических факторов [Methods and methods of processing phosphogypsum into conditioned raw materials, taking into account environmental factors]. Econ Ecol Territ Form 1:82–84

Tayibi H, Choura M, López FA, Alguacil FJ, López-Delgado A (2009) Environmental impact and management of phosphogypsum (review). Available at: http://digital.csic.es/bitstream/10261/45241/3/Environmental%20impact%20and%20management%20of%20phosphogypsum.pdf

Thoms GE, Green K (2008) Improved method of capturing carbon dioxide and converting to carbonate anions and then combining with calcium cations to form calcium carbonate. From Granted Innovation Pat. (Australia), AU 2007101174 A4 20080131

Tovazhnyanskij LL, Kapustenko PA, Buhkalo SI, Perevertajlenko AYu, Havin GL, Arsen'eva OP (2009) К вопросу повышения энергоэффективности комплексных технологий конверсии фосфогипса [On the issue of improving the energy efficiency of complex phosphogypsum conversion technologies]. Integr Technol Energy Conserv 1:3–8

Villa M, Mosqueda F, Hurtado S, Mantero J, Manjón G, Periañez R, Vaca F, García-Tenorio R (2009) Contamination and restoration of an estuary affected by phosphogypsum releases. Sci Total Environ 408(1):69–77

Wolicka D (2008) Biotransformation of phosphogypsum in wastewaters from the dairy industry. Bioresour Technol 99(13):5666–5672

Wolicka D, Kowalski W (2006) Biotransformation of phosphogypsum on distillery decoctions (Preliminary results). Pol J Microbiol 55(2):147–151

Yakovlev AS, Kaniskin MA, Terekhova VA (2013) Ecological evaluation of artificial soils treated with phosphogypsum. Eurasian Soil Sci 46(6):697–703

Yurkova R Ye, Dokuchaeva LM (2016) Влияние способов и доз внесения фосфогипса на физические свойства почв комплексного покрова [The influence of the methods and doses of phosphogypsum making on the physical properties of the soils of the complex cover]. Sci J Russ Sci Res Inst Land Improv Probl 3(23):102–115

Zhuang WQ, Fitts JP, Ajo-Franklin CM, Maes S, Alvarez-Cohen L, Hennebel T (2015) Recovery of critical metals using biometallurgy. Curr Opin Biotechnol 33:327–335

Can Circular Economy Tools Improve the Sustainable Management of Industrial Waste?

Mariana de Souza Silva Rodrigues and Alvim Borges

Abstract The circular production model indicates a restorative and regenerative system, in which the streams of materials and products take place in a cyclic way, recovering the value of waste by reducing disposal at most, creating sustainable production chains. Considering social pressures, major industrial enterprises perceived the need for readjusting their production chains according to circular chains, which are more sustainable and also consider the generated waste. In this research we aim to determine factors for sustainable waste management in major industrial enterprises based on the circular economy approach. We consider primary data on the availability of waste in a major industrial company. Approaches of circular economy, such as fault tree analysis and others, are applied to determine the conditions for the implementation of a circular process to industrial waste, especially those of lower value that have greater difficulties in being processed. The results show that the generating company alone cannot create a closed cycle for processing their waste. It is necessary to broaden the scope of participation in the production chain of industrial waste.

Keywords Circular economy · Sustainable management · Industrial waste · Sustainable chains · Circular economy tools

1 Introduction

Industrial activities generate different types of waste, with the most diverse characteristics. Industrial waste are quite varied, and can be represented by process residues, residues from pollution-control or decontamination operations, adulterated materials, materials and substances resulting from activities for contaminated soil remediation, wastes from the purification of raw materials and products, ashes, mud, oil, alkaline

M. de Souza Silva Rodrigues (✉) · A. Borges
Master of Science Program on Engineering and Sustainable Development, Federal University of Espírito Santo, Av. Fernando Ferrari, 514 – Goiabeiras, Vitória, Espírito Santo, Brazil
e-mail: marianade.silva@aluno.ufes.br

A. Borges
e-mail: alvim@pobox.com

© Springer Nature Switzerland AG 2020
W. Leal Filho et al. (eds.), *International Business, Trade and Institutional Sustainability*, World Sustainability Series,
https://doi.org/10.1007/978-3-030-26759-9_11

or acid waste, plastics, paper, wood, fiber, rubber, metal, slag, glass, and ceramic. It is estimated that, in Brazil, about 97,655,438 tons of industrial waste are generated per year (IPEA 2012). In Brazil, the *Política Nacional de Resíduos Sólidos* (Brazilian Solid Waste Policy) provides for obligations for the productive sector, in such a way that in addition to environmental benefits, the appropriate management of waste plays an important role in the proper expansion of the economic and social infrastructure of the country (IPEA 2012).

In our research, we addressed the Reverse Logistics of what we call unavoidable waste, which are generated despite all the actions directed to the best use and distribution of materials, which are inherent in industrial activity. The reverse logistics, within the context of Circular Economy, involves the management of returns of products, followed by the activities of end-of-life processing and recovery of products such as repair, reuse, reform, remanufacturing, and recycling (Bernon et al. 2018). By-products and wastes are potentially valuable inputs in various industrial processes. Markets are being developed to capitalize the wastes, recognizing the value of the use and reuse of these materials as inputs, creating the Reverse Logistics as a research area (Corbett and Kleindorfer 2001).

The Circular economy approach is an economic strategy that suggests innovative ways to transform the current predominantly linear system of consumption into a circular system, while achieving economic sustainability with the necessary economy of materials. The Circular Economy promotes high-value material cycles in addition to recycling, and develops system approaches for cooperation of producers, consumers, and other actors of society towards sustainable development (Korhonen et al. 2018). Hence, the circular production model indicates a restorative and regenerative system, in which the streams of materials and products take place in a cyclic manner, recovering the value of wastes, reducing disposal at most, and creating sustainable production chains.

Considering social pressures, major industrial enterprises should promote a readjustment of their linear production chains for circular chains, which are more sustainable and also include the generated waste (Vlajic et al. 2018). Thus, we need instruments that promote the sustainable management of waste of industrial companies. Our objective is to identify and assess the conditions for use of circular economy tools in the management of industrial waste.

2 Circular Economy of Industrial Waste

The linear production system, according to which the flow of energy and materials take place in a single extraction-production-use-disposal direction, has been causing dissatisfaction, since it is problematic concerning sustainability. Hence, Circular Economy suggests a new alternative approach to the achievement of a more sustainable development, with a cyclic and regenerative flow (Korhonen et al. 2018).

Circular Economy, unlike simple recycling, is an approach that emphasizes the reuse of products, components and materials, remanufacturing, reforms, repairs, and upgrade. It also considers the potential for sustainable energy sources, such as solar, wind, biomass, and energy derived from waste, used throughout the value chain of the product and using a life-cycle cradle-to-cradle approach. At a later time, the Circular Economy, when fully developed, should promote high-value material cycles rather than recycling only low-value raw materials such as is the case of traditional recycling. Thus, the concept of Circular Economy is not only a production system, but it also aims to develop a sustainable consumption in addition to sustainable production, promoting and implementing the sharing economy approach. According to Circular Economy, consumer groups share the function and the service provided by the physical product to replace current patterns of consumption based on individual property, in such a way that more value is extracted from physical resources within the economy (Korhonen et al. 2018). A successful circular economy should contribute to all three dimensions of sustainable development: social, economic, and environmental dimensions. Circular economy seeks to limit the flow of production to a level nature can tolerate, and uses ecosystem cycles in economic cycles, respecting their natural production rates.

After Industrial Revolution, disposable products with the explicit purpose of being discarded after use (built-in obsolescence) give voice to the age of fashion and style, stimulating the disposable mentality nowadays known as linear consumption behavior. Successively, problems of environmental pollution and landfills have become important leaders worldwide to start programs of waste reduction and recycling. More recently, economic opportunities of the circular economy were established by the European Union, emphasizing the advantages to the industrial sector such as cost reduction on materials or larger profit groups (Lieder and Rashid 2016).

3 Methodology

This exploratory research started with a visit to the Center of Discarded Materials (*Centro de Materiais Descartados*—CMD) of a major industrial company to which we will refer to as Company A. Not all wastes are disposed in the more sustainable way, and many of them remains deposited at the CMD for a long time. From this visit, we listed the most problematic unavoidable wastes concerning sustainable disposal, and we gathered data on generation and disposal of such wastes, as we can observe in Tables 3 and 4.

Then, we surveyed in the literature the main tools of Circular Economy, and we highlighted two reports from the Ellen MacArthur Foundation. The first report, *Growth within: A circular economy vision for a competitive Europe* (EMF 2015a), presents the ReSOLVE framework based on six business initiatives for companies and countries that aim to move to the circular economy: *Regenerate, Share, Optimize, Loop, Virtualize, and Exchange* (ReSOLVE). This framework is used to prioritize the 20 major economic sectors in the European Union for different actions. The second report, *Delivering the Circular Economy: A toolkit for policymakers* (EMF 2015b), provides a step-by-step methodology at the higher level of the policy maker to accelerate the transition to circular economy. Based on these reports, Kalmykova et al. (2018) created a framework in which they separate tools for implementing the circular economy according to the parts of the value chain (see Table 3). Data were analyzed considering these tools, and we verified their applicability to unavoidable waste generated by Company A and created a framework of application possibilities. Lieder and Rashid (2016) also propose a series of tools for implementing the circular economy on a large scale. As well as Kalmykova et al.(2018), they suggest an approach that works through public institutions from top to bottom, and through the industry from bottom to top. Lieder and Rashid (2016) suggest tools at the business-to-business level, such as the product passport concept, the network of initiatives for industrial symbiosis, or sustainable supply patterns. For manufacturing companies, the development of new innovative business models that fit on the circular economy context is vital. This involves reconsidering partnerships to enable new collaborative business models. In practice, this may mean incorporating activities, such as the remanufacture, that are generally deemed economically beneficial, but are currently regarded as parallel business, often outsourced and operated by outsourced companies. A step towards new partnerships within the context of circular economy would, consequently, promote a new perspective on resources in terms of value management and new business models (Lieder and Rashid 2016).

4 Identification of Tools for Waste Management According to the Circular Economy

To provide a framework of tools with applicability to sustainable management of unavoidable wastes of Company A, we searched the study of Kalmykova et al. (2018), from which we extracted a chart with definitions of the main tools of the Circular Economy according to its position in the value chain (Table 1).

4.1 Materials Sourcing

Within the value chain of the wastes from Company A, the *Materials Sourcing* position features tools intended for the supply of materials, whether for manufacturing (raw materials), or for use, which enable reducing or performing the correct disposal of waste. Within such position, we have many tools that are not suitable to the waste of Company A, such as:

- *Energy production/Energy autonomy;*
- *Life Cycle Assessment (LCA);*
- *Material substitution;*
- *Taxation;*
- *Tax credits and subsidies;*

Table 1 Circular economy tools according to parts of the value chain

1—Materials sourcing	
Diversity and cross-sector linkages	Establishment of industry standards to promote cross-sector collaboration through transparency, financial and risk-management tools, regulation and infrastructure development, and education
Energy production/energy autonomy	Energy production from by-products and/or residual/process/waste heat recovery to support facility operation
Green procurement	A process whereby public authorities/companies choose to procure goods and services with the same primary function, but lower environmental impact as measured, for example, by LCA-based comparison of goods and services
Life Cycle Assessment (LCA)	LCA is a structured, comprehensive, and internationally standardized method. It quantifies all relevant emissions, consumed resources, related environmental and health impacts, and resource depletion issues that are associated with any goods or services (EC 2015)
Material substitution	Replacing materials with those more abundant/renewable, hence making the production process more resilient to price fluctuations and resource scarcity
Taxation	Taxes on technologies, products, and inputs that are associated with negative externalities
Tax credits and subsidies	Reducing the tax on resources, for example on bio-based materials and products
2—Design	
Customization/made-to-order operation	Products are tailor-made to meet the needs and preferences of the customer. Such operation can reduce waste and prevent over-production. Customers who are satisfied with the products will return to the manufacturer to extend the service life of the products and keep their preferred features. Customer loyalty to the manufacturer is built-in
Design for disassembly/recycling	Design that considers the need to disassemble products for repair, refurbishment, or recycling

(continued)

Table 1 (continued)

Design for modularity	Products composed of functional modules in such a way they can be upgraded with newer features and/or functionalities. Modules can be individually repaired or replaced, thereby increasing longevity of the product core
Eco design	Product design with a focus on its environmental impacts during the whole life cycle
Reduction	Design and manufacturing involving reduction in the use of materials and elimination of harmful substances use
3—Manufacturing	
Energy efficiency	Providing the required services with reduced energy input, which can be achieved by reduced consumption and energy efficient processes
Material productivity	At the company level: the amount of economic value generated by a unit of material input or material consumption. At the economy-wide level: GDP per material input/consumption
Reproducible and adaptable manufacturing	A transparent and scalable production technology that can be emulated at other places using indigenously available resources and skills
4—Distribution and sales	
Optimized packaging design	Efficient packaging design strategies abiding regulations and using end-of-life of packaging materials
Redistribute and resell	Resale extends the product life by secondhand use. Therefore, fewer products, which serve for the same purpose, have to be produced. The complete products or their components can be resold
5—Consumption and use	
Community involvement	The voluntary involvement of the community and different stakeholders in organizing sharing platforms and providing guidance on product repair and replacement
Eco-labelling	A voluntary environmental protection certification of proven environmental preference of a product/service within its respective category. Credible and impartial labelling of product/service is usually overseen by public or private third parties

(continued)

Table 1 (continued)

Product as a service or Product Service System	The ownership of the product rests with the producer who provides design, usage, maintenance, repair, and recycling throughout the lifetime of the product. The customer pays a rent for the time of its usage
Product labelling	Aimed to guarantee that consumers have full information on the constituents, origin of raw materials etc. to enable them making informed decisions. It indicates no environmental or otherwise preference for certain products, in contrast to #11 Eco-labelling
Reuse	Direct secondary reuse extends the product life by secondhand use. Therefore, fewer products, which serve for the same purpose, have to be produced. The complete products or their components can be reused
Sharing	Shared use/access/ownership, for example, of space and products and sharing platforms enabling the shared use. Multipurpose space
Socially responsible consumption	A socially responsible consumer purchases products and services that are perceived to have less negative influence on the environment, and/or which support businesses that also have positive social impact
Stewardship	Taking responsibility in protecting the resource through conservation, recycling, regeneration, and restoration. A common good is considered, for example, a natural resource, in contrast to #16 Extended Producer Responsibility
Virtualize	Dematerialization. For example, electronic books/CDs, online shopping, and use of telecommunication to decrease the use of office space and travelling
6—Collection and disposal	
Extended Producer Responsibility (EPR)	"Extended Producer Responsibility is as an environmental policy approach in which a producer's responsibility for a product is extended to the post-consumer stage of a product's life cycle" (OECD 2015)
Incentivized recycling	A method for rewarding consistent and repeated recycling of recyclable materials, for example, a deposit refund

(continued)

Table 1 (continued)

Logistics/infrastructure building	Facilities to promote cost-effective, time-saving, and environmentally safe post-consumer collection and disposal. Solutions that render optimum collection
Separation	Biological constituents should be separated from technical or man-made/inorganic constituents. Technical nutrients ought to be used for remanufacture, and biological nutrients are used to be naturally restored or degraded
Take-back and trade-in systems	Efficient take-back systems ensure that products are recovered from the consumer after end-of-life and proceed to be remanufactured. Take-back systems may ensure a continuous flow of material for remanufacture
7—Recycling and recovery	
By-products use	Byproducts from other manufacturing processes and their corresponding value chains are used as raw materials for manufacturing new products
Cascading	Materials and components used through different value streams after end-of-life. The embedded extraction, labor, and capital are conserved through the cascade
Downcycling	It is the process of converting used products into different new products of lower quality or reduced functionality
Element/substance recovery	The process of recovering metals, nonmetals, and other reusable substances from a material waste stream
Energy recovery	The conversion of waste materials into useable heat, electricity, or fuel through a variety of waste-to-energy processes, including combustion, gasification, pyrolysis, anaerobic digestion, and landfill gas recovery
Extraction of biochemicals	Conversion of biomass into low-volume, but high-value chemical products, thereby generating heat, power, fuel, or chemicals from biomass
Functional recycling	Process of recovering materials for the original purpose or for other purposes, excluding energy recovery

(continued)

Table 1 (continued)

High-quality recycling	The recovery of materials in pure-form without contamination, to serve as secondary raw materials for subsequent production of the same or similar quality products
Industrial symbiosis	Exchange and/or sharing of resources, services, and by-products between companies
Restoration	Also known as composting. Process where biological nutrients are returned to the soil after breakdown by microorganisms and other species
Upcycling	Converting materials into new materials of higher quality and increased functionality
8—Remanufacture	
Refurbishment/remanufacture	Rebuilding a product by replacing defective components by reusable ones
Upgrading, maintenance, and repair	The most efficient way to retain or restore equipment to the desired level of performance is maintenance. Moreover, after-sales service is considered key for competitive advantage and business opportunity. Maintenance is also carried out as repair. To eradicate product obsolescence or extend the useful life of the product, services, such as upgrading, are required
9—Circular inputs	
Bio-based materials	Resource inputs or materials that last for longer than a single life cycle and which can be easily regenerated

Source Adapted from Kalmykova et al. (2018)

- *Extraction of biochemicals;*
- *Functional recycling;*
- *High quality recycling;*
- *Industrial symbiosis;*
- *Restoration;*
- *Upcycling.*

These tools cannot be used in the case of Company A, because the internal processes of the company should be modified, with the exception of *Taxation* and *Tax credits and subsidies* tools, which are within the government sphere. In the *Materials Sourcing* position, we highlight the following tools as the most suitable for the Company A's chain:

- *Diversity and cross-sector linkages*—in which industry standards are established to promote inter-sector collaboration through transparent, financial, and risk-management tools, regulation and development of infrastructure and education;

- *Green procurement*—in which authorities and public companies opt for purchasing goods and services with the same primary function, but which have less environmental impact. Overall, this tool is mainly applied by companies and government agencies; however, it can also be applied to the waste of Company A, considering its size and relevance as a major industrial company in the region where it is established;
- *Industrial symbiosis*—in which there are exchange and/or sharing of resources, services and/or by-products among companies. In this sense, cooperation between companies can lead to "win-win" situations, in which there is a competitive advantage for both companies;
- *Life Cycle Assessment (LCA)*—which is a structured, comprehensive, and internationally standardized method that quantifies all issues, relevant emissions and consumed resources, related environmental and health impacts, and resource depletion issues associated with any goods or services. LCA considers all environmental impacts of products or services from extraction, production, distribution, use, maintenance, remanufacturing, recycling, and disposal of raw materials.

Specifically, in the case of the wastes of Company A, these tools could be applied to the chain in which it is inserted from the creation of a structure that provides training for waste customers such as the supplier development programs. In these programs, major companies or sectors create training programs and certifications for smaller companies who want to be their suppliers.

4.2 Design and Manufacturing

In the *Design and Manufacturing* positions, the proposed tools do not contemplate issues related to the wastes of Company A considering the nature of the business and the characteristics of wastes and production.

4.3 Distribution and Sales

In this position, the *Optimized packaging design* strategy is not suitable for the reality of Company A, since both the company's products and the characteristics of the waste end up limiting the specificities of packaging, because it is an industry that produces and exports commodities.

On the other hand, the *Redistribute and resell* tool can be applied, even if with some exceptions, because resale increases the shelf life of secondhand products. However, the waste surveyed in our study have been presenting a major difficulty in being resold, because the market is unprepared to absorb, in a sustainable way, the generated material and the stream in which it is produced.

4.4 Consumption and Use

Regarding the *Consumption and use* position, the *Community involvement, Eco-labelling, Product as a service or Product Service System, Product labelling, Reuse, Socially responsible consumption, Stewardship,* and *Virtualize* tools do not apply production to the waste generated by Company A because it is not a material sold to the final consumer.

4.5 Remaining Positions

Regarding the *Collection and disposal, Recycling and recovery, Remanufacture,* and *Circular inputs* positions, we did not find tools that were suitable to the needs for developing a sustainable management of Company A's waste, mainly because such encounters difficulties in the characteristics of the generated materials and production processes of the company.

We consolidated our results in Table 2, relating the surveyed Circular Economy tools with the possibilities of application and the potential benefits of their implementation, considering the unavoidable wastes of major industrial companies. Overall, we can perceive the importance of creating networks of training and sharing between waste-generator companies and consumers.

5 Circular Economy Tools for the Waste Management of Company A

Regarding waste generation, in 2017, the Company A generated, according to its waste inventory, 30,222,562.15 kg of residues divided into 15 groups, namely: batteries; rubber and tires; rubble of work; mud, sludge, and dust from treatment systems; wood; metals; nonmetals; oily; paper and cardboard; plastics, polymers, fabrics, and canvas; mercury waste; household waste; special waste; and mixed waste; glass. In Table 3 we can observe how the wastes are distributed regarding the generated amounts.

Among the groups of wastes aforementioned, we have some that are not complying with the sustainable destination in a reasonable time, and remain stored in CMD for quite some time. This happens due to lack of appropriate solutions that meet not only the legislation, but also the values linked to the company's sustainability.

The residues selected for our research were: batteries; expired residues of chemicals (laboratory); waste from several lamps; IBC (Intermediate Bulk Container) scrap, uncontaminated (tote bins, cylinders, spray cans); rubber scrap (wrap and strap); residues from filters of off-road trucks; scrap of abrasives (grinding wheels, sandpaper, and abrasive discs); scrap of optical fiber cables; scrap of glass boards and

Table 2 Possibilities of application of circular economy tools

Tools	Application	Benefits
Diversity and cross-sector linkages	Establishing structures for training and development of customers (for instance, the supplier development programs) from the creation of a network of partnerships that develops standards and certifications for waste-purchasing companies	Reducing the risks regarding waste disposal. Creating a waste-consumer market. Stimulating innovation in the waste market
Green procurement	Creating a policy that prioritizes the hiring of services and the purchase of raw materials and consumables of environmentally responsible companies, which prove to cause less adverse impact on the environment	Reducing the impacts of the chain as a whole. Quality control of unavoidable waste
Life Cycle Assessment (LCA)	Creating programs for the development of suppliers and customers of waste that promote trainings and certifications according to the LCA	LCA allows identifying problematic areas within the supply chain, alternatives, and the most appropriate option to be chosen
Industrial symbiosis	Establishing a network of waste sharing between companies generating, processing, and consuming wastes	Reducing the use of resources, dependence on non-renewable sources, pollutant emissions, and waste disposals. Reducing the costs of inputs, production, and management of waste and generating additional income due to the added value to by-products and waste streams
Redistribute and resell	Creating programs for the development of suppliers and customers of waste that promote trainings and certifications guaranteeing the purchase of the produced material	Resuming the value of the materials. Reuse can increase profits by reducing the costs of materials and energy

Source Developed by the authors

Table 3 Solid waste generation, 2017—Company A—CMD

Solid waste generation in the year 2017		
Groups residues		2017 (kg)
Group 1	Batteries	103,557.70
Group 2	Rubber and tires	2,056,544.10
Group 3	Rubble of work	632,069.80
Group 4	Mud, sludge, and dust from treatment systems	54,820.00
Group 5	Wood	1,456,924.50
Group 6	Metals	14,688,612.50
Group 7	Nonmetals	1,607,460.00
Group 8	Oily	1,848,110.00
Group 9	Paper and cardboard	183,171.80
Group 10	Plastics, polymers, fabrics, and canvas	1,044,637.25
Group 11	Mercury waste	4,861.70
Group 12	Household waste	1,587,890.00
Group 13	Special waste	31,990.00
Group 14	Mixed waste	4,911,391.70
Group 15	Glass	10,521.10
	Total	30,222,562.15

Source Solid waste inventory of Company A, 2017

pieces; scrap of toner cartridge; scrap of polyurethane (conveyor belts and flippers); residues from filters of locomotives; and wastes from aluminum thermal welding of rails. Some of the listed wastes are not part of the inventory of wastes from 2017, and therefore will not be considered in our study, such as: resides from truck filters; wastes from filters of locomotives; scrap of optical fiber; and wastes from aluminum welding of rails (Table 4).

As we can observe in Table 4, among the selected unavoidable wastes, some of them have a significantly smaller disposal than generation, and other were not even disposed. This reflects a difficulty in absorption on the part of the local waste market, making Company A to store this material indefinitely.

More specifically, when applying the tools that we highlighted towards wastes individually, we have the following scenario (Table 5).

6   Conclusions

The application of circular economy tools to the waste of a major company enables us to observe that major industrial companies are not able to perform a transition from a linear production system to a circular economy without the aid of other companies. Our results showed the need to broaden the scope of participation in the production

Table 4 Waste generation and disposal in 2017

Waste	Generated 2017 (kg)	Disposed 2017 (kg)
Batteries waste (several and locomotives)	4550.20	0.00
Expired chemicals residues	5100.00	0.00
Abrasive scraps	16,990.00	0.00
Residues from several lamps	4861.70	2780.00
IBC scrap	19,320.00	18,900.00
Rubber scrap (wrap and strap)	495,182.00	480,640.00
Scrap from glass boards and pieces	10,521.10	6650.00
Toner cartridge scrap	102.30	0.00
Polyurethane scrap	18,480.00	12,250.00
Truck filter residues	Unrelated	
Residues from filter of locomotives	Unrelated	
Scrap of optical fiber	Unrelated	
Wastes from aluminum thermal welding of rails	Unrelated	

Source Inventory of wastes of Company A, 2017

Table 5 Application of circular economy tools to the wastes of Company A

Waste	Diversity and cross-sector linkages	Green procurement	Life Cycle Assessment (LCA)	Industrial symbiosis	Redistribute and resell
Batteries waste (several and locomotives)	Partnering with organizations that collect the residue along with the community	Purchasing only from companies that take responsibility for the collection and disposal of waste	Evaluating the entire life cycle of the residue and defining the ideal disposal to be chosen	Partnering to collect the residue in order to generate the volume required for processing	Not applicable
Expired chemicals residues	Not applicable	Purchasing only from companies that take responsibility for the collection and disposal of waste	Evaluating the entire life cycle of the residue and defining the ideal disposal to be chosen	Not applicable	Not applicable

(continued)

Table 5 (continued)

Waste	Diversity and cross-sector linkages	Green procurement	Life Cycle Assessment (LCA)	Industrial symbiosis	Redistribute and resell
Abrasive scraps	Developing a structure to qualify waste-purchasing companies	Purchasing only from companies that take responsibility for the collection and disposal of waste	Evaluating the entire life cycle of the residue and defining the ideal disposal to be chosen	Creating waste exchanges composed of qualified companies for processing of wastes	Conducting auctions in waste exchanges composed of qualified companies
Residues from several lamps	Partnering with organizations that collect the residue along with the community	Purchasing only from companies that take responsibility for the collection and disposal of waste	Evaluating the entire life cycle of the residue and defining the ideal disposal to be chosen	Partnering with other organizations to collect the residue in order to generate the volume required for processing	Not applicable
IBC scrap	Developing a structure to qualify waste-purchasing companies	Purchasing only from companies that take responsibility for the collection and disposal of waste	Evaluating the entire life cycle of the residue and defining the ideal disposal to be chosen	Creating waste exchanges composed of qualified companies for processing of wastes	Conducting auctions in waste exchanges composed of qualified companies
Rubber scrap (wrap and strap)	Developing a structure to qualify waste-purchasing companies	Purchasing only from companies that take responsibility for the collection and disposal of waste	Evaluating the entire life cycle of the residue and defining the ideal disposal to be chosen	Creating waste exchanges composed of qualified companies for processing of wastes	Conducting auctions in waste exchanges composed of qualified companies
Scrap from glass boards and pieces	Developing a structure to qualify waste-purchasing companies	Not applicable	Evaluating the entire life cycle of the residue and defining the ideal disposal to be chosen	Creating waste exchanges composed of qualified companies for processing of wastes	Conducting auctions in waste exchanges composed of qualified companies

(continued)

Table 5 (continued)

Waste	Diversity and cross-sector linkages	Green procurement	Life Cycle Assessment (LCA)	Industrial symbiosis	Redistribute and resell
Toner cartridge scrap	Partnering with organizations that collect the residue along with the community	Purchasing only from companies that take responsibility for the collection and disposal of waste	Evaluating the entire life cycle of the residue and defining the ideal disposal to be chosen	Partnering to collect the residue in order to generate the volume required for processing	Not applicable
Polyurethane scrap	Developing a structure to qualify waste-purchasing companies	Not applicable	Evaluating the entire life cycle of the residue and defining the ideal disposal to be chosen	Creating waste exchanges composed of qualified companies for processing of wastes	Conducting auctions in waste exchanges composed of qualified companies
Truck filter residues	Developing a structure to qualify waste-purchasing companies	Purchasing only from companies that take responsibility for the collection and disposal of waste	Evaluating the entire life cycle of the residue and defining the ideal disposal to be chosen	Creating waste exchanges composed of qualified companies for processing of wastes	Conducting auctions in waste exchanges composed of qualified companies
Residues from filter of locomotives	Developing a structure to qualify waste-purchasing companies	Not applicable	Evaluating the entire life cycle of the residue and defining the ideal disposal to be chosen	Creating waste exchanges composed of qualified companies for processing of wastes	Conducting auctions in waste exchanges composed of qualified companies

(continued)

Table 5 (continued)

Waste	Diversity and cross-sector linkages	Green procurement	Life Cycle Assessment (LCA)	Industrial symbiosis	Redistribute and resell
Scrap of optical fiber	Developing a structure to qualify waste-purchasing companies	Not applicable	Evaluating the entire life cycle of the residue and defining the ideal disposal to be chosen	Creating waste exchanges composed of qualified companies for processing of wastes	Conducting auctions in waste exchanges composed of qualified companies
Wastes from aluminum thermal welding of rails	Developing a structure to qualify waste-purchasing companies	Not applicable	Evaluating the entire life cycle of the residue and defining the ideal disposal to be chosen	Creating waste exchanges composed of qualified companies for processing of wastes	Conducting auctions in waste exchanges composed of qualified companies

Source Developed by the authors

chain of industrial wastes. This can be accomplished with the creation of a network for training and development of smaller companies in such a way they become able to absorb the waste and make its disposal circular, in which the value is maintained or recovered.

Among barriers to the implementation of tools that require cooperation between companies for the implementation of a circular economy, overall, the lack of awareness and sense of urgency regarding the development of sustainable chains is included (Masi et al. 2018). Similarly, the implementation of circular economy tools seems to be driven by economic behaviors, and not environmentally aware, with a marked preference for practices that generate an economic return in the short term. Masi et al. (2018) also highlight the companies' preference for practices at the company level instead of those at the supply-chain level, in line with the theory of supply chain management.

Acknowledgements The authors would like to thank VALE S.A. company for financial support.

References

Bernon M, Tjahjono B, Ripanti EF (2018) Aligning retail reverse logistics practice with circular economy values: an exploratory framework. Prod Plan Control. https://doi.org/10.1080/09537287.2018.1449266

Corbett CJ, Kleindorfer PR (2001) Environmental management and operations management: introduction to part 1 (manufacturing and ecologistics). Prod Oper Manage 10(2):107–111

EC (2015). Single market for green products initiative. https://ec.europa.eu/environment/eussd/smgp/

EMF (2015a) Delivering the circular economy a toolkit for policymakers. https://www.ellenmacarthurfoundation.org/

EMF (2015b) Growth within: a circular economy vision for a competitive Europe. Available from: https://www.ellenmacarthurfoundation.org/. ESA 2013. Going for growth: a practical route to a Circular Economy from https://www.esauk.org/esa_reports/Circular_Economy_Report_FINAL_High_Res_For_Release.pdf

IPEA – Instituto de Pesquisa Econômica Aplicada (2012) Diagnóstico dos Resíduos Sólidos Urbanos Diagnóstico dos Resíduos. IPEA, Brasília, 77 p

Kalmykova Y, Sadagopan M, Rosado L (2018) Circular economy—from review of theories and practices to development of implementation tools. Resour Conserv Recycl 135:190–201

Korhonen J, Nuur C, Feldmann A, Eshetu S (2018) Circular economy as an essentially contested concept 175:544–552. https://doi.org/10.1016/J.RESCONREC.2017.10.034

Lieder M, Rashid A (2016) Towards circular economy implementation: a comprehensive review in context of manufacturing industry. J Clean Prod 115:36–51. https://doi.org/10.1016/J.JCLEPRO.2015.12.042

Masi D, Kumar V, Garza-Reyes JA, Godsell J (2018) Towards a more circular economy: exploring the awareness, practices, and barriers from a focal firm perspective. Prod Plan Control 29(6):539–550. https://doi.org/10.1080/09537287.2018.1449246

OECD (2015) Extended producer responsibility. https://www.oecd.org/env/waste/extendedproducerresponsibility.htm

Vlajic JV, Mijailovic R, Bogdanova M (2018) Creating loops with value recovery: empirical study of fresh food supply chains. Prod Plan Control 29(6):522–538. https://doi.org/10.1080/09537287.2018.1449264

Mariana de Souza Silva Rodrigues is an economist, graduated from the Federal University of Espírito Santo, Master of Science Program on Engineering and Sustainable Development from the Federal University of Espírito Santo. She currently conducts research in the field of reverse logistics of industrial waste.

Alvim Borges is Ph.D. in *Sciences de Gestion* from the Aix-Marseille Université, Centre de Recherche sur le Transport et la Logistique (CRET-LOG), and Ph.D. in Energy from the Institute of Energy and Environment (IEE), University of São Paulo. Mechanical Engineer, specialist in Finance from PUC/Rio, Master in Production Engineering from the Federal University of Santa Catarina. Researcher of the Professional Master's Program in Engineering and Sustainable Development and professor at the Department of Administration of the Federal University of Espírito Santo. He currently works with research in finance and logistics, focusing on the energy industry and sustainable development.

Decision Model for Selecting Advanced Technologies for Municipal Solid Waste Management

Douglas Alcindo da Roza, Guilherme Teixeira Aguiar, Edson Pinheiro de Lima, Sergio Eduardo Gouvea da Costa and Gilson Oliveira Adamczuk

Abstract The National Solid Waste Policy (PNRS) was implemented in 2010; it aims to prevent and reduce waste generation, eliminate the so-called 'landfills' and implement municipal solid waste management plans. Waste selective collection and recycling can reduce the amount of waste destined to landfills, reduce transportation frequencies and minimize overall disposal costs. However, the conventional approach currently adopted in Brazil is inefficient considering economic, social and environmental aspects. There is an urgent need to improve municipal solid waste management (MSWM) by proposing alternatives, which cover product-service systems (PSS) and IoT based smart trash dustbins. This study provides an assessment through multicriteria analysis whether the implementation of scenarios using IoT smart trash cans is to be considered a sustainable operational strategy for the municipal public administration. The criteria used in the study was extracted from a systemic literature review; AHP and TOPSIS Fuzzy Methods were applied to achieve the best solution according to the selected criteria, which is the research main objective.

Keywords Selective waste collection · Procknow-C · AHP · Fuzzy-TOPSIS

D. A. da Roza (✉) · G. T. Aguiar · E. Pinheiro de Lima · S. E. Gouvea da Costa · G. O. Adamczuk
Industrial and Systems Engineering, Federal University of Technology—Parana, Via do Conhecimento, Km 1, Pato Branco, Paraná, Brazil
e-mail: douglas.alcindo@hotmail.com

G. T. Aguiar
e-mail: guilhermea@alunos.utfpr.edu.br

E. Pinheiro de Lima
e-mail: pinheiro@utfpr.edu.br

S. E. Gouvea da Costa
e-mail: gouvea@utfpr.edu.br

G. O. Adamczuk
e-mail: gilson@utfpr.eu.br

E. Pinheiro de Lima · S. E. Gouvea da Costa
Pontifical Catholic University of Parana—PUCPR, R. Imaculada Conceicao, 1155, Curitiba, Paraná, Brazil

© Springer Nature Switzerland AG 2020   201
W. Leal Filho et al. (eds.), *International Business, Trade and Institutional Sustainability*, World Sustainability Series,
https://doi.org/10.1007/978-3-030-26759-9_12

1   Introduction

The Law 12305, which deals with the National Solid Waste Policy (PNRS), was implemented in 2010. It aims to eradicate the so-called 'dumps', prepare and implement municipal solid waste plans. Thinking in terms of the circular economy, Da Silva (2018) argues that investments in public policies of environmental education, sectoral and innovation policies are necessary to reorganize chains, turning a problem into an opportunity for municipalities.

An alternative to selective waste collection is to transform the traditional service (static) approach into a Product-Service-System (PSS). Thus, a service component (waste collection) is improved by a product component, i.e., technological alternatives to monitor waste, allowing the management of variable and dynamic waste streams (Elia et al. 2018). These technical options are part of a new way of thinking productive systems through smart cities. Díaz-Díaz et al. (2017) analyzed and compared business models in Intelligent Cities. Their results indicated that municipal services using smart technologies generally present a value proposition focused on service efficiency, which in consequence reduces environmental impact and lower costs.

Sustainability in operations is a necessity associated with waste management activities due to the complexity and amount of produced waste. Alternative systems and technologies for waste management have been researched as a way to solve or improve conventional systems, using dumpsters with sensors and Internet of Things (IoT), for example (Misra et al. 2018; Wen et al. 2017; Yerraboina et al. 2018).

However, these alternatives are not only technological. Others are related to paying schemes for the produced waste, by either weight or volume (Dahlen et al. 2010). These options can be very efficient when associated with waste bins technologies with RFID sensors tags. These bins can assist in charging through the measurement of weight and volume as well as in the inspection of the collection, transport and final destination (Wen et al. 2017).

Considering this context, the main objective behind the study is to structure a model to evaluate the implementation of sustainable operations and innovative technologies in the MSWM. A systematic review of the literature was carried out. The opinion of experts was used to select criteria and more adequate alternatives adjusted to the setting of a municipality in the Western region of the Brazilian state of Santa Catarina.

2   Theoretical Background

The management of solid waste by municipalities is crucial for public health, environmental protection and avoid visual pollution. It is necessary to properly manage all activities involving solid waste, from collection to final disposal (Al-Khatib et al. 2007). Hlatka et al. (2018) point out that residue separation is significantly influenced by the conditions of households for solid waste sustainable management. Topaloglu et al. (2018) emphasized that waste management must be environmentally sound, economically viable and socially acceptable.

According to Coban et al. (2018), solid waste management depends on the composition of waste produced by the population; it is strongly influenced by socioeconomic factors, seasons and family size. Considering an urban development with inefficient infrastructure and management, waste issues become increasingly complex. Authorities require effective tools to select appropriate technologies that meet the needs of the local infrastructure. Coban et al. (2018) state that multicriteria decision tools (MCDM) stand out as a group of techniques to evaluate MSWM scenarios. The authors also indicates that MCDM methods have gained popularity over the last decade in the area of MSWM, since complex and integrated processes involving distinct dimensions such as environmental, social and economic are very solvable with the use of MCDM. Among the MCDM tools proposed by several authors, TOPSIS has been the most prevalent, because of its ease of use and consistency of results. Additionally, Coban et al. (2018) show that using a single MCDM method to rank alternatives may lead to proposed solutions susceptible to uncertainties.

The uncertainties arise from qualitative parameters, better known as linguistic variables, collected in the study, which are essential for the decision-making. Thus, the fuzzy method proposes the solution by converting linguistic terms into diffuse numbers (Topaloglu et al. 2018). In the present study, two MCDM were used: AHP and fuzzy TOPSIS. The integration of these methods is explained and discussed in the research methodology section.

3 Research Design

This paper presents an evaluation model for the selection of practices and technologies for solid waste management (Fig. 1). The model initially consists of a systemic literature review for the identification and hierarchization of the evaluation criteria. The opinion of experts allows the selection of the most relevant strategy for solid

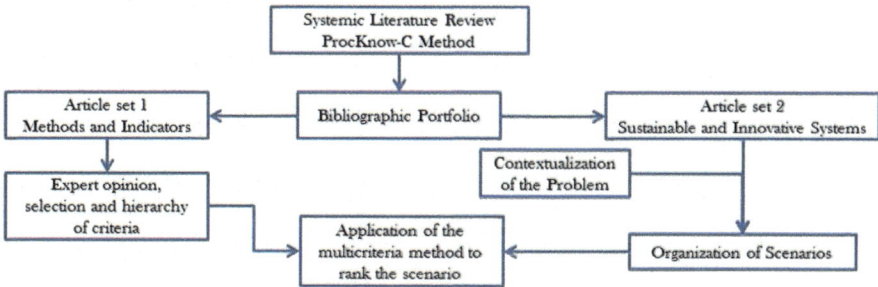

Fig. 1 Selection model of practices and technologies for waste management. *Source* The authors

waste management. Practices of sustainable and innovative operations, which associated with a specific context, may support technological and organizational alternatives for the construction of scenarios were identified through the literature review. Multicriteria decision support methods guide the selection of the best practices and technologies for waste management based on the selected criteria and organization of scenarios.

Using a systemic review of literature based on the method proposed by Ensslin et al. (2010), the Knowledge Development Process-Constructivist (Procknow-C), a 'Paper Set' was organized to analyze the content of the papers. It aimed to identify evaluation criteria and characteristics of innovative and sustainable operations for waste management. A questionnaire based on these results was developed (Table 1) and filled out by experts, selecting criteria relevant to the implementation of waste management sustainable and innovative operating systems.

The Analytic Hierarchy Process (AHP) analyzes the criteria judgments, performed by the experts, through the correlation between the criteria, using the classification shown in Table 2.

The TOPSIS methodology is based on the principle that there are 'n' criteria and 'm' alternatives. The selected alternative has a minimum distance from the ideal positive solution and a maximum distance from the ideal negative solution (Gupta and

Table 1 Questionnaire for selection of criteria

Public acceptance	Quality of collected waste
Political support	Amount of collected waste
Infrastructure capacity	Revenue
Capacity of innovation	Power recovery
Products lifecycle	Recovery of raw materials
Public awareness	Reduction of recyclables in landfills
Creation of new jobs	Income from recyclables sold
Cost with equipment	Environmental risks/impacts
Cost with qualified staff	Noise
Maintenance costs	Safety and hygiene
Transportation costs	System sustainability
Investment costs	Size of the population to be served
Operational costs	Discard rate
Unemployment	Recycling rate
Availability of collection points	Collection time
System efficiency	Type of waste to be collected
Atmospheric emissions	Vehicle traffic
Aesthetics	Waste treatment
Odor	Operational feasibility

Source The authors

Table 2 Numerical classification

Value	Definition	Explanation
1	Equal importance	Identical contribution
3	Low importance	Slightly higher judgment
5	Strong importance	Judgment strongly in favor
7	Very strong importance	Recognized domain
9	Absolute importance	Proven domain
2, 4, 6, 8	Intermediate values	Doubt

Source Adapted from Saaty (2008)

Barua 2018). The ideal positive solution is the solution that maximizes benefit criteria and minimizes cost criteria. The ideal negative solution is the solution that maximizes cost criteria and minimizes benefit criteria (Mesquita 2014). Chen (2000) extended TOPSIS as triangular Fuzzy Numbers (FN). The researcher introduced a vertex method to calculate the distance between two triangular FN. If $\tilde{x} = (a1, b1, c1)$, $\tilde{y} = (a2, b2, c2)$ are two triangular FN (1).

$$d(\tilde{x}, \tilde{y}) := \sqrt{\frac{1}{3}\left[(a1 - a2)^2 + (b1 - b2)^2 + (c1 - c2)^2\right]} \qquad (1)$$

In the study, TOPSIS Fuzzy procedure was applied as per the instructions by Nădăban et al. (2016). Step 1 is the assignment of rating to the criteria and alternatives, assuming there is a decision group with K members. The Fuzzy classification of the decision makers kth about alternatives A_i w.r.t. criterion C_j is denoted $\tilde{x}_{ij}^k = (a_{ij}^k, b_{ij}^k, c_{ij}^k)$ and the weight of the criterion C_j is denoted $\tilde{w}_{kj} = (w_{kj1}, w_{kj2}, w_{kj3})$. In Step 2, the aggregate diffuse classifications for the alternatives (Table 3) and diffuse weights aggregated for the criteria are calculated (Table 2).

The aggregated fuzzy classification $\tilde{x}_{ij} = (a_{ij}, b_{ij}, c_{ij})$ of ith alternative w.r.t. jth. The criterion is obtained as per Eq. (2).

Table 3 IVIFS linguistic values for linguistic terms

Linguistic term	IVIFS
Low (L)	(0.0; 0.1; 0.3)
Reasonably low (RL)	(0.1; 0.3; 0.5)
Medium (M)	(0.3; 0.5; 0.7)
Reasonably high (RH)	(0.5; 0.7; 0.9)
High (H)	(0.7; 0.9; 1.0)

Source Adapted from Nădăban et al. (2016)

$$a_{ij} = \frac{\min}{k}\{a_{ij}^k\}, \, b_{ij} = \frac{1}{K}\sum_{k=1}^{k} b_{ij}^k, \, c_{ij} = \frac{\max}{k}\{c_{ij}^k\}. \qquad (2)$$

The aggregate weight fuzzy $\tilde{w}_j = (w_{j1}, w_{j2}, w_{j3})$ for the criterion C_j is calculated by the formulas:

$$w_{j1} = \frac{\min}{k}\{w_{j1}^k\}, \, w_{j2} = \frac{1}{k}\sum_{k=1}^{k} w_{kj2}, \, w_{j3} = \frac{\max}{k}\{w_{j3}^k\}. \qquad (3)$$

The normalized fuzzy decision matrix is calculated in Step 3. The normalized fuzzy decision matrix is $\tilde{R} = [\tilde{r}_{ij}]$, (4) and (5).

$$\tilde{r}_{ij} = \left(\frac{a_{ij}}{c_j^*}, \frac{bc_i}{c_j^*}, \frac{cc_i}{c_j^*}\right) \text{e} \, c_j^* = \frac{\max}{i}\{c_{ij}\} (\text{benefit criterion}) \qquad (4)$$

$$\tilde{r}_{ij} = \left(\frac{a_j^-}{c_{ij}}, \frac{a_j^-}{b_{ij}}, \frac{a_j^-}{a_{ij}}\right) \text{e} \, c_j^- = \frac{\min}{i}\{a_{ij}\} (\text{cost criterion}) \qquad (5)$$

The weighted normalized fuzzy decision matrix is calculated in Step 4. The weighted normalized fuzzy decision matrix is $\tilde{V} = (\tilde{v}_{ij})$, where $\tilde{v}_{ij} = \tilde{r}_{ij} \times w_j$. In Step 5, the Ideal Positive Diffuse Solution (FPIS) (6) and the Ideal Fuzzy Negative Solution (FNIS) (7) are determined. FPIS and FNIS are calculated as per Eqs. (6) and (7):

$$A^* = (\tilde{v}_1^*, \tilde{v}_2^*, \dots, \tilde{v}_n^*), \text{ where } \tilde{v}_j^* = \frac{\max}{i}\{v_{ij3}\}; \qquad (6)$$

$$A^- = (\tilde{v}_1^-, \tilde{v}_2^-, \dots, \tilde{v}_n^-), \text{ where } \tilde{v}_j^- = \frac{\min}{i}\{v_{ij1}\}; \qquad (7)$$

The distance from each alternative to FPIS and FNIS is determined (Step 6). Compute the distance from each alternative Ai to FPIS and FNIS, respectively (Eq. 8).

$$d_i^* = \sum_{j=1}^{n} d(\tilde{v}_{ij}, \tilde{v}_j^*), \, d_j^- = \sum_{j=1}^{n} d(\tilde{v}_{ij}, \tilde{v}_j^-) \qquad (8)$$

In Step 7, the closeness coefficient CC_i for each alternative is determined. For each alternative A_i, the closeness coefficient CC_i is calculated as per Eq. (9).

$$CC_i = \frac{d_i^-}{d_i^- + d_I^*} \qquad (9)$$

Finally, in Step 8, the alternatives are classified. The alternative with the highest closeness coefficient represents the best alternative. The TOPSIS Fuzzy method was applied using a spreadsheet program.

4 Results

The results are described in four subsections. The first subsection was a systematic review of the literature with content analysis on the methods, evaluation criteria and characteristics related to the topic 'waste management'. The second involved interviews with experts for the criteria selection and hierarchization. The third subsection presents the contextualization of a real problem and organization of alternatives for a possible solution. The fourth subsection describes the application of the multicriteria method to select the best alternative.

5 Systemic Literature Review

The application of the Procknow-C methodology starts with the definition of keywords. A list of 23 research terms was divided into three research axes (Table 4).

The collection of papers was performed in the Web of Science™ and Scopus® databases through combinations of the search terms and axes. The search was limited to the last 10 years (only papers). Resulting references were inserted in the Mendeley® software; duplicated papers were excluded. A total of 21,040 papers was obtained. The process continued with the analysis of the titles, which resulted in 503 papers aligned with the research theme.

In the scientific recognition analysis, 228 papers with more than five citations passed through the analysis of abstracts and 88 were regarded as aligned with the

Table 4 Axes and terms used in the research

Axes	Diagnosis	Sustainable operations	Waste management
Search terms	Diagn*	Sustainab*	"waste management"
	Audit*	"Triple bottom line"	"Solid waste*"
	Evaluat*	"Value creation"	"Municipal waste*"
	Analy*	"Business model"	"Zero waste*"
	Perform*	"Smart Cit*"	
	Assess*	IOT	
	Manag*	"Internet of things"	
	Means*	Recy*	
		Upcyc*	
		Reduc*	
		Reus*	

Source The authors

research theme. The authors of these papers composed a data set of 254 authors. A list of 275 papers with less than five citations went to the reanalysis process; other 22 were selected. The sum of the selected references resulted in 110 items, after being thoroughly read, the set of papers was organized and composed. Then, the paper set was divided into two other sets. The first aim was to identify waste management methodologies and evaluation criteria in different contexts. The second aim was to identify sustainable and innovative waste management systems. The references and evaluation methodologies of the first set can be verified in Table 5.

Table 5 shows that the TOPSIS evaluation method was the most used as seen in Arıkan et al. (2017), Coban et al. (2018), Hlatka et al. (2018), Jovanovic et al. (2016), Mir et al. (2016), Pires et al. (2011) and Topaloglu et al. (2018). The criteria were identified (Frame 1). Nevertheless, due to different contexts for different indicators, it was decided to group similar criteria. For instance, Jovanovic et al. (2016) uses particulate matter, emission of gases (CH_4, CO_2 and N_2O); Stefanović et al. (2016) used and classified emissions of greenhouse gases (CO_2) and emissions of acid gases (NO_x and SO_2) as environmental indicators. In the current study, all of those are considered in the atmospheric emissions criterion. The second set of papers encompassed 13 references (Table 6). Content analysis aimed to identify solid waste management sustainable and innovative systems.

As shown in Table 6, the IoT information technologies are the most prevalent features due to the number of papers that address them within the set. They have been studied by Díaz-Díaz et al. (2017), Elia et al. (2015, 2018), Misra et al. (2018), Wen et al. (2017) and Yerraboina et al. (2018).

6   Selection and Hierarchization of Criteria

Three experts were selected. All of them graduated in Environmental Engineering; one has a master's degree in Building Engineering and is a lecturer in the subject of solid waste management. The other two have master's degrees in Environmental Engineering, with experience in municipal waste management. The experts were requested to indicate relevant criteria for the implementation and operations of sustainable and innovative waste management systems, using a questionnaire with closed questions. Each of the experts received a questionnaire to evaluate the criteria, individually and without any consultation with the other interviewees. The 11 selected criteria (Table 7) are observed in the literature and considered relevant by the experts. For the application of multicriteria methods, the selected criteria were divided into three categories (environmental, economic and social).

The experts, according to the AHP methodology and Saaty's classification (Table 2), performed peer comparison. Table 8 shows the results of the weights for each criterion after the judgement by the experts through the AHP method.

Table 5 Methods identified in the first set of papers

Authors	Methods													
	AHP	ASPID	TOPSIS	ELECTRE	VIKOR	ANP	MCDA-C	DEA	PROMETHEE	GAIA	WARM	FUZZY	SAW	Authorial
Arıkan et al. (2017)			X						X			X		
Banar et al. (2010)				X		X								
Coban et al. (2018)			X						X					
Deus et al. (2016)											X			
Herva and Roca (2013)	X								X	X				
Hlatka et al. (2018)			X											
Huang et al. (2011)								X						
Jovanovic et al. (2016)			X										X	
Khalili and Duecker (2013)				X										
Lolli et al. (2016)									X			X		
Makan et al. (2013)									X					
Milutinović et al. (2014)	X													
Milutinovic et al. (2016)	X													

(continued)

Table 5 (continued)

Authors	Methods													
	AHP	ASPID	TOPSIS	ELECTRE	VIKOR	ANP	MCDA-C	DEA	PROMETHEE	GAIA	WARM	FUZZY	SAW	Authorial
Mir et al. (2016)			X		X									
Pires et al. (2011)	X		X											
Rigamonti et al. (2016)														X
Rodrigues et al. (2018)							X							
Sarra et al. (2017)								X						
Simões et al. (2012)								X						
Stefanović et al. (2016)	X	X												
Topaloglu et al. (2018)			X									X		
Vucijak and Silajd (2015)	X				X			X						

Source The authors

Table 6 Sustainable and innovative features identified in the second set of papers

References	Features
Da Silva (2018)	The Circular Economy as a new way of thinking about current issues of urban planning and management, creating opportunities
Dahlen et al. (2010)	Scheme of payments as per waste collection rate *Pay as you throw* based on weight and volume
Díaz-Díaz et al. (2017)	Intelligent city business models based on the benchmarking of eight urban services rendered in the city of Santander; waste management was highlighted due to a 20% reduction in the cost of providing the service yearly
Elia et al. (2015)	It proposes a holistic framework for designing and managing PAYT systems applied to MSWM services through intelligent technology solutions
Elia et al. (2018)	Dynamic collection schemes for Electrical and Electronic Equipment Waste through IoT technology
Gelbmann and Hammerl (2015)	Appropriate business model for the establishment of new (sustainable) systems of products and services (SPSS) for reuse in social enterprises with labor integration ecologically oriented practices (ECO-WISEs)
Manni and Runhaar (2014)	Scheme for the payment of a waste collection rate *Pay as you throw* as an incentive to reduce waste
Misra et al. (2018)	It features an intelligent waste collection system based on an IoT trash can, which measures the level of materials and presence of gases; it sends this information to a cloud server for storage and processing over the Internet
Rada et al. (2013)	Optimization of the selective collection with the implementation of a system based on Geographic Information Systems (Web-GIS)
Rebehy et al. (2017)	Social innovation proposed through a sustainable and inclusive business model, with intensive use of information technology and logistics
Tseng and Bui (2017)	Business management through eco-innovation, and industrial symbiosis to achieve win-win status in supply chain networks
Wen et al. (2017)	Studies the implementation and evaluation of a sensor-based IoT network technology to improve waste management of restaurant food in the city of Suzhou, China
Yerraboina et al. (2018)	Prototype of an IoT trash can called "*Smart Garbage Bin*"

Source The authors

Table 7 Indicators observed in the literature and selected by the experts

Category	Criteria
Social	C1—Public acceptance
	C2—Awareness
	C3—Safety and hygiene
Environmental	C4—Amount of waste collected
	C5—Environmental risks and impacts
	C6—Reduction of recyclables in landfills
Economic	C7—Equipment costs
	C8—Investment costs
	C9—Operational costs
	C10—Income of recyclables
	C11—System efficiency

Source The authors

Table 8 Weights of the criteria

	Criteria	Weights			Average
		Spec. 1	Spec. 2	Spec. 3	
Social	C1	0.106	0.283	0.106	0.165
	C2	0.633	0.643	0.633	0.637
	C3	0.260	0.074	0.260	0.198
Environmental	C4	0.455	0.748	0.633	0.612
	C5	0.091	0.071	0.106	0.089
	C6	0.455	0.180	0.260	0.298
Economic	C7	0.053	0.370	0.056	0.160
	C8	0.057	0.370	0.084	0.171
	C9	0.362	0.151	0.216	0.243
	C10	0.143	0.073	0.154	0.124
	C11	0.385	0.035	0.490	0.303

Source The authors

7 Contextualization of the Solid Waste Management in a Municipality in the Western Region of the Brazilian State of Santa Catarina to Build Possible Scenarios

According to the Environmental Department of a municipality located in the Western region of the Santa Catarina state, the city does not have landfills or its own machinery for garbage collection. This service is the responsibility of a private company. The municipality pays a fixed amount for the collection of recyclable waste and a variable

Table 9 Description of scenarios

Scenario	Description of the recycling collection scenarios
A1	Current scenario without any changes, door-to-door collection without a recyclable waste identification system and without financial incentives or fines
A2	Door-to-door collection, with the use of identification systems by colored packs/bags without any use of technology or other financial incentives
A3	Door-to-door collection, using identification systems by colored packs/bags without using technology, with the application of fines for those who fail to separate waste
A4	Door-to-door collection with the use of GIS tools, route classification and priority locations, identification of recyclable materials and application of fines for those who fail to perform the separation
A5	Recyclable waste collection points in the municipality and neighborhoods centers placed at strategic points without using technology or financial incentives or fines
A6	Collection points for recyclable waste using IoT trash cans in the center of the municipality, in the center of the neighborhoods, in strategic locations, without a discount in the property taxes or other similar financial incentives
A7	Collection points for recyclable waste using IoT trash cans in the center of the municipality, in the center of the neighborhoods, in strategic locations, with a discount in the property taxes or other similar financial incentives

Source The authors

rate, according to the amount of residue. Decreasing the amount of recyclable organic matter mixed with residue decreases the value of the variable rate to be paid. Thus, the issue can be summarized in the following question: What are the systems alternatives for the optimization of solid waste management?

Considering the results observed in the second set of papers in the portfolio (Table 6), the selected indicators and context, seven possible scenarios were developed. They are used to compare and apply the multicriteria methodology and select a possible ideal scenario. These scenarios are described in Table 9.

8 Analysis Using the TOPSIS Fuzzy Methodology

Through the analysis of the three experts, seven scenarios or alternatives for the collection of recyclable solid waste were evaluated. The context of the city was studied, considering the 11 selected criteria and their respective weights. It should be indicated that the classifications, environmental, economic and social, have equal weights in the study. Hence, the sum of the environmental criteria has the same weight as the sum of the social criteria, which in turn is equal to the sum of the economic criteria.

For the application of the TOPSIS Fuzzy methodology, the experts filled out a spreadsheet with linguistic variables (Table 3), relating the scenarios to the criteria. Tables 10, 11 and 12 show the linguistic judgments regarding the performance of the alternatives.

Table 10 Expert assessment matrix 1 on the performance of alternatives

Alternatives	C1	C2	C3	C4	C5	C6	C7	C8	C9	C10	C11
A1	H	L	L	H	H	L	L	L	L	L	L
A2	RH	RL	RL	RH	RH	L	L	L	L	L	L
A3	RL	RL	RL	M	M	M	L	L	L	L	RL
A4	RL	M	M	RL	M	M	RL	RL	RL	M	RH
A5	M	M	M	RL	M	RH	M	M	M	M	RH
A6	RL	M	RH	L	RL	H	RH	RH	M	H	H
A7	M	M	RH	L	RL	H	RH	RH	M	H	H

Source The authors

Table 11 Expert assessment matrix 2 on the performance of alternatives

Alternatives	C1	C2	C3	C4	C5	C6	C7	C8	C9	C10	C11
A1	M	RL	L	RL	RH	L	L	L	RL	RL	RL
A2	M	M	RL	M	M	RL	RL	RL	RL	RL	M
A3	L	RH	M	M	RL	RL	M	RL	M	M	M
A4	RL	H	H	RH	L	M	RH	RH	M	RH	RH
A5	M	RH	RH	RH	RL	RH	RH	RH	RH	RH	RH
A6	RH	RH	RH	H	L	H	H	H	H	H	H
A6	H	H	H	H	L	H	H	H	H	H	H

Source The authors

Table 12 Expert assessment matrix 3 on the performance of alternatives

Alternatives	C1	C2	C3	C4	C5	C6	C7	C8	C8	C10	C11
A1	H	L	L	H	H	L	M	M	RH	L	RL
A2	M	RL	RL	H	RH	RL	RH	RH	H	RL	RL
A3	RL	RH	M	H	RL	RH	RH	RH	H	RH	RH
A4	RL	RH	M	H	RL	RH	RH	RH	H	RH	RH
A5	L	M	RH	L	M	M	M	M	RL	M	M
A6	L	M	RH	M	M	M	M	H	M	RH	M
A7	M	RH	H	RH	L	RH	M	H	M	H	RH

Source The authors

Table 13 Result with the ranking of alternatives

Alternatives	Ranking expert 1	OP	Ranking expert 2	OP	Ranking expert 3	OP	Final ranking	DG average
A1	7	0.509	7	0.513	7	0.476	7	0.488
A2	5	0.526	6	0.528	6	0.478	6	0.494
A3	2	0.531	4	0.534	3	0.522	4	0.526
A4	1	0.533	1	0.552	3	0.522	2	0.532
A5	4	0.527	5	0.534	2	0.526	3	0.528
A6	6	0.525	3	0.534	5	0.514	5	0.521
A7	3	0.528	2	0.552	1	0.548	1	0.549

Source The authors

The values presented in the tables were converted into fuzzy numbers, and the normalized results were then multiplied by the respective weight of each criterion. The ideal positive and negative solutions were calculated according to (4) and (5). Using the method according to (6), (7), and (8), the distances between the values and the ideal positive solutions (FPIS) and the negative (FNIS) were determined. Using (9), the closeness coefficient (CCi) was calculated. Table 13 lists the rankings of the alternatives and their respective Overall Performance (OP) according to the judgment of each expert. Table 13 shows a final ranking, that is, the result of a weighted average of the results.

It is possible to see that the Alternative A7, that is, recyclable garbage collection points using IoT trash cans in the center of the municipality, the center of the neighborhoods, in strategic locations, with a discount in property taxes or other similar financial incentives, has the highest ranking positions. This means that the collection of recyclable solid waste comes closest to the ideal positive solution; it is also the furthest from the ideal negative solution.

9 Conclusion

The study achieved its main goal, i.e., structuring a model to evaluate the implementation of MSWM sustainable operations and technologies. Based on these results, it is not yet possible to ensure that technological and innovative systems are a final solution. However, it is clear, considering the alternatives selected by experts, current waste management methods are not the more adequate. The interpretation of the indicators by experts is regarded as a limitation related to this investigation. The experts selected the criteria based on their experiences. As a future agenda is possible to move beyond the replication of the search, choosing other criteria. In organizing scenarios and selecting possible alternatives, the MCDM approach should also consider the managers' opinions in the criteria selection and hierarchy.

References

Al-Khatib IA, Arafat HA, Basheer T, Shawahneh H, Salahat A, Eid J, Ali W (2007) Trends and problems of solid waste management in developing countries: a case study in seven Palestinian districts. Waste Manage 27(12):1910–1919. https://doi.org/10.1016/J.WASMAN.2006.11.006

Arıkan E, Şimşit-Kalender ZT, Vayvay Ö (2017) Solid waste disposal methodology selection using multi-criteria decision making methods and an application in Turkey. J Clean Prod 142:403–412. https://doi.org/10.1016/j.jclepro.2015.10.054

Banar M, Özkan A, Kulaç A (2010) Choosing a recycling system using ANP and ELECTRE III techniques. Turk J Eng Environ Sci 34(3):145–154. https://doi.org/10.3906/muh-0906-47

Brazil. Law nº. 12,305, of August 2 (2010) Institutes the national policy on solid waste. Available in http://www.planalto.gov.br/ccivil_03/_ato2007-2010/2010/lei/l12305.htm

Chen CT (2000) Extensions of the TOPSIS for group decision-making under fuzzy environment. Fuzzy Sets Syst 114(1):1–9. https://doi.org/10.1016/S0165-0114(97)00377-1

Coban A, Ertis IF, Cavdaroglu NA (2018) Municipal solid waste management via multi-criteria decision making methods: a case study in Istanbul, Turkey. J Clean Prod 180:159–167. https://doi.org/10.1016/j.jclepro.2018.01.130

Da Silva CL (2018) Proposal of a dynamic model to evaluate public policies for the circular economy: scenarios applied to the municipality of Curitiba. Waste Manage 78:456–466. https://doi.org/10.1016/j.wasman.2018.06.007

Dahlen L, Lagerkvist A, Dahlén L, Lagerkvist A (2010) Pay as you throw Strengths and weaknesses of weight-based billing in household waste collection systems in Sweden. Waste Manage 30(1):23–31. https://doi.org/10.1016/j.wasman.2009.09.022

Deus RM, Battistelle RAG, Silva GHR (2016) Scenario evaluation for the management of household solid waste in small Brazilian municipalities. Clean Technol Environ Policy 19(1):205–214. https://doi.org/10.1007/s10098-016-1205-0

Díaz-Díaz R, Muñoz L, Pérez-González D (2017) Business model analysis of public services operating in the smart city ecosystem: the case of SmartSantander. Future Gener Comput Syst Int J Esci 76:198–214. https://doi.org/10.1016/j.future.2017.01.032

Elia V, Gnoni MG, Tornese F (2015) Designing Pay-As-You-Throw schemes in municipal waste management services: a holistic approach. Waste Manage 44:188–195. https://doi.org/10.1016/j.wasman.2015.07.040

Elia V, Gnoni MG, Tornese F (2018) Improving logistic efficiency of WEEE collection through dynamic scheduling using simulation modeling. Waste Manage 72:78–86. https://doi.org/10.1016/j.wasman.2017.11.016

Ensslin L, Ensslin SR, Lacerda RTO, Tasca JE (2010) Bibliometric analysis process. Technical file with patent pending registration with INPI, Brazil

Gelbmann U, Hammerl B (2015) Integrative re-use systems as innovative business models for devising sustainable product-service-systems. J Clean Prod 97:50–60. https://doi.org/10.1016/j.jclepro.2014.01.104

Gupta H, Barua MK (2018) A framework to overcome barriers to green innovation in SMEs using BWM and fuzzy TOPSIS. Sci Total Environ 633:122–139. https://doi.org/10.1016/j.scitotenv.2018.03.173

Herva M, Roca E (2013) Ranking municipal solid waste treatment alternatives based on ecological footprint and multi-criteria analysis. Ecol Ind 25:77–84. https://doi.org/10.1016/j.ecolind.2012.09.005

Hlatka M, Stopka O, Chovancova M (2018) The solution of the sorted waste collection using the methods of multi-criteria decision-making 164–170

Huang Y-T, Pan T-C, Kao J-J (2011) Performance assessment for municipal solid waste collection in Taiwan. J Environ Manage 92(4):1277–1283. https://doi.org/10.1016/j.jenvman.2010.12.002

Jovanovic S, Savic S, Jovicic N, Boskovic G, Djordjevic Z (2016) Using multi-criteria decision making for selection of the optimal strategy for municipal solid waste management. Waste Manage Res 34(9):884–895. https://doi.org/10.1177/0734242X16654753

Khalili NR, Duecker S (2013) Application of multi-criteria decision analysis in design of sustainable environmental management system framework. J Clean Prod 47:188–198. https://doi.org/10.1016/j.jclepro.2012.10.044

Lolli F, Ishizaka A, Gamberini R, Rimini B, Ferrari AM, Marinelli S, Savazza R (2016) Waste treatment: an environmental, economic and social analysis with a new group fuzzy PROMETHEE approach. Clean Technol Environ Policy 18(5, SI):1317–1332. https://doi.org/10.1007/s10098-015-1087-6

Makan A, Malamis D, Assobhei O, Loizidou M, Mountadar M (2013) Multi-criteria decision aid approach for the selection of the best compromise management scheme for the treatment of municipal solid waste in Morocco. Int J Environ Waste Manage 12(3):300–317. https://doi.org/10.1504/IJEWM.2013.056197

Manni LA, Runhaar HAC (2014) The social efficiency of pay-as-you-throw schemes for municipal solid waste reduction: a cost-benefit analysis of four financial incentive schemes applied in Switzerland. J Environ Assess Policy Manage 16(1). https://doi.org/10.1142/S146433321450001X

Mesquita JM (2014) Method of evaluation of the sustainability level of rural electrification programs with individual photovoltaic systems. Dissertation industrial and systems engineering, Federal University of Technology, Parana, Pato Branco, Brazil. Available in: http://repositorio.utfpr.edu.br/jspui/bitstream/1/869/1/PB_PPGEE_M_Mesquita%2C%20Jos%C3%A9%20Manuel_2014.pdf

Milutinović B, Stefanović G, Dassisti M, Marković D, Vučković G (2014) Multi-criteria analysis as a tool for sustainability assessment of a waste management model. Energy 74(C):190–201. https://doi.org/10.1016/j.energy.2014.05.056

Milutinovic B, Stefanovic G, Kyoseva V, Yordanova D, Dombalov I (2016) Sustainability assessment and comparison of waste management systems: the Cities of Sofia and Niš case studies. Waste Manage Res 34(9):896–904. https://doi.org/10.1177/0734242X16654755

Mir MA, Ghazvinei PT, Sulaiman NMN, Basri NEA, Saheri S, Mahmood NZ, Jahan A, Begum RA, Aghamohammadi N (2016) Application of TOPSIS and VIKOR improved versions in a multi criteria decision analysis to develop an optimized municipal solid waste management model. J Environ Manage 166:109–115. https://doi.org/10.1016/j.jenvman.2015.09.028

Misra D, Das G, Chakrabortty T, Das D (2018) An IoT-based waste management system monitored by cloud. J Mater Cycles Waste Manage 20(3):1574–1582. https://doi.org/10.1007/s10163-018-0720-y

Năḍăban S, Dzitac S, Dzitac I (2016) Fuzzy TOPSIS: a general view. Procedia Comput Sci 91(Itqm):823–831. https://doi.org/10.1016/j.procs.2016.07.088

Pires A, Chang N-B, Martinho G (2011) An AHP-based fuzzy interval TOPSIS assessment for sustainable expansion of the solid waste management system in Setúbal Peninsula, Portugal. Resour Conserv Recycl 56(1):7–21. https://doi.org/10.1016/j.resconrec.2011.08.004

Rada EC, Ragazzi M, Fedrizzi P (2013) Web-GIS oriented systems viability for municipal solid waste selective collection optimization in developed and transient economies. Waste Manage 33(4):785–792. https://doi.org/10.1016/j.wasman.2013.01.002

Rebehy PCPW, Costa AL, Campello CA, de Freitas Espinoza D, Neto MJ (2017) Innovative social business of selective waste collection in Brazil: cleaner production and poverty reduction. J Clean Prod 154:462–473. https://doi.org/10.1016/j.jclepro.2017.03.173

Rigamonti L, Sterpi I, Grosso M (2016) Integrated municipal waste management systems: an indicator to assess their environmental and economic sustainability. Ecol Ind 60:1–7. https://doi.org/10.1016/j.ecolind.2015.06.022

Rodrigues AP, Fernandes ML, Rodrigues MFF, Bortoluzzi SC, da Costa SE, de Lima E (2018) Developing criteria for performance assessment in municipal solid waste management. J Clean Prod 186:748–757. https://doi.org/10.1016/j.jclepro.2018.03.067

Saaty TL (2008) Relative measurement and its generalization in decision making: why pairwise comparisons are central in mathematics for the measurement of intangible factors. Rev Real Acad Ciencias 102(2):251–318. Retrieved from http://www.rac.es/ficheros/doc/00576.PDF

Sarra A, Mazzocchitti M, Rapposelli A (2017) Evaluating joint environmental and cost performance in municipal waste management systems through data envelopment analysis: scale effects and policy implications. Ecol Ind 73:756–771. https://doi.org/10.1016/j.ecolind.2016.10.035

Simões P, Carvalho P, Marques RC (2012) Performance assessment of refuse collection services using robust efficiency measures. Resour Conserv Recycl 67:56–66. https://doi.org/10.1016/j.resconrec.2012.07.006

Stefanović G, Milutinović B, Vučićević B, Denčić-Mihajlov K, Turanjanin V (2016) A comparison of the Analytic Hierarchy Process and the Analysis and Synthesis of Parameters under Information Deficiency method for assessing the sustainability of waste management scenarios. J Clean Prod 130:155–165. https://doi.org/10.1016/j.jclepro.2015.12.050

Topaloglu M, Yarkin F, Kaya T (2018) Solid waste collection system selection for smart cities based on a type-2 fuzzy multi-criteria decision technique. Soft Comput 22(15):4879–4890. https://doi.org/10.1007/s00500-018-3232-8

Tseng ML, Bui TD (2017) Identifying eco-innovation in industrial symbiosis under linguistic preferences: a novel hierarchical approach. J Clean Prod 140:1376–1389. https://doi.org/10.1016/j.jclepro.2016.10.014

Vucijak B, Silajd I (2015) Multicriteria decision making in selecting best solid waste management scenario: a municipal case study from Bosnia and Herzegovina. J Clean Prod. https://doi.org/10.1016/j.jclepro.2015.11.030

Wen Z, Hu S, De Clercq D, Beck MB, Zhang H, Zhang H, Fei F, Liu J (2017) Design, implementation, and evaluation of an Internet of Things (IoT) network system for restaurant food waste management. Waste Manage. https://doi.org/10.1016/j.wasman.2017.11.054

Yerraboina S, Kumar NM, Parimala KS, Aruna Jyothi N (2018) Monitoring the smart garbage bin filling status: an IoT application towards waste management. Int J Civ Eng Technol 9(6):373–381. Retrieved from https://www.scopus.com/inward/record.uri?eid=2-s2.0-85049254292&partnerID=40&md5=ad40ccee9eda35018c4e3566a6ea3473

Diagnostic Model in Sustainable and Innovative Operations for Municipal Solid Waste Management

Douglas Alcindo da Roza, Edson Pinheiro de Lima
and Sergio Eduardo Gouvea da Costa

Abstract Studies and assessments of municipal solid waste management systems are of fundamental importance to public authorities. Public management strategies and systems should be based on the design of integrated approaches to ensure sustainability requirements to mitigate negative impacts on the environment, economy and society. Innovative organizational and technological systems of waste management have been studied as alternative solutions to the inefficiency of conventional systems. They can be designed to guarantee quality and efficiency and understanding these new technologies is necessary. Appropriate systems for measuring sustainability performance and supporting decisions should be based on criteria and aspects relevant to the implementation of these new innovative and sustainable waste management concepts. This study aims to (1) propose a model of analysis and diagnosis for sustainable and innovative operations in the management of solid urban waste; (2) support the development of criteria and inclusion of aspects relevant to the construction of scenarios and projection of future waste management systems. The model draws from a systemic review of literature and interviews with experts in conventional waste management. The results highlight a great concern with financial factors and population awareness to obtain good results in the management systems and technologies.

Keywords Solid waste management · Innovative and sustainable operations · Systematic literature review

D. A. da Roza (✉) · E. Pinheiro de Lima · S. E. Gouvea da Costa
Industrial and Systems Engineering, Federal University of Technology—Parana, Via do Conhecimento, Km 1, Pato Branco, Paraná, Brazil
e-mail: douglas.alcindo@hotmail.com

E. Pinheiro de Lima
e-mail: pinheiro@utfpr.edu.br

S. E. Gouvea da Costa
e-mail: gouvea@utfpr.edu.br

E. Pinheiro de Lima · S. E. Gouvea da Costa
Pontifical Catholic University of Parana—PUCPR, R. Imaculada Conceicao, 1155, Curitiba, Paraná, Brazil

© Springer Nature Switzerland AG 2020   221
W. Leal Filho et al. (eds.), *International Business, Trade and Institutional Sustainability*, World Sustainability Series,
https://doi.org/10.1007/978-3-030-26759-9_13

1 Introduction

The 'National Solid Waste Policy' aims to eliminate landfills, implement municipal solid waste plans, selective collections and composting systems (Brazil 2010).

According to the 'National Solid Waste Policy', waste management should encompass the following order of priorities: non-generation of waste ('zero' waste), waste reduction, reuse, recycling, treatment and adequate environmentally final disposition of waste. The final disposal in landfills is not prohibited, but it is not recommended.

Many municipalities do not comply with the legislation as observed by Deus et al. (2016). Small municipalities that do not dispose of waste in landfills compatible with regulatory standards can seek inter-municipal cooperation and adopt integrated waste management programs.

Other less complex alternatives can improve conventional systems. Hlatka et al. (2018) propose alternatives to increase the effectiveness of the selective collection and reduce the cost of waste management in Pyšely (municipality of the Czech Republic). Partial results of the analysis of the current state of the system, classification of municipal waste and results of a study conducted through a questionnaire with the municipality residents were used in the study. Based on these analyses, proposed alternatives were evaluated using multicriteria decision-making methods.

These alternatives become even more interesting by using the perspective of innovation and sustainability to improve efficiency in the use of municipal resources. In order to improve the quality of life through efficiency, smart cities have been studied and planned based on the integration of information technologies for urban planning and management of municipal resources and services.

Inefficient infrastructures and services in today's cities justify the need for sustainable and innovative intelligent systems. Nonetheless, because it is a recent issue, there are difficulties in measuring and defining criteria for the implementation of intelligent management systems, which are sometimes unknown to even experienced and highly specialized professionals.

Although the potential of innovative organizational technologies and systems seems to be evident, designing these complex systems becomes a difficult task. Topaloglu et al. (2018) suggest a multi-criteria methodology to evaluate and classify alternative waste collection systems in an intelligent city environment. This environment involves design concepts based on emerging information and communication technologies (ICTs). The researcher uses the Fuzzy method TOPSIS Type 2 in a real case study in Eskisehir in Turkey; the evaluation criteria is determined, considering the local needs of the municipal experts.

Nevertheless, Da Silva (2018) argues that management systems are not enough. There must be a public policy oriented towards education, awareness of the importance of recycling and separating waste, resulting in decreased costs of disposal. Moreover, it should involve local development by structuring the value chain for socioeconomic benefits, creation of work and income, environmental gains through the reduction of the extraction of new raw materials, cultural perception gains with collective action in the community and political gains due to collective efficiency triggered by social action.

The current study aims to propose a model to analyze sustainable and innovative operations for solid waste management. The model encompasses criteria and characteristics for the implementation of innovative and technological systems for waste management, identified in the literature and based on expert opinion.

2 Theoretical Background

The topic 'sustainable development' has been present in the management of operations in several segments of the economy in light of the urban occupation and population growth. According to the Brazilian National Solid Waste Policy principles, the polluter pays and shares responsibility. The concepts of reduction, reuse and recycling, among others described in the standard, characterize the concern with sustainability.

Sustainability in operations is a necessity for waste management activities due to the complexity and amount of produced waste. Thus, alternative systems and technologies for waste management have been researched as a way of solving or improving conventional systems (Misra et al. 2018; Wen et al. 2017; Yerraboina et al. 2018).

New technologies driven by the concepts of smart cities and IoT stimulate innovation and new business models based on 'Product Service System' for waste management and other economy sectors (Díaz-Díaz et al. 2017; Elia et al. 2015, 2018). Nonetheless, innovation can also be social, creating economic, environmental and social value through 'Sustainable Products Services Systems' (Gelbmann and Hammerl 2015).

Considering this context, this study begins with a systemic review of the literature based on the discussed concepts of sustainable and innovative operations for solid waste management.

3 Research Design

This study deals with the organization of a conceptual model to analyze scenarios of sustainable and innovative operations for the management of urban solid waste through literature review, interviews and data analysis. The model structuring is approached qualitatively; it has descriptive and explanatory purposes.

The work starts with a systemic review of the literature, aiming to organize a 'Bibliographic Portfolio', using the process and procedures proposed by Ensslin et al. (2010). The Knowledge Development Process-Constructivist (ProKnow-C) is defined by a sequence of steps, which begins with the definition of research axes and keywords (search terms), until the organization of the Bibliographic Portfolio as seen in Fig. 1.

In the systematic analysis of the papers, which compose the bibliographical portfolio, assessment indicators in case studies reports for different contexts and sustainable and innovative management practices for solid waste management are identified.

The results obtained in the literature review are presented to experts on conventional national waste management systems. Semi-structured interviews are performed with the objective of identifying barriers and impacts related to technologies of sustainable and innovative organizational systems for the collection of urban solid waste, and criteria for the implementation of innovative and technological systems.

The interviews were divided in two stages, the first part was carried out in a group with a sequence of dialogues, following the script of questions:

I. What is the knowledge of the interviewees about the presented innovative characteristics, their opinion and potential application?
   Opening for dialogue among interviewees.
II. Intelligent waste collection systems can be regarded as a solution for solid waste management. Because …
   Opening for dialogue among interviewees.
III. Could other innovative forms of management also be applied as a solution?
   Opening for dialogue among interviewees.

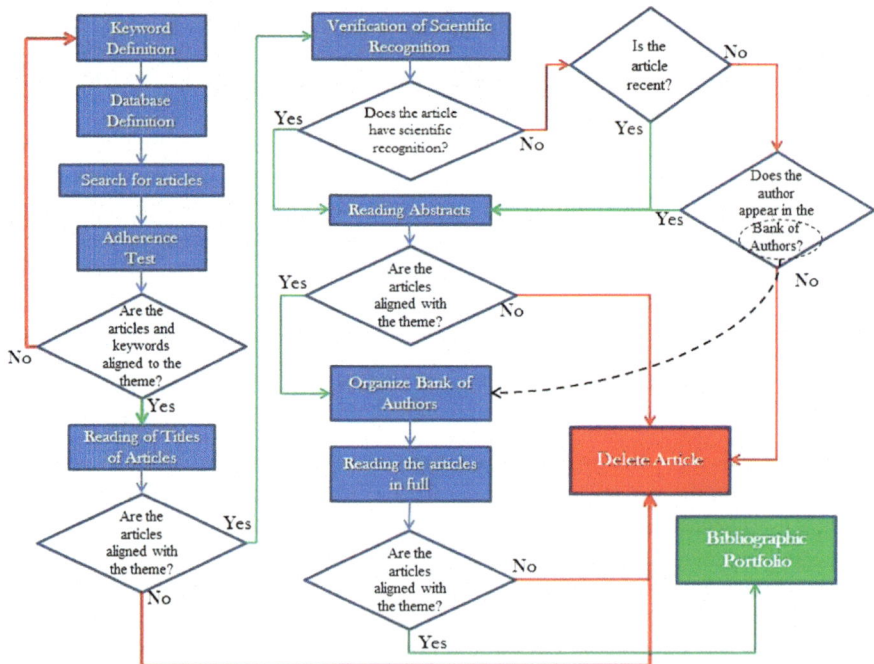

Fig. 1 Summary of ProKnow-C steps. *Source* The authors

Table 1 Waste management criteria

Public acceptance	Quality of waste
Political support	Amount of waste
Infrastructure capacity	Revenue
Innovation	Power recovery
Products lifecycle	Recovery of raw materials
Public awareness	Reduction of waste in landfills
Creation of new jobs	Income from sold recyclables
Costs with equipment	Environmental risks/impacts
Qualified staff	Noise
Maintenance costs	Safety and hygiene
Transportation costs	System sustainability
Investment costs	Size of the population to be served
Operational costs	Discard rate
Unemployment	Recycling rate
Availability of collection points	Collection time
System efficiency	Type of waste to be collected
Atmospheric emissions	Vehicle traffic
Aesthetics	Waste treatment
Odor	Operational viability

Source The authors

IV. What are the main difficulties in the implementation of smart and innovative technologies for municipal waste management and which (potentially) positive and negative impacts would affect the population and municipalities.
Opening for dialogue among interviewees.

In the second stage, carried out individually, the interviewees were required to indicate, through a questionnaire, which criteria are relevant to the implementation and operation of innovative and sustainable solid waste management systems. The questionnaire is shown in Table 1.

The criteria shown in Table 1 is organized according to the results of the first bibliographic portfolio of literature review.

4   Results and Discussions

In this section, the results of the study are described in three stages. The first stage involves a systematic review of literature in papers, content analysis on the evaluation criteria and characteristics related to the research topic. In the second stage, the interviews with experts and the characteristics of sustainable and innovative operations are discussed. In the last stage, the relevant criteria for the implementation and management of these systems is selected as per expert opinion.

5 Systemic Review of Literature

After defining the keywords/search terms, a list of 23 search terms was divided into three research axes as presented in Table 2. The research was carried out in the Web of Science™ and Scopus® databases. Combinations of the search terms between the axes were performed. The search was limited to 10 years and only journal articles, the resulting references were inserted in the Mendeley® software. Duplicated and/or repeated references were excluded, resulting in 21,040 references. The process continued with the analysis of the titles, resulting in 503 papers aligned with the research theme.

Scientific recognition is accomplished through an analysis of the citations in Google Scholar®, the 228 papers with more than five citations passed through an analysis of the abstracts and 88 were considered aligned with the research topic. The authors of these selected papers composed a data set of 254 authors.

The 275 papers with less than five citations passed through a process called reanalysis, which involves the verification if the paper is recent (less than two years of publication) or if the author(s) is/are in the data set of 254 authors. If one of these requirements is met, the abstract is read and can be selected or not. In the reanalysis process, 22 references were selected.

The sum of the selected references resulted in 110 papers, which were read thoroughly. This allowed the organization of two bibliographical portfolios.

The first portfolio is composed by 22 papers that address solid waste management in different contexts, the authors and identified criteria can be observed in Table 3.

Since different contexts encompass different indicators, similar criteria were grouped. For instance, Jovanovic et al. (2016) use particulate materials, gases emissions (CH_4, CO_2 and N_2O), Stefanović et al. (2016) use and classify emissions of greenhouse gases (CO_2) and emissions of acid gases (NO_x and SO_2) as environmen-

Table 2 Research axes and terms used in the research

Axis	Search terms
Diagnostic	Diagn*; Audit*; Evaluat*; Analy*; Perform*; Assess*; Manag*; Means*
Sustainable operations	Sustainab*; "Triple bottom line"; "Value creation"; "Business model"; "Smart Cit*"; IoT; "Internet of things"; Recy*; Upcyc*; Reduc*; Reus*
Waste management	"waste management"; "Solid waste*"; "Municipal waste*"; "Zero waste*"

Source The authors

Table 3 Criteria identified in the literature

Authors	Criteria											
	Public acceptance	Political support	Infrastructure capacity	Innovation	Products lifecycle	Public awareness	Creation of new jobs	Costs with equipment	Maintenance costs	Costs of operation	Transportation costs	Investment costs
Arıkan et al. (2017)			X				X			X	X	X
Banar et al. (2010)	X						X	X	X			
Coban et al. (2018)			X						X		X	X
Deus et al. (2016)					X							
Herva and Roca (2013)												
Hlatka et al. (2018)										X		X
Huang et al. (2011)												

(continued)

Table 3 (continued)

Authors	Criteria											
	Public acceptance	Political support	Infrastructure capacity	Innovation	Products lifecycle	Public awareness	Creation of new jobs	Costs with equipment	Maintenance costs	Costs of operation	Transportation costs	Investment costs
Jovanovic et al. (2016)										X		
Khalili and Duecker (2013)							X			X		
Lolli et al. (2016)	X											
Makan et al. (2013)	X						X		X	X		X
Milutinović et al. (2014)										X		X
(Milutinovic et al. 2016)										X		X
Mir et al. (2016)		X			X	X				X		X
Pires et al. (2011)										X		X

(continued)

Table 3 (continued)

Authors	Criteria											
	Public acceptance	Political support	Infrastructure capacity	Innovation	Products lifecycle	Public awareness	Creation of new jobs	Costs with equipment	Maintenance costs	Costs of operation	Transportation costs	Investment costs
Rigamonti et al. (2016)										X		
Rodrigues et al. (2018)										X		X
Sarra et al. (2017)												
Simões et al. (2012)										X		
Stefanović et al. (2016)	X						X			X		X
Topaloglu et al. (2018)				X								
Vucijak and Silajd (2015)	X						X			X		

(continued)

Table 3 (continued)

Authors	Criteria											
	Availability of collection points	System efficiency	Atmospheric emissions	Aesthetics	Odor	Qualified people	Waste quality	Quantity/volume of waste	Revenue	Power recovery	Material recovery	Reduction of the amount of waste in landfills
Arikan et al. (2017)		X	X							X	X	
Banar et al. (2010)						X	X		X			
Coban et al. (2018)						X						
Deus et al. (2016)			X							X		
Herva and Roca (2013)			X									X
Hlatka et al. (2018)	X					X						
Huang et al. (2011)								X				

(continued)

Table 3 (continued)

Authors	Criteria											
	Availability of collection points	System efficiency	Atmospheric emissions	Aesthetics	Odor	Qualified people	Waste quality	Quantity/volume of waste	Revenue	Power recovery	Material recovery	Reduction of the amount of waste in landfills
Jovanovic et al. (2016)			X									X
Khalili and Duecker (2013)								X				
Lolli et al. (2016)												
Makan et al. (2013)			X	X								
Milutinović et al. (2014)			X					X	X			
(Milutinovic et al. 2016)			X					X	X			
Mir et al. (2016)			X		X							
Pires et al. (2011)									X			X

(continued)

Table 3 (continued)

Authors	Criteria											
	Availability of collection points	System efficiency	Atmospheric emissions	Aesthetics	Odor	Qualified people	Waste quality	Quantity/volume of waste	Revenue	Power recovery	Material recovery	Reduction of the amount of waste in landfills
Rigamonti et al. (2016)										X	X	
Rodrigues et al. (2018)												
Sarra et al. (2017)												
Simões et al. (2012)								X				
Stefanović et al. (2016)			X					X	X			
Topaloglu et al. (2018)		X		X								
Vucijak and Silajd (2015)											X	X

(continued)

Table 3 (continued)

Authors	Criteria												
	Income from sold recyclables	Environmental risks/impacts	Noise	Safety and hygiene	System sustainability	Size of the population to be served	Monetary discounting/collection rate	Recycling rate	Time/Frequency of collection	Type of waste to be collected	Vehicle traffic	Waste treatment	Operational viability
Arıkan et al. (2017)		X								X			
Banar et al. (2010)		X		X					X		X		
Coban et al. (2018)		X											
Deus et al. (2016)													
Herva and Roca (2013)		X								X			
Hlatka et al. (2018)									X				
Huang et al. (2011)						X	X		X				

(continued)

Table 3 (continued)

Authors	Criteria												
	Income from sold recyclables	Environmental risks/impacts	Noise	Safety and hygiene	System sustainability	Size of the population to be served	Monetary discounting/collection rate	Recycling rate	Time/Frequency of collection	Type of waste to be collected	Vehicle traffic	Waste treatment	Operational viability
Jovanovic et al. (2016)		X											
Khalili and Duecker (2013)													
Lolli et al. (2016)		X											
Makan et al. (2013)		X	X										
Milutinović et al. (2014)		X						X					
(Milutinovic et al. 2016)		X					X						
Mir et al. (2016)													
Pires et al. (2011)		X		X			X						

(continued)

Table 3 (continued)

Authors	Criteria												
	Income from sold recyclables	Environmental risks/impacts	Noise	Safety and hygiene	System sustainability	Size of the population to be served	Monetary discounting/collection rate	Recycling rate	Time/Frequency of collection	Type of waste to be collected	Vehicle traffic	Waste treatment	Operational viability
Rigamonti et al. (2016)													
Rodrigues et al. (2018)		X										X	
Sarra et al. (2017)							X			X			
Simões et al. (2012)													
Stefanović et al. (2016)					X								
Topaloglu et al. (2018)													X
Vucijak and Silajd (2015)	X			X									

Source The authors

tal indicators. In this study, all of them are considered in the atmospheric emissions criterion.

It is worth mentioning that the number of indicators varies according to the paper and its context. Pires et al. (2011) use the AHP and TOPSIS methods to classify 18 alternatives for improving the sustainability of solid waste management in the Setubal region of Portugal. The criteria was divided into technical, environmental, economic and social aspects, all of which used 14 indicators. Khalili and Duecker (2013) used the ELECTRE III methodology to select the best yeast waste treatment project, the indicators included costs to the company, waste amount and job creation potential.

The second set is constituted by 13 references. The selected references are used to identify characteristics of sustainable operations or potentially innovative systems for solid waste management. Table 4 lists the references of the second bibliographic portfolio and its characteristics.

The main objective behind the review of the selected papers was to identify sustainable and innovative operations and features related to solid waste management. According to Tseng and Bui (2017), companies gain competitive advantages in business management through eco-innovation. Furthermore, industrial symbiosis contributes to achieving a win-win status in supply chain networks. The authors identify main attributes of eco-innovation to improve the performance of industrial symbiosis.

In addition to waste management eco-efficiency characteristics, social innovation can be observed in Gelbmann and Hammerl (2015) and Rebehy et al. (2017).

Rebehy et al. (2017) proposes an inclusive and decentralized institutional arrangement for selective collection with logistic technology and organization of collectors in micro- cooperatives or Temporal Accumulation Points. According to the model, collectors will use electric transportation vehicles equipped with GPS to follow predetermined collection routes. Moreover, there will be an Application (App) to allow communication with the route residents, providing specific calendars and bags for waste disposal. Gelbmann and Hammerl (2015) study a business model of sustainable systems of products and services for reuse, through socially oriented social integration companies, called ECO-WISES.

Systems with IoT technologies were the prevalent results in the portfolio (Díaz-Díaz et al. 2017; Elia et al. 2015, 2018; Misra et al. 2018; Wen et al. 2017; Yerraboina et al. 2018).

Yerraboina et al. (2018) and Misra et al. (2018) focus on smart cities with IoT technology. The first authors developed a prototype named "Smart Garbage Bin" with experimental studies of operations, but without large-scale applications. The second study presents a system that measures the level of waste and presence of foul gases, sending the information to a server over the Internet.

Díaz-Díaz et al. (2017) also use an intelligent cities rationale to study the city of Santander in Spain and eight urban services provided through IoT technology. According to the authors, the city has developed a service of collecting recyclable waste with sensors. These sensors monitor the containers and provide real-time information on location, dump characteristics and filling levels. This allows route and time

Table 4 Authors and characteristics identified in the literature

References	Features
Da Silva (2018)	The Circular Economy as a new way of thinking about current issues of urban planning and management, creating opportunities
Dahlen et al. (2010)	Scheme of payments as per waste collection rate *Pay as you throw* based on weight and volume
Díaz-Díaz et al. (2017)	Intelligent city business models based on the benchmarking of eight urban services rendered in the city of Santander; waste management was highlighted due to a 20% reduction in the cost of providing the service yearly
Elia et al. (2015)	It proposes a holistic framework for designing and managing PAYT systems applied to MSWM services through intelligent technology solutions
Elia et al. (2018)	Dynamic collection schemes for Electrical and Electronic Equipment Waste through IoT technology
Gelbmann and Hammerl (2015)	Appropriate business model for the establishment of new (sustainable) systems of products and services (SPSS) for reuse in social enterprises with labor integration ecologically oriented practices (ECO-WISEs)
Manni and Runhaar (2014)	Scheme for the payment of a waste collection rate Pay as you throw as an incentive to reduce waste
Misra et al. (2018)	It features an intelligent waste collection system based on an IoT trash can, which measures the level of materials and presence of gases; it sends this information to a cloud server for storage and processing over the Internet
Rada et al. (2013)	Optimization of the selective collection with the implementation of a system based on Geographic Information Systems (Web-GIS)
Rebehy et al. (2017)	Social innovation proposed through a sustainable and inclusive business model, with intensive use of information technology and logistics
Tseng and Bui (2017)	Business management through eco-innovation, and industrial symbiosis to achieve win-win status in supply chain networks
Wen et al. (2017)	Studies the implementation and evaluation of a sensor-based IoT network technology to improve waste management of restaurant food in the city of Suzhou, China
Yerraboina et al. (2018)	Prototype of an IoT trash can called "*Smart Garbage Bin*"

Source The authors

planning. Moreover, it has a system that allows citizens to be aware of the level of disposal points, routes and schedules as well as report of any incidents.

Wen et al. (2017) study the implementation and assessment of a system to improve the management of organic food waste from restaurants in the city of Suzhou in China. Their main objective was to achieve inspection improvements and decrease the generation of food waste. The IoT-based system encompasses the generation, collection, transportation and final disposal. It uses sensors to measure the level of waste and RFID tags to record information such as name, address, company turnover, weight, capacity and number of collection boxes. This information is processed automatically by a reader installed in the collection truck and sent in real-time to the on-board vehicle driver, along with the collection time and weight of the collected waste.

Another form of innovative technology application is observed by Elia et al. (2015). Their study aims to propose a holistic structure to design and manage PAYT (Pay As You Throw) systems in a municipality solid waste management services with organizational and technological solutions. It emphasizes the application of IoT technologies to measure waste volume and weight and identify users through different technologies such as bar code, QR code and radiofrequency technologies RFID. According to the researchers, even though they are more expensive, they also guarantee better reliability in user identification and faster read times.

IoT technology is also studied in Elia et al. (2018). It analyzes schemes of dynamic collection of electrical and electronic equipment waste, using data obtained in real-time, through these technologies. This scheme makes it possible to collect waste as demand demands it. Using a similar rationale of technology in waste collection, four case studies are presented by Rada et al. (2013). The authors also discuss aspects related to the implementation of systems based on Web-GIS (Geographic Information Systems), using GPS, RFID tag technologies at collection points for user identification and monitoring the collecting vehicle by a control room with a cartographic database.

Another feature found in the portfolio of papers is related to the 'polluter pays' principle designated as 'Pay As You Throw' or PAYT (Dahlen et al. 2010; Manni and Runhaar 2014; Elia et al. 2015).

PAYT systems can be practiced to guarantee a fairer rate to the user, according to the amount of produced waste as well as encourage the reduction of waste generation and segregation at the source. Dahlen et al. (2010) focused on the payment system by weight and volume. According to Manni and Runhaar (2014), the literature indicates that these financial incentive schemes are effective, but are criticized for raising costs for society. The authors indicate that PAYT systems have a great potential to reduce management costs and the amount of generated municipal solid waste.

Similarly, the concepts of the 'Circular Economy' also present conflicts of interest. Da Silva (2018) analyzes different scenarios for the city of Curitiba, considering trade-offs between investment in new landfills and policies to increase recycling rates. The author points out that investing in public policies of environmental education is important, but it is not enough. Sectoral policies and innovation are necessary in association with environmental policies, so that the circular economy can turn the waste problem into an opportunity for municipalities.

6 Interviews with Experts

Three experts were selected. All of them graduated in Environmental Engineering; one has a master's degree in Building Engineering and is a lecturer in the subject of solid waste management. The other two have master's degrees in Environmental Engineering, with experience in municipal waste management.

The experts were gathered and research objectives, used methodologies and results of the systemic literature review were presented. Moreover, the concepts of each innovative characteristic observed in the papers of the second portfolio were described (Table 4).

After the presentation, the interviewees were asked questions and from their answers, discussions were held until a consensus was obtained. As basic guidance for the interviews, the following questions were asked to start the dialogue. The first question was related to the interviewees' knowledge about the presented innovative characteristics and their opinion on potential applications. The second question inquired whether intelligent waste collection systems represent a solution to solid waste management. The third question related to other innovative forms of management that could also be applied as a solution. The fourth question inquired about the main difficulties in implementing smart and innovative technologies in municipal waste management and, potentially, which positive and negative impacts would affect the population and municipalities.

Regarding the first question, only one of the respondents had prior knowledge about industrial symbiosis. They all agreed and indicated there is great potential in applying the innovative forms described in the presentation, especially intelligent collection systems using IoT dumps and payment rates according to waste generation/provision.

In the second question, about intelligent waste collection systems, the answers were unanimous: they represent a solution for waste management. The interviewees also cited the association of smart technologies with collection or tax exemption rates to encourage selective collection and separation at the source. One of the interviewees said that the application of a rate for the amount of collected waste would reduce the generation of waste. Better results would be obtained through the application of fines for those who do not separate materials or higher rates for those who generate more waste.

Regarding the third question on other innovative forms of management as solutions, one of the interviewees spoke about tools such as Geographic Information Systems. These systems map generation points and waste amounts. Thus, it would be possible to map priority points for the collection and improve routing. The interviewee said that "many municipalities do not take into account the amount of generated waste or priority areas to trace the route". The experts also cited a case of a local government, where a worker at the very beginning of the waste collection implementation defined a route and taught the same route to those who succeeded him and so on for years. Furthermore, the existence of a route planning would reduce fuel costs and number of truck displacements.

Referring to the difficulties to implement innovative technologies (fourth question), the focus of the answers was mainly related to the difficulty of making it economically feasible and lack of user awareness. The interviewees cited other issues such as financial stimulus, inspection and lack of information. In terms of impacts, the experts indicated it could positively affect awareness, enhance public interest in proper waste disposal and generation or reallocation of jobs. In terms of negative impacts, it could increase the amount of residue in irregular areas and inappropriate disposal when dealing with environmentally uneducated citizens.

7   Selection of Indicators by Experts

After the interview and dialogue on innovations for waste management, the criteria and indicators identified in the literature were described and presented (Table 4). The experts were requested to select relevant criteria for the implementation and operations of waste management sustainable and innovative systems. Each of the experts received a spreadsheet to mark the criteria (according to Table 1). This step was performed individually and without any consultation among interviewees. The 11 criteria selected as shown in Table 5 are observed in the literature and regarded as relevant by the experts.

During the interview, experts highlighted as main difficulties for the implementation of sustainable and innovative technologies: population awareness and economic viability. This was observed in the choice of criteria as a confirmation of the opinions expressed in the interview. The experts selected costs with equipment, investments and operational costs in addition to income from sold recyclables, public awareness and acceptance.

According to the selected criteria, it is also inferred that there is a concern with the quality of the service to be provided, by choosing the system efficiency crite-

Table 5 Indicators observed in literature and selected by the experts

Criteria	
Public acceptance	Amount of collected waste
Public awareness	Reduction of waste in landfills
Costs with equipment	Income from sold recyclables
Investment costs	Environmental risks/impacts
Operational costs	Safety and hygiene
System efficiency	

Source The authors

rion, amount of collected waste and reduction of waste in landfills. Concerns with potential harm to people and environment were highlighted by the selection of the environmental risks and impacts criterion and safety and hygiene criterion.

It is often difficult to propose projects for the implementation of waste management, so that innovative and sustainable ideas can support the construction of potential scenarios. These in turn can be evaluated by multicriteria decision support methods, considering the criteria identified in the literature and selected by the experts.

8 Conclusion

This study aimed to analyze sustainable and innovative operations for waste management. The results present criteria that can be considered in the analysis and implementation of these systems.

The interview with experts shows the lack of knowledge of professionals specialized in the conventional management of municipal solid waste, regarding new technologies and sustainable alternative systems of waste management.

The interpretation of the indicators by the experts is regarded as a limiting factor in the current study. During the interview, the concepts of the criteria for different contexts identified in the literature were not described. Thus, the experts selected the criteria based on their experiences and empirical knowledge.

In future studies, aiming to identify sustainable and innovative characteristics of waste management and selection of indicators by experts; it is possible to build scenarios oriented to answer the concerns of these professionals. This can be achieved through multicriteria decision support methodologies such as TOPSIS by classifying best practices for a specific context.

References

Arıkan E, Şimşit-Kalender ZT, Vayvay Ö (2017) Solid waste disposal methodology selection using multi-criteria decision making methods and an application in Turkey. J Clean Prod 142:403–412. https://doi.org/10.1016/j.jclepro.2015.10.054

Banar M, Özkan A, Kulaç A (2010) Choosing a recycling system using ANP and ELECTRE III techniques. Turk J Eng Environ Sci 34(3):145–154. https://doi.org/10.3906/muh-0906-47

Brazil. Law n°. 12,305, of August 2 (2010) Institutes the national policy on solid waste. Available in: http://www.planalto.gov.br/ccivil_03/_ato2007-2010/2010/lei/l12305.htm

Coban A, Ertis IF, Cavdaroglu NA (2018) Municipal solid waste management via multi-criteria decision making methods: a case study in Istanbul, Turkey. J Clean Prod 180:159–167. https://doi.org/10.1016/j.jclepro.2018.01.130

Da Silva CL (2018) Proposal of a dynamic model to evaluate public policies for the circular economy: scenarios applied to the municipality of Curitiba. Waste Manage 78:456–466. https://doi.org/10.1016/j.wasman.2018.06.007

Dahlen L, Lagerkvist A, Dahlén L, Lagerkvist A (2010) Pay as you throw strengths and weaknesses of weight-based billing in household waste collection systems in Sweden. Waste Manage 30(1):23–31. https://doi.org/10.1016/j.wasman.2009.09.022

Deus RM, Battistelle RAG, Silva GHR (2016) Scenario evaluation for the management of household solid waste in small Brazilian municipalities. Clean Technol Environ Policy 19(1):205–214. https://doi.org/10.1007/s10098-016-1205-0

Díaz-Díaz R, Muñoz L, Pérez-González D, Perez-Gonzalez D (2017) Business model analysis of public services operating in the smart city ecosystem: the case of SmartSantander. Future Gener Comput Syst Int J Esci 76:198–214. https://doi.org/10.1016/j.future.2017.01.032

Elia V, Gnoni MG, Tornese F (2015) Designing Pay-As-You-Throw schemes in municipal waste management services: a holistic approach. Waste Manage 44:188–195. https://doi.org/10.1016/j.wasman.2015.07.040

Elia V, Gnoni MG, Tornese F (2018) Improving logistic efficiency of WEEE collection through dynamic scheduling using simulation modeling. Waste Manage 72:78–86. https://doi.org/10.1016/j.wasman.2017.11.016

Ensslin L, Ensslin SR, Lacerda RTO, Tasca JE (2010) Bibliometric analysis process. Technical file with patent pending registration with INPI, Brazil

Gelbmann U, Hammerl B (2015) Integrative re-use systems as innovative business models for devising sustainable product-service-systems. J Clean Prod 97:50–60. https://doi.org/10.1016/j.jclepro.2014.01.104

Herva M, Roca E (2013) Ranking municipal solid waste treatment alternatives based on ecological footprint and multi-criteria analysis. Ecol Ind 25:77–84. https://doi.org/10.1016/j.ecolind.2012.09.005

Hlatka M, Stopka O, Chovancova M (2018) The solution of the sorted waste collection using the methods of multi-criteria decision-making 164–170

Huang Y-T, Pan T-C, Kao J-J (2011) Performance assessment for municipal solid waste collection in Taiwan. J Environ Manage 92(4):1277–1283. https://doi.org/10.1016/j.jenvman.2010.12.002

Jovanovic S, Savic S, Jovicic N, Boskovic G, Djordjevic Z (2016) Using multi-criteria decision making for selection of the optimal strategy for municipal solid waste management. Waste Manage Res 34(9):884–895. https://doi.org/10.1177/0734242X16654753

Khalili NR, Duecker S (2013) Application of multi-criteria decision analysis in design of sustainable environmental management system framework. J Clean Prod 47:188–198. https://doi.org/10.1016/j.jclepro.2012.10.044

Lolli F, Ishizaka A, Gamberini R, Rimini B, Ferrari AM, Marinelli S, Savazza R (2016) Waste treatment: an environmental, economic and social analysis with a new group fuzzy PROMETHEE approach. Clean Technol Environ Policy 18(5, SI):1317–1332. https://doi.org/10.1007/s10098-015-1087-6

Makan A, Malamis D, Assobhei O, Loizidou M, Mountadar M (2013) Multi-criteria decision aid approach for the selection of the best compromise management scheme for the treatment of municipal solid waste in Morocco. Int J Environ Waste Manage 12(3):300–317. https://doi.org/10.1504/IJEWM.2013.056197

Manni LA, Runhaar HAC (2014) The social efficiency of pay-as-you-throw schemes for municipal solid waste reduction: a cost-benefit analysis of four financial incentive schemes applied in Switzerland. J Environ Assess Policy Manage 16(1). https://doi.org/10.1142/S146433321450001X

Milutinović B, Stefanović G, Dassisti M, Marković D, Vučković G (2014) Multi-criteria analysis as a tool for sustainability assessment of a waste management model. Energy 74(C):190–201. https://doi.org/10.1016/j.energy.2014.05.056

Milutinovic B, Stefanovic G, Kyoseva V, Yordanova D, Dombalov I (2016) Sustainability assessment and comparison of waste management systems: the Cities of Sofia and Niš case studies. Waste Manage Res 34(9):896–904. https://doi.org/10.1177/0734242X16654755

Mir MA, Ghazvinei PT, Sulaiman NMN, Basri NE A, Saheri S, Mahmood NZ, Jahan A, Begum RA, Aghamohammadi N (2016) Application of TOPSIS and VIKOR improved versions in a multi criteria decision analysis to develop an optimized municipal solid waste management model. J Environ Manage 166:109–115. https://doi.org/10.1016/j.jenvman.2015.09.028

Misra D, Das G, Chakrabortty T, Das D (2018) An IoT-based waste management system monitored by cloud. J Mater Cycles Waste Manage 20(3):1574–1582. https://doi.org/10.1007/s10163-018-0720-y

Pires A, Chang N-B, Martinho G (2011) An AHP-based fuzzy interval TOPSIS assessment for sustainable expansion of the solid waste management system in Setúbal Peninsula, Portugal. Resour Conserv Recycl 56(1):7–21. https://doi.org/10.1016/j.resconrec.2011.08.004

Rebehy PCPW, Costa AL, Campello CA, de Freitas Espinoza D, Neto MJ (2017) Innovative social business of selective waste collection in Brazil: cleaner production and poverty reduction. J Clean Prod 154:462–473. https://doi.org/10.1016/j.jclepro.2017.03.173

Rada EC, Ragazzi M, Fedrizzi P (2013) Web-GIS oriented systems viability for municipal solid waste selective collection optimization in developed and transient economies. Waste Manage 33(4):785–792. https://doi.org/10.1016/j.wasman.2013.01.002

Rigamonti L, Sterpi I, Grosso M (2016) Integrated municipal waste management systems: an indicator to assess their environmental and economic sustainability. Ecol Ind 60:1–7. https://doi.org/10.1016/j.ecolind.2015.06.022

Rodrigues AP, Fernandes ML, Rodrigues MFF, Bortoluzzi SC, da Costa SE, de Lima E (2018) Developing criteria for performance assessment in municipal solid waste management. J Clean Prod 186:748–757. https://doi.org/10.1016/j.jclepro.2018.03.067

Sarra A, Mazzocchitti M, Rapposelli A (2017) Evaluating joint environmental and cost performance in municipal waste management systems through data envelopment analysis: scale effects and policy implications. Ecol Ind 73:756–771. https://doi.org/10.1016/j.ecolind.2016.10.035

Simões P, Carvalho P, Marques RC (2012) Performance assessment of refuse collection services using robust efficiency measures. Resour Conserv Recycl 67:56–66. https://doi.org/10.1016/j.resconrec.2012.07.006

Stefanović G, Milutinović B, Vučićević B, Denčić-Mihajlov K, Turanjanin V (2016) A comparison of the Analytic Hierarchy Process and the Analysis and Synthesis of Parameters under Information Deficiency method for assessing the sustainability of waste management scenarios. J Clean Prod 130:155–165. https://doi.org/10.1016/j.jclepro.2015.12.050

Topaloglu M, Yarkin F, Kaya T (2018) Solid waste collection system selection for smart cities based on a type-2 fuzzy multi-criteria decision technique. Soft Comput 22(15):4879–4890. https://doi.org/10.1007/s00500-018-3232-8

Tseng ML, Bui TD (2017) Identifying eco-innovation in industrial symbiosis under linguistic preferences: a novel hierarchical approach. J Clean Prod 140:1376–1389. https://doi.org/10.1016/j.jclepro.2016.10.014

Vucijak B, Silajd I (2015) Multicriteria decision making in selecting best solid waste management scenario: a municipal case study from Bosnia and Herzegovina. J Clean Prod. https://doi.org/10.1016/j.jclepro.2015.11.030

Wen Z, Hu S, De Clercq D, Beck MB, Zhang H, Zhang H, Fei F, Liu J (2017) Design, implementation, and evaluation of an Internet of Things (IoT) network system for restaurant food waste management. Waste Manage. https://doi.org/10.1016/j.wasman.2017.11.054

Yerraboina S, Kumar NM, Parimala KS, Aruna Jyothi N (2018) Monitoring the smart garbage bin filling status: an IoT application towards waste management. Int J Civ Eng Technol 9(6):373–381. Retrieved from https://www.scopus.com/inward/record.uri?eid=2-s2.0-85049254292&partnerID=40&md5=ad40ccee9eda35018c4e3566a6ea3473

The Development Role of Customers in the Reverse Logistics of Industrial Waste

Tamiris de Oliveira and Alvim Borges

Abstract Waste management in large companies is becoming increasingly important. Waste resulting directly or indirectly from production process has very different values, importance, and characteristics, all of which need to be properly processed. The lack of sufficient processors to deal with a wide range of wastes has been shown to limit the sustainability levels of this reverse logistics chain. Thus, this research aims at evaluating the conditions for the development of a waste customers chain of large industrial companies. This research builds on theories and studies such as the Theory of Planned Behavior and the approach to Supplier Development, hitherto used upstream to be now applied downstream. The research is based on the primary data of a large industrial company, the survey of environmental legislation for some types of industrial waste, interviews with environmental agencies and other important stakeholders in the structuring of this logistics chain. The results show there are still no consolidated logistic chains for the industrial waste market. The following are the factors and recommendations for the industrial waste logistics to become more comprehensive and sustainable.

Keywords Waste management · Reverse logistics · Industrial waste · Development of customers

1 Introduction

According to data presented by the United Nations, the world population is estimated to be 9.6 billion people in 2050, and it would take three planets to provide the required resources to maintain the same conditions of consumption as today. The depletion of

T. de Oliveira (✉) · A. Borges
Master of Science Program on Engineering and Sustainable Development, Universidade Federal do Espírito Santo - UFES, Avenida Fernando Ferrari, 514, Goiabeiras, Vitoria, ES, Brazil
e-mail: tamiris.oliveira.02@aluno.ufes.br

A. Borges
e-mail: alvim@pobox.com

© Springer Nature Switzerland AG 2020
W. Leal Filho et al. (eds.), *International Business, Trade and Institutional Sustainability*, World Sustainability Series,
https://doi.org/10.1007/978-3-030-26759-9_14

natural resources and the negative impacts of environmental degradation exacerbate the list of challenges humanity must face (ONU 2018).

Inadequate disposal of waste from manufacturing processes can damage to the environment from its generation, collection, and treatment to its final disposal, making the proper management of such wastes a necessity from the sanitary, economic, environmental, and social points of view (IEMA 2018b). With the waste collection by the population, originated from public awareness, laws started to be created in several countries aimed at charging generators the responsibility for their products at the end of lifespan, bringing concepts such as reverse logistics (RL) and closed-loop supply chains (CLSC) as a revenue opportunity for companies, rather than as a cost minimization approach (Govindan et al. 2014).

In Brazil, there are already laws and decrees towards this goal, such as the National Solid Waste Policy, which recognizes the reusable and recyclable solid wastes as an economic and social good, generating work and income and promoting citizenship, and determines that the waste generator is responsible for its proper destination (LEI N° 12.305/2010 2010).

Most industrial production companies generate a considerably greater volume of potentially reusable waste when compared to any other sector of society, giving them an important role in achieving recycling and reusing goals (Levänen 2014). The greatest challenge for these companies is then facing the trade-off between achieving economic performance and, at the same time, guaranteeing their performance in social and environmental dimensions (Goebel et al. 2018).

The waste produced by these large industrial companies is highly relevant to sustainability. They can be reused through paths such as the open-loop chain—where they give rise to other types of products—or closed loop—giving rise to similar products. However, the greatest difficulty in inserting these residues as substitutes for raw materials in a manufacturing process is the lack of predictability regarding volume generated, quality control, delivery deadline, among other basic competitive characteristics that are guaranteed by suppliers when raw materials are purchased (González-Benito 2007).

To minimize or even avoid such difficulties, an integrated system for the management of solid sectoral and municipal waste should be implemented. The development of this system requires knowledge on characteristics of the industrial wastes and their status within the companies: their quantity, quality, and destination (IEMA 2018c).

Considering the needs of generating companies and their possibilities of waste management, this research aims at identifying the main difficulties faces by companies for the sustainable destination of industrial waste, through the application of a case study. From the results, one can present recommendations that contribute to the solution of such problems.

2   Industrial Waste Management

The waste management approach has some basic elements, such as sustainability, reverse logistics, and the closed-loop supply chain.

Broadly defined, sustainability is the balance between social, environmental, and economic criteria. In 1998, Elkington called this combination "triple bottom line (TBL)" (Goebel et al. 2018).

Reverse Logistics (RL) is defined as the process of planning, implementing, and controlling the efficient and economical flow of raw materials, process inventories, finished products, and related information from the point of consumption to the point of origin, to recover the value of proper disposal. In a more comprehensive way, the RL begins with the final use of the product, when they are gathered with the purpose of directing them to one of the different possible paths, such as recycling, remanufacturing, repairing, or elimination of some parts (Govindan et al. 2014).

The closed-loop supply chain is defined as the "design, control, and operation of a system to maximize value creation throughout the life cycle of a product, with dynamic value recovery of different types and return volumes over the time" (Govindan et al. 2014).

3   Research Methodology

This research was carried out using primary data from a mining company, where the data collection and the analysis on the quantitative data of waste management for 2017 were performed.

To better design the scenario studied, semi-structured face-to-face interviews were conducted with representatives of the company, the State management, and the waste management body (IEMA), which works directly with the subject studied. In addition, research was conducted on legislation and other studies related to the subject.

With a qualitative and exploratory approach, information and data were analyzed to identify the main waste management difficulties in the case studied, to then structure and propose measures to facilitate the development of customers for the waste produced. To ensure confidentiality, the name of the company and its partners will not be disclosed—fictitious names shall be used to identify them throughout the study.

4   Waste Management Regulation

In the case studied, the regulation on waste disposal is carried out by the State Institute of Environmental and Water Resources (IEMA), whose purpose is to plan, coordinate, execute, supervise, and control activities involving the environment, state water and natural resources, whose management has been delegated by the Union (IEMA

2018a). The IEMA Solid Waste and Sanitation Coordination (CRSS) is responsible for, among other activities, authorizing, licensing and maintaining the environmental control of activities or undertakings related to solid waste, environmental sanitation, drainage systems and their interference (IEMA 2018c).

In this context, in 2013 the IEMA proposed to establish and maintain the Information System and Inventory of Solid Waste of the State of Espírito Santo—Sinir/ES. This system aims at computerizing the data provided to the IEMA by the generators, transporters, and users of solid waste located in the State of Espírito Santo, formulating a database with the information on generators, collection, transportation, treatment, disposal, and reuse of solid waste. This instrument will allow data spatialization, allowing the visualization of waste areas and streams generated and destined in Espírito Santo, thus enabling a discussion of public policy propositions based on real data and information. Its implementation is scheduled for March 2019 (IEMA 2018b).

5 Collected Data

According to the information obtained, the Federal Law 10,650/2003 ensures access to data and procedural information held by the IEMA. These documents contain, among other information, data on the waste generation of enterprises in the state, such as: volume, type of waste, destination etc. Such information is important to describe the waste scenario in the state and to identify its possibilities. Currently, to access this data, one must follow the guidelines established by the SEAMA/IEMA/AGERH Joint Ordinance No. 08 of July 31, 2017, and one needs to reapply it for each company process one wishes to consult, turning this into an irksome and unproductive procedure.

The Delta company, where this study was applied, is a mining company present in several countries around the world and plays an important role in the waste trade in Brazil. It has a temporary waste storage area (Center for Miscellaneous Materials—CMD) for each complex, where all waste generation of this complex units is sent for temporary storage until its final destination, when the waste is weighted and accounted for in the system.

In 2017, 24 companies acquired waste generated by the Delta company. As the generator of such waste, the company has the responsibility to prove to IEMA, the inspection body, its correct destination. For this reason, purchasing companies must be approved by Delta's Environment Board so they can be included in the list of possible purchasers of waste from the company. The required documentation must also be presented (destination certificate) so that the intended use of such waste is actually proven, thus ensuring that it was in fact destined for what was informed at the time of purchasing.

Table 1 presents the data on the commercialization of industrial waste in 2017. It includes data on the type of waste and volume, as well as the percentage of each

Table 1 Waste marketed in relation to the total generated in 2017

Buyer	Waste	Amount (kg)	Percentage (%)
A	Soy/corn/bran reject	422,880.00	94.80
	Rests of food	180,920.00	43.63
B	Iron and steel scrap	9,703,250.00	98.67
	Scrap metal and steel powder	1,536,480.00	100.32
	Railway track scrap	576,000.00	99.93
	Unleaded brake shoe scrap/asbestos	170,600.00	97.03
C	Rubber scrap in blanket and strip	212,490.00	42.91
	Conveyor belt scrap	1,295,800.00	97.20
D	Refractory material scrap	913,970.00	99.74
E	"Residual" contaminant with oil and grease	0.80	0.00
	"Residual" fuel oil and altered	0.80	0.30
F	Ore-residue with impurity	3,653,670.00	99.94
G	Copper scrap	5,590.00	120.47
	Electrical and electronic component scrap	85,250.00	30.50
H	Paper and cardboard scrap	151,690.00	84.12
	Plastic scrap	232,520.00	46.74
	Uncontaminated big bags scrap	593,360.00	116.49
	Uncontaminated IBC scrap (TOT BIN)	17,440.00	90.27
I	Scrap copper wire/cable	14,940.00	14.28
J	Ore-residue with impurity	2,040.00	0.06
K	Wood scrap packaging waste	593,810.00	84.99
	Wood scrap not recyclable	30,930.00	23.19
L	"Waste" acid lead battery	99,290.00	100.29
M	Rail axis scrap	200,770.00	100.01
N	Tire scrap miscellaneous vehicles	61,450.00	143.80
O	Wood scrap packaging waste	20,320.00	2.91
P	Railway wheel	2,395,670.00	99.99
Q	Stainless steel scrap	30,850.00	85.32
	Aluminum scrap	88,580.00	106.94
	Bronze scrap	9,060.00	111.44
	Scrap copper wire/cable	92,590.00	88.48
R	"Waste" used lubricant oil	482,269.97	99.27
S	Wood scrap packaging waste	54,170.00	7.75
	Wood in Tora exotic tree	268,360.00	42.94
T	"Waste" ETE sludge	3,692.00	24.24
	"Residue" contaminated with oil and grease	970,292.00	77.13

(continued)

Table 1 (continued)

Buyer	Waste	Amount (kg)	Percentage (%)
	"Residue" fuel oil and altered	244.00	90.37
	Plastic scrap	107,380.00	21.59
	"Residue" lamp (dangerous substance)	2,780.00	57.18
	Oil and fat residue plant and animal	42.00	10.77
	"Residue" cont. non-oily hazardous substance	15,524.00	58.21
U	Electrical and electronic component scrap	139,440.00	49.89
V	Plastic scrap	320.00	0.06
	Uncontaminated IBC scrap (TOT BIN)	230.00	1.19
W	Stainless steel scrap	5,990.00	16.57
X	Rubber scrap in blanket and strip	249,090.00	50.30
	Conveyor belt scrap with steel	84,370.00	161.78
	Mill coating scrap	208,430.00	156.28

ETE Effluent treatment station
Source Prepared by the Author

purchase compared to the total volume generated by the Delta company in the same year.

As shown in Table 2, most (74.84%) wastes acquired by purchasing companies were destined for recycling, sales, or reprocessing. External and internal landfills combine to account for 9.74% of waste destinations. Remaining destinations include internal storage, co-processing, composting, re-refining, and so on.

Table 2 Volume by destination

Waste allocation	Weight (kg)
Recycling/Sale/reprocessing	23,249,301.60
External landfill	1,880,245.40
Internal Landfill	1,144,810.00
Co-processing	1,166,424.00
Re-refining	482,269.97
Analysis and testing/transformation	370,670.00
Exploitation transfer/carbonization	369,410.00
Blending	283,910.00
Marketing/reuse	195,771.00
Composting	607,740.00
Decontamination	2,780.00
Storage/stock	1,312,151.5
Total volume	31,065,483.47

Source Prepared by the Author

In 2016, the total volume of waste stored in the CMD was 815,089.20 kg, and in 2017 this volume grew to 1,312,151.50 kg. When comparing the volumes of December 2016 and December 2017, of the 75 residues generated, 24 had no stock, 20 had smaller volumes compared to December 2016, but still had a volume in stock, and 31 had a larger volume at the end of 2017 than in 2016.

The mean time of waste storage in the CMD is 12 months. However, it is possible to find cases in which the waiting time for an adequate and sustainable destination is nearing three years.

According to the interview with sector managers, the main problem the company faces in this management is the difficulty of finding companies in the market that are interested in the purchasing waste, considering the requirements in the homologation stage and the proof of proper destination. In addition, many companies have a very low volume and/or frequency buying interest in waste removal when compared with the destination requirements demanded by Delta.

6 Analysis of Data and Results

As observed in our data, the average of items purchased per company is 2; in other words, buying companies are interested in a low variety of waste. In addition, 41.33% of the items presented a growth in the volume in stock between 2016 and 2017, resulting in an increased total volume of waste in inventory of approximately 61%.

Despite the significant commercialization volume of Delta's waste this year (75% of the total volume), the company still had a stock of 1312.15 tons of waste occupying a large storage area that could be used for another purpose, besides generating costs related to the control and maintenance of the site, in addition to the loss of value due to the depreciation of these residues by the stored time. Another major difficulty identified in the interview was that, in general, buyers have a frequency and/or volume of waste removal below the Delta's needs.

In considering all these problems raised, one may identify that the common point between them is the difficulty in developing this market, due to the clients' limitations. For this reason, with the objective of zeroing the waste inventories generated in its production process and providing a suitable and sustainable destination, considering the particularities of waste management, it is proposed that the company uses the Development of Clients, structured on the approaches of supplier development and the Theory of Planned Behavior (TPB).

Supplier development is defined as any effort of a company to increase the performance and/or capabilities of its supplier to meet short and/or long-term supply demands. These efforts can range from limited efforts to performance improvement requests (Krause and Krause 1997). Studies show that buyers' actions typically result in improvements in supplier performance and capabilities (Humphreys et al. 2004). Such measures are important especially when the improper business practices of a supplier can significantly affect the image and, consequently, the financial results of the purchasing company, as the latter generally has greater visibility and collection by supervisory bodies and the community (Goebel et al. 2018).

Considering the similarities between the purchasing company (supplier development) and the company that generates waste (customer development)—such as the company's responsibility, control, and collection, as well as interest in increasing sales—adaptations are proposed for the development of suppliers in a way that can be used as a reference for structuring the customer development. The waste generator plays a pivotal role in the chain, becoming active in the behavior of its customers (current and future) through actions that can contribute to develop the interest of potential buyers, as well as increase performance and the current customer capabilities, making them frequent waste consumers. Some examples are shown in Table 3, where a comparison is made between actions of supplier development and customer development.

According to TPB, attitudes, in addition to aspects related to perceived social pressure (subjective norm), available infrastructure, and the ability of individuals (degree of perceived behavioral control), are predictive of behavioral intentions concerning a specific behavior, that is, the greater one or more of these constructs are, the greater the behavioral intentions of an individual (Heidemann et al. 2012). The term 'attitude' is defined as the willingness to respond favorably or unfavorably to an object, person, institution, or event; i.e., the evaluations the person makes on a particular subject that lead them act positively or negatively. Personal traits are much more resistant to transformations (Ajzen 1991). Such constructs can define behav-

Table 3 Supplier development actions × customer development actions

Supplier development	Customer development
Closer, more efficient communication to solve problems and make adjustments in less time	Instant communication to avoid difficulties and facilitate the continuous flow of waste acquisition
Provides data that proves the advantages in making adjustments and changes to meet certain buyer requirements	Provides data that shows the advantages of replacing the new raw material with waste in the production processes
Offers interchange of employees as a way to exchange experiences and knowledge for mutual growth	Promotes the interchange of ideas as a way to improve processes and knowledge for mutual growth
Develops long-term strategic orientations to achieve mutual gains	Develops long-term economic, social, and environmental assessments to demonstrate mutual and societal gains
Control and inspection to guarantee the requirements due to the collections and inspections on the buyer	Control and inspection to proper destination due to the charges and inspections on the generator
Transmits the sustainability measures applied to the purchasing company to the companies upstream of the chain	Promotes a permanent exchange of experiences among the various waste customers to improve processes throughout the production chain

Source Prepared by the Author

ioral intentions but can also be used for interventions to change behavior, provided there is room for change in this construct (Ajzen 2006). The TPB is seen as a theoretical and methodological support in the development of research aimed at measuring and interpreting attitudes, in addition to contributing in the planning of intervention programs (Heidemann et al. 2012).

As the existing relationship between the generating company and its current customers can offer important information, such as:

- the difficulties encountered by them during the homologation process and the proof of destination;
- the variables that influence such companies to become buyers of industrial waste;
- the intentions that guided its insertion in the waste market;
- difficulties to substitute raw materials for waste in manufacturing processes.

TPB complements the framework for structuring the client development program, minimizing the risks of data collection and analysis, and enabling understanding and influencing the behavior of companies that are not yet part of the industrial waste market.

An additional contribution that may be significant in the customer development process is that of the IEMA through the Sinir system, as it offers important information in an easy and integrated way to buyers interested in identifying other companies that generate waste and can complement their demand, mainly when only one company is unable to supply it, either by volume, frequency of generation, or any other factor requiring a partnership between two or more companies.

7 Conclusions

The results show that, although the company trades a significant volume of waste generated, there is still a considerable volume of materials in stock, consuming resources while losing their value due to the difficulty of being inserted in the market. Customer Development, structured with the support of the Theory of Planned Behavior and guided by studies and experiences of the supplier development application, come as a solution to the demand presented by the company, through application actions that stimulate the interest of other companies in consuming/processing industrial waste in their processes.

By expanding the partners in the industrial waste market, even if they still are interested in only a small variety of waste, or if they do not show significant frequency or volume, the amount purchased by a larger group of buyers will be able to absorb the volume of waste generated by the company without it having to be stored for a long period. As a result, management costs and storage time will be reduced, generating a revenue from this marketing and, more importantly, returning those materials to the production chain rather than simply discarding them. It is expected that, in obtaining the first satisfactory results of such actions, other generating companies with similar problems will follow the same example, which is important for the waste market development.

Acknowledgements The authors would like to thank the VALE S.A. company for financial support, and Marcos Paulo Valadares de Oliveira for his relevant contributions.

References

Ajzen I (1991) The theory of planned behavior. Organ Behav Hum Decis Process 179–211
Ajzen I (2006) Behavioral interventions based on the theory of planned behavior, January 2006
Goebel P, Reuter C, Pibernik R, Sichtmann C, Bals L (2018) Purchasing managers' willingness to pay for attributes that constitute sustainability. J Oper Manage 62:44–58
González-Benito J (2007) A theory of purchasing's contribution to business performance. J Oper Manage 25(4):901–917. https://doi.org/10.1016/j.jom.2007.02.001
Govindan K, Soleimani H, Kannan D (2014) Reverse logistics and closed-loop supply chain: a comprehensive review to explore the future. Eur J Oper Res 603–626. http://dx.doi.org/10.1016/j.ejor.2014.07.012
Heidemann LA, Araujo IS, Veit EA (2012) Um referencial teórico-metodológico teórico para o desenvolvimento de pesquisas sobre atitude: a Teoria do Comportamento Planejado de Icek Ajzen. Revista Electrónica de Investigación En Educación En Ciencias
Humphreys PK, Li WL, Chan LY (2004) The impact of supplier development on buyer–supplier performance. Omega Int J Manage Sci 32:131–143. https://doi.org/10.1016/j.omega.2003.09.016
IEMA (2018a) Competências. Retrieved 5 Nov 2011, from https://iema.es.gov.br/historia
IEMA (2018b) Gestão de Resíduos Sólidos. Retrieved 5 Nov 2018, from https://iema.es.gov.br/gestao_de_residuos_solidos/sinires
IEMA (2018c) Gestão de Resíduos Sólidos - Logística Reversa. Retrieved 5 Nov 2018, from https://iema.es.gov.br/gestao-de-residuos-solidos
Krause DR, Krause DR (1997) Supplier development: current practices and outcomes. Int J Purchasing Mater Manage 12–19
LEI Nº12.305/2010 (2010) Política Nacional de Resíduos Sólidos
Levänen J (2014) Ending waste by law: institutions and collective learning in the development of industrial recycling in Finland. J Clean Prod 87:542–549. https://doi.org/10.1016/j.jclepro.2014.09.085
ONU (2018) Transformando Nosso Mundo: A Agenda 2030 para o Desenvolvimento Sustentável Declaração. Retrieved 28 Oct 2018, from https://nacoesunidas.org/pos2015/agenda2030/

Tamiris de Oliveira is a Master's student in Engineering and Sustainable Development at the Federal University of Espírito Santo. Bachelor in Civil Engineering from the FAESA University Center.

Professor Alvim Borges holds a Ph.D. in Management Sciences from the Aix-Marseille Université at the Center de Recherche sur le Transport et la Logistique (CRET-LOG) and a Ph.D. in Energy from the Institute of Energy and Environment (IEE) at the University of São Paulo. Mechanical Engineer, Specialist in Finance from PUC/Rio, Master in Production Engineering from the Federal University of Santa Catarina. Researcher of the Professional Master's in Engineering and Sustainable Development and professor in the Administration Department of the Federal University of Espírito Santo. He currently works with finance and logistics research, focusing on the energy industry and sustainable development.

Challenges in Reducing Construction and Demolition Waste Generation in Construction Sites in Curitiba

Leilane Kusunoki, Eduardo Felga Gobbi and Patricia Charvet

Abstract Currently, construction and demolition wastes (CDW) represent more than 50% of the urban solid waste share in Brazil. It is a primary environmental concern for the construction industry and also affects city waste management plans. The CDW final disposal often occurs on improper areas, which combined with the urban expansion, postpone the sustainable city development. This study analyzed the class A waste generation leading causes, with multiple statistical regression, assessing the most significant factors that influenced the CDW generation. The class A CDW can be recycled and reused in construction sites, according to national regulations. Thirty construction sites in Curitiba composed the database. Eleven variables with high significance were diagnosed, and the main challenges were discussed. The results revealed expected and unexpected correlations, such as the more experienced and trained was the labor force, the less CDW were produced, and the higher the quality standard of the building, the more CDW were generated. The lack of a public, easy to access database and data treatment were also challenges that had to be overcome. Knowledge about the construction waste generation leading causes is essential to ensure the correct destination, a more sustainable construction industry, good life quality and a sustainable city.

Keyword Construction waste management · Multiple linear regression · Quantification methodology · Waste segregation · Reuse

L. Kusunoki (✉) · E. F. Gobbi · P. Charvet
Programa de Mestrado em Meio Ambiente Urbano e Industrial (PPGMAUI), Universidade Federal do Paraná (UFPR), Centro Politécnico - Av. Cel. Francisco H. dos Santos, 100, Curitiba, PR, Brazil
e-mail: leikusunoki@hotmail.com

E. F. Gobbi
e-mail: eduardo.felga@gmail.com

P. Charvet
e-mail: pchalm@gmail.com

© Springer Nature Switzerland AG 2020
W. Leal Filho et al. (eds.), *International Business, Trade and Institutional Sustainability*, World Sustainability Series,
https://doi.org/10.1007/978-3-030-26759-9_15

255

1 Introduction

CDW comes from materials such as ceramics, concrete, mortar, soil, gravel, sand, cement, lime, plastic, conductors, paper, and paperboard, glass, gypsum, styrofoam, metals, cement bags, paint cans, paints, solvents, oils, contaminated materials, asbestos, among others.

The factors that contribute to the waste to be pointed as an environmental issue due to the subclasses being categorized as hazardous waste (CONAMA n° 307 2002), by the accumulated volume, they represent more than 50% of the total urban solid residues generated annually in Brazil and inadequate final disposition, which can often be noticed on undeveloped terrain, pedestrian walkways, river streams and permanent preservation area (PPA), causing soil and water contamination, human health risk and environmental damage.

The first step to an effective construction waste management is to estimate which residue subclasses will be generated, in which steps of the constructive process it can happen, and the final volume accumulated (Wu et al. 2014). However, the scarcity of methodology and data for the development of estimates means that the forms of management that need to be elaborated for formalization before public agencies are difficult to apply in practice.

The construction waste management assists in the quality control of the construction site and enables verification of improvement actions implementation, such as training and occasional failures. All the advantages of applying this tool tend to reduce the waste of material, contain the aggression to the environment and generate budget savings.

Some publications sought to prove the economic return on applying the CDW management (Zanna et al. 2017), while others tried to develop methodologies to estimate CDW generation considering several factors with high significance to produce the residue. The methodology developed by Nagalli (2012, 2014) considered the factors of training and experience of the workforce, the construction process, history of specific losses, internal monitoring and the risk of delay in the schedule according to the activities developed.

Nagalli and Carvalho (2018) used the linear regression technique to obtain a regression model with the variables constructed area, masonry type, size of the team related to the needs, monitoring and workforce.

Data from eighteen buildings constructed between 2008 and 2013, in Porto Alegre and Metropolitan Region of Rio Grande do Sul State, were used in the study conducted by Dias (2013). By the linear regression, the regression equation was obtained, which calculates the total waste generated in construction sites.

The dependent variable used by Dias (2013) and Kern et al. (2015) is the total CDW generation of the project. While the independent variables tested in the model were: total area; floor area of each apartment; the number of floors; construction system; CDW reuse; the number of apartments on the floor/number of floors; density of the wall; EIC, which is the economic compaction index of the project. This equation is for the exclusive use of vertical buildings with reinforced concrete structure (Dias 2013; Kern et al. 2015).

Caetano et al. (2018) increased the number of sample elements used by Dias (2013) and Kern et al. (2015). One data excluded by Dias (2013) presented a constructive system in structural masonry (concrete). The authors added three more data of structural masonry (concrete), which influenced the variable "constructive system" and added a total of twenty-two sample elements.

The three data added were tested in the model obtained and presented a smaller error than the equation of Dias (2013), considering that the objective of the study was to model a regression equation to estimate the residues generated in vertical buildings with structural masonry (concrete) construction.

Thus, this work aims to identify the main perceived challenges to estimating CDW production at construction sites, to the applicability of the construction waste management at the construction site and also to the effective waste management in Curitiba.

2 Method

All data obtained to this work development happened by contacting construction companies, environmental companies, and public agencies, by e-mail, telephone or in person, from December 2017 to June 2018. The names of the contacts, companies, and enterprises will not be mentioned in this work and will be renamed by ordinal numerals.

After the database was reassembled, it was defined the tool for the analysis of the influences of the class A CDW generation. The multivariate analysis predicts the relation between two or more variables by statistical techniques. The variables are random and interrelated, however, it does not present significance in the results when studied individually (Hair et al. 2009).

Linear regression is the best option among multivariate techniques, when the survey has a dependent variable and several independent ones. In other words, when a characteristic, also called a dependent variable, suffers variation because it is being influenced by other characteristics, called independent variables. Linear regression is very common in business decision-making, and it is indicated when results are presented in metric scale. After weighing the independent variables, the equation or regression model, also called the statistical variable (Hair et al. 2009), is formed.

Thus, the multivariate regression technique was considered the best tool for the analysis of the influential variables in the CDW generation. The data processing was performed by the IBM SPSS Statistics program, version 2.3 (IBM 2015).

After all the linear regression technique verifications were checked (normality, outliers, significance, residues, the determination coefficient, Durbin-Watson, variance and collinearity), the equation was obtained (Field 2009). Then, the sample data used in the model were tested in the obtained equation, the results were compared with the data collected and the error was calculated, in percentage.

This work analyzed only class A waste due to the high representativeness of this class when compared to the total CDW generated at construction sites. According to the National Industrial Learning Service (SENAI 2014), approximately 70% of the CDW accumulated in construction sites are class A waste. Class A waste can be recycled and reused in the same or another construction site. However, soil residues (that are class A CDW) were disregarded from the study, as they are related to the location of the property, topography and foundation design.

The enterprises used as sample data in the study are located in Curitiba, Parana State, in South of Brazil and have followed the current legislation. The buildings could be for residential and/or commercial purposes. Infrastructure construction sites were not considered in this work.

The partial Construction and Demolition Waste Management Report (CDWMR) data were discarded from the study because they did not fully represent the construction site. Only reports from finished buildings were used as data. The sample elements considered were judged to be reliable and close to reality.

3 Results and Discussion

The first stage of the database survey is the definition of the independent and dependent variables. The dependent variable is the generation of class A CDW in buildings constructed in Curitiba and the independent variables are presented in Table 1.

Thus, the database was composed of thirty sample elements provided by a public agency, an environmental area company and five construction companies operating in the municipality of Curitiba. The quantitative variables of the database are presented in Table 2.

Table 3 describes the variables X11–X20.

Partial residue graphs or residuals dispersion diagrams represent the effect that the independent variable exerts on the dependent variable.

The definitions of learning curve and economies of scale can be a possible explanation for the variables X3 and X7 are negatively related to the generation of residues, the higher the number of blocks or units/floors in the project, the lower the Class A CDW generation.

Economics of scale refers to the evolution in the ability to perform a certain activity at a low cost because it is carried out on a large scale for a certain period. In turn, the learning curve attributes cost reduction to accumulated experience, independently of the activity (Besanko et al. 2013).

The repetition of pavements or buildings tends to train the workforce, causing the reduction of CDW generation to occur along the work schedule. Figure 1a, b show the decreasing trend an accumulation of residues as higher the number of buildings or units/floors.

The variables area/unit (X6), units (X5) and schedule (X8) had a strongly positive relation, that is, higher the area/unit, the number of units in the project and the period of construction, the tendency is that the CDW generation grow.

Table 1 Description of all variables analyzed

Variable	Definition	Type	Description
X1	Constructed área (m^2)	Quantitative	Total constructed área
X2	Floors	Quantitative	Number of floors
X3	Buildings	Quantitative	Number of buildings
X4	Floors/building	Quantitative	Number of floors per number of buildings
X5	Units	Quantitative	Number of units
X6	Constructed área per unit (m^2)	Quantitative	Constructed area per number of units
X7	Units/floor	Quantitative	Number of units per number of floors
X8	Schedule (months)	Quantitative	Time to finish the construction
X9	Constructed área/month	Quantitative	Constructed area per time of construction
X10	CDW reuse	Quantitative	CDW volume reuse
X11	Residencial/Commercial building	Qualitative	1-Commercial; 2-Both; 3-Residential
X12	Masonry type	Qualitative	1-Concrete blocks; 2-Ceramic blocks
X13	Constructive standard	Qualitative	1-Luxury; 2-Medium; 3-Low standard
X14	Monitoring	Qualitative	1-Permanent; 2-Regular; 3-Not happen
X15	Organization	Qualitative	1-Great; 2-Regular; 3-Poor
X16	CDW segregation	Qualitative	1-By subclass; 2-By class; 3-Not happen
X17	Green certification	Qualitative	1-Yes; 2-No
X18	Workforce experience	Qualitative	1-Experienced; 2-Not experienced
X19	Workforce needed	Qualitative	1-Compatible with the need; 2-Smaller
X20	Raw material storage	Qualitative	1-Great storage; 2-Bad storage
*	Class A CDW generation (m^3)	Dependent variable	

The area/unit and the schedule show positive growth because as long the execution period and the area executed, the tendency is for the waste to be generated proportionally. The exception can occur when punctual actions are applied at the construction site. However, in this study's database, this proportion appears to be growing.

The number of units is showing positive growth and it can be explained by the requirement of execution of more dividing walls in the delimitations of each unit.

Figure 2a–c demonstrate the increasing trend the accumulation of residues the

Table 2 Quantitative variables database

Data	Variable Dependent CDW_A (m³)	X1 Area (m³)	X2 Floors (un)	X3 Buildings (un)	X4 Floors/Buildings	X5 Units (un)	X6 Area/Units (m²)	X7 Units/Floor	X8 Schedule (months)	X9 Area/Month (m²/month)	X10 Reuse (m³)
1	29.00	4388.41	11	1	11	31	141.56	2.82	22	199.47	1.00
2	50.00	3740.14	14	4	3.5	64	58.44	4.57	19	196.85	10.00
3	95.00	3512.32	8	1	8	96	36.59	12.00	13	270.18	0.00
4	123.41	2945.93	9	1	9	36	81.83	4.00	24	122.75	123.41
5	157.50	5592.06	12	2	6	62	90.19	5.17	23	243.13	59.60
6	160.00	4069.14	5	1	5	22	184.96	4.40	24	169.55	160.00
7	172.00	16,503.75	56	14	4	224	73.68	4.00	20	825.19	130.00
8	181.50	12,382.06	16	1	16	16	773.88	1.00	24	515.92	0.00
9	225.00	9620.65	17	1	17	105	91.63	6.18	45	213.79	0.00
10	228.00	7582.01	10	1	10	46	164.83	4.60	27	280.82	40.00
11	277.75	9793.15	11	2	5.5	64	153.02	5.82	29	337.69	0.00
12	316.80	10,976.81	36	9	4	144	76.23	4.00	22	498.95	83.20
13	328.00	7353.96	32	8	4	128	57.45	4.00	15	490.26	0.00
14	386.00	2808.39	7	1	7	28	100.30	4.00	32	87.76	0.00
15	390.60	24,773.35	112	28	4	448	55.30	4.00	18	1376.30	0.00
16	433.55	4620.60	8	1	8	30	154.02	3.75	19	243.19	0.00
17	533.00	33,879.77	27	1	27	240	141.17	8.89	48	705.83	30.00
18	550.00	14,031.48	15	1	15	244	57.51	16.27	44	318.90	0.00
19	600.00	6157.05	11	1	11	14	439.79	1.27	44	139.93	15.00

(continued)

Table 2 (continued)

Data	Variable Dependent CDW_A (m³)	X1 Area (m³)	X2 Floors (un)	X3 Buildings (un)	X4 Floors/Buildings	X5 Units (un)	X6 Area/Units (m²)	X7 Units/Floor	X8 Schedule (months)	X9 Area/Month (m²/month)	X10 Reuse (m³)
20	634.20	7485.78	9	1	9	30	249.53	3.33	40	187.14	0.00
21	639.00	5933.76	10	1	10	29	204.61	2.90	18	329.65	0.00
22	712.40	22,588.81	26	2	13	288	78.43	11.08	28	806.74	0.00
23	712.52	13,653.99	21	1	21	64	213.34	3.05	40	341.35	0.00
24	735.00	7419.23	20	2	10	14	529.95	0.70	44	168.62	20.00
25	795.00	3950.78	9	1	9	35	112.88	3.89	40	98.77	0.00
26	1018.85	34,593.05	23	2	11.5	539	64.18	23.43	40	864.83	0.00
27	2035.00	48,331.48	51	2	25.5	390	123.93	7.65	49	986.36	80.00
28	2257.50	16,890.75	10	2	5	112	150.81	11.20	44	383.88	0.00
29	2280.42	10,594.50	13	1	13	38	278.80	2.92	26	407.48	0.00
30	3339.87	73753.76	42	3	14	705	104.62	16.79	35	2107.25	0.00

Table 3 Qualitative variable database

Data	Dependent Variable CDW_A (m³)	X11 Comercial/Residential	X12 Masonry type	X13 Constructed standard	X14 Monitoring	X15 Organization	X16 CDW segregation	X17 Green certification	X18 Work force experience	X19 Work force needed	X20 Raw material storage
1	29.00	2	2	2	2	2	2	2	2	1	1
2	50.00	3	1	3	1	2	2	2	1	1	1
3	95.00	3	1	2	1	1	1	2	1	1	1
4	123.41	3	2	2	1	2	1	2	1	1	1
5	157.50	3	2	3	1	1	2	2	1	2	2
6	160.00	3	2	2	1	1	2	2	1	1	1
7	172.00	3	1	3	1	1	2	2	1	1	1
8	181.50	1	1	1	1	1	1	1	1	1	1
9	225.00	1	2	2	1	1	2	2	2	1	1
10	228.00	2	2	2	1	2	1	2	1	1	1
11	277.75	3	1	2	2	2	1	2	1	2	1
12	316.80	3	1	3	1	1	2	2	1	1	1
13	328.00	3	1	3	1	1	1	2	1	1	1
14	386.00	3	2	1	1	2	2	2	1	1	2
15	390.60	3	1	3	1	1	1	2	1	1	1
16	433.55	1	1	1	1	1	2	2	1	1	1
17	533.00	1	1	1	1	1	1	1	1	1	1

(continued)

Table 3 (continued)

Data	Variable										
	Dependent	X11	X12	X13	X14	X15	X16	X17	X18	X19	X20
	CDW_A (m^3)	Comercial/Residential	Masonry type	Constructed standard	Monitoring	Organization	CDW segregation	Green certification	Work force experience	Work force needed	Raw material storage
18	550.00	1	2	2	1	1	2	2	1	1	1
19	600.00	3	2	1	2	2	2	2	1	1	1
20	634.20	3	2	1	2	2	2	2	2	1	1
21	639.00	3	1	3	1	1	2	2	3	1	1
22	712.40	2	2	2	1	1	2	2	1	1	1
23	712.52	3	2	1	1	1	2	2	1	1	1
24	735.00	3	2	1	2	2	2	2	1	1	1
25	795.00	3	2	1	1	2	2	2	2	1	1
26	1018.85	2	2	2	1	1	2	2	1	1	1
27	2035.00	2	1	1	1	1	1	1	2	1	1
28	2257.50	3	2	2	1	1	2	2	1	1	1
29	2280.42	1	2	1	1	1	2	2	1	1	1
30	3339.87	2	2	2	1	1	2	2	1	1	1

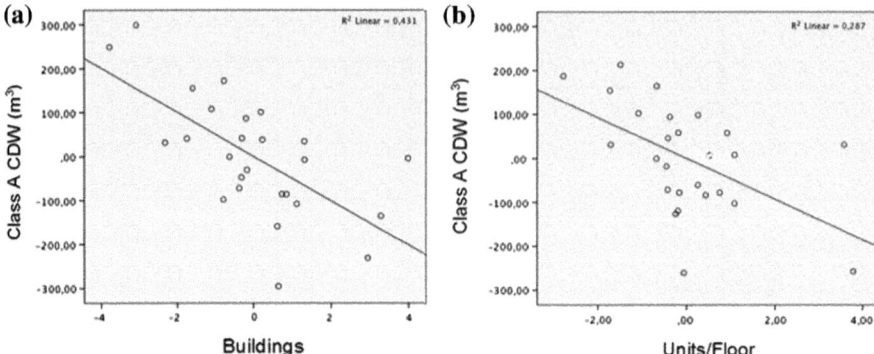

Fig. 1 **a** Buildings, **b** units/floor

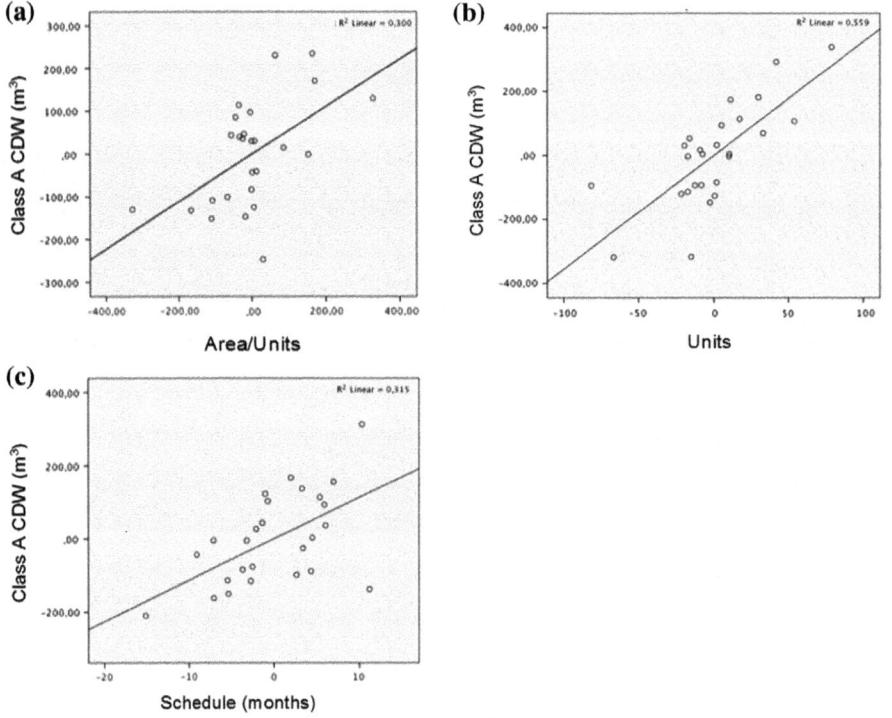

Fig. 2 **a** Area/units, **b** units and **c** schedule

greater the number of units, months of execution schedule or area per unit built in the enterprise.

The construction system (X16) was classified in structural masonry (1) and conventional masonry (2). The publicity of the structural masonry numbers as one of its positive points the CDW smaller generation when compared to the conventional masonry. However, the dispersion of the residues had the opposite result.

Perhaps the explanation for this decrease is that of the eleven data that used structural masonry, eight are projects of residential use, which would represent an increase of walls building and partitions and linear area executed, possibly generating an increase in the volume of CDW generated.

The constructive standard variable (X13) presented a strongly negative relation, demonstrating that as higher the project standard, more class A CDW is generated at the site. One hypothesis for this characteristic may be the low concern with the cost of the inputs installed in the building, because the enterprise is considering the high investment for the execution of the service, or, for the greater diversity of materials in the construction site.

The classification of variable X13 followed the standard luxury or high standard (1), normal or average standard (2) and low or popular standard (3).

The monitoring variable (X14) presented a strong negative relation. When the inspection took place permanently (1), the enterprise generated more waste than when inspection occurred sporadically (2). The option that there was no inspection at the site (3) was not answered.

The fact that the inspection is constantly being carried out can represent a greater control and, consequently, the data of residues generation can present larger volume than the sites that are not being actively supervised.

Surveillance is just a tool to verify that the workforce is performing the task correctly. The actions that are taken after the inspection verify problems in the execution is what can improve the reduction of accumulated CDW volume.

Figure 3a–c show the decreasing trend in residues accumulation as higher the constructional standard when using concrete blocks masonry instead of conventional masonry and when the supervision is permanent.

The option for the variable building use (X11) occurred because of the difficulty of the companies to provide the linear meter of walls executed of each enterprise. Thus, it was decided to use a nonmetric variable.

The variable building use was classified according to the purpose, that is, commercial (1), residential and commercial (2) and residential (3). The distribution presented an increasing characteristic. Residential properties usually have more dividing walls, consequently a tendency to generate more CDW.

The variable green certification is another not metric one, which comprised two possibilities of classification, the goal for certification (1) and projects that did not aim at it (2). The distribution shows that the enterprises that had some green certification generated less CDW at the construction site.

The workforce experience variable presented increasing growth. If the team was trained and experienced (1), the CDW generation was smaller than when the team did not have experience or training (2).

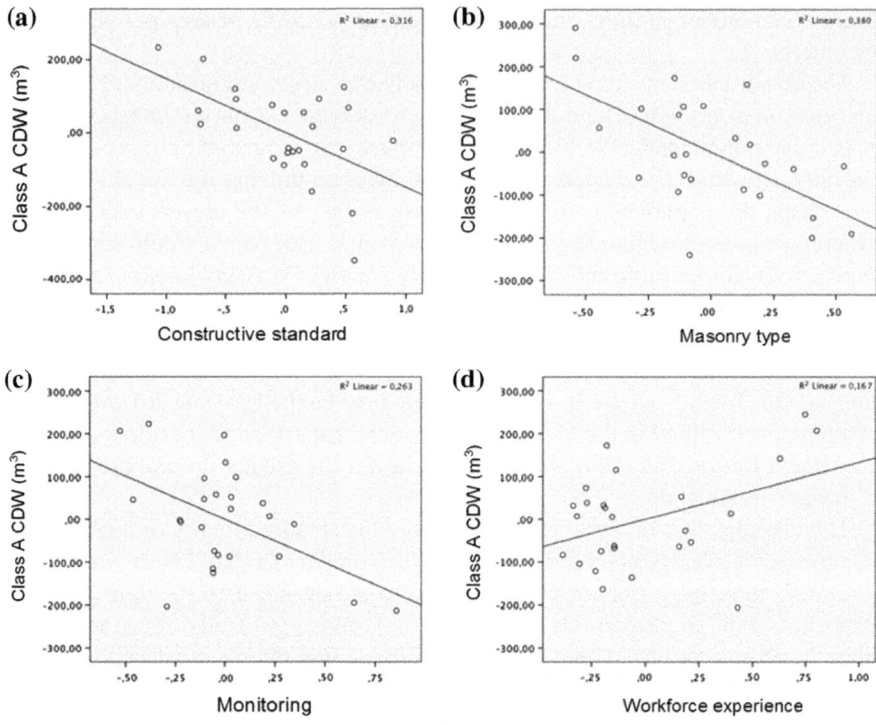

Fig. 3 **a** Constructive standard, **b** masonry type, **c** monitoring, **d** workforce experience

Figures 3d and 4a, b demonstrate the increasing trend in wastes accumulation when labor is not trained and/or experienced, the enterprise has the purpose of residential use and when the enterprise does not seek to obtain green certification.

Table 4 presents a summary of the variables effectively used in the regression

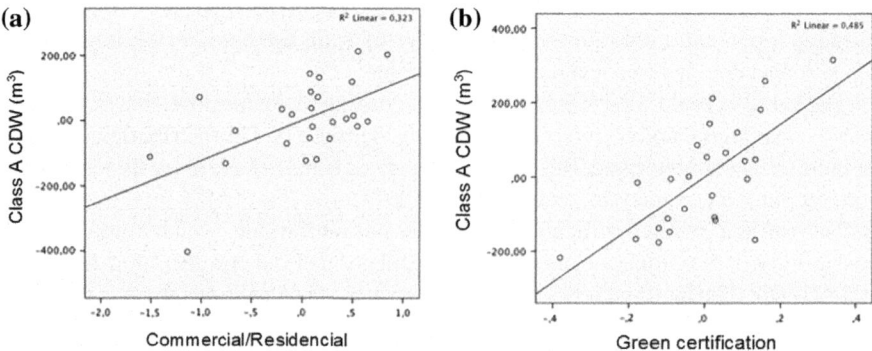

Fig. 4 **a** Comercial/Residencial and **b** green certification

Table 4 Analysis of variables and their CDW generation influence

Growth	Variable	Tendency
Much more	Number of buildings (X3)	Smaller the CDW generation
Much more	Number of units (X5)	Bigger the CDW generation
Much more	Area/units (X6)	Bigger the CDW generation
Much more	Units/floor (X7)	Smaller the CDW generation
Much more	Schedule, in months (X8)	Bigger the CDW generation
Much more	Residential use (X11)	Bigger the CDW generation
Much more	Ceramic blocks masonry (X12)	Bigger the CDW generation
Much more	Luxury standard (X13)	Bigger the CDW generation
Much more	Monitoring permanent (X14)	Smaller the CDW generation
Much more	Green certification (X17)	Smaller the CDW generation
Much more	Experienced workforce (X18)	Smaller the CDW generation

model and their respective trends.

After the verification of normality, the absence of outliers, determination coefficient, variance, collinearity, significance, Durbin-Watson, correlations and residues, the coefficients were obtained and the model that presented the best fit to the requirements of the multivariate regression is presented in Eq. 1.

$$
\begin{aligned}
\text{Class A CDW} = {}& -600{,}697 + (-45{,}963 \cdot X3) + 3352 \cdot X5 \\
& + 0.564 \cdot X6 + (-41{,}741 \cdot X7) + 12{,}655 \cdot X8 \\
& + 126{,}934 \cdot X11 + (-270{,}283 \cdot X12) + (-159{,}505 \cdot X13) \\
& + (-226{,}445 \cdot X14) + 711{,}483 \cdot X17 + 103{,}032 \cdot X18,
\end{aligned} \quad (1)
$$

In which:

"Class A CDW" is the estimated volume of waste to be generated in a construction site;

"X3" is the total of buildings;

"X5" is the total number of units constructed;

"X6" is the total area constructed per unit, in m2;

"X7" is the total number of units per floor;

"X8" is the schedule, in months;

"X11" is the use of the enterprise [1 is commercial use, 2 is commercial and residential use (mixed) and 3 is residential use];

"X12" refers to the constructive system of the enterprise, [1 is structural masonry (concrete blocks) and 2 is conventional masonry (ceramic blocks)];

"X13" is the constructive standard of the enterprise, (1 is high standard or luxury, 2 is normal or average standard and 3 is popular or low standard);

"X14" is the occurrence of inspection at the construction site, (1 is permanent, 2 is sporadic and 3 is when there is no inspection);

"X17" refers to the search for green certification for the enterprise, (1 is when the enterprise seeks to certify and 2 is when there is no interest in certification);
"X18" refers to the manpower team, (1 is trained and experienced and 2 is untrained and inexperienced).

The model was used to estimate the class A CDW generation of the sample data used in the regression model. Table 5 presents the results calculated with Eq. 1.

Table 5 Real values X estimated values

Data	Real	Estimated	Mistake (%)
1	29.00	68.06	−135
2	50.00	187.15	−274
3	95.00	177.42	−87
4	123.41	213.84	−73
5	157.50	47.14	70
6	160.00	203.01	−27
7	172.00	299.83	−74
8	185.00	236.90	−28
9	225.00	488.57	−117
10	228.00	178.59	22
11	277.75	341.05	−23
12	316.80	290.60	8
13	328.00	194.36	41
14	386.00	434.53	−13
15	390.60	360.09	8
16	433.55	358.49	17
17	533.00	480.94	10
18	550.00	353.76	36
19	600.00	632.07	−5
20	634.20	576.06	9
21	639.00	494.08	23
22	712.40	653.61	8
23	712.52	759.98	−7
24	735.00	658.44	10
25	795.00	695.38	13
26	1018.85	1104.03	−8

4 Discussion

The regression model obtained presented seven sample data with a difference of less than 10% between the values calculated and provided by the companies, whereas twelve data showed a difference smaller than 17% and eighteen data showed a difference smaller than 36%, that is, 70% of the elements.

By observing the data in Table 5, we can note the data in bold, which are those that presented greater variance between the actual and calculated values. From these data, the calculated values were higher than the real values, with the exception of data 5.

The coefficients that contributed to making the final value smaller than the real are blocks (X3), unit/pavement (X7), construction standard (X13), monitoring (X14) and masonry type (X12). The characteristics of the data 5 for these variables tend to intensify the reduction in the final value obtained with the regression equation since the data of variables X3, X7, X12, X13, and X14 are among the highest in the database.

The effective variables of the model have high significance, so it is suggested that the set of variables influenced the discrepancy between the actual and calculated values. However, the variable green certification (X17) must have been the one that most influenced the calculated values well above the real ones, considering that they have the highest coefficient and tend to increase the calculated value.

The green certification variable is the one that causes the greatest difference in the final values when the coefficients are analyzed since the characteristic of searching (1) or not the certificate (2) is multiplied by the coefficient of highest value (711.483). This context contributes so that the calculated values are above the expected values, of some enterprises.

In addition to the individual analysis of each model variable and the regression model tests, it is important to highlight the main challenge faced in the development of this work, which was the elaboration of the database, especially considering that the legalization of buildings in Curitiba is conditioned to the delivery of documents with CDW management data, in physical ways. The lack of a public database and digital format can be seen as a point to be improved in waste management at the municipal level.

5 Conclusion

When the construction companies ensure a correct CDW management, possibly the need for raw materials will reduce, the CDW can be recycled and reused, and the CDW that cannot be recycled would have proper disposal, which is essential for a more sustainable city, industry, and for good life quality.

The first step to have an effective CDW management is to have an electronic historic of all construction sites by every company and then by city, compound a database that can be easily accessed. It would help to estimate and identify which stage can be improved.

Finally, this work enables to analyze some significant factors in the generation of waste, to obtain an equation that estimates the CDW volume and to point out the main difficulty in the elaboration of the database, hence the goal was satisfactorily achieved.

Acknowledgements The authors thank the Master's Program in Urban and Industrial Environment (PPGMAUI-UFPR), a partnership between the Federal University of Paraná (UFPR), the University of Stuttgart and the National Service of Industrial Apprenticeship (SENAI-PR).
We are thankful for the suggestions given by professors M.Sc. Sandra Mara Pereira de Queiroz and Dr. André Nagalli. Our gratitude also to the companies and institutions that shared the data used in this study.

References

Besanko D, Dranove D, Shanley M, Schaffer S (2013) Economics of strategy, 6th edn. Wiley, Hoboken

Caetano MO, Fagundes AB, Gomes LP (2018) Modelo de regressão linear para estimativa de geração de RCD em obras de alvenaria estrutural. Ambiente Construído 18(2):309–324 (in Portuguese)

Dias MF (2013) Modelo para estimar a geração de resíduos na produção de obras residenciais verticais. Master in Civil Engineering, University of Vale do Rio dos Sinos, São Leopoldo (in Portuguese)

Field A (2009) Descobrindo a estatística usando o SPSS, 2nd edn. Artmed, Porto Alegre (in Portuguese)

Hair JF, Black WC, Babin BJ, Anderson RE, Tatham RL (2009) Análise Multivariada de Dados (Multivariate data analysis), 6th edn. Bookman, Porto Alegre

IBM (2015) IBM SPSS statistics, version 23. IBM

Kern AP, Dias MF, Kulakowski MP, Gomes LP (2015) Waste generated in high-rise buildings construction: a quantification model based on statistical multiple regression. Waste Manage 39:35–44

Nagalli A (2012) Quantitative method for estimating construction waster generation. Electron J Geotech Eng 17:1157–1162

Nagalli A (2014) Gerenciamento de Resíduos Sólidos na Construção Civil, 1st edn. Oficina de Textos, São Paulo (in Portuguese)

Nagalli A, Carvalho KQ (2018) Model for estimating construction waste generation in masonry building. Waste Resour Manage 1–9

SENAI – PARANÁ (Serviço Nacional de Aprendizagem Industrial do Paraná) (2014) Relatório Técnico – Plano de Logística Reversa Resíduos de Construção Civil. Curitiba (in Portuguese)

Wu Z, Yu ATW, Shen L, Liu G (2014) Quantifying construction and demolition waste: an analytical review. Waste Manage 34:1683–1692

Zanna CD, Fernandes F, Gasparine JC (2017) Solid construction waste management in large civil construction companies through use of specific software—case study. Acta Scientiarum Technology 39(2):169–176

A Methodology for Sewage Network Maintenance Toward the Fulfillment of Sustainable Development Goals

Luciano Rodrigues Penido, Karen Juliana do Amaral,
Regina Maria Matos Jorge, Jörg Wolfgang Metzger, Jefferson Skroch
and Rafael Cabral Gonçalves

Abstract Domestic effluents collected from urban areas are transported along extensive sewer networks that are assumed to operate as watertight systems. Such operability is ensured by maintenance routines, both preventive and corrective, that are capable of detecting weak points where loss of effluents might occur and affect bodies of water. Since urban rivers are the destination of such organic loads, they may be used as indicators of potential weak points along the sewage network. The implementation of initiatives whose goal is to avoid such input of pollutant loads is in line with at least three of the 17 Sustainable Development Goals from the 2030

L. R. Penido
Industrial and Urban Environment, Federal University of Paraná (UFPR), Curitiba, Paraná, Brazil
e-mail: luciano.penido@gmail.com

Remote Sensing, INPE, São José dos Campos, Brazil

L. R. Penido · J. Skroch · R. C. Gonçalves
Sanepar Sanitation Company, Rua Engenheiro Rebouças, 1376, Curitiba, Paraná, Brazil
e-mail: jeffersons@sanepar.com.br

R. C. Gonçalves
e-mail: rafaelcg@sanepar.com.br

K. J. do Amaral (✉) · J. W. Metzger
Institut für Siedlungswasserbau, Wassergüte- und Abfallwirtschaft, Universität Stuttgart,
Stuttgart, Büsnau, Germany
e-mail: karenjamaral@gmail.com

J. W. Metzger
e-mail: joerg.metzger@iswa.uni-stuttgart.de

K. J. do Amaral · R. M. M. Jorge · J. W. Metzger
Urban and Industrial Environment, Federal University of Paraná, Av. Coronel Francisco Heráclito
dos Santos, 100, Curitiba, Paraná, Brazil
e-mail: rjorge@ufpr.br

R. M. M. Jorge
Department of Chemical Engineering, Federal University of Paraná (UFPR), Curitiba, Paraná,
Brazil

Food Engineering, Federal University of Paraná (UFPR), Av. Francisco Hoffman dos Santos, s. n.,
Curitiba, Paraná 81530-900, Brazil

© Springer Nature Switzerland AG 2020   271
W. Leal Filho et al. (eds.), *International Business, Trade and Institutional
Sustainability*, World Sustainability Series,
https://doi.org/10.1007/978-3-030-26759-9_16

Agenda. The assumption that rivers may be used as indicators forms the basis of the Urban River Revitalization Program (PRRU) carried out since 2011 by the Water Resources Management division of the Sanitation Company of Paraná—Sanepar. Since its inception, over 3410 weak points have already been located and repaired within the Metropolitan Region of Curitiba, which covers about 98 river basins including the Belém River, one of the most important, located completely inside the urban area. In its northern basin 309 weak spots have been corrected, resulting in improved appearance and odor, as well as higher levels of dissolved oxygen.

Keywords Sewage collection network · Sustainable development goals · Maintenance · Effluent leakage

1 Introduction

Basic sanitation is a key factor for achieving a high level of quality in environmental and social management, in which improvements in the environment are most notably achieved by sanitary sewage systems (SSS) that capture and quickly transport domestic effluents produced in inhabited areas to Sewage Treatment Plants (STP) in order to reduce their potential for contamination, allowing them to be subsequently discharged into bodies of water.

In the absence of a SSS, the impact on the environment is perceived mainly in rivers that run through urban and industrial areas, in which the water receives effluents of all types. Further negative impact arises from the inadequacy of programs that aim to improve the environment, which are usually incapable of pinpointing and monitoring pollutants at their point of origin.

The preferred sewage collection system in Brazil is that of a separate sewage system, in which domestic effluents are managed in a context that is kept separate from that of rainwater management. Whereas the former are collected by a sewage sanitation system, rainwater is diverted directly into the river network by means of rainwater galleries, receiving no treatment. Each system is meant to be watertight, the public service water and rainwater remaining separate.

In accordance with the concept of universalization of sanitation, sewage sanitation systems should collect as much of the effluents produced by residential, commercial and institutional zones as is possible, as well as those produced in the sanitation installations of industrial units.

Another important issue concerning sewage disposal is the support from public policies. According to Giller et al. (2018), the SDGs focus the attention of public policies on resolving worldwide challenges, even though full control has not yet been achieved over the formulation of indicators that are capable of measuring the progress that has been made.

Tratabrasil (2018) observed that in Brazil, sewage treatment is the least universalized indicator, since 44 of its 100 most populous municipalities collect less than 80% of the sewage they produce, and no more than 20% of the collected sewage is treated. According to Malik et al. (2015), this discordant state is due to a high rate of urban growth, to an infrastructure for sewage treatment whose dimensions are too small to handle the total amount of the sewage it receives, or else to financial resources that are inadequate for maintaining the operational capacity of the sewage treatment plant.

According to Malik et al. (2015), untreated sewage affects the quality of both river water and human health, resulting in environmental impacts such as the rapid growth of algae, the eutrophication of lakes, the poisoning of fish, and the contamination of the water itself by drugs and other chemical substances.

Although universalization has the advantage of consolidating the extensive sewage collection and transportation pipes from the domestic sewage networks, it must be kept in mind that civil works have a limited useful lifespan and are vulnerable to various types of harmful forces, such as wear, fatigue and events that give rise to weak points, irregularities in the Wastewater Collection System (WCS) reported by DSD (1995), such as breaches in the walls of the pipelines and sections such as the manholes (MHs), that compromise the watertight quality of the system, resulting in effluent leakage.

Figure 1 shows the most commonly occurring weak points as catalogued by Amick and Burgess (2000), including joints between the parts (1), connection of the branches with the MH (2), internal walls of the MH (3), as well as longitudinal cracks in the sewer main (4) and fissures due to lateral forces (5).

Bertrand-Krajewski et al. (2005) define the weak points as infiltration, leakage, and overflow. Infiltration occurs when rainwater or water from the water table enters the ducts, diluting the effluents and thus hindering treatment by the STP. Leakage occurs when there is an underground loss of effluents through defects in the col-

Fig. 1 Occurrences of weak points in the WCS structures. *Source* Adapted from Amick and Burgess (2000)

lection structure. Overflow occurs when blocked or undersized mains result in the accumulation and above-ground leakage of the effluent.

Such weak points reduce the efficiency of the sewage collection network by "allowing" domestic effluents to remain in inhabited areas, thus frustrating the main goal of sanitation, which is to quickly remove sewage from inhabited areas. This may result in unsanitary conditions in these inhabited areas, thus counteracting at least three of the United Nations Sustainable Development Goals (SDGs) from the 2030 Agenda for Sustainable Development (UN 2015)—"Good Health and Well-Being", "Clean Water and Sanitation", and "Life Below Water".

For example, the issue of "maintenance" of the collection network, is not a heading that occurs frequently among basic sanitation indicators, despite the fact that the existence of weak points, in conjunction with the concept of universalization, makes necessary the fomentation of protocols for the monitoring and maintenance (both predictive and corrective) of the WCS, including the provision of funds and resources in order to carry out necessary interventions.

The current work relates to the Urban River Revitalization Program (PRRU), developed by the Sanitation Company of Paraná (Sanepar), presenting it as a paradigm shift in the management of the sewage collection network in that it takes the quality of the river environment as its point of reference for evaluating the integrity of the WCS, making use of qualitative parameters of the river environment as guides in the search for undesirable weak points.

2 Methodology

The Urban River Revitalization Program (PRRU) was developed at the company SANEPAR in the Water Resources Management division (GHID), which is part of the Environmental Management and whose goal is to provide an accurate evaluation of the sewage collection network, based on the analysis of the rivers that permeate the urban area (Skroch 2013).

Effluents that improperly escape the sewage collection network percolate through the soil by underground drainage until they reach the nearest waterbody, contaminating the waters upon their arrival. In view of this, a river that carries an organic load may indicate a discharge of domestic effluents upstream from the river basin (Skroch 2014).

Thus, this same author considers the river basin to be the main realm of action of the PRRU, with the understanding that the analyses must be carried out on location during initial visits to sections of its main river in order to perform a qualitative evaluation of the river water and measure its Dissolved Oxygen (DO). The DO readings, taken along the course of the river, provide a descriptive curve for this parameter, known as the River Profile (Fig. 2); this provides significant assistance in identifying the presence of an undesirable organic load in its waters.

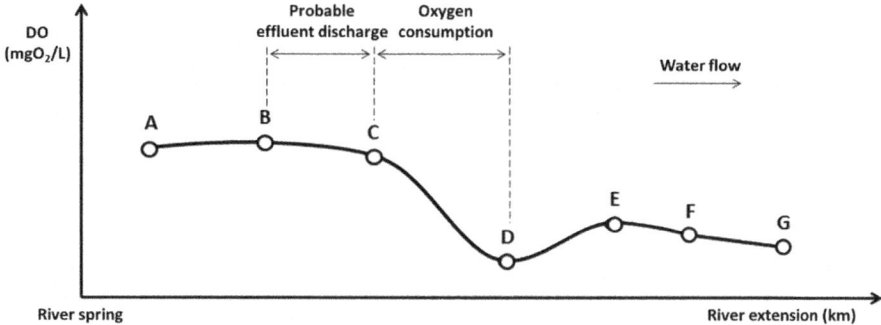

Fig. 2 Variation of oxygen along the course of the river, from its source to its mouth. *Source* Adapted from Skroch (2014)

The analysis proposed by the PRRU can help in the WCS maintenance taking into account the presence of organic load in the water body and its probable origin in the basin (Penido et al. 2017).

According to Skroch (2014), an abrupt variation in DO along the flow of the river may indicate an influx of organic load from effluents upstream, potentially as a result of a weak point in the WCS. In Fig. 2, the descending section of the curve (C-D) suggests a gradual DO consumption, showing a self-purifying behavior in the river of its organic load.

According to the same author, the organic load detected in the river provides the final vertex of the trajectory of the effluents, and all that is needed is to find the initial vertex, indicating the origin of the polluting load. In the case shown in Fig. 2, the discharge probably occurs in the area whose waters drain into the river between sections 'b' and 'c,' thus providing a field of research as deduced from the drainage topography of the waters that flow into this segment of the river (Fig. 3).

2.1 Premises

The analyses conducted by the PRRU are based on the premise that the urban river was a static system during the field research period, exhibiting stable organic load concentrations. Thus, the result of the samplings from the river segments is interpreted as being descriptive of a single point in time, providing a River Profile that functions as a snapshot. For this purpose, the following hypotheses are assumed:

I.   The polluting source produces a continuous disturbance in the river water;
II. The system under analysis exists in a static state;
III. The parameters being measured are in a steady state during the period of time in which the samples are collected.

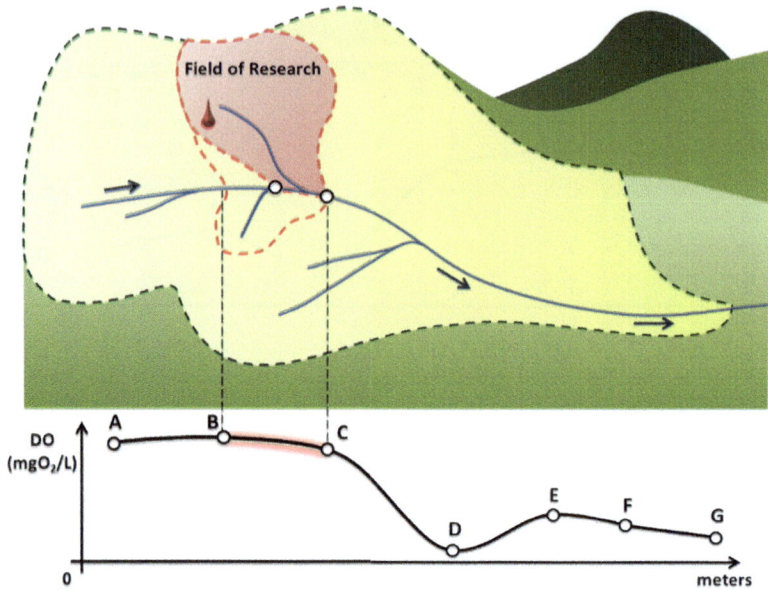

Fig. 3 Field of research as defined by the decline in DO. *Source* Adapted from Skroch (2013)

2.2 Operating Cycle

According to Skroch (2014), the method used by the PRRU is made up of the following steps, as shown in Fig. 4 and described below: "preliminary analysis," "investigation of causes," "maintenance," "verification" and "follow-up."

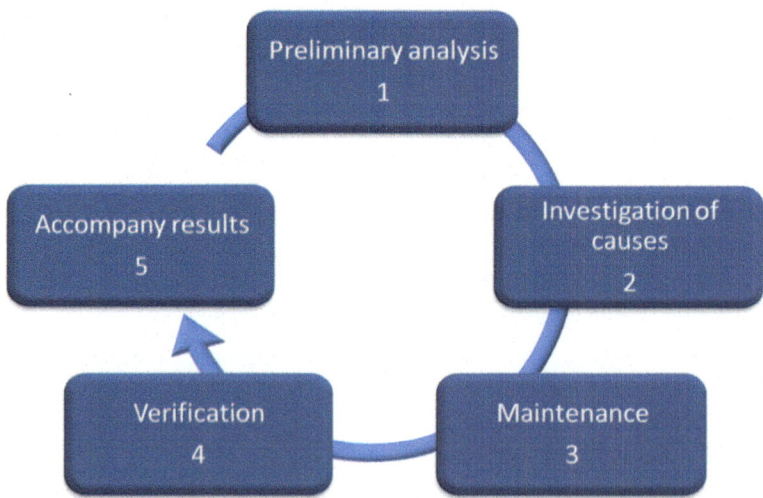

Fig. 4 Cycle of processes of the PRRU. *Source* Adapted from Skroch (2012)

2.2.1 Step 1—Preliminary Analysis

The PRRU begins with the recognition of its area of activity in the river basin, with the aid of the field map containing the WCS, the hydrography and respective sections of the river to be visited. Field activities are then carried out using an automotive vehicle, EPIs, an optical sensor oximeter, deionized water and devices for photographic recording and localization, as well as steel rods (known as "hooks") adapted to fit the manhole cover for its removal and repositioning. The employees should never be allowed to enter the bodies of water, storm water pipes, or the internal manhole chamber.

A DO reading is taken at each section visited, any value inferior to 5 mg O_2/L being reported as "low DO", and all values superior to this as "high DO". The 5 mg O_2/L threshold has been established in Brazil (2005) as the minimum desirable level for rivers of class 2, to be used for human consumption after conventional treatment, the protection of aquatic communities and other more demanding uses.

Of equal importance is the observation of the organoleptic properties of color, turbidity and odor indicated on the field map (Table 1), which are classified as being either "good" or "poor." A river classified as "good" does not contain elements that indicate a significant presence of organic loads. In contrast, a river classified as "poor" has a grayish color and exhibits an unpleasant odor.

Table 1 presents the manner in which the organoleptic aspects observed at the various segments of the river are recorded on the map, the letter "O" indicating the presence of a small quantity of matter particles in suspension, whereas "OO" indicates that the water is transparent. The letter "X" indicates the presence of visible pollution, whereas "XX" indicates turbid or opaque waters.

In connection with the field analyses in this first step of Preliminary Analysis, it should be noted that heavy rains raise the DO levels in the body of water; therefore, DO readings should be take only after 48 h have elapsed after a heavy rain.

Table 1 Classification of the organoleptic aspects observed in the river

Condition	Visual appearance	Description	Smell	Notation on map
Good	Natural	Good, natural crystal waters, river bed is visible	Odors	(OO)
			Foul odor	(O)
Poor	Gray	Grayish, opaque, or turbid water exhibiting a foul odor	Odors	(X)
			Foul odor	(XX)

Source Penido (2014)

2.2.2 Step 2—Investigation of Causes

Once the DO measurements have been concluded and a drop in DO has been identified, a field of research is marked off in the areas located upstream from the sections in which the drop in DO was observed, according to the water drainage topography, previously illustrated in Fig. 3.

The focus of this step is to determine the distance to the polluting event. The solution may be obtained from the DO-aspect binomial, which is an integrated analysis of the organoleptic parameters and the DO, and which is interpreted according to the rules presented in Table 2. Its application imparts agility to the activities of this cause integration step.

Possible causes of the DO variations observed in the river must be explored by visually inspecting the internal chambers of the manholes in order to determine their physical integrity and observe the flow of effluents in order to infer potential weak points in the adjacent branch, which if found will result in the calling of the appropriate maintenance crews.

The information regarding the relative distance of the source of pollution provides the field crew with a parameter that provides guidance concerning which WCS branches should be investigated, that is, which manholes will be inspected. The purpose of the manhole inspection is to identify alterations in the intensity of the flow of effluents by visual comparison.

The comparison of the flow in two adjacent manholes is conducted by removing the manhole cover from the manhole in order to observe with the naked eye the flow of effluents through the half-pipe located at the bottom of the internal chamber (Fig. 5).

As shown in Fig. 6, the effluents flow from the upper manhole (MH_1) to the lower manhole (MH_2) in such a manner that under normal operational conditions the same rate of flow is observed for MH_2 as that observed for MH_1. The comparative analyses of the flow in adjacent MHs may result in one of three outcomes:

Table 2 Interpretation of the DO-aspect binomial

DO-aspect binomial	Notation on map	DO	Analysis
Good condition, associated with low DO	(OO)	≥ 5 mg O_2/L	The polluting event is further away. It is a recovery zone, where the matter has already been consumed, but DO has not yet been recovered
	(O)		
Poor condition, associated with high DO	(X)	<5 mg O_2/L	The polluting event is near. It is a decomposition zone, with visible effects, but without a significant reduction in DO

Source Adapted from Skroch (2014)

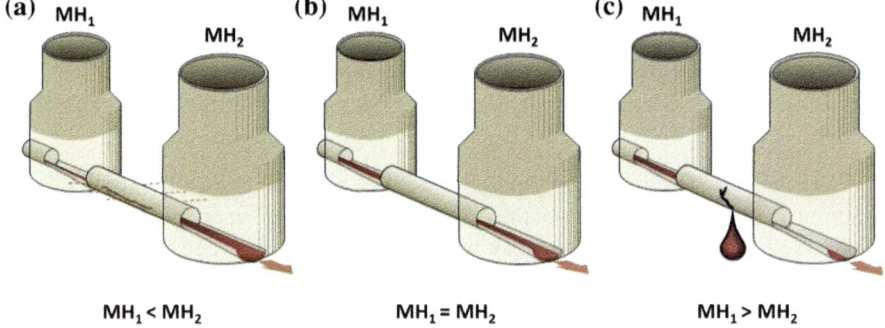

Fig. 5 Flow through adjacent manholes with inflow from subterranean or pluvial water (**a**), under normal conditions (**b**) and with loss of effluents (**c**). *Source* Penido (2014)

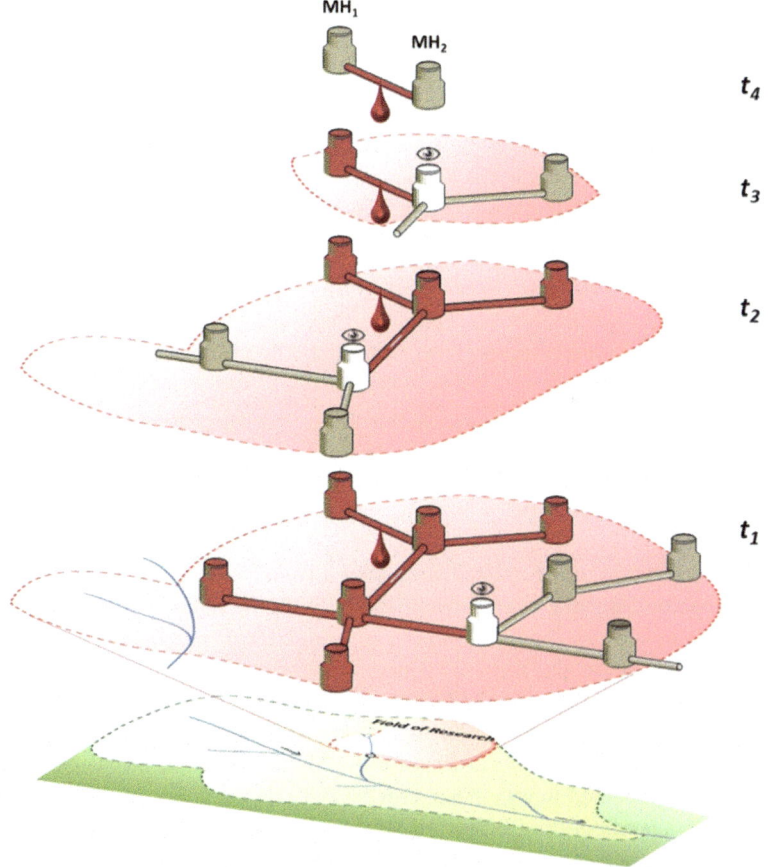

Fig. 6 The diminishing field of research as each MH is accessed. *Source* Penido (2014)

- the flow in MH$_1$ is inferior to that of MH$_2$: the main between the two MHs receives inflow from rainwater, treated water or underground infiltration;
- MH$_1$ and MH$_2$ exhibit equal rate of flow: normal condition;
- the flow in MH$_1$ is superior to that in MH$_2$: there is a loss of flow between the two MHs.

The loss of flow may be confirmed by adding a colorant to the effluent in order to function as a physical tracer from the current in the manhole upstream. If the colorant is observed outside the WCS, a weak point in the branch between the manholes has been identified.

Figure 6 illustrates the process for analyzing the sewer flow and demonstrates how the field of research diminishes as the investigation proceeds through the WCS. By placing the focus on the bifurcation under analysis, while paying attention to the passage of the effluents. Each moment in time t_1, t_2, t_3 and t_4 represents a further step through the WCS structure. The colors given to the MHs correspond to the following designations:

- white: bifurcation in which the flow analysis is conducted;
- beige: the sector of the WCS that has been discarded after having exhibited flow compatible with what is expected for the respective area serviced by the sewer collection;
- red: a sector exhibiting a flow inferior to what is expected for the serviced area.

At the end of each flow analysis, the sector of the WCS appearing in beige is discarded, while the red sector is maintained for analysis. The white manhole is also discarded. Each manhole cover is replaced immediately in order to avoid accidents.

If new weak points have been identified in the collection network, it is concluded that the polluting event is not located in the WCS and must instead come from structures that are not connected to the sewage collection network.

2.2.3   Step 3—Maintenance

Maintenance services are requested whenever a weak point is identified or deduced, or when other undesirable events are observed with the naked eye during the manhole inspection, such as cracked walls or irregular sewer or rainwater connections. Swamped manholes are also reported, since they reveal an obstruction in the branches downstream.

Communication is carried out by means of succinct reports known as sketches, which contain descriptive observations for the identification of the weak point.

Appropriately trained maintenance crews verify the presence of the "weak point" event, and the corresponding branch receives corrective measures. A further investigative method such as tele diagnosis may be applied in order to identify with certainty the branch that must be repaired. Once a corrective action has been carried out, it is reported to the PRRU team. These corrective actions represent concrete advances in item 3.9 of SDG Good Health and Well Being.

2.2.4 Step 4—Verification

The PRRU then returns to the location in order to verify the effectiveness of the corrective measures, and a new field visit is carried out for evaluation of the area that had been affected by the polluting load. Ideally, the cessation of the inflow of organic load should be reflected in a reading of "high DO associated with a good condition." If the results are not satisfactory, a new River Profile survey is carried out in search of polluting events that have not yet been located.

2.2.5 Step 5—Follow-up

A further visit must be made to the location in which the weak point was identified and corrected in order to predict new River Profile surveys, with the understanding that waters of higher quality indicate the effectiveness of the maintenance service. However, it is possible to conclude whether a further polluting event exists upstream, giving rise to a new survey.

According to Sloane and Pröbstl-Haider (2019), volunteers have always been a part of foundations, clubs, organizations, and other institutions related to the conservation of nature and the world environment. Therefore, the strategy adopted for this step of the method was that of forming a team of volunteers; thus, the "Follow-up" step must also rely on the voluntary participation of people who live in the river basin, who are invited to monitor the environmental conditions of the rivers in order to identify potential problems that may occur even after the repairs have been made. By engaging the efforts of the local community in the maintenance of the quality of both the rivers and the WCS, the PRRU broadens the scope of the capillarity of the sanitation company in a manner that is consistent with SGD 6.6b. 4.

Studies involving the participation of the society in environmental issues were carried out by Feszterova and Jomova (2015), using the participation of students inserted in different activities and work in different areas of the environment, in order to expand and develop their knowledge and information regarding the environment. The main strategy used in this case was to the perception and status of the environment in society by popularizing scientific activities among the students, with an emphasis on the awareness of the specific role of the young generation to care for the environment.

Active participation in environmental issues may also extend to local communities, such as in the initiative implemented by the PRRU, in which potential volunteer monitors were invited after a careful screening, and directed by emails that provided information concerning organoleptic properties of the river waters, thus ensuring a follow-up conducted in the form of participative monitoring by citizens who act as environmental agents, in line with SDG 6.6b. Hermes et al. (2004, p.3) states that these volunteers are '*entrusted with responsibilities and obligations that they accept voluntarily in order to carry out the monitoring of the quality of multiple-use water, in the absence of any employment relationship*'. According to Souza and Marques Junior (2001), the participation of the population promotes Environmental

Fig. 7 Seal of participatory monitoring (8 cm diameter), translation: "Urban river revitalization program—you are important for maintaining the water quality of this river basin". *Source* Adapted from Sanepar (2014a)

Education initiatives, since the community begins to participate and interact with the environment in which it lives.

The function of the volunteer is to remain attentive to 'his' river, reporting any alterations in its appearance to the manager of the collection network by email. For this purpose, the seal containing the email address of the Program shown in Fig. 7 was created. It is distributed in the form of a magnetic card so that the volunteer can place it where it may be easily referenced.

The volunteers are carefully selected by Socio-environmental Education agents and must show a visible interest in the recovery and maintenance of the environmental conditions of the river, demonstrate zeal for preservation of the environment in which they reside, have a positive attitude with regard to their environment, be capable of observing the river daily and have access to the Internet in order to report pollution events in the river. The screening of these reports is also a responsibility of the Socio-environmental Education agents.

3   Results and Discussion

The methodology used by the PRRU was conceived by Sanepar, and began to be implemented in the year 2011, at first on a trial basis; however, after providing such objective results with great agility and little financial investment, the methodology was extended to the maintenance divisions and integrated into their routines.

Over the course of the PRRU activity, more than 3410 weak points have been identified in approximately 98 river basins encompassed by the Metropolitan Region of Curitiba. In the Belem river alone, a water body that permeates the central region of the municipality of Curitiba and is densely urbanized along nearly the entire extent of its river basin, more than 309 weak spots have been repaired, producing such surprising results as an elevation of DO levels from less than 1.0 mg O_2/L to a threshold superior to 5.0 mg O_2/L in the northern segment of its course. Figure 8 shows the extent to which this achievement may be perceived visually.

After repairing the weak spots on the basis of the work that has already taken place in Curitiba, an appreciable improvement in the environmental quality of several bodies of water has been observed, both in terms of appearance and smell and in terms of maintenance of DO levels. In some cases, such as in that of the Uvú river, a tributary of the Barigui that encompasses almost the entirety of the restaurants in the Santa Felicidade neighborhood, the river continued exhibiting clear waters and DO superior to 5.0 mg O_2/L for several consecutive months after the weak points had been corrected.

Since heavy rains elevate DO levels in the rivers, the effectiveness of the results of the PRRU is measured during dry periods, as was the case with the Uvú River.

One factor that must be considered is that it is not always possible to maintain the improvements that have been achieved after repairing the weak points signaled by the PRRU. This may be attributed to the potential appearance of weak points as a result of the permanent wear on the structures. The Belém River itself reached

Fig. 8 The arrival of the waters from the Belém River to the lake in São Lourenço Park: on the left, before the PRRU; on the right, after the PRRU. *Source* Sanepar (2014b)

DO levels of at least 5.0 mg O$_2$/L from its source to its mouth during a dry period; however, this condition did not last beyond a one-month period.

An important facet of the PRRU to bear in mind is that the method allows the identification of only the main discharges during the field visit. Once these points are repaired, they may not immediately exhibit improvement due to the existence of further weak points that contribute significant polluting loads. Therefore, further field research may still be necessary in order for the recovery of a body of water to be observable.

In view of the above, the field surveys carried out by the PRRU may be likened to examinations that provide X-rays of the WCS, useful for locating the main areas that require intervention by the maintenance crews.

Among the irregularities reported by the PRRU, 8% involve underground water loss from the public water supply.

It must be noted that the diagnoses obtained using the PRRU methodology may be easily interpreted and located on the field, allowing the maintenance crews to direct themselves to the location while knowing beforehand the precise collection branch that must be serviced.

When reporting the services provided at the request of the PRRU, the maintenance crews identify the type of procedure that was carried out. This provides knowledge regarding the practical significance of the weak points reported by the Program, most of which involve cracks in the connection between the manhole and the sewage collection main. Solicitations for unblocking the sewage collection network are also common. Complex operations generally make up a small percentage of the corrections undertaken by the maintenance crews.

An additional advantage arising from the support provided by the PRRU in the maintenance of the WCS has been the unblocking of ducts before they can overflow, thus reducing their occurrence.

4 Conclusion

The PRRU is innovative in its use of the river as an indicator of the quality and efficiency of the RCI, reducing the possible gap between the sewage collection network infrastructure and the sewage that is actually removed, and bringing positive sanitation results for the serviced urban area. Moreover, the participatory monitoring aspect involves the local population in the effort to achieve improvements in the quality of the environment.

The analysis proposed by the PRRU has been demonstrated to be adequate for the maintenance of the WCS, finding in the rivers the answers regarding a significant presence of organic load and its probable point of origin in the river basin.

The Program transcends the various areas of a sanitation company—management of the collection network, water resources, and environmental education. Its initiative in using participative monitoring brings citizens closer to the environment in which they live and entrusts them with the responsibility for caring for the river in their river basin.

The main strength of the PRRU is its providing a diagnosis of the WCS that is supported by swift field work and is capable of locating the main weak points in the WCS and/or irregular discharges in the river basin while operating in heavily urbanized environments.

WCS diagnoses have proved to be an important resource for correcting potential weak points. In conjunction with this observation, it must be noted that the indicators used to measure the universalization of the sewage treatment system must also evaluate the conditions of its transportation along the collection networks. Thus, in addition to assess the collection and final treatment capacity of the effluents, the total transit of the effluents must be taken into account, from the point of their collection to their arrival at the treatment plant. This would lead to the implementation of maintenance programs directed toward the avoidance of effluent loss in transit, which could potentially contaminate water bodies within inhabited areas.

In a broader sense, incentive must be provided for the performance of maintenance on the collection network in order to improve the quality of the WCS and its effects on the sustainable development goals as established by the 2030 Agenda. In this regard, the contributions made by the PRRU may be thus described: participation of the community in pinpointing weak points, the reduction of effluent loss, a diminished influx of pollutant loads into urban rivers, less environmental degradation, and, consequently, the recovery of the quality of river water. When an urban area located in a river basin is involved, a further benefit is that of water security in the gross provision of water for the public water system.

The administration of the collection network, which should be watertight and provide fluidity in the transportation of effluents, is even more relevant in cities that are undergoing an intense process of expansion and densification, since a WCS is envisioned to serve a specific population which may increase at a later date, whether through expansion into rural areas or through the occupation and verticalization of urban areas. In such a case, an evaluation of the importance of maintenance of the collection network must be included in urban planning.

With reference to the SDVGs of the UN (2015), the advances in sustainability achieved by the PRRU may be understood according to SDGs 3, 6 and 14, as described below:

- Good Health and Well-Being (SDG 3), item 3.9, involving the reduction of deaths and illnesses from water pollution and contamination by 2030, including hazardous chemicals, air and soil contamination. The repair of the weak points accomplished by the PRRU results in a greater portion of the effluents being removed from inhabited areas, in line with the greatest objective of the sanitation system, which is their rapid removal from urban areas, thus avoiding greater concentrations of pathogenic microorganisms in water bodies located within the urban area, minimizing the presence of disease carriers such as rats and cockroaches;
- Clean Water and Sanitation (SDG 6): Item 6.2 envisions that all should have access to "adequate hygiene and sanitation", whereas item 6.3 involves the reduction of

pollution and suppression of discharges of "untreated residual waters". These items assume that sewage collection system should be equipped with effluent treatment systems, minimizing discharges of untreated effluents into rivers.

In WCSs such as those in the Metropolitan Region of Curitiba that are already serviced by Sewage Treatment Plants, the PRRU has been ensuring the transportation and treatment of the collected sewage, thus avoiding improper disposal due to the appearance of weak points in the collection structure.

Item 6.6.b aims to foment the participation of local communities in order to improve "the administration of the water and sanitation," an item that is wholly achieved through the incorporation of citizens into the Participative Monitoring program, who are committed to the periodic evaluation of the organoleptic properties of the river water. The strategic placement of these agents along the urban rivers allows the PRRU to act wherever the effects of effluent discharges have been perceived.

In treating the losses due to the depreciation of the collection network, the PRRU anticipates problems by adhering to a preventative maintenance schedule, conducting repairs in hidden underground areas and thereby improving the quality of life of urban areas. This is even more relevant in urbanized areas near springs, a recurring feature of the Brazilian topography, resulting in benefits to the raw water and water security and reducing the amount of chemicals that must be applied during treatment processes.

ODS 6 does not cite "losses," but it must be noted that 8% of the occurrences identified by the PRRU relate to the loss of treated water in the distribution network, a hidden loss that breaches the WCS and rainwater drainage structures.

- Life Below Water (SDG 14): The activities of the PRRU result in the reduction of irregular sewage discharges, which reduces the damage to marine life, both in urban or coastal areas, in the case of transportation by the river water of effluents to urban areas closer to the coast.

The PRRU weaknesses include its incapability of diagnosing all weak points in the WCS, requiring successive work fronts, as well as its incapability of operating effectively for at least 48 h after heavy rainfall in the river basin, due to the resulting dilution of the polluting loads in the rivers.

Over the course of the River Profile survey, the DO readings must be taken within short periods of time, in agreement with the premise that the DO parameter is static. In practice, it has been shown that the values from the DO readings exhibit the least fluctuation in the morning, whereas in the afternoon organic discharges may be detected more easily, as a result of the greater fluctuation in the DO reading.

The PRRU has proved to be capable of becoming an integral part of the solutions conventionally used for WCS maintenance, such as tele diagnosis and smoke tests. The benefit of such integration is the ability to avoid weak points before they are firmly established and result in losses in the quality of the environment of inhabited areas, as well as of lentic and fluvial environments.

The concepts and premises that guide the PRRU lead one to desire that the indicators for universalization of sanitation, which currently are grounded in the extent of the collection network present or in the population served, may eventually incorporate broader parameters, so as to express the efficiency with which the WCS is actually accomplishing its goal of effective removal of effluents.

References

Amick RS, Burgess EH (2000) Exfiltration in sewer systems. National Risk Management Laboratory, Cincinatti, Ohio

Bertrand-Krajewski J-L, Baer E, Cardoso MA, De Bénédittis J, Ellis B, Franz T, Frehmann T, Giulianelli M, Gujer W, Karpf C, Kohout D, Kracht O, Krebs P, Metelka T, Pliska Z, Pollert J, Prigiobbe V, Princ I, Pryl K, Revitt M, Rieckermann J, Rutsch M, Vanecek S (2005) Assessing infiltration and exfiltration on the performance of urban sewer systems (APUSS)—final report. Leon, France. Retrieved from: https://pdfs.semanticscholar.org/b57c/df743f71b8b6c9e2e2bcfad07dc5b82748d3.pdf

Brasil (2005) Resolução CONAMA N° 357. Dispõe sobre a classificação dos corpos d'água

(DSD) Drainage Services Department (1995) Sewerage manual—key planning issues and gravity collection system—part 1. Hong Kong, China. Retrieved from: https://www.dsd.gov.hk/EN/Files/Technical_Manual/technical_manuals/Sewerage_Manual_1_Eurocodes.pdf

Feszterova M, Jomova K (2015) Character of innovations in environmental education. Procedia Soc Behav Sci 197:1697–1702. In: 7th world conference on educational sciences (WCES-2015), 05–07 Feb 2015, Novotel Athens Convention Center, Athens, Greece. https://doi.org/10.1016/j.sbspro.2015.07.222

Giller KE, Drupady IM, Fontana LB, Oldekop JA (2018) Editorial overview: the SDGs—aspirations or inspirations for global sustainability. Curr Opin Environ Sustain 34:A1–A2

Hermes LC, Fay EF, Buschinell CCDA, Silva AS, França E Silva ÊF (2004) Participação comunitária em monitoramento da qualidade da água. Circular técnica n° 8, Embrapa. Jaguariúna, São Paulo. pp 1–8

Malik OA, Hsu A, Johnson LA, Sherbinin A (2015) A global indicator of wastewater treatment to inform the Sustainable Development Goals (SDGs). Environ Sci Policy 48:172–185

Penido LR (2014) Avaliação de metodologia de apoio à manutenção de rede coletora de esgotos. Dissertação (Mestrado em Meio Ambiente Urbano e Industrial) Setor de Tecnologia. Universidade Federal do Paraná, Curitiba

Penido LR, Amaral KJ, Matos Jorge RM, Metzger JW (2017) Análise Crítica da Metodologia PRRU para Recuperação de Rede Coletora Fragilizada. In: Amaral KJ et al (eds) Meio Ambiente Urbano e Industrial: desafios, tecnologias e soluções. Editora do Setor de Tecnologia da UFPR, Curitiba, pp 188–212

Sanepar - Companhia de Saneamento do Estado do (2014a) Selo do monitoramento participativo do PRRU. Curitiba, Paraná

Sanepar - Companhia de Saneamento do Estado do Paraná (2014b) Registros fotográficos das ações do PRRU. Curitiba, Paraná

Skroch J (2012) Gestão da rede coletora de esgoto: Revitalização de rios urbanos. Prêmio Nacional da Qualidade em Saneamento – PNQS. Inovação da Gestão em Saneamento – IGS. IX Seminário da Inovação em Gestão do Saneamento. Fortaleza

Skroch J (2013) Monitoramento de coletores de esgoto sanitário em áreas de reservatório de água para abastecimento público. Curso de Capacitação de Facilitadores para Qualidade. PUC, Curitiba

Skroch J (2014) Procedimentos e conceitos do Programa de Revitalização de Rios Urbanos [Personal message]. Message received by: lpenido@sanepar. com.br, 15/5/2014

Sloane GMT, Pröbstl-Haider U (2019) Motivation for environmental volunteering—a comparison between Austria and Great Britain. J Outdoor Recreation Tourism 25:158–168

Souza PABF, Marques Junior S (2001) A importância da Educação Ambiental na formação de profissionais de engenharia relacionado ao setor de transportes urbanos. Congresso Brasileiro de Educação em Engenharia. Retrieved from: http://www.abenge.org.br/cobenge/arquivos/18/trabalhos/EMA027.pdf

Tratabrasil - Instituto Trata Brasil (2018) Ranking do Saneamento Instituto Trata Brasil 2018. São Paulo. 2018. Retrieved from: http://www.tratabrasil.org.br/images/estudos/itb/ranking-2018/realatorio-completo.pdf

United Nations General Assembly (UN) (2015) Resolution A/RES/70/1. Transforming our world: the 2030 agenda for sustainable development. Resolution adopted by the General Assembly on 25 September 2015

Global Sustainable Food Systems and Agricultural Markets

Internet of Things: The Potentialities for Sustainable Agriculture

Tehmina Khan

Abstract Internet of Things (IoT) technology has a substantial role to play in promoting sustainable agriculture and food production. SMEs are playing a leading role in this space. Tools that allow continuous monitoring and decision making for example regarding the use of fertilisers or water consumption are being applied in various parts of the world by SMEs as in Vietnam and Australia, to undertake sustainable agriculture and food production. The aim in this article is to provide an overview of the latest technology which has been developed and applied in this space. Through the consideration of examples of applications, the resulting implications and impacts for example relating to water and energy use are provided. The potential for reducing environmental impacts while creating sustainable agriculture and food production from the perspective of technological applications by SMEs is discussed with broader environmental impact considerations.

Keywords Internet of things · Sustainable agriculture · SMEs

1 Introduction

Although by 2050 there will be approximately nine billion people to feed, critical challenges faced in the agricultural sector including droughts, floods and changing rainfall patterns continue to pose a threat to sustainable agriculture; that is climate change continues to negatively impact global food supply (Turner 2018). Agriculture, prior to the adoption of IoT has faced numerous challenges which smart agriculture IoT is addressing. These challenges have included: lack of reliable data relating to for example weather conditions, soil quality, crop growth, staff performance and equipment efficiency. Additional factors include: lack of control over internal processes, production risks and the lack of ability to foresee output and production, weaknesses relating to cost management and high waste generation, reduced business efficiency and low product quality and volumes (Aleksandrova 2018). These factors have had a negative impact on revenue in the Agriculture sector (ibid.).

T. Khan (✉)
School of Accounting, RMIT University, Melbourne, Australia
e-mail: Tehmina.khan@rmit.edu.au

© Springer Nature Switzerland AG 2020
W. Leal Filho et al. (eds.), *International Business, Trade and Institutional Sustainability*, World Sustainability Series,
https://doi.org/10.1007/978-3-030-26759-9_17

Internet of Things (IoT) is offering various opportunities for improved processes and impacts that have the potential to reduce resource utilisation through continuous monitoring and the availability of in-depth data in the Cloud. Internet of Things is the concept of connecting any device (with an on/off switch) to the Internet and to each other (Morgan 2014). It is also a network and relationship of connectedness between people, people and things (ibid.). The use of IoT is resulting in reduced costs and is positively impacting the bottom line for farmers. In this Chapter the key applications of IOT for sustainable agriculture are discussed together with examples (cases) of the applications of IoT in the agriculture sector in Vietnam and Australia; to highlight the benefits attained for sustainable agriculture. Prior literature has looked at the business benefits and poverty reduction potential of IoT (Dlodlo and Kalezhi 2015) and the technical aspects of IoT (see Patil et al. 2012; Mohanraj et al. 2016) relating to the agricultural sector. This Chapter provides a critical contribution to sustainable agriculture literature from the perspective of focusing on the potentials and benefits of IOT, from SMEs perspective, relating to the adoption of IoT in the agriculture sector.

2 Internet of Things-Potential for Continuous Monitoring by SMEs

The term Internet of Things implies the interconnection of devices through the Internet to send and receive data. The elements which can be connected through the Internet can include computing devices, mechanical and digital machines, objects, animals or people with unique identifiers and data transfer between these elements can take place without human-to-human or human-to-computer interactions (Rouse 2018). IoT has the potential to impact daily lives and work functions (Morgan 2014). The main reasons for the increasing use of the Internet for business activities include: the wider availability of broadband Internet, decreasing costs of connection, greater WiFi capabilities and use of sensors built in the devices, smartphone penetration and reduced technology costs (ibid.).

The need for IoT is highlighted in multiple benefits which can be attained through the use and implementation of the technology. Organisations in multiple industries are using IoT to improve efficiency, to promote business understanding of processes and customers, to enhance decision making and it is being used to increase business value (Rouse 2018). IoT is an extension of SCADA (supervisory control and data acquisition) used for processes control and for real time data gathering from remote locations for controlling conditions and equipment (ibid.).

SCADA technology is hardware oriented and suffers from numerous problems for example vendor specific protocols which could be mutually incompatible and the technology is unable to anticipate problems (Zambrano 2018). IoT asset managed systems on the other hand have been designed to support proactive integration strategies and are more flexible and less costly than SCADA (ibid.). IoT systems operate on open standard protocols which facilitate the incorporation of SCADA input, resulting in the extending of the value of legacy systems (Zambrano 2018).

3 Continuous Monitoring and the Role of IOT (SME and Agriculture Focus)

There are capabilities offered through IoT technologies such as smart farming which allow farmers to reduce waste and increase productivity (Writer 2018). Smart farming involves high investment in capital and technology which promotes sustainability in production for mass consumption. The technology aspect of farming involves continuous monitoring including the use of sensors to detect various climatic conditions and the automation of the irrigation system (Writer 2018). Writer (2018) has stressed that IoT based smart farming is not only applicable for large commercial applications; it can also be used for small scale farming such as organic farming. It promotes highly transparent farming. According to Writer (2018) various applications of IoT in agriculture include precision farming, agricultural drones, livestock monitoring and smart greenhouses.

3.1 Precision Farming

Precision farming involves managing variations in the field in an accurate manner to increase food production with less use of resources and less production costs (CEMA 2018). Precision farming involves the use of cost-effective monitors, controllers and the integration of data in a single management system (ibid.). Other tools include GPS guidance, control systems, sensors, robotics, drones, autonomous vehicles, variable rate technology, GPS-based soil sampling, automated hardware, telematics and software (Schmaltz 2017). Practice of precision takes place in the following areas: soil preparation, seeding, crop management and harvesting (CEMA 2018). The concept of precision farming commenced in the 1990s and the value of precision agriculture is expected to be \$43.4 billion by 2025 (Schmaltz 2017). Precision ergonomics comprises of the following elements:

1. Variable rate technology which enables variable application of inputs and control over the amount of inputs. The technology requires the use of a computer, software, controller and global positioning system (DGPS).
2. GPS soil sampling: Soil samples are collected and data is used for input regarding variable rate applications to optimise seeding and fertilizer use.
3. Computer based applications: These are used to create precise farm plans, crop scouting and yield maps. Pesticides, herbicides and fertilizers are applied more precisely; expenses are reduced resulting in higher yields and environmentally friendly operations.

4. Remote sensing technology: it is used for monitoring and managing land, water and other resources. Drones and satellites are used to provide a wide range of data; that is big data.

(Schmaltz 2017).

3.2 Agricultural Drones

GPS assisted unmanned aerial vehicles (UAV's) with various features including thermal imaging for example help growers and agronomists with cost reductions, increased efficiencies and time savings compared to traditional monitoring techniques such as labourers (National Drones 2018). Data captured through drones helps map the health of the crops, helps monitor sunlight absorption and plant physiology (ibid.). Benefits of accurate data collection through drones include: real time information on diseases, weeds and pests, better yields and immediate remedial or preventative action (National Drones 2018).

3.3 Livestock Monitoring

The main issue faced by farmers that negatively impacts the sustainability of livestock and profit generating ability is animal illness (Telit 2018). IoT facilitates livestock management through the following means:

Wearable connected sensors in livestock which help monitor heart rate, blood pressure, respiratory rate, temperature, digestion and other vital systems' functioning,
The transfer of data from the sensors to the Cloud allows farmers to identify illnesses and feeding problems,
Farmers use IoT to monitor reproductive cycles, calving process and sensors are also used to track sick animals' location and their grazing patterns.

(Telit 2018).

Reese (2016) has addressed the use of IoT for cows specifically as follows: to monitor cattle movement, to detect health and fertility of the cows through sensors, to detect cows' activity throughout the day, for robotic milking production based on cows' preferences relating to when they would like to be milked, use of IDs and transponders to identify times when the cows would produce more milk, utilisation of thousands of data points and detailed analysis of the data and early identification of illnesses in the cows. Another area of concern with livestock is missing cattle. IoT provides solutions through tracking collars and smartphone applications to locate missing animals in addition to its use to regulate grazing areas to maximise milk yield while reducing erosion and soil pollution (IoT Solutions 2016).

The ultimate impact of the use of IoT smart tracking devices which cost less than U.S. $10 per cattle includes large cost savings in the long term, reduced losses from predators, animal illnesses or escaped animals and savings in labour costs for farm maintenance (Tournier 2017). Tournier has identified substantial cost benefits including reduction in device and services costs, increased battery life and coverage or reach through the use of Low Power Wide Area (LPWA) technologies. These technologies require less processing power, less memory, low current, longer battery life, less processing, greater coverage and less risk of losing track of the asset through reliable connection even in rough weather conditions.

3.4 Smart Greenhouses

The main problem with greenhouses is that they are energy intensive with high electricity costs (Andrews 2017). With smart greenhouses, there is potential to increase growth by more than 20% and they require less water (ibid.). Smart greenhouses automate farmers' work, protect plants from extreme weather conditions and include the use of automatic drip irrigation (Kodali et al. 2017). Smart greenhouses allow optimal water use, optimal nutrition use through drip fertigation techniques and facilitate the provision of required wavelength light during night periods (ibid.). Temperature and air humidity are controlled through sensors (Kodali et al. 2017). Data is uploaded to the Cloud and can be shared with key stakeholders for example purchasing partners.

Immense benefits of smart greenhouses include insecticide and pesticide free crops, potential for alternative sources of income, cultivation of numerous types of crops, massive reduction of water requirements, increase in yield and rate of growth and direct connection with consumers (Kodali et al. 2017).

4 Methodology: Vietnamese and Australian SMEs Case Studies

The following brief coverage of examples (case studies) captures the benefits of implementation of IoT in the agricultural sector in Australia and Vietnam. They highlight the benefits of the use of IoT in the agriculture sector and the promotion of greater sustainability of the sector in-spite of increasing challenges against the sector.

4.1 Vietnam

In-spite of industrialisation occurring in Vietnam, the availability of a large amount of agricultural land and the dominance of the agricultural sector in Vietnam, provides opportunity for the utilisation of IoT to promote sustainable agriculture in Vietnam. Vietnam has 13.8 million farmer households and 78 million fragmented fields raising the necessity of agricultural restructuring (Hanoi Times 2018). According to Professor, Dr. Chu Pham Ngoc Son in Science-IT (2018) opportunities associated with IOT including the use of IOT apps on cheap mobile devices and wide availability of the Internet in Vietnam (as at the beginning of 2018, Vietnam has around 64 million Internet users; 67% of the total population), support the advancement of technologies and there is government's support in agricultural restructuring. These factors are facilitating successes in the application of IoT in agriculture in Vietnam (Hanoi Times 2018). Other factors promoting the success of IoT in Vietnamese agriculture include servers not being required due to the availability of the Cloud. IoT implementation in the agricultural sector is an exciting opportunity for clean agricultural production in Vietnam. Clean produce is of increasing demand in Vietnam as food security is of major concern in the country. There is a call in Vietnam to develop agricultural businesses which would ensure business safety by promoting innovative agricultural technologies and which would address the issue of overuse of pesticides.

4.2 Case Study 1 FPT Akisai Farm and Vegetable Factory (Sources: Fujitsu 2015; Fresh Plaza 2016)

1. Facility	Fujitsu—FPT Akisai Farm and Vegetable Factory
2. Inauguration	Wednesday, February 24, 2016
3. Location	Trau Quy town, Gia Lam district, Hanoi city, Vietnam
4. Size	403 m^2, including a greenhouse cultivation zone (259 m^2) a vegetable factory cultivation zone (15 m^2) and a presentation zone
5. Cultivated crops	Greenhouse cultivation: medium-sized tomatoes grown with Imec$^®$ Film Farming Vegetable factory cultivation: low-potassium leaf lettuce

Fujitisu has two production facilities in Vietnam that employ greenhouse cultivation and vegetable factory cultivation. The facilities have introduced high value added vegetables. The key features of the collaborative project between FPT Akisai Farm and Fujitsu include:

Green house cultivation which employs greenhouse horticulture, detailed real time environmental information including information relating to temperature, humidity, sunlight, CO_2, rain and wind, to autonomously control the facilities. The vegetable factory is a completely closed plant factory. It also uses detailed real time environmental information collected via multiple sensors. The FPT-Fujitsu Akisai Farm and Vegetable project applies Cloud computing technology and smart agriculture. This is the first project of its kind where Japanese technology solution has been applied in Vietnamese agricultural sector.

4.3 Case Study 2 Intel and Cao Dat Farm (Sources: Reuters 2016; Foo 2016)

The farm is based in Dalat and utilises a system of sensors, weather stations, and robots to enable the management of farm operations via the Cloud. The farm partnered with Intel and imported expertise and hardware. It is also serving as a platform for other agricultural businesses to join the network to implement IoT in their operations and processes. Cao Dat Farm has aimed to establish an agricultural database, the first of its kind in Vietnam to connect value chains, farmers, agribusiness companies, retailers, experts, and users.

4.4 Australia

In Australia, National Farmers' Federation (NFF)'s aim is to accelerate the growth of Australian agriculture to $100 billion in farm output by 2030 (Poole et al. 2018). One of the key themes that have emerged to achieve this goal is the adoption of new technology including IoT. Accordingly, by 2030, every farm in Australia should be connected to the Internet of Things using either traditional or emerging networks (ibid.).

The main factors that are enabling IoT for Australian agriculture are:

Decreasing cost of intelligent sensors and actuators (components of a system which control mechanisms),
Connectivity: a wide range of IoT connectivity solutions and technologies including wired and wireless devices being tested for geographical dispersion, mobility and energy requirements,
Low Powered Wide Area Network (LoRaWan) and Bluetooth networks specifically designed for IoT connectivity,
Large reductions in costs of computer processing power, storage and cloud computing with greater capacity to process and analyse big data and to identify trends for predictive decision making,
Interoperability: the ability for devices to connect and communicate with each other through open source standards,
A well-developed machine to machine (M2M) communications network including process specific sensors which collect real time data to send back to smart devices

and to stakeholders including regulators, banks, customers, suppliers and insurance providers (Koch 2017).

As Koch (2017) has pointed out connectivity is a critical issue for Australia's rural areas and it requires fast and effective action towards low cost networks in Australia. Precision agriculture is worth $3.7 Billion in Australia in 2018 (McPhail 2018). It involves the use of physical devices such as computers, sensors and phones and network connection to measure soil and air quality, humidity and temperature, livestock numbers, satellite imagery (ibid.). Combined with other data such as weather patterns and pesticide updates, it is helping farmers make smart and informed decisions around when to fertilise, irrigate, harvest and transport as well as to undertake long term planning.

The Australian government is investing $200 million in research grants for advancing agricultural technology. But reliable regional networks are also required to achieve a connected (IoT) based smart and precision agriculture in Australia.

4.5   Case Study 1 Full Profile

Full Profile comprises of a team of software engineers, farmers and agribusiness professionals. It is the world's first block chain enabled commodity management platform (Full Profile 2015). Full Profile's AgriDigital is a Cloud based commodity management platform and it apparently makes supply chain easier and secure from farmer to consumer. It allows the management of contracts, deliveries, stock, invoices, orders and payments through one medium and in real time (AgriDigital 2018). 2.97 million tonnes and economic value of $594 million of inventory has been transacted through the platform. The key features of the platform relevant to farmers include: digital sales contracts, real time access to contract quantities and status, tracking of harvest to delivery locations and detailed quantity related data, invoice management and simplified financial reporting through digital sales and payments reports, as well as automatic end point collections and payments. Other features include live SMS notifications, live dash boarding of contracts etc. and data export for analysis.

The platform provides benefits including monitoring of storage conditions, real time data on contracts and delivery, detailed customer data, total commodity received and sold, for storage operators as well. The platform is also available for traders with transactions details, net positions, prices etc.

The founders of the platform have captured the value of IoT for their business in the following words: "We all knew there was a better way; and that by using emerging technologies we could create cost-effective, efficient and world-leading agri-commodity management and supply chain solutions that would have global impact" (Watson 2018). And

> Supply chains are really complex with multiple participants, lots of double-entry, and manual handling. It's hard to match payments with title and asset transfers. There's a real lack of transparency and trust. We set about designing a platform that solved these problems,

managing grain from farm to consumer should not be so complex; we work with our customers to simplify and increase the efficiency of their individual operations; we can then incrementally improve the whole of supply chain experience. (McKay and Reid 2018)

4.6 Case Study 2 The Yield

The Yield provides technological solutions for agriculture and aquaculture. The main focus is on storage of big data in the Cloud and the data is identified as: personal, farm (collected via farm sensors), aggregated data: collected from multiple farms and anonymized. Data is transformed and is publicly provided through government bodies such as through the Bureau of Meteorology (The Yield 2018a). The main benefits that The Yield aims to provide through its products include increased yield, reduced waste, risk and cost mitigation and environmental sustainability (The Yield 2018b). The business recognises that food production demand is increasing in the presence of resource constraints and climate change. The system offered by Yield measures rainfall, temperature, air pressure, leaf wetness, soil moisture, humidity and converts raw data through analytics and AI to help deliver predictions which provide numerous benefits. These benefits include: accurate decision making around harvesting, irrigating, planting, feeding and crop protection and proactive responses to changing weather conditions (The Yield 2018c). For aquaculture purposes Yields' sensors measure air pressure, wind speed and direction, air temperature, rainfall, sea tide height, salinity, dissolved oxygen, water temperature and water depth. Data is converted to a three day harvest area forecast (The Yield 2018d).

With both agriculture and aquaculture data collection and analytical technologies, partnerships with larger companies including with Microsoft and Bosch play a critical role in the successful provision of the products and services by Yield.

The key benefits of Yield's technology are captured in the quotes from their clients: "We had a dry period early last year, which led to us increasing water. A lot of that decision was made on gut feel. But Sensing+™ will be able to tell us more accurately. Having that ability during growing season is critical" (Polley, Pooley's Wines.)

Also in relation to Aquaculture,

What we need is information to help us make decisions, and that's where The Yield fits in. Their solution gives us a roadmap of what is going on in the water, which has always been difficult to quantify. Goc, Barilla Bay Oysters. (The Yield 2017a, b)

All the examples and case studies provided by Yield highlight the importance of accurate data collection and analysis for forecasting and resource allocation purposes by Yield's clients. The impact from regulator's perspective is also considered for aquaculture as harvesting periods and non-harvesting periods are more accurately determined based on the detailed data which is available through sensors.

5   Discussion and Conclusions

As the above examples from Vietnam and Australia have suggested there is immense potential and benefits associated with the use of IoT to undertake sustainability supporting agriculture. Sustainability supporting agriculture is critical as land and food resources pressures continue to increase (Hermes 2017). In recognition of these pressures governments around the world such as in Australia and Vietnam are undertaking financial investments in innovative agricultural practices such as IoT based smart and precision farming. IoT is encouraging a new wave of agricultural practices which can facilitate practices which have reduced environmental impacts, increased productivity and reduced use of precious resources such as water.

The embedding of IoT in agriculture is represented quite succinctly as: drones (ground and aerial based) for field monitoring, Internet (Cloud) connected farming equipment and machines, soil, temperature, humidity etc. monitoring through sensors, and minimisation of resource use including water management (Data Flair 2018). In relation to precision farming, variable rate irrigation (VRI) maximises profitability in relation to irrigated fields by improving yields and enhancing water use efficiency (ibid.). The critical role of IoT in livestock management has also been looked at and the vast potential relating to cost reduction, livestock loss to sickness, death, to predators and to other factors has also been addressed. Smart Greenhouses provide the potential for remote access and control of various conditions in the greenhouse including temperature control; they minimise or eliminate the use of pesticides, and insecticides and they help promote organic farming and reduce water consumption.

From an economic perspective, increasing adoption and implementation of IoT based agriculture can be determined from these statistics: the market value of smart agriculture is forecasted to be 27 billion U.S. by 2020, precision farming is estimated to be worth 2.42 billion U.S. by 2020 (Statistica 2018). The main point to keep in mind is that IoT based agricultural practices are not a phenomenon that is restricted to economically developed countries. Internet connected technologies are cost effective and practically available to farmers around the world, including in economically developing countries. Further research in the economic impacts of the use of IoT in agricultural practices using a case study approach or even interviewing farmers who have implemented IoT can further highlight the potential benefits of IoT in agriculture. The barriers to adoption or the lack of achievement of benefits can be considered as well in order to understand limitations of IoT, which can pave the way for improvements and the optimal use of the technology to benefit the agriculture sector. Real time, comprehensive information, monitoring, analysis and relevant action are important to support sustainable agriculture, to meet increasing food production requirements and these factors are readily enforceable through IoT implementation in agricultural practices.

References

Agridigital (2018) About us. Available at https://www.agridigital.io/about

Aleksandrova M (2018) IoT in agriculture: 5 technology use cases for smart farming (and 4 challenges to consider). Available at https://easternpeak.com/blog/iot-in-agriculture-5-technology-use-cases-for-smart-farming-and-4-challenges-to-consider/. Accessed 12 Nov 2018

Andrews R (2017) Smart' greenhouses can grow crops and generate electricity simultaneously. Available at http://www.iflscience.com/environment/smart-greenhouses-grow-crops-generate-electricity-simultaneously/

CEMA (2018) Precision farming. Available at https://www.cema-agri.org/precision-farming. Accessed 15 Nov 2018

Data Flair (2018) IoT applications in agriculture—4 best benefits of IoT in agriculture. Available at https://data-flair.training/blogs/IoT-applications-in-agriculture/

Dlodlo N, Kalezhi J (2015) The internet of things in agriculture for sustainable rural development. In: 2015 international conference on emerging trends in networks and computer communications (ETNCC), pp 13–18. IEEE

Foo V (2016) Seedcom-backed Cau Dat Farm brings smart technology into Vietnam agricultural production. Available at https://www.reuters.com/brandfeatures/venture-capital/article?id=1725

Fresh Plaza (2016) Japan invests in Vietnamese agriculture. http://www.freshplaza.com/article/154171/Japan-invests-in-Vietnamese-agriculture

Fujitsu (2015) Fujitsu and FPT implement smart agriculture in Vietnam. Available at http://www.fujitsu.com/global/about/resources/news/press-releases/2015/1208–01.html

Full Profile Reimagining Agriculture (2015) Available at https://www.fullprofile.com.au/. Accessed 15 Nov 2018

Hanoi Times (2018) Prospects for IoT development in Vietnam's agriculture. http://www.hanoitimes.vn/economy/agriculture/2018/06/81e0c7d0/prospects-for-IoT-development-in-vietnam-s-agriculture/

Hermes J (2017) Farmers must get from farm to fork faster; IoT will facilitate. Available at https://www.environmentalleader.com/2017/10/174558/

IoT Solutions (2016) Livestock monitoring. Available at https://www.IoTsolution.eu/en/agriculture-livestock/livestock-monitoring/

Koch A (2017) IoT in agriculture-how is it evolving and which policy areas need addressing to facilitate its uptake. Aust Farm Inst Q Newsl 14(1). Available at file:/rmit.internal/USRHome/el1/E75321/Downloads/Feb17_newsletter_web.pdf

Kodali R, Jain V, Karagwal S (2017) IOT based smart greenhouse. Available at https://www.researchgate.net/publication/316448621_IoT_based_smart_greenhouse

McKay B, Reid B (2018) Passionate about agriculture and technology, we're bringing trust and transparency to agricultural supply chains. Available at https://www.agridigital.io/about

McPhail (2018) Precision agriculture: IoT in the field. Available at https://www.vodafone.com.au/red-wire/IoT-agriculture

Mohanraj I, Ashokumar K, Naren J (2016) Field monitoring and automation using IOT in agriculture domain. Procedia Comput Sci 93:931–939

Morgan J (2014) A simple explanation of the 'Internet of Things'. Forbes. Available at https://ieeexplore.ieee.org/document/7906846/www.forbes.com

National Drones (2018) Drones in agriculture. Available at https://nationaldrones.com.au/agriculture/

Patil VC, Al-Gaadi KA, Biradar DP, Rangaswamy M (2012) Internet of things (IoT) and cloud computing for agriculture: an overview. In: Proceedings of agro-informatics and precision agriculture (AIPA 2012), India, pp 292–296

Poole R, Delden B, Liddell P (2018) Talking 2030: growing agriculture into a $100 billion industry. Available at https://home.kpmg.com/au/en/home/insights/2018/03/talking-2030-growing-australian-agriculture-industry.html

Reese H (2016) IoT for cows: 4 ways farmers are collecting and analyzing data from cattle. Available at https://www.techrepublic.com/article/IoT-for-cows-4-ways-farmers-are-collecting-and-analyzing-data-from-cattle/

Reuters (2016). Seedcom-backed Cau Dat Farm brings smart technology into Vietnam agricultural production. Available at https://www.reuters.com/brandfeatures/venture-capital/article?id=1725

Rouse M (2018) Internet of things (IoT). Available at https://internetofthingsagenda.techtarget.com/definition/Internet-of-Things-IoT. Accessed 14 Nov 2018

Schmaltz R (2017) What is precision agriculture? AgFUNDER. Available at https://agfundernews.com/what-is-precision-agriculture.html

Science-IT (2018) IoT to be applied to agriculture. Available at http://english.vietnamnet.vn/fms/science-it/192265/IoT-to-be-applied-to-agriculture.html

Statistica (2018) Smart agriculture—statistics & facts. Available at https://www.statista.com/topics/4134/smart-agriculture/

Telit (2018). Crop and livestock management. Available at https://www.telit.com/industries-solutions/agriculture/crop-livestock-monitoring/

The Yield (2017a) Customer stories: meet Pooley wines. Available at https://www.theyield.com/post/pooley-wines

The Yield (2017b) Customer stories: Barilla Bay Oysters. Available at https://www.theyield.com/post/barilla-bay

The Yield (2018a) Our approach. Available at https://www.theyield.com/our-company/the-yield-story

The Yield (2018b) The Yield story. Available at https://www.theyield.com/products/free-growers-app

The Yield (2018c) Sensing+ for agriculture. Available at https://www.theyield.com/products/sensing-plus-for-aquaculture

The Yield (2018d) Sensing+ for aquaculture. Available at https://www.theyield.com/products/sensing-plus-for-aquaculture

Tournier B (2017) Tracking devices for livestock increase farm profits. Available at https://www.sierrawireless.com/IoT-blog/IoT-blog/2017/10/tracking_devices_for_livestock_increase_farm-_profits/

Turner G (2018) Food and agriculture: what are the sustainability challenges? The Guardian. Available at https://www.theguardian.com/sustainable-business/food-sustainability-agriculture-challenges-opportunities

Watson E (2018) Passionate about agriculture and technology, we're bringing trust and transparency to agricultural supply chains. Available at https://www.agridigital.io/about

Writer G (2018) IoT applications in agriculture. IoT for all. Available at https://www.IoTforall.com/IoT-applications-in-agriculture/

Zambrano S (2018) Taking SCADA to the next level with IoT. Available at http://www.b2b.com/-take-scada-to-the-next-level-with-iot

Geographic Information Systems as a Tool to Display Agribusiness and Human Development Synergy

Rodrigo Martins Moreira

Abstract Agribusiness is one of Brazilian economy's main pillars. In this scenario, soybean was the most exported agricultural commodity in 2017. International markets have postulated requirements demanding that this commodity way of production be adapted to environmentally friendly best practices. Brazilian producers have positively responded to these demands by implementing best practices in soil and water resources' usage, producing more by exploiting less. This work explores how Geographic Information System, free and open source software, can be used to graphically present agricultural and human development synergy. Data regarding soybean production, Human Development Index (HDI), from 2000 and 2010 were compared. QuantumGis software was used to graphically express changes in soybean production and HDI at Goiás mesoregions. This work's main findings show that municipalities that increased soybean production also increased its HDI; the use of GIS software is an effective and low-cost tool to present agribusiness information to stakeholders.

Keywords Environmental management · Moran index · Spatial social data distribution

Agribusiness is one of Brazilian economy's main pillars. In this scenario, soybean was the most exported agricultural commodity in 2017. International markets have postulated requirements demanding that this commodity way of production be adapted to environmentally friendly best practices. Brazilian producers have positively responded to these demands by implementing best practices regarding work conditions, soil and water resources' usage, producing more by exploiting less. Goiás state is one of the Brazil's main agricultural producers. By integrating Goiás State municipalities' data in a geographic database within QuantumGis and TerraView softwares, free and open source GIS platforms, decision makers can identify where to implement strategic policies, plans and programs to tackle life quality enhancement through an agribusiness that considers the synergy between environmental fragility, economic growth and human development. This work explores how Geo-

R. M. Moreira (✉)
Department of Environmental Engineering, Federal University of Rondônia, Ji-Paraná, Brazil
e-mail: moreirarmt@gmail.com

W. Leal Filho et al. (eds.), *International Business, Trade and Institutional Sustainability*, World Sustainability Series,
https://doi.org/10.1007/978-3-030-26759-9_18

303

graphic Information System, free and open source software, can be used to graphically present agricultural and human development synergy. Data regarding soybean production, Human Development Index (HDI), from 2000 and 2010 were statistically compared and clustered. QuantumGis and TerraView softwares were used to graphically display changes in soybean production and HDI at Goiás mesoregions. This work's main findings show that municipalities that increased soybean production also increased its HDI; the use of GIS software is an effective and low-cost tool to present agribusiness information to stakeholders.

1 Background

Agribusiness holds a notorious place in global economy due to farm related production, processing and distribution of goods such as food, cereals, livestock, crops for biofuels and biotechnology, and livestock nutrition. These activities generate jobs and move into a continuous economic flux throughout the world, intra and inter countries. According to Food and Agricultural Organization of the United Nations, the Gross per Capita Index Number for 2016 in a global scale was International $111.70, for South America Int. $117.40 and for Brazil, Int. $122.19. In this sense, sustainable agribusiness is key to global social and economic aspects (FAOSTATS 2018).

FAO presents an approach for enhancing agricultural management by using a results-based method, which "is an approach that integrates strategy, people, resources, processes and measurement to improve decision-making while increasing transparency and accountability". This method is based on "measuring performance against goals, learning from experience and adjusting to new conditions, reporting outcomes, and achieving objectives" (FAO 2018).

In this scenario, agribusiness is one of Brazilian economy's main pillars; soybean was the most exported agricultural commodity in 2017. International markets have postulated requirements demanding that this commodity way of production be adapted to environmental friendly best practices. Brazilian producers have positively responded to these demands by implementing best practices in soil and water resources usage, producing more by exploiting less, enhancing work environment quality for employees.

Geographic Information Systems are low cost set of technological tools for the upraised data efficient display, enhancing decision making by presenting indexes graphically, in an easily understandable approach (Bateman et al. 2002; Paterson and Boyle 2002; Swetnam et al. 2011). Brazil has enrolled most of its social, environmental and economic information on online databases, with time and spatial scale, providing management tools for states, basin regions, and municipalities. Nevertheless, lack of trained human resources to integrate these information databases and free GIS platforms, to express them in a geographical approach, in a municipal and regional scale is still a challenge.

By integrating Goiás State municipalities' data in a geographic database within QuantumGis software, a free and open source GIS platform, decision makers can identify where to implement strategic policies, plans and programs to tackle life quality enhancement through an agribusiness that considerate the synergy between environmental fragility, economic growth and human development (Hettinga 2014; Eloy et al. 2016; Gosch et al. 2017; Parente and Ferreira 2018).

Goiás state is one of the Brazil's main agricultural producers, and to the researcher knowledge, neither work explores how its human development and life quality enhancement synergy with agriculture, nor, how agriculture in this specific state follows the path of sustainability by incorporating good practices for a more environmental friendly production.

This chapter explores the relationship between Goiás State mesoregions' human development index and the synergy with agribusiness improvement displaying data using a free and open-source GIS platform.

1.1 GIS as a Tool to Support Decision-Making

The availability of Global Information Systems (GIS) technologies presented an increase in the past decades, with a wide variety of applications on data presentation and analysis, these innovative approaches have been applied at environmental, social and economic dimensions data display, always moving towards improving management and decision-making processes (De Almeida Bressiani et al. 2015; Sahu 2015; Cegan et al. 2017; Hiloidhari et al. 2017; Ramadas and Samantaray 2018).

Global Information Systems applied to agribusiness provides producers and distributors the possibility of aggregating key information regarding:

- Soil quality monitoring (Pierce and Lal 2017);
- Physical and chemical soil characteristics (Santos 2017);
- Productivity indexes (Glennie and Anyamba 2018);
- Crops yields (Seynave et al. 2018);
- Vegetation sanity (Du and Noguchi 2017; Del Grosso et al. 2018);
- Productivity potential zoning (Corbelle-Rico et al. 2018);
- Energy savings by optimized Vehicles Rout Management (Rodias et al. 2017).

Regarding social and economic GIS applications, applicability of these technologies to represent data has been observed, with the capability of been applied in a time scale by expressing changes throughout time, and in spatial scale by comparing countries, regions and cities. Some application examples are:

- Behavior tendencies (Balram and Dragićević 2005);
- Life quality (Spinney et al. 2009);
- Crime rate zones (Ratcliffe and Chainey 2013);
- Representation of several indices such as Human Development Index, and Gross Domestic Product (Huang and Jiang 2017);

- Health related data, such as epidemiology and disease spreads (Maksimov et al. 2017; Zambrano et al. 2017);
- Transportation planning (Mokashi et al. 2017);

Li and Weng (2007), in their research developed a methodology to integrate census data and Geographic Information System tools to measure life quality in the city of Indianapolis, Indiana, United States. The framework integrates environmental variables such as temperature, greenness, land use and socioeconomic variables such as "population density, housing density, median family income, median household income, per capita income, median house value, median number of rooms, percentage of college above graduates, unemployment rate and percentage of families under poverty level."

In this sense, GIS software by integrating and building time and space information database composed by social and economic indicators can provide decision makers an additional management tool to tackle social, environmental and economic issues, and also the development of policies, strategic plans and programs.

1.2 Best Practices in Soybean Production at Goiás State

Brazilian policy agenda regarding agriculture has a strong focus on moving towards a more sustainable path, improving agribusiness systems to address agricultural and life quality enhancement. The 2017/2018 Brazilian Agricultural and Livestock Plan focuses on overcoming challenges such as transport and logistic infrastructure deficits, and implementation of the Low Carbon Agriculture program, a set of requirements for a line of credit that provides financial incentives for producers that seek to invest on:

> Restoration of degraded areas and pastures, implementation and extension of crop-livestock-forestry integration systems, correction and fertilization of soils, implementation of soil conservation practices, implementation and maintenance of commercial forests, such as palm oil, açaí, walnut and olive trees, implementation of organic agriculture, restoration of permanent preservation areas or legal reserve, and other practices that involve sustainable production and culminate in low emission of greenhouse gases.

Soybean is an oleaginous plant that represents almost half of Brazil's grains cultivated area and is one of the main sources of protein for animal feed such as swine, birds and bovines (IBGE 2018). Hence, agribusiness directly employs 95,402 people in the state. It accounts for 1.5% of all the state population.

Figure 1 presents the historic evolution of soybean-cultivated area for Brazil and for Goiás State, where clearly can be seen that the studied state follows the country's tendency on increasing its production.

The main environmental sustainability certification for soy production in Goiás state is the Round Table on Responsible Soy Association, that in 2018 certified 49 producers in the state, corresponding to 72,667 ha of cultivated area, and 290,742 tons of soy produced. This is a pro-active approach to meet the national and international

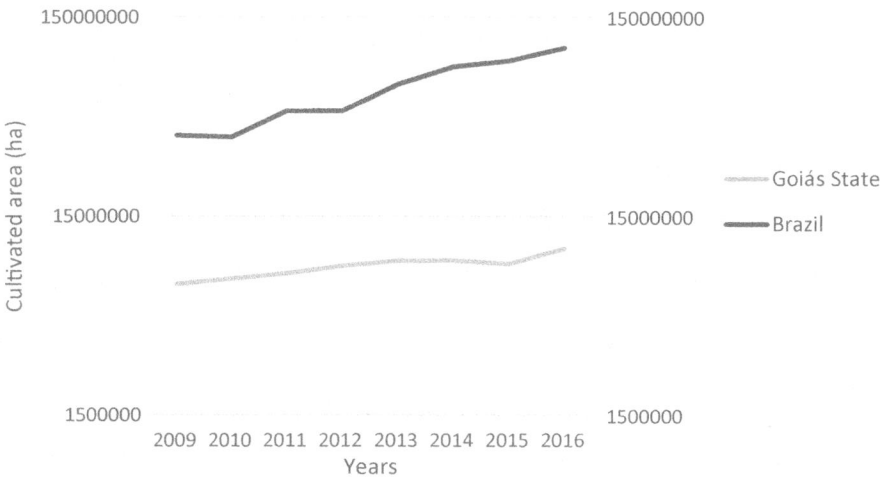

Fig. 1 Historic evolution of soybean cultivated area for Brazil and Goiás State

markets and society demands and tendencies for an environmentally friendly product, causing less negative social and environmental impacts on communities, and sustainably promoting economic growth assuring financial results for the producer.

The Round Table on Responsible Soy Association (RTRS) is a civil organization that aims at enhancing soy production, processing and fostering environmental responsibility. According to RTRS' mission is:

> Encourage current and future soybean production in a responsible manner to reduce social and environmental impacts while maintaining or improving the economic status of the producer through:

> • The development, implementation and verification of a global standard;
> • The commitment of the stakeholders involved in the value chain of soybean.

RTRS' vision is that soy "helps to meet social, environmental and economic needs of the present generation without compromising the resources and the welfare of future generations and allowing the development of a better world."

RTR's objectives:

• Facilitate a global dialogue on soy that is economically viable, socially equitable and environmentally sound.
• Reach consensus among key stakeholders and players linked to the soy industry.
• Act as a Forum to develop and promote a standard of sustainability for the production, processing, trading and use of soy.
• Act as an internationally recognized forum for the monitoring of global soy production in terms of sustainability.
• Mobilize diverse sectors interested in participating in the Round Table process.

2 Methods

2.1 Area of Study

Goiás State is in the Midwest region of Brazil. In the past decades, the state has presented an increase in agribusiness. Its main products are soybean, cotton, sugarcane, and livestock. Figure 2 presents the location of the studied area and mesoregions' division used for this work (IBGE 2018).

2.2 Data Collection

Human Development Index and Soybean Production data were acquired from the Brazilian Institute of Geography and Statistics (IBGE Cities 2018).

QuantumGis software was used to graphically express changes in soybean production and HDI at Goiás mesoregion.

2.3 GIS Methods

This work explores how Geographic Information System, free and open source software, can be used to display agricultural and human development synergy.

Human Development Index and Soybean Production data, for the years of 2000 and 2010, were integrated to QuantumGis software, used to graphically express changes in soybean production and Human Development Index in Goiás mesoregions.

Goiás State vectoral data, in ESRI shape-file format, were acquired from the Department of the Brazilian Institute of Geography and Statistics Maps (IBGE Cities 2018).

Human Development Index data and Soybean Production tables were integrated to the Goiás state vectoral shape-file archive to generate the Geographical Database within the QuantumGis software using the "join" tool. Data were clustered using Jenks Natural Breaks method (Chen et al. 2013)

TerraView GIS version 5.1.5, a free and open source software, was used to develop the spatial statistical analysis (TerraView 2010).

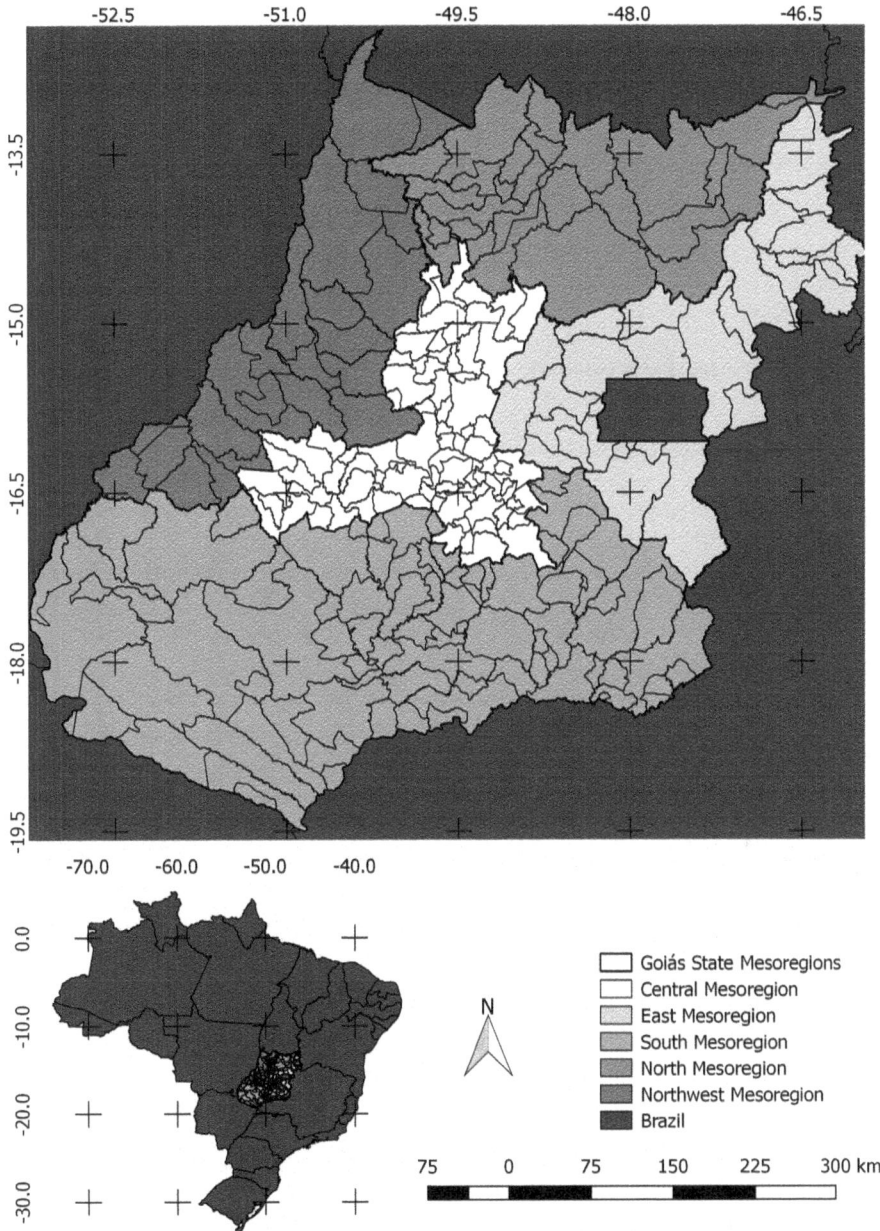

Fig. 2 Goiás state, study area location and mesoregions division

2.4   Statistical Analysis

Tukey test with 5% significance was used to compare Soybean Production for the years of 2000 and 2010, and for Human Development Index for 2000 and 2010. Statistical Package for Social Sciences (SPSS) software was used to conduct statistical tests.

Clusters were created using the Moran method for spatial analysis; the indicator used was the Local Indicator of Spatial Association (LISA) (Anselin 1995).

3   Results and Discussion

Table 1 presents the data comparison and statistical significance for Soybean Production for the years of 2000 and 2010, and for Human Development Index for 2000 and 2010. The statistical analysis present significant increase in three mesoregions of Goiás State for production variable, and two display no significant change for production variable. Regarding HDI, all mesoregions present significant change for the analyzed time period. de Andrade et al. (2004) conducted a work using the Tukey statistical test to HDI index studies for Paraná state, Brazil, and the statistical test was suitable for HDI analysis. Maciel and Khan (2009), used the Tukey statistical test to conduct analyses regarding agribusiness in Brazil, the test was suitable for the research purposes.

As shown in Table 1, all mesoregions present an increase in Soybean production; thus, Center, South and Northwest mesoregions present statistical significance regarding this increase. The mesoregion that presented the highest increase was the Northwest region with a mean production of 100 tons in the year 2000 and 2767 in the year 2010, followed by the Center region with a mean production of 546 tons in the year 2000 and 11,758 in the year 2010, and South region with a mean production of 44,915 tons in the year 2000 and 70,097 in the year 2010.

Table 1 Mean scores for soybean production for the years of 2000 and 2010, and for human development index for 2000 and 2010 for Goiás mesoregions' data comparison and statistical significance

Regions	Mean score		Significance	Mean score		Significance
	Production (ton)			HDI		
	2000	2010		2000	2010	
Center	546	1758	*	0.56	0.70	***
East	9855	33,874	Non-significant	0.51	0.67	***
South	44,915	70,097	*	0.59	0.71	***
North	1899	7896	Non-significant	0.51	0.66	***
Northwest	100	2767	*	0.54	0.68	***

Figure 3 displays the data mentioned in the last paragraph, it can be observed a pattern of production redistribution, where in the year 2000 there was a more widespread production, in the year 2010 this production is compiled in certain municipalities that in the year 2000 already displayed a notorious production number. In the year 2000, it can be seen four main producers, distributed in the center, east, and south mesore-

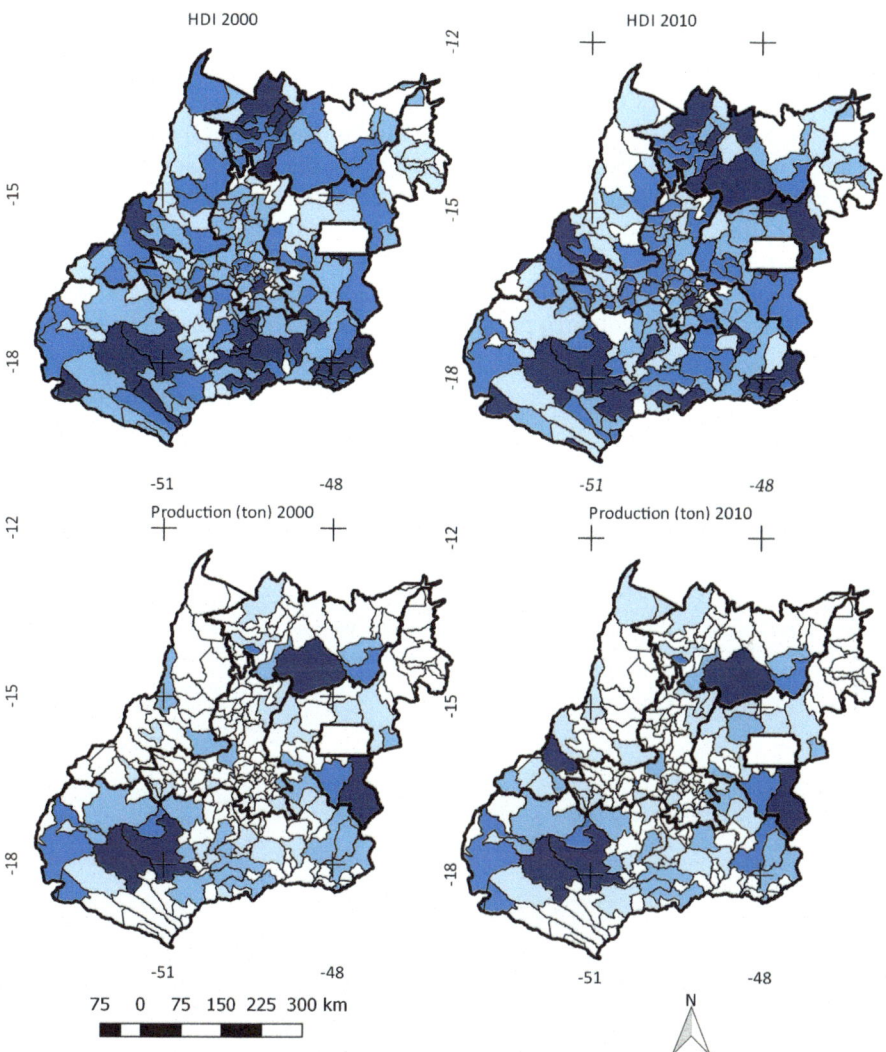

Fig. 3 Geographical display of information regarding human development index for the years of 2000 and 2010 and soybean production (ton) for the years of 2000 and 2010. Legend can be found on Fig. 4

gions. In the year 2010, it can be observed that the north and northwest mesoregions added municipalities with the highest yield of production.

The maps display increase for Human Development Index for the studied Goiás State mesoregions. The maps display that certain municipalities decreased soybean production, indicating that the increase was focused on certain strategically positioned cities. Constraints that influence this production migration are related to logistic and transportation issues, as to agricultural zoning, water availability, and soil potential for agriculture, this data can be found in the state focused efforts on conducted studies that identified optimum locations to soybean production, as the Agricultural Zoning for Goiás State. As seen in Fig. 4.

Figure 5 presents studied data using the Moran method for spatial distribution, it can be observed that clusters are formed showing the municipalities that present similar yield of soy production, and similar level of human development. The areas that present dark green color (3) have a strong statistical similarity, they have strong spatial correlation with p-value $= 0.001$, areas with moderate dark green color (2) present p value $= 0.01$, areas with light green color (1) present p value $= 0.05$, and areas with bright green color (0) present no statistical correlation.

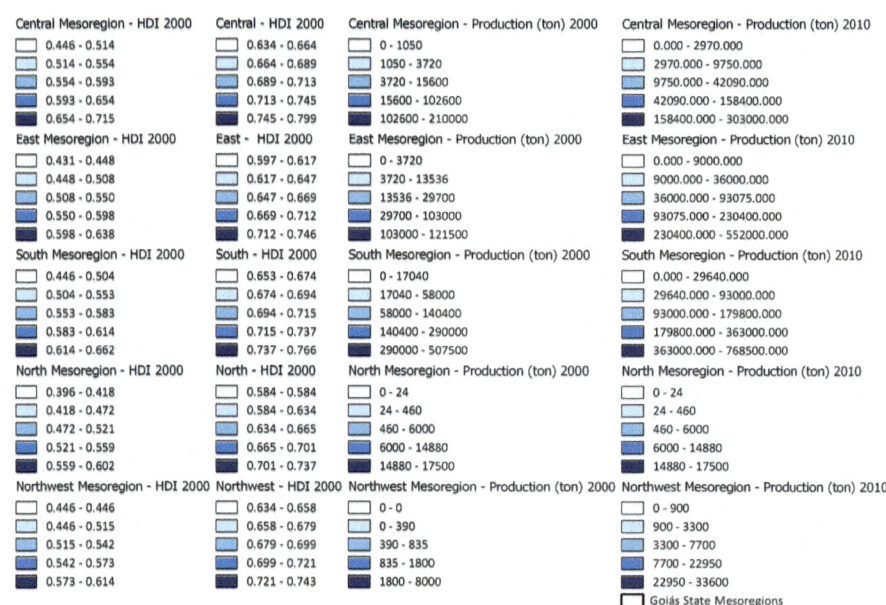

Fig. 4 Values for human development index for the years of 2000 and 2010 and soybean production (ton) for the years of 200 and 2010 of the aforementioned maps. The figure shows the first column the HDI index for year 2000 of the five studied mesoregions, in the second column the HDI index for year 2010 of the five studied mesoregions, in the third column is presented data regarding soybean production (ton) for year 2000 of the five studied mesoregions, and in the fourth column is presented data regarding the soybean production (ton) for year 2010 of the five studied mesoregions

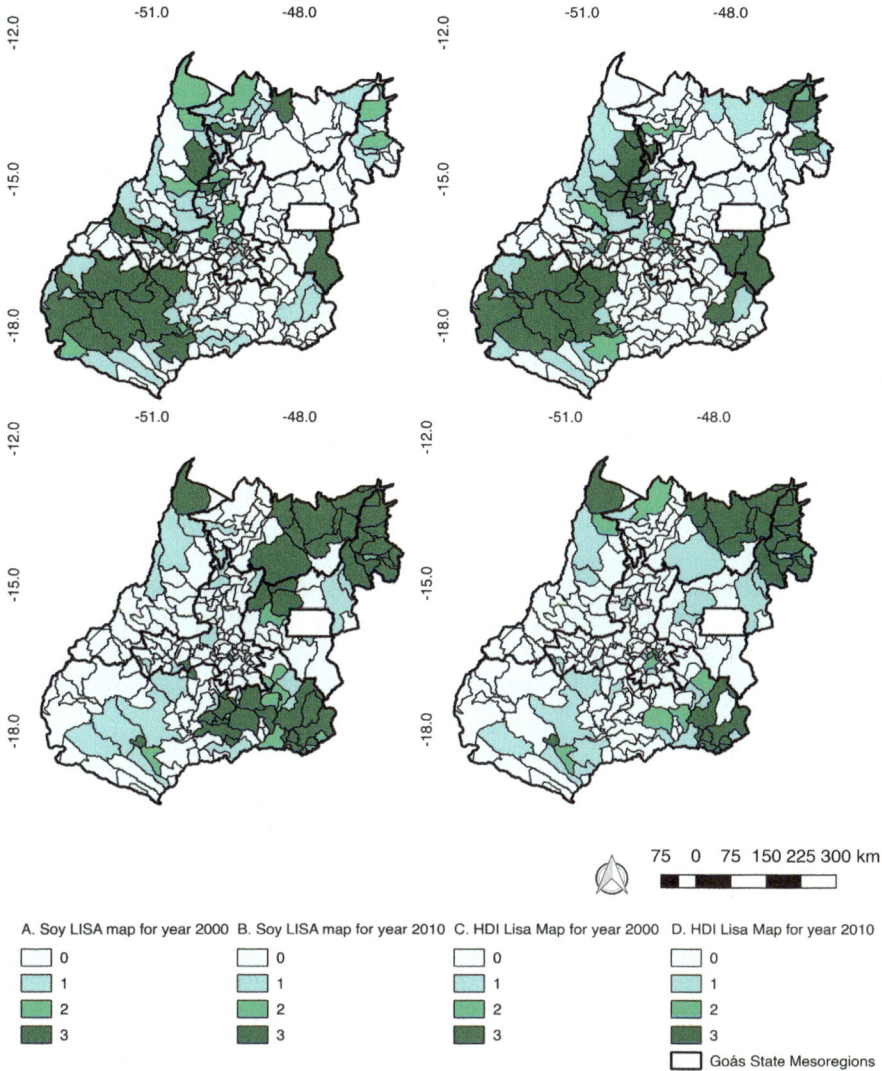

Fig. 5 Local index spatial association map for soy production and HDI values, for the year 2000 and 2010

The map shows that soy production has evolved to be more centralized in all regions, from the year 2000 to year 2010. Regarding the HDI, it showed an increase in all municipalities. The figure displays clearly that certain regions are highly influenced by soy production increase, for example, the extreme East micro region within South mesoregion increased its soy production, as well as human development, the extreme northeast micro region of Goiás State presents a centralization of soy production, causing a cluster of municipalities with high HDI index in the same region.

Table 2 Goiás State human development index divided by its composing aspects

Aspect	SHDI		
	1991	2000	2010
Education	0.278	0.439	0.646
Longevity	0.668	0.773	0.827
Income	0.633	0.686	0.742

The maps display Human Development Index for Goiás State mesoregions; it can be observed that 98% of all 246 municipalities presented increase with statistical significance, as seen in the aforementioned Table 1.

Even though the increase in Soybean Production was not a reality for all Goiás State Municipalities, this increase reflects directly on Human Development indexes, the municipality Gross Domestic Product increased, as well as population growth was a reality. When we observe municipality services such as health, labor and education, it is clear that there was also an increase in investment on these variables. The Goiás State general Human Development Index has risen from 0.49 in 1991 to 0.75 in 2010.

The State Human Development Index (SHDI) for Goiás State can be seen in Fig. 5. Regarding educational aspects, Goiás State has increased its literacy from 89% in 2000 to 93% in 2010, also the state has increased its general index from 0.28 in 1991 to 0.06 in 2010, for Longevity, from 0.67 in 1991 to 0.8 in 2010, and for Income, from 0.63 in 1991 to 0.7 in 2010, as shown in Table 2.

3.1 Social, Economic, and Environmental Synergy with Agribusiness for Sustainable Development

Environmentally friendly Agribusiness best practices reflect directly on surrounding community life quality by elevating the municipalities Gross Domestic Product, employment and per capita income growth, access to good quality health services, food security, and education (Tersoo 2013; Mellor 2017). Specifically to soybean production, studies have shown that the increase in productivity is related to decrease in poverty, better schools and improvement in HDI (Richards et al. 2015).

Goiás municipalities' Human Development Index increase can be reinforced when analyzing Fig. 6, which graphically shows the increase in Education, Health and Labor expenditures by mesoregions. The figure also shows the Agricultural Municipal income. All the data present an increase trend.

This increase can also be related to the agribusiness system complex in developing cities and regions, due to investments, attractiveness of new trades and promotion of employment, new industries and investors. It also influences the service sector and commerce. It is a resourceful relationship of synergy (Fig. 7).

Countries that rely on agribusiness present a perspective of incorporating good practices in agriculture, moving towards a more sustainable system. These practices

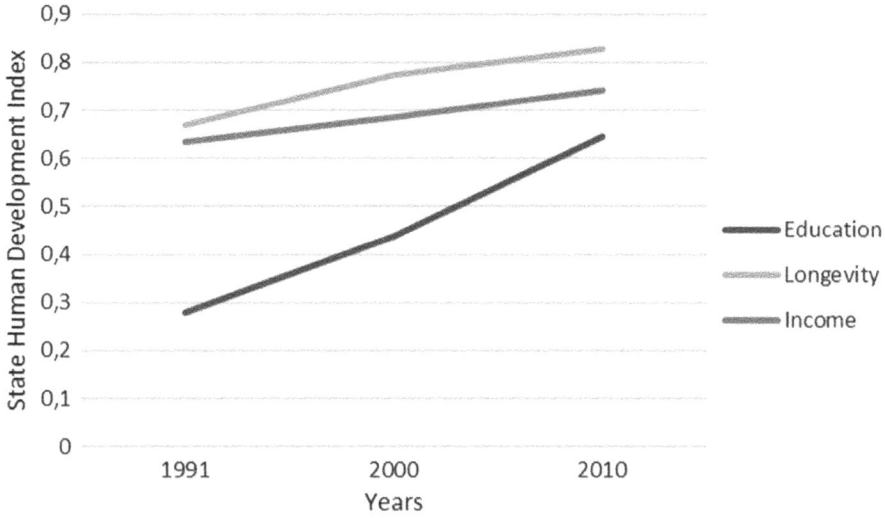

Fig. 6 Goiás State human development index presenting education, longevity and income aspects for the years of 1991, 2000 and 2010

unfold a chain reaction on several dimensions such as social, with creation of new job positions such as environmental managers, environmental by adopting soil, water and air impact minimizing technologies and monitoring initiatives; cultural by including it in undergrad and grad courses using innovative approaches towards agricultural sustainable development; and economic by creating new markets for organic food and environmental certifications.

Advantages of good sustainable practices for livestock and agriculture synergy with human development can be cited as the example of farm effluents treated by bio digesters, that present some characteristics as follow (Lansing et al. 2008):

- Great low cost source of macronutrients such as Nitrogen, Phosphorous, and Potassium,
- Vegetal development improvement that can be used for livestock nutrition,
- Innovative entrepreneurship approach that attracts investors,
- Biogas can be used as any other inflammable gas,
- It has the advantage of being produced within the farm,
- It's environmentally friendly when compared to other main electricity sources of the region such as coal, and fossil fuels,
- It provides electricity for small communities without access to the main grid.
- Reduces gender inequality.

Some of the challenges raised by the researchers of the Federal University of Ceará, in the Brazilian northeast region that implanted several biodigesters for treatment and reuse as fertilizers and electricity generation out of livestock waste were: (1) difficulty regarding lack of trained human resources to install and provide main-

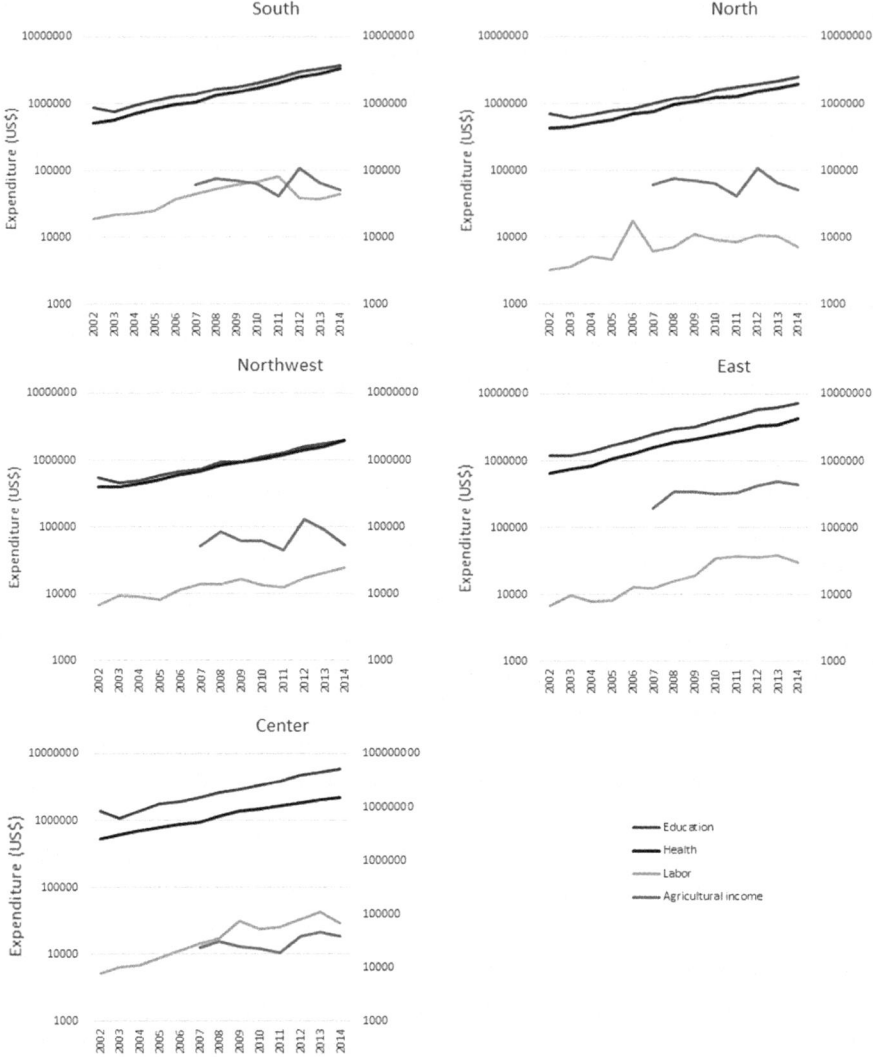

Fig. 7 Municipal expenditure in education, health and labor from 2002 to 2014, and agricultural municipal income from 2007 to 2014

tenance, (2) lack of ecological mindset and (3) lack of politic continuity and policy related tools (2002).

Brazil follows this path by incorporating in this system, by demand or proactively, best practices of sustainable production, replicating and innovating diversely on its many types of production, such as livestock, crops, and cereals. Also, implementing these practices in all logistic steps of the production, distribution and consumption chain reduces cost and environmental impacts (Onwude et al. 2016). Practices such as using agricultural waste to produce energy or to feed livestock are spreading throughout Brazil. In Goiás state, producers see the potential of this practice on its daily costs by reducing energy and food consumption (Ferreira-Leitão et al. 2010).

4 Conclusions

This work's main findings show that municipalities that increased soybean production also increased its HDI, although several variables might influence the increase in life quality, agribusiness is known to be one of the main drivers; the use of GIS software is an effective and low-cost tool to present agribusiness information to stakeholders.

Brazilian production is following international market requirements demanding that these commodity methods of production be adapted to environmentally friendly best practices. Brazilian producers have positively responded to these demands by implementing best practices in soil and water resources usage, producing more by exploiting less, increasing the work environment quality of employees, and life quality to surrounding communities, fostering the allocation of basic educational, health and economic resources. Brazil and Goiás State present a trend in incorporating an agribusiness system, sustainable best practices for production, distribution and consumption of agricultural resources, and waste reuse.

Global Information System (GIS) technologies' usage in social, environmental and economic sciences presented an increase in the past decades, enhancing data display and analysis by geographically presenting this information, always moving towards improving management and decision-making processes.

GIS is a key tool for agribusiness decision-making. By integrating Goiás State social and economic data in a geographic database within QuantumGis and TerraView softwares, free and open source GIS platforms, decision makers can identify, with statistical support and graphical display, where to implement plans and programs to promote agricultural development considering the synergy between environmental fragility and economical aspects, hence, fostering economic growth and human development.

References

Anselin L (1995) Local indicators of spatial association—LISA. Geogr Anal 27:91–115
Balram S, Dragićević S (2005) Attitudes toward urban green spaces: integrating questionnaire survey and collaborative GIS techniques to improve attitude measurements. Landscape Urban Plan 71(2–4):147–162. https://doi.org/10.1016/J.LANDURBPLAN.2004.02.007
Bateman IJ et al (2002) Applying Geographical Information Systems (GIS) to environmental and resource economics. Environ Resour Econ 22(1/2):219–269. https://doi.org/10.1023/A:1015575214292

Cegan JC et al (2017) Trends and applications of multi-criteria decision analysis in environmental sciences: literature review. Environ Syst Decisions 37(2):123–133. https://doi.org/10.1007/s10669-017-9642-9

Chen J et al (2013) Research on geographical environment unit division based on the method of natural breaks (Jenks). https://doi.org/10.5194/isprsarchives-xl-4-w3-47-2013

Corbelle-Rico E, Santé-Riveira I, Crecente-Maseda R (2018) A decision support system for farmland preservation: integration of past and present land use. Springer, Berlin, pp 173–192. https://doi.org/10.1007/978-3-642-37896-6_8

de Almeida Bressiani D, Gassman PW, Fernandes JG, Garbossa LHP, Srinivasan R, Bonumá NB, Mendiondo EM (2015) Review of soil and water assessment tool (SWAT) applications in Brazil: challenges and prospects. Int J Agric Biol Eng 8(3):9–35

de Andrade SM, Matsuo T, Soares DA, de Souza RKT, de Freitas Mathias TA, Iwakura MLH, Zequim MA (2004) Condições de vida no Estado do Paraná: análise ecológica com base em variáveis do Censo Demográfico de 2000. Semina: Ciências Biológicas e da Saúde 25(1):73–80

Del Grosso SJ et al (2018) Simple models to predict grassland ecosystem C exchange and actual evapotranspiration using NDVI and environmental variables. Agric Forest Meteorol 249:1–10. https://doi.org/10.1016/J.AGRFORMET.2017.11.007

Du M, Noguchi N (2017) Monitoring of wheat growth status and mapping of wheat yield's within-field spatial variations using color images acquired from UAV-camera system. Remote Sens 9(3):289. https://doi.org/10.3390/rs9030289

Eloy L et al (2016) On the margins of soy farms: traditional populations and selective environmental policies in the Brazilian Cerrado. J Peasant Stud 43(2):494–516. https://doi.org/10.1080/03066150.2015.1013099

FAO. Food and agriculture organization (2018) FAOSTATS. Available at: http://www.fao.org/faostat/en/#data. Accessed: 12 July 2017

FAO. Food and agriculture organization (2018) Reviewed strategic framework. Available at: http://www.fao.org/3/a-ms431reve.pdf. Accessed: 15 July 2017

Ferreira-Leitão V et al (2010) Biomass residues in Brazil: availability and potential uses. Waste Biomass Valorization 1(1):65–76. https://doi.org/10.1007/s12649-010-9008-8

Glennie E, Anyamba A (2018) Midwest agriculture and ENSO: a comparison of AVHRR NDVI3g data and crop yields in the United States corn belt from 1982 to 2014. Int J Appl Earth Obs Geoinf 68:180–188. https://doi.org/10.1016/J.JAG.2017.12.011

Gosch MS, Ferreira ME, Medina GDS (2017) The role of the rural settlements in the Brazilian savanna deforesting process. J Land Use Sci 12(1):55–70. https://doi.org/10.1080/1747423x.2016.1254687

Hettinga S (2014) Spatially optimizing the distribution of sustainable sugarcane production in Goiás, Brazil, based on GHG-emissions and costs. Available at: https://dspace.library.uu.nl/handle/1874/295981. Accessed: 25 May 2018

Hiloidhari M et al (2017) Emerging role of Geographical Information System (GIS), Life Cycle Assessment (LCA) and spatial LCA (GIS-LCA) in sustainable bioenergy planning. Bioresour Technol 242:218–226. https://doi.org/10.1016/J.BIORTECH.2017.03.079

Huang G, Jiang Y (2017) Urbanization and socioeconomic development in inner Mongolia in 2000 and 2010: a GIS analysis. Sustainability 9(2):235. https://doi.org/10.3390/su9020235

IBGE. Geography and statistics brazilian institute. Soybean production report. Available at: https://sidra.ibge.gov.br/pesquisa/lspa/tabelas. Accessed: 04 Apr 2017

IBGE Cities. Geography and statistics brazilian institute. Cities. Available at: https://cidades.ibge.gov.br/. Accessed: 04 Apr 2017

Lansing S, Víquez J, Martínez H, Botero R, Martin J (2008) Quantifying electricity generation and waste transformations in a low-cost, plug-flow anaerobic digestion system. Ecol Eng 34(4):332–348

Li G, Weng Q (2007) Measuring the quality of life in city of Indianapolis by integration of remote sensing and census data. Int J Remote Sens 28(2):249–267. https://doi.org/10.1080/01431160600735624

Maciel HM, Khan AS (2009) O impacto do programa de microcrédito rural (agroamigo) na melhoria das condições de vida das famílias beneficiadas no estado do Ceará: um estudo de caso. Revista de Economia e Agronegócio 7(1)

Maksimov P et al (2017) GIS-supported epidemiological analysis on canine Angiostrongylus vasorum and Crenosoma vulpis infections in Germany. Parasites Vectors 10(1):108. https://doi.org/10.1186/s13071-017-2054-3

Mellor JW (2017) The subsistence farmer in traditional economies. In: Subsistence agriculture and economic development, pp. 209–226. Routledge

Mokashi MS, Perdita Okeke P, Uma Mohan U (2017) Study on the use of Geographic Information Systems (GIS) for effective transport planning for Transport for London (TfL). Springer, Singapore, pp 719–728. https://doi.org/10.1007/978-981-10-1678-3_69

Onwude DI et al (2016) Mechanisation of large-scale agricultural fields in developing countries—a review. J Sci Food Agric 96(12):3969–3976. https://doi.org/10.1002/jsfa.7699

Parente L, Ferreira L (2018) Assessing the spatial and occupation dynamics of the Brazilian pasturelands based on the automated classification of MODIS images from 2000 to 2016. Remote Sens 10(4):606. https://doi.org/10.3390/rs10040606

Paterson RW, Boyle KJ (2002) Out of sight, out of mind? Using GIS to incorporate visibility in hedonic property value models. Land Econ 78(3):417–425. https://doi.org/10.2307/3146899

Pierce FJ, Lal R (2017) Monitoring the impact of soil erosion on crop productivity. Routledge, London, pp 235–263. https://doi.org/10.1201/9780203739358-10

Ramadas M, Samantaray AK (2018) Applications of remote sensing and GIS in water quality monitoring and remediation: a state-of-the-art review. Springer, Singapore, pp 225–246. https://doi.org/10.1007/978-981-10-7551-3_13

Ratcliffe J, Chainey S (2013) GIS and crime mapping. Wiley, Hoboken. Available at: https://books.google.com.br/books?hl=pt-BR&lr=&id=FUEh9TUVNagC&oi=fnd&pg=PT11&dq=crime+rate+zone+gis&ots=jKAJ7V83Uz&sig=10-_fyfndVuRBUDd7NxRpuTe7Ho#v=onepage&q=crimeratezonegis&f=false. Accessed: 24 May 2018

Richards P, Pellegrina H, VanWey L, Spera S (2015) Soybean development: the impact of a decade of agricultural change on urban and economic growth in Mato Grosso, Brazil. PLoS one 10(4). https://doi.org/10.1371/journal.pone.0122510

Rodias E et al (2017) Energy savings from optimised in-field route planning for agricultural machinery. Sustainability 9(11):1956. https://doi.org/10.3390/su9111956

Sahu BK (2015) A study on global solar PV energy developments and policies with special focus on the top ten solar PV power producing countries. Renew Sustain Energy Rev 43:621–634

Santos KEL, Bernardi ADC, Bettiol GM, Crestana S (2017) Geoestatística e geoprocessamento na tomada de decisão do uso de insumos em uma pastagem/geostatistics and gis in the decision making of the use of inputs in a pasture. Revista Brasileira de Engenharia de Biossistemas 11(3):294–307

Seynave I et al (2018) GIS Coop: networks of silvicultural trials for supporting forest management under changing environment. Ann Forest Sci 75(2):48. https://doi.org/10.1007/s13595-018-0692-z

Spinney JEL, Scott DM, Newbold KB (2009) Transport mobility benefits and quality of life: a time-use perspective of elderly Canadians. Transp Policy 16(1):1–11. https://doi.org/10.1016/J.TRANPOL.2009.01.002

Swetnam RD et al (2011) Mapping socio-economic scenarios of land cover change: a GIS method to enable ecosystem service modelling. J Environ Manag 92(3):563–574. https://doi.org/10.1016/J.JENVMAN.2010.09.007

TerraView 5.1.5. INPE, São José dos Campos, SP (2010)

Tersoo P (2013) Agribusiness as a veritable tool for rural development in Nigeria. Mediterr J Soc Sci. https://doi.org/10.5901/mjss.2013.v4n12p21

Zambrano LI et al (2017) Estimating and mapping the incidence of dengue and chikungunya in Honduras during 2015 using Geographic Information Systems (GIS). J Infect Public Health 10(4):446–456. https://doi.org/10.1016/j.jiph.2016.08.003

The Effects of Climatic Variations on Agriculture: An Analysis of Brazilian Food Exports

Gisele Mazon, Beatrice Maria Zanellato Fonseca Mayer,
João Marcelo Pereira Ribeiro, Sthefanie Aguiar da Silva,
Wellyngton Silva de Amorim, Larissa Pereira Cipoli Ribeiro,
Nicole Roussenq Brognoli, Ricardo Luis Barcelos, Gabriel Cremona Parma,
Jameson Henry McQueen, Issa Ibrahim Berchin
and José Baltazar Salgueirinho Osório de Andrade Guerra

Abstract Studies on climate change's impact on agricultural production including soybean and corn crops in Brazil require subnational analysis due to the large expanse of the country and its diverse climate, geographies, biomes and availability of natural resources. Accordingly, this study aims to understand the effects of climatic

G. Mazon (✉)
University of Vale do Itajaí (UNIVALI), Itajaí, Brazil
e-mail: gisamazon@gmail.com

G. Mazon · B. M. Z. F. Mayer · J. M. P. Ribeiro · W. S. de Amorim · L. P. C. Ribeiro ·
G. C. Parma · I. I. Berchin
Research Centre on Energy Efficiency and Sustainability (GREENS), Rio de Janeiro, Brazil
e-mail: beatrice.mayer@unisul.br

J. M. P. Ribeiro
e-mail: joaomarceloprdk@gmail.com

W. S. de Amorim
e-mail: wellyngton.amorim@gmail.com

L. P. C. Ribeiro
e-mail: larissapcr@gmail.com

G. C. Parma
e-mail: gabriel.parma@unisul.br

I. I. Berchin
e-mail: issa.berchin@gmail.com

B. M. Z. F. Mayer · J. M. P. Ribeiro · S. A. da Silva · W. S. de Amorim · N. R. Brognoli ·
R. L. Barcelos · J. B. S. O. de Andrade Guerra
University of Southern Santa Catarina (UNISUL), Tubarão, Brazil
e-mail: sthefanie.sads@hotmail.com

N. R. Brognoli
e-mail: nicoleroussenq@gmail.com

R. L. Barcelos
e-mail: ricardo.barcelos@unisul.br

© Springer Nature Switzerland AG 2020   321
W. Leal Filho et al. (eds.), *International Business, Trade and Institutional
Sustainability*, World Sustainability Series,
https://doi.org/10.1007/978-3-030-26759-9_19

variations on Brazilian agriculture and to understand how the expansion of land dedicated to crop production is influenced by production for export. Accordingly, four hypotheses were created and tested. The descriptive analysis of the data showed that soybean production for exportation is heavily concentrated in the last four months of the year, which indicates the seasonality of its production. The cluster analysis of all the variables (climate and soybean and corn production) shows that there is no relationship between climate variables and grain production variables, since both sets of elements are in separate groups. The non-relationship of climate variables and grain production are also confirmed when the data are analyzed by the principal components model, either by the Kaiser criterion or by the scree plot criterion. In this case, it is possible that crop yields are influenced by irrigation technologies, which reduces the dependence on rainfall patterns, but this data was not considered in this study.

Keywords Climate change · Agriculture · Food security · Sustainable development · International trade

1 Introduction

The creation of the World Trade Organization (WTO) facilitated international trade by reducing trade barriers and providing a platform for dialogue, fair trade, and conflict resolution. In Brazil, international trade gained prominence with a process of economic liberalization and government push to encourage an "exportation culture" within the country (Caron 2009). This process was supported by the creation of the Mercosul, which aimed to increase economic and trade integration in South America (Amann et al. 2014).

In the past few decades, Brazil has increased its share in international trade, particularly of agricultural commodities. This is due, in part, to its comparative advantages regarding the abundance of natural resources and land, which contribute to large agricultural production and exports. Thus, the Brazilian exportations of

J. B. S. O. de Andrade Guerra
e-mail: baltazar.guerra@unisul.br

R. L. Barcelos
PROFORME and GESEG Groups, UNIVALI University, Itajaí, Brazil

G. C. Parma
Federal University of Santa Catarina (UFSC), Florianópolis, Brazil

J. H. McQueen
Whittier College, Whittier, USA
e-mail: jameson.mcqueen@gmail.com

J. B. S. O. de Andrade Guerra
Department of Land Economy, Cambridge Centre for Environment, Energy and Natural Resource Governance (C-EENRG), University of Cambridge, Cambridge, UK

agricultural commodities was favored by the increasing prices of the commodities and an exchange rate, which placed the Brazilian currency at a far lower worth than the US dollar (Medeiros 2013; Serrano and Summa 2014).

The Brazilian preference for specializing in agricultural commodities raise questions regarding the actual benefits brought to Brazil's own internal development. Some authors argue (Gala 2006; Bresser-Pereira and Marconi 2008; Oreiro and Feijó 2010) that investing in more technological sectors within the industry would be more beneficial for the economy. Moreover, they argue that a given country would experience advantages regarding national development by having a positive coefficient of income elasticity, better education, infrastructure, and by producing goods with higher added value. Thus, this specialized and more technological/industrialized production would result in better development and economic growth rates (Sachs and Warner 1995; Dalum et al. 1999; Rodrik 2006).

However, the international exportation of agricultural commodities detects more stability in demand and prices. This is accurate in periods of economic crises, such as observed in 2008 (Brahmbhatt and Canuto 2010). Thus, the valorization of the Brazilian real in relation to the US dollar had a negative effect on the industry, whilst the increasing international demand for natural resources and agricultural commodities benefited this sector.

In 2017, Brazil was the largest exporter of soybeans, corresponding to 42.5% of the total amount traded globally (Bradesco 2017a, b). Brazil's 2017 soybean harvest produced 114 million tons, from which over 50% was exported (USDA 2017). This data highlights the economic importance of soybean exportation for Brazil in the international trade market.

Brazil's National Supply Company (Conab) estimates that the 2017/2018 grain harvest will reach 225.6 million tons, from which the majority of the production (88%) are corn and soybean. Brazil is one of the largest soybean and corn exporting country in the world (FAO 2018).

International debates on food security urge for a fair trade of agricultural commodities (Vasconcelos et al. 2014), which are susceptible to climate and environmental conditions (Pellegrino et al. 2007). Overall, agricultural production is highly dependent on favorable climate, particularly rain-dependent agriculture and largescale monoculture (Drabo 2017). Climate change, therefore, affects agriculture and the international trade of these commodities (Rong 2010; Drabo 2017), by which the majority of production is localized to a few countries, such as Brazil and the United States of America who globally export to several countries (USDA 2017; FAO 2018).

Both soybean and corn production are seasonal crops. While soybeans start its production in May and its harvest in October, the corn starts its production in April and its harvest in September. This seasonality is due to the crop's high sensitivity to temperature and precipitation. This can also explain the discrepancy in the production within different regions of Brazil. In 2016, the Middle-West region produced 45.8% of all Brazilian soybean, whilst the South region produced 36.9% (Bradesco 2017a, b).

Brazil is responsible for 9.2% of the global corn production. However, its exportations responds to 22.4% of the world trade of this commodity (FIESP 2018). From the total amount of corn produced in 2017/2018 harvest (85 million tons), 30% was exported. Corn production in Brazil is also centralized in the Middle-West (42.5%) and the South (34.7%). The OECD-FAO (2017) Agricultural Outlook 2017–2026 projects an increase of both soybean (2.6% per year) and corn (1.7% per year) production in Brazil, mainly due to an increase of the planted area and the productivity.

Studies on climate change's impact on agricultural production in Brazil, for instance soybean and corn, require subnational analysis due to the diversity of climate, geographies, biomes and natural resources available. For this purpose, we aim to analyze the effects of climatic variations in Brazil, focusing on the country's agricultural exports. As a response, we will seek to analyze the impacts of climate change in Brazil, including the impacts of climatic variations on agriculture during 11 years of studies.

2   The Challenges of Climate Change in Brazil

Due to its fragile ecosystems and very large and diverse area, Brazil is prone to experience hard effects from climate change (Rong 2010). Particularly, it is very susceptible to suffer from more intense and longer droughts (Falloon et al. 2007), which besides affecting human health, community wellbeing and food production, also influences the environment and its ecosystems (Lucena et al. 2010).

Brazilian food production suffers from different impacts depending on the producing region. For example, the future scenarios of climate change for soybean production shows greater risk of climate change effects in the Northeast, extreme South and part of the Middle-West Brazil, whilst in the rest of the country the production remains viable (Pellegrino et al. 2007). Considering that the size of the planted area may be a factor in the food production, this study proposes the following hypothesis to be tested:

H1: The larger the planted area, the larger the production of the states.

The future water availability and precipitation rates also depend on the location of the Brazilian states, such as possible droughts in the eastern Amazon, the transformation of the Northeast semiarid to an even more arid area and a systematic increase of rainfall in South Brazil (Marengo 2008).

In the Northeast region, which is already characterized by high temperatures and an irregular precipitation pattern (Cunha et al. 2014), there is a risk for a large-scale migration of its population to other regions (Fearnside 1995; Barbieri et al. 2010) due to even lower precipitation rates and increased temperature which intensify regional droughts (Roland et al. 2012; Fearnside 1999). This context, stimulant on the worst-case scenario for water distribution and vulnerability, have negative socioeconomic impacts, including food insecurity and damages in crop production and exportation (Confalonieri et al. 2014).

Southern Brazil will face an increase in precipitation pattern due to climate change (Hirata and Conicelli 2012). This may change production seasonality, which may benefit irrigated crops like rice (Walter et al. 2014). It will also imply challenges for disaster management, coastal environmental, livelihood protection and food production (Figueiredo 2013; Vasconcelos et al. 2013; Marengo 2008). In general, expectations are for an increase in the frequency and intensity of rainfall, due to the subtropical climate (PBMC 2015), and temperature increases up to 6 °C, affecting soybean, corn and wheat production (Alberto 2006).

In the Southeast, climate change impacts can be perceived in many ways. Coffee crops, which are largely produced in São Paulo and Minas Gerais, will be drastically reduced in the next 100 years due to climate change, restricting production to a few municipalities and single adaption to increases in temperature and decrease in precipitation is likely, crop migration is also projected to be necessary (Assad et al. 2004).

The Middle-West Brazil must face the occurrence of extreme events, ranging from years of extreme rainfall patterns and years of severe drought periods, relying on the complementary irrigation for crop production (Marin and Nassif 2013). As for the Northeast, the Middle-West will face an increase in temperature that would decrease its soybean production capacity (Lucena et al. 2009).

As a large producer of food commodities, the Middle-West Brazil risks having a reduction in its production capacity causing economic impacts for the country and on food security (Vasconcelos et al. 2014).

The factors that may influence it is about the quantity, quality and regulation of food distribution. Under this perspective, it will necessary to use more fertilizers that restore nitrogen to the crop so that such a productivity loss can be reversed (Watanabe et al. 2012).

The second possible impact on food production can be described by the second hypotheses:

H2a: The higher the temperature, the larger the production volume, the larger the exportation of the states.
H2b: The higher the volume of rainfall, the larger the volume of production, the larger the exportation of the states.

The Brazilian Atlantic Forest, which extends along the entire coast of the country, will also suffer strong effects of climate change. Such hotspots for biodiversity and environmental conservation should be protected from deforestation and environmental/climate impacts by local governments and development programs (Scarano and Ceotto 2015; Guariguata et al. 2008).

The intensification of global climate changes, due to anthropogenic causes (pollution, overexploitation of resources and high emissions of greenhouse gases) threatens sustainable development and human wellbeing by rising sea levels, warming the atmosphere and oceans, including its acidification, increasing the intensity of droughts, floods and extreme weather events (IPCC 2013).

In addition to these events, the world population has grown dramatically in the past decades, whilst increasing the demand for more natural resources and land (Butler 2016). Trade is also negatively affected by climate change, which has a direct impact on food production. Governments and entrepreneurs are adopting new strategies and investing in new technologies to balance climate adaptation with the need to increase production (Ojima et al. 2008; Ribeiro 2014).

The third possible impact of climate change on food production is expressed by:

H3a: The higher the temperature, the higher the exportation of the states.
H4b: The higher the precipitation rate, the higher the exportation of the states.

The influence of climate change on international trade is inevitable as the environmental impacts on agriculture changes both the price of commodities and consumption patterns. Therefore, governments and international organizations such as the WTO addresses climate change as a main challenge for international trade, also developing strategies to mitigate the consequences of climate change. Among these strategies, are: Carbon credits, sustainable products (Ribeiro 2014) and tariff adjustments for countries that do not follow sustainable production standards (Matias 2013; Drabo 2017).

Brazil is one of the world's largest commodity exporters and it has a key role in facing the challenges of ensuring global food security. Climate models indicate that water resources will remain available for agriculture in Brazil, which enable an increase of agriculture production (Camargo ct al. 2017). This, however, does not signify that Brazil faces a simple road ahead. Several challenges, such as poverty, inequality, climate change and environmental issues persist to be on the frontline as national issues (OECD-FAO 2015).

The impact of food production on exportation by state are described as:

H4a: The higher the commodities' value, the higher the volume of production of the states.
H4b: The higher the commodities' value, the lower the exportation volume of the states.

3   Methods

3.1   Analysis and Statistical Data Modeling

The analysis to develop the statistical modeling used a univariate description of the data, which were collected and organized by month in a period of 11 years (from 2007 to 2017). This univariate organization of the data was used to define monthly averages, standard deviations, amplitude and monthly histograms of each variable. With the histograms and averages, the months of greater influence of the variables were observed, thus identifying the months of major and minor productions and the different climatic conditions analyzed.

The unidimensional descriptive measures allow the identification of the indicators of central tendencies and the measures of variabilities of each element. Thus, allowing the analysis of the distribution of each variable from the histogram, in this case grouped by month, to observe if its distribution is regular (Larson and Farber 2010).

Subsequently, a simple linear regression model was implemented to the variables of soybean and corn, aiming to understand the relationships between production for domestic consumption and for exportation. Thus, another simple linear regression was used among the export variables of corn and soybean, resulting in weak or very weak correlations between the variables analyzed two-to-two.

The descriptive analysis of the multivariate data used the hierarchical grouping and the principal component method to analyze the interrelations between all the variables. Which indicated that the relationships between grain and climate variables were not related to each other.

The clustering analysis is a statistical/mathematical model that groups individuals or variables in clusters. The model observes the greatest similarities between the elements of the groups and the greatest differences between different groups (Santos et al. 2007). This study used the Ward relationship method and the Euclidean distance.

The principal components method was applied to verify how the variables were regrouped in the new components, according to their correlations and inter-correlations, in order to obtain the scores of the variables in each of the principal components analyzed.

Once the univariate and multivariate descriptive analysis of the data was done, the multivariate modeling was performed by the multiple linear regression model, which used grains (soybean, corn) and climate variables as independent variables. This analysis demonstrated little or minimal influence of climate variables on the corn and soybean exportation in the analyzed regions.

4   Results: Univariate and Multivariate Descriptive Data Analysis

The descriptions of the data are presented by histograms, means and standard deviations of the main variables to analyze their behavior over the period analyzed.

Figures 1, 2, 3, 4, 5, 6 and 7 show the sets of histograms, modeled by months. They present the analysis of the main variables. For instance, Fig. 1 shows the constant distribution of the soybean production for domestic consumption, during the 12 months of the year. Figure 2 shows the histogram of the soybean production for exportation by month, in which it is observed that in the domestic production, there is a strong concentration of production in the last four months of the year.

Figures 3 and 4 shows the histograms of both internal and external corn production by month. In the case of domestic production, as in the case of soybeans, a fairly constant distribution is observed over the months; same observation of normal distribution can be made for monthly production for exportation over the year.

Figures 5, 6 and 7 shows the histograms for the climate variables, separated by each region analyzed. Through comparing these three histograms, we observe the climatic similarities between the Southeast and Middle-West regions and their

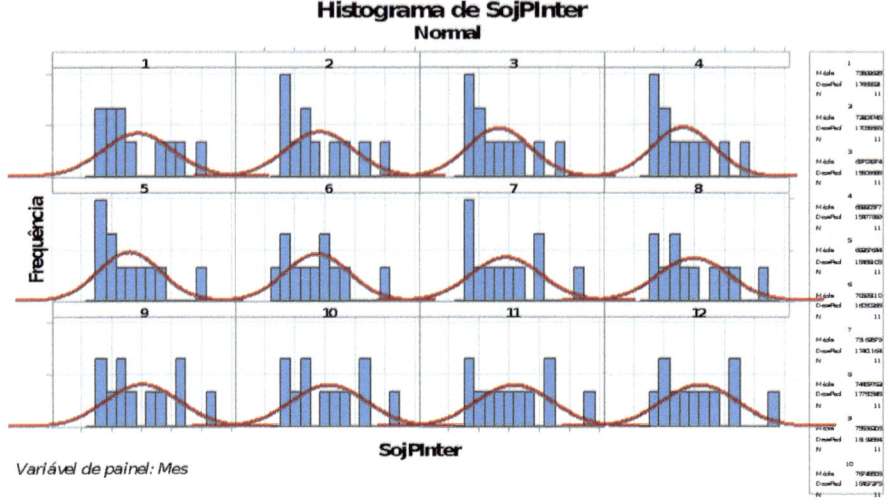

Fig. 1 Histogram of the soybean production for domestic consumption by month

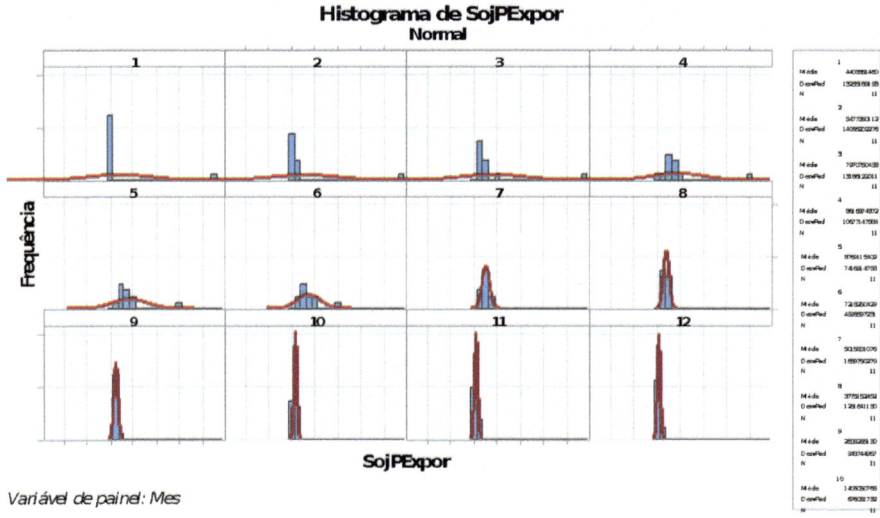

Fig. 2 Histogram of the soybean production for exportation by month

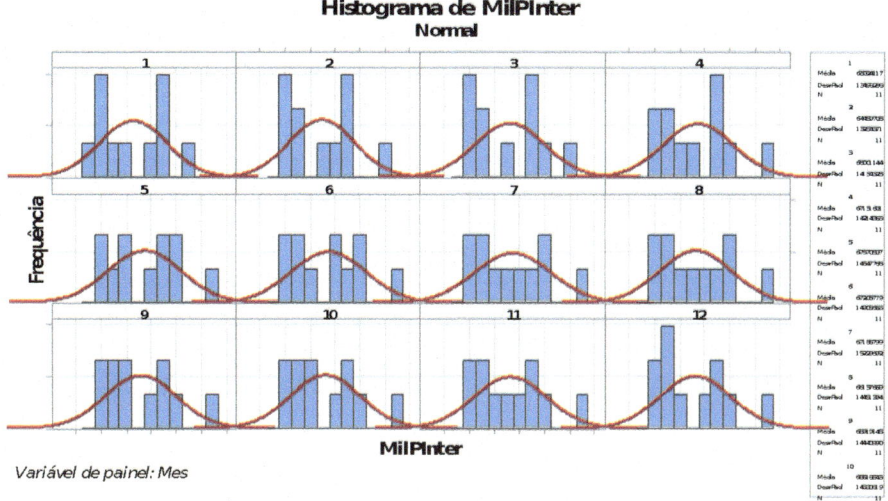

Fig. 3 Histogram of the soybean production for domestic consumption by month

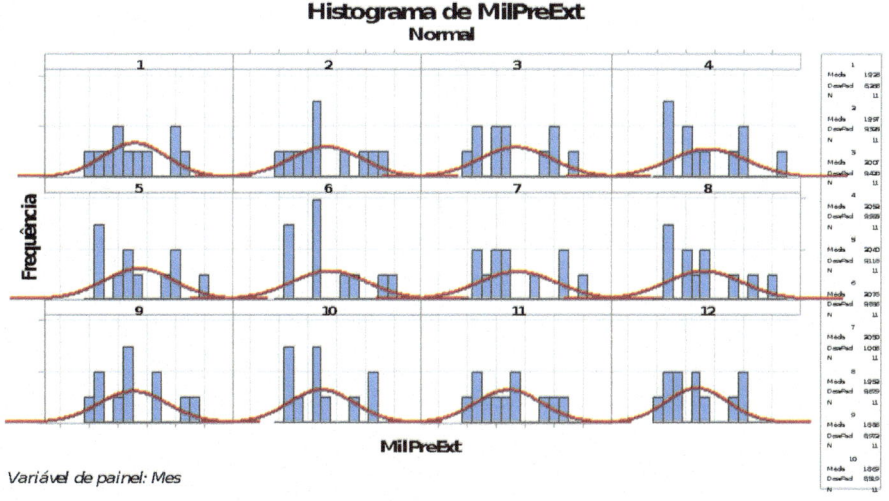

Fig. 4 Histogram of the corn production for exportation by month

differences with the climatic conditions of the Southern region. This fact will be observed in the analysis by hierarchical groupings and their dendrograms.

The simple linear regression was used between the variables of soybean and corn production for internal consumption and exportation, as well as the simple linear regression between corn and soybean exports, in order to analyze if there is any relationship between them in period analyzed.

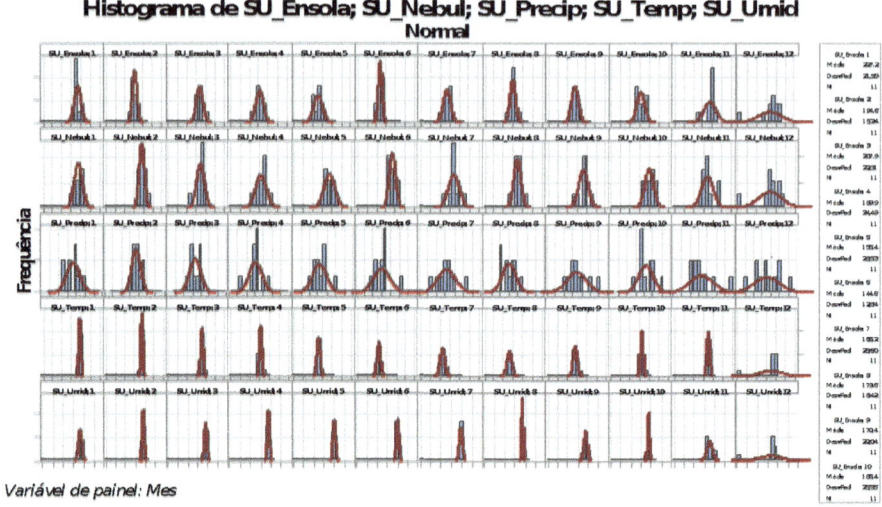

Fig. 5 Histograms of the climate variables of the South region by month. Variables: solar irradiation, cloudiness, precipitation, temperature, humidity

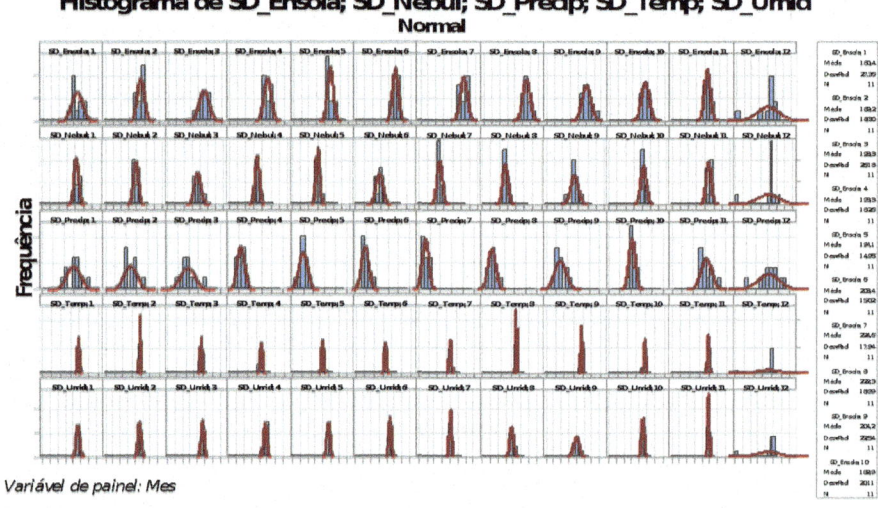

Fig. 6 Histograms of the climate variables of the Southeast region by month. Variables: solar irradiation, cloudiness, precipitation, temperature, humidity

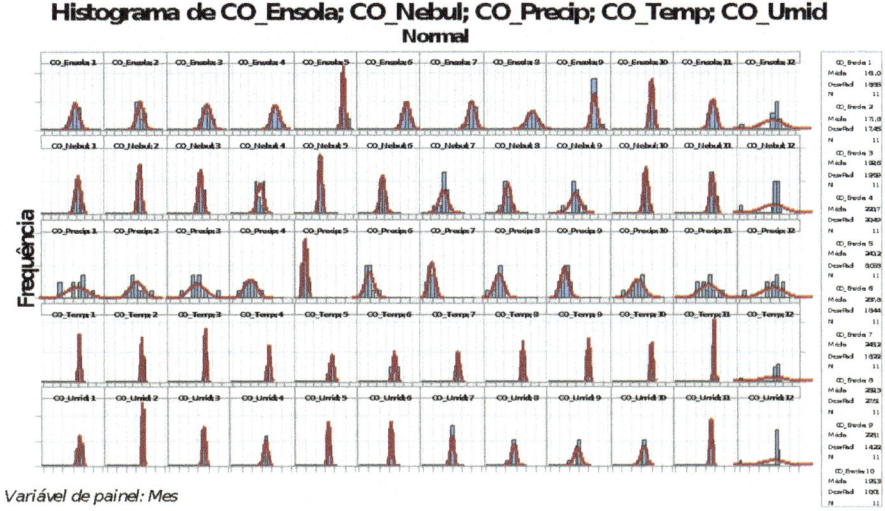

Fig. 7 Histograms of the climate variables of the Middle-West region by month. Variables: Solar irradiation, cloudiness, precipitation, temperature, humidity

Figures 8, 9 and 10 shows the graphs of linear regressions, the regression equation of the R^2 of each of the relationships made.

Figure 8 shows the simple linear regression between corn production for export and domestic consumption, it also shows that the relationship between the two variables is very low ($R^2 = 20\%$), which indicates that variability of domestic production only demonstrates a 20% variability of production for export. This low relation

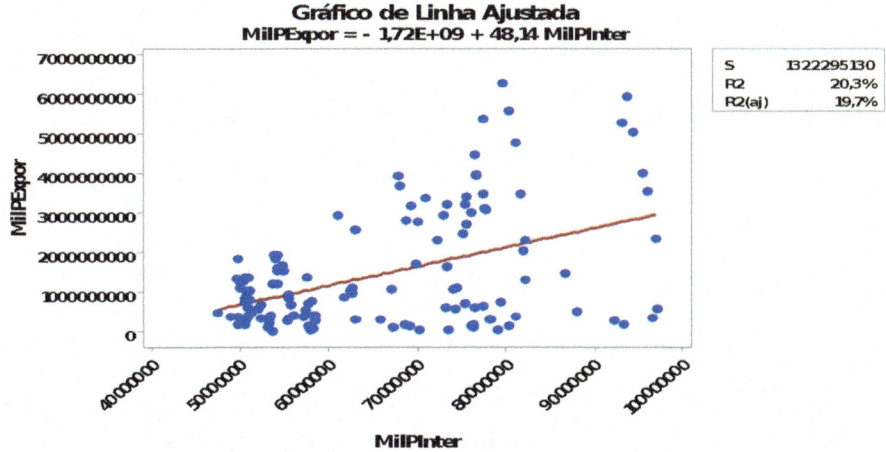

Fig. 8 Simple linear regression between corn production for domestic use (x) and for exportation (y)

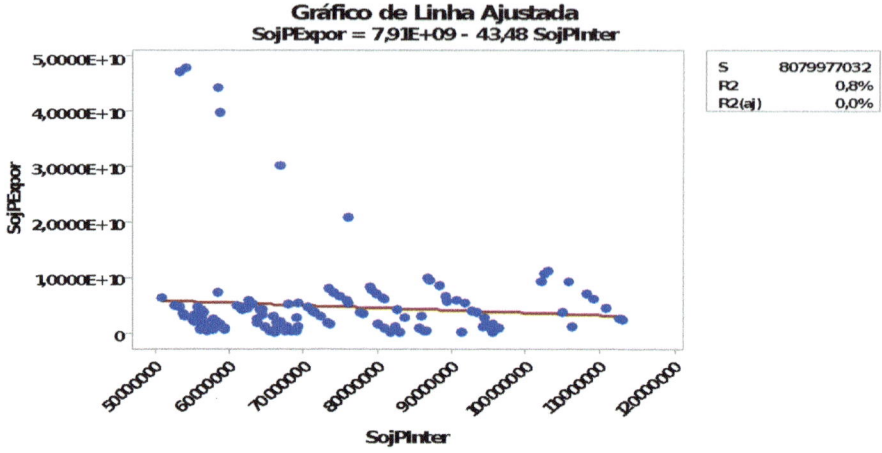

Fig. 9 Simple linear regression between soybean production for domestic use (x) and for exportation (y)

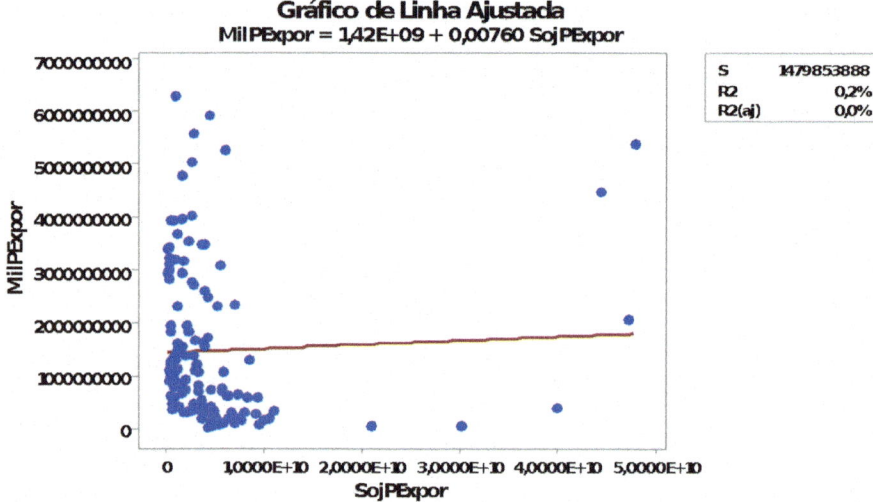

Fig. 10 Simple linear regression between the production for exportation of the soybean (x) and the corn (y)

expressed by the R2 indicates that there are other causes of the variability of production for exportation, in addition to domestic production.

The simple linear regression between the production for exportation of the soybean and the corn (Fig. 9) shows an independence among the variables (with a R2 = 0.8%). Thus, the regression line is almost horizontal, indicating that it is not statistically significant. This situation was verified by the regression hypothesis tests.

In both graphs (Figs. 8 and 9) discrepant points are observed, corresponding to different years from the normal production average. Figure 10 shows the lack of correlation among the production of both soybean and corn for exportation, as shown by the extremely low value of the R2 (R2 = 0.2%), which denotes the lack of relationships among these products.

This lack of relationship between the products allows the separate study of corn and soybean exports, relating them to other variables of each production and to the climate variables. As part of the descriptive analysis of the data, the multivariate statistical method was also applied by hierarchical grouping/clustering to identify the different levels of similarities among the variables analyzed. The clustering model was developed with the Ward method of binding and the absolute correlation coefficient distance.

Figure 11 shows the dendrogram for the variables related to the production of soybean and corn, in which five groups with a high level of similarity can be observed. From these five groups, the first identifies the internal and external prices of both grain products; the second group includes the domestic production of corn and soybeans, the third group is made up of corn production for exportation, the fourth group with soybean production for exportation and the fifth group shows the planted areas of both grains.

The analysis of this dendrogram indicates that the productions for exportation were isolated, which means that they have a very low level of similarity in relation to the other variables and to each other. This result corroborates the simple linear regression analysis performed in the bivariate analysis process, which did not presented correlations between the indicated variables.

As for the clustering analysis carried out with climate data, different groups can be observed in Fig. 12, among which it is possible to identify similarities, mainly, between the climatic variables of the Middle-West and Southeast. Generally, the

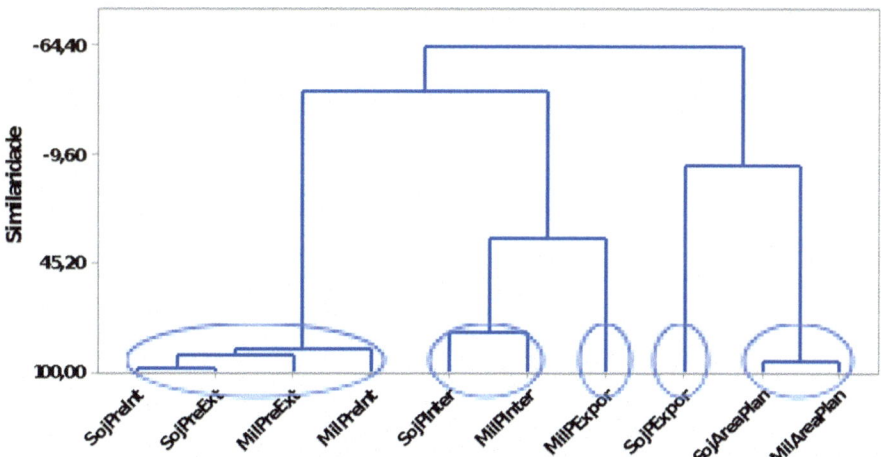

Fig. 11 Dendrogram of the production variables, with its five clusters

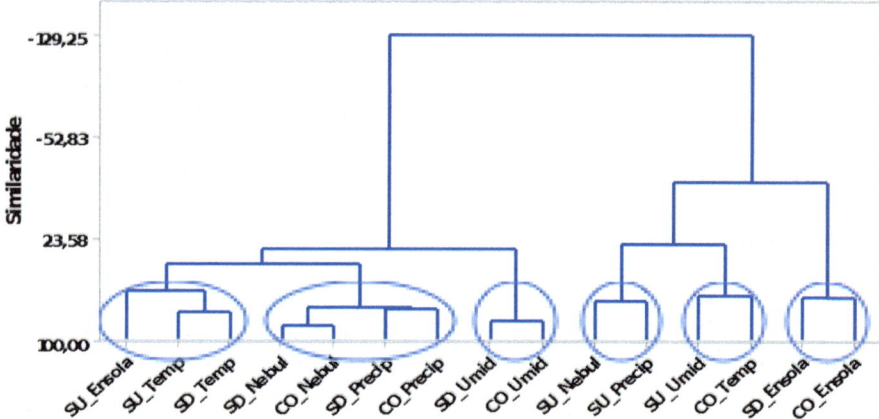

Fig. 12 Dendrogram of the climate variables, with its six clusters

climate variables of the South region are isolated from the other or grouped independently.

Following the multivariate descriptive analysis, Fig. 13 presents the joint dendrogram of the climate variables and the production of soybean and corn, which shows that they in fact do not relate to each other. The left-side of the dendrogram

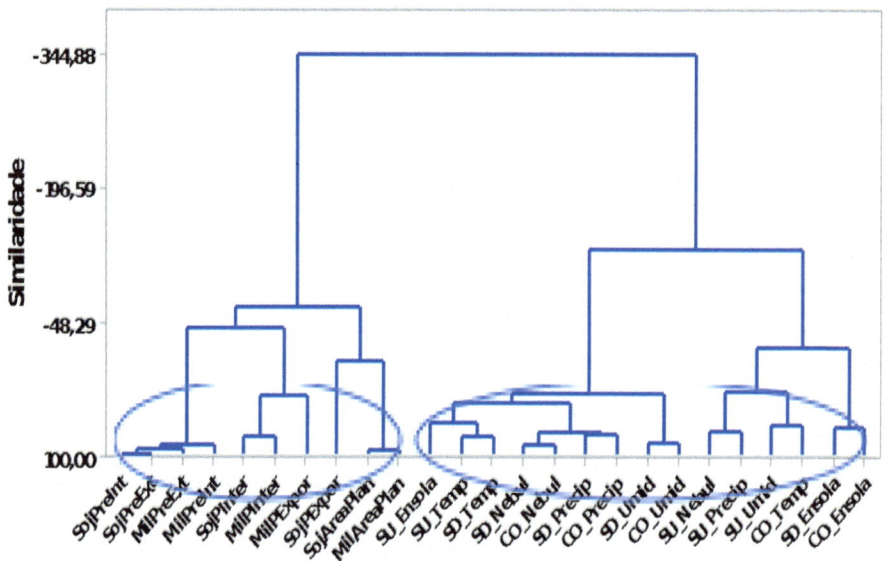

Fig. 13 Joint dendrogram of the production and climate variables, which demonstrates the independence of the variables of production and climate in relation to each other

presents the variables related to the production, whilst the right-side group the climate variables.

To test these results showing an independence between production and climate variables, Table 1 shows the results of the principal components analysis of the variables.

According to Kaiser's criterion, to determine the number of principal components that can replace the initial set of data, the eigenvalues of the major components must be greater than or equal to the units. So, according to the eigenvalue table, seven major components would be needed, which show 89% of the data variability, a highly favorable value (Table 1).

When analyzing the factorial load of each variable in the main components (Table 2), it is observed that in the first principal component, five climate variables are predominant, most of them from the Southeast and Middle-West. In the second principal component, the variables of domestic and exportation prices of soybean and corn are highlighted. Whilst in the third, the climate variables, mainly from the Middle-West, are again predominant. In the fourth principal components, the planted area of corn is the main variable that makes up this principal component. In the fifth, the climate variable of the South predominates, while in the sixth, the production of maize for exportation is the most relevant and, in the seventh, the production of soybean for exportation is the most relevant.

In the first two principal components, which have the most variability of the variance (56% as shown in Table 1), the predominant variables have no relation with the production for exportation, as well as in the last two selected variables, which are those that least demonstrate the variability of the variance (10%). The following variables predominate production for export of corn and soybean, showing that these two variables do not have a strong relationship with the climate, as it was observed in clustering analysis.

When the four principal components are analyzed according to the scree plot criterion (Fig. 14). It can be observed that these components are sufficient for this dataset, which demonstrate 73% of variance variability, sufficient value considering that part of the variables are climate data.

The analysis of principal components shows that there is no significant relationship among the production variables for corn and soybean exports with the other variables.

4.1 Multivariate Analysis of Soybean Data

The analysis of the data showed that there is no relation between the export of soybean and corn. Accordingly, this supported a multivariate multiple regression analysis of the soybean production for exportation combined with the other 19 variables, including climate variables and those related to the production of soybean.

The multivariate multiple regression analysis used the stepwise method, which removes and adds terms to the model in order to identify a statistically significant subset of variables. The 19 elements selected as independent variables are candidates

Table 1 Eigenvalues and proportion of variability of individual and accumulated variance

Eigenvalue	7.44	5.33	3.64	1.80	1.52	1.45	**1.00**	0.8786	0.5014	0.2431	0.2203	0.1848
Proportion	0.30	0.21	0.15	0.07	0.06	0.06	0.04	0.035	0.020	0.010	0.009	0.007
Accumulated	0.30	0.51	0.66	**0.73**	0.79	0.85	**0.89**	0.922	0.942	0.952	0.961	0.968
Eigenvalue	0.1638	0.1227	0.1191	0.0887	0.0639	0.0602	0.0454	0.0398	0.0392	0.0251		
Proportion	0.007	0.005	0.005	0.004	0.003	0.002	0.002	0.002	0.002	0.001		
Accumulated	0.975	0.980	0.984	0.988	0.990	0.993	0.995	0.996	0.998	0.999		
Eigenvalue			0.0144				0.0082			0.0054		
Proportion			0.001				0.000			0.000		
Accumulated			0.999				1.000			1.000		

Table 2 Eigenvectors or factor loads of for each principal component

Variables	CP1	CP2	CP3	CP4	CP5	CP6	CP7
SojPreInt	0.07	**0.38**	−0.09	0.08	−0.10	−0.19	0.23
SojPreExt	0.07	**0.39**	−0.09	0.01	−0.12	−0.16	0.17
SojPInter	−0.10	−0.28	−0.04	0.23	0.05	−0.33	0.21
SojPExpor	−0.06	−0.12	0.15	−0.15	0.10	0.32	**0.54**
SojAreaPlan	−0.07	−0.23	0.08	−0.51	−0.11	−0.30	0.16
MilPreInt	0.09	**0.36**	−0.10	−0.10	−0.15	−0.17	0.19
M.ilPreExt	0.07	**0.37**	−0.09	0.04	−0.10	−0.16	0.29
MilPInter	−0.12	−0.26	0.02	0.29	0.11	−0.13	0.52
MilPExpor	0.01	−0.21	−0.11	0.33	−0.06	**−0.38**	0.01
MilAreaPlan	−0.06	−0.15	0.07	**−0.61**	−0.13	−0.30	0.07
SU_Ensola	0.26	0.01	−0.13	−0.05	0.46	−0.18	−0.03
SU_Nebul	0.15	−0.05	**0.35**	0.08	−0.44	0.02	−0.07
SU_Precip	0.12	−0.13	0.19	0.16	**−0.56**	−0.02	0.00
SU_Temp	**0.32**	−0.12	−0.03	−0.05	0.06	−0.10	0.00
SU_Umid	0.11	0.07	0.46	0.06	−0.02	0.10	0.07
SD_Ensola	−0.02	0.19	0.33	−0.06	0.24	−0.24	−0.20
SD_Nebul	**0.33**	−0.11	−0.06	0.05	−0.03	−0.01	−0.02
SD_Precip	0.27	−0.15	−0.21	−0.03	−0.10	−0.00	−0.03
SD_Temp	**0.31**	−0.03	0.16	0.04	0.08	−0.24	−0.18
SD_Umid	0.28	−0.01	0.18	−0.03	0.13	0.24	0.26
CO_Ensola	−0.12	0.16	**0.39**	0.05	0.16	−0.12	−0.04
CO_Nebul	**0.34**	−0.07	−0.09	−0.01	−0.02	0.06	0.04
CO_Precip	0.29	−0.07	−0.20	−0.08	−0.05	0.02	−0.02
CO_Temp	0.21	−0.05	**0.34**	0.11	0.17	−0.22	0.06
CO_Umid	**0.33**	0.04	0.05	−0.13	0.09	0.19	0.13

for the final model. Table 3 shows the 19 variables used as input of the stepwise multiple regression model.

After the execution of the model; out of the 19 candidate variables for predictors, only 8 of them were defined as statistically significant (p-value < 0.05), with R2 = 51%, R2 adjusted = 47% and R2 prediction = 20%, the regression equation was translated into the mathematical expression:

$$\text{SojPExpor} = 21{,}523{,}905{,}902 - 160{,}094{,}676\,\text{SojPreExt} - 2107\,\text{SojPInter}$$
$$+ 156{,}236\,\text{SojAreaPlan} - 6{,}386{,}687{,}120\,\text{SD\_Nebul}$$
$$+ 998{,}531{,}167\,\text{SD\_Umid} - 136{,}423{,}004\,\text{CO\_Ensola}$$
$$+ 1{,}216{,}251{,}797\,\text{CO\_Temp} - 514{,}150{,}101\,\text{CO\_Umid}$$

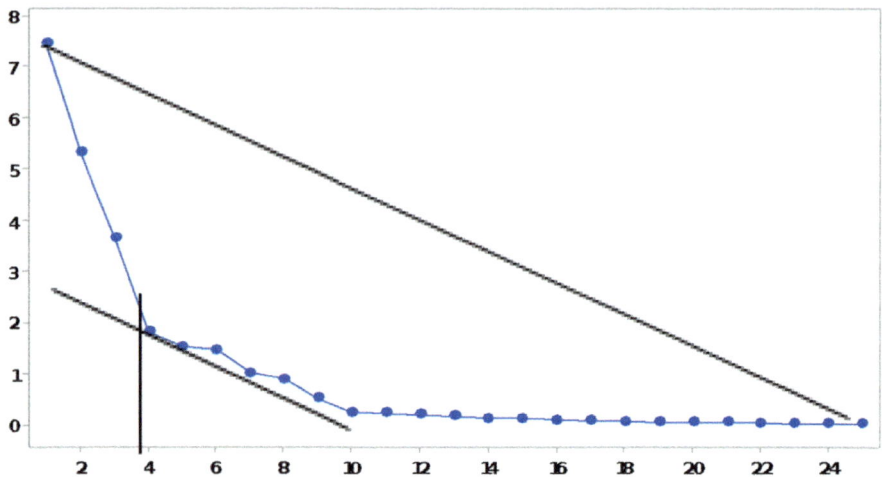

Fig. 14 Scree plot of major components × eigenvalues

Table 3 Predictor or regressor variables for soybean

Predictor	Average	Standard deviation
SojPreInt	41	15
SojPreExt	44	16
SojPInter	73,294,518	16,752,551
SojAreaPlan	25,252	8985
SU_Ensola	187	39
SU_Nebul	5	1
SU_Precip	146	60
SU_Temp	19	4
SU_Umid	76	8
SD_Ensola	194	30
SD_Nebul	5	1
SD_Precip	110	64
SD_Temp	22	3
SD_Umid	68	8
CO_Ensola	205	38
CO_Nebul	5	1
CO_Precip	127	73
CO_Temp	25	3
CO_Umid	67	10

Table 4 Variance analysis of the coefficients of the regression equation

Source	GL	Contribution (%)	F value	P value
Regression	8	51	15.76	0.000
SojPreExt	1	12	14.19	0.000
SojPInter	1	10	27.43	0.000
SojAreaPlan	1	2	5.74	0.018
SD_Nebul	1	4	51.53	0.000
SD_Umid	1	10	36.81	0.000
CO_Ensola	1	3	29.36	0.000
CO_Temp	1	3	12.03	0.001
CO_Umid	1	6	13.89	0.000
Error	123	49		

Table 4 shows the analysis of variance, which indicates that all p-values were calculated as being statistically significant, as a result of the F test.

Finally, the model generates the statistical inference on the coefficients of the regression equation, the standard errors, confidence intervals, p-value by the t-test and the multicollinearity VIF (variance inflation factors) test obtained for each of the coefficients, as shown in Table 5.

The coefficients also showed a statistically significant p-value (p-value < 0.05), the VIF test that measures how much the variance of an estimated regression coefficient increases if its predictors are correlated are, in some of the coefficients, greater than 5, which is the limit value for the nonexistence of collinearity.

4.2 Multivariate Analysis of Corn Data

Similar to soybean, and for the same reasons, a multivariate multiple regression analysis of corn production for exports was used with the other 19 variables under analysis (climate and others related to soybean production).

Accordingly, as for the soybean analysis, the stepwise method was selected for the multiple linear regression model, in which the 19 variables selected as independent or regressive variables are candidates for the final model. All variables were standardized by the different units involved in the process. Table 6 shows the 19 variables used as inputs to the stepwise multiple regression model.

After running the model, from the 19 candidate variables for predictors, only 6 were defined as statistically significant by the model (p-value < 0.05), with R2 = 66%, R2 adjusted = 64% and R2 prediction = 60%, the regression equation is:

$$\text{MilPExpor} = 318{,}938{,}749 + 4512 \, \text{MilPInter} - 46{,}550 \, \text{MilAreaPlan}$$
$$+ \, 412{,}714{,}391 \, \text{SD\_Nebul} + 9{,}233{,}866 \, \text{SD\_Precip}$$

Table 5 Inference on the coefficients and the T and VIF tests

Term	Coefficients	Coefficient's EP	IC 95%	T value	P value	VIF
Constant	4,718,501,278	510,149,013	(3,708,692,586; 5,728,309,970)	9.25	0.000	–
SojPreExt	−2,557,492,473	678,835,149	(−3,901,205,028; −1,213,779,919)	−3.77	0.000	1.76
SojPInter	−3,529,459,158	673,841,242	(−4,863,286,581; −2,195,631,735)	−5.24	0.000	1.73
SojAreaPlan	1,403,809,121	586,179,365	(243,503,009; 2,564,115,233)	2.39	0.018	1.31
SD_Nebul	−7,518,463,903	1,047,370,515	(−9,591,669,618; −5,445,258,188)	−7.18	0.000	4.18
SD_Umid	8,414,188,880	1,386,810,782	(5,669,081,969; 11,159,295,790)	6.07	0.000	7.33
CO_Ensola	−5,152,521,703	950,987,871	(−7,034,943,884; −3,270,099,523)	−5.42	0.000	3.45
CO_Temp	3,077,163,692	887,052,592	(1,321,297,476; 4,833,029,907)	3.47	0.001	3.00
CO_Umid	−5,080,051,896	1,363,219,556	(−7,778,461,423; −2,381,642,370)	−3.73	0.000	7.09

Table 6 Predictors or regressors variables for corn

Predictor	Average	Standard deviation
MilPreInt	20	9
MilPreExt	20	9
MilPInter	66,047,256	13,809,957
MilAreaPlan	14060	4329
SU_Ensola	187	39
SU_Nebul	5	1
SU_Precip	146	60
SU_Temp	19	4
SU_Umid	76	8
SD_Ensola	194	30
SD_Nebul	5	1
SD_Precip	110	64
SD_Temp	22	3
SD_Umid	68	8
CO_Ensola	205	38
CO_Nebul	5	1
CO_Precip	127	73
CO_Temp	25	3
CO_Umid	67	10

$$- 158,363,519 \, SD\_Umid + 251,027,155 \, CO\_Temp$$

Table 7 shows the results of the variance analysis, which indicates that all variables were statistically significant.

Finally, Table 8 shows the results of the inference on the coefficients of the regres-

Table 7 Variance analysis of the coefficients of the regression equation

Source	GL	Contribution (%)	F value	P value
Regression	6	66.0	40.36	0.000
MilPInter	1	20.3	62.12	0.000
MilAreaPlan	1	0.3	6.40	0.013
SD_Nebul	1	7.0	5.77	0.018
SD_Precip	1	4.9	13.22	0.000
SD_Umid	1	24.8	119.78	0.000
CO_Temp	1	8.7	31.89	0.000
Error	125	34.0		

Table 8 Inference on coefficients and T and VIF tests

Term	Coefficient	Coefficient's EP	IC 95%	T value	P value	VIF
Constant	1,459,090,382	76,713,154	(1,307,265,529; 1,610,915,236)	19.02	0.000	–
MilPInter	623,143,337	79,065,423	(466,663,052; 779,623,622)	7.88	0.000	1.05
MilAreaPlan	−201,514,010	79,664,276	(−359,179,499; −43,848,521)	−2.53	0.013	1.07
SD_Nebul	485,850,989	202,238,080	(85,596,728; 886,105,249)	2.40	0.018	6.90
SD_Precip	593,256,774	163,193,991	(270,275,611; 916,237,937)	3.64	0.000	4.49
SD_Umid	−1,334,460,656	121,930,898	(−1,575,777,037; −1,093,144,276)	−10.94	0.000	2.51
CO_Temp	635,108,329	112,465,106	(412,525,926; 857,690,732)	5.65	0.000	2.13

sion equation, the standard errors, confidence intervals, p-value for the t-test, and the multicollinearity VIF test.

The coefficients showed a statistically significant p-value, the VIF test was greater than 5 in one of the coefficients, which is the limit value for non-collinearity.

When the multiple linear regression modeling between the production variables of each type of grain, its production characteristics and the climate variables, relating them to the production for exportation of the grains, as dependent variables, it can be inferred that in the case of soybean production for exportation, from the 19 variables analyzed only 8 are statistically significant by an adequate degree to be included in the multiple linear regression model. However, the regression model obtained is only able to demonstrate 51% (R2) of variability of the dependent variable, which means that there are still 49% of unexplained causes.

The model showed a low coefficient of determination for the prediction (20%), indicating that the values of production for exportation obtained by the model can be different from the reality. Thus, when the statistical inference on all coefficients of the regression model was performed, although the p-value was adequate (<0.05), the value of the VIF tests for each of them had two coefficients higher than 5, which indicates the existence of collinearity among some of the variables. This may be the reason for the low R2 of the model.

In the case of corn production for exportation, from the 19 variables analyzed only 6 are statistically significant to be incorporated into the multiple linear regression model. The regression model obtained covers 66% (R2) of variability of the dependent variable, which means that there are still 49% of unexplained causes. The model also presents an average coefficient of determination for the prediction (60%), indicating that the values of production for exportation obtained by the model, may be different from the reality.

When the statistical inference on all the coefficients of the regression model was performed, although the p-value was adequate (<0.05), the value of the VIF tests for each of them present in one of the coefficients values higher than 5, indicating the existence of collinearity among some of the variables, which may be the reason for the low R2 of the model, although higher than in the case of soybean.

5 Conclusion

The descriptive analysis of the data showed that soybean production for exportation has a strong concentration in the last four months of the year, which indicates the seasonality of its production. The Middlesouth-West and Southeast regions have similar climates, and both regions differ from the South. The correlation between soybean production for domestic and exportation markets is low, with a small coefficient of determination (R2 < 1%).

The production of corn for domestic consumption and exportation presents a very low correlation with a coefficient of determination (R2) of 20%, which indicates that only 20% of the exportation variability can be explained by the variability of the domestic production. When analyzing the relation between the production for exportation of both grains (soybean and corn), it can be concluded that there is no correction among them, since the coefficient of determination (R2 $\cong$ 0) allows to do the multivariate study for each type of grain analyzed.

The clustering analysis of the variables related to grains showed an independence of soybean and corn production for exportation, corroborating the regression analysis. The dendrogram brings the clustering analysis of climate variables, which shows some similarities between the climate variables of the Middle West and Southeast. Data for grain storage in silos was not considered for this study. Grain storage is a common practice among producers that study the financial market to conduct exportation transactions at the best prices.

The cluster analysis of all the variables (climate and soybean and corn production) shows that there is no relationship between climate variables and grain production variables, since both sets of elements are in separate groups. The non-relationship of climate variables and grain production are also confirmed when the data are analyzed by the principal components model, either by the Kaiser criterion or by the scree plot criterion. In this case, it is possible that there is an influence of irrigation technologies, which reduces the dependence on rainfall patterns, but this data was not considered in this study.

Future research could replicate this model in other regions, with other types of grains. In addition, it is suggested that regional analyses would allow for the wide variation in climate, the seasons of grain production, and the unique natural resources of each region. This will lead to a more accurate calculation of the regression model and will allow it to be used for better production planning.

Acknowledgements This study was conducted by the Centre for Sustainable Development (Greens), from the University of Southern Santa Catarina (Unisul), in the context of the project BRIDGE - Building Resilience in a Dynamic Global Economy: Complexity across scales in the Brazilian Food-Water-Energy Nexus; funded by the Newton Fund, Fundação de Amparo à Pesquisa e Inovação do Estado de Santa Catarina, Coordenação de Aperfeiçoamento de Pessoal de Nível superior (CAPES), National Council for Scientific and Technological Development (CNPq) and the Research Councils United Kingdom (RCUK).

References

Alberto NASCM (2006) Simulação do impacto da mudança climática sobre a água disponível do solo em agroecossistemas de trigo, soja e milho em Santa Maria, RS. Ciência Rural 36(2)

Amann JC et al (2014) Brasil e Mercosul: aspectos econômicos e a relevância do bloco para o país, vol 39. Revista Estudos do CEPE, Santa Cruz do Sul, pp 107–138

Assad ED et al (2004) Impacto das mudanças climáticas no zoneamento agroclimático do café no Brasil. Pesquisa Agropecuária Brasileira 39(11):1057–1064

Barbieri AF, Domingues E, Queiroz BL, Ruiz RM, Rigotti JI, Carvalho JA, Resende MF (2010) Climate change and population migration in Brazil's Northeast: scenarios for 2025–2050. Popul Environ 31(5):344–370

BRADESCO and DEPEC (2017a) Departamento de Pesquisas e Estudos Econômicos. Soja. Junho de 2017. Available at: https://www.economiaemdia.com.br/EconomiaEmDia/pdf/infset_soja.pdf. Accessed on: 13.07.2019

BRADESCO and DEPEC (2017b) Milho. Junho de 2017. Available at: https://www.economiaemdia.com.br/EconomiaEmDia/pdf/infset_milho.pdf. Accessed on: 13.07.2019

Brahmbhatt M, Canuto O (2010) Natural resources and development strategy after the crisis. World Bank, Washington, DC (PREM notes, Economic Policy, no 147)

Bresser-Pereira LC, Marconi N (2008) Existe Doença Holandesa no Brasil? Fórum De Economia Da Fundação Getúlio Vargas, 4, São Paulo

Butler CD (2016) Planetary overload, limits to growth and health. Curr Environ Health Rep 3:360–369

Camargo FAO, Silva LS, Merten GH, Carlos FS, Baveye PC, Triplett EW (2017) Brazilian agriculture in perspective: great expectations vs reality. Adv Agron 141:53–114

Caron B (2009) A evolução do comércio exterior brasileiro, as ações, modificações e adaptações internas necessárias e a importância do respeito à cultura e aos aspectos culturais dos mercados externos nas negociações das empresas brasileiras. Faculdade Opet: Revista eletrônica, 1

Confalonieri UE, Lima ACL, Brito I, Quintão AF (2014) Social, environmental and health vulnerability to climate change in the Brazilian Northeastern region. Clim Change 127(1):123–137

Cunha DAD, Coelho AB, Féres JG, Braga MJ (2014) Effects of climate change on irrigation adoption in Brazil. Acta Scientiarum Agron 36(1):01–09

Dalum B, Laursen K, Verspagen B (1999) Does specialization matters for growth? Ind Corp Change 8(2):267–288

Drabo A (2017) Climate change mitigation and agricultural development models: primary commodity exports or local consumption production? Ecol Econ 137:110–125

Falloon P et al (2007) Climate change and its impact on soil and vegetation carbon storage in Kenya, Jordan, India and Brazil. Agr Ecosyst Environ 122(1):114–124

FAO (2018) Food balance sheets. Food and Agriculture Organization of the United Nations, Rome. Available at: http://www.fao.org/economic/ess/fbs/en/. Accessed on: 13.07.2019

Fearnside PM (1995) Potential impacts of climatic change on natural forests and forestry in Brazilian Amazonia. For Ecol Manage 78(1):51–70

Fearnside PM (1999) Plantation forestry in Brazil: the potential impacts of climatic change. Biomass Bioenerg 16(2):91–102

FIESP (2018) Informativo Junho de 2018. Safra Mundial de Milho 2018/19 – 2o Levantamento do USDA. Available at: file:///C:/Users/Tom/Downloads/file-20180613175924-boletimmilhojunho2018.pdf. Accessed on: 13.07.2019

Figueiredo SA (2013) Modelling climate change effects in southern Brazil. J Coastal Res 2(65):1933

Gala PSOS (2006) Política Cambial e Macroeconomia do Desenvolvimento. Escola de Administração de Empresas, Fundação Getúlio Vargas, São Paulo

Guariguata MR, Cornelius JP, Locatelli B, Forner C, Sánchez-Azofeifa GA (2008) Mitigation needs adaptation: tropical forestry and climate change. Mitig Adapt Strat Glob Change 13(8):793–808

Hirata R, Conicelli BP (2012) Groundwater resources in Brazil: a review of possible impacts caused by climate change. An Acad Bras Ciênc 84(2):297–312

IPCC (2013) Climate change 2013: the physical science basis. In: Stocker TF, Qin D, Plattner G-K, Tignor M, Allen SK, Boschung J, Nauels A, Xia Y, Bex V, Midgley PM (eds) Contribution of Working Group I to the fifth assessment report of the Intergovernmental Panel on Climate Change. Cambridge University Press, Cambridge, 1535 pp

Larson R, Farber B (2010) Estatística aplicada. Pearson Prentice Hall, São Paulo

Lucena AFP et al (2009) The vulnerability of renewable energy to climate change in Brazil. Energy Policy 37(3):879–889

Lucena AFP, Szklo AS, Schaeffer R, Dutra RM (2010) The vulnerability of wind power to climate change in Brazil. Renew Energy 35(5):904–912

Marengo JA (2008) Água e mudanças climáticas. Estudos Avançados 22(63)

Marin F, Nassif DS (2013) Mudanças climáticas e a cana-de-açúcar no Brasil: Fisiologia, conjuntura e cenário futuro. Revista Brasileira de Engenharia Agrícola e Ambiental 17(2):232–239

Matias EFP (2013) Políticas Climáticas e Livre Comércio. Consulex 404(12):1–2

Medeiros CA (2013) Padrões de investimento, mudança institucional e transformação estrutural na economia chinesa. In: CGEE. Padrões de Desenvolvimento Econômico, vol 2. América Latina, Ásia e Rússia. CGEE, Brasília

OECD-FAO (2015) Agricultural outlook 2015. Disponível em: http://www.oecd-ilibrary. org/docserver/download/5115021ec005.pdf?expires=1521657595&id=id&accname=guest& checksum=217DE2FCA672CF5E443F2A0C990DCE46. Acesso em: 19.03.2019

OECD-FAO (2017) Perspectivas Agrícolas 2017–2026, Éditions OCDE, París. Available at: http:// dx.doi.org/10.1787/agr_outlook-2017-es. Accessed on: 12.07.2019

Ojima ALR et al (2008) (Nova) Riqueza das Nações: Exportação e Importação Brasileira da Água Virtual e os Desafios Frente às Mudanças Climáticas. Revista Tecnologia & Inovação Agropecuária 1(1):64–73

Oreiro JL, Feijó CA (2010) Desindustrialização: conceituação, causas, efeitos e o caso brasileiro. Revista de Economia Política 30(2):219–232

PBMC (2015) Relatório de Avaliação Nacional: Base Científica das Mudanças Climáticas. Painel Brasileiro de Mudanças Climáticas

Pellegrino GQ, Assad ED, Marin FR (2007) Mudanças climáticas globais e a agricultura no Brasil. Revista Multiciência, Campinas 8:139–162

Ribeiro E (2014) As consequências das mudanças climáticas sobre o comércio internacional e as ações do Brasil. In: Public forum (WTO), vol 13, pp 1–2

Rodrik D (2006) What is so special about China's exports? National Bureau of Economic Research, Cambridge (NBER working paper series, 11947)

Roland F, Huszar VLM, Farjalla VF, Enrich-Prast A, Amado AM, Ometto JPHB (2012) Climate change in Brazil: perspective on the biogeochemistry of inland waters. Braz J Biol 72(3):709–722

Rong F (2010) Understanding developing country stances on post-2012 climate change negotiations: comparative analysis of Brazil, China, India, Mexico, and South Africa. Energy Policy 38(8):4582–4591

Sachs JD, Warner AM (1995) Natural resource abundance and economic growth. National Bureau of Economic Research, Cambridge

Santos PT et al (2007) Análise Multivariada de Dados Ecológicos da Baía de Guanabara- RJ, com Base em Foraminíferos Bentônicos. Anuário do Instituto de Geociências - Ufrj 30(1):109–115

Scarano FR, Ceotto P (2015) Brazilian Atlantic forest: impact, vulnerability, and adaptation to climate change. Biodivers Conserv 24(9):2319–2331

Serrano F, Summa R (2014) Demanda agregada e a desaceleração do crescimento econômico brasileiro de 2011 a 2014. Center for Economic and Policy Research

USDA (2017) Agricultural statistics. U.S. Government Printing Office, Washington, DC. Available at: https://www.nass.usda.gov/Statistics_by_State/Washington/Publications/Annual_Statistical_ Bulletin/. Accessed on: 12.09.2019

Vasconcelos ACF, Bonatti M, Schlindwein SL, D'Agostini LR, Homem LR, Nelson R (2013) Landraces as an adaptation strategy to climate change for smallholders in Santa Catarina, Southern Brazil. Land Use Policy 34:250–254

Vasconcelos ACF, Schlindwein SL, Lana MA, Fantini AC, Bonatti M, D'Agostini LR, Martins SR (2014) Land use dynamics in Brazilian La Plata Basin and anthropogenic climate change. Clim Change 127(1):73–81

Walter LC, Streck NA, Rosa HT, Ferraz SET, Cera JC (2014) Mudanças climáticas e seus efeitos no rendimento de arroz irrigado no Rio Grande do Sul. Pesquisa Agropecuária Brasileira 49(12):915–924

Watanabe M, Ortega E, Bergier I, Silva JSV (2012) Nitrogen cycle and ecosystem services in the Brazilian La Plata Basin: anthropogenic influence and climate change. Braz J Biol 72(3):691–708

Gisele Mazon Ph.D. in Administration and Tourism at the University of Vale do Itajaí (UNI-VALI). Researcher at the Research Centre on Energy Efficiency and Sustainability (GREENS) and Group of Research in Entrepreneurship and Management of Micro and Small Companies—GRU-PEM. Brazil.

Beatrice Maria Zanellato Fonseca Mayer Professor of international business at University of Southern Santa Catarina (UNISUL). Researcher at the Research Centre on Energy Efficiency and Sustainability (GREENS). Brazil.

João Marcelo Pereira Ribeiro Master in Administration at the University of Southern Santa Catarina (UNISUL). Bachelor in International Relations (UNISUL). Researcher at Group on Energy Efficiency and Sustainability (GREENS). Brazil.

Sthefanie Aguiar da Silva Student in International Relations at University of Southern Santa Catarina (UNISUL). Brazil.

Wellyngton Silva de Amorim Master student in Environmental Sciences at University of Southern Santa Catarina (UNISUL). Bachelor in International Relations. Researcher at the Research Centre on Energy Efficiency and Sustainability (GREENS). Brazil.

Larissa Pereira Cipoli Ribeiro Researcher at the Research Centre on Energy Efficiency and Sustainability (GREENS). Brazil.

Nicole Roussenq Brognoli Bachelor in International Relations at the University of Souther Santa Catarina (UNISUL). Brazil.

Ricardo Luis Barcelos Professor in the postgraduate program in Corporate Strategic Management and graduation in Management Processes at the SENAC Faculty; and postgraduate professor in Logistics Management and graduation of the disciplines of logistics for the Administration axis at UNISUL university. Researcher in the PROFORME and GESEG groups at UNIVALI University and in the groups Active Materials and GREENS at UNISUL University. Brazil.

Gabriel Cremona Parma Ph.D. in Civil Engineering in the Federal University of Santa Catarina (UFSC). Researcher at the Research Centre on Energy Efficiency and Sustainability (GREENS). Argentina.

Jameson Henry McQueen Student in Social Work at the Whittier College. United States of America.

Issa Ibrahim Berchin Director of Graduate Studies at Faculdade Anclivepa. Professor at Faculdade Anclivepa. Researcher at the Research Centre for Energy Efficiency and Sustainability (GREENS). Brazil.

José Baltazar Salgueirinho Osório de Andrade Guerra Professor at the Graduate programmes in Management and Environmental Sciences at the University of Southern Santa Catarina (UNISUL). Director of the Centre for Sustainable Development (GREENS). Fellow at the Cambridge Centre for Environment, Energy and Natural Resource Governance (C-EENRG), Department of Land Economy, University of Cambridge, United Kingdom.

Public Policy on Sustainable Food and Agricultural Markets: Legal Perspective from Nigeria

Fatimah M. Opebiyi

Abstract Worldwide, there is a need to increase food production and storage to meet human consumption requirements. This need is rendered more important because developing countries struggle to produce sufficient food to feed their population. In Nigeria, the local food and agricultural market suffers from inadequate local production. Consequently, the state faces food shortage. In addition to insufficient food and agricultural production, majority of the food available do not meet the minimum requirements relevant to qualifying such food as being fit for human consumption. Nonetheless, the prevailing poverty makes consumers nonchalant about the need for producers/merchants to abide by best practices in food production/sale, especially when such food is available at extra charges. This paper focuses on the role of public policies on measures taken by the public sector to improve sustainable food and agricultural markets, and the societal responses to the policies in Nigeria.

Keywords Sustainable food · Agricultural markets · Law, policy on food importation

1 Introduction

The problems of insufficient food for local consumption and export faced by Nigeria are attributable to poor financing, labour deficit, inefficient system for setting and enforcing food quality standards, insufficient food testing facilities, a weak inspectorate system, and poor knowledge of target markets.[1] In addition to this, post-harvest

---

[1]In Nigeria, the local demand for rice, wheat and fish daily are 6.3, 4.7 and 2.7 Million Metric Tonnes (MMT) respectively, while the locally produced volume do not exceed 2.3, 0.06 and 0.8 MMT see note 3 below p. 9; Federal Ministry of Agriculture and Rural Development (FMARD) 2016. Accessed, 8 September, 2018. 5.

F. M. Opebiyi (✉)
School of Law, University of Manchester, Manchester, UK
e-mail: fatimah.opebiyi@postgrad.manchester.ac.uk

© Springer Nature Switzerland AG 2020
W. Leal Filho et al. (eds.), *International Business, Trade and Institutional Sustainability*, World Sustainability Series,
https://doi.org/10.1007/978-3-030-26759-9_20

losses and illegal food imports deprive farmers of market opportunities.[2] Environmental enthusiasts and nutritionists agree on the imperativeness of preference for sustainable food and agricultural markets notwithstanding the attendant immediate high costs of such over cheaper and unhealthy alternatives.[3] Conversely, the decision is not always simply made by consumers, especially in societies where poverty is prevalent and the standard of living is low.[4]

This chapter discusses the drive towards achieving sustainable food and agricultural markets in Nigeria and how difficult and distant its achievement has been, especially, when the prevalent poverty in the country is combined with the poor purchasing power of the majority of the consumers. The attainment of sustainable food and agricultural markets involves the provision of access to healthy and nutritious foods for current and future local consumption needs.[5] The underlying theory of this chapter is that to achieve sustainable food and agricultural markets, there is the need for collaboration; a public/private partnership (PPP) between policy makers and producers/merchants. PPP will occasion the reduction of the gap between the costs of sustainable food and agricultural markets and their unsustainable alternatives.

The methodology adopted to achieve the aims of this chapter is a textual analysis of policy documents and relevant journal articles. Considering that the achievement of a sustainable food and agricultural markets requires overt acts by the Nigerian government, and the attainment of the sustainability goal is dependent upon efficient implementation of relevant policies, the use of a qualitative methodology will best achieve the aim of this discourse. Due to time and resources constraints, this chapter does not conduct an independent empirical research into the specific inputs of local producers to local food consumption, rather, the chapter relies upon recent statistics made available by the Food and Agricultural Organisation and data provided in Nigeria policy documents relating to the topic.

In achieving its aims, this chapter is divided into six major parts. Part 1 introduces the chapter, while the Part 2 explores the factors that determine the choices made by food consumers in purchases and how these choices impact on the drive towards attainment of sustainable food and agricultural markets in Nigeria. Part 3 examines the global policy on sustainable food and agricultural markets. This part focuses on creating a background for the purpose of measuring the position Nigeria occupies in the global drive towards sustainable food and agricultural markets. Part 4 discusses the government policies on sustainable food and agricultural markets in Nigeria and how effective their implementation has been. Part 5 examines how a public private partnership can aid in achieving sustainable foods and agricultural market. Part 6 concludes the work by stating that the achievement of sustainable food and agricultural markets in Nigeria is dependent upon an improved economy achievable through a public private partnership, and also the implementation of policies that discourage producers/merchants from providing foods of poor quality.

---

[2]Ibid. 8.
[3]Drewnowski and Darmon (2005).
[4]Ibid.
[5]Andrés and Delvaux (2018).

2 Conceptual Clarification

2.1 Public Policy

According to Merriam Webster dictionary, policy refers to 'a course or principle of action adopted or proposed by an organisation or individual'.[6] Iwuchukwu and Anor defined public policy to be a guideline consisting of principles and rules governing the behavior of persons in an organisation.[7] Public policy is the proposed action plan of a government which is not backed by law. The absence of law backing for public policies aids in ensuring administrative flexibility in public policy implementation.

2.2 Sustainable Food and Agricultural Market

The word 'sustain' refers to supporting, maintaining something or to persist in making an effort over a long period.[8] Similarly, Merriam-Webster Dictionary explains the adjective 'sustainable' to be 'relating to, or being a method of harvesting or using resources in a manner which ensures that such resource is not depleted or permanently damaged.'[9] In the words of Banuri, sustainability refers to '… a bridge [between] environment and development, North and South, government, business, and civil society, present and future, long term and short term, science and policy, and efficiency, equity, and participation.'[10] Sustainability not only ensures preservation for continuity and future uses, but it also takes into consideration the provision for immediate uses. 'Sustainable food and agricultural market' as it is being used in the context of this paper refers to the ability of a society to make provisions for agricultural outputs and the medium of exchange which satisfies the current healthy food and market needs of the society without jeopardising the ability of the later generations to have access to better means of satisfying their local food needs.

2.3 Public Private Partnership (PPP)

PPP is a framework which engages the private sector i.e. entrepreneurs, non-governmental organisations and peer groups, while acknowledging the role of gov-

---

[6]Merriam-Webster Dictionary https://en.oxforddictionaries.com/definition/policy Accessed 4 January 2018.

[7]Iwuchukwu and Igbokwe (2012).

[8]Garner (2009).

[9]Merriam-Webster Dictionary (2018).

[10]Cited in Alkon (2012), 20. *JSTOR*. www.jstor.org/stable/j.ctt46nh81.6. Accessed 11 August 2018.

ernment in ensuring that social obligations are met.[11] According to the Food and Agriculture Organisation (FAO), Public Private Partnership (PPP), for the purpose of agricultural oriented business development, refers to… 'a formalised partnership between public institutions and private partners designed to address sustainable agricultural development objectives, where the public benefits anticipated from the partnership are clearly defined, investment contributions and risks are shared, and active roles exist for all partners at various stages throughout the PPP project lifecycle.'[12]

3 Global Policy on Sustainable Food and Agricultural Markets

Agricultural production in Africa is acclaimed to be one of the worst performances among other continents in the world.[13] Conversely, Africa has the fastest growing population in the world.[14] A combination of the steady population increment and insufficient food production in Africa drags the goal of the world to attain sustainable agricultural production behind.[15] Consequently, hunger and poverty are endemic in Africa.[16] Due to the shortfall in food production occasioned as a result of Africa's complicity, amongst other factors, commentators are in agreement on the need for a significant increase in food production now, and in the future.[17]

Notwithstanding the Africa related issues in food production, the argument in developed countries has shifted from current food shortages to the need to make food and agricultural markets sustainable. Therefore, the aim of developed countries is to ensure that the current food consumption and market needs are being achieved without scuttling the ability to attain the provision for future needs. Although the drive towards increased production and storage for future use seems to encourage major agricultural trading countries to increase output of food safe for consumption at the expense of environmental concerns,[18] global policy seems to be in favour of increasing food production and where possible, protecting the environment. The best approach is to note the importance of increasing world agricultural production in order make sufficient food available and, at the same time, encourage the protection of the environment. In addition to the above, the global policy on sustainable food and agricultural markets recognises the need to focus on improved productivity while minimising externalities.[19]

---

[11]FAO (2016). 4.

[12]Ibid. viii.

[13]Pretty et al. (2011).

[14]United Nations (2019).

[15]Pretty et al. (2011, n 13).

[16]Ibid. 6.

[17]Ibid.7.

[18]Buller and Morris (2004), 1066.

[19]Pretty et al. (2011, n 13) 7.

Externalities refer to one of the inefficiencies in the market system, whereby the price of a product, in the instant case an agricultural product, does not reflect all the cost associated with the production of that product.[20] Externalities can imply negative implications on the environment encountered in the diminished ability of the same parcel of land to continue to produce optimally where this is brought about by the use of that land to produce some crops presently or in the past. Currently, one of the best approaches to reducing externalities is by sustainable intensification. Sustainable intensification refers to the ability to improve agricultural outputs by increasing yields per acre achieved through the use of new and improved agricultural managements.[21] In which case, each piece of land will be optimally utilised.

Furthermore, sustainable food production and agricultural markets will be attained when the maximum agricultural production safe for consumption is achieved with the minimum negative impact on the environment. In fact, the limit to growth argument posits that economic growth, which can be achieved by increase in food production, would protect the environment because profits realised through this means can be taxed and applied to improving the environment.[22] Therefore, growth in food production and agricultural market can translate to better environment protection.

4   Food and Agricultural Markets in Nigeria

One of the major problems confronting the actualization of sustainable food and agricultural markets in Nigeria relates to inadequacy of local sources from which food and agricultural products come from. In Nigeria, food production and the agricultural markets cannot satisfy local food demands.[23] Admittedly, there is the need for sufficient agricultural products, and also the products available must be healthy and nutritious.[24] According to the claims made by the Nigerian Government, particularly, the Minister for Agriculture, Nigeria is capable of producing enough food crops to satisfy local consumption, export, as well as future storage.[25] However, the above statement best describes an ambitious thinking by the Nigerian government.

The Nigeria food sector is majorly import based as the agricultural markets heavily depend on importation to supply local needs.[26] Statistics for the year 2016 reports that about $3–$5 billion worth of food, especially wheat, rice, fish and sundry items including fresh fruits, are imported annually.[27] Rice, being one of the most consumed staple food items, will be used to measure the success of the Nigeria government in

---

[20]Morgan and Yeung (2007) 19.

[21]Pretty et al. (2011, n 13) 9.

[22]Alkon (2012), 18. *JSTOR*. www.jstor.org/stable/j.ctt46nh81.6. Accessed 11 December 2018.

[23]FMARD (2016, n 1) 4.

[24]Andrés and Delvaux (2018, n 5).

[25]Babatunde (2014).

[26]FMARD (2016, n 1) 3.

[27]Ibid. 4.

adequately meeting local demands. Although Nigeria is one of the largest producers of rice in Africa, she is equally the country with the highest rice consumption in Africa. Consequently, the local production of rice cannot satisfy local consumption.[28] Due to her large population, Nigeria is one of the largest importers of rice in the world. Inasmuch as rice consumption exceeds its production locally, there will continue to be the need to source the difference through importation.[29]

As at 2011, 44.7% of the Nigeria land area is cultivated,[30] hence the problems associated with insufficient food production is not about lack of efforts, but the insufficiency and inefficiency of the efforts vested in agricultural production.[31] Although Nigeria produces a modest part of the food consumed locally, harvest and post-harvest problems reduces further the outputs. Out of the foods grown locally, a substantial amount falls to wastage due to poor processing/preservation techniques.[32] Going by the estimates for year 2016, for tomatoes alone, from the local production of 1.5MMT, 0.7MMT fell to wastage post-harvest due to factors ranging from poor transportation to inefficient preservation.[33] Therefore, a large part of agricultural products do not get to the market.

Other factors responsible for agricultural wastage range from poor outdated harvest mechanisms to poor storage systems.[34] The absence of adequate motivation to farmers is an important factor which prevents the interest of young entrants into the agricultural businesses. Consequently, the availability of subsidies for the cost of production and assurance on the part of the government that market will be made available for produced agricultural products will go a long way in encouraging agricultural producers and new entrants into the agricultural businesses.

Furthermore, other impediments to the attainment of a sustainable food and agricultural markets are the absence of a legislated agricultural extension policy, non-standardised policy in the sector, grossly inadequate and untimely funding, poor risk management, poor leadership and coordination, low private sector participation, a very weak supply driven approach to production, infrastructure deficit, absence of appropriate local agricultural technology and neglect of the rural development.[35] The drive by the Nigeria government to improve local supply of food items is thwarted by misjudgment, as against dedicated action plan capable of achieving sustainable agriculture. The above assertion is rendered more readily apparent from the move by the Nigeria government to restrict importation of staple food items such as rice and frozen foods at a time that local food production could not cater for local consumption. Due to the insufficiency of food items available locally, they are smuggled

---

[28]FOA, 'FAO in Nigeria; Nigeria at a glance', www.fao.org/nigeria/fao-in-nigeria/nigeria-at-a-glance/en/. accessed: January 7, 2019.

[29]Matemilola and Elegbede (2017).

[30]Akanoa et al. (2018), 2.

[31]Matemilola and Elegbede (2017, n 29) 9.

[32]FMARD (2016, n 1) 4.

[33]Ibid. 9.

[34]Matemilola and Elegbede (2017, n 29) 13.

[35]Ajani and Igbokwe (2014) 245. 241, FMARD (2016, n 1) 11.

into the country. In majority of the cases where food products are smuggled into the country, their quality are not examined and certified by the relevant agencies charged with these duties. Hence, the quality of food consumed locally cannot be guaranteed to be of ideal international standards for safe and healthy consumption. Furthermore, due to the failure to pay import duties and requisite taxation, smuggled food products are available in the market at cheaper prices when compared with locally produced food products.[36] The cheaper prices of the food products sourced through smuggling makes them the preferred choice of consumers. Even where the requirements for ideal standards for food production are enforced on locally produced food products, the porous borders and customs system make the achievement of ideal food consumption difficult.

In line with the restrictions on the importation of staple foods in Nigeria which was born out of health concerns over the importation of food unfit for human consumption, rice importation witnessed a drastic reduction in 2014.[37] Conversely, the increase recorded in the neighboring Benin Republic alludes to the fact that local consumption of imported rice has not reduced to the level reflected by the reduction in rice importation. Consequently, what occurred was an alteration of the route through which rice and other staple foods are imported into Nigeria.[38] This turn of events creates a more dangerous situation because food smuggled into the country would not be presented for government certification before they are made available at the market for local consumption. By analogy, the same results are applicable to other consumable agricultural products such as frozen fish, turkeys and chickens.

To return to the previous point, the Nigeria Information Minister identified rice importation as one of the greatest setbacks to local rice production in Nigeria.[39] Similarly, the Minister of State for Agriculture also acknowledged smuggling as a threat to government's effort and success recorded in the agricultural sector.[40] Furthermore, considering that the local production industry has been incapable of satisfying local consumption of rice; there is a shortfall which is being satisfied by smugglers. The above assertion by the Minister of State for Agriculture is rendered more readily apparent by the fact that smuggled rice, as well as other food products, come at a cheaper price, endears it to consumers—smuggled rice sells for between N1, 000 and N13, 000 per 50 kg bag, while Nigeria processed rice sells for between N14, 500 and N15, 000 per 50 kg bag.[41] Therefore, in attempting to account for smuggled food products, rice imports exceed $1 billion/annum.[42] Given that the current local food consumption needs cannot be satisfied by local production, provision for the

---

[36]Matemilola and Elegbede (2017, n 29) 9.

[37]Awojulugbe (2018).

[38]Republic of Benin recorded an increase in rice imports from Thailand, from 805,765 mt in 2015 to 1,647,387 mt as at November 2017. ibid.

[39]News Agency of Nigeria (NAN) (2018).

[40]Onusi (2018).

[41]At the prevailing exchange rate of 360 NGN to 1 USD the price varies from 30.5 to 36 USD for smuggled rice and 40 to 41 USD for locally processed rice.

[42]FMARD (2016, n 1) 3.

future is secondary. From the foregoing, it can be safely assumed that the Nigerian food and agricultural market has failed to function in a sustainable manner, hence the burden falls on government to fund and encourage sustainable production and trade in agricultural products.

To achieve sustainable food and agricultural markets, the Nigerian government has made several efforts to increase the acceptance of local foods. In a bid to improve the acceptance of locally produced food and local agricultural markets, there has been a recent focus on sensitisation of the populace as a means of discouraging consumption of smuggled food products. Sensitisation, alone, might not be helpful in discouraging people from consuming smuggled foods. Undeniably, the importance of knowledge of the ills attendant to the consumption of uncertified food products cannot be overemphasised. However, in addition to sensitisation, a higher purchasing power by the consumer will aid in the consumer's decision to opt for the healthier choice between smuggled uncertified food products and locally produced food products.

The problems and health concerns associated with smuggling of staple food items are unlikely to stop in the foreseeable future except the Nigerian government takes at least the following major decisive steps. First, there is the need to eliminate the market created for smuggled goods. Nigerians patronise these food products because they are cheaper. The failure of Nigerian consumers to consider the health implications of consuming such food items which are made available in the market past their shelve lives boarders on the availability of the food products at cheaper rate. If sufficient local alternatives are available at prices which are almost at par with the smuggled versions, it is likely that the market demand for smuggled items will reduce.

Second, creating a tighter border security along the unmanned boarders all over the country boundaries will further discourage the smuggling of these food items. Tighter boarder security creates some sort of scarcity which results is higher prices for the smuggled food items. This does not detract from the need to employ more immigration personnel and sufficiently motivate these immigration officers in order to discourage collusion with smugglers. Third, creating a friendly environment and incentives for marketers of locally produced agricultural/food items will help in diverting the interest of smugglers into taking the legitimate route. Undeniably, smugglers are first business men who are profit oriented, except for a few, majority of smugglers are willing to divert to alternative legal businesses.[43]

---

[43]In an interview conducted with 11 rice smugglers in Ogun State, Nigeria, the author was made to understand that 80% of the smugglers are willing to take the normal route of importation if the government allows importation of food items because the risks attendant to the business at times outweigh the profits made.

5 Public Policy on Sustainable Food and Agricultural Markets

Understanding that there exists the need for the local agricultural production and processing sectors to feed the agricultural markets is imperative to having a vibrant food and agricultural markets. Dating back to the 1970s, government intervention in the sector was a common phenomenon. In the past, the Nigerian government tried to solve the food shortage problems through the introduction of several programmes and initiatives. The central purpose of a majority of these programmes was the need to increase food production for local consumption, export and storage for future purposes. It is implicit that if production gaps are bridged through increasing and improving yield, then, resultant costs of production per unit will also reduce, thereby, helping reduce food costs and ultimately, inflation.[44] Several policy pronouncements were made for the purpose of achieving increased patronising of locally processed foods. The next section discusses some of the steps by the Nigerian Government to improve upon the capacity and quality of the local food and agricultural markets.

5.1 Actions Taken on Food Importation

In 2015, the Nigerian Government, in an effort to tighten up her foreign exchange policy and improve local food production, rolled out a new initiative whereby some items were banned from importation into the country. The government composed an exclusive list of goods that could be procured on a lower exchange rate of the local currency—Naira with the American Dollar.[45] Rice and frozen poultry products feature on the list at Items 1 and 7 respectively.[46] These two food items are food items mostly consumed by the largest part of the population on a daily basis. In addition to the above foreign currency exchange policy, the Nigerian Customs Service pamphlet includes live, dead or frozen poultry as items banned for importation.[47] The central aim of these policies is to improve the economy by encouraging the preference for local produce. The resultant effect of this policy would improve the consumption of fresher and healthier products. Nevertheless, because of the failure to improve upon the local agricultural market and food industry for the purpose of making the food items available at cheaper rates, the policies have not achieved the expected results. It was discovered that cheap imported foods compete favourably with their local alternatives.

Aside from overt acts of restricting the availability of cheaper Dollar-to-Naira rates to importers, the Nigerian government, through the Information Minister, warned

---

[44]Ibid. 8.
[45]Central Bank of Nigeria (2015), Banjo (2018).
[46]Ibid.
[47]Nigeria Custom Services (2018).

Nigerians to desist from consuming imported rice because of the possibility of it being imported and sold beyond its shelf life.[48] According to the Information Minister, Mohammed, '[the Nigerian Government] could not guarantee the healthy status of the rice [imported] having spent months on the high seas and warehouse'.[49] The policy on restriction of access to foreign exchange for importation of food items has not achieved much success in restricting its importation. Rather, it has created a wider gap for profit margins for the merchants. Consequently, the higher prices and resultant health implication associated are passed on to the end purchasers, since the cost of purchasing smuggled rice, for instance, is still cheaper than locally produced rice, even when the exchange restrictions are factored into its costs.

5.2   Credit Schemes

In addition to the other moves, at several points in time, Nigerian Government implemented initiatives capable of making credits available to farmers and agricultural producers. The initiatives were necessitated by the inability of farmers to have access to credit facilities offered by financial institutions. Due to the fact that majority of farmers in Nigeria are predominantly poor, they are not predisposed to possessing the required assets which will make them qualified to access credit facilities.[50] About 90% of farmers in rural arrears engage in subsistence farming[51] and in order to encourage a significant increase in their outputs, the Nigerian government created the Agricultural Credit Guarantee Scheme fund (ACGSF) to act as guarantee for credit facilities extended to farmers by banks up to 75% of the amount aside from the any security supplied by the borrower.

The Central Bank of Nigeria manages the fund in order to ensure that there is an absence of a long line of official bottlenecks.[52] The scheme was established in 1977 and it became operational in 1978.[53] Up till 1987, the scheme was successful as was evident in the increase recorded in agricultural production recorded during its operation. However, due to the deregulation of banks and the attendant risks of non-recovery of the total loan given by the banks, the scheme suffered a setback.[54] Several other similar initiatives such as Agricultural Credit Support Scheme, and more importantly Commercial Agricultural Credit Support Scheme were established by the Nigerian government. The latter credit scheme targets agricultural producers who engage in production, processing, storage and marketing of agricultural products in commercial quantities. The objectives of the scheme include the following:

---

[48] News Agency of Nigeria (NAN) (2018).

[49] Ibid.

[50] Zakaree (2014).

[51] Ibid.

[52] Central Bank of Nigeria (2018).

[53] Ibid.

[54] Ibid.

1. Fast-tracking the development of the agricultural sector of the Nigerian economy by providing credit facilities to large-scale commercial farmers at a single digit interest rate;
2. Enhance national food security by increasing food supply and effecting lower agricultural produce and products prices, thereby promoting low food inflation;
3. Reduce the cost of credit in agricultural production to enable farmers exploit the untapped potentials of the sector; and
4. Increase output, generate employment, diversify Nigeria's revenue base, raise the level of foreign exchange earnings and provide input for manufacturing and processing on a sustainable basis.[55]

Jointly managed by the Central Bank of Nigeria and the Federal Ministry of Agriculture and water resources, the scheme offers loan to qualified entities at a reduced interest rate of 9% per annum. The scheme is financed by N200 billion bond raised by the Debt Management Office.[56]

5.3 Non-credit Based Programmes and Initiatives

Aside from extending credits to farmers to increase output, the Nigerian government has tried several non-credit based programmes and initiatives in the past, such as the Special Programme for Food Security (SPFS); Fadama I and II Programmes; Fertilizer Revolving Fund; Presidential Initiative on Cassava, Rice, Vegetable Oil, Tree Crops and Livestock; restructuring and recapitalisation of the Nigerian Agricultural, Operation Feed the Nation of 1976 (OFN); the Value Added Tax Exemption for locally produced agricultural inputs such as fertilizer; and agricultural machinery, storage and processing facilities, agricultural development and marketing initiatives. The major purpose of OFN was to significantly increase food production in the country so as to satisfy local needs. Under the OFN, mass sensitisation was conducted to encourage engage in farming, at least, at a subsistence level. The probable rationale of this drive, if successfully keyed into, was that when most people grow what they eat, agriculture would be mastered by the majority of the people up to a level whereby there will be more food than what is required for local consumption and agricultural produce exports would be boosted. Secondly, if food items are produced in abundance, there will be a reduced demand in the market, hence a fall in prices would be recorded.

Apparently, the OFN failed in achieving its targeted objective; forty two years after its commencement, the country is still being fed majorly by imported food, notwithstanding that there are sufficient natural resources to see sufficient local agricultural production to fruition. A writer attributed the failure of the OFN to the organisational and operational structure targeted at non subsistence farmers.[57] Undeniably,

---

[55]Ibid.

[56]Ibid.

[57]Arua (1982), 101.

the structure of OFN lost sight of the goals achievable by focusing on subsistence farmers, but instead directed its attention at large scale production by individual corporations. A lot could have been achieved if more people engaged in agriculture at a subsistence level as more output can be guaranteed as against when fewer people engage at an industrial level, particularly where storage and preservative methods are still at the infancy stage of their development in Nigeria and mastery of business management skills is at its lowest ebb among entrepreneurs and administrators in the country. The Programme was doomed to fail before its commencement.

5.3.1 Agricultural Transformation Agenda

While discouraging the populace from patronising non-sustainable food and agricultural markets, the Nigerian government has over the years tried to encourage farmers through the provision of finance either directly from the budget or mandating banks to offer cheap credits. The aim of the Nigerian government was to ensure that sustainable food and agricultural markets are prioritised in the country. Amongst other initiatives, the Nigerian Government introduced the Agricultural Transformation Agenda (ATA) to protect and improve local production of foods and other agricultural products.[58] Under the agenda, a tariff regime on rice was proposed to achieve an increase in the output produced for local consumption and export. The focus of this programme is to achieve the above aims through the liberation of seed and fertilizer supply. While understanding that achieving optimum food production is important, the initiative operated without losing sight of the importance of limiting negative impacts to the environment. For the purpose of deferring global warming, ATA includes carbon sequestration and soil restoration as non-tangible benefits of the programme.[59]

5.3.2 The Green Alternative

Currently, the Nigerian government, in an attempt to build upon the achievements recorded by the ATA, introduced the Agricultural Promotion Policy 2016–2020 (APP); otherwise called the Green Alternative.[60] The APP focuses on enterprises development which spans over successive stages of value chain in agricultural production. Another focus of the APP is stimulating agricultural production on a sustainable basis, and stabilising prices of agricultural produce through market led price stabilisation mechanisms.[61] The initial crop prioritisation of the APP between the period of 2016 and 2018 are staple crop foods such as rice, wheat, maize, soya beans

---

[58] Ajani and Anor (n 35).

[59] Federal Ministry of Agriculture and Rural Development (FMARD) (2018).

[60] Ibid.

[61] Ibid.

and tomatoes for local consumption.[62] While for the purpose of exports, the initial food products of interest are cocoa, cassava, oil palm, sesame and gum Arabic. However, from 2018, the export focus will be on bananas, avocado, mango, fish and cashew nuts.[63]

Furthermore, the APP focuses on sustainable foods and agricultural markets. Its aim is to achieve optimal food production realised through a sustainable use of natural resources with future generations in mind, while at the same time generate an increase in agricultural production, marketing and other human activities in the agricultural sector.[64] The trust of the programme, as it touches upon green agriculture initiatives, relates to creating public awareness. The policy paper made mention of aiming to attain sustainable foods and agricultural markets, by focusing on stimulating agricultural production on a sustainable basis through ensuring that supply outweighs demand. Currently, the consumption demands of food products such as rice, palm oil and fish exceed local supply. Notwithstanding the plan outlined in the policy document, there have been no significant change in the gap between demand and supply of the food products and the current prices of staple food items are observed to have been on the increase. At the same time, the volume of unhealthy goods smuggled into the country has not decreased.

Considering that several policies of the Nigerian government on improving the volume of food produced locally have failed in achieving their aims, the quest to ensure sustainable foods and agricultural market requires more than a policy paper to be successful.[65] In addition to the will power of the government, private businesses and consumers, there is the need to draw an achievable blueprint which displays how the policy would be achieved. A declaration by the Nigerian government that she will boost public awareness through advertisement of importance of encouraging climate smart agriculture will not be sufficient because the target audience must be able to sufficiently benefit from such awareness. The major precursor to achieving consumer education, which will subsequently influence their choices, is the need to resolve the question of the particular audience that needs to key into the initiative of climate friendly agriculture and how such sensitisation can be achieved.

Further thrusts of the APP police are as follows:

i.   Boosting public awareness through advertising of importance of climate smart agriculture The management of land, water, soil and other natural resources will be improved

ii. Institutional linkages and partnerships will be strengthened for ensuring climate smart agricultural governance, policies, legislations and financial mechanisms

iii. Environmental impact assessment will be carried out on major agricultural projects

---

[62]Ibid.

[63]Ibid.

[64]Ibid.

[65]The situation has not shifted from the status quo, the food grown and produced in Nigeria is still not sufficient to satisfy local consumption.

iv. The use of renewable energy will be promoted with the involvement of private sector

v. Broad public and stakeholder awareness on Climate Smart Agriculture will be created

vi. Government will facilitate soil map to improve land use and management practices

vii. Government will increase the adoption of global best practices on climate change, including the aspects of adaptation, mitigation and carbon credit.[66]

Admittedly, according to the APP Policy document, "Nigeria suffers from policy instability driven by high rate of turnover of programmes and personnel, which in turn has made the application of policy instruments unstable."[67] However, the APP does not seem to learn from this observation. Even though the initiative is acclaimed to be a continuation and a better version of the ATA policy, the modus operandi makes it different from the ATA policy in operation before 2011. For there to be development which translates to sustainable food and agricultural markets in Nigeria, a level of stability is required policy wise. A lesson to be learnt from above is that majority of actors, who are not the initiators of these policies could merely see the policies as placeholders pending the change in administration, since every administrator/minister seeks to do something different from their predecessors in order to be seen as performing. A better position would be for policy makers not to discontinue earlier policies, if there is the urge to create a new policy, new policy and existing one can exist side by side so as to engender continuity.

Generally, the failure recorded in ensuring sustainable food and agricultural markets in Nigeria cannot be divorced from the basic weaknesses of the numerous agricultural policies in Nigeria, coupled with the incompetence of succeeding administrators to solve the insufficient food production in Nigeria.[68] The policies discussed above are, by no means, exhaustive of the agricultural policies made by the Nigerian government. A general theme linking the policies together is that most of them perform overlapping roles while conjunctively, they have all failed to stimulate the primary purpose of producing food sufficient for local consumption.

6   Towards a Public/Private Partnership (PPP) Between Policymakers and Producers/Sellers

Turning to how best a sustainable food and agriculture markets can be achieved, a partnership between the government and the agricultural/food producers is a precursor to attaining a sustainable food and agricultural market. Private producers and sellers are better suited to take the risks involved for the purpose of driving the agri-

---

[66]FMARD (2016, n 1).

[67]Ibid 10.

[68]Iwuchukwu and Igbokwe (2012, n 7).

cultural economy, while the larger capital infrastructure related projects should be provided by the government so as to create an enabling environment for the agricultural producers and merchants to thrive.[69] Due to the level of management required in overseeing business oriented entities, private individuals are more relevant in overseeing and ensuring agricultural businesses runs smoothly.

Similarly, Ajani & anor are of the view that public intervention in agricultural transformation should be focused on infrastructure development related to 'rehabilitation of existing rural and agricultural infrastructures and establishing new ones to help in mitigating impacts of climate change, restore soil fertility, remove barriers to domestic trade and flows of food.'[70] Furthermore, to record success in attaining sufficient food and agricultural production, there is the need for an increase in budgetary allocation to the agricultural sector.

Recognising the importance of collaboration between the government and private businessmen, the APP policy paper states that "the private sector will remain in the lead while government facilitates, as well as provides supporting infrastructure, systems, control processes, and oversight."[71] Profit oriented organisations are relevant in the agricultural sector of the economy to foster a desirable development in the local and export sectors of the food and agricultural markets. The APP policy document evidences the intention of government to contribute to sustainable agriculture through the creation of an enabling environment.

Core among the factors contributing to insufficient agricultural production is inadequate labour. It is basic economics that an increase in labour combined with available capital, in the form of land, will increase agricultural output. It is believed that if Africa, and by extension Nigeria, can leverage on their high youth population, there is the likelihood of an improvement in agricultural production. However, the first priority should be on ensuring agricultural personnel are well trained and skilled before such improvement can be guaranteed.

In addition, the APP policy identified that its success depends primarily upon the levels at which market participants, farmers, states, investors, financial institutions, research laboratories, Nigeria Customs Service, donors and communities are engaged.[72] The government aims to utilise its oversight functions in ensuring that farmers and investors operate in a safe, competitive and capable of enabling wealth creation in the future. It is not sufficient for the government to enter into a partnership with private merchants/farmers, the success of a PPP is dependent on the existence of an enabling environment and adequate infrastructure.[73] The land allocation procedure as evident in the practice among the states of the federation appears to be too cumbersome and discriminatory. This ends up discouraging entrepreneurs who intend to venture into agriculture business.

---

[69] Ibid 3.
[70] Ajani and Anor (n 35).
[71] Ibid.
[72] Ibid.
[73] FAO (2016), x.

PPP has been tried on small scales by state and local governments. A successful instance of PPP in agriculture can be seen in the Shonda case where the Kwara State government entered into a partnership with five banks in Nigeria. The PPP recorded success in the generation of 4000 new jobs, investment in infrastructure (around US$20 000 for roads improvement and schools provision), and personnel development of over 500 school leavers. There was also technology transfer from 13 Zimbabwean farmers to the local farmers.[74] In a similar vein, Lagos State represents one of the states that have keyed into the PPP initiative successfully in Nigeria. The Lagos State government also entered into a partnership with farmers from Kebbi State. Amongst others, the purpose of the partnership was to crash the price of rice paddy, and create food security. According to the Lagos State Commissioner for Agriculture, one of the major aims of the partnership was to "… bring about the cultivation of 32,000 ha of farm land to produce rice paddy, equating to an estimated 130 million Kg of processed rice per year which is an equivalent of 2.6 million 50 kg bags of rice."[75]

The Lake Rice initiative resulted in an increase in the population of new entrants into rice farming, either as primary or secondary occupation in Lagos State.[76] It was further declared by the commissioner that the Lagos State government intends to utilize the partnership to establish a 32 tonnes per hour rice mill with AG ("Bühler"), a leading rice-mill producer in the world.[77] For obvious reasons, Lake Rice partnership appears to be a perfect move in the direction of achieving a sustainable food and agriculture marker. First, through the partnership, Lake Rice was made available when the price of rice in the country was skyrocketing. Second, the partnership created a healthier alternative to the rice paddy available in the market. Also, the PPP was taken improved to further ensure that Lake Rice has an even market penetration, the Lagos State Government not only entered into a partnership agreement with distributors to ensure that Lagos residents are able to purchase Lake Rice from designated sales centres, but also from the open market at the official government price.[78] Under the agreement, the role of the distributors include the transportation, marketing, engaging in equitable distribution of the product by avoiding hoarding and other sharp practices and ensuring all year round continuous and equitable distribution of Rice.[79]

In addition to the benefits derived from the Lagos-Kebbi partnership, the Lake Rice is a fresh healthy alternative to the imported rice that is freely available at the market, given that the imported rice available has a minimum storage of five to six years storage life span.[80] The major role played by the Lagos State government, aside from financing the partnership, was to demand the acceptable standard of rice

---

[74]Mickiewicz et al. (2018).

[75]The Nation (2018).

[76]Ibid.

[77]Ibid.

[78]Vanguard, (2017).

[79]Ibid.

[80]Vanguard (2016).

produced by the partnership. If enabling policies for the creation of initiatives similar to that of Lagos-Kebbi State partnership can be replicated by other states, the volume of food produced in Nigeria will be significantly improved upon. Through similar agreements, the output quality can be predetermined by the agreements to be that which underscores the requirements of what is capable of ensuring sustainable food and agricultural markets.

7 Conclusion

Food, as a basic need of humans for the purpose of survival, needs special attention on the part of the policy makers. This chapter examines the state of food production and agricultural markets in Nigeria. The fact that the Nigerian agricultural market is import driven does not portend a promising future for the coming generations. Overt and determined efforts are required on the part of the government in providing sustainable and agricultural markets. While agreeing that the making and implementation of effective agricultural policies will go a long way in filling the gaps in local needs and production, collaboration between the federal, state and local government is required for a successful implementation of these policies.

Farmers are key actors in the drive towards sustainable food and agricultural markets. To enable farmers to fulfil their part in the making sustainable food and agricultural markets a reality, there is the need to train the farmers and grant the access to some knowledge and information. According to Pretty and Ors, key knowledge and information needs of farmers include 'information on systems that sustain agricultural production, information on current and new technology, and its performance in real farm settings; business management advice; information on markets, including an ability to investigate.'[81]

Conclusively, for there to be sustainable food and agricultural markets, it is essential to improve upon the efficiency of the supply outlets. A systematic combination of economic, social, environmental and technological strategies will ensure that a sustainable food and agricultural market is present in Nigeria.[82] This can be partly executed by increasing better implementation of relevant policies and increasing the percentage of budgetary allocation to agriculture be brought to be above what is obtainable in countries that produces sufficient food to satisfy local food needs and export. It is preposterous that the government expects to achieve laudable achievements with a budgetary allocation of 2% while the Maputo Declaration prescribes nothing less than 10%.[83] There should be friendly laws and reduced bottlenecks on land allocation to farmers at cheap rates, in order to encourage farming. Sustainable

---

[81]Pretty et al. (2011, n 13) 16.

[82]Matemilola and Elegbede (2017, n 29).

[83]Bassey (2018), 517. www.researchgate.net/publication/326405700_AGRICULTURAL_ EXPENDITURE_MAPUTO_DECLARATION_TARGET_AND_AGRICULTURAL_ OUTPUT_A_CASE_STUDY_OF_NIGERIA/download. Accessed 16 January 2019.

food and agricultural markets cannot be achieved by the public authorities acting in isolation, but through a Public Private Partnership. There is the need for collaboration between the government through the provisions of infrastructure, certifications and enabling laws on the one hand, and the merchants/farmers, through undertaking the risks of doing businesses and cultivating and on the other hand.

References

Ajani EN, Igbokwe EM (2014) A review of agricultural transformation agenda in Nigeria: the case of public and private sector participation. Res J Agric Environ Manag 3(5):238

Akanoa O, Modirwaa S, Yusuf A, Oladele O (2018) Making smallholder farming systems in Nigeria sustainable and climate smart. www.ifsa2018.gr/uploads/attachments/121/Theme3_Akano.pdf. Accessed 16 Jan 2019

Alkon AH (2012) Understanding the green economy. Black, White, and Green: Farmers Mark, Race, Green Econ 16. www.jstor.org/stable/j.ctt46nh81.6. Accessed 15 Aug 2018

Andrés P, Delvaux G (2018) Access to common resources and food security: evidence from national surveys in Nigeria. Food Secur 10:121

Arua EO (1982) Achieving food sufficiency in Nigeria through the operation 'feed the nation' programme. Agric Adm 9(2):91. ISSN: 0309-586X

Awojulugbe O (2018) INFOGRAPHICS: Thailand's rice export to Nigeria drops dramatically—but increases in Benin Rep Rice exports from Thailand to Nigeria dropped from 1.23 million metric tonnes in 2014 to 23, 192 mt as at Nov 2017. www.thecable.ng/infographic-drastic-drop-thailands-rice-export-nigeria/amp. Accessed 20 June 2018

Babatunde J (2014) Nigeria has enough rice to feed its population-Adesina. Vanguard Newspaper, April 4. www.vanguardngr.com/2014/04/nigeria-enough-local-rice-feed-population-adesina/. Accessed Aug 4 2018

Banjo T (2018) Full list of 40 imported items banned from Nigeria's Forex market. www.nigerianmonitor.com/full-list-of-40-imported-items-cbn-banned-from-nigerias-forex-market/. Accessed 17 July 2018

Bassey E (2018) Agricultural expenditure, Maputo Declaration Target and Agricultural Expenditure, Maputo Declaration Target and Agricultural Output : A Case Study of Nigeria

Buller H, Morris C (2004) Growing goods: the market, the state, and sustainable food production. Environ Plan A 36:1065–1084

Central Bank of Nigeria (2015) Inclusion of some imported goods and services on the list of items not valid for foreign exchange in the Nigeria Foreign Exchange Markets. www.cbn.gov.ng/Out/2015/TED/TED.FEM.FPC.GEN.01.011.pdf. Accessed 10 July 2018

Central Bank of Nigeria (2018) Agricultural credit guarantee scheme fund (ACGSF). www.cbn.gov.ng/Devfin/acgsf.asp. Accessed 28 June 2018

Drewnowski A, Darmon N (2005) Symposium: modifying the food environment: energy density, food costs, and portion size portion sizes and the obesity epidemic. The J Nutr 900, 903. jn.nutrition.org. Accessed 16 Jan 2019

FAO (2016) Public–private partnerships for agribusiness development—a review of international experiences. In: Rankin M, Gálvez Nogales E, Santacoloma P, Mhlanga N, Rizzo C, Rome, Italy

Federal Ministry of Agriculture and Rural Development (FMARD) (2016) The agriculture promotion policy (2016–2020) building on the successes of the ATA, closing key gaps. Policy and strategy document, 2016. https://fscluster.org/sites/default/files/documents/2016-nigeria-agric-sector-policy-roadmap_june-15-2016_final1.pdf. Accessed 15 Aug 2018

Federal Ministry of Agriculture and Rural Development (FMARD) (2018) The green alternative. https://fmard.gov.ng/the-green-alternative/. Accessed 15 Aug 2018

Garner BA (ed) (2009) Black's Law Dictionary, 9th edn 1585

Iwuchukwu and Igbokwe (2012) Lessons from Agricultural Policies and Programmes in Nigeria. J Law, Policy Glob 5:11. http://www.iiste.org/Journals/index.php/JLPG/article/viewFile/2334/2335. Accessed 16 Jan 2019

Matemilola S, Elegbede I (2017) The challenges of food security in Nigeria. 04 OALib 1:9. www.oalib.com/paper/pdf/5291809. Accessed 16 Jan 2019

Merriam-Webster Dictionary (2018) https://www.merriam-webster.com/dictionary/sustainable. Accessed 15 Aug 2018

Mickiewicz TM, Ifedolapo T, Olarewaju A (2018) Evolution in transaction costs and ownership of public-private partnerships: Shonga farms. Acad Manag Proc 2018(1)

Morgan B, Yeung K (2007) An introduction to law and regulation; text and materials. Cambridge University Press

News Agency of Nigeria (NAN) (2018) Beware of imported rice—FG warns Nigerians. www.thecable.ng/beware-imported-rice-fg-warns-nigerians/amp. Accessed 20 June 2018

Nigeria Custom Services (2018) Import prohibition list. www.customs.gov.ng/ProhibitionList/import.php. Accessed 12 July 2018

Onusi A (2018) Smuggling, threat to FG's efforts in agriculture- Lokpobiri. https://fmard.gov.ng/smuggling-threat-to-fgs-efforts-in-agriculure-lokpobiri/. Accessed 15 August, 2018

Pretty J, Toulmin C, Williams S (2011) Sustainable intensification in African agriculture. Int J Agric Sustain 9, 5

The Nation (2018) Lagos spends N1.04billion on Lake Rice. http://thenationonlineng.net/lagos-spends-n1-04billion-on-lake-rice/. accessed 17 Sept 2018

United Nations (2019) Population. www.un.org/en/sections/issues-depth/population/. Accessed 16 January 2019

Vanguard (2016) Ambode, Bagudu launch LAKE rice in Lagos. www.vanguardngr.com/2016/12/ambode-bagudu-launch-lake-rice-lagos/. Accessed 17 Sept 2018

Vanguard (2017) LAKE rice: Lagos signs agreement with major rice distributors. www.vanguardngr.com/2017/12/lake-rice-lagos-signs-agreement-major-rice-distributors/. Accessed 17 Sept 2018

Zakaree SS (2014) Impact of agricultural credit guarantee scheme fund (ACGSF) on domestic food supply in Nigeria. Br J Econ, Manag Trade 4(8):1273–1284, p 1275

Some Ways of Environmentally Sustainable Agriculture Production in the Context of Global Market and Natural Barriers

Eugeny V. Krasnov, Galina M. Barinova, Dara V. Gaeva and Timur V. Gaev

Abstract This paper aims to describe the different types of market barriers in relation to the trade in agricultural goods and the possibilities of its overcoming through environmentally friendly technologies. The main reasons and features of both positive and negative effects of a market economy in connection with environmentally sustainable approaches for agricultural production will be considered. Some perspectives for sustainable agriculture production are observed in the context of global challenges faced by humanity. This article will be focused on the relationships between small and middle agro-firms, on the one hand, and also holdings, on the other hand, in connection with global and regional obstacles including climate change and market barriers. More important questions for now are multifunctional contradictions in a production technology, legal conditions, market pressure, pricing policy and other risk factors. Nevertheless, many small agrarian firms are often more variable, more mobile in relocation, etc., and have supply options for more environmentally friendly products. But, in different countries and regions, there are negative consequences for agrarian business such as bankrupting and the reduction of small and middle companies. Finally, we would like to consider cooperation as a more perspective option for survival not only for small farmers.

E. V. Krasnov (✉) · G. M. Barinova
Institute of Environmental Management, Urban Development and Spatial Planning (IEMUD&SP), Immanuel Kant Baltic Federal University, Kaliningrad, Russia
e-mail: ecogeography@rambler.ru

G. M. Barinova
e-mail: GBarinova@kantiana.ru

D. V. Gaeva
Service for Organization of Scientific Research Activity, Immanuel Kant Baltic Federal University, Kaliningrad, Russia
e-mail: DGaeva@kantiana.ru

T. V. Gaev
Department of Animal Husbandry, Kaliningrad Branch of Saint-Petersburg State Agrarian University, Kaliningrad, Russia
e-mail: timurgaew@rambler.ru

© Springer Nature Switzerland AG 2020
W. Leal Filho et al. (eds.), *International Business, Trade and Institutional Sustainability*, World Sustainability Series,
https://doi.org/10.1007/978-3-030-26759-9_21

369

Keywords Agriculture · Market and natural barriers · Obstacles · Globalization · Small farms · Risks and opportunities · Sustainable agriculture

1 Introduction

Agriculture accounts for the major share of land use and over two-thirds of water use. Agricultural soils and livestock are the largest sources of nitrogen emissions. In order to fulfill today's environmental requirements and demand in food, raw materials in the context of global changes, agriculture must break a number of external market barriers. Market barriers affect not only the economics of countries and regions, but also indirectly affect the environmental situation. Finding a way for environmentally sustainable agriculture requires equilibrium between biodiversity, human well-being, ecosystem services and economic profitability.

Liberalized international agrifood trade and domestic markets globally shift towards differentiated products. More and more often consumers demand product quality, safety, certification and labels reflecting these grades and standards (G&S). However, developing countries in Africa, Asia, Eastern Europe, and Latin America are not ready for such conditions. This phenomenon causes increasing in trade with the developed world and foreign direct investments by multinational companies. So, the local rural economies become increasingly linked to both urban and export economies (Reardon et al. 1999).

In order to meet rising demands, grain production must increase approximately by 25–70% from recent levels. Sustaining of average annual yield growth rate until 2050 will require the widespread intensification of fertilizer, pesticide, and irrigation regimes. While maintaining functioning ecosystems, meeting food demand will require an updated sustainable intensification (SI) strategy which will couple new production goals with required environmental targets (Hunter et al. 2017).

2 Modern Agriculture Across Market Barriers

To analyze the barriers, in this study were compared some modern concepts and views on this issue (Table 1).

In the context of global changes, solving the problem of accessibility of agricultural product markets for farmers has ceased to be a task just for economists.

Modern interdisciplinary research indicates the ambiguity of the impact on the economic, social and environmental situations of various countries by prohibitions and restrictions on the goods importation or the full integration of markets.

The study reveals the significant variety of market barriers caused not only by society and economics, but also by natural factors. In countries and regions with different levels of technological development, overcoming of barriers requires the use of various organizational, institutional and legal measures. For developed countries,

Table 1 Limiting factors to market access and some ways to solve them in modern research and development

	Type of limiting factors	Ways to solve	References
1.	Mental: (a) By customer (b) By consumer	New business models Legal instruments Cultural evolution	Starr et al. (2003)
		Legal defense Education Cultural development	Maertens and Swinnen (2015)
2.	Corporative: (a) Price unification (b) Monopolization	Legal instruments Knowledge	Wegren and Elvestad (2018)
		Legal instruments Open audit Documentation control	Schmutz et al. (2018)
3.	Governmental: (a) Tariff policy (b) Non-tariff limits	Press minimization	Edwards (2018), Swinbank (2017), Wetherill and Gray (2015)
		Cooperation co-agreement	Hunter et al. (2017), Scott (2015)
4.	Natural: (a) Weather (b) Climate change	Events Prevention risk analysis Different scenarios'	Baldos and Hertel (2015)
		Long-term modeling	Redlich et al (2018)

this is primarily the introduction of innovative approaches, and, at the same time, for developing countries, reforms in education and management are needed.

In international trade, tariffs or taxes on goods are used to protect domestic industries. In opposition to trade liberalization, governments are increasingly using non-tariff barriers to protect some industries of their countries. Such barriers have several basic forms, such as, firstly, quantitative restrictions or barriers limiting the quantity of exported or imported goods, secondly, some laws, regulations, policies or procedures for the impeding of international trade, and "buy local" campaigns (Morschett et al. 2015).

Based on a 19th century database from Austria historical studies of the relationships between market integration and culture show that even regional religious differences can influence on an increase in grain prices. Road infrastructure development is also negatively correlated with religious diversity (Walker 2018).

Long-term restrictions on the import of agricultural products are ambiguous. They have a positive effect on the development of agricultural production in importing countries, but it often leads to the loss of sustainable trading partners—suppliers of agricultural products. For example, after the Russia's food embargo in 2014 its domestic production has increased and Russia has become more food self-sufficient since then. But, some western food trading partners have been replaced by trading partners from Asia and Central Asia. Consequently, in the future the Russian food market can be lost for Western exporters (Wegren and Elvestad 2018).

In the period from 2014 to 2016, the embargo caused a decrease by 90% in the import of some food products to Russia (Fig. 1). On the one hand, this led to some reduction in the variety of food products for Russians; on the other hand, it provided an opportunity for local farmers (including small ones) to develop their own processing of milk, meat and other agriculture products.

For example, in Russia, cheese production increased by more than 100 thousand tons from 2014 to 2016 (Rosstat 2019). At the present time the main way to overcome market barriers at the global level is to create common markets, customs and currency unions, and free trade zones. All these agreements are accompanied by both benefits and losses (environmental, economic and social) for the Allied states. For example, the North American Free Trade Agreement between the USA, Mexico and Canada (NAFTA, Eng. North American Free Trade Agreement), the Eurasian Economic Union (EAEU), the economic and monetary union of the EU, etc.

According to Swinbank (2017), the EU has influenced on the UK food supplies and prices, the profitability of farm businesses, the rural environment and land use in a number of ways, for example, through agricultural subsidies and a highly protective trade regime. After Brexit, the taxpayer support to the farm sector is likely to be reduced and become more focused on environmental goals.

The weakening of trade barriers can be an economically profitable option for many countries. However, for example, the liberalization of trade in so-called "environmental goods", such as biofuels, may adversely affect the ecological conditions of the territory. Due to an increase in the area of arable land under corn, sugar cane and other crops, soil fertility as well as the area under less-profitable crops decreases. In the late 2000s, the biofuels boom provoked the greater cultivation of corn and soybeans in the United States. The crop expansion resulted in a significantly landscape transformation with the conversion of long-term unimproved grasslands and land that had not been previously used for agriculture. Corn was the most common crop planted directly on new land. The cropland expansion occurred most rapidly on land that was less suitable for cultivation which could adverse environmental and economic costs (Lark et al. 2015). According to the FAO, the area under rape crops in the world for the period from 1961 to 2016 increased from 6.2 to 33.7 million ha, and soybeans from 23.8 to 121.5 million ha (Fig. 2).

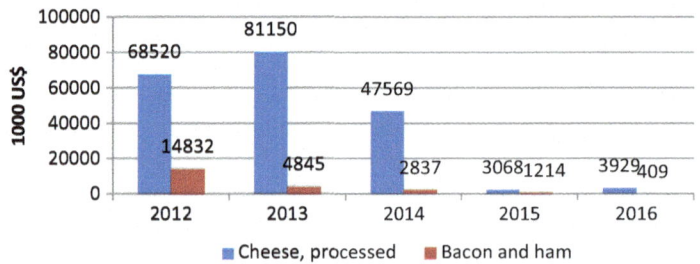

Fig. 1 Dynamics of Russia's imports of certain food products, 1000 USS. *Source* Authors, according to FAO (2018a)

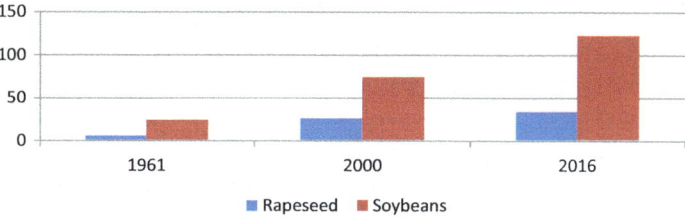

Fig. 2 Area harvested of rapeseed and soybeans world, million ha. *Source* Authors, according to FAO (2018b)

According to FAO data (2018b), in Europe from 1961 to 2016, the area under permanent meadows and pastures decreased by 12.4%, while the area under rapeseed increased by 7.5 million hectares over the same period.

The transport biofuels of the first generation, for example, biodiesel based on oilseed rape or ethanol from wheat/maize, are far less efficient use of resources. A broader mix of crops can reduce environmental impacts. For example, perennial crops (energy grasses or short rotation willow plantations) can enhance ecosystem services provided by farmland, such as flood prevention and water filtration (Popp et al. 2014).

In order to increase competitiveness in the global market, states provide some support to farmers in the form of subsidies. But, in some cases, subsidies can harm the ecosystem. Subsidies cause overproduction which draws lower-quality farmlands into active production. Additionally, areas that can be used for parks, forests, grasslands, and wetlands become used for agriculture. Lands which are used for pasture or grazing has been shifted into crop production lands. Because of subsidizing, producers on marginal lands tend to use more fertilizers and pesticides which can cause environmental problems (Edwards 2018).

3   Non-tariff Barriers in Agriculture Trade: Environmental and Social Effects

As an exception to the general rules of trade, non-tariff trade restrictions can be introduced to protect the environment, to combat animal and plant diseases, to prevent negative impacts on the lives and health of citizens, etc. But, these restrictions lead to financial losses for states under various restrictive measures.

"The effects of the EU non-tariff measures (NTM) on beef, vegetables, and fruits from the U.S. are significant. The AVE (a tariff estimated as a percentage of the price—the ad valorem equivalent) effect of the U.S.' NTMs on the EU exports ranges from 37% for vegetables to 45% for fruits. The AVE effect by the EU restrictions on beta agonists and trichinae for pork is estimated as 81%" (Arita et al. 2015).

A significant number of different sustainable initiatives have emerged to improve sustainability and inclusion of small farmers in global supply chains. Farmers have improved their profits by receiving better payment conditions, reducing their production costs and maintaining or even increasing their yields. In addition, positive environmental impacts are achieved while reducing or even stopping the use of chemical pesticides and fertilizers. Company-based projects have an easier market access because they have their own supply chains and market connections. Development-NGO projects are relevant to assure small farm livelihoods and take into consideration local characteristics and needs (Fayet and Vermeulen 2014).

Some national governments seek to support domestic producers, protect themselves from increasing in food imports for the food supply sustainability, and protect domestic food markets. But, in some cases, such measures challenge the sustainability of food supply by decreasing food availability and the quality of food products, and cause dietary changes, and threaten the food security of the country (Erokhin 2017).

For example, South Korea has the most supported agricultural sector among member countries of the Organization for Economic Cooperation and Development (OECD) and has nearly prohibitive tariffs in rice, meat, and dairy sectors. The state also has high production subsidies in many other sectors and significant non-tariff trade barriers on many commodities, including administrative barriers (import monopolies) and sanitary restrictions. Such self-sufficiency measures are detrimental to (poor) consumers and inconsistent with the food security as "access to food for all" proposed by the World Food Summit of the Food and Agriculture Organization (Beghin and Bureau 2015).

4   Local Small Farming in a Strategy for Food Security

The presence of competition both within the state and at the international level is the generally recognized and necessary attribute of a competitive economy. However, due to various deviations and restrictions on competition, it is necessary to implement government policies aiming to maintain an optimal level of competition. Thus, in the EU, the special attention of the state is directed at the activities of large TNCs controlling most of the market (Zaytseva 2013). Transnational corporations (TNCs) are important players in the food and agriculture globalization and, in order to promote their approach of improving the food system, they are actively engaged in relations with public. Similar actions are needed to be done to offer alternatives to TNCs, such as small-scale farming systems. To solve this problem, the multiple environmental discourses and growing awareness's about the health costs associated with poor diets must be brought together as a response to TNCs (Scott 2015).

Nelles (2015) suggests that rural small-holder rights to adequate, healthy, safe, nutritious food, which is depending from land, seeds, waters, and knowledge, are often ignored, and inadequately supported by governments. A small farmers' ability to earn a decent, socially responsible and ecologically sustainable livelihood may be broken by governments and large multinational companies through unjust policies and laws, or corporate criminal behavior.

The country's mode of integration into the global economy accelerates its population's dependence on imported products and processed food from transnational corporations. The global fast-food expansion in China, India, and Russia illustrates how fast food directly shapes food availability and food options in the global market, strengthening and expanding dietary dependence on imported, processed, and fast-food varieties (Christian and Gereffi 2018).

According to Dodds et al. (2014), the increasing popularity of local food consumption can be attributed to the heightened awareness of food safety concerns, carbon emissions produced from food transportation, and the understanding of ways large corporations' obtain their food supplies. Using the case study of Toronto, Canada, they indicate that the main purpose of visiting farmers' markets is not solely to fulfill grocery needs. The quality of products offered and the ability to support the local community are the primary motivators to visit the markets.

As a Sweden example by Sivertsson and Tell (2015) shows, micro-small companies with low technological production face many barriers to business model innovation such as the high costs of fixed assets, a lot of government regulations, weather conditions, and traditions. In addition, small farmers have supplementary barriers as the farming mentality and culture. Farmers don't think of themselves as business managers despite having most of the same issues and concerns that all business managers have.

An important method, which can help rural farm-households in developing countries to benefit from agrifood exports, is contract-farming with exporters or overseas buyers. Successful contract-farming involves technology and capital transfers. In many cases, to make these value chains function, it requires farm assistance programs which help to overcome constraints on domestic firms with a limited access to capital and technology (Maertens and Swinnen 2015).

A study by Croft et al. (2016) from Kenya shows that the most common markets' barrier is an access to capital. However, informal vendors, more than formal vendors, perceive that as a major problem. Overall, 97% of vendors say that they see the vegetable market grow, which suggests that this market is still expanding. To improve urban nutritional security, the most important change policymakers can enact is to increase an access to capital and improve infrastructure to connect rural growers with urban consumers.

Beyond encouraging farmers along the avenue of direct marketing, a number of policy recommendations can help farmers and restaurants make connections. For example, a straight forward policy change on a state level mandating that state institutions purchase major Colorado crops (potatoes, onions, apples) from Colorado farmers will be the single most powerful market-based support for them. Price alone is not a significant factor in purchasing decisions. Food buyers prioritize quality as their top purchasing criterion but they are not aware that local farmers can provide higher quality, and that institutions are interested in buying locally. Moreover, small farms can offer comparable or higher quality production and service. So, farmers need to show buyers the quality of production and service they can provide (Starr et al. 2003).

In terms of urbanization, urban and suburban agriculture is of major socio-economic and ecological importance. However, even in countries where urban agriculture is common, such as in the US, the most urban farmers do not receive the same level of financial support from the government as rural commodity farmers, as they grow non-commodity crops. The other problems include a lack of low-cost land, resources for irrigation and training programs in urban agriculture. All these obstacles place urban farmers at an increased financial risk (Castillo et al. 2016).

Zelinskaya et al. (2015) identifies the main problems of the state regulation of farm enterprises in the Krasnodar Territory as the difficulty of obtaining loans, limited addressees' product sales, difficulties in access to land resources, and the lack of indicators demonstrating accounting activities of farmers.

In the development of agriculture (in particular, urban and suburban agriculture), an important factor is the acceptance of agrolandscape and the form of agriculture management of the local population and stakeholders. The results of survey among 386 urban participants in Berlin, Germany, show that more than 80% of the respondents prefer having accessible systems such as public green spaces, intercultural gardens, and rooftop gardens. More than 50% are willing to buy horticultural products, but they reject products from intensive production systems and animal-farming mechanisms, with more than 70% rejection for animal products. The main insight is that the highest degree of acceptance is reached for multifunctional urban agriculture which combines commercial goals with ecological and social (Specht et al. 2016).

Urban agriculture (UA) is spreading within the Global North, largely for food production, ranging from household individual gardens to community gardens which boost neighborhood regeneration. Additionally, UA is also being integrated into buildings, such as urban rooftop farming (URF). Urban agriculture is largely perceived as a social activity rather than a food production initiative. However, several stakeholders highlight the potential to increase urban fertility through URF by occupying currently unused spaces. A lot of UA-related stakeholders (e.g., food co-ops, NGOs) prefer soil-based UA. Newer stakeholders (e.g., architects) highlight the economic, social and environmental opportunities of local and efficient food production through innovative URF (Sanyé-Mengual et al. 2016).

The results from the sustainability impact assessment (SIA) by Schmutz et al. (2018) in London on the example of different types of urban and peri-urban 'short food supply chains' (SFSCs) show that 'community supported agriculture' (CSA) is regarded as delivering the highest overall social, economic and environmental benefits, followed by urban gardening (both commercial and self-supply) and direct sales (off-farm).

Thus, the acceptance of local products by the population is connected, among other things, with the side social or environmental benefits that the ecosystem provides.

The next study (Belfrage et al. 2015) in Sweden, across six small (<50 ha) and six large farms (>135 ha), shows that small farms rather than large farms have significantly higher on-farm landscape heterogeneity. Strong positive relations between on-farm landscape heterogeneity and a number of breeding birds, butterflies, and herbaceous plant species are found. The study indicates that, to increase biodiversity, farm size should be taken into consideration.

A perceptions study by Wetherill and Gray (2015) about FM (farmers' markets) foods and barriers to FM use in Oklahoma among 64 Participants Supplemental Nutrition Assistance Programs beneficiaries receiving Temporary Assistance for Needy Families shows that few participants eat fresh products regularly and mostly appreciate the convenience of shopping at a supermarket. The results show that farmers' markets are not perceived as available or accommodating to shopping needs and affordability and acceptability concerns are expressed. Therefore, it is very important to simplify the access of the population to local products by the optimization of the location of farmers' markets, an increase in advertising of local products and an explanation of their advantages through the media.

5 Balance of Interests in Food Production and Marketing

A market is an exchange place between sellers and buyers where an exchange is an entry point to a market system which takes a product from initial production to consumption by a final consumer. Each market sector has its own product standards that can be varied between and within countries. The industrial market system for food has changed rapidly. The competition for efficiency of the industrial system is said to be not between operators of a chain at the same level like in a more traditional approach of competition, but between chains. As a result of the quest for chains efficiency, the coordination of various components of the chain and the relationships between them has become increasingly important (Ramsay and Morgan 2009).

However, overcoming trade barriers is important for countries with high-risk agriculture as a guarantee of food security. The populations in Sub-Saharan Africa and South Asia are relatively more vulnerable to the long-term food security impacts from climate change. But, when markets are fully integrated, there is significant moderation in the range of malnutrition which highlights the importance of international trade in mitigating the long-term food security impacts of climate change (Baldos and Hertel 2015).

6 Few Words on Contract Farming

The way of reducing the farmers' risks and continuing the food specialization of farms is contract farming with cooperation of small and medium-sized farms. Contract farming is agricultural production carried out according to an agreement between a buyer and farmers, which establishes conditions for the production and marketing of a farm product or products. Mostly, the farmer should agree with the agreed provision quantities of a specific agricultural product. These products should meet the quality standards of the purchaser and be supplied at the time determined by the purchaser.

In turn, the buyer should purchase the product and, in some cases, support production through supply of farm inputs, land preparation and the provision of technical advice and other actions (FAO 2017).

The increasing need for vertical coordination is not limited to transactions between farmers and traders, processors or retailers, but it results in overall integration of (smallholder) farmers with agricultural value chains. Contract Farming (CF) is very effective for implementing strict value chain coordination. Whereas the traditional reasons for adopting CF, such as failing markets for farm credit and seeds, fertilizers, and crop protection, are important, the need of strengthening vertical coordination throughout the agrifood value chain has recently become more vital, according to FAO (2013).

7 Perspectives of Sustainable Agriculture in the Face of Global Contradictions

According to Geertsema et al. (2016), ecological intensification of agriculture (EI) aims to conserve and promote biodiversity and the sustainable use of associated ecosystem services in order to support resource-efficient production. In this case, science should be based on four principles, such as: (1) biodiversity conservation for the delivery of ecosystem services; (2) management of ecosystem services benefit from a landscape-scale approach; (3) articulated ecosystem service of trade-offs and synergies; (4) EI must be associated with social dynamics involving farmers, governments, researchers, and related institutions.

Jouzi et al. (2017) shows that the most significant advantages of organic farming (OF) are environmental protection and higher resilience to environmental changes, increasing an income of farmers and reducing an external input cost, enhancing a social capacity and increasing employment opportunities. Food security can be enhanced by increasing the food purchasing power of local people. The main challenges of this food production system are lower yields in comparison with conventional systems, difficulties with soil fertility management, market barriers, and the educational needs of small-holders.

A new "commercial ecological agriculture" should unite sustainable agricultural practices and must be coordinated among all the stakeholders via learning and adaptation with time. In order to provide agricultural sustainability, it should be least-disturbing, resilience-building, and efficient in energy and nutrient use, site-specific, labor and skill-intensive, low-input, diversified and integrated, and closely nature harmonized (Srivastava et al. 2016). Such types of agriculture management can increase ecological benefits and yields. Redlich et al. (2018) shows that increased crop diversity can lower the dependence on insecticides while enhancing yield stability through the ecological intensification of farming. A simple diversification approach, for example, growth of nectar-producing plants around the rice fields, can contribute to the ecological intensification of agricultural systems. Such an inexpensive intervention significantly reduces populations of key pests as well as insecticide applications by 70%. Moreover, it increases grain yields by 5% and delivers an economic advantage of 7.5% (Gurr et al. 2016).

Due to an increase in competition and lower prices for agricultural products, many farms are forced to expand their sown area or switch to growing of energy monocultures. To reduce the negative impact of agriculture on the ecosystem, the EU applies agri-environment measures which are designed to encourage farmers to protect and enhance the environment on their farmlands. They provide payments for farmers in return for a service of carrying out agri-environmental commitments which mean more than the application of usual good farming practices (European Commission 2005). However, for farms oriented to the production of commercial crops with a high proportion of arable land, it may be unattractive to use agri-environmental measures (Gailhard and Bojnec 2015).

8 Conclusion

Due to many barriers of various origins like trade (price) or mental (behavioral, educational and cultural) barriers, the "invisible hand" of the market increasingly fails to cope with the need to provide the normal functioning of a system. The most common barriers are associated with the seller's domination over the buyer, overpricing of goods and services, monopolization of market conditions by large transnational corporations, and random restrictions on market access for small-scale producers.

In Poland, meat processing companies repeatedly bind meat producers by preconditions (for example, limit the slaughter weight of pork's they purchase), that can damage small and medium-sized farms, especially (Pepliński 2016). In French winemaking, wholesalers confront farmers to supply more wines of a low price category in order to compete with cheap wines from Spain (Mustacich 2017). But, in France, wine prices historically vary depending on the region in which the grapes are grown. It can be very damageable for the quality of French wine and profitability of the whole winemaking branch, while even small farms can be highly profitable, regardless of the size of the plantation (Delord et al. 2015). And such examples around the world are far from unique.

So, in the EU countries the cooperation of agricultural producers with unique regulation and application of competition rules are still welcome. Companies associated with the formation of national markets for agricultural products are sometimes excluded from the general ban of anti-competitive activities. The practice of subsidizing agrarians is widespread in the EU. To overcome such barriers, closer cooperation of science, production and trade organizations, reduction of the number of intermediaries between farmers and end users of agricultural products and a number of other measures are proposed. But, the most promising are such innovative approaches as diversification of production, increasing the quality of products through the introduction of organic farming and higher technologies in land reclamation, recycling of industrial waste, involvement of new types of natural raw materials in the agricultural production.

In many countries, the share of small and medium-sized agricultural enterprises is declining. However, in some countries of the EU it reaches up to 70% of farms with an area of less than 5 ha, for example, in Italy. In Russia, from 2006 to 2016 the number of agricultural cooperatives of a moderate size increased by seven times.

On the whole, the picture is rather contradictory: only small enterprises do not withstand competition and they are forced out of the market by larger ones. Only by uniting into cooperatives they can not only survive, but also successfully compete with agro-holdings.

Although TNCs and small farms are often opposed to each other, still, not only competition, but also partnerships can and must exist between them. Thus, in the Gulf countries (Gulf Group States), the economics transits from dependent to more diversified, using other types of mineral and biological raw materials, including the "green", in order to spread and successfully develop the recreational and tourist business to sea coasts and islands. In many developed countries, TNCs are attracting the closest attention of law enforcement. Even the most ingenious corruption schemes are revealed and this gradually clears the business community of criminal structures. However, in this matter, it is impossible to rely only on state or even international legal bodies. In underdeveloped and developing countries, the formation of education systems and an increase in the general culture of the population is of paramount importance in overcoming mental barriers. With the Internet many countries in Africa, Asia and Latin America are rapidly becoming involved in various types and methods of education and in a short time overcome barriers of underdevelopment in agribusiness and related industries (bio energy, etc.). In the markets of the Scandinavian countries, the relations of buyers and sellers have reached a high degree of trust: there is a kind of consensus between them which is largely related to the information literacy of the population and it is contributing to a more transparent and honest business. Due to the open reports of leading Swedish farms, the access to a larger piece of information for Swedish housewives has become almost 100%.

The knowledge of several foreign languages is an excellent prerequisite for innovations development and inventions in all areas of business and, foremost, in agro-industrial production, the profitability of which is revealed much faster than of industry.

Summarizing this work the main conclusion can be made: in agribusiness there are still many markets and other barriers, but with a systematic approach and matrix management of all components of market relations, these barriers are quite climbable, although not in different countries and regions simultaneously.

References

Arita S, Mitchell L, Beckman J (2015) Estimating the effects of selected sanitary and phytosanitary measures and technical barriers to trade on US-EU agricultural trade (No. 212887). United States Department of Agriculture, Economic Research Service

Baldos ULC, Hertel TW (2015) The role of international trade in managing food security risks from climate change. Food Security 7(2):275–290

Beghin JC, Bureau JC (2015) The cost of food self-sufficiency and agricultural protection in South Korea. Iowa Ag Rev 8(1):2

Belfrage K, Björklund J, Salomonsson L (2015) Effects of farm size and on-farm landscape hetero-geneity on biodiversity—case study of twelve farms in a Swedish landscape. Agroecol Sustain Food Syst 39(2):170–188. https://doi.org/10.1080/21683565.2014.967437

Castillo SR, Winkle CR, Krauss S, Turkewitz A, Silva C, Heinemann ES (2016) Regulatory and other barriers to urban and peri-urban agriculture: a case study of urban planners and urban farmers from the greater Chicago metropolitan area. J Agric, Food Syst, Community Dev 3(3):155–166

Christian M, Gereffi G (2018) Fast-food value chains and childhood obesity: a global perspective. In: Pediatric obesity. Humana Press, Cham, pp. 717–730

Croft MM, Marshall MI, Hallett SG (2016) Market barriers faced by formal and informal vendors of African leafy vegetables in Western Kenya. J Food Distrib Res 47(3)

Delord B, Montaigne É, Coelho A (2015) Vine planting rights, farm size and economic performance: do economies of scale matter in the French viticulture sector? Wine Econ Policy 4(1):22–34

Dodds R, Holmes M, Arunsopha V, Chin N, Le T, Maung S, Shum M (2014) Consumer choice and farmers' markets. J Agric Environ Ethics 27(3):397–416

Edwards C (2018) Agricultural subsidies. from https://www.downsizinggovernment.org/agriculture/subsidies

Erokhin V (2017) Self-Sufficiency versus Security: how trade protectionism challenges the sustain-ability of the food supply in Russia. Sustainability 9(11):1939

European Commission (2005) Agri-environment measures. Overview on general principles, Types of measures, and application. Directorate General for Agriculture and Rural Development. Unit G-4—Evaluation of Measures applied to Agriculture, Studies. From https://ec.europa.eu/agriculture/sites/agriculture/files/publi/reports/agrienv/rep_en.pdf

FAO (2013) Contract farming for inclusive market access. Rome. From http://www.fao.org/3/a-i3526e.pdf#page=33

FAO (2017) Legal fundamentals for the design of contract farming agreements by Caterina Pultrone (Rural Infrastructure and Agro-industries Division, FAO), with contributions from Carlos A. da Silva (Rural Infrastructure and Agro-industries Division, FAO) and Carmen Bullón Caro (FAO Legal Office). From http://www.fao.org/3/a-i8059e.pdf

FAO (2018a) Crops and livestock products. From http://www.fao.org/faostat/en/#data/TP

FAO (2018b) Crops. From http://www.fao.org/faostat/en/#data/QC

Fayet L, Vermeulen WJ (2014) Supporting smallholders to access sustainable supply chains: lessons from the Indian cotton supply chain. Sustain Dev 22(5):289–310

Gailhard İU, Bojnec Š (2015) Farm size and participation in agri-environmental measures: farm-level evidence from Slovenia. Land Use Policy 46:273–282

Geertsema W, Rossing WA, Landis DA, Bianchi FJ, Van Rijn PC, Schaminée JH, … Van Der Werf W (2016) Actionable knowledge for ecological intensification of agriculture. Front Ecol Environ 14(4):209–216

Gurr GM, Lu Z, Zheng X, Xu H, Zhu P, Chen G, … Villareal S (2016) Multi-country evidence that crop diversification promotes ecological intensification of agriculture. Nat Plants 2(3):16014

Hunter MC, Smith RG, Schipanski ME, Atwood LW, Mortensen DA (2017) Agriculture in 2050: recalibrating targets for sustainable intensification. Bioscience 67(4):386–391

Jouzi Z, Azadi H, Taheri F, Zarafshani K, Gebrehiwot K, Van Passel S, Lebailly P (2017) Organic farming and small-scale farmers: main opportunities and challenges. Ecol Econ 132:144–154

Lark TJ, Salmon JM, Gibbs HK (2015) Cropland expansion outpaces agricultural and biofuel policies in the United States. Environ Res Lett 10(4):044003

Maertens M, Swinnen J (2015) Agricultural trade and development: a value chain perspective (No. ERSD-2015-04). WTO Staff Working Paper

Morschett D, Schramm-Klein H, Zentes J (2015) Market barriers, global and regional integration. In: Strategic international management. Springer Gabler, Wiesbaden, pp 147–169

Mustacich S (2017) Young, French and angry: winegrowers in Southern France Are Struggling https://www.winespectator.com/webfeature/show/id/Winegrowers-in-Southern-France-Are-Struggling

Nelles W (2015) The right to organic/ecological agriculture and small-holder family farming for food security as an ethical concern. In: Hongladarom S (ed) Food security and food safety for the twenty-first century. Springer, Singapore

Pepliński B (2016) Factors affecting changes in the population of sows in Poland. Regional analysis. ACTA SCIENTIARUM POLONORUM 75

Popp J, Lakner Z, Harangi-Rakos M, Fari M (2014) The effect of bioenergy expansion: food, energy, and environment. Renew Sustain Energy Rev 32:559–578

Ramsay G, Morgan B (2009) Barriers to market entry, poor livestock producers and public policy. Pro–Poor Livestock Policy Initiative (PPLPI) Working Paper (46)

Reardon T, Codron JM, Busch L, Bingen J, Harris C (1999) Global change in agrifood grades and standards: agribusiness strategic responses in developing countries. The Int Food Agribus Manag Rev 2(3–4):421–435

Redlich S, Martin EA, Steffan-Dewenter I (2018) Landscape-level crop diversity benefits biological pest control. J Appl Ecol

Rosstat (2019) Federal'naya sluzhba gosudarstvennoy statistiki. Regiony Rossii. Sotsial'no-ekonomicheskiye pokazateli—2017. From http://www.gks.ru/bgd/regl/b17_14p/Main.htm

Sanyé-Mengual E, Anguelovski I, Oliver-Solà J, Montero JI, Rieradevall J (2016) Resolving differing stakeholder perceptions of urban rooftop farming in Mediterranean cities: promoting food production as a driver for innovative forms of urban agriculture. Agric Hum Values 33(1):101–120

Schmutz U, Kneafsey M, Kay CS, Doernberg A, Zasada I (2018) Sustainability impact assessments of different urban short food supply chains: examples from London, UK. Renew Agric Food Syst 33(6):518–529

Scott CM (2015) The role of transnational food and agriculture corporations in creating and responding to food crises—synthesis paper. Can Food Stud/La Revue canadienne des études sur l'alimentation 2(2):146–151

Sivertsson O, Tell J (2015) Barriers to business model innovation in Swedish agriculture. Sustainability 7(2):1957–1969

Specht K, Weith T, Swoboda K, Siebert R (2016) Socially acceptable urban agriculture businesses. Agron Sustain Dev 36(1):17

Srivastava P, Singh R, Tripathi S, Raghubanshi AS (2016) An urgent need for sustainable thinking in agriculture–an Indian scenario. Ecol Ind 67:611–622

Starr A, Card A, Benepe C, Auld G, Lamm D, Smith K, Wilken K (2003) Sustaining local agriculture barriers and opportunities to direct marketing between farms and restaurants in Colorado. Agric Hum Values 20(3):301–321

Swinbank A (2017) World trade rules and the policy options for British agriculture post-Brexit. Briefing paper, 7

Walker S (2018) Cultural barriers to market integration: evidence from 19th century Austria. J Comp Econ

Wegren SK, Elvestad C (2018) Russia's food self-sufficiency and food security: an assessment. Post-Communist Econ 1–23

Wetherill MS, Gray KA (2015) Farmers' markets and the local food environment: identifying perceived accessibility barriers for SNAP consumers receiving temporary assistance for needy families (TANF) in an urban Oklahoma community. J Nutr Educ Behav 47(2):127–133

Zaytseva AL (2013) Agroprodovol'stvennyy kompleks YES kak oblast' realizatsii konkurent-noy politiki. Chelovecheskoye izmereniye mirovoy politiki i ekonomiki. Mirovoye razvitiye. 9:141–152

Zelinskaya MV, Plotnikova EV, Zayceva MV, Chueva TI, Osmolovskaya MS (2015) The problem of development of farm enterprises in the Krasnodar region. Mediterranean J Soc Sci 6(5 S3), 164

Indicators for Assessing Sustainability Performance of Small Rural Properties

Aleriane Zanetti Vian, Dalmarino Setti and Edson Pinheiro de Lima

Abstract Sustainability development requires the use of strategies that involve all actors of the socioeconomic system and society in the pursuit of a common goal: to sustain current demands without compromising resources for future generations. These strategies generally address economic, social, and environmental performance, framed and named as the Triple Bottom Line. One of the fundamentals for developing adequate organizational strategies is the development of models to assess sustainability that contemplate the Triple Bottom Line of companies, organizations, processes and economic activities, such as agricultural activities. In this way, the objective of this work is to identify the economic, social and environmental indicators that could be used in sustainability assessment models of small rural properties. For that purpose, it is used the ProKnow-C methodology for literature review and bibliometrics analysis, and after that, it is submitted the developed conceptual model to experts' review. This work develops a conceptual model, and the obtained results involved the identification of the main adopted sustainability indicators and their respective data sources, contributing to the development of new models for assessing the sustainability performance of small rural properties.

Keywords Sustainability performance · Assessment model · Small rural properties

A. Z. Vian (✉) · D. Setti · E. P. de Lima
Industrial and Systems Engineering, Federal University of Technology—Paraná (UTFPR), Via do Conhecimento, Km 1, Pato Branco, Parana, Brazil
e-mail: alerianezanetti@hotmail.com

D. Setti
e-mail: dalmarino@utfpr.edu.br

E. P. de Lima
e-mail: pinheiro@utfpr.edu.br

E. P. de Lima
Pontifical Catholic University of Parana—PUCPR, R. Imaculada Conceicao, Curitiba, Parana 1155, Brazil

© Springer Nature Switzerland AG 2020
W. Leal Filho et al. (eds.), *International Business, Trade and Institutional Sustainability*, World Sustainability Series,
https://doi.org/10.1007/978-3-030-26759-9_22

1 Introduction

Sustainability and sustainable development are gaining more and more space in society and in the economic sphere of organizations, requiring mechanisms that also contemplate the demands of future generations and monitor performance over time (Kruger 2017; Rodrigues et al. 2018). Svensson et al. (2018) observe that by adopting the Triple Bottom Line, companies could develop competitive advantages, greater market share, and optimize shareholder value.

In this way, as pointed out by Louette (2007), business organizations, both for profit and non-profit, aim to find mechanisms that help to transform concepts such as social responsibility and sustainable development into management practices. Several studies isolated approach the dimensions of sustainability, arising from the need to broaden discussions and understandings, to overcome these limitations and encourage assessments that together portray economic, social and environmental sustainability (Dias 2016; Kruger 2017). Therefore, the approach in question is justified by intended to cover the Triple Bottom Line, that is, the dimensions of economic, social and environmental.

In addition, Kruger (2017) argues that there is still a gap with regard to the application of management indicators in rural areas that make it possible to perform performance an assessment from a sustainable perspective, despite the various discussions on the theme. Thus, the importance of assessing sustainability performance as a tool to help decision making by rural properties is highlighted, and for this, the use of indicators can be an efficient mechanism.

Therefore, the proposed research question is: "Which indicators can be used to assess the sustainability performance of small farms?" In order to respond to the problem in question, the objective of the research is to identify the economic, social and environmental indicators, used in models for assessing sustainability identified in the literature, to subsidize the structuring of a model for small rural properties. After that, three sustainability review the proposed model in order to produce a first revised version.

This work develops a conceptual model, and the expected results involve the identification of the main indicators used and the data sources necessary to obtain their information, contributing to the development of new models for evaluating the sustainability performance of small farms. As for the structuring, it is presented as follows: first, the issues of the literature on the identification of indicators for performance evaluation in sustainability are first evidenced. Subsequently the methodological procedures performed for the construction of the database, the results obtained and the conclusions on the studied topic.

2   Identifying Indicators for Assessing Sustainability Performance

The sustainable development can be portrayed a formula that maintains the economic, social and environmental dimensions, being that many terminologies in conceptualizing sustainability, approach the Triple Bottom Line contemplating these dimensions, that aid the management process and are differentiated by the level of concentrated efforts to achieve the desired goal (Louette 2007; Rodrigues et al. 2018; Svensson et al. 2018).

In addition, sustainable development is also addressed when issues of social responsibility are discussed, with a view to their union with the economic and environmental aspects of the environment, thus, laws, standards, certifications, principles and guidelines become the direction to reach the objectives, and the greater the applicability of the various existing mechanisms and tools, the greater the possibility to assess risks, opportunities, successes and improve management (Louette 2007).

Therefore, the demand for food and other natural resources is increasing, and at the same time, the demands for improving sustainability, reducing impacts and performance efficiency, and pushing farmers to innovate in sustainability aspects to meet the qualified needs of consumers (Rodrigues et al. 2010; Ruviaro et al. 2012; Rótolo et al. 2015).

In this line of thinking, using indicators it is possible to assess how close or distant the organization is to sustainability, enabling the measurement of the stage or degree of achievement of goals and objectives (Passel et al. 2007; Van der Werf et al. 2007). In addition, as mention by Rodrigues et al. (2010), indicators can express a measure of the changes brought about by agricultural practices and trends over time, guiding corrective management actions.

In this way, by analyzing the literature, some perceptions and considerations can be extracted, for example: some studies use more than one method in their research seeking to verify the relations and bottlenecks, as is the case of Siche et al. (2008) assess the environmental sustainability of nations with the ESI (Environmental Sustainability Index), EF (Ecological Footprint) and EMPIs (Emergy Performance Indices). It is evident that the methods have benefits and limitations, each one with its particularities, scope and complexity, but that allow to obtain an estimate of the sustainability performance in a general way, being able to provide patterns for similar results assessment.

Similarly, Van der Werf et al. (2007) investigated the methods: LCA (Life Cycle Assessment), EF (Ecological Footprint), EMA (Environmental Management for Agriculture), FarmSmart and DIALECTE, aiming at assessing the differences of the results obtained by each method and the characteristics involved in the process. As well as, De Olde et al. (2016) compared the RISE (Response Inducing Sustainability Evaluation), SAFA (Sustainability Assessment of Food and Agriculture Systems), PG (Public Goods Tool) and IDEA (Indicators of Sustainability of Farms) to assess sustainability in practice and verify the perception of farmers. Highlighting the importance attributed in queries such as ease of use, complexity, language, context and representativeness of the tool for the designer and its user.

Still, there are researches that use methods that are essentially similar, such as LCA (Life Cycle Assessment) to assess environmental impacts, with a view to sustainable performance (Van der Werf et al. 2007; Ruviaro et al. 2012; Mohammadi et al. 2015; Wang et al. 2016, Mousavi-Avval et al. 2017; Khanali et al. 2018). From this perspective, each study may resemble the basis, in the context, and may approach or distance itself as it traverses the intended goal.

It is worth remembering that in this structuring process, there are works with methods that are composed of indicators or that directly approach the indicators. Paracchini et al. (2015) show the SOSTARE model, which is made up of indicators, covering the following aspects: economic, agronomic and ecological, to analyze sustainability in farms, providing farmers with a diagnostic tool and performance assessment. Rodrigues et al. (2010) used the "System for Weighted Assessment of the Environmental Impact of Rural Activities"—APOIA-NovoRural, which involved sixty-two indicators, aiming to provide an integrated assessment of agricultural sustainability in the environmental management of activities, enabling a diagnostic tool for managers.

Funes-Monzote et al. (2009) used fifteen agroecological and financial indicators to assess the conversion of specialized dairy farming systems into sustainable mixed agricultural systems. And, Zorn et al. (2018) calculated seventeen financial ratios to evaluate economic sustainability in dairy farms, formed by the analysis of profitability, liquidity, financial efficiency, stability, solvency and payment capacity. Emphasizing that the use of financial ratios can be an incentive factor for farmers to understand the effects of sustainability assessments in the sector.

Also, there are studies that combine methods that in fact generated new approaches for assessing sustainability performance. Blasi et al. (2016) sought to assess the environmental impact and the relationship with economic development in an agricultural production system using EFA (Ecological Footprint Accounting) that involved the EF (Ecological Footprint) metrics measuring the ecological demand of human activities and BC (Biocapacity) capacity of the ecosystem to offer ecological goods and services, evidencing the importance of the care with economic and environmental impacts. Rótolo et al. (2015) explored how land allocation and technological innovation affect agricultural sustainability, with SUMMA (Sustainability Multimethod Multiscale Assessment) combining MFA (Material Flow Accounting), CEA (Cumulative Energy Accounting), EMA (Emergy Accounting) and CML2 baseline 2000, each with its indicators and with essence in the life cycle to address the performance of the agricultural system.

In addition, approaches may include one or several dimensions. Since only the environmental dimension (Van der Werf et al. 2007; Ruviaro et al. 2012; Rótolo et al. 2015; Mousavi-Avval et al. 2017; Khanali et al. 2018), the social (Sidhoum 2018) or economic (Zorn et al. 2018). The three dimensions, economic, social and environmental in a single study (Passel et al. 2007; De Olde et al. 2016; Azevedo et al. 2018) or go beyond these, with environmental, economic, social, ethical and cultural approaches, for example (Siche et al. 2008).

Considering the analysis of the literature, it is observed that several studies are being developed seeking to contemplate the demands that involve sustainability, through different terminologies, methods, indicators and sectors. With similarities when assessing, analyzing, diagnosing, portraying, assisting, providing feedback or measuring performance, aiming to meet a common goal, "sustainability."

3 Methodological Procedures

In order to meet the proposed objective of identifying economic, social and environmental indicators used in sustainability assessment models of small farms, a structured data collection and analysis procedure was adopted, including ProKnow-C (Knowledge Development Process—Constructivist), to perform the literature review and bibliometrics analysis. Lacerda et al. (2012), Dutra et al. (2015) and Ensslin et al. (2017) could be consulted for further details on the SLR method. Figure 1 illustrates the delimitations of the search axes and the keywords in each axis, along with their combinations, in order to select the raw paper database.

The axes and keywords presented in Fig. 1 made it possible to execute the following combinations: *(sustain* OR environment OR "sustainable development") AND ("performance assessment" OR "performance measurement" OR kpi) AND ("family farming" OR "small farm*" OR "agricultural production system*")* and *(sustainability OR environment OR "sustainable development") AND ("performance assessment" OR "performance measurement" OR kpi) AND ("family farming" OR "small farm" OR "agricultural production system")*. Being executed one combination or another, depending on the possibility of running in the database.

The databases used were the *Scopus, Science Direct, Emerald, Taylor and Francis* and the *Springer*. In each database the combination was executed only once, so that the return of papers meets a manageable ratio. The search was carried out in the period from 03 to 25 September 2018 obtaining a gross return of 438 papers. Subsequently, the papers and papers of review were selected, restricting the 326 papers, with these,

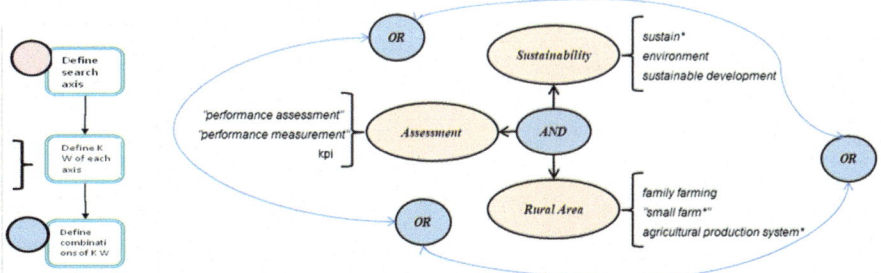

Fig. 1 Definition and combination of the search keywords and axes. *Source* Prepared by the authors

it was tested the adherence of the search terms, that answered the researched context, not needing new inclusions.

With the paper set of 326 papers gross, filtering began, exporting to the Mendeley program to assist in portfolio management. Duplicates were eliminated, leaving 324 papers that read the titles. After that, 45 papers were left that formed the paper set of non-repeated papers with aligned titles. Later, with Google Scholar, we verified the number of citations of the 45 papers, it was ordered by the number of citations and participation in percentage, the representability was fixed in five citations. Separating into two different data sets, the set "A" composed of papers with confirmed scientific acknowledgment, and set "B", with scientific recognition not yet confirmed, as evidenced in Fig. 2.

Figure 2 shows the composition of the sample for a representativeness of five citations, of which 29 cited papers representing 98% of citations make up the set "A" and 16 papers representing 2% of citations and make up the set "B". Following, the summaries of the "A" paper set were read, if they were aligned, they would remain in the database, with 27 remaining papers that formed the "A" paper set and a set of authors.

It was filtered the "B" paper set that had 16 papers and that had not yet confirmed scientific acknowledgment. So, if the paper were from the last two years would remain in the set to be read the summaries, read 15 papers. One paper did not meet the period of the last two years, it was verified if the author was in the set of authors formed by the set of papers "A", as it was stated, was again considered, going through the reading of the summary. Of the 16 papers, 13 had abstracts aligned forming the "B" paper set.

The "A" and "B" paper sets formed the "C" repository containing 40 papers. With the "C" repository, the whole paper was read to verify if it met the research

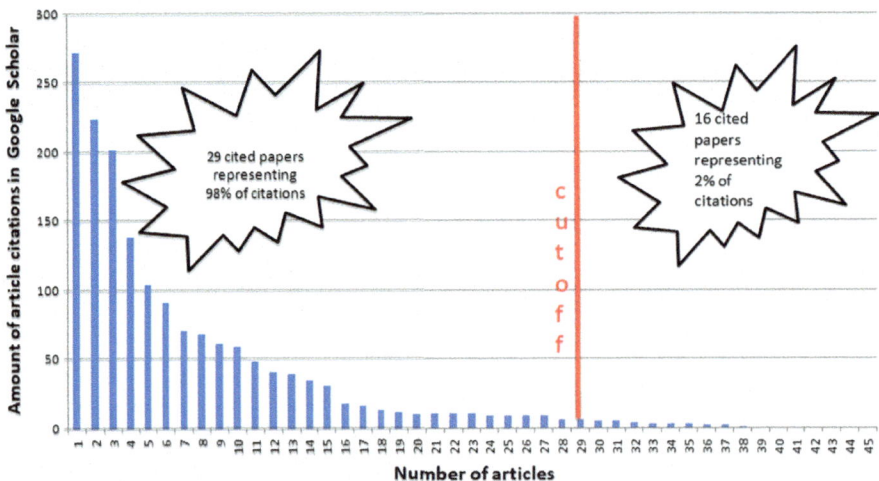

Fig. 2 Alignment of scientific recognition. *Source* Prepared by the authors

Passel *et al*. (2007)	*Measuring farm sustainability and explaining differences in sustainable efficiency*
Van der Werf *et al*. (2007)	*Environmental impacts of farm scenarios according to five assessment methods*
Tipraqsa *et al*. (2007)	*Resource integration for multiple benefits: Multifunctionality of integrated farming systems in Northeast Thailand*
Siche *et al*. (2008)	*Sustainability of nations by indices: Comparative study between environmental sustainability index, ecological footprint and the emergy performance indices*
Funes-Monzote *et al*. (2009)	*Conversion of specialized dairy farming systems into sustainable mixed farming systems in Cuba*
Rodrigues *et al*. (2010)	*Integrated farm sustainability assessment for the environmental management of rural activities*
Ruviaro *et al*. (2012)	*Life cycle assessment in Brazilian agriculture facing worldwide trends*
Mohammadi *et al*. (2015)	*Joint Life Cycle Assessment and Data Envelopment Analysis for the benchmarking of environmental impacts in rice paddy production*
Paracchini *et al*. (2015)	*A diagnostic system to assess sustainability at a farm level: The SOSTARE model*
Rótolo *et al*. (2015)	*How land allocation and technology innovation affect the sustainability of agriculture in Argentina Pampas: An expanded life cycle analysis*
De Olde *et al*. (2016)	*Assessing sustainability at farm-level: Lessons learned from a comparison of tools in practice*
Wang *et al*. (2016)	*Integrated analysis on economic and environmental consequences of livestock husbandry on different scale in China*
Blasi *et al*. (2016)	*An ecological footprint approach to environmental–economic evaluation of farm results*
Mousavi-Avval *et al*. (2017)	*Use of LCA indicators to assess Iranian rapeseed production systems with different residue management practices*
Azevedo *et al*. (2018)	*The influence of collaboration initiatives on the sustainability of the cashew supply chain*
Sidhoum (2018)	*Valuing social sustainability in agriculture: An approach based on social outputs' shadow prices*
Zom *et al*. (2018)	*Financial ratios as indicators of economic sustainability: A quantitative analysis for Swiss dairy farms*
Khanali *et al*. (2018)	*Life cycle assessment of canola edible oil production in Iran: A case study in Isfahan province*

Fig. 3 Paper portfolio. *Source* Prepared by the authors

objective, leaving 18 papers that were analyzed through the bibliometric analysis. Figure 3 shows the selected papers.

Figure 3 depicts the papers used to obtain indicators that were used to construct the conceptual model, which aims to contribute to the development of new models for assessing the sustainability performance of small rural properties. Thus, after the literature review, the analysis of bibliometrics was carried out, extraction of the present indicators evidenced in the essence of the portfolio and after submission of the developed conceptual model to experts' review.

In order to refine the proposed model, a questionnaire was sent to three sustainability experts, verifying their perceptions about: "which data sources could be consulted to obtain the information regarding the group of indicators", "how important is the group of indicators for assessing the sustainability of small farms" and "whether there were any set of indicators or indicators that should be excluded or included in the survey". This stage of the process aimed at identifying whether the proposed indicators, in the opinion of the experts, would serve the context or would they need other indicators, as well as the level of importance and the sources of consultation of the indicators.

4 Results

The analysis of bibliometrics allowed the deepening of the portfolio with a focus on perceiving its characteristics. Figure 4 shows the structure that included the analysis of bibliometrics, and the papers of the portfolio and their references were verified regarding the relevance of journals and authors, the scientific recognition of papers and the use of keywords.

Figure 4 showed the 18 papers that compose the portfolio under analysis and the 1258 papers that are present in the references of this portfolio. Among the papers that compose the bibliographic portfolio, we analyzed its relevance, using the number of citations in Google Scholar and Scopus, according to Fig. 5.

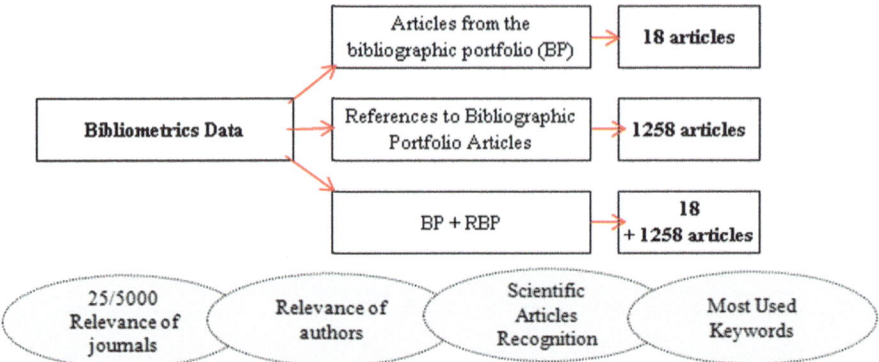

Fig. 4 Structure of bibliometry. *Source* Prepared by authors

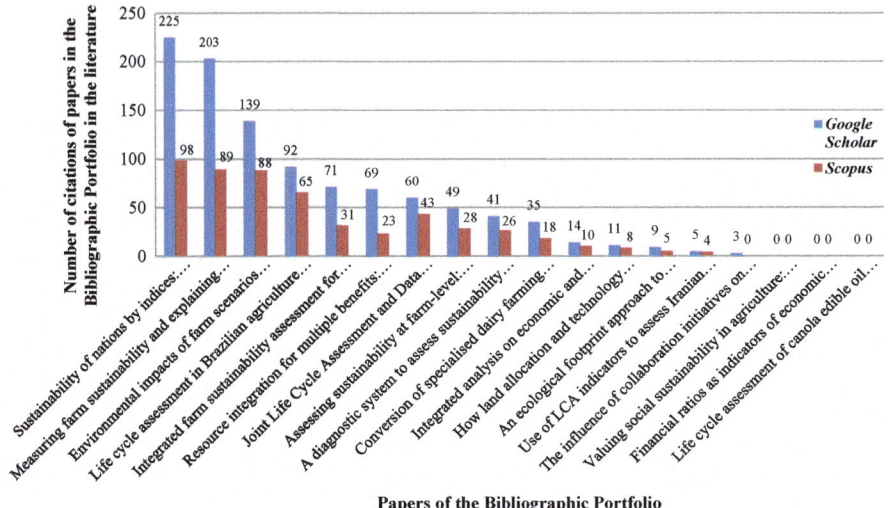

Fig. 5 Relevance of papers in the bibliographic portfolio. *Source* Prepared by the authors

Figure 5 shows the number of times the portfolio item was cited in a certain time period of the research execution. The most cited paper in both consultations was the "Sustainability of nations by indices: Comparative study between environmental sustainability index, ecological footprint and the emerging performance indices" by Siche et al. (2008) containing 225 citations in Google Scholar and 98 citations in Scopus. It should be noted that the papers with the fewest citations were the most recent, thus have not yet been so cited. Also, according to Fig. 6 the relevance of the papers in the references of the bibliographic portfolio was analyzed.

According to Fig. 6, the paper "Environmental impacts of farm scenarios according to five assessment methods" by Van der Werf et al. (2007) was evidenced three

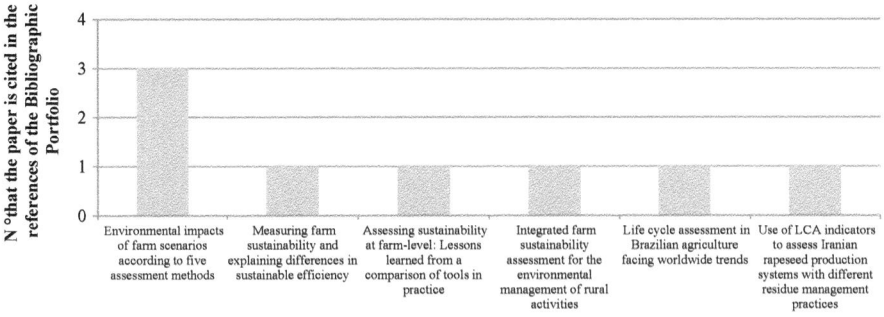

Papers of the Bibliographic Portfolio

Fig. 6 Relevance of papers in bibliographic portfolio reference. *Source* Prepared by the authors

times in the papers of the references and another five papers were evidenced only once. Figure 7 shows the most commonly used keywords in bibliographic portfolio papers.

According to Fig. 7, the most used keyword was Life Cycle Assessment (LCA), used in five papers, reporting life cycle assessment analyzes (Van der Werf et al. 2007; Mohammadi et al. 2015; Wang et al. 2016; Mousavi-Avval et al. 2017; Khanali et al. 2018). Also, the relevance of journals of the portfolio and their references were analyzed according to Fig. 8.

In the analysis of the relevance of the journals of the portfolio and their references, which were jointly presented through Fig. 8, the most outstanding journal was the Journal of Cleaner Production, evidenced in 5 papers, of the 18 that compose the

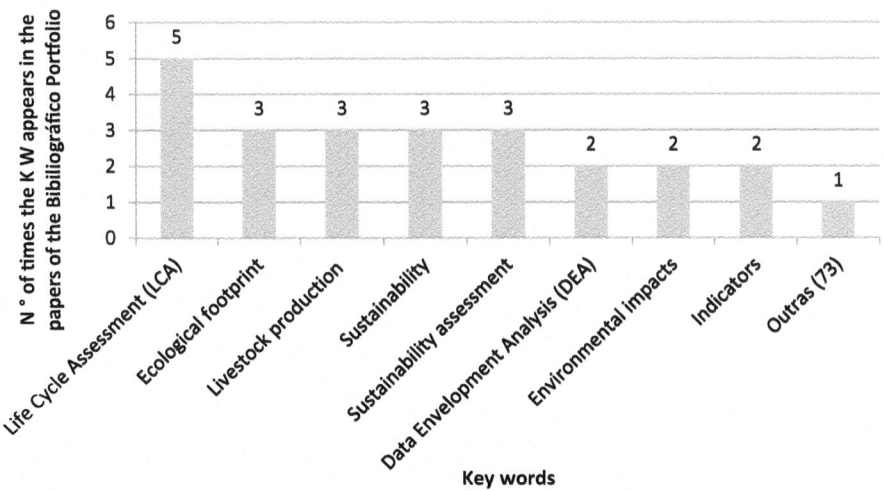

Fig. 7 Most used keywords. *Source* Prepared by the authors

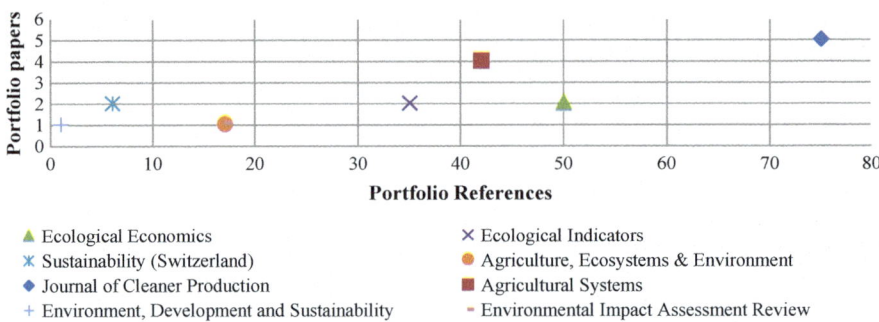

Fig. 8 Relevance of journals of the portfolio and their references. *Source* Prepared by the authors

portfolio and in 75 papers of the references that appear in the papers of this portfolio. Figure 9 shows the authors who stood out most in both bases (portfolio and their references).

According to Fig. 9, the author Rafiee, S. was the most outstanding, present in 2 portfolio papers and 22 portfolio reference papers, subsequently Khanali, M. in 2 portfolio papers and 3 papers of references and Sharifi, M which on both bases is evidenced 2 times.

After the analysis of the portfolio regarding its structure, the search began for aspects that contemplated the assessments surrounding sustainability. Each study is formed by its terminologies and structure, therefore, faced with the diversification of the terminologies found, we sought to analyze the essence of the paper and at the same time meet the proposal of this research. The evidence extracted from the literature was structured, in the three dimensions of the Triple Bottom Line: economic, social and environmental, and also considering the relationships between dimensions, stakeholders and context.

Given the above, the results obtained in the economic dimension involved aspects that formed "groups of indicators" referring to: economic-financial performance, degree of indebtedness, costs, infrastructure and diversification of business and rev-

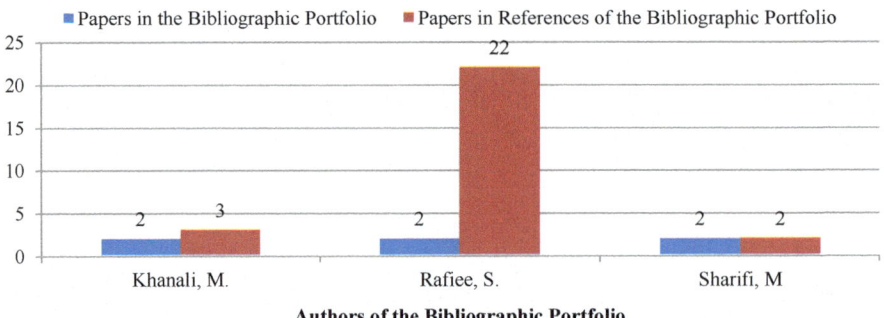

Fig. 9 Most cited authors. *Source* Prepared by the authors

enues. Through the group of indicators corresponding to economic-financial performance, it was possible to show indicators such as economic profit based on land productivity; economic profit based on labor productivity; net profit; family income; gross revenue; return on assets—ROA and return on equity—ROE (Passel et al. 2007; Tipraqsa et al. 2007; Funes-Monzote et al. 2009; Rodrigues et al. 2010; Mohammadi et al. 2015; Paracchini et al. 2015; Blasi et al. 2016; Wang et al. 2016; Mousavi-Avval et al. 2017; Azevedo et al. 2018; Zorn et al. 2018).

The degree of indebtedness, with indicators such as the independence of subsidies, debt index and the ability to pay (Rodrigues et al. 2010, Paracchini et al. 2015, Zorn et al. 2018). The costs relate to the total costs of production, such as fixed and variable (Funes-Monzote et al. 2009; Blasi et al. 2016; Wang et al. 2016). The group of infrastructure indicators, depicting indicators such as quality and quantity of housing, equipment, vehicles and external accessibility (Tipraqsa et al. 2007; Rodrigues et al. 2010). And, the diversification of business and revenues, for diversity between agricultural businesses and sources of revenue (Rodrigues et al. 2010; Paracchini et al. 2015).

To refine the model obtained from the literature review, experts were asked if there were any "group of indicators" or "indicators" that should be excluded or included in the research. Figure 10 shows the proposal of a model for sustainability performance assessment of small farms in the economic dimension, based on the analysis of the literature that was structured in the groups of indicators and indicators, and the highlights made in blue contemplate the opinions of the specialists.

In the social dimension, groups of indicators related to education, socio-cultural values and social inclusion, health and safety, food security, job, rural exodus and continuity, management and administration were highlighted. The group of education indicators aims to contemplate the level of training, with indicators such as higher education, access to education and local opportunity for higher qualification (Passel et al. 2007; Rodrigues et al. 2010; Azevedo et al. 2018). Sociocultural values/social inclusion, referring to indicators such as access to public services, for example, sports and leisure; preservation of the historical legacy; fair standards of living and consumption; social inclusion of genders and beliefs; assistance to the development of communities and social aspect (Rodrigues et al. 2010; Azevedo et al. 2018; Sidhoum 2018).

The group of health and safety indicators, with indicators such as human vulnerability; nutritional status and diseases related to the environment and health and safety at work (Siche et al. 2008; Rodrigues et al. 2010). Food Safety, with indicators on the diversity of food products available and agricultural products safe for health (Tipraqsa et al. 2007; Sidhoum 2018). The job is reported to indicators of quality of employment and labor force that represents the amount of labor available (Tipraqsa et al. 2007; Rodrigues et al. 2010). The rural exodus/continuity, covering the migration of people from rural areas to urban areas and the successor on the farm (Passel et al. 2007; Tipraqsa et al. 2007). And, the management and administration, being the profile of the manager and dedication; marketing conditions; sharing decisions and choices (Rodrigues et al. 2010; Azevedo et al. 2018).

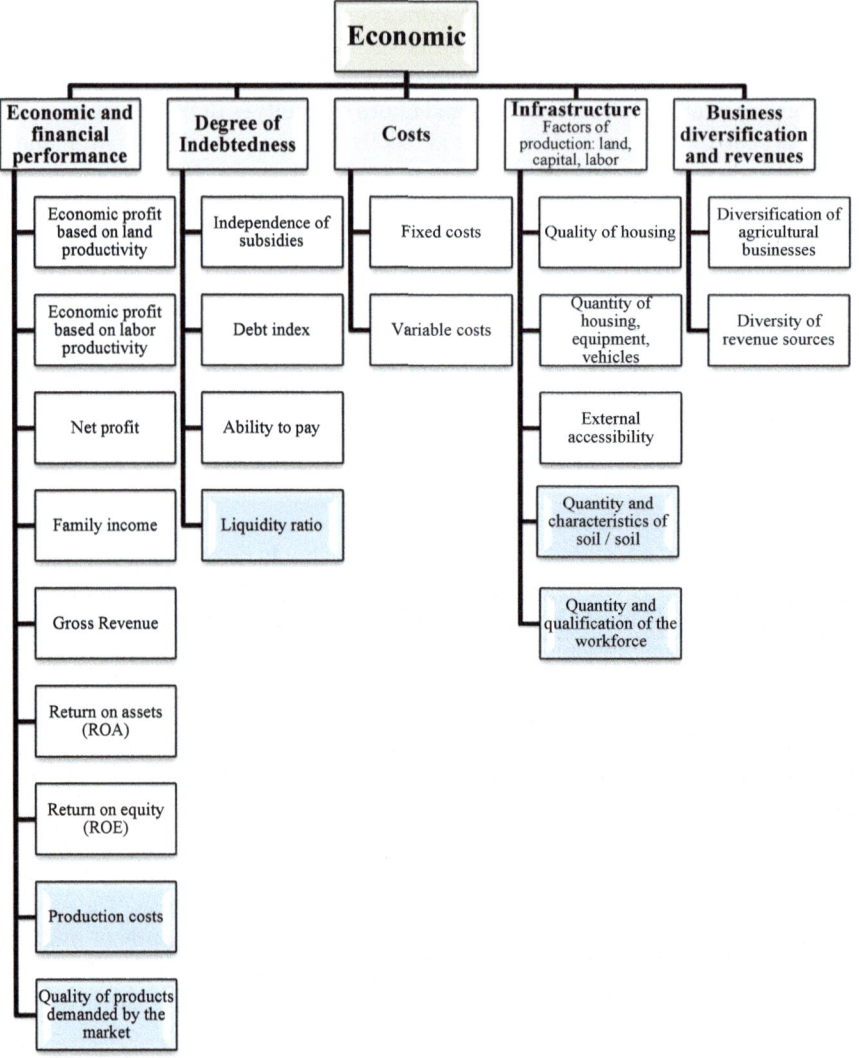

Fig. 10 Assessment model for economic dimension. *Source* Prepared by the authors

For the social dimension, experts were also asked if there were any "group of indicators" or "indicators" that should be excluded or included in the research and Fig. 11 shows the structure of the groups of indicators and indicators formed by the literature analyzes and in blue the opinions of the experts.

The environmental dimension involved the groups of indicators: environmental pollution, water, soil, biodiversity/land use. Environmental pollution had the frequency of pesticide treatment as indicators; the use of fertilizers, herbicides, fungicides and insecticides; particles/smoke; emission of greenhouse gases and the man-

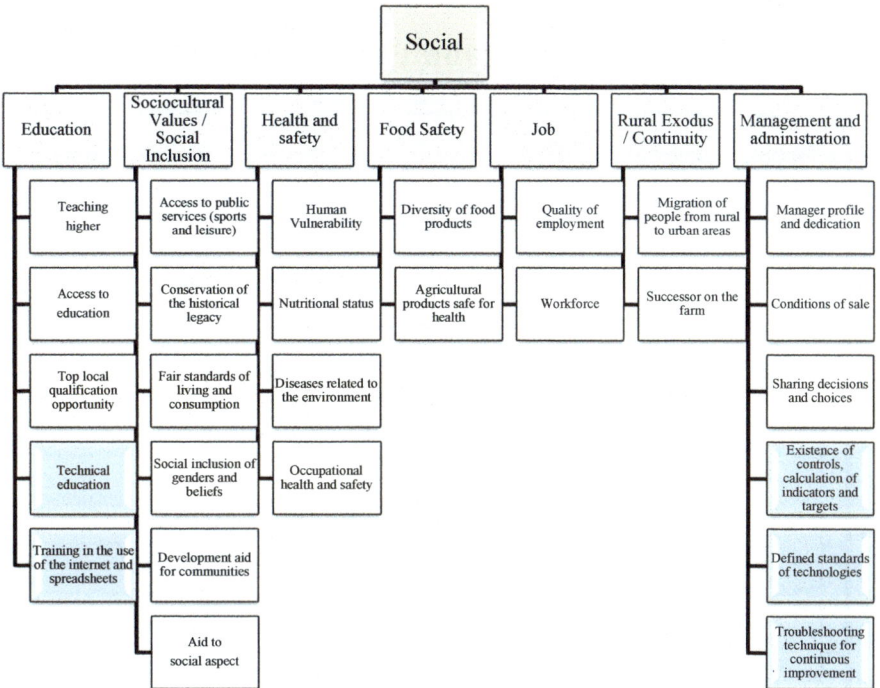

Fig. 11 Assessment model for social dimension. *Source* Prepared by the authors

agement of waste, inputs and chemical residues (Van der Werf et al. 2007; Rodrigues et al. 2010; Ruviaro et al. 2012; Mohammadi et al. 2015; Paracchini et al. 2015; Rótolo et al. 2015; Blasi et al. 2016; Mousavi-Avval et al. 2017; Azevedo et al. 2018; Khanali et al. 2018).

In the group of water indicators, the indicators were: water consumption; water quality and water efficiency, or their availability (Van der Werf et al. 2007; Siche et al. 2008; Rodrigues et al. 2010; Mohammadi et al. 2015; Paracchini et al. 2015; Rótolo et al. 2015; De Olde et al. 2016; Azevedo et al. 2018).

The soil encompassing the indicators quality and soil management; soil fertility management; preservation of organic matter and soil erosion (Tipraqsa et al. 2007; Van der Werf et al. 2007; Siche et al. 2008; Funes-Monzote et al. 2009; Rodrigues et al. 2010; Paracchini et al. 2015; Rótolo et al. 2015). And finally, biodiversity/land use referring to biodiversity conservation indicators; maintenance of healthy habitats; species richness; diversity of production and reforestation; and land occupation (Van der Werf et al. 2007; Funes-Monzote et al. 2009; Rodrigues et al. 2010; Mohammadi et al. 2015; Paracchini et al. 2015; Azevedo et al. 2018). Likewise, Fig. 12 shows the structure of the groups of indicators and indicators formed by the analyzes of the literature and in blue the opinions of the specialists.

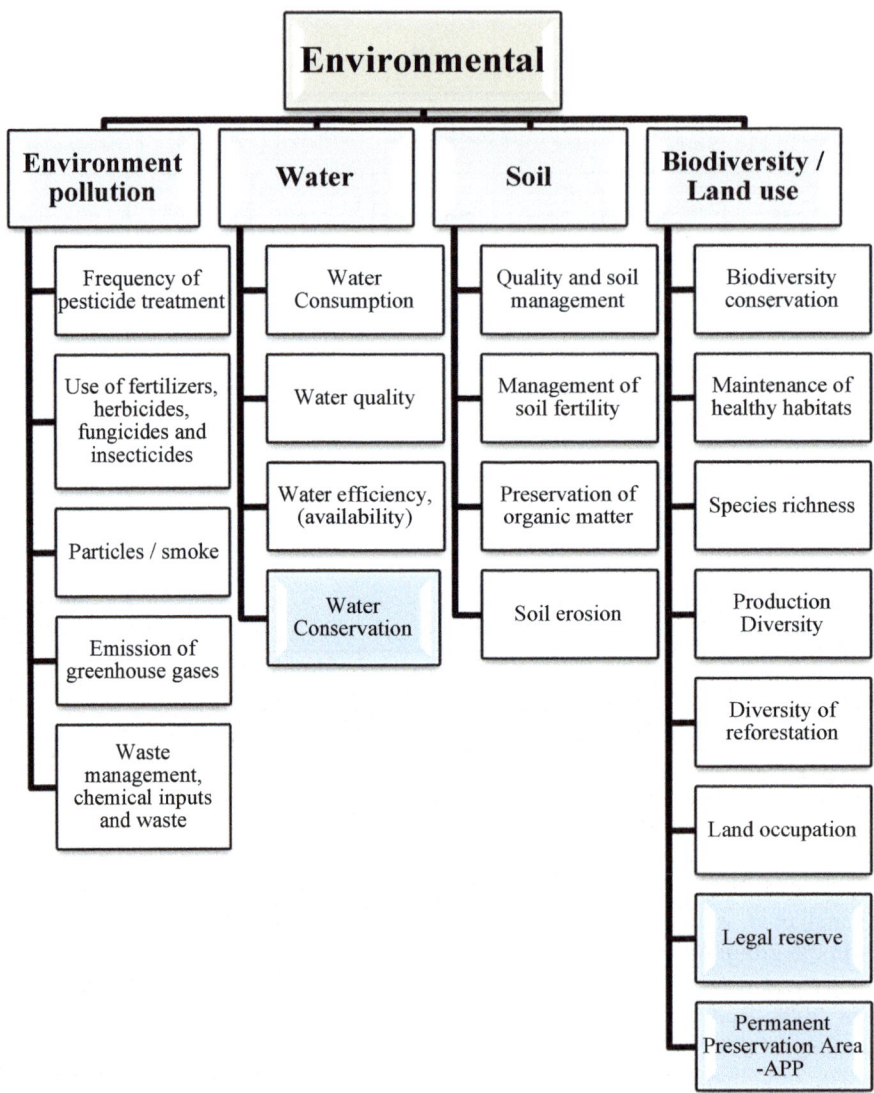

Fig. 12 Assessment model for environmental dimension. *Source* Prepared by the authors

Also, the experts were asked: What sources of data could be consulted to obtain information on the "indicator group"? The experts suggested searching the properties, locally, on sites such as Brazilian Institute of Geography and Statistics—IBGE, Instituto Paranaense de Development Economic and Social—IPARDES, Ministry of Agrarian Development—MDA, through other specialists, and other bibliographic sources, such as books and manuals, are also consulted.

Subsequently, it was questioned: How important is the "indicator group" to assess the sustainability of small farms? (Very Low, Low, Medium, High, Very High). In order to make possible the analysis of the results, the linguistic scale was transformed into numbers, so in the indicators that were not answered by the specialists, scores were assigned = 0, and the others were set in Very Low = 2, Low = 4, Average = 6, High = 8 and Very High = 10, to consolidate the information, the average of the three specialists' answers was calculated for each group of indicators. Figure 13 shows the importance attributed.

Figure 13 shows that in the economic dimension, the group of indicators "economic and financial performance" was the group that obtained the most prominence, with the average opinion among specialists of 8.7. The social dimension, the group of indicators "education" obtained a higher score with 9.3. And, the environmental dimension the groups of indicators "soil" and "biodiversity land use" were the groups of indicators that obtained the highest score with an average of 9.3. Based on the experts' considerations, it was possible to perceive the importance of the proposed indicators and also that sustainability is broad.

Among the lessons learned in the development of the proposed model, it was evidenced that the assessments involving sustainability have different characteristics and contexts and the level of complexity depends on the degree of deepening of

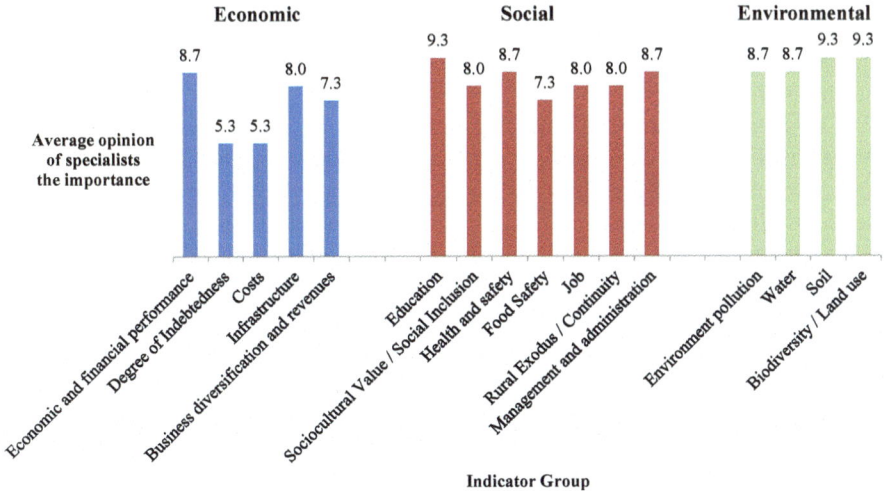

Fig. 13 Average importance attributed to the group of indicators. *Source* Prepared by the authors

the study. In addition, they have different terminologies and mechanisms aimed at meeting the objective, if resembling the essence of sustainability.

The Triple Bottom Line covers the economic, social and environmental dimensions.

It is worth mentioning that the union of these dimensions allows analysis in relation to: economic and social, socioeconomic, social and environmental, socioenvironmental, environmental and economic, ecoefficiency, which can also be portrayed when addressing the issues of sustainable development that encompasses people, the planet and profit. Sustainable demands require the involvement of all of society in the same direction, carrying out present actions to guarantee a sustainable future for our descendants.

5 Conclusion

This study aimed to answer the question about which indicators can be used to assess the sustainability performance of small farms. For this, the ProKnow-C methodology (Knowledge Development Process—Constructivist) was used to perform literature review and bibliometric analysis.

Among the main results of the bibliometrics analysis, the most used keyword was Life Cycle Assessment (LCA), the most cited paper in the scholarly portfolio in Google Scholar and Scopus was the "Sustainability of nations by indices: Comparative study between environmental sustainability index, ecological footprint and the emerging performance indices" and the paper of bibliographic portfolio that was most cited in the portfolio references was "Environmental impacts of farm scenarios according to five assessment methods". As for the relevance of the periodicals of the portfolio and their references, the journal that stood out the most was the Journal of Cleaner Production, and among the authors who stood out most in both bases was the author Rafiee, S.

In relation to the proposed model with indicators to assess the sustainability performance of small rural properties, the structuring evaluated the "three dimensions" that contemplate the Triple Bottom Line: economic, social and environmental, within the dimensions the "groups of indicators", as a general approach and the "indicators" being the breakdown of the group. The "groups of indicators" included: economic dimension, economic and financial performance, degree of indebtedness, costs, infrastructure and diversification of business and revenues; in the social dimension, education, socio-cultural values/social inclusion, health and safety, food safety, job, rural exodus/continuity, management and administration and in the environmental dimension, environmental pollution, water, soil, biodiversity/land use.

In addition, in order to refine the model obtained from the literature review, sustainability specialists were consulted, seeking to identify: what were the sources of data, which could be consulted to obtain the information regarding the "group of indicators", so the respondents showed the possibility of obtaining the information, directly on the property, through websites, books and manuals. As for the question about the importance of the "group of indicators" to assess the sustainability of small farms, average scores were good. And if there were any "indicator group" or "indicator" that should be excluded or included in the survey, opinions were about changes in structure and inclusions, but "group of indicators" or "indicators" were not excluded.

This research aimed to identify the economic, social and environmental indicators used in sustainability assessment models of small farms, so it was not intended to exhaust the discussions on this subject, but to contribute to the evaluations. Thus, for future research, it is suggested the application of the proposed model, obtained by the analysis of the literature and opinions of the specialists, that can be done obtaining the information of the indicators as highlighted in the opinion of the specialists and differentiating according to the scope and availability of the evaluator.

Through the results of this research, it was hoped to identify the main indicators used and the data sources necessary to obtain its information and thus contribute to the development of new models for assessing the sustainability performance of small properties. Making it possible to conclude that sustainability is something dynamic and broad, requiring evaluations in all sectors of the economy, with a view to performance and help in decision making.

References

Azevedo SG, Silva ME, Matias JCO, Dias GP (2018) The influence of collaboration initiatives on the sustainability of the cashew supply chain. Sustain (Switz) 10:6
Blasi E, Passeri N, Franco S, Galli A (2016) An ecological footprint approach to environmental–economic evaluation of farm results. Agric Syst 145:76–82
De Olde EM, Oudshoorn FW, Sorensen CAG, Bokkers EAM, De Boer IJM (2016) Assessing sustainability at farm-level: lessons learned from a comparison of tools in practice. Ecol Indic 66:391–404
Dias JMA (2016) Avaliação da sustentabilidade de países em nível mundial. Dissertação. Mestrado em Ciências Ambientais, Universidade Federal de Alfenas. Alfenas/MG, 2016, 87f (in Portuguese)
Dutra A, Ripoll-Feliu VM, Fillol AG, Ensslin SR, Ensslin L (2015) The construction of knowledge from the scientific literature about the theme seaport performance evaluation. Int J Prod Perform Manag 64(2):243–269
Ensslin L, Rolim S, Dutra A, Nunes NA, Reis C (2017) BPM governance: a literature analysis of performance evaluation. Bus Process Manag J 23(1):71–86
Funes-Monzote FR, Monzote M, Lantinga EA, van Keulen H (2009) Conversion of specialised dairy farming systems into sustainable mixed farming systems in Cuba. Environ Dev Sustain 11(4):765–783
Khanali M, Mousavi SA, Sharifi M, KeyhaniNasab F, Chau K-W (2018) Life cycle assessment of canola edible oil production in Iran: a case study in Isfahan province. J Clean Prod 196:714–725
Kruger SD (2017) Conjunto de indicadores para avaliação da sustentabilidade da produção suinícola. Tese (doutorado) - Universidade Federal de Santa Catarina, Centro Sócio-Econômico, Programa de Pós-Graduação em Contabilidade, Florianópolis, 2017, 226p (in Portuguese)
Lacerda RTDO, Ensslin L, Ensslin SR (2012) Uma análise bibliométrica da literatura sobre estratégia e avaliação de desempenho. Gestão & Produção 19(1):59–78 (in Portuguese)

Louette A (2007) Compêndio para a Sustentabilidade: Ferramentas de Gestão de Responsabilidade Socioambiental. São Paulo: Antakarana 2007(1):184p (in Portuguese)

Mohammadi A, Rafiee S, Jafari A, Keyhani A, Dalgaard T, Knudsen MT, Hermansen JE (2015) Joint life cycle assessment and data envelopment analysis for the benchmarking of environmental impacts in rice paddy production. J Clean Prod 106:521–532

Mousavi-Avval SH, Rafiee S, Sharifi M, Hosseinpour S, Notarnicola B, Tassielli G, Khanali M (2017) Use of LCA indicators to assess Iranian rapeseed production systems with different residue management practices. Ecol Ind 80:31–39

Paracchini ML, Bulgheroni C, Borreani G, Tabacco E, Banterle A, Bertoni D, De Paola C (2015) A diagnostic system to assess sustainability at a farm level: the SOSTARE model. Agric Syst 133:35–53

Passel S Van, Nevens F, Mathijs E, Huylenbroeck G Van (2007) Measuring farm sustainability and explaining differences in sustainable efficiency. Ecol Econ 62(1):149–161

Rodrigues AP, Fernandes ML, Rodrigues MFF, Bortoluzzi SC, da Costa SEG, de Lima EP (2018) Developing criteria for performance assessment in municipal solid waste management. J Clean Prod 186:748–757

Rodrigues GS, Rodrigues IA, Buschinelli CC, de Almeida CC, de Barros I (2010) Integrated farm sustainability assessment for the environmental management of rural activities. Environ Impact Assess Rev 30(4):229–239

Rótolo GC, Montico S, Francis CA, Ulgiati S (2015) How land allocation and technology innovation affect the sustainability of agriculture in Argentina Pampas: an expanded life cycle analysis. Agric Syst 141:79–93

Ruviaro CF, Gianezini M, Brandão FS, Winck CA, Dewes H (2012) Life cycle assessment in Brazilian agriculture facing worldwide trends. J Clean Prod 28:9–24

Siche JR, Agostinho F, Ortega E, Romeiro A (2008) Sustainability of nations by indices: comparative study between environmental sustainability index, ecological footprint and the emergy performance índices. Ecol Econ 66(4):628–637

Sidhoum AA (2018) Valuing social sustainability in agriculture: an approach based on social outputs' shadow prices. J Clean Prod 203:273–286

Svensson G, Ferro C, Hogevold N, Padin C, Varela JCS, Sarstedt M (2018) Framing the triple bottom line approach: Direct and mediation effects between economic, social and environmental elements. J Clean Prod 197:972–991

Tipraqsa P, Craswell ET, Noble AD, Schmidt-Vogt D (2007) Resource integration for multiple benefits: multifunctionality of integrated farming systems in Northeast Thailand. Agric Syst 94(3):694–703

Van der Werf HMG, Tzilivakis J, Lewis K, Basset-Mens C (2007) Environmental impacts of farm scenarios according to five assessment methods. Agr Ecosyst Environ 118(1):327–338

Wang X, Wu X, Yan P, Gao W, Chen Y, Sui P (2016) Integrated analysis on economic and environmental consequences of livestock husbandry on different scale in China. J Clean Prod 119:1–12

Zorn A, Esteves M, Baur I, Lips M (2018) Financial ratios as indicators of economic sustainability: a quantitative analysis for Swiss dairy farms. Sustain (Switz) 10:8

Analysis of Sustainability Indicators in Irrigated Rice Production in the South of Santa Catarina, Brazil

Tiago Comin Colombo and Melissa Watanabe

Abstract This study aims to analyze the degree of sustainability in irrigated rice cultivation in the South of Santa Catarina state. Qualitative research with deductive method was carried out with 10 farmers from the study region. The SAFA (Sustainability Assessment of Food and Agriculture) method developed by FAO was searched for the measurement. The results show a convergence between the research participants, who classified their properties as having a high degree of sustainability in the areas of social welfare and economic resilience and performing poorly in relation to environmental integrity and governance. When validating the aspects found with specialists in the productive chain, they also had similar orientations to farmers, except for the government agency, which justified the development of sustainability policies for the rice production chain. These policies are translated into the sector as actions that promote economic gains, thus achieving an optimized result when applying their policies in the field. The authors conclude that the degree of sustainability of the irrigated rice in the South of Santa Catarina state has relevant aspects when it directly affects the producer in relation to the legislation or the economic result; on the other hand, it is considered low in relation to the indicators that they value for transparency in the market.

Keywords SAFA · FAO · Sustainable production · Sustainable chain

1 Introduction

Projections suggest that by 2050, the world population will be 9.3 billion people and this will present major food changes. Consequently, such demand will need 60% more food in the world (FAO 2014). In the past, technology, innovation, and research

T. C. Colombo · M. Watanabe (✉)
UNESC—University of the Extreme South of Santa Catarina, Universitaria Avenue, 1105, 88806-000 Criciuma, Santa Catarina, Brazil
e-mail: melissawatanabe@unesc.net

T. C. Colombo
e-mail: tiagocolombo@gmail.com

© Springer Nature Switzerland AG 2020
W. Leal Filho et al. (eds.), *International Business, Trade and Institutional Sustainability*, World Sustainability Series,
https://doi.org/10.1007/978-3-030-26759-9_23

403

improvements have led to significant gains in agricultural production and productivity, bringing the Green Revolution to a threefold increase in world agricultural output in 50 years, with only 12% growth in the agricultural production area (FAO 2014).

Despite such advancement, the greater need for food in the future will tend to demand a larger land area for agricultural production, as well as other inputs (FAO 2014). However, food production on land and aquatic systems already dominates much of the Earth's surface and has major negative impacts on ecosystems on the planet, while rural areas still house the majority of the world's poor citizens and populations in vulnerable situations, which rely heavily on the "natural capital" for their subsistence (FAO 2014).

FAO (2018) also states that increasing productivity needs to be achieved using methods that ensure the conservation of natural resources and that continually seek efficient ways to use these resources. Based on this statement it is possible to say that agricultural production needs to become ecologically correct and of high efficiency in the use of its resources. Nordhaus, in 1991, stated that agriculture was the part of the economy most sensitive to climate change, and recent studies indicate that the environment is an important determinant of human activity and simultaneously affects society and its economic activity. Thus, these studies bring the unison search of the knowledge of nature for the field of economic analysis (Roberts 1991; Nobel Prize 2018).

The US Department of Agriculture (USDA) indicates that embracing the quest to meet food demand to climate change while respecting social and environmental limits is an important and necessary step towards a more sustainable agricultural production. To this end, academia, government and private sector increasingly need to create an appropriate environment to discuss and promote dialogues aimed at sustainable improvements in food production (Brown et al. 2015).

Therefore, one can observe that the strategy of using more efficient ways possible, in order to maximize the results for a greater gain, brings with it the concept of sustainability (Fischer-Kowalski et al. 2016). When observing this context, sustainability studies are necessary on the main agricultural products of various regions of the world, especially in those regions where a given agricultural production can contribute to the local socioeconomic development. FAO (2013) recommends that for a study on sustainability in agriculture, the SAFA (*Sustainability Assessment of Food and Agricultures*) system works in a compiled format of other systems and this particular model allows both micro and macro evaluation. As an example, one can evaluate from a small property, as well as, the entire agrifood value chain (FAO 2014).

The main themes of this indicator are Governance (G), Environmental Integrity (E), Economic Resilience (C) and Social Well-being (S). These themes link to the method and are divided into 21 variables, allowing application in national, regional, productive chains or rural properties and companies. This allows one to unfold these points in practical results to measured and analyze them, making possible management decisions. The themes are subdivided into 58 subthemes, allowing the identification of gaps, risks and even problems that at first glance would not be directly

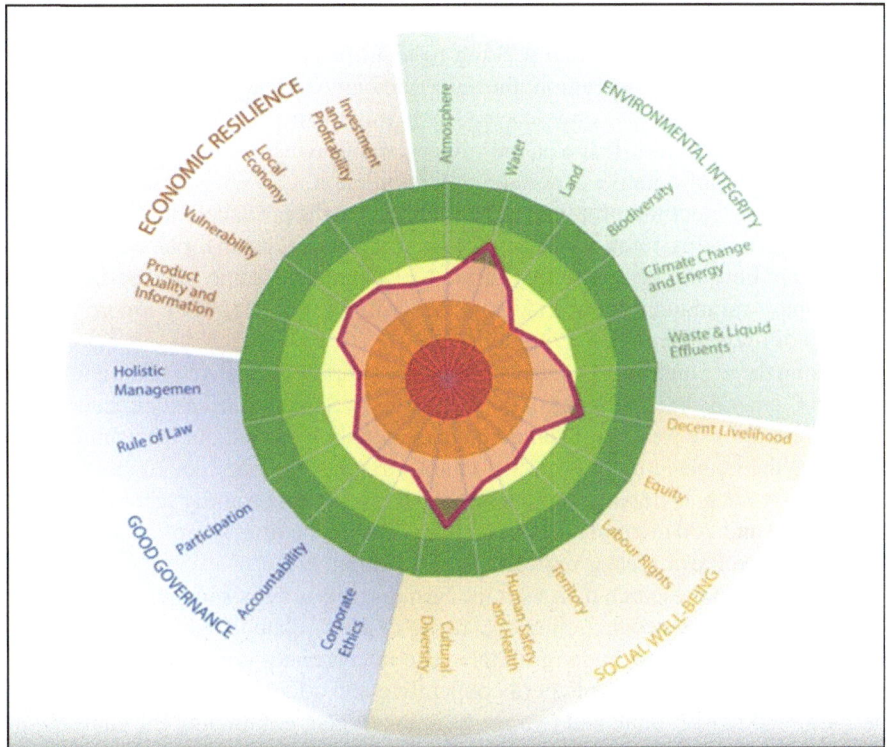

Fig. 1 SAFA holistic model. *Source* FAO (2013)

involved. For global reach, the subtopics are divided into 116 measurable indicators that measure the actual performance of the whole, as shown in Fig. 1 (FAO 2013).

Based on this premise, the present work aims to analyze the degree of sustainability in the production of irrigated rice in the South of Santa Catarina state from SAFA (*Sustainability Assessment of Food and Agriculture systems*) method.

2 Materials and Methods

For the research of the SAFA system, the characteristics subdivided by dimensions with their respective themes, sub-themes and indicators were searched. The information was compiled into an electronic spreadsheet system, which was used to organize and process the data.

The authors established the research procedure, which gives a score of 0–5 to the 116 indicators, 0 for nonexistent and 5 for complete aspects about the variable searched (FAO 2014).

As the character of the research is not to know the reality of each individual but to be a comprehensive research seeking to identify in general the practices linked to the sustainability of the region, the selection can be made by judgment criterion. According to Kish (1965) and Aaker (1995) this method is common by choosing people with a great knowledge on the subject, so these research representatives gave a typical and representative panorama of a population.

In order to select the properties, the authors sought producers in each of the 10 main rice-producing municipalities of southern Santa Catarina, placed in order of production: Forquilhinha, Meleiro, Turvo, Nova Veneza, Jacinto Machado, Tubarão, Jaguaruna, Araranguá, Imaruí and Ermo, according to IBGE (2017). Figure 2 shows the map of the region with the municipalities.

Among these municipalities, the authors also proposed to select the individuals by means of general characteristics of the properties in Santa Catarina state according to Conab (2015). The properties should have the planting carried out in swampy areas, including the irrigated mode. Another characteristic is the presence of machinery in the property such as tractors and harvesters, partial or total family labor and having between 50 and 200 ha of planted area, thus, seeking the diversity of data among the properties to achieve greater regional representation.

To validate the research data with the farmers, the authors sought representatives of the irrigated rice production chain, being experts related to the stakeholders of the system according to Ludwig (2004). The interviewees consisted of: part of the productive chain; input suppliers (a cooperative and a local trade); financing agents (a traditional market bank and two credit unions); and, completing the analysis, the

Fig. 2 Map of Santa Catarina state. *Source* Adapted from IBGE (2017)

agency of the government of the State of Santa Catarina linked to agriculture, which were interviewed in the first half of 2017. The authors recorded and transcribed the interviews, which took in average 45 min.

Table 1 shows an item as example within the governance dimension, to present the splits and adaptations made in the SAFA method until the questioning of the interviewees. All inquiries made to farmers come from the issues raised in FAO (2013) and the adaptations were made in Colombo (2018).

After tabulating the results of the 10 interviewees, the study sought to organize in radar format graphics, which express how much each point converges or diverges to the main subject.

Table 1 Explanatory table of the dimensions and questionings made to the interviewees

Dimension	Theme	Sub-theme	Indicator	Questions
Governance	Corporate ethics	Mission	Explicitness of mission	Is the company's mission articulated in all reports and understood by all members of the organization?
			Mission targeting	Does the company's mission converge with the development policies adopted by the government?
		Due diligence	Due diligence	Does the company measurable adopt the policies so that decisions can be taken on impacts in the area of sustainability?
	Accountability	Holistic audits	Holistic audits	Does the company use any international instrument model to measure its sustainability indicators?
		Responsibility	Responsibility	Is the company able to demonstrate through documents the convergence between its mission and all stakeholders?
		Transparency	Transparency	Does the company have a process management policy carried out by the company with access to interested parties?
	Participation	Interaction among stakeholders	Identification of stakeholders	Is the company describing the entire process by identifying all the materials needed to carry out the verification procedures?

(continued)

Table 1 (continued)

Dimension	Theme	Sub-theme	Indicator	Questions
			Interaction among stakeholders	Does the company use different means to communicate with different stakeholders in its business?
			Barriers to engagement	Is the company aware that some stakeholder groups do not have access to the information? Is the company working to improve it?
			Effective participation	Is the company able to describe current *Participation* stakeholders (including "less powerful" ones), what is its impact on decision-making and how has this impact been *Interested*?
		Complaint procedures	Complaint procedures	Is the company able to describe the complaint procedures for each group of stakeholders? How they are disclosed (especially with "less powerful") and how is their current use?
		Conflict resolution	Conflict resolution	Is the company able to identify conflicts between stakeholders so as examples of how some kind of conflict has been resolved based on dialogue and mutual understanding?
	Rules and legislation	Legitimacy	Legitimacy	Does the company's policy or code of practice explicitly require that all applicable laws and regulations and adopted or existing standards be communicated to the government, their members or staff, and regularly reviewed for consistency with the mission?

(continued)

Table 1 (continued)

Dimension	Theme	Sub-theme	Indicator	Questions
		Repair, restoration or compensation	Repair, restoration or compensation	Is the company able to show evidence of a prompt and responsible response to legal, regulatory, human rights and voluntary violations of the code, including how the violation has been corrected, how the effects of the violation will be restored or compensated, and the policies and procedures instituted to prevent further violations?
		Civic responsibility	Civic responsibility	Civic responsibility: within its sphere of influence, does the company proactively and transparently support the improvement of the legal and regulatory situation in all four dimensions of sustainability in order to avoid the impact of human rights or sustainability norms or regulation?
		Appropriation of resources	Free, prior and informed consent	Is the company aware of the stakeholders' interests? Pre-existing access to land, water and resources has mapped this to the satisfaction of all affected stakeholders and has agreed not to reduce access until it has fully informed stakeholders on equal terms and has provided mutually agreed sustainable livelihoods?
			Right of possession	Does the company have all documents proving its full establishment and operating conditions?

(continued)

Table 1 (continued)

Dimension	Theme	Sub-theme	Indicator	Questions
	Holistic management	Sustainability management plan	Sustainability management plan	Does the company have a sustainability plan approved by its governing body (or members of the association or contractors), which provides a sustainability vision of the company and covers each of the dimensions: environmental, economic, social and governance, including references to the mission and demonstration of progress against the plan, or how the plan has driven specific decisions and results?
		Complete cost accounting	Cost accounting	Is the company's business success measured and reported to stakeholders, taking into account the direct and indirect impacts on the economy, society and on physical environment?

Source Adapted from SAFA (FAO 2013)

3 Result Analysis

The study sought, by means of the SAFA system, to conduct a research with 10 farmers from the South of Santa Catarina state in order to check the sustainability degree in the irrigated rice cultivation. During the research, one can notice that all the interviewees have already heard about the topic, but little did know in fact that it could be divided into three macro aspects and none of these interviewees knew about the existence of an economic indicator within this concept. For the respondents the word sustainability was linked only to environmental issues.

When analyzing the four themes: Governance, Environmental Integrity, Economic Resilience and Social Well-being, one can notice the interviewees were surprised, paid attention to the questions and realized the importance of sustainability as a whole and not only as an environmental issue. After ten interviews, one can have a parameter to analyze the indicators, which made clear that farmers in this market segment are concerned with the compliance with established market regulations and with financial gains.

Rice production chain shows a constant technological evolution of rice production in Brazil, according to Conab (2015), from a national level of 1.4 thousand kilos per hectare to 5.5 thousand kilos per hectare. In Santa Catarina, this number went from 2.5 thousand kilos to more than 7 thousand kilos per hectare over a period of four decades, which corroborates with the theory of Boserup (1965), which shows that there are technological leaps caused by a necessity, not allowing the population collapsed.

One can realize that in the South of Santa Catarina State there is no intentional concern with sustainability conditions, as per result of the research shown in Fig. 3.

The research as a whole shows a small concern directly related to sustainability, but when evaluating the four major themes one can noticed that, there are themes with a perception of greater importance to the detriment of others, in the interviewees' point of view.

Within governance, for instance, the issues of rules and legislation that must be fulfilled stands out, as well as when talking about social welfare, basic needs items linked to people that must be fulfilled by force of Law and that can cause legal damages such as fines, imprisonment, among others, in case of noncompliance.

In the field of environmental integrity, producers gave too little value on this subject, as the inspection and benefits do not seem to be enough for producers to invest.

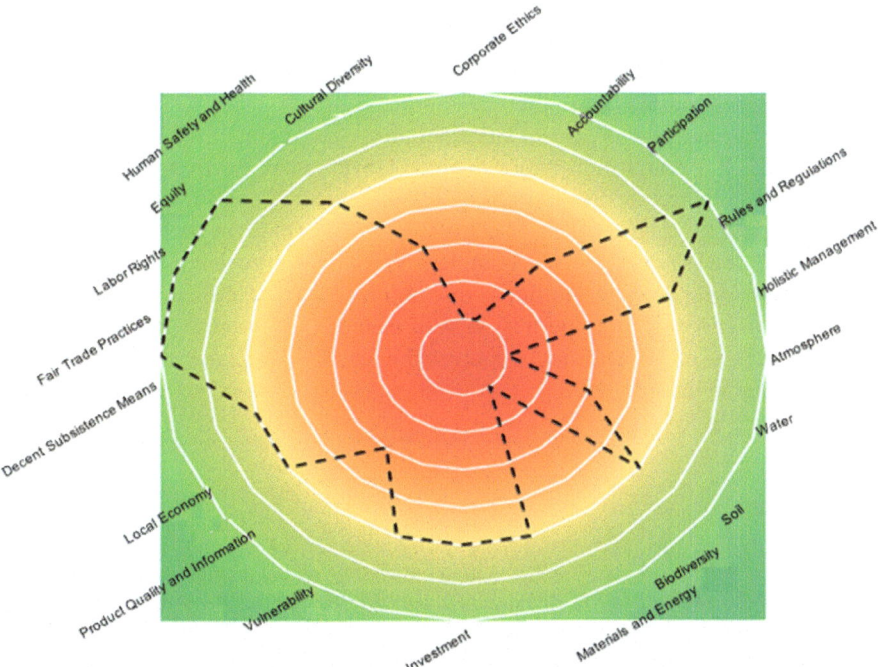

Fig. 3 SAFA global indicators. *Source* Research data

In the last theme, which explores economic resilience, there is a constant concern with most of the indicators, since it is crucial for the accomplishment of agricultural activities.

During data validation by specialists, one can notice that the sustainability subject is something propagated in the strategic scope of the companies as a matter of relevance. However as each company has its specific segment to check they end up not communicating and taking care of their environment specifically.

Only one of the stakeholders interviewed—the government through its rural development company—seemed concerned with the whole during the interview. This company, in its turn, understands the longings of the farmer and the chain and seeks to translate relevant environmental and social issues in another way so that the producer can buy the idea and see it as something feasible to implement.

4   Conclusions

According to De Oliveira et al. (2012), society has been transforming itself technologically for centuries, from the first industrial revolution to the present moment with the genetic improvement of living beings. The present study analyzed the SAFA (Sustainability assessment of food and agriculture) system as a possible indicator that seeks to measure a degree of sustainability for the cultivation of irrigated rice in the South of Santa Catarina state.

In order to assess the degree of sustainability, a tool called SAFA was developed by FAO, which, through four major themes: Governance, Environmental Integrity, Economic Resilience and Social Welfare, 21 variables, 58 sub-themes and 116 indicators translate the degree of sustainability of a given chain, or region.

Although the use of this tool is complex, one can observe that it is feasible to use it in the field, with farmers. As future research it is suggested the extension of such instrument to more rural properties, as well as the application in other links of the productive chain of rice in order to analyze the feasibility and possible challenges to be transposed in its use.

References

Boserup E (1965) The conditions of agricultural growth: the economics of agrarian change under population pressure. London: G. Allen and Unwin. http://www.biw.kuleuven.be/aee/clo/idessa_files/boserup1965.pdf. Accessed 10 Jan 2016

Brown ME, Antle JM, Backlund PW, Carr ER, Easterling WE, Walsh M, …. Dancheck V (2015) Climate change and global food security: food access, utilization, and the US food system. In: AGU fall meeting abstracts. http://www.usda.gov/oce/climate_change/FoodSecurity2015Assessment/FullAssessment.pdf. Accessed 30 Nov 2018

Colombo TC (2018) Análise dos indicadores de sustentabilidade na produção de arroz irrigado no Sul de Santa Catarina. Thesis – Universidade do Extremo Sul catarinense – UNESC. http://repositorio.unesc.net/handle/1/5652. Accessed 30 Nov 2018

CONAB, Companhia Nacional de Abastecimento (2015) A cultura do arroz. Brasília: Conab, http://www.conab.gov.br/OlalaCMS/uploads/arquivos/16_03_01_16_56_00_a_cultura_do_arroz_-_conab.pdf. Accessed 19 Nov 2016

De Oliveira LR, Medeiros RM, de Bragança Terra P, Quelhas OLG (2012) Sustentabilidade: da evolução dos conceitos à implementação como estratégia nas organizações. Production 22(1):70–82. Available: http://dx.doi.org/10.1590/S0103-65132011005000062. Accessed 30 Nov 2018

FAO, Food and Agriculture Organization of the United Nations (2013) SAFA Sustainability Assessment of Food and Agriculture systems indicators. Rome. http://www.fao.org/fileadmin/templates/nr/sustainability_pathways/docs/SAFA_Indicators_final_19122013.pdf. Accessed 23 Dec 2015

FAO, Food and Agriculture Organization of the United Nations (2014) Building a common vision for sustainable food and agriculture: principles and approaches. On-line Report. http://www.fao.org/3/919235b7-4553-4a4a-bf38-a76797dc5b23/i3940e.pdf. Accessed 29 Apr 2016

FAO, Food and Agriculture Organization of the United Nations (2018) Our Strategic objectives: Make agriculture, forestry, and fisheries more productive and sustainable. http://www.fao.org/about/what-we-do/so2/en/. Accessed 30 Nov 2018

Fischer-Kowalski M, Mayer A, Schaffartzik A, Reenberg A (2016) Ester Boserup's legacy on sustainability: orientations for contemporary research. Austrian Science Fund (FWF). Springer Open

IBGE. Instituto Brasileiro de Geografia e Estatística (2017) Cidades@. http://www.cidades.ibge.gov.br/cartograma/tabDadosUfCsv.php?lang=&idtema=18&codv=V16&coduf=42. Accessed 3 Feb 2017

Ludwig VS (2004) Agroindústria processadora de arroz: Um estudo das principais características organizacionais e estratégias das empresas líderes gaúchas. Universidade Federal do Rio Grande do Sul, Porto Alegre

Nobel Prize (2018) Economic growth, technological change, and climate change. https://www.nobelprize.org/uploads/2018/10/advanced-economicsciencesprize2018.pdf. Accessed 10 Oct 2018

Roberts L (1991) Academy panel split on greenhouse adaptation: its conclusion that the United States can adapt relatively painlessly to global warming draws two vigorous dissents. Science 253(5025):1206. Academic OneFile, https://link.galegroup.com/apps/doc/A11360605/AONE?u=googlescholar&sid=AONE&xid=2f73786e. Accessed 30 Nov 2018

Study of the Antioxidant and Nutraceutical Properties of Strawberry *Fragaria x ananassa* Organically Grown as an Option of Sustainable Agriculture

Juan Carlos Baltazar-Vera, Ma del Rosario Abraham-Juárez,
Iovanna Consuelo Torres-Arteaga, Nancy Karina Vargas-Ramos,
Gilberto Carreño-Aguilera, Cesar Ozuna-Lopez,
Ana Isabel Mireles-Arriaga and María Elena Sosa-Morales

Abstract The development of farming technologies that allow the production of agricultural products free of chemical compounds that are harmful to the agricultural environment as well as the consumers is a challenge that several research groups have addressed. In several case studies the effect of various treatments of organic fertilization on quality variables of the agricultural product to be produced has been analyzed. The above is of vital importance for the development of agriculture that allows the sustainable development of the communities. The objective of this work was to study the antioxidant and nutraceutical properties of samples at 5 degrees of maturation of the variety Fragaria x ananassa Camino Real organically grown and compared with a commercial strawberry sample as a control; The parameters to be considered were: Superoxide dismutase activity, amount of anthocyanidins and total reducing sugars. The results indicate that the strawberry organically grown presents competitive nutritional properties and in some parameters superior to the commer-

J. C. Baltazar-Vera
Departamento de Ingeniería en Minas, Metalurgia y Geología, Universidad de Guanajuato, Ex Hda. de San Matías s/n. Col. San Javier, Guanajuato, Gto 36020, Mexico

M. R. Abraham-Juárez (✉) · N. K. Vargas-Ramos · C. Ozuna-Lopez · M. E. Sosa-Morales
Departamento de Alimentos, Universidad de Guanajuato, Ex Hacienda El Copal k.m. 9; Carretera Irapuato-Silao; A.P. 311, Irapuato, Gto 36500, Mexico
e-mail: mrosarioabdiciva@gmail.com

I. C. Torres-Arteaga
Departamento en Ingeniería en Agrotecnología, Universidad Politécnica del Bicentenario, Carretera Estatal Silao-Romita km 2. Col. San Juan de los Durán, Silao, Gto 36283, Mexico

G. Carreño-Aguilera
Departamento de Ingeniería en Geomática e Hidraúlica, Universidad de Guanajuato, Juárez 77. Col. Zona Centro, Guanajuato, Gto 36000, Mexico

A. I. Mireles-Arriaga
Departamento de Ingeniería en Agronomía, Universidad de Guanajuato, Ex Hacienda El Copal k.m. 9; Carretera Irapuato-Silao; A.P. 311, Irapuato, Gto 36500, Mexico

© Springer Nature Switzerland AG 2020
W. Leal Filho et al. (eds.), *International Business, Trade and Institutional Sustainability*, World Sustainability Series,
https://doi.org/10.1007/978-3-030-26759-9_24

415

cially grown strawberry. In this work we can elucidate the benefits of developing sustainable agriculture techniques which allow the production of food free of toxic chemical agents for human health, as well as contribute to the preservation of the environment of the various communities.

Keywords Sustainable agriculture · Human health · Environment · Communities

1 Introduction

The exponential growth of the human population has resulted in the development of food production technologies in mass form, these practices (large scale agriculture) have generated diverse imbalances in the flora and fauna of the media where they develop as well as problems of health in humans, this is due to the handling of pesticides and chemical-based supplements (Salgado Sánchez 2015). This is why the development of sustainable agriculture techniques is fundamental to achieve an adequate environmental and social balance.

Currently strawberry consumption has attracted the attention of various research groups due to the relationship it has with respect to the decrease in the incidence of chronic and degenerative diseases due to its antioxidant properties (Carvajal et al. 2012). The strawberry is native of diverse zones with temperate climates of the planet and this fruit is cultivated as much on a large scale as for local consumption by small farmers (SAGARPA 2012).

Several constituents of strawberry have been studied as chemo preventive agents at different levels of the process of cancer maturation, avoiding oxidation of DNA, promoting cell repair, inhibiting the formation of carcinogenic compounds and reducing tumor proliferation (Zheng et al. 2007; Crecente et al. 2012; Luo et al. 2011). The varieties of commercially grown strawberries are mostly hybrids specifically *Fragaria x ananassa* (Bielinski and Henner 2009). Among the most used varieties in Mexico are foreign varieties (mainly from California and Florida, marketed by Eurosemillas) among them: "Festival", "Sweet Charlie", "Galexia" from the University of Florida; as well as those from the University of California: "Camino Real", "Albión", "Camarosa", "Aromas", "Ventana" y "Diamante" (CONAFRE 2009).

Among the constituent agents of strawberry are: Vitamin C, tannins, flavonoids, anthocyanins, catechin, quercetin and kaempferol, organic acids (citric, malic, oxalic, salicylic and ellagic) and minerals (K, P, Ca, Na y Fe), In addition to pigments and essential oil (Restrepo et al. 2009). The objective of this work was to determine the nutraceutical and antioxidant properties of organically grown strawberry and compare with a commercial fresh sample.

2   Materials and Methods

2.1   Location of the Sample

Organic strawberry samples, variety "Camino Real", were grown in the community Las Malvas (Revolución), Municipio de Irapuato, Guanajuato (20° 40' 47.20''N, 101° 18' 13.50''W, elevation 1,734 m above sea level). The commercial fruits employed for comparison were bought from the Hidalgo market of the City of Irapuato, Guanajuato, Mexico.

2.2   Collection of Biological Material and Lyophilization of the Sample

Strawberry fruits were collected in different degrees of maturation: Green, Mature Green, Rose, Ripe, Over Ripe and Strawberry Control. For the lyophilization, aliquots were taken from the different samples, which were placed in poly paper bags and inmersed in liquid nitrogen, using a freeze-dryer brand Labconco model Freezone$^{2.5}$ at a temperature of $-47\ ^{\circ}C$ and at a pressure of 0 mBar for 3 days.

2.3   Determination of Antioxidant Capacity

The ABTS method is based on the quantification of the discoloration of the radical $ABTS^{+}$. The cationic radical $ABTS^{+}$ is a chromophore that absorbs at a wavelength of 734 nm and is generated by an oxidation reaction of ABTS with potassium persulfate, the results are expressed as TEAC (trolox equivalent antioxidant capacity) values by constructing a standard curve using as an antioxidant Trolox (Mesa et al. 2010). For the determination of antioxidant capacity, the Antioxidant Assay Kit, of Sigma was used. The standard curve of Trolox was prepared using solutions in a range of 0–0.42 mM concentration.

2.4   Determination of Superoxide Dismutase (SOD) in Strawberry and Anthocyanidins

Superoxide dismutase (SOD) is a group of isoenzymes that function as scavengers of the superoxide radical. These enzymes protect the cells from oxidative stress by catalyzing the dismutation of the superoxide radical to hydrogen peroxide and oxygen. For the determination of superoxide dismutase, the SOD Assay Kit was used.

For the determination of the content of anthocyanidins, the spectrophotometric method of Wrolstad was used for the pelargonidin 3-glucoside, The anthocyanin that predominates in the fruit of the strawberry, a Lambda spectrophotometer was used XLS/XLAS reading at a wavelength of 520 nm (Vargas 2013).

2.5 Determination of Total Reducing Sugars (ART). Method of Lane and Eynon

The method of Lane and Eynon (1923) is made to react cupric sulfate with reducing sugar in alkaline medium, forming cuprous oxide (Del Ángel et al. 2013). The measurement of the total reducing sugars was carried out by acid hydrolysis with HCl and subsequently by titration with 2 M NaOH for the reduction of copper by: $CuSO_4 \cdot 5H_2O$ a 0.27 M y $KNaC_4O_6 \cdot 4H_2O$, 1.2 M.

2.6 Statistic Analysis

A completely randomized design (CRD) was carried out for the following variables: antioxidant capacity, superoxide dismutase activity, content of anthocyanidins and reducing sugars. Five degrees of maturation, one extract and three repetitions were considered. Statistical analysis was carried out using the statistical package SAS (Statistical Analysis System) Version 8, using Duncan's multiple-interval test.

3 Results

Figure 1 shows the antioxidant capacity of Trolox dilutions based on Trolox units (mM) determined by the ABTS technique. The calibration curve for the Trolox shows an acceptable correlation between the absorbance and the concentration at different dilutions, with a value of $R^2 = 0.9372$ which is very close to 1.

Figure 2 shows the antioxidant capacity of strawberry fruits in their different degrees of maturation, expressed in Trolox equivalents (mM) per 100 mg of lyophilized sample. In the case of the Green (GS) and Mature Green (MGS) samples, the antioxidant capacity was 0.855 and 0.841 mM of Trolox, respectively. For the cases corresponding to strawberry pink (PS), mature (MS), over ripe (ORS) and strawberry control (CS), the antioxidant activity was 0.806, 0.819, 0.795 and 0.796 mM of Trolox, respectively. According to the above, when comparing the fruit grown organically with the commercial product in the same phase of maturation

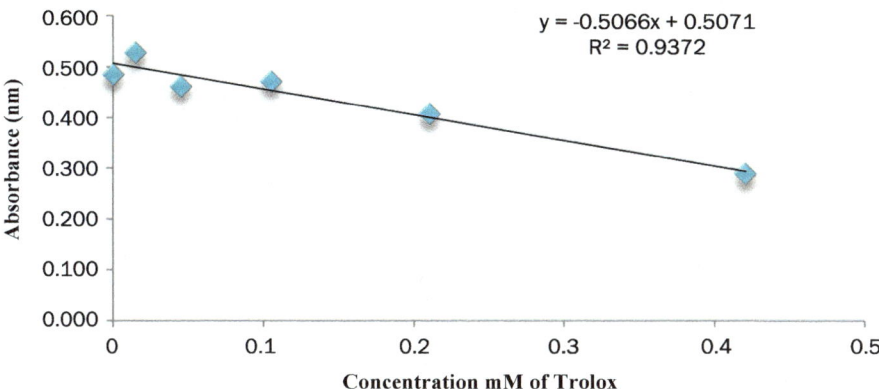

Fig. 1 Trolox standard curve for the calculation of antioxidant capacity

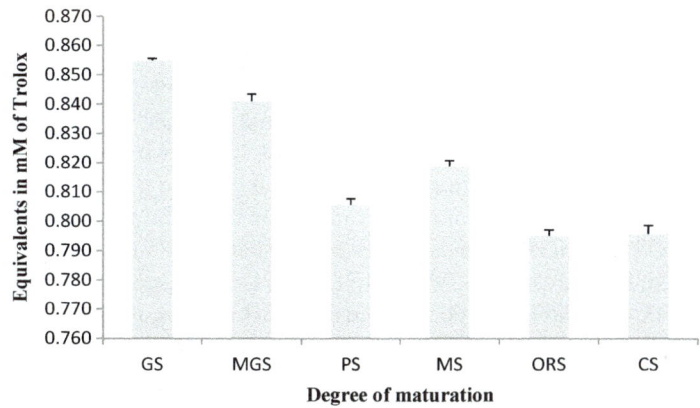

Fig. 2 Antioxidant capacity of strawberry fruits in different degrees of maturation. Degree of maturation: Green (GS), mature green (MGS), pink (PS), mature (MS), over ripe (ORS), control strawberry (CS)

(PS), we have that the organic product has an increase in the antioxidant capacity corresponding to 2.89% with respect to the commercial fruit.

In this sense it can be observed that antioxidant capacity is higher in green fruits than in fruits with coloration; the Green strawberry has the highest antioxidant capacity. There was a significant difference between the means of the samples of strawberry fruits in degree of maturation GS, MGS and MS ($p < 0.05$), while for the ORS and CS cases there was not difference ($p < 0.05$).

Figure 3 shows the activity of superoxide dismutase (SOD) in percentage of inhibition rate for the different degrees of strawberry ripening, for the cases of GS and MGS, values of 88.71 and 75.09% respectively were obtained, for PS, MS, ORS and CS samples, values of 103.07, 92.25, 154.78 and 113%, respectively, were recorded.

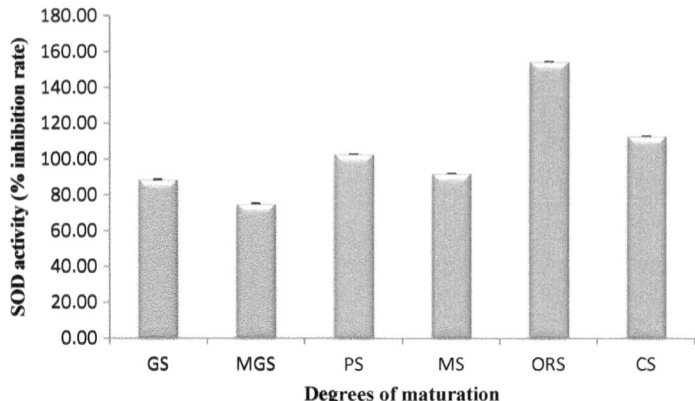

Fig. 3 SOD activity in strawberry fruits in different degrees of maturation. Degree of maturation: Green (GS), mature green (MGS), pink (PS), mature (MS), over ripe (ORS), control strawberry (CS)

Taking the case with the highest inhibition rate was ORS. When comparing the strawberry organically grown with the commercial product, the latter has to provide 20.75% more in relation to the inhibition rate. There was significant difference between the degrees of maturation ORS, MGS and GS ($p < 0.05$).

Figure 4 shows the content of anthocyanidins in the different degrees of maturation, for the GS and MGS samples, values of 22.04 and 19.85 mg/100 g respectively were obtained. For the cases PS, MS, ORS and CS were registered amounts of 21.92, 56.02, 107.20 and 30.42 mg/100 g respectively. The above indicates that the content of anthocyanidins is 84.15% higher in the organic product with respect to the

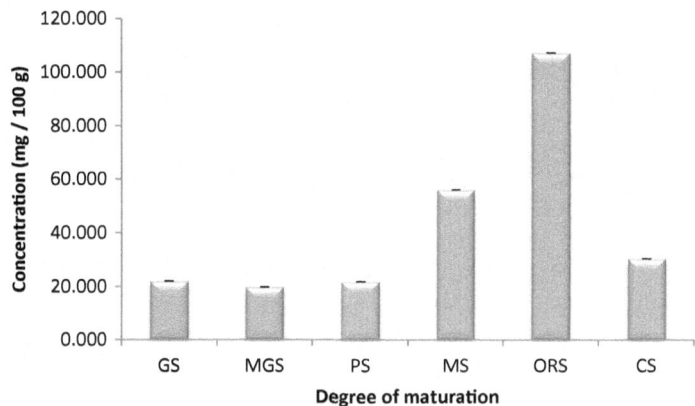

Fig. 4 Content of anthocyanins in strawberries in different degrees of maturation. Degree of maturation: Green (GS), mature green (MGS), pink (PS), mature (MS), over ripe (ORS), control strawberry (CS)

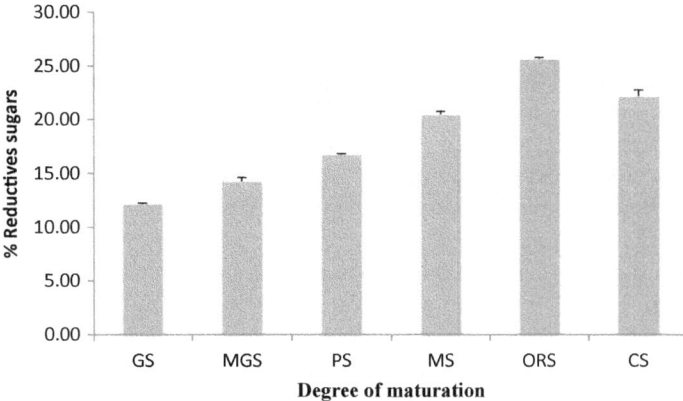

Fig. 5 Reductive sugars in strawberry fruits in different degrees of maturation. Degree of maturation: Green (GS), mature green (MGS), pink (PS), mature (MS), over ripe (ORS), control strawberry (CS)

commercial fruit for the same degree of maturation (MS). The results indicate that the greater the degree of maturation, the greater the concentration of anthocyanidins (Zhao 2007), which suggests that this parameter is directly related to the amount of pigment in the fruit. The statistical analysis indicates that there are significant differences between the degrees of maturation ORS, MS and CS, while for GS, PS and MGS do not exist.

Figure 5 shows the content of strawberry reducing sugars at different degrees of maturity, for the GS and MGS cases values of 12.14 and 14.23% respectively are reported in 100 g, for the samples corresponding to PS, MS, ORS and CS, results of 16.72, 20.44, 25.64 and 22.14% were obtained in 100 g, respectively. When comparing the organic fruit with respect to the commercial one in the same degree of maturation (MS), it is observed that the commercial product presents a content of reducing sugars higher by 1.7%. As in the case of the concentration of anthocyanidins, the content of reducing sugars increases as the degree of maturation increases (García et al. 1998). The CDR using a value of ($p < 0.05$) shows that there is a significant difference between the different degrees of strawberry ripening.

4 Conclusion

The obtained results show that the antioxidant capacity is benefited when the strawberry is harvested organically since said parameter was higher in all the degrees of maturation of the samples organically grown with respect to the control sample. In terms of the superoxide dismutase (SOD) activity, the over ripe strawberry organically grown has the highest value.

In the case of quantity of anthocyanidins the sample of strawberry over ripe, organically grown presents the highest content of this parameter, which is consistent since there are reports that the production of anthocyanidins is favored in advanced stages of maturation (Zhao 2007). In relation to the presence of reducing sugars, the state over ripe organic was the one that presented a greater quantity of this parameter, his coincides with previous studies that report that the amount of reducing sugars increases with the degree of maturation (García et al. 1998). When comparing the strawberry organically grown with respect to the commercial product it was obtained that for the antioxidant capacity and the content of anthocyanidins the organic product was 2.89 and 84.15% higher respectively. Regarding the SOD and the content of reducing sugars, the commercial product reported values of 20.75 and 1.7% higher respectively.

Comparative results show that organic strawberry cultivation is a viable option since the antioxidant capacity, SOD, quantity of anthocyanidins and reducing sugars showed values very close to the control sample for the same phase of maturation, which is indicative of obtaining of a good product when growing organically. These results are consistent with studies that report an increase in nutraceutical and antioxidant properties of organically grown products (Martínez et al. 2014), which is encouraging since it allows elucidating the advantages of this culture technique.

In this work it can be observed that organic strawberry cultivation provides a fruit with competitive antioxidant and nutraceutical properties with respect to the commercial product, in this sense it can be elucidated that the implementation of this type of production is attractive to generate sustainable agriculture systems.

Acknowledgements The authors thank the University of Guanajuato and Universidad Politécnica del Bicentenario for the academic support in the realization of this project.

References

Bielinski MS, Henner AO (2009). Prácticas culturales para la producción comercial de fresas en florida. Documento HS1160. Horticultural Sciences, Servicio de Extensión Cooperativa de la Florida, Instituto de Alimentos y Ciencias Agrícolas,. Universidad de la Florida

Carvajal L, El Hadi C, Cartagena R, Peláez C, Gaviria C, Rojano C (2012) Antioxidant capacity of two *Fragaria x ananassa* (Weston) Duchesne (strawberry) varieties subjected to variations in vegetal nutrition. Revista Cubana de Plantas Medicinales 17(1):37–53

CONAFRE (2009) Sistema Producto Fresa, Plan Rector Nacional Convergente. 13

Crecente J, Nunes M, Romero M, Vázquez M (2012) Color, anthocyanin pigment, ascorbic acid and total phenolic compound determination in organic versus conventional strawberries (*Fragaria x ananassa* Duch, cv. Selva). J Food Compos Anal (28) 23–30

Del Ángel A, Isaac M, Martínez A (2013) Manual de prácticas de bromatología: gravimetría y volumetría. Universidad de Guadalajara 79

García M, Martino M, Zaritzky N (1998) Plasticized starch-based coatings to improve strawberry (*Fragaria x ananassa*) quality and stability. J Agric Food Chem 46:3758–3767

Lane J, Eynon L (1923) Determination of sugars by Fehling solution with methylene blue as indicator. J Soc Chem India 42:32–34

Luo Y, Tang H, Wang X, Zhang Y, Liu Z (2011) Antioxidant properties and involved compounds of strawberry fruit at different maturity stages. J Food Agric Environ 9(1):166–170

Martínez G, Plaza P, Romero R, Garrido A (2014) Analytical approaches for the determination of pesticide residues in nutraceutical products and related matrices by chromatographic techniques coupled to mass spectrometer. Talanta J 118:277–291

Mesa A, Gaviria A, Cardona F, Sáez J, Blair S, Rojano A (2010) Actividad antioxidante y contenido de fenoles totales de algunas especies del género *Calophyllum*. Revista Cubana de Plantas Medicinales. 15(2):13–26

Restrepo A, Cortés M, Rojano B (2009) Determinación de la vida útil de fresa (*Fragaria ananassa* Dutch.) fortificada con vitamina E. Revista DYNA, 159:163–175

SAGARPA (2012) Fondo Sectorial de Investigación en materia Agrícola, Pecuaria, Acuacultura, Agrobiotecnología y Recursos Fitogenéticos, ANEXO B. DEMANDAS DEL SECTOR 2012-3. Pág.2. http://www.conacyt.gob.mx/fondos/FondosSectoriales/SAGARPA/201203/Demanda_especifica-2012-3.pdf

Salgado Sánchez R (2015) Agricultura sustentable y sus posibilidades en relación con consumidores urbanos. Estudios Sociales 23(45):113–140

Vargas N (2013) Determinación de la actividad enzimática de la superóxido dismutasa y del poder antioxidante de frutos de fresa (*Fragaria x ananassa* Dutch.), Tesis Licenciatura, Universidad de Guanajuato

Zhao Y (2007) Berry fruit: value-added products for health promotion. CRC Press 1:25–28

Zheng Y, Wang S, Wang C, Zheng W (2007) Changes in strawberry phenolics, anthocyanins, and antioxidant capacity in response to high oxygen treatments. LTW- Food Sci Technol 40:49–67

Indicators for Assessing Sustainable Operations in a Poultry Slaughterhouse, Considering Industry 4.0 Perspective

Débora de Souza Soares, Marcelo Gonçalves Trentin and Edson Pinheiro de Lima

Abstract Corporate Social Responsibility is being highlighted and demanded by consumers, government and non-governmental entities, and many other organizations. Companies' environmental concern has become a matter of survival in the market. With the 4th Industrial Revolution many changes are happening in companies, new technologies are being used, innovation is present all the time, actually it is being required to innovate in a sustainable way, so innovation and sustainability are two fundamental issues for companies to remain competitive. Some organizations have already understood that becoming sustainable can bring many benefits and have identified the importance of balancing economic, social and environmental requirements. Brazil has a great worldwide participation in the production and export of poultry meat, which makes it a direct competitor of international companies, so it is necessary to worry about sustainable issues, so that it does not become a barrier to grow on new customers and to maintain the existing ones. In this paper, a systematic review of the literature was performed on economic, social and sustainable pillar indicators in agroindustry companies and in industry 4.0. The objective of the present work is to identify the indicators for assessing sustainable operations in a poultry slaughterhouse and to propose a model for assessing sustainable operations from the perspective of digital transformation. Three indicators were suggested for each pillar of the Triple Bottom Line (economic, social and environmental) as a model for assessing sustainable poultry slaughterhouse operations under the perspective of digital transformation.

D. de Souza Soares (✉) · M. G. Trentin · E. P. de Lima
Industrial and Systems Engineering, Federal University of Technology—Parana (UTFPR), Via Do Conhecimento, km 1, Pato Branco, Parana, Brazil
e-mail: deborasouzasoares@yahoo.com.br

M. G. Trentin
e-mail: marcelo@utfpr.edu.br

E. P. de Lima
e-mail: pinheiro@utfpr.edu.br

E. P. de Lima
Pontifical Catholic Univeristy of Parana—PUCPR, R.Imaculada Conceição,1155, Curitiba, Parana, Brazil

© Springer Nature Switzerland AG 2020
W. Leal Filho et al. (eds.), *International Business, Trade and Institutional Sustainability*, World Sustainability Series,
https://doi.org/10.1007/978-3-030-26759-9_25

Keywords Performance management · Industry 4.0 · Digital transformation · Slaughterhouse · Sustainability

1 Introduction

The technological advance has made it possible to monitor and synchronize information between all the areas of a company, besides allowing machines to be always connected, making companies much more efficient and productive. This tendency to use available technologies more effectively is turning traditional companies into smart ones, inserting these manufactures into a generation known as Industry 4.0 (Lee et al. 2015).

The reality of the food industries shows a great variation of quality of the final product, due to the unpredictable supply of the raw materials of the manufactures of this branch of activity and due to the different requirements of customers (Dora et al. 2014). As the system being introduced by the fourth industrial revolution is concerned with customer satisfaction, long-range integration of production and sustainable development, this can help in solving problems experienced by slaughterhouses manufactures.

Sustainable development was defined by the World Commission on Environment and Development (WCED) as "development that meets the needs of the present without compromising the ability of future generations to meet their own needs." Although many companies focus on environmental sustainability, it is necessary to understand that the concept is broader, in which it covers measures of "quality of life" and "corporate social responsibility", worrying about financial, social and environmental issues, known as Triple Bottom Line (Wilkinson et al. 2001; Carrol 1991 and Van der Wiele et al. 2001).

There are not many studies that propose indicators for assessing of sustainable operations in poultry slaughterhouses, the main contribution is found for farms, food industries and agro-industries, branches not so specific to the poultry slaughtering industry. Kruger et al. (2015), Skunca et al. (2015, 2018) returned to their studies to identify indicators to evaluate the sustainable operations of chicken production, similar to the purpose of this paper.

Related to Industry 4.0, Watanabe et al. (2017) and Chaim et al. (2018) returned their studies to identify indicators to evaluate sustainable operations of industries that are undergoing the 4th Industrial Revolution. No papers were found suggesting indicators for poultry slaughterhouses from the perspective of Industry 4.0.

For an effective implementation of sustainability in poultry slaughterhouses, it is necessary to introduce efficient strategic management tools that allow the identification and monitoring of the sustainable operations of these companies, and these management tools can be represented by the implementation of sustainability performance assessment through the use of sustainability indicators (Whitehead 2016).

The paper is structured and organized as follows: will be presented a theoretical reference on industry 4.0, sustainability and performance assessment. Next, the methodology used in the research will be presented, and the results and conclusions will be presented.

2 Industry 4.0, Sustainability and Performance Assessment

In the last decades, the need to obtain competitive advantages has driven the organizations in the search for new technologies and forms of manufacture that minimize the cost of production. In this context, emerged the term Industry 4.0, which describes a strategy proposed by Germany to accelerate productivity return, characterized by Cyber Physical Systems (CPS) and dynamic data processes that use intelligent units and data (Sirkin et al. 2015).

These changes were already visible at the beginning of the 21st century, with the development of the Internet of powerful sensors, software and hardware enhancement and the increase the capacity of machines that learn and teach. This modification became known to teachers Erilk Braynjolfsson and Andrew McAfee of the Massachusetts Institute of Technology as the second age of the machine, however, in 2011, industry 4.0 was speaking at the Hannover Industrial Fair, in Germany (Dreher 2018).

With the transformations caused by the 4th Industrial Revolution, sensors, machinery, parts and systems of Information Technology (IT) will be connected along the value chain of manufactures, interacting with each other with the help of standard Internet-based protocols, which will facilitate data analysis, fault prevention, change, production of high-quality goods and cost savings (Rüßmann et al. 2015).

Industry 4.0 is based on 9 pillars, they are: simulation, additive manufacturing, cybersecurity, cloud computing, internet of things, big data, autonomous robots, system integration and augmented reality (Rüßmann et al. 2015).

In addition to worrying about the insertion of new technologies, with the increasing influence of various entities pressing companies due to environmental issues, it is necessary for the organization to begin to consider sustainable issues for decision-making (Castka and Balzarova 2008).

The parameters for pursuing a fully sustainable industry are hampered due to economic factors, because many of the tools that are used to have this sustainable growth within the enterprise become momentarily infeasible. However, the return to an industry that seeks to be sustainable must be seen in the long run, since the environment degraded by industrial liabilities is slow to be seen and noticed by society (Kruger et al. 2015).

The result of the implementation of sustainability in the productive process of an industry entails a better visualization in the external market, this makes more investors commit to the industry, generating higher results at the end of each cycle (Kruger et al. 2015).

For the implementation of sustainability in the industry, it is important to be based on 4 items in its development: (I) man made capital (financial resources, machine and tool acquisition.), (II) human capital (skills and characteristics of a group of people), (III) natural capital (natural resources) and (IV) social capital (interpersonal relations). Being able to interact these four items, sustainable development in industry will take place naturally (Kruger et al. 2015).

(I) Man-made capital:

The tools used, machinery purchased, and structure used, all this enters this item, since without structure there is no industry that can function correctly. It is important to analyze the tools and machinery that offer a good job, but that balance in the economic factor of the industry, being this one of the pillars of the sustainability.

(II) Human capital:

The skills of the group of people who will work for the sustainability of the industry should be requested in the smallest detail, since a well-qualified team can deal with any anomaly recorded in the process.

(III) Natural capital:

The natural resources acquired to complete the process of industry must be "hand-picked", because a quality raw material saves time and money in the process. However, one must always be within the pillars of sustainability, not extrapolating the economic barrier in the process.

(IV) Social capital:

The ability to interrelate with people in the present day is the great factor that leverages the productive process. If you can align this factor of sustainable development with the process, you have halfway gone.

Companies also need to assess the performance of their sustainable operations to assess whether the measures taken are having an effect. Companies are accustomed to using data to assess the performance of organizations in many different areas, such as the use of indicators to assess financial health and overall efficiency, and sustainability indicators are being inserted by companies to assess sustainable performance (Chaim et al. 2018).

Glenn and Pannell (1998) define indicators as quantitative measures of certain aspects of the expected performance of a management policy or strategy that can be assessed. Indicators are addressed with the aim of improving decision-making in an organization (Pannell and Glenn 2000).

In order to be able to efficiently assess the sustainability of a system, the indicators should be compared with reference values, and these can be determined by identifying a minimum value for each indicator that represents the minimum acceptable level of sustainability that a system should achieve (Van der Wiele et al. @@@2001), or through benchmarking results between two systems, different in spatial or time scale, to assess which is the most sustainable (Van Passel and Meul 2012).

3   Research Methodology

For the development of this paper, a systematic review of the literature was conducted to identify research related to models of assessment of sustainable operations in poultry slaughterhouses and/ or food industries and models of assessment of sustainable operations in Industry 4.0. The methodology used was an adaptation of the Process of Development of Knowledge—Constructive (Proknow-C). This methodology was chosen due to its structured and scientific process of search (Bortoluzzi 2013), shown in Fig. 1.

Two researches were carried out, the first involved the following research axes and strings: Industry 4.0 (Industry 4;0, Industrial Internet of Things, connected manufacturing, smart manufacturing e integrated manufacturing), Assessment (assessment, performance measurement, performance evaluation, measure*, performance management, performance measurement syste*, key performance indicator e KPI) and Sustainability (sustainable development, triple bottom line, sustain*, environment, shared value, HSE, HSMS, health and safety e management system). In the other research carried out, the research axes Assessment and Sustainability were main-

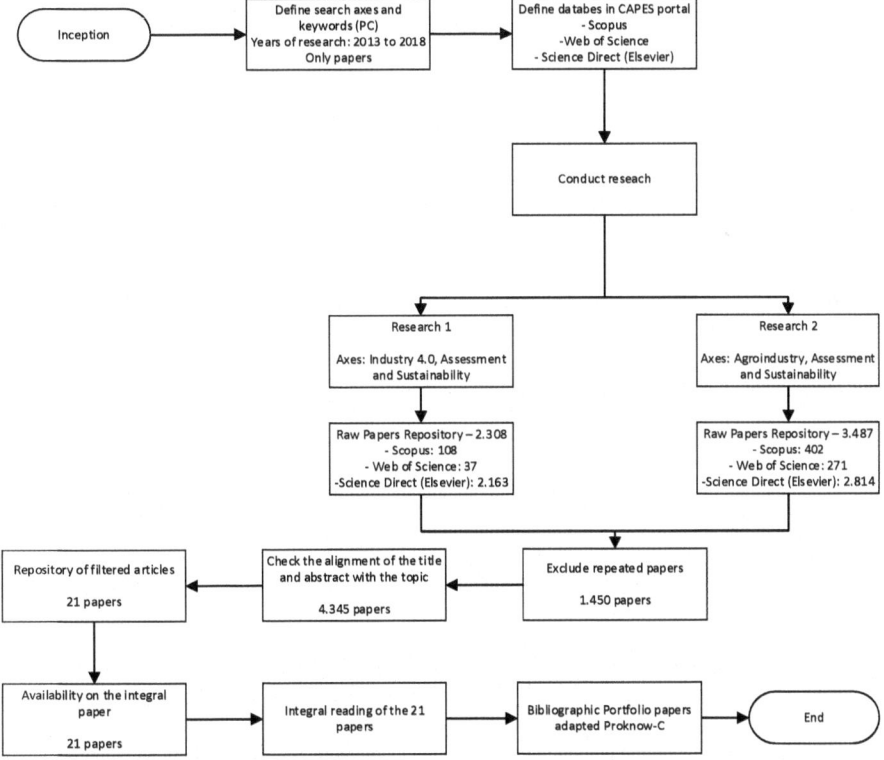

Fig. 1 Selection process of the documents that compose the Bibliographic Portfolio. *Source* Authors

tained and were combined with the Agorindústria axis, with the following strings: agroindustr*, agro-industr*, agricultural cooperativ*, poultry e slaughterhouse. The search terms chosen to carry out the study were defined based on the subject of the paper. The symbol (*) has the function of including the variations in the searched terms, such as words in the plural. The search axis and the strings used can be seen in Table 1.

The Scopus, Web of Science and Science Direct (Elsevier) databases were defined as 510,308 and 4977 papers, respectively, considering the two surveys and using the time filter for the last 5 years. At the end of the database search, Mendeley Desktop® software was used to organize and manage the papers.

Of the 4345 remaining papers, all titles were read to assess whether they were consistent with the topic. The paper abstract was read for the papers that generated doubts about the pertinence of the title to the theme, allowing the paper to be discarded or not. With the reading of the titles and some abstracts, 4324 papers were eliminated, leaving 21 papers to be complete reading the texts. When reading the papers in full, 9 papers were eliminated because they were not in accordance with the theme proposed in this paper, therefore, 12 documents compose the bibliographic portfolio.

With the portfolio formed, the performance indicators used were mapped, mounting tables and radar charts it was possible to visualize which indicators were most used by the authors. Based on the most representative indicators in the documents analyzed and discussion with experts through an informal and unstructured interview, it was possible to suggest 3 indicators from each pillar (social, economic and

Table 1 Axes and strings used

	Industry 4.0	Assessment	Sustainability	Agroindustry
Strings	Industry 4.0	Assessment	Sustainable development	Agroindustr*
	Industrial internet of things	Performance measurement	Triple bottom line	Agro-industr*
	Connected manufacturing	Performance evaluation	Sustain*	Agricultural cooperativ*
	Smart manufacturing	Measur	Environment	Poultry
	Integrated manufacturing	Performance management	Shared value	Slaughterhouse
		Performance neasurement syste*	HSE	
		Key performance indicator	HSMS	
		KPI	Health and safety management system	

Source Authors

environmental) to be used as a tool to evaluate sustainable operations in poultry slaughterhouses from a perspective of digitization.

In the interview with the experts, was discussed on the possible indicators to be used in the poultry slaughtering industry, characterized as key to transforming sustainable operations. Items such as water consumption, energy, raw materials, air emissions and waste generation were more effectively addressed. Although the economic and social pillars were treated with the experts, the environmental pillar received more attention.

The water indicator was presented with greater relevance, the experts pointed out that this is charged more frequently by supervisory bodies. Measures that reduce the impacts caused by the removal of water from the environment and tools that help in the reuse of water were explained by the experts.

4 Results and Discussion

The documents that make up the bibliographic portfolio can be visualized in Table 2.

Based on data collected from the papers that compose the portfolio of this research, in relation to the indicators suggested in Industries 4.0, it is possible to visualize them in Tables 3, 4 and 5, corresponding to the social, economic and environmental pillar, respectively, and in Fig. 2.

As many indicators were found in the literature, the indicators were grouped, thus creating an indicator encompassing several other indicators. Through Table 3 it is possible to identify the main indication and the indicators involved found in

Table 2 ProKnow-C bibliographic portfolio

Portfolio articles—ProKnow-C	
1	Marcis et al. (2018)
2	Petit et al. (2018
3	Tsolakis et al. (2018)
4	Mahon et al. (2018)
5	Kruger et al. (2015)
6	Carrasquer et al. (2017)
7	Skunca et al. (2015)
8	Djekic et al. (2016)
9	Castellani et al. (2017)
10	Skunca et al. (2018)
11	Watanabe et al. (2017)
12	Chaim et al. (2018)

Source Authors

Table 3 Indicators of the social pillar related to Industry 4.0

Pillar	Indicators name	Indicators involved	Authors	Representativeness
Social	Community Responsibility	Development (Community)	Chaim, O., Muschard, B., Cazarini, E., and Rozenfeld, H	4
		Justice (Community)	Chaim, O., Muschard, B., Cazarini, E., and Rozenfeld, H	
		Community Suggestions	Watanabe, E. H., da Silva, R. M., Tsuzuki, M. S., Junqueira, F., dos Santos Filho, D. J., and Miyagi, P. E	
		Product Responsibility	Chaim, O., Muschard, B., Cazarini, E., and Rozenfeld, H	
		Customer Rights	Chaim, O., Muschard, B., Cazarini, E., and Rozenfeld, H	
		Customer Complaints	Watanabe, E. H., da Silva, R. M., Tsuzuki, M. S., Junqueira, F., dos Santos Filho, D. J., and Miyagi, P. E	
	Customer Satisfaction	Customer Satisfaction	Chaim, O., Muschard, B., Cazarini, E., and Rozenfeld, H	4
		Health and Safety (Customer)	Chaim, O., Muschard, B., Cazarini, E., and Rozenfeld, H	
		Development (Employee)	Chaim, O., Muschard, B., Cazarini, E., and Rozenfeld, H	

(continued)

Table 3 (continued)

Pillar	Indicators name	Indicators involved	Authors	Representativeness
		Work Days Lost	Watanabe, E. H., da Silva, R. M., Tsuzuki, M. S., Junqueira, F., dos Santos Filho, D. J., and Miyagi, P. E	
		Satisfaction Employee	Chaim, O., Muschard, B., Cazarini, E., and Rozenfeld, H	
	Employee Satisfaction	Job Satisfaction	Watanabe, E. H., da Silva, R. M., Tsuzuki, M. S., Junqueira, F., dos Santos Filho, D. J., and Miyagi, P. E.	8
		Health and Safety (Employee)	Chaim, O., Muschard, B., Cazarini, E., and Rozenfeld, H	
		Employee Suggestions	Watanabe, E. H., da Silva, R. M., Tsuzuki, M. S., Junqueira, F., dos Santos Filho, D. J., and Miyagi, P. E	
	Labour Training in Sustainability	Labour Accidents Rate	Watanabe, E. H., da Silva, R. M., Tsuzuki, M. S., Junqueira, F., dos Santos Filho, D. J., and Miyagi, P. E	
		Labour Productivity Rate	Watanabe, E. H., da Silva, R. M., Tsuzuki, M. S., Junqueira, F., dos Santos Filho, D. J., and Miyagi, P. E	
	Sustainability Award	–	Watanabe, E. H., da Silva, R. M., Tsuzuki, M. S., Junqueira, F., dos Santos Filho, D. J., and Miyagi, P. E	1

(continued)

Table 3 (continued)

Pillar	Indicators name	Indicators involved	Authors	Representativeness
		–	Watanabe, E. H., da Silva, R. M., Tsuzuki, M. S., Junqueira, F., dos Santos Filho, D. J., and Miyagi, P. E	1
	Sustainability Reports	–	Watanabe, E. H., da Silva, R. M., Tsuzuki, M. S., Junqueira, F.,dos Santos Filho, D. J., and Miyagi, P. E	1

Source Authors

Table 4 Indicators of the economic pillar related to Industry 4.0

Pillar	Indicators name	Indicators involved	Authors	Representativeness
Economic	Profit	Reused Material Economy	Watanabe, E. H., da Silva, R. M., Tsuzuki, M. S., Junqueira, F., dos Santos Filho, D. J., and Miyagi, P. E	4
		Disposal Waste Economy	Watanabe, E. H., da Silva, R. M., Tsuzuki, M. S., Junqueira, F., dos Santos Filho, D. J., and Miyagi, P. E	
		Traditional Energy Reduction	Watanabe, E. H., da Silva, R. M., Tsuzuki, M. S., Junqueira, F., dos Santos Filho, D. J., and Miyagi, P. E	
		Recycled Material Economy	Watanabe, E. H., da Silva, R. M., Tsuzuki, M. S., Junqueira, F., dos Santos Filho, D. J., and Miyagi, P. E	

(continued)

Table 4 (continued)

Pillar	Indicators name	Indicators involved	Authors	Representativeness
	Benefit	Renewable Energy Benefit	Watanabe, E. H., da Silva, R. M., Tsuzuki, M. S., Junqueira, F., dos Santos Filho, D. J., and Miyagi, P. E	2
		Carbon Footprint Benefit	Watanabe, E. H., da Silva, R. M., Tsuzuki, M. S., Junqueira, F., dos Santos Filho, D. J., and Miyagi, P. E	
		Material Cost	Watanabe, E. H., da Silva, R. M., Tsuzuki, M. S., Junqueira, F., dos Santos Filho, D. J., and Miyagi, P. E	
	Cost	Energy Costs	Watanabe, E. H., da Silva, R. M., Tsuzuki, M. S., Junqueira, F., dos Santos Filho, D. J., and Miyagi, P. E	4
		Operational and Capital Costs	Watanabe, E. H., da Silva, R. M., Tsuzuki, M. S., Junqueira, F., dos Santos Filho, D. J., and Miyagi, P. E	
		Labour Costs	Watanabe, E. H., da Silva, R. M., Tsuzuki, M. S., Junqueira, F., dos Santos Filho, D. J., and Miyagi, P. E	

Source Authors

the literature. The indicators with the symbol (-) in the indicators involved column, means that it has not been grouped into any other indicators.

Through Tables 3, 4 and 5 and Fig. 2 it is possible to observe that the 3 most representative indicators of the Social Pillar are employee satisfaction, followed by customer satisfaction and community responsibility.

In relation to the economic pillar, the indicators were grouped into 3 groups, with the most representative of the profit and cost indicators, followed by the benefits indicator.

Table 5 Indicators of the environmental pillar related to Industry 4.0

Pillar	Indicators name	Indicators involved	Authors	Representativeness
Environmental	Water	Resource Consumption Water	Chaim, O., Muschard, B., Cazarini, E., and Rozenfeld, H	4
		Waste Water Discharged	Watanabe, E. H., da Silva, R. M., Tsuzuki, M. S., Junqueira, F., dos Santos Filho, D. J., and Miyagi, P. E	
		Water Reused	Watanabe, E. H., da Silva, R. M., Tsuzuki, M. S., Junqueira, F., dos Santos Filho, D. J., and Miyagi, P. E	
		Water intensity	Watanabe, E. H., da Silva, R. M., Tsuzuki, M. S., Junqueira, F., dos Santos Filho, D. J., and Miyagi, P. E	
		Resource Consumption— Energy	Chaim, O., Muschard, B., Cazarini, E., and Rozenfeld, H	
	Energy	Total Energy Consumption	Watanabe, E. H., da Silva, R. M., Tsuzuki, M. S., Junqueira, F., dos Santos Filho, D. J., and Miyagi, P. E	2
		Biodiversity	Chaim, O., Muschard, B., Cazarini, E., and Rozenfeld, H	

(continued)

Table 5 (continued)

Pillar	Indicators name	Indicators involved	Authors	Representativeness
	Habitat Management	Habitat Management	Chaim, O., Muschard, B., Cazarini, E., and Rozenfeld, H	3
		Natural Habitat Conservation	Chaim, O., Muschard, B., Cazarini, E., and Rozenfeld, H	
	Packaging	Packaging Materials Discarded	Watanabe, E. H., da Silva, R. M., Tsuzuki, M. S., Junqueira, F.,dos Santos Filho, D. J., and Miyagi, P. E	2
		Packaging Materials Reused	Watanabe, E. H., da Silva, R. M., Tsuzuki, M. S., Junqueira, F.,dos Santos Filho, D. J., and Miyagi, P. E	
	Solid waste	–	Chaim, O., Muschard, B., Cazarini, E., and Rozenfeld, H	1
	Air quality		Watanabe, E. H., da Silva, R. M., Tsuzuki, M. S., Junqueira, F., dos Santos Filho, D. J., and Miyagi, P. E	
	Effluent emission	–	Chaim, O., Muschard, B., Cazarini, E., and Rozenfeld, H	1
	Other Pollutant (Pollution)	–	Chaim, O., Muschard, B., Cazarini, E., and Rozenfeld, H	1

(continued)

Table 5 (continued)

Pillar	Indicators name	Indicators involved	Authors	Representativeness
	Waste energy emission	–	Chaim, O., Muschard, B., Cazarini, E., and Rozenfeld, H	1
	Hazardous Substances (Pollution)	–	Chaim, O., Muschard, B., Cazarini, E., and Rozenfeld, H	1
	Resource Consumption—Material	–	Chaim, O., Muschard, B., Cazarini, E., and Rozenfeld, H	1
	Resource Consumption—Land	–	Chaim, O., Muschard, B., Cazarini, E., and Rozenfeld, H	1
	Waste Material Discarded	–	Watanabe, E. H., da Silva, R. M., Tsuzuki, M. S., Junqueira, F., dos Santos Filho, D. J., and Miyagi, P. E	1
	Reused/Recycled Materials	–	Watanabe, E. H., da Silva, R. M., Tsuzuki, M. S., Junqueira, F., dos Santos Filho, D. J., and Miyagi, P. E	1

Source Author

Finally, in the environmental pillar, the water indicator was more representative, followed by the Habitat Management indicator, and thirdly, the indicators of packaging and energy.

In the agribusiness sector, the number of indicators found was higher. In Table 6 it is possible to visualize the indicators and the aspects involved that represent the environmental pillar, and in Fig. 3 it is possible to visualize the radar chart, which allows the quick visualization of the indicators with greater representativeness.

Through Table 6 and Fig. 3 it is possible to observe that in agroindustries the most cited indicator is water, the second is air emission and the third is between waste and energy. In Table 7 it is possible to visualize the indicators and the aspects involved that represent the economic pillar in agroindustries, and in Fig. 4 it is possible to

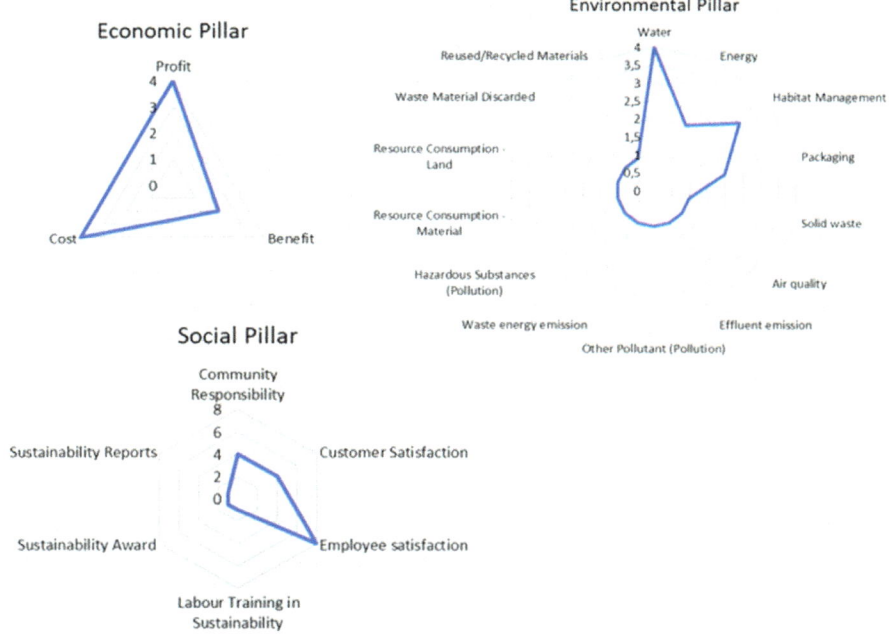

Fig. 2 Radar chart of the indicators related to Industry 4.0. *Source* Authors

visualize the radar chart of the indicators of the economic pillar, which allows the quick visualization of the indicators with greater representativeness.

Through Table 7 and Fig. 4 it can be observed that in agribusinesses the most cited economic indicator is cost and the most cited indicators and that assume the second position are profit, production performance and investments.

In Table 8 it is possible to visualize the indicators and the aspects involved that represent the social pillar in agroindustries, and in Fig. 5 it is possible to visualize the radar graph of the indicators of the social pillar, which allows the quick visualization of the indicators with greater representativeness.

Through Table 8 and Fig. 5 it is possible to observe that in agroindustries the most cited social indicator is employee satisfaction. The second and third most cited indicators were farmers satisfaction and community responsibility.

Table 6 Indicators of the environmental pillar related to Agroindustry

Pillar	Indicators name	Indicators involved	Authors	Representativeness
Enviromental	Water	Water losses	Petit, G., Sablayrolles, C., and Yannou-Le Bris, G	17
		Water consumption/ kg of product	Kruger, S. D., Petri, S. M., Ensslin, S. R., and dos Santos Matos, L	
		Consumption of Resources—Water	Tsolakis, N., Anastasiadis, F., and Srai, J	
		Water consumption	Skunca, D., Tomasevic, I., and Djekic, I	
		Wastewater	Skunca, D., Tomasevic, I., and Djekic, I	
		Water use efficiency	Mahon, N., Crute, I., Di Bonito, M., Simmons, E. A.,& Islam, M. M	
		Preservation of water	Marcis, J., Bortoluzzi, S. C., de Lima, E. P., and da Costa, S. E. G	
		Management of water	Djekic, I., Blagojevic, B., Antic, D., Cegar, S., Tomasevic, I., and Smigic, N	
		Management of wastewater	Djekic, I., Blagojevic, B., Antic, D., Cegar, S., Tomasevic, I., and Smigic, N	
		Water footprint	Carrasquer, B., Uche, J., and Martínez-Gracia, A	

(continued)

Table 6 (continued)

Pillar	Indicators name	Indicators involved	Authors	Representativeness
		Water consumption	Carrasquer, B., Uche, J., and Martínez-Gracia, A	
		Impact of water use	Carrasquer, B., Uche, J., and Martínez-Gracia, A	
		Monitoring of water quality by the Environment Agency	Mahon, N., Crute, I., Di Bonito, M., Simmons, E. A., and Islam, M. M	
		Concentration of crop protection chemicals in waterways	Mahon, N., Crute, I., Di Bonito, M., Simmons, E. A., and Islam, M. M	
		Presence of national polices on water quality	Mahon, N., Crute, I., Di Bonito, M., Simmons, E. A., and Islam, M. M	
		Degree of water reuse	Carrasquer, B., Uche, J., and Martínez-Gracia, A	
		Emissions to water	Castellani, V., Sala, S., and Benini, L	
		Consumption of Resources— Energy	Tsolakis, N., Anastasiadis, F., and Srai, J	
		Energy consumption	Skunca, D., Tomasevic, I., and Djekic, I	
		Amount of renewable energy generated	Mahon, N., Crute, I., Di Bonito, M., Simmons, E. A., and Islam, M. M	

(continued)

Table 6 (continued)

Pillar	Indicators name	Indicators involved	Authors	Representativeness
	Energy	Energy use efficiency	Mahon, N., Crute, I., Di Bonito, M., Simmons, E. A., and Islam, M. M	8
		Management of energy	Djekic, I., Blagojevic, B., Antic, D., Cegar, S., Tomasevic, I., and Smigic, N	
		Electricity consumption	Castellani, V., Sala, S., and Benini, L	
		Heat consumption	Castellani, V., Sala, S., and Benini, L	
		Energy consumption	Carrasquer, B., Uche, J., and Martínez-Gracia, A	
		Particulate matter in the air	Mahon, N., Crute, I., Di Bonito, M., Simmons, E. A.,& Islam, M. M	
		Greenhouse gas emission	Mahon, N., Crute, I., Di Bonito, M., Simmons, E. A., and Islam, M. M	
		Amount of carbon sequestered on farm	Mahon, N., Crute, I., Di Bonito, M., Simmons, E. A., and Islam, M. M	

(continued)

Table 6 (continued)

Pillar	Indicators name	Indicators involved	Authors	Representativeness
	Air emission	Preservation of air	Marcis, J., Bortoluzzi, S. C., de Lima, E. P., and da Costa, S. E.G	9
		Greenhouse gases emission	Castellani, V., Sala, S., and Benini, L	
		Emissions to air	Castellani, V., Sala, S., and Benini, L	
		CO_2 emission	Carrasquer, B., Uche, J., and Martínez-Gracia, A	
		carbon dioxide emission	Carrasquer, B., Uche, J., and Martínez-Gracia, A	
		Carbon footprint	Carrasquer, B., Uche, J., and Martínez-Gracia, A	
	Waste	Excessive waste generated by animals	Kruger, S. D., Petri, S. M., Ensslin, S. R., and dos Santos Matos, L	8
		Volume of waste/ kg of product	Kruger, S. D., Petri, S. M., Ensslin, S. R., and dos Santos Matos, L	
		Waste Management	Tsolakis, N., Anastasiadis, F., and Srai, J	

(continued)

Table 6 (continued)

Pillar	Indicators name	Indicators involved	Authors	Representativeness
		Plastic and paper waste	Skunca, D., Tomasevic, I., and Djekic, I	
		Biowaste	Skunca, D., Tomasevic, I., and Djekic, I	
		Volume of waste produced on farm	Mahon, N., Crute, I., Di Bonito, M., Simmons, E. A., and Islam, M. M	
		Efficient use of the waste	Marcis, J., Bortoluzzi, S. C., de Lima, E. P., and da Costa, S. E.G	
		Management of waste	Djekic, I., Blagojevic, B., Antic, D., Cegar, S., Tomasevic, I., and Smigic, N	
		Emissions of metals to soil	Castellani, V., Sala, S., and Benini, L	
	Soil	Preservation of soil	Marcis, J., Bortoluzzi, S. C., de Lima, E. P., and da Costa, S. E.G	2
	Packaging	Packaging (materials, quantity)	Tsolakis, N., Anastasiadis, F., and Srai, J	3
		Recyclability potential	Carrasquer, B., Uche, J., and Martínez-Gracia, A	
		Packaging materials	Skunca, D., Tomasevic, I., and Djekic, I.	

(continued)

Table 6 (continued)

Pillar	Indicators name	Indicators involved	Authors	Representativeness
	Biodiversity	Biodiversity	Petit, G., Sablayrolles, C., and Yannou-Le Bris, G	2
		Presence of national polices on	Mahon, N., Crute, I., Di Bonito, M., Simmons, E. A.	
	Climate change	Climate change	Petit, G., Sablayrolles, C., and Yannou-Le Bris, G	
		Presence of international polices on climate change	Mahon, N., Crute, I., Di Bonito, M., Simmons, E. A., and Islam, M. M	
		Presence of national polices on climate change	Mahon, N., Crute, I., Di Bonito, M., Simmons, E. A., and Islam, M. M	3
	GMO feed ratio/formula	–	Petit, G., Sablayrolles, C., and Yannou-Le Bris, G	1
	Production area	–	Kruger, S. D., Petri, S. M., Ensslin, S. R., and dos Santos Matos, L	1
	Eco-Production Practices	–	Tsolakis, N., Anastasiadis, F., and Srai, J	1
	Cleaning agents	–	Skunca, D., Tomasevic, I., and Djekic, I	1
	Presence of subsidies to encourage more environmentally sensitive farming	–	Mahon, N., Crute, I., Di Bonito, M., Simmons, E. A., and Islam, M. M	1

(continued)

Table 6 (continued)

Pillar	Indicators name	Indicators involved	Authors	Representativeness
	Efficient use of the product	–	Marcis, J., Bortoluzzi, S. C., de Lima, E. P., and da Costa, S. E. G	1
	Environmental management	–	Marcis, J., Bortoluzzi, S. C., de Lima, E. P., and da Costa, S. E. G	1

Source Authors

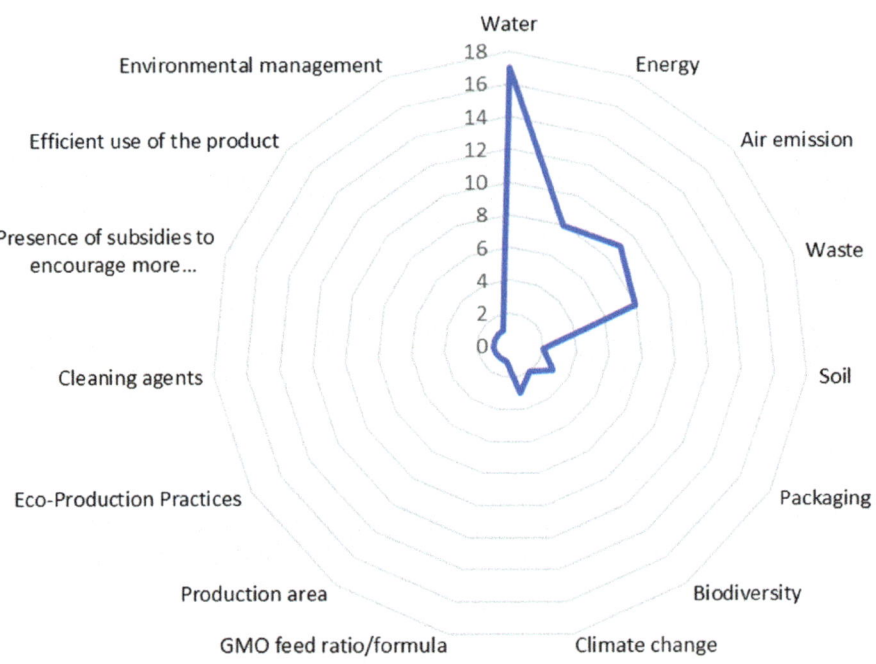

Fig. 3 Radar Chart of the indicators of the environmental pillar related to Agroindustry. *Source* Authors

Table 7 Indicators of the economic pillar related to Agroindustry

Pillar	Indicators name	Indicators involved	Authors	Representativeness
Economic	Cost	Additional cost paid to the farmer	Petit, G., Sablayrolles, C., and Yannou-Le Bris, G	9
		Variation in labor cost	Petit, G., Sablayrolles, C., and Yannou-Le Bris, G	
		Variation of cost-to-make	Petit, G., Sablayrolles, C., and Yannou-Le Bris, G	
		Production cost	Carrasquer, B., Uche, J., and Martínez-Gracia, A	
		Cost	Marcis, J., Bortoluzzi, S. C., de Lima, E. P., and da Costa, S. E. G	
		Cost of production	Mahon, N., Crute, I., Di Bonito, M., Simmons, E. A., and Islam, M. M	
		Ability of farmers to negotiate fair prices for what they produce	Mahon, N., Crute, I., Di Bonito, M., Simmons, E. A., and Islam, M. M	
		Economic benefits for its members	Marcis, J., Bortoluzzi, S. C., de Lima, E. P., and da Costa, S. E. G	
		Cost of production, materials and labor	Tsolakis, N., Anastasiadis, F., and Srai, J	

(continued)

Table 7 (continued)

Pillar	Indicators name	Indicators involved	Authors	Representativeness
		Short-term investment	Petit, G., Sablayrolles, C., and Yannou-Le Bris, G	
		Long-term investment	Petit, G., Sablayrolles, C., and Yannou-Le Bris, G	
	Investment	Amount of financial capital invested in farm	Mahon, N., Crute, I., Di Bonito, M., Simmons, E. A., and Islam, M. M	5
		Age of farm infrastructure	Mahon, N., Crute, I., Di Bonito, M., Simmons, E. A., and Islam, M. M	
		Investiments	Tsolakis, N., Anastasiadis, F., and Srai, J	
		Profitability	Marcis, J., Bortoluzzi, S. C., de Lima, E. P., and da Costa, S. E. G	
		Distribution of net income	Marcis, J., Bortoluzzi, S. C., de Lima, E. P., and da Costa, S. E. G	
	Profit	Total income from subsidies	Mahon, N., Crute, I., Di Bonito, M., Simmons, E. A., and Islam, M. M	5
		Presence of farm income from nonagricultural activities	Mahon, N., Crute, I., Di Bonito, M., Simmons, E. A., and Islam, M. M	
		Profit estimate	Marcis, J., Bortoluzzi, S. C., de Lima, E. P., and da Costa, S. E. G	

(continued)

Table 7 (continued)

Pillar	Indicators name	Indicators involved	Authors	Representativeness
		Energy performance	Carrasquer, B., Uche, J., and Martínez-Gracia, A	
	Energy Performance	Energy and exergy utilization	Carrasquer, B., Uche, J., and Martínez-Gracia, A	3
		Amount of renewable energy generated	Mahon, N., Crute, I., Di Bonito, M., Simmons, E. A., and Islam, M. M	
		Financial measures	Marcis, J., Bortoluzzi, S. C., de Lima, E. P., and da Costa, S. E. G	
	Financial measures	Return on assets	Marcis, J., Bortoluzzi, S. C., de Lima, E. P., and da Costa, S. E. G	4
		Return on equity	Marcis, J., Bortoluzzi, S. C., de Lima, E. P., and da Costa, S. E.G	
		Return on investment	Marcis, J., Bortoluzzi, S. C., de Lima, E. P., and da Costa, S. E.G	
	Production Performance	Production valorization (loss rate)	Petit, G., Sablayrolles, C., and Yannou-Le Bris, G.	5

(continued)

Table 7 (continued)

Pillar	Indicators name	Indicators involved	Authors	Representativeness
		Lean muscle percentage	Petit, G., Sablayrolles, C., and Yannou-Le Bris, G	
		Waste and losses rate	Petit, G., Sablayrolles, C., and Yannou-Le Bris, G	
		Product performance	Carrasquer, B., Uche, J., and Martínez-Gracia, A	
		Traceability	Tsolakis, N., Anastasiadis, F., and Srai, J	
		Carcass pH	Petit, G., Sablayrolles, C., and Yannou-Le Bris, G	
	Quality	Sensory evaluation score	Petit, G., Sablayrolles, C., and Yannou-Le Bris, G	2
	Consumer preference	–	Mahon, N., Crute, I., Di Bonito, M., Simmons, E. A., and Islam, M. M	1
	Water utilization efficiency	–	Carrasquer, B., Uche, J., and Martínez-Gracia, A	1
	Farm Income	–	Tsolakis, N., Anastasiadis, F., and Srai, J	1

Source Authors

5 Proposal of Environmental, Economic and Social Indicators for Poultry Slaughterhouse Operations from the Perspective of Industry 4.0

In Table 9 it is possible to visualize the indicators suggested in this paper for assessing of sustainable operations of poultry slaughtering industries from the perspective of Industry 4.0.

Fig. 4 Radar Chart of the indicators of the economic pillar related to Agroindustry. *Source* Authors

According to the indicators found in the literature and through an informal and unstructured interview with specialists, it was possible to determine indicators that fit the poultry slaughterhouse under the perspective of digitization, thus enabling a better view of the brand by customers and investors.

In relation to the environmental pillar, indicators such as water, air emission and energy are suggested. The water and energy indicators, in addition to being very representative in the literature review in what concerns agroindustry, also had great representability in the surveys carried out regarding Industry 4.0.

Water, a primordial element for life in any resort, is also seen as a priority in the industrial environment. In relation to slaughterhouse, where it fits in the branch of transformation industries, the consumption of water represents the third greater use, with urban supply and irrigated agriculture in first and second place respectively (ANA 2017). At all stages of the industrial process in a poultry slaughterhouse, water becomes necessary, for cleaning the plant, transporting by-products, cooling the carcass, cleaning products, among others.

It is necessary that, not only in poultry slaughtering industries, but in all industries, be aware of water consumption, consumption without reducing the quality of the product, therefore, it is necessary to establish water consumption targets, and this goal can be determined by water/poultry consumption. According to Cristóbal et al. (2018), the consumption of water/kg of product, stipulated as a reference indicator for the process, stresses the need to avoid waste, and even if a predetermined quantity of water per finished product, all the staff will worry about these wastes, thus avoiding higher costs with water treatment.

Table 8 Indicators of the social pillar related to Agroindustry

Pillar	Indicators name	Indicators involved	Authors	Representativeness
Social	Employee satisfaction	Rh policies (benefits)	Marcis, J., Bortoluzzi, S. C., de Lima, E. P., and da Costa, S. E. G	11
		Rh policies (Remuneration)	Marcis, J., Bortoluzzi, S. C., de Lima, E. P., and da Costa, S. E. G	
		Rh policies (Job Plan)	Marcis, J., Bortoluzzi, S. C., de Lima, E. P., and da Costa, S. E. G	
		Rh policies (Salary Plan)	Marcis, J., Bortoluzzi, S. C., de Lima, E. P., and da Costa, S. E. G	
		Health and safety	Marcis, J., Bortoluzzi, S. C., de Lima, E. P., and da Costa, S. E. G	
		Welfare (employees and cooperative member)	Marcis, J., Bortoluzzi, S. C., de Lima, E. P., and da Costa, S. E. G	
		Quality of life (employees and cooperative member)	Marcis, J., Bortoluzzi, S. C., de Lima, E. P., and da Costa, S. E. G	
		Education and qualification (employees and cooperative member)	Marcis, J., Bortoluzzi, S. C., de Lima, E. P., and da Costa, S. E. G	
		Employee welfare	Petit, G., Sablayrolles, C., and Yannou-Le Bris, G	

(continued)

Table 8 (continued)

Pillar	Indicators name	Indicators involved	Authors	Representativeness
		Max transport without pause	Petit, G., Sablayrolles, C., and Yannou-Le Bris, G	
		Staff training	Tsolakis, N., Anastasiadis, F., and Srai, J	
		Farm worker happiness	Mahon, N., Crute, I., Di Bonito, M., Simmons, E. A., and Islam, M. M	
		Farmer welfare	Petit, G., Sablayrolles, C., and Yannou-Le Bris, G	
		Educational level of famers	Mahon, N., Crute, I., Di Bonito, M., Simmons, E. A., and Islam, M. M	
	Famers satisfaction	Farmer access to multiple sources of information	Mahon, N., Crute, I., Di Bonito, M., Simmons, E. A., and Islam, M. M	7
		Farmer's training	Petit, G., Sablayrolles, C., and Yannou-Le Bris, G	
		Affordability of rural housing	Mahon, N., Crute, I., Di Bonito, M., Simmons, E. A., and Islam, M. M	
		Age of farmers	Mahon, N., Crute, I., Di Bonito, M., Simmons, E. A., and Islam, M. M	
		Employment	Marcis, J., Bortoluzzi, S. C., de Lima, E. P., and da Costa, S. E. G	
		Project related to community	Marcis, J., Bortoluzzi, S. C., de Lima, E. P., and da Costa, S. E. G	

(continued)

Table 8 (continued)

Pillar	Indicators name	Indicators involved	Authors	Representativeness
	Community Responsibility	Localness	Petit, G., Sablayrolles, C., and Yannou-Le Bris, G	7
		Number of hires	Petit, G., Sablayrolles, C., and Yannou-Le Bris, G	
		Food safety	Tsolakis, N., Anastasiadis, F., and Srai, J	
		Number of people in agricultural employment	Mahon, N., Crute, I., Di Bonito, M., Simmons, E. A., and Islam, M. M	
		Income generation for farmers	Kruger, S. D., Petri, S. M., Ensslin, S. R., and dos Santos Matos, L	
	Exclusive Suppliers	–	Tsolakis, N., Anastasiadis, F., and Srai, J	1
	Certification	–	Tsolakis, N., Anastasiadis, F., and Srai, J	1

Source Authors

The reuse of wastewater is also an activity that favors the water indicator, because if the effluent is well treated, the time demanded by the environment to be regenerated the impact caused by the impurities present in the water decreases.

The water indicator was also strongly suggested for industry 4.0, then this indicator can be used from the perspective of industry 4.0.

Giving continuity to the indicators of the environmental pillar, energy is the other indicator suggested. Energy consumption is another factor that influences both environmentally and economically in industry, this factor is able to align, with feasibility, these two factors of sustainability. Industries that invest in generations of sustainable energy have better visibility in the market, and this if well executed, it becomes an asset to your brand.

Investments in energy generators that do not pollute the environment as Djekic et al. (2016) explains, the renewable energies, will be a great work front for the industries, that however they demand high investment, in the long run they give economic return to the industry.

Fig. 5 Radar Chart of the indicators of the social pillar related to Agroindustry. *Source* Authors

Table 9 Suggested Indicators

Pillar	Indicator
Economic	Cost Investments Profit
Social	Employee satisfaction Farmers satisfaction Community responsability
Environmental	Water Air emission Energy

Source Authors

Energy consumption in the poultry slaughtering industry is large, especially in industries that are more automated. As automation demands great energy consumption and this is the basis of Industry 4.0, this indicator was cited for use in Industries 4.0, both for the environmental and economic pillar.

And finally, air emission enters as a strong indicator candidate for an industry that aims at sustainability. Air pollutants are strongly linked to rising temperatures of the globe, this because they help in the destruction of the ozone layer. An industry that cares about its atmospheric emissions and invests in technologies that help the environment to absorb these pollutants has a great advantage in terms of sustainability.

Emissions generated by the boilers, the so-called "hearts of industries", has great representativity in the total degradation by means of air emissions of the planet. For the treatment of these emissions, the investment is high, which often makes the project unfeasible if top management does not support the company's sustainable issues. One of the remedial measures is the sequestration of carbon through reforestation areas, or greenbelts, which eventually absorb part of this emission of pollutants. In addition to absorbing these air pollutants, the green belts also absorb the wastewater generated in the industries, through fertigation projects.

The air emission indicator was one of the most representative in Industry 4.0, so it can be said that with this indicator the poultry slaughtering industry is in accordance with the principles of Industry 4.0.

Finally, the industries that have a high level of sustainability implemented in their processes should be concerned with their generations of abrasives to the environment, because it is from him that the industry takes its raw material to generate the products, jobs and a socioeconomic and environmental balance for its community involved.

As a proposal of economic indicators for poultry slaughterhouses the following is suggested: cost, investments and profit. Only 3 indicators were cited in the industry 4.0, benefits, costs and profit, which are in line with the indicators proposed in agroindustries.

Cost and investment are important economic indicators, the decision to become sustainable will depend very much on these indicators. High investments and long-term financial returns are not seen by companies, but this reality is changing with regard to sustainable issues, because companies are realizing the need to become sustainable and are investing in this issue (Epstein et al. 2015).

The cost indicator is important for the company to remain competitive in the market. It is necessary for the company to know its costs, know the cost of the finished product, so that it can form the selling price of the product/ service, so that you make a profit on your sales and have a competitive price. In addition, with the cost indicator, it is possible for companies to know their waste and allow intervention to reduce cost and waste.

The third indicator proposed is profit, which presents great importance, because without profit, the company does not hold. Profitability can be measured as a percentage of profit or R $/finished product, the important thing is for the company to have this well-structured and easily understood indicator. Profit and cost are indicators that run parallel, the profit depends on the cost, the lower the cost of the company, the greater your profit.

Sustainability helps reduce company costs by reducing waste, reducing waste, reducing energy and water consumption. On the economic side, sustainability brings great benefits.

Investments are necessary and need to be measured, the company needs to know the investments made and their return, and it is long-term or if it is short-term, and often, if these investments will be paid. Many investments are needed to improve the company's visible to investors and stakeholders, for a better view of the brand. In aspects of Industry 4.0 the investments are fundamental, without the investment in intelligent systems, the company will not be able to make the transition from the third to the fourth industrial revolution. Although many poultry slaughtering industries are quite automated, a very large investment is still needed in this respect, so this indicator is fundamental for poultry slaughtering industries from the perspective of Industry 4.0.

Regarding the social indicators for poultry slaughterers, it is suggested: employee satisfaction, community responsibility and farmer satisfaction. The first two indicators were also recommended in Industry 4.0, so it can be stated that these indicators evaluate sustainable operations in a poultry slaughtering industry from the perspective of Industry 4.0.

It is necessary to evaluate employee satisfaction, because the more satisfied the employee is, the more productive he becomes, besides being necessary to take care of the health and safety of the worker, so that it is possible to avoid possible deviations, in this way, to maintain the intellectual capital of the company.

Another suggested indicator, farmer satisfaction, is extremely important because it depends on the farmer's management to obtain better meat production indexes. Once the management is carried out correctly, production performance will be greater, animal welfare management will be carried out and the industry will obtain quality meat products, so that it can meet the most demanding customers who pay the best for the finished product.

Finally, the last suggested indicator is community responsibility. This indicator is also very important because industries need to benefit the community and take responsibility for it. Businesses need to benefit the community by generating jobs, not polluting rivers, soil and air. The location of a poultry slaughtering industry needs to be planned and away from cities so as not to harm the population with possible odors, and it is necessary that companies be open to suggestions from the community.

6 Conclusion

It is necessary to worry about the assessment of sustainable operations in poultry slaughterhouses, since it is an industry that presents a great consumption of water, there are significant air emissions, high energy use and waste generation, both in manufacturing and in the poultry production.

Many are the indicators used to evaluate the poultry production chain, and in Industries 4.0 the question of assessment of sustainable operations is still being inserted. There are still few documents found regarding this subject in this new reality of the industries because it is a relatively new topic, and this represented a limitation of this paper.

The objective of proposing indicators for evaluating sustainable operations in poultry slaughtering industry from the perspective of Industry 4.0 has been reached, it was suggested to use three indicators for the social pillar, three indicators for the economic pillar and three indicators for the environmental pillar.

It is seen that if any of the indicators are inserted in a chain of processes, it will have some gain, and whatever the sustainability pillar, since these indicators act in a way to reduce the impacts generated and still save in stages of the process. However, something that comes up against the idea of sustainability is the high investments needed to minimize negative impacts on the environment, thus generating a difficulty in adopting sustainable practices in the organization.

This paper contributes to the academic productions, since there is a gap in the surveys that involve adequate indicators for the reality of poultry slaughterhouses that have an expectation regarding Industry 4.0, it was only possible to find studies on these subjects separately.

It is suggested as future paper the survey of indicators of the entire supply chain of poultry slaughtering industries, since the activity of farmers results in high rates of soil and air pollution. If a company wants to become sustainable, it needs to worry about its entire production chain.

As it was not possible to apply this sustainable operations assessment model, another future paper proposal is to apply the model to assess whether it is suitable for poultry slaughtering industries and assess whether it is in line with the principles proposed by Industry 4.0.

References

ANA (2017) Estudo da Agência Nacional de Água aborda uso da água no setor industrial. Retrieved from http://www3.ana.gov.br/portal/ANA/noticias/estudo-da-agencia-nacional-de-aguas-aborda-uso-da-agua-no-setor-industrial. (in Portuguese)

Bortoluzzi SC (2013) Proposta teórico-metodológica fundamentada na avaliação de desempenho multicritério para a gestão do relacionamento de arranjo produtivo local (APL) e suas empresas individuais. Tese de Doutorado em Engenharia de Produção, Florianópolis 2013:553p (in Portuguese)

Carrasquer B, Uche J, Martínez-Gracia A (2017) A new indicator to estimate the efficiency of water and energy use in agro-industries. J Clean Prod 143:462–473

Carroll AB (1991) The pyramid of corporate social responsibility: toward the moral management of organizational stakeholders. Business Horizons

Castellani V, Sala S, Benini L (2017) Hotspots analysis and critical interpretation of food life cycle assessment studies for selecting eco-innovation options and for policy support. J Clean Prod 140:556–568

Castka P, Balzarova MA (2008) ISO 26000 and supply chains—On the diffusion of the social responsibility standard. Int J Prod Econ 111(2):274–286

Chaim O, Muschard B, Cazarini E, Rozenfeld H (2018) Insertion of sustainability performance indicators in an industry 4.0 virtual learning environment. Procedia Manufacturing 21:446–453

Cristóbal J, Castellani V, Manfredi S, Sala S (2018) Prioritizing and optimizing sustainable measures for food waste prevention and management. Waste Manag 72:3–16

Dora M, Van Goubergen D, Kumar M, Molnar A, Gellynck X (2014) Application of lean practices in small and medium-sized food enterprises. British Food Journal 116(1):125–141

Dreher A (2018) The smart factory of the future—Part 1. Belden News. Retrieved from https://www.belden.com/blog/industrial-ethernet/the-smart-factory-of-the-future-part-1

Djekic I, Blagojevic B, Antic D, Cegar S, Tomasevic I, Smigic N (2016) Assessment of environmental practices in Serbian meat companies. J Clean Prod 112:2495–2504

Epstein MJ, Buhovac AR, Yuthas K (2015) Managing social, environmental and financial performance simultaneously. Long Range Plan 48(1):35–45

Glenn NA, Pannell DJ (1998) The Economics and Application of Sustainability Indicators in Agriculture. In:42nd Annual conference of the australian agricultural and resource economics society, University of New England, Armidale, pp 19–21

Kruger SD, Petri SM, Ensslin SR, dos Santos Matos L (2015) Performance evaluation of poultry production sustainability: international mapping regarding this issue. CEP 89:060

Lee J, Bagheri B, Kao H (2015) A cyber-physical systems architecture for industry 4.0-based manufacturing systems. Manufact Lett 3:18–23

Mahon N, Crute I, Di Bonito M, Simmons EA, Islam MM (2018) Towards a broad-based and holistic framework of sustainable intensification indicators. Land Use Policy 77:576–597

Marcis J, Bortoluzzi SC, de Lima EP, da Costa SEG (2018) Sustainability performance evaluation of agricultural cooperatives' operations: a systemic review of the literature. Environ Develop Sustain 1–16

Pannell DJ, Glenn NA (2000) A framework for the economic evaluation and selection of sustainability indicators in agriculture. Ecol Econ 33:135–149

Petit G, Sablayrolles C, Yannou-Le Bris G (2018) Combining eco-social and environmental indicators to assess the sustainability performance of a food value chain: a case study. J Clean Prod 191:135–143

Rüßmann M, Lorenz M, Gerbert P, Waldner M, Justus J, Engel P, Harnisch M (2015) Industry 4.0: the future of productivity and growth in manufacturing industries. Boston Consulting Group, 9

Sirkin HL, Zinser M, Rose JM (2015) Why advanced manufacturing will boost productivity. Boston Consulting Group, 2

Skunca D, Tomasevic I, Djekic I (2015) Environmental performance of the poultry meat chain–LCA approach. Procedia Food Science 5:258–261

Skunca D, Tomasevic I, Nastasijevic I, Tomovic V, Djekic I (2018) Life cycle assessment of the chicken meat chain. J Clean Prod 184:440–450

Tsolakis N, Anastasiadis F, Srai J (2018) Sustainability performance in food supply networks: Insights from the UK industry. Sustainability 10(9):3148

Van der Wiele T, Kok P, McKenna R, Brown A (2001) A corporate social responsibility audit within a quality management framework. J Bus Ethics 31(4):285–297

Van Passel S, Meul M (2012) Multilevel and multi-user sustainability assessment of farming systems. Environ Impact Assess Rev 32:170–180

Watanabe EH, da Silva RM, Tsuzuki MS, Junqueira F, dos Santos Filho DJ, Miyagi PE (2017) Assessment of sustainability for production control based on petri net and cyber-physical cloud system. IFAC-PapersOnLine 50(1):12985–12990

WCED—World Commission on Environment and Development. Our Common Future, Chairman's Foreword. Retrieved from http://www.un-documents.net/our-common-future.pdf

Whitehead J (2016) Prioritizing sustainability indicators: using materiality analysis to guide sustainability assessment and strategy. Bus Strategy Env 26(3):399–412

Wilkinson A, Hill MR, Gollan P (2001) The sustainability debate. Int J Oper Prod Manag 21(12):1492–1502

The Use of the Seed Germination Test to Evaluate Phytotoxicity in Small-Scale Organic Compounds: A Study on Scientific Production and Its Contributions to Goals 2 and 12 of the UN 2030 Agenda

Isael Colonna Ribeiro, Walessa Nunes Barcellos, Isabella Maria de Castro Filogônio, Poliana Daré Zampirolli Pires, Jacqueline Rogéria Bringhenti, Sheila Souza da Silva Ribeiro and Adriana Marcia Nicolau Korres

Abstract Composting generates a suitable product to be used as a substrate for seed germination, contributing to resilient agricultural practices and progressively improving soil quality. Considering the growth of this compound use in home gardens and knowing that it may contain phytotoxic substances, which prevent seed development, the goal of this work was to evaluate the scientific production on the effectiveness of the use of the germination test in organic compound analyzes. The bibliographical research was developed through Capes' Periodicals Portal, using the

I. C. Ribeiro (✉) · W. N. Barcellos · I. M. de Castro Filogônio · P. D. Z. Pires · J. R. Bringhenti
Federal Institute of Education, Science and Technology of Espírito Santo, 1729 Vitoria Ave.
Vitoria, Espírito Santo 29040-780, Brazil
e-mail: isaelcolonna@gmail.com

W. N. Barcellos
e-mail: wabarcellos@hotmail.com

I. M. de Castro Filogônio
e-mail: isabella_mcf@hotmail.com

P. D. Z. Pires
e-mail: poliana.pires@ifes.edu.br

J. R. Bringhenti
e-mail: jacquelineb@ifes.edu.br

S. S. da Silva Ribeiro
Laboratory of Microbiology, Federal Institute of Education, Science and Technology of Espírito
Santo, 1729 Vitoria Ave. Vitoria, Espírito Santo 29040-780, Brazil
e-mail: sheilasouza@ifes.edu.br

A. M. N. Korres
Sanitary and Environmental Engineering Department, Federal Institute of Education, Science and
Technology of Espírito Santo, 1729 Vitoria Ave. Vitoria, Espírito Santo 29040-780, Brazil
e-mail: adrianak@ifes.edu.br

© Springer Nature Switzerland AG 2020
W. Leal Filho et al. (eds.), *International Business, Trade and Institutional
Sustainability*, World Sustainability Series,
https://doi.org/10.1007/978-3-030-26759-9_26

following descriptors: solid waste composting, toxicity, germination, fertilizers and coffee grounds, in both English and Portuguese languages, over a 20-year time horizon. Nine researches focused on the germination procedure, of which five of them were dissertations. The work evidences the effectiveness of the germination test in phytotoxic analyzes and contributes to more sustainable cultivation methods and improving the quality of ecosystems, what can achieve safety regarding the use of these compounds as fertilizers, on a domestic scale. In addition, composting contributes to consumption and production in a more responsible way, reducing food losses in production chains and waste, both encompassed by the sustainable development goals of the UN 2030 Agenda.

Keywords Organic waste · Composting · Organic compost · Germination · Phytotoxicity

1 Introduction: Composting and Toxicity Analysis Methods of Final Compost

Composting is a natural process in that organic residues are transformed in stable compounds. This technique involves transformations of biochemical nature, promoted by microorganisms that use organic matter as a source of energy, mineral nutrients and carbon (Souza and Rezende 2014). It is a way of providing adequate conditions to the microorganisms so that they degrade the organic matter and generate a final product rich in nutrients. In this way, composting is an easy, economical and natural process of biodegradation, which allows the recovery of organic waste (Kadir et al. 2016).

The use of the final compound as fertilizer has grown in small-scale systems such as community and home gardens, small flowerbeds, plant pots, among others. This alternative is feasible from the environmental point of view, since the organic residues generated on site can be reused (Stock et al. 2018). In addition, this initiative provides support for more sustainable cultivation methods, which are important as they are related to organic production and consumption reduction, what is addressed in the Agenda 2030, more specifically in Sustainable Development Goals 2 and 12 (UNDP 2015).

Some studies have emphasized the use of a particular organic residue, the coffee grounds, since it is generated in great quantity in domestic and institutional environments (Korres et al. 2013 and Costa et al. 2016). However, the use of this residue is controversial due to its physicochemical characteristics. The constitution of the coffee grounds may vary according to several factors, for example, the mode of production, the addition of sugar, and compounds such as caffeine, tannins, and polyphenols found in this material which can confer a certain toxicity, leading to environmental pollution problems (Pandey et al. 2000).

Due to this, and also seeking ways to improve the storage of material, academic research has described pre-treatment methodologies of this residue, however, there also seems to be no consensus of the most appropriate techniques to be used. The compost generated from the composting of the coffee grounds combined with other residues should have a chemical composition capable of allowing the availability of nutrients so that the plants grown can develop without, however, suffer some kind of toxicity.

According to Torrentó et al. (2008), due to its characteristics, the quality of the compost has been proven due to its use in soil applications, resulting in improvements in water infiltration and retention, decrease of temperature variations, reduction of erosion, improvement of crop health and nutrient supply to the growth of plants. However, the absence of toxicity of this compound is not guaranteed by monitoring only its physicochemical aspects. In this context, there are specific tests for the analysis of phytotoxicity of a compound, the most usual being germination and seed growth tests (Batista and Batista 2007).

The germination tests are aimed at proving the phytotoxic effects of the organic matter to be analyzed, that is, to verify if the material is toxic on the germination and/or development of seedlings and rootlets rootlet and of some vegetal species. This test consists in determining, by comparison with a control (blank test), the percentage and/or other germination parameters in the presence of a sample of compound (Cordeiro 2010).

The study of caffeine toxicity on seed germination is ancient. Ransom (1911) reported caffeine as a reducing agent for the germination of seeds of different plants, such as Endive, Onion, and Lettuce. It is believed that caffeine may act in the disintegration of some proteins of both animals and vegetables.

A literature search on the toxicity of organic compounds found a lack in the definition and specification of techniques, measurements and, mainly, a variety of types of seeds used in the experiments. When researching the use of coffee grounds as part of the original composting material, the literature reports the presence of compounds potentially harmful to seed germination but the results concerning toxicity are inconclusive or not clear. The goal of this research was to carry out a literature review on the validation of germination tests as a measure for the phytotoxicity analysis of compounds from composting. If they do not present toxicity to the plant species, they can be used as organic fertilizer in the environment where small-scale composting and cultivation practices are established. This validation of compost use encourages local sustainable practices and, from the gradual increase of these small actions, an impact on global sustainable agriculture is generated.

2 Sustainable Practices of Composting as Tools for the Achievement of the Sustainable Development Goals

In September 2015, at the UN Summit on Sustainable Development, the 17 Sustainable Development Goals (SDG) were announced, under the name Agenda 2030

(UNDP 2015). These proposals were built between 2013 and 2015 and should guide national policies and international cooperation activities over the next fifteen years, succeeding and updating the Millennium Development Goals (MDG), as well as strengthening and contributing to the discussion on environmental issues.

The MDG were established in the year 2000 and included eight anti-poverty targets to be achieved by the end of 2015. Since then, progress has been made, with global poverty continuing to decline, more children are attending primary schools, child deaths have fallen dramatically, access to drinking water has expanded, among others.

It is noted that the MDG have translated into positive impacts on people's lives, showing that targets are efficient to reduce poverty. From this, the United Nations defined the SDG as part of a new sustainable development agenda that should complement the work of the MDG. This agenda was launched in 2015 during the Sustainable Development Summit, where Member States and civil society negotiated their contributions.

Agenda 2030 reflects the new development challenges and is related to the outcome of the conference Rio + 20, which was held in June 2012 in Rio de Janeiro, Brazil. This agenda is composed of 17 goals and 169 targets, and these decisions will determine the global course of action to end poverty, promoting prosperity and well-being for all, protecting the environment and tackling climate change. Among the objectives included in Agenda 2030, the activities reported in this research are related to the number 2 and 12.

Goal 12 is focused on ensuring sustainable production and consumption patterns. Some of the targets for achieving this goal include reducing food losses along production and supply chains; achieve environmentally sound management of chemicals and significantly reduce their release to air, water and soil; substantially reduce the generation of waste through the prevention, reduction, recycling and reuse; ensure that people have relevant information and awareness for sustainable development and lifestyles in harmony with nature, among others.

Goal 2 focuses on ending hunger, achieving food security and improving nutrition and promoting sustainable agriculture. One of the targets to achieve this goal includes the implementation of sustainable food production systems and resilient agricultural practices that progressively improve land and soil quality. The present work can contribute directly to the incentive to these practices, since it is related to composting and the validation of its use for the increase of agricultural crops.

These targets demonstrate how composting is related to the goals, since it is a practice that promotes the reuse of organic waste, transforming them into a stable final product that can be used as a natural fertilizer, reducing the need for chemical fertilizers that attack the environment. In addition, this practice promotes environmental awareness and encourages a lifestyle in balance with the environment.

It is also worth mentioning the relationship between the germination tests to evaluate the toxicity of organic composts and the SDG mentioned, whereas the implementation of this technique contributes to the determination of the maturation of the composts and, therefore, can provide the knowledge of the feasibility of applying them to the soil.

The application of stabilized and matured composts in soils can beneficially modify their physical, chemical and biological properties. However, agricultural application of immature products can cause damage to germination and plant development (Gajalakshmi and Abbasi 2008). In this sense, due to the contamination risks associated with immature composts, it is key to adopt strategies to know the maturity of the compound.

3 Method

The methodology includes bibliographic search in the Portal of Periodicals of the Coordination of Improvement of Higher Level Personnel (CAPES) using the following key words: "solid waste", "composting", "compost", "coffee grounds", "mutagenicity tests", "germination" and "fertilizer", in English, and, in Portuguese, "residuos sólidos", "compostagem", "composto", "borra de café", "toxicidade", "germinação" and "fertilizante".

The most accessed search databases were Science Direct (Elsevier), Scielo and the CAPES Thesis Bank. The searches were restricted to a time horizon of 20 years (1998–2018). We also used bibliographic suggestions made available by Mendeley according to the works saved and organized in the authors' virtual library.

The bibliographic materials collected from the research were classified and analyzed according to the characteristics of the publication, which includes type of research, year of publication and type of residue used in composting.

For the organization and compilation of the works raised during the search in the research portals, the criteria adopted by Luo et al. (2018) for data ordering were used, adding the variables: number of replications, incubation temperature, plaque size, and type of statistical analysis used.

4 Results and Discussions

The present research resulted in a survey of nine papers, five dissertations in Portuguese (four from Portugal and one from Brazil) and four scientific articles, three in English and one in Portuguese.

The papers differed greatly in relation to the material used in composting. Green residues, municipal solid waste, paper waste, coffee residues, among others were used. In addition, some parameters analyzed on the germination test and the statistical analysis used by the authors also differ.

In a recent literature review, Luo et al. (2018) reviewed the use of the seed germination test to evaluate the toxicity of the compound, giving focus to its rules, problems and proposals. These authors analyzed variables such as primary feedstock of compost, sample to water ratio, shake, centrifugation, filtration, the seeds of plant species used in the test, and number of seeds and volume of extract per Petri dish, incubation time and operational definition of germination. Based on these variables, the criteria are presented in the Table 1.

Analyzing the methodologies used by the authors, according to Table 1, it is observed that on obtaining the aqueous extract, 5 of the 9 works cited the dilution of 1:10 or 10% dilution in water. Three of these papers performed the test using only this dilution and the other two reported the combination with other dilutions. Other types of dilutions employed were: 20, 25, 30, 40, 50 and 60%. Dilutions varied according to type of water was used. Three cited the use of distilled water; one double distilled water; two, deionized water; one, sterile water (the authors did not specify the water purification) and two works did not specify anything about the liquid used in dilution. This variation in the degree of water purification and the dilution level may interfere with the results of the tests.

The negative control used in the tests varied between the studies reported. This test is of paramount importance to ensure the reliability of the results and still to be used as a benchmark. Three papers cited the use of distilled water; four cited the use of deionized water; two did not make clear what the control parameter was.

Other parameters that comprise obtaining the aqueous extract are the type of homogenization, if it was by agitation, centrifugation and filtration. Analyzing the selected papers it was possible to conclude that there is no standardization regarding these analyzes, with large variation in the methodologies used.

Over the shaking, three of the nine works reported no stirring; four mention an hour of agitation; one paper used 30 min and another, three hours. Regarding centrifugation, four of the total works do not mention any use of centrifugation and the others vary on their approaches. Concerning filtration, two works did not mention anything; five used paper filters (a blue band) of different porosities and sizes; and two used membrane filters. This step is essential for the removal of larger particulates that may interfere with test results. These variations depend on the type of extract to be obtained, the characteristics of the material used, and the level of filtration required.

When considering the plant species used, in the tests, one of the most frequent was cress (*Lepidium sativum* L.), cited by six of the nine studies analyzed. In five of these studies, only this species was used and in one of them other types of seeds were also used. Another well-quoted seed species was lettuce (*Lactuca sativa* L.). Four of the total studies used this species, one of them isolated and three along with others. This demonstrates that *Lepidium sativum* L. and *Lactuca sativa* L. are the most usual for this type of test, due to its characteristics of rapid germination, cress 4–10 days and lettuce 4–7 days (Brasil 2009). Moreover, they are species that are easily used in home or community gardens and are useful as indicators of the toxicity of the compound to be added to the soil.

Another criterion on seed germination is the amount of seeds used per Petri dish. This parameter varied between seven and 50 seeds, besides the variation between the size of the plates used. Two papers cited the use of seven seeds per plate; other paper used eight; three used 10; one cited 20; one quote 40 and one quote 50 seeds per Petri dish. The most commonly found values are close to or equal to 10. These

Table 1 Bibliographical survey of scientific publications that address seed germination tests to evaluate the toxicity of compounds, in the time horizon from 1998 to 2018

Composting materials	Aqueous extract				Seed germination					Statistical analysis	References
	Dilution sample to water ratio (w/v)	Shake	Centrifugation	Filtration	Species of seed	Number of seeds per Petri dish (Ø mm)	Volume of extract (mL) per Petri dish	Replicates	Incubation time (d)		
Soybean residues and leaves	1:10 (dry weight basis with double distilled water)	1 h	10,000 rpm	0.45 μm membrane filters	Cress (*Lepidium sativum* L.)	10	5	3	2 (20–25 °C/dark)	The least significant difference test at P = 0.05	Wong et al. (2001)
Compost made from "alperujo"	Samples (4 g dry) were moistened at 60%; after stand for 30 min, more deionized water (54 ml) was added	30 min	Stand for 30 min	0.45 μm Whatman filter papers	Cress (*Lepidium sativum* L.)	8 (100)	1	10	2 (27 °C/dark)	Analysis of variance and the least significant difference at P < 0.05	Alburquerque et al. (2006)
Wastepaper	Dilutions (25 and 50% v/v) of the extract of the sample previously moistened at 90% for 3 h	–	–	–	Cress (*Lepidium sativum* L.)	40	2	3	1 (28 °C incubation hot house)	Mean and analysis of variance	Felícia (2009)

(continued)

Table 1 (continued)

Composting materials	Aqueous extract				Seed germination					Statistical analysis	References
	Dilution sample to water ratio (w/v)	Shake	Centrifugation	Filtration	Species of seed	Number of seeds per Petri dish (Ø mm)	Volume of extract (mL) per Petri dish	Replicates	Incubation time (d)		
Green waste	Dilution 30% of the aqueous extract of the sample (with sterilized water) previously moistened at 60%	–	Stand for 30 min	0.2 μm Nalgene filter papers	Cress (*Lepidium sativum* L.), lettuce (*Lactuca sativa*) and tomato (*Lycopersicum esculentum* L.)	7	1 and 2 (cress); 3 (lettuce; tomato)	15	1 and 2 days (cress); 5 days (lettuce) and 6 days (tomato)—(27 °C greenhouse)	Analysis of variance at P = 0.05 and Tukey test	Cordeiro (2010)
Kitchen bio-waste, aerobic and anaerobic sludge	Compost—deionized water extracts (1:1, v/v)	1 h at 180 rpm	5000 rpm for 15 min	Pre-rinsed filter paper (Whatman 4, Ø15 cm)	Cress (*Lepidium sativum* L.)	20 (90)	5	3	2 (25–27 °C/dark)	Averages and standard deviations; T-test (p < 0.05) and Bernoulli test	Himanen and Hänninen (2011)
Coffee dregs and *Acacia dealbata* L. wastes in composting	Aqueous extracts of the composts by the dilutions of 10, 20, 30 and 40% in distilled water	3 h (20 °C, 200 rpm)	–	0.45 μm membrane filters	*Lactuca sativa* L., *Sorghum sudanense* L. and *Acacia dealbata* L.	50 (125)	5	3	Observe until no more germination occurs (Lovibond 23 °C)	Analysis of variance and Tukey test at P = 0.05	Fonseca (2012)

(continued)

Table 1 (continued)

Composting materials	Aqueous extract				Seed germination					Statistical analysis	References
	Dilution sample to water ratio (w/v)	Shake	Centrifugation	Filtration	Species of seed	Number of seeds per Petri dish (Ø mm)	Volume of extract (mL) per Petri dish	Replicates	Incubation time (d)		
Mixtures of Urban Solid Waste compost and raw coffee grounds	1:10 (sample with distilled water)	1 h	3500 rpm for 20 min	Filter blue band	Cress (Lepidium sativum L.)	7 (70)	1	10	1 (27 °C/dark)	Analysis of variance and the least significant difference at 5%	Lima (2016)
Sludge from fish industrialization with rice hulls	1:10 (sample in distilled water)	1 h	–	Whatman filter paper 3, two times	Lettuce (Lactuca sativa), cucumber (Cucumis sativus) androoster-tail (Celosia plumosa)	10 (90)	10	5	2 (25 °C)	Normality by the W test; Hartley test; analysis of variance by the F test ($p \leq 0.05$).	Fonseca (2017)
Coffee grounds, eggshells and rice	10, 50 and 100%	–	–	–	Lettuce (Lactuca sativa)	10	5	3	5 (20 °C ± 2)	Averages, with their respective amplitudes (minimum-maximum)	Kim et al. (2018)

Source Prepared by the authors (2018)

variations interfere with the statistical analyzes, because the higher the number of repetitions, the greater the precision of the experiment (Banzatto and Kronka 2013).

Four of nine papers did not specify the size of the Petri dish used. One of the nine papers reported the use of dish with diameter of 100 mm; two used dishes of 90 mm diameter; one of 70 mm and one of 125 mm diameters. The size of the plaque is related to the size and quantity of the seeds and the amount of extract to be applied in the test.

The volume of extract per Petri dish varied, in most cases, between one and five milliliters. Two papers cited one ml; one reported the use of two ml; one reported the use of one, two and three, according to the seed used; four cited the use of five ml and one used 10 ml. Note that this variation may be related to the size of the Petri dish, the number of seeds applied, the type of seed, since some species have larger seeds and require more or less moisture to germinate, to the type of paper used as liner (when used) and how much it absorbs water, among other factors.

The number of repetitions ranged from 3 to 15, with five of the nine studies addressing the use of 3 replicates. One paper cited 5 repetitions; two cited 10 reps, and one used 15 repetitions on the experiments. This variation is related to the chosen experimental design and the amount of treatments adopted (Banzatto and Kronka 2013).

The incubation period for the determination of germination ranged from one to six days, with two papers quoting one day; four quoted two days; one paper cited one, two, five and six days, depending on the type of seed. One paper reported the time of five days incubation and one observed the Petri dishes until no further germination occurred. The majority (seven) of the authors cited a period of one to two days of incubation of the seeds in the extracts, what may be considered short, allowing a rapid return of test results.

The incubation temperature ranged from 20 to 28 °C. Four of the nine studies mentioned the incubation occurred in the absence of light. According to Brasil (2009), the temperatures necessary for the germination test with the cress species are: 20–30; 20 and 15 °C, an isolated number means constant temperature and two numbers separated by dash means alternating temperatures. Following the same analogy, Brasil (2009) points to the following temperatures in relation to the lettuce: 20 and 15 °C.

The temperature variation reported by the papers found by this work was from 20 to 28 °C, while Luo et al. (2018) reported a range between 20 and 25 °C. Both range of temperature extrapolate a bit the ideal temperature recommended by Brasil (2009) for germination of cress (15–30 °C) and lettuce (15–20 °C).

After the incubation period, the works varied over the measured and calculated parameters used. Eight papers performed germination and root growth test; three did growth test; one did test for vegetation in pots (EV). Three works calculated the relative germination percentage (% GR) and one presented this percentage as a function of time. Six papers measured germination index (GI). Still on this aspect, two papers measured the growth index (GI); one paper calculated the ratio of germination and seedling length/shoots weight and control.

Another parameter cited by Luo et al. (2018) is the operational definition of germination. One of the papers reported the accuracy of 1 mm. The seeds were considered germinated when the rootlet was at least 1 mm long. The other papers did not make it clear if these considerations were considered.

Regarding the statistical analysis employed in the different studies, it should be summarily noted that very similar and usual analyzes were performed. The authors cited analyzes where the mean, standard deviation, variance analysis, mean comparison test using the least significant difference test. Seven of the nine papers cited the use of tests for a significance level of 5%.

The software used to carry out the statistical analyzes varied among the works surveyed. Among the software mentioned by the authors were SPSS 11.0; Statistic 7.0; STATISTIX 7.0 (Analytical Software, Tallahassee, USA); R (version 2.8.1); SAS statistical package; Microsoft Office 2007 and MS Office Excel 2003.

To evaluate and conclude about the toxicity of the material used in the germination test, the works differed according to the reference used. Luo et al. (2018) cited the GI value, not less than 80%, what usually means that compost has no phytotoxicity (Tiquia et al. 1996). Table 2 presents the conclusions about the toxicity obtained in the germination tests carried out in the selected studies.

Analysis of the selected studies indicated the germination index and hence the toxicity varied according to the plant species, the dilution and, especially, the maturation stage of the compost. It was verified that the more mature the compost, the lower the toxicity indexes and, therefore, the higher the germination indexes. Thus, for small-scale composting, the safety of the compost in terms of toxicity can be guaranteed by the extent of complete. In general, the maturity of the compost can be determined when it presents absence of odor and uniform granulometry, respecting a composting time between 45 and 90 days.

Germination seed test as indicative of safety and toxicity of organic composts obtained from composting contribute for the Circular Economy, the Zero Waste Strategy, as well as the sustainable agriculture, all goals pursued by Sustainable Development Goals. When disposing organic solid waste, including coffee grounds, for small-scale composting, the safety of the use of the final product in small gardens and properties must be ensured and the seed germination test is a feasible tool for that.

5 Conclusion and Recommendations

The literature review performed by this work revealed seed germination tests for compost toxicity is very different from paper to paper concerning many parameters, as composting materials, dilution sample to water ratio, types of seed and number of seeds per Petri dish.

Despite their great applicability, the methodologies are not consolidated worldwide, and the standardization of the technique may collaborate with a way of mon-

Table 2 Conclusions about the toxicity obtained in the germination tests carried out in the selected studies

Parameter for measuring toxicity	Conclusions	References
Non-phytotoxic: GI > 50%	Non-phytotoxic	Wong et al. (2001)
No defined GI	Acceptable degree of maturity and stability	Alburquerque et al. (2006)
Non-phytotoxic: GI > 60% for dilution at 25% (v/v)	Most of treatments were phytotoxic	Felícia (2009)
Non-phytotoxic: GI > 60%	For lettuce, phytotoxic For cress and tomato, non-phytotoxic	Cordeiro (2010)
Non-phytotoxic: Control: GI > 95% Treatments: no statistically significant difference (p > 0.05) between the parameter (germination or seedling length) in compost extract and in the control	Non-phytotoxic	Himanen and Hänninen (2011)
Non-phytotoxic: Control: GI > 90% Treatments: not specified	Control: For lettuce and sorghum, non-phytotoxic For mimosa, phytotoxic For treatments, not conclusive	Fonseca (2012)
Mature compost: IC > 50%	Immature composts	Lima (2016)
Non-phytotoxic: GI > 60%	Phytotoxic	Fonseca (2017)
Relative index value were classified into three categories: (1) Inhibition of root elongation—toxic (2) No significant effect (3) Root elongation stimulus—nontoxic	Dilution of 10 and 50%: non-toxic Dilution of 100%: toxic	Kim et al. (2018)

Source Prepared by the authors (2018)

itoring the quality of the compound obtained. Also, the repeatability and reliability of the method may be reflected in quality and safe in use of the final product.

Stimulating small-scale composting and composting in home gardens is an effective way to achieve sustainability with the vision of acting locally and thinking globally as these actions are easier and simpler to implement. The use of this biotechnological technique for the use of organic waste in a practical, agile and easily replicated way has the potential to contribute to a culture that reintegrates man and the environment, based on Zero Waste and Circular Economy strategies.

The implementation of local sustainable actions is a determining factor for the success and the establishment of measures to build a more sustainable planet. In this context, higher education institutions play a fundamental role in the construction of sustainable development, since they can act as living laboratories, supporting the implementation of sustainable development goals.

References

Alburquerque JA, Gonzálvez J, García D, Cegarra J (2006) Measuring detoxification and maturity in compost made from "alperujo", the solid by-product of extracting olive oil by the two-phase centrifugation system. Chemosphere 64(3):470–477

Banzatto DA, Kronka SDN (2013) Experimentação agrícola. Jaboticabal: Funep, 4, 237p

Batista JGF, Batista ERB (2007) Compostagem: Utilização de compostos em horticultura. Universidade dos Açores, Centro de Investigação de Tecnologias Agrárias, 252p

Brasil (2009) Ministério da Agricultura, Pecuária e Abastecimento. Secretaria de Defesa Agropecuária. Regras para análise de sementes. http://www.agricultura.gov.br/assuntos/ insumos-agropecuarios/arquivos-publicacoes-insumos/2946_regras_analise__sementes.pdf. Last Accessed 07 Jan 2019

Cordeiro NM (2010) Compostagem de resíduos verdes e avaliação da qualidade dos compostos obtidos-caso de estudo da algar SA. Master's Dissertation in Environmental Engineering—Environmental Technologies, Instituto Superior de Agronomia, Lisboa. https://www.repository.utl.pt/ bitstream/10400.5/3353/1/TESE.pdf. Last Accessed 07 Jan 2019

Costa PM, Bringhenti JR, Korres AMN Faé C (2016) Awareness and practice of solid waste selective collect for vermicomposting: case study in an educational institution. In: XXXV Congreso de AIDIS y 59° Congreso Internacional ACODAL, 2016, Cartagena, Colombia. Anais do XXXV Congreso de AIDIS y 59° Congreso Internacional ACODAL. AIDIS, México

Felícia DG (2009) Estudo do comportamento do resíduo papel no processo de compostagem. Master's Dissertation in Environmental Engineering, University of Aveiro. https://ria.ua.pt/bitstream/ 10773/664/1/2010000400.pdf. Last Accessed 07 Jan 2019

Fonseca JPQB (2012) Efeito da adição de borras de café sobre a compostagem de resíduos de Acacia dealbata L. (mimosa). Master's Dissertation in Agronomic Engineering, University of Trás os Montes e Alto Douro Vila Real, 85p

Fonseca CB (2017) Compostagem de lodo da industrialização de pescado com casca de arroz em diferentes volumes. Master's Dissertation in Food Science and Technology, Federal University of Pelotas. http://dctaufpel.com.br/ppgcta/manager/uploads/documentos/teses/dissertacao_ fonseca,_camilo_bruno.pdf. Last Accessed 07 Jan 2019

Gajalakshmi S, Abbasi A (2008) Solid waste management by composting: state of the art Critical Rev Environ Sci Technol 38:5, 311–400

Himanen M, Hänninen K (2011) Composting of bio-waste, aerobic and anaerobic sludges–Effect of feedstock on the process and quality of compost. Biores Technol 102(3):2842–2852

Kadir AA, Ismail SNM, Jamaludin SN (2016) Food waste composting study from Makanan Ringan Mas. IOP Conf Series Mater Sci Eng 136:1

Kim JKF et al (2018) Análise de toxicidade do adubo orgânico, a partir de borra de café, casca de ovo e arroz, na germinação de sementes de alface (Lactuca sativa L.). In: Congresso Nacional de Meio ambiente, XV, 2018, Poços de Caldas-MG. Anais...Poços de Caldas-MG: Ifes campus Muzambinho. 5p

Korres AMN, Bringhenti JR, Costa PM Filogônio IMC (2013) A sensibilização e envolvimento da comunidade escolar sobre a prática da coleta seletiva de resíduos sólidos orgânicos e a compostagem como forma de destinação final de material orgânico. In: XXVII Congresso Brasileiro de Engenharia Sanitária e Ambiental, Goiânia

Lima PMMF (2016) Estudo como corretivo orgânico de misturas de composto de RSU com borras de café. Master's Dissertation in Agronomic Engineering, Higher Institute of Agronomy, University of Lisbon. https://www.repository.utl.pt/bitstream/10400.5/12177/1/TeseCorretivoOrganicoMisturasCompostoRSUBorrasCafe-C%C3%B3pia.pdf. Last Accessed 07 Jan 2019

Luo Y, Liang J, Zeng G, Chen M, Mo D, Li G, Zhang D (2018) Seed germination test for toxicity evaluation of compost: Its roles, problems and prospects. Waste Manag 71(2018):109–114

Pandey A, Soccol CR, Nigam P, Brand D, Mohan R, Roussos S (2000) Biotechnological potential for coffee pulp and coffee husk for bioprocesses. Biochem Eng J 6:153–162

Ransom F (1911) The Effects of Caffeine upon the germination and growth of seeds. Biochem J 151–155. https://www.ncbi.nlm.nih.gov/pmc/articles/PMC1276408/pdf/biochemj01218-0003.pdf. Last Accessed 30 Nov 2018

Souza JL, Rezende P (2014) Manual de horticultura orgânica. Viçosa, Minas Gerais: Aprenda Fácil, 2 ed

Stock MA, Santiago MR, Rosa AM, Dotto AR F., Da Silva MCF (2018) Composteira caseira como alternativa para o tratamento de resíduos orgânicos. Anais do Salão Internacional de Ensino, Pesquisa e Extensão, 9(1)

Tiquia SM, Tam NFY, Hodgkiss IJ (1996) Effects of composting on phytotoxicity of spent pig-manure sawdust litter. Environm Pollution 93(3):249–256

Torrentó MS, Martínez ML, Huerta O (2008) Antecedentes y fundamentos del proceso de compostaje. In Compostaje. Mundi Prensa Libros SA, pp 75–92

UNDP (2015) Objetivos de Desenvolvimento Sustentável. Programa das Nações Unidas para o Desenvolvimento. https://nacoesunidas.org/pnud-explica-transicao-dos-objetivos-do-milenio-aos-objetivos-de-desenvolvimento-sustentavel/. Last Accessed 17 Nov 2018

Wong JWC, Mak KF, Chan NW, Lam A, Fang M, Zhou LX, Liao XD (2001) Co-composting of soybean residues and leaves in Hong Kong. Biores Technol 76(2):99–106

Innovations in Agriculture: The Important Role of Agroforestry in Achieving SDG 13

Marcia Fajardo Cavalcanti de Albuquerque

Abstract It is noteworthy that agriculture is a key element for achieving several SDG, such as fight against poverty, zero hunger, biodiversity protection and sustainable agriculture, production and consumption. Agriculture is also able to contribute to climate change mitigation and adaptation. Agroforestry stands out in face of other agricultural practices. It can contribute to both climate change mitigation and adaptation through enhanced carbon sequestration, vulnerability reduction, production diversification and smallholder's capacity building to adapt to climate change. This paper makes an analysis of the main international instruments that promote the adoption of agroforestry practices to tackle climate change. Nevertheless, its adoption may be constrained by some adverse national policies. Thus, this paper also aims to present some legal and policy elements underpinning successful agroforestry regimes.

Keywords SDG · Climate change · Sustainable agriculture · Agroforestry · Legislation · Public policy

1 Introduction

Agriculture is part of the human foundation and no other human activity has caused as many positive and negative impacts on our planet and for the people inhabiting it. The agricultural sector is one of the largest contributors to greenhouse gas emissions and biodiversity loss.[1] But agriculture is also able to enhance biodiversity levels and promote climate change mitigation and adaptation depending on the practices implemented.

Climate change is not a new phenomenon. It has already caused extreme weather events, sea level rise, changes in the quality and quantity of water and in agricultural

M. F. C. de Albuquerque (✉)
University of Paris, Sorbonne, Paris, France
e-mail: Marcia.Fajardo-Cavalcanti-Albuquerque@etu.univ-paris1.fr

Mackenzie University of São Paulo, São Paulo, Brazil

[1] «Globally, agriculture, land-use change and forestry are responsible for 19–29% of greenhouse gas (GHG) emissions» (CGIAR, CCAFS, Why climate-smart agriculture?).

© Springer Nature Switzerland AG 2020
W. Leal Filho et al. (eds.), *International Business, Trade and Institutional Sustainability*, World Sustainability Series,
https://doi.org/10.1007/978-3-030-26759-9_27

productivity. In 2015, at a United Nations Summit on Sustainable Development, countries adopted the 2030 Agenda for Sustainable Development and the 17 Sustainable Development Goals (SDGs). These goals address five key areas: humanity, planet, prosperity, peace and partnerships. SDG 13 urges countries to take urgent action to combat climate change and its impacts, specially through (1) the strengthening resilience and adaptive capacity to climate-related hazards and natural disasters in all countries; (2) the integration of climate change measures into national policies, strategies and planning; and (3) the improvement of education, awareness-raising and human and institutional capacity on climate change mitigation, adaptation, impact reduction and early warning. In addition, in order to fight against climate change, developed countries should mobilize financial and human resources to assist developing countries in the implementation of mitigation and adaptation actions (UN, SDG).

The Sustainable Development Goals core aim is to build a resilient[2] future. Thus, urgent action to tackle climate change is essential for the successful implementation of all SDG.

As regards to agriculture, climate is a crucial factor for crop production, once it determines nature and characteristics of vegetation and crops. So, how to implement agricultural systems resilient to climate change? One can go further, is agriculture a viable approach to reach climate change adaptation and mitigation?

Agroforestry is the answer. It can contribute to both climate change mitigation and adaptation through enhanced carbon sequestration, vulnerability reduction, production diversification and smallholder's capacity building to adapt to climate change. However, agroforestry systems encompass complex practices and if not managed correctly they may not fulfill its potential of sustainability and climate change mitigation.

This paper presents an analysis of the main international instruments that promote the adoption of agroforestry practices as a means to fight climate change. Nevertheless, its adoption may be constrained by some adverse national policies. This paper also aims to bring forward some legal and policy elements underpinning successful agroforestry regimes.

2 Is Agriculture Capable to Address Climate Change? the Agroforestry Case

Agriculture is both affected by climate change and has considerable influence on it. Yet, agro-ecological practices are considered as potential mitigation practices, which promote the sequestration of carbon from the atmosphere. In this sense, the climate

---

[2]According to Bizikova et al., «resilience is the capacity of a socioecological system to absorb or withstand perturbations and other stressors such that the system remains within the same regime, essentially maintaining its structure and functions. It describes the degree to which the system is capable of self-organization, learning and adaptation» (Bizikova et al. 2019).

smart agriculture approach (CSA) was developed at the 2010 FAO Conference on Agriculture, Food Security and Climate Change in Hague. It is an approach that seeks synergy between mitigation and adaptation, aimed at redirecting agricultural production systems towards sustainable development. According to the CSA guide organized by the Consultative Group on International Agricultural Research (CGIAR) and the CGIAR Research Program on Climate Change, Agriculture and Food Security (CCAFS), «CSA is sustainable agriculture that incorporates resilience concerns while at the same time seeking to reduce greenhouse gas emissions» (CGIAR and CCAFS).

This approach has been developed in order to sustainably increase agricultural productivity and incomes; adapt and build resilience to climate change; reduce greenhouse gas emissions when possible; and promote the achievement of national food security and development goals (FAO, Climate-smart agriculture). CSA approaches are capable to address food security, misdistribution and malnutrition, the relationship between agriculture and poverty and the relation between climate change and agriculture (CGIAR and CCAFS). They are based on sustainable intensification, adaptation, resilience, ecosystem services maintenance and mitigation specially through soil and tree management that «maximizes their potential to act as carbon sinks and absorb CO_2 from the atmosphere» (CGIAR and CCAFS).

The implementation of such an approach is based on five points of action: broadening the knowledge base, encouraging supportive policy frameworks, strengthening national and local institutions, improving funding options, and improving implementation of practices in the field (FAO 2019). CSA assists stakeholders in identifying agricultural strategies that are adapted to their local conditions (FAO 2018).

Agroforestry can be considered as a climate-smart agriculture practice in the sense that it promotes the production of diverse ecosystem services (such as supply services such as food production), the presence of trees in agricultural fields enhances resilience to natural hazards, reduces deforestation and increases carbon sequestration (CGIAR and CCAFS).

According to the definition given by the World Agroforestry Center, agroforestry is

> a collective name for all land-use systems and practices in which woody perennials are deliberately grown at the same land management unit as crops and/or animals. This can be either in some form of spatial arrangement or in a time sequence. To qualify as agroforestry, a given land-use system or practice must permit significant economic and ecological interactions between the woody and non woody components. (Clarke and Thaman 1993, p. 9)

King and Chandler complement this definition stating that agroforestry is «a sustainable land management system which increases the overall yield of the land, combines the production of crops (including tree crops) and forest plants and/or animals simultaneously or sequentially, on the same unit of land, and applies management practices that are compatible with the cultural practices of the local population» (King and Chandler 1978, p. 2).

Agroforestry system may comprise different land management practices such as «crop diversification, long rotation systems for soil conservation, homegardens, boundary plantings, perennial crops, hedgerow intercropping, live fences, improved fallows or mixed strata agroforestry» (Mbow et al. 2014). Agroforestry practices are able to mitigating non-point source pollution, controlling soil erosion, creating wildlife habitat and improving resilience to rapid ecological change (Mbow et al. 2014). According to Mbow et al., «if well managed (success hinges essentially upon proper implementation), agroforestry can play a crucial role in improving resilience to uncertain climates through microclimate buffering and regulation of water flow» (Mbow et al. 2014).

Agroforestry systems tend to sequester much higher quantities of carbon than agricultural systems without trees, once it stores carbon both in vegetation and in soils. Agroforestry can contribute to climate change mitigation through enhanced carbon sequestration, micro climate and macro-climate improvement (Uthappa et al. 2017). Trees on farms are able to positively influence the micro climate through radiation flux, air temperature and wind speed (Uthappa et al. 2017, p. 4).

Agroforestry is capable of enhancing the uptake of CO_2 or reduce its emission and of removing CO_2 from the atmosphere «if the trees are harvested, accompanied by replanting of same and/or other area, and sequestered carbon is locked through non-destructive use of such wood» (Newaj et al. 2016, p. 6). This potential vary depending on the species combination, age of trees, local geographic and climatic factors and management practices (Newaj et al. 2016, p. 7). Recent studies have estimated the carbon sequestration potential of tropical agroforestry systems between 12 and 228 Mg ha-1 with a mean value of 95 Mg ha-1 (Newaj et al. 2016, p. 7).

It also contributes to climate change adaptation, once the presence of trees in agricultural fields improves resilience to natural hazards, reduces vulnerability, diversifies production and income sources and improves livelihoods, building the capacity of smallholders to adapt to climate change (Uthappa et al. 2017). Tree-based systems are able to maintain productivity over dry and wet seasons, thus «diversifying the production system to include a significant tree component may buffer against income risks associated with climatic variability» (Newaj et al. 2016, p. 8).

Agroforestry's potential of climate change mitigation and adaptation has already been recognized by some international instruments, as it will be analyzed next.

3   The International Recognition of Agroforestry's Adaptation and Mitigation Potential

The United Nations Framework Convention on Climate Change (UNFCCC) is the main international legal instrument to address climate change, through the «stabilization of greenhouse gas concentrations in the atmosphere at a level that would prevent dangerous anthropogenic interference with the climate system. Such a level should be achieved within a time frame sufficient to allow ecosystems to adapt naturally to climate change, to ensure that food production is not threatened and to enable economic development to proceed in a sustainable manner» (Article 2, UNFCC).

Both the UNFCCC and the Intergovernmental Panel on Climate Change (IPCC) recognize agroforestry practices as a component of climate-smart agriculture (Buttoud, 2013, p. 4). In 2008, the secretariat of the UNFCC elaborated a technical paper on challenges and opportunities for mitigation in the agriculture sector. This paper recognized that agroforestry practices have great potential to promote carbon sequestration and carbon reduction sector at non-prohibitive costs:

> expanding the role of agroforestry offers the potential for synergies between mitigation programs and adaptation to climate change. In many instances, improved agroforestry systems can reduce the vulnerability of small-scale farmers to inter-annual climate variability and help them adapt to changing condition. (UNFCCC. FCCC/TP/2008/8, p. 23)

The same technical paper affirms that the adoption of agroforestry practices can contribute to biodiversity and wildlife habitat promotion, climate change adaptation and poverty reduction (UNFCCC. FCCC/TP/2008/8, p. 48). Furthermore, a CGIAR and CCAFS paper, supplement to the UNFCCC NAP Technical Guidelines, about the ten best innovations for agriculture, acknowledges the essential role of agroforestry in diversifying farms and enhance resilience (Dinesh et al. 2017).

Agroforestry's potential has also been recognized by the United Nations Convention to Combat Desertification (UNCCD). Parties to the Convention stated the object of the Asian Thematic Program Network on Agroforestry and Soil Conservation as the promotion of agroforestry and soil conservation in Asia in the context of combating desertification and mitigating the effects of drought through enhancing local, national, sub-regional regional and international cooperation (UNCCD. Decision New Delhi, India, 15/03/2000).

FAO have equally recognized agroforestry as a powerful tool to globally promote climate change mitigation and adaptation, because «when applied strategically on a large scale agroforestry enables agricultural lands to withstand weather events, such as floods and drought, and climate change» (Buttoud 2013, vii). Additionally, National Adaptation Plans of Action (NAPAs) and Nationally Appropriate Mitigation Actions (NAMAs) consider agroforestry as an important component in agricultural sector actions (Buttoud 2013, p. 4).

In 2012, a Memorandum of Understanding MoU was signed between Biodiversity International, the International Center for Tropical Agriculture (CIAT), the Center for International Forestry Research (CIFOR) and the International Centre for Research Agroforestry (ICRAF) as partners in the CGIAR Research Program on Forests, Trees and Agroforestry: Livelihoods, Landscapes and Governance, together with the CBD Secretariat to make it easier to implement activities related to the research program and implementation of the CBD Program of Work on Forest Biodiversity under the Strategic Plan for Biodiversity 2011–2020 and the Aichi Targets, specially 3, 5, 7, 14, 15 and 18 for the period 2012–2016 (CGIAR and CBD, 2012).

The agreement aims to improve the livelihoods of smallholders by supporting and improving ecosystem services through the development and application of knowledge on the use of trees to diversify farming systems. It has also recognized agroforestry as a practice that improves different aspects of agriculture, reconciling livestock, crop farming and forestry, and represents an alternative to the problems of environmental degradation and loss of soil productivity. Associated with ordinary agricultural practices. The benefits of adopting agroforestry practices include increasing agricultural and forest biodiversity, diversification of food production and food security (CGIAR and CBD, 2012).

In 2016, Decision 3 of CBD COP 13 also affirmed agroforestry's potential to help achieve the Aichi Biodiversity Targets. The 20 Aichi Biodiversity Targets were adopted within the CBD's Strategic Plan for Biodiversity for the period from 2011 until 2020. Regarding the fight against climate change, the Aichi Target 15 is aimed at ecosystem resilience enhancement and biodiversity conservation as carbon stocks to contribue to climate change mitigation and adaptation and to combat desertification.

Agroforestry's potential to fight climate change has been internationally recognized over the years. This potential has been both constrained by adverse national policies and harnessed by innovative policies.

4   Constrains and Opportunities

Agroforestry systems encompass complex practices and if not managed correctly they may not fulfill its potential of sustainability and climate change mitigation. They could be constrained by local customs, institutions and national policies, lack of information and technical assistance, such as capacity building, extension and research programmes.

FAO's Agroforestry Working Paper identifies some major obstacles to the adoption of successful agroforestry practices. Such as (Buttoud 2013):

(1) agricultural policies can discourage farmers from practicing agroforestry and incentives are often given to commercial agriculture;
(2) perverse subsidies are often provided to specific inputs associated with commercial agriculture like fertilizer, inducing higher usage of fertilizers whilst simultaneously discouraging the adoption of more sustainable practices, such as agroforestry;
(3) markets for the produce derived from agroforestry are often underdeveloped;
(4) there is often a lack of appreciation of the advantages of agroforestry;
(5) the legal status of land and tree resources is often unclear;
(6) forestry legislation can constrain tree growing on farms through restricting or prohibiting harvesting, cutting or selling activities;
(7) restrictions placed on multifunctional land management and entangled taxation frameworks can inhibit agroforestry development;
(8) lack of coordination between relevant government sectors, coupled with policy conflicts and omissions, create gaps or adverse incentives;
(9) problems in the germ plasm sector often frustrate the implementation of agroforestry practices. The same working paper list a wide range of solutions to overcome these obstacles:

(1) policy support and financial incentives: policies must eliminate legal and institutional constraints on agroforestry and support positive outcomes;
(2) certification of wood products and better integration into the carbon market;
(3) clear and secure property rights: this accords farmers the necessary confidence to make long-term investments;
(4) community-based forest management: enabling local communities to organize themselves and develop their own rules to regulate agroforestry initiatives;
(5) inter-sectoral coordination: between the environmental, agricultural and forestry sectors through the creation of consultative bodies, adoption of cross-sectoral strategies to foster collaboration; and the establishment of participatory approaches, including wide stakeholder consultation and decentralized governance structures;
(6) strengthen farmers' access to markets regarding tree products;
(7) accountable and transparent decision-making.

Financial incentives granted through government subsidies or through private contracts are a very important tool to promote agroforestry systems, since its implementation or the transition from conventional agriculture to agroforestry may be costly.

Agroforestry practices could be the subject of various market-based instruments, such as REDD+.[3] The REDD+ is an incentive developed under the UNFCCC to financially reward developing countries for their greenhouse gas emission reductions from deforestation and forest degradation, conservation of forest carbon stocks, sustainable management of forests and increase of forest carbon stocks. (Proforest 2014). CGIAR and CCAFS highlight that agroforestry can contribute to REDD+ as part of REDD+ itself depending on the definition of forest in a given country; and as part of a strategy for achieving REDD+ in landscapes (Minang et al. 2011). According to Minang et al. (2013), «agroforestry (the deliberate management of trees on farms) is not explicitly mentioned as part of REDD+ or any current United Nations Framework Convention for Climate Change (UNFCCC) Mechanism. However, considering the UNFCCC definition of forest, a great deal of existing agroforestry systems worldwide could qualify to be an integral part of a REDD+ mechanism.»

There is not yet a formal international mechanism for REDD implementation, but several countries are implementing voluntary REDD projects, such as: Envirotrade's Sofala Community Carbon Project in Southern Africa and offset credits are available from both the Agroforestry and REDD activities (CBD 2010) and the reforestation-based carbon-offset project on the collective lands of Ipetí-Emberá (Panama) where mixed-species agroforests or timber-only plots were established by a subset of community members under voluntary carbon-offset agreements with a private client (Holmes et al. 2017). According to the ASB Partnership for the Tropical Forest Margins, «countries should consider giving agroforestry a special place in the REDD+ and NAMAs strategies given the great potential benefits from emission reductions as well as the biodiversity and livelihoods benefits that can be generated» (ASB 2011).

---

[3]Reducing emissions from deforestation and forest degradation.

In Brazil, Environmental Reserve Quotas (CRA) were regulated in 2008, allowing rural property owners to negotiate forest assets, including the intercropping of native and exotic species, in an agroforestry system (Decree 9.640/2018), with the condition of the vegetation cover is not decharacterised and does not impair the conservation of the native vegetation (article 22, I, of the Brazilian Forest Code). A CRA buying and selling market is being gradually established. The state of Rio de Janeiro, for example, has created an environmental stock market that allows, through market operations, producers and rural owners to earn money by preserving the native vegetation of their properties (Oeco 2015). The negotiation of environmental quotas promotes not only the preservation of natural resources, but also supports the transition to a low carbon economy.

Another successful financial initiative has been adopted in Vietnam. The Agroforestry Partnership Fund invests in smallholder producers in Vietnam to help improve their efficiency, a kind of "climate-smart" investment fund that aims to reduce greenhouse gas emissions, conserve important natural resources, assure food security and improve rural livelihoods (UNFCC).

5 Conclusion

The adoption of the SDG is a clear effort to set the blueprint to achieve a better and more sustainable future for all and the fight against climate change is a core goal.

Regarding agriculture, it is both affected by climate change and has considerable influence on it, but some agro-ecological practices are considered to be a good tool to fight the phenomenon. In this sense, climate smart agriculture approaches are been developed to «transform and reorient agricultural systems to effectively support development and ensure food security in a changing climate» (FAO, Climate-Smart Agriculture). Agroforestry has been internationally recognized as a CSA approach. At the farm level, agroforestry combine mitigation and adaptation to enhance the socio-environmental resilience to natural hazards. Besides promoting climate change adaptation and mitigation agroforestry is able to ensure food security, having great potential to provide several ecosystem services essential to both man and nature.

However, in order to be sustainable, it requires learning of advanced cultivation methods and support to ensure its adoption. Furthermore, agroforestry implementation may be constrained by local customs, institutions and national policies, lack of information and technical assistance, such as capacity building, extension and research programmes. Nevertheless, some important key insights have been presented to overcome these obstacles, such as market based instruments and government support. REDD projects, PES schemes and investments funds are been successfully implemented worldwide in order to promote agroforestry's adoption as a strong instrument to address climate change.

References

Bizikova L, Larkin P, Mitchell S, Waldick R (2019) An indicator set to track resilience to climate change in agriculture: a policy-maker's perspective. Land Use Policy 82:444–456. https://www.sciencedirect.com/science/article/pii/S0264837717312358. Last accessed 22 Feb 2019

Buttoud G (2013) Agroforestry working paper n. 1. Advancing agroforestry on the policy agenda. FAO. http://www.fao.org/3/a-i3182e.pdf. Last accessed 23 Feb 2019

CBD (2010) REDD plus and biodiversity e-newsletter. vol. 12. https://www.cbd.int/forest/newsletters/redd-12.htm. Last accessed 23 Feb 2019

CGIAR, CCAFS (2018) Climate-smart agriculture guide. Retrieved from https://csa.guide. Last accessed 23 Feb 2019

Clarke WC, Thaman RR (1993) Agroforestry in the pacific islands: systems for sustainability. United Nations University Press, Tokyo, New York, Paris

Dinesh D, Campbell B, Bonilla-Findji O, Richards M (eds) (2017) 10 best bet innovations for adaptation in agriculture: a supplement to the UNFCCC NAP Technical Guidelines. CCAFS Working Paper no. 215. Wageningen, The Netherlands: CGIAR Research Program on Climate Change, Agriculture and Food Security (CCAFS). www.ccafs.cgiar.org. Last accessed 22 Feb 2019

FAO (2018) Climate Smart Agriculture Sourcebook. http://www.fao.org/climate-smart-agriculture-sourcebook/concept/module-a1-introducing-csa/chapter-a1-2/en/. Last accessed 24 Feb 2019

FAO Climate-Smart agriculture. http://www.fao.org/climate-smart-agriculture/en/. Last accessed 24 Feb 2019

Holmes I, Kirby KR, Potvin (2017) Agroforestry within REDD+: experiences of an indigenous Emberá community in Panama. Agroforestry systems, Springer 91(6):1181–1197

King KF, Chandler NT (1978) The wasted lands: the program of work of the International Council for Research in Agroforestry (ICRAF). Nairobi, Kenya

Mbow C, Smith P, Skole D, Duguma L, Bustamante M (2014) Achieving mitigation and adaptation to climate change through sustainable agroforestry practices in Africa. Curr Opinion Environ Sustain 6:8–14

Minang PA, Bernard F, van Noordwijk M, Kahurani E (2011) Agroforestry in REDD+: opportunities and challenges. ASB Policy Brief 26. ASB Partnership for the Tropical Forest Margins, Nairobi, Kenya, http://www.asb.cgiar.org/PDFwebdocs/ASB_PB26.pdf. Last accessed 22 Feb 2019

Newaj R, Chaturvedi OP, Handa AK (2016) Recent development in agroforestry research and its role in climate change adaptation and mitigation. Indian J. Agroforestry 18(1):1–9. https://www.researchgate.net/publication/304879691_Recent_development_in_agroforestry_research_and_its_role_in_climate_change_adaptation_and_mitigation. Last accessed 23 Feb 2019

Oeco (2015) O que são Cotas de Reserva Ambiental (CRAs).https://www.oeco.org.br/dicionario-ambiental/28921-o-que-sao-cotas-de-reserva-ambiental-cras/. Last accessed 25 Feb 2019

Proforest (2014) Interactions FLEGT-REDD+ Information note 12. http://www.euredd.efi.int/documents/15552/154912/Linking+FLEGT+and+REDD%2B/7152b991-8ae6-4c8a-8679-02c1fbb1765e. Last accessed 24 Feb 2019

UN Sustainable Développement Goals. https://www.un.org/sustainabledevelopment/sustainable-development-goals/. Last accessed 24 Feb 2019

UNCCD (2000) Decision New Delhi, India. http://www.unccd.int/Lists/SiteDocumentLibrary/Regions/Asia/meetings/regional/TPN2_3_2000/decision.pdf. Last accessed 22 Feb 2019

UNFCCC (2008) FCCC/TP/2008/8 Technical Paper. Challenges and opportunities for mitigation in the agricultural sector.http://unfccc.int/resource/docs/2008/tp/08.pdf. Last accessed 22 Feb 2019

UNFCC. Agroforestry Partnership Fund. https://unfccc.int/climate-action/momentum-for-change/activity-database/momentum-for-change-agroforestry-partnership-fund Last accessed 25 Feb 2019

Uthappa et a. (2017) Agroforestry-A sustainable solution to address climate change challenges In: Gupta SK, Panwar P, Kaushal R (eds) Agroforestry for increased production and livelihood security. Chapter: 1, pp 1–22. https://www.researchgate.net/publication/313011619_Agroforestry-_A_Sustainable_Solution_to_Address_Climate_Change_Challenges. Last accessed 23 Feb 2019

Marcia Fajardo Cavalcanti de Albuquerque received her Law Degree from de University Center of Brasilia, in 2009. In 2011, she received her «Diplôme d'Université» in Sustainable development and land management from the University of Montpellier 1. In 2012, she received her Masters Degree in Environmental law from the Universities of Paris 1 and 2-Panthéon/Sorbonne. Still in 2012, Marcia was an intern at IUCN Environmental Law Center. Currently, Marcia is a doctoral student under a co-supervision regime between the University of Paris 1-Sorbonne and Mackenzie University of São Paulo. Marcia's research is focussed on law and agroecokogy.

Organic Foods in Brazil: A Bibliometric Study of Academic Researches Related to Organic Production Before and After the Conceptual Law

Juliana Fatima de Moraes Hernandes and Viviane Beraldo Rosolen

Abstract In Brazil, a series of activities and public policies aimed at regulating the production of organic food have been observed in the last decades, as well as an increase in the farmers' interest to adopt this form of production. In 2003 the brasilian Law no. 10.831 de 23 de dezembro de 2003 defined organic agriculture. Therefore, this article has as general objective to verify the behavior of academic production regarding the topic of organic food and four related terms, which are food security, food education, sustainability and agroecology before and after the year 2003. For this purpose, the volume of scientific production was searched in the portal of scientific articles Web of Science and Biblioteca Digital Brasileira de Teses e Dissertações—BDTD, in the fifteen years before and after the enactment of the afore mentioned law. It was verified that the previous production was not very expressive for all the terms. In the years after 2003, there was a significant increase in the number of publications in all that terms and in their combinations. Based on the data it was observed that there was a simultaneous growth of publications and production and demand for organic products, after the year 2003, in Brazil.

Keywords Organic foods · Researches in organic foods · Bibliometric studies

1 Introduction

Organic agriculture is gaining ground every year, both in the fields of production and cultivation, as well as in consumer demand. In Brazil, Law no. 10831/2003 defines organic agriculture as being

> (…) all those in which specific techniques are adopted, optimizing the use of available natural and socio-economic resources and respect for the cultural integrity of rural communities,

J. F. de Moraes Hernandes (✉) · V. B. Rosolen
School of Business Administration, State University of Londrina, Rod. Celso Garcia Cid, Km 380, Londrina, Brazil
e-mail: hernandes.sul@gmail.com

V. B. Rosolen
e-mail: viviane_brosolen@yahoo.com.br

© Springer Nature Switzerland AG 2020
W. Leal Filho et al. (eds.), *International Business, Trade and Institutional Sustainability*, World Sustainability Series,
https://doi.org/10.1007/978-3-030-26759-9_28

aiming at economic and ecological sustainability, maximizing social benefits, the minimization of non-renewable energy dependence, using as far as possible cultural, biological and
mechanical methods, as opposed to the use of synthetic materials, the elimination of the
use of genetically modified organisms and ionizing radiation at any stage of the production,
processing, storage, distribution and marketing, and the protection of the environment.

In this type of production is adopted practices that aim at the preservation and
responsible use of soil, water and air, in order to reduce the forms of contamination
and waste of natural resources, which results in a more sustainable system. In addition to minimizing the impacts to the environment, using organic inputs, the use of
agrochemicals and genetically modified organisms, for which scientific clarity is not
yet available, is ruled out. From the text of the law, presented above, five more constructs were extracted, as well as organic foods, which were used as themes related to
this work, being: sustainability, food safety, food education, Agroecology and food
sovereignty. In view of this scenario, we ask: in view of the growth of organic production in Brazil, has academic scientific production increased and been significant
on organic agriculture? Is there scientific academic production of related topics such
as food security, food sovereignty, food education, sustainability and Agroecology?

Thus, this work will have as a general objective to verify the behavior of academic
production regarding the topic of organic and other related topics, described in the
question above. The research was based on the year 2003, when the concept of organic
agriculture was defined by Law no. 10831, dated December 23, 2003, cited above. To
reach this objective, two databases of relevance for scientific publications in Brazil
will be consulted: the Web of Science portal and the Brazilian Digital Library of
Theses and Dissertations (BDTD).

2   What Does Literature Tell Us?

(a) The production of organic—In Brazil the literature on organic foods has been
     growing in the last decades. The analysis of the academic production frame-
     work on organic foods makes possible the characterization of the support that
     the university and the scientific knowledge have exerted for the evolution and
     development of this field.

The understanding of research activity as a dynamic, social and historical process—and not only as a resource for the improvement of scientific doing—is evidenced in Lloyd (1995, p. 25) arguments, when he considers that a scientific domain
is a resource for theoretically constituting objects as well as "to incorporate and do
justice to the history of science in its accumulation of knowledge."

As for the commercialization and market, Dias et al. (2015) noted that commercialization of certified organic agriculture (OA) has spread to over 130 countries
worldwide and the demand for organic products is driven by the belief that they are
healthier, tastier and more environmentally friendly than conventional products.

The Ministry of Agriculture, Livestock and Food Supply states that the area destined to organic production in Brazil surpassed 750 thousand hectares in 2016, a growth of 15% compared to 2015. In just three years, the number of organic production units jumped from 6700 to 15,700—that is, between 2013 and 2016, more than double the growth of this type of planting in Brazil. In the ranking of the regions that produce the most organic food, the Southeast comes first, totaling 333 thousand hectares and 2729 records of producers in the National Register of Organic Producers (CNPO). Following, the North (158 thousand hectares), Northeast (118.4 thousand), Midwest (101.8 thousand) and South (37.6 thousand) regions. 75% of the CNPO registered farmers were family farmers. In 2016, the organic market earned more than RS 3 billion in domestic market, besides RS 145 million in exports. It is estimated today that Brazil is the largest producer of organic rice in Latin America.

Within Latin America, Chile, which signed an agreement with Brazil this year for the organic food trade, presented in 2016 an area of 684 552 thousand hectares of organic production (ODEPA 2018). For Argentina, in the year 2016, the data presented a harvest area of 83,754 ha, reaching this figure a maximum in production since the beginning of the organic activity in this country (SENASA 2017).

In the United States of America organic production in 2016 earned the figure of $ 7.6 billion in certified organic products (NASS), US Department of Agriculture (USDA) 2017). In Europe, for its part, in 2016 the market reached US $ 35 billion, with Germany and France being the countries with the most prominence in production. Segundo Dias et al. (2015)

> The rise of the market for natural and organic products follows a worldwide trend of increasing demand for products and services that provide health and well-being. Added to this is the growing distrust of some sectors of society over modern industry, which has brought a number of facilities to everyday life, but has also significantly increased the handling of persistent chemicals in the environment, with serious consequences for human health and for natural ecosystems.

In Brazil, organic production is closely linked to family agriculture and, in turn, the development of family organic agriculture in Brazil is directly linked to the National Policy on Agroecology and Organic Production (PNAPO). PNAPO was launched by the Federal Government, with the issuance of decree 7794, of August 20, 2012, as an important step for the expansion and implementation of actions to promote sustainable rural development. One of the main instruments of this policy is the National Agroecology and Organic Production Plan (PLANAPO), also known as Agroecological Brazil. The first PLANAPO ended in 2015 and benefited 678 449 family farmers, organic producers, indigenous peoples and traditional peoples and communities, technicians and extensionists. In 2016 a new planning cycle for this theme began, with the launch of PLANAPO 2016–2019. In order to assist organic food producers in the country, the Ministry of Agriculture still has the Organic Production Commissions (CPOrgs), spread by numerous units of the Federation, which are responsible for coordinating actions to promote organic agriculture. In all, they are made up of 578 public and private entities.

There is a series of activities and public policies in Brazil aimed at consolidating organic production, as well as an increase in the willingness of farmers to adopt this form of production.

(b) Approach of related topics—In the scope of this research, having as presupposition the Law no. 10831/2003, which conceptualizes what are organic products in Brazil, some issues related to the "organic" construct are listed. These themes are the constructs of food security, food sovereignty, sustainability, food education and Agroecology. It is understood that these constructs, are present between the lines of the Law in question and that, therefore, verify the academic production on the construct "organic", completes itself verifying what the Academy has produced on these related subjects. At first, then, it is necessary to understand and conceptualize each of these themes. Beginning with food security, we have that hunger on the global stage began to receive more attention after the First World War, when many European economies were deeply shaken. However, it was from the Second World War, with the worsening of the problem of hunger on the planet, that some measures began to be taken to get around the problem. One of the most important measures was the creation of the Food and Agriculture Organization of the United Nations (FAO) in 1945 as the main initiative for the elaboration and planning of strategies to combat hunger worldwide, 2014). Since its inception, FAO has been raising discussions on hunger, drawing attention primarily to issues related to food and nutritional security and food sovereignty. With the emergence of these concepts, there was also a need for clarification as to its meaning. Food security is defined by the FAO as the production or purchase of food sufficient to meet the daily needs of the people, so as to enable them to lead an active and healthy life. Already, food sovereignty is defined as a national strategy of the countries, once the importance of self-sufficiency of food has been perceived, especially in times of international conflicts. (Maluf and Menezes 2015).

With the strengthening of the concepts of food security and sovereignty, the need arose for the establishment of a food education. Food education also comprises a strategy that reinforces the concepts of food security and sovereignty, acting as a field of knowledge aimed at promoting the autonomous and voluntary practice of healthy eating habits. With this, some challenges arise regarding different food cultures, the strengthening of regional habits, the reduction of food waste and the dimensions related to sustainability (MDS—Ministry of Social Development 2018).

These dimensions of sustainability have begun to gain importance mainly with the aggravation of concerns about sustainable development. In this sense, sustainability, in a logical sense, refers to the capacity of something to sustain or to remain forever (Mikhailova 2004). This means that, using natural resources as an example, they could be used, but in a way and speed that would allow their renewal and not cause their exhaustion. In view of this, together with the challenge related to food education to strengthen regional habits, the alternative forms of agriculture began to receive more attention, with emphasis on Agroecology, the last subject related to organic, which will be discussed in this paper.

Agroecology "represents a way of approaching agriculture that incorporates special care related to the environment, social problems and the ecological sustainability of production systems" (Moreira and do Carmo 2004, p. 45).

Once the concepts of the themes of importance for this study have been defined, we also discuss their close relationship with the Brazilian Law, mentioned above, which conceptualizes organic production. Based on the understanding of food sovereignty as a strategy to have food, it is verified in the Law in question, indications of procedure for the acquisition of organic food and this respecting natural resources and maximizing the benefits to the populations. This also means applying the concept of sustainability. When the Act refers to the elimination of the use of "genetically modified organisms and ionizing radiation at any stage of the process of production, processing, storage, distribution and marketing, and the protection of the environment" draws attention to a need for education food and all its contents suggests a concern for food security, understood as the provision of food in order to provide an active and healthy life.

With this, it is understood the conceptualization and meaning of the constructs related to the organic theme, present in the Law that defines and regulates the activity in Brazil, and seeks to understand the relevance of these themes to the activity itself, as well as the relational process in each case.

(c) Knowledge management, a support theory—To understand how a market field develops, it is interesting to observe how academic-level research on the desired topic and related topics is being developed. According to Rodrigues (2011)

> The growing importance of knowledge in organizational models suggests not only that it is an asset that can be stored, retrieved and transferred to third parties in an eyedropper, but also that it is increasingly possible to transform intangible knowledge into specific products

Knowledge has been gaining increasing attention in organizations, since it represents a strategic character in obtaining results. In this way, its management has been analyzed in order to understand how to use the knowledge in favor of the achievement of the organizational objectives.

According to Nonaka (2008), knowledge is traditionally defined as "justified true belief". It differs from information for three reasons: Knowledge first concerns beliefs and commitments, unlike information; second, it relates to action, different from information and; third, "knowledge as information, refers to meaning. It is context-specific and relational. "The authors also affirm that knowledge comes from two dimensions: the epistemological and the ontological. The first one covers tacit and explicit knowledge, the first being personal, specific and contextual, and the second one transmissible through formal and systematic language. The ontological dimension concerns what is created by the individual, emphasizing that "an organization can not create knowledge without individuals".

Nonaka (2008) stated that the secret of the success of the most successful Japanese companies lies in managing the creation of new knowledge. This specific case presents evidence of the importance of knowledge for organizational success, which may indicate a path to the formulation of market-oriented strategies.

According to Lincório et al. (2013), the knowledge management concept presented by American Productivity and Quality Center (2006), is characterized by the "set of strategies and processes for identifying, and leveraging knowledge to enhance competitiveness". As can be observed, the author links business competitiveness to knowledge management, addressing "knowledge as a competitive differential of organizations" (Lincório et al. 2013, p. 6).

Terra (2005), through empirical research, addresses seven dimensions of organizational practice for the understanding of Knowledge Management, namely: Strategic factors and the role of top management; Culture and organizational values; Organizational structure; Human resource Management; Information systems; Measurement of results and; Learning with the environment. The author emphasizes that underlying most of these dimensions is human capital, formed by individual and organizational values and norms, as well as by the skills and attitudes of each employee, is the driving force behind the generation of knowledge and value in companies (Terra 2005).

In this way, one can observe the relevance of knowledge management understanding to analyze the functioning of organizations and markets. Or turn, academic production can give subsidy to the development of the market. Once there is research on organic foods, the market can seize these contents to develop strategies, which would help boost their production and consumption.

3   Materials and Methods

This article is characterized as a bibliometric, quantitative research and in order to reach the objective it proposes, two databases of relevance in terms of scientific publication were consulted: the Web of Science portal and the Brazilian Digital Library of Theses and Dissertations Portal (BDTD). Portals were searched for publications in the general search database and title search. The search was done in two phases: (i) the first phase looked for publications with the name of the individual constructs. (ii) the second phase sought the two-to-two combination of each construct, three to three and a combination of six.

The period of analysis was divided into two time series: (a) From 1988 to 2002 (b) 2003 to 2017. It was decided to break the first series in the year 2002 due to the validity in 2003 of Law no. 10831/2003 that defines and defines the organic activity in the country. To better understand the form of the research, the following Table 1 was set up:

Table 1 Bibliometric indicators used

Elementos da Pesquisa	Indicador
Elements of research	Indicator
Portals	Web of Science Brazilian Digital Library of Theses and Dissertations
Filters	Item Search Title
Publication characteristic	Series Constructs of Interest

Source Prepared by the authors (2018)

4 Results and Discussion

The search in the portals chosen by this research showed two well-defined and different time series in terms of the number of publications. The first time series, from 1988 to 2002, in the general search and by title shows expressive results for the constructs in question, as we can see in Graphs 1 and 2.

For the combination between the main theme and the other related themes the volume of publications is inexpressive and in many cases of combination, it is null, within the time series 1988–2002, as shown in Fig. 2, below. For the purpose of legend the constructs are determined as (1) Sustainability; (2) Food security; (3) Nutrition Education; (4) Organics Food/"Organic Foods"; (5) Agroecology/(6) Food Sovereignty/"Food Sovereignty".

The 2003–2017 time series, however, shows more robust publication data within the same search pattern, that is, general and title search. Even with respect to the related researched, of a combination of the constructs, the results are more expressive,

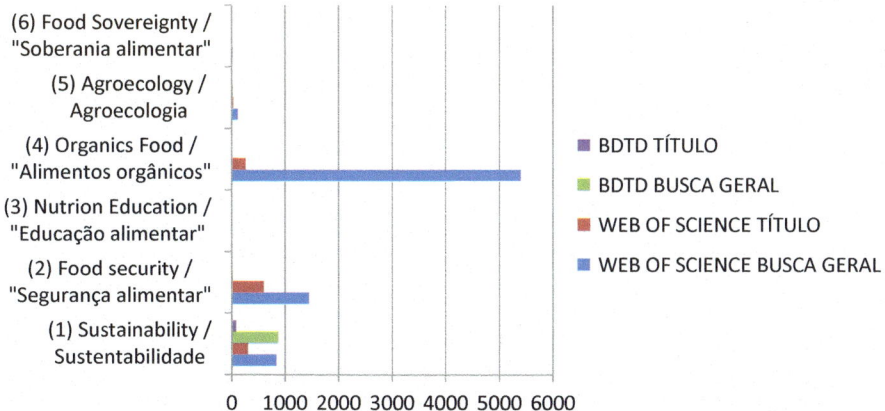

Graph 1 Number of publications of the main theme and related, in general search and by title, in the time series 1988–2002. *Source* Prepared by the authors (2018)

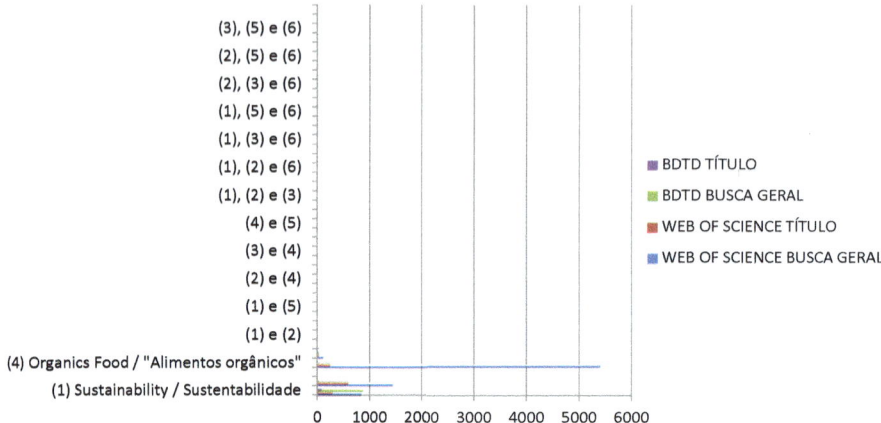

Graph 2 Number of publications with the combinations between main theme and related, in search of gral and by title, in the time series 1988–2002. *Source* Prepared by the authors (2018)

within the portal of the Brazilian Digital Library of Theses and Dissertations, as shown in Graphs 3 and 4.

For the purposes of legend, the constructs continue to be determined as (1) Sustainability; (2) Food security; (3) Nutrition Education; (4) Organics Food/"Organic Foods"; (5) Agroecology/(6) Food Sovereignty/"Food Sovereignty".

A total of 9997 publications regarding the first time series were found using the constructs organic food, food security, food sovereignty, sustainability, food education and Agroecology, adding the results of the search in the "general search" and "title" items in the portals used in research. In this same pattern, the result for the second time series was 197 043 publications. We thus see a significant increase in the number of publications in the second time series, i.e. in the period 2003–2017.

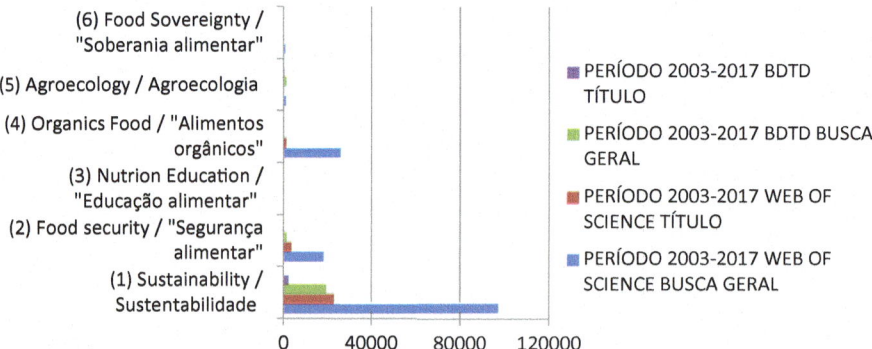

Graph 3 Number of publications of the main theme and related, in general search and by title, in the time series 2003–2017. *Source* Prepared by the authors (2018)

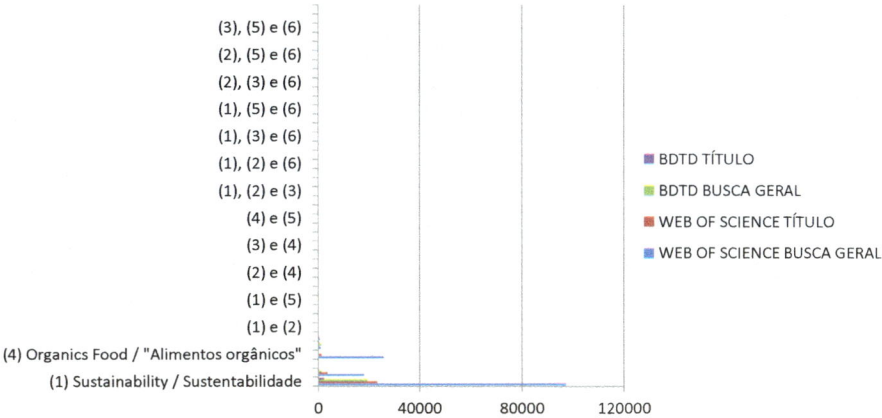

Graph 4 Number of publications of the combinations between main theme and related, in search of gral and by title, in the time series 2003–2017. *Source* Prepared by the authors (2018)

The data found showed the classification of articles in order of relevance within the entire time series of this research, i.e., the period from 1988 to 2017, indexed in the Web of Science portal. The ten most relevant articles are listed, in order, in Table 2.

The results show variety of research within the theme of "organic food". There are articles that present their object of study directly related to organic foods, but there is also an article on organic market, on organic production and the relation of products with health, among others. Hence we can infer that there is variety in the field of research, both on the variations of the topic and in the areas of knowledge with interest in the research.

It was also sought to verify, in order of relevance, the first ten theses and dissertations of the portal of the Brazilian Digital Library of Theses and Dissertations—BDTD according to Table 3.

The results show a higher presence of dissertations X theses and points to a role of the Federal University of Santa Catarina, in the number of publications on the topic listed among the ten most relevant.

Thus, this research verified that in view of the growth of organic production in Brazil, academic scientific production increased since 2001 and has been expressive on the subject of organic agriculture and also on related topics, food security, food sovereignty, education food, sustainability and Agroecology.

It is not possible to infer the direct relation between market increase and increase of academic production. The data presented here only show that the growth of publications increased significantly, from 203, the year in which Law no. 10831 that defines and conceptualizes organic in Brazil came into force.

In a next moment of this work one can look for a more direct relation on the influence of the Academy in the Market and vice versa.

Table 2 The ten articles on "organic food" of greater relevance published between 1988 and 2017 in the Web of Science portal

Article	Author	Source
Organic Baby Food	BEAL, Judy A	The American Journal of Maternal Child Nursing
Predicting intentions to purchase organic food: the moderating effect as of organic food prices	LIANG, Rong-Da	British Food Journal
Organic Food Market in Romania	SPERDEA, Natalia Maria; VLADU, Marius	SGEM
Organic Foods Come Up Rosy	ANONYMOUS	Science
Food Safety and Organic Meats	VAN LOO, Ellen J.; ALALI, Walid; RICKE, Steven C.	Annual Review of Food Science and Technology
Organic Food: Panacea for Health?	ANONYMOUS	LANCET
Organic Acids as Food antimicrobials	GURTLER, Joshua	Abstracts of Papers of the American Chemical Society
Food & Agriculture Tropical Organic Farming	BURKE, Maria	Chemistry & Industry
Organic Food Awareness in Turkey	DERMITAS, Bekir; PARLAKAY, Oguz; TAPKI, Nuran	Emirates Journal of Food and Agriculture
Consumers Attitude Towards Organic Food	BASHA, Mohamed Bilial; MASON, Cordelia; SHAMSUDIN, MiladAbdelnabi	International Accounting and Business Conference

Source Prepared by the author

5   Conclusion

In the search for the panorama of publications and observation compared to 2003, there was a significant academic production in the portals surveyed, mainly in the period between 2003 and 2017.

The articles related to studies on the sustainability of organic production, the profile of the consumer and on the market of these products, appear as the themes of greatest number of publication in the universe researched.

Organic agriculture and Agroecology have an essential approach linked to a more environmentally friendly and more sustainable production.

Among the publications we must highlight the high number of theses and dissertations on the constructs of this research in its temporal universe. There are 23 058 theses and dissertations on all constructs.

There is thus an academic interest in the subject and it is hoped that intellectual knowledge can serve as a basis for a practical application in society.

Table 3 The ten theses/Dissertations on "organic foods", of greater relevance published in the period 1988–2017 in the portal of the Brazilian Digital Library of Theses and Dissertations—BDTD

THESIS (T)/DISSERTATION(D)	Author	Institution of teaching
Alimentos orgânicos na alimentação escolar (D)	Silverio, Gabriela de Andrade	Universidade Federal de Santa Catarina—UFSC
Supermercados e alimentos orgânicos no Brasil (D)	Guivant, Julia Silvia	Universidade Federal de Santa Catarina—UFSC
Demanda domiciliar por alimentos orgânicos no Brasil (D)	Ferreira, Alberes Sousa	Universidade Federal de Viçosa—UFV
Disponibilidade domiciliar de alimentos orgânicos no Brasil (D)	EdinéiaDottiMooz	Universiade Estadual de São Paulo—USP
Organismos geneticamente modificados e alimentos transgênicos (D)	Caus, Cesar Antônio	Universidade Federal de Santa Catarina—UFSC
Alimentos e funções orgânicas: uma situação de estudo (D)	Vescovi, Elisete Coser	Universidade do Vale do Taquari—UNIVATES
Na teia do alimento orgânico no Espírito Santo (D)	GOUVEIA, K. A. N.	Universidade Federal do Espírito Santo
Estratégias de marketing junto ao mercado de consumo para aquisição de alimentos orgânicos (D)	Nava, Evandro Jackson Redivo	Universidade Federal de Santa Catarina—UFSC
Motivações e envolvimento dos consumidores de alimentos orgânicos no Brasil (D)	BARROS, José Eduardo de Melo	Universidade Federal Rural do Pernanbuco—UFRPE
Alimentos orgânicos: perfil dos consumidores e variáveis que afetam o consumo (D)	Tavares, Victor de Souza	Universidade Federal de Viçosa—UFV

Source Prepared by the authors

References

Camara dos Deputados do Brasil (2018) PROJETO DE LEI N° 6.299, DE 2002

Conselho Nacional de Segurança Alimentar e Nutricional (2017) Produção de Orgânicos Cresce no Brasil e mostra que é um bom negócio. Recuperado em 17 outubro, 2018, de http://www4.planalto. gov.br/consea/comunicacao/noticias/2017/dezembro/organicos-tambem-sao-um-bom-negocio

Dias VV, Schultz G, Schuster MS, Talamini E, Revillion JP (2015) Ambiente & Sociedade. O Mercado de Alimentos Orgânicos: um Panorama Quantitativo e Qualitativo das Publicações Internacionais, 18:161–182. Recuperado em 17 outubro, 2018, de http://www.scielo.br/pdf/asoc/ v18n1/pt_1414-753X-asoc-18-01-00155.pdf

IPEA—Instituto de Pesquisa Econômica Aplicada (2014) A trajetória histórica da Segurança Alimentar e Nutricional na agenda política nacional: projetos, descontinuidades e consolidação. Rio de Janeiro. Recuperado em 17 outubro, 2018, de http://repositorio.ipea.gov.br/bitstream/11058/3019/1/TD_1953.pdf

Licorio AMO, Siena O, Almeida MRA (2013) Gestão do conhecimento: Análise Bibliométrica de Produção Científica no Período de 1990 a 2012. Convibra. Recuperado em 17 novembro, 2018, de http://www.convibra.org/upload/paper/2013/31/2013_31_6950.pdf

Lloyd C (1995) As estruturas da história. Zahar

MDS—inistério do Desenvolvimento Social (2018) Princípios e Práticas para Educação Alimentar e Nutricional. Recuperado em 17 outubro, 2018, de https://www.mds.gov.br/webarquivos/arquivo/seguranca_alimentar/caisan/Publicacao/Educacao_Alimentar_Nutricional/21_Principios_Praticas_para_EAN.pdf

Maluf RS, Menezes F (2015) *Caderno* "Segurança Alimentar". Recuperado em 17 outubro, 2018, de https://www.researchgate.net/publication/266884132_Caderno_'Seguranca_Alimentar'

Mikhailova I (2004). Sustentabilidade: Evolução dos conceitos teóricos e os problemas da mensuração prática. Revista Economia e Desenvolvimento, n.16. Recuperado em 17 outubro, 2018, de http://w3.ufsm.br/depcie/arquivos/artigo/ii_sustentabilidade.pdf

Moreira RM, Do Carmo MS (2004) Agroecologia na construção do Desenvolvimnto Rural Sustentável. Agric. São Paulo. São Paulo, v. 51, 2, 37–56. Recuperado em 17 outubro, 2018, de http://www.iea.sp.gov.br/out/publicacoes/pdf/asp-2-04-4.pdf

NASS—United States of America. S. National Agricultural Statistics Service; USDA—U.S. Department of Agriculture (2017) Os números da produção orgânica nos EUA em 2016. Recuperado em 18 outubro, 2018, de http://www.organicsnet.com.br/2017/10/os-numeros-da-producao-organica-nos-eua-em-2016/

Nonaka I (2008) A empresa criadora de conhecimento. In: Takeuchi H, Nonaka I (eds) Gestão do conhecimento. Tradução Ana Thorell. Porto Alegre, Bockman

ODEPA—Oficina de Estudios y Políticas Agrarias do Chile (2018) Estadísticas Productivas. Recuperado em 18 outubro, 2018, de https://www.odepa.gob.cl/estadisticas-del-sector/estadisticas-productivas

Presidência da República do Brasil. Casa Civil (2003) Subchefia para assuntos Jurídicos. Lei no. 10.831 23/12/2003. Recuperado em 17 outubro, 2018, de http://www.planalto.gov.br/ccivil_03/LEIS/2003/L10.831.htm

Recuperado em 18 outubro, 2018, de https://www.camara.gov.br/proposicoesWeb/prop_mostrarintegra?codteor=1654426&filename=Tramitacao-PL+6299/2002

Rodrigues SB (2011) De fábricas a lojas de conhecimento: as universidades e a desconstrução do conhecimento sem cliente. Gestão Estratégica do Conhecimento, Atlas

SENASA—Servicio Nacional de Sanidad y Calidad Agroalimentaria (2017) Situación de la Producción Orgánica en la Argentina durante el año 2016. Buenos Aires. Recuperado em 18 outubro, 2000, de file:///C:/Users/LG/Desktop/segurancaalimentar/200_Estadistica_SENASA_2016%20ARGENTINA.pdf

Terra JCC (2005) Gestão do conhecimento: o grande desafio empresarial. Elsevier

Global Responsible Mining and Energy

Integrating Corporate Social Responsibility and Sustainability for Deep Seabed Mining

Yao Zhou

Abstract Deep seabed mining is a high-risk activity due to its economic, environmental, social, and technical challenges. As the only international organization that has the authority to enact and enforce regulations for activities in the international seabed (Area), the International Seabed Authority (ISA) has established a number of exploration contracts with member States and their sponsored firms. These sponsoring States, including both developing and developed countries, should adhere to the same standards for ensuring responsible mining. There are trends in the linkage between corporate social responsibility (CSR) and sustainability for addressing social and environmental concerns on the global arena. Although entities that intend to conduct seabed mining can be regarded as triggers of environmental and biological changes, they are generally equipped with capacities to solve relevant problems, such as filling gaps in laws, regulations, and practices. The CSR framework can be a model for governing common resources to promote sustainability. However, the implementation of CSR varies among contractors in the Area, and how CSR can contribute to all humanity is still unknown. This chapter investigates practices of existing contractors and explores the role of corporations in sustainable development of the Area. It investigates the contract-based system, including the present status of contracts, the rights and responsibilities of contractors and sponsoring States, and a parallel system for benefit-sharing. It also provides some feasible approaches to integrating CSR and sustainability pertaining to the Area.

Yao Zhou, SJD, is a Research Assistant at the Morgridge Institute for Research in the Ethics group. She received her SJD from the University of Wisconsin Law School in the United States, and her master's degree in international law from Shanghai Jiao Tong University in China. For studying governance mechanisms on international seabed, she has conducted qualitative research interviews and ethnographic observations in Jamaica, China, and America, and presented at the International Underwater Mining Conference, the Deep-Sea Biology Symposium, and the Law and Society Association Annual Meeting. The author is thankful for comments received from presentation participants at the 2018 Midwest Law and Society Retreat and for editing help from Mariah Watts and Paulo Brito with regard to this chapter.

Y. Zhou (✉)
Morgridge Institute for Research, Madison, WI, USA
e-mail: yzhou@morgridge.org

© Springer Nature Switzerland AG 2020   499
W. Leal Filho et al. (eds.), *International Business, Trade and Institutional Sustainability*, World Sustainability Series,
https://doi.org/10.1007/978-3-030-26759-9_29

Keywords Deep seabed mining · Corporate social responsibility · Sustainability · International seabed (Area)

1 Introduction

The United Nations Convention on the Law of the Sea (UNCLOS) and the 1994 Agreement relating to the Implementation of Part XI of the Convention (1994 Agreement) established a legal regime for the international seabed (Area) governed by the International Seabed Authority (ISA). The ISA is an autonomous international organization through which the State Parties to the Convention organize and control activities in the Area. The ISA is comprised of three major organs—the Assembly, the Council, and the Secretariat—as well as two subsidiary expert groups—the Legal and Technical Commission (LTC) and the Finance Committee. The ISA is currently developing the Mining Code and has signed a number of contracts for exploring mineral resources in the Area.

Due to economic, environmental, social, and technical challenges, mining in the Area is a high-risk activity. It requires effective protection of the marine environment, human health and safety, equitable sharing of financial and other economic benefits derived from activities in the Area, and fully integrated participation of developing countries. One major challenge of deep seabed mining is the lack of adequate information about its implications. The ISA is now working on gathering information through establishing and developing databases of scientific and technical information.[1] This provides an opportunity to obtain a better understanding of the Area and to evaluate the potential impacts of mining activities.

Regardless of whether or not mining in the Area will occur, there is a need to set high standards of practice and guidelines. Several delegates to the ISA have expressed that the Mining Code cannot be completed until the accompanying standards and guidelines are developed.[2] Setting standards can promote stakeholder engagements and enhance both the competition and quality of the mining activities. As the only international organization that has the authority to enact and enforce regulations for mining in the Area, the ISA is responsible for ensuring that developers adhere to the best practices of the industry.

Although entities that intend to conduct deep seabed mining can be regarded as triggers of environmental and biological changes, they are generally equipped with capacities to address relevant problems, such as filling gaps in the legal framework. The contractors have been working with the ISA to organize a series of workshops on technical and environmental issues, provide training programs for personnel from

---

[1] Report of the Secretary-General of the International Seabed Authority under Article 166, Paragraph 4, of the United Nations Convention on the Law of the Sea (2018). Retrieved from https://www.isa.org.jm/document/isba24a2.

[2] ISA-24 Part 2 Highlights (18 July 2018). Retrieved from https://www.isa.org.jm/document/enb-iisd-bulletin-isa-24-part-2-2.

developing countries, and launch initiatives in local communities of the sponsoring States.

This chapter investigates practices of existing contractors and explores the role of corporations in sustainable development of the Area. As an emerging governance framework, corporate social responsibility (CSR) is increasingly being used as a tool for enforcement by regulators.[3] For example, the United Nations Global Compact was established to address corporate conduct built on a "social constructivist" approach (Ruggie 2017). It considers not only traditional power and interests, but also the role of ideas, norms, human agency, and institutional learning. Several major mining companies have adopted unilateral and collaborative voluntary initiatives for sustainable development (Dashwood 2014). Some scholars have demonstrated the potential role of CSR in business regulation, law, and ethics (McBarnet et al. 2007). However, the implementation of CSR varies among contractors in the Area, and how CSR can contribute to all humanity is still unknown. Thus, the study of a CSR framework pertaining to the Area can help to build a model for governing common resources to promote sustainability.

2   Research Methodology and Objectives

This is, to the author's knowledge, the first study of a CSR framework regarding the Area. It is based on data gathered from a literature review, present practices, and lessons learned as relayed in an empirical study of governance mechanisms for the Area (Zhou 2016). It examines the contract-based system for activities in the Area, including the current status of contracts, the rights and responsibilities of contractors and sponsoring States, and a parallel system for benefit-sharing. It also investigates contractors' training commitment, studies the first private contractor in the Area and its initiatives, and provides some feasible approaches to integrating CSR and sustainability pertaining to the Area.

3   A Contract-Based System

Before conducting exploration or mining in the Area, any interested entity must sign a legally binding contract with the ISA. In accordance with the UNCLOS and relevant regulations, the entity should submit an application with a formal written plan of work, which will form part of the contract. The LTC is mandated to review the work plans, in which applicants spell out the Area-related activities they intend to perform during a limited time period. Then, the Council decides whether to approve

---

[3] Examples include the United Nations Global Compact, the Organization for Economic Cooperation and Development's Guidelines for Multinational Enterprises, and the Global Reporting Initiative. Discussed *infra*, Sect. 5.

the application or not. This section examines the current status of contracts, the rights and responsibilities of contractors and sponsoring States, and a parallel system for benefit-sharing established under the UNCLOS.

3.1 Current Status of Contracts

To date, no real deep seabed mining has occurred in the Area. The present contracts pertain to each of the three mineral resources for which the ISA has adopted exploration regulations. The ISA has awarded 29 exploration contracts for polymetallic nodules, polymetallic sulphides, and cobalt-rich ferromanganese crusts by public and private entities (contractors).[4] The exploration contract is limited to 15 years. Before the expiration date, the contractor must either apply for a mining plan of work or seek an extension of the exploration contract. Seven contractors have extended their exploration contracts for an additional five years.[5]

Recently, several contractors have proposed to conduct mining tests in the Area. For instance, the German Federal Institute for Geosciences and Natural Resources and the Global Sea Mineral Resources submitted environmental impact assessments for the testing of collector components in their contract areas. The ISA will review the environmental impact assessments to determine their completeness, accuracy, and statistical reliability.

However, due to an increasing workload, the ISA needs more human and financial resources to complete its tasks. For example, the LTC had insufficient time to complete its agenda in 2012 and 2013.[6] Since 2013, an additional preparatory meeting of the LTC has been held several months prior to the main session to work on confidential documents, such as the contractors' reports, and to carry out other collaborative work in preparation for the main session. This allows the LTC to focus more efficiently on decision-making during the main session. However, the LTC still needs more time to deal with pressing matters, such as the operationalization of the Enterprise.[7]

---

[4]Deep Seabed Minerals Contractors. Retrieved from https://www.isa.org.jm/deep-seabed-minerals-contractors.

[5]*Id.*

[6]Information Note on Matters before the Legal and Technical Commission at the Twentieth Session of the International Seabed Authority (2014). Retrieved from http://www.isa.org.jm/files/documents/EN/20Sess/LTC-InfoNote.pdf.

[7]Discussed *infra*, Sect. 3.3.

3.2 Rights and Responsibilities of Contractors and Sponsoring States

While contractors have exclusive rights of exploration or mining in the Area for a limited time period, they should take precautions, such as ensuring adequate financial and technical capacity to carry out their plans of work. By signing the exploration contracts, they commit to reporting annually to the ISA on their activities in the Area. The annual reports should include their actual and direct exploration expenditures. The LTC reviews their work and then reports to the Council regarding the extent to which contractors have met their commitments under the plans of work. Furthermore, contractors should pay the fees for processing applications and overhead expenses for the administration and supervision of their contracts. Once mining becomes profitable, contractors will pay royalties to the ISA, which will distribute these receipts equitably by taking into account the interests and needs of developing countries. When disputes arise, the Council has the power to issue emergency orders, including suspension or termination of operations. The ISA can impose monetary penalties upon contractors under certain circumstances. In addition, contractors should acquire baseline data on existing environmental conditions and monitor seabed activities that might cause serious environmental harm. They are encouraged to conduct activities in close collaboration with the ISA.

Regarding the obligations of sponsoring States, sponsoring States must take all necessary and appropriate measures to ensure compliance by contractors under the UNCLOS. Upon a request by the ISA, the Seabed Dispute Chamber of the International Tribunal for the Law of the Sea delivered an advisory opinion on the limits of State liability when a contractor that a State is sponsoring to explore or exploit the Area causes damage or harm. The Seabed Dispute Chamber found that the measures should include domestic laws, regulations, and administrative measures, which are at least as stringent as those adopted by the ISA (Seabed Dispute Chamber of the Internal Tribunal for the Law of the Sea 2011, para. 241). The Chamber also set the highest standards of due diligence, or "obligation to ensure," for sponsoring States and identified other "direct obligations," such as applying a precautionary approach, best environmental practices, and conducting an environmental impact assessment (*Id.*, para. 145). As these obligations apply to both developed and developing countries, contractors would need to help their sponsoring States to develop appropriate law and enforcement capacity (Anton et al. 2011; Rayfuse 2011).

Drawing on the advisory opinion, the Council requested the Secretary-General to invite sponsoring States to provide information and prepare a report on the laws, regulations, and administrative measures with respect to activities in the Area. After consideration of the report submitted by the Secretary-General in 2012, the Council made the matter a standing item on its agenda and requested the Secretary-General to prepare an updated report annually. Moreover, the Secretariat has launched an online database incorporating the updates to national laws.

As deep seabed mining may have widespread and long-lasting impacts on the marine environment with implications for human health, any activities or use of special equipment should be well assessed by experts. In addition to conducting an analysis of existing regulations and international treaties, sponsoring States should hold expert and public consultations regarding their environmental impact assessments. Furthermore, national laws should make it illegal to conduct experimental projects without certain plans of work in marine areas beyond national jurisdiction.

3.3 A Parallel System for Benefit-Sharing

Under the Common Heritage principle, the Area regime aims to realize a just and equitable international economic order that considers the interests and needs of all humanity. This reflects both inter- and intra-generational equity underlying the concept of sustainability. To achieve these goals, the UNCLOS established a parallel system for benefit-sharing. One essential element of the parallel system is a "site bank" mechanism. This means that any interested entity that desires to extract minerals in the Area must propose two mining sites of equal estimated commercial value to the ISA. One site will be reserved for the ISA to be developed by the Enterprise or developing countries, and then the original applicant will work on the non-reserved site. As of February 6, 2018, the Council has approved six applications for exploration of polymetallic nodules in reserved areas (International Seabed Authority 2018).

However, during the review of the applications by the Nauru Ocean Resources and the Tonga Offshore Mining Limited (TOML), several members of the Council expressed concerns about the issue of "effective control" to ensure real benefits for developing countries. In 2014, the Secretariat delivered a note to the Council for its consideration to analyze the relevant provisions.[8] The note indicated guidance as to "effective control" by sponsoring States and concluded that: (1) no single definition of "effective control" exists, and its meaning depends upon the context and the purpose for which the test of efficacy is being applied; (2) conditions and standards which define "effective control" fall under the competence of the State that exercises it; and (3) the Area regime follows the same approach as the law and practice relating to the flagging of vessels and civil aviation. Thus, the test is on a case-by-case basis. In spite of this, the ISA could set some standards for the identification of "effective control" by developing countries to promote efficiency.

As an operational arm of the ISA, the Enterprise is intended to extract mineral resources from the Area directly. Since the 1994 Agreement, the Enterprise should operate under sound commercial principals and pursue its initial deep seabed mining through joint ventures. The current functions of the Enterprise are carried out by the Secretariat. In 2012, Nautilus Minerals, a Canadian company, submitted a proposal

---

[8]Analysis of Regulation 11.2 of the Regulations on Prospecting and Exploration for Polymetallic Nodules and Polymetallic Sulphides in the Area (2014). Retrieved from https://www.isa.org.jm/document/isba20ltc10.

for a joint venture operation with the Enterprise for several reserved sites in the Area.[9] Upon consideration, the Council concluded that it was premature for the Enterprise to function independently at that time. In 2018, the Government of Poland submitted another application for a joint venture arrangement.[10] The African Group also submitted a proposal for the independent functioning of the Enterprise to consider and respond to the proposal made by the Government of Poland.[11] Therefore, it is urgent for the ISA to elaborate the conditions under which joint ventures may operate with the Enterprise's participation.

4 Practices of Existing Contractors

As mentioned above, contractors in the Area should carry out their plans of work approved in the form of contracts. A plan of work for exploration is comprised of a general description and a schedule of the proposed programs, including an initial five-year program of Area-related activities, a program of oceanographic and environmental baselines studies, a preliminary environmental impact assessment of the proposed activities, and a schedule of expected yearly expenditures.

Current monitoring mechanisms include annual reports submitted by contractors and periodic reviews of the plans of work for exploration. In particular, the contractor is required to submit an annual report to the Secretary-General covering its activities in the exploration area and work with the Secretary-General to jointly undertake a periodic review at intervals of five years. For the review, the Secretary-General may require the contractor to submit additional data and information. Following the review, the contractor should make any necessary adjustments to its plan of work and submit its program of activities for the next five years, including a revised schedule of expected yearly expenditures.

The ISA has taken several measures to improve efficiency in the reporting process, including the establishment of a contract management unit within the Secretariat, an annual meeting of contractors, and the launch of a new database. Different organs of the ISA take respective responsibilities in terms of reporting. A sponsoring State should provide details of the measures taken to ensure compliance under a contract. Furthermore, an appropriate inspection mechanism is necessary, particularly in exploitation regulations.

Noncompliance denotes a failure or refusal to comply with a regulatory requirement and may have serious consequences for contractors. Based on the review of annual reports, the issues referred to as noncompliance include: (1) the failure to sub-

---

[9]Considerations relating to A Proposal by Nautilus Minerals Inc. for A Joint Venture Operation with the Enterprise (2013). Retrieved from https://www.isa.org.jm/documents/isba19c6.

[10]Considerations relating to A Proposal by the Government of Poland for A Possible Joint-Venture Operation with the Enterprise (2018). Retrieved from https://www.isa.org.jm/document/isba24c12.

[11]Request for Consideration by the Council of the African Group's Proposals for the Operationalization of the "Enterprise" (2018). Retrieved from https://www.isa.org.jm/document/statement-algeria-obo-african-group-1.

mit annual reports on time; (2) the failure to follow recommended reporting formats and methodologies; and (3) reported delays in advancing activities under the plan of work (International Seabed Authority 2018). Noncompliance may lead to the suspension or termination of a contract or monetary penalties. To date, no enforcement action has been taken, and no monetary penalties have been imposed with respect to any contractor.

However, there is a general concern that in many cases, contractors' actual expenditure is lower than their expected expenditure. In 2016, reported expenditure was lower than planned expenditure in 12 cases, and the percentages varied from 3 to 99 percent lower.[12] The contractors should provide more information about the reasons for reductions in expenditures. Moreover, as some contractors have not advanced environmental objectives, several delegates urged the Council to develop a concrete recommendation on noncompliance regarding expenditures.[13]

This section further investigates the contractors' training commitment and provides a case study of Nautilus Minerals and its initiatives for integrating CSR and sustainability.

4.1 Training Commitment

The importance of international technical and scientific cooperation with regard to activities in the Area, including training personnel of the ISA and developing countries, is recognized in Articles 144 and 148 of the UNCLOS and Sect. 5 of the Annex to the 1994 Agreement. The legal obligations of contractors with regard to training are required under Article 15 of Annex III to the UNCLOS and regulations adopted by the ISA relating to prospecting and exploration. The LTC also issued recommendations for the guidance of contractors and sponsoring States relating to training programs under their plans of work for exploration. According to the standard clauses for exploration contracts, prior to the commencement of exploration, each contractor should submit to the ISA proposed programs for training personnel of the ISA and developing countries, including the participation of such personnel in all of the contractor's activities under the contract, for approval. Each contract includes a schedule of practical training programs organized by the contractor in cooperation with the ISA and the sponsoring State.

The goal of training programs is to benefit the trainee, the nominating country, and members of the ISA, especially developing countries. The contractor in cooperation with the ISA and the sponsoring State should ensure that training contributes to the training and capacity development needs of the participants' country of origin. The provision of training must be afforded the same priority as other exploration activities

[12]Information relating to Compliance by Contractors with Plans of Work for Exploration (2018). Retrieved from https://www.isa.org.jm/document/isba24c4.

[13]ISA-24 Part 2 Highlights (18 July 2018). Retrieved from https://www.isa.org.jm/document/enb-iisd-bulletin-isa-24-part-2-4.

in terms of time, effort, and financing. The ISA and sponsoring States must ensure that the benefits to the trainee and the country's involvement in activities are related to the Area.

Regarding the content of training programs, contractors must provide training which is practical, focused on Area-related activities, and run for the full term of contracts. According to the recommendations by the LTC, each contractor should provide training for at least 10 trainees during every five-year period of the contract (International Seabed Authority Legal and Technical Commission 2013). The contractor should seek viable alternatives to address issues beyond the control of potential trainees, such as language barriers. The recommendations also emphasize the participation by sponsoring States in the negotiation of the scope and financing of training programs. In particular, if the sponsoring State is a developing country, it may inform the Secretariat when it needs any training which its contractor may not be able to satisfy under a bilateral agreement.

A reporting system on training activities is essential to achieve the objectives of accountability and transparency. The contractor should include information about its completed training in its annual reports and make any changes to training programs in its plans of work. The Secretariat is mandated to monitor the performance of training and to carry out a regular evaluation. Since 2013, 45 trainings have been provided by nine contractors.[14] The types of training programs include at-sea training, engineering training, fellowship training, education programs, and workshop internships. These programs provide personnel from developing countries an opportunity to gain expertise in conducting activities of the Area.

Furthermore, different types of training should be tailored to the needs of the ISA and developing countries. Contractors in collaboration with the ISA and sponsoring States should take into account the shortage of any skills and requirements of the industry. They should consider how to provide training to under-represented groups, such as female scientists. Recently, the ISA and Canada have co-organized an event on enhancing the role of women in the Area's marine scientific research. The Secretary-General emphasized the ISA's commitment to increasing the participation of women in activities of the Area through capacity-building opportunities, such as contractor training programs.[15] For instance, the ISA could give preference to qualified female applicants from developing countries through the training programs.

4.2 Nautilus Minerals and Its Initiatives

A subsidiary of Nautilus Minerals, TOML, is the first private sector organization to explore mineral resources in the Area. Although it is controlled by the Kingdom of

---

[14]Contractor Training. Retrieved from https://www.isa.org.jm/contractor/training-activities.

[15]Canada and the ISA Address Challenges Facing Women in Deep-Sea Marine Scientific Research. Retrieved from https://www.isa.org.jm/news/canada-and-isa-address-challenges-facing-women-deep-sea-marine-scientific-research.

Tonga, TOML can call on the technical and financial expertise of Nautilus Minerals. Teck and Anglo American, the world's leading resource companies, are the two largest shareholders of Nautilus Minerals.[16] With the sponsorship of the Kingdom of Tonga (a developing country), TOML was granted a reserved area of the ISA. Nautilus Minerals has also been awarded the world's first deep seabed mine, Solwara 1, in Papua New Guinea.

Nautilus Minerals recognized two types of approval for seabed mining: one based on relevant laws and regulations and the other which involves obtaining a social license to operate from concerned stakeholders (Smith 2011). Nautilus Minerals has launched several CSR initiatives for the local communities in Tonga. For example, the CARES program is intended to ensure local support through building capacity, community partnerships, collaborations, and initiatives in terms of infrastructure, health, and education.[17]

Furthermore, the company has explored stakeholder engagement opportunities through holding workshops on environmental and social impact assessment, environmental management plans, and data sharing and collaboration on studies. It also works on the monitoring program and community awareness project, including consultations with national and provincial governments. Although these initiatives can help the company build general acceptance of its project, commentators have argued that it lacks the credibility of possessing a social license because the identification of social actors is uncertain (Filer and Gabriel 2016).

As the Area is a common heritage owned by all humanity, it is difficult to settle the question of which actors should be involved to make judgment on a social license. To date, TOML has provided two at-sea training programs for personnel from developing countries.[18] Regardless of whether or not the company can be granted a social license, the ISA is mandated to ensure equitable sharing of benefits derived from the Area, and negotiate terms, such as training requirements, with contractors. It could develop guidance for initiatives in the local communities of sponsoring States, particularly for developing countries, to promote the integration of CSR and sustainability.

5 Approaches to Integrating CSR and Sustainability

There have been trends in the linkage between CSR and sustainability for addressing social and environmental concerns in the global arena (Jenkins and Yakovleva 2006). The World Business Council for Sustainable Development (2002) defines CSR as "the commitment of business to contribute to sustainable economic development,

---

[16]Decision of the Council relating to A Request for Approval of A Plan of Work for Exploration for Polymetallic Nodules Submitted by Tonga Offshore Mining Limited (2011). Retrieved from https://www.isa.org.jm/document/isba17c15.

[17]Nautilus Cares. Retrieved from http://www.nautilusminerals.com.

[18]Tonga At Sea Training Programme. Retrieved from
    https://www.isa.org.jm/scientific-activities/tonga-sea-training-programme.

working with employees, their families, the local community and society at large to improve their quality of life" (p. 229). Recently, the United Nations Human Rights Council has adopted the Guiding Principles on Business and Human Rights to build a three-pillar framework—protect, respect, and remedy (United Nations 2011). It indicates that States have a duty to protect against human rights abuses by third parties; business enterprises have an independent responsibility to respect human rights; and States and enterprises have a role to play in enabling individuals' access to effective remedy when their human rights are harmed.

Based on practices of existing contractors, this section provides several feasible approaches to integrating CSR and sustainability pertaining to the Area.

5.1 Develop Industry-Wide Norms and Standards

Industry-wide norms and standards can help to fill gaps in the legal framework of the Area. Several international professional societies have adopted codes of conduct and guidelines for deep seabed mining. For example, the International Marine Minerals Society developed the Code for Environmental Management of Marine Mining to provide its experience in the implementation and development of marine mining and associated environmental practices, particularly for environmental restoration and monitoring.[19] The ISA published the Code to educate interested entities and individuals. Although such a code is not legally binding, contractors should meet the standards or the best practices of the industry.

To date, the LTC has issued a number of recommendations on the protection of the environment, which will be incorporated into mining contracts. One could argue that the ISA's discussion on general issues, particularly for issues regarding the marine environment, should be open to the public. In 2014, the LTC held its first open meeting to review the status of the implementation of the environmental management plan for the Clarion-Clipperton Zone.[20] Furthermore, the ISA should work with contractors and States to ensure that all relevant stakeholders, especially under-represented groups, can get access to the discussion and articulation of these meetings.

In addition, partnership between the ISA and contractors in environmental protection offers a means for integrating CSR and sustainability to achieve environmental management objectives. For instance, the ISA has collaborated with contractors to develop environmental management plans (Lodge et al. 2014; Durden et al. 2017), particularly for organizing workshops on the development of a Regional Environ-

---

[19]The International Marine Minerals Society (2010). Code for Environmental Management of Marine Mining. Retrieved from http://www.immsoc.org/IMMS_downloads/LTC_&_ISA_presentations_final_summary.pdf.

[20]Clarion-Clipperton Zone is the best-known "nodule-rich" region of the Area, which is named for the two E-W-trending fracture zones it lies between.

mental Management Plan (REMP).[21] REMPs can help prevent, reduce, and control pollution and other hazards to all the areas where the ISA has issued contracts by protecting the natural resources and preventing damage to the flora and fauna of the marine environment. In addition, through partnership with the ISA for these activities, corporations could meet their CSR objectives for sustainability.

5.2   Launch Broader Initiatives in Local Communities

Corporations, especially those sponsored by developing countries, could launch broader initiatives in local communities to support their participation in Area-related activities. These initiatives can help to build the ability of more people in the world, particularly of those from developing countries, to be involved in activities to promote social progress. The local initiatives give practical examples of how the goal of sustainability can be embedded in CSR.

Walker and Howard (2002) have provided reasons why CSR voluntary initiatives are important for the mining industry: (1) public opinion of the industry as a whole is poor; (2) pressure groups have consistently targeted the industry at local and international levels, challenging its legitimacy; (3) the financial sector increasingly focuses on the industry from both risk management and social responsibility perspectives; (4) many companies have invested significantly in improved environmental and social performance, yet cannot demonstrate significant added value; and (5) maintaining "a license to operate" is a constant challenge. Therefore, launching initiatives in local communities can help corporations gain public trust and support.

To date, the ISA has been organizing sensitization seminars in partnership with contractors to offer technical assistance. The purpose of these seminars is to inform countries and different groups of entities about UNCLOS and established institutions. They also aim to address current issues relating to deep seabed mining. Representatives of governments and various entities can discuss scientific research on marine minerals and propose mechanisms for improving cooperation in scientific and marine mineral development at these seminars. Furthermore, the ISA could work with contractors and States to provide a variety of programs, such as infrastructure and health initiatives, to help ensure real benefits to developing countries.

5.3   Work on Data Collection and Sharing

One main challenge of deep seabed mining is the lack of data to understand its implications. Contractors can play a significant role in data collection and sharing.

---

[21] Workshop on the Development of A REMP for Cobalt-Rich Ferromanganese Crusts in the Northwest Pacific Ocean (2018). Retrieved from https://www.isa.org.jm/workshop/workshop-development-remp-cobalt-rich-ferromanganese-crusts-northwest-pacific-ocean-26-29.

Based on their plans of work, contractors are required to perform cruise activities and submit environmental, geological, and technological data to the ISA. The ISA reviews the submitted data to monitor and evaluate the contractors' activities in the Area. In 2016, the ISA proposed a data management strategy.[22] The new infrastructure for retrieval of Area-related data will include contractor templates and website functions. Corporations should be active in the discussion and collect both environmental and geological data to (1) establish baselines to assess natural change and likely effects of activities in the exploration areas; (2) monitor and evaluate methodologies and programs; and (3) designate impact and preservation reference zones (International Seabed Authority Legal and Technical Commission, 2011, para. 41). Moreover, a reporting system is helpful to promote transparency and accountability.

No individual scientist or institution has the capability to take on the challenge or bear the burden of deep seabed mining alone. Knowledge and data sharing can help to reduce production costs, facilitate new product developments, promote team performance, and improve firm innovation capabilities and performance (Wang and Noe 2010; Cummings 2004; Hansen 2002; Lin 2007; Mesmer-Magnus and DeChurch 2009; Arthur and Hutley 2005; Collins and Smith 2006). Cooperation and collaboration promote innovation to benefit the entire world. Therefore, it is essential to foster such a culture in the Area.

While data sharing can promote further research and enable new researchers to explore projects in the Area, data is usually a vital part of an exploration or mining project. In some cases, the only source of data may be a contractor that is conducting its own project in the Area. The contractor may consider its data to be an important proprietary resource. Under such cases, it is difficult to achieve integrity and transparency in managing, using, and sharing data.

Some feasible approaches can be used to balance transparency and confidentiality in sharing data derived from the Area. There are two approaches to sharing data: open and controlled. Data sharing can be fully open to the public when the information is available on an unrestricted website, easily discoverable, or without any access restriction, while some data is only available to qualified data users under a controlled-access procedure (Ossorio 2011). Therefore, an open and collaborative approach can help to create transparency and cooperation, particularly for sharing biological data to protect the marine environment. With regard to some proprietary data, such as information about mineral resources in the mining areas, a controlled access approach can be used to allow the ISA to get access to those data and monitor contractors' activities.

Last but not least, the ISA and other international organizations, such as the Intergovernmental Oceanographic Commission of UNESCO,[23] could provide some tools or legal templates to enable data sharing under the least restrictive terms. For instance, data users can be treated differently depending on the field of use: standard data use agreements can be crafted for users from industry and academia respectively.

---

[22]Data Management Strategy of the International Seabed Authority (2016). Retrieved from https://www.isa.org.jm/document/isba22ltc15.

[23]Intergovernmental Oceanographic Commission. Retrieved from http://ioc-unesco.org.

6 Conclusion

The integration of CSR and sustainability can help to address environmental and social concerns about deep seabed mining. Based on analysis of present practices, this chapter demonstrates that corporations for exploration or mining in the Area could work with the ISA and sponsoring States to fill gaps in the legal framework through various approaches. For example, they could develop industry-wide norms and standards, launch broader initiatives in local communities, and contribute to collecting and sharing data derived from activities in the Area.

As a preexisting governance institution in the Area, the ISA could provide guidance for the establishment of a CSR framework to ensure sustainable management of the Area. The framework could be built on practices of existing contractors in the Area and serve the changing needs and times. It could also be a model for governing other common resources, such as biological resources in marine areas beyond national jurisdiction, to promote sustainability.

References

Agreement relating to the Implementation of Part XI of the United Nations Convention on the Law of the Sea of 10 December 1982 (1994) S. Treaty Doc. No. 103-39, 1836 U.N.T.S. 3

Anton DK, Makgill RA, Payne CR (2011) Advisory opinion on responsibility and liability for international seabed mining (ITLOS case no. 17): international environmental law in the Seabed Disputes Chamber. Environ Policy Law 41:60–65

Arthur JB, Huntley CL (2005) Ramping up the organizational learning curve: assessing the impact of deliberate learning on organizational performance under gainsharing. Acad Manag J 48:1159–1170

Collins CJ, Smith KG (2006) Knowledge exchange and combination: the role of human resource practices in the performance of high-technology firms. Acad Manag J 49:554–560

Cummings JN (2004) Work groups, structural diversity, and knowledge sharing in a global organization. Manage Sci 50:352–364

Dashwood HS (2014) Sustainable development and industry self-regulation: developments in the global mining sector. Bus Soc 53:551–582

Durden JM, Murphy K, Jaeckel A, Van Dover CL, Christiansen S, Gjerde K, Ortega A, Jones DOB (2017) A procedural framework for robust environmental management of deep-sea mining projects using a conceptual model. Marine Policy 84:193–201

Filer C, Gabriel J (2016) How could Nautilus Minerals get a social licence to operate the world's first deep sea mine? Marine Policy 95:394–400

Hansen MT (2002) Knowledge network: explaining effective knowledge sharing in multiunit companies. Organ Sci 13:232–248

International Seabed Authority. Regulations on Prospecting and Exploration for Polymetallic Nodules in the Area, ISBA/6/A/18 (July 13, 2000), amended by ISBA/19/A/9 and ISBA/19/A/12 (July 25, 2013) and ISBA/20/A/9 (July 24, 2014)

International Seabed Authority. Regulations on Prospecting and Exploration for Polymetallic Sulphides in the Area, ISBA/16/A/12/Rev.1 (Nov. 15, 2010), amended by ISBA/19/A/12 (July 25, 2013) and ISBA/20/A/10 (July 24, 2014)

International Seabed Authority. Regulations on Prospecting and Exploration for Cobalt-rich Ferro-manganese Crusts in the Area, ISBA 18/A/11 (Oct. 22, 2012), amended by ISBA/19/A/12 (July 25, 2013)

International Seabed Authority (2010) The international marine minerals society's code for environmental management of marine mining

International Seabed Authority (2014) Summary report of the chair of the legal and technical commission on the work of the commission during the twentieth session of the international seabed authority

International Seabed Authority (2018) Status of contracts for exploration in the area

International Seabed Authority Legal and Technical Commission (2009) Recommendations for the guidance of contractors for the reporting of actual and direct exploration expenditures as required by Annex 4, Section 10, of the regulations on prospecting and exploration for polymetallic nodules in the area

International Seabed Authority Legal and Technical Commission (2010) Recommendations for the guidance of contractors for the assessment of the possible environmental impacts arising from the exploration for polymathic nodules in the area

International Seabed Authority Legal and Technical Commission (2011) Environmental management plan for the clarion-clipperton zone

International Seabed Authority Legal and Technical Commission (2013) Recommendations for the guidance of contractors and sponsoring states relating to training programmes under plans of work for exploration

International Seabed Authority Legal and Technical Commission (2018) Draft regulations on exploitation of mineral resources in the area

Jenkins H, Yakovleva N (2006) Corporate social responsibility in the mining industry: exploring trends in social and environmental disclosure. J Clean Prod 14:271–284

Lin HF (2007) Knowledge sharing and firm innovation capability: an empirical study. Int J Manpower 28:315–332

Lodge M, Johnson D, Gurun GL, Wengler M, Weaver P, Gunn V (2014) Seabed mining: international Seabed Authority environmental management plan for the Clarion-Clipperton Zone: a partnership approach. Marine Policy 49:66–72

McBarnet D, Voiculescu A, Campbell T (eds) (2007) The new corporate accountability: corporate social responsibility and the law. Cambridge University Press, New York

Mesmer-Magnus JR, DeChurch LA (2009) Information sharing and team performance: a meta-analysis. J Appl Psychol 94:535–546

Ossorio PN (2011) Bodies of data: genomic data and bioscience data sharing. Social Research 78:907–932

Rayfuse R (2011) Differentiating the common? The responsibilities and obligations of States sponsoring deep seabed mining activities in the Area. German Yearbook Int Law 54:459–488

Ruggie JG (2017) The social construction of the UN guiding principles on business and human rights. Harvard Kennedy School Corporate Responsibility Initiative Working Paper, 67, 1–23

Seabed Dispute Chamber of the Internal Tribunal for the Law of the Sea (2011) Responsibilities and obligations of states sponsoring persons and entities with respect to activities in the area, Advisory Opinion

Smith S (2011) Deep ocean seafloor mineral extraction —environmental and social responsibility for a new industry. In: IEEE Conference United Nations Convention on the Law of the Sea (1982), 1833 U.N.T.S. 397

United Nations (2011) Guiding principles on business and human rights: implementing the United Nations "Protect, respect and remedy" framework

Walker J, Howard S (2002) Finding the way forward: how could voluntary action move mining towards sustainable development? International Institute for Environment and Development and WBCSD Report

Wang S, Noe RA (2010) Knowledge sharing: a review and directions for future research. Human Res Manage Rev 20:115–131

World Business Council for Sustainable Development Executive Committee (2002) The business case for sustainable development: making a difference towards the Earth Summit 2002 and beyond. Corporate Environ Strat 9:226–235

Zhou Y (2016) Governance in the International Seabed: technology diffusion and its implementation. SJD Dissertation, University of Wisconsin Law School

Integrated Bio-cycle System for Rehabilitation of Open-Pit Coal Mining Areas in Tropical Ecosystems

Cahyono Agus, Enggal Primananda and Malihatun Nufus

Abstract Tropical rainforests are often called the "lungs of the planet" because they produce oxygen, which helps regulate carbon dioxide levels in the atmosphere. Open-pit coal mining in tropical forests exposes carbon and heavy metals and thus is a main cause of severe local, regional, and global environmental damage. The paradigm shift from mining to natural resources empowerment provides a new opportunity to change from red and green economic concepts to blue, which is smarter, wider, global, deeper, focused, sustainable, and futuristic. Tropical natural resources have the highest productivity in the world but still have a relatively low economic value. The net primary production in tropical ecosystems is more supported by the rapid bio-cycling than the almost infertile, weathered, acidic soil, which is caused by the high rainfall, moisture, temperature, and light intensity over a period of one year. Genetic engineering using fast-growing species (exotic and indigenous) and site engineering through land preparation and soil amendment facilitate the rehabilitation of post-coal mining areas and add value to the environment, economy, sociocultural setting, and health. An integrated bio-cycle system (IBS) is an almost natural ecosystem based on landscape ecological management for managing land resources (i.e., soil, water, mineral, air, and microclimate) and biological resources (i.e., fauna, flora, and human) under one integrated area; it is an important strategy for sustainable productivity in tropical ecosystems.

Keywords Genetic and site engineering · Integrated bio-cycle system · Open-pit coal mining · Sustainable development · Tropical ecosystem

C. Agus (✉)
Faculty of Forestry, Universitas Gadjah Mada, Yogyakarta 55281, Indonesia
e-mail: cahyonoagus@gadjahmada.edu

E. Primananda
Center for Plant Conservation, Indonesian Institute of Sciences, Bogor, Indonesia
e-mail: enggal.primananda@lipi.go.id

M. Nufus
Faculty of Agriculture, Universitas Sebelas Maret, Surakarta, Indonesia
e-mail: malihatunufus@staff.uns.ac.id

© Springer Nature Switzerland AG 2020
W. Leal Filho et al. (eds.), *International Business, Trade and Institutional Sustainability*, World Sustainability Series,
https://doi.org/10.1007/978-3-030-26759-9_30

1 Introduction

Indonesia's natural resources of minerals and mining materials significantly contribute to its national income (Agus et al. 2017a, b). The mining sector contributes 11.78% to the country's gross domestic income (Anonim 2013). One of the strategic mining materials is coal, whose total reserve is 21.131 billion tons of the total 105.187 billion tons of natural resources (Anonim 2011). The use of the coal in Indonesia increased significantly from 13.2 million tons in 1997 to 45.3 million tons in 2007 to satisfy the energy demands of coal-fired steam power plants, industries, and households. Consequently, the production capacity of coal increased from 77 million tons in 2000 to 466.307 million tons in 2012 (Iswanto 2013). In 2013 the Central Bureau of Statistics noted that there was a significant annual increase (14.45%) of coal production in Indonesia in the period of 2010–2012 (Iswanto 2013). Mining has contributed most to the development of Indonesia's economy for more than 30 years (Manaf 2009; Agus et al. 2014). There have been 833 mining activities in Indonesia, including open-pit mining activities, and a total mining area of 36 million ha; these activities have contributed to deforestation and land degradation in Indonesia.

The increase in coal production via the open-pit mining technique leads to a wider opening of vegetated land (deforestation) (Widyati and dan Kresno 2007). Subsequently, the deforestation results in degraded and derelict lands, representing unstable and highly eroded landform condition with high sedimentation and serious groundwater and soil nutrient deficits because of the loss of topsoil and subsoil (Puspaningsih et al. 2010). There are 1.3 million ha of degraded land resulting from the open-pit coal mining (Widyati and dan Kresno 2007) conducted in the coal production areas of several Indonesian islands such as Sumatera, Kalimantan, and Sulawesi.

Heavy metal contamination and the low pH land condition are serious problems found in the post-coal mining lands. The toxicity of the heavy metals and the low pH land hamper the land revegetation process, which is an important post-mining activity for ecosystem recovery. The heavy metals Cu, Zn, Cd, and Pb have been reported to be the main contaminants found in the post-mining areas (Mokhtari et al. 2014). Environmental manipulation is thus necessary to overcome the problem of the low pH land and the presence of pyrites in the post-mining lands and to support vegetation growth for faster succession process.

Open-pit coal mining in tropical ecosystems will in long term cause increasingly serious environmental degradation. Muhdar (2015) reported that coal mining in East Kalimantan covers 7.2 ha of land, representing 25% of the total territory, and it causes serious deforestation and environmental degradation. Organizing an optimal and sustainable rehabilitation that creates a balance between the economy and environmental sustainability is required. Therefore, land degradation must be minimized. Rehabilitation can be realized by first improving soil chemical, physical, and biological fertilities by growing plants through organic media. The use of organic materials in pots is a method to improve the soil properties by increasing microbial activities (Agus 2018). The method is useful for successful land rehabilitation. Microorganisms such as bacteria and fungi have been widely used because of their capability to decompose pollutants. Imanudin (2010) mentioned that many heavy metals are found in post-mining lands, and they pollute the environment and are toxic human beings. Microorganisms should be used to decompose such pollutants to prevent further land degradation or environmental pollution.

It is necessary to reclaim and revegetate post-mining lands to recover the ecosystem and microclimate, soil fertility, and water reserve function. The characteristics of reclaimed lands are open condition, high intensity of sunlight, high temperature and extreme fluctuation, low pH, and reduced number of flora and fauna species and ground microorganisms (Rahmawaty 2002). The general and predominant characteristic of the post-coal mining land is serious degradation, which causes erosion, diminishment or loss of top oil, hard-to-process solid soil, and soil properties (structure, texture, porosity, and bulk density) that do not support roots development and hamper plant growth. As a result, only certain plant species can grow in post-mining lands.

Post-mining sites can be improved using ameliorants such as organic materials and ectomycorrhizae fungi. One of the methods to recover the environmental condition is to eliminate contaminants using microorganisms such as ecofriendly fungi (Widyati 2008). Additionally, organic materials that produce decomposed humus can improve soil physical and chemical fertilities (Agus 2018). Humus improves the water-binding capability of soils and increases aggregate granulation for more stable soils. A good soil aggregation will enable air and water circulation, which enables organisms' activities and increases the availability of soil nutrients. Moreover, the capability of the soil to retain water for longer periods will influence the temperature and the moisture around the rooting area; the root tips and root hairs will grow well and function optimally. The addition of organic natural ameliorants in organic pots (growing media) provides initial sites that support rapid development of plant roots (Sumardi 2008).

Furthermore, the application of adaptive mycorrhiza also plays an important role in the success of post-mining land rehabilitation because of its capability to remedy the land (Wulandari et al. 2014) and minimize the contamination of the heavy metals (Long et al. 2010). Adaptive mycorrhiza such as *Glomusclarum* and *Gigasporadecipiens* are able to support the growth of *A. saman* and *M. paniculatus* in post-coal mining lands (Wulandari et al. 2014).

The integrated bio-cycle farming system (IBFS) is a prospective system for sustainable management of tropical ecosystems; it harmoniously combines agricultural sectors (agriculture, horticulture, plantation, animal husbandry, fisheries, forestry) with non-agricultural aspects (settlements, agro-industry, tourism, industry), and both are managed based on landscape ecological management under one integrated area (Agus 2018). Integrated bio-cycle system (IBS) is expected to improve land productivity, environmental conservation, and rural development programs.

2 Tropical Coal Mining

Coal is petrified prehistoric plants that initially accumulated in swampy and peat moss areas. It is a fossil fuel and is thus combustible. It comprises sediments of organic materials, especially carbon, hydrogen, and oxygen. Coal strata were formed from the consolidation of coal with other stone strata and subsequent transformation by subjection to high pressure and heat for millions of years (Anonim 2007).

The Indonesian thermal coal production is about 36% of international market share. In 2015, Indonesian coal production was 392 million tons, and about 319.87 million tons (81.6%) were exported to Japan, China, India, and South Korea. Meanwhile, domestically consumed coal was 72.13 million tons (18.4%). The domestic consumption of the coal increased by 9.83% (from 61 to 67 million tons) between January and September 2015. Coal is still the most long term reliable form of energy, and hence, there is huge potential of increase in coal mining in Indonesia.

The statistical data show that there is a constant increase in coal utilization over time. It increased from 41 million tons in 2005 to 67 million tons in 2015. Kalimantan is the biggest coal producing area in Indonesia. It contributes 92% to the national coal production. However, South Sumatera has a huge coal reserve. East Kalimantan has 22 billion tons of coal, and its annual production is about 40 million tons. The ongoing coal mining will in the long run cause excessive deforestation.

The mining activities that are conducted by opening forest and open mining will contribute to deforestation and land degradation in Indonesia (Agus et al. 2014). The mean deforestation rate in 2009–2013 was 1.13 million hectares/year. This is consistent with the data of the Ministry of Forestry showing that between 2008 and March 2013, a total of 2.98 million ha of forest area was leased to the mining industry. Thus, mining activities are one of the causal factors of the deforestation in Indonesia (Anonim 2015b).

Coal mining activities can be conducted using two methods: (i) surface/shallow mining, including open-pit mining, in-line mining, and hydraulic mining and (ii) sub-surface/deep mining. The common mining system in Indonesia is open-pit mining, meaning that ground layers are opened until the coal deposit is found. The most serious impact of the open-pit mining is the emerging phenomena of acid mine drainage and acid rock drainage as a result of oxidized sulfur-containing minerals (Untung 1993). Acid mine drainage has a series of interrelated impacts, including decrease in soil pH, distortion of the availability and balance of soil nutrients, and increase in the dissolvability of micro elements, which are generally metal elements and are highly toxic to plants (Havlin et al. 1999). Additionally, the soil becomes solid and exhibits an unstable structure, slow permeability, and poor aeration (Sidik et al. 1995). The condition of post-mining lands requires serious improvement.

3 Degradation of Mining Environment

In 2013, there were 1722 registered mining units and a total mining area of 8.8 million ha. The exploitation of forest area caused problems, and the potential loss from those problems in seven provinces of Indonesia was predicted to cost almost IDR 273 trillion. Of the 1722 units, 223 operated in East Kalimantan, with a total mining area of 774.5 thousand ha (Anonim 2015c). The mining activities include opening forest and open mining, which has significantly contributed to deforestation and land degradation in Indonesia (Agus et al. 2014).

The majority of the natural resources exploitation activities have significant impact on environmental degradation owing to the presence of the mineral wastes of poorly managed mining. There are three aspects of environmental degradation: natural resources depletion; decrease in forests width and in the life-supporting capability of the environment (Ljubojev et al. 2001; Sengupta 1993); and water, land, and air pollution (Stojadinovic et al. 2013; Singh 2005). However, the most serious impact of natural resources exploitation activities is the decrease in forest width (deforestation) and environmental degradation (Lawrence 2003; Gupta et al. 2006).

The accumulated materials comprise massive solid soil fragments consisting of clay stones, siltstones, sandstone, and lignite coal shales. Whitemore et al. (2011) mentioned that blue clay layer contains 2:1 minerals (smectite and interlayered A1) with sulfate salts such as Jerosite $KFe_3^{3+}(SO_4)_2(OH)_6$, epsomite/magnesium sulfate heptahydrate ($MgSO_4 \cdot 7H_2O$), hanksite/sodium-potassium chlorocarbonate sulfate ($Na_{22}KCl(CO_3)_2(SO_4)_9$), pyrite/ferrous sulfide (FeS_2), and carbonate. The salts play a main role in increasing soil acidity and degradation of the 2:1 minerals. Widjaja (1993) confirmed that the blue clay first undergoes reaction to become alkaline (pH 7.0–8.0); once they have been oxidized, the constituent pyrite forms salt and sulfate acid, which drastically lowers the pH of the land on which the materials resulting from the excavating activities are accumulated. Widyati (2006) stated that the sulfate content of the ground resulting from the excavation in the PT Bukit Asam mining operation in South Sumatera is 60,000 ppm.

The remaining minerals (both the overburden and the minerals contained in the excavated ground) are rapidly oxidized when exposed to air and water and produce sulfate acid, a very strong acid that greatly lowers soil pH. The measurement results of the excavated ground of PT Bukit Asam mining show that the soil pH ranges from 2.8 to 3.2, while the pH of water ranges from 1.6 to 5.2 (Widyati and dan Kresno 2007).

High rainfall and the minerals in the excavated ground, including sulfidic minerals, are the main causal factors of highly acidic mine water in Indonesia. The people around mining locations are not allowed to use such water. The water is characterized by a change in color to orangish red because of the oxidation of ferrous iron (Fe^{2+}) in the pyrite to ferric iron (Fe^{3+}). The results of the measurement of the Fe, Mn, Zn, and Cu metals in the excavated ground of PT. Bukit Asam mining show that the metals are above the existing allowed threshold, while the Fe and the Mn in water are above the existing allowed threshold (Widyati and dan Kresno 2007). This result is consistent with that of Neculita et al. (2007), who stated that the acidic mine water is characterized by very low pH, high sulfate concentration, and the accumulation of heavy metals. Other minerals that are generally found in the post-coal mining lands in addition to pyrite (FeS) include sphalerite (ZnS), galena (PbS), millerite (NiS), greenockite (CdS), covellite (CuS), and chalcopyrite (CuFeS) (Costello 2003). The highly acidic mine water of post-coal mining lands requires serious management, especially to lower the acidity and lessen the accumulated heavy metals.

4 Reclamation of Mined Land

The Regulation of the Government of the Republic of Indonesia Number 76 of 2008 on forest rehabilitation and reclamation requires each mining company to organize revegetation of the critical post-mining land. The revegetation may be organized by growing reclamation vegetation in the post-mining locations even though the mining activities as a whole still persist. The reclamation aims at recovering the mined land into its condition before the mining (Elaw 2010). Revegetation is a part of the reclamation activity. It represents an effort of reforesting the post-mining land. Generally, it is conducted in three steps: by growing cover crops, fast-growing species, and then climax species (Darmawan and Irawan 2009).

The soil quality of post-mining lands can be improved through revegetation of pioneering fast-growing legumes that can produce organic litter and humus and quickly reactivate the carbon cycle (Agus et al. 2014). The organic materials contain nitrogen that will in turn be decomposed and transformed into nitrate and ammonium necessary for plants. Moreover, the revegetation not only aims at covering the post-mining land, but also at improving the land to become more prospective and productive in the future. Pioneering fast-growing and adaptive plants for the revegetation of the post-coal mining land include sengon (*A. chinensis*), acacia (*Acacia mangium* Willd.), sungkai (*Peronema canescens*), *Gmelina arborea* tree, angsana (*Pterocarpus indicus*), castor oil plant, and legume cover crops. These species have a more significant impact on the improvement of the organic C, total N, and soil pH of post-coal mining lands than the baseline of wet tropical forest (Agus et al. 2014).

Pongamia pinnata (Linn.) Pierre is one of the species grown for the vegetation of post-mining lands because it can associate with *Rhizobium* bacteria to bind nitrogen, and it can grow very well and produce a lot of biomass since it lasts up to 2–18 years (Bohre et al. 2013, 2014). The presence of the N-binding bacteria also plays an important role in N availability in soil. Agus et al. (2004) showed that the N-mineralizing capability of the species is high; it can supply 3–5 times the available N in the soil. Meanwhile, the use of land-covering legumes has been able to supply 9–27 times the N available in the soil (Agus et al. 2003).

The final condition of the reclaimed land can be very close to the condition before the mining or other stipulated conditions (Rachmanadi 2009). The restoration of seriously degraded ecosystem has three objectives, which are protective, productive and conserving ones (Rahmawaty 2002). One of the main keys to successful revegetation is selecting suitable species of trees based on their adaptability, their fast-growing capability, applicable silviculture, availability of planting materials, and capability to establish mutual symbiosis with mycorrhiza. The *Pongamia pinnata* species are capable of establishing mutual symbiosis with the biggest quantity of spores and colonies of vesicular-arbuscular mycorrhiza (VAM), compared to *Acacia nilotica*, *Acacia catechu*, *Acacia indica*, and *Leucaena leucocephala* (Kumar et al. 2010).

The revegetation of the post-mining land is conducted by considering the species of plants and also by applying organic pot technology and soil microbes. The VAM plays an important role to help the plants absorb nutrients, especially phosphate (Suhardi et al. 2006), in the nitrogen- and phosphorus-deficit lands (Booze-Daniels et al. 2000; Bucking and dan Shachar-Hill 2005). The application of mycorrhiza is important, as the species can serve as an ecofriendly biological fertilizer (Rahmawaty 2002; Widyati 2008).

Endomycorrhiza inoculation can improve N and P absorptions and increase the biomass of *Albizia saman* and *Mallotus paniculatus* (Wulandari et al. 2014). Thus, the endomycorrhiza is effective for improving nutrients absorption and accelerating the growth of plants that support the revegetation and sustainability of the post-coal mining areas. Meanwhile, endomycorrhiza inoculation in pongamia seedlings can accelerate the development the plants shoots and roots by 92 and 48%, respectively (Anonim 2015a). Additionally, pongamia can establish mutual symbiosis between VAM and the highest quantity of spores and colonies, compared to other species (Kumar et al. 2010). The application of the endomycorrhiza on pongamia in red soil has shown that there are as many colonies as those found in black soil (Venkatesh et al. 2009). Revegetation by fast-growing species of *Pongamia pinnata* and application of Arbuscular Mycorrhyza Fungi (AMF) has been found to extremely improve soil fertility, supporting rehabilitation of post-coal mining areas (Agus et al. 2017a).

Mycorrhiza plays an important role in maintaining the flora diversity by transferring nutrients from the roots of a plant to those of neighboring plants (hypha bridge). Mycorrhiza also plays an important role in biogeochemical cycle; it can accelerate natural succession in extremely disturbed natural habitat by improving the tolerance of its hosts to heavy metals (Orcutt and dan Nilsen 2002; Aghababaei et al. 2014). Mycorrhiza increases the sensitivity and the tolerance of its hosts to heavy metals through some mechanisms such as filtering toxic heavy metals in hyphal sheaths or Hartig nets, reducing the transfer of the metals to the trunks of the plants (shoots), absorbing the metals in miselium, modifying the absorption system of plasmalemma, intracellular detoxification, and chelation by humic acid and other substances it produces (Godbold 2004). The presence of the heavy metals Zn, Cu, Cd, and Ni hampers and limits the amount of phosphorus, potassium, and iron extant in root system. Phosphorus as a nutrient in plants accelerates the growth of roots, development of meristem tissue, flowering, and conception, and serves as building materials of nucleus, fat, and protein (Sarief 1993). The protection mechanism against heavy metals and toxic elements takes place through the filtration effect, chemical deactivation, or accumulation of the toxic elements in the hypha of the fungi. The application of the mycorrhiza in the plants grown in heavy metal-contaminated media can reduce the content of the metals Pb, Cu, and Zn (Rossiana 2009).

The direct uses of the mycorrhiza are to improve plants resistance to drought, absorb P nutrient in lands with P deficit and too high or too low pH, improve plants endurance against highly toxic metals such as Al, Fe, and Mn (Sieverding 1991), control pathogenic infection, and stimulate useful microorganism activities. The indirect uses of the mycorrhiza are to improve soil structure and soil aggregation, smoothen the cycle of minerals, and enable weathering of essential materials (Santoso et al. 2007).

Organic materials are important components for forest rehabilitation activities, land remediation, and ecofriendly soil amelioration, and they are easy to apply in seriously degraded lands. Some basic principles in manipulating the growing environment of plants include the application of organic materials to increase moisture regime, temperature regime, and nutrient regime (Agus 2018). However, when organic materials are applied to the soil, especially in the degraded post-mining areas, the possibility of leaching should be considered. Therefore, in post-coal mining lands, organic materials should be applied using organic pots. The use of organic pots is suitable in nutrient-deficit dry lands because they are compact and can retain water, absorb heat, and reduce evaporation, and the gradual decomposition of the organic pots can provide plants with nutrients as needed (Agus 2018). The application of ameliorants in a molded media can accelerate the establishment of the root systems of plants, especially in the initial planting period, because the roots can use the water and the nutrients extant in the organic planting media. Once the root system has been established and optimally functions, the plants can grow and develop very well, and in the next period they can absorb the water and the nutrients in more distant locations (Nugroho and Sumardi 2010).

The ideal quantity of the organic materials is about 3–5% (194). The organic materials improve soil properties and support the growth of plants via the following ways: (i) They improve soil structure; (ii) they serve as a source of soil nutrients, including N, P, and S (iii) they improve the water-retaining capability of the soil; (iv) they improve the nutrient-retaining capability of the soil (increasing cation exchange capability); and (v) they serve as a source of energy and nutrients for organisms. A soil with a fine texture is more capable of protecting organic materials than that with a coarse texture so that the former has more organic material content than the latter (Hassink et al. 1997).

5   Integrated Bio-Cycle System (IBS)

Tropical natural resources have the highest productivity in the world but still have a relatively low economic value. The net primary production in tropical ecosystems is more supported by the rapid bio-cycling than the almost infertile, weathered, acidic soil, which is caused by the high rainfall, moisture, temperature, and light intensity over a period of one year. Genetic engineering using fast-growing species (exotic and indigenous) and site engineering through land preparation and soil amendment facilitate the rehabilitation of post-coal mining areas and add value to the environment, economy, sociocultural setting, and health. Biotechnology through bio-artificial and functional nanotechnology will make the realization of integrated bio-cycle management in tropical regions more successful (Agus et al 2011; Agus2018). This will be a good contribution in terms of a sustainable economy, the environment, and the sociocultural setting.

A red economy that is only economy-oriented results in the degradation of the environment and life. A green economy that is oriented toward environmental and healthy values is expensive. Gunter Pauli of the Zero Emissions Research and Initiatives Foundation in 2009 developed the concept of the blue economy, which focuses on accelerating natural cycle processes by empowering land resources (i.e., soil, water, mineral, air, and microclimate) and biological resources (i.e., fauna, flora, and human); this adds value in terms of economy, environment, and sociocultural aspect (Agus et al. 2016; Agus 2018). The blue economy facilitates and stimulates efficient investment, innovation, funding, job creation, social capital development, and entrepreneurship. Waste and abandoned goods are converted to food, energy, and employment opportunities, thus turning poverty to sustainable development, and scarcity to availability. The blue economy has provided new, innovative, green, and creative opportunities.

The IBFS is a prospective system that harmoniously combines agricultural and non-agricultural sectors, based on landscape ecological management. The cycles of materials, organic matter and carbon, energy, water, nutrient, production, crop, and money are managed through 9R (reuse, reduce, recycle, refill, replace, repair, replant, rebuild, and reward) for the farmer, community, and agriculture and global environment to gain optimal benefits. The system has multiple functions and includes multiples products (food, feed, fertilizer, fuel, wood, water, oxygen, pharmacy, edutainment, and ecotourism). IBS can provide additional benefits through the recycling of organic waste into high-value renewable resources, such as organic fertilizer (liquid and solid), animal feed, and bio-energy (Agus 2018).

The IBFS has a comparative comprehensive advantage compared to other integrated farming systems such as low-input farming, organic farming, bio-dynamic farming, and agroforestry system (Table 1). The key characteristics of IBS are as follows: (i) integration of agriculture and non-agriculture sectors, (ii) added value of environment, esthetics, and economics, (iii) rotation and diversification of plants, (iv) nanotechnology, artificial and functional biotechnology and probiotics, (v) closed organic cycle and integration in an integrated area, (vi) ecosystem health management and integrated bio-protection (vii) agropolitan concept and landscape ecological management, (viii) specific management of plants, and (ix) holistic and integrated system (Agus 2018).

6 Conclusion

Open-pit coal mining in tropical forests exposes carbon and heavy metals and thus is a main cause of severe local, regional, and global environmental damage. The paradigm

Table 1 Key characteristics of various types of sustainable agricultural systems

Low input/integration	Organic farming	Bio-dynamic agriculture	Agroforestry	Integrated bio-cycle
Integration of advantageous natural processes	Integration of land, environment, and human health	Management of organisms that optimize quality of land, plants, animals, and human health	Integration of Wood and herbal plants	Integration of agriculture and non-agriculture sectors
Adding environmental values	Natural fertilizer Environmental values	Economic values	Environmental values	Value of environment, esthetics, and economics
Plant rotation	Plant rotation, diversification, and ideal space	Plant rotation, diversification, and ideal space	Spatial diversitas tipe crop	Rotation and diversity of plants
Impact of minimum land management	Adequacy of N through N fixation	Adequacy of N through N fixation, special preparation for improving land quality and living plants	Plant variation and pastoral system	Artificial and functional biotechnology, nanotechnology, and probiotics
The use of chemical fertilizer	Prohibition on treatment of plants and fertilizer	Prohibition on treatment of plants and fertilizer	Fertilization of agricultural plants; the use of cycle in forest plants	Management of closed organic cycle and integration of crop, moisture, nutrient and pest management
The use of pesticide	Management of traditional animals	Management of traditional animals		Management of integrated bio-protection and ecosystem health management
General principle	Principle of grouping units	Principle of grouping units	General principle	Landscape ecological management, agro-politan concept
Specific management of plants	Specific management of plants	Specific management of plants	Specific management of plants	Specific management of plants

(continued)

Table 1 (continued)

Low input/integration	Organic farming	Bio-dynamic agriculture	Agroforestry	Integrated bio-cycle
Semi-traditional	Natural	Integrated	Traditional	Holistic and integrated
Stockdale and Cookson (2003), Chan (2006)	IFOAM (1998)	Koepf et al. (1976)	Stockdale and Cookson (2003)	Agus (2013)

Source Agus (2018)

shift from mining to natural resources empowerment provides a new opportunity to change from red and green economic concepts to blue, which is smarter, wider, global, deeper, focused, and futuristic for sustainable development. Genetic engineering using fast-growing species (exotic and indigenous) and site engineering through land preparation and soil amendment facilitate the rehabilitation of post-coal mining areas and add value to the environment, economy, sociocultural setting, and health. The IBS is an important strategy for sustainable productivity in tropical ecosystems.

Acknowledgements Table 1 was reprinted by permission from Springer Nature. Agus (2018), https://doi.org/10.1007/978-3-319-73028-8_9.

References

Aghababaei F, Raiesi F, dan Hosseinpur A (2014) The significant contribution of mycorrhizal fungi and earthworms to maize protection and phytoremediationin cd-polluted soils. Pedobiologia J Soil Ecol 57:223–233

Agus C (2013) Management of tropical bio-geo-resources through integrated bio-cycle farming system for healthy food and renewable energy sovereignty: sustainable food, feed, fiber, fertilizer, energy, pharmacy for marginalized communities in Indonesia. In: Proceedings of 2013 IEEE Global Humanitarian Technology Conference (GHTC). www.ieeeghtc.org. San Jose, California USA October 20–23, 2013

Agus C (2018) Development of blue revolution through integrated bio-cycles system on tropical natural resources management. In: Leal Filho W, Pociovălişteanu D, Borges de Brito P, Borges de Lima I (eds) Towards a sustainable bioeconomy: principles, challenges and perspectives. World sustainability series. Springer, Cham, pp 155–172

Agus C, Karyanto O, Hardiwinoto S, Haibara K, Kita S, Toda H (2003) Legume cover crop as a soil amendment in short rotation plantation of tropical forest. J For Environ 45(1):13–19

Agus C, Karyanto O, Kita S, Haibara K, Toda H, Hardiwinoto S, Supriyo H, Na'iem M, Wardana W, Sipayung M, Khomsatun, Wijoyo S (2004) Sustainable site productivity and nutrient management in a short rotation *Gmelina arborea* plantation in East Kalimantan, Indonesia. New For J 28:277–285

Agus C, Sunarminto BH, Suhartanto B, Pertiwiningrum A, Wiratni Setiawan I, Pudjowadi D (2011) Integrated bio-cycles farming system for production of bio-gas through GAMA DIGESTER GAMA PURIFICATION and GAMA COMPRESSING. J Jpn Inst Energy 90(11):1086–1090

Agus C, Pradipa E, Wulandari D, Supriyo H, Saridi, Herika D (2014) Role of revegetation on the soil restoration in rehabilitation areas of tropical coal mining. J Manusia dan Lingkungan 21(1):60–66 (in Indonesian)

Agus C, Putra PB, Faridah E, Wulandari D, Napitupulu RNP (2016) Organic carbon stock and their dynamics in rehabilitation ecosystem areas of post open coal mining at tropical region. Procedia Eng 159:329–337

Agus, C, Wulandari D, Primananda E, Hendryan A, Harianja V (2017) The role of soil amendment on tropical post tin mining area in Bangka island Indonesia for dignified and sustainable environment and life. IOP Conf Ser: Earth Environ Sci 83:012030

Agus CE, Primananda E, Faridah D Wulandari, Lestari T (2017a) Role of *arbuscular mycorrhizal fungi* and *Pongamia pinnata* for revegetation of tropical open-pit coal mining soils. Int J Environ Sci Technol (IJEST) 15(11):1–11. https://doi.org/10.1007/s13762-018-1983-5

Agus C, Hendryan A, Harianja V, Faridah E, Atmanto WD, Cahyanti PAB, Wulandari D, Pertiwiningrum A, Suhartanto B, Bantara I, Hutahaean BP, Suparto B, Lestari T (2017b) The role of soil organic amendment of humus paramagnetic and compost for remediation of post tin mining tailing media and their growth of *Reutealis trisperma* seedling. Int J Smart Grid Clean Energy (IJSGCE) (in Printing)

Anonim (2007) Analisis Dampak Lingkungan Hidup (AMDAL) Peningkatan Produksi Batubara Dari 2 juta ton/tahun menjadi 5 juta ton/tahun PT. Berau Coal Site Binungan Kecamatan Sambaliung Kabupaten Berau Provinsi Kalimantan Timur. PT. Berau Coal

Anonim (2013) Pendapatan Nasional Indonesia 2009–2012. Badan Pusat Statistik. Indonesia. Jakarta. Sumber: http://www.bps.go.id/hasil_publikasi/pend_nas_2009_2012/index3.php?pub=Pendapatan%20Nasional%20Indonesia%20Tahun%202009-2012. 13 Apr 2018

Anonim (2012) Statistic of coal. Source:www.esdm.go.id/statistik/data-sektor-esdm/cat_view/58-publikasi/240-statistik/337-statistik batubara.html. 6 March 2014

Anonim (2015a) Biodiesel crops as candidates for the rehabilitation of degraded lands in India (International Crops Research Institute for the Semi-Arid Tropics) [Online]. Tersedia: www.icrisat.org.html. 22 Mar 2018

Anonim (2015b) Nasib Hutan Alam Indonesia, Tambang Kontributor Dominan Hilangnya Tutupan Hutan, Intip Hutan Indonesia. Halaman 4–13. Forest Watch Indonesia. Bogor. http://fwi.or.id/publikasi/mau-dibawa-kemana-nasib-hutan-kita/.html. 4 Jan 2018

Anonim (2015c) Consession Area PT Berau Coal. http://www.beraucoalenergy.co.id/id/profil-kami-2/operasi/pt-berau-coal/operasional.html. 2 Jan 2018

Bohre P, Chaubey OP, dan Singhal PK (2013) Biomass production and carbon sequestration by Cassia siamea Lamk in degraded ecosystem. In Chaubey OP, Prakash R (ed) Sustainable biodiversity conservation in the landscape. Aavishkar publishers, distributors, Jaipur, India, pp 15–34

Bohre P, dan Chaubey OP (2014) Restoration of degraded lands through plantation forests. Glob J Sci Front Res Biol Sci 14(1)

Booze-Daniels JN, Daniels WL, Schmidt RE, Krouse JM, dan Wright DL (2000) Establishment of low maintenance vegetation in highway corridors. In: Barnhisel RI (eds) Reclamation of drastically disturbed lands. American Society of Agronomy (41):887–920

Bucking H, dan Shachar-Hill Y (2005) Phosphate uptake, transport and transfer by arbuscular mycorrhizal fungus is increased by carbohydrate availability. New Phytol 165(3):889–912

Chan GL (2006) Integrated farming system. http://www.scizerinm.org/chanarticle.html. 5 Sep 2018.

Costello C (2003) Acid mine drainage: innovative treatment technologises. U.S. environmental protection agency office of solid waste and emergency response technology innovation office, Washington, DC.www.clu-in.org

Darmawan A, Irawan MA (2009) Reklamasi Lahan Bekas Tambang Batubara PT Berau Coal, Kaltim. Workshop IPTEK Penyelamatan Hutan Melalui Rehabilitasi Lahan Bekas Tambang Batubara. Banjarmasin

Elaw (2010) Guidebook for evaluating mining project EIAs. Environmental law alliance worldwide. Eugene, USA

Godbold DL (2004) Mycorrhizae. Encyclopedia of forest science, vol 4. Elsevier Academic Press, Oxford

Gupta H, Gupta K, Singh P, dan Sharma R (2006) A sustainable development and environmental quality management strategy for Indore. Environ Qual Manage 15(4):57–68

Hassink J, Matus FJ, Chenu C, Dalenberg JW (1997) Interactions between soil biota, soil organic matter and soil structure. Adv. Agroecol (in press)

Havlin JL, Beaton JB, Tisdale SL, dan Nelson WL (1999) Soil fertility and fertilizers. An introduction to nutrient management. Prentice Hall, New Jersey

IFOAM (International Federation of Organic Agriculture Movements) (1998) Basic standard for organic production and processing. Tholey-Thelei.Germany, IFOAM

Imanudin T (2010) Pemanfaatan Mikroorganisme sebagai Salah Satu Alternatif untuk Mengurangi Pencemaran Logam Berat di Perairan. Katalog. PDII. http://LIPI.go.id/index.php/searchkatalog/ .../6889/6890.pdf. 30 May 2018

Iswanto D (2013) Statistik Pertambangan Non Minyak dan Gas Bumi 2010–2012. Editor: Sodikin dan Gumelar. Badan Pusat Statistik. Jakarta. http://www.bps.go.id/hasil_publikasi/stat_non_ migas_2008_2012/index3.php?pub=Statistik%20Pertambangan%20Non%20Minyak%20dan% 20Gas%20Bumi%202008-2012. 6 Mar 2018

Koepf HH, Pettersson DD, Schaumann DD (1976) Biodynamic agriculture. Anthroposophic Press, Spring Valey New York

Kumar A, Raghuwanshi R, Upadhyay RS (2010) Arbuscular mycorrhizal technology in reclamation and revegetation of coal mine spoils under various revegetation models. Eng J 2:683–689

Lawrence D (2003) Environmental impacts assessment: practical solutions to recurrent problems. Wiley

Ljubojev M, Popovic R, Bogdanovic D (2001) Determination of period for surface stabilization for the civil engineering use. In: Proceedings of the Third International Symposium on Mining and Environmental Protection MEP-01, Belgrade, pp 415–418

Long LK, Yao Q, Guo J, Yang RH, Huang YH, Zhu HH (2010) Molecular community analysis of arbuscular mycorrhizal fungi associated with five selected plant species from heavy metal polluted soils. Eur J Soil Biol 46:288–294

Manaf MH (2009) The environmental impact of small scale mining in Indonesia. www.gemeed.cl. 5 May 2009

Mokhtari AR, Rodsari PR, Cohen DR, Emami A, Bafghi AAD, Ghegeni ZK (2014) Metal specification in agricultural soils adjacent to the Irankuh Pb–Zn mining area, central Iran. J Afr Earth Sci 101:186–193

Muhdar M (2015) Aspek Hukum Reklamasi Pertambangan Batubara pada Kawasan Hutan di Kalimantan Timur. Mimbar Hukum 27(3):472–486

Neculita CM, Zagury GJ, danBussiere B (2007) Passive treatment of acid mine drainage in bioreactors using sulfate-reducing bacteria. J Environ Qual 36:1–16

Nugroho AW, Sumardi S (2010) Ameliorasi Tapak untuk Pemapanan Cemara Udang (Casuarina equisetifolia Linn.) pada Gumuk Pasir Pantai. Jurnal Penelitian Hutan dan Konservasi Alam 4(7):381–397

Orcutt DM, dan Nilsen ET (2002) Physiology of plant under stress: soil and biotic factors. Wiley, New York

Puspaningsih N, Murtilaksono K, Sinukaban N, Jaya INS, Setiadi dY (2010) Estimasi umur harapan pencapaian keberhasilan reforestasi kawasan pertambangan PT. INCO, Sorowako, Sulawesi Selatan. Jurnal Forum Pascasarjana 33(4): 275–283

Rachmanadi D (2009) Upaya Reklamasi Lahan Pascatambang Batubara di Kalimantan Selatan. Prosiding Workshop IPTEK Penyelamatan Hutan Melalui Rehabilitasi Lahan Pascatambang Batubara. Balai Besar Penelitian Dipterokarpa. Samarinda, pp 46–52

Rahmawaty (2002) Restorasi Lahan Bekas Tambang Berdasarkan Kaidah Ekologi. [Online]. http:// library.usu.ac.id/download/fp/hutan-rahmawaty5.pdf.html. 6 July 2018

Rossiana N (2009) Penurunan Kandungan Logam Berat dan Pertumbuhan Tanaman Sengon (Paraserianthes falcataria) Bermikorisa dalam Medium Limbah Lumpur Minyak Hasil Ekstraksi.

Lab. Mikrobiologi dan Biologi Lingkungan UNPAD. Bandung. http://pustaka.unpad.ac.id/wpcontent/uploads/2009/04/penurunan_kandungan_logam_berat_dan_pertumbuhan_tanaman_sengon.pdf. 3 Mar 2018

Santoso E, Turjaman M, Irianto RSB (2007) Aplikasi Mikoriza untuk Meningkatkan Kegiatan Rehabilitasi Hutan dan Lahan Terdegradasi. www.dephut.go.id/files/Erdy.pdf. 10 July 2018

Sarief S (1993) Kesuburan dan Pemupukan Tanah Pertanian. Pustaka Buana, Bandung

Sengupta M (1993) Environmental impacts of mining: monitoring, restoration and control. Lewis Publishers, p 494

Sidik H, Tala'ohu S, Moersidi S, dan Gunawan S (1995) Sifat Fisiko-Kimia Tanah Timbunan di PTBA Tanjung Enim, Sumatera Selatan. Hal. 39–48 dalam Prosiding Pembahasan Penelitian Tanah dan Agroklimat, Bidang Fisika, Konservasi Tanah dan Air dan Agroklimat, Bogor, 26–28 September 1995. Pusat Penelitian Tanah dan Agroklimat, Bogor

Sieverding E (1991) Vesicular-arbuskular mycorrhiza management in tropical agrosystem. Technical Coorporation Federal Republik of Germany

Singh O (2005) Mining environment, problems and remedies. Regency Publications, New Delhi, p 278

Stojadinovic S, Zikic M, Pantovic R, Svrkota I, dan Petrovic D (2013) High slope waste dumps—a proven possibility. Acta Montanistica Slovaca 18(1):40–51

Stockdale EA, Cookson WR (2003) Sustainable farming systems and their impact on soil biological fertility-some case studies. In: Abbott LK, Murphy DV (eds). Soil biological fertility: A key to sustainable land use in agriculture. Kluwer Ac. Pub, Dordrecht, pp 225–239

Suhardi, Faridah E, Iskandar E, Rahayu S (2006) Mycorrihizal formation and growth of *Shorea leprosula* in Bukit Suharto after using charcoal and rockphosphate. In: Plantation technology in tropical forest science. Springer, Tokyo

Sumardi (2008) Prinsip Silvikultur Reforestasi dalam Rehabilitasi Formasi Gumuk Pasir di Kawasan Pantai Kebumen. Seminar Nasional Silvikultur Rehabilitasi: Pengembangan Strategi untuk Mengendalikan Tingginya Laju Degradasi Hutan. Wanagama I, 24–25 November 2008. Yogyakarta

Untung SR (1993) Dampak Air Asam Tambang dan Upaya Pengelolaannya. Pusat Penelitian Tambang Batubara dan Mineral. Bandung (Unpublished)

Venkatesh L, Naik ST, dan Suryanarayana V (2009) Survey for occurrence of arbuscullar mycorrhizal fungi associated with *Jatropha curcas* (L). and *Pongamia pinnata* (L), Pierre in three agroclimatic zones of Karnataka. J Agric Sci 22(2):373–376

Widjaja A IPG (1993) Penjajagan Hara/Kendala Tanah Berbagai Lapisan di PTBA Tanjung Enim. Laporan Akhir Reklamasi, Penelitian dan Pengembangan Sumber daya Lahan serta Pelatihan. Kerja Sama PTBA dengan Pusat Penelitian Tanah dan Agroklimat. Bogor. (tidak dipublikasikan)

Widyati E (2006) Bioremediasi Tanah Bekas tambang Batubara dengan Sludge Industri Kertas untuk Memacu Revegetasi Lahan. Disertasi Doktor, Sekolah Pascasarjana IPB, Bogor

Widyati E (2008) Peranan Mikroba Tanah pada Kegiatan Rehabilitasi Lahan Bekas Tambang. Info Hutan 5(2):151–160

Widyati E, dan Kresno BS (2007) Bioremediasi air asam tambang di beberapa KPL PT. Bukit Asam dengan bakteri pereduksi sulfat (Unpublished)

Wulandari D, Saridi, Cheng W, Tawaraya K (2014) Arbuscular mycorrhizal colonization enhanced early growth of *Mallotus paniculatus* and *Albizia saman* under nursery condition in East Kalimantan, Indonesia. Int J For Res 2014, Article ID 898494

Best Practice for Responsible Small Scale Aggregates Mining in Developing Countries

P. Schneider, K.-D. Oswald, W. Riedel, A. Le Hung, A. Meyer, I. Nolivos and L. Dominguez-Granda

Abstract Small scale aggregates mining includes artisanal as well as small scale industrial mining, summarised as artisanal and small-scale mining (ASM). Aggregates mining forms the backbone of the urbanisation process in developing countries as it is the source of construction materials for the infrastructural development process. Responsible mining is a mining process that uses cleaner production strategies, and that respects the social responsibility towards the employees, generally aiming on a long term stable economic growth. As such, the best practice for responsible small-scale aggregate mining addresses sustainability, particularly the following Sustainable Development Goals: end poverty, good health and well-being, decent work and economic growth, industry, innovation and infrastructure, sustainable cities and communities as well as responsible consumption and Production. This contribution

P. Schneider (✉) · A. Meyer
University of Applied Sciences Magdeburg-Stendal, Breitscheidstr. 2, 39114 Magdeburg, Germany
e-mail: petra.schneider@hs-magdeburg.de

A. Meyer
e-mail: andreas.meyer@hs-magdeburg.de

K.-D. Oswald · W. Riedel
C&E Consulting und Engineering GmbH, Jagdschänkenstr. 52, 09117 Chemnitz, Germany
e-mail: klaus-dieter.oswald@cue.gmbh

W. Riedel
e-mail: wriedel24@gmail.com

A. Le Hung
Industrial University of Ho Chi Minh City, No. 12 Nguyen van Bao, Ward 4, Ho Chi Minh City, Vietnam
e-mail: lh.anh.9@googlemail.com

I. Nolivos · L. Dominguez-Granda
Escuela Superior Politécnica del Litoral (ESPOL), Campus Gustavo Galindo, Km. 30.5 vía Perimetral, Guayaquil, Ecuador
e-mail: inolivos@espol.edu.ec

L. Dominguez-Granda
e-mail: ldomingu@espol.edu.ec

© Springer Nature Switzerland AG 2020
W. Leal Filho et al. (eds.), *International Business, Trade and Institutional Sustainability*, World Sustainability Series,
https://doi.org/10.1007/978-3-030-26759-9_31

illustrates best practice using examples from Vietnam (South East Asia), Uganda (East Africa), and Ecuador (Latin America).

Keywords Artisanal and small-scale mining (ASM) · Small scale aggregates mining · Responsible mining · Sustainability

1 Introduction

Mineral raw materials are one of the most important foundations of our economy and lifestyle. As these raw materials are important for many areas of production, they form the basis for mankind's prosperity. The global demand for most of the raw materials is increasing, making it harder and more expensive to boost supplies. And both the mining and product's use, come along with its environmental damage. In addition to biological diversity, water, soil or clean air, minerals are natural resources. Raw materials are unprocessed or only slightly processed substances or mixtures that can be used in production processes (German National Environmental Agency 2014). Usually, a distinction between primary and secondary raw materials is made: primary raw materials are taken from nature and secondary raw materials originate from recycling, and secondary raw materials originate from recycling. For many primary raw materials, the term mineral resources is used, including in legal regulations on mining such as the EU Extractive Waste Directive (2006/21/EC; European Commission 2006): "*'mineral resource' or 'mineral' means a naturally occurring deposit in the earth's crust of an organic or inorganic substance, such as energy fuels, metal ores, industrial minerals and construction minerals, but excluding water.*"

Abiotic raw materials are all raw materials that are not sourced or derived from living organisms (Drielsma et al. 2016). These include ores and other mineral resources, construction minerals such as sand, gravel, stones and industrial minerals such as quartz sand, potash salts (Drielsma et al. 2016). Fossil fuels like coal and oil are also referred to as abiotic raw materials, although they are originally derived from biomass. Yet, they only developed to their present state after millions or billions of years by slow geological processes (diagenesis). These abiotic raw materials are also referred to as non-renewable raw materials contrary to renewable raw materials derived from agriculture and forestry. These are animal and plant substances also referred to as renewable raw materials (German National Environmental Agency 2014). Usually, the environmental impact of raw material extraction is estimated through the Life Cycle Assessment method, based on a Material Flow Analysis, using the Material Footprint as sum parameter (Drielsma et al. 2016; Wiesen and Wirges 2017).

Raw materials are important to produce goods and thus are the basis of a society's prosperity. As the world's population still grows and emerging and developing economies become increasingly wealthy, the world economy's demand for raw material rises and an increasing amount of natural resources need to be extracted.

According to the German Ministry of Environment, Nature Protection and Nuclear Safety (2018), the global raw material consumption has already reached 90 billion tons in 2017, while high-income countries currently consume 10 times more raw materials per capita than low-income countries. In 2010, for example, more than 70 billion tons of raw materials were used worldwide, about one third more than in 2000, two thirds more than in 1990 (United Nations Environment Programme 2016). For comparison, in the late 1970s, the global raw materials extraction was 22 billion tons (United Nations Environment Programme 2016). Aggregates for construction purposes, like sand and gravel, are the world's most commonly mined raw materials with 11 billion tonnes mined in 2010 (World Economic Forum 2017). The Asia-Pacific region was the largest producer, followed by Europe and North America.

Increasing raw material demand is a major challenge from an economic as well as an ecological and social point of view. From an economic point of view, constantly increasing or fluctuating commodity prices can be considered a burden. In addition, it is problematic when the supply of raw material is not ensured. Some important raw materials, such as petroleum, cobalt and rare earth metals, are getting less accessible from readily available sources. From an ecological point of view, the increasing use of raw materials is worsening global environmental problems which need to be assessed through a comprehensive Life Cycle Assessment, considering particularly the material flows (Drielsma et al. 2016; Wiesen and Wirges 2017). The possible environmental impact concerns the entire value chain, from extraction to disposal. They range from the release of greenhouse gases, pollutants into the air, water and soil, soil degradation to the increasing loss of biodiversity, especially in eco-logically sensitive areas. Overall, the use of natural resources already exceeds the earth's ability to regenerate today.

Raw material extraction intervenes with nature and the environment in a variety of ways and occurs over a long time. Typically, a raw material extraction project goes through several stages of its life cycle. The first phase is the exploration of a deposit. Subsequently, it is developed and usually mined over many years. The raw materials are processed and refined. At the end, the deposit is closed and ideally there is a follow-up. In each of these phases, different environmental impacts occur, sometimes in different places—for example, when the raw material is not processed at the deposit. Further, there are effects through access roads, which are built in previously untouched areas. Of importance are interferences with the water balance and the natural environment, which, among other things, can affect the biodiversity. Other uses of nature by humans can be affected, for example, the extraction of groundwater or the use of agricultural land. In addition, the extraction of raw material consumes energy and pollutants might be released into water, soil and air. From an ecological point of view, it is also important how mining waste is handled.

Nature and extent of the environmental impact of raw material extraction can vary considerably. They depend on the raw material, the technologies used and the ecological conditions. Other factors are the political and social conditions. For example, a weak government is barely able to fully en-force compliance with environmental and social standards in mining. The step from exploration to development and closure can often increase the effect on the environment. Provided that effective closure processes take place, they usually decline during the closing and rehabilitation phase of the raw material extraction site. These effects are described as "ecological back-pack" of the raw material obtained. For instance, in Germany, the yearly extraction of aggregates sums up to 520 million tons, causing another "ecological backpack" of 100 million tons (Statistical Office of Germany 2017).

Many raw materials are mined in open-pit mining, particularly construction minerals such as sands, gravels, stones and mineral industrial raw materials. Often, the open pit has serious consequences for the surrounding environment: the degradation of natural stones on mountain slopes affects the vegetation, sand and gravel mining intervene with the soil structure and the groundwater balance. Often, the landscape is fundamentally changed, and the land is irretrievably lost for uses like agriculture and forestry. Even after the end of mining, opencast mines are a challenge. Due to the changes in the soil structure, later use for settlements is hardly possible, and for agriculture becomes difficult. Natural soil fertility is usually not fully restored. Northey (2018) summarised that there is a strong incentive to improve the understanding of the effects of mining so that informed decision making, improved mine-site management, and better societal outcomes can be achieved. Further, it was recognised that there are needed criteria that can be used to assess the progress and contribution of the mining industry towards meeting sustainable development goals (Fonseca et al. 2013; Moran et al. 2014).

This contribution refers to artisanal and small-scale mining (ASM), with a particular relevance to developing countries. According to Hentschel et al. (2003), ASM refers to mining by individuals, groups, families or cooperatives with minimal or no mechanization, often in the informal sector of the market. In ASM activities, mineral raw materials are mined from deposits which are unsuitable for industrial mining, mined with simple means and with little use of machinery. Artisanal miners often work self-organized or under the control of local patrons, but also illegally. The extraction of raw materials often entails precarious working conditions, without regard to labour, children or women's rights. Often, miners earn only a few dollars daily, while profits go to middlemen and local patrons. The causes of these problems are often weak institutions, insufficient legal framework conditions and the lack of regulation of the sector. According to the World Bank, ASM occurs in approximately 80 countries worldwide, is largely poverty-driven and a part of the rural economy (Labonne 2014).

Small-scale mining forms one of the biggest challenges in the extractive sector. It is often illegal, sometimes catastrophic and causes substantial environmental damage (Villegas et al. 2012). Often, basic human rights are ignored, such as the rights of women and children or health protection. Informal small-scale mining is labour-intensive and depending on the price of raw materials, it employs between 15 and 30 million people worldwide (Villegas et al. 2012). Considering family members and downstream supply chains, 80–100 million people globally depend on small-scale mining, according to estimates by the World Bank. Thus, small-scale mining is an important source of income and substantially contributes to the livelihood of the local population (Villegas et al. 2012).

2 Methodology

This investigation is based on an in-depth literature review and field studies in Vietnam (South East Asia), Uganda (East Africa) and Ecuador (Latin America). The field study in Vietnam was performed in the regions of the Hanoi and Hoa Binh provinces as well as the Ho Chi Minh City in November 2017, the field study in the south west region of Uganda in March 2018 and the field studies in the south western region between Salinas in the lowlands and Cuenca in the highlands in Ecuador in July and September 2018. These countries were selected due to their status as low- or middle-income countries with a strong economic development and a fast urbanisation, still having a substantial proportion of ASM. Evaluation of case study countries was done qualitatively and semi-quantitatively through literature review and site visits, providing a national overview of the country context, its ASM sector and prevailing policy and legal frameworks.

For these countries, the basic data relevant for the investigation was collected, as there is population development, and data per capita, like brut income, biocapacity, Ecological Footprint (EF) as well as CO_2 emissions. Unsustainability occurs if the area's EF exceeds its biocapacity, where the EF is defined as the impact of a person or community on the environment, expressed as the amount of land required to sustain their use of natural resources (given in global hectares [gha]).

3 Case Study Sites: Aggregates Mining in Vietnam, Uganda and Ecuador

3.1 Small-Scale Aggregate Mining in Vietnam

Vietnam has a vast mineral resource potential. There are approximately 4200 mining projects by 2000 businesses active in Vietnam (Thanhniennews 2012). The country hosts all dimensions of mining, from small artisanal mines to huge multi-million-dollar mining operations (Fong-Sam 2013). According to Fong-Sam (2015), Vietnam has 930 quarries and mines under legal exploitation, out of which 433 produce construction aggregates, 88 produce clays, 81 produce cobblestone and sand and 52 produce coal (anthracite). The remaining 276 quarries and mines produce ferrous and nonferrous minerals. Aggregates play a fundamental role in the urbanisation and industrialisation activities of a country. Most of the aggregate companies are small-scale industrial facilities, while also informal artisanal sand mining does exist. In Vietnam, the urbanisation rate developed from 28.5% in 2007 to 34.9% in 2017 (Statistica 2017). The UN Human Development Index for Vietnam is 0.683 (place 115 out of 188 countries), the Gini coefficient (expression of inequality) is 34.8, expressing a high percentage of inequality.

With approximately 92.7 million inhabitants (2017) on an area of about 332,000 km^2, Vietnam ranks place 14 of the largest states in the world according to population and place 40 in terms of population density. Table 1 gives an overview of the economic and ecologic capacity of the country.

The establishment of Vietnam's minerals policy began during the Doi Moi reformation policy, and in 1987 Vietnam's government created the Law on Foreign Investment. Originally, the Mineral Law passed in 1996 and has been amended twice, most recently in November 2010 (fully effective since July 2011). Vietnam's government established an environmental regulation beginning in 1992, when the first environmental agency, the Ministry of Science, Technology, and the Environment, was established (Nguyen and Harrison 2017). The Law on Environmental Protection (LEP) was established in 1993 and it requires all investment properties to submit Environmental Impact Assessments (EIA). Further, in art. 4(1) the LEP requires that *"environmental protection must be in harmony with economic development and assure social advancement for national sustainable development."* Thus, environmental protection is understood as a policy that must support economic development and national security, not having as a priority the protection of environmental health. Shortly after the LEP was issued, a government directive was issued to strengthen environmental protection during the industrialization and modernization process of the country. Further, the Vietnamese Constitution (articles 17–18) requires the people of Vietnam's societal responsibilities to the environment such as ownership of the land, water, and other natural resources and prohibits *"all acts"* that result in the depletion or destruction of the environment by individuals, organizations or the government. The 2010 Mineral Law addresses the *"Protection of Environment, and Use of Land, Water and Infrastructure in Mineral Activities"* which requires mining entities to use environmentally-friendly equipment and materials, minimize any adverse impact on the environment and rehabilitate the environment according to law.

Megatrends of urbanization and industrialization are driving Vietnam's construction industry forward. In Vietnam, the construction industry has accounted for between 5 and 6% of the nominal GDP in recent years. The gross value added by the construction industry ranged between $9 and 10 billion and is projected to double by 2020. Ministry surveys in 2015 revealed that up to 50–60 million m^3 of sand

Table 1 Economic development in Vietnam (*source* National Footprint Accounts 2018 edition, data year 2016; World Development Indicators, The World Bank, 2016; U.N. Food and Agriculture Organization; https://data.footprintnetwork.org, the unit gha refers to global hectares)

Year	1971	1981	1991	2001	2011	2014	2017
Inhabitants (Mio)	42.7	53.7	66.0	78.6	87.9	90.7	92.7
Brut income per capita (US$)	n.d.	n.d.	130	430	1390	1900	2050
Biocapacity per capita [gha]	1.0	0.8	0.7	0.9	1.0	1.0	n.d.
EF per capita [gha]	0.8	0.7	0.7	1.0	1.7	1.7	n.d.
Built up land [gha]	0.04	0.04	0.07	0.09	0.11	0.11	n.d.
CO_2 emission per capita (t)	0.7	0.3	0.3	0.8	1.8	1.7	1.7

n.d. No data available

are necessary to meet the annual demand of on-going construction projects in the country. A government-issued report from the Department of Construction Materials in the Vietnam Ministry of Construction in August 2017, based on statistics from 49 provinces and cities, indicated that by the end of 2016, sand and gravel there have been issued mining permits for the mining of 691 million m^3 of sand and gravel have been issued. Ministry surveys in 2015 revealed that up to 50–60 million m^3 of sand are necessary to meet the annual demand of on-going construction projects in the country. According to information from the Ministry of Construction in Vietnam, the domestic demand for construction sand between 2016 and 2020 is estimated at around 2.3 billion m^3, while the country's total sand reserves are just about 2 billion m^3. Continuing the current sand consumption in Vietnam, the country will run out of building sand as building material by the year 2020. Gravel and sand in Vietnam are usually extracted from river beds (Fig. 1), while other aggregates are extracted from quarries (Fig. 2).

Fig. 1 Small scale industrial aggregates mining through river sand dredging in Vietnam; left: Son La river in Hoa Binh, right: Mekong delta

Fig. 2 Small scale industrial aggregates mining through hard rock and soil cover cutting in Hoa Binh province in Vietnam

3.2 Small-Scale Aggregates Mining in Uganda

Located in the Great Lakes Region in East Africa at the Equator, Uganda is a low-income country with a relatively solid economic growth. Having ambitions to achieve middle-income status by 2020, the mineral sector plays a key role in the nation's economic transformation, as was outlined as one of Uganda's top five priorities in the Second National Development Plan (NDP II) 2015–2020. The country lies in the East African highlands with heights of 1000–1500 m and is traversed by the East African ditch system. It has a size of 240,000 km^2, out of which almost 20% are taken up by large inland lakes. Uganda's population with over 40 million people, is of great ethnic diversity. The UN Human Development Index for Uganda is 0.493 (place 163 out of 188 countries), the Gini coefficient (expression of inequality) is 44.3, expressing a high percentage of inequality.

Uganda's industrial sector is underdeveloped. While small amounts of copper ore and gold are mined, the most important industries are tobacco processing, food, textile, wood, metal and building materials production as well as the chemical industry. Main trading partners are Kenya, Great Britain and Germany. The Government of Uganda started to plan massive infrastructure improvements in order to achieve industrialization, job creation, robust economic growth and, ultimately, to fulfil the nation's ambitions of middle-income status by 2020 as laid down in the NDP II. Public investments in infrastructure coupled with private demands for housing and other key products provide opportunities for Uganda's Development Minerals sector (Hinton 2009; UBOS 2017). At 6% annual growth, the construction sector continues to be strong and becomes a market for a broad range of aggregates including sand, clay, limestone, marble, kaolin and sources of stone aggregate (Hinton et al. 2018). In Uganda, the urbanisation rate developed from 13.6% in 2007 to 16.8% in 2017 (Statistica 2017). Table 2 gives an overview of the economic and ecologic capacity of the country.

Main mineral resources found in Uganda include common clays, specialty clays (including kaolin and bentonite), sand, a range of rocks used for stone aggregate and dimension stone production, limestone and marble, pozzolanic ash, gypsum, semi-

Table 2 Economic development in Uganda (*source* National Footprint Accounts 2018 edition, data year 2016; World Development Indicators, The World Bank, 2016; U.N. Food and Agriculture Organization; https://data.footprintnetwork.org, the unit gha refers to global hectares)

Year	1971	1981	1991	2001	2011	2014	2017
Inhabitants (Mio)	9.4	12.5	17.4	24.8	35.1	38.8	41.5
Brut income per capita (US$)	n.d.	n.d.	320	240	630	670	660
Biocapacity per capita [gha]	1.7	1.1	1.0	0.9	0.6	0.5	n.d.
EF per capita [gha]	2.1	1.9	1.7	1.5	1.3	1.2	n.d.
Built up land [gha]	0.04	0.04	0.04	0.05	0.04	0.04	n.d.
CO_2 emission per capita (t)	0.2	0.2	0.2	0.1	0.1	0.1	0.1

n.d. No data available

precious gemstones as well as other agro-minerals and industrial minerals, such as vermiculite and phosphate (Hinton et al. 2018). Due to the common occurrence of clays, for instance as laterite, ASM laterite soil mining is very common for instance for the home production of normal bricks and Interlocking Stabilized Soil Bricks (ISSB) (Fig. 5). In Uganda, ASM is summarised under the term small scale operation and defined as mining operations whose expenditure does not exceed 10 million Uganda shillings (about $2526) or use special-ized technology (INTOSAI—WGEI 2018).

Due to the location at the Equator and the related rainfalls, Ugandan clays are of sedimentary origin and well-suited for moulding and production of ceramics such as bricks, drainage pipes and floor and wall tiles and even as "fire clays" suitable for production of refractory (high temperature resistant) bricks (Hinton et al. 2018). ASM extraction of clays takes place particularly near urban centres. Ac-cording to Hinton et al. (2018), within a 150 km radius around the capital city Kampala 576 active and abandoned clay sites and 346 active and abandoned sand extraction sites exist (Hinton et al. 2018). Other areas of highly concentrated activity are found in the southwestern (Ntungamo and Bushenyi Districts) and western regions of the country (Mityana and Mubende Districts). Figure 3 illustrates a laterite clay deposit, a sample from the Mbarara region in Uganda, used for artisanal brick production.

Other aggregate production in Uganda includes stone aggregates and dimension stones. Stone aggregates are crushed stones with a specified size suitable as construc-tion material, mainly for use with cement and sand in the production of concrete. ASM operations avoid the dimension stone sector, as they prefer slightly softer, weathered rock (Fig. 4), whereas large-scale extraction (Fig. 5), which uses more sophisticated technology, favours harder rock (Hinton et al. 2018).

In accordance with the urbanisation strategy, aggregate quarries are located mainly within and around densely populated urban centres. According to Hinton et al. (2018), at least 316 stone quarries (both industrial and ASM) can be found in a 150 km radius of Kampala. Some temporary industrial quarries are opened throughout the country to meet the needs of specific infrastructure projects (Fig. 5). Dimension stones refer to slabs or blocks produced from natural stones that meet basic dimension requirements (length, width, thickness, shape) (Hinton et al. 2018).

Fig. 3 Artisanal brick production from local laterite clay in the Mbarara region in Uganda

Fig. 4 Artisanal aggregates mining in the Kisoro region in Uganda

Fig. 5 Small scale industrial aggregates mining in the Kisoro region in Uganda

The mining sector in Uganda is regulated by the Constitution of the Republic of Uganda, the Mineral Policy (2001), the Mining Act (2003, amended 2016) and the Mining Regulations (2004). There are also other laws that effect the mining sector, such as the National Environment Act, the Income Tax Act and the Land Act (Saferworld 2017). The regulatory framework to guide the extraction and development of mining in Uganda was set up with the draft of the new mineral law and the Draft Mining & Mineral Policy for Uganda in 2016, replacing the Mining Act of 2003. The policy aims on steering the establishment of a mineral exploration department to guide industry developments. According to the Ministry of Energy, the goal of the new policy is to develop the industry through increased investment, value addition, national participation and revenue generation to contribute to socio economic transformation and poverty eradication (The Monitor 2016). On the health and safety regulations, there are concerns that the current standards are not comprehensive and are not specific to mining. In terms of licences, in 1999 there were 66 licences issued for exploration and mining combined; by the beginning of 2010 there was a total of 517 licences issued (The Monitor 2016).

3.3 Small-Scale Aggregates Mining in Ecuador (Latin America)

The Republic of Ecuador covers an area of $271{,}000$ km^2 (including Galapagos Islands) and increased from 14.5 (national census performed in 2010 by Instituto Nacional de Estadística y Censos INEC) to 17.1 million inhabitants (national INEC-projections for 2018), of which about 2.6 million people are in the capital region and economic capital of Quito (2018 INEC projection per municipality). With a gross domestic product (GDP) of $97.8 bn in 2016, Ecuador is one of the middle-income countries. The informal sector is overstretched in Ecuador. More than one-third of all middle and low-income families are micro-enterprises, contributing to one quarter of the GDP. Only about a quarter of the companies are registered. According to the National Development Plan, measures such as strengthening social systems and redistributive policies have substantially reduced multidimensional poverty in the country between 2009 and 2016 from 51.5 to 35.1%. However, 22.9% still live below the national poverty line. The population and economy of Ecuador are also suffering from the consequences of the climate variability events, as El Niño Southern Oscillation (ENSO), investments in infrastructure projects are subsequently required and politically prioritized. In addition, new legislative frameworks, increased investment in low-population areas of the country and in the industrial sector (especially mining, oil production and tourism) should stimulate economic growth. To this end, the Institute for the Promotion of Exports and Investments *Pro Ecuador* created an investment catalogue in October 2016 with projects totalling more than $40 bn. In Ecuador, the urbanisation rate developed from 62.1% in 2007 to 64.2% in 2017 (Statistica 2017). Table 3 gives an overview on the economic and ecologic capacity of the country. The UN Human Development Index for Ecuador is 0.739 (place 89 out of 188 countries, expresses a high human development), the Gini coefficient (expression of inequality) is 45.9, expressing a high percentage of inequality.

In the 1980s, mining had a contribution of only 0.7% to the GDP and employing around 7000 people (eResearch Corporation 2018). Mineral extraction occurred in

Table 3 Economic development in Ecuador (*source* National Footprint Accounts 2018 edition, data year 2016; World Development Indicators, The World Bank, 2016; U.N. Food and Agriculture Organization; https://data.footprintnetwork.org, the unit gha refers to global hectares

Year	1971	1981	1991	2001	2011	2014	2017
Inhabitants (Mio)	6.1	7.9	10.2	12.8	15.2	15.9	16.4
Brut income per capita (US$)	520	2140	1370	1550	4900	6150	5820
Biocapacity per capita [gha]	5.3	4.2	3.4	2.8	2.3	2.1	n.d.
EF per capita [gha]	1.4	1.9	1.8	2.0	2.1	2.0	n.d.
Built up land [gha]	0.0	0.0	0.0	0.0	0.0	0.07	n.d.
CO_2 emission per capita (t)	0.7	1.7	1.6	1.8	2.6	2.8	2.8

n.d. No data available

regions with little to no access, thus exploration activities were hindered. Only in the last decade, mining activities were developed on a larger scale in Ecuador. In 2008, mining was declared a key strategic sector by the Government of Ecuador and refers to the National Development Plan of the Mining Sector (PNDSM). Currently, Ecuador's mining sector employs 3700 people, estimated to rise to about 16,000 by 2020 (Mining.com 2017). In the context of a growing market and largely untapped resources, Ecuador currently implements 17 major mining projects focusing on copper, gold, silver and molybdenum. The country dependents on exports and as a result relocates its investments into crude oil production, accounting to about 30% of exports (Mateo and García 2014). A new mining law was passed by the Congress in June 2013, making Ecuador much more lucrative for foreign investors. Yet, mining tends to have a negative connotation, especially among indigenous peoples. Environmental damage, conflicts over land rights and neglect of occupational safety and health are ongoing challenges that call for context-specific solutions. Therefore, several indigenous groups are in opposition to large-scale mining, planning to take their cases to the international courts.

The Mine Promotion Law of 1974 was the first law to clearly address the legalization of ASM in Ecuador (UNEP 2012). This law defines small-scale mining as activities with a production less than 1500 tons per month of ore or 50 tons daily of mineralized material for placer or alluvial deposits. Since 1999, new companies in Ecuador must apply for a permit before registering any facility that has potential environmental impacts. For the new registrations, an environmental assessment (comparable to EIA) must be presented. After registration, companies must submit every 6 months a new report on liquid and solid waste and emissions into the atmosphere to the competent environmental authority. The current Mining Law of Ecuador was established in 2009 and is characterized by neo-liberal reforms, which were supported by the World Bank between 1991 and 2008 (UNEP 2012). The legislation aimed at establishing large-scale construction projects without substantial state participation and neglected social and environmental aspects. As a result, the current Mining Act strengthens the role of the state in the regulation and implementation of mining projects (UNEP 2012). In September 2008, the new Constitution for Ecuador established that non-renewable natural resources are the inalienable property of the state. The Constitution grants the state the right to regulate non-renewable resources to guarantee a good life quality for the population while keeping sustainable development, cultural diversity, biodiversity and the capacity for ecosystems regeneration (Wacaster 2014).

The contribution of aggregate mining to the mineral production from mines and quarries accounted for 8% in 2014 (Wacaster 2014), out of which 73% originate from the Provinces of Guayas and Pichincha where mainly clay for brick production and sand is mined (Fig. 6). Limestone and clays are produced basically in the Imbabura Province (Wacaster 2014). Ecuador is quite advanced in terms of ASM: it has 1349 artisanal operations out of which 1069 are gold operations (UNEP 2012).

In 2017, the Ministry of Mining in Ecuador informed that the government had recently awarded 414 artisanal mining licenses and 36 small-scale mining titles, assuming the creation of more than 9000 jobs. At the beginning of September 2018,

Fig. 6 Artisanal brick production from local clay in Guayas Province (left) and artisanal sand extraction (right) in Santa Elena Province in Ecuador

Fig. 7 Small scale industrial aggregates mining in Azuay Province in Ecuador

Ecuador's Ministry of Mining announced that it expected the country's mining industry to produce $500 million in revenue in 2018. The National Mining Company announced to open 33 mining sites in the sub-district of Muyuyacu, Azuay, connected with the creation of an added 1800 jobs (Midwest 2018). Figure 7 shows a small scale industrial aggregates mining site in Azuay Province in Ecuador.

4 Results of the Data Collection and Discussion

4.1 General Considerations

Three types of gaps were found that lead to reduced responsibility in ASM: operational gaps, institutional gaps, and organisational gaps (Schneider et al. 2018). Each type of identified gap is somehow related to the relatively small scale of the mining operations and affects the companies' operational performance, their resource efficiency potential, as well as the connected environmental and social impacts. The operational gaps concern non-compliance with Best Practice technologies in open cast mining. Institutional gaps underline that mines are unable to implement state-of-the-art preparation and extraction technology due to licences for overly small extraction volumes and mine surfaces. The small-scale organisations and the short lifespans of the quarries result in organisational gaps which make it difficult to generate revenue for investment in technology and equipment. The result is a harmful cycle: non-efficient operations cause low revenues, which hinder investments to improve the existing situation, and the mines' resulting low productivity makes it furthermore challenging to obtain loans for investment (Schneider et al. 2018).

4.2 The Role of the Legal Framework

All investigated case studies have legal regulations for the mining sector, which have been revised by the countries over the past decade and hold a broad environmental legislation and a focus on economic and infrastructural development. Therefore, the problems of the sector are often not in the absence of laws, but in their implementation or in their implementation control. One of the main problems with implementation, especially in the control and supervision mechanisms, is the lack of human and financial capacity. Authorities are often poorly equipped and lacking the knowledge and capacity to effectively implement and monitor (e.g. Schneider et al. 2018). These problems are often compounded by the lack of legislation and insufficient institutional capacity to fight corruption, but national and international non-governmental organizations (NGOs) could become important supervisory and supervisory bodies (Rüttinger and Scholl 2017).

Ecuador has a specific regulatory framework for ASM but has also integrated ASM as part of the overall mining sector regulation in the Mining Law, General Mining Regulations and Environmental Mining Regulations (UNEP 2012). Generally, and considering Ecuador as an exception in terms of ASM, the informal sector is not regulated and environmental standards and regulations are often ignored. Usually, no aftercare is considered: reclamation, renaturation or a remediation of abandoned mining sites is often missing. For that reason, on a global scale, there are strong governmental intentions for the formalisation of the informal mining sectors (Marshall and Veiga 2017; Salo et al. 2016).

As Marshall and Veiga (2017) pointed out, in the formalisation process a) the legislation needs to be changed to directly correlate taxes with minerals production and b) formalization cannot exist without education. As Marshall and Veiga (2017) expressed: "*Without training and the organization of artisanal miners into groups, cooperatives or associations, governments cannot expect that these rudimentary miners will be converted into educated, responsible miners, who are willing to adopt cleaner, more sustainable practices*".

In Vietnam no regulation for ASM exists. In Uganda, the Mineral Policy (2001) has the scope to regulate and improve artisanal and small-scale mining. According to the Uganda Mineral Policy, the government shall apply facilitated regulation in small-scale mining, provide information on available production and marketing facilities, provide extension services to the small-scale miners through their associations, and carry out awareness campaigns targeting the artisanal and small-scale miners (INTOSAI—WGEI 2018).

4.3 The Role of Technology

Schneider et al. (2018) investigated the role of technology for the Vietnamese artisanal and small-scale aggregate mining. Typical for ASM, almost each of the visited open cast mine shows a non-compliance with the Best Practice in open cast mining technologies. Instead of a regular technology with exploratory and pre-processing works for the installation of slopes, berms and ramps, dangerous and hazardous exploitation without a continuous mineral extraction is carried out. The manual exploitation through climbing and blasting is dangerous for the workers, as they often transport the drilling equipment and explosives uphill without protection equipment (Schneider et al. 2018). A similar situation was seen for the ASM in Uganda: hard rocks are extracted by fire-setting. Another reason for non-compliance are discrepancies between the approved operational design and the practiced mining technology.

As Schneider et al. (2018) pointed out, the small-scale mining operations in Vietnam (50% less than 5 ha) lead to small annual extraction volume (less than 50,000 m^3/a; for comparison: the average in German aggregate mines is 300,000 m^3/a). This volume is not enough to generate enough financial resources for investments in technology and equipment. Further, the approved quantities suggest a relatively short operation time of the quarries. The mining operators can't invest to improve the existing situation due to the small revenues from a non-efficient operation, and the fact that they have difficulties to receive loans due to the low productivity. According to Hotelling's rule (Hotelling 1931), which defines the net price path as a function of time while maximising economic rent in the time of fully extracting a non-renewable natural resource, a maximum rent cannot be obtained with the low-tech equipment applied in ASM because the stock resource cannot be fully used.

4.4 The Role of Geographical Location

Increasingly, raw materials are being extracted and produced in remote, ecologically sensitive or politically unstable regions with little or insufficiently implemented environmental and social standards (Rüttinger and Scholl 2017). In this regard, a responsible mining approach as well as a sensitive approach to the natural capital is mandatory.

4.5   The Role of Governance

According to UNEP (2012) *"the advanced status of small-scale mining in Ecuador is a product of several favourable regulatory dispositions, international cooperation support and national acceptance of the sector. Another factor that may have played an important role is that Ecuador never had a considerable medium and large-scale mining sector"*. Consequently, some relevant conclusions can be drawn from this Ecuadorian situation:

> All mining title-holders are recognized and regulated in terms of their relationships and these provisions can enable 'good neighbour' behaviour. This integrated approach also creates the opportunity for partnership among the different operations and allows flexibility for development that reflects the diversity and complexity of the geology and the economic reality of the country. UNEP (2012)

Likewise, in Uganda ASM governance plays a large role. An example is the „Buy Uganda Build Uganda" (BUBU) initiative that represents policy reforms aimed at poverty eradication and improving the income levels of Ugandan citizens (Ministry of Trade, Industry and Cooperatives Uganda 2014). This initiative is regarded as one of the contributors to reducing poverty levels from 31% in 2008 to 19.7% in 2013 (Hinton et al. 2018). BUBU policy is premised on existing Government policies that support and encourage the consumption of locally produced goods and services. Part of the policy is to provide guidance to policy makers to ensure that promotion is carried out (Hinton et al. 2018).

A fundamental role in enabling ASM plays also the accessibility of microfinancing options and the development of local value chains, particularly for aggregates, as they do not have a high intrinsic material value. A feasible framework for local value chains might be provided through the Spaces—Practises—Goods Nexus, which refers to a sustainable consumption of locally produced goods representing the regional identity (Schneider and Popovici 2019).

4.6   The Role of Environmental Responsibility and Economic Aspects

Environmental issues, particularly in developing and emerging economies, cannot be seen isolated from social and economic aspects, as the resource sector, apart from its negative effects, also holds great opportunities for societal and economic development. The environmental impact of extracting raw materials has often had a direct and far-reaching effect on the livelihoods of the local population. Large parts of the population in developing countries, and indigenous people, depend on agriculture as a livelihood and means of gainful employment. They depend on functioning ecosystems such as rainforests, water or wetlands and their ecosystem services (Rüttinger and Scholl 2017). According to Rüttinger and Scholl (2017), the environmental and social impacts of the mining sector have acted as a primary driver of conflict in some cases, while in other cases they worsened existing conflicts. In most of the case studies, environmental and social impacts interacted as conflict drivers. To avoid conflicts, institutional control and supervisory mechanisms are needed. It should also be noted that the intensity of environmental impacts is related to social impacts, and greater negative environmental impacts might also lead to greater negative societal impacts.

4.7 The Role of Social Responsibility

A major social setback are the health effects of miners and the local population: for miners, there is a direct risk of accidents, rockfalls and landslides. Especially in the ASM, mining often takes place in unsecured locations and without proper security measures (Schneider et al. 2018 for Vietnam). Other common health effects for miners and the surrounding population included respiratory, skin and eye diseases. In this context, social responsibility plays a role that must not be underestimated.

Social responsibility has two dimensions:

(a) Social responsibility of the members of the society
(b) Corporate Social Responsibility (CSR).

Social responsibility is a duty that every individual must fulfil to keep a balance between the economy and ecosystems, which concerns the whole society consisting of producers and consumers. Corporate Social Responsibility (CSR) means the conscious and voluntary striving of companies to reconcile social and environmental goals with economic activity. The categories for sustainability schemes in mining in relation to the core subjects of social responsibility are framed in the ISO 26000 (Table 4; Kickler and Franken 2017). Further, international reference standards from institutions like the International Labor Organization (ILO) and the Organisation for Economic Co-operation and Development (OECD) are helpful.

5 Initiatives and Recommendations for Best Practices

Responsible mining requires that exploration, development and operation must be planned from the outset about the environmental effects respecting the precautionary principle. In addition, the so-called aftercare needs to be considered and implemented. This includes the handling of waste as well as the re-cultivation and restoration of freshwater ecosystems used for aggregate extraction after the end of the mining. Careful and efficient use of natural resources is a key competence of sustainable societies. In the past decade, several initiatives have been established to promote responsible mining, for example the Alliance for Responsible Mining (ARM), which is a global initiative founded in 2004 with the aim of transforming ASM into a socially

Table 4 Categories for sustainability schemes in mining in relation to the core subjects of social responsibility according to ISO 26000 (according to Kickler and Franken 2017)

ISO 26000 seven core subjects to social responsibility

Human Rights	Labour practices	Community involvement and development	The Environment	Fair operating practices	Organisational governance	Consumer issues
1. Human and workers rights		2. Social welfare	3. Use of natural resources	4. Emissions and land reclamation	5. Company governance	–
Serious human rights abuses		Community rights	Land use and biodiversity	Closure and land rehabilitation	Business practices	–
Employment conditions		Value added	Water use	Mine wastes and waste water	Management practices	–
Occupational health and safety			Energy use	Air emissions and noise		–
				Material use		

Identified the five categories and fourteen subordinate issues

and environmentally responsible activity that improves the quality of life of artisanal miners and their environment.

A particular role plays the "International Conference on Artisanal and Small-scale Mining & Quarrying" (ASM Conference) initiative, which was an activity of the ACP-EU Development Minerals Programme, organised by the African Caribbean and Pacific Group of States, European Union (EU), United Nations Development Programme (UNDP), and The Government of Zambia, with the support of The World Bank (WB), The African Union (AU), Organisation for Economic Co-operation and Development (OECD), International Conference on the Great Lakes Region (ICGLR), The Intergovernmental Forum on Mining, Minerals, Metals and Sustainable Development (IGF) and German Development Cooperation (GIZ). 547 delegates, representing 72 nations assembled in Livingstone, Zambia in September 2018 for the ASM Conference (International Conference on Artisanal and Small-scale Mining & Quarrying 2018) and adopted the 'Mosi-oa-Tunya Declaration', the first declaration of its kind since a decade, based on the earlier ASM conference declarations from Harare (1993), Washington (1996) and Yaoundé (2002). The key aspects in terms of Best Practice approaches of the 'Mosi-oa-Tunya Declaration' based on the former declarations are:

- *Upscaling ASM through multi-stakeholder partnerships for synergized action,*
- *Governments shall take primary responsibility to improve the legal and regulatory conditions for artisanal mining and ensuring the appropriate institutions to carry out this mandate,*
- *Integration of ASM policy into the Poverty Reduction Strategy Paper (PRSP) process with linkages to other rural sectors, and the development of a strategic framework for PRSPs,*
- *Identification and dissemination of Best Practice regulations,*
- *Mineral Resources Development, and a transparent, equitable and optimal exploitation of mineral resources to underpin broad-based sustainable growth and socio-economic development,*
- *Enhancing the capacity of public mineral institutions; developing mineral exploration and geoscientific information systems; developing the small- and medium-scale mining sectors; reducing adverse social and environmental impacts; improving energy and transport infrastructure related to mining and enhance mineral-based industrialisation,*
- *Curbing the illegal exploitation of natural resources through a (1) Regional Certification Mechanism; (2) Harmonization of National Legislation; (3) Regional Database on Mineral Flows, (4) Formalization of the Artisanal Mining Sector; (5) Promotion of the Extractive Industry Transparency Initiative (EITI) and a (6) Whistle Blowing Mechanism,*
- *Fighting against Illegal Exploitation of Natural Resources,*
- *Recalling the OECD Due Diligence Guidance for Responsible Supply Chains of Minerals,*
- *Acknowledgement of the role of Development Minerals in sustainable development to overcome environmental, social, labour and other challenges,*

- *Recognition that the health of the natural environment is critical to the livelihoods and the health of everyone, and that mining practices that maintain the integrity of rivers, oceans and forests are important both in their own right, and for ecosystem services such as firewood, food, and clean drinking water,*
- *Recognition that seasonality has led many miners and farmers to practice both agriculture and mining as part of livelihood diversification strategies,*
- *Restating the primary role of governments is to protect human rights as set forth in the UN Guiding Principles for Business and Human Rights,*
- *Promotion of transparency in Mineral Supply Chains.*

Further initiatives are: Fair Stone e.V.; the Global Reporting Initiative (GRI), Initiative for Responsible Mining Assurance (IRMA), and Xertifi X e.V. The XertifiX Association has developed two seals, which are awarded when clear standards are met: XertifiX and XertifiX plus. Since 2006, XertifiX has been inspecting quarries and natural stone companies in India, since 2014 in China and Vietnam (XertifiX 2017). The controls are used to check compliance with the XertifiX standard. As a result, XertifiX ensures that no child labour or slavery is carried out, that all ILO core labour standards are obeyed, the working conditions of adult workers are progressively improved and basic environmental protection measures are obeyed (XertifiX 2017). In addition to processing companies all quarries are assessed. The production sites are controlled twice a year and one of these control visits must be unannounced. XertifiX e.V. provides the XertifiX PLUS-Label when the following aspects are met:

- Compliance with ILO core labor standards,
- Fair working hours (ILO Conventions No. 1 and No. 14),
- Payment of statutory minimum wages according to national legislation,
- Provision and use of personal protective equipment (e.g., boots, helmets, eye protection, ear protection, mouthguards).

A specific approach to traceability of material from mine to retail is the "closed pipe" supply chain used by some companies. This entails a greater control and transparency of the supply chain and minerals' origins provide incentives for mining companies such as long-term partnerships, better contracting or pricing conditions which can complement traditional assurance mechanisms for ensuring compliance (Kickler and Franken 2017). The main recommendations concluded from this contribution for an ASM Best Practices Framework are:

- Respect the relevant ISO 26000 core subjects of social responsibility (human rights, labour practices, community involvement and development, environment, fair operating practices, and organisational governance),
- Development of (preferably local) value chains and their life cycle assessment leading to Certified Trading Chains (CTC) (Kickler and Franken 2017),
- Strengthening primary and secondary raw material awareness,
- Systematic assessment of waste streams,
- "Good neighbour" behaviour as base for a partnership among the different operations,
- If applicable and feasible: formalisation of the (often) informal activities,

- Defining and enforcing requirements for the extraction of raw materials and the supply chains for raw materials. Compliance with environmental, social and transparency standards on the part of consumers. Establishment of a Regional Mineral Certification Mechanism (RCM) (Kickler and Franken 2017),
- A preventive and recovery-oriented waste management, including the integration of Industrial Ecology thinking into the management of mining waste (Lèbre and Corder 2015).

According to Laurence (2011) the International Council on Mining and Metals (ICMM) principles *"ignore one important dimension that distinguishes mining from all other activities: a focus on the mineral resource itself"* and underlined the need for recovery-oriented waste management of mineral resources. Since the use of natural resources is closely related to economic performance and thus to wealth, it is the policy of policy makers to decouple economic growth from the use of resources. This means that fewer natural resources are used and the environmental impact is to be reduced with the same or increasing economic performance. This approach is often referred to as increasing resource efficiency or "decoupling".

Even though the ecological impacts of ASM are well documented on aquatic and terrestrial ecosystems (e.g. Costea 2018; Ncube-Phiri et al. 2015; Ahmed and Oruonye 2016), published studies also demonstrate the after-life effects caused by mining activities if no proper closure procedure is adopted (e.g. Warra and Prasad 2018; Swab et al. 2017; Oyewo et al. 2018). Loss of ecosystems services due to deforestation as well as habitat degradation on streams and rivers (e.g. riparian deterioration, impaired water quality) are among the major challenges for ecosystem reclamation. Proper management and follow up is needed after the end of mining activities, especially in developing countries, where legal and institutional gaps can overlook environmental liabilities. Current environmental engineering techniques for land remediation and effluent treatment can be valuable tools to mitigate the negative effect of ASM, improving the quality of life of the communities involved (e.g. Kemp and Owen 2013; Majer 2013; Hinton et al. 2003). A guideline for the responsible mining and processing of aggregates and the mining sites after-use was recently published by Oswald et al. (2018) for Vietnam, being applicable also for other countries.

6 Summary

In conclusion, the best practice for responsible small-scale aggregate mining addresses sustainability, particularly the Sustainable Development Goals, end poverty, Good health and well-being, Decent work and economic growth, industry, innovation and infrastructure, Sustainable Cities and Communities and Responsible Consumption and Production. Responsible small-scale mining can open new, environmentally friendly livelihoods for the poor and future generations.

To reduce the burden on the environment in the extraction of raw materials to an acceptable level, various strategies are pursued. This includes:

- to reduce the environmental impact directly in the extraction of raw materials, in production and consumption to use raw materials more economically and efficiently,
- to promote the use of secondary raw materials from waste and thus to reduce the use of raw materials (primary raw materials) taken directly from nature,
- In addition, the best available techniques should be used.

For instance, the German Federal Nature Conservation Act stipulates that major adverse effects on nature and the landscape are to be avoided and minimized as a matter of priority. Unavoidable impairments must be compensated. Either through compensation or replacement measures or, if this is not possible, through compensation payments. This approach could be also transferable to other countries.

To promote responsible mining strategies, the international "Extractive Industries Transparency Initiative" (EITI) was founded. Among other approaches, this initiative aims to combat corruption in connection with commodity transactions. Consumers, too, must come into direct contact with the question of the conditions under which raw materials are promoted. Further, a responsible mining governance, for example, promotes better environmental and social standards in the mining sector as part of a raw material diplomacy and development cooperation. At the same time, commodity-importing companies also must review their supply chains and to ensure compliance with environmental and social standards along the entire value chain.

References

Ahmed YM, Oruonye ED (2016) Socioeconomic impact of artisanal and small scale mining on the Mambilla Plateau of Taraba State, Nigeria. https://doi.org/10.22158/wjssr.v3n1p1

Carleton University, Partnership Africa Canada, Development Research and Social Policy Analysis Centre (2013) Women in artisanal and small-scale mining. https://olc.worldbank.org/sites/default/files/WB_Nairobi_Notes_4_RD3_0.pdf. Accessed 9 Sept 2018

Costea M (2018) Impact of floodplain gravel mining on landforms and processes: a study case in Orlat gravel pit (Romania). Environ Earth Sci 77(4):119. https://doi.org/10.1007/s12665-018-7320-y

Drielsma JA, Allington R, Brady T, Guinée J, Hammarstrom J, Hummen T, Russell-Vaccari A, Schneider L, Sonnemann G, Weihed P (2016) Abiotic raw-materials in life cycle impact assessments: an emerging consensus across disciplines. Resources 5(1):12. https://doi.org/10.3390/resources5010012

eResearch Corporation (2018) Initiating report gold. https://www.baystreet.ca/articles/research_reports/eresearch/CoreGold_021918-P.pdf. Accessed 27 Oct 2018

European Commission (2006) Directive 2006/21/EC of the European Parliament and of the Council of 15 March 2006 on the management of waste from extractive industries and amending Directive 2004/35/EC

Fong-Sam Y (2013) The mineral industry of Vietnam. U.S. Geological Survey Minerals Yearbook (2013). https://www.usgs.gov/centers/nmic/asia-and-pacific#vm

Fong-Sam Y (2015) The mineral industry of Vietnam. U.S. Geological Survey Minerals Yearbook (2015). https://www.usgs.gov/centers/nmic/asia-and-pacific#vm

Fonseca A, McAllister ML, Fitzpatrick P (2013) Measuring what? A comparative anatomy of five mining sustainability frameworks. Miner Eng 46–47:180–186. https://doi.org/10.1016/j.mineng. 2013.04.008

Fonseca A, McAllister ML, Fitzpatrick P (2014) Sustainability reporting among mining corporations: a constructive critique of the GRI approach. J Clean Prod 84:70–83. https://doi.org/10. 1016/j.jclepro.2012.11.050

German Ministry of Environment, Nature Protection and Nuclear Safety (2018) Verantwortungsvolle Nutzung von nicht erneuerbaren Rohstoffen. Available online: https://www.umwelt-im-unterricht.de/hintergrund/verantwortungsvolle-nutzung-von-nicht-erneuerbaren-rohstoffen/. Accessed 3 Sept 2018

German National Environmental Agency (2014) Ecofriendly abiotic resource extraction. Available online: https://www.umweltbundesamt.de/en/topics/waste-resources/resource-conservation-in-the-manufacturing/ecofriendly-abiotic-resource-extraction#textpart-1. Accessed 8 Sept 2018

Hentschel T, Hruschka F, Priester M (2003) Artisanal and small-scale mining: challenges and opportunities, Russell (Nottingham) (this report was already published in 2002 as "Global Report on Artisanal & Small-Scale Mining" by the International Institute for Environment and Development (IIED), report Nr. 70)

Hinton J (2009) National strategy for the advancement of artisanal and small scale mining in Uganda. Report to Ministry of Energy and Mineral Development (MEMD), 144 pp

Hinton J, Veiga MM, Beinhoff C (2003) Women and artisanal mining: gender roles and the road ahead. In: Hilson G (ed) The socio-economic impacts of artisanal and small-scale mining in developing countries, Balkema (Amsterdam)

Hinton J, Lyster O, Katusiime J, Nanteza M, Naulo G, Rolfe A, Kombo F, Grundel H, MacLeod K, Kyarisiima H, Pakoun L, Ngonze C, Franks DM (2018) Baseline assessment of development minerals in Uganda, vol 2, Market Study and Value Chain Analysis, ACP-EU Development Minerals Programme. Implemented in Partnership with UNDP, © 2018 by the United Nations Development Programme. Available online: http://www.ug.undp.org/content/dam/uganda/docs/ UNDPUg18%20-%20DevMinBaseLineUganda-Vol.2.pdf. Accessed 3 Sept 2018

Hotelling H (1931) The economics of exhaustible resources. J Political Econ 39(2):137–175

International Conference on Artisanal and Small-scale Mining & Quarrying (2018) Mosi-oa-Tunya declaration on artisanal and small-scale mining, quarrying and development. In: International Conference on Artisanal and Small-scale Mining & Quarrying, 11–13 September 2018, Livingstone, Zambia. https://asmconference.org/pdf/Mosi-oa-Tunya_Declaration_EN.pdf. Accessed 27 Oct 2018

INTOSAI—WGEI (2018) Artisanal mining in Uganda, International Organisation of Supreme Audit Institutions (INTOSAI), Working Group on Audit of Extractive Industries (WGEI). Available online: http://www.wgei.org/mining/artisanal-mining-in-uganda/. Accessed 27 Oct 2018

Kemp D, Owen JR (2013) Community relations and mining: core to business but not "core business". Resour Policy 38(4):523–531. https://doi.org/10.1016/j.resourpol.2013.08.003

Kickler K, Franken G (2017) Sustainability schemes for mineral resources: a comparative overview. Federal Institute for Geosciences and Natural Resources. ISBN: 978-3-943566-94-9

Labonne B (2014) Who is afraid of artisanal and small-scale mining (ASM)? Extr Ind Soc 1:121–123

Laurence D (2011) Establishing a sustainable mining operation: an overview. J Clean Prod 1:278–284

Lèbre E, Corder G (2015) Integrating industrial ecology thinking into the management of mining waste. Resources 4:765–786. https://doi.org/10.3390/resources4040765

Majer M (2013) The practice of mining companies in building relationships with local communities in the context of CSR formula. J Sustain Min 12(3):38–47. https://doi.org/10.7424/jsm130305

Marshall BG, Veiga MM (2017) Formalization of artisanal miners: stop the train, we need to get off! Extr Ind Soc 4(2):300–303

Mateo JP, García S (2014) El sector petrolero en Ecuador. Problemas del Desarrollo. Revista Latinoamericana de Economía 45(177):113–139

Midwest (2018) Ecuador's mining industry projected to produce $1 billion in revenue in 2018. Available online: http://blog.midwestind.com/ecuador-mining-revenues. Accessed 3 Sept 2018

Mining.com (2017) Ecuador anticipates $4 billion in mining investments by 2021. Available online: http://www.mining.com/ecuador-anticipates-4-billion-in-mining-investments-by-2021. Accessed 27 Oct 2018

Ministry of Trade, Industry and Cooperatives Uganda (2014) Buy Uganda Build Uganda, Ministry of Trade, Industry and Cooperatives; The Republic of Uganda, September 2014. Available online: www.mtic.go.ug/images/policies/bubu.pdf. Accessed 27 Oct 2018

Moran CJ, Lodhia S, Kunz NC, Huisingh D (2014) Sustainability as it pertains to minerals and energy supply and demand: a new interpretative perspective for assessing progress. J Clean Prod 84:16–26. https://doi.org/10.1016/j.jclepro.2014.09.016

Ncube-Phiri S, Ncube A, Mucherera B, Ncube M (2015) Artisanal small-scale mining: potential ecological disaster in Mzingwane District, Zimbabwe, Jàmbá. J Disaster Risk Stud 7(1), Art. #158, 11 pp. http://dx.doi.org/10.4102/jamba.v7i1.158

Nguyen HT, Harrison D (2017) The international comparative legal guide to mining law in 2018, chapter 32: Vietnam, A practical cross-border insight into mining law, 5th edn. Ashford Colour Press Ltd. ISBN 978-1-911367-74-1, ISSN 2052-5427

Northey SA (2018) Assessing water risks in the mining industry using life cycle assessment based approaches. Ph.D. thesis, Monash University, Department of Civil Engineering, Faculty of Engineering

Oswald K-D, Schneider P, Riedel W (2018) Technical solutions for mining and processing of aggregates and the mining sites after-use: a cleaner production guideline for Vietnam; Issue 3. ISBN 978-3-933053-39-8, Open Access Publication CC BY-NC-ND-SA 3.0 DE, IÖR Eigenverlag Dresden

Oyewo OA, Agboola O, Onyango MS, Popoola P, Bobape MF (2018) Current methods for the remediation of acid mine drainage including continuous removal of metals from wastewater and mine dump. In: Bio-geotechnologies for mine site rehabilitation, pp 103–114

Rüttinger L, Scholl C (2017) Verantwortungsvolle Rohstoffgewinnung? Herausforderungen, Perspektiven, Lösungsansätze, Zusammenfassung der Ergebnisse des Forschungsvorhabens "Ansätze zur Reduzierung von Umweltbelastung und negativen sozialen Auswirkungen bei der Gewinnung von Metallrohstoffen (UmSoRess)", Schriften des Umweltbundesamtes FKZ 3712 94 315, UBA-Texte | 67/2017

UBOS (2017) Uganda national household survey. http://www.ubos.org/onlinefiles/uploads/ubos/pdf%20documents/UNHS_VI_2017_Version_I_%2027th_September_2017.pdf. Accessed 8 Dec 2017

Saferworld (2017) Mining in Uganda—a conflict sensitive analysis

Salo M, Hiedanpää J, Karlsson T, Ávila LC, Kotilainen J, Jounela P, García RR (2016) Local perspectives on the formalization of artisanal and small-scale mining in the Madre de Dios gold fields, Peru. Extr Ind Soc 3(2016):1058–1066

Schneider P, Popovici LD (2019) Approaches for the implementation of water-related cultural ecosystem services in teaching programs on sustainable development. In: Handbook of sustainability and humanities: linking social values, theology and spirituality towards sustainability. Springer, Cham, pp. 267–289. https://doi.org/10.1007/978-3-319-95336-6

Schneider P, Oswald K-D, Riedel W, Meyer A, Schiller G, Bimesmeier T, Pham Thi VA, Nguyen Khac L (2018) Engineering perspectives and environmental life cycle optimization to enhance aggregate mining in Vietnam. Sustainability 10:525. https://doi.org/10.3390/su10020525

Statistica (2017) Statistica—the statistics portal. Available online: https://www.statista.com/statistics. Accessed 27 Oct 2018

Statistical Office of Germany / Statistisches Bundesamt (2017) Umweltnutzung und Wirtschaft: Tabellen zu den Umweltökonomischen Gesamtrechnungen. Available online: http://www.destatis.de. Accessed 3rd Sept 2018

Swab RM, Lorenz N, Byrd S, Dick R (2017) Native vegetation in reclamation: improving habitat and ecosystem function through using prairie species in mine land reclamation. Ecol Eng 108:525–536

Thanhniennews (2012) Vietnam's lack of control over mining wastes resources: house committee. 17 Aug 2012. http://www.thanhniennews.com/index/pages/20120817-vietnam-loose-control-onmining-puts-resources-to-waste.aspx

The Economist (2009) Bauxite bashers. Available at http://www.economist.com/node/13527969. 3 Apr 2009

The Monitor (2016) Uganda: new mineral law in offing. Available online: https://allafrica.com/stories/201610030016.html. Accessed 3 Sept 2018

United Nations Development Programme UNEP (2012) Analysis of formalization approaches in the artisanal and small-scale gold mining sector based on experiences in Ecuador, Mongolia, Peru, Tanzania and Uganda—Ecuador Case Study

United Nations Environment Programme (2016) Global material flows and resource productivity—Assessment Report for the UNEP International Resource Panel. United Nations Environment Programme. ISBN 978-92-807-3554-3

Villegas C, Weinberg R, Levin E, Hund K (2012) Artisanal and Small-scale Mining in Protected Areas and Critical Ecosystems Programme (ASM -PACE), Working together towards Responsible Artisanal and Small Scale Mining—A Global Solutions Study. ASM-PACE: Global Solutions Study, p 1, © Estelle Levin Ltd. and WWF, September 2012

Wacaster S (2014) The mineral industry of Ecuador. U.S. Geological Survey Minerals Yearbook (2014). https://minerals.usgs.gov/minerals/pubs/country/2014/myb3-2014-ec.pdf

Warra AA, Prasad MN (2018) Artisanal and small-scale gold mining waste rehabilitation with energy crops and native flora—a case study from Nigeria. In: Bio-geotechnologies for mine site rehabilitation, pp 473–491

Wiesen K, Wirges M (2017) From cumulated energy demand to cumulated raw material demand: the material footprint as a sum parameter in life cycle assessment. Energ Sustain Soc 7:13. https://doi.org/10.1186/s13705-017-0115-2

World Economic Forum (2017) We're heading for a global sand crisis. Available online: http://theconversation.com/the-world-is-facing-a-global-sand-crisis-83557. Accessed 4 Apr 2019

XertifiX (2017) The XertifiX PLUS-Label. Available online: http://www.xertifix.de/en/siegel/. Accessed 3 Sept 2018

Assessing Sustainability in Mining Industry: Social License to Operate and Other Economic and Social Indicators in Canaã dos Carajás (Pará, Brazil)

Thiago Leite Cruz, Valente José Matlaba, José Aroudo Mota,
Celso de Oliveira Júnior, Jorge Filipe dos Santos, Leon Nazaré da Cruz
and Eduardo Nicolau Demétrio Neto

Abstract This research assessed sustainability in the mining industry in the municipality of Canaã dos Carajás, Brazilian Amazon, home to the world's largest iron ore mine. It analysed the mining Social License to Operate (SLO) and constructed Economic and Social Sustainability indicators. The SLO level was identified: Acceptance (3.62; Likert scale), the second level in a four-levelled scale. Procedural fairness, company-community interactions and improvement in infrastructure favoured the SLO, whereas the increasing cost of living and environmental impacts harmed it. Variables related to UN's SDG were selected from secondary sources. A Principal Component Analysis identified the most significant variables to calculate their weights, constructing the indicators. Positive influence of mining in the long term has been proved in both economic and social dimensions, from 2004/2008 to 2017. Benefits were observed through growth in jobs, taxes and production; improvements in education; social investments from the mining companies; increasing female par-

T. L. Cruz (✉) · V. J. Matlaba · J. A. Mota · C. de Oliveira Júnior · J. F. dos Santos ·
L. N. da Cruz · E. N. D. Neto
Instituto Tecnológico Vale, R. Boaventura da Silva, 955, Belém, PA 66055-090, Brazil
e-mail: thiago.james@live.com

V. J. Matlaba
e-mail: valente.matlaba@itv.org

J. A. Mota
e-mail: jose.aroudo.mota@itv.org

C. de Oliveira Júnior
e-mail: celso.oliveira@itv.org

J. F. dos Santos
e-mail: jorge.filipe@itv.org

L. N. da Cruz
e-mail: leonncruzz@gmail.com

E. N. D. Neto
e-mail: eduardodemetrio1@hotmail.com

© Springer Nature Switzerland AG 2020
W. Leal Filho et al. (eds.), *International Business, Trade and Institutional Sustainability*, World Sustainability Series,
https://doi.org/10.1007/978-3-030-26759-9_32

ticipation in mining jobs; and safeguard of staff's rights to unionization and collective negotiation. This research developed economic and social indicators to assess sustainability, and was the first to quantitatively assess the SLO in Brazil. These findings can guide actions to improve sustainability in the mining industry in accordance with the UN's SDG.

Keywords Social license to operate · Mining · Amazon · Sustainability · Indicators

1 Introduction

Despite globally publicized negative impacts, the importance of mining in daily life and for global development is undeniable. Notwithstanding, pressures for sustainability in mining are transcending the academic field and affecting international trade and law. There are already processes and certifications promoting trade of minerals from conflict-free areas (EU 2017; Mancini and Sala 2018). Mining companies are increasingly being taken to courts due to operational and environmental concerns. Civil society is struggling to have its demands addressed (Conde 2017). Social dissatisfaction with mining increases social risk, prompting protests that harm the activities of companies, which lead to financial losses and negative effects on their public image (Prno and Slocombe 2012).

The region of Carajás, located in the state of Pará in the Brazilian Amazon, is home to the greatest iron ore mine in the world (Duddu 2018). However, its activities are frequently interrupted by protests against the mining industry (G1 PA 2016; Survival International 2016). Despite many economic and social benefits provided by mining, the industry has caused issues for people, such as: inflation, real estate speculation, greater inflow of migrants and environmental impacts (Cabral et al. 2011).

Boutilier and Thomson (2011), Prno and Slocombe (2014) and Moffat and Zhang (2014) noticed that fulfilling all legal requirements does not guarantee successful operations. It demands engaging with stakeholders and managing impacts in multiple dimensions. Thus, there is an urgent need for developing sustainability indicators in mining to better understand its impacts and benefits in multiple aspects of daily life (Mota et al. 2017).

Gallopin (1996) highlights some universal requirements for indicators: measurability, availability, standardization, technical-financial feasibility in data mining, and political acceptance by decision-makers. Indicators allow comparison of different scenarios and identify specific characteristics and trends, adequately approaching sustainability (Liu et al. 2018).

Frameworks like the United Nations' Sustainable Development Goals (SDG), Global Reporting Initiative (GRI) and Social Life Cycle Assessment (SLCA) influenced the development of sustainability indicators. They guide good practices regarding economic, social and environmental impacts. Their goals involve matters such as human rights, health, education, land use etc., which can be assessed using secondary sources such as global reports, national censuses etc. (Mancini and Sala 2018).

Based on these frameworks, Mota et al. (2017) suggested key factors to construct sustainability indicators for the mining industry. In the social dimension: health and safety; education; diversity and opportunities; freedom of association; child labor; forced labor; indigenous and traditional peoples' rights; community management; bribe and corruption; investments in social infrastructure; stakeholders' engagement. In economics: financial contributions to the government; tax payment; job creation.

On the other hand, the Social License to Operate (SLO) is an indicator that reflects communities' and other stakeholders' opinions and feelings towards companies and their activities. It is tacit and changes over time, but it can be qualitatively and quantitatively assessed (Boutilier and Thomson 2011).

The SLO is independent of operational and environmental licenses regulated by governments, being informally granted by society. Based on public opinion regarding companies' performances, it requires engaging with stakeholders' needs and expectations. Thus, it may be considered a tool of social empowerment and an indicator of sustainability in mining (Prno and Slocombe 2014; Moffat and Zhang 2014).

Local communities are the most influential and most affected stakeholders in the mining industry but groups such as politicians, journalists and NGOs also have influence (Prno and Slocombe 2012). Different social dynamics demand specific approaches to acquire and maintain an SLO. It is crucial to understand each local reality and how different variables may affect each location (Prno and Slocombe 2014; Moffat and Zhang 2014).

Considering the importance of the municipality of Canaã dos Carajás for the world mining industry and the undeniable impacts of mining in the region, it is necessary to understand its social dynamics and the effects experienced locally to guide industry and government in adopting effective sustainable practices. This research aimed to assess the sustainability of mining activities in Canaã dos Carajás—PA—Brazil, by measuring its Social License to Operate and analysing economic and social variables to construct indicators of sustainability.

This paper is divided into five sections. The second section reviews the literature on SLO, the third explains the methodology, the fourth presents and discusses the results and, finally, the concluding remarks are found in the fifth section.

2   The Social License to Operate

At a World Bank Conference in 1997, the term Social License to Operate (SLO) was first coined by Jim Cooney. He recognized that the mining industry needed to engage with local communities in order to get a social license to avoid financial losses. Boutilier and Thomson (2011) were the first to theoretically analyse it as a conceptual model, describing the SLO as a pyramid consisting of four different levels (Fig. 1).

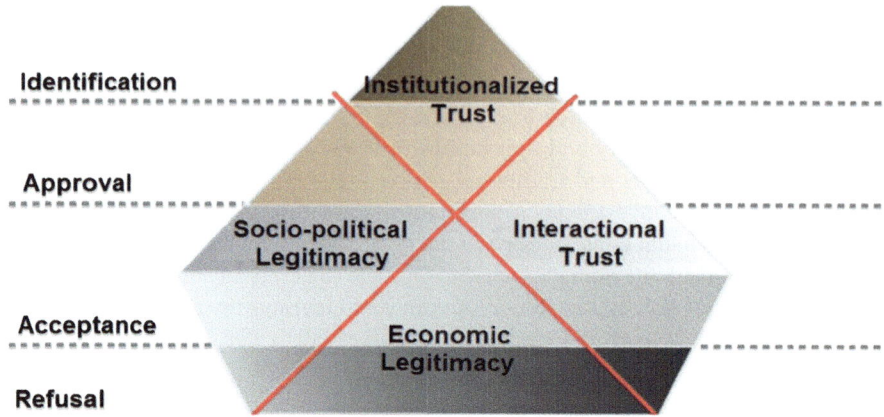

Fig. 1 Levels of the SLO and its factors (adapted from Boutilier and Thomson 2011)

Refusal is present when stakeholders deny any kind of support or sympathy for the company. Acceptance indicates tolerance and consent to the company's activities. Approval indicates satisfaction towards the industry. Identification is the highest level, when stakeholders enthusiastically support the companies and their activities, feeling it is an essential part of their lives. To reach higher levels, the company should build economic and socio-political legitimacy, as well as interactional and institutionalized trust in its relationship with the different stakeholders.

Trust in the industry and the perceived legitimacy of the company and its activities are key factors in the SLO. Moffat and Zhang (2014) confirmed this assumption, reckoning that although trust is harmed by negative impacts on social infrastructure, it is favoured by positive perceptions on procedural fairness and good-quality interactions between company and communities. Trust also positively influences acceptance and approval of the company's activities (Fig. 2).

Their study was conducted in Australia, in a region of CSG exploration. Using online surveys, 123 people were interviewed, and one year later the interview was repeated with 142 people. Using the Likert Scale, the participants answered questions regarding each analysed factor. Participants were all selected using the company's database, thus they may not have been representative of the community's diversity. Bivariate correlations and Path Analysis assessed the relationships between the variables. However, the study did not assess the influence of environmental and economic impacts, suggesting that future studies should investigate these areas. In fact, Prno and Slocombe (2014) coined the term *Cross-scale Effects*, noticing that negative experiences with mining in the past or in other regions may harm popular perceptions of mining companies anywhere.

To avoid biases and ensure the achievement of goals such as the UN's SDG, metrics and indicators must consider influences in multiple scales. Liu et al. (2018) stressed that analyses should articulate diverse impacts and benefits of mining activities in an integrated manner. Such an approach identifies trade-offs and influences that would not be otherwise acknowledged.

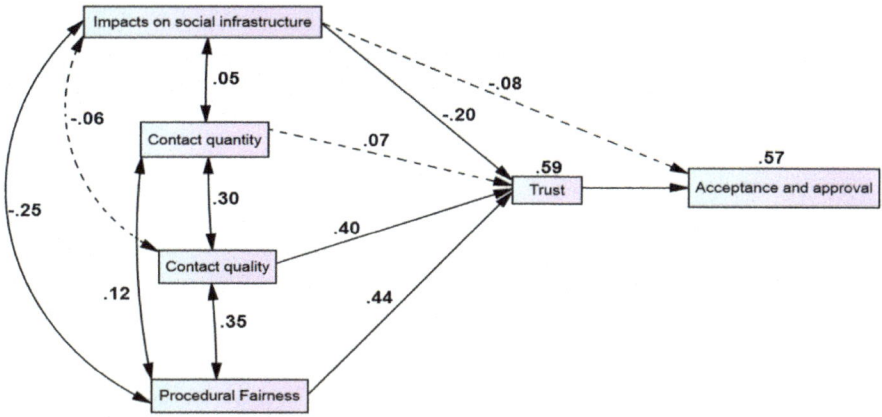

Fig. 2 Factors identified by Moffat and Zhang (2014). Block lines indicate statistically significant relationships, dashed lines indicate non-significance. Beta weights indicates the strength of relationships. The explained variance of trust, acceptance and approval is represented by estimated coefficients above them

In Brazil, very few studies mention the SLO to date. The concept is already known, but there are not yet specific procedures associated with the license (FGV et al. 2008; Angloamerican 2012). Only one research study analysed the SLO of a social project conducted by a mining company in Brazil, using just qualitative data (Santiago and Demajorovic 2016). The SLO has not yet been measured in a Brazilian mining project, and which factors may influence it remain unknown. Likewise, Mota et al. (2017) highlighted that few studies have analysed indicators for sustainability in mining, regarding the realities of Brazil or other developing countries.

3 Methodology

3.1 Location

Canaã dos Carajás, located in the state of Pará in the Brazilian Amazon (Fig. 3), was a small farming settlement, where agriculture and livestock had shares of 55 and 52% in the municipality's GDP in 1999 and 2002, respectively. The discovery of its mineral potential in 2004 and the immediate implementation of the first exploration project changed the region quickly. Nowadays, the locality is home to the world's largest iron ore mine (Duddu 2018). The mining companies invest generously in the region (DNPM 2018). The industry share in the GDP evolved from 9 to 84%, between 2002 and 2013. Its population grew from 11,139 inhabitants in 1996 to 33,632 in 2015.

Fig. 3 Location of Canaã dos Carajás. *Source* Created by the authors, using ArcGIS version 10.4, with data collected from IBGE (2018) and satellite imagery

Cabral et al. (2011) noticed that mining had positive effects in the municipality, such as: job generation, higher income, and construction of education and health facilities. Higher local government revenues due to mining taxes led to increased investments in infrastructure and public services.

However, Palheta da Silva et al. (2014) and Cabral et al. (2011) uncovered some negative impacts: company-community conflicts, real estate speculation, increased cost of living, and environmental impacts. The local government was unable to properly manage all the abrupt changes, and benefits and impacts were not fairly distributed among the population, which contributed to social dissatisfaction.

3.2 Assessing the Social License to Operate

190 people participated in this study. They belonged to seven social groups that reflected local diversity: students, teachers, people involved in NGOs and trade unions, journalists, clergy people, politicians, public service and private sector workers.

128 (67.4%) were men and 62 (32.6%) were women. The average age and its standard deviation were 31.4 and 11.7, respectively. Age varied from 18 to 78 years. The average years of education and its standard deviation were 10.6 and 3.3, respectively.

Score	Level
1.00–2.49	Withdraw
2.5–3.99	Acceptance
4.0–4.49	Approval
4.5–5.0	Identification

Table 1 Scores and corresponding levels of the SLO (Adapted from Boutilier and Thomson 2011)

17.4% of the participants completed elementary school, with 8 years of education; 17.4% reached high school, with 11 years; and 20% had completed a bachelor's degree, with 15 years. Only 3.2% completed more than 15 years of education.

Data collection took place from June to August 2016. People were randomly approached in the streets and asked to rate 34 topics using a five-point Likert scale (1 = strongly disagree; 5 = strongly agree). The first 15 were adapted from Boutilier and Thomson (2011) to measure the SLO. The mean average score indicated the level of SLO (Table 1):

1- We can benefit from a relationship with the companies.
2- We need the cooperation of the companies to reach our most important goals.
3- The companies keep the promises.
4- We are very pleased with our relationship with the companies.
5- The presence of mining projects is a benefit to us.
6- The companies listen to us.
7- In the long term, mining will contribute to the well-being of the whole region.
8- The companies treat everyone fairly.
9- The companies respect our way of life.
10- Our group (teachers, students, journalists...) and the companies have a similar vision for the future of the region.
11- The companies give support to those who they negatively affect.
12- The companies share decision-making with us.
13- The companies' decisions are fair and take into account our interests.
14- The companies are concerned with our interests.
15- The companies openly share relevant information with us.

The next 19 topics assessed people's perceptions regarding procedural fairness of the companies, company-community interactions, improvements on social infrastructure, and mining's impacts on the economy and environment. Each topic was rated using a five-point Likert scale (1 = strongly disagree; 5 = strongly agree). Improvements on social infrastructure were rated according to people's perception (1 = very negative; 5 = very positive), higher rates indicating improvements on this factor. Topics regarding impacts on economy and environment were inversely rated (1 = very positive; 5 = very negative), with higher rates meaning worse perceptions (Table 2).

The scores for each topic were averaged for each participant, for each group, and, finally, for the entire sample. A Path Analysis (a sophisticated Multiple Regression) was conducted to evaluate the relationships between the variables using the IBM

Table 2 Survey topics and analysed variables (Adapted from Moffat and Zhang 2014)

Survey topics	SLO variables
- 6, 9, 12, 13, 14 and 15	Procedural fairness
16- Our contact with the companies is frequent 17- Our contact with the companies and their staff has positive outcomes 18- Our contact with the companies and their staff is pleasant	Company-community interactions
19- We trust the companies 20- We have goodwill towards mining and the companies 21- The companies act responsibly	Trust
22- Access to medical and health facilities 23- Access to education 24- Quality of the urban infrastructure (e.g.: paving of streets and roads) 25- Sanitation	Improvements in social infrastructure
26- Cost of living 27- Job creation 28- Income opportunities	Impacts on economy
29- Water resources 30- Forest resources 31- Soil quality 32- Air quality (dust; pollution)	Impacts on environment
33- We accept mining companies 32- We approve the mining companies	Acceptance and approval

SPSS Amos 24 software (IBM Corp. 2015). This method tests hypotheses' coherence with the existing data. The following hypotheses were tested (Fig. 4):

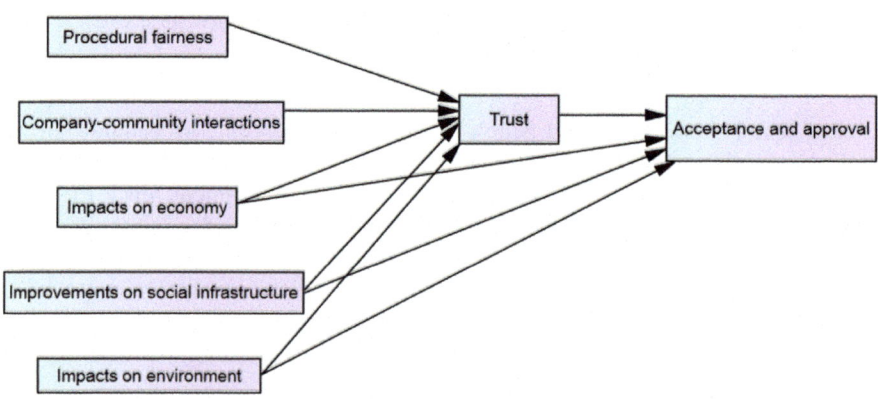

Fig. 4 Hypothetical model tested by the Path Analysis method

(i)   Procedural Fairness and company-community interactions have positive and direct influences on trust in the companies and their projects;

(ii) Impacts on economy and environment have negative and direct influences on trust, acceptance and approval; whereas impacts on social infrastructure have positive and direct influences on trust, acceptance and approval;

(iii) Trust positively and directly affects acceptance and approval of the companies and their activities.

The Path Analysis evaluates the goodness-of-fit of the hypothesised models to check whether the proposed relationships are adequate using the chi-square test, the Comparative Fit Index (CFI), the Normed Fit Index (NFI), and the Root Mean Square Error of Approximation (RMSEA). A satisfactory fit was indicated by a non-significant chi-square test, CFI $\geq$ .95, NFI $\geq$ .95, and RMSEA $\leq$ .06 (Hu and Bentler 1999; Kenny and McCoach 2003; Moffat and Zhang 2014).

3.3   Indicators of Economic and Social Sustainability

Based on suggestions by Mota et al. (2017), this paper constructed Indicators of Economic and Social Sustainability using variables (Table 3) from data collected from secondary sources, whose access is either publicly available or provided by official institutions such as Brazilian Ministry of Labour (RAIS 2018), Research Institute of Applied Economics (Ipea Data 2018), IBGE (2018), Compara Brasil (2018), GRI and Sustainability Reports (Vale 2018).

Data regarding each variable were statistically assessed using Principal Component Analysis (PCA) to estimate weights and values for each, in economic and social dimensions (Marôco 2014):

$$P_1 = a_{11}X_1 + a_{12}X_2 +, \ldots, +a_{1k}X_k$$
$$P_2 = a_{21}X_1 + a_{22}X_2 +, \ldots + a_{2k}X_k$$
$$\ldots$$
$$\ldots$$
$$\ldots$$
$$P_k = a_{k1}X_1 + a_{k2}X_2 +, \ldots + a_{kk}X_k$$

The equation of the components indicates how each variable is weighted to estimate the Indicators of Sustainability. Thus, the proposed indicators follow the equation:

$$V_i = f(X_1, X_2, X_3, \ldots X_n),$$

V_i value represents the indicator, whereas X_i indicates the independent variables.

Table 3 Variables for indicators of economic and social sustainability (Mota et al. 2017)

Sustainability dimensions	
Economic[a]	Social[b]
CFEM distributed to municipality/Total taxes received in the municipality (CFEM_Imp_Tot)	Training hours/Hours worked at mining companies (per year) (IVG_H_Training)
Distributed CFEM/GDP of the municipality (CFEM_GDP)	Scope of collective agreement in Trade Unions (IVG_Freedom_Assoc)
Distributed CFEM/Total Revenue of the Municipality (CFEM_Revenue)	% Women employed in mining (IVG_Women)
ICMS + ISS/Total taxes received in municipality (ICMS_ISS_Tot_Tax)	% Women in leading roles in mining (IVG_Women_Lead)
ICMS + ISS/Total Revenue of the Municipality (ICMS_ISS_Revenue)	% Women in mining jobs (IOL_Women)
Jobs in mining and related areas/Total jobs in the municipality (Jobs_Mining)	% Local mining employees who completed at least High School (IOL_High_School)
Salary amount in Mining and related areas/Total salary amount in the municipality (Sal_Amount_Mining)	Social investments by mining companies in the municipality/Total estimated revenue (IVL_Invest_Social_FAT)
Total taxes received in the municipality/Total Revenue of the municipality (Tot_Tax_Revenue)	Social investments by mining companies in the municipality/Total social investments by mining companies in the State (IVL_Invest_Soc_Munic_State)
Industry GDP/Total GDP of the Municipality (GDP_Industry)	Infrastructure investments by mining companies in the municipality/Total social investments by mining companies in the municipality (IVL_Invest_Infra)
	Investments in education by mining companies in the municipality/Social investments by mining companies in the municipality (IVL_Invest_Educ)

[a]CFEM is a Brazilian taxation for mining, paid by companies to surrounding municipalities. ICMS and ISS are general taxes for Industry and Services
[b]Some data have been collected and grouped in the following sets:
IVG—Global Vale Information: Information provided by Vale company, and made available in its sustainability reports (Vale 2018)
IVL—Local Vale Information: Information made available in the company's databases
IOL—Official Local Information: Data collected in official secondary sources (RAIS 2018)

To identify which variables were relevant for constructing the indicators, each one had their statistical significance checked using bivariate and anti-image correlations, KMO and Bartlett tests, and weighted means estimative matrixes. The weight of the significant variables was assessed through PCA; then, the Indicators of Economic and Social Sustainability were constructed.

4   Results and Discussion

4.1   SLO Level and Influencing Factors

The current SLO level of the mining industry in Canaã dos Carajás (Fig. 5) is **Acceptance** (3.63 points, standard deviation: 0.87), indicating that the stakeholders recognize the benefits provided by the mining industry and acknowledge its economic legitimacy. However, any trouble could generate crises and harm the public opinion, as the goodwill towards mining companies is still weak (Boutilier and Thomson 2011).

Businessmen should engage with the communities in transparent and reciprocal dialogue, with mutual respect, acknowledgement of communities' demands and exceeding expectations. This enhances social-political legitimacy and trust, leading to **Approval** and eventually to the **Identification** level.

People involved with NGOs, Public Servants and Industry workers showed the highest SLO level. The companies engage in partnerships and sponsorships with local NGOs to enhance social development. The workers were also very pleased with job and income opportunities provided directly and indirectly by mining.

Surprisingly, farmers declared satisfaction with mining. In the past, they were indemnified to move their farms and crops to other locations due to mine implementations and bought larger pieces of land. On the other hand, journalists severely criticized mining for its perceived negative impacts. Among politicians, opinions were divided; some praised mining and some were very critical. Some participants, though, acknowledge that the companies respect the laws and pay the taxes, but they declared that local politicians do not manage the economic resources adequately. This suggests that governance and mistrust of politicians may play an important role in SLO in Canaã dos Carajás, which should be investigated in future studies.

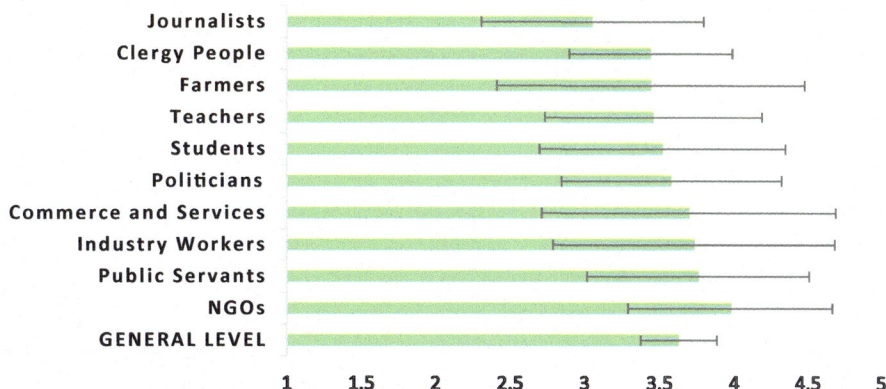

Fig. 5 Levels and standard deviation (thin bars) of the SLO in Canaã dos Carajás, for each stakeholder

As the SLO is granted by a large society, not only by some groups, open dialogue with different stakeholders is essential. The companies currently engage in dialogue with specific institutions and social leaders, leaving behind ordinary citizens and other stakeholders. Many people who were approached for this study were unable to respond to the survey because they were unfamiliar with the companies, only aware of their existence.

Among the participants, 40% complained that they had received little to no information about the projects implemented by the companies due to the lack of frequent and good-quality contact with the companies' staffs. The companies should be present at schools, churches and in the daily life of all ordinary citizens who are somehow influenced and affected by their operations.

The Path Analysis confirmed the hypotheses (Fig. 6). The nonsignificant Chi-square (χ^2[2df] = .467; p-value = .792) indicated that the hypothesized covariance matrix did not differ from the actual one. The other parameters confirmed the good fit of the model to the data (CFI = 1.0; NFI = .999 and RMSEA = .000). The hypothesized model explained 71.7% of the variance in Trust and 51.7% in acceptance and approval of mining, indicating a good index (Hu and Bentler 1999; Kenny and McCoach 2003; Moffat and Zhang 2014).

It was confirmed that *procedural fairness, company-community interactions, improvements on social infrastructure, trust* and *acceptance and approval* are positively correlated with each other. *Economic* and *environmental impacts* are positively

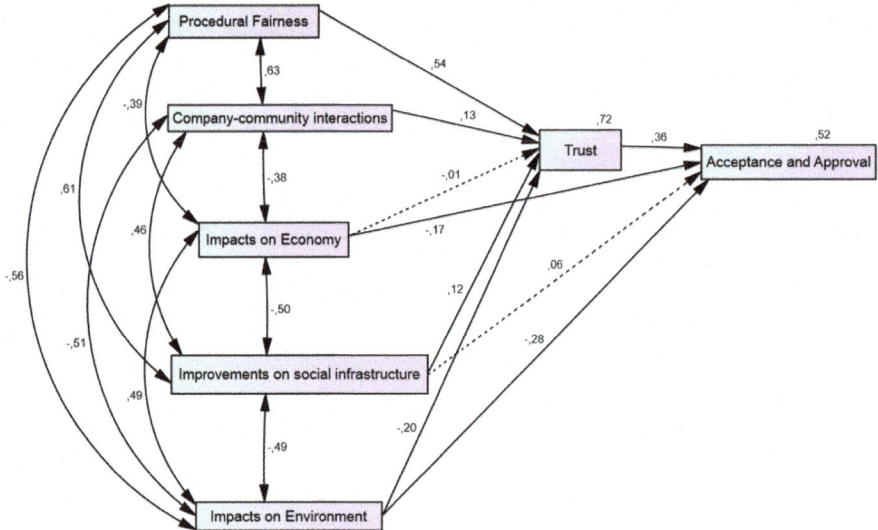

Fig. 6 Path Analysis explaining the relationships between the influencing factors of the SLO in Canaã dos Carajás (adapted from Cruz 2017). Block lines indicate statistically significant relationships, dashed lines indicate nonsignificant ones. Beta weights indicate the strength of the relationships. The explained variance of trust, acceptance and approval is indicated by the estimated coefficients above them

correlated with each other and negatively with the other factors, including *trust* and *acceptance and approval* of mining.

The greatest positive effect on *trust* by *procedural fairness* ($\beta = .54$; p-value = .000) highlights the importance of companies being honest and honouring promises to the community, listening and corresponding to the community's expectations. Likewise, good perceptions of *company-community interactions* ($\beta = .13$; p-value = .014) and *improvements in social infrastructure* ($\beta = .12$; p-value = .028) relate to higher *trust*.

The results suggest that it is more important to engage actively with the stakeholders in transparent and good-quality interactions, although the company should continue to mitigate its negative impacts in the region. The companies currently invest in education, health and infrastructure by building hospitals, schools, training labour forces and maintaining roads, but involving the community in the decision-making process and making them feel included is essential. Most companies invest more in mitigating negative impacts and creating infrastructure, disregarding this important factor.

Negative perceptions of *economic impacts* significantly influenced *acceptance and approval* ($\beta = -.17$; p-value = .007), but not *trust* ($\beta = -.01$; p-value = .897). Despite the job generation by mining, factors such as intense inflation and speculation contributed to popular dissatisfaction regarding the *economic impacts* of mining. In fact, 61% of the participants attributed the worst rate to the topic **Costs of Living** in the survey. However, some respondents attributed the inflation to the national political-economic crisis that took place in 2016, rather than mining. Prno and Slocombe (2014) had already discussed the *cross-scale effects*, noticing how perceptions of mining are influenced by factors of different spatial and time scales.

Such effects also influenced perceptions of *environmental impacts*, which negatively and significantly affect both *trust* ($\beta = -.20$; p-value = .000) and *acceptance and approval* ($\beta = -.28$; p-value = .000). Although recent mining projects are eco-friendly and less polluting, people are unaware of this (lack of dialogue and transparency) and suspicious due to previous negative experiences with mining across the country (Palheta da Silva et al. 2014). The break of a dam in a mining venture located in the southeastern Brazilian state of Minas Gerais in 2015 caused damage that caused serious harm to the environment and to people's lives and identities (Phillips 2016).

During periods of crisis, companies should engage with communities in transparent relationships, strengthening *trust* to enhance people's identification with the mining industry. In fact, *trust* was confirmed as an important predictor of *acceptance and approval* of mining companies and their activities ($\beta = .36$; p-value = .000), which is in line with previous studies (Moffat and Zhang 2014; Boutilier and Thomson 2011). Stakeholders accept and even support projects that present harmful impacts when they *trust* the decision-making process, considering it fair and transparent (Moffat and Zhang 2014).

4.2 Indicator of Economic Sustainability

The indicators were based on data from 2004 to 2017. The variables from Table 3 were tested for significance using bivariate and anti-image correlations, KMO and Bartlett tests, and weighted means estimative matrixes. The statistically relevant variables were:

- Distributed CFEM/Total Revenue of the Municipality (**CFEM_Revenue**)
- ICMS + ISS/Total Revenue of the Municipality (**ICMS_ISS_Revenue**)
- Jobs in mining and related areas/Total jobs in the municipality (**Jobs_Mining**)
- Industry GDP/Total GDP of the Municipality (**GDP_Industry**).

The baseline for this analysis was the period between 2004 and 08. Principal Component Analysis assessed fitness and weight of each factor (Table 3) for the municipality's economic dynamics.

The factor **CFEM_Revenue** (0.880) explains the first part of the component (C1 = 0.620), **ICMS_ISS_Revenue** (0.897) explains its second part (C2 = 0.567), **GDP_Industry** (0.984) its third part (C3 = 0.328) and **Jobs_Mining** (0.698) its fourth part (C4 = 0.432).

Using the results, the Indicator of Economic Sustainability was constructed through the formula:

$$IESt = [(C1)2.Rt + (C2)2.It + (C3)2.Pt + (C4)2.Et].$$

In which:
IES = Indicator of Economic Sustainability
Ci (i = 1, ..., 4) = Part of the Principal Component (weight)
t = year

✓ R = CFEM_Revenue
✓ I = ICMS_ISS_Revenue
✓ P = GDP_Industry
✓ E = Jobs_Mining

Data from 2004 to 2008 were used as the baseline (100) in the formula of the Indicator:

$$IES = \frac{IES_t}{Mean\ Average\ (IES_{2004-08})}$$

We noticed a strong correlation between mining and the economic dynamics in the municipality (Fig. 7). Price variations of iron ore influence the local economy, contributing to the fluctuations in the graph.

One result was below the baseline in 2009 because the increase in tax collection rate (52%) was lower than the municipality's revenue growth rate (99%). The implementation of the eco-friendly and largest mine in the region was reflected in higher

Fig. 7 Indicator of economic sustainability in Canaã dos Carajás—PA—Brazil

results between 2010 and 2015, due to job generation and increase in GDP and tax collection. Many constructing companies left the region in 2015, explaining the fall of the index. Nevertheless, the indicator grew again in 2017 when operations actually started.

A strong correlation was found between mining jobs and the municipality's GDP (0.777). Taxation is also influenced by mining, as 19% of local's GDP comes from CFEM and other taxes. During that period, the municipality's revenue increased 6 times, whereas ICMS and ISS increased 15 and 4 times, respectively. Such increases may also be related to job generation. Regression analyses indicated that for every R\$1.00 produced in mining salaries, R\$5.00 are generated in ICMS and ISS.

4.3   Indicator of Social Sustainability

Due to data availability, the indicators were based on data from 2008 to 2017. The variables in Table 3 were tested for significance using bivariate and anti-image correlations, KMO and Bartlett tests, and weighted means estimative matrixes. The statistically relevant variables were:

- % Women employed in mining companies (**IVG_Women**)
- % Local mining employees who completed at least High School (**IOL_High_School**)
- Social investments by mining companies in the municipality/Total social investments by mining companies in the State (**IVL_Invest_Soc_Munic_State**).

The baseline for this analysis was the year 2008. Principal Component Analysis assessed fitness and weight of each factor (Table 3) for the municipality's social dynamics.

The factor **IOL_High_School** (0.922) constitutes the first part of the component (C1 = −0.623), **IVG_Women** (0.976) its second part (C2 = 0.478) and **IVL_Invest_Soc_Munic_State** (0.878) its third part (C3 = 0.619).

Thus, Indicator of Economic Sustainability was constructed using the formula:

$$ISS_t = \left[(C1)^2 . En_t + (C2)^2 . Mu_t + (C3)^2 . In_t \right]$$

where:
IES = Indicator of Economic Sustainability
Ci (i = 1, ..., 4) = Part of the Principal Component (weight)
t = year

✓ En = IOL_High_School
✓ Mu = IVG_Women
✓ In = IVL_Invest_Soc_Munic_State

Data from 2008 were used as the baseline (100) in the formula of the Indicator of Social Sustainability:

$$ISS = \frac{ISS_t}{\text{Mean Average } (ISS_{2008})}$$

The Indicator remained positive (Fig. 8), especially in 2015, when the world's largest iron ore mine was launched. The construction of infrastructure and the generation of jobs, profits and taxes contributed to social sustainability in the region, as both the local government and mining companies could invest in social infrastructure and education, and promote gender equality.

Tax collection from mining companies promotes local government's investments in Education, and companies directly invest in labour-force training. Around 65% of the mining employees in the municipality have finished High School, between 2008 and 2016. This rate is higher than among workers in other sectors (55%). Regarding gender issues, female participation in the job market has increased 179% in Canaã dos Carajás, from 2008 to 2017, solely in mining jobs. Direct social investments from mining companies benefited health, sanitation, leisure and infrastructure. Compa-

Fig. 8 Indicator of social sustainability in Canaã dos Carajás—PA—Brazil

nies invested R$85.65 million in social areas in 2015; R$18.23 million in 2016 and R$13.21 million in 2017.

Mota et al. (2017) suggested other important factors regarding social sustainability, but they could not be quantitatively assessed due to their natures (binary variables) or unavailability. Some of these factors are indigenous rights, community management, child and/or forced labour, bribery and corruption, and stakeholder engagement. However, local mining companies promote good practices on social justice, such as committing to international agreements regarding Child and Forced Labour (Vale 2018).

5   Concluding Remarks

This research assessed sustainability in the mining industry, focusing on the municipality of Canaã dos Carajás, where the world's largest iron ore mine is located. It was the first study to quantitatively and qualitatively analyse the mining Social License to Operate in Brazil. It also constructed and assessed indicators of social and economic sustainability.

Under the current SLO level, **Acceptance** (3.62 in a Likert scale), mining companies are allowed to operate, but they lack social support and goodwill. Reaching higher levels (Approval and Identification) is crucial to lower social risks and to avoid losses. Achieving these goals demands engaging with the communities and involving all stakeholders in the decision-making processes. People are prone to support companies, despite their negative impacts, if the decision-making is considered fair and transparent. Strengthening and keeping an SLO demands attention to social, economic and environmental issues, and encourages social empowerment, thus promoting sustainability in multiple dimensions.

The economic and social indicators analysed here suggest that, despite some harmful impacts, mining activities are socially and economically sustainable in the region and they improve social and economic welfare, as the indicator values are equal or higher than the baseline. Increases in local GDP, tax collection and job generation benefited economic sustainability; improvements in education, gender issues and social investments benefited social sustainability. Such findings help to manage the local impacts of mining.

Negative experiences across time-space scales, governance and political distrust influenced social perceptions regarding mining. Mining can generate different impacts and benefits depending on geographical, cultural and political scenarios. Further studies should assess the influence of different factors in different scenarios to understand the dynamics involving mining sustainability using tools such as SLO and other indicators. This may contribute to empowering communities and stimulate companies to increase investment in social responsibility.

6 Biographical Notes

Thiago Leite Cruz has Bachelor's degrees in International Relations and Biology from Universidade da Amazônia and Universidade Federal do Pará—Brazil, and a Professional Master's degree in Sustainable Development. He has a postgraduate certificate in Business Management (London School of Business and Finance). This paper was based on his Master's thesis at Instituto Tecnológico Vale (Vale Institute of Technology—Brazil), the most complete research involving SLO in a Brazilian mining context. Sustainability, SLO and CSR are his areas of expertise.

Prof. Valente Matlaba (Ph.D. in Economics, University of Waikato—New Zealand), Prof. José Aroudo Mota (Ph.D. in Sustainable Development, Universidade de Brasília—Brazil), Celso Oliveira Júnior (Master's degree in Economics, Universidade Federal do Pará—Brazil), Prof. Jorge Filipe dos Santos (Ph.D. in Geographical Engineering, Universidade de Coimbra—Portugal), Leon Nazaré Cruz (Master's degree in Sustainable Development, Instituto Tecnológico Vale—Brazil), and Eduardo Demétrio Neto (Bachelor in Forest Engineering, Universidade Federal do Pará—Brazil) study sustainable development and mining impacts in the Socioeconomic Research Group at Instituto Tecnológico Vale. They helped gather and analyse data for this paper.

References

Angloamerican (2012) Relatório à sociedade 2012. http://relatoriosociedade.angloamerican.com.br/2012/equilibrio-comunidade.php. Accessed 01/12/2016

ArcGIS. Version 10.4. www.esri.com. Accessed 30 July 2019

Boutilier RG, Thomson I (2011) Modelling and measuring the social license to operate: fruits of a dialogue between theory and pratice. International Mine Management Conference. University of Queensland, Australia. https://socialicense.com/publications/Modelling%20and%20Measuring%20the%20SLO.pdf. Accessed 01/09/2018

Cabral ER, Enríquez MA, Dos Santos DV (2011) Canaã dos Carajás - do leite ao cobre: transformações estruturais do município após a implantação de uma grande mina. In: Fernandes FRC, Enríquez MARS, Alamino RC (eds) Recursos Minerais & Sustentabilidade Territorial: Grandes Minas. CETEM/MCTI, Rio de Janeiro, pp 39–69

Compara Brasil (2018) Painel de Arrecadação Municipal. http://comparabrasil.com/comparabrasil/municipios/Paginas/sobre.aspx?m=1. Accessed 07/12/2018

Conde M (2017) Resistance to mining. A review. Ecol Econ 132:80–90

Cruz TL (2017) A licença social de operação em Canaã dos Carajás como instrumento de sustentabilidade do Projeto Ferro Carajás S11D (Master's Dissertation). Programa de Mestrado Profissional em Uso Sustentável de Recursos Naturais em Regiões Tropicais/Instituto Tecnológico Vale, Belém-PA-Brasil. http://www.itv.org/wp-content/uploads/2018/03/Disserta%C3%A7%C3%A3o-Thiago-Cruz.pdf. Accessed 30 July 2019

DNPM (2018) CFEM. https://sistemas.dnpm.gov.br/arrecadacao/extra/Relatorios/arrecadacao_cfem.aspx. Accessed 08/01/2018

Duddu P (2018) The world's biggest iron ore mines. https://www.mining-technology.com/features/featurethe-worlds-11-biggest-iron-ore-mines-4180663/. Accessed 12/05/2018

EU—European Union (2017) Regulation (EU) 2017/821 of the European Parliament and of the Council of 17. May 2017 laying down supply chain due diligence obligations for Union importers of tin, tantalum and tungsten, their ores, and gold originating from conflict-affected and high-risk areas. https://eur-lex.europa.eu/legal-content/EN/TXT/?uri=CELEX%3A32017R0821. Accessed 08/19/2018

FGV/CES, FUNBIO, ALCOA (2008) Juruti sustentável. Uma proposta de modelo para o desen-volvimento local. GVCes, São Paulo. http://bibliotecadigital.fgv.br/dspace/bitstream/handle/10438/18488/GVces_Juruti%20sustent%C3%A1vel.pdf?sequence=1&isAllowed=y. Accessed 01/09/2018

G1 PA (2016) Em protesto, camponesas impedem acesso à mineradora no Pará. http://g1.globo.com/pa/para/noticia/2016/03/em-protesto-camponesas-impedem-acesso-mineradora-no-para.html. Accessed 10/15/2018

Gallopin GC (1996) Environmental and sustainability indicators and the concept of situational indicators. A system approach. Environmental Modelling & Assessment 1:101–117

Hu L, Bentler P (1999) Cutoff criteria for fit indexes in covariance structure analysis: conventional criteria versus new alternatives. Struct Equ Model 6:1–55

IBGE (2018) Canaã dos Carajás. https://cidades.ibge.gov.br/brasil/pa/canaa-dos-carajas/panorama. Accessed 05/17/2018

IBM Corporation (2015). IBM SPSS statistics for Windows, version 24.0. IBM Corporation, Armonk

Ipea Data (2018) Municípios. http://www.ipeadata.gov.br/Default.aspx. Accessed 11/22/2018

Kenny D, McCoach D (2003) Effect of the number of variables on measures of fit in structural equation modelling. Struct Equ Model 10:333–351

Liu J, Hull V, Charles H, Godfray J, Tilman D, Gleick P, Hoff H, Pahl-Wostl C, Xu Z, Chung MG, Sun J, Li S (2018) Nexus approaches to global sustainable development. Nat Sustain 1:466–476

Mancini L, Sala S (2018) Social impact assessment in the mining sector: review and comparison of indicators frameworks. Resour Policy 57:98–111

Marôco J (2014) Análise Estatística com o SPSS Statistics, 6a edn. Manuel Barbosa & Filhos, Lisboa

Moffat K, Zhang A (2014) The paths to social licence to operate: an integrative model explaining community acceptance of mining. Resour Policy 39:61–70

Mota JA, Maneschy MC, Souza-Filho PWM, Torres VFN, Siqueira JO, Santos JF, Matlaba V (2017) Uma nova proposta de indicadores de sustentabilidade na mineração. Sustentabilidade em debate 8(2):15–29

Palheta da Silva J, Da Silva C, Chagas C, Medeiros G (2014) Geography and mining in Carajás/Pará (Northern Region of Brazil). Int J Geosci 5:1426–1434

Phillips D (2016) Samarco dam collapse: one year on from Brazil's worst environmental disas-ter. The Guardian. https://www.theguardian.com/sustainablebusiness/2016/oct/15/samarco-dam-collapse-brazil-worst-environmental-disaster-bhpbilliton-vale-mining. Accessed 08/02/2017

Prno J, Slocombe DS (2012) Exploring the origins of 'social license to operate. Resour Policy 37:346–357

Prno J, Slocombe DS (2014) The paths to social licence to operate: an integrative model explaining community acceptance of mining. Resour Policy 39:61–70

RAIS (2018) Índice de PDET Microdados. ftp://ftp.mtps.gov.br/pdet/microdados/RAIS/. Accessed 05/21/2018

Santiago A, Demajorovic J (2016) Social license to operate: a case study from a Brazilian mining industry. Lat Am J Manage Sustain Dev 3(1)

Survival International (2016) Tribo bloqueia ferrovia em protesto contra grande miner-adora. Disponível em: http://www.survivalinternational.org/ultimas-noticias/11331. Accessed 10/15/2016

Vale. Relatórios (2018). http://www.vale.com/PT/investors/information-market/quarterly-results/ResultadosTrimestrais/Forms/AllItems.aspx. Accessed 06/05/2018

A Review on Multi-criteria Decision Analysis in the Life Cycle Assessment of Electricity Generation Systems

José Guilherme de Paula do Rosário, Rodrigo Salvador, Murillo Vetroni Barros, Cassiano Moro Piekarski, Leila Mendes da Luz and Antonio Carlos de Francisco

Abstract Strategic planning of energy matrices have become more complex with the increase in electricity demand and the need for reducing the impacts generated by human activities. Thus, the techniques of Life Cycle Assessment (LCA) and Multi-Criteria Decision Analysis (MCDA) have been used to support sustainable decisions. By means of a systematic search in the literature a research gap on this subject has been identified. Therefore, the objective of the present paper is to identify how LCA and MCDA are being used in the choice of technologies to be used in the expansion of electricity generation capacity. To that end, it was conducted a literature review to assess the state-of-the-art literature. The *Methodi Ordinatio* was used to rank the main studies in the addressed theme. The articles were analyzed according to 5 criteria: Multi-Criteria model, Technical and Sustainability Indicators, Location, Application and Life Cycle Impact Assessment methods. Results show the most used MCDA methods, the LCA assumptions, the most addressed approaches, as well as the locations where the studies were conducted and the technologies that showed better performance. The methods and topics researched suggest that the concern with sustainability of each technology and life cycle thinking has been being incorporated to the analyses and planning of energy systems. Moreover, this study may be helpful

J. G. de Paula do Rosário · R. Salvador (✉) · M. V. Barros · C. M. Piekarski · L. M. da Luz · A. C. de Francisco
Sustainable Production Systems Laboratory (LESP), Federal University of Technology—Parana (UTFPR), Monteiro Lobato Av., no number, Km 04, Ponta Grossa, Parana 84016-210, Brazil
e-mail: salvador.rodrigors@gmail.com

J. G. de Paula do Rosário
e-mail: jrosario@alunos.utfpr.edu.br

M. V. Barros
e-mail: murillo.vetroni@gmail.com

C. M. Piekarski
e-mail: piekarski@utfpr.edu.br

L. M. da Luz
e-mail: leila.mendesdaluz@gmail.com

A. C. de Francisco
e-mail: acfrancisco@utfpr.edu.br

© Springer Nature Switzerland AG 2020
W. Leal Filho et al. (eds.), *International Business, Trade and Institutional Sustainability*, World Sustainability Series,
https://doi.org/10.1007/978-3-030-26759-9_33

to decision-makers on promoting environmentally more sustainable actions. Nevertheless, this piece of research contributes to the Sustainable Developments Goal number 7, proposed by the United Nations, providing insights and investigating paths for an "affordable and clean energy".

Keywords Multi-criteria analysis · Life cycle assessment · Electricity generation · Renewable energies

1 Introduction

Considering the world's increasing demand for electricity, it is inevitable to seek transformation in the electricity generation systems to allow a cleaner and less socially impacting production. Therefore, researchers and decision makers around the world have been working on and assessing technologies and future scenarios in order to invest in alternatives that contribute to a sustainable development of electricity generation.

Nonetheless, the variables involved in the electricity grid's decision process do not make this process easy, since there might be conflicting choices and there is no technology that has a best performance on all aspects. That makes the assessment of technology and generation scenarios highly complex, there being many trade-offs to choose the one causing the least impacts whereas bringing the greatest benefits according to decision makers' preferences, without harming the less important variables that still need to be considered due to political, technical, social, environmental or economic factors, for it to be considered a sustainable alternative.

On the one hand, in the long term, fossil fuels such as oil will become scarce and more expensive (Vis and Ursavas 2016). On the other hand, renewable sources such as wind and solar have been showing great potential for reduction of greenhouse gases (GHG). In addition, renewable sources can significantly contribute to decreasing the electricity generation costs in off-grid systems, which produce power independently of the utility grid (Maleki and Askarzadeh 2014) and the price per watt installed for photovoltaic and wind systems has been decreasing over time (Kolhe et al. 2015).

With the need of indicators that represent the technologies and scenarios' impacts and performances, it is common for researchers to use the Life Cycle Assessment (LCA) (ISO 2006a, b), which is a technique that allows for a robust assessment of the aspects and potential impacts observed throughout the life cycle of a product, service or activity. However, many times such indicators are in different measurement units from the ones commonly used to support decisions in projects, making difficult their comparison with economic and political indicators.

As a means to minimize the differences among the indicators and translate them into the same unit, Multi-Criteria Decision Analysis (MCDA) methods are widely used for such problems. They allow to normalize the data and to assess several alternatives giving different weights to each one, resulting in an order of preference for those alternatives. A study by Zhou et al. (2006) shows that the main areas where MCDA is applied are energy policies, electricity planning and project assessment.

MCDA methods are a specific alternative of using the various aspects involved in decision-making (i.e., environmental, economic and social). It is also worth highlighting that MCDA provides, in a structured and objective way, assistance for decision-making, even when there is a significant amount of variables to be considered. Therefore, this importance can be noted when energy is the subject, for there are various sustainability criteria involved.

Among the existing methods, there is the multi-attribute value theory (MAVT), originating from the utility theory for multi-objective problems, presenting efficiency in modeling problems with high levels of uncertainty, providing a capacity for identifying and correcting parameters generated by decision-makers' mistakes. It is capable of encompassing the three sustainability spheres providing an excellent analysis of each one. This method is an excellent tool when used in political issues, which always involve the three spheres of sustainability. In this context sustainability can be understood as a basis to the triple bottom line approach and the three dimensions must be considered - environmental, economic and social. This study focused strongly on the environmental aspect, due to the consideration of LCA, however it does not disregard the other two dimensions.

The literature presents some studies on this subject. Noorollahi et al. (2016) analyzed a multi-criteria decision support system to define wind energy resources in western Iran. Løken (2007) provided an overview of some of the most important MCDA methods that have been proposed over the years. Troldborg et al. (2014) assessed the sustainability of renewable energy technologies using multi-criteria analysis. Volkart et al. (2017) worked on the MCDA of energy system transformation pathways.

Notwithstanding, yet a considerable problem for decision-makers is that even with such tools and techniques to assess and create indicators, for a fair comparison, it is necessary to use the same method for analysis for both the LCA and MCDA. It becomes a problem once there are several methods available and the decision-makers have their preferences, resource availability, policies and other variables that affect the problem at hand.

In summary, the world needs cleaner energy sources and production methods and in this regard, there are many variables (read, criteria) involved, all which need to be accounted for. There are methods for multi-criteria decision analysis, however there is a need to investigate which are better suited to assess the environmental aspects of electricity generation systems. On top of it, the LCA is the most used and complete tool for environmental assessment. Therefore, the objective of the present paper is to identify how LCA and MCDA are being used in the choice of technologies to be used in the expansion of electricity generation capacity. Although few studies have addressed the use of LCA and MCDA for electricity generation, to the best of the authors' knowledge, there have been no studies such as the one presented here, assessing this theme, thus, this study presents an unprecedented proposal using a systematic review of the literature, presenting and discussing how two techniques (LCA and MCDA) have been being used by researchers on the generation of electricity worldwide. In addition, this piece of work can also contribute to promoting cleaner electricity generation.

To that end, a search in three databases (ScienceDirect, Scopus and Web of Science) was conducted considering high impact articles published during the period 2013–2017 to identify and discuss the main aspects of each study. The dimensions of sustainability and indicators of each dimension in the studies from the literature were analyzed. The most relevant studies were identified using an adaptation of the *Methodi Ordinatio* (see Pagani et al. 2015).

This paper is organized as follows. This first section presented the initial considerations and objective of the study. In the sequence, it is presented the methods used to conduct this systematic review. The following section fulfills this study's aim presenting the trends on the addressed body of literature, assisted by the use of tables. The second last section shows the highlighting issues on LCA and MCDA of electricity generation systems. Finally, the last section draws on this study's final considerations and further investigations.

2   Methods

To conduct this research, the *Methodi Ordinatio* (Pagani et al. 2015) was used. The method allows conducting a systematic literature review and helps identify and rank publications according to their publication year, number of citations and journal impact factor. The method can be summarized in 9 steps: (i) Establishing the intention of research, (ii) Preliminary exploratory search of keywords in databases, (iii) Definition and combination of keywords and databases, (iv) Final search in the databases, (v) Filtering procedures, (vi) Identifying impact factor, year and number of citations, (vii) Ranking the papers using the *InOrdinatio* coefficient, (viii) Finding the full papers, (ix) Reading and systematic analyzing the papers.

According to the steps by the *Methodi Ordinatio*, after establishing the intention of research, it was conducted an exploratory search to analyze the keywords used in articles that fit the scope of this piece of research. It was, then, defined the keywords to be used in the search and their combinations, as well as the databases where the search would be conducted.

It was analyzed articles published between January 2013 and December 2017 in peer-reviewed journals, in 3 databases: (i) ScienceDirect, (ii) Scopus and (iii) Web of Science. Thus, Table 1 presents the keyword combinations (using Boolean operators and wildcards) used in the three databases searched, which covered from 2013 to 2017.

The search resulted in 54 articles: 26 from the ScienceDirect, 9 from Scopus and 19 articles from the Web of Science. These documents were exported to the reference manager software EndNote, where it was possible to exclude duplicates and articles published in sources other than peer-reviewed journals. A total of 34 articles was excluded. Thereafter, title and abstract filters were applied to identify which articles

Table 1 Combinations used in the search on the databases

Keywords
"LCA" OR "LCIA" OR "LCI" OR "LIFE CYCLE ASSESSMENT"
"MULTI-CRITERI*" OR "MULTI CRITERI*" OR "MCDA" OR "MCDM"
"ELECTRICITY GENERATION" OR "POWER GENERATION SYSTEM*" OR "ELECTRICITY PRODUCTION" OR "ELECTRICITY MIX"

Source Authors (2018)

were aligned with the intent of the present research. Therefore, a total of 12 articles remained for full analysis.

The search in databases, impact factor and number of citations was conducted in August 27, 2018. The impact factor of the journals was retrieved from the Journal Citation Reports® (JCR) and the number of citations was retrieved from Google Scholar.

For the 12 remaining articles, the (vii) step from the *Methodi Ordinatio* was applied, where the *InOrdinatio* coefficient (Eq. 1) was calculated. The method was then used to rank the articles. Thereafter, it was possible to sort the articles, as shown in Table 2, according to the article's reference number, author, title, journal, impact factor, number of citations (obtained with Google Scholar) and the *InOrdinatio* coefficient.

$$InOrdinatio = (\text{IF}/1000) + \alpha \times [10 - (\text{Research Year} - \text{Publish Year})] + \left(\sum c_i\right) \tag{1}$$

- IF: impact factor;
- C_i: number of citations;
- α: ponderation factor (1–10) attributed by the researcher.

In the analysis the ponderation factor (α) assigned was 5.

Table 2 reports on the characteristics of the 12 articles found most relevant in the literature, informing the reference, article title, journal where it was published, journal impact factor, number of citations, and ranking of *InOrdinatio*.

When searching for the documents, only the last one listed in Table 2 was not found available, as it was the one with the lowest *InOrdinatio* coefficient value, it was ruled out.

Thus, 12 articles were analyzed on this theme, however, the discussion was based on articles published in journals with a high impact factor, besides, many of the studies have a high number of citations.

Thereafter, for the remaining 12 articles, the factors that were taken into consideration in the assessment were: (i) Multi-Criteria model, (ii) Criteria, (iii) Location, (iv) Application and the (v) LCA Methods, which are presented in Sect. 3 (Results).

In this context, LCA prepares policy makers and decision-makers to support adequate and environmentally sound energy supply systems (Vilcekova and Burdova

Table 2 Selected articles' classification

ID	Reference	Title	Journal	Impact factor	No. of citations	InOrdinatio
1	Santoyo-Castelazo and Azapagic (2014)	Sustainability assessment of energy systems: Integrating environmental, economic and social aspects	Journal of Cleaner Production	5.651	193	223
2	Troldborg et al. (2014)	Assessing the sustainability of renewable energy technologies using multi-criteria analysis: Suitability of approach for national-scale assessments and associated uncertainties	Renewable & Sustainable Energy Reviews	9.184	92	122
3	Atilgan and Azapagic (2016)	An integrated life cycle sustainability assessment of electricity generation in Turkey	Energy Policy	4.039	45	85
4	Noori et al. (2015)	A macro-level decision analysis of wind power as a solution for sustainable energy in the USA	International Journal of Sustainable Energy	0	40	75
5	Klein and Whalley (2015)	Comparing the sustainability of U.S. electricity options through multi-criteria decision analysis	Energy Policy	4.039	40	70
6	Theodosiou et al. (2015)	Integration of the environmental management aspect in the optimization of the design and planning of energy systems	Journal of Cleaner Production	5.651	30	65
7	Barba et al. (2016)	A technical evaluation, performance analysis and risk assessment of multiple novel oxy-turbine power cycles with complete CO_2 capture	Journal of Cleaner Production	5.651	20	60

(continued)

Table 2 (continued)

ID	Reference	Title	Journal	Impact factor	No. of citations	InOrdinatio
8	Hong et al. (2014)	Nuclear power can reduce emissions and maintain a strong economy: Rating Australia's optimal future electricity-generation mix by technologies and policies	Applied Energy	7.900	16	56
9	Volkart et al. (2017)	Multi-criteria decision analysis of energy system transformation pathways: A case study for Switzerland	Energy Policy	4.039	9	54
10	Volkart et al. (2016)	Interdisciplinary assessment of renewable, nuclear and fossil power generation with and without carbon capture and storage in view of the new Swiss energy policy	International Journal of Greenhouse Gas Control	4.078	10	50
11	Atilgan and Azapagic (2017)	Energy challenges for Turkey: Identifying sustainable options for future electricity generation up to 2050	Sustainable Production and Consumption	0	3	48
12	Štreimikienė (2013)	Assessment of energy technologies in electricity and transport sectors based on carbon intensity and costs	Technological and Economic Development of Economy	3.244	7	32

Source Authors (2018)

2014), supported by the use of detailed input and output parameters operating within the limits of the designated system (Kadiyala et al. 2017).

To identify the LCA and the MCDA methods and their interrelationship, as well as the technologies used in the expansion of electricity generation capacity the full reading of the articles in the final portfolio was conducted, building the necessary connections and organizing the content in Tables 3 and 4, for further assessment in the results section.

3 Results

The results section is divided into two subsections, the first presents the criteria used in the studies, the results found in the analysis conducted and what sustainability indicators are being used and, the other subsection presents Table 4, which contains the aspects of each study analyzed.

3.1 Technical and Sustainability Indicators

After the documents' analysis, it was possible to identify what indicators are being addressed in recent studies and with what frequency each one has been being approached. Table 3 shows the sustainability dimension (Environmental, Economic and Social) and Technical where the indicator fits, as well as the indicator itself and which studies are using certain criterion. The studies follow the same sequence of the identification (ID) in Table 2.

Together with the life cycle approaches, it is normal that the indicators take into consideration not only the implementation or operation costs, but also the costs throughout the whole project or investment life, therefore the most used indicator, among the reviewed articles, for the economic dimension was the indicator "Levelized Costs".

The indicator most commonly used for the environmental dimension among the articles analyzed was "Greenhouse gas emissions", which measures the technology's GHG emissions (in CO_2 equivalent) throughout its life cycle, the second most used environmental indicator was "Ecosystem Quality".

For the social dimension, the highlighting indicator, present in all studies analyzed was the "Total employment (direct + indirect)", which determines the number of jobs generated due to a project for electricity generation, result from an action or a proposed scenario, being that out of the 6 studies using such indicator, 5 assess technologies, not scenarios.

In the technical dimension, only 8 out of the 12 studies accounted for such dimension. The ones often present in the analyzed studies were "Efficiency", which indicates how much of its capacity or resource consumption the technology is converting into energy, and; "Availability of energy sources", which measures the necessary

Table 3 Indicators and studies

Dimension	Indicator	ID						
Environmental	Metal depletion	9	10					
	Abiotic resource depletion potential (elements)	1	4	11				
	Abiotic resource depletion potential (fossil fuels)	1	4	9	10	11		
	Global warming potential	1	4	11				
	Acidification potential	1	4	11				
	Eutrophication potential	1	4	11				
	Fresh water aquatic ecotoxicity potential	1	4	11				
	Marine aquatic ecotoxicity potential	1	4	11				
	Ozone layer depletion potential	1	4	11				
	Photochemical oxidants creation potential	1	4	11				
	Terrestrial ecotoxicity potential	1	4	11				
	Human Health	1	4	6	9	11		
	Ecosystem Quality	2	3	6	8	9	10	
	Resources	3	6					
	Greenhouse gas emissions	2	3	7	8	9	10	12
	Area requirements	2	3	5	8			
	Nuclear energy depletion	10						
	Water withdrawal	5	8					
	Carbon footprint	5						
	Solid waste	3						
	Chemical waste	9	10					

(continued)

Table 3 (continued)

Dimension	Indicator	ID	1	2	3	4	5	6	7	8	9	10	11	12
	Radioactive waste	10												
Economic	Fuel sensitivity	10												
	Capital costs	1				4			7		9	10	11	
	Total annual costs	1				4						10	11	
	Levelized costs	1		2	3	4		6	7	8	9	10	11	12
Social	Direct employment	4					5						11	
	Total employment (direct + indirect)	2				4	5		7	8			11	
	Injuries	4										10	11	
	Fatalities due to large accidents	4								8	9	10	11	
	Imported fossil fuel potentially avoided	4					5						11	
	Diversity of fuel supply mix	4											11	
	Security and diversity of supply	1		2							9	10		
	Public acceptability	1		2										
	Health and safety	1										10		
	Conflict potential	9										10		
	Percentage use of RES	6												
Technical	Efficiency	2							7	8		10		
	Availability of energy sources	2					5	6			9			
	Safety costs	3												
	Dispatchability	10												

Source Authors (2018)

584   J. G. de Paula do Rosário et al.

Table 4 Studies' aspects

ID	Reference	Study's location	LCIA method	MCDA	Dimensions	Application
1	Santoyo-Castelazo and Azapagic (2014)	University of Manchester, UK	CML 2001	Multi-attribute value theory (MAVT)	Environmental, Economic and Social	Electricity Generation Scenarios
2	Troldborg et al. (2014)	The James Hutton Institute, UK	Review of Research and Reports	PROMETHEE with Monte Carlo simulation	Environmental, Economic, Technical and Social	Electricity Generation Technologies
3	Theodosiou et al. (2015)	Aristotle University of Thessaloniki, Greece and University of Wolverhampton, UK	Eco Indicator 99	Multiobjective optimization	Environmental, Economic, Technical and Social	Electricity Generation Technologies
4	Noori et al. (2015)	University of Central Florida, USA	Review of Research and Reports	Analytic Hierarchy Process (AHP) with Monte Carlo simulation	Environmental, Economic and Social	Electricity Generation Technologies
5	Klein and Whalley (2015)	University of Maine, USA	Review of Research and Reports	Weighted Sum Model (WSM)	Environmental, Economic, Technical and Social	Electricity Generation Technologies
6	Hong et al. (2014)	The University of Adelaide, Australia	Review of Research and Reports	Weighted Sum Model (WSM) with DELPHI	Environmental, Economic and Technical	Electricity Generation Scenarios
7	Barba et al. (2016)	Cranfield University, UK	Not specified	Decision Evaluation in Complex Risk Network Systems (DECERNS)	Environmental, Economic and Technical	Electricity Generation Technologies
8	Atilgan and Azapagic (2016)	University of Manchester, UK	CML 2001 Update 2010	Multi-attribute value theory (MAVT)	Environmental, Economic and Social	Electricity Generation Technologies

(continued)

Table 4 (continued)

ID	Reference	Study's location	LCIA method	MCDA	Dimensions	Application
9	Atılgan and Azapagic (2017)	The University of Manchester, UK	CML 2001 Update 2010	Multi-attribute value theory (MAVT)	Environmental, Economic and Social	Electricity Generation Scenarios
10	Volkart et al. (2016)	Paul Scherrer Institut, Switzerland	CED and ReCiPe	Pairwise Outpermentance Aggregation (POA) and Weighted Sum Approach (WSA)	Environmental, Economic, Social and Technical	Electricity Generation Technologies
11	Volkart et al. (2017)	Paul Scherrer Institut and University of Basel, Switzerland	CED and ReCiPe	Weighted Sum Approach (WSA)	Environmental, Economic, Social and Technical	Electricity Generation Scenarios
12	Štreimikienė (2013)	Vilnius University, Lithuania	Review of Research and Reports	TOPSIS	Economic and Environmental	Electricity Generation Technologies

Source Authors (2018)

resources availability for generating energy by means of a certain technology, considering both the quantity available and location.

3.2 Analysis of Aspects

The studies were assessed and Table 4 was built indicating the location where the study was conducted, the LCA method, MCDA method, the dimensions addressed by the indicators in the study and whether the study refers to some technology for electricity generation or possible scenarios.

Thus, Table 4 shows other aspects of the 12 articles, such as, the institution where the research was conducted, the Life Cycle Impact Assessment (LCIA) method, the MCDA method, as well as the dimensions and area of application.

The countries where the 12 studies were developed (following the final portfolio of this research) were: 6 studies were carried out in United Kingdom (UK) being developed in partnership with a Greek university; 2 pieces of work in United States of America (USA) and 2 in Switzerland. Lithuania and Australia presented 1 study each.

Regarding the LCA methods, the highlight is in the studies that used data from review of reports and research, being that 5 of the 12 studies used this as input method in their models, followed by "CML 2001 update 2010", seen in 3 studies, "ReCiPe" and "CED" seen in 2 studies, and "Eco Indicator 99" seen in 1 study.

With regard to MCDA, studies used a mixed approach, using 2 MCDA methods or combined with another simulation method, such as Monte Carlo. In general, the most used method was the Weighted Sum Model (WSM), seen in 4 studies, followed by Multi-attribute Value Theory (MAVT) seen in 3 studies. Other methods observed only once were Pairwise Outpermentance Aggregation (POA), PROMETHEE, TOPSIS, DELPHI and Multi-objective algorithms.

Among the studies analyzed, 6 of them addressed the three dimensions of sustainability (environmental, social and economic), 5 addressed the three dimensions of sustainability and the technical dimension, and only one addressed only the social and environmental dimensions.

Furthermore, regarding the applications, 8 studies conducted analyses with focus on listing technologies of electricity generation, and 4 studies conducted analyses of different energy scenarios.

A study conducted by Strantzali and Aravossis (2016) confirms the tendency in the utilization of combined methods with the purpose of facilitating its application, as well as it gives results more reliability and precision. Just as this study identified 4 studies using mixed methods applied together. So, the use of tools such as LCA and MCDA can contribute to the decision maker in the option in terms of reducing the potential environmental impact of each source of electricity.

4 Highlighting Issues on LCA and MCDA of Electricity Generation Systems

This piece of research identified how LCA and MCDA are being used in the choice of technologies to be used in the expansion of the electricity generation capacity and a few characteristics of this relationship. By means of the review of the state-of-the-art literature, given the procedures and settings presented in Sect. 2, supported by the *Methodi Ordinatio*, a few conclusions can be drawn:

- The authors who stood out in the field were Atilgan B., Azapagic A. and Volkart K.;
- Most studies concentrated on two research institutes, the University of Manchester (United Kingdom) and the Paul Scherrer Institut (Switzerland);
- The vast majority of studies were based on reviews of research and reports;
- The most used methods used were CML, CED and ReCipe.

  In relation to size and dimensions:

- For the environmental dimension, most studies focused on the indicators: Greenhouse Gas Emissions, Ecosystem Quality, Human Health and Abiotic Resource Depletion Potential (fossil fuels);
- For the economic dimension, the indicators focused on Levelized costs and Capital costs;
- Most studies consider in the social dimension: Employment Indicators (direct + indirect) and Fatalities due to large accidents;
- In the technical dimension, the indicators of Efficiency and Availability of Energy Sources were more relevant.

  In addition, most of the high impact studies have been conducted in Europe, with a small participation of developing countries. However, most studies address the three dimensions of sustainability, so that this thinking will add value to organizations in the future.

5 Final Considerations

There is evidence that the body of literature on the joint use of MCDA and LCA on the assessment of electricity generation systems has been receiving contributions globally, however, it is not enough to affirm that there are solidly established trends. Thereupon, the development of studies on the theme are welcome to help identify advances in LCA and MCDA. On those grounds, further studies can be expected in the analysis of different impact categories and indicators for the triple bottom line.

Future prospects on the theme include the wider consideration of social and technical aspects when assessing the sustainability of electricity generation systems, for the technical dimension is of essential consideration for the feasibility of such systems and even though a certain technology may be feasible in the social and environmental dimensions, it will not be possible to implement if not technically feasible. On top of it, there is room for further assessment on the implications of the compensatory nature of the criteria in the four dimensions.

Notwithstanding, this study does not claim to be exhaustive nor exempt from any limitations, for the search conducted was limited to the terms and databases used and described in Methods (Sect. 2). Further limitations on the possible choice of technologies using MCDA and LCA in electricity generation systems are (i) the small number of studies in the current literature that allow for the consideration of relevant tested/validated sustainable criteria; (ii) the broadness of both fields of LCA and MCDA, which are constantly expanding and evolving, thus, challenging research to always be up-to-date on their state-of-the-art, which leads to (iii) the time delimitation used in this study, which covered from 2013 to 2017, aiming to cover the last five years of research previous to this piece of work.

To the best of the authors' knowledge, there are no studies assessing the joint use of MCDA methods and the LCA in the analysis of electricity generation systems, neither the expansion of such systems, hence, there it lies the novelty and originality of this study.

Acknowledgements The present study was developed under the financial support of the Fundação Araucária Paraná/Brazil, the Coordination of Improvement of Higher Education Personnel (CAPES) and the National Council for Scientific and Technological Development (CNPq).
The authors declare that they have no conflict of interest.

References

Atilgan B, Azapagic A (2016) An integrated life cycle sustainability assessment of electricity generation in Turkey. Energy Policy 93:168–186

Atilgan B, Azapagic A (2017) Energy challenges for Turkey: identifying sustainable options for future electricity generation up to 2050. Sustain Prod Consumption 12:234–254

Barba FC, Sánchez GMD, Seguí BS, Darabkhani HG, Anthony EJ (2016) A technical evaluation, performance analysis and risk assessment of multiple novel oxy-turbine power cycles with complete CO_2 capture. J Clean Prod 133:971–985

Hong S, Bradshaw CJA, Brook BW (2014) Nuclear power can reduce emissions and maintain a strong economy: rating Australia's optimal future electricity-generation mix by technologies and policies. Appl Energy 136:712–725

International Organization for Standardization (ISO) (2006a) Environmental management—life cycle assessment—principles and framework, 2nd edn. ISO 14040:2006; ISO, Geneva

International Organization for Standardization (ISO) (2006b) Environmental management—life cycle assessment—requirements and guidelines, 1st edn. ISO 14044:2006; ISO, Geneva

Kadiyala A, Kommalapati R, Huque Z (2017) Characterization of the life cycle greenhouse gas emissions from wind electricity generation systems. Int J Energy Environ Eng 8(1):55–64

Klein SJW, Whalley S (2015) Comparing the sustainability of US electricity options through multi-criteria decision analysis. Energy Policy 79:127–149

Kolhe ML, Ranaweera KMIU, Gunawardana AGBS (2015) Techno-economic sizing of off-grid hybrid renewable energy system for rural electrification in Sri Lanka. Sustain Energy Technol Assess 11:53–64

Løken E (2007) Use of multicriteria decision analysis methods for energy planning problems. Renew Sustain Energy Rev 11(7):1584–1595

Maleki A, Askarzadeh A (2014) Optimal sizing of a PV/wind/diesel system with battery storage for electrification to an off-grid remote region: a case study of Rafsanjan, Iran. Sustain Energy Technol Assess 7:147–153

Noori M, Kucukvar M, Tatari O (2015) A macro-level decision analysis of wind power as a solution for sustainable energy in the USA. Int J Sustain Energ 34(10):629–644

Noorollahi Y, Yousefi H, Mohammadi M (2016) Multi-criteria decision support system for wind farm site selection using GIS. Sustain Energy Technol Assess 13:38–50

Pagani RN, Kovaleski JL, Resende LM (2015) Methodi Ordinatio: a proposed methodology to select and rank relevant scientific papers encompassing the impact factor, number of citation, and year of publication. Scientometrics 105(3):2109–2135

Santoyo-Castelazo E, Azapagic A (2014) Sustainability assessment of energy systems: integrating environmental, economic and social aspects. J Clean Prod 80:119–138

Strantzali E, Aravossis K (2016) Decision making in renewable energy investments: a review. Renew Sustain Energy Rev 55:885–898

Štreimikienė D (2013) Assessment of energy technologies in electricity and transport sectors based on carbon intensity and costs. Technol Econ Dev Econ 19(4):606–620

Theodosiou G, Stylos N, Koroneos C (2015) Integration of the environmental management aspect in the optimization of the design and planning of energy systems. J Clean Prod 106:576–593

Troldborg M, Heslop S, Hough RL (2014) Assessing the sustainability of renewable energy technologies using multi-criteria analysis: suitability of approach for national-scale assessments and associated uncertainties. Renew Sustain Energy Rev 39:1173–1184

Vilcekova S, Burdova EK (2014) Multi-criteria analysis of building assessment regarding energy performance using a life-cycle approach. Int J Energy Environ Eng 5(2–3):83

Vis IFA, Ursavas E (2016) Assessment approaches to logistics for offshore wind energy installation. Sustain Energy Technol Assess 14:80–91

Volkart K, Bauer C, Burgherr P, Hirschberg S, Schenler W, Spada M (2016) Interdisciplinary assessment of renewable, nuclear and fossil power generation with and without carbon capture and storage in view of the new Swiss energy policy. Int J Greenhouse Gas Control 54:1–14

Volkart K, Weidmann N, Bauer C, Hirschberg S (2017) Multi-criteria decision analysis of energy system transformation pathways: a case study for Switzerland. Energy Policy 106:155–168

Zhou P, Ang BW, Poh KL (2006) Decision analysis in energy and environmental modeling: an update. Energy 31(14):2604–2622

Towards More Sustainable Extractive Industries: Study of Simulation of Efficient Ventilation Systems in the Emission Reduction of Gases for the Development of Mine Works

Pablo Vizguerra-Morales, Rosa Isela Lopez-Mejia,
Juan Carlos Baltazar-Vera, Joel Everardo Valtierra-Olivares,
Roberto Ontiveros-Ibarra, Carolina de Jesús Rodríguez-Rodríguez,
Juan Esteban García-Dobarganes Bueno, Gilberto Carreño-Aguilera
and Alberto Florentino Aguilera-Alvarado

Abstract Mining has been throughout history one of the basic activities for economic and technological development for humanity; In this sense, an important challenge for the mineral extraction industry is to achieve production processes more friendly to the environment; that is why the ventilation in contemporary mine tries to solve the problem of the emission gases as well as to minimize the consumption of energy resources for this operation, contributing solutions in this topic for a sustainable development of the mining communities. In this project, a simulation comparative study was carried out using three models for a system without extraction versus extraction in an underground mine development work, this in order to reduce CO and CO_2 concentrations which are emission gases harmful to the environment; as well as obtaining a model that allows to vary parameters related to the best use of energy resources. The simulation work was carried out using the CFD software of ANSYS. The results show that in all cases the emission gases concentrations were reduced to values below the limits established by the standards (NOM-010-STPS-1999).

P. Vizguerra-Morales · R. I. Lopez-Mejia · J. C. Baltazar-Vera (✉) · J. E. Valtierra-Olivares · R. Ontiveros-Ibarra · C. de J. Rodríguez-Rodríguez · J. E. García-Dobarganes Bueno
Departamento de Ingeniería en Minas, Metalurgia y Geología, Universidad de Guanajuato, Ex Hda. de San Matías s/n. Col. San Javier, Gto. 36020 Guanajuato, Mexico
e-mail: jc.baltazarvera@ugto.mx

G. Carreño-Aguilera
Departamento de Ingeniería en Geomática e Hidraúlica, Universidad de Guanajuato, Juárez 77. Col. Zona Centro. Guanajuato, Gto. 36000 Guanajuato, Mexico

A. F. Aguilera-Alvarado
Departamento de Ingeniería Química, Universidad de Guanajuato, Noria Alta s/n. Col. Noria Alta, Gto. 36050 Guanajuato, Mexico

© Springer Nature Switzerland AG 2020   591
W. Leal Filho et al. (eds.), *International Business, Trade and Institutional Sustainability*, World Sustainability Series,
https://doi.org/10.1007/978-3-030-26759-9_34

Keywords Sustainable extractive industries · Ventilation · Underground mine · Simulation · Sustainable development

1 Introduction

Mining over time has been one of the basic activities for the economic and techno-logical development of humanity; this is due to the evolution of humans and mining as part of the technological development of society (Hartman 1992). At present, well-known worldwide efforts have been carried out with the purpose of generating sustainable extractive industries; that is why a commitment that must be taken into account to achieve this goal is the concept of sustainable development proposed by the Brundtland report of the World Commission on Environment and Development which states: "achieve the needs of the present without compromising the possibili-ties of future generations to achieve their own needs" (United Nations 1987, p. 43).

The sustainability of the mining industry is supported in three aspects of utmost importance: economic, environmental and social (Richards 2002); These aspects must be considered in an integral manner to obtain a sustainable solution. In terms of the environmental aspect, the resources that must be treated in a sustainable manner are: hydric, earth and atmospheric. According to the above, this work focuses on the study of an important parameter concerning the atmospheric resource (ventilation in underground mine).

Underground mining involves the extraction of minerals from the rocky massifs of the earth's crust through excavations of considered depths. The extracted min-eral is transported to the surface to carry out the extraction process of materials of commercial value (Armstrong and Menon 1998).

The air in its natural state is formed by a mixture of gases approximately 20.9% volume of Oxygen (O) and 79% of nitrogen, in addition to 1% of carbon anhydrides and a tiny amount of water vapors, which depend on the temperature and atmospheric pressure. Ventilation is a constant flow of air that circulates through the underground mining works developed for this activity such as, counterpoises, robins, ramps, levels, etc.; which occurs due to pressure differences (Beard 1894). There are two types of ventilation in underground, natural and forced, the first is to maintain a constant flow of air (due to the difference in temperatures) this difference in temperature generates a specific weight difference between fresh air and stale air (outgoing) which allows the eviction of the latter, maintaining the mining work with a constant clean air flow (Hartman 1961). The forced ventilation consists of the implementation of extractors and/or air injectors to generate an air flow in the ventilation circuit, decreasing the amount of harmful gases produced by the different activities that are carried out in the interior of the mine (Quevedo 2013); This type of ventilation is the most used in underground mining works and is due to the use of the equipment required for this purpose, that is why the generation of analyzes that allow studying the energy requirements as well as the final disposal of toxic gases is very important for the development of sustainable processes in underground mine.

To carry out jobs, operations and services in underground mining there is legislation and regulations in this regard, which has been applied over the years in an ordinary way for the importance that this entails in terms of safety. Therefore, it is important to highlight the need to monitor compliance with the provisions of the Official Mexican Standards NOM-121-STPS-1996 which establishes the ventilation requirements for the work carried out in the mines (Normas Oficiales Mexicanas STPS 1996) and NOM-023-STPS-2003 which indicates the characteristics that a ventilation system in an underground mine should contain (Secretaría de Trabajo y Previsión Social 2012). In this work, a simulation study of a ventilation system for the extraction of CO and CO_2 by Computational Fluids Dynamics (CFD) is carried out in order to generate projections of the effects on mass and energy transfer phenomena. air flows and pollutants; This will allow the development of ventilation systems with greater energy efficiency and control over the emission gases in order to create ventilation processes in sustainable underground mines.

According to what has been written in previous paragraphs, generating a model that allows the design of sustainable ventilation systems in terms of the three mentioned aspects is of vital importance. In terms of environmental impact, obtaining a model for the design of a ventilation system will allow to know the behavior of the emission gases of the underground mine, which is important in the implementation of trap and treatment systems for these gases, minimizing the emissions to the surface. In the economic and environmental field, having an efficient model for ventilation design is a powerful tool to minimize energy, which reduces costs and the consumption of the energetic resource. Finally, having an efficient ventilation system and therefore reducing the environmental impact produced by the emission gases from underground mines improves the quality of life of mine workers as well as the families that live in these communities.

2 Methodology

For the study of mass and energy transfer of the contaminants present in an underground mine the $k - \varepsilon$ estándar (Wilcox 1994; Fluent Inc. 2006), Transition k, kl-omega (3 ecn), (Herruzo 2015) and Reynolds Stress (RMS) 7 ecn (Sarkar and Balakrishnan 1990) models were used. For the elaboration of the mesh of the mine work, a method of construction of upper–lower geometry in GAMBIT was used, using meshing units of the mixed type with tetrahedral and hexahedral cells with a total of 250,000 units (see Figs. 1, 2 and 3) using the following meshing dimensions:

- Dimensions of 7 m × 7 m with a length of 150 m.
- Ventilation sleeve is formed by a concentric pipe of 0.5 m and a length of 120 m, located in the upper right.
- Extraction pipe has a diameter of 1 m with a length of 75 m, located in the center of the mine.
- Combustion generator has a diameter of 0.08 m with a height of 1 m.

Fig. 1 Scheme of mine work

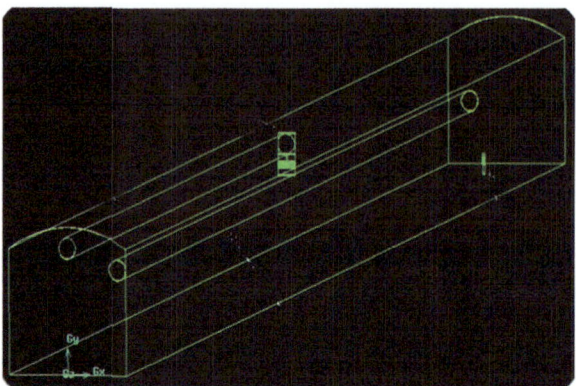

Fig. 2 Mine work mesh

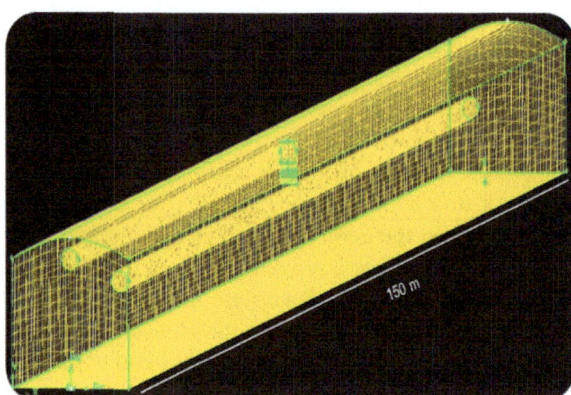

Fig. 3 Internal mesh of mine work

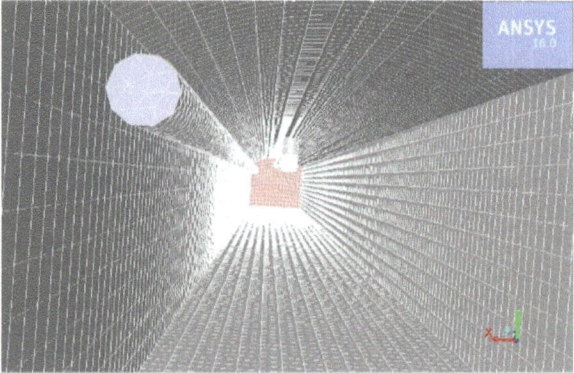

To calculate the amount of air required, the following specifications were considered: Use of 1 power Scoop 278.94 HP and a low profile power truck 409.01 HP; it was considered that the mining work is at 2012 meters above sea level and a group of 8 people working in the development work; an extractor-injector ATC axial tube injector with injected aluminum propeller and/or glass fiber reinforced polypropylene with 7 blades at an angle of 45° was used. The properties of air mixture and emission gases, as well as the boundary conditions necessary to perform the simulation are shown below (Tables 1 and 2).

The simulation of the phenomenon of mass and energy transfer during 500 s was carried out, considering the combustion reaction that is generated in the motors of the work equipment main sources of emission gases, these reactions are shown below.

$$C_{10}H_{22} + 10.5O_2 \rightarrow 10CO + 11H_2O \tag{1}$$

$$2CO + 2H_2O + O_2 \rightarrow 2CO_2 + 2H_2O \tag{2}$$

Table 1 Properties of the air mixture and emission gases in underground mine

Border condition	Type of condition	Condition values	
		Velocity (m/s)	Temperature (K)
Mine entrance	Output pressure	–	298
Extractor inlet	Output pressure	15	300
Fresh air	Mass flow input	16.82	298
Burner air	Mass flow input	–	300
Combustion Air	Mass flow input	–	315
Right wall	Wall	0	300
Left wall	Wall	0	300
Ceiling	Wall	0	300
Front of work	Wall	0	310

Table 2 Boundary conditions for the simulation in the development work of the underground mine

Border condition	Type of condition	Condition values
Interior default	Inside	Inside
Lighting_200_5_Interior_superior	Inside	Inside
Lighting_200_5_Interior_superior	Wall	Stationary wall
Upper shaft_3_ of the intermediate shaft	Wall	Stationary wall
Wall	Wall	Stationary wall
Output pressure	Output pressure	177,608.933 pa
Tank	Mass flow input	16.82 m/s

Once the simulations with the three proposed models have been carried out, profiles and outline schemes are generated to elucidate the differences in the desired parameters between each one.

3 Results

3.1 Effect of CO Concentrations in the Case Study (Model KL-Standard Epsilon)

Figure 4 shows the profile of carbon monoxide in the development work of the underground mine without implementing the ventilation model, in this figure it can be seen that at a distance of 20 m there is a concentration of 68 PPM of CO; at a distance of 50 m there is a value of 68.1 PPM of CO; corresponding to the central distance (75 m) the 68.1 PPM of CO is recorded; later at the distance 120 m a concentration of 68.2 PPM of CO is observed and finally for the distance corresponding to the front of the development work of the mine (150 m) we have 68.2 PPM of CO. These data show that in the whole section of the underground work values are obtained above the NOM-010-STPS-1999 that specifies that the maximum permissible limits of CO exposure in a working day of 8 h is 50 PPM.

Figure 5 shows the contour of CO distribution using the standard KL-epsilon model. The concentration recorded tends to increase in relation to the distance where the lowest concentration is 68 PPM corresponding to the first 20 m, as well as on the front of the development work of the mine the maximum concentration is recorded which is 68.3 PPM.

Figure 6 shows the profile of carbon monoxide along the underground mine (150 m) when a ventilation circuit is applied, in the first 20 m a CO concentration of 49.4 PPM is recorded, which is maintained until the 50 m; at a distance of

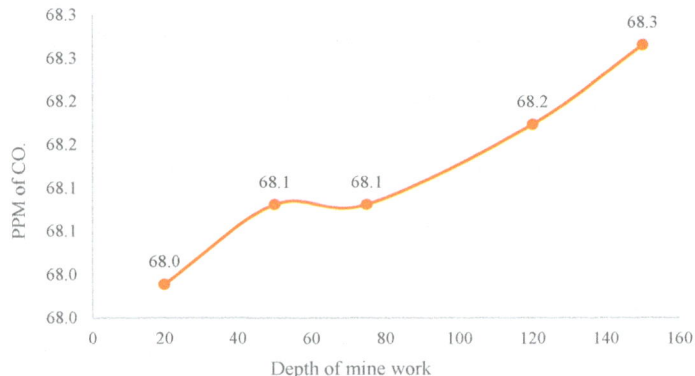

Fig. 4 Carbon monoxide profile without extraction using the KL-Epsilon standard model

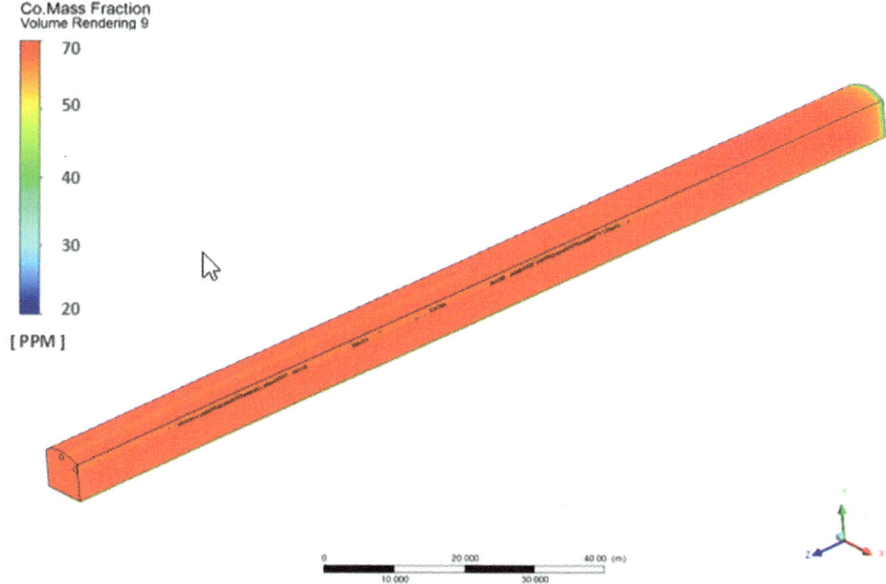

Fig. 5 Contour of carbon monoxide without extraction using the KL model-Epsilon standard

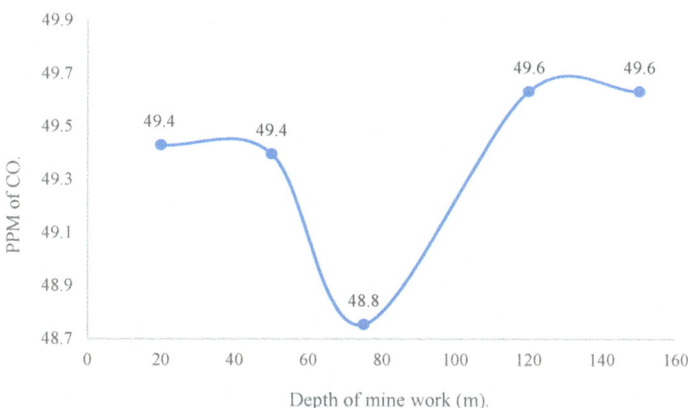

Fig. 6 Carbon monoxide profile with extraction using the KL-Epsilon standard model

75 m the concentration tends to decrease considerably until 48.8 PPM; This concentration tends to rise to a distance of 120 m which remains constant up to 150 m
giving a value of 49.6 PPM. These records indicate when supplying the amount of
fresh air keeping the flow constant, the CO concentration will be lower. As it can
be observed when applying this model of mass and energy it is necessary that the
concentration of CO reaches levels below the NOM-010-STPS-1999.

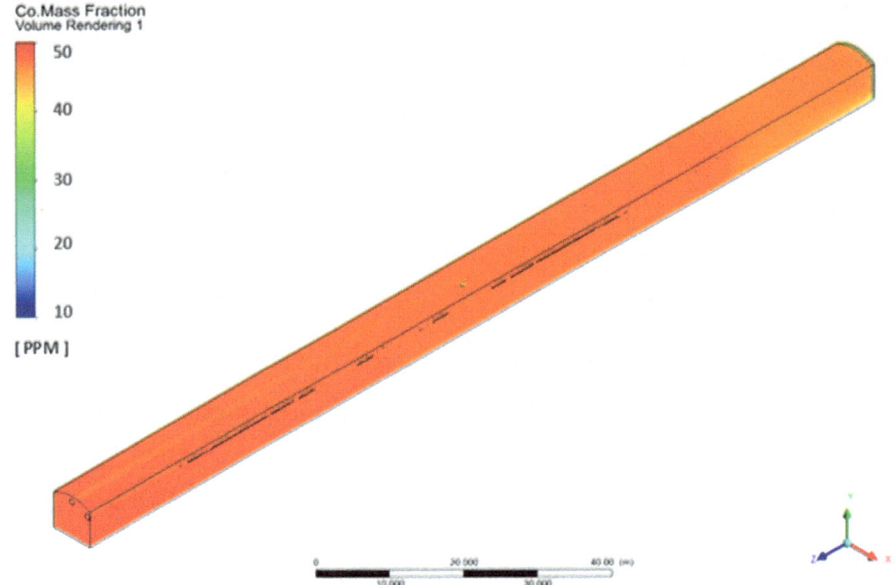

Fig. 7 Carbon monoxide contour with extraction using the KL-Epsilon standard model

The contour corresponding to the concentration of carbon monoxide generated during the simulation in a period of 500 s with the application of a ventilation system (see Fig. 7), shows that from the entrance of the mine to 50 m exists the same concentration of CO, which tends to decrease in the middle part of the total length of the mine (48.8 PPM); in the last meters the length increases to 49.6 PPM.

3.2 Effect of CO Concentrations in the Case Study (Model K-KL Omega 3 Ec.)

Figure 8 shows the carbon monoxide profile along the development work of the underground mine, in this figure it can be seen that at a distance of 20 m there is a concentration of 68 PPM of CO; as regards the distance of 50 m, a value of 68 PPM of CO is reported; corresponding to the central distance (75 m) the 68.1 PPM of CO is recorded; subsequently at a distance of 120 m, a concentration of 68.1 ppm of CO is observed; and finally, for the distance corresponding to the front of the development work of the mine (150 m), the maximum concentration of 68.2 PPM of CO is recorded.

Figure 9 shows the carbon monoxide contour. The figure shows the distribution of CO registered throughout the mine development work without the application of a ventilation system. The minimum concentration is at the entrance of the development

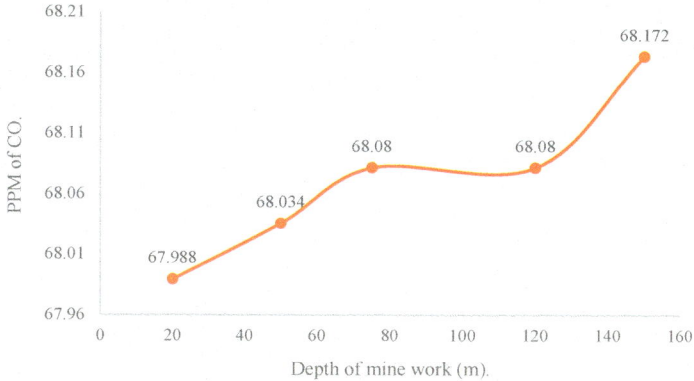

Fig. 8 Carbon monoxide profile without extraction using the model K-KL omega 3 ecn

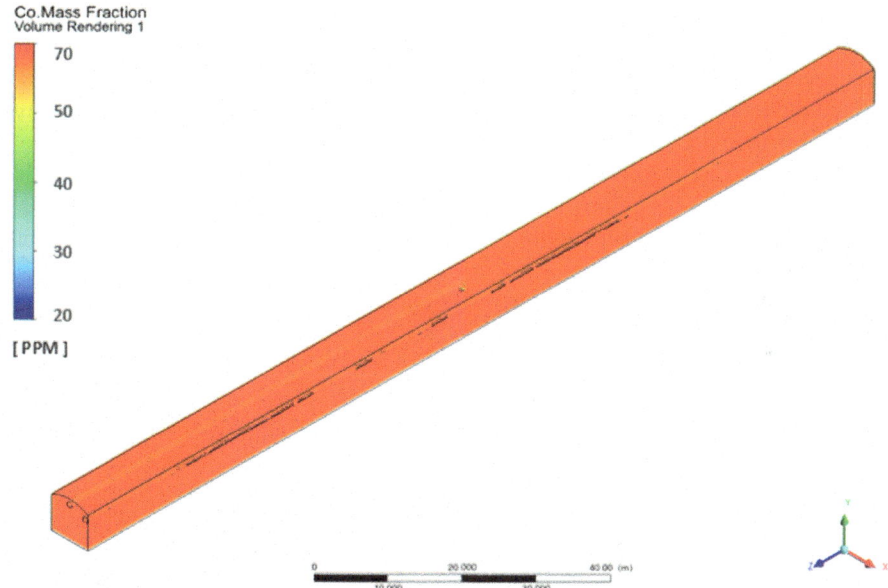

Fig. 9 Carbon monoxide contour without extraction using the model K-KL omega 3 ecn

work of the mine, later this is increasing until registering the maximum concentration on the front of the development work of the mine (150 m).

Figure 10 shows the profile of carbon monoxide when applying a ventilation system. In this figure it can be noted that at a distance of 20 m the concentration of CO is 48.7 PPM and this is maintained up to 50 m; corresponding to the central distance (75 m) a concentration of 48 PPM is recorded due to the extraction that is being carried out; subsequently at a distance of 120 m a concentration of 48.8

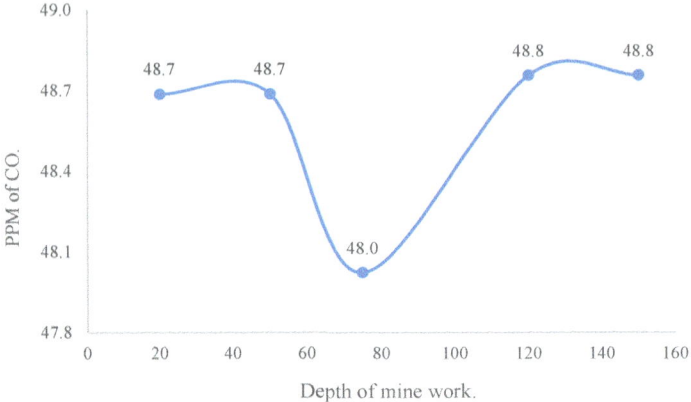

Fig. 10 Carbon monoxide profile with extraction using the model K-KL omega 3 ecn

PPM is observed and finally for the distance corresponding to the mine front (150 m) the concentration remains constant, since that is where the working machinery is located (generation of the combustion gases incomplete). In the same way as in the previous system it can be observed that the registered CO concentrations are below the permissible limits in NOM-010-STPS-1990.

Figure 11 shows the behavior of carbon monoxide during the simulation of 500 s, which tends to descend from the mine front to the entrance of. In the middle part of

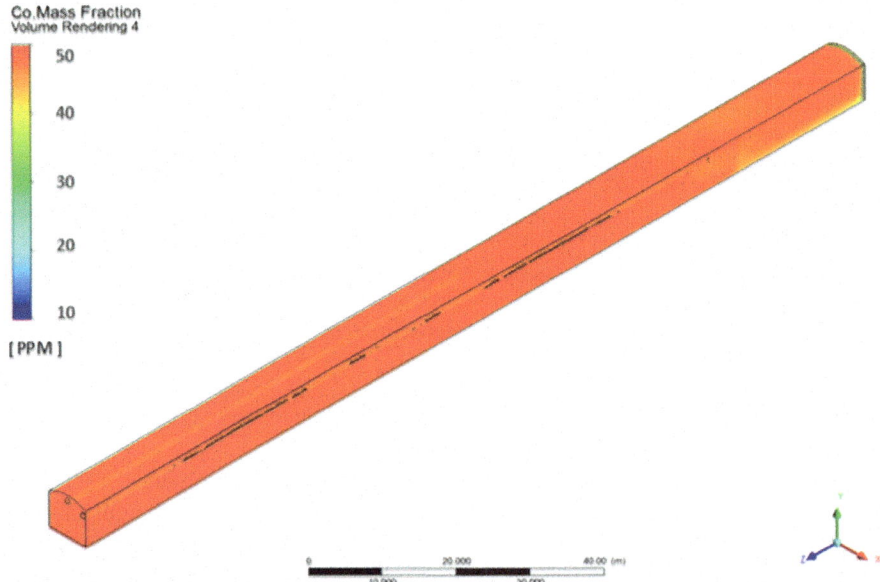

Fig. 11 Carbon monoxide contour with extraction using the model K-KL omega 3 ecn

the total length where the extractor is located, it is the point where there is a lower concentration of CO (48 PPM), so it can be deduced that the implementation of a ventilation circuit decreases gas concentrations, in the mine front the concentration is higher than in the other points (20, 50 and 75 m).

3.3   Effect of CO Concentrations in the Case Study (Reynolds Stress 7 Ec. Model)

Figure 12 shows the profile of carbon monoxide in underground mine with the application of the ventilation system, in the figure it is observed that in the first 20 m of the concentration of CO is 48.8 PPM and this increases 0.1 PPM to 50 m away; at a distance of 75 m the concentration decreases giving a record of 48 PPM; with respect to the distance 120 m it is observed that the concentration of CO increases to 49.4 PP; As regards the 150 m, the concentration is maintained. As in the previous models, it is observed that the concentrations recorded throughout the development work in the underground mine are below that established in NOM-010-STPS-1999.

Figure 13 shows the contour with the behavior of carbon monoxide at different points to the total length of the underground mine, in the figure it can be seen that the highest concentration of CO is registered in the front of the mine and it tends to decrease until 20 m where you have the first CO record.

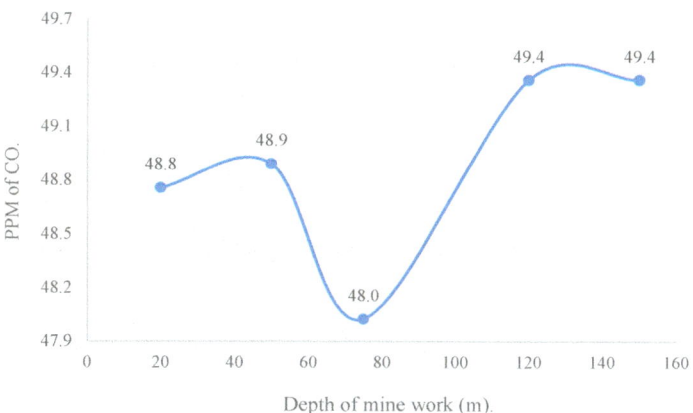

Fig. 12 Carbon monoxide profile with extraction using the Reynolds Stress 7 ecn model

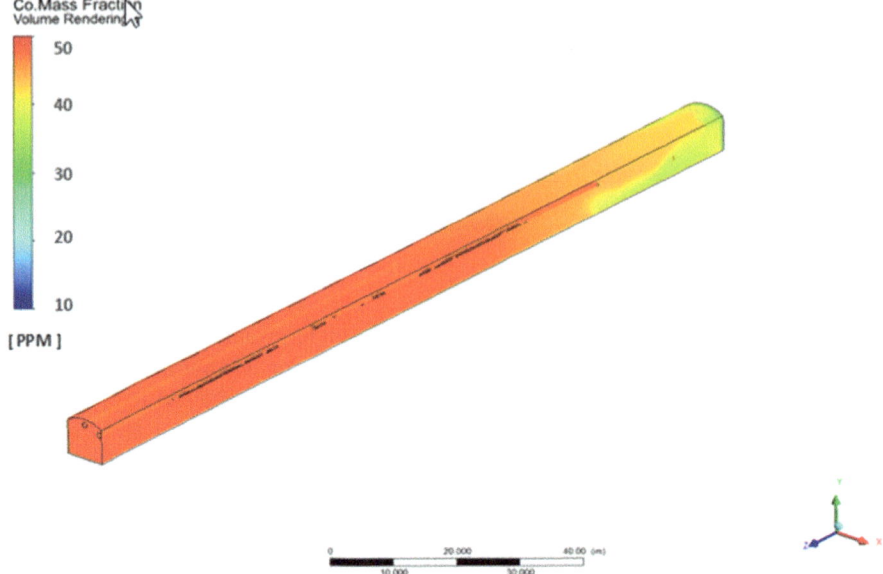

Fig. 13 Carbon monoxide contour with extraction using the Reynolds Stress model 7 ecn

3.4 Analysis of the Three Mass and Energy Transfer Models for CO Concentration

Figure 14 shows the comparison of the carbon monoxide concentration results obtained in the simulation with the implementation of a ventilation model versus

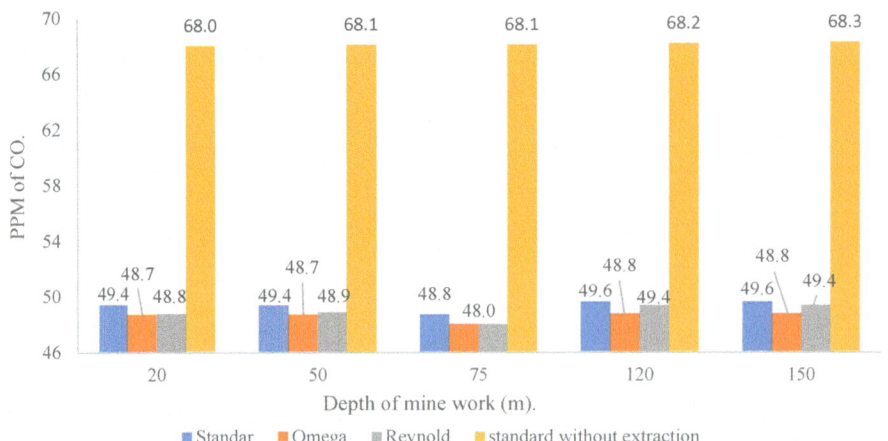

Fig. 14 Analysis with the three models used in the simulation for the concentration of CO

the model without the extraction that recorded the highest concentration of CO in the development work of the mine. underground. The figure shows that the concentration of CO in the first 20 m of distance is 18.6, 19.2, 19.3 PPM lower respectively to the Epsilon-Standard, Reynolds Stress, K-KL Omega models in relation to the model that registered the highest concentration of CO without the extraction; at the distance of 75 m (position of the extractor) the value of the concentration decreases 19.3, 20.1, 20.1 PPM respectively with respect to the results obtained with the model without application of a ventilation circuit; the maximum CO concentration records obtained at the front of the development work (150 m) in the Epsilon-Standard, Reynolds Strees, K-KL Omega models are 18.6, 18.9, 19.5 PPM lower than those obtained in the model without the removal. These results show that with the K-KL Omega model lower CO concentrations are obtained and therefore it is the most promising model to generate adequate conditions in the ventilation of the development work of the underground mine.

3.5   Effect of CO_2 Concentrations in the Case Study (Model KL-Standard Epsilon)

Figure 15 shows the carbon dioxide profile without the implementation of a ventilation circuit. In this figure it can be seen that at a distance of 20 m there is a CO_2 concentration of 5005 PPM; with respect to the distance of 50 m there is a concentration of 5071 PPM of CO_2, while at 75 m distance the concentration is higher (5203 PPM); corresponding to the distance of 120 m, there is a concentration of 5280 PPM. The maximum concentration of CO_2 recorded is at the front of the development work of the mine, which corresponds to 5291 PPM.

Figure 16 shows the carbon dioxide contour along the development work of the

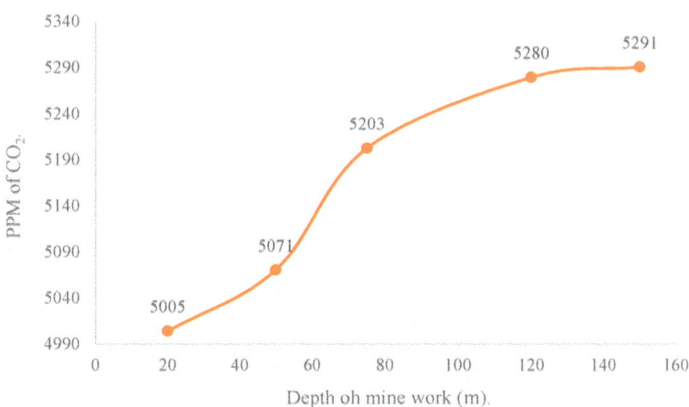

Fig. 15 Carbon dioxide profile without extraction using the standard KL-Epsilon model

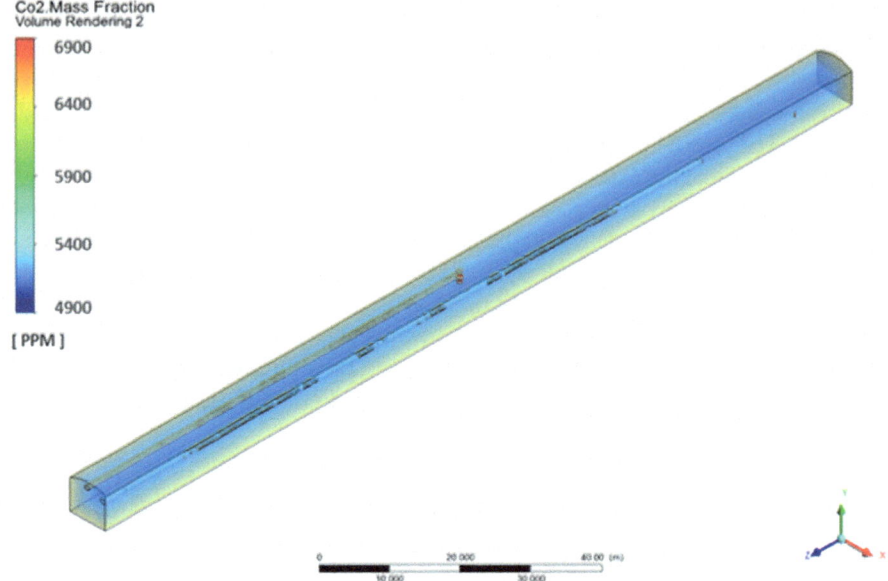

Fig. 16 Carbon dioxide contour without extraction using the KL-Epsilon standard model

underground mine with a simulation of 500 s, in which a concentration in a range of 5000 and 5300 PPM can be observed. In this outline it is shown that the highest concentration of CO_2 occurs at 150 m due to the machinery that is in operation.

Figure 17 shows the carbon dioxide profile in the underground mine with the implementation of a ventilation model, in which the trend and concentration of CO_2 can be observed, at a distance of 20 there is a concentration of 4390 PPM; in relation to a distance of 50 m they have a concentration of 4623 PPM; in the central part

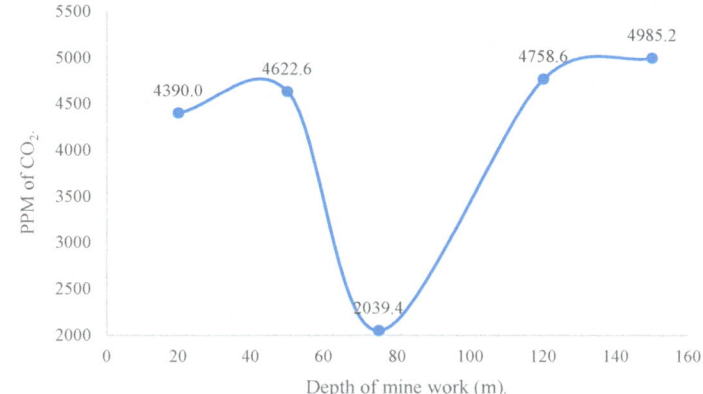

Fig. 17 Carbon dioxide profile with extraction using the standard KL-Epsilon model

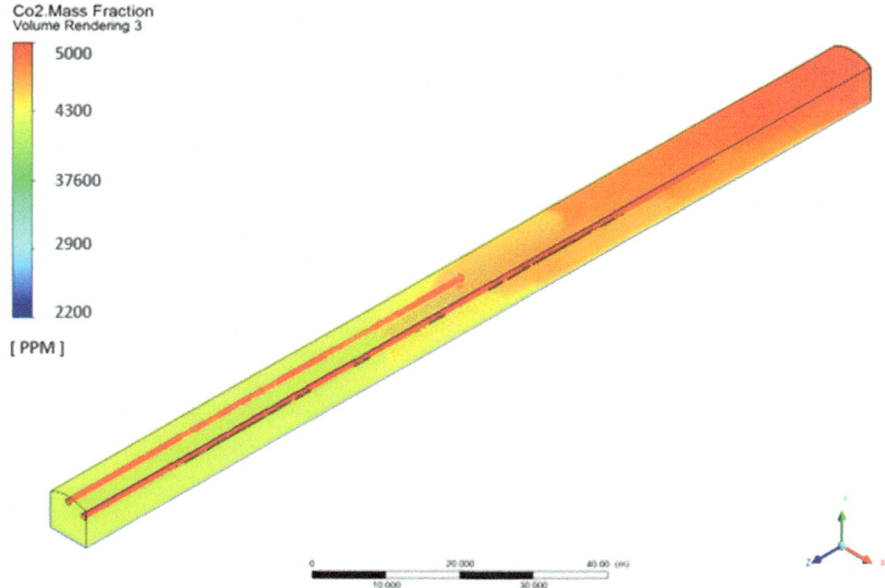

Co2.Mass Fraction
Volume Rendering 3

5000

4300

37600

2900

2200

[PPM]

Fig. 18 Carbon dioxide contour with extraction using the model KL-Epsilon standard

of the total length of the mine the concentration decreases 2584 PPM due to the extraction that is being made at that point; in the distance corresponding to 120 m the recorded concentration is 4759 PPM; finally at 150 m there is an increase in the previous concentration to 4985 PPM. The results obtained from CO_2 indicate that the concentrations are below the concentration established in NOM-010-STPS-199 corresponding to 5000 PPM.

The contour shown in Fig. 18 shows the variation in the concentration of CO_2 in the different points of the underground mine by implementing the standard KL-epsilon model, where the minimum concentration is 4390 PPM at a distance of 20 m and the maximum concentration is 4985 PPM recorded at 150 m.

3.6 Effect of CO_2 Concentrations in the Case Study (Model K-KL Omega 3 Ec.)

Figure 19 shows the behavior of carbon dioxide in the development work of the underground mine generated by the diesel combustion machinery. Where the lowest concentration of CO_2 is found at 20 m from the entrance of the mine: at a distance of 50 m there is a concentration of 5233 PPM; in relation to the middle part of the total length (75 m) there is a concentration of 5239 PPM; for the distance of 120 m the concentration increases 2 PPM (5241 PPM); to the 150 m that correspond to the front of the development work of the mine, there is the highest CO_2 concentration

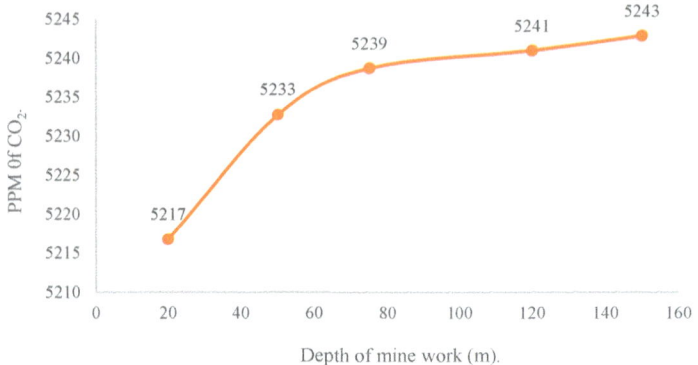

Fig. 19 Carbon dioxide profile without extraction using the model KL-omega 3 ecn

of 5243 PPM. These records indicate that fresh air circulation is not present or is minimal throughout the development work of the mine.

Figure 20 shows the contour of carbon dioxide in the development work of the underground mine using the model K-KL omega. The simulation shows the variation in the concentration of CO_2 at different points in the total length of the development work of the mine; The minimum concentration is 5217 PPM of CO_2, the maximum concentration is 5243 PPM corresponding to the 150 m distance.

Figure 21 shows the results of concentration of carbon dioxide obtained in the

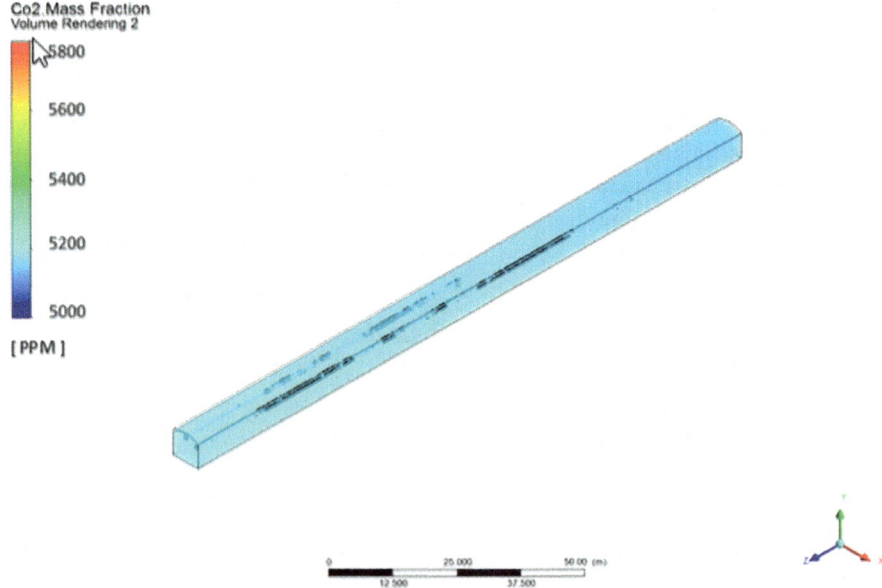

Fig. 20 Carbon dioxide contour without extraction using the K-KL omega model

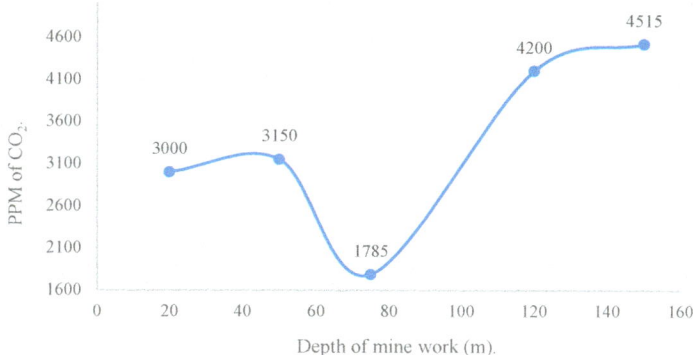

Fig. 21 Carbon dioxide profile with extraction using the model KL-omega 3 ecn

simulation carried out by applying a ventilation circuit. In it, it is observed at a distance of 20 m, there is a concentration of 3000 PPM, while at 50 m distance the concentration increases 150 PPM, this concentration decreases considerably at the distance of 75 m (1785 PPM) since at this point is the extractor; in relation to the distance of 120 m, the concentration is 4200 PPM, which tends to increase until reaching a concentration of 4515 PPM (in front of the mine). As well as the results obtained in the previous model, the concentrations registered with the application of this model are below the maximum permissible CO_2 established in NOM-010-STPS-1999.

Figure 22 shows the CO_2 contour where its behavior is observed, as well as the different concentrations of it. The variation in the most representative concentration of CO_2 in the underground mine is shown in the section from 50 to 75 m where having 3150 PPM decreases 1365 PPM.

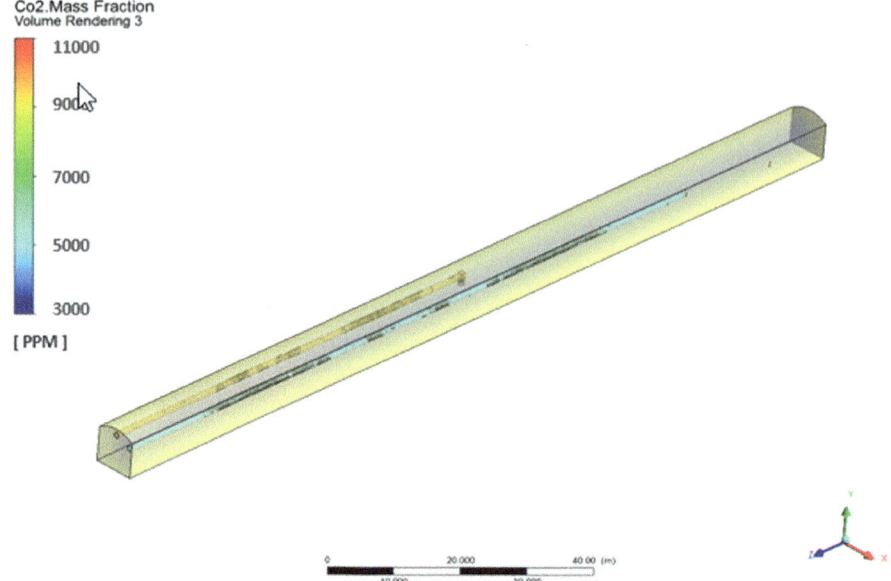

Fig. 22 Carbon dioxide contour with extraction using the model K-KL omega

3.7 Effect of CO₂ Concentrations in the Case Study (Reynolds Stress 7 Ec. Model)

Figure 23 shows the profile of carbon dioxide in the development work of the underground mine without the implementation of a ventilation model. In the figure it can be observed that the behavior of CO_2 oscillates between 5200 PPM along the depth

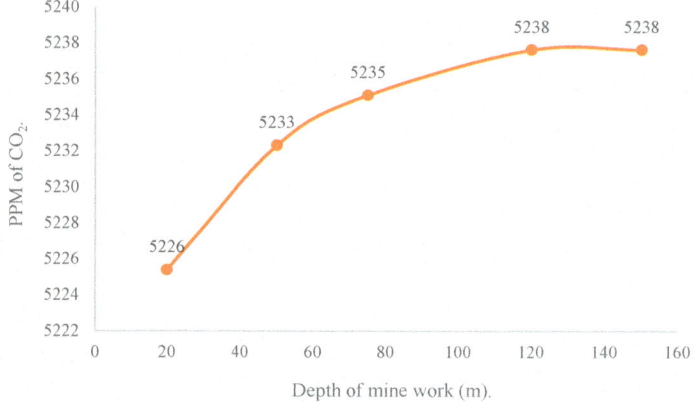

Fig. 23 Carbon dioxide profile without extraction using the Reynolds Stress model

of the development work of the mine, at 20 m there is a concentration of 5226 PPM; with respect to the distance of 50 m the concentration is 5233 PPM; in relation to the distance of 75 m to concentration of 5235 PPM of CO2 is registered; at 120 m in length of the concentration of carbon dioxide increases 3 PPM; On the front of the development work of the mine that corresponds to a distance of 1501 m, the concentration of 5238 PPM is conserved than in the 120 m.

Figure 24 shows the behavior of carbon dioxide at different points in the total length of the development work of the underground mine. The variation in the concentration of CO_2 from the different points of the mine is minimal because it increases 2 or 3 PPM from 50 m from the entrance of the mine. In the section of 20–50 m the concentration has a greater variation (around 7 PPM) compared with the other distances.

Figure 25 shows the profile of carbon dioxide, as well as its behavior along the total length of the underground mine. The figure shows that the maximum concentration of CO_2 is recorded at a distance of 150 m, while the minimum concentration of CO_2 is at 75 m from the entrance of the mine; in the distances of 20 and 50 m the concentration of CO_2 is shown in a range of 4800 and 4700 PPM. The recorded results agree with the position of the extractor (75 m) as well as the working machinery (general toxic gases). These results show in the same way as in the previous models that the concentrations registered throughout the development work in the underground mine are below what is established in NOM-010-STPS-1999.

Figure 26 shows the variation in the results of the concentration of CO_2. In contour shows that from the distance where the extractor was placed the concentration of CO_2

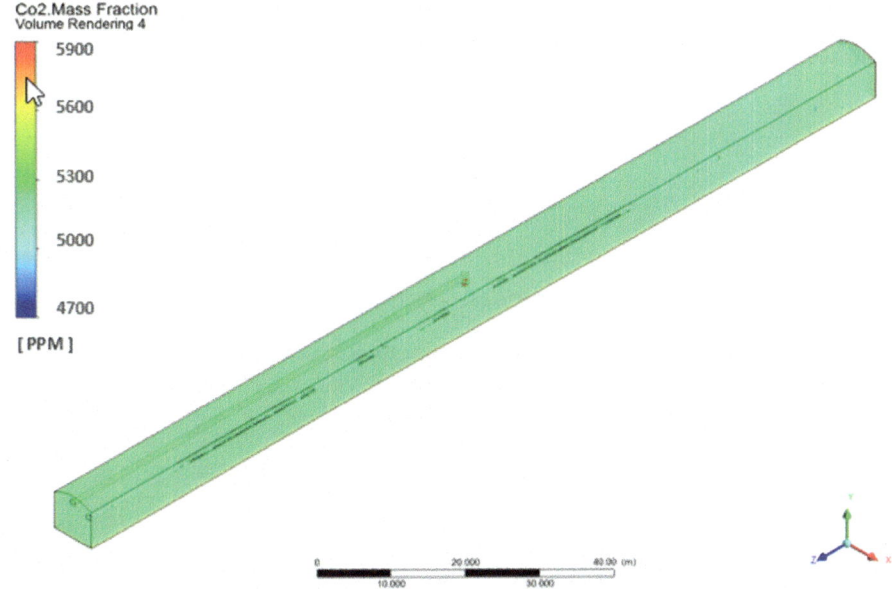

Fig. 24 Carbon dioxide contour without extraction using the Reynolds Stress model

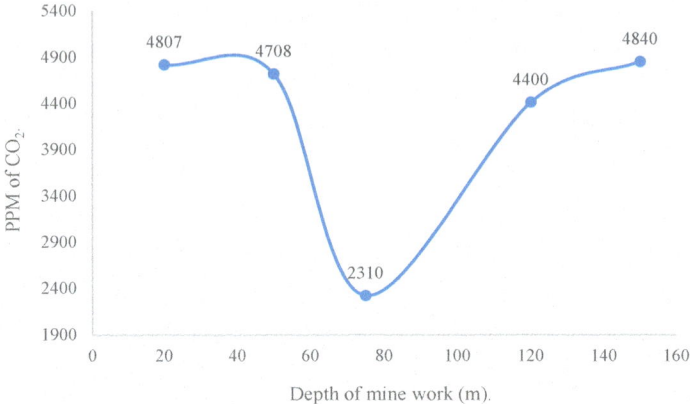

Fig. 25 Carbon dioxide profile with extraction using the Reynolds Stress model

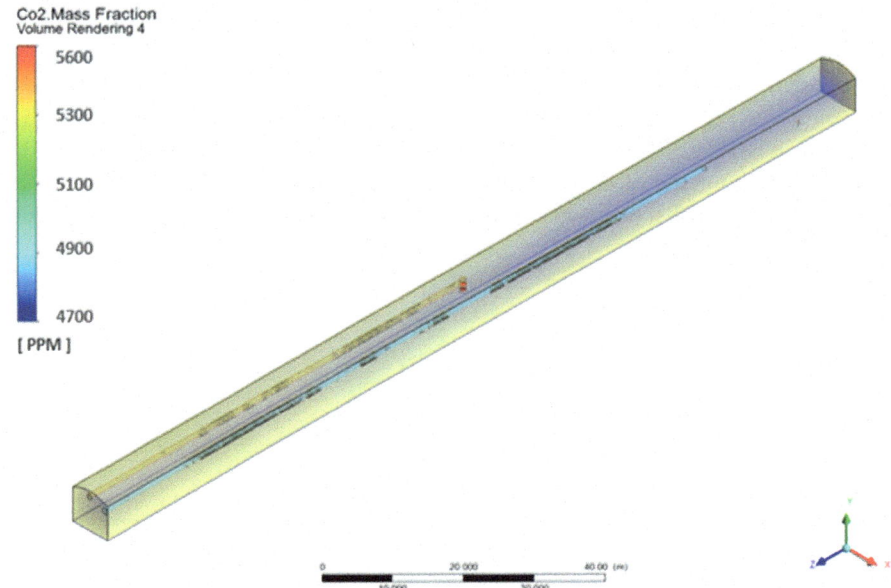

Fig. 26 Carbon dioxide contour with extraction using the Reynolds Stress model

tends to decrease because this is generating the necessary ventilation circuit so that the concentrations of CO_2 inside the underground mine decrease.

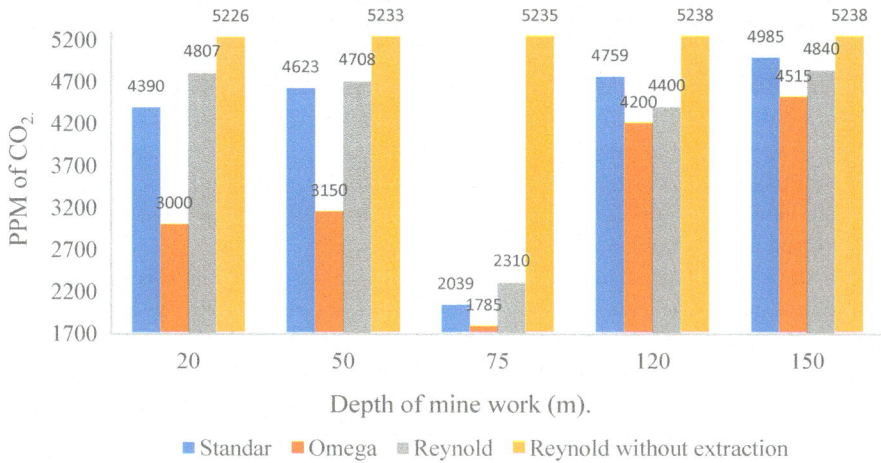

Fig. 27 Comparative analysis of three models of mass and energy transfer used in the simulation for the concentration of CO_2

3.8 Analysis of the Three Models of Mass and Energy Transfer for the Concentration of CO_2

Figure 27 shows the comparison of the concentration of carbon dioxide with the three models implemented for ventilation versus the results of the model that presented the highest concentrations without extraction. At a distance of 20 m the concentration of CO_2 decreases 419, 836, 2226 PPM in the Reynolds Stress, Epsilon-Standard, K-KL Omega models, respectively in relation to the model that generated the highest concentrations of CO_2 without extraction; corresponding to the distance of 75 m, the concentration is 2925, 3196, 3450 PPM respectively to the Reynolds Stress, Epsilon-Standard, K-KL Omega models with respect to the concentrations obtained in the model without extraction; at 150 m distance the concentrations decrease to 253, 398, 723 PPM respectively to the Epsilon-Standard, Reynolds Stress, K-KL Omega models, in relation to the maximum concentrations recorded in the model without extraction. With the results obtained from the concentration of CO_2 in the different points of the underground mine with the three models applied; The one that showed the least presence of CO_2 with the extraction is the K-KL Omega model.

4 Conclusions

With the application of the three models used (Epsilon-Standard, K-KL Omega, Reynolds Stress) in the simulation of ventilation of 500 s for each one of the factors (CO and CO_2) that take part daily in the works carried out in the different works of development in underground mine, results were obtained below the maximum permissible established within the norm NOM-010-STPS-1999 with each of the models, which present a great difference between the maximum and minimum records with the simulations without the implementation of a ventilation circuit. These results show that the three models applied in the simulations represent an option to ventilate a certain area in underground mines. However, when implementing the Epsilon-Standard and Reynolds Stress models, there could be a probability that the ventilation applied to a site will generate results that are within the limits established by the standards.

On the other hand, if the model K-KL Omega is applied in a ventilation, the concentrations of the gases present in the work works are below the established in the standards considering a safety factor for each of these. Likewise, this method could be adapted for any type of work that needs a ventilation circuit taking into account the characteristics of the site (elevation, temperature, humidity, ventilation works, etc.); In the same way, the type of fan that is needed so that the development work is ventilated could be selected, as well as the position of it to generate better working conditions and therefore determine the time it takes for the area to be ventilated (decreasing downtime). This model presents a possibility of saving in terms of the electrical energy consumed and therefore the ventilation processes will be economized.

In general, this project represents a good option in the study of sustainable ventilation systems because, based on these simulations, it is possible to generate implementation projections (prior validation), taking into account extremely important parameters such as the reduction of resources energy, as well as the reduction of emission gases in underground mines, which creates conditions for improving the health of workers and contributes to the reduction of toxic gases in the environment where these types of activities are carried out.

Acknowledgements The authors thank the support for the realization of this research project to the Support Program for the Incorporation of NPTC-PRODEP UGTO-PTC-633.

References

Armstrong J, Menon R (1998) Minas Y Canteras. Industrias basadas en recursos, vol 1, 4th edn. Naturales, pp 74–75

Beard J (1894) Ventilation of mines, 1st edn. USA, pp 1–21. https://archive.org/details/ventilationmine00beargoog/page/n2

Brundtland Report of the World Commission on Environment and Development (1987) United Nations, p 43. https://www.are.admin.ch/are/en/home/sustainable-development/international-cooperation/2030agenda/un-_-milestones-in-sustainable-development/1987–brundtland-report.html

Fluent Inc. (2006) FLUENT 6.3 getting started guide. Reaction Design Inc., PathScale Corporation, USA, p 1

Hartman H (1961) Mine ventilation and air conditioning. Vanderbilt University, The Ronald Press Company, USA

Hartman H (1992) SME mining engineering handbook, vol 2, 2nd edn. USA, pp 1273–1275

Herruzo J (2015) Diseño, cálculo y fabricación de un sistema de escape para motor 250 cc monocilíndrico. Dep. de Ingeniería Mecánica y Fabricación Universidad de Sevilla, España, p 1. https://www.springer.com/cda/content/document/cda_downloaddocument/Key_Style_Points_BasicRef.pdf?SGWID=0-0-45-1330668-0

Normas Oficiales Mexicanas STPS (1996) Norma Oficial Mexicana NOM-121-STPS-1996, Seguridad e Higiene para los trabajos que se realizan en Minas. Estados Unidos Mexicanos. http://legismex.mty.itesm.mx/normas/stps/stps121.pdf

Quevedo C (2013) Sistema de ventilación de diez kilometros del túnel de conducción de la Central Hidroeléctrica Huanza. Dissertation, Universidad Nacional Mayor de San Marcos, pp 28–29

Richards J (2002) Sustainable development and the mineral industry. SEG Newsletter, Society of Economic Geologists, Canadá, p 48

Sarkar S, Balakrishnan L (1990) Application of a Reynolds-stress turbulence model to the compressible shear layer. ICASE Report 90-18, NASA CR 182002, USA, p 1. https://apps.dtic.mil/dtic/tr/fulltext/u2/a227097.pdf

Secretaría de Trabajo y Previsión Social (2012) NORMA Oficial Mexicana NOM-023-STPS-2012, Minas subterráneas y minas a cielo abierto - Condiciones de seguridad y salud en el trabajo. Estados Unidos Mexicanos. http://www.stps.gob.mx/bp/secciones/dgsst/normatividad/p-a-nom-023-stps-2012-19.06.13.pdf

Wilcox D (1994) Turbulence modeling for CFD, vol 1. DCW Industries, Inc., USA, pp 1–17

Affordable and Clean Energy: A Study on the Advantages and Disadvantages of the Main Modalities

Pablo Carpejani, Érica Tessaro de Jesus,
Bárbara Luzia Santor Bonfim Catapan, Sergio Eduardo Gouvea da Costa,
Edson Pinheiro de Lima, Ubiratã Tortato, Carla Gonçalves Machado
and Bernardo Keller Richter

Abstract The purpose of this study is to identify several renewable and clean energy sources and investigate their accessibility. The structuring of the energetic sources was outlined to display the advantages and disadvantages of their use. To do so, the literature review method and the snowball method were used as the research methodology. Previous results had determined the elements of a sample of the main existing forms of energy. Therefore, this research analysed the following sources: solar energy, wind energy, hydroelectric power, thermoelectric energy with renewable fuels, tidal energy, biogas energy, geothermal energy and hydrogen energy. The results discuss the benefits of using sustainable energies, such as being helpful to the environment, as well as the implementation obstacles that, in this case, are stripped down to the high financial cost of initial investment. Because no previous research provided a structure to compare different energy forms, this study is expected to act as an initial guide for researchers and professionals in the field. As a limitation and recommendation for future research efforts, it is suggested to discover and verify mechanisms capable of reducing the high initial investment costs associated with sustainable energies.

P. Carpejani (✉) · É. T. de Jesus · B. L. S. Bonfim Catapan · S. E. Gouvea da Costa ·
E. Pinheiro de Lima · B. K. Richter
Department of Industrial and Systems Engineering, Pontifical Catholic University of Parana (PUCPR), Imaculada Conceição—Prado Velho, Curitiba 1155, Brazil
e-mail: pablo.carpejani@pucpr.edu.br

S. E. Gouvea da Costa · E. Pinheiro de Lima
Department of Industrial and Systems Engineering, Federal University of Technology - Parana (UTFPR), Via do Conhecimento—Fraron, Pato Branco, Brazil

U. Tortato
Department of Business, Pontifical Catholic University of Parana (PUCPR), Imaculada Conceição—Prado Velho, Curitiba 1155, Brazil

C. G. Machado
Department of Technology Management and Economics, Chalmers University of Technology, Chalmersplatsen 4, 412 96 Göteborg, Sweden

© Springer Nature Switzerland AG 2020   615
W. Leal Filho et al. (eds.), *International Business, Trade and Institutional Sustainability*, World Sustainability Series,
https://doi.org/10.1007/978-3-030-26759-9_35

Keywords Clean energy · Advantages · Disadvantages · Sustainable development goals · Goal seven

1 Introduction

Presently, the quest for self-sufficiency in energy generation, together with a diversification of the energy matrix, keeps increasing the demand for different sources of alternative energy. Organizations and countries are increasingly seeking clean energies and renewable resources to supply their own demand. This inclination has been amplified due to the trend of fossil fuel shortages. The environmental, social and economic changes resulting from this reality are gradually reaching the spotlight of the academic community's agenda, which continually explores these factors to ensure society's progress. In addition to governments and the academic world, the demand for clean and renewable energy is also part of the reality of NGOs, consumers, private companies and other institutions.

The use of clean and renewable energies is regarded as a priority in our contemporary climate, and the development of their use is increasingly receiving financial and research resources. Pacheco (2006) defines clean energy as a source of energy not produced by fossil fuels. These energy sources are also commonly known as green energies, because they do not pollute the atmosphere and contribute to the greenhouse gas effect. Renewable energy is any type of energy sourced in natural cycles of solar radiation conversion; in other words, it is the product of natural processes, such as sunlight and wind (Bhattacharya et al. 2016) and includes almost every source of energy naturally available on Earth. The term "renewable" means that the main source is inexhaustible; therefore, it does not harmor influence the planet's thermal balance. Another trait is that renewable energy sources have no usage limit, as long as they are properly cared for (Marques 2007; Pacheco 2006).

In September 2015, following the favourable trend of these energy forms, the United Nations approved a global agenda with 17 goals and 169 targets for the implementation of initiatives capable of making the planet more sustainable, resilient and socially fair The objectives and targets are intended to stimulate action during the next 15 years in fields deeply important for humanity and the planet (United Nations 2018a). Specifically, among these objectives, goal no. 7, which is the object of this research, seeks to establish universal, reliable, sustainable, modern and affordable access to energy services for everyone by 2030, thus significantly increasing the share of renewable energy in the global energy matrix, doubling the overall rate of energy efficiency improvement, and bolstering international support for the promotion of access, research and investments in energy infrastructure and clean energy technologies (United Nations 2018a). These goals are an extension of the actions of the Brundtland Report, since 1987, has consubstantiated the importance of sustainable development (United Nations 2018b).

Nevertheless, in order for these goals to be accomplished, specificpremises have to be accepted; for instance, for global progress, each nation has to comply with the requirement to use renewable energy (Yüksel 2010). The push is to have the energy scene progress toward 100% sustainable use; in other words, the world will be responsible for using fully renewable energies by the year 2050 (Greenpeace 2018; Fiesp 2018). As an example of forward progress in this direction, Brazil intends to increase energy usage from renewable sources from 28 to 33% by the year 2030 (Brasil 2016).

To accomplish these goals and reap the many benefits of relying on clean, renewable energy, the study of energies is a pivotal element for human evolution, since its results are needed by present and future generations. Therefore, understanding and distinguishing between different types of energy is a decisive step toward magnifying the strengths and neutralizing the weaknesses of each energy matrix. In this regard, the objective of this paper's research is the following: to review the advantages and disadvantages of the main sources of renewable energy and establish a comparative table to engage and support the debate on the accessibility of each energy type. Therefore, this work is justified from both a practical and a theoretical standpoint. Regarding the practical side, this document will serve as a tool to help in the achievement and development of the United Nation's Goal 7 while also being a guide for corporations and governments that intend to collaborate in the work of sustainable development. As a theoretical contribution, the article will be the starting point for future research that requires a list of the main types of existing clean energy and a comparison of their accessibility (using advantages and disadvantages).

2 Methodology

The literature review method was used to identify the main sources of existing clean energies and the main modalities.

The following databases were consulted: Web of Science, Google Scholar and Scopus. The initial research strings were based on the following word combinations: "clean energy", "renewable energy", "energy matrix", "advantages" and "disadvantages". To assess the existence of potential and new modalities, the *snowball* technique was used. This technique uses the references of the initial articles as a kind of network, allow in researchers to discover new information through articles not encountered by the initial searches (Baldin and Munhoz 2011). Thus, the references of the first articles were analyzed in order to find other similar documents.

3 Results

Based on the findings of the literature review—in the main results found—this research investigates the following energy sources: solar energy, wind energy, hydroelectric power, thermoelectric energy with renewable fuels, tidal energy, biogas energy, geothermal energy and hydrogen energy. Following the investigation, the primary results are presented, and the research protocol is explained. Table 1 presents a summary of the results.

4 Solar Energy

Solar energy is one of the renewable sources with greater potential to be used in the future. Solar energy is one of the best options to address future energy demands, because it is an inexhaustible, clean, affordable and inexpensive source compared to other renewable energy sources (Kabir et al. 2018).

According to Hernandez et al. (2014), of all renewable energies, solar energy has one of the lowest impacts on climate change due to its low emissions of pollutant substances. Furthermore, one of the main advantages of solar systems is the generation of energy near the place of usage, which reduces costs and losses in its diffusion. This happens mostly in urban areas with the use of walls and roofs in homes to capture energy.

Nonetheless, the collection and concentration of solar energy for consumption requires vast amounts of space, which can cause negative impacts on natural ecosystems and biodiversity (Hernandez et al. 2014; Northrup and Wittemyer 2013). Kabir et al. (2018) declare that the system requires high initial expenditures for the installation of this energy's capturing method. Also, the efficiencies of most domestic solar panels account for 10–20%, and more efficient solar panels have higher costs; these higher costs are the deficiency of solar technology. Other shortcomings identified are associated with the maintenance of systems, such as the shortage of manpower to address the growing demands related to the installation, maintenance, repair and evaluation of solar energy systems (Kabir et al. 2018).

5 Wind Energy

Although most people think of "wind energy" as synonymous with "wind", wind energy comes from the heat released by the sun (Ilkiliç et al. 2011). The sun is responsible for the Earth's heat release, causing different temperature ratios in the most diverse geographical areas of the planet. The surface response is to absorb, retain, reflect and release the captured heat, phenomena that result in the origin of wind (Ilkiliç and Türkbay 2010). Aerogenerators capture the kinetic movement of the wind and transform it into electrical energy (Dincer 2011).

Wind energy is a source of free, clean and inexhaustible energy (Alsaad et al. 2013). The use of this type of energy was expanded due to society's growing environmental concerns in the last decades, resulting in an increased awareness of green energy (Ilkiliç and Türkbay 2010; Tabassum et al. 2014).

Benefits include cleanliness, low cost and the abundance of wind in any part of the globe. The technology that converts wind energy into other types of energy is

Table 1 Types of renewable energy, advantages and disadvantages

Affordability

Type of renewable energy	Advantages	Disadvantages
Solar	It does not pollute, and it requires minimal maintenance. Installation is easy, and it does not require a high investment for its diffusion	It requires extensive urban areas for energy capturing and depends on daily solar irradiation
Wind	It is a free, clean, unlimited source of energy and also one of the cheapest sources of energy; it does not require a fuel supply	Production depends on weather conditions; it changes the usual behaviour of wild animals while also contributing to noise pollution
Hydropower	It does not emit polluting substances and is inexhaustible; the production cost is low	There are significant costs related to installation and deactivation; it requires large water reservoirs and contributes to soil erosion that has a negative effect on local vegetation
Thermoelectric with renewable fuels	The thermoelectric energy conversion system does not present obstacles; it is compact, soundless, highly reliable and does not harm the environment	The conversion process may be inefficient, reaching a level deemed low when there is interference with other energy sources
Tidal wave	A source of clean energy that does not present major environmental hazards, it is an inexhaustible resource, and the energy-generating equipment has low maintenance costs	Production depends on sea conditions; the installation of equipment for tidal power generation requires a significant investment
Biogas	It is regarded as an efficient, sustainable and low-cost alternative; the transformation of organic waste into biogas has environmental advantages	Generates a high concentration of methane gas, which contributes significantly to environmental pollution
Geothermal	Low levels of pollution and environmental impact but economically viable	If not properly extracted, it damages the planet. The implementation has significant costs, and geothermal plants produce noise pollution
Hydrogen	Provides renewable and sustainable energy while also producing electricity and hot water steam; has a high potential for technological development	A high cost when compared to conventional energy sources; highly hazardous and significant costs related to the transportation and distribution of hydrogen gas

Source the authors (2018)

cheaper than other conversion systems. The wind is commonly available in nature, and through wind turbines and mechanical conversion, wind energy can be transformed into electrical energy (Ilkiliç and Türkbay 2010). To operate, wind turbines do not need fuel. Wind power generation does not produce emissions like carbon dioxide, sulphur dioxide, mercury, particulates or any kind of air pollution (Alsaad 2013; Fang 2014).

When compared to traditional means, wind energy has high costs (Alsaad 2013). The negative impacts derived from wind power projects manifest locally; in other words, they mostly have to do with the dislodgment of wild animals in the surroundings, as well as visual and sound effects (Slattery et al. 2012).

6 Hydroelectric Power

The cornerstone of hydroelectric power is the conversion of water's potential force into energy. Hydro turbines transform hydraulic pressure into energy resources for processing machines, which is the same as what happens with electricity generators (Frey and Linke 2002; Yuksek et al. 2006). Even though it is feasible to merge the usage, the turbines used in the process have two classifications: impulse and reaction. The energy associated with water pressure is converted into kinetic energy, in which the movements of the items, required for the process, change the flow of the water stream, thereby generating energy initiation (Balje 1981; Bartle 2002; Date et al. 2013). Water is pressurized as it moves through the production process (Date et al. 2013).

Currently, hydroelectric power is the most used renewable resource for electricity generation (Frey and Linke 2002; Yuksek et al. 2006; Kendir and Ozdamar 2013). Since the most important component for viability is proper water management, it is deemed sustainable in the long term (Yuksek et al. 2006).

Hydroelectric is still the most efficient way to generate electricity, because modern turbines can convert approximately 90% of the energy available into electricity (Yuksek et al. 2006). In comparison, fossil fuel plants are only 50% efficient (Frey and Linke 2002). Its usage contributes to energy independence for countries without fossil fuel resources (Yuksek et al. 2006). Its benefits are associated with human well-being, including a safe water supply, irrigation for food production, flood control and social improvements, such as increased navigation, the development of fishing activities, home-based industries and more (WCD 2000; Yuksek et al. 2006). These advantages can be associated with electric power or related to the environment's development (OUD 2002). The indirect benefits are also better than fossil-based energy generation. In 1997, the estimates indicated that hydroelectric power prevented the production of emissions equivalent to the pollution of every car on the planet (IHA et al. 2000).

This type of energy is highly renewable (over 90%) and has low energy costs, high longevity (200 years), a short amortization period (5 and 10 years) and extremely low operational costs (0.20 cents/kWh). Moreover, it does not depend on external energy sources and incurs no fuel costs (Kendir and Ozdamar 2013).

Proper geographic conditions are required to support the creation of hydroelectric plants. Developing countries have great potential to develop hydroelectric (Iea 2002; Koch 2002; Yuksek et al. 2006). In comparison to thermal power generation, hydroelectric plants require more time for business forecasting. For example, their approval needs to overcome more hurdles, construction can take significantly longer and they require a higher initial investment, which means that recovering the capital invested entails more time (Yuksek et al. 2006).

7 Thermoelectric Energy with Renewable Fuels

Thermal generation directly transforms thermal energy into electricity using thermoelectric conversion materials, like oil and its derivatives, natural gas and mineral coal, among others (Chen et al. 2012). The steam generated is used to handle the plant's turbines. Generally speaking, the thermoelectric power conversion system does not pose major obstacles, it is compact, quiet, highly reliable and environmentally friendly (Kim et al. 2006; Bell 2008; Gou et al. 2010). Nevertheless, renewable fuels such as ethanol, biodiesel and biomass may also be used.

Thermoelectric devices (thermoelectric modules) maintain the ideal temperature. Therefore, the energy is transformed by the thermoelectric generator, which is represented by a single heat engine, safeguarding the loading, transportation and fluid work services (Riffat and Ma 2003). As long as the temperature remains controlled, the devices connected to the module will generate energy (Riffat and Ma 2003).

Thermoelectric plants are essentially silent when compared to other energy options. Their operation is reliable, as it includes several fuels, such as petroleum, for the execution and generation of electrical energy.

Nonetheless, the conversion process is inefficient, attaining a level deemed low when there is interference from other energy sources (Gou et al. 2010). In thermoelectric power, energy generators are limited by their low efficiency (Rowe and Min 1998. Riffat and Ma 2003).

8 Tidal Energy

Tides are being increasingly applied in different forms. Presently, ocean exploration is giving birth to new concepts and shapes. Particularly, the kinetic energy of tidal flows can generate relevant levels of electricity (O'Rourke et al. 2010; Neill et al. 2016). Waves are generated through several factors, such as the interaction between the Earth, the moon and the sun. Waves, whenever they are influenced by the moon, cause an imbalance between the forces acting on a particle due to the gravitational attraction, which triggers a centrifugal force due to Earth's rotation on the centre of gravity (Neill et al. 2016). With the movements deriving from the unevenness of the weather, and also from the sea, it is possible to generate power through dams and turbines (Devine-wright 2011).

The energy of the tides is acknowledged by society as an important potential source of clean energy, because it is a structure in which kinetic energy can be extracted from tidal currents. The forecasts indicate that this type of energy will acquire more notoriety and space in power extraction within the next decades, reaching a great percentage of generation of total electricity production (Atwater and Lawrence 2011).

The method's potential expansion is one of the key future benefits, as is the high index of seas to be explored (O'Rourke et al. 2010; Neill et al. 2016). The turbines of tidal streams have a high-performance percentage between cycles, and there are none of the interruptions that occur with other devices that capture renewable energy (Fagan et al. 2016).

Many energy conversion systems for tidal streams do not have the capacity to actively respond to direction changes. They rely on nearby bidirectional flows or energy conversion systems that are relatively insensitive to small directional changes (Harding and Bryden 2012). There are also regulatory and economic issues, which, in all likelihood, will limit the generation of tidal energy to levels significantly lower than the maximum values that are physically possible (Atwater and Lawrence 2011). Generally speaking, to date, few empirical studies are available on the subject's specificities and complexity (Devine-wright 2011).

9 Biogas Energy

Biogas is a renewable energy source derived from the anaerobic digestion of biomass, which can be a product of organic waste, animal manure, sewage and industrial effluents. It often consists of 40–75% methane, 20–45% carbon dioxide and other compounds (Budzianowski 2012).

According to Deublein and Steinhauser (2008), anaerobic digestion has great potential for energy production, yet the process also relies on energy to maintain itself. Therefore, energy has to be used to heat new substrates and to make up for the loss of heat in digesters and pipes. This energy can be made from the biogas combustion itself. The amount of energy required depends on different factors, such as the temperature of the digester, the temperature of new substrates, the room temperature and the insulation of digesters and pipes to reduce their heat loss. Furthermore, pumps, agitation equipment and monitoring are required. The replacement of fossil fuels by biogas also reduces the greenhouse gas emissions (Morken and Sapci 2013).

On the other hand, there are some obstacles to using biogas for electrical power generation, particularly the lack of suitable technology for its conversion into energy (Coelho et al. 2004). The lack of information on biogas production, as well as market studies on the energy to be produced, discourage potential investors. Likewise, project technology assessments, safety, legislation, capital costs, operation and maintenance are extremely important, but they still lack a reliable body of research and coverage in the scholarly literature (Coelho et al. 2004).

10   Geothermal Energy

Geothermal energy is produced by the Earth's internal heat, and it can be regarded as a solution to the current issues of energy shortage and combustion of fossil fuels (Arboit et al. 2013). According to Rabelo et al. (2002), the ideal temperature for the use of this heat ranges between 35 °C and 148° for residential, industrial and agricultural purposes, but it must be higher than 300 °C for the generation of energy.

Geothermal energy can be economically viable. Furthermore, it is also a source of clean energy available in different regions of the planet. If robust and viable technologies are developed for its capture and use, an energy revolution may take place, and humanity's dependence on fossil fuels may be reduced (Vichi and Mansor 2009).

According to Campos et al. (2017), there are some cons relating to geothermal energy's operating conditions. For instance, plants can only be established in propitious geological zones, which are located in only 10% of the planet. Additionally, it has a high initial cost fora plant's well drilling, study and establishment, and there is significant noise pollution during plant construction.

11   Hydrogen Energy

Hydrogen energy is obtained through the combination of hydrogen and oxygen, causing vapours capable of releasing energy (Zhang et al. 2016). Hydrogen can be produced from different means, such as fossil fuels, nuclear energy, renewable biomass and renewable electricity using thermal energy and biochemicals (Mazloomi and Gomes 2012; Zhang et al. 2016). Presently, the majority of hydrogen production is derived from fossil fuels (Zhang et al. 2016). The hydrogen produced can be stored in compressed or liquid gases, which are widely used in the industry. Hydrogen technology has the potential to change energy infrastructure and current lifestyles (Zhang et al. 2016). Hydrogen is an important *driver* of energy storage that benefits renewable and sustainable energy (Zhang et al. 2016). Hydrogen energy provides an energy vector needed for storage and for energy exchange between smart centres (Maroufmashat et al. 2016). Hydrogen is an efficient energy carrier, because it registers a low number of losses during transportation when compared to the traditional grid (Mazloomi and Gomes 2012; Walker et al. 2016).

There are limitations in energy generation, and the gaseous storage requires quite large tanks. In addition, liquefaction requires great initial workforce and storage tanks and major insulation efforts, while a metal hydride may absorb only small amounts of hydrogen (Mazloomi and Gomes 2012; Kanoglu et al. 2016). The hydrogen energy method is criticized for its high energy consumption and carbon emissions (Kanoglu et al. 2016).

12 Discussion and Conclusion

This research intended to identify the advantages and disadvantages of several sustainable energy sources. In the future, this article is expected to be the initial instrument for theoretical and practical analyses that will precede any other form of application.

Due to the conceivable scarcity of non-renewable resources, like coal and oil, as well as the damage that their extraction and usage do to human health and the environment, the search for renewable energy forms is increasing. Therefore, it is important to have a general background of the possible forms of energy generation in order to enable a systematic view of the decision-making process.

The main benefit of renewable and clean energy sources is the reduction of the environmental impact due to the lowering of greenhouse gas emissions and the use of non-finite resources. The use of different clean energy generation technologies can reduce the consumption of resources and the emission of polluting gases into the environment, while also aiding in human efforts to tackle climate change, acid pollution and photochemical pollution (Hongtao and Wenjia 2018). Nevertheless, the main disadvantage identified in this research is that, given the few viable technologies and the lack of government incentives, renewable sources demand a high investment (Lima 2012).

The proposed methodology produced a list of eight different types of energy (solar energy, wind energy, hydroelectric power, renewable energy, tidal energy, biogas energy, geothermal energy and hydrogen energy). It was possible to identify energy forms that are not usually discussed in comparison with other types of renewable energy—for example, geothermal energy, which does not have thorough coverage in academic literature. On the other hand, other energy forms, like solar, wind and hydroelectric, have a wide theoretical body of literature, favouring research and implementation.

Generally speaking, because different types of energy have different generation costs, the production of clean and renewable energy demands considerable financial resources (Hongtao and Wenjia 2018). Likewise, a significant investment in infrastructure may take several years (besides having to rely on geographic assistance, asis the case with hydroelectric, wind and geothermal energies). Regardless of the cost and difficulty of execution, these types of energy forms will write off the investment in the long run, not only due to their benefits to the planet, but also due to their ability to save resources. The most commonly used renewable energy sources are hydroelectric and thermoelectric plants with renewable fuels, whereas solar, wind and biogas energies have attained considerable growth. Nonetheless, each form has its own pros and cons, and it is important to verify the existing investment conditions, scenarios and available local resources.

Counterweighting the aforementioned movement, only the long-term perspective creates the need to analyse the investment in this type of energy in the future. The conscious search for energy also ends up supporting economic development, since the implementation costs offset the amount invested in the long run. The ideology behind sustainable energy has moved from a reflective idea to a worldwide need of governments, companies and society. It is possible to benefit the environment while also acquiring great economic advantages (Lima 2012).

Finally, this research supports the development of goal no. 7 of the UN's Sustainable Development Goals, highlighting the advantages and disadvantages of the eight types of energy deemed sustainable. Although goal no. 7 significantly stimulates the use of renewable energies in the global energy matrix, this research identifies the initial price of investing in clean energy as a primary hindrance to its adoption. As previously seen, it is a major obstacle to society's broad adoption of sustainable energy. Therefore, the suggestion for future research efforts is to find methods to popularize clean and renewable energies.

Acknowledgements This study was financed in part by the Coordenação de Aperfeiçoamento de Pessoal de Nível Superior—Brasil (CAPES)—Finance Code 001.

References

Alsaad M, El-Suleiman A, Nasir A (2013) An assessment of wind energy resource in north central Nigeria, Plateau. Sci J Energy Eng 3:13–17

Alsaad M (2013) Wind energy potential in selected areas in Jordan. Energy Convers Manag 65:704–708

Atwater J, Lawrence G (2011) Regulatory, design and methodological impacts in determining tidal-in-stream power resource potential. Energy Policy 39(3):1694–1698

Arboit NKS et al (2013) Potencialidade de utilização de energia geotérmica no Brasil—umarevisão de literatura. Revista do Departamento de Geografia—USP 26:155–168

Baldin N, Munhoz BME (2011) Educação Ambiental Comunitária: umaexperiência com a técnica de pesquisa snowball (bola de neve). Revista Eletrônica do Mestrado em EducaçãoAmbiental, 27

Balje OE (1981) Turbo-machines, a guide to design, selection and theory. Wiley

Bhattacharya M, Paramati RS, Ozturk I, Bhattacharya S (2016) The effect of renewable energy consumption on economic growth: evidence from top 38 countries. Energy Appl 162:733–74

Bartle A (2002) Hydropower potential and development activities. Energy Policy 30(14):1231–1239

Bell L (2008) Cooling, heating, generating power, and recovering waste heat with thermoelectric systems. Science 321(5895):1457–1461

Brasil (2016) em: http://www.brasil.gov.br/meio-ambiente/2015/11/com-proposta-mais-ambiciosa-Brasil-chega-a-COP21-como-importante-negociador-mundial-do-clima (Last accessed 10 June 2016)

Budzianowski WM (2012) Sustainable biogas energy in Poland: prospects and challenges. Renew Sustain Energy Rev 16:342–34

Chen Z, Han G, Yang L, Cheng L, Zou J (2012) Nanostructured thermoelectric materials: current research and future challenge. Prog Nat Sci: Mater Int 22(6):535–549

Campos AF, Scarpati CBL, Santos LT, Pagel UR, Souza VHA (2017) Um panorama sobre a energia-geotérmica no Brasil e no Mundo: Aspectosambientais e econômicos. RevistaEspacios 38(1):8. https://www.revistaespacios.com/a17v38n01/a17v38n01p08.pdf

Coelho ST, Velazquez SMSG, da Silva OC, Varkulya A, Pecora V (2004) Programa de uso racional de energia e fontes alternativas. Enc Energ Meio Rural, 5

Date A, Date A, Akbarzadeh A (2013) Investigating the potential for using a simple water reaction turbine for power production from low head hydro resources. Energy Convers Manag 66:257–270

Devine-Wright P (2011) Enhancing local distinctiveness fosters public acceptance of tidal energy: a UK case study. Energy Policy 39(1):83–93

Deublein D, Steinhauser A (2008) Biogas from waste and renewable resources: an introduction. Wiley-VCH Verlag, Weinheim. https://doi.org/10.1002/9783527621705

Dincer F (2011) The analysis on wind energy electricity generation status, potential and policies in the world. Renew Sustain Energy Rev 15(9):5135–5142

Fagan E, Kennedy C, Leen S, Goggins J (2016) Damage mechanics based design methodology for tidal current turbine composite blades. Renew Energy 97:358–372

Fang H (2014) Wind energy potential assessment for the offshore areas of Taiwan west coast and Penghu Archipelago. Renew Energy 67:237–241

FIESP (2018) Federação das Industrias do Estado de São Paulo. http://www.fiesp.com.br/noticias/matriz-100-renovavel-no-brasil-em-2050-e-viavel-mostra-greenpeace-na-fiesp. (Last accessed 17 July 2018)

Frey G, Linke D (2002) Hydropower as a renewable and sustainable energy resource meeting global energy challenges in a reasonable way. Energy Policy 30(14):1261–1265

Gou X, Xiao H, Yang S (2010) Modeling, experimental study and optimization on low-temperature waste heat thermoelectric generator system. Appl Energy 87(10):3131–3136

Greenpeace (2018) http://www.greenpeace.org/brasil/pt/Blog/15-graus-celsius-na-veia-100-energia-renovvel/blog/55060/. (Last accessed 01 Oct 2018)

Harding S, Bryden I (2012) Directionality in prospective Northern UK tidal current energy deployment sites. Renew Energy 44:474–477

Hernandez RR et al (2014) Environmental impacts of utility-scale solar energy. Renew Sustain Energy Rev 29:766–779

Hongtao L, Wenjia L (2018) The analysis of effects of clean energy power generation. Energy Procedia 152:947–952

IEA (2002) International energy agency. World Energy Outlook, IEA/OECD, Paris

IHA (2000) International Hydropower Association, ICOLD (International Commission on Large Dams), IAHTP IEA (Implementing Agreement on Hydropower and Programmes, IEA), CHA (Canadian Hydropower Association), 2000. Hydropower and world's energy future, pp 1–14

Ilkılıç C, Aydın H, Behçet R (2011) The current status of wind energy in Turkey and in the world. Energy Policy 39(2):961–967

İlkiliç C, Türkbay İ (2010) Determination and utilization of wind energy potential for Turkey. Renew Sustain Energy Rev 14:2202–2207

Kanoglu M, Yilmaz C, Abusoglu A (2016) Geothermal energy use in absorption precooling for Claude hydrogen liquefaction cycle. Int J Hydrogen Energy 41(26):11185–11200

Kabir E, Kumar P, Kumar S, Adelodun AA, Kim KH (2018) Solar energy: potential and future prospects. Renew Sustain Energy Rev 82:894–900

Kendir T, Ozdamar A (2013) Numerical and experimental investigation of optimum surge tank forms in hydroelectric power plants. Renew Energy 60:323–331

Kim W, Zide J, Gossard A et al (2006) Thermal conductivity reduction and thermoelectric figure of merit increase by embedding nanoparticles in crystalline semiconductors. Phys Rev Lett 96:4

Koch F Hydropower (2002) The politics of water and energy: introduction and overview. Energy Policy 30(14):1207–1213

Lima AR (2012) A produção de energias renováveis e o desenvolvimento sustentável: uma análise no cenário da mudança do clima. Revista Eletrônica de Direito Energia 5:4

Maroufmashat A, Fowler M, SattariKhavas S et al (2016) Mixed integer linear programing based approach for optimal planning and operation of a smart urban energy network to support the hydrogen economy. Int J Hydrogen Energy 41(19):7700–7716

Marques S (2007) Energias fosseis versus energias renováveis: proposta de intervenção de educação ambiental no 1° ciclo do Ensino básico. Dissertação (Mestrado) em Estudos da Criança, Universidade do Minho, Braga. http://hdl.handle.net/1822/7275

Mazloomi K, Gomes C (2012) Hydrogen as an energy carrier: prospects and challenges. Renew Sustain Energy Rev 16(5):3024–3033

Morken J, Sapci Z (2013) Evaluating biogas in Norway—bioenergy and greenhouse gas reduction potentials. Agric Eng Int CIGR J 15(2):13

Neill S, Hashemi M, Lewis M (2016) Tidal energy leasing and tidal phasing. Renew Energy 85:580–587

Northrup JM, Wittemyer G (2013) Characterising the impacts of emerging energy development on wildlife, with an eye towards mitigation. Ecol Lett 16(1):112–125

O'Rourke F, Boyle F, Reynolds A (2010) Tidal energy update 2009. Appl Energy 87(2):398–409

Oud E (2002) The evolving context for hydropower development. Energy Policy 30(14):1215–1223

Pacheco F (2006) Energias renováveis: breves conceitos. Conjuntura e Planejamento 149:4–11

Rabelo JL, de Oliveira JN, de Rezende RJ, Wendland E (2002) Aproveitamento da energia geotérmica do sistema Aqüífero Guarani: estudo de caso. In: XII Congresso Brasileiro De Águas Subterrâneas. Florianópolis

Riffat S, Ma X (2003) Thermoelectrics: a review of present and potential applications. Appl Therm Eng 23(8):913–935

Rowe D, Min G (1998) Evaluation of thermoelectric modules for power generation. J Power Sources 73(2):193–198

Slattery M, Johnson B, Swofford J, Pasqualetti M (2012) The predominance of economic development in the support for large-scale wind farms in the U.S. Great Plains. Renew Sustain Energy Rev 16 (6):3690–3701

Tabassum A, Premalatha M, Abbasi T, Abbasi SA (2014) Wind energy: increasing deployment, rising environmental concerns. Renew Sustain Energy Rev 31:270–288

United Nations (2018a) https://nacoesunidas.org/pos2015/. (Last accessed: 01 Sept 2018)

United Nations (2018b) http://www.onu.org.br/rio20/documentos/. (Last accessed: 01 June 2018)

Vichi FM, Mansor MTC (2009) Energia, meio ambiente e economia: o Brasil no context mundial. Quim Nova 32(3):757–767. https://doi.org/10.1590/S0100-40422009000300019

Walker S, Mukherjee U, Fowler M, Elkamel A (2016) Benchmarking and selection of power-to-gas utilizing electrolytic hydrogen as an energy storage alternative. Int J Hydrogen Energy 41(19):7717–7731

WCD (World Commission on Dams) (2000) Dams and development—A new framework for decision making. Earthscan, London

Yuksek O, Komurcu M, Yuksel I, Kaygusuz K (2006) The role of hydropower in meeting Turkey's electric energy demand. Energy Policy 34(17):3093–3103

Yüksel I (2010) Energy production and sustainable energy policies in Turkey. Renew Energy 35(7):1469–1476

Zhang F, Zhao P, Niu M, Maddy J (2016) The survey of key technologies in hydrogen energy storage. Int J Hydrogen Energy 41:14535–14

Is Energy Planning Moving Towards Sustainable Development? A Review of Energy Systems Modeling and Their Focus on Sustainability

Pedro Gerber Machado, Dominique Mouette, Régis Rathmann, Edmilson dos Santos and Drielli Peyerl

Abstract An "energy system" is encompasses the "combined processes of acquiring and using energy in a given society or economy, from primary energy to its final use. For its important role in economic development, energy systems models started to appear in the 70s following the oil crisis. However, in the 90s and 00s, environmental issues as climate change and local pollution have taken the center of the stage in energy modeling due to the high share of negative impacts related to energy production. Today, these models move towards sustainability in general, but are not there quite yet. Having that in mind, this chapter analyzes how energy models are moving towards the sustainable development goals, through a review of 442 publications encompassing 34 modeling tools. Results show that 65% of the studies are concentrated in 2 sustainability goals, with much room for geographical redistribution of models, focusing on less developed countries and tackling other important sustainable development goals related directly or indirectly to energy.

Keywords Energy systems planning · Energy modeling · Sustainable development goals · Review

P. G. Machado (✉) · R. Rathmann · E. dos Santos · D. Peyerl
IEE—Institute for Energy and the Environment, University of São Paulo, Av. Prof. Luciano Gualberto, 1289 São Paulo, SP, Brazil
e-mail: pgerber@usp.br

R. Rathmann
e-mail: regisrat@hotmail.com

E. dos Santos
e-mail: edsantos@iee.usp.br

D. Peyerl
e-mail: driellipeyerl@gmail.com

D. Mouette
School of Arts, Sciences and Humanities, University of São Paulo, Rua Arlindo Béttio, 1000 São Paulo, SP, Brazil
e-mail: dominiquem@usp.br

© Springer Nature Switzerland AG 2020
W. Leal Filho et al. (eds.), *International Business, Trade and Institutional Sustainability*, World Sustainability Series,
https://doi.org/10.1007/978-3-030-26759-9_36

1 Introduction: Energy and Sustainability

Sustainable development was conceived from the need to address ecological and social issues in a world with increasing economic growth to the detriment of the quality of life of society and the health of the environment (Buch-Hansen 2014). The famous concept advocates that development must be planned in order to "meet the necessities of the present generation without harming the future generation's capacity to meet their own" (Brundtland 1987, p. 54). The current concepts of sustainable development are continuously evolving, moving beyond the environmental, economic and social development concerns, towards impacting people's well-being in the form of 5 new pillars: People, Planet, Prosperity, Peace and Partnership (United Nations 2015). In this context, science needs to be at the service of politics, as well as dealing with governments and the multitude of requests when faced with the challenge of accomplishing sustainable development (Gusmão Caiado et al. 2018).

Within global economy, the energy sector plays an important part in developing and maintaining the growth expected by any capitalist system (Gozgor et al. 2018). Also, energy is linked to better quality of life and social development (Pasten and Santamarina 2012). The emergence of new challenges related to climate change, security of energy supply, and economic recessions have pushed energy research towards developing models, which became more available for analyzing energy systems or their sub-systems. In the early 1970s, concerned with a number of purposes, such as demand forecasting, energy—environment and energy—economy interactions (Bhattacharyya and Timilsina 2010; Connolly et al. 2010), energy modeling tools began to come to light. However, energy modeling has yet to address the sustainability goals in a direct and coordinated way.

In order analyze the energy modeling researchers' concern; this study analyzes 442 papers from 6 energy modeling tools to identify their focus and relationship with the 17 UN sustainable development goals (SDG).

2 Literature Review: Previous Reviews

An "energy system" is defined as the process chain from the extraction of primary energy to the final use of energy to supply services and goods. In other words, an energy system encompasses the "combined processes of acquiring and using energy in a given society or economy" (Pfenninger et al. 2014). An energy system is modeled through several tools or software. This section describes previous reviews published and their focus, shown in Table 1. In this table, models are considered the set of premises used, while approach is how the modeler chose to treat a certain case, such as top-down or bottom-up approaches. Finally, tools are the platforms used to describe the model. Previous reviews are listed as follows:

Following the chronological order, there are some recent broad overviews of energy models analyzed, such as Jebaraj and Iniyan (2006), which contains a list

Table 1 Recent energy systems models reviews, their focus and coverage

Publication	Mentions sustainability?	Coverage
Jebaraj and Iniyan (2006)	Yes	6 groups of models
Connolly et al. (2010)	No	37 tools
Bhattacharyya and Timilsina (2010)	No	5 approaches and 10 tools
Bazmi and Zahedi (2011)	No	277 publications
Suganthi and Samuel (2012)	No	12 methods
Keirstead et al. (2012)	Yes	219 publications
Gargiulo and Gallachóir (2013)	No	18 tools
Capros et al. (2014)	No	7 models
Pfenninger et al. (2014)	No	4 groups of models
Allegrini et al. (2015)	No	3 approaches and 4 tools
Martinez Soto and Jentsch (2016)	No	12 models and tools
Hall and Buckley (2016)	No	110 publications
Laha and Chakraborty (2017)	Yes	15 tools and models
Wagh and Kulkarni (2018)	Yes	3 groups of models

of models published up to 2005, ranging from demand-focused models to planning, policy, and operation models. The authors found that the models used technology, efficiency, supply, demand, employment and resource availability as constraints, and observed that the behavioral or econometric models used reflected the characteristics of energy supply and consumption, oriented towards forecasting. The authors mention papers related to energy sustainability and list two models built for sustainability assessment. However, the review is not based on sustainability parameters. In 2010, Connolly et al. (2010) conducted a review of the different computer tools that can be used to analyze the integration of renewable energy into energy matrices, analyzing a total of 37 tools, providing the necessary information to identify a suitable one for different objectives. The authors do not mention sustainability, or analyze how models treat sustainability in their mathematics.

In the same year (2010), Bhattacharyya and Timilsina (2010) compared tools and approaches for energy modeling. Although the authors call them models, tools and approaches based on criteria defined by the authors such as geographical coverage, activity coverage, level of disaggregation and data and skill needs are actually analyzed. They conclude that the tools are incapable of reflecting the specific features of energy systems of developing countries, leaving out important issues, such as rural-urban divide, traditional energies, informal economies, technological diversities and inequity in these countries. The authors do not mention how tools deal with sustainability issues. Bazmi and Zahedi (2011) focus on the power sector and optimization models in particular, looking at models ranging from plant operation, power distribution, consumption in residential and industrial buildings, to larger-scale system models for policy and planning. They report that modeling and optimization are proved to be effective and useful tools for problem solving in the power and in the supply sector. The authors, however, do not inform how models are linked to sustainable development. From 2012, two publications are highlighted. First, Suganthi and Samuel (2012) focus on traditional methods such as time series, regression, econometric models, ARIMA as well as soft computing techniques, such as fuzzy logic, genetic algorithm, neural networks and bottom-up tools. The authors focus on demand-side forecasting and describe methods and explore publications using each of the 12 methods to determine energy demand. Second, Keirstead et al. (2012) study urban energy system models by analyzing 219 papers to identify the major approaches used in urban energy systems. Only the second modeling review analyzes how models deal with sustainability.

In the next year, another review of tools was conducted by Gargiulo and Gallachóir (2013), who analyzed half of the tools previously analyzed by Connolly et al. (2010), with a focus similar to the latter. This publication does not focus on sustainability issues. In 2014, another publication by Pfenninger et al. (2014) studied groups of models divided into energy systems optimization models, energy systems simulation models, power systems and electricity market models and qualitative and mixed-methods scenarios. Capros et al. (2014) also studied models, but focused on European decarbonisation pathways, based on 7 models, describing them and providing key quantitative information from the models, such as GHG emissions, energy and climate policies included and European power generation mix. None of the publications directly mentions s sustainability s, or analyze the developments in model theory related to sustainability.

Other tool reviews come again in 2015 by Allegrini et al. (2015), but interested in district-scale energy systems modeling. In 2016, Martinez Soto and Jentsch (2016) study a mixture of models and tools for determining energy demand in the residential sector of a country. Again in 2016 Hall and Buckley compare available models using a classification scheme defined to characterize the 110 publications and models used. The authors state that policy driven papers are scarcer than academic papers, and that there are several tools not utilized either in policy or academia. These tools are also hardly compared, since models are unclear regarding their characteristics. None of the publications mentioned sustainability. In 2017, Laha and Chakraborty (2017) discussed the usefulness of energy models focused on the Indian energy system.

Finally, (Wagh and Kulkarni 2018) analyzed 3 groups of models: energy planning models, energy forecasting models, and renewable energy integration models, aiming to study the integration of renewable energy. They concluded that most of the energy planners find renewable energy resources the most important energy alternative for conventional fuels, normally considering maximum penetration in the grid for their planning. These two publications, from 2017 and 2018, present the highest levels of sustainability discussions, as compared to the other reviews discussed.

From these previous reviews, one notices the lack of direction towards the sustainable goals, and a few reviews mention sustainability and analyze how sustainability is dealt with within the tools and models analyzed.

3   Materials and Methods

This study analyzes 442 publications regarding 34 energy modeling tools to identify research gaps in energy system planning using energy modeling tools regarding the sustainability goals. To find these gaps, the publications analyzed were categorized based on the SD goal they were most related to. First, the tools were selected based on their importance according to the literature, followed by the selection of publications, using the online scientific search engines Scopus[1] and Web of Science.[2] The following sections describe the selection of tools and publications.

4   Selection of Tools

The first step in the selection of publications was to filter the tools to be analyzed. The 14 authors in Table 1 were screened to serve as a basis of the tools relevant for our review. These reviews screened a large number of models, going from simple 3-tool comparisons to complex and extensive 96 models/tools analysis, and investigated the most important models; they include a wide range of model applications and coverage, distributed over 12 years (from 2006 to 2018). From this screening, a total of 157 tools and models were selected.

Those tools appearing in at least 2 of the review publications were taken into consideration and are shown in Fig. 1.

In a second screening, 22 tools were excluded. HOMER, WASP, EMCAS, WILMAR, SIVAEL, RAMSES, ProdRisk, PLEXOS, ORCED, GTMax, EMPS, ELMOD, AEOLIUS, EnergyPRO, BCHP, RETScreen and HYDROgems were excluded due to their particular focus on one energy source sector or on project feasibility, which falls outside the scope of this review. E4Cast, COMPOSE and

---

[1]Scopus is an abstract and citation database from the Elsevier publishing company, which allows searching for articles, proceedings, books and book chapters.

[2]Web of Science is a scientific citation indexing service based on online subscriptions.

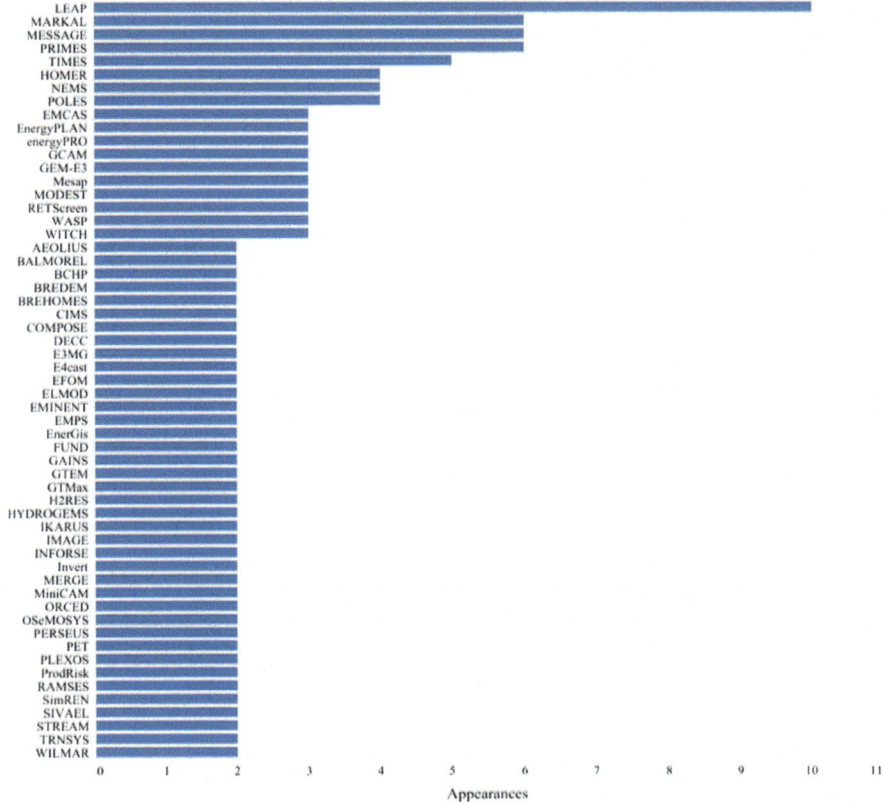

Fig. 1 First screening of tools and models

INFORSE were excluded because no publications applying these tools were found. Finally, PET was excluded since it is a regionalization of TIMES. Table 2 gives an overview of the 34 remaining tools included in this study.

Table 2 Overview of the modeling tools in this study

Model family	Tools	Primary focus
Optimization models	MARKAL, TIMES, MODEST, PERSEUS, IKARUS, GAINS	Normative scenarios
Simulation models	LEAP, MESSAGE, PRIMES, POLES, NEMS, WITCH, MESAP, GEM-E3, GCAM, EnergyPLAN, STREAM, SimREN, MiniCAM, GTEM, E3MG, BALMOREL	Forecasts, predictions
Mixed models	OSeMOSYS, MERGE, INVERT, IMAGE, H2RES, FUND, EnerGis, EMINENT, DECC, CIMS, BREHOMES, BREDEM	Exploratory scenarios

5 Selection of Publications

In energy system modeling, the tool represents a computational environment with a built-in algorithm to represent any energy system that falls within the scope of that tool. When a specific system is described in any given tool, it is called a model. In this sense, each tool can be used several times to construct different models of different energy systems. For that reason, our study analyzes publications that model various energy systems, using the tools described above. Publications were selected using Scopus and Web of Science (WoS). Every tool was searched using the words "TOOL_NAME energy model", with TOOL_NAME being the acronym for each tool. Publications were next filtered to exclude the results unrelated to energy modeling. The publications were then randomly chosen for the analysis. Based on the population size of SCOPUS and Web of Science results, a margin of error of 4.1% was calculated for the total sample.

Note that none of the publications selected were explicit about the SDG they refer to. This does not mean, however, that SDGs are not addressed regarding energy. For example, Dada and Mbohwa (2018) deal with waste as a source of energy, and how it could contribute to SDG 7. Büyüközkan et al. (2018) developed a multi-criteria model to select energy options based on the SDG. Schwerhoff and Sy (2017) also tackle the importance of renewable energy for the SDG, using Africa as a case study.

6 Sustainable Development Goals: Connection with Energy Modeling

In order to analyze how energy models relate to sustainability, the publications were divided into groups, which represent each of the SDGs. Since virtually none of the publications were explicit regarding the SDG they were most related to, they had to be classified based on pre-defined criteria. Next, the SDG will be presented along with the objectives in the publications selected that were most related to the SDG.

SDG 1: No poverty
Publications related to improving income.

SDG 2: Zero hunger
Publications related to food production or food prices.

SDG42: Good health and well-being
Publications related to health impacts and health benefits.

SDG 4: Quality education
Publications related to improving education.

SDG 5: Gender equality
Publications related to women's rights, gender equality, or gender pay gap.

SDG 6: Clean water and sanitation
Publications related to water use, water quality and hydropower.

SDG 7: Affordable and clean energy
Publications related to energy price, renewable energy and energy costs.

SDG 8: Decent work and economic growth
Publications related to economic growth, macroeconomic impacts, equilibrium and better jobs.

SDG 9: Industry, innovation and infrastructure
Publications related to technology development, technology diffusion and new technologies.

SDG 10: Reduced inequalities
Publications related to overall inequality.

SDG 11: Sustainable cities and communities
Publications related to urban, urban planning and local energy systems.

SDG 12: Responsible consumption and production
Publications related to waste and end-product efficiency.

SDG 13: Climate action
Publications related to Greenhouse Gases emissions, CO_2 prices and climate policy.

SDG 14: Life below water
Publications related to aquatic biodiversity.

SDG 15: Life on land
Publications related to land use and land use change.

SDG 16: peace, justice and strong institutions
Publications related to government.

SDG 17: peace, justice and strong institutions
Publications related to multilateral collaborations.

7   Results: Energy Models and Their Relation with the SDG

Figure 2 shows the evolution of publication over the years. The series starts with MARKAL, TIMES and with a smaller participation of LEAP in the early 1980s, up to the late 1990s. In this period, MESSAGE started to have some importance in the field around 1995. In the early 2000s, LEAP started to gain strength, along with other free tools, such as MESSAGE, especially in developing countries. POLES and

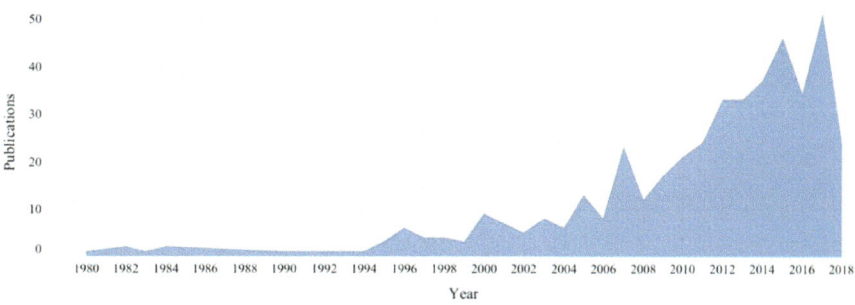

Fig. 2 Number of publications per year

PRIME entered the publication circuit in the late 1990s, along with BREHOMES, FUND, IMAGE, MESAP and MODEST.

The publication of energy systems and energy planning analyses gained force after the Kyoto Protocol in 1997, with special boost after its entering into force in 2005. The period after 2010 showed the highest rate of increase of publications, following the continued interest in energy planning and the environmental implications of energy supply and use, as will be seen in the next sections.

The initial purpose of modelers in the 1980s was model construction, without any particular objective other than present the model (Egberts 1981; Bergendahl and Bergström 1982; Sailor and Rath-Nagel 1982) and exploring energy demand (de L. Musgrove 1984), evolving to a majority of environmentally-directed publications.

In 2005, the Paris Agreement entered into force. This agreement was signed at the 21st Conference of the Parties (COP-21) of the United Nations Framework Convention on Climate Change (UNFCCC) and the participating countries needed to define their mitigation actions through intended Nationally Determined Contributions (iNDCs). Many countries up to 2015 used energy modeling to create energy and emissions scenarios that were the basis for their iNDCs.

When it comes to the SDG, energy modeling has shown importance for SDG 13, and represents 41% of all the publications. Figure 3 shows the distribution of publications regarding the SDG.

Climate change, climate policy, CO_2 prices, emission reductions and related subjects compose the publications related to climate action. This is a reflection of the importance of the energy sector for the anthropogenic carbon emissions and consequent climate change. Overall, the authors of publications on energy systems modeling analyze how energy systems can guarantee emissions in the medium and long term to keep the globe temperature increase under the 2 °C mark (UN 2018).

Following climate action (SDG 13), comes clean and affordable energy (SDG 7). In most cases, publications intrinsically related to both SDG 13 and 7. However, those classified as related to SDG 7 focus on energy costs and other environmental impacts other than climate change, or climate change is a secondary impact studied. Either way, it shows the importance given to the environmental aspect of sustainability, be it in SDG 13 or in SDG 7. Together, they represent 65% of all the publications studied.

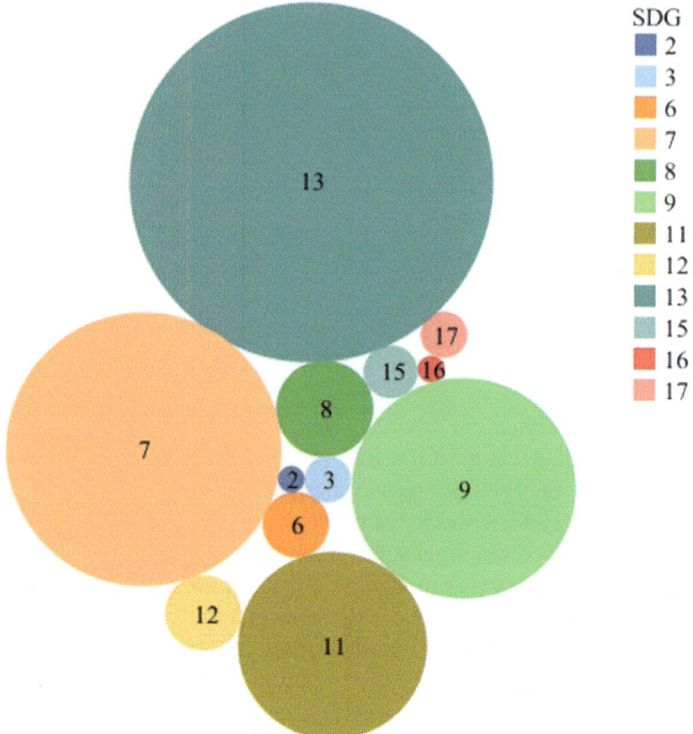

Fig. 3 Publications and their relationship to the SDG

In the case of SDG 9 (innovation and industry), the description of technologies and their analysis was in the forefront of the objectives of the publications in this group. Some of them studied how these new technologies could reduce emissions. Publications focusing on a particular industry sector were also placed in this group, as is the case of Murphy, Rivers and Jaccard (2007) that, despite study emissions, focus on the industry sector in Canada.

Other SDGs might be harder to connect with energy production. The relationship between energy and sustainability is mostly perceived on the environmental side of SD. However, energy is an important part of development and efforts should be made to include other areas of SD in the area of energy systems modeling. If the SDGs were analyzed by country, interesting outcomes could be shown. Figure 4 shows the number of publications by country of each SDG, excluding those countries and SDGs appearing only once. Publications on the global or at the European level represent 24% of all the publications. In total, 34 countries were studied plus Asia, Africa, Europe and Global regions, leaving out at least 127 countries to be analyzed in the world.

There is a clear geographical inequality among studies, with a concentration on Europe, North America and China. China comes first, with 8% of the publications,

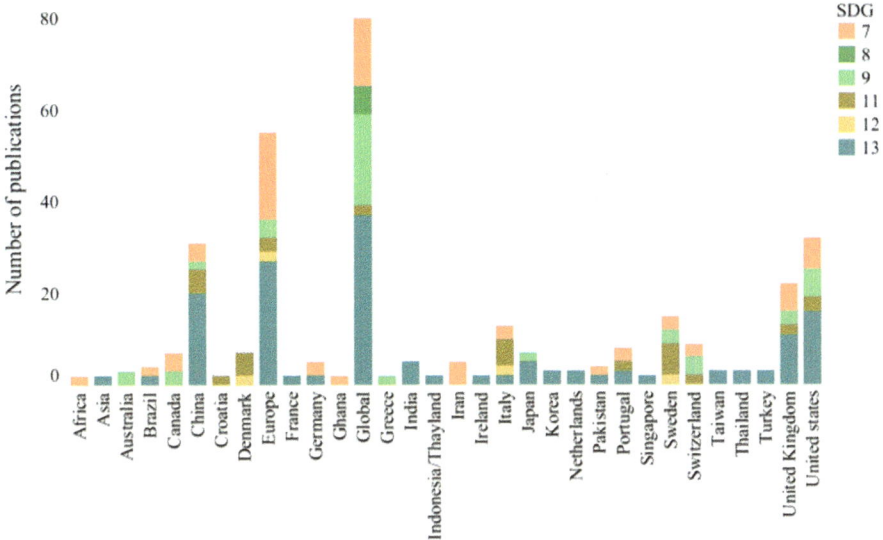

Fig. 4 SDG-related publications by country and/or region

followed by the USA, with 7.9%, the United Kingdom with 5% and Sweden with 3.8%, regardless of the SDG.

The relationship between energy and social development has long been studied and proven. Therefore, energy systems should be expected to be planned to help the social development of underdeveloped societies. Yet the application of models seems to be proportional to development (highly developed countries have more studies than less developed ones), most likely due to the required modeling skills and volume of data, which requires strong capacity and high costs for the modeling work. This represents a gap in knowledge, even for social sciences and for the analysis of the energy-society nexus. One way to reduce this inequality is to have bilateral (or multilateral) agreements to develop capacities and modeling abilities between developed and less developed countries. Some examples are Luukkanen et al. (2015), in which Finnish researches analyze Cambodia and Laos, or Taliotis et al. (2016) a study conducted by Swedish researchers in cooperation with Ethiopian researchers about the electricity sector in Africa.

Another aspect that calls for attention is that SDG 13 is a priority even in developing countries. Although countries such as Thailand could benefit more from the analysis of other SDG, their efforts are also focused on energy modeling for climate change action.

From the tool point of view, LEAP, MARKAL and TIMES are the most widely used. Their focus lies mostly on climate action and clean energy (goals 7 and 13) and technology analysis (goal 9). Other tools have smaller reach, mostly focused on limited regions or objectives. None of them, however, treats all SDGs, or are able to derive proper results from each SDG (Fig. 5).

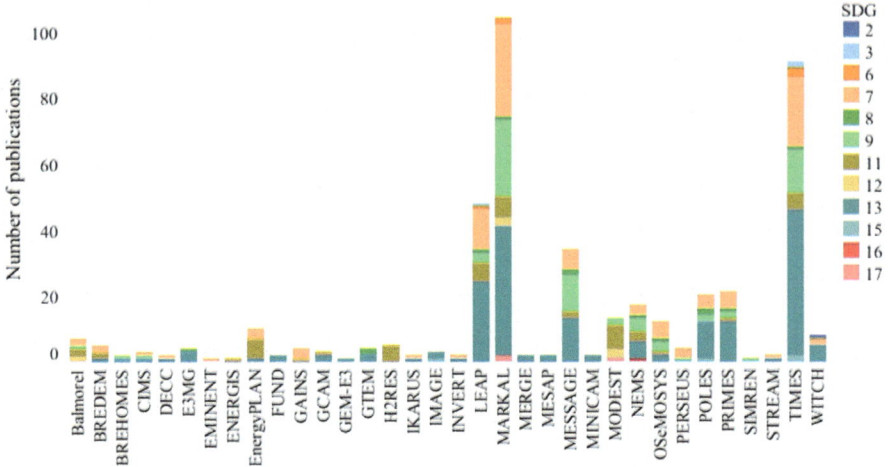

Fig. 5 SDG-related publications by tool used

For sustainable cities (goal 11), H2RES and EnergyPLAN are the most suitable and specialized models, since they were built to analyze local energy systems. Due to its basis on general equilibrium theory, GTEM is the tool with closest relation to goal 8, since it is able to analyze economic growth from energy-related decisions. Modest has focused on regional modeling and, therefore, is more useful to goals 11 and also 12.

8 Discussions and Conclusions

This study focuses on how the existing models are not adequate when it comes to the geographic coverage of sustainable development goals, with several gaps and inequalities in research regarding energy generation, use and impacts.

Regarding geographic distribution, there is clear geographical inequality among the studies, with a focus on Europe, North America and China. There is considerable space for modeling energy systems in developing countries; although they represent a large part of the world's energy production, there is a lack of studies concerned with better understanding the energy systems in Africa and in South America. Even though few of these countries are major energy suppliers, they present opportunities for renewable energy (with interesting experiences from the point of view of SDG 13) and also have very complex issues, related to social SDG, such as social inequalities (SDG 10) and regional differences (interesting from the research point of view).

Environmentally-oriented publications related to SDG 13 and 7 account for 65% of all the publications studied, showing that modelers have greater interest in studying issues related to climate change mitigation than in other areas of sustainability. The application range of models and tools show the room for research on extremely important social issues, such as health and SDG 3, hunger and SDG 2, poverty and SDG 1 and inequalities and SDG 10. The linkages between these SDG and energy might not me straightforward, but they still require attention when it comes to energy modeling. Diminishing any SDG could induce a snowball effect. If one only considers SDG 13 for example, there could be effects on personal income at the regional level, which could represent a threat to SDG 1.

There is also considerable space for modeling energy systems at the national scale, in a number of countries. In fact, the 442 publications studied are focused on 34 countries, leaving out at least 127 countries to be modeled.

Within the theme of climate change mitigation, most of the publications show interest in the long-term, while short-term mitigation actions are necessary, because the advances in recent years are still occurring at a very slow pace, requiring urgent and greater participation of the private sector and governments.

From the results found and from the data crossing, a few suggestions are made for the future modelers, based on the research gaps found:

- Tackling energy demand in transport considering the changes in technology and behavioral changes in the market: electrification of light duty fleet, self-driving cars, consumers' preference for public transportation, changes in modals due to increase in income, etc. This issue is much related to goal 11 and goal 12, with a particular interest in the transport sector, due to its major challenges regarding energy and energy sustainability, since it is a sector with heavy dependency on oil;
- Tackling other impacts other than environmental ones, such as economic development (using hybrid macroeconomic models), related to SDG 8, local development and poverty related to SDG 1, using social indicators and other social impact categories to address inequalities, health and education and how energy can help develop those fields, also including social factors, such as job creation, income generation, gender inequality, which are much related to SDG 2, 3, 4, 5 10 and 16.
- In environmental impacts, in studies related to SDG 7 and 13, expanding the model through modular build-up in water use, land use and carbon capture and storage, especially in developing countries to address SDG related to the environment in an integrated way, to avoid externalities;
- Investigating the emerging technologies for renewable energy supply, focusing not only on new sources of energy, but also on efficiency increase due to technological developments and innovation and infrastructure improvements;
- Forecasting energy demand for different macroeconomic set-ups, modeling economic crisis and how energy-related subjects change in a scenario to avoid energy dependency and to guarantee affordable and clean energy;
- Modeling African, Latin American and Southeast Asian countries, at the national or regional level, taking their individual characteristics into account, to increase the accuracy of models in southern countries, helping their sustainability and developing sustainable energy systems;
- Investigating non-energy sectors, such as industry and urban buildings and disaggregating their energy consumption. Investigating the non-renewable sectors and

the room for improvements in energy efficiency and energy supply within these sectors.

Although the ideal model should take all SDG into consideration and choose the best way to supply energy, its practicality is nearly impracticable. For that reason, further suggestions are made for future research directly linked to the SDG:

- Goal 1: No poverty

How much energy would be required to give everyone enough income to leave poverty?

- Goal 2: Zero Hunger

How much energy do we need to feed everyone?

- Goal 3: Good health and well-being

What are the best options for energy generation to guarantee the fewest impacts on human health?

- Goal 4: Quality education

Can energy generation help improve education? Can energy supply be linked to skilled labor?

- Goal 6: Clean water and sanitation

How much energy is required to offer clean water and sanitation to everyone? And how does energy generation impact water quality and availability?

- Goal 14 and 15: Life below water and on land

What are the impacts of energy production on biodiversity and on animal life?

- Goal 17: Partnerships for the goals

How can bilateral and multilateral agreements for technology exchange help achieve the other SDG goals?

References

Allegrini J et al (2015) A review of modelling approaches and tools for the simulation of district-scale energy systems. Renew Sustain Energy Rev 52:1391–1404. https://doi.org/10.1016/j.rser.2015.07.123 (Elsevier Ltd.)

Bazmi AA, Zahedi G (2011) Sustainable energy systems: role of optimization modeling techniques in power generation and supply—a review. Renew Sustain Energy Rev 15(8), pp. 3480–3500. https://doi.org/10.1016/j.rser.2011.05.003 (Elsevier Ltd.)

Bergendahl P-A, Bergström C (1982) Long-term oil substitution—the IEA-Markal model and some simulation results for Sweden. In: The impact of rising oil prices on the world economy. Palgrave Macmillan, London, UK, pp 97–112. https://doi.org/10.1007/978-1-349-06361-1_7

Bhattacharyya SC, Timilsina GR (2010) A review of energy system models. Int J Energy Sect Manage 4(4):494–518. https://doi.org/10.1108/17506221011092742

Brundtland GH (1987) Our common future (The Brundtland Report), Report of the World Commission on Environment and Development: Our Common Future (The Brundtland Report). https://doi.org/10.1080/07488008808408783

Buch-Hansen H (2014) Capitalist diversity and de-growth trajectories to steady-state economies. Ecol Econ 106:167–173. https://doi.org/10.1016/j.ecolecon.2014.07.030 (Elsevier)

Büyüközkan G, Karabulut Y, Mukul E (2018) A novel renewable energy selection model for United Nations' sustainable development goals. Energy 165:290–302. https://doi.org/10.1016/j.energy.2018.08.215 (Pergamon)

Capros P et al (2014) Description of models and scenarios used to assess European decarbonisation pathways. Energy Strategy Rev 2(3–4):220–230. https://doi.org/10.1016/j.esr.2013.12.008 (Elsevier Ltd.)

Connolly D et al (2010) A review of computer tools for analysing the integration of renewable energy into various energy systems. Appl Energy 87(4):1059–1082. https://doi.org/10.1016/j.apenergy.2009.09.026 (Elsevier Ltd.)

Dada O, Mbohwa C (2018) Energy from waste: a possible way of meeting goal 7 of the sustainable development goals. Mater Today: Proc 5(4):10577–10584. https://doi.org/10.1016/j.matpr.2017.12.390 (Elsevier)

Egberts G (1981) 'MARKAL—Ein LP-Modell für die Internationale Energieagentur. OR Spektrum 3(2):95–100. https://doi.org/10.1007/bf01720101 (Springer-Verlag)

Gargiulo M, Gallachóir BÓ (2013) Long-term energy models: principles, characteristics, focus, and limitations. Wiley Interdisc Rev: Energy and Environ 2(2):158–177. https://doi.org/10.1002/wene.62

Gozgor G, Lau CKM, Lu Z (2018) Energy consumption and economic growth: new evidence from the OECD countries. Energy 153:27–34. https://doi.org/10.1016/j.energy.2018.03.158 (Pergamon)

Gusmão Caiado RG et al (2018) A literature-based review on potentials and constraints in the implementation of the sustainable development goals. J Cleaner Prod 198:1276–1288. https://doi.org/10.1016/j.jclepro.2018.07.102 (Elsevier)

Hall LMH, Buckley AR (2016) A review of energy systems models in the UK: prevalent usage and categorisation. Appl Energy 169:607–628. https://doi.org/10.1016/j.apenergy.2016.02.044 (Elsevier Ltd.)

Jebaraj S, Iniyan S (2006) A review of energy models. Renew Sustain Energy Rev 10(4):281–311. https://doi.org/10.1016/j.rser.2004.09.004

Keirstead J, Jennings M, Sivakumar A (2012) A review of urban energy system models: approaches, challenges and opportunities. Renew Sustain Energy Rev 16(6):3847–3866. https://doi.org/10.1016/J.RSER.2012.02.047 (Pergamon)

de L. Musgrove AR (1984) A linear programming analysis of liquid-fuel production and use options for Australia. Energy 9(4):281–302. https://doi.org/10.1016/0360-5442(84)90100-2 (Pergamon)

Laha P, Chakraborty B (2017) Energy model—a tool for preventing energy dysfunction. Renew Sustain Energy Rev 95–114. https://doi.org/10.1016/j.rser.2017.01.106 (Pergamon)

Luukkanen J et al (2015) Long-run energy scenarios for Cambodia and Laos: Building an integrated techno-economic and environmental modelling framework for scenario analyses. Energy 91:866–881. https://doi.org/10.1016/J.ENERGY.2015.08.091 (Pergamon)

Martinez Soto A, Jentsch MF (2016) Comparison of prediction models for determining energy demand in the residential sector of a country. Energy Build 128:38–55. https://doi.org/10.1016/j.enbuild.2016.06.063 (Elsevier B.V.)

Murphy R, Rivers N, Jaccard M (2007) Hybrid modeling of industrial energy consumption and greenhouse gas emissions with an application to Canada. Energy Econ 29(4):826–846. https://doi.org/10.1016/J.ENECO.2007.01.006 (North-Holland)

Pasten C, Santamarina JC (2012) Energy and quality of life. Energy Policy 49:468–476. https://doi.org/10.1016/J.ENPOL.2012.06.051 (Elsevier)

Pfenninger S, Hawkes A, Keirstead J (2014) Energy systems modeling for twenty-first century energy challenges. Renew Sustain Energy Rev 33:74–86. https://doi.org/10.1016/j.rser.2014.02.003 (Elsevier)

Sailor VL, Rath-Nagel S (1982) Markal, a computer model designed for multi-national energy system analyses. In Energy modelling studies and conservation. Elsevier, p 635. https://doi.org/10.1016/b978-0-08-027416-4.50057-0

Schwerhoff G, Sy M (2017) Financing renewable energy in Africa—key challenge of the sustainable development goals. Renew Sustain Energy Rev 75:393–401. https://doi.org/10.1016/j.rser.2016.11.004

Suganthi L, Samuel AA (2012) Energy models for demand forecasting—a review. Renew Sustain Energy Rev 16(2):1223–1240. https://doi.org/10.1016/j.rser.2011.08.014 (Elsevier Ltd.)

Taliotis C et al (2016) An indicative analysis of investment opportunities in the African electricity supply sector—using TEMBA (The electricity model base for Africa). Energy for Sustain Dev 31:50–66. https://doi.org/10.1016/J.ESD.2015.12.001 (Elsevier)

UN (2018) Climate action—United Nations sustainable development. Available at: https://www.un.org/sustainabledevelopment/climate-action/ (Accessed: 14 Nov 2018)

United Nations (2015) Transforming our world: the 2030 agenda for sustainable development. https://doi.org/10.1007/s13398-014-0173-7.2

Wagh MM, Kulkarni VV (2018) Modeling and optimization of integration of renewable energy resources (RER) for minimum energy cost, minimum CO_2 emissions and sustainable development, in recent years: a review. Mater Today: Proc 5(1):11–21. https://doi.org/10.1016/j.matpr.2017.11.047 (Elsevier Ltd.)

Sustainable Development Goals as a Tool to Evaluate Multidimensional Clean Energy Initiatives

Karen L. Mascarenhas, Drielli Peyerl, Nathália Weber, Dominique Mouette, Walter Oscar Serrate Cuellar, Julio R. Meneghini and Evandro M. Moretto

Abstract Great contributions are expected by the public from science and technology (S&T), especially now, given the highly complex environmental and social challenges such as climate change, cleaner and affordable energy provision and economic crises, among others. These problems require a multidimensional approach involving all stakeholders working together: government, non-governmental organizations, funding agents, companies, academy and society. Based on the Sustainable Development Goals (SDGs) launched in 2015, actions have been carried out to develop capacities for the 2030 Energy Agenda in Brazil. Hence, this work aims to present a case study of the actions being developed in the analysis and integration of

K. L. Mascarenhas (✉)
Research Centre for Gas Innovation, Institute of Psychology, University of São Paulo and Fundação Getúlio Vargas, São Paulo, Brazil
e-mail: karenmascarenhas@usp.br

D. Peyerl · N. Weber
Research Centre for Gas Innovation, Institute of Energy and Environment, University of São Paulo, São Paulo, Brazil
e-mail: driellipeyerl@gmail.com

N. Weber
e-mail: nathaliaweber.m@gmail.com

D. Mouette · E. M. Moretto
Faculty of Arts, Sciences and Humanities, Research Centre for Gas Innovation, Institute of Energy and Environment, University of São Paulo, São Paulo, Brazil
e-mail: dominiquem@usp.br

E. M. Moretto
e-mail: evandromm@usp.br

W. O. S. Cuellar
Research Centre for Gas Innovation, University of São Paulo, São Paulo, Brazil
e-mail: oscar.serrate@usp.br

J. R. Meneghini
Research Centre for Gas Innovation (RCGI), Escola Politécnica, University of São Paulo, São Paulo, Brazil
e-mail: jmeneg@usp.br

© Springer Nature Switzerland AG 2020
W. Leal Filho et al. (eds.), *International Business, Trade and Institutional Sustainability*, World Sustainability Series,
https://doi.org/10.1007/978-3-030-26759-9_37

clean energy initiatives employing natural gas, biogas, hydrogen and Carbon Capture and Storage of CO_2 (CCS). Particularly, this work focus on the experiences of the Research Centre for Gas Innovation (RCGI), created in 2016 by the University of São Paulo, Shell and the São Paulo Research Foundation (FAPESP). This integration of strategic partnerships that incorporates academy, a private company and a research funding agency is evolving through 46 research projects that aim to contribute with sustainable solutions for the Energy agenda. The Centre is also contributing with the construction of the Social License to Operate for implementing clean energies initiatives and the development of a database and indicators that enable an evaluation of the Centre's performance, to be made available to all the stakeholders in the framework of the SDGs.

Keywords Sustainability · Public perception · Social license to operate · Carbon capture and storage · Sustainable development goals

1 Introduction

At the beginning, it was only R for Research, by the end of the 20th century, it became R&D to emphasize the connection between Research and Development. A few years later, it became RD&I, because Innovation was the challenge and the method for the new Era of accelerating demands. Our proposition is that it is time for another change: RDI&S. S for Social, because now it is not possible to launch a new technology without a close encounter with its potential beneficiaries. Citizens want to understand our scientific labs and papers. Communities and countries expect a transparent and immediate delivery of results from academia, businesses, and governments.

There are innumerous examples that include the 'public perception' issue, as well as 'media' or 'political' issues. People demand to participate, in the process of building the Social License to Operate (SLO). The inclusion has not been that simple, because it implies a paramount change. The new "S" implies migrating from the old analog world to the new digital reality, from a top-down learning approach to a more creative bottom-up method of ideas and interactions. It means choosing between closed or open-ended approaches. Choosing between ego and eco, shifting from focusing only on a single individual's point of view to considering the viewpoint of all the other players in the eco-systems (Scharmer 2018). The old world is not yet completely gone, and the new horizons are not yet completely clear, but that is precisely the most exciting challenge for the Science and Technology (S&T) communities: the possibility to work jointly with and for the society that created them. It is also a promising path for businesses and governmental organizations, which were originally established aiming their users or citizens.

Therefore, the 'public' is not an obstacle for project development. People are the most important partners in any initiative that benefits them. However, as any good partner, the 'public' needs to be involved from the very beginning of the idea in order to obtain full transparency and to build solid trust among all parties that take part in the process (Ashworth et al. 2009a, b; Seigo et al. 2014; Karimi and Toikka 2018). That is the 'state of the art' in successful academic, private or public projects. However, such issue becomes even more important when Energy is at stake. Many cases of strong initiatives or companies collapsed because they could not obtain or maintain their SLO (Feenstra et al. 2011). Those cases are widely known, because they are highlighted by the media or by specialized channels, diminishing the potential to advance the project from the planning stage (Feenstra et al. 2011; Van Egmond and Hekkert 2015). Nonetheless, lately there are several examples of success and win-win partnerships. They all have in common the message that successful RD&Is require the S, becoming RDI&S (Rooney et al. 2014; Asayamaa and Ishiib 2017; Ashworth et al. 2009a, b).

Whenever partnerships are required, good associations are verified to happen when the parties find a shared purpose, a common language and joint processes. Successful partners build a vehicle with 4 wheels (RDI&S) and drive it together in the same direction. In this approach, the Sustainable Development Goals (SDGs), agreed by all members of the United Nations, have been playing an important role in establishing a common purpose (Büyükozkan et al. 2018; Callan and Ashworth 2004), resulting in a variety of negotiated agreements. SDGs also have been useful by providing sets of indicators that allow following up their progress. Moreover, it has gradually become a common language to support projects and to create capacities for their implementation and diffusion.

Therefore, the objective of this paper is to closely examine the case study of the Research Centre for Gas Innovation (RCGI) hosted at the University of São Paulo (USP), Brazil, and to analyse the perspective of including the S factor into its research projects. The proposal is to present, given the current state of the art, options for incorporating the SDGs as a supporting and monitoring tool, and to align actions for the roadmap of 'The Brazilian 2030 Agenda for Sustainable Development'. The working hypothesis is that it is possible to design and to build a sustainable future, by developing capacities to sustain innovations incorporating the essential S factor.

2 University and SDGs: Innovative Approaches to Support the Sustainable Development Goals in Brazil

The 1992, after the Earth Summit held in Rio de Janeiro City, Brazil, under the umbrella of the United Nations, a new paradigm was opened integrating economic, social and environmental visions and launched a common Sustainable Development Program for the 21st century. Ten years later, consensus and progress were achieved again with the framework of the Millennium Development Goals (MDGs) approved in South Africa in 2002. In 2012, the United Nations (U.N.) Conference on Sustainable Development (UNCSD or "Rio + 20" in Brazil) brought back the theme and new challenging goals.

The newly globally accepted programme is known as 'The *2030 Agenda for Sustainable Development*', and claims that "this Agenda is a plan of action for people, planet and prosperity" (U.N. Transforming our world... 2015: 3). It includes 17 Sustainable Development Goals (SDGs) and 169 targets that came into effect in January 2016 with the approval of all U.N. member countries which "demonstrate the scale and ambition of this new universal Agenda. [...]. The Goals and targets will stimulate action over the next 15 years in areas of critical importance for humanity and the planet" (Transforming our world... 2015: 3). To ensure its legitimacy "the U.N. conducted the largest consultation exercise in its history to ensure wide ownership of the goals" (Kroll 2015: 12).

Indeed, the implementation of a broad and diverse agenda as the one mentioned above represents a challenge for all sectors from local to national levels. It is noteworthy that the progress in the implementation of the SDGs is achieved by integrating the three dimensions—economic, social and environmental—which is already widely accepted, and was reiterated both in Agenda 2030 and in the document *The Future We Want*, of the United Nations Conference on Sustainable Development (Rio + 20) (UNEP 2015). The Paris Agreement (effective in November 4th, 2016) also reinforces this understanding by recognizing the value of sustainable development in mitigation measures.

To illustrate the connection between universities and the goals, RCGI, hosted at the University of São Paulo was selected as a case study. For that, in general, the technologies under development at the RCGI have a potential to contribute towards at least two goals: SDG #7—clean energy; as well as SDG #13—climate change, with spillovers in many other goals, such as: sustaining industry innovation and infrastructure (SDG #9), indirectly helping to reduce poverty (SDG #1) and hunger (SDG #2), as well as providing better health and well-being (SDG #3), to mention just a few. For meeting the goals, partnerships (SDG #17) are to be established among government, industry, society, Non-Governmental Organization (NGOs), academy and interested stakeholders.

For the implementation of the SDGs, the key roles universities are playing are: (1) "providing the knowledge and solutions to underpin the implementation of the SDGs"; (2) "creating current and future SDGs implementers"; (3) "embodying the principles of the SDGs through organisational governance, operations and culture"; and (4) "providing cross-sectoral leadership in implementation" (SDSN Australia/Pacific 2017: 8).

Base on SDSN Australia; Pacific (2017), the key benefits for universities still include: (1) validating university and academic impact; (2) capturing demand for SDG-related education; (3) building new external and internal partnerships; (4) accessing new funding streams; and (5) adopting a comprehensive and globally accepted definition of a responsible and globally aware university. More suggestions for enhancing universities actions include:

- Provide students with knowledge, skills, and motivation to understand and to address the challenges of the SDGs
- Empower and mobilise young people and researchers

- Provide in-depth academic or vocational training to implement SDGs solutions
- Enhance opportunities for capacity building of students and professionals from developing countries to address challenges relating to the SDGs (SDSN Australia/Pacific 2017: 12).

In this sense, this paper shows the relation between the theory discussed above, the application of SDGs in the RCGI projects and, lastly, the role of the University of São Paulo in the SDGs.

3 The SDGs as a Tool to Communicate and to Build the SLO in Brazilian Sustainable Energy Initiatives

The RCGI, through the development of 46 projects focused on engineering, physical-chemistry, energy policies and economics, CO_2 abatement and the recently created geophysics programme, expects to contribute to the SDGs. These projects are centered on developing solutions that enable the transition from fossil to cleaner energy, incorporating fuels in the transitional period, such as natural gas, and later, the use of biogas, solar, wind and hydrogen, in a renewable energy society. As the energy needs are higher than the current capacity to fulfill them only from renewable sources, natural gas, as one of the lowest-emission fossil fuel energy sources, will be necessary to support this transition, offering the potential to deliver cleaner and affordable energy to a large number of people in Brazil and enable economic sustainable growth in the decades to come.

The specific geological formation of the pre-salt basin on the Brazilian coast in the states of São Paulo, Rio de Janeiro and Espírito Santo, allows extracting large quantities of oil and natural gas, with a high concentration of Carbon Dioxide (CO_2) (Lima 2009). The pre-salt has specific characteristics that enable the creation of salt caverns capable of storing high quantities of CO_2. Technologies are being developed to separate Methane (CH_4), Carbon Dioxide (CO_2) and other gases in caverns through a gravimetric method and other innovative technologies, keeping the CO_2 captured without the need of extracting and then reinjecting it back in, thus preventing its release to the atmosphere. Other technologies in a renewable area, such as the capture of CO_2 liberated by the fermentation process in the production of ethanol, are creating a condition for Bioenergy Carbon Capture, Usage and Storage (BECCS), which can lead to a process with a negative CO_2 footprint, since the emissions from the ethanol combustion are neutralized by sugarcane plantations.

In addition to these technologies, the RCGI is also researching how to avoid emissions through energy planning, analyzing the use of lower or negative carbon emissions as the natural gas and biogas on both, the transformation sector, and its use by the economy. The efficiency of new technologies is also being considered by evaluating the use of natural gas and biomethane in the transport sector.

RCGI was founded and began operating in January 2016. Its initial founders were FAPESP (São Paulo Research Foundation) and Shell, together with the University of São Paulo, which characterizes a triple helix innovation model (Etzkowitz and

Leydesdorff 2000; Etzkowitz 2008) as it is established by the partnership among government (funding agency), industry and academy. Other representations should also be included, as the technologies resulting from the research projects are aimed to solve concrete problems that usually require a SLO (Ashworth et al. 2009a, b, 2015). Ashworth et al. (2009a, 2015) stressed the need to understand public perception and include the voices of society, policy makers and NGOs to jointly build an agreement on the best ways to deploy any technology that affects people's lives and the country's economy.

Therefore, after the first three years of operation, the RCGI recognizes the need to extend its multidisciplinary groups to be even more comprehensive and to discuss beyond technical solutions, reaching out to the social sciences for insights and the framework to implement the new cleaner energy technologies. This requires education, communication, sharing information with government's legislators, leaders, industries, communities and the general public.

Lessons learned by studying and reviewing the literature in SLO (Rooney et al. 2014; Gough et al. 2018), public perception (Ashworth et al. 2009a, b; Whitmarsh et al. 2011; Scruggs and Benegal 2012), communication and dissemination of knowledge (Ashworth et al. 2009a, b, 2015; Gardner et al. 2009; Callan and Ashworth 2004; Kubota and Shimota 2017; Vercelli et al. 2017; Asayamaa and Ishiib 2017) were reinforced by visiting and exchanging information with scientists involved in Carbon Capture and Storage (CCS) demonstration projects in Tomakomai, Japan (Mabon et al. 2017), and Otway, Australia, denominated CO_2CRC (Sharma et al. 2006). These projects carry some similarities with the CO_2 Abatement Programme projects being developed in the RCGI, as they involve CO_2 capture, transport and storage, as well as technologies for detecting leakages. The major difference is that those demonstration plants are onshore or close to the shore facilities while the RCGI projects aim to work in caverns in the pre-salt layer of the Brazilian basin, far from the coast and in very deep waters. Considering the biogas projects and the cycle of ethanol production, carbon capture usage and storage, the conditions could be more akin to the demonstration plant in Australia, as the location is onshore. The question that arises is how to involve and to create knowledge about these technologies within Brazil and internationally?

Our hypothesis is that the SDGs would be the language to communicate with different stakeholders and to create a common ground to follow the path toward cleaner energy (Mc Dade 2015) and the mitigation of climate change. This could contribute to generating wealth and economic development, through innovation and inclusion. SDGs could help to develop the tools that enable society, industry, academy, government and NGOs partnering to implement consensual energy innovations that incorporate carbon emission mitigation as an emblematic solution to the challenges of climate change that locally and globally affect the sustainable future.

This should be a collective effort, which identifies leaderships and enables them to understand alternatives such as CCS (Carbon Capture and Storage) and CCUS (Carbon Capture Storage and Usage), their costs and benefits, and to involve them in the decision making. The model under which the SDGs were developed, through public consultation, shows that considering a universal approach can lead to even better sustainable solutions. Hence, this would be carried out by: (1) disseminating qualified knowledge about energy solutions and mitigation of emissions of Greenhouse Gases (GHG); and (2) developing individual and collective capacities that enable the Social Licenses to Operate cleaner energy systems and to ensure the sustainability of innovations in the medium and long term.

4 Capacity Development

As presented herein, the RCGI proves to be a driving force for the development of cleaner energy technologies, logistics and regulatory innovations for sustainable development. However, innovations, in the broad sense, consist not only in creating new ideas, but also in implementing and disseminating them in economic and social systems (Fagerberg 2018). Some technologies have suffered resistance from society at its various levels, be it by the Not In My Backyard (NIMBY) effect, by the perceived disadvantageous cost-benefit and even by the general lack of understanding (Huijts et al. 2012). It is this kind of public resistance that may affect negatively the implementation of these cleaner energy technologies, making it difficult to achieve the SDGs.

Obtaining SLO is, in many cases, essential for developing projects aimed at reducing GHG emissions. From the origin of the term SLO in the 1990s, at that time related to the mining industry (Gunningham et al. 2004), the bulk of the literature on the subject raises central questions about the industry and community relationship, highlighting corporate citizenship, social sustainability, reputation and legitimacy (Owen and Kemp 2012). At the level of local public acceptance, there are individual psychological factors that go beyond perceived risks and benefits, such as trust in the stakeholders and the perception of fairness in the process of implementing a technology (Huijts et al. 2012). Also, access to information is essential, since the public acceptance of CCUS, for example, has as its fundamental premise, the recognition of the anthropogenic contribution to climate change and the need to reduce CO_2 emissions (Ashworth et al. 2015).

The SLO and the capacity to implement cleaner energy technologies are not only dependent on interactions with local communities, but also on national and international contexts, structured by relationships with other stakeholders—governments, NGOs, business, civil society, funding agencies, universities, and the press, among others. Cooperation and coordination among these actors are responsible for the evolution of the technological and institutional environment capable of providing the necessary legitimacy (Pellegrino and Lodhia 2012) for disseminating innovation towards SDGs and a low carbon economy.

Stakeholder governance challenges, identified as central points of sustainability-related issues (Bowen et al. 2017), include the need for agent empowerment, engagement, incorporation of values and synergies, in order to drive the achievement of goals in the scale and effectiveness required, considering the three dimensions of sustainability—social, economic and environmental—in an integrated approach. The success of the transition to a low carbon economy may depend on building new inclusive and transparent institutional arrangements.

In this sense, we identified the importance of capacity building, inspired by the Capacity for Agenda 2030 of the U.N., towards an environment favorable to implementing innovations in cleaner energies, with the proposal of contributing to the development of the capacities described in Fig. 1.

GLOBAL CAPACITIES

- Participation in networks and in the interchange with entities related to the U.N. System dedicated to SDGs.

STRATEGIC CAPACITIES

- Elaboration of a model of the Brazilian energy history, with future projections to 2030 and 2050, to plan the roadmaps that guide the collective action, relating them to long-term scenarios planning strategies as those used by Shell and other organizations.

INSTITUTIONAL CAPACITIES

- Building partnerships with scientific and economic decision-making organizations for sustainable development aiming at catalytic effects with current and future resources.

SOCIAL CAPACITIES

- Development of leaders and construction of a virtual network of clean energies and carbon mitigation for promoting SLO and ensuring their maintenance and growth.

PROFESSIONAL CAPACITIES

- Dissemination of knowledge, ethical values, continuing education, promotion of new curricula and teaching methods, articles, magazines, books and social publications, including "social media", and construction of demonstrative projects accessible to specialists and to the public.

MANAGEMENT CAPACITIES

- Construction of a management information dashboard with technical, financial, reputational and environmental content and with data, indicators and trends for immediate and timely decision making, for monitoring the progress of innovations with mitigation.

ASSEMBLY OF THE COMMUNICATION SYSTEM

- To enrich and to integrate all the above tools into a single coherent system, articulated and presented according to the needs of each actor requesting training or information, always evaluating, with statistical methods, the public perception of the themes interest.

Fig. 1 Capacity building. *Source* Author's own

5 RCGI Emblematic Projects

Brazil was the ninth largest oil producer in the world in 2017 (EIA 2018), impacting the lives of about 200 million inhabitants. Through the most different channels, people daily follow the evolution of energy supply, investments, prices, or technologies under development.

The Brazilian national oil company, Petrobras, for a long time since its creation, enjoyed one of the best SLOs in the world. Trust diminished, however, and now all the efforts are being made to regain public confidence, which is also required for better financial and shares value. Similar challenges will certainly impact other players joining Petrobras for the development of its gigantic O&G reserves in the pre-salt basin. For them, and all the service industry, trust is not an option, it is a pre-condition for any operation. The same lessons learned also apply for the social and climate change issues.

Offshore gas production in Brazil represents approximately 91% of the total production (ANP 2018). Three states are currently receiving most of the direct economic benefits from the O&G close to their shores. The same for the neighboring municipalities, which are gradually developing because of royalties, job creation and economic growth. This evolution eventually impacts nature and human life. Therefore, it is of greatest importance to develop technological and social solutions simultaneously. Together with the technological innovations, the development of social engagement and the application of the indicators provided by the SDGs are the most useful tools for coordination, negotiation and follow up of these processes.

CCS is an emblematic solution. By storing CO_2, it is possible to increase fossil fuel production and simultaneously minimize greenhouse gases emissions. The amount is yet to be determined, but EIA (U.S. Energy Information Administration 2018) estimates that from the emissions mitigated until 2050, probably around 14% of them will have been captured by the CCS projects throughout the world. Brazilian exploitation will be one of the most important initiatives, given the enormous potential of the offshore reserves.

Onshore, biogas has contributed with 119 MW to generate electricity in 2016, considering an average annual growth of 66% since 2007 (EPE 2017). CCUS, which also includes the U for Usage, and the BECCUS (Bioenergy Carbon Capture Usage and Storage), based on the residual waste from the biofuel production, are part of RCGI priorities. Dealing with almost two hundred producing plants in the state of São Paulo, the Brazilian Energy Research Enterprise (EPE 2017) estimates that biogas has the potential to substitute approximately 50% of the Diesel consumed in the agriculture sector by the end of 2026. ABiogás (2018), the organization representing biogas producers, estimates that the unused biogas available could potentially have provided an equivalent to 25% of the country's energy in 2016. Some regions in Brazil would have the possibility of joining the areas with a negative carbon footprint.

6   The RDI&S Process

As observed, it was made clear that these technologies have a great potentiality to positively impact Brazil and other countries. SLOs need to become a regular routine, permanently developing the capacities described above. For universities, and other partners, the processes will need to establish renewed RDI&S. Learning from RCGI experiences and others, the following are an example of an action-oriented approach:

- **Research**

a. Focusing the activities at the accelerated speed required by the demanding transition towards a low carbon world.
b. Emphasizing the need to contribute in the exploration and exploitation of oil and gas, by learning, adapting and improving technology.
c. Redefining priorities, based on agreements with federal, state, municipal, universities and social organizations.
d. Building partnerships with public and private industry and benefitting from local and international experiences.
e. Assuring that the projects contribute to one or more of the goals set by the SDGs.

- **Development**

a. Strengthening national and international cooperation.
b. Building partnerships with governmental agencies, businesses, other universities and civil society.
c. The emblematic solutions require an extensive industry and service network.
d. Developing capacities allowing navigation in global and long-term scenarios, which include new institutions and networks and, above all, deliver citizen participation that sustain the SLOs.
e. Supporting the Environmental Impact Assessment (EIA) and other legal steps required by the projects. SLOs are not papers, they represent trust that legitimize and goes beyond legality while also including it.
f. The development of new Technologies will be based on the human capital to be educated and informed to ensure sustainability for decades to come.

- **Innovation**

a. All innovations have their own specific roadmap as for instance, CCUS and BECCUS. At the RCGI, the following innovations list is just an overview of what is being studied:

    i. Storage System.
    ii. Gas separation.
    iii. Emissions reduction.
    iv. Transportation.
    v. Acoustic Monitoring.
    vi. Salt Caverns and other Geological underground formations.

vii. Leakages detection.
viii. Reservoirs mapping.
ix. New products such as synthetic fuels.
x. Infrastructure.

- **Dissemination**

a. In modern times, knowledge dissemination is achieved mainly through digital means, including 'social media'.
b. Producers and consumers of knowledge are multiplying exponentially in all fields, a situation that increases, more than ever, the value of trustworthy, relevant and opportune qualified information.
c. Information is of the highest the most important value in interactions with stakeholders
d. The participants sharing knowledge include teachers, students, managers, workers, companies, markets, scientists, politicians, public servants, journalists and others.
e. It is important to exchange verifiable and useful data and support true and ethical values.
f. It is essential to promote horizontal and vertical valuable exchanges, promoting freedom of speech and mutual respect, exploiting the potential of modern communication technologies.
g. Recognizing that face-to-face interactions are the best way to create understanding, events bringing national and international researchers together is another essential mechanism for knowledge sharing.
h. Establishing showcase units and public open labs may also contribute to educating current and future generations.
i. Synchronizing the communication systems is essential to follow the principle of 'one objective, one message'.
j. Training people to replicate these approaches as extensively as possible.

7 Conclusions

The case study of the RCGI and the actions being developed in the analysis and integration of clean energy initiatives employing CCS are presented in this paper. Based on the SDGs, actions are underway to develop capacities for the 2030 Energy Agenda in Brazil, as an example of framework for other centres to align their efforts. Aspects of how this Centre is contributing to the construction of the Social License for implementing clean energy initiatives and the development of indicators for evaluating the Centre's performance in the context of the SDGs is also outlined.

Acknowledgements The authors gratefully acknowledge the *Escola Politécnica* and the Institute of Energy and Environment, both from the University of São Paulo and also the support from FAPESP and Shell, through the Research Centre for Gas Innovation—RCGI (FAPESP Grant Proc. 2014/50279-4). Drielli Peyerl thanks especially the current financial support of grant Process 2017/18208-8, São Paulo Research Foundation (Fundação de Amparo à Pesquisa do Estado de São Paulo—FAPESP).

References

ABiogás—Associação Brasileira de Biogás e Biometano (2018) Produção de energia elétrica a partir do biogás cresce 14% em 2017. Available in: www.abiogas.org.br/. Accessed in: Nov 2018

ANP—Agência Nacional do Petróleo, Gás Natural e Biocombustíveis (2018) Boletim da Produção de Petróleo e Gás Natural. Available in: http://www.anp.gov.br/. Accessed in: Dec 2018

Asayamaa S, Ishiib A (2017) Selling stories of techno-optimism? The role of narratives on discursive construction of carbon capture and storage in the Japanese media. Energy Res Soc Sci 31:50–59

Ashworth P, Wade S, Reiner D, Liang X (2015) Developments in public communications on CCS. Int J Greenhouse Gas Control 40:449–458

Ashworth P, Carr-Cornish S, Boughen N, Thambimuthu K (2009a) Engaging the public on carbon dioxide capture and storage: does a large group process work? Energy Procedia 1:4765–4773

Ashworth P, Boughen N, Mayhew M, Millar F (2009b) An integrated roadmap of communication activities around carbon capture and storage in Australia and beyond. Energy Procedia 1:4749–4756

Bowen KJ, Cradock-Henry NA, Koch F, Patterson J, Häyhä T, Vogt J, Barbi F (2017) Implementing the "sustainable development goals": towards addressing three key governance challenges—collective action, trade-offs, and accountability. Curr Opin Environ Sustain 26–27:90–96

Büyükozkan G, Karabulut Y, Mukul E (2018) A novel renewable energy selection model for United Nations' sustainable development goals. Energy. 165:290–302

Callan V, Ashworth P (2004) Working together: industry and VET provider training partnerships. Adelaide, NCVER

EIA—United States Energy Information Administration (2018) What countries are the top producers and consumers of oil? Available in: https://www.eia.gov/. Accessed in: Nov 2018

EPE—Empresa de Pesquisa Energética (2017) Impactos da participação do biogás e do biometano na matriz energética. In: IV Fórum do Biogás. São Paulo: Oct 2017

Etzkowitz H (2008) Triple helix innovation: industry, university, and government in action. Routledge, London/New York

Etzkowitz H, Leydesdorff L (2000) The dynamics of innovation: from National Systems and "Mode 2" to a Triple Helix of university–industry–government relations. Res Policy 29(2):109–123

Fagerberg J (2018) Mobilizing innovation for sustainability transitions: a comment on transformative innovation policy. Res Policy 47(9):1568–1576

Feenstra CFJ, Mikunda T, Brunsting S (2011) What happened in Barendrecht? Case study on the planned onshore carbon dioxide storage in Barendrecht, the Netherlands. Global CCS Institute. Available in: https://www.globalccsinstitute.com/. Accessed in: Oct 2018

Gardner J, Dowd A-M, Mason C, Ashworth P (2009) A framework for stakeholder engagement on climate adaptation. CSIRO Climate Adaptation Flagship Working paper No. 3. Available in: http://www.csiro.au/resources/CAF-working-papers.html. Accessed in: Oct 2018

Gough C, Cunningham R, Mander S (2018) Understanding key elements in establishing a social license for CCS: an empirical approach. Int J Greenhouse Gas Control 68:16–25

Gunningham N, Kagan R, Thornton D (2004) Social license and environmental protection: why businesses go beyond compliance. Law Soc. Inq 29:307–341

Huijts N, Molnia E, Steg L (2012) Psychological factors influencing sustainable energy technology acceptance: a review-based comprehensive framework. Renew Sustain Energy 16:525–531

Karimi F, Toikka A (2018) General public reactions to carbon capture and storage: does culture matter? Int J Greenhouse Gas Control 70:193–201

Kroll C (2015) Sustainable development goals: are the rich countries ready? Bertelsmann Foundation, Gütersloh

Kubota H, Shimota A (2017) How should information about CCS be shared with the Japanese public? Energy Procedia 114:7205–7211

Lima PCR (2009) O pré-sal e o aquecimento global. Câmara dos Deputados. Estudo Técnico, Brasília, pp 1–20

Mabon L, Kita J, Xue Z (2017) Challenges for social impact assessment in coastal regions: a case study of the Tomakomai CCS demonstration project. Marine Policy 83:243–25

Mc Dade S (2015) SDG 7 and sustainable energy development in Latin America and the Caribbean, vol LII, no 3. UN Chronicle

Owen J, Kemp D (2012) Social licence and mining: a critical perspective. Resour Policy 38:29–35

Pellegrino C, Lodhia S (2012) Climate change accounting and the Australian mining industry: exploring the links between corporate disclosure and the generation of legitimacy. J Clean Prod 36:68–82

Rooney D, Leach J, Ashworth P (2014) Doing the social in social license. Soc Epistemology 28(3–4):209–218

Sharma S, Cook PJ, Robinson S, Anderson C (2006) Regulatory challenges and managing public perception in planning a geological storage pilot project in Australia. In: Proceedings of the 8th international conference on greenhouse gas control technologies, 19–22 June 2006, Trondheim, Norway

Scharmer O (2018) Axial shift: the decline of trump, the rise of the greens, and the new coordinates of societal change. Available in: https://medium.com/presencing-institute-blog/axial-shift-the-decline-of-trump-the-rise-of-the-greens-and-the-new-coordinates-of-societal-b0bde2613a9e. Accessed in: Nov 2018

Scruggs L, Benegal S (2012) Declining public concern about climate change: can we blame the great recession? Glob Environ Change 22:505–515

SDSN Australia/Pacific (2017) Getting started with the SDGs in universities: a guide for universities, higher education institutions, and the academic sector. Australia, New Zealand and Pacific Edition. Sustainable Development Solutions Network—Australia/Pacific, Melbourne

Seigo SL, Dohle S, Siegrist M (2014) Public perception of carbon capture and storage (CCS): a review. Renew Sustain Energy Rev 38:848–863

UNEP—United Nations Environmental Programme (2015) The three dimensions of sustainable development is an integrated approach beyond our reach? Available in: http://web.unep.org/ourplanet/march-2015/unep-work/three-dimensionssustainable-development. Accessed in: Nov 2018

United Nations (2015) Transforming our world: The 2030 agenda for sustainable development—A/RES/70/1. Available in: https://sustainabledevelopment.un.org/. Accessed in: Nov 2018

Van Egmond S, Hekkert M (2015) Analysis of a prominent carbon storage project failure—the role of the national government as initiator and decision maker in the Barendrecht case. Int J Greenhouse Gas Control 34:1–11

Vercelli S, Lombardi S, Modesti F, Tartarello MC, Finoia MG, De Angelis D, Bigi S, Ruggiero L, Pirrotta S (2017) Making the communication of CCS more "human". Energy Procedia 114:7367–7378

Whitmarsh L, Seyfang G, O'Neill S (2011) Public engagement with carbon and climate change: To what extent is the public 'carbon capable'? Glob Environ Change 21:56–65

Appraising Services to the Ecosystem: An Analysis of Itaipu Power Plant's Water Supply in Energy Generation

Fabrício Baron Mussi, Ubiratã Tortato and Aline Alvares Melo

Abstract Among the approaches derived from the interface between sustainability and economic theory, the pursuit of appraising the economic value of natural resources, based on the evaluation of services to the ecosystem as a mechanism for environmental assessment and as a support tool for investments in preservation both by government and businesses is emphasized. The objective of this study is to analyze the geographic reach and distribution of Itaipu power plant's investments in environmental programs, among which their following actions: (i) sediment monitoring; (ii) micro-pollutants; (iii) water quality; (iv) vegetation management of the reservoir protection range; and (v) management of river basins, considering the area of influence and the water contributions of the municipalities upstream of its reservoir. A quantitative research approach was used to collect information on investments and water contributions from the analysis of water supply. It was observed that 94.6% of the investments were concentrated in regions whose water contribution to the generation of energy was less than 4%. From a results perspective, opportunities were identified for expanding investments in environmental preservation to other areas with greater relevance to maintenance of power generation

Keywords Environmental programs · Water supply · Ecosystem services

F. B. Mussi (✉)
Itaipu Binacional, Pontifícia Universidade Católica do Paraná, Imaculada Conceição, 1155, Curitiba 80215901, Brazil
e-mail: fabricio_mussi@hotmail.com

U. Tortato
Pontifícia Universidade Católica do Paraná, R. Imaculada Conceição, 1155, Curitiba 80215901, Brazil
e-mail: ubirata.tortato@pucpr.br

A. A. Melo
Federal University of Maranhão, Pontifícia Universidade Católica do Paraná, Imaculada Conceição, 1155, Curitiba 80215901, Brazil
e-mail: alinemelo19@yahoo.com.br

© Springer Nature Switzerland AG 2020
W. Leal Filho et al. (eds.), *International Business, Trade and Institutional Sustainability*, World Sustainability Series,
https://doi.org/10.1007/978-3-030-26759-9_38

1   Introduction

In sustainability studies, we see an increasing number of technical analyses on the socio-environmental impacts caused by major corporations in the area of power generation (Morimoto 2013; Liu et al. 2013; Jiang et al. 2016). Moreira et al. (2015) and Jiang et al. (2016) observed that the themes related with the issues hydroelectric plants face—among which those on environment management and sustainability— have become more attractive than the transformation technology itself, and that there is an interdisciplinary trend in research.

Due to the size of these projects and their impact on the environments where they are installed, hydroelectric actions related to sustainability are often discussed. From these, Jabbour et al. (2012) highlight the actions of a reactive nature, resulting from judicial demands, the pressure of stakeholders and surrounding communities; actions of a preventive nature, resulting from decisions to monitor environmental issues sensitive to the project (vegetation edges along the reservoir, water quality, etc.); and actions of a proactive nature, resulting from the strategic orientation of the corporations involved.

Still in regards to sustainability and hydroelectric dams, most studies contemplate economic, social and environmental assessments (Liu et al. 2013; Kumar and Katoch 2016). Local impact, the constraints these ventures can generate in ecosystems and nearby communities, changes in the dynamics of the aquatic and terrestrial habitat, the deposition of sediments in riverbeds, and others are also considered (Yuksel 2010; Zao et al. 2012).

It is recognized that corporations cause impact in the surroundings of their operation, and that cautionary attention to these impacts becomes absolutely necessary, particularly in the social, economic, and environmental spheres. In this context, much has been discussed about the need for corporations to address the subject of sustainability, not only as an accessory matter, but incorporating it on strategic decision-making processes (Engert et al. 2016; Moreira et al. 2015). It is in this scenario that many hydroelectric plants develop their sustainability programs, oriented by socio-environmental and economic issues. However, the definition of how to prioritize investments in these programs, as well as the magnitude of these investments still represents a challenge, so that one of the alternatives for this endeavor consists in the assessment of the ecosystem utility services (Garcia and Romeiro 2015). By evaluating the utility services provided by the environment economically, governments and businesses can establish references for their expenditure with preservation, support actions to mitigate the degradation of natural resources, and in the case of need for payments for environmental utility services (Motta 2011).

In face of the above, the objective of this study is to analyze the geographic reach and distribution of Itaipu power plant's investments in environmental programs, among which their following actions: (i) sediment monitoring; (ii) micro-pollutants; (iii) water quality; (iv) vegetation management of the reservoir protection range; and (v) management of river basins, considering the area of influence and the water contributions of the municipalities upstream of its reservoir.

This subject constitutes a research opportunity, since in preliminary literature review attempts to evaluate the ecosystem services of the reservoirs of the Brazilian hydroelectric plants were not identified, despite the fact that this type of plants represents the main source of energy in the country (MME 2017). In addition, the information generated through the evaluation can be useful for licensing processes, negotiation of the conditions for installation and operation, and environmental compensations. The article begins with the theoretical reference, and in sequence, the methodological procedures and findings are presented. Finally, the final considerations are discussed.

2 Theoretical References

2.1 Evaluating Ecosystem Utility Services

It is argued that part of poverty and social inequalities is due to the lack of preservation of ecosystems, or their use in an unsustainable way, especially when looking for the needs of future generations (MEA 2005). In this sense, attention has been focused on the economic benefits of maintaining biodiversity (TEEB 2010). Given its relevance, in the academic and business circles, the economic evaluation of the benefits these ecosystems bring to society is what has been sought. Environmental evaluation consists of the process by which one seeks to estimate the economic value of natural resources by determining the equivalence of other available resources in the economy (Castro 2015).

The following Table 1 presents the definition of ecosystem utility services.

Recognizing the consensus about the concept, the challenge is to verify how to measure it from an economic perspective. In this context, Turner et al. (1998), and Motta (2011) defend that although there are limits to economic calculation, recog-

Table 1 Definition of ecosystem utility services

Concept of ecosystem utility services	Author
Conditions and processes by which natural ecosystems provide support to human wellbeing	Daily (1997)
Benefits human populations directly or indirectly obtain from ecosystem functions	Costanza (1997)
Components of nature directly harnessed to maintain human wellbeing	Boyd and Banzhaf (2007)
Benefits people obtain from ecosystems	Wallace (2007)
Aspects of ecosystems used (directly or indirectly) for maintenance of human wellbeing	Fisher et al. (2009)
Benefits people obtain from ecosystems	MA (2005)

Source Prepared based on the literature review

nizing that not everything is subject to significant monetary valuation, measurement can play a significant role in the process of environmental policy evaluation. In a study by Groot et al. (2002), the understanding about the quantification of ecosystem utility services was still identified as a problem with no definitive answer, besides the different interpretations of the implications of these calculations for the academic and business perspective (Spangenberg and Settele 2010; Turner et al. 2010).

Turner and Daily (2007) point out the main challenges of defining a structure to evaluate ecosystem utility services, focusing on detailing information at relevant decision-making scales; practical *know-how* in the process of institutional design and implementation; and the presentation of compelling models of success in which economic incentives are aligned with preservation. In the words of these authors:

> Despite growing general awareness of conserved ecosystem benefits, detailed information at scales useful for decision makers on how people benefit from specific services remains deficient. This "information failure" is one reason why conservation investment finance is still too low and sometimes ineffective. (Turner and Daily 2007, p. 27).

Among other shortcomings identified by Turner and Daily (2007), we can mention (i) "*institutional failures*", in the sense that the beneficiaries of ecosystem utility services are often different and distant from those that gain with the transformation of the ecosystem. Local socio-ecological contexts, including property rights and institutions, often overlooked conservation programs; (ii) the "*market failures*" that derive from public characteristics of many benefits and their lack of prices. Markets typically reward short-term values of natural resources (overvaluing preservation opportunity costs) in detriment of long-term ecological health.

The arguments that support the evaluation exercise are based on the following points:

i.   With the absence of market prices and definitions of property rights over particular natural resources, the evaluation of ecosystem utility services is sometimes not considered (Turner et al. 2010);
ii. However, the evaluation has opened new spaces for the debate on environmental policies, including in areas where dialogue on modes of preservation occurred in an abstract and imprecise way (Turner et al. 2010);
iii. In addition, proving how valuable an ecosystem utility service is can help in the projections for the economic development of countries (Motta 2011);
iv. Evaluation can help prioritize organizations' investments in environmental preservation (Fu et al. 2014);
v.   It can help in the construction of policies aimed at paying for environmental utility services in a precise way (Fu et al. 2014);

As an exception to the exercise of the evaluation of ecosystem utility services, it is argued that the aggregation between the different functions provided by a given ecosystem should be restricted, due to the double counting risks. It is necessary to address possible incompatibilities between different evaluation measures (such as opportunity costs, consumer surplus, and market prices) (Turner et al. 1998), and the risk of problems related to the overlapping of ecosystem utility services and ambiguities in the interpretation of these services (Ojea et al. 2012). Finally, the problem of lack of consensus on the commonly used methodologies, with respect to their efficiency to fulfil the intended purpose should be looked at (Nogueira et al. 2018).

Discussing the challenges of defining the value of the ecosystem utility service of water supply, Garcia and Romero (2015, pp. 73–74) state that:

> "pricing" water is not a trivial task. Firstly, because there may not be sufficient information to allow an adequate valuation; secondly, because it is possible that situations occur where the adequate price cannot be fully charged to end-users [...] Adequate pricing must be understood as allowing for the maintenance of the "production" conditions, in terms of quantity and quality of the water resource.

3   Interfaces with Hydroelectric Plants

Among the evaluation models, we highlight the pioneering work of Costanza et al. (1997), which seeks to connect the processes and functions of the ecosystem with results of goods and services to which one can then assign economical value. Bryan et al (2010) sought to design an evaluation model in which it will be possible for decision-makers to establish investment priorities. Keller et al. (2012), in turn, present a model of evaluation of ecosystem services of regulation of water quality considering cause-effect relationships, concomitantly identifying the ecological pathways.

In an attempt to evaluate ecosystem utility services derived from the construction of reservoirs, Fu et al. (2014) argue that dams cannot completely replace the water conservation function of the ecosystem reservoir, and present high economic and environmental costs which must be paid. Compensation for water conservation services should become a basis for the ecological compensation owed by the hydroelectric plant. It is from this perspective that many of the initiatives materialized in sustainability programs developed by hydroelectric plants, and in payment programs for environmental services are justified. The former envisage preserving the environment affected by the hydroelectric plant, by acting in socio-environmental and economic initiatives, while the second consists in the remuneration of the agents that ensure the preservation of the environment. In both cases, understanding ecosystem utility services and their corresponding manner of assessment helps to define values, establish priorities, and evaluate results. Still according to these authors:

> Hydropower development is an important way to solve the energy demand in developing countries. In the global context of climate change, its importance is more prominent. But unscientific hydropower development causes great negative impact on the environment, thereby affecting the region's sustainable development. This requires stakeholders of hydropower development to correctly understand the relationship between protection and development, to fully consider the influence of ecosystem services on hydropower benefits, and to change the performance from passive compensation for environmental damage to active participation in watershed protection, so as to reduce the impact on the environment (Fu et al. 2014, p. 345).

Jager and Smith (2008), and Brauman et al. (2007) affirm that the reservoirs of large hydroelectric plants operate in systems that seek to maximize revenue based on the sale of energy, respecting some permits for use of the reservoir. Notwithstanding, these optimization systems do not usually consider the health of the aquatic ecosystem. To these authors, both situations must be reconciled, discarding a *trade-off* between the maximization of generation revenue and the preservation of the reservoir, so that harmonizing generation efficiency with environmental preservation becomes more plausible, based on the valuation of water supply.

In this way, in addition to deepening the debate on potential mitigating actions of risks of environmental degradation, as well as initiatives for preservation, it is still possible to remunerate agents—by paying for environmental services—who participate in initiatives of this nature, either by mitigating environmental impacts (Chan et al. 2006), soil erosion reduction (Lu and Li 2006), or the cost of opportunities resulting from the fact that local residents give up their crops to preserve river springs (Brinkman 2001).

In Brazil, approximately 65% of energy generation comes from hydroelectric plants, representing the main power source, followed by thermoelectric plants (coal, natural gas, and biomass) with 15%, wind power (4%), nuclear (3%), and other sources with 13% (MME 2017). Considering the relevance of this source to energy supply, the main ecosystem services and their potential interfaces with hydroelectric plants are presented below (Table 2).

In this research, we considered the ecosystem utility service that is water supply as a priority, once it represents the input for generation, both for the total quantity provisioned and for the change in flow patterns. The existence of the enterprise, in turn, can also affect the availability of this ecosystem utility service to third parties. Besides, it is understood that the different uses of the soil in the basin, with the presence of greater or lesser degree of vegetation, potentially affects the availability of water for the generation system (GVces 2018).

4 Methodological Procedures

The proposal for the integration of the themes is illustrated here in the form of Itaipu hydroelectric plant and its particular *Sustainability Program*. The choice was intentional, once the corporation is considered the largest power generating hydroelectric plant in the world, with its sustainability program in force for over a decade, being internationally recognized for its contribution to the socio-economic development of the western region of Parana state, for its participation in the supply of energy to Brazil (approx. 18% of the country), and for its water management and conservation practices. The operation of this plant began in 1984. Its reservoir is 170 km long, and 20 power units generating 700 MW each were installed.

In addition, it is a company whose sustainability is supported by strategic planning, and whose vision shows. "Until 2020, ITAIPU Binational will consolidate as the best performance generator of clean and renewable power, with the best oper-

Table 2 Main ecosystem services and interfaces with hydroelectric plants

Ecosystem utility services	Concept	Relation with hydroelectric plants
General provision	Production of tangible goods (food or inputs) that generate wellbeing	Fish supply/monitoring
Water supply	Contribution in terms of quantity of water	Dependence for generation/impact on downstream users
Water quality regulation	Water quality control	Influence turbine operation
Regulation of assimilation of liquid effluents	Capacity of the ecosystems to dilute a pollutant load	Upstream third-part effluents may influence the plant
Regulation of global climate	Influence on emissions of relevant greenhouse gases	Maintenance and restauration of surrounding areas
Regulation of soil erosion	Role of ecosystems in the control of soil erosion processes	Control and monitoring, depending on the impact on the life of the reservoir
Leisure and tourism	Role of ecosystems in relaxation and leisure	Influence in touristic activities
Cultural services	Natural benefits	Modification of landscapes and interaction with ecosystems

Source Adapted from TeSE—GV'ces (2017)

ative performance and best practices of sustainability in the world, impelling the sustainable development and regional integration" (ITAIPU 2018).

The research was conducted assessing documents such as the *Ten-Year Energy Plan* (Brazil 2017), the plant's annual sustainability reports, other documents such as technical reports, and energy auction notices. The purpose of this stage of data collection was to obtain information about the actions contemplated by the sustainability program of this plant, as well as the localities benefited, and the data related to water consumption, energy generation, and revenues. Subsequently, we performed the valuation exercise for the water supply utility service using both the reposition cost and market prices methods (Fu et al. 2014; GVces 2018). With this information, we analyzed:

i. The expenses with environmental actions (in the period of 2010–2017) that may contribute for the maintenance of said ecosystem utility service;
ii. The history of energy generation, water consumption, and revenues obtained (in the same period: 2010–2017);
iii. The costs of replacing this source, in the eventuality of interruption of water supply;
iv. The localities where such investments were made, considering the map of the water contributions for the generation of energy.

Finally, the data obtained was presented to the managers of the environmental actions analyzed, and to the Itaipu power plant's superintendent of environmental management, in order to obtain validation in face of the results.

5 Presenting the Findings

The following table presents the environmental actions considered in this evaluation, taking in consideration its objectives, the justification for the development and the disbursement made during the period considered: 2010 to 2017.

The environmental programs selected were the following (Table 3).

Regarding the history of energy generation and consumption, the following table presents the information that denotes dependency of water as an input, once its decline causes a decrease in generation. In this case, it is suggested that the economic evaluation reflects the loss of equivalent billing (GVces 2018) (Table 4).

For the evaluation of the costs of replacing the energy source, the disbursement of replacing water was not examined, as it did not apply to the scenario. As an alternative (GVces 2018), the replacement costs were used to deliver the same amount of energy to the Brazilian electric system. For this analysis, the lowest and the highest price of the energy auctions for an alternative source, thermal energy, from the last auctions of the chamber of commercialization of electric energy were considered (CCEE 2017). In this scenario, it was verified that the disbursement for replacement of water would transit in a spectrum of 69–151% superior to Itaipu's revenue (Table 5).

For the conditions described in the first table above, the water supply services would be estimated by the market price method in approx. US$29 trillion dollars. For the conditions on the second table, using the method of costing replacement of water supply for power generation, the valuation would be in a minimum US$50 trillion dollars.

Recognizing this gradient as an estimate for the valuation of water supply, there is a strong possibility the analyzed power plant should expand its investments to include costs for ecologic compensation and environmental preservation (Fu et al. 2014) in order to ensure continuity in water supply for power generation. It is recognized, however, that the application of more than one method for the evaluation suggests a significant breadth of values, representing still a challenge (Turner and Daily 2007), as much as the impossibility of charging end-users an adequate value (Garcia and Romero 2015), either because of resource constraints or because of the absence of clear public policies to support these evaluations.

From the aspect of environmental sustainability (Motta 2011), this exercise would support actions to protect natural resources, such as water and micro-pollutants monitoring, besides management of the vegetation at the edge of the reservoir—and actions to mitigate degradation, such as those acting on water and soil, with the objective of reducing the ingress of sediments in the reservoir. Finally, we evaluated the localities where investments in environmental actions were carried out, considering the map of water contributions for the generation of energy, according to the following Fig. 1.

Table 3 Distribution of investments in environmental actions

Environmental action	Objective	Justification	Investments between 2010–2017
Sediment monitoring	Determine solid discharges and estimate the production of sediment in the water contribution basin, in order to estimate the sedimentation and guide the conservation actions. Operate the sedimentation measurement stations to estimate the life of the reservoir	Accelerated erosion that has been occurring in the soil, particularly in agricultural areas, has become increasingly critical and difficult to be contained. The lack of erosion control practices—called conservation practices—generates serious social, environmental, and economic impacts such as impoverishment of soil fertility, deposition of sediment in reservoirs (diminishing life), compromising their multiple uses	US$378,000.00
Water quality monitoring	Monitor the quality of the water in the reservoir, affluent streams, micro-basins, and groundwater. Provide technical subsidies for the management of the hydrous body	Artificial eutrophication of reservoirs occurs due to the release of nutrients from different origins, such as: domestic, industrial, and/or agricultural effluents. This type of eutrophication is responsible for the "premature aging" of the aquatic ecosystem, where deep physical, chemical, and biological mutations then occur. This phenomenon can compromise water supply and generation of energy due to the proliferation of (macrophyte) aquatic plants. This monitoring allows also for the recommendation of preventive and/or corrective sanitation measures that eventually can be adopted to improve water quality for different uses	US$1,670,430

(continued)

Table 3 (continued)

Environmental action	Objective	Justification	Investments between 2010–2017
Micro-pollutants monitoring	Know the technical factors that can limit the commitment of the productive and edaphic environment to establish rational programs of management and recommendation, whose more efficient use promotes the increase of harvests and reduces the costs and risks of environmental damage. Identify, quantify, and evaluate the main micro-pollutants in the cross-border region (BR-PY), in diverse matrices of environmental relevance (water, soil, nourishment and living organisms), seeking to understand the spread of these dynamics in the environment, and their relation with biodiversity	Monitoring agrotoxics and their metabolites in different matrices, and understanding their dynamics and influence in relation to biodiversity is a necessity of the region belonging to the area of influence of the Itaipu power plant. This type of study contributes to the development of a region in various aspects, such as the development of management activities aimed at increasing productivity, and reducing environmental impacts	US$202,680.00
Management of vegetation in the protected area	Preserve and recover the protected areas belonging to the Itaipu power plant, guaranteeing their biological integrity and compliance with legal precepts, contributing to the preservation of regional biodiversity	Itaipu protected areas, comprising the reservoir protection strip, reserves, and biological refuges require recovery actions, forest maintenance and monitoring, as well as legal regulations for their allowed multiple uses within sustainability criteria. The production of forest seedlings, foreseen in this action, aims to attend the reforestation programs for the areas belonging to the corporation, and the recovery of the permanent preservation areas in the micro-basins	US$1,000,128.00
Management of river basins	Implement a set of water and soil management activities for environmental monitoring of the micro-basins affected by the Itaipu reservoir	Reduce the contribution of sediments to the reservoir so that water is available with quality and quantity sufficient for energy production, and other uses	US$6,034,300.00

Source Elaborated by the authors based on collection of secondary data

Table 4 History of water generation and consumption

Year	Generation (m³)	Cooling (m³)	Total (m³)	Total (GWh)	Revenue (US$)	m³/US$
2010	302.097.254.400	365.868.058	302.463.122.458	85.303	3.450.500.000	87.65
2011	325.706.832.000	365.868.058	326.072.700.058	91.523	3.384.400.000	96.34
2012	344.470.233.600	365.868.058	344.836.101.658	97.533	3.703.500.000	93.11
2013	349.168.579.200	365.868.058	349.534.447.258	97.878	3.760.100.000	92.95
2014	308.814.940.800	365.868.058	309.180.808.858	87.165	3.680.400.000	84.00
2015	314.462.476.800	365.868.058	314.828.344.858	88.575	3.680.800.000	85.53
2016	369.632.851.200	365.868.058	369.998.719.258	102.335	3.811.500.000	97.07
2017	336.110.688.000	365.868.058	336.476.556.058	95.682	3.729.703.000	90.21

Source Based on secondary data

Table 5 Comparatives by source of energy (lowest and highest rate)

Year	Production MWh	Itaipu revenue	Thermal min. price (US$67.12/MWh)	Thermal max. price (US$92.96/MWh)	(Itaipu—P. min) (Itaipu—P. max)
2010	85.303.000.000,00	3.450.500.000,00	5.771.134.259,26	7.992.025.925,93	67–132%
2011	91.523.000.000,00	3.384.400.000,00	6.143.905.092,59	8.508.249.259,26	82 – ⬭151%⬭
2012	97.533.000.000,00	3.703.500.000,00	6.547.354.166,67	9.066.956.666,67	77–145%
2013	97.878.000.000,00	3.760.100.000,00	6.570.513.888,89	9.099.028.888,89	74–142%
2014	87.165.000.000,00	3.680.400.000,00	5.851.354.166,67	8.103.116.666,67	⬭59⬭–120%
2015	88.575.000.000,00	3.680.800.000,00	5.946.006.944,44	8.234.194.444,44	61–124%
2016	102.335.000.000,00	3.811.500.000,00	6.869.710.648,15	9.513.364.814,81	80–150%
2017	95.682.000.000,00	3.729.703.000,00	6.422.175.840,00	8.894.598.720,00	72–138%

Source Based on secondary data

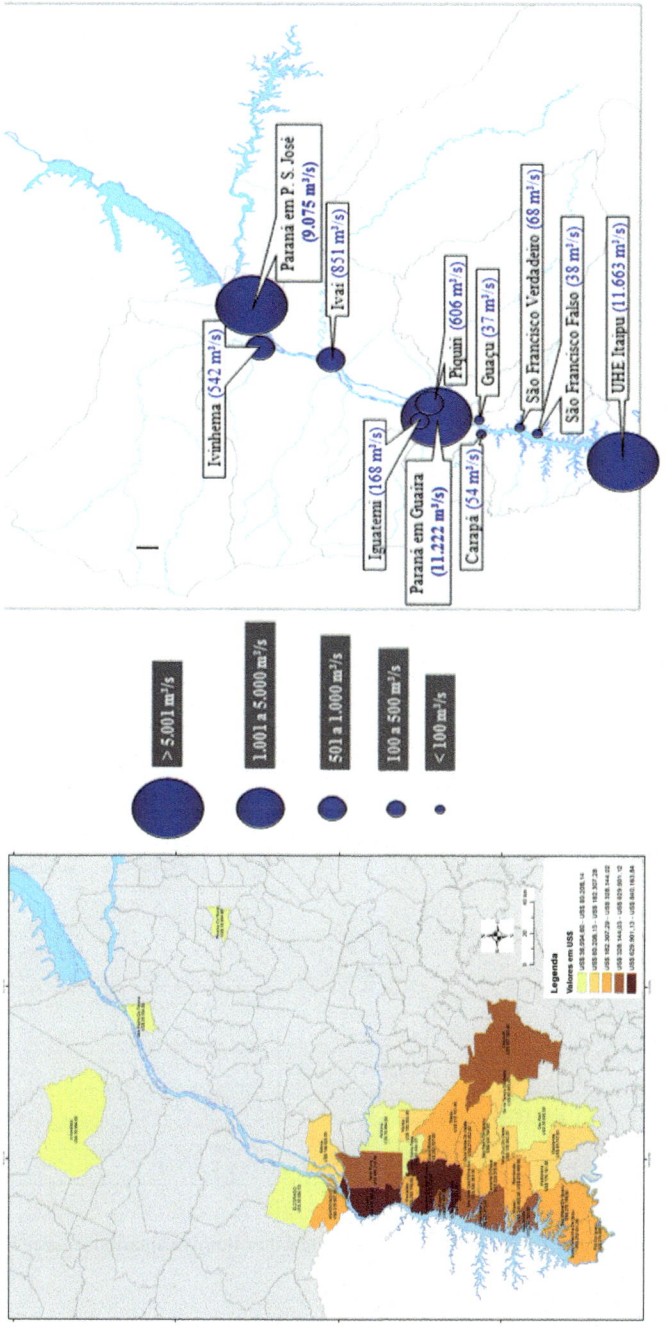

Fig. 1 Spatial distribution of the investments X water supply valuation. *Source* Elaborated by the authors

With the information on the evaluation of the ecosystem utility services of water supply for the Itaipu hydroelectric power plant, the water contributions for its reservoir, the spatial distribution of the investments in environmental actions aiming to maintain useful life in the reservoir, we next examined the following questions: (i) *Do the current investments spatially contemplate the regions that most contribute to the supply of water for power generation, i.e., the regions that provide the ecosystem service of water supply?* (ii) *What is the value of water provision for power generation?* The following map, validated by the representatives of the management board of the power plant in question, illustrates the questionings.

From the analysis of Fig. 2, it can be affirmed that the region that receives more investment in environmental preservation, through actions directed to sediments monitoring, micro-pollutants monitoring, water quality monitoring, management of vegetation in the reservoir's protective edge, and micro-basins management represents, from the perspective of evaluating ecosystem utility water supply services, the one that contributes with only 4% to the water resources used for power generation. Almost all of the input for power generation is in fact provided by the most upstream region.

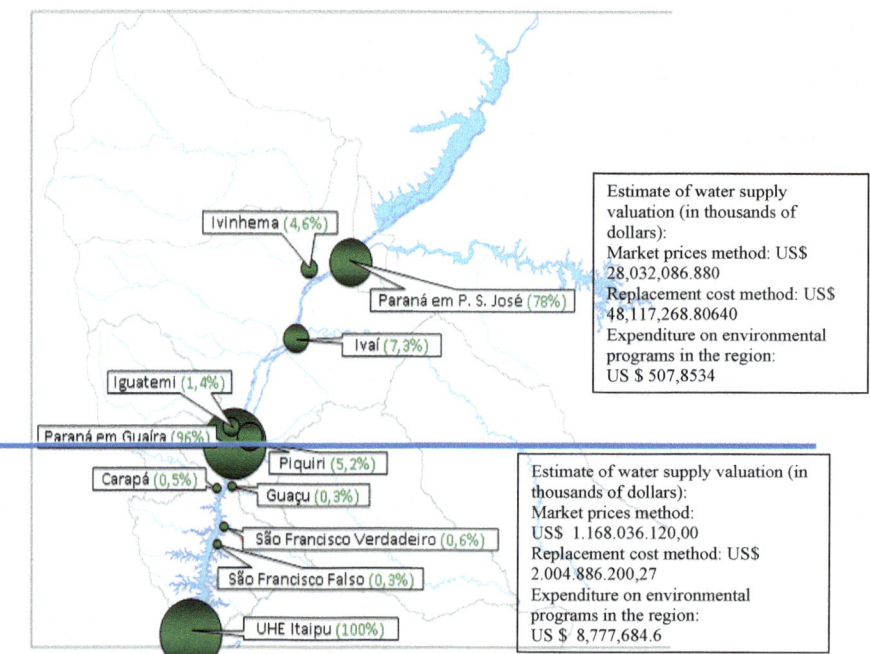

Fig. 2 Spatial distribution of the investments considering water supply valuation. *Source* Elaborated by the authors

6 Conclusions

The objective of this work was to analyze the geographic reach and distribution of Itaipu hydroelectric power plant's investments in environmental programs, among which their following actions: (i) sediment monitoring; (ii) micro-pollutants; (iii) water quality; (iv) vegetation management of the reservoir protection edge; and (v) management of river basins, considering the area of influence and the water contributions of the municipalities upstream of its reservoir. Information on water consumption was used to assess the provision of this input used in power generation, and concurrently the spatial distribution of investments was also observed.

It was found that approximately 5.4% of the investments in environmental preservation is concentrated in the region that contributes with approximately 96% of the water used for power generation, whereas that 94.6% of the investments are made in the region that contributes with only 4% of the water used. This evaluation exercise proves the relevance of environmental preservation actions, and provides support to the power plant under analysis, for the increase of investments to be made on the field of sustainability and the flow of these to specific regions. The evaluation exercise can also support new proactive and precautionary initiatives (Jabbour et al. 2012) turned to the preservation of the reservoir.

It should be noted that the power plant under analysis carries out innumerable other environmental actions which, due to their scope, were not considered in the current analysis, although they all are immensely relevant for the preservation of aquatic and terrestrial biodiversity. It should also be noted that, from a technical point of view, there are other factors that also contribute for energy generation, such as upstream and downstream water levels, availability of other generating units, and market demand for energy and streaming.

As a limitation of this research, we should mention the difficulties stemming from the non-replicability of some of the valuation, given the unique characteristics of the power plant under analysis, the fragmentation of the information required for the evaluation, the limited time perspective utilized (only 8 years), and the absence of comparatives, which could corroborate the analyzes performed in this research. Lastly, future research is suggested, undertaking similar studies in other large-scale hydroelectric power plants, and the re-evaluation of other environmental programs, from the perspective of their contributions to other localities supplying water for energy generation.

References

Boyd J, Banzhaf S (2007) What are ecosystem services? The need for standardized environmental accounting units. Ecol Econ 63(2–3):616–626

Bryan BA, Raymond CM, Crossman ND, Macdonald DH (2010) Targeting the management of ecosystem services based on social values: where, what, and how? Landscape Urban Plann 97:111–122. https://doi.org/10.1016/j.landurbplan.2010.05.002

Brauman KA, Daily GC, Duarte TK, Mooney HA (2007) The nature and value of ecosystem services: an overview highlighting hydrologic services. Ann Rev Environ Resour 32:67–98

Brinkman W (2001) Innovative financing mechanisms for conservation and sustainable forest management. Eur Trop Forest Res Netw. Retrived from: https://eldis.org/document/A12986

Câmara de Comercialização de Energia Elétrica (CCEE) (2017). Relatório Anual de Administração

Castro ALG (2015) Serviços ambientais: remoção de insetos em ambiente natural e de cultura. Dissertação (mestrado). Universidade Federal de São João Del Rei, Programa de Pós Graduação em Ciências Agrárias. 48 f

Chan KMA, Shaw MR, Cameron DR, Underwood EC, Daily GC (2006) Conservation planning for ecosystem services. PLoS Biol 4:2138–2152

Costanza R, d'Arge R, de Groot R, Farber S, Grasso M (1997) The value of the world's ecosystem services and naturalcapital. Nature 387:253–260

Daily GC (1997) Nature's services: societal dependence on natural ecosystems. Island Press, Washington, DC

De Groot RS, Wilson M, Boumans R (2002) A typology for the description, classification and valuation of ecosystem functions, goods and services. Ecol Econ 41(3):393–408

Engert S, Rauter R, Baumgartner RJ (2016) Exploring the integration of corporate sustainability into strategic management: a literature review. J Clean Prod 112(4):2833–2850. https://doi.org/10.1016/j.jclepro.2015.08.031

Fisher B, Turner KR, Morling P (2009) Defining and classifying ecosystem services for decision making. Ecol Econ 68(3):643–653

Fu B, Wang YK, Xu P, Yan K, Li M (2014) Value of ecosystem hydropower service and its impact on the payment for ecosystem services. Sci Total Environ 472(15):338–346

Garcia JR, Romeiro AR (2015) Valoração e cobrança pelo uso da água: uma abordagem econômico-ecológica. In: Tôsto SG, Belarmino LC, Romeiro AR, Rodrigues CAG (eds) (2015). Valoração de serviços ecossistêmicos: metodologias e estudos de caso/Brasília. Embrapa Monitoramento por Satélite, DF, pp 71–90

Itaipu (2018). Annual Sustainability Report. Retrived from: https://www.itaipu.gov.br/responsabilidade/relatorios-de-sustentabilidade

Jabbour CJC, Silva EM, Paiva EL, Santos FCA (2012) Environmental management in Brazil: is it a completely competitive priority? J Clean Prod 21(1):11–22. https://doi.org/10.1016/j.jclepro.2011.09.003

Jager HI, Smith BT (2008) Sustainable reservoir operation: can we generate hydropower and preserve ecosystem values? River Res Appl 24(3):340–352. https://doi.org/10.1002/rra.1069

Jiang H, Qiang M, Lin P (2016) A topic modeling based bibliometric exploration of hydropower research. Renew Sustain Energy Rev 57(3):226–237. https://doi.org/10.1016/j.rser.2015.12.194

Keeler BL, Polasky S, Brauman ka, Johnson ka, Finlay JC, O'Neill A, Kovacs K, Dalzell B (2012) Linking water quality and well-being for improved assessment and valuation of ecosystem services. PNAS 109 (45):18619–18624. Retrieved from: www.pnas.org/cgi/doi/10.1073/pnas.1215991109

Kumar D, Katoch SS (2016) Environmental sustainability of run of the river hydropower projects: A study from western Himalayan region of India. Renewable Energy 93:599–607. https://doi.org/10.1016/j.renene.2016.03.032

Liu J, Zuo J, Sun Z, Zillante G, Chen X (2013) Sustainability in hydropower development—a case study. Renew Sustain Energy Rev 19:230–237. https://doi.org/10.1016/j.rser.2012.11.03

Lu X, Li H (2006) Payment for environmental services: an approach to sustainable watershed management. In: Sustainable sloping lands and watershed management conference, Luang Prabang, Lao PDR

Millenium Ecosystem Assessment (MEA). (2005). Ecosystems and human well-being: current state and trends, vol 1. Island Press, Washington

Ministério de Minas e Energia (MME) (2017) Plano Decenal de Expansão de Energia 2026. MME/EPE, Brasília. Retrieved from: www.epe.gov.br/pt/publicacoes-dados-abertos/publicacoes/Plano-Decenal-de-Expansao-de-Energia-2026

Moreira JM, Cesaretti MA, Carajilescov P, Maiorino JR (2015) Sustainability deterioration of electricity generation in Brazil. Energy Policy 87(December):334–346. https://doi.org/10.1016/j.enpol.2015.09.021

Morinoto R (2013) Incorporating socio-environmental considerations into project assessment models using multicriteria analysis: a case study of Sri Lankan hydropower projects. Energy Policy 59(c):643–653. https://doi.org/10.1016/j.enpol.2013.04.020

Motta RS (2011) Valoração e precificação dos recursos ambientais para uma economia verde. Política Ambiental, Belo Horizonte 8(june):179–190

Nogueira EM, Yanai AM, Vasconcelos SS, Graça PMLA, Fearnside PM (2018) Carbon stocks and losses to deforestation in protected areas in Brazilian Amazonia. Regional Environmental Change 18(1):261–270. https://doi.org/10.1007/s10113-017-1198-1

Ojea E, Ortega JM, Chiabai A (2012) Defining and classifying ecosystem services for economic valuation: the case of forest water services. Environ Sci Policy 19–20 (May–June), 1–15

Spangenberg JH, Settele J (2010) Precisely incorrect? Monetising the value of ecosystem services. Ecol Complex 7(3):327–337

The Economics of Ecosystems and Biodiversity Project (TEEB) (2010) The economics of ecosystems and biodiversity ecological and economic foundations. Pushpam Kumar. Earthscan, London/Washington. Retrieved from: http://www.teebweb.org/our-publications/teeb-study-reports/ecological-and-economicfoundations/#.Ujr1xH9mOG8

GVces (2018) Aplicação das Diretrizes Empresariais para Valoração Econômica de Serviços Ecossistêmicos (DEVESE) e das Diretrizes Empresariais para valoração não econômica de Serviços Ecossistêmicos Culturais (DESEC) 1 para hidrelétricas. Centro de Estudos em Sustentabilidade da Escola de Administração de Empresas de São Paulo da Fundação Getulio Vargas. São Paulo, 52 p

Turner RK, Adger N, Brouwer R (1998) Ecosystem services value, research needs and policy relevance: a commentary. Ecol Econ 25:61–65

Turner RK, Daily GC (2007) The ecosystem services framework and natural capital conservation. Environ Resour Econ 39(1):25–35. https://doi.org/10.1007/s10640-007-9176-6

Turner RK, Morse-Jones S, Fisher B (2010) Ecosystem valuation, a sequential decision support system and quality assessment issues. Ann N Y Acad Sci 1185:79–101

Wallace KJ (2007) Classification of ecosystem services: problems and solutions. Biol Cons 139(3–4):235–246

Yüksel I (2010) Hydropower for sustainable water and energy development. Renew Sustain Energy Rev 14(1):462–469. https://doi.org/10.1016/j.rser.2009.07.025

Zhao X, Liu L, Liu X, Wang J, Liu P (2012) A critical analysis of the development of China's hydropower. Renew Energy 44(1):1–6. https://doi.org/10.1016/j.renene.2012.01.005

Using GIS to Map Priority Areas for Conservation Versus Mineral Exploration: Territorial Sea of Espírito Santo State, Brazil, Study Case

Viviane K. Bisch, Valeria S. Quaresma, João B. Teixeira and Alex C. Bastos

Abstract This paper applied a GIS platform (ArcGis software) as a tool to measure the conflicting use between priority areas for conservation and potential mineral and oil and gas exploitation sites. The study has as goal to analyze and quantify the distribution of each process stage, to determine the current and future multiple uses of the study area in order to understand the impact that such activities may have on the preservation areas. It's a tool that contributes to reach the Sustainable Development Goals, specifically number 14, about Conservation and sustainably use of the oceans, seas and marine resources. The study area encompasses the territorial sea of Espírito Santo State. The software was used to produce maps and to analyze the overlap between the different uses. Large numbers of authorization and research requirements were found in Priority Areas for Conservation (PAC) in order to extract substances such as salts of potassium and limestone in larger quantities in the PAC. The study provides a basis for conflict identification for the sustainable management of the exploitation of marine mineral resources, which should include the management of extraction activities aiming at the sustainability of the ecosystem, valuing the existence of areas with greater sensitivity.

Keywords GIS · Priority areas for conservation · Sustainability

V. K. Bisch (✉) · V. S. Quaresma (✉) · J. B. Teixeira (✉) · A. C. Bastos (✉)
Oceanography and Ecology Department, Universidade Federal do Espírito Santo, Av. Fernando Ferrari, 514, Campus Goiabeiras, 29060-900, Vitória, ES, Brazil
e-mail: vivikorres@gmail.com

V. S. Quaresma
e-mail: valeria.quaresma@ufes.br

J. B. Teixeira
e-mail: jboceano@gmail.com

A. C. Bastos
e-mail: alex.bastos@ufes.br

© Springer Nature Switzerland AG 2020
W. Leal Filho et al. (eds.), *International Business, Trade and Institutional Sustainability*, World Sustainability Series,
https://doi.org/10.1007/978-3-030-26759-9_39

1   Introduction

1.1   *Environmental Conflict Management and Spatial Decision Support Systems*

The growth and concentration of human activities are followed by the reduction of natural resources, along with conflicts between stakeholders of the target areas related to conservation and development. Most of the Earth's population, more than half, and a large part of world's economic output is connected with coastal and oceanic areas (Cicin-Sain and Belfiore 2005).

The sustainable management of coastal and marine resources is linked to the constant technical development, continuous monitoring of management and resolution of conflicts. Long-term conflicts can be expensive and reduce the probability of sustainable development policies to be applied. The conflict management process relies on research on the management process and environmental techniques that support decision-making, but these generally do not consider a spatial perspective within the context of ecological and human management systems (Mazor et al. 2014).

In order to identify the problems, is fundamental to acknowledge the several interdisciplinary interdependencies among the different perspectives on the resources and how these interact to generate dispute (Susskind et al. 1999; Wondolleck and Yaffee 2000). In this regard, landscape scanning and establishment of the most vulnerable areas to future disputes is an approach to mitigate environmental impacts.

Methods of conflict resolution are based on choosing one among the multiple scenarios proposed, and indicate a solution that is mutually consented. One of the methods is the Multiple Criteria Decision Making (MCDM) that has been used to choose the best alternative. The MCDM is mainly useful for resolving area conflicts for a specific use, such as power line and pipeline routes (Brody et al. 2004). The mentioned study is only one of several examples of the literature on multi-criteria approaches to solving environmental problems (Hipel et al. 1997; Ridgley et al. 1997; Agrell et al. 1998; Tecle et al. 1998; Hamalainen et al. 2000).

Starting in the 90's, researchers began to focus on conflict in terms of their relationship with physical areas, which came to be understood as coming from different perspectives on how the resources should be used. Based on this research, a new approach is developed that integrates the MCDM techniques with the emerging Geographic Information Systems (GIS) technology to generate a SDSS (Spatial Decision Support Systems). SDSS is a system for storing and manipulating information with geographic reference, such as points and polygons that identify a conflict environment. This approach facilitates the visualization and interpretation of problems by stakeholders (Jankowski and Nyerges 2001), and is useful for facilitating the integration of information by decision makers. Thus, managers observe geographic differences when assessing the region, and consider how conflicts are situated in space. All these factors are important in a negotiation situation to generate the division, delimitation or spatial union of the interests of the various parts (Godschalk 1992).

There are a few frameworks that guide the better understanding of the processes that take place in the marine environment, such as human activities, spatial and temporal overlap and their objectives, communication problems when different sectors of authority manage different areas as well as between the sea resources uses and the population that depends of them, added to the lack of protection of more sensitive marine areas (DEFRA 2007).

Another approach that have been very used in the past years is the ecosystem-based management, where there is not defined just as an area, but the natural ecosystem that will be the focus and the key to management. Studying all the processes that affect its natural state and how it changes a certain species or concern (Crowder et al. 2006).

Even though most of initiatives of study were about supporting decision of implementation of MPAs, marine spatial plan have been expanding the focus to managing the multiple uses of marine spaces as whole, as in countries as Belgium, Germany, the Netherlands and the UK. In this case MPAs stays as an important factor to be considered in the context of the larger area where they're inserted, defining strategies to manage the whole area and the uses outside of the MPAs, providing economic growth as well.

1.2 Coastal and Marine Protected Areas as a Source of Spatial Environmental Conflict

In 2004 the Brazilian Ministry of Environment (MMA) published the study "Priority Areas for Conservation" (PAC). This study presents the delimitation of the key areas for the implementation of biodiversity conservation, sustainable use, benefit sharing, research, species recovery and economic valuation of biodiversity projects (MMA 2007). The study resulted in maps of priority areas (high, very high or extremely high priority), with the objective of subsidizing the governmental decision related to the licensing of ventures with potential impact on biodiversity, such as hydrocarbon mining and exploration activities.

As seen in the Biodiversity Convention, Agenda 21 and other international acts, referred in Martins et al. (2013), Marine Protected Areas (MPAs) are important because they represent the environment that must be protected in a way that allow it to remain efficient (Prates 2003). Unlike a spatial plan for the use of marine resources, marine conservation plans aim to protect biodiversity (Agardy 2010). Thus, exploration activities cause interferences on this environment, such as: emission of sound pulses through seismic surveys; wells drilling; production and transport of the derived compounds, threatening the marine biota both directly and indirectly. In this way, PACs can be potential sources of conflict, mainly because the delimitation of a geographic area for conservation purposes may overlap areas previously used for other activities or propitious to their development. Exactly for this reason, protected coastal and marine areas are a very representative case for the study of environmental conflict and decision making, especially considering that they are areas with a described location within the coastal and marine environment and thus act as a spatially-defined management tool (National Research Council 2001).

The State of Espírito Santo (ES) was chosen for the GIS tool for application of the conflict identification as a test case. The region was chosen for being an area of recognized biodiversity and critical habitats and, presenting many interests from different sectors including environmental conservation, commercial fishing, port activity, research and structural development. And although the region is relatively less anthropized, it is subject to future population growth and is a good indicator of latent conflicts.

Conflicts were examined in discrete areas located within a geopolitical boundary (Territorial Sea), using GIS in order to map potential competitions between protected areas established in the state of Espírito Santo and oil and mineral exploration activities. Using the method of overlapping interests in space, it will be possible to identify areas with potential of interaction and environmental conflicts. This approach works as a quick tool where it is feasible to identify specific areas of opportunities to maximize joint gains while avoiding threats to the purpose of the protected area.

Thus, the main objective of this study is to evaluate the use of the GIS tool to identify the quantity and types of use of the continental shelf area, mainly in relation to mineral and oil exploration and its possible use conflicts.

1.3 GIS Measurement of Protected Areas in the Context of Global Biodiversity Targets

Protected areas can be seen as indicators of the implementation and achieving of global goals for sustainability, as they play a role in the Millennium Development Goal 7 and Target 9, concerning the topics of ensuring environmental sustainability, integrate the principles of sustainable development into country policies and programs and reverse the loss of environmental resources, respectively.

Protected areas are important indicators of the achieving global sustainability targets, but they only represent an area, so it is necessary to consider how human environment interacts with the space, in order to fulfill conservation objectives of the MPA, and not simply the idea of commitment with the environment (Chape et al. 2005). It is fundamental to analyze that point in order to understand the concerns raised in the development agendas.

The study provides basis for the identification of conflicts for the sustainable management of the exploitation of marine mineral resources, which should include the management of extraction activities aiming at the ecosystem sustainability, acknowledging the existence of areas with greater sensitivity. It is a tool that contributes to reach the Sustainable Development Goals, specifically number 14, about Conservation and sustainable use of the oceans, seas and marine resources.

Conservation activities that balance the needs of both humans and nature will require more information, in much finer detail. The goal of this paper is to analyze the integration of conservation policies and economic interest, by the use of a technique that supports conservation worldwide. Highlighting the global significance of sites for biodiversity can empower local conservation efforts and encourage regional support with the management of the areas.

Science has brought information about the changes in world natural areas and it became clear the need to measure potential use of the environment by human activity. The change in the oceans has been discussed within the Millennium Ecosystem Assessment, which presents the condition of marine ecosystems and analyzes trends of change, as well as in the context of other assessments such as UNEP's Global Environmental Outlook. These analyses are good, albeit incomplete indicators as they do not take into consideration the role of socioeconomic needs when they interact with the engagement of the sustainable development (Stojanovic and Farmer 2013).

The lack of a plan-based approach to management of the spatial ocean areas, the sectorial analysis and permissions of the exploration of marine resources, present challenges in order to deal with the conflicts of interest. The implementation of marine spatial planning would be a strategic, integrated and forward-looking framework for all uses of the sea, as stated by the UK Working Group (2005), and that helps to set the best way to achieve the goals of sustainable development.

Regarding the marine protection and resources, the Chap. 17 of Agenda 21 (1992) has several propositions to help build the structure to management, such as the integrated management instead of sectorial, that will help to reach sustainable use and conservation of marine living resources under national jurisdiction, strengthening international and regional cooperation and coordination. The Convention on Biological Diversity—CBD (2004) and the principles of the Jakarta Mandate also present a perspective of various aspects of marine spatial planning that include the role of MPAs in this process.

2 *Methods:* Study Area

The study area is the marine region adjacent to the State of Espírito Santo comprising the Territorial Sea (region under state management). Under the United Nations Convention on the Law of the Sea (UNCLOS), the territorial sea comprises a sea band adjacent to the State with a dimension of up to 12 nautical miles from the baselines (SOUZA 1999).

2.1 *Prospecting of Data*

A search was made for officially available public data. The georreferenced data, from mineral exploration (open processes) were obtained from the National Department of Mineral Production (DNPM 2017). The georreferenced files with blocks, wells and pretensions of oil exploitation, were obtained on the website of the National Petroleum Agency (ANP 2017). In addition, a map with the Brazilian priority areas for conservation was obtained from the Ministry of Environment website (MMA 2007).

2.2 Overlap Analysis

The ArcGis software (ArcMap 10.1) was used to overlap economic use intentions and the priority conservation areas. Once the databases were gathered, the shapefiles referring to the areas of interest were imported into ArcGis; the APCs that counted in the attribute table with the characterization in high, very high or extremely high; the mineral exploration areas, which had the attributes of the stage of the DNPM processes, the substance to be explored and its use; the oil blocks; and the extension of the Territorial Sea of the ES.

From these files a clip was made, so that only those processes that are within the boundary of the territorial Sea would be represented. Then, an overlapping analysis was performed, through the "tabulate area" function, which resulted in tables with the APC versus phase of DNPM and phases versus substance, showing the result in the form of areas.

3 Results

The Fig. 1 shows the map of multiple uses, superimposed on the map of conservation priority areas. In order to better visualize the processes, we divided the map into three regions that will be called south, central and north only to increase resolution and facilitate visualization (Fig. 2). The interpretation of current and intended uses were made from the available and mapped data of already active extraction areas and those that are still in the process of being activated.

The overlapping areas in the form of tables, DNPM phase with the Priority Areas for Conservation and the PAC with the substances or products to be explored, from which the graphs (Figs. 3 and 4) were assembled, obtained through the map by exporting ArcGIS results into Excel tables.

The studied area has a small representation of oil exploration blocks, which occupies an area of approximately 6 km^2 located in the northern portion of the state as a field of production. As for the phases of the open mining processes, there is a greater amount of areas with authorization for research (approximately 1636 km^2), followed by research request (approximately 405 km^2). The southernmost portion has a high concentration in terms of phases of mineral exploration process, counting on many areas with authorization and request for research, application and mining concession. It is important to point out that there is an extremely high priority area, almost entirely

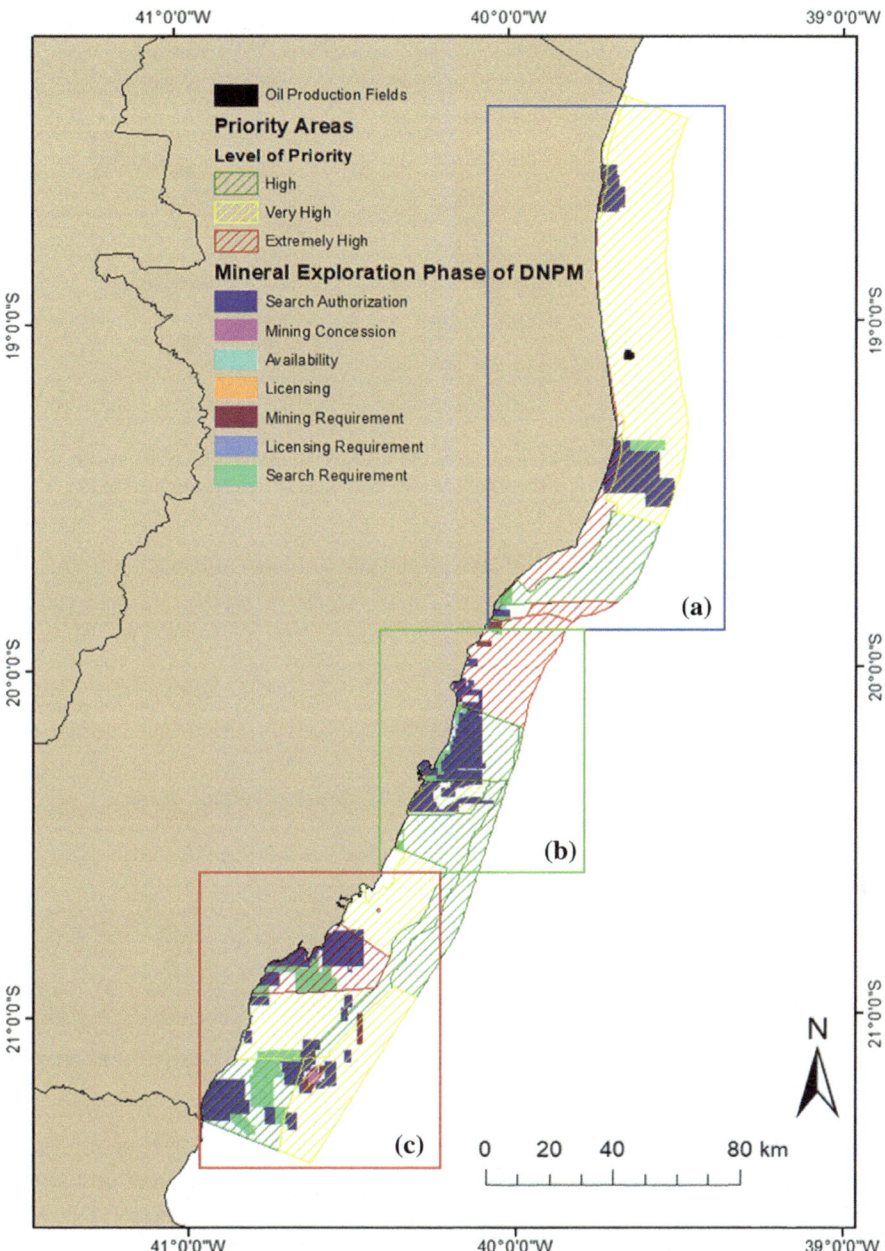

Fig. 1 Map of current and intended Multiple Uses of the ES Territorial Sea focusing on the northern region (**a**), the central region (**b**) and the southern region of state (**c**)

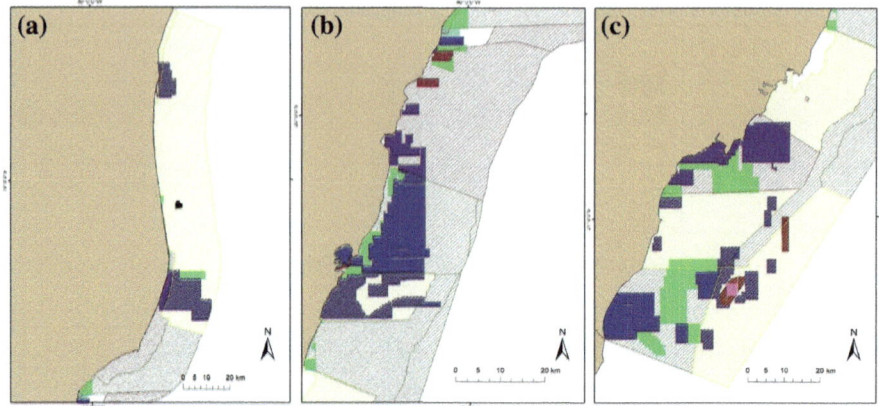

Fig. 2 Map of current and intended Multiple Uses of the ES Territorial Sea focusing on the northern region (**a**), the central region (**b**) and the southern region of state (**c**), as highlighted in Fig. 1

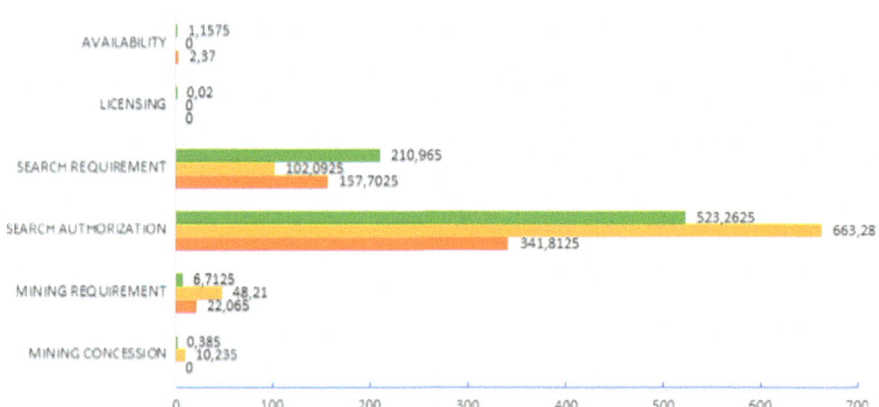

Fig. 3 Correlation between priority areas for conservation and mineral exploration phase DNPM

occupied by regions of pretension of use, with request and authorization of research in the vicinity of Anchieta and Piúma. There are also concessions for mining in a very high priority area at a distance of 25 km from the coast (Fig. 2c).

The central portion (Fig. 2b) presents authorization and research request, mining requirement and availability, being part of an area of extremely high priority. Although small (less than 1%) there is an already licensed area close to the coast, encompassing a high priority area. The availability of coral limestone lies in a place that is partially lacking Conservation Priority Areas, while another part is

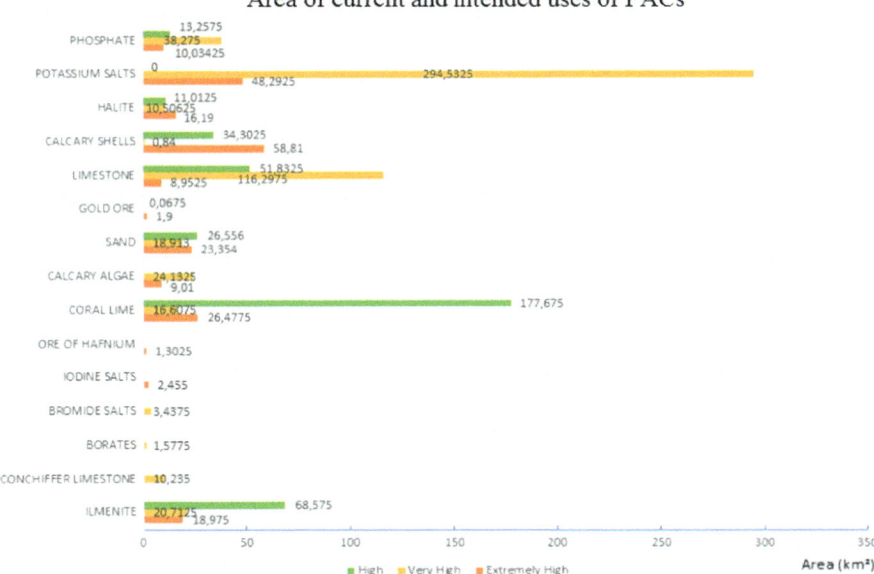

Fig. 4 Correlation of priority areas for conservation with resource to be explored

on an extremely high priority area. There are also research authorization and mining requirements within a region with extremely high priority.

The northern portion of the state (Fig. 2a) also presents authorization and research requirements, as well as an oil field. All of them are located in a very high priority area.

As stated above, the area occupied by the availability and licensing phases are relatively small compared to the others, so that they hardly appear on the chart.

It was observed that in the high priority areas there are more than 200 km^2 with research requests, and around 500 km^2 with research authorization, as well as a small area where there are already licensing processes going on.

The very high priority areas have fewer research requirements, but more than 600 km^2 of research authorization, and are leaders in terms of application and concession of mining, which demonstrates a greater advance of the process. The extremely high priority areas also have application and authorization of research, mining requirements and a small availability.

It is important to highlight a large area that is intended for the exploration of potassium salts in very high priority areas. Also, bioclasts aggregates and rhodolith extraction is observed in high priority areas.

There are several fields with search authorization for potassium salts for industrial use in the northern portion, (Fig. 2a) next to two blocks with research requests. These fields range from the coast to the 12-mile limit, with an area of over 300 km^2 in extremely high priority areas (near the coast) and very high priority areas.

In the northern part of the state (Fig. 2a) there is a concentration of blocks with authorization to search for gem salt for industrial use near the coast, close to 4 smaller areas where there are requirement for the production of bromine and iodine salts as well as borates.

In the southern portion of the state, there are concession for the development of rhodoliths, at approximately 25 km from the coast, with an exploitation area of 10,000 m^2, and concession of ilmenite mining in a high priority area of approximately 400 m^2 near the coast.

The licensing stage is underway for the exploration of sand for civil construction purposes, as well as mining requirements for this product with the same purpose inside the Vitória Bay and near the Tubarão port complex. This region is within the high priority classification, and yet there are several requirements and research authorizations related to sand for civil and industrial construction. Between 3 and 5 km distance from the coast (Fig. 1) there are several research authorizations of bioclasts aggregates for industrial use within high conservation priority areas.

This situation is repeated close to Aracruz and also near Marataízes (Fig. 2b), where there are several authorizations of sand research for civil construction and industrial use, together with ilmenite and gold ore for industrial activity in an extremely high priority and high area.

From Aracruz to the southern end of the state, the adjacent marine region shows an area covering over 200 km^2 having a requirement for phosphate research for use in fertilizers (Fig. 2c). In this same region several fields present search permits for limestone to produce quicklime, occupying approximately 105 km^2 in areas of high and very high priority.

4 Discussion

As expected, the proposition for spatial mapping of the intended mineral exploration sites and PAC in the coastal region demonstrated a high level of spatial conflict, reaching a critical point. Since mineral and oil exploration have great capacity to interact negatively with the environment, they generate significant adverse environmental impacts, altering the natural environment and causing losses to biodiversity. The coastal zone thus becomes a hotspot of conflicts of interest.

It is emphasized that this study is not directed to offer a support system to choose the location of PACs or MPAs but rather to offer a technique that scans the area of interest and identifies the potential conflicts within the landscape. Using GIS to spatially predict these environmental conflicts, as well as locating the best points of intervention, reduces the chances that initial management proposals will evolve into intractable disputes.

Although this study was carried out for the Espírito Santo Territorial Sea, its approach can be applied anywhere in the world, since it is a technique limited only by the spatial data available, not depending on jurisdictional or political boundaries. The technique can be used on a regional scale as in this study but also at global or local scales, depending on the objective. For example, authors such as Brody et al. (2004) performed it for Bay of Matagorda—Texas, where GIS was used to identify conflicts in the Bay region, focusing on conflicts only in the bay region, where it is useful for port implementation decisions or for land use as done by Hamalainen et al. (2001).

When different objectives coincide in the same area, along with the conflict environment-user happen there are also user-user conflicts. Maes et al. (2005) classified the conflicts between users as "manageable in time, space and overlap" or "mutual exclusion", revealing that the approach used in this paper does not represent the whole picture. The paper study (Maes, op cit) does not consider the negative interactions between the uses. Sectorial management states that in each sector the regulation takes place in particular activities or projects considered as independent at a particular location, while the management of areas considers that in a particular area, both sustainable development and use will be established for all activities taking the whole area as interdependent.

The use of the GIS tool was especially interesting in highlighting the concentration of exploration blocks in certain areas that were mainly the purpose of this exploration. It was also crucial to highlight exploration claims within Federal Conservation Units, which have described their objectives, indication of authorization for exploration of goods and compensation for environmental impact through Decree 4.340/2002. The mapping tool makes more evident the need to apply the current legislation, including the need to give the direction of what should be the concern through the data of the resource to be explored.

Understanding exactly the key hotspots where environmental conflicts are most likely to occur in response to the PAC proposal, serves to alert those responsible for formulating environmental policies. This avoids the use of these areas, as well as helps to develop alternative solutions to disputes.

The technique used has limitations such as not including stakeholder views and public opinion regarding the real interest in using the scanned areas; the method depends on the availability of data by public agencies that may have fast-outdating data or may not represent processes if they were not processed by these public agencies. The need for marine spatial planning is related directly with the need for a comprehensible framework allowing decision makers and planners to manage the increasing demand for ocean space, avoiding gaps in the ecologically responsibility of the use of the sea.

The use of local GIS facilitate planning processes that mitigate the interaction conflicts in the marine area, and it is an alert, indicating which companies or people should be listed for surveys or personal interviews to validate the probability of occurrence of conflicts, and how they could be solved, adding usefulness of the mapping techniques.

5 Conclusion

This paper has emphasized the importance of a comprehensive approach to the use of protected areas and it's interaction with economic interests as an indicator for meeting global biodiversity targets. Measurements of numbers and extent of MPA must be combined with the intent of their use, in order to achieve meaningful results.

The GIS tool is useful for highlighting and mapping areas of potential conflict at large scale, facilitating proactive planning of environmental policy management and implementation processes geared towards the mitigation of conflicts. Using this technique to map conflicts in priority areas supports planning in the Territorial Sea and other systems, and is particularly useful in view of future population growth and development.

The tool was very interesting as it allowed the easy visualization of the existing conflicts, considering that the literature review showed that there are difficulties in creating and expanding preservation areas, especially when there are interests in multiple uses of the region. However, the GIS tool is incomplete with respect to giving value to these resources, since the geographic analysis only gives indication of where there are potential for conflicts, without measuring if there is actual interest of a particular group for that area.

The study represents an innovative method while it uses available data using well-known software, facilitating its replication. To replicate the study in other sites it is only necessary to access the database including the polygon boundary information for sites, which can be derived from a broad range of sources. These sources include official government systems, used in this study due to the easy access of such information in Brazil, and reliable secondary sources that include NGOs working on sites or in specific countries.

References

Agardy T (2010) Ocean zoning: making marine management more effective: earthscan. UK, London., p 207

Agenda 21, Chapter 17 (1992) United Nations conference on environment and development

Agrell PJ, Lence BJ, Stam A (1998) An interactive multicriteria decision model for multipurpose reservoir management: the Shellmouth reservoir. J Multicriteria Decis Anal 7:61–86

ANP Agência Nacional de Petróleo, Gás Natural e Biocombustíveis. http://www.anp.gov.br (accessed Oct 2017)

Brody SD, Highfield W, Arlikatti S, Bierling DH, Ismailova RM (2004) Conflict on the coast: using geographic information systems to map potential environmental disputes in Matagorda Bay, Texas. Environ Manage 34(1):11–25

Chape S, Harrison J, Spalding M, Lysenko I (2005) Measuring the extent and effectiveness of protected areas as an indicator for meeting global biodiversity targets. R Soc 360:443–455

Cicin-Sain B, Belfiore S (2005) Linking marine protected areas to integrated coastal and ocean management: a review of theory and practice. Ocean Coast Manag 48:847–868

Convention on Biological Diversity (CBD) (2004) Decision VII/5, App. 3 of the Conference of the Parties to the Convention on Biological Diversity, Kuala Lumpur, 9–20 February. Elements of a Marine and Coastal Biodiversity Management Framework. Marine and Coastal Biological Diversity

Crowder LB, Osherenko G, Young OR, Airamé S, Norse EA, Baron N, Langdon SJ (2006) Resolving mismatches in US ocean governance. Science 313(4):617–618

DEFRA (2007) A sea change. A marine bill white paper. In: Presented to parliament by the secretary of state for environment, food and rural affairs by command of Her Majesty. London

DNPM, Sistema de informações geográficas da mineração. Departamento nacional de produção mineral. http://www.dnpm.gov.br (accessed Oct 2017)

Godschalk D (1992) Negotiating intergovernmental development policy conflicts: practice-based guidelines. J Am Plann Assoc 58:368–378

Hamalainen RP, Lindstedt M, Sinkko K (2000) Multi-attribute risk analysis in nuclear emergency management. Risk Anal 20:455–468

Hamalainen RP, Kettunen E, Ehtamo H (2001) Evaluating a framework for multi-stakeholder decision support in water resources management. Group Decis Negot 10:331–353

Hipel KW, Kilgour DM, Fang L, Peng X (1997) The decision support system GMCR in environmental conflict management. Appl Math Comput 83:117–152

Jankowski P, Nyerges T (2001) GIS-supported collaborative decision making: results of an experiment. Ann Assoc Am Geogr 9:48–70

Maes F, De Batist M, Van Lancker V, Leroy D, Vincx M (2005) Towards a spatial structure plan for sustainable management of the sea. Belgian Science Policy

Martins CCA, Andriolo A, Engel MH, Kinas PG, Saito CH (2013) Ocean & coastal management identifying priority areas for humpback whale conservation at Eastern Brazilian Coast. Ocean Coast Manag 75:63–71. https://doi.org/10.1016/j.ocecoaman.2013.02.006

Mazor T, Possingham HP, Edelist D, Brokovich E, Kark S (2014) The crowded sea: incorporating multiple marine activities in conservation plans can significantly alter spatial priorities. PLoS ONE 9:e104489

Ministério do Meio Ambiente—MMA (2007) Áreas Prioritárias para Conservação, Uso Sustentável e Repartição dos Benefícios da Biodiversidade Brasileira http://www.mma.gov.br/portalbio

National Research Council (2001) Marine protected areas: tools for sustaining ocean systems. National Academy Press, Washington, D. C

Prates APL (2003) Recifes de Coral e Unidades de Consevação Costeiras e Marinhas no Brasil: uma análise da representatividade e eficiência na conservação da biodiversidade. Ph. D. thesis. Universidade de Brasília, Brasília, p 159

Ridgley MA, Penn DC, Tran L (1997) Multicriteria decision support for a conflict over stream diversion and land-water reallocation in Hawaii. Appl Math Comput 83:153–172

Souza JM (1999) Mar Territorial, Zona Econômica Exclusiva ou Plataforma Continental? Revista Brasileira de Geofísica 17(1)

Stojanovic TA, Farmer CJQ (2013) The development of world oceans and coasts and concepts of sustainability. Marine Policy 42:157–165

Susskind L, Mckearnan S, Thomas-Larmer J (eds) (1999) Consensus building handbook: a comprehensive guide to reaching agreement. Sage Publications, California

Tecle A, Shrestha BP, Duckstein L (1998) A multiobjective decision support system for multiresource forest management. Group Decis Negot 7:23–40

UK-MSP Working Group (2005) Added value of marine spatial planning. County Agencies. Interagency MSP Working Group, United Kingdom

Wondolleck J, Yaffee S (2000) Making collaboration work: lessons from innovation in natural resource management. Island Press, Washington, D.C

Challenges and Opportunities Due to Energy Access in Traditional Populations: The Quilombo Ivaporunduva Case, Eldorado—SP

Rodolfo Pereira Medeiros and Célio Bermann

Abstract The aim of our study is to contribute to the achievement of the seventh goal of the Sustainable Development Goals, access to universal, affordable, and clean energy, based on a bottom-up and quantitative-qualitative methodological approach, little used in energy studies. We performed a case study in the Quilombo Ivaporunduva population, in Eldorado—SP, located in the largest continuous area of the remnant of Atlantic Forest in Brazil, a biome with high biodiversity index. Our objective is to understand how the electrification process took place in this traditional population and which are the consequences for the life in the community through productive activities and its culture. Still under development, the preliminary results were produced in the field work, elaborated based on the technique of direct observation and application of a mixed questionnaire. The collected preliminary data indicate that the models of the electrification programs applied to the community over the years did not consider their specificity as a culturally differentiated community, resulting in financial difficulties, misuse of electricity, and unsatisfactory service delivery by the distributor. Thus, we emphasize the importance of rethinking the current electrification program for the constitution of a more effective public policy.

Keywords Energy access · Sustainable development goals · Traditional population

1 Introduction: On Energy Access

According to the International Energy Agency, 1.1 billion people worldwide lack access to electricity and other 2.8 billion people make use of polluting fuels for cooking, which reflects in 3.5 million deaths per year potentiated by indoor air pollution in homes. Of the polluting fuels, 120 million use kerosene, 170 million use coal and 2.5 billion, traditional biomass (IEA 2017).

R. P. Medeiros (✉) · C. Bermann
Institute of Energy and Environment, University of São Paulo—USP, São Paulo, Brazil
e-mail: medeiros.rodolfop@gmail.com

C. Bermann
e-mail: cbermann@iee.usp.br

© Springer Nature Switzerland AG 2020
W. Leal Filho et al. (eds.), *International Business, Trade and Institutional Sustainability*, World Sustainability Series,
https://doi.org/10.1007/978-3-030-26759-9_40

The use of traditional biomass corresponds to the inefficient burning of solid fuels, usually firewood, for generating thermal energy in order to meet the demand for energy services for cooking and heating. This inefficient combustion is performed through inadequate technology and entails a series of health and environmental problems (Karekesi et al. 2005).

Energy access, according to the same agency, refers both to electricity access and to clean cooking facilities, for example, efficient stoves that do not emit pollutants, whether they are of LPG or the modern biomass stoves (IEA 2017).

Thus, these two fields of activity are the basic principle for the establishment and minimum provision of energy services that a population needs.

However, advances in the energy inclusion of these people have been slow and complex, especially regarding clean cooking facilities.

Regarding electric power, the results are positive and significant, as we indicate in Fig. 1.

However, when observing the number of people without access to clean cooking facilities, the scenario is negative. Despite the incorporation of a significant number of people who gained access to clean facilities, the population growth in these regions and the persistent use of polluting fuels made the total number of people without access to clean cooking facilities to remain the same among the years 2000 and 2015. We can perceive this in the following graph (Fig. 2).

Both initiatives of energy access comprise the various benefits that this energy transition entails to contemplated populations. As already indicated in several studies, they directly influence economic, social and environmental issues, for instance: reduction in air pollution; increased efficiency of workload; reduction in physical effort and injuries in productive or collection activities; increase in the level of medical care due to refrigeration and adequate packaging of medicines; reduction in costs

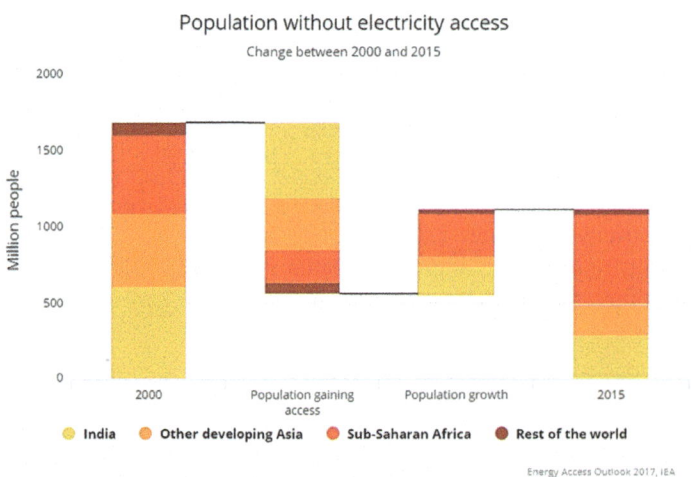

Fig. 1 Population without access to electricity: changes between 2000 and 2015 (IEA 2017)

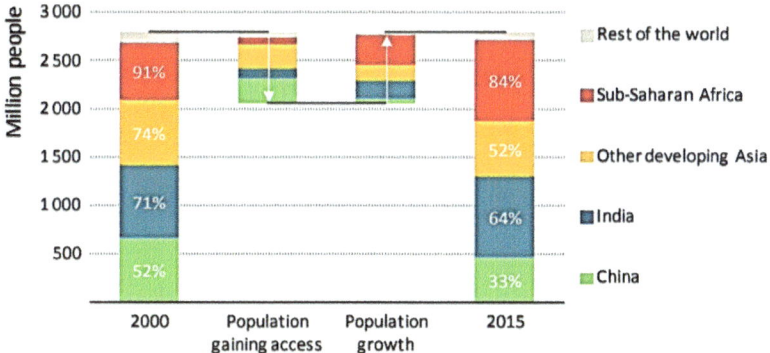

Fig. 2 Number of people without access to clean cooking between 2000 and 2015 (WEO 2017)

in less efficient liquid or solid fuels; reduction in mortality and morbidity; alleviation of gender inequality; reduction in domestic tasks; access to education; increase in the time available for other activities for women; and environmental improvement through the reduction in the emission of greenhouse gases with the use of firewood and deforestation for the collection of firewood, as in the case of Africa. (Sovacool 2012; Kumar 2018; Bhattacharyya 2012; Sokona et al. 2012; WEO 2017). In Chart 1 we present these and other benefits in a synthetic way.

Within this context, it seems undeniable that energy transitions always have a positive effect on the lives of those contemplated by the access to them.

However, when focusing on specific case studies, we observe that not everything works in the harmonious way as expected.

Geographical issues, social organization, economic and cultural dynamics tend to make the energy inclusion task increasingly complex. Stretching a transmission line and connecting it to a residence, for example, does not seem such a complicated task, requiring at most some engineering and economical adjustments, explaining how it will take place and who will pay for the expansion.

However, realizing how the beneficiary reacts to this access, how this is reflected in their day-to-day life and if it really incorporates all the imagined benefits is an arduous task. First, energy comprises moral and cultural values, topics rarely addressed in studies on the energy area.

Sovacool (2014) made a survey on production areas, followed methods and the background of authors who wrote in the main journals on the energy area between 1999 and 2013, demonstrating that 85% of the studies are related to technology and decision-making, mainly supported by quantitative methods (58%) and implementation of models. Only 4.3% of the studies focused on the users' perceptions and behaviors. Therefore, there is a lack in social studies on energy, leaving a huge gap in neglected issues that are fundamental when thinking about sustainable development goals and expected results.

Chart 1 Scope and consequences of the benefits of access to energy

Consequences	Scope			
	Poverty and productivity	Health	Gender and education	Environment
Benefits	Mitigates poverty scenarios, positively influences productivity	Reduction in internal pollution and GHG emissions, reduction in injuries during collection of water and firewood, conditioning and sterilization of hospital supplies, hygiene care	Education of youth and adults, extension of study hours, reduction in absenteeism due to diseases, gender equality	Reduction in deforestation and GHG emissions, increasing soil fertility by reducing the use of food and its waste for combustion
Ways	Economy with fuel, collection equipment, processing and product conditioning	Use of efficient fuels, water pumping, refrigeration, information and communication (radio and television)	Artificial lighting, access to information (radio, television and the Internet), reduction or cessation of water and firewood collection activities (performed by women and children)	Use of modern fuels for cooking and heating

Analyses considering a wider range of areas of knowledge, in addition to key questions such *how? what for? whom for?* (Bermann 2007) should be incorporated and guide the studies within the seventh sustainable development goal.

2 From Energy Access to the Cultural Incorporation of Modern Energy

In this section, as indicated by its title, we sought the change of perspective of the paradigm of energy access. This change will be guided by the transfer of the analysis focus from the top-down approach to the bottom-up, that is, henceforth the focus of analysis will be the actors involved, the study community and not the institutional perspective that seeks to impose its way of being or acting on the population.

This is because every change really occurs only when it is incorporated by the actors, by the culture. Nothing that is imposed on a population will truly result in what was expected. In addition, a more effective policy concerning its goals only succeeds when we recognize its target audience, virtue aspects and vulnerabilities.

Thus, the energy transition concept will be based on the multiple fuels or energy stack perspective, rather than on the traditional energy ladder concept.

The energy ladder comprises a linear and positivistic view of the energy transition in which as the economic condition of a population increases, it tends to migrate to efficient and modern fuels, abandoning other forms of energy use (van der Kroon et al. 2013). This type of approach is simplistic and disregards the receptivity or not of the actors involved in the process, cultural aspects and the choice of each actor, tending to impose a world view and a majority practice on the others.

For top-down models the energy ladder can be well-accepted and can usually satisfy the problems encountered in this order of magnitude. However, for bottom-up models, it is insufficient and fails to explain locally-mediated events by actors and culture, often contradictory to hegemonic reasoning.

The energy stack approach can explain that although income influences the choices of the used energy or fuels, there are other factors that prevent the transitions from being linearly demarcated. This approach comprises a more dynamic perspective on energy transitions, which are locally defined by different criteria that may indicate the return to a type of fuel no longer used, for example due to some cultural or socioeconomic factor (Masera 1997) (Fig. 3).

Working from the perspective of the energy stack model, understanding the various non-economic factors that influence the decisions of end uses of energy is paramount when it comes to specific experiences of energy access and their consequences to the affected population.

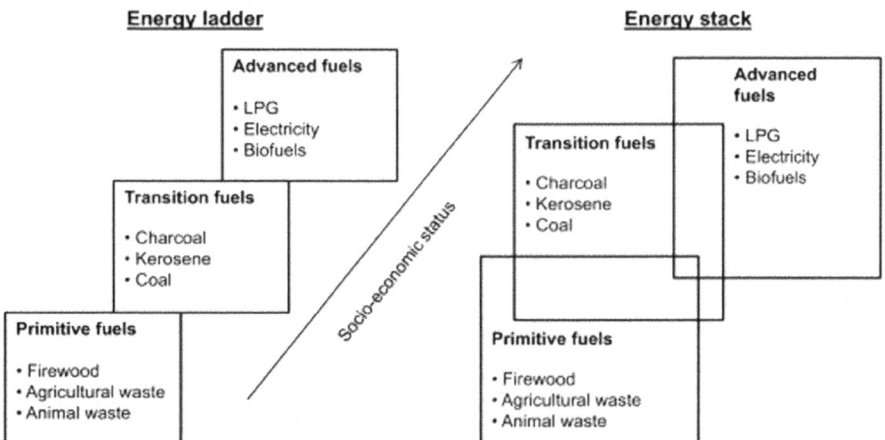

Fig. 3 Differences between the energy ladder and the energy stack models (van der Kroon et al. 2013)

In order to better measure the effects of top-down policies on energy access at the daily level of populations, we must follow qualitative or quantitative methods that grasp the repercussions of these energy transitions on the communities.

Thus, more than the call for the expansion of non-economic social studies on the field of energy, we must also adopt qualitative methods, characteristic of the human sciences, which allow the analysis of cultural factors and social organization in the populations contemplated by the policies on access to energy.

Therefore, unlike the unrestricted beneficial effect on the populations contemplated by such policies, we seek to better discriminate the subtle nuances identified in the contradictions of this process that is also marked by adversity.

3   Traditional Culturally Differentiated Populations and the Incorporation of Electrical Energy

Traditional populations consist in communities that have historically been distant from the urban or rural development idealized by the values of modernity. Thus, they can be understood as populations of rustic cultures, with sharing of knowledge based on orality, differentiated in the way of relating to nature, according to which man integrates nature and such is governed by cosmology (Arruda 1999).

As nature is the very essence of its existence, the places inhabited by these populations are environmentally conserved, and their mode of social reproduction is perhaps the closest to what the contemporary society seeks as a sustainable development.

Traditional populations differ among themselves concerning their ethnic origin and where they live, thus varying some of their cultural practices, but without losing this close relationship with the environment. Hence, in Brazil there are several traditional populations that inhabit its territory, ranging from *caiçaras* (near the coast), *ribeirinhos* (near the rivers), and *quilombolas* (descendants of slaves), among others.

Throughout their existence, and more frequently in recent decades, these populations and their territories have been subjected to political, ideological and economic pressures. Discourses to invalidate their history, way of living and territory tackle from legal questions about the possession of the land, to ideological ones such as their economic unproductiveness, from the hegemonic perspective. They have even been pressured by environmental organizations, which have conceived their existence as a threat to preservation units up to date (Silva Pimentel and Ribeiro 2016).

Without legal support for a long time, these traditional populations needed to get politically organized to face these constant threats to their culture and territory, and to claim their rights. To illustrate that, the decree on traditional peoples and communities was only enacted in Brazil in 2007, which establishes the National Policy for the Sustainable Development of Traditional Peoples and Communities, and its principles and objectives are strongly linked to the recognition and "guarantee of their territorial, social, environmental, economic and cultural rights, with respect and valuing to their identity, their forms of organization and their institutions" (Brasil 2007).

Among these rights, one is the right to energy, as indicated by Camargo (2015), which demonstrates its recognition of as a public service indispensable to the population.

As a result, traditional populations began to be electrified due to public policies, by the federal program *Luz para Todos* [Light for All], which aims to guarantee universal access to energy in Brazil.

However, accordingly, the Light for All program is characterized as a top-down policy and, therefore, it may produce contradictory effects in its achievement.

Hence, our study comes from a Master's research in which we seeks to understand the repercussions of the incorporation of electrical energy in the Ivaporunduva *quilombola* community, from Eldorado—SP, Brazil. The results are based on the field work carried out in the community by direct observation methods and the application of a mixed questionnaire (quanti-qualitative).

Direct observation of the field was recorded from a field diary and photographs, and we aimed to witness the daily life of the community concerning its energetic uses, evidencing and trying to respond to the categories: (i) subject—*who?*; (ii) scenario—*where?*, and (iii) behavior—*how?*.

The mixed questionnaires were applied to a sample of 26% of the community residences, spread across the various sectors (areas) that compose Ivaporunduva. The structured questionnaire aimed at obtaining quantitative data on equipment, prices and frequency of use. The open questionnaire focused on obtaining qualitative data regarding the behavioral or cultural changes observed by the actors themselves during the electrification period.

4   The Quilombo Ivaporunduva

The Quilombo Ivaporunduva rural traditional community is located in Southeastern Brazil, in the Vale do Ribeira region, composed of the states of São Paulo and Paraná. Its creation dates from the end of the seventeenth century to the beginning of the eighteenth century, when abandoned slaves of a deceased farmer and other fugitive slaves from the vicinity gathered in the region (Fig. 4).

The Vale do Ribeira region has two outstanding characteristics: first, it is the poorest region of the state of São Paulo, with low rates of economic and social development. Second, it is the region with enormous social and environmental diversity, both due to the largest continuous area of the Atlantic Forest in the country, accounting for 23% of the total remaining, and to the cultural diversity of the traditional populations that live there.

The river *Ribeira de Iguape*, which names the valley, has already been the target of hydroelectric projects that would affect all of this socio-environmental wealth, especially the *Tijuco Alto* project, which had been considered by environmental organizations since the end of the 1980s and was only actually closed in November 2016 (ISA 2016).

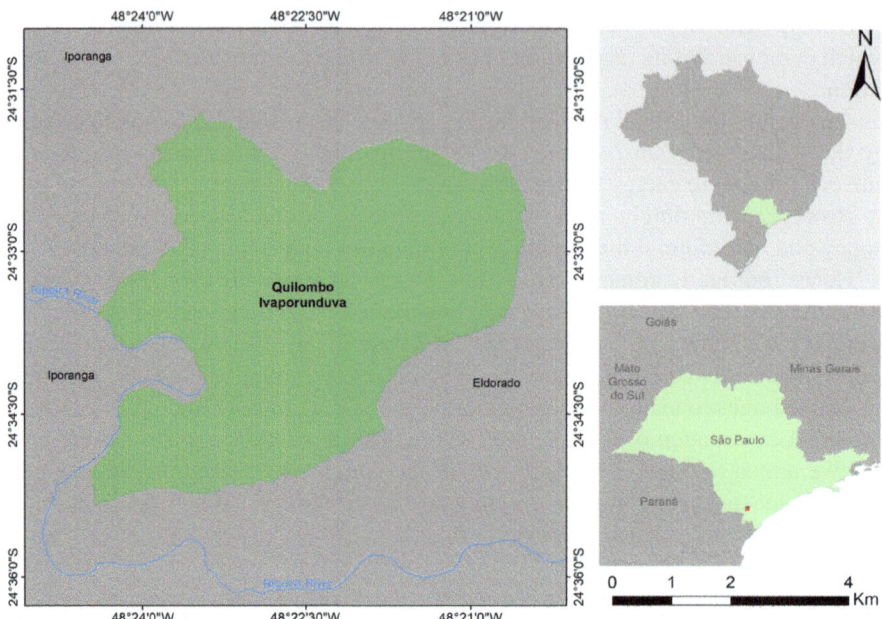

Fig. 4 Location and area of Quilombo Ivaporunduva (Luís Campanha, 2019)

During all this time, the quilombola communities of the region became politically involved against the project, and this struggle resulted in the creation of the
Movement of the Threatened by Dam (*Movimento dos Ameaçados por Barragem—*
MOAB). This articulation united the communities, conferring political unity and
strengthening their organization, which started to claim their rights, mainly those
concerning identification, recognition, delimitation and titling of the occupied lands.

Land titling is a fundamental step for these communities, which in turn will have
greater legal support to claim their other rights.

The recognition of remaining quilombola communities was only ceded by ITESP[1]
in 1998, being its first titling of lands (672 ha) from 2003. In 2010, INCRA[2] completed
the titling of lands registering another 2032 ha (Frizero 2016).

Still, according to Frizero (2016), there are 400 residents in Ivaporunduva, divided
into 110 families, occupying an area of 2754.36 ha. All these lands are in the name of
the *Associação do Quilombo Ivaporunduva* [Quilombo Ivaporunduva Association]
and belong to all the residents.

Electric power, according to the residents, arrived in the community in 1989, being
all its structure of posts and materials brought by themselves, with ferry through the
Ribeira river. This first electrical installation only benefited houses located around

---

[1] *Instituto de Terras do Estado de São Paulo* [Land Institute of the State of São Paulo]. State agency
responsible for the recognition of quilombos and land regularization.

[2] *Instituto Nacional de Colonização e Reforma* Agrária [Brazilian Institute of Colonization and
Agrarian Reform]. Federal agency responsible for agrarian reform and land tenure.

the chapel, central landmark of the community, and was financed by the city hall itself.

Between 1989 and 2005, there were occasional initiatives of electrification in the community by public policies such as the *Luz da Terra* [Light of the Land] program of the state government of São Paulo. This program subsidized the purchase of the necessary equipment for electrification, although still charging the population for such service.

Until 2005, less than 30% of households had access to electricity, the date in which the Light for All program was implemented in the community. Through the initiatives of such program there was universal access to electric power in the community.

Thus, electrification in the community was carried out over a long period of time and stimulated by several agents, which confers the analysis a comparison characteristic.

Electrification carried out by the city hall reached a very small number of residences, although there was no cost to the beneficiaries. It was an isolated action and did not have continuity, which is very common to initiatives of political interest, used at the end of official terms in Brazil.

Electrifications carried out by the Light of the Land and Light for All programs consist in public policy initiatives.

The Light of the Land was one of the first electrification programs in the country with a social inclusion bias, perceiving energy as a right. Until then, initiatives had a strictly economic perspective, both to the beneficiary and to the company (Ribeiro 2003).

However, despite the provided subsidy, beneficiaries themselves paid for the equipment, in significant amounts for this low-income population. Thus, in addition to the energy bill that arrived at the end of the month, the installments of equipment were added, which heavily weighted in the family budget.

Light for All is still the main electrification program in the country and has served 16 million people since its beginning in 2003 (MME 2017). It is linked to other social government programs and perceives energy as a fundamental right. During its implementation, beneficiaries have no cost of equipment or installation, which is fully subsidized by the federal government and the company.

It works from the perspective of sunk costs, according to which the State takes responsibility for the eventual economic deficit generated by the expansion/return ratio.

Most of electrification in the community was done through this program, but recently, there was a lot of delay in the service for new connections. This delay, in addition to the urgency of applicants to have electricity, makes some people prefer to bear the costs of the equipment and ask the company only to install it.

Overall, the Light for All program was the one that best served the community's longings both for serving all places, even the most remote ones, and for the gratuity of the service. This became clear during the experience and conversation with the residents during the field work. In the questionnaire sample, the following result was obtained regarding those responsible for electrification and the need for having their own resource for the installation (Figs. 5 and 6).

Fig. 5 Responsable for
electrification. *Source*
Authors

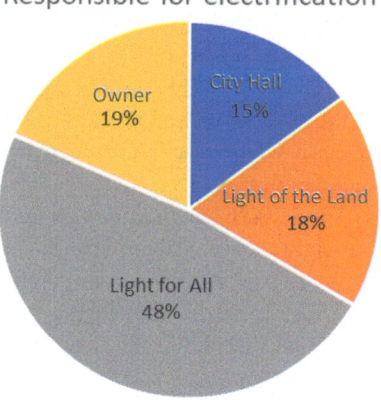

Fig. 6 Residence
responsible for the
installation payment. *Source*
Authors

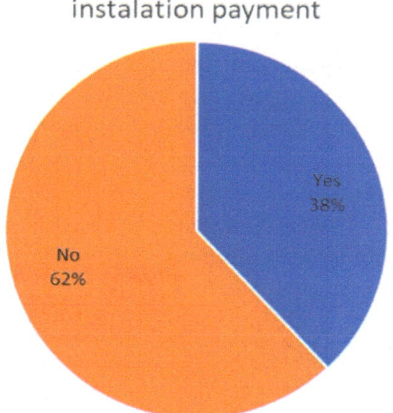

Regarding the access to clean cooking facilities, the use of the wood stove is still common.

Traditional mud houses, built of wood, liana and clay, remain attached to the new masonry residences. In such houses there are wood stoves, which, for cultural reasons, continue to be used mainly for cooking hard grains, such as beans, and more elaborate and traditional meals claimed to be tastier when cooked in these stoves.

LPG stoves, for families that prefer to use firewood stoves, are only used for quick meals, such as breakfast, or to heat some food, as well as on rainy or cold days, and residents do not need to leave the masonry house. In the next figure we show the preference of the stove type for meal preparation (Fig. 7).

Fig. 7 Energy stove
preference. *Source* Authors

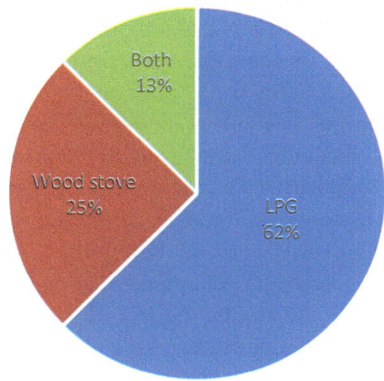

Recently, with the successive and significant increase[3] in the prices of LPG bottles in Brazil, due to an economic policy, many families said they would resume the use of wood stove, since it does not result in increases in the budget of the house, only being necessary to collect firewood in the near forests for its operation.

5   The Challenges

The weak point of all electrification programs, especially the Light for All, due to its scope, is that many of its works are done with wooden posts, which, by the current legislation, does not comply with the new technical standard, according to which such work is done with concrete posts. Thus, in the eventual needs of calling the company to carry out maintenance or solve any problem, such may refuse to attend residents due to the irregular situation of the property. Many residents, in order to regularize their situation, are buying concrete posts and having to bear these costs.

Another challenge faced by the residents with the implementation of electrical energy is related to the cultural condition.

Since this is a population that has always had their resources in nature, they have never had to pay bills for provided services. That is, the consumed water, the wood removed from the woods, the place of habitation, nothing was paid for.

With an electric power input, customers had direct access to an individual energy payment policy, generating much confusion about the individual bills as well as situations of economic defaults that suspended the electricity service.

---

[3]Between September 2017 and November 2018, there was a 30% increase in LPG refineries. The transfer of the increase to consumers is around 15%. These increases are related to the adoption of Petrobras pricing policy to keep pace with the dollar and oil prices in the international market.

Furthermore, habits of use and the efficiency of electronic devices were not considered in the electrification of the community, and a great energy waste was noted in the residences. The lights and televisions are switched on all day and are observed and reported as usual by the residents.

As for the light issue, the situation of public lighting provided by the distributor, which is low and of poor quality, should be mentioned. Thus, it has become a practice among residents to let outdoor lamps in their homes switched on all the time in order to meet this demand. Such practices are detrimental to the residents.

Electricity has also included households within the world of consumption, which are now stimulated by all modern media to consume all types of products, creating new needs, causing them to lose their budgets.

Regarding reductions that can be applied to tariffs, low-income families, which are in the unique register of social programs, can adhere to the social tariff of electrical energy. Such grants discounts, depending on the range of consumption of the residence, can be free of charge in the case of quilombolas with consumption of up to 50 kWh/month (Brasil 2010). The field survey indicated that only 28% of respondents are within this category.

Residents are unaware of their right to recover tariffs at the end of the month when there is a suspension of the energy service, as well as compensation for damages caused to equipment or loss of products due to this (ANEEL, RN. 414, Articles 152 and 204 2010).

During the rainy season, the community reports power interruptions that last from two to three days. Such events are usual due to the lack of maintenance and the delay of the company personnel service; however, they are not considered in the final value of the bill.

Finally, concerning wood stoves, it is very difficult to change the population's habit of using it because it is related to the history and culture of the community, tied to subjective issues of personal taste. In addition, logistic issues in the distribution and economic policies on LPG tend to make the transition to this fuel even more complex.

6 Opportunities

Access to electricity created several opportunities. It enabled to create a primary school and a healthcare center in the community.

In addition, it enabled to access information through television and, more recently, through the Internet. The latter, along with smartphones, favored the communication among the residents through instant messaging applications. The mobile network does not cover the quilombo area and the community has only two fixed lines, one in the central region and another at the *Centro de Visitantes* [Visitors' Center], with communication totally dependent on electricity. Reports of improvement in communication are recurrent, especially by residents of more remote sectors who indicated important health issues, reports of death and possible accidents or transportation problems within the community.

From the economic point of view, it enabled the diversification and improvement of the quality of products and provided services. The community's main products are bananas and heart of palm. With the electrical energy and resources provided by public bids, the community built a Banana Factory and a Visitor Center, guaranteeing other sources of income.

The banana factory, although not yet in full operation, already ensures autonomy and improvement in the production of banana by the air conditioning, equipment that accelerates the process of the fruit maturation. Thus, according to residents, it was possible to increase the value of transferring to the producer by 1000%, disregarding the action of *atravessadores*.[4]

The Visitor Center allowed the environmental tourism in the community, guaranteeing lodging and food for tourists as well as employing residents as environmental monitors, cooks, janitors, among other functions. According to reports, tourism has been the activity that guarantees greater economic return to the community, more than banana production itself.

Wood stoves, during the period in which the price of LPG has increased, consisted in a viable technology option, alleviating the household budget. Thus, improving its construction to better yields and to generate less pollutants seems to be a good alternative. Hence, its cultural value will also be preserved. Eventual constructive restructurings could take advantage of the heat generated in the process to warm the bath water, reducing the demand of the main consumer of energy in Brazilian homes, the electric shower.

7 Final Considerations

Traditional populations, unlike other types of rural communities, have their own cultural characteristics and are susceptible to the significant changes caused by the process of energy access.

Despite the benefits sought by the policies on access to electricity, or the insertion of clean cooking facilities, our case study demonstrates that this process presupposes a series of difficult elements to be considered in their accomplishment, which may eventually have an opposite effect than the expected.

All the issues we mentioned suggest problems the residents must face themselves, making them socially and economically vulnerable.

Cultural mediation during the energy transition process of the community, conducted by specialists who understand their culture and demands, would make this process more coherent and successful.

---

[4]*Atravessadores* are marketing agents who act as intermediaries between producers and consumers, paying very low prices to producers because of their inability in bringing their products to the market.

Without intending to superficially criticize the results achieved by the Light for All program, which played a fundamental role in guaranteeing the right to electricity access for the population, our research sought to add to the studies and goals of the sustainable development by indicating points for reconsideration in the public policies on access to energy, in such a way they can be more assertive in their implementation.

References

ANEEL—National Agency for Electric Energy—Normative Resolution No. 414. Establishes the General Conditions for the Provision of Electric Power in an updated and consolidated manner (2010)

Arruda R (1999) Traditional populations and the protection of natural resources in protected areas. Environ Soc, Year II, no. 5

Bermann C (2007) Impasses and controversies of hydroelectricity. Adv Stud 21(59)

Bhattacharyya SC (2012) Energy access programs and sustainable development: a critical review and analysis. Energy Sustain Dev 16(3):260–271

Brasil (2007) Decree No. 6040. Institutes the national policy for the sustainable development of traditional peoples and communities

Brasil (2010) Law 12.212. Provides for the social electricity tariff

Camargo EJS (2015) Light for all program: in search of a self-sustaining state policy. Thesis of Doctorate in Science, PPGE. IEE/USP, São Paulo

Frizero MG (2016) Quilombo Ivaporunduva. Belo Horizonte: FAFICH 16 p.: il. (Quilombos lands) Based on scientific technical report of the Ivaporunduva Community, by Cleyde Rodrigues Amorim

IEA (2017) © OECD/IEA 2017, Energy access outlook: from poverty to prosperity. IEA Publishing. Licence www.iea.org/t&c

ISA (2016) Socioenvironmental Institute. Ribeira Valley population is free of Tijuco Alto. November 2016. Available from https://www.socioambiental.org/en/noticias-socioambientais/populacao-do-vale-do-ribeira-esta-livre-de-tijuco-alto. Accessed on 26 Dec 2017

Karekesi et al (2005) Status of biomass energy in developing countries and prospects for international collaboration. In GFSE-5 Enhancing international cooperation on biomass. Background paper. Austria

Kumar A (2018) Justice and politics in energy access for education, livelihoods and health: how socio-cultural processes mediate the winners and losers. Energy Res Soc Sci 40:3–13

Masera O, Navia J (1997) Fuel Switching or multiple cooking fuels? Understanding inter-fuel substitution on patterns in rural Mexican households. Biomass Bioenergy 12(5):347–361

MME—Ministry of Mines and Energy (2017) National program for the universalization of access and use of electric power operation manual for the period 2015 to 2018 rev. 1

Ribeiro FS et al (2003) "Light of the land" program—participatory rural electrification model. Annals of the 3rd energy meeting in the rural environment

Silva Pimentel MA, Ribeiro WC (2016) Traditional populations and conflicts in protected areas. Geousp—Space Time (Online) 20(2):224–237

Sokona Y et al (2012) Widening energy access in Africa: towards energy transition. Energy Policy 47, Supplement 1:3–10

Sovacool BK (2014) Diversity: energy studies need social Science. Nature 511:529–530

Sovacool BK (2012) The political economy of energy poverty: a review of key challenges. Energy Sustain Dev 13(3):249–388

van der Kroon et al (2013) The energy ladder: theoretical myth or empirical truth? Results from a meta-analysis. Renew Sustain Energy Rev 20:504–513

WEO (2017) © OECD/IEA 2017, World energy outlook. IEA Publishing. Licence www.iea.org/t&c

Global Sustainable Transportation, Construction and Infrastructure

Collaborative Outsourcing
for Sustainable Transport Management

Marzenna Cichosz, Katarzyna Nowicka and Barbara Ocicka

Abstract The chapter aims to investigate vertical collaboration focused on making transport of chemical products more sustainable and safer by shifting from road to multimodal. This collaboration on transport planning, energy and emission management is supported by technological tools (Intermodal Links Planner and CO_2 calculator) developed within ChemMultimodal Project. The research problem is analyzed based on literature review and empirical research on collaborative outsourcing in chemical industry. The main findings show that close collaboration between chemical companies and LSPs has a strategic importance for the sustainable transport development. On the one hand, the role of LSPs, as architects of transport processes as well as integrators of transport modes and connections, is crucial for rising eco-efficiency. On the other hand, chemical companies view their LSPs as critical partners in business development and supply chain management and are willing to share their plans to achieve competitive advantage. Both parties are expected to invest in the collaborative relationships that is strongly supported by technological tools for sustainable transport management. The preliminary results show that based on close collaboration companies are able to reduce CO_2 emission in transport activities almost by 60%.

Keywords Collaboration · Outsourcing · Sustainability · Multimodal transport · CO_2 emission · Chemical logistics

M. Cichosz
Institute of Infrastructure, Transport and Mobility, SGH Warsaw School of Economics, Warsaw, Poland
e-mail: marzenna.cichosz@sgh.waw.pl

K. Nowicka · B. Ocicka (✉)
Department of Logistics, SGH Warsaw School of Economics, Warsaw, Poland
e-mail: barbara.ocicka@sgh.waw.pl

K. Nowicka
e-mail: katarzyna.nowicka@sgh.waw.pl

© Springer Nature Switzerland AG 2020
W. Leal Filho et al. (eds.), *International Business, Trade and Institutional Sustainability*, World Sustainability Series,
https://doi.org/10.1007/978-3-030-26759-9_41

1   Introduction

Sustainability has been spread worldwide and is currently considered to be one of the most important themes to emerge not only in production, but also in other parts of supply chain management, i.e. in logistics operations. The key logistics process, which creates time and place utility, is transport management. Due to globalization, in most of the cases transportation expands to international or global routes what results in the growth of the distance as well as the volume of freight transport and causes problems when it comes to accommodating product flow in an efficient and sustainable way.

One of the biggest challenges is a very high share of road transport within inland flows. According to Eurostat, in European Union countries in 2014 (EU-28) almost 75% of total inland freight transport was done via road, i.e. four times more than via rail (18.3%) (Eurostat 2016). The European Commission (EC) warns that this imbalanced modal split with road haulage domination has its negative effects, which cost the EU more than 250 billion EUR annually (Woxenius and Barthel 2013), of which half relates to congestion and longer delivery times. Additional causes of this problem are environmental deterioration (by CO_2 and NOx emissions) and the inefficient use of energy, as well as social costs of road accidents and noise (Piecyk and McKinnon 2007).

According to a recent global Annual Third-Party Logistics Study, transport is the most frequently outsourced logistics activity (Langley 2018). Moreover, the study shows that logistics service providers (LSPs) continue to move away from primarily transactional relationships towards meaningful partnerships. Thus, collaborative relationships between LSPs and their customers, when designing, implementing and controlling eco-efficient and socially responsible transport network, are central to the sustainable transport management success of both, LSPs and their customers.

The chapter aims to investigate vertical collaboration focused on making transport of chemical products more sustainable and safer by shifting from road to multimodal. This collaboration on transport planning, energy and emission management is supported by technological tools (here: Intermodal Links Planner and CO_2 calculator developed within ChemMultimodal Project). This research problem combines literature review on logistics outsourcing, and collaboration with empirical research on collaborative outsourcing in the chemical industry.

2   A Framework for the Collaborative Logistics Outsourcing Development

Logistics outsourcing, termed also as "third-party logistics" (3PL) or "contract logistics", refers to the organizational practice of contracting out part of or all logistical activities, which had previously been performed in-house, to specialist providers

(Bowersox 1990; Lieb 1992; Virum 1993; Skjoett-Larsen 2000; Selviaridis and Spring 2007).

The degree of logistics outsourcing varies, and outsourced activities differ in complexity. Delfmann et al. (2002) identify three types of LSPs regarding their service offerings. The first group provides standardized and isolated logistics services such as e.g. transportation, warehousing. The second group provides bundled service, i.e. the standard core logistics services combined with value-added logistics services according to the customers' demand. The third group offers customized logistics solutions responding specific customer's requirements. An example of advanced customizing LSP is the fourth-party logistics (4PL) provider, who assembles the resources, capabilities and technology of its own organization and other organizations to design, plan and run comprehensive supply chain solution (van Hoek and Chong 2001).

Contemporary logistics outsourcing arrangements are based on formal (both short- and long-term) contractual relations as opposed to spot purchases of logistics services (Murphy and Poist 1998). Companies entering into logistics outsourcing face the choice of developing a transaction-based or a more cooperative relationship-based arrangement. The choice of latter is typically made because of perceived benefits, which include the synergy gained through shared expertise and resources, exchange of information, better planning and support, and joint problem solving (Stank et al. 1999).

Daugherty (2011) investigates the spectrum of logistics outsourcing relationships and in accordance with channel literature places them on the governance spectrum. At one extreme she puts arm's length transactions (market-governed), at the other —vertical integration (hierarchical-governed). In between the two there is coopera- tive relationship (hybrid-governed). For the cooperative relationship Lambert et al. (2004, p. 24) use the term partnership and divide it into three types of partnerships according to partners' involvement into coordinating and planning activities. The authors emphasize that "because each relationship has its own set of motivating fac- tors driving its development as well as its own unique operating environment, the duration, breadth, strength and closeness of partnership will vary from case to case and over time." Świtała and Klosa (2015) also grade inter-firm cooperative relation- ships on a scale from cooperation, through coordination, up to collaboration when collaborating companies treat each other as an "extension" of their organizations.

Recently, more and more researchers investigating supply chain relationships, apply the term of collaboration for "a high level of integration between two companies working together to create a competitive advantage and higher profits than could be achieved by operating alone" (Soosay and Hyland 2015, p. 161). Hartmann and de Grahl (2011) emphasize that collaboration could also be considered as a unique dynamic capability. Fawcett et al. (2011) see collaboration as a strategy.

The focus of this chapter is on collaboration between an LSP and its client within outsourcing relationship. The competences of LSPs make them an attractive partner for logistics collaboration (Cichosz 2017). As firms collaborate, they develop con- nections, especially through information sharing for joint planning (Cichosz et al. 2018) and problem solving what finally has a positive effect on the supply chain

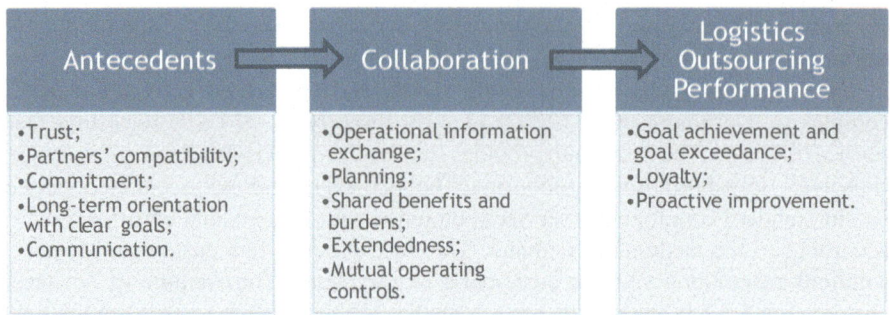

Fig. 1 Framework for collaborative logistics outsourcing

performance. Min et al. (2005), based on personal interviews, have developed conceptual model of supply chain collaboration covering antecedents of collaborative relationships, collaboration itself, and consequences.

We used the framework (antecedents-collaboration-performance) to structure our analysis in logistics outsourcing (Fig. 1). We took interorganizational vertical collaboration perspective i.e. 3PLs and their customers. Base on the literature reviews, we formulated the list of the antecedents embracing: trust (e.g. Moor 1998), partners' compatibility (Whipple and Frankel 2000), commitment (e.g. Wallenburg et al. 2011), long-term orientation with clear goals (e.g. Hofer et al. 2012), and communication (e.g. Rollins et al. 2011).

After Hofer et al. (2009) we applied set of five behavioural dimensions, which institute successful LSP-client collaboration:

- operational information exchange;
- planning;
- shared benefits and burdens;
- extendedness; and
- mutual operating controls.

And finally, regarding the logistics outsourcing performance, we stressed goal achievement and goal exceedance (e.g. Deepen et al. 2008; Hartmann and de Grahl 2011), loyalty (e.g. Wallenburg et al. 2011), and proactive improvement (e.g. Hofer et al. 2012) aspects.

Additionally, there are different reasons for logistics collaboration. According to Cruijssen (2012, p. 31), LSPs collaborate vertically and horizontally as well, mainly for cutting costs, enhancing customer service and improving efficiency. To achieve eco-efficiency in transport companies work on: transport planning, reducing empty movements, increasing loading factor, stimulating multi-modality, reducing transport externalities such as greenhouse gas emissions.

3 Service Capability and Performance of LSPs in the Chemical Industry

Before addressing specific transport market trends and tailored solutions for the chemical companies, it will be helpful to outline what is currently going on between supply and demand in the global chemical industry. World chemicals turnover reached the value of 3360 billion EUR in 2016 (Cefic 2017, p. 5). A growth of 4.5% per year up to 2030 is predicted for the global chemical industry (Leker et al. 2018, p. 17). Most important geographical regions of chemical sales are nowadays: China (1331 billion EUR), NAFTA countries (528 billion EUR), and the EU (507 billion EUR) (Cefic 2017). Among different industries, the chemical sector generates significant demand for transport services globally, that might be characterised by long distances, huge cargo volumes and regular links in supply chain networks. There is a need to manage trade-offs between different issues in transport operations, like e.g. reducing CO_2 and simultaneously, to optimize transport costs and ensure expected service level in supply chain management.

Following the vision and trends of the chemical industry development till 2020, logistics management in this sector will face significant challenges, like: longer and complex supply chains from suppliers via producers to customers, higher logistics costs and constant pressure on transport capacity, shifting power between shippers and LSPs, increased regulation, focused on emission reduction and improvements in safety and security standards, demands for a responsible and sustainable approach to business (Cefic and Deloitte 2001, p. 3). The opportunity for chemical production and trade flows growth is unlikely to reach its full potential without addressing challenges in transport processes and infrastructure. Based on the U.S. research findings, logistics shortcomings across primary transport modes will greatly affect the chemical industry and determine high costs, including: a cost of 22 billion USD in working capital because of excess inventories held due to transportation delays, increase of capital expenditures by 23 billion USD for equipment and infrastructure required to handle congestion and delays, increase of operating costs by 29 billion USD over a 10-year period due to logistical inefficiencies (PwC 2017, p. 2).

Chemical manufacturers and distributors increasingly outsource logistics functions to third- and fourth-party logistics (3PL, 4PL) providers. According to a 2015 report by London-based Technavio Research, logistics services market for chemical companies will reach between 450 and 500 billion USD by 2019 in comparison to 350 billion USD in 2014 (Inbound Logistics 2016). The global chemical logistics market will grow close to 14% annually in the period 2018–2022, in comparison to 5.48% over the period 2013–2018 (Technavio 2018a, b). The following companies were mentioned in this report: Agility, BDP International, CH Robinson, DB Schenker, Deutsche Post DHL Group. The global chemical warehousing and storage market is expected to grow at the annual rate of around 11% in the period 2016–2020 (Technavio 2016). Globally, APAC is the biggest chemical warehousing and storage region (42%) because of dynamic urbanization and industrialization processes, in comparison to percentage shares of North America (25%) and Europe (21%). The

European chemical warehousing and storage market is expected to grow at a rate of close to 10%. The top vendors in the global chemical warehousing and storage market noticed in the Technavio report are: BDP International, Agility, Americold, DB Schenker and DHL. The global cross docking facility market in the chemical logistics industry will grow over 5% annually from 2018 to 2022. Main players noticed in the report are: A.P. Moller-Maersk, C.H. Robinson, CEVA Logistics, Deutsche Bahn, Deutsche Post DHL Group, Kuehne + Nagel.

Based on the Cefic and Deloitte report (2001), chemical shippers must rethink and reconfigure their supply chains and logistics network models to cope with increased volatility and uncertainty in a global environment (p. 31). Close collaboration with LSPs might help chemical companies to adapt to the changes driven by several trends in the 21st century. Kannegiesser (2008) noted the following trends as most important for the chemical industry development: globalization, consolidation, commoditization and margin pressure, innovation, legislation and sustainability (pp. 72–73). Firstly, LSPs can advise scenarios of changes to manage global material flows and finally, to compete with new rivals from emerging markets. Next, LSPs are partners with expertise regarding supply chain processes optimization and volumes consolidation, essential to achieve logistics costs reduction and enhance costs competitiveness of the chemical companies operating under margin pressure. Furthermore, the importance of innovative technologies should be stressed especially as a source of competitive advantage. Besides developing and managing innovation in production technology, there is an enormous potential to improve effectiveness of logistics processes (transport, warehousing, handling operations) through the usage of Industry 4.0 technologies, like e.g. automation, robotics, mobile technologies, cloud computing or big data analytics. LSPs are leaders in the development of logistics technologies globally. Besides that, they implement process innovations driven by customers' supply chain strategies. Moreover, chemical supply chain management is highly complex due to legislation requirements. Firms are confronted with legislation targets, documentation and test procedures requiring professional support. In a non-core competence like logistics, chemical companies might rely on core competences of external partners. Sustainability is one of the imperatives for chemical supply chain management. Chemical companies have already started to translate this purpose into the industry practice with respects to products development and production, but the sustainability of their logistics processes is underdeveloped. Chemical shippers outsource logistics processes and engage LSPs mostly to achieve costs reduction and quality of deliveries in terms of punctuality and safety. Because of LSPs are owners of the logistics processes, the achievement of sustainable objectives is highly dependent on their practices in contract logistics. LSPs can develop new supply chain models to deliver carbon emission reduction in the face of new regulations focused e.g. on carbon taxes and/or emission trading schemes through switching to greener transport modes and reducing the speed of transport (Cefic and Deloitte, p. 27).

4 ChemMultimodal Project—A Systematic Approach to Develop and Manage Sustainable Transport

4.1 Project Objectives

The main objective of ChemMultimodal Project is the promotion of multimodal transport of chemical goods. The project aims to achieve this goal by coordinating and facilitating cooperation between chemical companies, specialized logistics service providers, terminal operators and public authorities in chemical regions mainly in Central Europe. The Project is carried out under the Interreg Central Europe Programme between June 2016 and May 2019. Project is conducted in seven countries or regions in Central Europe by fourteen Partners. All the Project's activities are performed simultaneously in all Partners' countries—Poland, Germany, Austria, Italy, Czech Republic, Slovakia and Hungary.

The ChemMultimodal Project presents a systemic approach to manage sustainable transport activities on a global scale. This is because it consists of four elements which complement each other to provide comprehensive solution based on cooperation in the outsourcing model. These elements are described in more details in the next section.

4.2 ChemMultimodal Toolbox

The toolbox consists of four elements (consulting services, planning guidelines, Intermodal Links platform and CO_2 calculator) and has been tested in 5 pilots with chemical companies in the partner countries to facilitate real modal shift. The toolbox elements are presented on Fig. 2.

Consulting Services for chemical companies to improve multimodal transport serves as a moderating framework for hosting workshops, bilateral meetings while

Fig. 2 ChemMultimodal toolbox

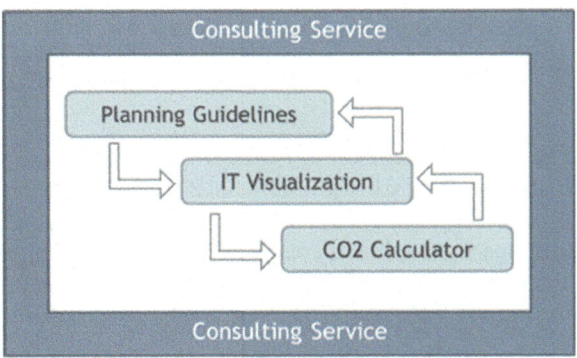

discussing the potential to shift unimodal transport to multimodal. The aim is to establish bilateral cooperation and to develop a database of contacts generated throughout the project. The main role of the project partners is providing information, engaging in discussion, facilitation of cooperation and networking between companies and logistics service providers.

Planning Guidelines for increasing multimodal transport are established to capture necessary transport facts. The planning guidelines serve as an output sheet whereby most important indicators related to multimodal transport are gathered. Such indicators are:

(a) product type to receive information about the products characteristics,
(b) volume to estimate if intermodal transport poses a suitable alternative way of transport,
(c) countries crossed along the route with respective driving and loading regulations,
(d) bundling options to achieve a more efficient use of capacities and
(e) the number of transport units based on the volumes (tons or liters) foreseen for the transport.

The Intermodal Links Planner allows the visualization of existing intermodal transport routes and provides information about frequency of departure, availability of LSPs and terminal operators, arranged feed, delivery of transports to and from different terminals. It fulfils the core requirements like European wide availability, high topicality of the routes and connections and integration of different transport modes. The Intermodal Links Planner is available at: intermodallinks.com/GetAccess.

The CO_2 emissions calculator was developed based on activity-based method. ChemMultimodal CO_2 calculator uses McKinnon's methodology and findings on average emission factors that serve as the basis for calculating transport related CO_2 emissions (Mc Kinnon and Piecyk 2009). It is useful for both, chemical companies and logistics operators, and provides the one-click calculation of CO_2 emissions of intermodal connections from the place of origin to the cargo destination with possibility to define freight characteristics. The value of CO_2 emissions is based on average emission factors. The calculator is available at: ifsl50.mb.uni-magdeburg.de/chemmultimodal/ and presented on Fig. 3.

5 Methodology Within the ChemMultimodal Project

The research methodology used involve qualitative and quantitative methods. The first phase of the ChemMultimodal Project concentrated on diagnose on how physical flows of chemical goods are managed and what kind of tools are used in supply chains to support smooth and continues flows. During this phase the survey research was conducted within 49 main market players in Poland by SGH Warsaw School of Economics and Polish Chamber of Chemical Industry. The research was conducted between August and September 2016. There were two groups of respondents invited to the study—chemical producers and LSPs that cooperate with producers or within

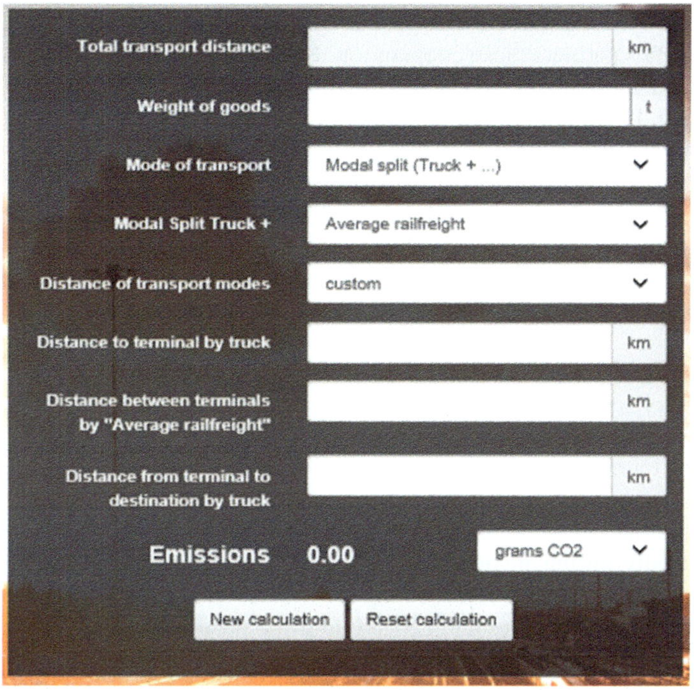

Fig. 3 ChemMultimodal CO_2 emissions calculator

chemical supply chains in Poland. Finally, 22 questionnaires were filled in—13 by chemical producers and 9 by LSPs.

Based on a detailed analysis of the needs for improving multimodal transport of chemical goods, the project developed a toolbox to support chemical companies and logistics service providers in their strategic and operational planning for increasing the share of multimodal transport. This toolbox was tested with 30 chemical companies in the partner countries to facilitate real modal shift. The preliminary assumptions of the pilot phase were to increase of multimodal transport by 10% and to reduce CO_2 footprint by 5% until the end of project duration. The final phase of the Project will concentrate on developing one common multimodal transport strategy and seven regional action plans to continue and intensify activities after the project end.

For the purpose of analyzing the environmental impact of transport modes shifts from road to multimodal solutions in the chapter, five companies with five chemical cargo transport routes tested during the first half of 2018 were chosen. Firstly, companies were asked to identify routes that are currently served by trucks and have a potential to shift to multimodal route. Multimodal transport starts to be an attractive solution whenever long distances are considered (in Europe it means that there is a distance of at least 300 km between first loading and last unloading places) and the volume of goods transported is high enough to use railway capacity potential. The specific potential routes were analyzed and chosen based on chemical companies' and their LPS's partners close cooperation. Both companies discussed the potential solutions with the help of Intermodal Links platform where different options of routes are available. However, the LSP as architect of transport processes and integrator of transport modes and connections was requested to advise the best solution not only in terms of time and cost but also eco-efficiency. In this phase the ChemMultimodal CO_2 emission calculator was a helpful tool used for potential saving estimation. The role of chemical company was to analyze the capacity for the production and its flexibility as the multimodal solution impacts also the production size and cycles. One of the most important part in this case was the agreement from the customer to change delivery schedule in terms of frequency and volume. Therefore, it must be underlined that besides of time of delivery and cost of transport also the quality of customer service and its potential changes have to be discussed. All these conditions had to be met before any change in transport timetable might be done.

After partners analyzed and selected routes for modal shifting they had started pilots' tests - new multimodal routes. In each of the cases pilots were conducted three times to eliminate any accidental events that might impact on results and conclusions.

Additionally, during the pilots' testing phase, individual or group interviews were conducted with managers representing both sides of the supply chain relations, namely chemical companies and LSPs. It allowed the authors to gather information helpful to transform their experience into knowledge considering success factors in collaborative outsourcing development.

6 Sustainable Transport Development Through Collaborative Outsourcing: Success Factors and Effects

6.1 Success Factors for Collaborative Outsourcing

It is clear that modal shifts can only be carried out in the pilots' testing phase in a situation of close cooperation between chemical companies and LSPs. In the light of research findings within the ChemMultimodal Project focused on collaboration practices in the chemical supply chains, the following success factors should be highlighted as worth following up in the development of multimodal transport solutions.

(a) Operational information exchange and collaborative planning
    Chemical companies share information on their business operations development (including transport needs, stock levels, production cycles and other interdependent plans concerning procurement, production or distribution) with LSPs, that stay up-to-date with development of their activities on supply and demand side of supply chains. LSPs plan transport operations tailored to the needs and requirements of service level of the customers. Furthermore, there is an ongoing cooperation between supply chain partners regarding changes in shipments

requiring flexibility in transport management. As a logistics manager representing a chemical manufacturer explained:

In an optimal scenario at a regular basis, LSP prepares plans of raw materials deliveries in accordance to our production plan for the whole week, including both road and intermodal shipments. But simultaneously, there is a crucial need for the high flexibility due to changes in our transport orders and requirements. It stems from the specificity of chemical supply chain management, including changes of orders placed by our clients considering volume or products diversity, then as a result increase (or decrease) and new structure of production volume and raw materials deliveries from suppliers. Planning of chemical transport operations is a dynamic process.

(b) Collaborative process designing and redesigning

The joint agreement of the transport modal shift between chemical companies and their LSPs was a critical factor for the pilots' phase realization within the ChemMultimodal Project. Most of the companies declared that the modal shift is strictly dependent on the engagement of the LSPs with whom they have developed long-term relations. As a representative of LSP stressed:

In case of planning new or changes in existing routes, we discuss together the needs and requirements of our client and check all possible scenarios to design or redesign transport process by usage different transport modes or multimodal solutions. Next, we prepare the costs analysis of the available transport options and outline trade-offs regarding logistics service (including comparison of total transshipment time, punctuality and quality of deliveries). Then, we realize tests of new scenarios and our client controls their effects. Finally, we are able to advise best scenario in the customer's supply chain management and agree on its realization.

(c) Mutual trust and loyalty based on operations reliability

The ChemMultimodal partners developed consulting services for chemical companies participating in the pilots' testing phase through the usage of Intermodal Links Planner and CO_2 calculator, then advising them possible low-carbon scenarios of transport development. Most of the companies discussed the proposed routes scenarios with logistics partners or even make their engagement an essential condition for modal shift. A logistics manager representing a chemical company described the character of relations with LSP as follows:

I would like to confirm that our company is interested in testing one of the routes generated by the Intermodal Links Planner and proposed within ChemMultimodal consulting services, but we would like to engage and collaborate with our logistics service partner within this process. We rely on its operational reliability and trust each other.

(d) Proactive improvement

Based on their logistics expertise and knowledge of the market, LSPs might propose proactively the improvement initiatives in transport management. The scope of the proactive improvement might include innovations, especially process and technological ones aiming at added value in economic and environmental performance. The experience of the chemical companies is highlighted in the following comments of key account and projects managers representing logistics service providers:

There are no universal technological platforms or tools available on the market, that support successfully planning of intermodal transport. We configure a tailored system suitable to our needs as LSP to effectively manage our assets and to the customers' requirements in transport management.

In my opinion, the most important factors in building LSP-client relationship in chemical industry are: experience, transparency, information sharing and innovations. I perceive the ChemMultimodal Project as a valuable initiative supporting green innovations such as e.g. green solutions responding to the EU transport policy.

(e) Sharing benefits and burdens

The collaborative outsourcing practices in the chemical sector confirmed that engaged partners share benefits from implementing good and best solutions as well as burdens and risk if something went wrong. As reflected in the following comment of a transport manager:

We had developed a multimodal solution, that was effective in economic and environmental dimensions in international transport services. It was valuable and beneficial solution for the chemical supplier and its client. Unfortunately, in practice, a trucker offering its services in the key intermodal terminal had proven to be unreliable and we gave up. We shared the risk and burdens of this situation with supply chain partners. Nowadays, we realize road transport on this route.

6.2 Effects of Sustainable Transport Development

After modal shifts have been carried out within the ChemMultimodal pilots' phase, the CO_2 emissions were calculated for each of the route to compare road and multimodal transport emissions. One can observe that even when the distance is not that long (350 km), CO_2 emission decreased dramatically—more than 53%. The higher railway share in multimodal transport, the lower CO_2 emission was produced. The detailed CO_2 emission reduction and tested routes details are presented in Table 1.

7 Conclusion

The chapter develops a systematic approach to sustainable transport management in the chemical industry in collaboration with logistics service providers. The findings are based on the qualitative and quantitative research carried out within the ChemMultimodal Project in 2018. Authors focused their attention on sustainable transport solutions in the chemical industry that enable to achieve great effects in

Table 1 Impact of modal shift on CO_2 emission reduction

Route no.	Shipped materials or goods	Quantity (per month) (t)	Road transport destinations and distance (in km)	CO_2 emitted by road transport (per month) (t)	Modal split (in %)	CO_2 emitted by multi-modal transport (per month) (t)	CO_2 reduction after transport mode shift
1	Plastic & rubber	22,800	Poland—Espania 2400	3392.6	Short-sea (87%) Rail (12%) Road (1%)	1892.8	1499.8 t **(44,2%)**
2	Raw materials	125	The Nederland—Poland 1172.5	9	Rail (93.5%) Road (6.5%)	3	6 t **(59,7%)**
3	Raw materials	900	Loading and unloading in Poland 651	36.3	Rail (59%) Road (41%)	22.3	14 t **(38,7%)**
4	Raw materials	100	Loading and unloading in Poland 349	2.2	Rail (83%) Road (17%)	0.99	1.2 t **(53,9%)**
5	Raw materials	360	Great Britain—Poland 1842	41.1	Rail (66%) Road (28%) Short-sea (6%)	21	20 t **(48,9%)**

CO_2 emission reduction through transport modal shift in collaboration with LSPs. Additionally, they highlighted the possibilities how to support collaborative outsourcing development by modern digital open platform. Case studies of modal shifts were supported by the usage of the Intermodal Links Planner and CO_2 calculator. The presentation of success factors for collaborative outsourcing is enriched with the professional comments of managers representing LSPs and chemical companies collected during the pilots' tests implementation.

Sustainable transport management highly depends on the collaboration between customers and LSPs responsible for planning and managing transport processes. In the chemical industry, companies can achieve reduction of CO_2 emissions by 60% in international transport management due to close collaboration with LSPs. LSPs usually play strategic role as outsourcing partners that are not only offering and managing transport services, but also (re)designing transport processes and developing innovations to continuously improve supply chains performance of the customers. Successful relationships between LSPs and chemical companies are based on operational information exchange, collaborative planning and process management, mutual trust and loyalty, proactive improvement, sharing benefits and burdens.

Acknowledgements The chapter is a result of ChemMultimodal Project implementation and is co-financed by Interreg Central Europe Programme.

Scientific work financed from funding for science in the years 2016–2019 granted for the implementation of the co-financed international project.

References

Bowersox DJ (1990) The strategic benefits of logistics alliances. Harvard Bus Rev 68(4):36–45

Cefic (2017) Facts & figures 2017 of the European chemical industry. Retrieved July 25th, 2018, from http://www.cefic.org/Facts-and-Figures/

Cefic, Deloitte (2001) Chemical logistics vision 2020. The next decade's key trends, impacts and solution areas, September. Retrieved July 26th, 2018, from http://www.cefic.org/Documents/IndustrySupport/Transport-and-Logistics/Chemical-Logistics-Vision-2020%20-190911-final.pdf

Cichosz M (2017) Collaborating on green logistics in chemical supply chains: insights from Poland. Bus Logistics Mod Manage 17:507–522

Cichosz M, Nowicka K, Ocicka B (2018) Efektywne planowanie transportu multimodalnego/Eco-efficient multimodal transport planning. Przemysł Chemiczny 97(9):1421–1425. https://doi.org/10.15199/62.2018.9.2

Cruijssen F (2012) Framework for collaboration: a CO_3 position paper. Collaboration concepts for co-modality

Daugherty PJ (2011) Review of logistics and supply chain relationship literature and suggested research agenda. Int J Phys Distrib Logistics Manage 41(1):16–31

Delfmann W, Albers S, Gehring M (2002) The impact of electronic commerce on logistics service providers. Int J Phys Distrib Logistics Manage 32(3):203–222

Deepen JM, Goldsby TJ, Knemeyer AM, Wallenburg CM (2008) Beyond expectations: an examination of logistics outsourcing goal achievement and goal exceedance. J Bus Logistics 29(2):75–105

Eurostat (2016) http://ec.europa.eu/eurostat/statistics-explained/index.php/Freight_transport_statistics_-_modal_split

Fawcett SE, Wallin C, Allred C, Fawcett AM, Magnan GM (2011) Information technology as an enabler of supply chain collaboration: a dynamic-capabilities perspective. J Supply Chain Manage 47(1):38–59

Hartmann E, de Grahl A (2011) Logistics outsourcing interfaces: the role of customer partnering behavior. In: Success Factors in Logistics Outsourcing. Gabler Verlag, pp. 53–78

Hofer AR, Knemeyer AM, Dresner ME (2009) Antecedents and dimensions of customer partnering behavior in logistics outsourcing relationships. J Bus Logistics 30(2):141–159

Hofer AR, Knemeyer AM, Murphy PR (2012) The roles of procedural and distributive justice in logistics outsourcing relationships. J Bus Logistics 33(3):196–209

Inbound Logistics (2016) Chemical logistics: formula for success, June 15. Retrieved August 1st, 2018, from https://www.inboundlogistics.com/cms/article/chemical-logistics-formula-for-success/

Kannegiesser M (2008) Value chain management in the chemical industry. global value chain planning of commodities. Physica-Verlag, Heidelberg

Lambert DM, Knemeyer AM, Gardner JT (2004) Supply chain partnerships: model validation and implementation. J Bus Logistics 25(2):21–42

Langley CJ Jr (2018) 2018 Third-party logistics study. The state of logistics outsourcing. Infosys Consulting, Penske, Korn Ferry, Penn State University

Leker J, Gelhard C, von Deft S (2018) Business chemistry. How to build and sustain thriving businesses in the chemical industry. Wiley, Hoboken

Lieb RC (1992) The use of third-party logistics services by large American. J Bus Logistics 13(2):29

Mc Kinnon A, Piecyk M (2009) Measuring and managing CO_2 emissions in European chemical transport. Edited by Cefic—The European Chemical Industry Council. Heriot-Watt University, Logistics Research Centre, Edinburgh. http://www.cefic.org

Min S, Roath AS, Daugherty PJ, Genchev SE, Chen H, Arndt AD, Glenn Richey R (2005) Supply chain collaboration: what's happening? Int J Logistics Manage 16(2):237–256

Moore KR (1998) Trust and relationship commitment in logistics alliances: a buyer perspective. Int J Purchasing Mater Manage 34(4):24–37

Murphy PR, Poist RF (1998) Third-party logistics usage: an assessment of propositions based on previous research. Transp J 37(4):26–35

Piecyk M, McKinnon AC (2007) Internalising the external costs of road freight transport in the UK. Heriot-Watt University, Edinburgh

PwC (2017) Transporting growth: delivering a chemical manufacturing renaissance, March. Retrieved August 2nd, 2018, from https://www.americanchemistry.com/Transporting-Growth-Delivering-a-Chemical-Manufacturing-Renaissance.pdf

Rollins M, Pekkarinen S, Mehtälä M (2011) Inter-firm customer knowledge sharing in logistics services: an empirical study. Int J Phys Distrib Logistics Manage 41(10):956–971

Selviaridis K, Spring M (2007) Third party logistics: a literature review and research agenda. Int J Logistics Manage 18(1):125–150

Skjoett-Larsen T (2000) Third party logistics–from an interorganizational point of view. Int J Phys Distrib Logistics Manage 30(2):112–127

Soosay CA, Hyland P (2015) A decade of supply chain collaboration and directions for future research. Supply Chain Manage Int J 20(6):613–630

Stank TP, Daugherty PJ, Autry CW (1999) Collaborative planning: supporting automatic replenishment programs. Supply Chain Manage Int J 4(2):75–85

Stefansson G (2006) Collaborative logistics management and the role of third-party service providers. Int J Phys Distrib Logistics Manage 36(2):76–92

Świtała M, Klosa E (2015) The determinants of logistics cooperation in the supply chain-selected results of the opinion poll within logistics service providers and their customers. LogForum 11(4):329–340

Technavio (2016) Global chemical warehousing and storage market 2016–2020, November

Technavio (2018a) Global chemical logistics market 2018–2022, April

Technavio (2018b) Global cross docking facility market in chemical logistics industry 2018–2022, April

Wallenburg CM, Cahill DL, Michael Knemeyer A, Goldsby TJ (2011) Commitment and trust as drivers of loyalty in logistics outsourcing relationships: cultural differences between the United States and Germany. J Bus Logistics 32(1):83–98

Whipple JM, Frankel R (2000) Strategic alliance success factors. J Supply Chain Manage 36(2):21–28

Woxenius J, Barthel F (2013) The future of intermodal transport. Edward Elgar Publishing Inc, Glos UK

van Hoek RI, Chong I (2001) Epilogue: UPS logistics–practical approaches to the e-supply chain. Int J Phys Distrib Logistics Manage 31(6):463–468

Virum H (1993) Third party logistics development in Europe. Logistics Transp Rev 29(4):355–362

Public Attitude Toward Investment in Sustainable Cities in Taiwan

Meng-Fen Yen and Yuh-Yuh Li

Abstract The purpose of this study was to investigate public attitude toward investment in sustainable infrastructure in Taiwan. We decomposed the value of sustainability development into three dimensions, namely, environmental, societal, and economic values, and then showed that there were relationships between public attitude toward investment in sustainable cities and these sub-values of sustainability development. To do so, we built scales to measure public attitude toward investment in sustainable infrastructure and these sub-values of sustainability development. We used a questionnaire to collect our data, interviewing 359 undergraduate students in June of 2018. Multiple regression models were then employed for statistical analysis. We first found that, after controlling for the students' gender and majors, the public attitude toward sustainable cities in Taiwan was correlated with environmental value but only partially related to economic value. Meanwhile, the relationship between the public attitude toward investment in sustainable cities and societal value was trivial. Moreover, we also found that the economic, societal, and environmental values of sustainability development are not mutually exclusive concepts but, rather, are compatible with each other in the overall concept of sustainable city design. We concluded that investment in sustainable cities in Taiwan was regarded by the public as mainly contributing to environmental value. These findings provide a theoretical contribution to the literature of environmental psychology. Furthermore, the conclusions of the study have methodological and practical implications for countries and firms involved in the public infrastructure development of sustainable cities.

Keywords Sustainable infrastructure · Public attitude · Sustainable cities · Environmental psychology

M.-F. Yen
Center for Research in Econometric Theory and Applications (CRETA), National Taiwan University, Taipei, Taiwan
e-mail: mengfyen@ntu.edu.tw

Y.-Y. Li (✉)
Research Center for Promoting Civic Literacy, National Sun Yat-sen University, Kaohsiung, Taiwan
e-mail: yyli@mail.nsysu.edu.tw

© Springer Nature Switzerland AG 2020
W. Leal Filho et al. (eds.), *International Business, Trade and Institutional Sustainability*, World Sustainability Series,
https://doi.org/10.1007/978-3-030-26759-9_42

1 Introduction

The sustainability practices of the public sector have been a focus of considerable academic attention recently (Bulkeley 2013; Falkena et al. 2002; Judyta 2016; Krause et al. 2014; Wang et al. 2012; Yen et al. 2019). In the recent trade-related literature, for example, increased government investments in sustainable cities are viewed as constituting a strategy for upgrading a country's trade status-quo (Francois et al. 2013). This strategy of public investment in sustainable cities is particularly important with respect to small- and medium-sized cities as it can provide opportunities to invest in green infrastructures and, thus, reduce the use of old energy technologies, which may in turn lead to positive changes in social development before social inequities become unsustainable. The results of the World Economic and Social Survey (WESS) (United Nations Department of Economic and Social Affairs 2013, p. 69) have suggested that, for cities in middle- and high-income countries, aspects of infrastructure related to capacity development and urban resilience are important for the growth and development of local firms. In particular, it is critical to focus on investing in the production and use of renewable sources of energy, as well as on the renovation of existing infrastructure, including the retrofitting of buildings and improved efficiency in the use of electricity and water. The environmental Kuznets curve (Dinda 2004; Stern 2004; Apergis and Ozturk 2015; Lau et al. 2018) suggests that economic development initially leads to environmental deterioration, but after a certain level of economic growth, a society begins to improve its relationship with the environment and the level of environmental degradation is reduced. Relatedly, it has been shown that as wealthier countries become increasingly willing to pay for environmental quality, international business relationships in a free trade environment allow for the free flow and spread of cleaner technologies across borders. More recent research further suggests that international business has not only facilitated access to clean technologies but has also led to more rapid adoption of these technologies.

At the country level, the concepts of sustainability have been spread worldwide and are considered to constitute one of the most important themes to have emerged in recent decades as pressing concerns for the future of the planet and human populations increase. The European Science for Environment Policy (2018) focused on eight categories of relevant indicators, namely, energy, buildings, transport, water, wasteland use, air quality, environmental governance, and CO_2. Relatedly, Jabareen (2006) stated that the concept of a sustainable city primarily includes, generally speaking, the following eight key concepts: compactness and density, mixed land use, social and housing, diversity, sustainable transportation, passive solar design, and integrating nature into the urban environment. These concepts of a sustainable city reveal positive economic, societal, and environmental outcomes of sustainable development in the long run.

The Ministry of the Interior in Taiwan established a green building assessment system in 1999 and began enforcement of the Special Chapter of Green Buildings to the Building Technical Regulations (https://pwbgis.kcg.gov.tw/Sustainable_en/main_2.aspx) in 2003. To comply with the international trend toward the setting of carbon reduction goals and reduce the Earth's environmental load, municipal governments are encouraged to formulate and adopt green building regulations that are based on more stringent carbon reduction standards than those in Taiwan's national regulations. To that end, the local government of Kaohsiung city initiated new architecture regulations while also providing developers with financial motivations to adopt various green-building innovations, with the aim of building so-called "Kaohsiung Buildings". The goal of the regulations is to adapt to the higher temperatures brought by global warming by developing various innovative building technologies based on the three perspectives of environmental sustainability, local relevance, and livability and health. Also, the emergence of the green economy has prompted the government of Taiwan to begin building infrastructure that will lay the foundation for the country's green energy sector, transform the nation into a nuclear-free society, and spur industrial innovation. In line with these objectives, Taiwan's government recently launched an eight-year (2017–2024) program called the Forward-looking Infrastructure Development Program (FIDP). The purpose of the FIDP is to build the necessary infrastructures for sustainable development of the country over the subsequent 30 years. Among the targeted projects of the FIDP are various types of infrastructure projects such as railway projects, water-related infrastructure projects, green energy projects, digital development projects, and urban and rural infrastructure projects. The debt of the FIDP to the 3-year-average GDP ratio for 2018 is projected to be 33.4%, which is within the legal limit, but the financial burden for the government is nonetheless quite large. The FIDP projects will also draw investments from leading international companies and encourage them to work with high-potential local firms, forming globally competitive teams that can boost Taiwan's industrial future and ensure the nation's sustainable development.

Sustainability has come to represent an important dimension of corporate strategy for business firms around the globe, with an increasing number of firms trying to measure, monitor, and improve the social and environmental impacts of their operations. Despite this explosion in interest, however, the effective incorporation of sustainability into business and management practices faces serious obstacles, raising the need for further research and implementation efforts. For example, by analyzing optimal investment in public infrastructure in one country when public investment by a trading partner is fixed, Yen et al. (2019) found that, *ceteris paribus*, greater public investment in commercial infrastructure raises general labor productivity, leading to gains in workers' real income.

One of the major obstacles is not that clean technologies are not available, but that they often come at a much higher cost than traditional technologies. In addition to such financial considerations, investments in the building of sustainable cities might not be fully supported by citizens, because that there is some degree of conflict between what is required to make a city sustainable and the livability of that city for its citizens (Judyta 2016; Leiserowitz et al. 2006; Levi et al. 2010). Judyta's study (2016) concluded that new public investments in urban infrastructure likely lead to numerous protests by the inhabitants of local communities relating to their concerns about the negative effects of the investments on the natural environment, local inhabitants, and property values. Leiserowitz et al. (2006) stated that values/attitudes often do not translate directly into actual behavior, with gaps between expressed values or attitudes and actual behaviors having been discovered. Three types of barriers associated with these gaps are mentioned by Leiserowitz, Kates, and Parris. The first type of barrier consists of the existence, direction, and strength of particular values (e.g., economic growth) that effectively counter the other values. The second type of barrier to sustainable behaviors include, for individuals, the lack of time, money, access, literacy, knowledge, skills, power, or perceived efficacy required to translate their values into action. The third type of barrier are structural barriers, including laws, available technology, and social norms. Levi et al. (2010) argued that even though sustainable cities have many environmental benefits in terms of land use, resources, and energy consumption, the negative impacts on people and their lifestyles prevent citizens from being willing to support the associated environmental changes. The specific negative impacts identified in their study were crowding, reduced private or personal spaces (such as backyards), crime and related issues (such as fear and lack of safety), and social conflict.

The present study was undertaken in light of recent developments in the study of environmental attitude (Heberlein 2012; Johnson et al. 2004; Macnaghten and Jacobs 1997; Kaiser and Fuhrer 2003; Milfont et al. 2010; Stern et al. 1999). There is little theoretical disagreement that environmental attitude (in particularly, the affect components of such attitude) should be regarded as one of the precursors of behavior. Furthermore, Dietz et al. (1998) and Stern et al. (1995) stated that environmental attitude (such as the attitude toward sustainable cities) plays a central role in a schematic causal model of belief and behavior. Through the attitude, a more sequential way to predict environmental behavior is seen in common. That is, on the one hand, behaviors, or behavioral commitments and intentions, are based on an individual's attitude toward a specific environmental topic. On the other hand, such an attitude toward a specific environmental topic has also been found to be associated with a hierarchical influence of cognitions relating to various general beliefs and values (Homer and Kahle 1988). A 2016 study conducted by Stjernborg and Mattisson in Sweden regarding the public's perceptions of infrastructure showed that public transport is regarded as an important factor in achieving other goals and other public values, particularly those related to economic and environmental issues, whereas the social dimension of such transport is not as strongly prioritized. Macnaghten and Jacobs (1997) concluded, meanwhile, that whether the public identifies with the concept of sustainable development and the analyses underlying it is the more meaningful topic in terms of the study of sustainable attitudes.

In this study, we propose an empirical method for measuring public opinions regarding investment in sustainable cities (ISC), as the support of citizens for ISC is a determining factor it the success of such projects in the social context of Taiwan. We developed a scale for the measurement of public attitudes toward ISC in order to examine public opinions of ISC. We predicted the support for or disapproval of ISC through the evaluation of three value dimensions (as shown in Fig. 1), which we subsequently termed the three sustainability value dimensions. It was expected that citizens' attitudes toward ISC would be positively correlated with these sustainability values. To measure the three sustainability values, namely, economic, societal, and

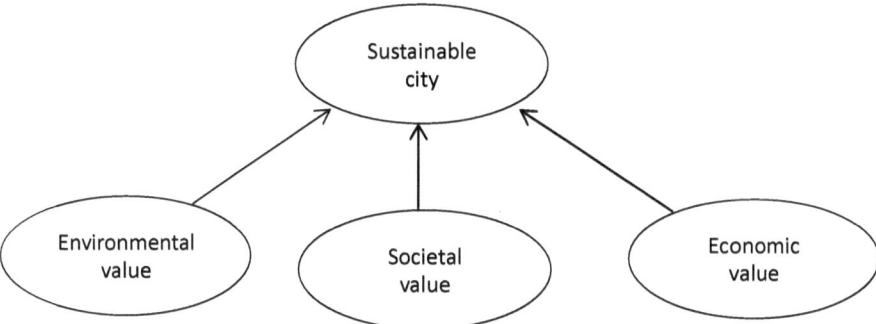

Fig. 1 Three value dimensions of sustainable cities

environmental values, we first developed tools to measure attitude toward ISC and to measure perceptions regarding the three values. Secondly, regression analysis was used to examine their relationships with one another in general and with specific investment targets. We expect that the results of these investigations will provide useful insights for Southeast Asian countries seeking to elicit firm innovation through public investment in ISC projects.

2 Methods

2.1 Participants

359 undergraduate students were interviewed from a national owned university located in the city, Kaohsiung, of Taiwan during June of 2018. These students were randomly selected from five classes that were all part of the general curriculum. The students who took these five courses were thus highly representative of the student population of the university in general, including students majoring in a variety of disciplines. More specifically (Table 1), the participating students included students

Table 1 Background information of study participants

Variable	Category	N	%
Gender	Male	239	67.1
	Female	120	32.9
Major	Science	78	21.7
	Engineering	129	57.7
	Business	61	17.0
	Social/humanities	91	25.3
	Total	359	100

from the freshman to senior classes (with ages ranging from 18 to 22 years old) with a range of different majors. 21.7% of them were science majors, 57.7% were engineering majors, 17.0% were business majors, and 25.3% were social sciences majors. In terms of gender, 67% of the students were male, and 32.9% were female.

3 Concept and Measure

3.1 Attitude Toward ISC

In this study, "attitude toward ISC" was defined as the behavioral intention to support governmental financial investment in the building of infrastructure to ensure sustainable cities. We then created a scale to measure students' attitude toward ISC. A report regarding the World Economic and Social Survey provided the expert validity required in developing this scale. The WESS report (United Nations Department of Economic and Social Affairs 2013) argued that sustainable cities should include the following: renewable energy sources; efficiency in the use of water and electricity; compactness in the design and implementation of city planning; the retrofitting of buildings; increased green areas; fast, reliable, and affordable public transportation; and improved waste and recycling systems. This study created an ISC scale comprised of the 7 items shown in Table 2. The ISC scale starts by asking participants the following question: "As a resident of Kaohsiung city, do you agree or disagree with the statement that the city government should spend money on infrastructure investment in the following categories?" We employed a 4-point Likert scale (with the response options being "strongly disagree," "neutral," "agree," and "strongly agree") in developing the ISC scale. Table 1 shows the respondents' preferences regarding the different types of infrastructure in sustainable cities. We found that recycling (mean = 3.44), the improvement of electric power distribution and water distribution networks (mean = 3.33), renewable energy (mean = 3.26), and mass

Table 2 Descriptive analysis of responses to different types of infrastructure investments

Different types of infrastructure	Range	Mean	S.D.	Factor loading
Recycling	1–4	3.44	0.599	0.767
Improvement of electric power distribution and water distribution networks	1–4	3.33	0.597	0.777
Renewable energy	1–4	3.26	0.703	0.657
Mass rapid transit system	1–4	3.21	0.668	0.718
Compact city	1–4	2.99	0.705	0.528
Bike trails	1–4	2.98	0.711	0.610
Light rails	1–4	2.45	0.907	0.354

Note 1 = strongly disagree, 4 = strongly agree

rapid transit (mean = 3.21) projects were the most preferred types of infrastructure projects. Meanwhile, compact city (mean = 2.99), bike trail (mean = 2.98), and light rail (mean = 2.45) projects were the least favored types of projects. The factor loadings of the construct validity ranged from 0.354 to 0.718. We removed the question regarding light rail projects from our scale because it had the lowest factor loading. The Cronbach's α of internal consistency reliability value for the scale with 6 items was 0.763 (Table 2).

3.2   Value Dimensions of Sustainability (VDOS)

We measured the "value dimensions of sustainability" using a VDOS scale. There were 7 indicators employed by this VDOS scale (please see below), with the different indicators representing a citizen's perceptions of sustainable development. We adopted an idea from a study by Levi et al. (2010) in constructing the VDOS scale. Levi et al. used the concept of livability to measure citizens' evaluations of infrastructure for sustainable cities. In developing our scale, we assumed that citizens' evaluations concerning sustainable infrastructure projects are evolved from several fundamental values of sustainability, namely, economic, societal, and environmental values. We used a 5-point Likert scale for the construction of the VDOS scale, with the indicators being evaluated by the respondents according to their choice from among five options regarding the given indicator. The five options were as follows: "this idea only creates problems and has no benefits," "this idea creates more problems than benefits," "this idea creates both problems and benefits equally," "this idea creates more benefits than problems," and "this idea creates only benefits and no problems." We also provided a brief definition of each indicator in the questionnaire itself.

The following list includes all the indicators used in the VDOS scale:
What do you think about using each of the following concepts in designing a city? Does the given concept bring more problems or more benefits?

1. *Compactness (description: urban areas should be limited in how much they can expand by preventing their boundaries from growing.)*
2. *Sustainable Transport (description: cities should support walking, cycling, and efficient public transport.)*
3. *Density (description: people should live in higher densities; there should be more people and dwelling units in a given area.)*
4. *Mixed Land Uses (description: compatible land uses, such as housing, commercial areas, and offices, should be located close to each other.)*
5. *Housing Diversity (description: there should be different types, styles, and densities of housing within urban areas.)*
6. *Social Diversity (description: there should be a variety of income-level groups and cultures within urban areas.)*

7. *Greening (description: more nature should be integrated into cities through parks, street trees, etc.)*

There are three sub-scales of VDOS that represent the three values of sustainability development, i.e., economic, societal, and environmental values. The concepts of sustainable city design that require compactness, density, and mixed land uses were categorized as relating to economic value. The concepts of housing diversity and social diversity were categorized as relating to societal value. The concepts of sustainable transport and greening were categorized as relating to environmental value. We employed factor analysis to examine the validity of these categorizations. Factor loadings represent the degree of construct validity for each sub-scale in a larger scale. We used principal component analysis as the extraction method and used Varimax with Kaiser Normalization as the rotation method. The Cronbach's α reliability index values indicated the internal consistency of the scale. Table 1 shows the validity and reliability values of the VDOS scale. The factor loadings of the 3 economic value sub-scale indicators were 0.708, 0.758, and 0.662, respectively. The Cronbach's α value for economic value was 0.547. The 2 items in the societal value sub-scale had factor loadings of 0.815 and 0.812, respectively. The reliability for societal value was 0.587. The factor loadings of the 2 environmental value sub-scale indicators were 0.829 and 0.756, respectively. The reliability for environmental value was 0.477.

3.3 Controlling Variables

Gender and academic backgrounds such as majors were controlling variables. For gender, female was set as the dummy variable equal to 1. We set up four other dummy variables for four different types of majors: science, engineering, business, and social/humanities majors.

4 Statistical Model

We used multiple regression models to test the hypothesized relationships between attitude towards ISC and the sustainability values. 7 models were estimated. Model 1 was a general model to test the relationships between three types of variables (gender, major, and values of sustainability) and the dependent variable, attitude towards ISC. In model 2, the dependent variable was infrastructure investment in renewable energy (RE). In model 3, the dependent variable was investment in mass rapid transportation (MRT). In model 4, the dependent variable was investment in bike trails (BT). In model 5, the dependent variable was investment in the improvement of

Table 3 Factor loadings and Cronbach's α values of VDOS scale

Indicator	Range	Mean	S.D.	Value 1: economic value	Value 2: societal value	Value 3: environmental value
1 Compactness	1–5	3.13	0.776	0.708		
2 Sustainable transport	1–5	3.96	0.679			0.829
3 Density	1–5	2.81	0.823	0.758		
4 Mixed land uses	1–5	2.95	0.807	0.662		
5 Housing diversity	1–5	3.55	0.745		0.815	
6 Social diversity	1–5	3.38	0.837		0.812	
7 Greening	2–5	4.20	0.605			0.756
Total variance explained	–	–	–	23.16%	21.70%	18.83%
Cronbach's α	–	–	–	0.547	0.587	0.477

electric power distribution and water distribution networks (AUS). In model 6, the dependent variable was investment in a compact city (CC). Finally, in model 7, the dependent variable was investment in recycling (RR). Meanwhile, the independent variables in models 2–7 were the same as those in model 1.

5 Regression Model

$$f(x) = a_0 + a_1 female + a_2 Engineering + a_3 Management + a_4 Social\_Humanity$$
$$+ a_5 Economic + a_6 Societal + a_7 Environmental + error$$

6 Results

6.1 Bi-Variate Correlations of Attitude Toward ISC and Sustainability Values

Table 3 shows the correlations between attitude toward ISC and the three sustainability values, as well the correlations between those three values. Attitude toward ISC was correlated with economic value ($r = 0.14$, p-value < 0.01), with societal value ($r =$

Table 4 Descriptive statistics and bi-correlations of attitude toward ISC and values of sustainable cities

	Attitude toward ISC	Economic value	Societal value	Environmental value
Attitude toward ISC	1.00	–	–	–
Economic value	0.14**	1.00	–	–
Societal value	0.14**	0.14**	1.00	–
Environmental value	0.26**	0.04	0.21**	1.00
Mean	3.20	2.96	3.46	4.07
S.D.	0.45	0.59	0.66	0.52

$*P < 0.05$, $**P < 0.01$, $***P < 0.001$

0.14, p-value < 0.01), and with environmental value ($r = 0.26$, p-value < 0.01). Meanwhile, economic value was correlated with societal value ($r = 0.14$, p-value < 0.01) but not with environmental value ($r = 0.04$, p-value > 0.05), while societal value was correlated with environmental value ($r = 0.21$, p-value < 0.01). Our analysis in Table 4 showed that attitude toward ISC was more highly correlated with environmental value than with societal value or economic value if we did not control for the confounding variables in the model. Also, we found that societal value was correlated with both economic value and environmental value. However, there was no significant correlation between economic value and environmental value.

7 Attitude Toward ISC Was Related to Environmental Value and Economic Value in the Young Generation in Taiwan

The results of the multiple regression analysis are shown in Table 5. The multiple regression analysis allowed us to show any partial relationship between two variables, *ceteris paribus*. Model 1 showed the general relationships between attitude toward ISC and the three sustainability values after controlling for gender, major, and other sustainability values in the same model. As indicated in Table 5, in model 1, economic value (beta = 0.121, p-value = 0.018) and environmental value (beta = 0.226, p-value < 0.001) had direct relationships with attitude toward ISC. However, in general, students' genders and majors did not affect their attitude toward ISC, even though the female students were slightly disapproving overall of ISC and the business majors were slightly approving overall of ISC.

Table 5 also shows the results regarding investments in some specific categories of sustainable infrastructure projects as determined by models 2 through 7, namely, renewable energy, mass rapid transportation, bike trails, the improvement of electric

Table 5 Multiple regression analysis of attitude toward ISC

	Model 1 ISC Beta (p-value)	Model 2 RE Beta (p-value)	Model 3 MRT Beta (p-value)	Model 4 BT Beta (p-value)	Model 5 AUS Beta (p-value)	Model 6 CC Beta (p-value)	Model 7 RR Beta (p-value)
Gender (male = 0)							
Female	−0.063(0.241)	0.054(0.328)	0.030(0.591)	−0.111(0.042)	−0.112(0.041)	−0.106(0.050)	−0.010(0.854)
Major (science = 0)							
Engineering	−0.065(0.334)	−0.084(0.223)	0.040(0.560)	−0.038(0.582)	−0.078(0.254)	−0.070(0.305)	−.033(0.627)
Business	0.098 (0.111)	0.047(0.453)	0.073(0.249)	0.127(0.043)	0.018(0.779)	0.078(0.216)	0.045(0.463)
Social & humanity	0.022(0.736)	0.027(0.684)	0.041(0.539)	0.095(0.150)	−0.047(0.477)	−0.036(0.586)	−0.043(0.966)
Values							
Economic	0.121(0.018)	0.114(0.031)	0.072(0.174)	0.016(0.753)	0.094(0.072)	0.205(0.000)	−0.024(0.641)
Societal	0.072(0.170)	0.026(0.626)	0.061(0.257)	0.048(0.369)	0.040(0.459)	0.068(0.201)	0.048(0.362)
Environmental	0.226(0.000)	0.131(0.014)	0.184(0.001)	0.175(0.001)	0.177(0.001)	−0.013(0.812)	0.290(0.000)
N	359	359	359	359	359	359	359
R square	0.107	0.057	0.055	0.071	0.066	0.073	0.099
Adj. R square	0.089	0.038	0.037	0.053	0.047	0.054	0.081

Note 1. *ISC* index of the *development of sustainable cities (sum of RE, MRT, BT, AUS, CC, and RR)*; *RE* Renewable energy; *MRT* Mass rapid transportation; *BT* Bike trails; *AUS* Advanced utilities supply; *CC* compact city; *RR* rubbish recycling. 2. Bold number refers to statistically significant at p-value < 0.05

power distribution and water distribution networks, compact city, and recycling. We found that there was a positive relationship between economic value and attitude toward investment in renewable energy (beta = 0.114, p-value = 0.031), as well as between economic value and attitude toward investment in a compact city (beta = 0.205, p-value < 0.001). There was also a positive relationship between environmental value and attitude toward investment in renewable energy (beta = 0.131, p-value = 0.014), between environmental value and attitude toward investment in mass rapid transportation (beta = 0.184, p-value = 0.001), between environmental value and attitude toward investment in bike trails (beta = 0.175, p-value = 0.001), between environmental value and attitude toward investment in the improvement of electric power distribution and water distribution networks, (beta = 0.177, p-value = 0.001), and between environmental value and attitude toward investment in recycling (beta = 0.290, p-value < 0.001).

8 Discussion and Conclusion

In this study, we examined students' attitude toward ISC. We also explored how this attitude toward ISC is correlated with the three key values of sustainability by building upon past studies in environmental psychology. This study contributes both methodological and theoretical advances to the study of sustainable city development. In methodological terms, it contributes to the literature by providing a new scale for measuring citizens' attitude toward ISC. It is important for both the public sector and private sectors to understand citizens' attitude toward ISC because ISC provides both environmental and economic benefits. The ISC scale used in this study was proven to be a valid tool for measuring attitude toward ISC. The one-dimensional construct of the ISC scale suggests that Taiwanese citizens' attitude toward investment in infrastructure could be regarded as a latent variable, that is, one that cannot be observed directly. This means that the approval or disapproval expressed by citizens toward different types of ISC projects, such as light rails and bike trails, is highly correlated with their underlying core attitude toward ISC. Relatedly, we can predict that if some types of ISC projects are more in line with the sustainability values favored by the public, then these projects stand a higher chance of being supported by the public.

Among the sustainability values, we found that there is a trivial but positive relationship between environmental value and economic value, a weak relationship between societal value and economic value, and a stronger relationship between societal value and environmental value. Two key implications of the findings are as follows. First, our study provides strong evidence that environmental sustainability and economic sustainability can coexist, whereas the traditional perspective was that the pursuit of economic development cannot coexist with the pursuit of environmental protection. In contrast with that traditional perspective, the result of this study showed that, in the view of Taiwan's citizens, economic development is not one of the barriers to environmental protection. Second, the small correlations revealed between the environmental, societal, and economic values of sustainability proved that the concept of sustainable development is very complex. The weak connections between these three key values of sustainability suggest that, in terms of how people think about sustainability values, these three sustainability values are independent of one another. This finding suggests, in turn, that there is a need to update the understanding of Taiwan's citizenry regarding the concept of sustainability so that it includes more holistic and cohesive relationships between three aspects of values.

The most important finding in our study is that attitude toward ISC is mainly related to environmental value. We found that there is a very strong positive relationship between the support for sustainable city infrastructure investment and environmental value in Taiwan. Our findings are compatible with the hypothesis of the environmental Kuznets curve (Dinda 2004; Stern 2004; Apergis and Ozturk 2015; Lau et al. 2018), which states that while economic development initially leads to environmental deterioration, once a certain level of economic growth is achieved, a society begins to improve its relationship with the environment and environmental degradation levels are reduced. Taiwan experienced rapid economic development during the period from 1960 to 1990 at the expense of the environment. Environmental value, nonetheless, really plays the most critical role in determining citizens' approval or disapproval of sustainability-related investment projects when economic development reaches a certain level. In other words, we proved the argument of the environmental Kuznets curve in this study.

In conclusion, the results of this study show that ISC in Taiwan is mainly regarded by the public as contributing to environmental values. Many of today's sustainability challenges, such as fossil fuel dependency, global warming, poverty, and social exclusion, are highly relevant for urban development. We suggest that the disapproval of public investment in infrastructure projects may be prevented by successful communication with the public. Citizens' appreciation of environmental value offers a promising means for overcoming such disapproval based on the findings of this study and the environmental Kuznets curve. However, at the same time, it should be noted that the perceived lack of economic and social values in ISC might be a possible barrier for the future development of sustainable infrastructure.

Acknowledgements The work was supported by all of the participating science centers and Taiwan's Ministry of Science and Technology under Grant 106-2511-S-110-007.

References

Apergis N, Ozturk I (2015) Testing environmental Kuznets curve hypothesis in Asian countries. Ecol Ind 52:16–22

Bulkeley H (2013) Cities and climate change. Routledge

Dietz T, Stern PC, Guagnano GA (1998) Social structural and social psychological bases of environmental concern. Environ Behav 30:450–471

Dinda S (2004) Environmental Kuznets curve hypothesis: a survey. Ecol Econ 49(4):431–455

Falkena HJ, Moll HC, Noorman KJ (2002) Towards a sustainable city: roles, behaviour and attitudes of citizens, local organisations and the authorities. WIT Trans Ecol Environ 54

Francois J, Manchin M (2013) Institutions, infrastructure, and trade. World Dev 46:165–175

Heberlein TA (2012) Navigating environmental attitudes. Oxford Scholarship Online: January 2013. https://doi.org/10.1093/acprof:oso/9780199773329.001.0001

Homer PM, Kahle LR (1988) A structural equation test of the value-attitude-behavior hierarchy. J Pers Soc Psychol 54(4):638

Jabareen Y (2006) Sustainable urban forms: their typologies, models, and concepts. J Plann Educ Res 26(1):38–52

Johnson CY, Bowker JM, Cordell HK (2004) Ethnic variation in environmental belief and behavior: an examination of the new ecological paradigm in a social psychological context. Environ Behav 36(2):157–186

Judyta W (2016) Urban infrastructure facilities as an essential public investment for sustainable cities–indispensable but unwelcome objects of social conflicts. Case study of Warsaw, Poland. Transp Res Procedia 16:553–565

Kaiser FG, Fuhrer U (2003) Ecological behavior's dependency on different forms of knowledge. Appl Psychol Int Rev 52(4):598–613

Krause RM, Feiock RC, Hawkins CV (2014) The administrative organization of sustainability within local government. J Public Adm Res Theor 26(1):113–127

Lau LS, Choong CK, Ng CF (2018) Role of institutional quality on environmental Kuznets curve: a comparative study in developed and developing countries. In: Advances in pacific basin business, economics and finance. Emerald Publishing Limited, pp 223–247

Leiserowitz AA, Kates RW, Parris TM (2006) Sustainability values, attitudes, and behaviors: a review of multinational and global trends. Annu Rev Environ Resour 31:413–444

Levi D, Casswell R, Gonzales U, Lopez A (2010) Attitudes towards sustainable cities: are sustainable cities liable cities? Focus J City Reg Plann Dept 7(7):33–35

Macnaghten P, Jacobs M (1997) Public identification with sustainable development: investigating cultural barriers to participation. Glob Environ Change 7(1):5–24

Milfont TL, Duckitt J, Wagner C (2010) A cross-cultural test of the value–attitude–behavior hierarchy. J Appl Soc Psychol 40(11):2791–2813

Science for Environment Policy (2018) Indicators for sustainable cities. In-depth Report 12. Produced for the European Commission DG Environment by the Science Communication Unit, UWE, Bristol. Available at http://ec.europa.eu/science-environment-policy

Stern DI (2004) The rise and fall of the environmental Kuznets curve. World Dev 32(8):1419–1439

Stern PC, Dietz T, Abel T, Guagnano GA, Kalof L (1999) A value-belief-norm theory of support for social movements: the case of environmentalism. Hum Ecol Rev:81–97

Stern PC, Kalof L, Dietz T, Guagnano GA (1995) Values, beliefs, and pro-environmental attitude formation toward emergent attitude objects. J Appl Soc Psychol 25:1611–1636. https://doi.org/10.1111/j.1559-1816.1995.tb02636.x

Stjernborg V, Mattisson O (2016) The role of public transport in society—a case study of general policy documents in Sweden. Sustainability 8(11):1120

United Nations Department of Economic and Social Affairs (2013) Chapter III Towards sustainable cities. In: From world economic and social survey 2013: sustainable development challenges. United Nations, NY

Wang X, Hawkins C, Lebredo N, Berman E (2012) Capacity to sustain sustainability: a study of U.S. cases. Public Adm Rev 72:841–853

Yen M-F, Wu R, Miranda MJ (2019) A general equilibrium model of bilateral trade with strategic public investment in commercial infrastructure. J Int Trade Econ Dev 28(6):712–731

Circularity in the Built Environment: A Focus on India

Usha Iyer-Raniga, Priyanka Erasmus, Pekka Huovila and Soumen Maity

Abstract The built environment operates in a linear way where large amounts of non-renewable resources are used to feed the growing and rapid city building activities taking place globally, particularly in the Asia Pacific and Latin American regions. Recent estimates from the World Bank (Ellen MacArthur Foundation 2017, p. 4) indicate that over half of the world's population will live in urban areas, whilst providing over 80% of global GDP generation. Building and construction uses 36% of energy consumption, produces 40% waste and estimated approximately 40% carbon dioxide emissions (GABC 2017). 'Achieving Growth' (Ellen MacArthur Foundation 2017, p. 12) has identified 115 billion euros investment opportunities in the built environment for designing and constructing buildings based on circular principles, closing the loop on building construction and demolition materials, and building circular cities. Not just creating jobs, application of circular principles supports resilience, reduces resource use and lowers overall emissions. This paper focuses on India as an emerging economy. It discusses the potential of placing the country on a path of circularity with reference to the built environment. Two case studies are used to demonstrate examples of as-yet untapped upscaling potential of integrating principles of circularity. It offers opportunities to increase knowledge in the sector, develop main-

U. Iyer-Raniga (✉) · P. Erasmus
School of Property, Construction and Project Management, RMIT University, Melbourne, Australia
e-mail: usha.iyer-raniga@rmit.edu.au

P. Erasmus
e-mail: priyanka.erasmus@rmit.edu.au

U. Iyer-Raniga · P. Erasmus · P. Huovila · S. Maity
Sustainable Buildings and Construction Programme, One Planet Network (United Nations 10 Year Framework), Paris, France

P. Huovila
Green Building Council Finland, Helsinki, Finland
e-mail: pekka.huovila@figbc.fi

S. Maity
Development Alternatives, New Delhi, India
e-mail: smaity@devalt.org

© Springer Nature Switzerland AG 2020
W. Leal Filho et al. (eds.), *International Business, Trade and Institutional Sustainability*, World Sustainability Series,
https://doi.org/10.1007/978-3-030-26759-9_43

streaming platforms from fragmented examples, and most importantly decoupling economic growth with resource consumption. By sharing these learnings, the key value drivers of increasing life cycles of the asset from multiple functional perspectives, increasing utilization and expanding regenerative potentials in an increasingly digitized world are highlighted.

Keywords Closed loop · Circularity · Waste · Non-linear · Built environment · Sustainability

1 Introduction

The built environment has an impact on resource use. Estimates are that for building and construction a significant proportion of world's resources are used. Globally, building and construction requires 40% of global energy, produces 40% waste, emits 30% of GHG emissions and uses 12% of the fresh water use while employing about 12% of the workforce (UN Environment 2017; UNEP-SBCI 2016). According to Seto et al. (2014), urban areas are estimated to consume between 67 and 76% of global energy and generate approximately three quarters of global carbon emissions, of which buildings and other infrastructure constitute a significant proportion as cited by Ness and Xing (2017). The developing economies are in a state of rapid city building as they are in the process of catching up with the developed world with respect to urbanization. It is anticipated that of the two tiger economies of Asia; China and India, will undergo massive urbanization. In China, by 2050, 2/3rd of its population will be housed in urban centers, many of which will be in megacities with a population over 5 million (McKinsey Global Institute 2009). In the case of India, 40% of the population or close to 600 million will be urbanized (McKinsey Global Institute 2010).

It is in the emerging economies in particular that a systemic approach can be taken to fully understand building life cycle operation and start of innovative models to explore the value chain fully. Circular economy may be defined as a 'regenerative system in which resource input and waste, emission and energy leakage are minimized by slowing, closing and narrowing material and energy loops. This can be achieved through long-lasting design, maintenance, repair, reuse, remanufacturing, refurbishing and recycling' (Geissdoerfer et al. 2017).

Adopting circular economy approaches in a high-growth, high-waste sector like the built environment presents a tremendous opportunity for businesses, governments and cities to minimise structural waste and thus realise greater value from built environment assets. In a circular economy, renewable materials are used where possible, energy is provided from renewable sources, natural systems are preserved and enhanced, and waste and negative impacts are designed out. Materials, products and components are instead managed in loops, maintaining them at their highest possible intrinsic value (Ellen Macarthur Foundation 2018). According to a report by UNEP (2016), in its most basic form, a circular economy can be loosely defined as one which balances economic development with environmental and resource protection and in this form; it appears to be inseparable from industrial ecology, and close to the three pillars (economic, environmental and social) of sustainable development (Murray et al. 2017, p. 373). However, circular economy needs some level of societal engagement to put into practice. Other barriers are financial, structural, operational, attitudinal and technological (Ritzen and Sandstrom 2017).

The European Union's vision for a circular economy is to move away from linear processes. The objective is to link production and consumption processes such that waste, and waste management can become resources for manufacturing and production leading to a system of non-linear, closed loop production and consumption (European Commission 2015). By-products created during manufacturing, re-usable products, and materials at the product's end-of-life are no longer considered as waste but as a valuable resource to be put back into the 'system'. Resources are kept within the economy as secondary raw materials, to be used again and again to create further value (Stahel 2016; Wijkman and Skanberg 2015).

Sustainable consumption and production processes and practices are therefore, essential to achieving the circular economy. The EU's plan for a circular economy is expected to bring net savings of euros 600 billion for EU businesses, which could then be passed on to consumers thus improving competitiveness (SWITCH-Asia 2017), and an estimated 170,000 jobs are expected to be created by 2035. The economic value underlying such approaches cannot be underestimated. In the EU alone by 2030, there is a potential boost to the economy by 1.8 trillion euros (Ellen MacArthur Foundation 2016).

A good place to start is where waste currently is not being used as resource (CoE-Resources 2016). The hierarchy where prevention of use of resources is at the top with the lowest priority given to recovering energy from waste, is considered to be the most ineffective use of waste. In between these scales are other options in lower order of priority: reduce the use of resources, find a new product to support second hand use, maintain and repair the product, refurbish the product where the product itself is improved, creating new product from second hand product through the process of re-manufacture, reuse of the product for a different purpose, otherwise known as repurposing, recycling where the reuse of the raw material of the product is considered. Bocken et al. (2017, p. 487) argue that even the lower order priority of the circular concept from a waste perspective such as recycling is not universal.

Waste generated by developing countries in Asia is expected to rise by 60% in the forthcoming decade. Significantly less than 70% of the waste is collected and the same amount of waste is dumped without treatment into the environment (SWITCH-Asia 2017). Hence, waste management needs to be set up as a precondition to advancing towards a circular and inclusive green economy. In nature, waste is a resource for another set of species down the chain.

This paper commences with a brief description of circularity principles, followed by an understanding of what circularity means for the built environment. Two case studies from India are presented to highlight the importance of circularity in the built environment in emerging economies. Discussions and conclusions follow.

2   Circularity Principles

The circularity approach moves away from current approaches for underpinning our linear economy. Circularity is restorative and regenerative in its approach. It decouples growth from finite resource consumption (Ellen MacArthur Foundation 2015, 2017). Three principles for circularity have been touted as essential to advancing away from current linear approaches (Ellen MacArthur Foundation 2015, 2016; Ellen MacArthur Foundation et al. 2016).

Principle 1 Preserving and enhancing natural capital by controlling finite stocks and balancing renewable resource flows.

Where possible, utilities are virtual (they may have physical impacts), and high/higher efficiencies are favored. Processes that use renewable or better performing resources are considered. Natural capital has the opportunity to regenerate itself and support systemic flows of materials and nutrients leading to balanced ecological cycles.

Principle 2 Optimizing resource yields by circulating products, components and materials at their highest utility at all times, in both technical and biological cycles. This approach ensures that materials, products and components are at their highest value at all times.

This principle focuses on supporting refurbishing, remanufacturing and recycling to keep all products, components, and materials circulating and contributing to the economy. Material consumption per capita has been growing in many developing economies, as a result of the burgeoning middle-class population. Per capita use of material resources in the developed world has always far surpassed that of the developing world. Therefore, it is critical, now more than ever, that circularity approaches are put in place to avoid lock-in.

Principle 3 Fostering system effectiveness by revealing and designing out negative externalities.

Negative externalities of economic activity include land degradation, air, water and noise pollution, release of toxic substances and greenhouse gas (GHG) emissions. By taking a systemic approach, such negative externalities can be reduced or eliminated. This principle ensures that natural capital is not eroded by ensuring that waste and pollution are designed out to support cleaner, greener approaches where possible.

3 Circularity in the Built Environment

Circularity in the built environment considers mimicking natural stocks and flows as much as possible. The stocks and flows will design, build and operate a built environment that will generate power and food by setting up cycles of water, nutrients, materials, and energy to replicate the natural cycles. Nutrients, especially food waste may be used as organic fertilizer supporting both rural and urban agriculture. Municipal solid waste may thus be reduced, and waste from energy will support circular systems. Energy systems that are efficient, resilient, renewable, localized, distributed and reduce costs, while having a positive impact on the environment will be used. Spaces within and outside buildings will be healthy, improve quality of life for the users and will be made with materials that minimize the use of virgin materials. The built forms will use efficient construction techniques with maximized utilization rates. Building components will only be replaced if they cannot be maintained or renewed.

The design of the built environment itself will encourage the creation of local value loops (Ellen MacArthur Foundation 2017). This means that local production, local use and local knowledge that are often entrenched in thousands of years of cultural underpinnings will flourish. Supported by collective resource banks and digital applications to put the demand and supply components of the value chain together, negative externalities dominated by current global supply chains will be reduced or eliminated.

Spaces connecting buildings and built forms will be supported by accessible, affordable and effective means. As far as possible, transport will be used by highly efficient, electric-powered vehicles operated by renewable energy and where the designs themselves support durability, efficiency, easy maintenance and resource efficient manufacturing or remanufacturing. The vehicles will be shared and automated. There will be no pollution to air or water as a result of vehicular traffic.

4 Circularity in Buildings and Construction in India: Untapped Potential

The housing and construction industry are the largest sources of employment in India, accounting for 60% of the working population. As a result, these two sectors consume vast amounts of raw materials. Due to rising population, housing food and mobility will bear the brunt of most impact (World Bank 2017; IGEP 2013). Indian construction industry is expected to see a growth of 6.5% annually (Sustainability Outlook 2015) heading into the near future.

In addition to employment, urbanization of 60% of India's population will put severe stress on cities. Like other cities across the world, Indian cities are also expected to contribute to 75% of carbon emissions while using approximately 75% of natural resources (UNEP-SBCI 2016) while at the same time cities in India are expected to contribute to 75% of GDP by 2030 (Ministry of Urban Development India 2015).

Indian construction sector is expected to become the third largest globally (KPMG 2016). While globally, the construction industry uses about 40% of raw materials (UN Environment 2017; UNEP-SBCI 2016), in India, the construction sector uses about 20% of materials (WEF 2016; GIZ 2016). Attendant waste generated by the building and construction sector is also 40% (UN Environment 2017) whereas in India, it is about a third. In terms of energy use, globally buildings use 40% (UN Environment 2017) and in India, this is 34% (UNDP 2015). By 2020, 200 million metric tons of steel, 454 million metric tons of cement, and 311 million metric tons of bricks are expected to be used; by 2030, these figures are 500, 723 and 589 million metric tons respectively (Sustainability Outlook 2015).

Unexplored opportunities at the city level

As cities get more urbanized, smart and effective urban planning to optimize land utilization and transport flows are needed. Similarly, infrastructure for effective nutrient and material cycles are needed to ensure that natural capital is not eroded and materials not wasted. Maximizing asset utilization as much as possible ensures that accessibility and affordability may be maintained. Currently, 15% of India's offices stand vacant (Indian Express 2016), and sharing spaces further reduces operational costs and emissions associated with building operations. If such opportunities for space utilization are used, the sheer volume of buildings that need to be constructed in the country would reduce.

Designing buildings that apply circular economy principles to construction, operation, and end of use has untapped opportunities impacting buildings across its whole life cycle. Buildings designed for water and energy efficiency can provide savings across the whole-of-life of buildings. If buildings are well designed bearing in mind the local climate and context, energy and water use in buildings may be reduced. Buildings designed well above the minimum energy standards can support good design and use of passive techniques. If buildings can generate their own energy on site, this supports the move towards zero energy buildings. Likewise, using water efficient fixtures and fittings, capturing rain water, using grey water recycling systems in buildings can support overall reductions in the use of water, particularly, potable water.

Design will need to consider deconstruction, reassembly and future flexibility. To support these design principles, information on cost, the condition of the material or product, resource productivity, life cycle data, ownership, warranty, traceability are all required. Collaboration to support shared incentivization, transparency, innovation across the scale and between scales, and amongst sectors will need to be considered, in addition to longer term business models. These will result in better operation where performance over ownership, better utilization of the product or materials and more options are presented to consumers. Assets will need to consider materials and products at higher value for longer period of time, total cost benefits along whole of life chains, and maintenance and replacement certainties are needed. Material security, waste reductions over the life cycle and open and closed loop solutions need to be considered from a waste perspective (Ellen MacArthur Foundation 2016; Ellen MacArthur Foundation et al. 2016).

Modular construction has a huge potential in India. Modular approaches can reduce time, cost, and material use while supporting high quality construction. The role of building information modelling (BIM) before, during and after the construction processes may support ongoing building performance to optimize energy and water use throughout the life cycle of buildings. Such approaches to construction can support the immediate needs, particularly housing in the country.

Selection of appropriate building materials will reduce resource use and also support the materials to be reused later when buildings need to be dismantled. Buildings used as material banks, where materials are reused after useful life of the building ensures recycling of building materials, and reduction or elimination in the use of virgin materials (BAMB 2018). Therefore, transparency of material composition is a fundamental shift that needs to occur in standard building and construction practices. Use of QR codes or other forms of tagging can provide knowledge of chemical composition and related intelligence to ensure that materials can be sourced back to be used in buildings or infrastructure by supporting the supply chain system. Use of alternative building materials such as bamboo and engineered clay has a huge potential, yet to be tapped in the country.

Recycling of construction waste, including using recycled aggregates supports circular approaches. Examples such as using rice husks as binder and thermal insulation supports innovative approaches where, discarded materials from one sector; in this instance from the agriculture sector can be redirected into the buildings and construction sector. Other alternatives to clay fired bricks are autoclaved concrete blocks (AAC), recycled aggregates in places of conventional aggregates: sand and gravel, and ordinary Portland cement to be replaced by blended cement using recycled fly-ash or slag (Sustainability Outlook 2015). AAC blocks in particular, is expected to have a potential market worth USD 5.5 billion by 2030, whilst the consumption of raw materials and energy in the manufacture is the lowest (400 kg/m^3 consumption of raw materials and 60 kWh/m^3 consumption of energy), compared to calcium silicate bricks, standard clay fired bricks and porous clay brick (Sustainability Outlook 2015).

Energy consumption by the residential building sector in India is particularly significant. Given the vast number of buildings yet to be built by 2030, attendant energy consumption is expected to increase more than eight-fold by 2050, with annual electricity use per household predicted to increase 4.5 times, from 650 kWh in 2012 to 2750 kWh by 2050 (GBPN 2014).

There are of course, technological, market and operational risks. Lack of data for product performance impacts on the operational cost of the asset itself. Currently products are deliberately designed for shorter life spans with planned obsolescence in mind. Therefore, current mindsets need to be changed and the business models underlying such changes also need to be made.

Bringing in participatory links

Social enablers to make circularity a reality need to be catalyzed. In 'advancing the circular economy in Asia', the role of local actors has been emphasized (SWITCH-Asia 2017). Connections with practices and daily needs of people is critical to support the needs and daily aspirations of the large numbers of middle and lower income citizens of the developing regions of Asia, Latin America and Africa. To support local citizens, it is essential to set up effective policies around who shall act, where, what needs to be done, when actions will take place and how initiatives will be pursued.

New knowledge and capacity building

Building circular economy knowledge and capacity is required (Ellen MacArthur Foundation 2012, 2014, 2017). A vision that embraces circular product design across all areas of day to day life, governance and business models and reverse logistics are needed. Innovation is critical, and investment directed towards pilot projects is required to support circular economy activities. Current inertia of following trends from the 'west' and overcoming socio-economic challenges are needed. Integrating circular economy principles into strategy and processes to create value proposition are needed. Incentives, particularly government incentives can support cross functional collaborations forming the beginnings of circular approaches. Using buildings as material banks actually supports recovery of materials in the long term, and therefore, long term visions are required to support such practices.

Governments can play a role

The role of governments for such approaches are critical. Integrating circularity in current policies and regulatory frameworks supports and encourages private and public investment in seeking new models. As advancing circularity requires moving away from standard business practices, governments can support transition models and spur private sector investments. Building an evidence base is required, as is also, skilling or reskilling the work force to support capacity in circularity. It may also need new models of education. Preparing the current and future generation of students to break away from current linear thinking requires shaping a different pedagogy that supports circular models of learning and engagement.

The construction sector in India has been seduced by the Prime Minister's *Make in India* (Make in India 2018) program, set up in 2014 and this is anticipated to play a huge role in the next few decades. *Make in India* facilitates investment, fosters innovation, enhances skill development to build the best manufacturing infrastructure while protecting intellectual property.

Multi-stakeholder platforms that address key issues are critical to achieve systemic changes moving away from current linear models. Collaboration across sectors and moving away from silo thinking is urgently needed. Coordination and sharing case studies and examples will reiterate and enhance mutually supportive alignment towards circularity. Governments themselves can set examples of supporting circular models through public procurement and infrastructure. By leading by example, governments can 'walk the talk'.

Economic potential of circularity in India

The potential of using circularity principles for the Indian economy is currently untapped. The country's economic growth has grown at an average of 7.4% a year in the last decade (Ellen MacArthur Foundation 2016). Taking a circular pathway for the Indian economy can bring India annual benefits of USD $624 billion in 2050, compared to the current development path. This is equivalent to 30% of India's GDP in 2050 and 11% of GDP in 2030. Perhaps, the greatest potential lies in the trajectory India can take to ensure linear models of infrastructure and services are not set up in the first place, avoiding locking-in mechanisms, thus reducing the shift to move to circularity in the more advanced economies. This would put the India market well ahead of other mature economies.

An estimated 700–900 million m^2 of new commercial and residential space is yet to be built in India (McKinsey Global Institute 2010). The areas of focus to achieve this are: cities and construction, food and agriculture, and mobility and vehicle manufacturing. Household expenditure typically focuses on housing, food and mobility in urban and rural areas of India. The circularity pathway will also lower greenhouse gas emissions for the country: 23% lower in 2030 and 44% lower in 2050 compared to current BAU development path. Water usage in the construction industry would be 19% lower in 2030 and 24% lower in 2050 compared to BAU scenarios. Use of virgin materials in cities and construction, food and agriculture and mobility and vehicle manufacturing would be reduced. It would be 19% lower in 2030 and 24% lower in 2050 (McKinsey Global Institute 2009, 2010).

Supporting circularity will deliver cheaper products and services and reduce congestion and pollution in India. Decreased costs will support initiatives such as the *Pradhan Mantri Awas Yojana* (Housing for All) where 38 million units are yet to be built to the year 2030 (IBEF 2012). The Smart Cities Mission also need to consider water, sanitation and waste services at scale, creating more effective urban stocks and flow cycles. Higher efficiency of life cycle operating costs from both buildings and infrastructure perspectives will support rising housing and infrastructure needs of the country. Seventy percent of the buildings expected to stand in India by 2030 are yet to be built, compared with 25% in mature economies such as the UK (CoE 2016; CSE 2015; NRDC-ASC 2012). Digital technology to support connectivity may be considered in the current highly capable IT sector in India. Mobility increases and digitally enabled sharing solutions to increase utilization of floor space in buildings has tremendous, yet untapped potential to make significant inroads in the buildings and construction sector. Mobility increases through growth in car ownership has been on the rise. If models encouraging increased mobility with low car ownerships are in place, this avoids locking-in car ownership and associated emissions.

Investment in the transition to circular economy will need to also involve not just a financial outlay, but also research and development, asset and stranded investments, and subsidy payments to promote market penetration. Governments can support through subsidies to catalyze uptake and support public expenditure for digital infrastructures (Ellen MacArthur Foundation 2016). The role of circularity in the informal economy is quite high in countries like India, but the management practices may not necessarily provide value to all or provide value to only a small section rather than integration across the whole supply chain. For instance, 60% of discarded plastics are recycled in India compared to only 6% in the US (Ellen MacArthur Foundation 2016). Ninety-five percent of this activity happens in the informal sector. Furthermore, as these practices occur at the end of the supply chain, much of the value is lost and the opportunities for circularity are also low. Therefore, long term thinking is required.

The informal sector is also quite critical, particularly in a country like India, and indeed in other emerging economies. A systemic understanding of manufacturing and the service industry may support the creation of cross functional opportunities, including reverse logistics. This requires coordination and alignment of actors across the entire supply chain. For example, in the current business as usual model in the building and construction industry is the split incentive approach; where the developer has no vested interest in investing in green technologies or features (often at a higher cost) in buildings to ensure optimized operational performance as there is no benefit to the developer in investing in such features and technologies. The benefit of these are passed on to the users of the building. However, for owner occupied buildings and in most residential cases, the occupier ends up paying for the costs of energy and water, so it makes logical sense in such a scenario to invest up-front ensuring green building outcomes are achieved. In cases where the developer on-sells the building, clients can demand such features and technologies as the developer is keen to recover funds to move on to the next project. Such options may be pursued from a circularity perspective but bearing in mind transition costs associated with breaking free from business as usual models to models of circular economy.

As the middle-income segment of India's population continues to grow, the standard of living of the middle class is also expected to improve; leading to lower opportunities for recycling moving towards the direction of more advanced economies. It is critical therefore, that a systemic approach to circularity is supported either through policies and regulatory frameworks and government leading by example to support a circular model of engagement across the supply chain.

An example of a systematic approach to circularity can be seen through two case studies from India presented below. The two case studies show that most of the construction and demolition (C&D) waste can be recycled in productive ways by replacing natural stone and sand in structural and non-structural applications. The construction systems used in the second case study are advanced yet easy to use, decentralized and environmentally friendly. A combination of the two case studies not only shows a systematic approach to circularity in the construction industry but also an efficient, profitable and easy to use system.

4.1 Construction and Demolition Study in India

A country wide study conducted by Development Alternatives in India on C&D waste revealed that India generates about 716 million tons of C&D waste per year, second highest in the world after China. The study showed that about 90% of C&D waste can be used in productive and profitable applications. Presently some of the value-added products like wooden door and windows, steel, glass etc. are recycled by the informal sector and is a thriving market. These are sometimes used as such or recycled again. C&D waste can replace natural stone and sand in structural and non-structural applications. Life cycle assessment of C&D waste processing showed a 21% saving of carbon dioxide (CO_2) as compared to natural stone processing. Business case study supported by industry revealed that products made with C&D are 10–20% cheaper, compared to conventional products with equivalent or even better properties. Technology was developed and transferred to industries for production and application. The research group developed a training module and provided training to about 40 municipal corporations across India on management of C&D waste. Interventions with industry led to green certification of first C&D waste-based paver block by GRIHA (Green Rating for Integrated Habitat Assessment), an Indian-based green rating system. Management and use of C&D waste is also supported by many national policies such as Construction and Demolition Waste Management Rules, 2016 & Indian Standard (IS) Code 383 (Caleb et al. 2017a, b). Support provided to Municipal Corporations led to regulations for preferential procurement of C&D waste-based building materials in Ahmadabad, a city in the state of Gujarat in India (Caleb et al. 2017a, b) (Fig. 1).

4.2 Development Alternatives Building-Efficient Resource Utilization

As indicated already, globally, the construction sector accounts for 30–40% of all material flows. Resource-efficient measures hold significant material-saving potential of more than 40%. About 50 billion tonnes of materials may be saved if all the housing demand were constructed using resource-efficient options by 2030 (IGEP 2013) (Fig. 2).

Fig. 1 C&D waste processing unit in Ahmedabad (*Source* Development Alternatives)

Fig. 2 Vertical sunshades to
minimize East-West sunrays
at Development Alternatives
Headquarters (New Delhi)
(*Source* Development
Alternatives)

The Development Alternatives World Headquarters building exemplifies the
social and environmental values that the term 'Sustainable Development' seeks
to demonstrate including principles of circularity where possible. Its construction
involves a wide range of resource conserving strategies that include conservation
measures that harvest, reduce, reuse, recycle and recharge using scarcest resources:
energy and water. The construction systems used are an inventory of innovative and
green building materials and techniques that are easy to replicate in both urban and
rural areas (Holcim 2008). These include:

- Production systems for easy-to-use, quality prefabricated elements for roofs, floors
  and wall which are eco-friendly.
- Use of decentralized and on-site production methods such as utilizing debris from
  demolished buildings to make stabilized earth mud blocks.
- Use of low carbon cement-based building materials which are durable and have
  better strength compared to normal cement based materials.
- Use of advanced, environment-friendly construction systems that conventional
  contractors can easily adopt, such as ferro-cement channels and fly-ash blocks.
- Bespoke building management systems for water, energy and waste that can be
  adopted by urban neighborhoods to reduce their ecological foot-print (Holcim
  2008) (Fig. 3).

Fig. 3 Ferro-cement roofing channels used to minimize consumption of steel reinforcements (*Source* Development Alternatives)

5   Discussions

The significant impact that buildings and the built environment have on cities, and the rise of urbanization in India requires some rethinking in terms of how cities are designed and built. Despite schemes by the Indian government such as *Make in India* and *Housing for All*, fundamental shifts are required in thinking and approaching problems to make serious inroads in use of resources and arresting associated emissions.

The Ellen MacArthur Foundation (2016, p. 31) has offered six opportunities to shape Indian cities and the construction industry, described in Sects. 3 and 4, and summarized here. These are:

- Urban planning to optimize land utilization and transport flows
- Infrastructure for effective nutrient and material cycles
- Sharing and multi-use of spaces
- Buildings designed for energy and water efficiency
- Modular construction
- Selection and looping of construction materials.

A systemic approach to circularity will support and sustain the growth of urban centres in India. Urban planning that carefully supports residential, commercial zones, and optimizes land transport patterns to create thriving, liveable environments where demand for reduced energy, water and reduction in virgin, non-renewable resources are followed, is required. When well planned, this will also result in cutting costs and contributing to balancing natural stocks and flows to support the natural environment. This needs to be based on an infrastructure that treats waste as a resource. The informal economy that currently exists needs to be recognized as an important part of the supply chain in such a scenario.

To lead the transition to circularity, there are some key considerations:

1. *Building capacity* to incentivize and ensure the models of governance and decision- making processes are right to ensure incentives for short, medium and long-term value creation and cross functional collaborations.
2. Collaborations across the spectrum with *policy makers, private businesses and the informal economy* to support cross value chain networks and allow the support of additional value creation.
3. *Investing in opportunities* for circular economy based on cultural and contextual factors are needed. Scaling back on linear models will also need effort.
4. The role of government in supporting circularity is essential. Governments have a commendable influence in the market and can support the long-term vision for circularity:

    a. Governments can also support *regulatory frameworks and remove policy barriers.*
    b. Governments can create platforms for *multi-stakeholder collaboration.*
    c. Governments can support and *lead by example, circular models of public procurement and infrastructure.*

5. *Education* can build further capacities if circularity is embedded in primary, through to tertiary education so that circular thinking is not 'new'.
6. *Private and public organizations*, including universities can play a key role in supporting the transition to a circular economy.
7. *Further stakeholder research and engagement* within the country is required to scale up and mainstream circular economy opportunities in India (Ellen MacArthur Foundation 2016).

The next section presents conclusions.

6   Conclusions

Keeping circularity in mind, the building and construction industry needs to move from short term thinking to long term thinking. It needs to consider design for deconstruction, be innovative both in terms of design and supply chain considerations including non-linear financial models, use business as usual tensions between flexibility and durability, utilize new models of consumption and production where collaboration is the underpinning platform for producing outcomes.

In particular, waste needs to be built into construction practices today. New models of construction where the waste becomes resource upstream or downstream are desired. The recognition that we have the one planet and need to live within the stocks and flows of this one planet needs to be consciously taken into account in our everyday decision making. A change in societal mindset is needed where there is greater collaboration and less fragmentation.

Such approaches require Indian businesses to take initiatives, which may be outside their comfort zones, though they are well placed to lead the way in the transition. They need to build circular economy knowledge and capacity. Indian businesses need to innovate to create new products and business models. To demonstrate their success, they need to integrate circular economy principles into strategy and processes, collaborate with other businesses, policymakers, and the informal economy and invest in circular economy opportunities.

Beyond the existing initiatives in India, the government can set direction for the transition and create the right enabling conditions by setting direction and showing commitment, creating and enabling regulatory frameworks and removing policy barriers, creating platforms for multi-stakeholder collaboration, supporting circular models through public procurement and infrastructure and embedding circular economy principles into education.

Other organizations can play important supporting roles in the transition to a circular economy. In the short term, further stakeholder engagement and research is needed to create and maintain mechanisms for stakeholder dialogue and identify knowledge gaps and build an evidence base so these may be used to convince a variety of stakeholders and build further to support policy initiatives.

References

Bocken NMP, Ritala P, Huotari P (2017) The circular economy-exploring the introduction of the concept among S&P 500 firms. J Ind Ecol 21:487–489

Buildings as Material Banks (BAMB) (2018) About BAMB. Retrieved on 27th October from https://www.bamb2020.eu/about-bamb/

Centre of Expertise on Resources (CoE-Resources) (2016) The circular economy and developing countries. Retrieved on August 14th 2018 from https://hcss.nl/sites/default/files/files/reports/CEO_The%20Circular%20Economy.pdf

Centre for Science and Environment (CSE) (2015) CSE regional dialogue on building sense: towards sustainable buildings and habitat. Retrieved on August 14th 2018 from https://www.cseindia.org/regional-dialogue-on-building-sense-towards-sustainable-buildings-and-habitat-6145

Caleb PR, Gokarakonda S, Jain R, Niazi Z, Rathi V, Shrestha S, Thomas S, Topp K (2017a) Policy brief I: decoupling energy and resource use from growth in the indian construction sector a potential analysis study. GIZ GmbH Environmental Policy Programme, Germany. Retrieved on August 14th 2018 http://devalt.org/images/L2_ProjectPdfs/(3)REDecouplingBaseline.pdf?Oid=161

Caleb PR, Gokarakonda S, Jain R, Niazi Z, Rathi V, Shrestha S, Thomas S, Topp K (2017b) Policy brief II: decoupling energy and resource use from growth in the indian construction sector a potential analysis study. GIZ GmbH Environmental Policy Programme, Germany. Retrieved on August 14th 2018 http://devalt.org/images/L2_ProjectPdfs/(1)REDecouplingStudyAnalysis.pdf?Oid=159

Ellen MacArthur Foundation & ARUP (2018) From principles to practices: first steps towards a circular built environment. Retrieved on October 31st 2018 from https://www.arup.com/perspectives/publications/research/section/first-steps-towards-a-circular-built-environment

Ellen MacArthur Foundation (2017) Cities in the circular economy: an initial explo-
    ration. Retrieved on August 14th 2018 from https://www.ellenmacarthurfoundation.org/assets/
    downloads/publications/Cities-in-the-CE_An-Initial-Exploration.pdf
Ellen MacArthur Foundation (2016) Circular economy in India: rethinking growth for long-term
    prosperity. Retrieved on August 14th 2018 from https://www.ellenmacarthurfoundation.org/
    assets/downloads/Summary_Circular-economy-in-India_5-Dec_2016.pdf
Ellen MacArthur Foundation, ARUP & BAM (2016) Circular business models for the built envi-
    ronment. Retrieved on August 14th 2018 from https://www.ellenmacarthurfoundation.org/assets/
    downloads/ce100/CE100-CoPro-BE_Business-Models-Interactive.pdf
Ellen MacArthur Foundation (2015) Towards a circular economy: business rationale for an accel-
    erated transition. Retrieved on November 8th 2018 from https://www.ellenmacarthurfoundation.
    org/assets/downloads/TCE_Ellen-MacArthur-Foundation_9-Dec-2015.pdf
Ellen MacArthur Foundation (2014) Towards the circular economy: accelerating the scale-
    up across global supply chains, vol 3. Retrieved on August 14th 2018 from https://www.
    ellenmacarthurfoundation.org/assets/downloads/publications/Towards-the-circular-economy-
    volume-3.pdf
Ellen MacArthur Foundation (2012) Towards the circular economy-economic and business
    rationale for an accelerated transition, vol 1. Retrieved on August 14th 2018 from https://www.
    ellenmacarthurfoundation.org/assets/downloads/publications/Ellen-MacArthur-Foundation-
    Towards-the-Circular-Economy-vol.1.pdf
European Commission (2015) Closing the loop—an EU action plan for the circular econ-
    omy. Retrieved on August 14th 2018 from https://eur-lex.europa.eu/resource.html?uri=cellar:
    8a8ef5e8-99a0-11e5-b3b7-01aa75ed71a1.0012.02/DOC_1&format=PDF
Geissdoerfer M, Savaget P, Bocken NMP, Hultink EJ (2017) The circular economy—a new sus-
    tainability paradigm. J Clean Prod 143:757–767
Global Alliance for Buildings and Construction (GABC) (2017) Towards a zero-
    emission, efficient, and resilient buildings and construction sector—global status report.
    Retrieved on August 14th 2018 from https://www.globalabc.org/uploads/media/default/0001/01/
    35860b0b1bb31a8bcf2f6b0acd18841d8d00e1f6.pdf
GIZ Federal Ministry of the Environment, Nature Conservation, Building and Nuclear Safety (2016)
    Material Consumption Patterns in India. Retrieved on August 14th 2018 from https://www.
    international-climate-initiative.com/fileadmin/Dokumente/2016/GIZBaselineReportSummary_
    SinglePages.pdf
Global Buildings Performance Network (GBPN) (2014) Residential buildings in India: energy use
    projections and savings potentials. Retrieved on 14th August 2018 from http://www.gbpn.org/
    sites/default/files/02.%20India2014_Briefing_0.pdf
Holcim (2008) Office building in India development alternatives world headquarters. Holcim
    Foundation for Sustainable Construction, Switzerland. Retrieved August 14th from https://src.
    lafargeholcim-foundation.org/dnl/7e006509-87e9-4894-a442-68f3c256875d/DA_India.pdf
India Brand Equity Foundation (IBEF) (2012) Affordable housing in India: budding, expanding,
    compelling. Retrieved on August 14th 2018 from https://www.ibef.org/download/Affordable-
    Housing-in-India-24072012.pdf
Indian Express (2016) City Biz: 20% of Mumbai office space vacant, says JLL. Retrieved on
    August 14th 2018 from https://indianexpress.com/article/cities/mumbai/city-biz-20-of-mumbai-
    office-space-vacant-says-jll/
Indo-German Environment Partnership (IGEP) (2013) India's future needs for resources.
    Retrieved on August 14th 2018 from http://asiapacific.recpnet.org/uploads/resource/
    17be709b7b666696526eaa0c2a6ed620.pdf
KPMG (2016) Promising opportunities: Urban Indian real estate. Retrieved on August 14th 2018
    from https://assets.kpmg.com/content/dam/kpmg/in/pdf/2016/08/Urban_Indian_real_estate.pdf
McKinsey Global Institute (2010) India's urban awakening: building inclusive Cities, sustaining
    economic growth. Retrieved on August 14th 2018 from https://www.mckinsey.com/~/media/

McKinsey/Featured%20Insights/Urbanization/Urban%20awakening%20in%20India/MGI_Indias_urban_awakening_full_report.ashx

McKinsey Global Institute (2009) Preparing for China's Urban Billion. Retrieved on August 14th 2018 from https://www.mckinsey.com/~/media/McKinsey/Featured%20Insights/Urbanization/Preparing%20for%20urban%20billion%20in%20China/MGI_Preparing_for_Chinas_Urban_Billion_full_report.ashx

Make in India (2018) Retrieved on August 14th 2018 from http://www.pmindia.gov.in/en/major_initiatives/make-in-india/

Ministry of Urban Development, Government of India (2015) Smart cities, report on mission statement & guidelines. Retrieved on August 14th 2018 from http://smartcities.gov.in/upload/uploadfiles/files/SmartCityGuidelines(1).pdf

Murray A, Skene K, Haynes Kathryn (2017) The circular economy: an interdisciplinary exploration of the concept and application in a global context. J Bus Ethics 140:369–380

Natural Resources Defence Council and Administrative Staff College of India (NRDC—ASC) (2012) Constructing change: accelerating energy efficiency in India's buildings market. Retrieved on August 14th 2018 from https://www.nrdc.org/sites/default/files/india-constructing-change-report.pdf

Ness DA, Xing Ke (2017) Towards a resource-efficient built environment—a literature review and conceptual model. J Ind Ecol 21(3):572–592

Ritzen S, Sandstrom GO (2017) Barriers to the circular economy—integration of perspectives and domains. Sci Direct Procedia CIRP 64:7–12

Seto KC, Dhakal S, Bigio A, Blanco H, Delgado GC, Dewar D, Huang L et al (2014) Human settlements, infrastructure and spatial planning. In: Climate change 2014: Mitigation of climate change. IPCC Working Group III Contribution to AR5. Cambridge University Press, Cambridge, UK

Stahel WR (2016) Circular economy. Nature 531:435–438

Sustainability Outlook (2015) Sustainable materials in the construction industry. Retrieved on August 14th 2018 from http://sblf.sustainabilityoutlook.in/file_space/0_MI%20PDFs/Sustainable%20Materials%20Construction/Sustainable%20Materials%20in%20Construction%20Industry%20Jan%202015.pdf

SWITCH-Asia (2017) Advancing the circular economy in Asia. Retrieved on August 14th 2018 from https://www.switch-asia.eu/fileadmin/user_upload/SCREEN_final_singlepages02.pdf

Wijkman A, Skanberg K (2015) The circular economy and benefits for society—jobs and climate clear winners in an economy based on renewable energy and resource efficiency. Retrieved on August 14th 2018 from https://www.clubofrome.org/wp-content/uploads/2016/03/The-Circular-Economy-and-Benefits-for-Society.pdf

World Bank (2017) World development indicators—employment in agriculture (% of total employment). Retrieved on 14th August 2018 from https://data.worldbank.org/indicator/sl.agr.empl.zs

World Economic Forum (WEF) (2016) Shaping the future of construction: a breakthrough in mindset and technology. Retrieved on August 14th from http://www3.weforum.org/docs/WEF_Shaping_the_Future_of_Construction_full_report__.pdf

United Nations Development Programme (UNDP) (2015) Implementing energy efficiency in buildings: a compendium of experiences from across the world. In: International Conference on Energy Efficiency in Buildings (ICEEB). New Delhi, pp 3–158, http://www.undp.org/content/dam/india/docs/ICEEB%202015_Compendium.pdf

UN Environment (2017) Buildings and construction sector grows: time is running out. Retrieved August 15th 2018 from https://www.unenvironment.org/news-and-stories/press-release/buildings-and-construction-sector-grows-time-running-out-cut-energy

United Nations Environment Programme's Sustainable Consumption and Production Branch (UNEP-SBCI) (2016) Cities and buildings—UNEP-DTIE sustainable consumption and production branch. Retrieved on August 14th 2018 from https://issuu.com/rodrigovelasquezangel/docs/cities_and_buildings-unep_dtie_init

Modern City in the Perception
of Students–Architects

Olga Melnikova

Abstract The article discusses the problems of modern urban environment formation. Modern architectural environment is presented not only as a space organized by the law "benefit-strength-beauty", but also as an environment that meets the needs and requirements of the modern user. All currently available terms "sustainable architecture" or "smart city" refer primarily to the economy and construction technology, but do not reflect the aesthetic, symbolic, artistic and psychological aspects of architectural activity. Along with the developed infrastructure and effective information technologies, the modern "smart city" is an architectural environment that largely determines human behavior; it is an environment filled with images, symbols and meanings that form the "image of the city" and should be understood by any user; it is an environment which is designed to promote harmonious existence and sustainable development of the individual. In this regard, the need for in-depth study of the environmental characteristics of the space is particularly important for a professional architect who will design a comfortable urban environment. The article presents the results of empirical studies of the "image of the city" formation among the future architects at the most important stage—the initial stage of professional development, as well as comparative and factor analysis of the features of the existing architectural environment perception on the basis of the semantic differential method. The researchers revealed similarities in the emotional assessments of certain city types among students-architects and students of other specialties; the differences in the perception of the proposed "image of the city" characteristics among students of different professional training. In conclusion, there are examples of solving various environmental problems in different countries.

Keywords Sustainable architecture · Smart city · Architectural environment · Semantic differential method · Image of the city (city image)

O. Melnikova (✉)
Faculty of Architecture, Saint Petersburg State University of Architecture and Civil Engineering (SPbGASU), 2-ya Krasnoarmeiskaya st., 4, 190005 St. Petersburg, Russia
e-mail: melnikova.gasu@yahoo.com

© Springer Nature Switzerland AG 2020   757
W. Leal Filho et al. (eds.), *International Business, Trade and Institutional Sustainability*, World Sustainability Series,
https://doi.org/10.1007/978-3-030-26759-9_44

1   Introduction

25 September 2015, the General Assembly adopted the summit final report of the United Nations which was dedicated to the post 2015 Development Agenda and was called "Transforming our world: the 2030 Development Agenda for Sustainable Development". The preamble of the document stated that "this Agenda is a plan of action for people, the planet and prosperity". 17 goals and 169 targets were further formulated; they should be "integrated and indivisible" and ensure "the balance of all three components of sustainable development: economic, social and environmental" (UN General Assembly Resolution 2015). Along with global challenges, such as the eradication of poverty worldwide, the Agenda includes such challenges as the realization of human potential, as well as the right to exist in a healthy environment. Highlighting the main areas of great importance for the humanity and the planet (people, planet, prosperity, peace, and partnership), the authors outlined the main ways to solve the existing problems.

Among the seventeen goals devoted to various spheres of human life, the Agenda highlighted goal 11, which refers to the urgent problems of cities—"ensuring openness, security, resilience and environmental sustainability of cities and human settlements." The range of tasks to be solved by specialists in the near future is quite wide and includes not only the tasks of providing the population with affordable and safe housing, but also strengthening measures for the protection and preservation of the world cultural and natural heritage.

By 2050, according to expert estimates, 70% of the world's population will live in cities (Munro and Grierson 2017), so the creation of a healthy, comfortable, environmentally friendly architectural environment is a priority for the modern urbanist (Crabbé et al. 2017). It is easy to see that the concept of sustainable development of the city is largely intersected with the concept of a "Smart city". Most experts note that information and communication technologies (ICT) play a crucial role today, as well as innovations such as intelligent transport systems (ITS), "smart" water use, "smart" energy and "smart" waste management (Zhao 2016). However, experts in different areas recognize the fact that the introduction of «smart» technologies into the existing urban structure, as well as the design of a «smart» sustainable city from the zero cycle is a complex task (Newman 1999). Solving these and many other problems requires active implementation of an integrated and interdisciplinary approach, closer cooperation of specialists from various fields, including psychologists, power engineers, economists, builders, environmentalists and, of course, properly trained architects.

The place of architecture in the rapidly changing aspects of human life is widely discussed nowadays. The concept of Sustainable Development has also affected this type of human activity, giving rise to the term "sustainable architecture", which is widespread but not unambiguous. Along with the term "sustainable architecture" such concepts as "green architecture", "sustainable construction", "ecological architecture", "low-cost architecture", "high-tech architecture", "bioclimatic architecture", "energy efficient and smart construction" are often used (Salmina and

Bystrova 2015). Many definitions have an economic, social or technological bias, but architecture is not limited to any of these areas. All presented concepts relate primarily to the economy and technology of construction, but do not reflect the aesthetic, symbolic, artistic and psychological aspects of architectural activity. In architecture there is artistry, history, culture, conceptual design; it has its own expressive means (Salmina and Bystrova 2015) which cannot be reduced to a set of only technical or economically sound indicators.

Recent urbanization is a global problem that has led to the development of large cities as a major source of transformation and pollution, as well as to a significant change in the urban space in the different countries (Ding 2008). Scientific and technological progress, coupled with modern information technologies, on the one hand, greatly facilitate the everyday side of human life, but, on the other hand, increasingly alienate people from more favorable lifestyle. This leads to the fact that a person neglects communication with friends, relatives, nature. As a result of industrialization and urbanization, the sensory environment has become aggressive for sensory organs historically adapted to more positive effects (Melnikova 2011). Modern multi-storey buildings of many megacities are designed with maximum compaction which deprives a person of chamber, intimate spaces, recreational areas in which a person feels comfortable.

2   Methods

The modern city is a multifunctional and multidimensional structure, where material, spiritual, financial and human resources are concentrated (Melnikova 2016). In the middle of the 20th century, the population of cities began to increase dramatically, and it became clear that the existing methods and approaches to the organization of architectural space are outdated, knowledge of urban space is clearly not enough to make a comprehensive analysis of the problems of people's existence in the urban environment and provide basic solutions (Melnikova 2014). This encouraged the developers of a new approach to the urban environment to try to analyze the structure of the city from its inhabitant's point of view. One such innovator was Kevin Lynch, an American Professor and consultant in urban planning. He was one of the first to reveal the features of the urban environment perception and offered his vision of the ways of the modern urban space development. K. Lynch proposed to introduce the following elements into the structure of the city: roads, districts, nodes and landmarks (Lynch 1982).

Paths are communication lines along which people can move occasionally or continuously (streets, railways, motorways, pedestrian roads).

Lynch offers to consider the two-dimensional part of the city of average size as a district, which the user enters from the inside. The main characteristics of the area are the identifiability and the presence of a certain character.

Nodes are those points in the city, in which the observer can freely get: these are some places, focusing points (intersections, places of maximum concentration of any functions, for example, a cafe on the corner, a closed area).

In the course of the study, which was conducted in Saint-Petersburg in 2014 among the students of the fourth and the fifth year of architecture and construction university (SPbGASU), an attempt was made to analyze the image of one of the small Russian cities, using Lynch's theory about the structural elements of the city (Melnikova 2014).

The purpose of this study was to determine the characteristics of the formation and emotional evaluation of the "city image" by architectural specialties students. The hypothesis that was tested in this study implies the presence of professional features in the «city image» perception and students-architects assessment in comparison with students of general construction disciplines.

In the first part of the study, students had to use the proposed photos of the four objects, selected in accordance with Lynch's classification, to describe the impressions about the city, to imagine its size, location, population, age and gender characteristics of its inhabitants. As a sample, a small town with a standard housing was proposed, founded in the middle of the 20th century according to the construction standards of the USSR of that time. In the second part of the study, students, using the method of semantic differential (SD), had to evaluate the same four objects on the proposed scales. In the third part they had to finish one of the selected objects with the missing, according to the students, elements of the urban space.

The procedure of the study. A modified version of the binary opposition (SD) method was used as the main method in the study. SD data processing included factor analysis by the principal component method with Varimax rotation and scale extension according to the factors selected after rotation. The experiment involved two groups of senior students, a total of 166 people. The first group included students of architectural discipline, 87 people. The second one included students of technical general construction disciplines, 79 people.

3 Results and Discussion

The results of the first part of the experiment: almost all respondents described the city negatively or neutrally as a city devoid of individuality, features of urban development, with a small population.

The results of the second part. Objects presented for the assessment: 1-Yard, 2-Street, 3-Square and 4-Panoramic view of the city, made from the several hundred meters height.

Factor analysis of the most significant characteristics of the city image assessment

		Ā	A
	Factor	Dispersion	
1	"Aesthetics assessment"	60%	52%
2	"Space assessment"	37%	33%
3	"Creativity assessment"	–	15%

The content of the factor can be considered as a set of its features. The leading factor, with 60% dispersion among the students of general construction disciplines (Ā) and with 52% dispersion among students-architects (A) received the name "Aesthetics assessment". The second factor, with 37% dispersion among students of general construction disciplines and with 33% of dispersion among students-architects, was named "Space assessment". The third factor which was obtained only for students-architects, having 15% dispersion, was conditionally called "Creativity". All three variables included in it do not belong to any characteristic (Melnikova 2014). It can be assumed that the emergence of this factor is due to the presence of students-architects personal professional experience. Being one step away from being a professional, they were able to note the shortcomings of the urban environment, from the aesthetic and psychological point of view.

Thus, the method of semantic scaling with subsequent factor analysis allowed determining the features of the «city image» formation in the students-architects minds and building a model of subjective space. The selected factors represent a three-dimensional semantic space of the «city image» perception by students-architects. The statistical analysis showed that these factors describe the most noticeable differences between students-architects and students of general construction disciplines in the perception of certain characteristics of the «city image». Consequently, such differences are one of the manifestations of the urban space perception's features among students-architects.

The third part of our experiment was devoted to the existing urban environment reconstruction where students-architects were asked to draw graphically those ideas that make this urban space insufficient to be comfortable for human life. Photos of different types of the city were divided into groups in accordance with the theory of K. Lynch: nodes, areas, paths and landmarks. Students were asked to choose the types of work that they consider the most attractive for themselves and to finish the elements that the urban space lacks.

The most popular pictures among first-year students (50%) were pictures of the area, while students of the third (25%) and the fifth (22%) courses chose them less willingly. This is due to the fact that the architectural profession for the first-year students is, first of all, ambitions and self-realization; and the public space of open access offers the best opportunities for self-realization.

Among senior students, the most popular were pictures of the district (courtyard) and paths (streets)—they were chosen by 30% of the third-year students and 36% of the fifth-year students. This is due to the fact that over the course of training students come to understand that architecture, among other things, is also a concern for people, their needs and desires.

The street attracted all the students (29%, 25% and 22% in the first, third and fifth years respectively) as an element of architectural space and which is familiar to urban residents and those who comes from villages. It should be noted that students do not consider it necessary to equip the street not only with parking lots, which immediately cuts off car-users, but also with bins or garbage containers, which threatens to turn the street into a garbage dump over time.

Photos of the nodes (crossroads) for editing attracted the least number of students. For the first course (3%) crossroads are, first of all, road crossing, which implies the crosswalk. And among the fifth-year students-architects there was not anybody interested in the reconstruction of the intersection.

The new construction and reconstruction of the existing facades were chosen as the main elements for the completion. Students willingly operated with the height of new buildings, as well as completed the existing facades with the new elements such as balconies, canopies, awnings, advertising signs and even graffiti. The next most popular were the elements of landscape design: trees, shrubs, benches, paving, decorative fence. The pictures showed a large number of people and animals that suggests that students understand the city not only as a subject-spatial environment, but also as an environment designed for a comfortable existence of various groups of subjects. The new city in the view of students became more developed, modern, stylish and even fashionable, but it should be noted that this is just the only one, although the important side of the architectural environment—aesthetic. The ecological side of the city life in the view of the students includes bicycle paths and garbage containers in the yards, and some students even offered to clean the street from trees, because they closed the view of the buildings facades. But the modern city is unthinkable without separate garbage collection, without public toilets, bins, benches under canopies, energy-saving lighting and all the other things that make life easier for the citizen. All these elements with a reasonable design approach can become a bright and unforgettable accent in the urban environment, and indispensable assistant in the rapid pace of human life.

4 Conclusions

Architectural space is a territory filled with physical objects. But it also involves a person; it is perceived by him and provides the conditions for his life. The urban environment not only delimits and intelligently organizes its components, but also provides information about the object. This allows the architectural environment along with aesthetic, sensory, cognitive and other functions to support certain human behavior (Melnikova 2011). To see the examples of some urban elements transformation into the artist built, you can refer to the experience of some Western countries.

The following project was developed especially for London. Bins Renew Bin are equipped with large LCD displays, thanks to which you can find the latest news, weather forecast, exchange rates, aircraft schedules and access the network via Wi-Fi.

The problem of public toilets lack, especially in the evenings, in busy places, appeared long ago. Therefore, many European countries have found a way out of this situation by installing night Elevator toilets Urilift in the central streets.

Elevator toilet is a stainless steel cabin which is equipped with a sewerage system, lighting and automatic flushing. Urilift also has a heating system in cold weather. And there are many examples of "smart" use of architectural space elements; one has only to be a little bolder both in their plans and in the ways of their implementation. This requires a qualitative shift in the thinking of those who are responsible for the modern comfortable urban environment. The quality of the designed space depends not only on the technical knowledge received by students-architects during their studies at the university, but also on the ideas the architect is influenced currently.

Our research has proved the presence of professional features in the perception, evaluation and formation of the "city image" among students-architects. It can serve as a basis for future theoretical and empirical work, indicating the vector for researches, the most relevant ones in our time. There is no doubt that modern architects with such powerful tools as rich imagination, subtle perception and the principles of humanism, safety and environmental friendliness as the purpose of their activities can and should form such an architectural environment that modern residents expect from them. And these universal values are absolutely correlated with the principles of Sustainable development.

In order to implement the principles of the Sustainable development concept, the architect needs to add a psychological component to his activities, in addition to technical and aesthetic. And the "Smart city" in this case is presented not only as an innovative information and high-tech space, but also as an environment that helps a person to realize his potential, as well as satisfying his right to exist in a comfortable, healthy environment. The modern architect, thus, becomes not just the author who dictates the conditions of existence in the designed environment, but the co-author of that environment, who takes into account both wishes and hopes of future users, and consequences of the activity for a wide range of people. At the same time, Professor MARHI, Doctor of Arts Vyacheslav Glazychev noted that the transition to new methods cannot be carried out in the near future; the same revolutionary shift in thinking requires long-term work of generations; and that is why it is not an abstract future, but the reality of today's choice for any specialist operating in the arena of the city (Glazychev 1998).

References

Crabbé A, Bergmans A, Craps M (2017) Participation in spatial planning for sustainable cities: the importance of a learning-by-doing approach. In: lifelong learning and education in healthy and sustainable cities, pp 69–85

Ding CKD (2008) Sustainable construction—the role of environmental assessment tools. J Environ Manag 86(3):451–464

Esaulov G (2014) Sustainable architecture—from principles to development strategy. Vestnik TGASU 6(9)

Glazychev V (1998) Selected lectures on municipal policy. Lecture 4, Environmental approach in the development of the city. Lecture course

Lynch K (1982) Image of the city. Translated from English VL Glazycheva; Comp. AV Ikonnikov. Under the editorship of AV Ikonnikov. Stroyizdat, Moscow, p 328

Melnikova O (2011) Concepts of subject-spatial environment in psychology and architecture. In: Actual problems of modern construction: 64th international scientific and technical conference of young scientists. Spbgasu, 3 h. h. III. SPb., p 264

Melnikova O (2014) Complex analysis of "city image" formation by students on the basis of semantic differential method. Bull Civil Eng 3(44):302–307

Melnikova O (2016) Ergonomic aspects of the subject-spatial environment perception. In: In the collection: architecture-construction-transport materials 72-nd scientific conference of professors, teachers, researchers, engineers and graduate students of the university, pp 248–251

Munro K, Grierson D (2017) Nature, people and place: informing the design of urban environments in harmony with nature through the space/nature syntax. In: Lifelong learning and education in healthy and sustainable cities, pp 105–125

Newman WGP (1999) Sustainability and cities: extending the metabolism model. Landscape Urban Plan 44(4):219–226

Our common future. Report of the world Commission on environment and development. www.un.org/ru/ga/pdf/brundtland.pdf

Salmina O, Bystrova T (2015) The principles of sustainable architecture creation. Academic Bulletin of the Ural Research Institute project RAACH (4)

UN General Assembly Resolution A/RES/70/1 (2015) Transforming our world: The 2030 agenda for sustainable development. September 18, 2015. United Nations Assembly Resolution, New York. http://www.un.org/ga/search/view_doc.asp?symbol=A/RES/70/1&Lang=E

Zhao H (2016) MSE News 2

Sustainable Logistics: A Case Study of Vehicle Routing with Environmental Considerations

Aline Scaburi, Júlio César Ferreira and Maria Teresinha Arns Steiner

Abstract Logistics is one of the sectors that requires large investments of resources and planning in an organization. At the same time, it also demands maximum operational efficiency. Because of its impacts on a company's processes, it is possible to realize quickly the benefits that sustainable initiatives enable. This paper aims to optimize the newspaper delivery process of a printing company located in the city of Curitiba, in the State of Paraná, Brazil. For this purpose, the mathematical model of the Vehicle Routing Problem (VRP) is applied, and then, based on the optimized deliveries proposal, the intention is to evaluate the reduction in greenhouse gas emissions with the help of the GHG Protocol calculation tool. This tool is widely used by companies and governments in GHG inventories because it is in line with the parameters of the Intergovernmental Panel on Climate Change (IPCC). With the application of the proposed methodology, companies can find a new solution for the distribution process, demonstrating its economic, social and environmental responsibility. With regard to the economic aspect, operational costs will be reduced. From a social viewpoint, there is a greater customer satisfaction. Furthermore, the number of respiratory diseases will be reduced due to fewer emissions of atmospheric pollutants. In terms of the environment, there is a reduction in GHG. All these factors correspond to competitive advantages over other competitors. The tools presented here can be replicated in logistics distribution processes of a wide variety of segments, and is intended to increase sustainability in this segment.

A. Scaburi (✉)
Industrial Engineering Program, Universidade Federal do Paraná, 100 Cel. Francisco H. dos Santos Ave., Curitiba, Paraná 81530-000, Brazil
e-mail: alinescaburi@ufpr.br

J. C. Ferreira · M. T. A. Steiner
Industrial Engineering and Systems Program, Pontifícia Universidade Católica do Paraná, 1155 Imac. Conceição St., Curitiba, Paraná 80215-901, Brazil
e-mail: ferreira.julio@pucpr.edu.br

M. T. A. Steiner
e-mail: maria.steiner@pucpr.br

© Springer Nature Switzerland AG 2020
W. Leal Filho et al. (eds.), *International Business, Trade and Institutional Sustainability*, World Sustainability Series,
https://doi.org/10.1007/978-3-030-26759-9_45

Keywords Vehicle routing problem (VRP) · Optimization in newspaper
distribution · Reduction of emissions of greenhouse gases (GHG) · GHG protocol

1 Introduction

Since the last century, scientists have observed that concentrations of carbon dioxide (CO_2) in the atmosphere have increased significantly compared to pre-industrial levels. In 2016, the average CO_2 concentration was approximately 40% higher than in the mid 1800 s. Among the many human activities that produce greenhouse gases, the use of energy in its different forms is by far the major source of emissions, corresponding to 68% of total CO_2 emissions. This number is far higher than agriculture, which accounts for 12% of emissions, or industrial manufacturing processes, which correspond to 7%. Other activities, such as waste storage and indirect and non-agricultural emissions, correspond to 14% (IEA 2017).

The energy sector includes direct emissions from fuel combustion (fossil and/or renewable) and "fugitive" emissions, which are intentional or unintentional releases of gases resulting from the production, processes, transmission, storage and use of fuels. Among all the energy sector composition, two sub-sectors, (1) electricity generation and (2) transport on their own produced two-thirds of global CO_2 emissions emanating from fuel combustion, and the transportation sector accounted for 24% of these emissions in the year 2015 (IEA 2017).

The perceived increase in transport activities for the distribution of goods and services in urban areas in recent years is responsible for generating not only environmental impacts but also social and economic impacts. These can be highlighted as problems originating from the transport activity, the intensification of traffic jams in road networks, noise emission, emission of atmospheric pollutants and greenhouse gases (GHG), as well as the reduction of the safety of the population, due to the increase in the number of accidents involving vehicles (Oliveira et al. 2018).

However, the need to transport goods between producer and consumer centers is inevitable. The logistic distribution process must therefore be seen as the link between the market and the supply center, and the main objective of this process within an organization is to plan and coordinate all the activities necessary to ensure the availability of products and or services to customers, when and where they are needed. In this sense, logistics operations, among which the physical transportation of goods stands out, must be carried out by organizations efficiently, with a view to customer satisfaction and acceptable and competitive operating costs. (Christopher 2012).

According to Vasconcelos (2015), the integration of logistics systems and efficient supply chain management can result in gains in the form of competitive advantages, which, when aligned with good environmental practices, contribute to the sustainability of the process. According to Bertaglia (2016), the use of technological tools enables efficient management of transportation operations and commonly employs vehicle programming and routing techniques to define these activities, aiming to eliminate waste and obtain effective gains in the process.

In this context, the objective of this work is to provide a reduction of greenhouse gas emissions during the logistic process of newspaper delivery by a company located in the city of Curitiba, Paraná State, Brazil, through the application of mathematical techniques adopted in vehicle routing problems, quantifying this reduction using the GHG Protocol calculation tool. With the proposal of new optimized delivery routes, it also enables an improved operating result for the company in question by reducing operational costs.

The contribution of this work is the use of exact mathematical procedures and the optimization approach used to solve the real problem of distribution, given the optics of data collection. Researchers usually apply optimization techniques (exact or heuristic) using the lowest Euclidean distances between the demand points. However, in this work, the minimum real times of displacement between the demand points obtained through Google Maps will be considered, constituting a matrix of asymmetric displacement times.

This work is organized as follows. The theoretical background presents the vehicle routing problem and its mathematical model, some work related to this study, a brief description of Greenhouse Gas Emission Inventories. To finalize the theoretical rationale, the Brazilian Program for the control of emissions of greenhouse gases is presented. This is followed by a description of the method and details of the case study in question, an analysis of the results and, finally, the conclusions.

2 Theoretical Background

2.1 Vehicle Routing Problem (VRP)

The Vehicle Routing Problem (VRP) was introduced by Dantzig and Ramser under the title "The Truck Dispatching Problem". This problem consists of determining how a group of consumers will be served by means of a fleet of vehicles, starting from an established point (called a warehouse) and returning to it after delivery, in order to minimize the total distance traveled or the time involved in the displacement (Dantzig and Ramser 1959).

According to Laporte (2009), the VRP is an extension of the well-known Traveling Salesman Problem (TSP). However, as it incorporates some other characteristics of the real process of distributing goods in relation to the fleet, such as capacity and demand, its solution is more difficult in relation to TSP. The VRP is part of the NP-difficult problem class.

The mathematical model of the VRP, according to Arenales et al. (2007), is presented as equations and inequalities, as shown below in (1)–(8), in which the VRP is completely represented by a directed graph, $G = (N, E)$, where $N = C \cup \{0, n + 1\}$, $C = \{1, ..., n\}$ is the set of nodes that represent customers, and $\{0, n + 1\}$ are the nodes representing the warehouse and $E = \{(i, j)\}: i, j \in N, i \neq n + 1, j \neq 0\}$ the arcs between the nodes that represent the connections between the demand points. The cost involved in passing through a given arc generally represents the distance between the nodes to accomplish this displacement.

In this model, where the decision variables are $X_{ijk} = 1$, if the arc (i, j) is traversed by the vehicle k, $i \neq j$ and $X_{ijk} = 0$, otherwise. The objective function in (1) seeks to minimize the total cost of transport, which is represented by the total distance or travel time. The constraints on (2) define that each customer must be visited only once by a vehicle. In (3) there are vehicle capacity constraints, where the total demand served on a route must be less than or equal to the capacity of the vehicle K. The other constraints (4), (5) and (6) ensure the flow conditions in networks, determining that each vehicle K leaves the warehouse (node 0) once and must return to the warehouse (node $n + 1$). It should be noted that the constraints (6) maintain the network structure. The constraints in (7) do not allow the formation of sub-routes and those in (8) present the domain of the decision variables.

$$\text{Min} \sum_{K \in K} \sum_{(i,j) \in E} c_{ij} X_{ijk} \tag{1}$$

Subject to:

$$\sum_{k \in K} \sum_{(i,j) \in E} X_{ijk} = 1 \quad \forall i \in C \tag{2}$$

$$\sum_{i \in C} di \sum_{j \in N} X_{ijk} \leq q_k, \quad \forall K \in K \tag{3}$$

$$\sum_{j \in N} X_{0ij} = 1, \quad \forall k \in K \tag{4}$$

$$\sum_{i \in N} X_{ihk} - \sum_{j \in N} X_{hjk} = 0, \quad \forall h \in C, \quad \forall k \in K \tag{5}$$

$$\sum_{i \in N} X_{i,n+1,k} = 1, \quad \forall k \in K \tag{6}$$

$$\sum_{i \in S} \sum_{j \in S} X_{ij} \leq |S| - 1, S \subset 1, 2 \leq |S| \leq \left[\frac{n}{2}\right], \quad \forall k \in K \tag{7}$$

$$X \in B^{K|E|} \tag{8}$$

where K total number of vehicles; $n =$ total number of customers; $c_{ij} =$ cost of the arc (i, j); $d_i =$ demand for node i; $q_k =$ vehicle k capacity.

There are several variants of VRP in the literature due to the addition of new conditions so that the approach is closer to the real situation. One may cite as some examples of these variants, the VRP with Time Windows when temporary delivery windows are defined and when there being more than one warehouse, one has the

Multi-Depot VRP (MDVRP). When there is an extension of the planning period for a given period of days, one has the Periodic VRP. When not all the information relevant to the planning of the routes is known a priori, or when relevant information to the routing process can be changed after the initial routes have been constructed, there is the Dynamic VRP (D-VRP). When, in addition to deliveries, there is also the need to schedule and make collections, there is the Pickup and Delivery Problem (Braekers et al. 2016).

When environmental aspects are included in the VRP, we can have, according to the chosen objectives, three different approaches, the Pollution Routing Problem (PRP), the Green VRP and the VRP in Reverse Logistics (VRPRL). The PRP seeks vehicle routing that produces a smaller amount of pollution, in particular, with the reduction of greenhouse gases emissions. The green VRP aims to optimize energy consumption in transport and also seeks to reduce fuel consumption. The VRPRL focuses on the aspects of reverse logistics distribution (Ferreira et al. 2018).

2.2 Some Related Work

The 9 most cited articles on the TSP and VRP are presented in Table 1 in descending order of the average number of citations per year. It is organized by authors (including the year of publication), article title, journal, number of citations obtained through Google Scholar and the average number of citations per year since its publication. The article with the highest average citations per year is the work of Deng et al. (2015), with a total of 64 citations. Considering only the number of citations instead of the average number of citations per year, the most relevant work in the base is Hashimoto et al. (2008), with 138 citations.

Deng et al. (2015) proposed a new strategy for the generation of the initial population of the Genetic Algorithm (GA) using the k-means algorithm. Subsequently, the TSP was solved through the application of the GA. As the strategy decreases, the best error value when compared to the random initial population method at 29.15% and that of the initial greedy population at 37.87%, it is a good alternative resolution for the TSP. The TSP Multi-Paths with Stochastic Travel Costs (stochastic displacement times) were addressed by Tadei et al. (2017), making a deterministic approximation of these times to enable resolution. The solution obtained, when compared to that obtained by the Monte Carlo simulation, shows both the precision and the efficiency of the deterministic approach, with an average percentage gap of around 2%.

Cacchiani et al. (2014) solved a periodic VRP using a hybrid optimization algorithm by Linear Integer Mixed Programming (LIMP), incorporating a heuristic and exact components. A generalization of the VRP with Time Windows, allowing travel times and costs to be time-dependent functions, was presented by Hashimoto et al. (2008). The authors proposed a local search algorithm to determine the vehicle routes, and the sub optimization problem in relation to the time windows was solved by Dynamic Programming.

Table 1 Most relevant articles considering the number of citations

Authors	Title	Journal	Google Scholar citations	Average citations per year
Deng et al. (2015)	An improved genetic algorithm with initial population strategy for symmetric TSP	Mathematical Problems in Engineering	64	21.3
Tadei et al. (2017)	The multi-path traveling salesman problem with stochastic travel costs	Euro Journal on Transportation and Logistics	21	21.0
Cacchiani et al. (2014)	A set-covering based heuristic algorithm for the periodic vehicle routing problem	Discrete Applied Mathematics	66	16.5
Hashimoto et al. (2008)	An iterated local search algorithm for the time-dependent vehicle routing problem with time windows	Discrete Optimization	138	13.8
Yang et al. (2008)	Ant colony optimization method for generalized TSP problem	Progress in Natural Science	126	12.6
Odili e Kahar (2016)	Solving the traveling salesman's problem using the African Buffalo optimization	Computational Intelligence and Neuroscience	21	10.5

(continued)

Table 1 (continued)

Authors	Title	Journal	Google Scholar citations	Average citations per year
Mavrovouniotis et al. (2017)	Ant colony optimization with local search for dynamic traveling salesman problems	IEEE Transactions on Cybernetics	10	10.0
Jun-man and Yi (2012)	Application of an improved ant colony optimization on generalized traveling salesman problem	Energy Procedia	54	9.0
Majumdar and Bhunia (2011)	Genetic algorithm for asymmetric traveling salesman problem with imprecise travel times	Journal of Computational and Applied Mathematics	61	8.7

Source Prepared by the authors (2018)

Yang et al. (2008) used a heuristic approach, the ant colony optimization method, to solve the TSP. The results showed that when the 2-opt local search mutation process is used, the results obtained are better. However, the authors showed that the results are satisfactory only when the scale of the problem is less than 200 cities.

Odili and Kahar (2016) solved the TSP by proposing a new meta-heuristic, African Buffalo Optimization (ABO), a population-based stochastic optimization technique. The results obtained were satisfactory when compared to other heuristics, but the authors recommended new studies to establish the effectiveness of the proposed heuristic. A Dynamic TSP was approached by Mavrovouniotis et al. (2017), where the weights (or travel times) between two municipalities are subject to change. The resolution was obtained through the application of a memetic version of the Ant Colony Optimization (ACO) algorithm, with the integration of a local search operator. The results demonstrated that the method maintains a good balance between computation time and the quality of the approximate solution obtained.

Likewise, Jun-man and Yi (2012) used the ACO algorithm to solve the TSP, but the problem addressed did not consider the dynamic changes as in the case of Mavrovouniotis et al. (2017). Majumdar and Bhunia (2011) dealt with a version of the Asymmetric TSP, where the travel time between each pair of cities is represented by a range of values (in which the actual travel time is expected to occur) instead of a fixed (deterministic) value, as in the classic Asymmetric TSP. The solution was found by proposing a modified GA. The tests performed by the authors show that the results of the alternative approach are satisfactory when compared to traditional GAs.

2.3 Greenhouse Gas Emission Inventories

The greenhouse effect is a natural phenomenon characterized by the thermal insulation of the planet due to the concentration of gases (CO_2—carbon dioxide, CH4—methane and N20—nitrous oxide) in the atmospheric layer. A part of the solar rays is absorbed by the planet and transformed into heat enabling the maintenance of a mean planet temperature. However, the natural concentration of these gases has increased, influenced by human action, mainly by the burning of fossil fuels, deforestation, agriculture and industrial production. The intensification of these emissions means that some rays do not return to space, causing a rise in the temperature of the planet that configures global, leading to climate change (Sirvinskas 2018).

Because of the impacts related to climate change, many government initiatives have sought measures to reduce GHG emissions through public policy actions and the promotion of programs to contain climate change (Santos and Paiva 2018).

The first step in the management of GHG emissions and contributing to the fight against climate change is to know the profile of emissions from the diagnosis made by the inventory and then establish strategies, plans and targets for emission reduction and management. The inventory of GHG emissions has been the tool used by the institutions to identify, quantify and manage the positive and negative GHG emissions of their processes to increase the efficiency of their operational activities, while mitigating their impact, taking into account public policies, legal obligations or market demand.

Estimates of GHG emissions depend basically on the multiplication of activity data and a respective emission factor. Activity data are defined as a function of the magnitude of human activity over a given period of time, while the emission factor is the average emission rate of a certain GHG for a given activity (Azevedo et al. 2018).

2.4 GHG Protocol and Brazilian Program

According to FGV (2016), the GHG Protocol Program is a tool created in 1998 in the United States by the WRI (World Resource Institute) to understand, measure and manage GHG emissions. This tool is currently used by companies and governments around the world in GHG inventories, since it conforms with IPCC (Intergovernmental Panel on Climate Change) parameters.

Even though it is a worldwide program, the GHG Protocol was adapted according to the characteristics and needs of Brazil, resulting in the Brazilian GHG Protocol Program. The implementation of the Brazilian GHG Protocol Program is an initiative of the Center for Sustainability Studies, Fundação Getúlio Vargas (FGV) and WRI, in partnership with the Ministry of the Environment (MMA), the Brazilian Business Council for Sustainable Development (CEBDS) and the World Business Council for Sustainable Development (WBCSD), which seeks to promote the corporate culture of measuring, publishing and voluntary management of GHG emissions in Brazil. This provides participants with access to international quality instruments and standards for accounting and drafting GHG inventories (FGV 2016).

It should be noted that there is a growing commitment by companies to the Brazilian GHG Protocol program, since companies that seek to keep growing and ensure their continuity need to improve their image and reputation with their stakeholders (Santos et al. 2018).

The GHG Protocol Calculation Tool is available in a *Microsoft Excel®* software file on the website of the Brazilian GHG Protocol Program, and covers scopes of direct and indirect GHG emissions. Direct emissions are those related to sources that are owned or controlled by the company. The indirect ones are related to company activities, issued by sources that belong to or are controlled by another company. The definition of the scope does not depend on the type of source (such as equipment that burns fuel), but on who owns the control of the source. To help define this operational limit, three scopes were established: Scope 1, direct emissions; Scope 2, indirect emissions from the use of purchased energy purchased from third parties; and Scope 3, indirect emissions related to goods and services purchased, or goods and services sold which in general represent a very significant proportion of GHG emissions from an activity (FGV 2016).

Figure 1 shows the navigation menu of calculus tool GHG Protocol, where the three scopes are addressed for the preparation of the inventories.

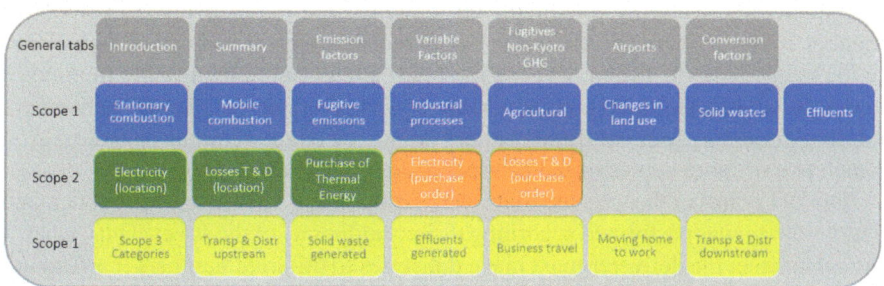

Fig. 1 Navigation Menu of the Brazilian Calculation Tool GHG Protocol Version 2018.1.4. *Source* GHG Protocol Tool Version 2018.1.4 (2018)

More specifically, in Scope 3, the item Transport and distribution (upstream), GHG emissions refers to the transportation and distribution of products purchased or acquired by the inventorying organization in the inventory year in vehicles and installations that are not owned or operated by organization. It also includes other outsourced transportation and distribution services (including outbound logistics for distribution and final product delivery).

3 Method—Case Study of the Newspaper Distribution Company

The newspaper distribution process analyzed here is carried out by a service provider to a print shop in the municipality of Curitiba. This process consists of the delivery (after weekly printing) of a single product. The printing company performs the printing process on Fridays, and delivery to subscribers and retail outlets (newsstands) occurs through distributors who retrieve newspapers from the printing company every Friday at 2 p.m. After the product has been collected, the distributor separates, labels, orders and distributes the printed newspapers. The delivery process begins around 9:00 pm on Friday and continues until 8:00 a.m. on Saturday.

The company in question delivers the newspapers to the municipalities of: Curitiba, Colombo, Almirante Tamandaré, Pinhais and São José de Pinhais, the last four being part of the metropolitan region of Curitiba. The distribution of deliveries is divided between two different distribution companies, one serving the regions located to the south and another serving the regions to the north. The average weekly delivery is to 18,000 subscribers and 1400 copies to newsstands.

In this analysis, we will consider the deliveries made by the distributor to the subscribers of the regions to the south. During the survey, this distributor made deliveries using 40 motorcycles with a capacity of 100 copies per trip. The distributor follows 89 different routes to make the deliveries.

In order to propose an optimized delivery route and analyze the reduction of GHG emissions, one of the 89 roadmaps will be discussed here, specifically Roadmap number 1010002 (name given by the company). It serves the suburb of Água Verde in the municipality Curitiba, where 479 copies are delivered by a single motorcycle courier to 194 different stop points. More copies of the newspaper are delivered at the stop points, as some of these demand points are condominiums or buildings, to which two or more copies of the newspaper are delivered.

The 194 points of the selected route were grouped into 35 macro points of demand. The grouping was performed according to the "midpoint" of the street. Thus, when in the same street the deliverer must make deliveries to different numbers (corresponding to different stops), we considered the passage of the same by the midpoint of the respective street, corresponding to a macro point. This grouping is shown in Fig. 2, where the point identified in red corresponds to the warehouse. In order to obtain the coordinates and times between the demand points, the Google Maps website was

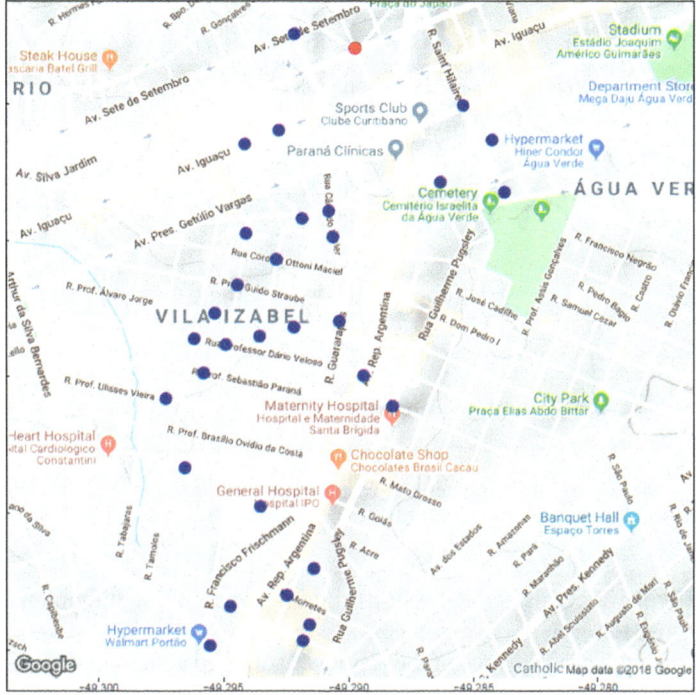

Fig. 2 Grouped demand points, with warehouse indicated in red. *Source* Prepared by the authors (2018)

used. The data regarding capacity and time of displacement were then compiled in matrices using Microsoft Excel software, and used in the interface with LINGO 12.0 (Language for Interactive Optimizer) software for the solution of the mathematical model of the VRP ((1) to (8) shown above), seeking to minimize the total time of deliveries while respecting the load capacity of the motorcycle on each route. As the capacity per trip is 100 copies of the newspaper and the demand to be met is 479 units, it was necessary to prepare five routes, departing from and returning to the warehouse.

4   Results and Discussion

From the LINGO software, the new routes obtained by minimizing the execution time of the deliveries were compiled and compared with the routes currently used by the distributor. Both (the current scenario and optimized scenario) are shown in Fig. 3. The proposed solution was obtained with 2 h of computational processing of

Fig. 3 Illustration of the current routes and proposed optimized routes. *Source* Prepared by the authors (2018)

the LINGO software, this being the stop criterion, where the solution obtained has a total of 87 min to serve the five routes.

To compare the performance of the proposed route in relation to the current route, Table 2 shows the travel time of each of the five routes, as well as the distance and total time involved in both the current situation and the proposed situation. The

Table 2 Comparison of the current situation and the proposed routing situation

Current scenario	Demand	Time (min)	Distance (km)
0-1-8-24-2-17-0	101	15	7.5
0-18-26-29-16-13-20-7-32-0	103	19	9.5
0-34-9-19-33-14-22-23-25-0	97	28	14
0-15-5-3-6-28-27-10-4-21-12-11-31-0	108	40	20
0-30-35-0	70	23	11.5
Totals	479	125	62.50
Optimized scenario			
0-2-13-20-16-35-22-14-7-0	96	17	8.5
0-8-34-25-6-15-5-3-24-19-9-0	99	23	11.5
0-17-1-0	97	6	3
0-26-29-18-32-33-23-0	88	13	6.5
0-28-4-21-31-12-11-30-27-10-0	99	28	14
Totals	479	87	43.50

Source Prepared by the authors (2018)

current deliveryman's routing is defined without any calculation tool, only through the deliverer's experience. In this way, the activity takes on average 120 min, surpassing in some cases the capacity of the motorcycle to load 100 copies of the newspaper per trip.

In the optimized proposal, there is a reduction of 55 min in the operation and 19 km less displacement. In addition, there is an improvement in the number of copies of the newspaper to be distributed per itinerary, and it is now possible to meet the capacity constraint of loading the motorcycle with 100 copies per trip.

The GHG emission calculation tool, GHG Protocol, Scope 3—Transport and Distribution Emissions category presents four different options for measuring emissions from the goods delivery process, with Option 1 (calculation by type and year of manufacture of the fleet of vehicles), Option 2 (calculation by fuel type), Option 3 (calculation by distance traveled and load weight in loaded vehicles) and Option 4 (calculation by distance traveled and age of the fleet) to calculate the emissions of this source. Using option 4, since there is a homogeneous fleet of motorcycles used by the distributor, which uses the same type of fuel, and making an extrapolation of the total distance traveled for the 89 different routes served by the southern distributor and for annual values (52 weeks), the total estimated distances of 289,250 km traveled in the current scenario and 201,318 km traveled in the optimized delivery scenario are obtained through simple multiplication. Based on these values of annual distances, the total CO_2 emissions are equivalent to 15,618 metric tons obtained for the current scenario and 10,672 metric tons for the optimized scenario.

5 Conclusions

In the present study, a methodology was adopted that can be replicated to several real problems of delivery and/or collection of goods, considering the shortest travel times (instead of distances) for the optimization of the operational process of newspaper distribution of a printing company located in the city of Curitiba, Paraná State, Brazil. Furthermore, through the GHG Protocol tool, the emission of carbon dioxide (CO_2) was measured both for the current delivery scenario and the evaluation of the proposed optimized scenario.

For the application of the proposed methodology, a data survey involving the demand points and quantities of newspapers to be delivered was performed. From the current situation of one of the distributors that serves 89 different delivery routes, one of these was selected and analyzed. On the selected route, deliveries of 479 copies are performed by a single deliveryman on a motorcycle at 194 different stopping points. To enable the application of the exact mathematical model of the VRP, the 194 points were grouped into 35 macro points of demand. From these 35 macro points of demand and considering also the warehouse, all the displacement times involved were collected, pairwise for all 1296 (36 × 36) elements of the matrix.

By solving the mathematical model of the VRP using LINGO software, it was shown that the deliveries are not performed in the best way and that, if the proposed optimized scenario were applied, it would be possible to obtain a reduction of 30.4% in time, as well as distances traveled for deliveries. When analyzing the potential for reducing the environmental impact of the newspaper delivery process, it can be seen that with the use of the optimized route, a reduction of 31.7% can be obtained in the annual levels of CO_2 emission, the main greenhouse gas.

Therefore, it can be seen that by applying mathematical procedures and Operational Research techniques, it is possible to contribute by reducing operational costs and employing sustainability actions. It is worth mentioning that the present methodology can be replicated in other sectors to achieve a reduction in the emission of greenhouse gases and the consequent global warming.

Acknowledgements This study was financed by the Coordination for the Improvement of Higher Education Personnel (Brazil) (CAPES; 1st. and 2nd. authors) and the National Council for Scientific and Technological Development (Brazil) (CNPq; 3rd. author).

References

Arenales M, Armentano V, Morabito R, Yanesse H (2007) Pesquisa Operacional para Cursos de Engenharia, 1ª edn. Elsevier, Rio de Janeiro

Azevedo TR, Ciniro CJ, Amintas BJ, Cremer MS, Piatto M, Kishinami R (2018) SEEG initiative estimates of Brazilian greenhouse gas emissions from 1970 to 2015. Sci Data 5:1–43

Bertaglia PR (2016) Logística e gerenciamento da cadeia de abastecimento. Ed. Saraiva, São Paulo

Braekers K, Ramaekers K, Van Nieuwenhuyse I (2016) The vehicle routing problem: State of the art classification and review. Comput Ind Eng 99:300–313

Cacchiani V, Hemmelmayr VC, Tricoire F (2014) A set-covering based heuristic algorithm for the periodic vehicle routing problem. Discrete Appl Math 163(1):SI, 53–64

Christopher M (2012) Supply chain logistics & management, 4th edn. Pearson, Edinburgh

Dantzig GB, Ramser JH (1959) The truck dispatching problem. Manage Sci 6:80–91

Deng Y, Liu Y, Zhou D (2015) An improved genetic algorithm with initial population strategy for symmetric TSP. Mathematical Prob Eng 2015:1–6

Ferreira JC, Steiner MTA, Canciglieri O Jr (2018) Um Survey sobre Otimização Multi-Objetivo para Problemas de Roteamento Green. In: 27th International workshop advances in cleaner production. Barranquilla

FGV, Fundação Getúlio Vargas (2016) Contabilização, Quantificação e Publicação de Inventários Corporativos de Emissões de Gases de Efeito Estufa. http://www.ghgprotocolbrasil.com.br/. Last accessed 11 Sept 2018

Hashimoto H, Yagiura M, Ibaraki T (2008) An iterated local search algorithm for the time-dependent vehicle routing problem with time windows. Discrete Optim 5(2):434–456

IEA, International Energy Agency (2017) Emissions from fuel combustion: highlights. https://webstore.iea.org/CO2-emissions-from-fuelcombustion-highlights-2017. Last accessed 11 Sept 18

Jun-Man K, Yi Z (2012) Application of an improved ant colony optimization on generalized traveling salesman problem. Energy Procedia 17(Part A):319–325

Laporte G (2009) Fifty years of vehicle routing. Transp Sci 43(4):408–416

Majumbar J, Bhunia AK (2011) Genetic algorithm for asymmetric traveling salesman problem with imprecise travel times. J Comput Appl Math 235(9):3063–3075

Mavrovouniotis M, Muller FM, Yang S (2017) Ant colony optimization with local search for dynamic traveling salesman problems. IEEE Trans Cybern 47(7):1743–1756

Odili JB, Kahar MNM (2016) Solving the traveling salesman's problem using the African Buffalo optimization. Comput Intell Neurosci 2016:1–12

Oliveira CM, Bandeira RAM, Goes GV, Schmitz DN, D'Agosto MA (2018) Alternativas sustentáveis para veículos utilizados na última milha do transporte urbano de carga: uma revisão bibliográfica sistemática. Revista Gestão e Sustentabilidade Ambiental 7(1):3–28

Santos JPP, Paiva I (2018) Brazilian external policy and climate change: analysis of the international acts signed by Brazil (1990-2017). Revista de Iniciação Científica em Relações Internacionais 5(10):112–134

Santos RO, Gomes SMS, Oliveira NC (2018) The impact of emission inventory (GHG) on the operational and financial performance of GHG participating companies. Rev Ambiente Contábil 10(2):266–284

Sirvinskas LP (2018) Manual de direito ambiental, 16ª edn. Saraiva, São Paulo

Tadei R, Perboli G, Perfetti F (2017) The multi-path traveling salesman problem with stochastic travel costs. Euro J Transp Logistics 6:1, SI, 3–23

Vasconcelos J (2015) Gestão da cadeia de suprimentos. Laureate, 1ª. edn. São Paulo

Yang J, Shi X, Marchese M, Liang Y (2008) Ant colony optimization method for generalized TSP problem. Prog Nat Sci 18(11):1417–1422

Green Supply Chain Management and the Contribution to Product Development Process

**Alda Yoshi Uemura Reche, Osiris Canciglieri Jr.,
Carla Cristina Amodio Estorilio and Marcelo Rudek**

Abstract A company that is developing a product must respect all phases: concept, research, analysis, development and launch. At the same time, the green supply chain management must be integrated with suppliers, production and distribution, along with the product development process. This study's aim is to present a model of product development process oriented to green supply chain management, resulting in a product that meets the environmentally sustainable standards. In order to search about the existing models in the literature and the research opportunities in the area, the following keywords were used: sustainability, product development process and green supply chain management; the search was conducted in CAPES Periodicals Portal (MEC). The year of publication was not limited, aiming to understand the evolution of all materials already published. The first section presents concepts about market demands, laws, regulations and green products. The second section discusses about green supply chain management and product development process, and the model presentation. As a result, it was discussed that the green supply chain management and the product development process have still been treated with a broad

A. Y. Uemura Reche (✉) · O. Canciglieri Jr. · M. Rudek
PPGEPS—Postgraduate Program in Production and System Engineering, PUCPR—Pontifical
Catholic University of Paraná, Rua Imaculada Conceição, 1155–Prado Velho, Curitiba, Paraná
80215-901, Brazil
e-mail: aldauemura@hotmail.com

O. Canciglieri Jr.
e-mail: osiris.canciglieri@pucpr.br

M. Rudek
e-mail: marcelo.rudek@pucpr.br

A. Y. Uemura Reche
SENAI—National Service for Industrial Training, Arapongas, Paraná, Brazil

C. C. A. Estorilio
PPGEM—Postgraduate Program in Materials Engineering, UTFPR—Federal Technological
University, Rua Deputado Heitor Alencar Furtado, 5000–Ecoville-Bloco M (amarelo)-3° andar,
Curitiba, Paraná 81280-340, Brazil
e-mail: amodio@utfpr.edu.br

© Springer Nature Switzerland AG 2020
W. Leal Filho et al. (eds.), *International Business, Trade and Institutional
Sustainability*, World Sustainability Series,
https://doi.org/10.1007/978-3-030-26759-9_46

approach, which made it possible to create a model that could guide the companies to develop a product aimed to the green supply chain management.

Keywords Green supply chain management · Green products · Product development process

1 Introduction: Green Supply Chain Management and Product Development Process

The topic "new product development" has been studied for decades and attracted attention to the engineering research, collaborative aspects and global teams. The development of new products focuses on providing a product for sale, in a small development cycle, with a view towards the market opportunities (Gmelin and Seuring 2014). The competitiveness in the market forces the companies to develop and launch new products in increasingly tight timeframes, as a consequence of shorter product life cycles (Cooper et al. 1997).

The environmental practices in companies have transformed the supply chain management, through changes ranging from the eco-product design until the concern with the use of materials, in order to reduce waste, reduce the use of hazardous materials, or replace virgin raw material with the reconstituted raw material or with those that can be reused (Vachon and Klassen 2008).

The environmental reflection must be integrated with supply chain management and product development through concepts that include product design, material selection, the production process, final product delivery to customers, as well as post-consumer reverse logistics, at the end of the product useful life. The environmental impacts must be considered without sacrificing quality, cost, reliability, performance or energy use (Simão et al. 2016), as well as the restriction of resources, laws, green supply chain concepts, product line design, supplier selection and logistics (Huang et al. 2016).

The environmental practices and supply chain, can drive a company to direct efforts for green supply chain strategies, that are practices to minimize the negative impact in supply chain. The negative impacts in supply chain are regarding to climate change, pollution, and non-renewable resource constraints (Mollenkopf et al. 2010).

The overall objective of the study is to present a product development model oriented to green supply chain management, so that the product meets the standards of environmental sustainability. Among the specific objectives, this study sought to identify existing models in the literature, published in international periodicals of high impact, peer-reviewed, with no restrictions about their years of publication and that deal with the green supply chain management and the product development process. This work will verify the evolution of the main proposals, in the models to be analyzed, that aim to contribute with the theme, as well as the research opportunities in this area.

2 Methodology

In order to search for product development models oriented to the management of the green supply chain, some articles were selected from periodicals, through a research in CAPES Periodicals Portal, a Brazilian database of the Coordination for the Improvement of Higher Education Personnel, subordinated to the Ministry of Education. The CAPES Periodicals Portal offers access to full texts available in more than 38,000 international and national periodicals, and 532 reference databases (including Cambridge Journals Online, Emerald Insight (Emerald), IEEE Xplore, Scopus (Elsevier), Science Direct, SpringerLink, Taylor and Francis and so forth), which bring together references and abstracts, from academic and scientific papers to technical standards, patents, theses and dissertations, among other types of material, covering all areas of knowledge. It also includes a selection of important sources of scientific and technological information of free access on the web. As a tool, this study used keywords for advanced search, such as sustainability, product development process and green supply chain management. No filters were established in relation to the years of publication; thus, the research found articles published until the year 2018, since the authors searched for all the publications previously made.

The research is exploratory because it is conducted on a research question or problem, about which there are few or no previous studies, so that all the information about the subject can be found (Collis and Hussey 2005). The search for the keywords was intended to verify the relationship between the searched keywords and the topic.

The research is qualitative, since it is a more subjective method; it examines and reflects on the perceptions and understanding of social and human activities (Collis and Hussey 2005). The qualitative research was conducted through the reading and analysis of articles published in periodicals on the themes described previously, as well as the selection of models on the development of products oriented to the green supply chain, published so far in international periodicals.

3 Product Development and Green Supply Chain Management—Market Demands, Laws, Regulations and Green Products

Since the beginning of 2000, a large number of authors have begun publishing on the importance of managing the supply chain systemically, rather than each company managing its business separately for individual goals. In this context, the supply chain alignment, which deals with common interests and aims to synchronize and coordinate processes, activities and decisions among supply chain partners, has been recognized as an important factor in aligning the product characteristics with the supply chain, thus emerging as an important topic of supply chain management research (Morita et al. 2015). Thus, when thinking about the product design, the marketing impact is emphasized, but it is also necessary to highlight the importance of aspects related to the supply of raw materials and components, production, distribution and even how the final products will be made available to customers at points of sale (Khan et al. 2012).

The supply chain integration allows the developing company to seek alternatives related to design, sales, and promotion and distribution plans, in order to meet the customer's expectations. However, we can highlight the green supply chain, as it moves towards sustainability, for it addresses environmental concerns, public pressure and environmental legislation. "The need for environmental protection and increasing demands for natural resources are forcing firms to reconsider their business models and restructure their supply chain operations (Wu and Pagell 2011)."

When the "green" component is added to the supply chain management, the supply chain and the natural environment are addressed together (Simão et al. 2016). About the green component, the interest toward sustainability has been found in both—academia and industry, specially in the cross-disciplinary field of green supply chain management (Schrettle et al. 2014).

Although dealing with green supply chain concepts, it is important to address the issues related to sustainability, considering the sustainable supply chain related to the environmental, social and economic dimensions of the supply chain management systems. The sustainability emerged as a concept, a strategic demand for the business scenario. While understanding the concept of sustainability is easy, its implementation is still a challenge for any business (Bhanot et al. 2017).

The environmental sustainability directs the company to develop environmental corporate activity and directs the green/sustainable supply chain management. The companies have adopted environmental practices such as environmental purchasing, green supply, green supply chain management and sustainable logistics strategies. The sustainable business changes enable the creation of new production and management systems, as well as creates difficulties for companies to achieve a balance between the traditional efficiency based on operational performance and the environmental benefits, which in turn influence the environmental management of business (Wu et al. 2014). The sustainability can advance firm financial performance, this is possible through the minimization business risks and maximization market opportunities (Fiore et al. 2017).

In order to mitigate the risks arising from the supply chain uncertainties, it is necessary to create analytical tools and performance indicators in order to structure and collaborate on the environmental topics related to the supply chain (Wu et al. 2014). This way, the scope of activities of the supply chain is extended, including the complete product life cycle, requiring the management of relations such as product design, manufacturing by-products, by-products produced during product use, product life extension, product end-of-life, and recovery processes at end-of-life, thus covering the domain of sustainable supply chain management (Bhanot et al. 2017).

Thus, terms such as green innovation or eco-innovation can be defined as the production and adoption of new technologies, which lead to the reduction of environmental risks, pollution and other negative impacts of resource use (including energy use), if compared to relevant alternatives. The number of academic studies on green innovation has grown in recent years and these studies investigate the factors that can sustain and adopt a green transformation in the economy (Castellacci and Lie 2017).

Recent publications show two different lines of study; the first one investigates the determinants of green innovation. However, the green innovation is undoubtedly a broad and complex phenomenon, covering many innovations related to renewable energy, new materials, carbon dioxide and pollution reduction, as well as recycling technologies. The topics are related to markedly different technological trajectories, requiring distinct managerial capacities and supportive policies, and asking how the various types of green innovation differ from one another, and what are the main drivers of green innovation. This line of study highlights the importance of investigating green innovation and distinguishing between different types (Castellacci and Lie 2017). A second line of study shows the importance of expanding the geographical scope of empirical studies, since many studies are focused on European countries, particularly Germany. Consequently, the empirical evidence on green innovation for non-European economies is still limited. The authors point out that the East Asian region is of great relevance as many nations have faced a process of rapid industrialization and have now reached the point where strategy and then environmental issues can no longer be applied. Concepts such as green innovation, green development, circular economy, close loop supply chain and the 3Rs (reduce, reuse and recycle) have already entered the political agendas of many East Asian countries. Particularly noteworthy is South Korea, with its success story of industrialization and the recent implementation of a national green growth strategy. This way, the authors highlight the need to ascertain the main patterns and determinants of the green innovation in East Asia, and how they differ from each other, in empirical patterns identified for European countries (Castellacci and Lie 2017).

The extraction need of earth's natural resources due to its population's increasing demand, have to be questioned face to the future generations. Many natural resources are consumed by supply chains, logistics and supply chain managers, so they have to be aware of 3R's integration (reduce, reuse and recycle) (Garg et al. 2015).

In their quest to meet demand, the companies also face internal and external pressures on sustainable development practices. The external pressures relate to government regulations and for-profit and not-for-profit organizations. In relation to the internal pressures, these are related to strategic objectives, managerial vision, health and safety at work, productivity and quality (Gunasekaran and Spalanzani 2012). The companies also highlight the difficulty of managing the sustainable supply chain, operating under different national regulations and social norms (Wu et al. 2014).

In order to meet the needs listed, an important consideration stands out in the supply chain, questioning whether or not the products can be manufactured to the desired specifications, to the correct materials and adequate supply, and whether the final product is packed and transported in the most efficient way. Analyzing the supply, production and distribution aspects, we realize that the product development process is an important precursor of supply chain decisions, with a view to better coordination of product development and supply chain. The need for the "design for the supply chain" is defined as part of the new product development (NPD) process, which is concerned with the product design, taking into account the performance and success of the supply chain. The good practices in supply chain activities will require that the integration of processes such as business, purchase and logistics, along with the product development, be geared towards the supply chain management (Khan et al. 2012).

The initiatives related to the product development process and the green supply chain management can involve all aspects, from the supplier to the end consumer. The early supplier involvement is a form of vertical cooperation where manufacturing companies engage suppliers, from the first stage of the product development, at the concept and design level. This reflects the ideas of concurrent engineering and offers a number of benefits, such as reducing product development time and improving product quality, as well as sharing technological expertise between the parties. The suppliers are an important resource to bring improvement to the product design, process and technical specialties (Khan et al. 2012).

Most industries consume a large amount of nonrenewable resources and, simultaneously, present waste generation and emission of pollutants. However, considering the growth of sustainable concerns, the companies have begun to work and direct their efforts to meet sustainable requirements and implement measures that fit into sustainable concepts. Also noteworthy are the initiatives related to the sustainable manufacturing, which is considered one of the most important aspects of the sustainable supply chain and which aims to develop the technological practices in the transformation of materials into finished products, aiming at reducing the energy consumption, emission of greenhouse gases, waste and use of non-renewable or toxic materials (Bhanot et al. 2017).

4   Models, Frameworks and Researches on Green Supply Chain Management and Product Development Process

Through the search for articles in the Periodicals Portal, with the keywords sustainability, product development process and green supply chain management, it was possible to analyze the models related to the keywords existing in the literature. Through this research, it was possible to analyze the studies and proposals of models and frameworks, allowing the identification and contribution of each model for the present study, as well as the relevant aspects and gaps, as will be discussed below.

The theme Green Supply Chain Management usually are related with green purchasing, internal environmental operations management, or green logistics. The authors reinforce the gaps existing in literature related to integration and whole supply chain approach. Many authors suggest that green supply chain coil move from subjectivities studies to experimental studies (Govindan et al. 2014)

The article "Antecedents to environmental supply chain strategies: The role of internal integration and environmental learning" (Graham 2018) presents a theoretical model (Fig. 1) based on the perspective of the NRBV natural resource-based

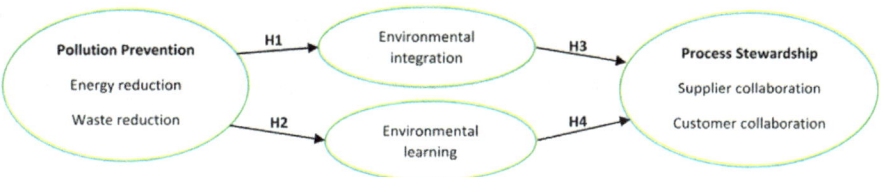

Fig. 1 Theoretical framework (Graham 2018)

view and dynamic capacities; therefore, the company strategy must be the internal implementation of pollution prevention (focus company), as it precedes the environmental integration and enables the environmental learning. A second key aspect that this framework presents is to consider the effects of the environmental integration and learning with a view towards the progression of pollution prevention, directing the implementation of strategies and internal environmental efforts to external levels of the supply chain, that is, the suppliers and customers.

The study has, as limitations, the context presentation of a single food industry and the data collection in a single region in the United Kingdom, limiting the context of the study to this area. The cross-sectional nature of the data does not allow them to be accounted for, since they are not assessed after the implementation of the framework in the company. The study contributes to the understanding of the progression of internal environmental strategies to more advanced strategies of the supply chain.

Ageron et al. (2012), presented, in the article "Sustainable supply management: An empirical study", a conceptual model of Sustainable Supply Management (SSM) (Fig. 2), which consists of 7 blocks that influence the SSM, namely: (1) reasons for sustainable SSM, (2) criteria employed for SSM, (3) greening supply chain, (4) characteristics of suppliers, (5) managerial approaches for SSM, (6) barriers for SSM and

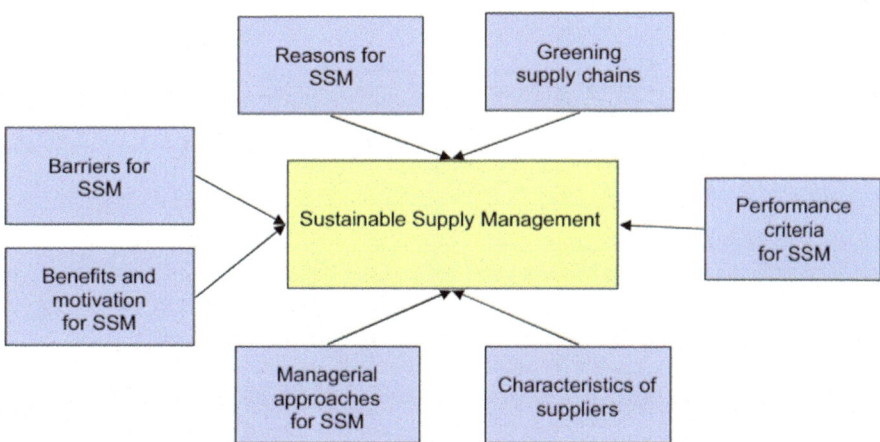

Fig. 2 Model for sustainable supply management (Ageron et al. 2012)

(7) benefits and motivation for SSM. "Greening supply chain" and "Characteristics of suppliers" stand out among the mentioned blocks. The model presented focuses on the development and management of sustainable business between suppliers and the focus company.

As a limitation of the study, we perceive the focus only on suppliers, that is, on the supply area of an industry; certain areas, such as production and distribution to customers, were not considered. As a contribution to the SSM conceptual model, it enables the study of the SCM upstream sustainability, considering the fact that the supply management (strategic alliances, supplier selection and its criteria) plays an important role in the supply chain management process.

In the article "Environmental sustainability in fashion supply chains: an exploratory case based research" (Caniato et al. 2012), the authors present a framework (Fig. 3) where it is possible to identify three factors: (i) drivers—they are considered internal to the company and are related to the company objectives and efficiency; (ii) market drivers—they are related to the customer's requirements, which may be final consumers or the industry; (iii) laws—related to regulations. In this context, the drivers encourage the companies to adopt green practices; these different practices can be used to improve environmental sustainability, and the environmental performance indicators can be used for industry measurement.

The study focuses on the analysis of incremental changes to product and process improvement in small businesses. The study highlights the remodeling of the supply chain in small companies, from the perspective of the internal and external logistics, where it is emphasized that, in practice, some large companies have difficulty in remodeling the supply chain, due to their scale.

Regarding the applicable practices, the study deals with product design and all the decisions related to the product characteristics: the process design, including the production process only in the focus company, and supply chain design, of decisions related to the supply chain—outsourcing, logistics channels, suppliers and distribu-

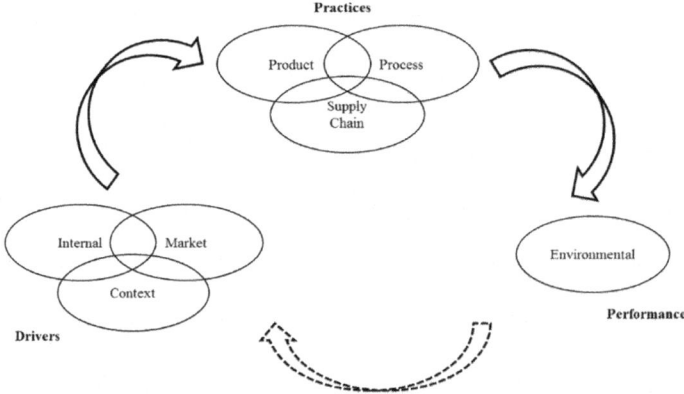

Fig. 3 Research framework (Caniato et al. 2012)

tors. As for performance, it relates materials, energy, water, biodiversity, emissions, effluents and waste, products and services, compliance, transport and business integration.

The contribution of the study is the presentation of examples of performance practices and measurement, as well as a new business model, applicable even to small businesses; these results can help companies identify the good practices to be adopted in the pursuit of sustainability goals.

The study's limitations include a small number of selected cases, two North American clothing companies operating globally, and three small companies operating in some regions of Italy, which have identified environmental sustainability as a key element to compete and survive in the market.

He et al. (2014), in the article "The impact of supplier integration on customer integration and new product performance: The mediating role of manufacturing flexibility under trust theory", explain that the integration of suppliers and the integration of customers have similar and different mechanisms in the development of a new product. Through integration, the companies can have external partners, such as interorganizational strategies, procedures and collaboration, with the purpose of adding value.

The model (Fig. 4) makes it possible to explore the complex relationship between integration and supplier, integration and customer and the performance of a new product through the mediation of roles between manufacturing flexibility and service capacity.

As limitations, we can emphasize that the model focuses on the study of product-based service, considering the information sharing with customers and suppliers, and the possibility of verifying the effects in the development of new products. Although the model presents a structure in relation to the supplier, production, service and client, the authors do not relate the environmental aspects.

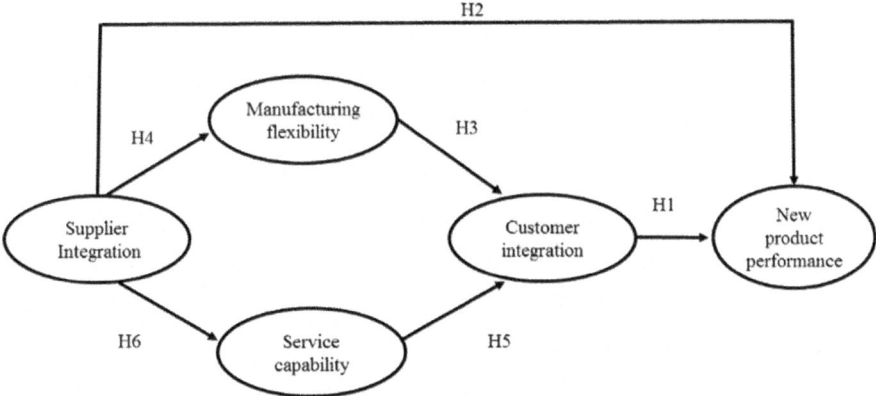

Fig. 4 The conceptual framework in this study (He et al. 2014)

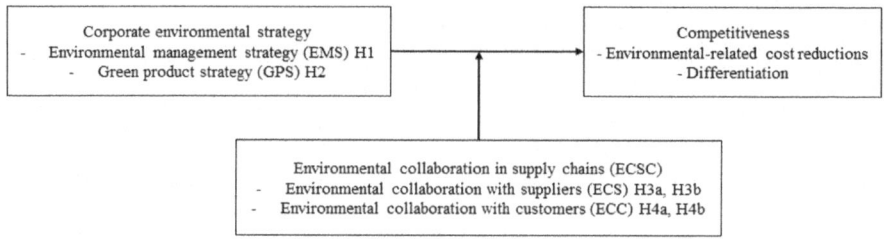

Fig. 5 The conceptual model (Chen et al. 2015)

Chen et al. (2015), in the article "Moderating effect of environmental supply chain collaboration", present a conceptual model from the perspective of five constructs: Environmental Management Strategy, Green Product Strategy, Competitiveness, Environmental collaboration with suppliers and Environmental Collaboration with customers. The study explores the approach of effects on performance, related to green initiatives in the companies, showing how the commitment and environmental effort lead to competitive and sustainable entrepreneurship.

For the model development (Fig. 5), the authors conducted a survey and regression models were used to analyze the data related to corporate environmental strategies, environmental management strategies and green products strategy, and how they affect the competitiveness of the company. The study also investigates the environmental collaboration at supply chain, supplier, and customer scopes, and how these collaborations are related to the environmental performance.

As per limitations, the study focuses on a limited number of relevant green initiatives; such initiatives must be analyzed together with the corporate social responsibility and innovations, focused on improving the company performance.

In relation to the models presented in this study, we can highlight that several articles with different models were observed during the selection process. The contribution of these articles to researchers and companies are worth mentioning, although other studies have not been presented. The selection made during the course of this study identifies the articles with the greatest potential to assist in the topic development, but does not rule out the importance of other publications.

5 Discussion of the Results of Models, Frameworks and Researches of Green Supply Chain Management and Product Development Process

When studying the development of products oriented to the green supply chain, we can emphasize that the focus is the relationship with suppliers, company focus and clients, although there are several subjects related to the subject that can aid in the development of the study. Although the green supply chain topic has recent

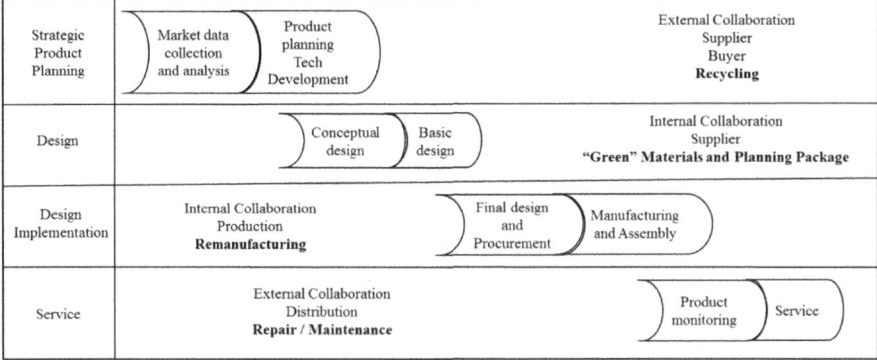

Fig. 6 The conceptual model—green supply chain management and the product development process

publications, they do not specifically address a model that guides companies in the product development process.

The models presented in this article deal with the topic on a stand-alone basis, without the integration of the supply chain in relation to sustainable issues. Therefore, the main objective of the study was to identify models and gaps in the literature. The product development process must be in the context of green supply chain management; when developing a product design, the integration of the supply chain should be considered as a whole, with a view towards the sustainability. This way, this article presents a preliminary model for the integration of the product development and the green supply chain management (Fig. 6).

In the model presented, we have the process of product development; at each stage, we can highlight the activities of the supply chain: supplies, production and distribution, as well as the recycling activities, "green" materials and packaging, remanufacturing, activities that are related to green supply chain management.

6 Final Considerations

This study aimed to present a model of product development oriented to green supply chain management, so that the product meets the standards of environmental sustainability. For this, it presented the models developed by authors of the area, and listed the limitations, applications and contributions of these models. Although there are a large number of published studies, there is a need for more in-depth studies and further research on the subject.

As a main difficulty, we can highlight the gaps in the literature, regarding a product development model that guides the green supply chain, in order to integrate the processes between suppliers, focus company and distribution (Uemura Reche et al. 2018).

Among the limitations of this study, we can point out that the research was structured to collect models and frameworks in articles that presented the keywords in the title, in the abstract and in the keywords set of the publications, that is, it is possible to have other studies that were not presented here.

Finally, we can conclude that the study of the development models of products oriented to the green supply chain is relevant, since it guides the companies to the development of more sustainable practices.

Acknowledgements This study was financed in part by the Coordination for the Improvement of Higher Education Personnel (Coordenação de Aperfeiçoamento de Pessoal de Nível Superior)—CAPES—Brazil—Finance Code 001.

References

Ageron B, Gunasekaran A, Spalanzani A (2012) Sustainable supply management: an empirical study. Int J Prod Econ 140:168–182. https://doi.org/10.1016/j.ijpe.2011.04.007

Bhanot N, Rao PV, Deshmuckh SG (2017) An integrated approach for analyzing the enablers and barriers of sustainable manufacturing. J Clean Prod 142:4412–4439. https://doi.org/10.1016/j.jclepro.2016.11.123

Caniato F, Caridi M, Crippa L, Moretto A (2012) Environmental sustainability in fashion supply chains: An exploratory case based research. Int J Prod Econ 135:659–670. https://doi.org/10.1016/j.ijpe.2011.06.001

Castellacci F, Lie CM (2017) A taxonomy of green innovators: empirical evidence from South Korea. J Clean Prod 143:1036–1047. https://doi.org/10.1016/j.jclepro.2016.12.016

Chen YJ, Wu YJ, Wu T (2015) Moderating effect of environmental supply chain collaboration – Evidence from Taiwan. Int J Phys Distrib Logistics 45:959–978. https://doi.org/10.1108/IJPDLM-08-2014-0183

Collis J, Hussey R (2005) Pesquisa em administração: um guia prático para alunos de graduação e pós-graduação, 2nd edn. Bookman, Porto Alegre

Cooper MC, Lambert DM, Pagh JD (1997) Supply chain management: more than a new name for logistics. Int J Logistics Manag 8(1):1–13. https://doi.org/10.1108/095740997

Fiore M, Silvestri R, Contò F, Pellegrini G (2017) Understanding the relationship between green approach and marketing innovations tools in the wine sector. J Cleaner Prod 142:4085–4091. https://doi.org/10.1016/j.jclepro.2016.10.026

Garg K, Kannan D, Diabat A, Jha PC (2015) A multi-criteria optimization approach to manage environmental issues in closed loop supply chain network design. J Cleaner Prod 100:297–314. https://doi.org/10.1016/j.jclepro.2015.02.075

Gmelin H, Seuring S (2014) Achieving sustainable new product development by integrating product life-cycle management capabilities. Int J Prod Econ 154:166–177. https://doi.org/10.1016/j.ijpe.2014.04.023

Graham S (2018) Antecedents to environmental supply chain strategies: The role of internal integration and environmental learning. Int J Prod Econ 197:283–296. https://doi.org/10.1016/j.ijpe.2018.01.005

Govindan K, Kaliyan M, Kannan D, Haq AN (2014) Barriers analysis for green supply chain management implementation in Indian industries using analytic hierarchy process. Int J Prod Econ 147:555–568. https://doi.org/10.1016/j.ijpe.2013.08.018

Gunasekaran A, Spalanzani A (2012) Sustainability of manufacturing and services: investigations for research and applications. Int J Prod Econ 140:35–47. https://doi.org/10.1016/j.ijpe.2011.05.011

He Y, Lai KK, Sun H, Chen Y (2014) The impact of supplier integration on customer integration and new product performance: the mediating role of manufacturing flexibility under trust theory. Int J Prod Econ 147:260–270. https://doi.org/10.1016/j.ijpe.2013.04.044

Huang Y, Wang K, Zhang T, Pang C (2016) Green supply chain coordination with greenhouse gases emissions management: a game-theoretic approach. J Clean Prod 112:2004–2014. https://doi.org/10.1016/j.jclepro.2015.05.137

Khan O, Christopher M, Creazza A (2012) Aligning product design with the supply chain: a case study. Supply Chain Manage: Int J 17(3):323–336. https://doi.org/10.1108/1359854121

Morita M, Machuca JAD, Flynn EJ, De los Ríos JLP (2015) Aligning product characteristics and the supply chain process—a normative perspective. Int J Prod Econ 161:228–241. https://doi.org/10.1016/j.ijpe.2014.09.024

Mollenkopf D, Stolze H, Tate WL, Ueltschy M (2010) Green, lean, and global supply chains. Int J Phys Distrib Logistics Manag 40:14–41. https://doi.org/10.1108/096000310

Simão LE, Gonçalves MB, Rodriguez CMT (2016) An approach to assess logistics and ecological supply chain performance using postponement strategies. Ecol Ind 63:398–408. https://doi.org/10.1016/j.ecolind.2015.10.048

Schrettle S, Hinz A, Scherrer–Rathje M, Friedli T (2014) Turning sustainability into action: explaining firms' sustainability efforts and their impact on firm performance. Int J Prod Econ 147:73–84. https://doi.org/10.1016/j.ijpe.2013.02.030

Uemura Reche AY, Canciglieri O Jr, Estorilio CCA, Rudek M (2018) How can green supply chain management contribute to the product development process? In: 7th International workshop advances in cleaner production. Barranquilla, Colombia

Vachon S, Klassen RD (2008) Environmental management and manufacturing performance: the role of collaboration in the supply chain. Int J Prod Econ 111:299–315. https://doi.org/10.1016/j.ijpe.2006.11.030

Wu Z, Pagell M (2011) Balancing priorities: decision-making in sustainable supply chain management. J Oper Manag 29:557–590. https://doi.org/10.1016/j.jom.2010.10.001

Wu T, Wu Y-CJ, Chen YJ, Goh M (2014) Aligning supply chain strategy with corporate environmental strategy: a contingency approach. Int J Prod Econ 147:220–229. https://doi.org/10.1016/j.ijpe.2013.02.027

Global Conservation and Sustainability Innovations, Investments, and Policies

Sharing Economy—Another Approach to Value Creation

Pawel Dec and Piotr Masiukiewicz

Abstract The paper concerns the sharing economy and its phenomenon in the world in recent years. The success of the sharing economy, especially in developed countries, which is a form of social reaction to crises, debt traps, high profits of large corporations, destruction of the environment or waste of raw materials and products, was the basis for the authors' research topic. Especially that the development of the current sharing economy indicates that a new period of market education and economic rationality in consumer behavior has begun. The dynamics of these types of services are considerable, hence the governments of individual countries will have to take a position on the regulation and taxation of transactions in the new market segment, which is the sharing economy. The main aim of the chapter is to answer the question about the future of the sharing economy and an attempt to indicate the value created by this phenomenon. The authors made a detailed analysis of the sharing economy on theoretical, quantitative, and practical grounds. In addition, thanks to the authors own research using a very large sample of research surveys with experts, it was pointed out that the sharing economy, on the one hand, will be subject to progressive commercialization processes, but on the other hand, many new services or products will be subject to the processes of sharing.

Keywords Sharing economy · Cooperation · Collaborative economy · Circular economy · Social lending · Generation Y

1 Introduction

Today's world is completely different from that of previous centuries, and this truism is not in principle subjected to any major criticism. But when we talk about global changes that occur significantly in shorter time horizons and even annually, it may be

P. Dec (✉)
Institute of Corporate Finance and Investment, Warsaw School of Economics, Warsaw, Poland
e-mail: paweldec@gmail.com

P. Masiukiewicz
Institute of Value Management, Warsaw School of Economics, Warsaw, Poland

© Springer Nature Switzerland AG 2020
W. Leal Filho et al. (eds.), *International Business, Trade and Institutional Sustainability*, World Sustainability Series,
https://doi.org/10.1007/978-3-030-26759-9_47

797

a surprise for many audiences. Meanwhile, the Internet, the openness of the minds of the new generations, and rapidly changing fashion have been contributions to the search for new solutions in both the social and economic spheres, a notable example of which is the development of the sharing economy. The economy of sharing (cooperation) is based not only on economic values but also on ethical values, such as empathy and twin aid, environmental saving, respect for work put into creating goods, and social responsibility. The phenomenon of the sharing economy is a phenomenon of recent years, especially in developed countries. It is a kind of reaction of societies to crises, debt traps, high profits of large corporations, destruction of the environment, or the waste of raw materials and products. A careful observation of the functioning of the sharing economy indicates that a new stage of market education and economic rationality in consumer behavior began. The first ventures of this type originated in the USA and are now spreading more and more throughout the world, even in these less developed countries, where it is mainly a method of saving costs incurred by customers for expenses, whether every day (car journeys) or for holidays (renting apartments). A typical example of the sharing economy is a company that provides a fleet of cars that everyone can use, paying only for how much the client really travels.

The key research questions are, therefore, whether and how the sharing economy will develop, what are the barriers and factors conducive to this activity, whether legal and tax regulations are needed and what are the risks for customers? Market peer-to-peer seems to have a big future, but it is a serious competitive threat to large commercial and service corporations. It is also difficult to predict today what the reactions of large companies will be in this matter. The main goal of the paper is to analyze the characteristics, opportunities, and threats of the sharing economy and its business utility and potential development. The authors put forward the thesis that the sharing economy segment will develop an important part of the circular economy concept, but this phenomenon will be subject to increasingly large commercialization processes that may distort the original assumptions. The circular economy is a new concept of a market more effective for the ecological and social environment. Such a concept as part of economic development is currently being created in the European Commission. Therefore, it is an extremely current research area, also for the authors. The methods used in this work are literature analysis, analysis of reports, desk research, and case studies, as well as the method of online individual surveys.

2 Features of the Sharing Economy—Double Creation of Value

Research on the sharing economy sector can be placed within the framework of Nano-finance theory, social enterprise theory, circular economy theory, business ethics, and also in the social economy (Arcidiacono et al. 2018; Murzyn and Pach 2018). The complexity of the sharing economy requires a broad and interdisciplinary approach to research needs, and ideally a holistic approach. Economic projects generally known as the sharing economy have different definitions; both broad and narrow (Botsman 2013; Poniatowska-Jaksch and Sobiecki 2017; Rudnicka 2018). There is no single official definition, and different names are used: *"sharing economy", "peer economy", "collaborative economy", "on-demand economy", "collaborative consumption"* are often being used interchangeably (Botsman 2015; Linne 2017; Petrini et al. 2017). An authentic manifestation of the sharing economy is free access to premises through couch surfing, to knowledge through bookcrossing or to a common pool of tools in the possession of the whole community, through the increasingly popular among American farmers tool-sharing. These ideas, however, do not break into the media, because they do not impress either the scale, generally local, or the profits they do not generate at all (Sharing economy 2016). In the European Commission document, the concept of the sharing economy was compared to business models whose activity is supported by cooperating platforms, creating open market space for periodic use of goods or services provided by private entities (European 2016). Transactions of this kind do not entail a change of ownership. However, the importance of access to the resource as a substitute for its property was emphasized. The EC document also points out that the potential of digital technology blurs the division into suppliers and consumers, employed and self-employed, professional and occasional suppliers. Bardhi and Eckhardt (2012) introduced the concept of the access economy as an adjustment to the concept of the sharing economy, used in relation to enterprises that have little to do with the real sharing of resources/products. According to some researchers, the sharing economy is also a more sustainable form of consumption; or the path to a decentralized, equitable, and sustainable economy (Martin 2016). The sharing economy is also unjustly identified with shared consumption—collaborative consumption (Belk 2014; Hamari et al. 2016). Services included in the sharing economy can be provided individually as part of the so-called informal economy or as part of registered activities, e.g. foundations or associations (Cusumano 2015).

The definition of the sharing economy, in narrow terms, proposed by the authors, is as follows: **sharing economy enterprises are organizational and legal entities or informal entities, organized by a group of people who provide services or resell used goods (second-hand assets or commodities of second hand) at low prices**. These are non-profit ventures and are self-help (Kathan et al. 2016; Dec and Masiukiewicz 2018). An important feature of the activity in this market segment is mutual assistance (self-help). In view of the above definition, they are so-called social enterprises, but you cannot include, for example, cooperative savings and credit unions, which are profit-oriented.

According to the authors, the most important features that distinguish economic entities offering sharing economy services are primarily professional and extensive customer service through social media and applications for smartphones and a high emphasis on trust as a basis for successful service provisioning. The occasional nature of the offered service or product is equally important (e.g. free space in a car on a popular route, or a free apartment in an attractive tourist place.) This is also clearly associated with flexible and, what is more important, attractive prices for renting/exchanging services or lending without overly complicated procedures.

Hence, low or zero marginal costs and low tax burden are characteristic for this type of enterprise in the sharing economy sector. On the other hand, the risk related to concluding a distance contract, often just for a word, is underestimated for the recipients of such services, via e-mail, text messages, or social media communication. The share economy sector creates double values—that is, value for the client and value for society—understood as an increase in rationality and macroeconomic efficiency—according to the rules: you do not have to buy, you can borrow, you do not have to have a new thing, you do not have to throw everything away. Certain values also get a service provider, i.e. financial (reimbursement). Rifkin (2016) has proposed a rule for marginal costs close to zero in this sector. The discussion also does not overlook the risk involved in sharing economy activities and regulatory issues related to this sector (Ferrari 2016; Miller 2016).

The authors believe that mutual services, provided free of charge, should not be taxed, because they are a manifestation of social activity and self-help. The awareness of people about having to pay tax for providing a service can be demotivating and limit creativity. The category of a free-of-charge service should not be distorted, because then it can cause its mistaken reception by potential users. On the other hand, a profitable business should be taxed. A clear definition of the demarcation line is needed here. An example of this is the organization by the local community of healthy food supplies directly from farmers. The food producer should pay taxes. A social organization set up to help the local community in providing food (transport, distribution, settlement)—providing free services should not be taxed. This question obviously requires further research. But for instance, Uber's entry into the international market has significantly changed the approach to taxi services, as well as the question of taxing this type of service (Wallsten 2015). This forced it to take actions in many countries to implement new solutions and regulations in the field of taxi services and to protect the domestic market from quite a predatory strategy of Uber.

3   The Scale and Borders of the Sharing Economy

The observation of the functioning of the sharing economy indicates that this segment of the market is developing in high- and medium-developed countries. In third-world countries, the trade in used assets and barter provision of services and exchange of goods has been known for a long time. The scope of the trade in goods and services is growing in this area, from clothing exchange to bitcoin money (Table 1). Unfortunately, there are already large companies appearing on the market, which under the sharing economy banner offer goods or services that are fully commercial and run a profit-oriented activity (Schor 2011, 2016). Services such as those addressed to travelers—Airbnb or dedicated to people looking for usable spaces—WeWork, are well suited to the needs of generation Y. This generation is characterized by a great need for independence, mobility, the desire to experience adventures, and a lack of financial stability and weak attachment to external signs of status (Benckendorff et al 2010; Sharing Economy 2016).

Table 1 Types of offers in the area of the sharing economy (risk estimation 0–5; estimation 5—the highest risk for the customer)

No.	Typical offers	Area of activity	Customer risk assessment
1.	Social lendings in internet platform	International or country	4
2.	Bitcoin—internet money	International	5
3.	Services of facilities car transport	International and local	2
4.	Flat facilities for holidays	International	2
5.	Library services	Local	0
6.	Change of second-hand clothes	Local	1
7.	Press/information local editors	Local	0
8.	Bicycles for rent	Local	1
9.	Household equipment for rent	Local	1
10.	Offers of digital work (remote work)	Local	2
11.	Rent of toys	Local	0
12.	Mutual change of services	Local	0

Own studies

A new, authentic manifestation of the sharing economy is free access to premises through couch surfing (Molz 2013), to knowledge through bookcrossing, to a common pool of tools owned by some local community, e.g. tool-sharing developing among American farmers. These ventures are still not very popular in the media (Sharing Economy 2016).

According to Fücks, with the spread of digital exchange models, such as borrowing, renting or exchanging, new forms of the transaction based on trust were created. It is saving money, energy, and raw materials. Thanks to virtual exchange exchanges on the web, the idea of shared use becomes more common; in Germany, for example, three students founded the Klederkreisel online exchange in 2009—three years later, the stock exchange numbered 380,000 members (Fücks 2013). However, when the co-users are transformed into corporations, they become observers to the entrepreneurs. These are some of the responses they have undergone, which may lead to disgraceful contamination and evasion of taxes (Domaradzki 2015). An essential factor is the use of close interactions between people. A suggestion to introduce special legal regulations for the new exchange of goods and services, among others, in terms of taxes and consumer protection rules, seems controversial—the principle of proportionality of regulations should apply here. Turnover in peer-to-peer transactions in the US market was around USD 100 billion a year (Rifkin 2016). By 2025, global revenue from the sharing economy will amount to USD 335 billion (Ekonomia

Fig. 1 Number of adult sharing economy users in the US in 2016–2021 in millions (Statista 2018a, b). *forecast

2016). Internet services based on the idea of sharing, offering, among other commodities; car transport and accommodation are becoming more and more popular in Poland: 26% of citizens actively use them. It seems that the sharing economy is a significant development potential for European economies. By 2019, the value of shared accommodation and hotel services will increase by 31% and transport by 23% (PwC Raport 2016). Published data on the number of sharing economy users indicates that in the case of the United States, there were almost 45 million of them in 2016, of which the most used Airbnb services (Gibbs et al. 2018; Priporas et al. 2017) and Uber (Baker 2014; Lee et al. 2018) (Fig. 1).

In turn, the forecast estimates assume that the number of these users will grow over the next few years, with the horizon up to 2021. In the years 2020–2021, the forecast indicates more than 80 million adult sharing economy users in the US, which indicates the huge development potential of this sector and the services and products offered there.

4 Social Lending—The Case Study

The social lending business model, which has been put into practice, combines the business models of several types of products, including loans:

- social lending—in which unknown entities lend each other financial resources,
- for online companies—where decisions are made mainly based on the borrower's financial performance and scoring,
- from friends—i.e. companies cooperating with each other, having relations e.g. belonging to the same business organization.

Implemented parameterized scoring algorithms and automatically generated documents based on templates improve the business process (Gonzalez and McAleer 2011). The platform administrator panel allows you to review, process, and manage the list of primary and secondary investors, loan applications, loan auctions, bonds and bills of exchange. Financial operations take place via online websites without the participation of financial institutions. First social lending was in Great Britain in

2005 and the first service of this type was ZOPA (2018). The business model consisted of borrowing and lending money directly between individuals. Investors, who had cash on hand, lent to those who needed financial resources. Anyone who decides to borrow from someone must first pass the registration and verification stage. The borrower is usually required to verify the account by making a small transfer from his bank account. In addition, the ID card, place of residence or source of income are checked. Most often, transactions take place in the form of an auction. Internet loan platforms differ, most often in the amounts and forms of collateral. Those interested in this form of money investment or those who want to use borrowing without intermediaries will find guides, information from forums, discussion groups, and blogs devoted to this subject. Blogs of experienced investors draw attention to the important issue of security and independent verification of borrowers. Currently, this is allowed by social media. Those who want to borrow propose a sum and offer a matching interest rate. They can also respond to ads placed by investors.

The vision of the world without banks looks debatable, but a competitive offer is offered by social lending platforms (Citi 2018). They write, among others about loan clubs, or platforms. From the point of view of those in need of cash, it is a private loan, from the point of view of holders of excess cash—an investment. The spread between the interest rate on the deposit and the loan in banks is much larger than in loan clubs. In highly developed countries, almost half of these types of social loans are used to refinance debts contracted in banks or to repay credit cards. This means that customers notice that the easiest way to get out of debt is by converting them into cheaper loans than can be taken out of the bank. In Poland, we already have a loan platform, which itself offers the loan interest rate to the investor based on scoring conducted based on similar parameters, such as banking. The scoring is the quality of the borrower, the loss of loans to this type of borrower and the expected risk premium.

In the case of the largest American Lending Club portal, the average loan value is USD 14,000, the average return for the investor is 5–9%, and the average interest rate is 13.2% per annum. The average "loss ratio" of clients does not exceed 5%. In another American social platform Prosper, the average interest rate is 15%, the average investor's profit is 9%, and the average "loss ratio" is 6–7% of loans. In the British ZOPA loan club, an average of 7500GBP was lent, the average interest rate did not exceed 8%, the investor's profit was on average 5%, and the percentage of loans not repaid on time—2%. In Asia, social loans occur not only in the online version but also offline, in which case an agent reaches the client. According to analysts, the social loans sector will grow at an unprecedented pace, according to forecasts of 2050 their value is expected to be close to USD one trillion (Statista 2018b).

5   New Opportunities and Threats of Sharing Economy

The participation of the young generation in those processes, which thanks to the unlimited possibilities of using social media and new technologies, became a reality in this area of the economy, contributed to the success of sharing economy. The sharing economy phenomenon should be viewed from the point of view of social benefits and threats (Baker 2015; Domaradzki 2015; Rifkin 2016; Frenken 2017). The benefits of this economy for pro-ecological development on a global scale are high. It is difficult to overestimate the effects of saving raw materials, materials, and energy. A social benefit from the development of the sharing economy as the official economy will probably be a reduction in the black market of trade, services, and finances. Also, sharing economy business models are and will be evaluated and will evolve in the future (Cheng 2016). The benefit for customers is, of course, the acquisition of goods and services at very low prices; including used products—often unique or rare (Fücks 2013). But consumers are exposed to several risks. Relatively frequent use of so-called oral contracts in transactions may pose some threat to customers in this market (e.g. lack of service delivery, poor quality); the risk may also apply to the prices of products or services offered (Masiukiewicz 2016b; Masiukiewicz 2017). A serious threat may be the development of unlicensed financial services under the guise of the sharing economy, especially the pyramid scheme. One should remember here about the bankruptcy of two bitcoin platforms in the US in recent years, which resulted in losses for clients amounting to USD 0.5 billion (Masiukiewicz 2016a). An important risk of borrowing funds on social lending platforms is the lack of loan repayment.

An opportunity for sustainable development is saving raw materials and limiting product waste, respecting human labor, thanks to which already used products have been created, creating jobs. A threat to the development of this market segment may be running the so-called undeclared activity, included in the black market, and, on the other hand, possible tendencies of state authorities to excessive legal regulations and the introduction of significant tax burdens for entities operating in the sharing economy market.

6 Sharing Economy and Experts View—In-Depth Own Research

To verify the thesis, the authors made individual online surveys with selected experts. The selection of experts who specialize professionally and scientifically in the area of the sharing economy followed the prior consultation and verification of potential participants in the pilot study. The survey was conducted globally because the experts came from all over the world. From the initial list of experts selected for the study, a group of 150 experts were finally selected, of which 74 expressed their willingness to participate in the study, i.e. the participation rate was nearly 50%, which can be described as very good. The survey was conducted in May-August 2018. The initial list of questions and research areas was limited from over 20 items to the 6 most important items from the point of implementation of the goals set in the article.

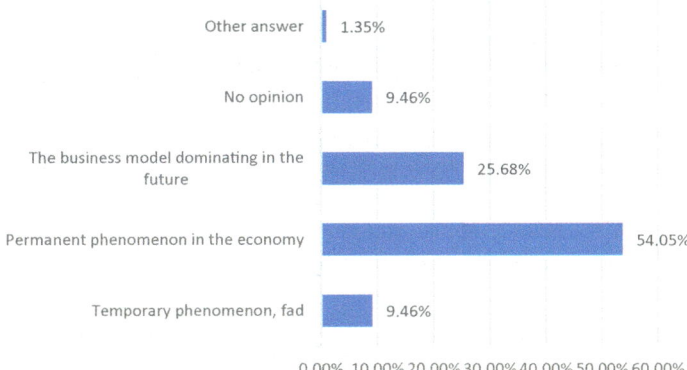

Fig. 2 The indication of the sharing economy

The first basic research problem was the self-qualification and diagnosis of sharing economy development in economic processes taking place in the world (Fig. 2).

According to almost 10% of the surveyed experts, the sharing economy is only a temporary phenomenon (fad), and the majority think that this is a constant phenomenon in economics (almost 55%). With more than 25% opting for it being the dominant business model in the future, it can be clearly stated that the sharing economy is permanently inscribed in the history of economics.

The responses to the second question regarding the biggest and most important challenges in the further development of the sharing economy in the future were already more diversified (Fig. 3).

Over 20% of experts pointed to the need for the state/regulator to provide a stable legal system relating to the functioning of the sharing economy. Thus, there is

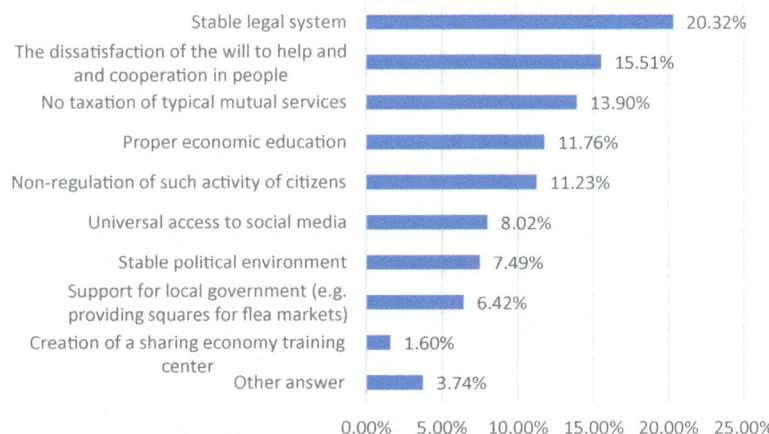

Fig. 3 Prerequisites for the further development of the sharing economy

a permanent lack of a permanent institutional and legal infrastructure for this phenomenon. On the other hand, 11% of experts appreciated certain regulatory freedoms of the area and the possibility of citizens flexing this type of activity. The second important issue pointed out by the surveyed was the uncertainty as to the continued willingness to help and cooperate with other people (over 15%). Thus, a key question arises as to how participants in sharing economy processes will continue to support and create selflessly, and not, for example, they have tried to commercialize these processes to a large extent. Another challenge that cannot be overlooked and which the experts interviewed rightly point out is the lack of regulation of tax issues in the scope of the sharing economy (almost 14% of responses). On the one hand, the lack of a rigid tax system allows the development of such services, and on the other, it lacks limits, to a certain degree, the tax revenues of individual countries. A large group of experts (almost 12%) pointed to the need for proper economic education of participants in sharing economy processes. This is undoubtedly important in terms of further development of such services and increasing their efficiency. Over 8% of responses concerned providing access to social media as an indicator of further development of the sharing economy. It is, of course, a fact that the development of the sharing economy was so high due to the huge power of portals, social applications, and a high level of having both internet access, as well as smartphones or laptops. Without these factors, the sharing economy force would certainly be smaller. Respondents also pointed to the necessity of fostering the political environment, both at the national and local level in terms of sharing economic development. It is largely from politicians, but also from urban or local activists and social activists who want to initiate a series of initiatives supporting or even initiating sharing economy processes. Among other indications raised by experts were those regarding risk management problems in the sharing economy area. Ensuring adequate security for participants in such processes, seem to be insufficiently stressed in the literature on the subject and may be an important factor limiting the further development of this sector.

The third problem raised by the authors in the study was directly related to earlier issues and directly referred to the possibility of commercialization of the sharing economy sector (Fig. 4).

According to almost three-quarters of the surveyed experts, sharing economy processes can be commercialized as much as possible, meaning that they can be sold directly for profit. Could this be the beginning of the end of sharing economy development and its main assumption about sharing, that of resources shared often selflessly and certainly not on commercial terms. This situation seems very likely, as only less than 7% of experts expressed the view that such processes will not be commercialized. Complementing the question about the commercialization of sharing economy processes is the question of determining who the largest beneficiary of this type of process is currently (Fig. 5).

According to more than 55% of surveyed experts, households and consumers are the largest beneficiaries of sharing economy processes. A large group of beneficiaries, according to respondents, are also internet companies and social media (almost 19% of responses), which, after all, not only ensure the flow of information about such services but also allow fast communication between interested parties. It is also not

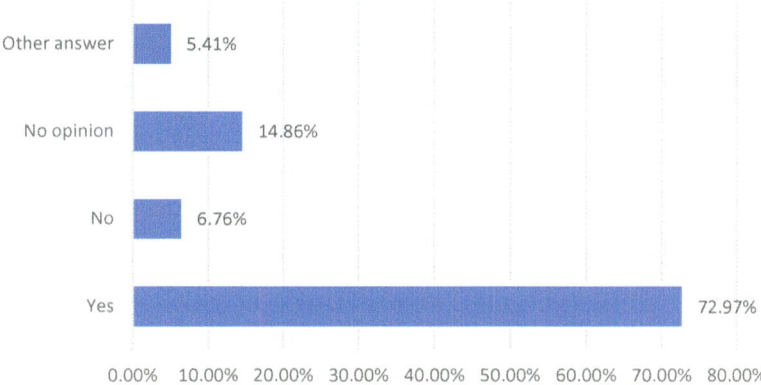

Fig. 4 The possibility of sharing the commercialization area economy

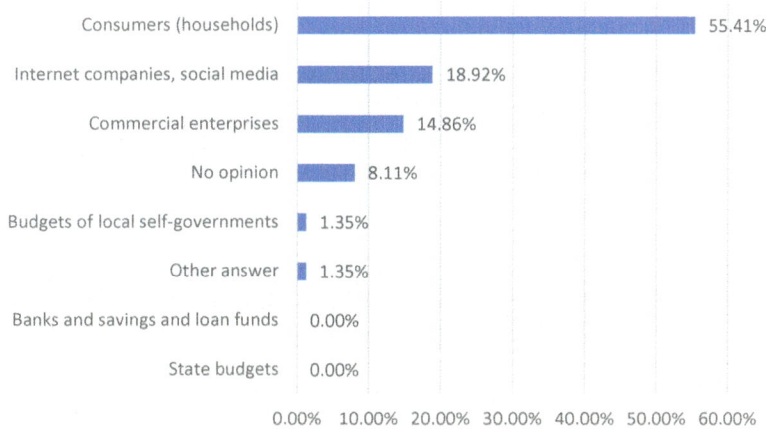

Fig. 5 The largest beneficiary of the sharing economy

surprising that companies in quite high positions in this list of private companies (15% of indications) have a business model based on sharing economy (such as rental of city bicycles, scooters). The experts clearly stated that the beneficiaries of sharing economy services are not the banking and finance sector, nor local or national budgets. Thus, one can see not only a threat to ensure tax revenues but also some independence of the sharing economy sector from bank financing, i.e. the ability to do business in sharing economy without the need to obtain support from the financial market. Finally, the authors tried to determine whether the need to regulate activities in the sharing economy should take place at the supranational or national/local level (Fig. 6).

According to almost 40% of respondents, such regulation of processes taking place in the sharing economy sector should take place at the supranational level, i.e.

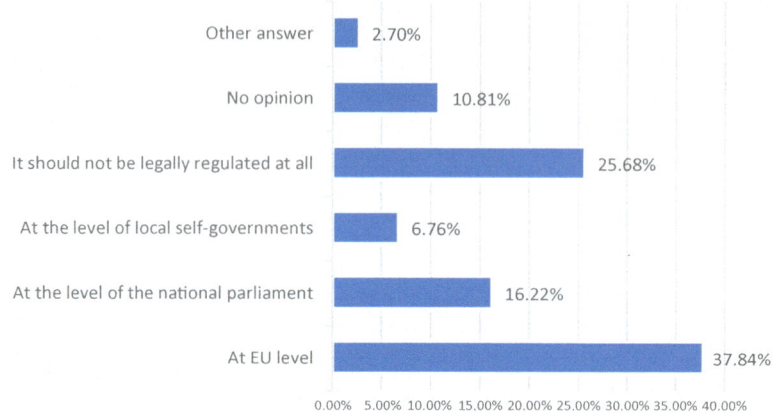

Fig. 6 Activities within the framework of the sharing economy should be legally/legally regulated

regulation at the level of the European Union. On the other hand, more than a quarter of experts stated that the sharing economy should not be regulated at any level at all. Undoubtedly, this is an argument in the context of the current development of the sharing economy based mainly on the lack of rigid regulations in this area. Over 16% of responses indicated the need to regulate such processes at the national level, and less than 7% at the local level. On the one hand, such decentralization may consider the specificity of a given market, but it may also limit competition in the sector (through the need to overcome barriers to entry by external providers of such services), which in the longer term may cause commercialization of these processes.

The last issue raised by the authors in the expert study was to indicate what kind of service or product in the sharing economy area can be mostly provided as part of the sharing economy and not sold by commercial companies in the next 10 years. The most experts (almost 10%) pointed to services related to renting apartments as well as transport services (car or two-wheeled car rental). It was quite an interesting indication about the possibilities of sharing care over elderly people or children, and looking at demographic trends in many countries seems to be the right direction. Among other indications, several experts gave the opportunity to share licenses for 3D printing services, use of video platforms or music on demand, or IT services. Single expert indications also concerned, for example, services or products such as: financial services, consulting services, everyday use goods, online education, know-how, open innovation, tourist equipment, social loans, renovation works, scientific books, energy from household RES, supplying home shopping, business models, legal and economic consulting, storage, medical services, housekeeping services. According to the authors, it is difficult to say unequivocally whether any of the services indicated has greater development potential in the sharing economy sector than others. Considering the monetization and commercialization processes of specific services or products, the sharing economy sector faces a difficult and turbulent period of further development.

7 Conclusions

The sharing economy is one of the most amazing and reflective phenomena that we face in the modern global economy. For many economists or experts, it is difficult to imagine and accept the scale of the phenomenon. The sharing economy is a phenomenon of recent years in developed countries; being a social reaction, among others, to crises, but very fast spreading and gaining hugely in other countries. Therefore, it also becomes the everyday life of Nano finances. The development of the sharing economy confirms the changes in attitudes and behavior of consumers, for whom tinsel and brand cease to be the most important, ethics and rationality become more important. According to the authors, on the one hand, the sharing economy escapes from classical economic theories, and on the other hand, the state tries to excessively regulate the phenomenon. This may cause the loss of the sense of sharing services or products, and bottom-up human initiatives will be condemned to failure in the long horizon.

The dynamic development of the sharing economy segment requires greater attention of researchers and new research, as it is not only a kind of breakthrough in traditional economics but it also constitutes a kind of competition for existing business models. The development of sharing economy is a social reaction of new generations to high profits of corporations, crises, and destruction of the natural environment. The political consequence of these changes is, in a way, the work of the European Union regarding the implementation of the circular economy concept. The governments of individual countries will have to take a position on the regulation and taxation of these transactions. It can be concluded that there are perspectives for the future development of the sharing economy, mainly due to new breakthrough technologies and ubiquitous digitization of life. The increase in turnover in this sector is also very real, which will attract the interest of many companies or institutions. And this, in turn, will be a challenge for big global companies, which because of sharing economy development are exposed to the loss of their previous, dominant position.

Acknowledgements The authors would like to express special thanks to professor Gabriel Główka and Piotr Staszkiewicz, Ph.D. (both from Warsaw School of Economics) for help with the research conducted in the text.

References

A European agenda for the collaborative economy, Communication from the Commission to the Parliament, the Council, the European Economic and Social Committee and the Committee of the Regions, Brussels, 2.06.2016, COM (2016) 356 final

Arcidiacono D, Gandini A, Pais I (2018) Sharing what? The 'sharing economy' in the sociological debate. Sociol Rev 66(2):275–288. https://doi.org/10.1177/0038026118758529

Baker D (2014) Don't buy the 'Sharing Economy' hype: Airbnb and Uber are facilitating ripoffs. The Guardian. Retrieved 15 June 2018 from http://www.theguardian.com/commentisfree/2014/may/27/airbnb-uber-taxes-regulation

Baker D (2015) The opportunities and risks of the sharing economy. Testimony before the Subcommittee on Commerce, Manufacturing, and Trade of the US House of Representatives Committee on Energy and Commerce. Washington, DC, September 29

Bardhi F, Eckhardt GM (2012) Access-based consumption: the case of car sharing. J Consum Res 39(4):881–898

Belk R (2014) You are what you can access: sharing and collaborative consumption online. J Bus Res 67(8):1595–1600

Benckendorff P, Moscardo G, Pendergast D (eds) (2010) Tourism and generation Y. Cabi. B2B loans platform. Retrieved 15 May 2018 from www.whitehill.eu

Botsman R (2013) The sharing economy lacks a shared definition. Fast Company 21

Botsman R (2015) The sharing economy: dictionary of commonly used terms. Collaborative Consumption 12

Cheng M (2016) Sharing economy: a review and agenda for future research. Int J Hospitality Manage 57:60–70

Citi (2018) Bank of the future. ABCs of Digital Disruption in Finance. Citi GPS: Global Perspectives & Solutions. Retrieved 12 June 2018 from http://www.vostokemergingfinance.com/content/uploads/2018/05/Citi-GPS-Bank-of-the-Future.pdf

Cusumano MA (2015) How traditional firms must compete in the sharing economy. Commun ACM 58(1):32–34

Dec P, Masiukiewicz P (2018) Sharing economy—new phenomenon. Bus Econ Res 8(2):1–10

Domaradzki K (2015) Oszczędzanie czy kantowanie. Retrieved 15 June 2018 from www.forbes.pl

Ekonomia współdzielenia to szansa dla Polski (2016). Retrieved 24 July 2018 from www.forsal.pl

Ferrari M (2016) Beyond uncertainties in the sharing economy: opportunities for social capital. Eur J Risk Regul 7(4):664–674. https://doi.org/10.1017/S1867299X00010102

Fücks R (2013) Intelligent Wachsen. Die grune Revolution. Carl Hanser Verlag

Frenken K (2017) Political economies and environmental futures for the sharing economy. Philos Trans R Soc A 375(2095):20160367

Gibbs C, Guttentag D, Gretzel U, Yao L, Morton J (2018) Use of dynamic pricing strategies by Airbnb hosts. Int J Contemp Hospitality Manage 30(1):2–20. https://doi.org/10.1108/IJCHM-09-2016-0540

Gonzalez L, McAleer K (2011) Online social lending: a peak at US Prosper and UK Zopa. J Account Finance Econ 1(2):26–41

Hamari J, Sjöklint M, Ukkonen A (2016) The sharing economy: why people participate in collaborative consumption. J Assoc Inf Sci Technol 67(9):2047–2059

Kathan W, Matzler K, Veider V (2016) The sharing economy: your business model's friend or foe? Bus Horiz 59(6):663–672

Lee ZWY, Chan TKH, Balaji MS, Chong AYL (2018) Why people participate in the sharing economy: an empirical investigation of Uber. Internet Res 28(3):829–850. https://doi.org/10.1108/IntR-01-2017-0037

Linne A (2017) What exactly is the sharing economy? Retrieved 25 July 2018 from www.weforum.org/agenda/2017/12/when-is-sharing-not-really-sharing/

Martin CJ (2016) The sharing economy: a pathway to sustainability or a nightmarish form of neoliberal capitalism? Ecol Econ 121:149–159. https://doi.org/10.1016/j.ecolecon.2015.11.027

Masiukiewicz P (2016a) Pyramids scheme—growth global problem, vol 8. Ekonomika i Organizacja Przedsiębiorstwa

Masiukiewicz P (2016b) Consumer protection in financial services. Social and Humanities, Prices Area, Eureka, p 4

Masiukiewicz P (2017) Konkurencja czy współpraca? vol 9. Nowoczesny Bank Spółdzielczy

Miller SR (2016) First principles for regulating the sharing economy. Harvard J Legislation 53(1):147–202

Molz JG (2013) Social networking technologies and the moral economy of alternative tourism: the case of couchsurfing. Ann Tourism Res 43:210–230

Murzyn D, Pach J (eds) (2018) Ekonomia społeczna. Między rynkiem, państwem a obywatelem. Difin, Warszawa

Petrini M, De Freitas CS, Da Silveira LM (2017) A proposal for a typology of sharing economy. Rev Administração Mackenzie 18(5):39–62

Poniatowska-Jaksch M, Sobiecki R (eds) (2017) Sharing economy (gospodarka współdzielenia). Oficyna Wydawnicza SGH, Warszawa

Priporas C-V, Stylos N, Rahimi R, Vedanthachari LN (2017) Unraveling the diverse nature of service quality in a sharing economy: a social exchange theory perspective of Airbnb accommodation. Int J Contemp Hospitality Manag 29(9):2279–2301. https://doi.org/10.1108/IJCHM-08-2016-0420

PwC Raport (2016) (Współ)dziel i rządź! Twój nowy model biznesowy jeszcze nie istnieje. Retrieved 11 May 2018 from www.pwc.pl

Rifkin J (2016) Społeczeństwo zerowych kosztów krańcowych: internet przedmiotów, ekonomia współdzielenia, zmierzch kapitalizm. Wydawnictwo Studio Emka, Warszawa

Rudnicka E (2018) Czym jest sharing economy i jak zmienia biznesowy krajobraz? Harvard Bus Rev Pol. Retrieved 10 June 2018 from www.hbrp.pl/b/YKrUu6Qu

Schor J (2011) True wealth: how and why millions of Americans are creating a time-rich, ecologically-light, small-scale, high-satisfaction economy. The Penguin Press, New York

Schor J (2016) Debating the sharing economy. J Self-Gov Manage Econ 4(3)

Sharing Economy (2016) Uber i Airbnb mają niewiele wspólnego z dzieleniem się czymkolwiek. Retrieved 27 June 2018 from www.gazetaprawna.pl/biznessharing-economy:Uber-i-Airbnb

Statista (2018a) Retrieved 15 Aug 2018 from https://www.statista.com/statistics/289856/number-sharing-economy-users-us

Statista (2018b) Value of global peer to peer lending from 2012 to 2025 (in billion U.S. dollars). Retrieved 30 Aug 2018 from https://www.statista.com/statistics/325902/global-p2p-lending

Wallsten S (2015) The competitive effects of the sharing economy: how is Uber changing taxis, vol 22. Technology Policy Institute

Zopa (2018) Retrieved 15 March 2018 from https://www.zopa.com/about

International Business, Trade and the Nagoya Protocol: Best Practices and Challenges for Sustainability in Access and Benefit-Sharing

Natalia Escobar-Pemberthy and Maria Alejandra Calle Saldarriaga

Abstract Biodiversity is critical for international trade, businesses and investment. However, the issue of their ownership and exploitation has been a matter of debate. Specifically, the access to biodiversity resources and the distribution of the associated benefits are at the core of the environmental economics debate. Since 1992, the Convention on Biological Diversity (CBD) established regulations for the access and benefits-sharing of genetic resources, which materialized in the 2010 Nagoya Protocol. In 2015, as part of the Sustainable Development Goals (SDGs), Goal 15 renewed countries' commitment to the promotion of appropriate access to genetic resources, and equitable and fair sharing of the associated benefits, calling for countries to adopt the policy and strategic frameworks to implement the ABS regime. Using examples from developing countries in Latin America and the Caribbean, this chapter explains how the ABS regime is being implemented in relation to international business and trade, and analyzes how it brings opportunities, best practices and challenges in order for these countries to improve the balance in the relationship between biodiversity conservation and economic activities. These new circumstances support the identification of actions that governments and international business actors can follow to effectively use the ABS regime as a policy instruments that contribute to sustainability through the implementation of economic and environmental regulations.

Keywords Biodiversity resources · Benefit-sharing · Nagoya Protocol · International business and trade · ABS regime

Nature and economics have always been related. However, the ongoing evolution of scientific research and technology in biological and genetic resources on one side, and of trade dynamics in the other have increased the relevance of this relationship, and its role in environmental law and governance, and in international trade and business. While economists see environmentalism as an obstacle in their quest for

N. Escobar-Pemberthy (✉) · M. A. Calle Saldarriaga
Universidad EAFIT, Medellin, Colombia
e-mail: nescoba3@eafit.edu.co

M. A. Calle Saldarriaga
e-mail: mcalle@eafit.edu.co

© Springer Nature Switzerland AG 2020
W. Leal Filho et al. (eds.), *International Business, Trade and Institutional Sustainability*, World Sustainability Series,
https://doi.org/10.1007/978-3-030-26759-9_48

growth and development, defenders of the environment perceive economic activities as insatiable causes of the planet's degradation (Clapp and Dauvergne 2011; Collier 2010). Historically, economists dominated this debate. However, the inclusion of legal and ethical elements into the economic approach and the definition of governance instruments that provide guidelines to combat unregulated access to biodiversity and biopiracy have changed the perception about the use of biodiversity resources in economic activities and international business. Three issues have particularly contributed to this. First, in 1968, Garrett Hardin (1968) argument of "The Tragedy of the Commons" demonstrated how it is possible for individuals, directed by their own self-interest, to deplete resources that are shared and limited. Second, the increasing recognition of the environmental responsibility of present societies with future generations raised awareness about the concept of sustainable development and the integrative nature of the link between economics, society and environment (Nunes et al. 2003; United Nations 2012; WCED 1987). Finally, the understanding of environmental resources as global public goods opened the door for new international cooperation regimes to regulate its management and the distribution of its economic value (Perrings and Gadgil 2003).

In pursuing this, however, two political economy obstacles persist: The determination of natural resources ownership and the definition of their real value. Both variables are required to determine the distribution of the associated benefits. By definition "natural assets have no natural owners" and states have the sovereign right to regulate the conditions for the use and access to their biodiversity based on their own environmental and development policies (Collier 2010; United Nations 1992a). This has fundamental economic implications, since ownership will determine the sovereign access to its exploitation and to the derived benefits (Boyce 2002). Environment valuation, the second obstacle, is even more complex. Natural resources are not only valuable because of its intrinsic monetary equivalency, but also because of the services they provide in terms of ecological balance, cultural and spiritual values, traditional knowledge, and identity, and their valuation is based both on their use and their conservation (CBD 2009; Pearce and Moran 1994; Secretariat of the Convention on Biological Diversity 2007; ten Kate and Laird 1999). Furthermore, debates persist about the extent to which environment valuation and market-based instruments can really contribute to support the conservation strategies and mechanisms established by governments and communities (Gómez-Baggethun and Muradian 2015; Iftekhar et al. 2017).

It is in this context that international environmental law and governance started addressing these issues and connected to international trade and business. Since 1972, the international community has tried to establish institutional and regulatory mechanisms to manage the environment. In 1992, when the United Nations Conference on Environment and Development (UNCED) was celebrated in Rio de Janeiro (Brazil), a new paradigm of sustainable development emerged and a general framework for biological diversity was established with the 1992 Convention on Biological Diversity (UNCED 1992). However, this international agreement—designed to regulate biodiversity conservation and its sustainable use—has not been absent of debate. Access to biodiversity and the distribution of the benefits emerging from their use are complex issues, required to develop sustainable paradigms for the conservation of biodiversity as common heritage of humankind (Ibisch et al. 2010; UNCTAD 2017). The definition of property and exploitation rights is also required in order for these resources to fulfill their economic role. The Nagoya Protocol, the instrument adopted in 2010 to further the objectives of the CBD in the context of benefit-sharing resulting from the utilization of genetic/biological resources (genetic and or biochemical composition of genetic resources) and associated traditional knowledge, addresses these issues.

Both agreements acknowledge the sovereign right of states over their natural resources and therefore to establish the conditions for access and equitable benefit-sharing arising from scientific research, commercial and other uses of genetic resources. This regulation is fundamental in the relationship between environment and trade, since it provides specific dispositions for controlling available resources and allows for their full economic use. In 2012, the Rio+20 United Nations Conference on Sustainable Development reinforced this agenda in the Sustainable Development Goals (SDGs)—among which biodiversity has a critical role (United Nations 2012). Specifically, SDG 15 Life on Land calls for the promotion of the "fair and equitable sharing of the benefits arising from the utilization of genetic resources" according to the international agreements established for this purpose, measuring the number of countries that adopt "legislative, administrative and policy frameworks" in this regard.

However, challenges persist on the implementation of these global regimes. States, as signatory parties of these international law instruments, are then called to adopt domestic policies, regulations and processes addressing issues such as the prior informed consent of the provider country, the negotiation of mutually agreed terms and the scope of benefit-sharing. In the case of the Nagoya Protocol, the monetary and non-monetary benefits associated with the use of genetic resources and traditional knowledge (including subsequent application and commercialization) shall be shared in accordance with the negotiated mutually agreed terms. Monetary benefits include royalty payments and joint ownership of intellectual property rights, while non-monetary benefits refer instead to research and development, training and education and transfer of technology (Schindel et al. 2015; Secretariat of the Convention On Biological Diversity 2011). In this context, signatories to the Nagoya Protocol are still struggling with the definition of legislation and policy mechanisms that guarantee the outcomes outlined by the agreement. Furthermore, obstacles persist in guaranteeing that new policy instruments and procedures don't make more difficult or onerous the access to genetic resources.

The purpose of this chapter is then to explain the extent to which the Nagoya Protocol is a critical environmental law instrument for international business and trade, characterizing the debate between environment and economics and the relationship between biodiversity, international trade and international business. By identifying the approach that developing countries have on the implementation of the ABS regime, it will be possible to outline the main challenges of this process and to offer valuable insights regarding the need to work with the private sector—with trade and international business actors—in the definition of the national legislation for access and benefit-sharing, and on the subsequent promotion of sustainability and sustainable development.

1 Biodiversity, Trade and International Business

Understood as the "variability among living organisms from all sources" including the diversity of genes, people, species, communities and ecosystems (United Nations 1992b), biodiversity (including genetic material and traditional knowledge) affects almost every human activity. Since 1992, the topic has been central to the international agenda. Factors such as the dimension of biodiversity resources, the legal control over them, their economic conservation and cultural value, the risk of species extinction, the localization and use of the resources, and the economic implications—costs and benefits of the use of biodiversity—have been some of the aspects discussed by international organizations and stakeholders (Adams et al. 2010; Collier 2010; IUCN 2016; MEA 2005; Pearce and Moran 1994; Rosendal 2000; UNCED 1992; United Nations 1992b).

In the process of reconciling biological resources with humanity's social and economic development, in 1988 UN Environment convened an Ad Hoc Working Group to explore the need for an international framework on biological resources, oriented to prepare a legal instrument for their conservation and sustainable use. This Working Group, later transformed in an Intergovernmental Negotiation Committee, culminated its task with the adoption, in 1992, of the Convention on Biological Diversity (Glowka et al. 1994; Rosendal 2000; UNCED 1992; United Nations 1992b). The CBD pursues three objectives: (1) the conservation of biological diversity, (2) the sustainable use of its components, and (3) the fair and equitable sharing of the benefits arising from the utilization of genetic resources (Oberthür and Rosendal 2000; Rosendal 2000; United Nations 1992b). Until today up to 196 countries have signed it (Secretariat of the Convention On Biological Diversity 2018b).

Since its adoption, the Convention has also aimed at raising awareness on economic practices at all levels (micro, small, medium and large enterprises) to make them compatible with biodiversity conservation objectives. This relationship, however is highly complex. Biodiversity sustains development thanks to its economic potential, but when lost it also exacerbates poverty since it deprives communities from materials for basic needs, livelihoods and subsistence through consumption, business and trade (Adams et al. 2010; Agrawal and Redford 2010; CBD 2009; Perrings and Gadgil 2003). Development on the other side is seen as a condition for countries to be able to protect the environment. Furthermore, these effects are not exclusively associated to its ecological implications—conservation of genetic information and the threats to the species from managed ecosystems—but also to the environmental services they provide. Ecosystems determine the development path of each community, and their international trade and business profiles, but at the same time said choice also affects the future of the ecosystem services available to each human group.

Biodiversity's benefits for human economic activities, human well-being, and poverty reduction are associated to five categories: Basic material for good life, health, good social relations, security and freedom of choice and action (MEA 2005). Genetic and species diversity provide communities with different products they can use for consumption, barter or trade. They also provide materials for other activities and serve as an input for industrial processes and for health services. Genetic diversity, specifically, allows agriculture to adapt to the changing conditions of the environment and to guarantee its subsistence (CBD 2009). Additionally, biodiversity supplies important ecological services and basic needs such as clean water, livelihoods, employment, health, nutrition and prevention of natural disasters, at the same time as it promotes economic activities such as agriculture, forestry, fisheries and ecotourism (Adams et al. 2010; Agrawal and Redford 2010; CBD 2009; MEA 2005). Many developing countries particularly rely on exports from these sectors and on natural resources, commodities, raw materials for industries and ecosystem services as their main sources of income. On the opposite side, some economic activities have direct and indirect impacts on the biodiversity, its conservation and sustainability. If not managed appropriately, industries such as mining, oil, gas, construction, infrastructure may represent fundamental biodiversity costs.

In addition, international business and trade are also the framework of activities in which market-based mechanisms can be established to fund conservation efforts (Arsel and Büscher 2012; Bishop and Pagiola 2012; Pagiola et al. 2002). As this type of instruments provide incentives for the conservation and effective management of ecosystems, they can also generate sustainable financing for communities with access to the related biodiversity resources. In the context of access to genetic resources, however, the definition of policy instruments for the implementation of the Nagoya Protocol does not guarantee the realization of the benefits from biodiversity use.

Governments then have the responsibility to guarantee the adequate management of all these sectors—both the ones resulting from biodiversity and the ones that affect biodiversity, recognizing their dependency on ecosystem services and their role on economic development and trade. This fundamental responsibility should be implemented under the framework of sustainability in order to balance three objectives: economic growth, environmental conservation and the principles of international trade and business. As stated by the Sustainable Development Agenda, the relation between nature and development goes beyond the basic protection of species and the reduction of biodiversity loss into the provision of basic needs, adequate livelihoods, and opportunities for economic growth (UN General Assembly 2015). Furthermore, improving environmental standards does not only guarantee sustainability, but it also contributes to achieve other objectives such as the eradication of hunger and the promotion of health and education.

This sustainability approach in global responses to biodiversity loss and conservation strategies specifically requires strengthening the management of the economic benefits derived from the world's protected areas and the traditional knowledge of indigenous communities. The implementation of different legal and economic mechanisms and the promotion of country leadership in these processes is then fundamental. Among them, the mechanisms established by the CBD constitute innovative political economy approaches, which if implemented effectively and efficiently will contribute to improve and consolidate environment's contribution to human well-being.

1.1 The Nagoya Protocol

The third objective of the Convention on Biological Diversity regarding the fair and equitable sharing of the benefits arising from the utilization of genetic resources is fundamental for different reasons. Most of the world's biodiversity is located in developing countries, and they need economic mechanisms that allow them to use these resources in a sustainable way (CBD 2009). However, despite the fact that in general the benefits of genetic resources outweigh the costs of their use, they are not perceived by these countries (Boyce 2002; Oberthür and Rosendal 2013; Rosendal 2000). That is why many of the dispositions established in the CBD require access and benefit-sharing. Articles 8j, 15, 16 and 19 are very specific in this matter and endorse the Convention's general recognition of the sovereign right of states over their biological resources, and the consequent authority of national governments to determine access to genetic resources (United Nations 1992b). It also obliges parties to facilitate access to genetic resources, subject to the prior and informed consent and on mutually agreed terms that promote a fair and equitable share of the benefits.

However, since the signing of the CBD, the implementation of the Access and Benefit-Sharing (ABS) regime proved to be a difficult task, particularly because of its close connection to international trade and to many other dispositions in the area of Intellectual Property Rights (IPRs). The failure to implement resulted in continuous misappropriation of genetic resources and traditional knowledge, and on permanent tensions between the CBD, the World Trade Organization (WTO) and the Trade-Related Aspects of Intellectual Property Rights (TRIPs) agreements (Cabrera Medaglia 2010). In this context, the implementation of the ABS regulation became an urgent matter. The problem was no longer exclusively about biodiversity loss, but equally much about the distribution of the benefits derived from using biological resources in economic and trade-related activities (Oberthür and Rosendal 2013; Rosendal 2000). Biodiversity loss does not only mean a reduction in the number of species, but also in the genetic richness. The implementation of institutional legal and economic mechanisms is still required for the actions of both supplier and user countries to guarantee an appropriate use of the resources and—under an optimal scenario—an equitable distribution of the benefits.

In 1998, the Fourth Meeting of the CBD Conference of the Parties—held in Bratislava (Slovakia)—established a panel of experts to clarify the main principles and concepts related to the ABS objective. In 2000, a Working Group was created with the mandate of defining a set of guidelines to assist the Parties on the implementation of the associated provisions of the Convention. The Bonn Guidelines on Access to Genetic Resources and Fair and Equitable Sharing of the Benefits arising out of their Utilization were adopted in 2002. Simultaneously, the 2002 World Summit on

Sustainable Development mandated the negotiation of an international regime on this matter, oriented to guarantee the effective implementation of the provisions in Articles 15 and 8(j) of the Convention and in general of all its dispositions (United Nations 2002). With that as a purpose, the Ad Hoc Open-ended Working Group on Access and Benefit-Sharing celebrated eleven meetings between 2005 and 2010 to negotiate the ABS regime. A draft protocol was accepted in March 2010 in Cali (Colombia). In 2010, the last meeting took place before the CBD Tenth Conference of the Parties. There, the Nagoya Protocol on Access to Genetic Resources and the Fair and Equitable Sharing of Benefits Arising from their Utilization to the Convention on Biological Diversity was adopted in October 29, 2010. Up to date, 107 are parties to the Protocol (see Fig. 1) (Secretariat of the Convention On Biological Diversity 2018c).

The Nagoya Protocol aims to establish appropriate mechanisms to share the benefits arising from the utilization of genetic resources in a fair and equitable way (Secretariat of the Convention On Biological Diversity 2010). This implies that each country has the sovereign authority to determine the appropriate mechanisms to guarantee such access, normally requiring prior informed consent (Oberthür and Rosendal 2013; Rosendal 2000; Secretariat of the Convention On Biological Diversity 2010). This access should also be complemented by benefit-sharing, implying that under mutually agreed terms, parties should establish the way in which the supplier country is recognized. Benefits can take financial, scientific, social and environmental forms, and there are different mechanisms to return them. Participation in the research using the resources, technology transfer and sharing the actual monetary benefits of the commercial use of the biological resources are the three instruments established by the protocol (Rosendal 2000; Secretariat of the Convention On Biological Diversity 2010). In general, the ABS regime moves away from the understanding of genetic resources as common heritage and create specific international political economy mechanisms to protect these resources from ownership claims, illegal trade, and biopiracy. The general obligations of the Protocol are summarized in Table 1.

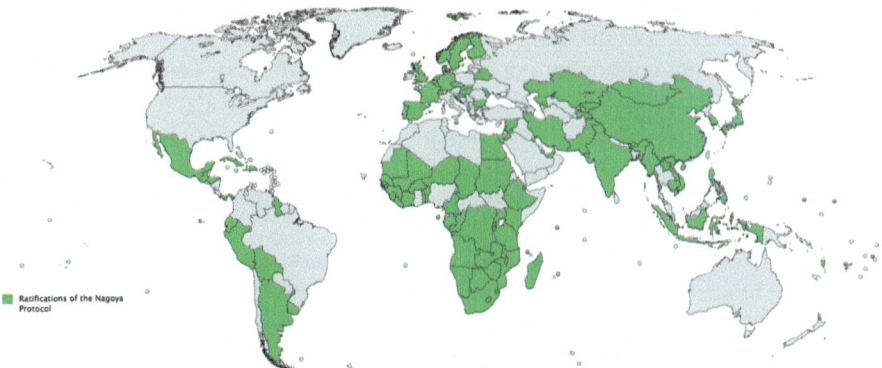

Ratifications of the Nagoya Protocol

Fig. 1 List of parties to the Nagoya Protocol

Table 1 Main obligations under the Nagoya Protocol ABS regime

Access	Benefit-sharing	Compliance
Domestic measures associated to access should: – Guarantee legal certainty, clarity and transparency – Provide fair and non-arbitrary rules and procedures – Establish mechanisms for prior informed consent and mutually agreed terms – Provide issuance of permits when access is granted – Create the conditions to promote research for biodiversity conservation and sustainable use – Consider measures for imminent emergencies that threaten human, animal or plant health – Consider the importance of genetic resources for agriculture and food security	Regarding benefit-sharing the regime establishes that: – Domestic measures should provide a fair and equitable sharing of the benefits arising from the utilization of genetic resources with the contracting party providing genetic resources – Utilization includes R&D and its subsequent application and commercialization – Sharing should be subject to mutually agreed terms. – Benefits can be monetary or non-monetary	Compliance obligations are an innovative mechanism to guarantee that user countries act according to the domestic legislation of the contracting party providing the genetic resources, and according the mutually agreed terms – Guarantee that genetic resources are accessed according to prior informed consent and mutually agreed terms – Ensure cooperation under mutually agreed terms in case of any violations or dispute – Take measures regarding access to justice – Take measures to monitor the utilization of genetic resources by the designation of effective checkpoints at the different stages of the value chain

Source Secretariat of the Convention On Biological Diversity (2010)

According to Article 8 of the Nagoya Protocol, the parties, should enable conditions to incentivize the sustainable use of biodiversity and to promote research and development (R&D) for non-commercial purposes, especially in developing countries (Nijar 2011; Schindel et al. 2015; Secretariat of the Convention On Biological Diversity 2010). Clarity, predictability and transparency are fundamental challenges in the implementation of the obligations of both the CBD and the Nagoya Protocol. Important trade and investment opportunities related to the sustainable use of bio-genetic resources may directly affect R&D and the commercial application of genetic and biotechnology discoveries with industrial application. One may think about the use of extracts of live or dead organisms of biological origin that can be used in the production of cosmetics, skincare, perfumery, pharmaceuticals and herbal/botanic medicine.

Access to genetic resources is also a necessary condition for advancing scientific research in biotechnology and the bio-industry in general. However, many economic sectors and industries can at some point of product development use bio-genetic resources. This is the case of the pharmaceutics and cosmetics. Concerns about misappropriation of genetic resources and associated traditional knowledge from developing countries and the potential monopoly of the profits derived from their use by developed countries can now be addressed with the implementation of the Nagoya Protocol (Nijar 2011). This agreement enables an international legal framework under which both developing and developed countries can regulate profit-sharing mechanisms between industrial manufacturers and suppliers of genetic resources (Cho 2017).

These obligations imply a multidimensional approach since they involve the participation of governments, private sector, intermediaries and communities. These measures also reflect three specific elements of the relation between environment and trade and international business. First, the recognition of the sovereignty of each state over its natural resources (Perrings and Gadgil 2003; UNCED 1992). Second, the increasing perceived value of biological resources due to international trade and the emergence of the green economy. And third, the uneven distribution of resources and the need to avoid the problems associated to this. Specifically, the definition of the benefits deserves special attention since they have both local and global dimensions. Contracting parties should take into consideration both the production and commercial functions of each resource, and the contribution of each species to the provision of economic and ecological services. In this context, the ABS regime brings innovative variables for environmental valuation, including social, environmental and ethical considerations (ten Kate and Laird 1999). The concept of traditional knowledge is also essential to the idea behind ABS. It refers to those innovations and practices that are the result of collective and ancestral experiences of communities in their relationship with their local biodiversity. Such knowledge has been adapted to local needs, cultures and environment, and transmitted from generation to generation. The importance of traditional knowledge is clear in the sense that innovations and practices of local communities can be a pool of knowledge in the identification of ancestral uses of genetic resources (plants, animals and other organisms). Without this knowledge, the development and commercialization of products elaborated with genetic material would be highly difficult (Secretariat of the Convention On Biological Diversity 2011). However, agreeing on these three elements and implementing them has not been an easy endeavor.

ABS emerged in 1970s as a policy concept intended to regulate the distribution and benefits derived from the use of bio-genetic resources (De Jonge and Korthals 2006; De Jonge and Louwaars 2009). If well implemented as a governance tool in access to genetic resources, the ABS mechanism is expected to incentivize the conservation and sustainable use of biodiversity, and thus to help addressing economic asymmetries between megadiverse developing countries providers of bio-genetic resources and developed countries, users of such resources. The definition of ABS was nonetheless absent in both the text of CBD, the Nagoya Protocol and also other derivative regional regulatory frameworks for genetic resources such as the one of the Andean Region. The ABS regime remains as another ambiguous key concept of International Environmental Law requiring further elaboration by states. De Jonge and Louwars identify six possible motivations behind such elusive concept: (1) The South-North imbalance in resource allocation and exploitation, (2) the need to conserve biodiversity, (3) biopiracy and the imbalance in property rights, (4) a shared interest in food security, (5) an imbalance between IP protection and the public interest, and (6) protecting the cultural identity of traditional communities (De Jonge and Louwaars 2009). The implementation of the measures regarding traditional knowledge, especially in the context of ABS and Intellectual Property Rights (IPRs), are not without difficulties. IPRs and its different categories, especially patents, breeders' rights, geographical indications (GIs) and trademarks (protection of innovation) are key for in the commercial use of bio-genetic resources, especially for biotechnology industries (e.g., pharmaceutical and agricultural). Specifically, the intersection between ABS and the IPR system is noticeable in the sense that both the CBD and the Nagoya Protocol attribute particular relevance to transparency on the origin, source, legal provenance and utilization of genetic resources and/or biochemical compounds and associated traditional knowledge. This in order to safeguard cultural, economic and moral interests of local communities (UNCTAD 2017).

2 ABS in Developing Countries: Practices and Challenges for Trade and International Business

The issue of national sovereignty over genetic resources was central to the position of developing countries during the negotiations of the CBD. Their main concern was to receive a compensation for the provision of natural resources and ecosystem services, a revenue that has been historically neglected to them since colonial times (Nijar 2011; ten Kate and Laird 1999). Additionally, a global response was required considering two fundamental challenges of developing countries. In first place they need technology to protect and to take full advantage of their biodiversity (Oberthür and Rosendal 2013; Rosendal 2000). An important part of the world's natural resources is located in the South but are the countries in the North the ones that have the technology for its exploitation and the markets and consumers for the products derived from them. Additionally, developing countries need to strengthen their domestic institutions to adopt the national conservation, access, benefit-sharing and compliance measures (Oberthür and Rosendal 2013; Perrings and Gadgil 2003; Rosendal 2000). Developing countries still have to work on the design and establishment of national incentive structures to avoid the market and public policy failures generated by biodiversity loss and promote its conservation, at the same time that they discourage perverse incentives against it.

The fact that most of the supplier countries are categorized as developing while most of the users are located in developed states reflects a broader trade policy debate. The commercial use of biodiversity and its importance in given economic sectors such as pharmaceutical, biotechnology, seed, crop protection, horticulture, cosmetic and personal care, fragrance and flavor, botanicals, and food and beverage industries, adds complexity to the implementation of the regime (Secretariat of the Convention On Biological Diversity 2008). The market value of these sectors and the economic power of some of the companies involved increase their bargaining power when negotiating with developing countries (Kamau et al. 2010; Nijar 2011). Additionally, the latter problematic of corruption and institutional weakness adds another layer to the threats to biological resources management. Ten Kate and Laird (1999) analyzed the position of both supplier countries and user companies to understand the different arguments and levels of negotiation power they have and how this can affect the position of developing countries as contracting parties of the ABS regime. These elements are summarized in Table 2.

ABS is also relevant for developing countries because its implementation can contribute to build a natural asset base essential for poverty reduction strategies. The democratization and regulation of access can be an important step in granting the lower-income communities the instruments they need to improve their situation. Investment, distribution, and appropriation of the benefits from biodiversity is required to build new natural capital or to improve the existing stocks, if the funding it generates gets transferred to the poor and they can have egalitarian use of these resources and abilities to capture their benefits (Boyce 2002). Reducing the disparities in terms of environmental wealth can undoubtedly contribute to sustainable development and to reduce other abuses that affect the evolution of the international economic and political system (Adams et al. 2010).

One final implication for developing countries is associated to traditional knowledge. Article 8(j) of the CBD establishes that each party should "subject to its national legislation, respect, preserve and maintain knowledge, innovations and practices of indigenous and local communities embodying traditional lifestyles relevant for the conservation and sustainable use of biological diversity and promote their wider application with the approval and involvement of the holders of such knowledge, innovations and practices and encourage the equitable sharing of the benefits arising from the utilization of such knowledge, innovations and practices" (United Nations 1992b). This represents a fundamental challenge since it is embedded in the broader issue of the recognition of indigenous communities and the support for their rights. According to the (UN 2009), even though these communities make up only 5% of

Table 2 Source countries and user companies positions regarding the ABS regime

Source countries	User companies
– Companies need genetic resources. These should not be considered valueless since they contribute to companies' profits. Source countries should then receive a compensation – Rationality. Developed countries' companies should share the royalties generated by biological resources – Legislation is flexible enough to open negotiation space for the conditions of prior informed consent and mutually agreed terms	– Research and development are what gives natural resources its value – Companies assume the risks associated to create patents – Being forced to pay a compensation for the access to genetic resources will affect the companies' competitiveness – Prices established in source countries are high – The legislation is still unclear at the national level. Some genetic resources can be obtained in other countries that are not parties of the CBD

Source ten Kate and Laird (1999)

the global population, they represent about one third of the world's rural poor, which evidences its vulnerability and dependence from biodiversity and traditional knowledge. Their contribution is central not only to the implementation of the ABS regime but also to the conservation of biodiversity. Developing countries face then the challenge of guaranteeing that these incentive mechanisms contribute to its well-being at the same time that they preserve their culture and traditional values.

The Biotrade initiative launched in 1997 by the United Nations Conference on Trade and Development (UNCTAD) addresses many of these challenges, promoting trade and investment in biological resources in support of sustainable development. This initiative—in line with the three objectives of the CBD—aims to provide incentives for the sustainable management of biodiversity while creating economic growth for rural populations in terms of business and investment opportunities (UNCTAD 2013). Latin American countries such as Colombia, Ecuador and Peru have been implementing UNCTAD Biotrade initiative for almost two decades. Biotrade covers a wide range of sectors including pharmaceuticals, cosmetics and personal care, fashion, ornamental flora and fauna, textiles, handcrafts, textiles and sustainable tourism (UNCTAD 2017). Due to its broad scope, Biotrade activities may result in the use of genetic resources, including R&D on the genetic and/or biochemical composition of such resources and therefore be governed by the obligations stemming from the Nagoya Protocol (UNCTAD 2017).

2.1   The Case of Latin American and the Caribbean

The status of ratifications of the ABS regime in Latin America and the Caribbean is diverse (see Fig. 2). Countries such as Argentina, Bolivia, Ecuador, Mexico and Peru were prompt to sign and ratify the agreement and to adopt the legislations in terms of the management and use of genetic resources. Nonetheless, they are still facing challenges on their implementation. Other countries are still to advance in this process. Megadiverse countries such as Brazil, Colombia, and Costa Rica signed the agreement but have not ratified it, which reflects on some of the practices and challenges that they have in the use of biodiversity. In other countries such as Chile, Haiti, Nicaragua, Paraguay, and Venezuela the signature of the Nagoya Protocol is still pending. However, the region evidences an important trend. Members of the Andean Community (Bolivia, Colombia, Ecuador and Peru) had already adopted other forms of domestic legislation that are consistent with the objectives of the CBD. Decision 391 of the Andean Community (the Common Regime on Access to Genetic Resources of the Andean Community)—adopted in 1996—is considered to be the first sub-regional response to the implementation of Article 15 of the CBD on access to genetic resources (Comunidad Andina de Naciones 1996).

The Andean Regime was regarded as regional response to the issues and procedures that were not addressed by the CBD, including sovereignty and ownership of genetic resources (Rosell 1997). The CBD and the Nagoya Protocol are ambiguous and key concepts such as "access" are not defined by such agreements. Divergent

Fig. 2 Status of ratifications for the Nagoya Protocol in Latin America and the Caribbean

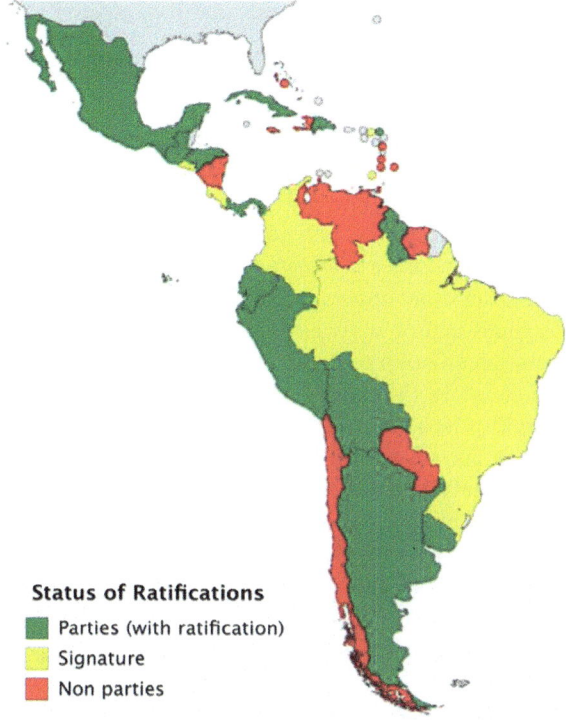

Status of Ratifications
- Parties (with ratification)
- Signature
- Non parties

interpretations on such concept may represent a challenge in the implementation of the obligations stemming from these international instruments, especially between "users" and "suppliers" of bio-genetic resources. However, these circumstances have not prevented some developing countries to consider that effective access happens when the resource is utilized. In contrast, developed countries (users of genetic resources) consider that access occurs in the moment that the genetic resource has crossed the borders of its country of origin (Bifani 2014). In spite of the above, regional instruments such as Decision 391 of the Andean Community defines access as "the obtaining and use of genetic resources conserved in situ and ex situ, of their by-products and, if applicable, of their intangible components, for purposes of research, biological prospecting, conservation, industrial application and commercial use, among other things" (Comunidad Andina de Naciones 1996). The preamble of the Andean Community legislation also highlights the multi-ethnic and pluricultural nature of the member states and the strategic value of their biological diversity, genetic resources, their endemism and rarity, and associated traditional knowledge (know-how, innovations and practices) of the native, Afro-American and local communities in the international context. The decision expressly refers to the importance of preserving and developing their biological and genetic heritage as well as the recognition of the historic contribution made by the native, Afro-American and local communities in the sustainable use of biodiversity (Comunidad Andina de Naciones 1996).

Colombia, one of the most megadiverse countries in the world,[1] signed the Nagoya Protocol in 2011. However, as mentioned above, it is still pending ratification. Nonetheless, the country has regulated access to genetic resources by implementing the CBD Convention (ratified by Law 165 of 1994) and Andean Decision 391. Under Colombian law, genetic resources are regarded as a dimension of biodiversity. They include genes, individuals, species, populations, ecosystems and landscapes. As these resources belong to the state and are considered inalienable, imprescriptible and unseizable, any natural or legal person who wishes to access these in the form of genes and/or derivative products requires a state's authorization. The Genetic Resources Group of the Colombian Directorate of Forest Biodiversity and Ecosystem Services (Ministry of Environment and Sustainable Development) is the legal authority in charge of authorizing access to genetic resources following a procedure in accordance with the provisions of the Andean Decision 391 (Ministerio de Ambiente Vivienda y Desarrollo Sostenible 2018).

The signature of an access contract between the Colombian state and the user (i.e., applicant) of the genetic resource or by-product is essential to the Andean procedure. According to Article 35 Andean Decision 391 when the contract also involves intangible components it needs to incorporate an annex stipulating the fair and equitable distribution of the profits from use of that component as a fundamental condition for the validity of the access contract (Comunidad Andina de Naciones 1996). This particular document shall be signed by the supplier of the intangible component and the applicant. Depending of the national legislation of the Andean Member, the annex of may also be signed by the Competent National Authority.

Also, access to genetic resources can be limited under certain circumstances. Under Article 45 of the Andean Decision 391, member states can implement regulations in order to limit such access in the cases that include: (a) Endemism, rarity or danger of extinction of species, subspecies, varieties or races or breeds; (b) Vulnerability or fragility of the structure or functioning of the ecosystems that could worsen as a result of access activities; (c) Adverse effects of access activities on human health or on elements essential to the cultural identity of nations; (d) Undesirable or not easily controlled environmental effects of access activities on the ecosystems; (e) Danger of genetic erosion caused by access activities; (f) Regulations on biosecurity; or (g) Genetic resources or geographic areas rated as strategic.

However, the definition of these legal and policy instruments does not guarantee—in the case of Colombia—the full implementation of the ABS regime. Some evaluations have found that the legal framework is still to be completed and is not supported by the required information and management systems. Bureaucracy, inefficiency, and lack of diffusion pose challenges not only to the use of biodiversity

---

[1] According to the country profiles established by the CBD, Colombia is one of the most megadiverse countries in the world, hosting about 10% of the biodiversity of the planet, registering more than 300 types of ecosystems. The country also ranks first in bird and orchid species diversity and second in plants, butterflies, freshwater fishes and amphibians (Secretariat of the Convention On Biological Diversity 2018a).

but also to research on genetic resources (Silvestri 2016). The resulting benefits, in some cases, are also non-monetary and conflicts between local communities and conservation authorities persist (Correa 2017; De Pourcq 2017). Previous consultations do not guarantee that communities are granted full access to the genetic resources extraction and use processes (Silvestri 2016). Furthermore, many of the measures established by Colombian and Andean legislation still have to fulfill the requirements of the Nagoya Protocol.

3 Conclusions

Natural resources in general and biodiversity in particularly are a fundamental part of the world assets and then are central to international trade and business. The close connection between environment and economics needs to recognize the value of natural assets and the need to guarantee its sustainability. Additionally, biodiversity's link to development is nowadays a central issue in the global strategies to improve human well-being. However, in this context the definition of the ownership and value of biological resources is central. The CBD was a fundamental step in this path. It demonstrated the need for international cooperation and regulation in order to guarantee environmental conservation and sustainable management. Specifically, the Convention established measures—in the Nagoya Protocol—to regulate the access and benefit-sharing associated to the utilization of biological resources.

Instruments as the ABS regime constitute fundamental steps to reconcile the economic dimension of the environment and its influence on international trade and business. However, the implementation of the ABS regime has not been as straightforward as required. The implications of the ABS regime on international trade and IPRs have been some of the obstacles. Additionally, its multidimensional approach has different implications that require further consideration. Stakeholders' engagement and the commitment of the private sector is fundamental to advance in regulation and measures that translate the international definitions of the ABS regime and reduce their ambiguity. Market incentives are also needed to improve the economic contribution of natural resources and to enhance its social, political and cultural value. Regulation on bio-genetic resources should work in ways that find a balance between the use and the conservation of the resources.

In the case of the developing countries, their position as suppliers of biological resources has to confront the economic interest of companies and multinational corporations of powerful sectors such as pharmaceuticals, cosmetics, and biotechnology, a situation that can seriously erode the former's negotiation power. The ABS regime has fundamental implications for developing countries not only in terms of national legislation but also in terms of a series of challenges they need to overcome to be able to fully use their biological wealth. Technology and institutions are two fundamental processes in this context, as well as the protection of traditional knowledge and indigenous rights. Some countries, nonetheless, have advanced in the regulation of this important dimension of international trade and businesses. Countries in Latin

America and the Caribbean—specifically the members of the Andean Community of Nations such as Colombia—have developed regional mechanisms to regulate the process. Other experiences in countries such as Brazil and Costa Rica—that also have not ratified the Nagoya Protocol—show different approaches, mechanisms, instruments and lessons that should be take into consideration. However, the existence of these legal frameworks does not guarantee their implementation or the contribution of biodiversity use and benefit-sharing to conservation and improved livelihoods.

Furthermore, a broader movement towards a more equitable distribution of the power represented by access and benefit-sharing is necessary to guarantee the promotion of environmental conservation. And the implementation of the Nagoya Protocol requires an active participation of developing states, so they can design the mechanisms that would eventually contribute to the collection of the expected benefits. Paths of action—that should be both practical and executable—include the use of trade negotiations and agreements as mechanisms to advance in the implementation of the Nagoya Protocol. In addition, a commitment by developed countries to fairly pay the developing countries for their natural resources and to extract them in ways that do not cause environmental degradation is essential for the developmental perspective. Otherwise, a valuable opportunity will be missed to protect natural resources and to give them a sustainable use, in ways that promote conservation and contribute to economic growth and development, without disincentivizing the role of investment, trade and international business in sustainability and sustainable development.

References

Adams WM, Aveling R, Brockington D, Dickson B, Elliott J, Hutton J, Wolmer W (2010) Biodiversity conservation and the eradication of poverty. In: Roe D, Elliott J (eds) The Earthscan reader in poverty and biodiversity conservation. Earthscan, London and Washington D.C., pp 18–26

Agrawal A, Redford K (2010) Poverty, development, and biodiversity conservation: shooting in the dark? In: Roe D, Elliott J (eds) The Earthscan reader in poverty and biodiversity conservation. Earthscan, London and Washington D.C., pp 42–47

Arsel M, Büscher B (2012) Nature™ Inc.: changes and continuities in neoliberal conservation and market-based environmental policy. Dev Change 43(1), 53–78

Bifani P (2014) Retos y perspectivas del Protocolo de Nagoya. ICTSD Puentes 15(9)

Bishop J, Pagiola S (eds) (2012) Selling forest environmental services: market-based mechanisms for conservation and development. Taylor & Francis

Boyce JK (2002) The political economy of the environment. Edward Elgar Publishing, Northampton

Cabrera Medaglia J (2010) The political economy of the international ABS regime negotiations: options and synergies with relevant IPR instruments and processes. International Centre for Trade and Sustainable Development, Geneva

CBD (2009) Biodiversity, development and poverty alleviation: recognizing the role of biodiversity for human well-being. Retrieved from Montreal: http://www.cbd.int/doc/bioday/2010/idb-2010-booklet-en.pdf

Cho A-Y (2017) Practical implementation issues for the convention on biological diversity and the Nagoya Protocol from a Korean perspective. Korean J Int Comp Law 5(1):61–82

Clapp J, Dauvergne P (2011) Paths to a green world: the political economy of the global environment. MIT Press, Cambridge

Collier P (2010) The plundered planet: why we must, and how we can, manage nature for global prosperity. Oxford University Press, Oxford and New York

Comunidad Andina de Naciones (1996) Common regime on access to genetic resources. Caracas, Venezuela. Retrieved from http://www.sice.oas.org/trade/JUNAC/decisiones/DEC391e.asp

Correa CM (2017) Access to and benefit sharing of Marine Genetic Resources Beyond National Jurisdiction: developing a new legally binding instrument. In: McManis CR, Burton O (eds) Routledge handbook of biodiversity and the Law. Routledge

De Jonge B, Korthals M (2006) Vicissitudes of benefit sharing of crop genetic resources: downstream and upstream. Dev World Bioeth 6(3):144–157

De Jonge B, Louwaars N (2009) The diversity of principles underlying the concept of benefit sharing. In: Genetic resources, traditional knowledge y the law: solutions for access and benefit sharing, pp 37–56

De Pourcq K et al (2017) Understanding and resolving conflict between local communities and conservation authorities in Colombia. World Development 93: 125–135

Glowka L, Burhenne-Guilmin F, Synge H, McNeely JA, Gundling L (1994) A guide to the convention on biological diversity. IUCN and Environmental Law Centre, Bonn and Gland

Gómez-Baggethun E, Muradian R (2015) In markets we trust? Setting the boundaries of market-based instruments in ecosystem services governance

Hardin G (1968) The tragedy of the commons. Science 162:1243–1248

Ibisch P, Vega EA, Hermann TM (2010) Interdependence of biodiversity and development under global change. Retrieved from Montreal, Canada

Iftekhar MS, Polyakov M, Ansell D, Gibson F, Kay GM (2017) How economics can further the success of ecological restoration. Conserv Biol 31(2):261–268

IUCN (2016) The IUCN red list of threatened species. Version 2016.3. Retrieved from http://www.iucnredlist.org

Kamau EC, Fedder B, Winter G (2010) The Nagoya Protocol on access to genetic resources and benefit sharing: what is new and what are the implications for provider and user countries and the scientific community. Law Environ Dev J 6:246

MEA (2005) Millennium ecosystem assessment. In: Ecosystems and human well-being: biodiversity synthesis. World Research Institute, Washington D.C.

Ministerio de Ambiente Vivienda y Desarrollo Sostenible (2018) Recursos genéticos. Retrieved from http://www.minambiente.gov.co/index.php/component/content/article/782-plantilla-bosques-biodiversidad-y-servicios-

Nijar GS (2011) The Nagoya Protocol on access and benefit sharing of genetic resources: analysis and implementation options for developing countries. South, Centre

Nunes P, Van Den Bergh J, Nijkamp P (2003) The ecological economics of biodiversity: methods and policy applications. Edward Elgar Publishing, Northampton

Oberthür S, Rosendal GK (2013) Global governance of genetic resources: access and benefit sharing after the Nagoya Protocol, Routledge

Pagiola S, Landell-Mills N, Bishop J (2002) Market-based mechanisms for forest conservation and development. Selling For Environ Serv Mark Based Mech Conserv Dev 1–13

Pearce DDW, Moran D (1994) The economic value of biodiversity. Earthscan Publications, London

Perrings C, Gadgil M (2003) Conserving biodiversity: reconciling local and global public benefits. In: Kaul I, Conceiçao P, Le Goulven K, Mendoza RU (eds) Providing global public goods: managing globalization (pp. 532–555). Oxford University Press, New York

Rosell M (1997) Access to genetic resources: a critical approach to decision 391'common regime on access to genetic resources' of the cartagena agreement. Rev Eur Commun Int Environ Law 6(3):274–283

Rosendal G (2000) The convention on biological diversity and developing countries. Kluwer Academic Publishers, Dordrecht

Schindel D, Bubela T, Rosenthal J, Castle D, du Plessis P, Bye R (2015) The new age of the Nagoya Protocol. Nat Conserv 12:43

Secretariat of the Convention on Biological Diversity (2007) An exploration of tools and methodologies for valuation of biodiversity and biodiversity resources and functions, pp 71. Technical Series No. 28

Secretariat of the Convention on Biological Diversity (2008) Access and benefit-sharing in practice: trends in partnerships across sectors, pp 140. Technical Series No. 38

Secretariat of the Convention on Biological Diversity (2010) Nagoya Protocol on access to genetic resources and the fair and equitable sharing of benefits arising from their utilization to the convention on biological diversity: text and annex. Montreal and United Nations

Secretariat of the Convention on Biological Diversity (2011) Introduction to access and benefit sharing. Montreal (Canada) Retrieved from https://www.cbd.int/abs/infokit/brochure-en.pdf

Secretariat of the Convention on Biological Diversity (2018a) Colombia country profile. Retrieved from https://www.cbd.int/countries/?country=co

Secretariat of the Convention on Biological Diversity (2018b) List of parties. Retrieved from https://www.cbd.int/information/parties.shtml

Secretariat of the Convention on Biological Diversity (2018c) List of parties for the Nagoya Protocol. Retrieved from https://www.cbd.int/abs/nagoya-protocol/signatories/default.shtml

Silvestri LC (2016) Acceso a recursos genéticos y distribución de beneficios en Colombia: desafíos del régimen normativo. Investigación Desarrollo 24(1)

ten Kate K, Laird SA (1999) The commercial use of biodiversity: access to genetic resources and benefit-sharing. Earthscan Publications, London

UNCED (1992) Agenda 21. Retrieved from New York: http://www.un.org/esa/sustdev/documents/agenda21/english/Agenda21.pdf

UNCTAD (2013) Guidelines for the sustainable management of BioTrade Products: resource assessment. Geneva, United Nations. Retrived from: http://unctad.org/en/Publicationslibrary/ditcted201041_en.pdF

UNCTAD (2017) BioTrade and access and benefit sharing: from concept to practice. In: A handbook for policymakers and regulators. Geneva and United Nations

UN General Assembly (2015) A/RES/70/1 Transforming our world: the 2030 Agenda for sustainable development. United Nations, New York. Retrived from: http://undocs.org/sp/A/RES/70/1

United Nations (1992a) A/CONF.151/26 Rio declaration on environment and development. In: United Nations conference on environment and development. Rio de Janeiro

United Nations (1992b) Convention on biological diversity. Retrieved from http://www.cbd.int/doc/legal/cbd-en.pdf

United Nations (2002) A/CONF.199/20 Johannesburg declaration on sustainable development. In: World summit on sustainable development

United Nations (2009) State of the wold's indigenous peoples. Retrieved fromhttp://www.un.org/esa/socdev/unpfii/en/sowip.html

United Nations (2012) A/RES/66/288 The future we want—outcome document from Rio+20. In: United Nations conference on sustainable development. Rio de Janeiro

WCED (1987) Our common future—the Brundtland report. Retrieved from http://www.un-documents.net/wced-ocf.htm

Global New Economy: Structure and Perspectives in Kaliningrad Region

Yulia Aleynikova

Abstract The emergence and development of innovative technologies have a mul-tivalued impact on the region social-economic sphere and environment. Obviously, it will move region economies entirely to next decade. To reveal the streams and features of "fresh" processes, it is extremely relevant to launch sustainable devel-opment at the first steps. And there are some prerequisites and stumbling blocks to deal with, they are important to be identified and researched. In this paper, you can find important methodology gap and its description: official statistic authorities follow traditional classification and new economy industries are complicated to be revealed, studied and supported. As a solution initiative research project was run. This chapter presents its key methods, stages and results briefly. New economy of Kaliningrad region (53 companies) is described as a scheme. Additionally you can find key participants from each new industry and their brief description. The most valuable part of the project is recommendations for authorities how make new econ-omy transparent for researches and effective government support. That is one stage of research series which is devoted to global innovation economy and sustainability.

Keywords New economy · Innovations · Sustainability · Baltic region

1 Goal of Research

The research is targeted to define approach for identification and study of 'new economy' phenomenon and its development tendencies in cities and regions. This project has been implemented in Kaliningrad region.

Y. Aleynikova (✉)
Department of Economics, Kaliningrad State Technical University, Kaliningrad, Russia
e-mail: j.aleynikova@icloud.com

© Springer Nature Switzerland AG 2020
W. Leal Filho et al. (eds.), *International Business, Trade and Institutional Sustainability*, World Sustainability Series,
https://doi.org/10.1007/978-3-030-26759-9_49

2 Definitions

In modern theory and practice global processes a variety of similar-root and similar-content definitions are used. That is the reason to pay special attention what 'new economy' is for researchers.

Till XXI century permanent and gradual changes in global production basis characterized 'postindustrial society' definition and its derivatives (for example, 'postindustrialization'). Cycle theories (by E. Toffler,[1] N. Kondratyev,[2] D. Bell[3] and others) underlie epoch periodization and general development of postindustrial society. Wealthy growth models for this society do not accent a number of employees and full employment as a favorite politicians' and macroeconomists milestone, but high professional and intellectual competences and skills payment. According to that last five decades income sources had been transforming slowly to more flexible ways and principles due to individual inventions, patents and contributions.[4]

However, first two decades of XXI century have performed an innovation mixer in global production which creates new and destroys old (traditional) industries on high-speed mode, and it has been occurring in ways with no precedents before. Ideas, technologies and knowledge diffuse free and fast due to the virtual architecture growing day-to-day and its intensive blending with real world. Very often it is defined 'digital revolution' or 'global digital transformation'. Then, 'postindustrialization' as an attribute of modern society development is displaced by 'new economy' definition (or 'innovation economy'). By itself it describes 'postindustrialization' mechanics and reflects the phenomenon in next stage of development and diffusion.

Therefore, we can make the conclusion that majority of postindustrialization predictions made by different scientific schools (generally about affection of knowledge and innovations to global production evolution) have come true. Nowadays new economy appears as a result of using the latest know-how and innovations technologies to transform traditional products, services, consumption habits and business-processes. But there is still no standard definition for 'new economy' and it exists as an abstract and rapidly-changing.

---

[1] Тоффлер (2004).
[2] Кондратьев (2002).
[3] Белл (2004).
[4] Алейникова (2015).

Specifically majority of researchers make accents on similar attributes of 'new economy' (A. Buzgalin,[5] S. Glazyev,[6] W. Deming,[7] Cambrigde school of Economics, business-society with internet platforms and others[8,9]):

- high proportion of intangible assets with developed infrastructure for their creation, storage, processing and monetization;
- key production factor is human capital;
- rapid gap from invention to innovation.

On our opinion, the best way to describe the phenomenon discussed is mega-trends scope which have been transforming the society core and image during last century and got the critical mass to re-design human values and habits, resources consumption, machines and technologies entirely. So using 'new economy' definition, we mean approach to global production with an attribute-set as described Table 1.

3 Research Scheme

Nowadays, there is obvious conflict in research fields: between official data sources which are based on traditional industries classification and pilot initiatives run by authorities, consulting companies, corporations and scientific centers. Actually, in advanced and developing countries last several years official statistic authorities has been adopting new approach to the data and now their databases consist of some newest industries, processes and accents, but still present not all objectives and are being fulfilled with time-lag. After public database and analytic sources study, we finally decided to follow traditional classification of industries and choose those of them where the transformation occurs materially (Scheme 1).

Kaliningrad region is an enclave region in Russian Federation located on Baltic Sea and surrounded by EU countries with no land connection to Russia. It directly affects the regional structure of economy. The region has been announced officially as the area with special preference status till 2045.[10] Such a long-term special status is the root for transformation and sustainable development, together with the neighboring countries (Scheme 1).[11]

---

[5]Бузгалин and Колганов (2007).

[6]Глазьев (2012).

[7]Deming (2000).

[8]Основные черты новой эпохи быстрых перемен [Электронный ресурс]. Режим доступа 01/12/2018: http://www.cecsi.ru/coach/new_economy.html.

[9]Understanding the Innovation Economy and Its Impact on Our World by philmckinney/ – Режим доступа 01/12/2018: https://philmckinney.com/understanding-innovation-economy-impact-world/.

[10]Портал Администрации ОЭЗ в Калининградской области. https://oez.gov39.ru/.

[11]Калининград рассчитывает привлечь инвесторов благодаря новым законам [Электронный ресурс]. Режим доступа 01/12/2018: http://tass.ru/ekonomika/4474221.

Table 1 Main attributes of 'new economy' ('innovation economy')

Attribute	New economy	Before
Human as a customer		
Values and priorities	Life quality, material and spirit values combination, individualization (customization)	Material values
Key factor for making decision	Time spent, environmental-friendly consumption, security in variety of scopes	Information about price and quality
Customer basket	High proportion of VR- or AR-goods (or VR/AR as a part of goods)	Material goods and supporting services
Market		
Competition	Global hyper-competition, cooperation and collaboration	Local market, the bigger is company the larger is market
Engine for growth	Innovation small-sized companies/teams using networking, based on horizontal management and intrapreneurship principles	Big-sized companies with fixed hierarchy and vertical management
Key competitive advantage	Decision speed and innovation technologies	Company size
Human as a production factor		
Employer philosophy	Human capital is major investment	Stuff means current expenses
Skills and competences	Cross-disciplinary, flexible, based on ethics	Standardized and focused
Employment	Flexible, intrapreneurs, based on emphatics	Stable, based on 'must have a work'
Company		
Engine for growth	Human capital, knowledge	Money and its derivatives
Priorities	Changes are regular way of things	Changes happen, adoption concept
Innovations	Permanent	Temporary, linear
Organizational structures	Decentralized, horizontal	Vertical, centralised, fixed
Production philosophy	Customization, decreasing of transaction costs based on collaboration, cooperation and digitalization	Mass-production, long-term investments, severe competition
Recourse consumption	Extensive growth based on resource-efficiency increase and resource-intensity reduction	Intensive growth (the more the better)

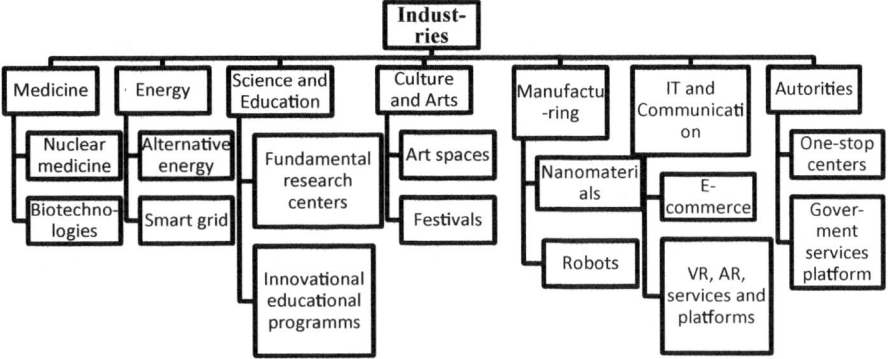

Scheme 1　Regional new economy (industries researched)

Key phases:

1. Teambuilding, research goal setting, data source determination. Workshop with business and government experts to specify hypothesis.

This initiative research was implemented by master students of KrausLab (master program for entrepreneurs at Baltic Federal University of I. Kant) and their business advisers (for support and methodology framework). The hypothesis was transformed to following agenda: there are some features and participants of new economy industries in the region but they are not transparent for government support and researchers because official statistics authorities followed traditional classifications (see above).

2. Public resources (national and international ones) study. Working-out of regional industries list to be researched.

Team started with targeted sub-research of public resources (consulting groups reports, articles and books) for collecting list of new industries. Then they organized workshop with authorities (Ministry of Industry in Kaliningrad region[12]) to match list points with traditional classification. For that team presented briefly the core of each new branch and asked experts whether they had dealt or heard about such kind of processes and products in Kaliningrad region. There were two key results. The first one was draft scheme of local new industries (scheme above). The second one was draft list of their participants (see chart below, updated after survey).

3. Questionnaire lists and interview framework compilation.

The next stage was devoted to study features of companies which have been launching new economy in region: their size, stuff, products and services, business conditions, perspectives and limits.

4. Survey and interview sessions.

---

[12]Портал Министерства промышленности Калининградской области. https://investinkaliningrad.ru/invest/otrasli.php.

Half of participants dealt with questionnaire by e-mail, team also took 11 interviews, and others gave no answer but some data were collected from public resources (media chronicles, articles).

5. Processing data, analysis.

Then team started with description of results. Unfortunately not full information was available for graphic analysis but interviews revealed key barriers and negative conditions for business development which were presented in final report.

6. Workshop with business and government experts: public presentation of results, discussion, recommendations. Final report.

4 Brief Results

The major result presents that the new economy is still an abstract definition, not obvious and clear for regional science centers, business and authorities. In this case, there are high risks to develop conditions and priorities needed and to launch transformation process efficiently based on sustainable approach.

According to data collected in the beginning of 2018, 53 projects set in the region can be defined as subjects of new economy and strategic valuable initiatives. Here are some of them.

In the energy sphere, the region became the first one in Russia where digital substation draft project was launched (smart grid solutions and engineering). Smart grids use IT and communication technologies for data collecting and processing to reduce resource consumption and increase energy efficiency. Also, there are some successful alternative energy projects: wind farms in Zelenogradsk and Kaliningrad, solar panels distribution company for individual houses and enterprise bases.

In manufacture sphere, regional new economy is presented by private technopark 'Technopolis GS' (GS Group) specialized in microelectronics and nanomaterials production and by companies which produce different industrial robots. Producers also collaborate with IT-companies and launch draft-projects to further automatization of their production process.

Alive System Institute (based at Baltic Federal University of I. Kant—BFU) provided a great shift for development of Kaliningrad region. It implements science co-working concept. Basically, it is specialized in genetics, bio-IT and neurobiology. It co-works with Medicine, Math, Physics and IT Institution (BFU) and launched researches and projects in nanomedicine. In 2019 the full-circle nuclear medicine center will be run with 'Rosatom" support.

In education, there were crated and launched several innovation programs and centers, such as Krauslab innovation master program, BFU Research Center, Technopark KGTU, Technopark 'Fabrica' BFU. They initiate their own projects and collaborate in IT and innovation studies. The new educational programs adopt learning-by-doing concept and blended education based on digital content and full-time sessions.

Some private regional initiatives developed good level IT-potential. Actually their activities can't be calculated clearly because usually they use outsourcing. But 11 residents note they are regional ones with investments in equipment and stuff. Their specialization is really flexible. Being on the new wave edge IT-companies adopt permanently to develop services for national and global market: games, applications, platforms. They create VR, some AR and Blockchain projects (government service platform, Golos platform and others). Additionally, two residents are focused on search technologies.

The research results approve that innovations also affect regional cultural environment and push it to find new commercial solutions. Creative spaces are new cultural phenomenon which have been evolving in region for last 5 years. S. Evans defines such spaces as a creative entrepreneurs communities which collaborate at the particular area.[13] In Kaliningrad, they are located at the Museum of Modern Arts, libraries, cinema hall and cafes. They also provide a lot of regional, national and international events and festivals for culture exchange and ideas diffusion.

Regional government and its authorities also launch digital initiatives to provide their public services faster and efficiently, transparent for society. During last 3 years they run several one-stop-centers and business center with co-working opportunities and telecommunication equipment; provided national and regional government services platform, several draft-projects on Blockchain.

Below the number of projects is structured in pie chart according to industry scheme defined (Fig. 1).

The leaders are IT and Communication (21%), Manufacturing (19%), Culture and Arts (17%) but it is less than 1% of total company number in the region. According to research results average net income from these pilot projects is less than 1–2 billion euro that also less than 1% Regional Gross Product.

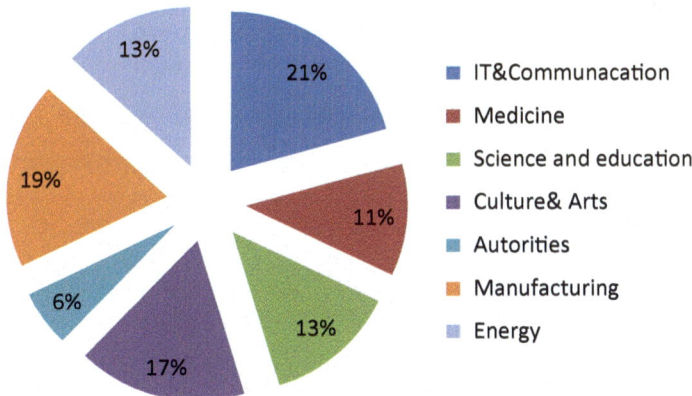

Fig. 1 Structure of new economy in Kaliningrad region

---

[13]Розамунд Дэйвис. Introducing the Creative Industries: From Theory to Practice [Текст] / Дэйвис Розамунд // Первая публикация. - 22 апреля 2013 г.

In this case active information and communication activities can be recommended to provide regional business and authorities with awareness of new economy tendencies and projects. It will be basis for privileges and support design and we can expect new economy sustainable growth. One of the practical solutions can be interactive map: official and free resource with regional participants of new economy and their actual brief information available. In prospect, it can be virtually enlarged to all Baltic area with number of projects to be observed, invested and industries to be supported; sustainable goals can be implemented by governments, business, scientific and educational centers collaboration.

References

Deming W (2000) The new economics for Industry, Government, Education, 3rd edn. The MIT Press, Boston

Алейникова Ю.А. Ключевые сценарии стратегии России и перспективы их реализации // Экономическая система современной России: пути и цели развития: Монография, под. ред. А.А.Пороховского (Электронное издание). М.: Экономический факультет МГУ им.Ломоносова, 2015. с.295–301

Белл Д (2004) Грядущее постиндустриальное общество. Опыт социального прогнозирования. М.: ACADEMIA, 790 с

Бузгалин АВ, Колганов АИ (2007) Глобальный капитал (2-е изд.) М: Едиториал УРСС, 560 с

Глазьев СЮ (2012) «Современная теория длинных волн в развитии экономики» // Экономическая наука современной России. № 2 (57) С.8–27

Основные черты новой эпохи быстрых перемен [Электронный ресурс]. Режим доступа 01/12/2018: http://www.cecsi.ru/coach/new_economy.html

Калининград рассчитывает привлечь инвесторов благодаря новым законам [Электронный ресурс]. Режим доступа 01/12/2018: http://tass.ru/ekonomika/4474221

Кондратьев НД (2002) Большие циклы конъюнктуры и теория предвидения / Сост. Ю. В. Яковец. М.: Экономика, 768 с

Розамунд Дэйвис. Introducing the creative industries: from theory to practice [Текст] / Дэйвис Розамунд // Первая публикация. - 22 апреля 2013 г

Тоффлер Э (2004) Глава 15. За пределами массового производства // Третья волна. Москва: АСТ, 781 с

Портал Администрации ОЭЗ в Калининградской области. https://oez.gov39.ru/

Портал Министерства промышленности Калининградской области. https://investinkaliningrad.ru/invest/otrasli.php

Understanding the Innovation Economy and Its Impact on Our World by philmckinney/ – Режим доступа 01/12/2018: https://philmckinney.com/understanding-innovation-economy-impact-world/

Foreign Direct Investment, Domestic Investment and Green Growth in Nigeria: Any Spillovers?

Akintoye V. Adejumo and Simplice A. Asongu

Abstract Globally, investments in physical and human capital have been identified to foster real economic growth and development in any economy. Investments, which could be domestic or foreign, have been established in the literature as either complements or substitutes in varying scenarios. While domestic investments bring about endogenous growth processes, foreign investment, though may be exogenous to growth, has been identified to bring about productivity and ecological spillovers. In view of these competing–conflicting perspectives, this chapter, examines the differential impacts of domestic and foreign investments on green growth in Nigeria during the period 1970–2017. The empirical evidence is based on Auto-regressive Distributed Lag (ARDL) and Granger causality estimates. Also, the study articulates the prospects for growth sustainability via domestic or foreign investments in Nigeria. The results show that domestic investment increases CO_2 emissions in the short run while foreign investment decreases CO_2 emissions in the long run. When the dataset is decomposed into three sub-samples in the light of cycles of investments within the trend analysis, findings of the third sub-sample (i.e. 2001–2017) reveal that both types of investments decrease CO_2 emissions in the long run while only domestic investment has a negative effect on CO_2 emissions in the short run. This study therefore concludes that as short-run distortions even out in the long-run, FDI and domestic investments has prospects for sustainable development in Nigeria through green growth.

Keywords Investments · Productivity · Sustainability · Growth

A. V. Adejumo (✉) · S. A. Asongu
Department of Economics, Obafemi Awolowo University, Ile-Ife, Nigeria
e-mail: adejumoakinvic@gmail.com; vadejumo@oauife.edu.ng

S. A. Asongu
Lead Economist, African Governance Development Institute, Yaounde, Cameroun

© Springer Nature Switzerland AG 2020
W. Leal Filho et al. (eds.), *International Business, Trade and Institutional Sustainability*, World Sustainability Series,
https://doi.org/10.1007/978-3-030-26759-9_50

1   Introduction

The continuous debate on the gains of foreign direct investment (FDI) and trade in the literature has made it impossible to isolate the effects of FDI on growth. While domestic investments are primarily geared towards the growth and development of local economies, the insufficiency of this form of investment has caused most developing economies to continually position themselves to attract FDI. However, the presence of FDI for local economic benefits has continuously attracted arguments in the literature. For instance, some studies have seen FDI and trade as catalysts for economic growth, augmenting physical and human capital and promoting efficiency in the production of goods and services (Feder 1983; Ram 1985; Salvatore and Hatcher 1991; Makki and Somwaru 2004). Conversely other studies have either seen FDI as a threat to resource allocation and the existence or development of industries within host economies (Boyd and Smith 1992; Bende-Nabende 2017). For instance, according to Smarzynska Javorcik (2004), Multinational Enterprises (MNEs) are mostly located strategically in highly productive industries; thereby, masking the genuine spillovers. It could be such that MNEs may force less productive domestic firms to exit and then increase their share of investment; thus, causing host economies to superficially pass the productivity test.

Theoretical discourse on the bearing of the FDI-growth nexus has been contested especially within the neoclassical growth doctrines. For instance, the exogenous growth theorists perceive FDI more as income-stimulating rather than stimulating long-run growth. Therefore, the exogenous growth theorist upheld that, if FDI will drive any long-run growth, it will be because it has affected population growth and technological progress positively (Solow 1957; De Mello 1997). However, the differing perspective of the endogenous growth theory posited that FDI can drive growth through spillover effects and positive externalities in outputs. Specifically, spillover possibilities via FDI gains within host economies could stem from FDI fostering innovations and entrepreneurship, building human capital through technological diffusion, as well as the introduction of new management and organizational systems. Thus, irrespective of the ideologies surrounding the FDI-growth nexus, there is a consensus on the plausibility of FDI fostering and generating high-growth opportunities. However, FDI stimulating growth remains contingent on the absorptive capacity of the host economy; as well as whether FDI causes or crowds out domestic investment.

Empirically, studies have ascertained the roles of FDI in either stimulating domestic investment (Makki and Somwaru 2004); or generating positive spillovers (Lall 1980; Rodriguez-Clare 1996; Markusen and Venables 1999; Lin and Saggi 2004). Also, some studies have examined the differential contributions of FDI and domestic investment in stimulating development. While some have found support for FDI (Borensztein et al.1998; Balasubramanyam et al. 1996); some others have adduced otherwise for domestic investments (Aitken and Harrison 1999). Irrespective of the direction of thought, the central focus, especially for African economies is development. And even more recently is the issue of sustainable development: that is, the development that thrives while engendering posterity. Thus, following the global emphasis on the post-2015 agenda, certain intrigues are brought to the fore which include: the extent to which FDI is desirable in an economy; the environmental effects of the presence of FDI; the threshold for the presence of FDI within a host economy; and the extent to which FDI is performing vis-à-vis domestic investments? (Agosin and Machado 2005; Halicioglu 2009; Lee, 2010; Asongu 2018).

Following these intrigues is the need to mainstream the FDI-growth nexus within the framework of sustainable development. This calls for a more challenging discourse beyond the traditional economic growth. Incidentally, the focus of sustainable development is tilted towards not just growth for development but green growth. These are growth processes that reduce environmental degradation and hazards to human posterity. While some strands in the literature are trending in this regard (Lee 2013; Omri et al. 2014; Omri and Khaoli 2014; Shahbaz et al. 2015; Asongu 2018), studies focusing on Nigeria to articulate the problem statement are scant. Therefore, despite these theoretical and empirical issues on the FDI-growth nexus, the direction of interest within this chapter is nipped within the sphere of sustainable development in Nigeria. Hence, the chapter conducts an assessment of the differential capacities of FDI and domestic investments to generate positive spillovers for sustainable development. Specifically, beyond the traditional emphasis on economic growth for development, this chapter hinges on which investments bring about sustained growth for sustainable development, otherwise known as green growth.

The rest of the chapter is structured as follows. The theoretical, empirical and Nigeria-centric stylized literature are covered in Sect. 2 while Sect. 3 discusses the data and methodology. The empirical results are disclosed in Sect. 4 whereas Sect. 5 concludes with implications and future research directions.

2 Literature Review

2.1 Theoretical Transfer Channels for Investment and Sustainable Development

Overtime, development economists have been positive on FDI driving economic growth; which in turn is capable of generating spillovers. Spillovers are said to occur in the case of FDI when the entry of MNEs brings about productivity increase to domestic firms and MNEs do not fully internalize the value of these benefits (Smarzynska Javorcik 2004). Therefore, Multinational Enterprises (MNEs) are seen as agents that increase competition in the host economy, transfer modern technology, and help achieve a more efficient allocation of resources. According to Blomström and Kokko (1998), dimensions to which FDI spillovers can occur include productivity spillovers and market access (export) spillovers and agglomeration (firm clusters). Basically, productivity spillovers occurs when local firms as a consequence of the presence of MNEs are more efficient in production.

According to Smarzynska Javorcik (2004), spillovers can take on different forms (See Fig. 1). It can occur through demonstration effects. Specifically, this is when domestic firms increase their efficiency by adopting technologies of MNEs operations in domestic firms, either as paid workers or through observation. Another form of spillover is apparent through linkage effects. The linkage spillover which could be backward or forward stems from the relationship that local firms establish. Local firms could work in consonance with MNE's as either subsidiaries or suppliers (backward linkages) or customers of intermediate inputs produced by MNEs subsidiaries (forward linkages) (Lall 1980; Rodriguez-Clare 1996; Markusen and Venables 1999; Lin and Saggi 2004). Also, positive spillovers could occur via competitive effects. Firstly, MNEs can act as competitive entrants with the aim of fostering more efficient utilization of domestic resources by local firms through superior technology (Wang and Blomstrom 1992; Blomström and Kokko 1998). Secondly, the presence and utilization of advanced technologies can stimulate domestic efforts to come up with ingenious innovations. Thirdly, given the need for technological absorption and

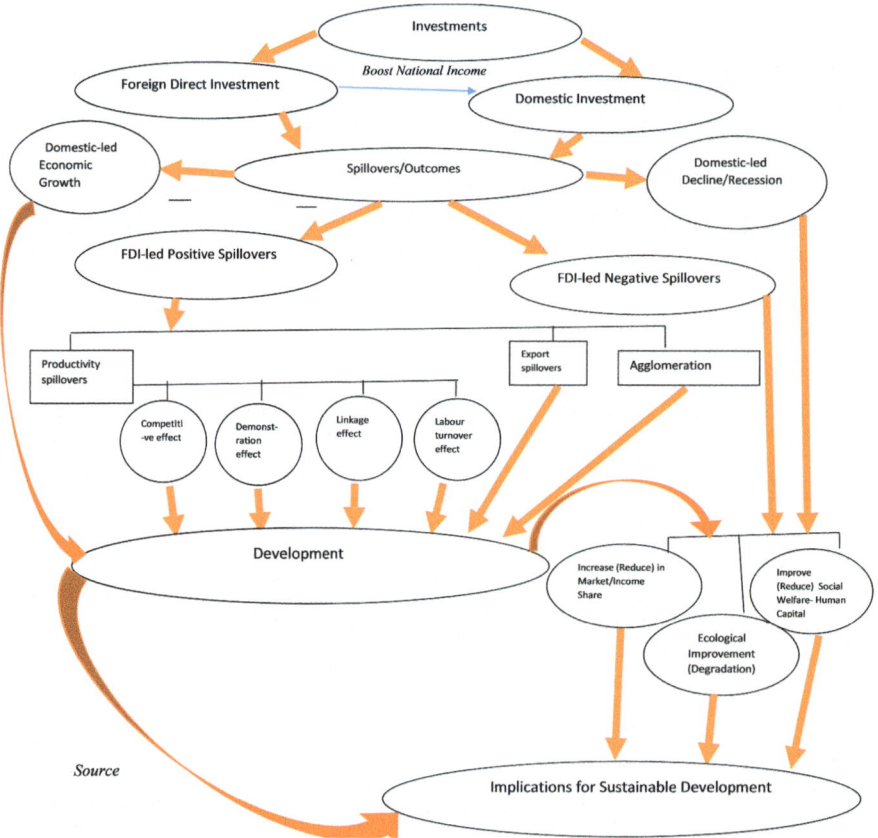

Fig. 1 Mechanisms of investments for sustainable development. *Source* Author

internalization, local firms will have to invest in human and physical capital; thereby raising productivity levels to match MNEs (Damijan et al. 2003; Crespo and Fontoura 2007). However, as appealing as this competitive effect sounds, there are challenges of loss of market shares for local firms which could equally impede the quantum of income retained within the local economy; as well as associated ecological issues from technological utilization; thereby questioning the benefits that could emanate from the presence of MNEs for local industries.

Therefore, following the spillover expectation of positive effects, the notion of development for sustainability raises issues on possible negative effects either from domestic financing or FDI. These issues are embedded in the spiral pollutant effect of production technologies on the environment; as well as its direct effects on market share and indirect effects on social and human development. For instance, Lee (2013) noted that FDIs are considered as one of the major factors that could lead to environmental degradation. Also, some other studies have equally advanced similar arguments (Smarzynska and Wei 2001; Xing and Kolstad 2002; Eskeland and Harrison 2003; He 2006; Zhang 2011). The notion of sustainability in this study amongst other components emphasizes green growth: a concept which denotes the means by which the current economy can make the transition to a sustainable economy while reducing pollution and greenhouse gas emissions, minimizing waste and inefficient use of natural resources, maintaining biodiversity, and strengthening energy security (OECD 2010). Theoretically, the *Environmental Kuznets Curve* (EKC) hypothesis expands on the green growth concept. The EKC advances income growth as a tool for achieving green growth and therefore posits a non-linear (inverted-U) relationship between the environment and growth (i.e. Kuznets shape nexus) (Panayotou 1993). With regard to FDI, other theories emphasizing green growth have been put forward. It is worthwhile to discuss some of them. First, the *Pollution Haven Hypothesis* explains how foreign investors can take advantage of governments of developing economies actions and inactions, especially when they downplay environmental issues through relaxed or non-enforced regulations (Copeland and Taylor 2004; Cole 2004). Second, the *Industrial Flight Hypothesis* postulates that foreign firms decide to move their capital to local economies where environmental issues are ignored in order to reduce production costs (Asghari 2013). Third, there is the *Pollution Halo Hypothesis* which advances the perspective that foreign firms prefer to function in economies that are environmentally conscious and apt (Zarsky 1999). Beyond these hypotheses is the Porter-Palmer's argument on environmental laws in domestic economies bringing about compliance cost; such that FDI could serve as a vehicle for stimulating innovations for clean technologies, especially when expected benefits outweigh costs (Porter 1991; Porter and Van der Linde 1995; Palmer et al. 1995).

The aim of investments, whether domestic or foreign, is to bring about positive outcomes such as growth, development and even sustainable development. From the context of Fig. 1, if domestic investments yield positive outcomes, economic growth will occur, which through appropriate income redistribution, development will occur. And if the components of sustainable development, which are income growth (economic), ecological sustainability (environment) and human welfare (social), are mainstreamed into the process of development, sustainable development will be assured. However, just like FDI, if domestic investments do not yield the required positive results, development and its sustainability will be affected; and this will be reflected via reduced income, ecological degradation and decreased human and social welfare. Thus, the implications of total investments for economic sustainability cannot be undermined given the differing possible outcomes.

2.2   Empirical Literature

Several papers in developed and developing economies have conducted studies with regard to FDI and economic growth, productivity growth and green growth. In addition to the outcomes of local activities and domestic investments bringing about economic growth, Hsiao and Shen (2003) explained how economic growth could act as a stimulant for attracting FDI in developing countries.

The role of FDI in bringing about economic growth and stimulating domestic investment or otherwise, has been continually discussed in the literature. For instance, Balasubramanyam et al. (1996) revealed the growth enhancing effects of FDI over domestic investments, which was made possible via the pursuit of export promotion policies by developing economies. For instance, several studies have shown that open economies grow faster(Dollar 1992; Sachs and Werner 1995; Rodriguez and Rodrik 1999; Lipsey 2003). Also, Barrel and Pain (1999) emphasized that positive effects of FDI and trade on economic growth may simply reflect the fact that FDI is attracted to countries that are expected to grow faster and follow open-trade policies. Furthermore, Borensztein et al. (1998), from a panel of 69 developing countries, asserted that FDI is an important vehicle of technology transfer, and that it contributes more to economic growth than domestic investment. Furthermore, Makki and Somwaru (2004) found a strong evidence for sixty-six developing economies, where a positive relationship between FDI and trade in promoting economic growth; as well as FDI stimulating domestic investment. The FDI-growth nexus in these economies was enhanced by human capital development, sound macroeconomic policies and institutional stability. Incidentally, Adams (2009) found an initial crowding out effect of FDI on domestic investment for Sub-Saharan African countries; however, this trend reverses in a latter period. Adams noted that for FDI to complement domestic investment and economic growth, there is the need for FDI to be targeted at specific sectors that require it. Also, just like Makki and Somwaru (2004), he emphasized the cooperation between government and MNEs to fostering mutual benefits, as well as the role of human capital especially as regards the absorption capacity of local firms. Beyond these, several panel studies established causality that is uni-directional (Lee, 2010); bi-directional (Choe 2003; Pao and Tsai 2011) and neutral (Herzer et al. 2008).

FDI, through spillover effects has been argued to basically generate productivity growth. In addition, FDI is seen to set the pace for domestic investment with regards to favourable investment climates, technology externalities, and learning effects (Feder, 1983; Ram 1985; Grossman and Krueger 1991, 1995). But Görg and Greenaway (2004) posit that empirical evidence to support positive spillovers are difficult and somewhat illusionary, given that foreign firms protect their assets. For instance, Girma et al. (2001) found no intra-industry spillovers. They also established that foreign firms were more productive than domestic firms. Also, Aitken and Harrison (1999) found that spillovers are limited to domestic firms where foreign investments are present within the Venezuelan economy while, Smarzynska Javorcik (2004) revealed that spillovers at firm levels are associated with projects of shared domestic and foreign ownership but not with fully owned foreign investments in the Lithuanian economy.

Quite a number of studies have also been conducted on the FDI-green growth nexus. We discuss some in what follows:

Grimes and Kentor (2003) noted that in developing economies, the presence of FDI in the energy sector is prominent. This is reflected in the significant effects to the growth of carbon dioxide emissions; while domestic investment has no significant effect on CO_2 emissions. Also, employing causality estimates on low, middle and high-income countries, Hoffmann et al. (2005) found a unidirectional causality from FDI to energy emissions in middle-income countries; and a uni-directional relationship from CO_2 emissions to FDI in low-income countries; while no causal relationship was apparent for high income countries, just like the Gulf Corporation Countries (GCC) (Al-mulali and Tang 2013). Similarly, Aliyu (2005) revealed that in OECD and non-OECD countries, while foreign outflows impacted the environment positively, foreign inflows impacted the environment negatively. Shahbaz et al. (2015) investigated the non-linear relationship between FDI and environmental degradation for high-, middle- and low-income countries. They found that the environmental Kuznets curve exists and FDI increases environmental degradation; thus validating the pollution heaven hypothesis (PHH) exists. Also, a bidirectional causality was seen between CO_2 emissions and foreign direct investment in the global panel. Similarly, Lee (2010) found a bi-directional relationship between FDI and energy pollutants for the Malaysian economy. The PHH was also found to be valid in China (Beak and Koo 2009; Bao et al. 2011; Cole et al. 2011), Taiwan and in the short-run in India (Beak and Koo 2009). In addition, Pao and Tsai found the evidence of EKC in BRIC (Brazil, Russia, India and China). However, in contrast to the PHH hypothesis, some studies analyzed the effects of FDI on CO2 emissions; and found that FDI improves environmental quality due to the use of energy efficient technology (Tamazian and Rao 2010; Al-mulali and Tang 2013).

For sub-Saharan African countries, Kivyiro and Arminen (2014) noted that FDI lead to increase in CO_2 emissions and causality runs from FDI to CO_2 emissions. Keho (2016) provides empirical evidence on ECOWAS countries which supports the environmental Kuznets curve for four countries (Cote d'Ivoire, Gambia, Mali and Niger). Also, economic growth and population contribute to environmental degradation. Incidentally, the effect of FDI on CO_2 emissions is contingent on trade openness. This effect is positive and increases with the degree of trade openness in Burkina Faso, Gambia and Nigeria, suggesting that trade and FDI are complementary in worsening environmental quality. The effect of FDI decreases with trade in Ghana,

Mali and Togo while in the cases of Benin, Niger, Senegal and Sierra Leone, FDI has no significant long-run effect on CO_2 emissions. Specifically in Nigeria, Alege and Ogundipe (2015) analyzed the impact of FDI on environmental sustainability. The findings were consistent with the PHH where FDI contributes to CO_2 emissions. This is attributed to the activities of resource- extracting industries which cause pollution in Nigeria. They also found that population growth leads to environmental degradation because most Nigerians are poor and depend on the environment for their livelihood thereby aiding depletion. Also, similar to the findings of Danladi (2013), Alege and Ogundipe (2015) found that growth in GDP spurs environmental sustainability, despite the low level of industrialization in Nigeria. Maku et al. (2018) has revealed that GDP has an insignificant positive influence on CO_2 emission while FDI and energy consumption also have an insignificant negative impact on CO_2 emission in Nigeria.

Overall, our literature review suggests that the empirical results of the previous studies are inconclusive. This inconclusiveness can be traceable to, *inter alia*: differences in the techniques of analysis and lack of adequate information on the direction of causality, especially for Nigeria. The causal segregation and comparison of these growths has not been quite articulated in the literature. This is pertinent because while drawing and implementing policies for development, the role of peculiar growth context stands different vis-à-vis investment feedbacks. For instance, while economic growth is geared towards income distribution, productivity growth concentrates on efficiency; while sustainable growth emphasizes on posterity. Therefore, the directions of influence of different growth indices vis-vis-vis investments are invaluable for policy decisions in Nigeria which is still developing. Also, an overview of previous studies concentrated mainly on FDI impact on the environment and downplays the effects that could equally emanate from domestic investments. Therefore, since both investments dictate the pace of economic activities within a local economy, they will be both considered as determinants of environmental sustainability in this study. This will also reduce the challenge of omitted variable bias.

2.3   Foreign Direct Investment and Domestic Investment in Nigeria

For the past two decades, FDI in form of foreign capital inflow to sub-Saharan African (SSA) countries has been ranked lowest; especially when compared to other regions. For instance, countries in SSA receive capital inflow that are 13 times lower than those flow to East Asia and Pacific in 2015 (UNCTAD 2018). In 2013, the *EY* report revealed that South Africa and Kenya, Nigeria were listed as the countries where FDI was highest. With investment projects in excess of more than 60 projects, Nigeria led Ghana in the West African region. The *EY* report noted that despite a minor decrease in FDI projects in 2013 in Nigeria, the country's remains an attractive place for investors especially given the size of the economy.

According to the United Nations Conference on Trade and Development (UNC-TAD) World Investment Report (2017), FDI flows to Africa fell by 3 percent- precisely from $61 billion in 2015 to $59 billion in 2016. The overall decline of FDI flows to the continent has been greatly attributed to weak commodity prices, especially oil to which Nigeria is mainly dependent. However, statistics revealed that in 2017, inflows into West Africa grew by about 12%. With a total of $11.4 billion foreign investments in the region, it was majorly driven by Nigeria's oil sales and Ghana's hydrocarbons and cocoa processing projects which amounted to $4.4 billion and $3.5 billion, respectively. Despite the fact that Nigeria remains among the first-three recipients of FDI in SSA, the capital inflow is not so reflected in the contributions of FDI to GDP in the country. From Figs. 2 and 3, the contribution of FDI to GDP when compared to domestic investments in Nigeria remains considerably low.

From Fig. 3, with an average contribution of 40–45%, domestic investment was a major contributor to the GDP in Nigeria especially during the early 1970s and 1980s. This is majorly attributed to the agrarian nature of the Nigerian economy then; which sustained and financed domestic investments within the economy before the oil boom.

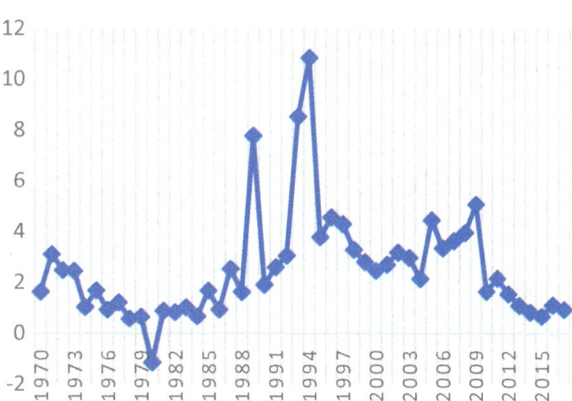

Fig. 2 FDI contribution to GDP. *Source* World Development Indicators (2018)

Fig. 3 Contribution of Domestic Investment to GDP. *Source* International Financial Statistics (2010) (for: 1970–1981). World Development Indicators (2018) (for: 1982–2017)

But a sharp decline is noticed in domestic investments from 1981 through, although with a slight rise in the mid-1980s, domestic investment's contribution remained low at an average of 15%; and even contributed as low as 10% in the mid-1990s. Although, there has been a slight increase since the early 2000s, with a contribution of about 20% in 2010; however, this contribution remains low when compared to the early 1970s.

The trend analysis of both investments in Nigeria reveals an interesting pattern. From the 1970s to 1985, while FDI contributed less than 3% to GDP, domestic investments contributed an average of 40%. With the exception of 1991 and 1997 where FDI fell sharply, FDI contributed as high as 8–10% between 1986 and 1999. By the year 2000 when FDI fell to 2%, domestic investments had started picking-up again to about 15% in 2010. In fact, except between 2003 and 2006 where FDI contributed about 3%, between 2009 and 2016, FDI has been dwindling to almost 0.5% in 2015; while domestic investment has risen gradually during that period in Nigeria. This pattern in contributions of FDI versus DI suggests a substitute-complementary type of relationship. However further empirical analysis is worthwhile in order to provide more causal insights. In all, the UNCTAD (2018) report has predicted increases in FDI inflows in Africa. This prediction is strengthened by the recent signing of the African Continental Free Trade Area (AfCFTA) by 44 African countries; and even the anticipation is higher in Nigeria, given an expectation in the improvement of commodity prices like oil.

3 Methodology

3.1 Data and Variable Definition

In order to assess the spillovers and investment outcomes in Nigeria, the paper employs the Auto-regressive Distributed Lag (ARDL) estimation procedure and causality tests. Specifically, in order to uncover the existence of a relationship and direction of causality between investments and sustainability in Nigeria, this paper utilized annual data covering the period 1970–2017 by disaggregating the time period into three-time horizons. The choice of the variables and time period is informed by the behaviour of the series of the focal variables: FDI and Domestic Investment (DI). It is worthwhile to note that all-time series variables are transformed into natural logarithms to avoid heteroskedasticity and spurious results. The sources and the description of the variables used in this paper are presented in Table 1.

Table 1 Description of variable

Variable	Definition	Sources
CO_2	Per capita CO_2 emission (measure in metric ton) is used as a proxy for environmental proxy	World Development Indicator (2017) CD ROM
RGDP	Real GDP in constant 2010 US dollar as proxy of economic growth	World Development Indicator (2017) CD ROM
FDI	Foreign direct investment as a percentage of GDP	World Development Indicator (2017) CD ROM
OPEN	Trade openness is the sum of export and import	World Development Indicator (2017) CD ROM
TFP	As a percentage of GDP Total Factor Productivity (measures productivity growth)	Author's Computation
ENG	Energy consumption per capita measure in kg of oil equivalent per capita	World Development Indicator (2017) CD ROM
GDP per capita	Income distribution	World Development Indicator (2017) CD ROM
ELECT	Data on Electricity Consumption	World Development Indicator (2017) CD ROM
DI	Gross fixed capital formation as a percentage of GDP is used to measure domestic investment	World Development Indicator (2017) CD ROM
URP	Urban Population Growth	World Development Indicator (2017) CD ROM

3.2 The ARDL Bound Testing

Overtime, a number of econometric techniques have been employed in different studies such as Enger and Granger (1987), Johansen (1988), Johansen and Juselius (1990), fully modified OLS of Phillip and Hansen (1990) and Johansen (1996), among others, to estimate the relationships between the variables. Thus, to explore the relationship between investment and sustainable development growth in Nigeria within the period of 1970–2017, this study utilizes the Auto-Regressive Distributed Lag Estimates (ARDL) bounds test approach proposed by Pesaran et al. (2001). This approach is justified to have numerous advantages over other cointegration approaches[1]. The unrestricted error correction model (UECM) version of the ARDL model is presented in Eq. (1) as follows:

---

[1]This approach is found to be applicable irrespective of the order of integration of variables, evades the need for pre-testing the integration order of variables, allows the variables to have different optimal lag length, possibility of deriving a dynamic unrestricted error correction model from the approach via a simple linear transformation and it integrates both the short run dynamics and long run dynamics together without loss of any long run information (see Halicioglu, 2008; Kohler, 2013; Sung et al. 2017 among others).

$$
\begin{aligned}
\Delta CO2_t = \rho_0 &+ \sum_{j=0}^{P} \alpha_j \Delta CO2_{t-j} + \sum_{j=0}^{q} \beta_j \Delta FDI_{t-j} + \sum_{j=0}^{r} \emptyset_j \Delta RGDP_{t-j} + \sum_{j=0}^{S} \psi_j \Delta DI_{t-j} \\
&+ \sum_{j=0}^{t} \omega_j \Delta ENG_{t-j} + \sum_{j=0}^{u} \varphi_j \Delta UPG_{t-j} + \sum_{j=0}^{v} \pi_j \Delta ELEC_{t-j} + \sum_{j=0}^{w} \Omega_j \Delta OPEN_{t-j} \\
&+ \lambda_1 CO2_{t-j} + \lambda_2 FDI_{t-j} + \lambda_3 RGDP_{t-j} + \lambda_4 DI_{t-j} + \lambda_5 ENG_{t-j} + \lambda_6 UPG_{t-j} + \lambda_7 ELEC_{t-j} \\
&+ \lambda_8 OPEN_{t-j} + \mu_t
\end{aligned}
\tag{1}
$$

Where Δ is the first difference operator, α, β, $\emptyset$, Ψ, ω, φ, π and Ω are the coefficient estimates of the chosen variables; μ is error term; p, q, r, s, t, u, v and w are the optimal lag lengths selected based on the optimal length selection criteria. Pesaran et al. (2001) suggest an F-test for joint significance of the coefficients of the lagged level of variables. For example, the null hypothesis of no long run relationship between the variables is tested against the alternative hypothesis of cointegration. Pesaran et al. (2001) computed two sets of critical values (lower and upper critical bounds) for a given significance level. A lower critical bound is applied if the regressors are I(0) and the upper critical bound is used for I(1). If the F-statistic exceeds the upper critical value, I(1), the null hypothesis will be rejected in favour of the alternative hypothesis and thus, we concluded that there is long run relationship. If the F-statistics falls below the lower critical bound, we cannot reject the null hypothesis of no cointegration. However, if the F-statistics lies between the lower and upper critical bounds, inference will be inconclusive.

3.3 Causality Test

To make this paper robust and increase its predictive power, this study equally determines the direction of causality among the variables using vector error correction model (VECM). Engle and Granger (1987) asserted that once there is existence of long run relationship between variables, then there must Granger causality in at least one direction. Though, they cautioned that Granger causality test conducted in the first difference variables by means of a VAR might be misleading in the presence of cointegration; thus, inclusion of an additional variable to the VAR system which is the error correction term will help to test the long run relationships between the variables. Thus, this paper determines the possible short run and long run causality among the variables using this technique which is the best alternative technique to capture this causality and the augmented form of the Granger causality test involving the error correction term is formulated in a multivariate pth order vector error correction model as follows.

It should be note that $(1 - L)$ is the lag operator used to explain the amount of lags include in the VAR and (EC_{t-1}) denotes the error correction term. Thus, this paper utilises the Granger causality derived for Eq. (2) above to check for the statistical significance of the lagged differences of the variables for each vector; which is a measure of short run casuality. While, the coefficient of the lagged error correction term represents the long run causality.

$$
\begin{bmatrix} CO_{2t} \\ TEP_t \\ DGP_t \\ FDI_t \\ DI_t \end{bmatrix} = \sum_{i=1}^{P} (1 - L) \begin{bmatrix} d_{11i}\ d_{12i}\ d_{13i}\ d_{14i}\ d_{15i}\ d_{16i} \\ d_{21i}\ d_{22i}\ d_{23i}\ d_{24i}\ d_{25i}\ d_{26i} \\ d_{31i}\ d_{32i}\ d_{33i}\ d_{34i}\ d_{35i}\ d_{36i} \\ d_{41i}\ d_{42i}\ d_{43i}\ d_{44i}\ d_{45i}\ d_{46i} \\ d_{51i}\ d_{52i}\ d_{53i}\ d_{54i}\ d_{55i}\ d_{56i} \end{bmatrix} \begin{bmatrix} CO_{2t-1} \\ TEP_{t-1} \\ DGP_{t-1} \\ FDI_{t-1} \\ DI_{t-1} \end{bmatrix}
$$

$$
+ \begin{bmatrix} \lambda_1 \\ \lambda_2 \\ \lambda_3 \\ \lambda_4 \\ \lambda_5 \end{bmatrix} [EC_{t-1}] + \begin{bmatrix} \Omega_1 \\ \Omega_2 \\ \Omega_3 \\ \Omega_4 \\ \Omega_5 \end{bmatrix} \tag{2}
$$

In all, the ARDL estimate was used to determine the type and degree of relationship that exist between FDI and green growth; and between DI and green growth. However, cutting down on the number of control variables in the ARDL model, the Eq. (2) which is the VECM-Granger model endogenizes five selected variables. This is with the aim of ascertaining if FDI stimulates DI; as well as ascertaining the prospects for sustainable development vis-à-vis growth channels which include economic growth (GDP), productivity growth (TFP) and green-growth (CO_2).

3.4 Model Stability

The issue in econometric techniques of testing the stability of estimated coefficient has denegrated into a controversial discussions among researchers with no consensus on the appropriate technique to determine the stability of estimated coefficients. Bahmani-Oskooee and Chomsisengphet (2002) argued that existence of a cointegration among the estimated long run coefficients is not a sufficient condition to conclude that the estimated coefficients are stable and different stability tests have been employed in the empirical literature such as Chow (1960), Brown et al. (1975), Hansen (1992) and Hansen and Johansen (1993). Hence, this paper utilises Brown et al. (1975) stability test which incorporates cumulative sum and cumulative sum of squares tests based on the recursive regression residuals. These tests also include the dynamics of the short-run to the long-run through the residuals with a graphical plot showing that the cumulative sum and cumulative sum of squares statistics fall inside the critical bounds of 5% significance which also provide information about the confirmation of the stability of the coefficients of the ARDL regression.

4 Empirical Evidence and Discussion

Prior to estimating the long run relationship between the investment spillovers in Nigeria, it is of paramount to check the stationarity of all variables, that is, to ascertain the order of integration of all the variables to avoid spurious results using Dickey and Fuller (1981) and Phillip-Perrons (1988) unit root testing procedures. Having verified the stationarity of all variables, this paper proceeds to confirm the existence of long run cointegration relationship among the variables using Pesaran et al. (2001) bounds test. It is imperative to first determine the optimal lag length using different selection criteria due to sensitivity of F-statistics to the numbers of lags. Table 2 presents the ARDL bounds test along with the error correction models. The results show that the computed F-statistics are greater than the upper critical bounds generated by Pesaran et al. (2001) at 5 and 10% significant levels which lead to the rejection of null hypothesis of no cointegration in favour of alternative hypothesis at both significant levels. This findings confirm the presence the of cointegration between the variables in the whole time periods, first and the third period but inconclusive results is reported only in the second period since the computed F-statistic fall within the upper and lower critical values at 5% level of significant. However, there is no evidence of cointegration between these variables in the second period since the computed F-statistic fall below the upper and lower critical values at both 5 and 10% significant levels.

Interestingly, the results of the error correction terms are negative and statistically significant in all the time periods and this implies that the error correction terms corroborates with the established cointegration results which lead us to conclude that changes in carbon emissions are corrected for at different significant levels for each time period. The results equally show that the magnitude of the adjustment coefficients reported in each time period varies with the fastest adjustment speed recorded in third period only.

Table 2 Estimated ARDL cointegration test results

Periods	Model	F-stat	ECM(-1)	Value 1%	Value 5%	Value 10%
Whole	1,1,3,0,1	4.61**	−0.58***	I(0) = 2.96	I(0) = 2.32	I(0) = 2.03
			(0.03)	I(1) = 4.26	I(1) = 3.5	I(1) = 3.13
First	1,0,0,1.1	3.23**	−0.62***	I(0) = 2.96	I(0) = 2.32	I(0) = 2.03
			(0.02)	I(1) = 4.26	I(1) = 3.5	I(1) = 3.13
Second	1,1,0,1,0	1.61	−0.39***	I(0) = 2.96	I(0) = 2.32	I(0) = 2.03
			(0.07)	I(1) = 4.26	I(1) = 3.5	I(1) = 3.13
Third	2,0,10,1	6.76**	−0.77***	I(0) = 2.96	I(0) = 2.32	I(0) = 2.03
			(0.01)	I(1) = 4.26	I(1) = 3.5	I(1) = 3.13

Note *** and ** represents significant at 5 and 10% levels respectively
Values in brackets represent the probability values

Table 3 ARDL estimates

Dependent variable CO_2	Whole coeff	First coeff	Second coeff	Third coeff
Long run				
FDI	−0.435**	0.14	−0.197**	−0.239*
DI	0.082	0.214	0.166	−1.130*
RGDP	−1.575**	1.013	0.112	−0.335**
OPEN	−0.267	−0.551	0.098	0.048
ELECT	−1.135**	−1.204	0.572	0.655
UPG	2.582***	−6.308	1.915**	6.287*
ENG	1.907	5.695	0.309	−2.850*
Short run				
FDI	−0.234	0.064	−0.262	−0.266
DI	0.139**	0.097***	0.221***	−0.258**
RGDP	0.261**	0.46***	0.014**	0.374**
OPEN	−0.238	−0.251	0.13	0.054
ELECT	−0.589	−0.547	0.761	0.731
UPG	1.341*	2.906	2.547	7.015*
ENG	2.572	2.591	0.411	3.18
Diagnostic				
Normality	0.087	0.3524	0.245	0.365
Serial	0.102	7.165	0.109	0.3826
ARCH	0.252	0.888	0.445	0.832
RAMSEY	2.86	0.063	0.432	0.322
CUSUM	STABLE	STABLE	STABLE	STABLE
CUSUMQ	STABLE	STABLE	STABLE	STABLE

Note *** and ** represent significant at 5 and 10% levels respectively

It is also important to note that the ECTs have the expected signs that are also within the acceptable theoretical interval (Asongu et al. 2016). Accordingly, whereas an ECT at equilibrium is zero, an ECT that is not zero implies that linkage pairs have deviated from the long term equilibrium. Therefore, the ECT helps to adjust and partially restore the long-run nexus. The underlying restoration is contingent on two main factors, notably, the ECT: (i) displays a negative sign and (ii) is within an interval of 0 and 1 which is necessary for the stability of the error correction mechanism (Asongu 2014a, b, c). Moreover, a positive ECT implies a deviation from the equilibrium. Hence, a negative sign reflects the restoration of the long term nexus after an exogenous shock. Within this framework, in determining the speed at which the equilibrium is reinstated, an ECT of 1 reflects full adjustment while an ECT of zero implies no adjustment.

The results of the long run and short run ARDL are presented in Table 3. The results show that the effect of FDI on CO_2 emission is negative and statistically significant in case of whole, second and third periods only but a positive insignificant relationship is reported in the first period. The empirical finding of a positive relationship between FDI and CO_2 emission is not surprising given the initial influx of most foreign investment in the primary sector during the oil boom; and most industries that dominated this sector produced highly pollution-intensive goods. However, the error correction estimates and subsequent periods reveal that the positive relations even out overtime. This tendency rejects the PHH position of downplaying green growth through inefficient technologies as far as CO_2 emissions are concerned. But, the findings show that domestic investment, though insignificant in the long-run, positively influences CO_2 in all the time periods with the exception of the third period only. This indicates that challenges to green growth are more, on the path of domestic investments. Although, it is expected that as FDI impacts the efficiency of domestic investments in the long-run; which may explain the negative effects of domestic investments on CO_2 in the third period.

The effect of real GDP on CO_2 is negative and statistically significant in the case of whole time period and third period only but an insignificant positive relationship is reported in first and second periods only. However, in the short-run, real GDP has significant positive influences on CO_2 emissions thus indicating the presence of EKC effects. The effect of trade openness and electricity consumption on CO_2 is positive and insignificant in second and third only but negative significant findings are reported for trade openness in first and whole period only. The estimated coefficient of urban population shows a positive and significant relationship with CO_2 emission in all the time periods but negative insignificant findings are recorded in first period only. Energy use shows a positive and insignificant association with CO_2 emission in the case of the whole, first and second periods only but a negative significant relationship is exclusively reported in third period.

Estimating the long run relationship between investment and sustainable growth without considering the short run dynamics between these variables is not sufficient enough. Thus, this study also considers the short run relationship between these variables over time. The impact of FDI on CO_2 is negative and statistically insignificant only in whole, second and third periods but a positive insignificant relationship is recorded in first period only; which is consistent with the long-run analysis. Also, just like in the long-run, the results of domestic investment and real GDP show that they are positively significant in influencing CO_2 emission in the short run in the case of all time periods but a negative and significant relationship is found between domestic investment and CO2 in the third period only. Similarly, trade openness and electricity generation are reported to be negative and statistically insignificant in whole and first period only but positive insignificant findings are reported only in the second and third periods. This indicates that trade openness and electricity generation have an insignificant but gradual increasing effect on green growth. The estimated coefficient of urban population growth has a positive and significant effect on CO_2 emission in whole and third period only but insignificant effects in first and second periods. The results also show that there is a positive and insignificant short run relationship between energy use and CO_2 emission in all the time periods.

The results of the diagnostic tests such as serial correlation, functional form, normality and Heteroskedasticity are also reported in Table 3. The results show that all the estimated coefficients are statistically significant and there are no serial correlations. Also, the stability of the model is confirmed with the test of CUSUM and CUSUMQ used to assess the recursive residue in the mean and variance respectively for the whole time period

Having confirmed that the existence of long run relationship between the variables, then, this paper examines the direction of causality between five variables in each time horizon using vector error correction model (VECM) by verifying the causal effect through the significance of the coefficient of the lagged error correction term and the joint significance of the lagged differences of the explanatory variables using the Wald test. The results of the causality test are presented in Table 4. In the case of the whole period, the results show that there are unidirectional Granger casual relationships in the short run: from GDP to CO_2 emission, from domestic investment to CO_2 emission, FDI to domestic investment and FDI to TFP. This finding implies that changes in GDP and domestic investment cause changes in the CO2 emission,

Table 4 Granger causality estimates

Periods	Dependent variable	CO_2	TFP	GDP	FDI	DI	ECT
		F-stat	F-stat	F-stat	F-stat	F-stat	t-stat
Whole periods	CO_2		0.771	0.1315**	0.081	0.194**	−0.19***
	TFP	1.022		2.349	3.135**	1.7554	
	GDP	7.603	5.422		3.407	2.199	−0.52**
	FDI	12.72	3.109	0.55		1.542	−0.78**
	DI	0.532	1.117	2.199	0.897***		
First	CO2		5.225***	4.574**	3.894	1.34	−0.36**
	TFP	3.055		2.643	10.305**	10.134**	
	GDP	2.323	0.2116		2.218	6.229	−2.75
	FDI	0.739	1.1963	2.287		0.975	
	DI	0.598	3.129	2.435	7.6563		−0.45***
Second	CO2		11.92***	2.603	14.42***	5.371	
	TFP	0.535		1.066	0.103	15.22	−0.54
	GDP	1.231	9.568		2.399	0.977	−0.25
	FDI	2.373	3.854	13.98		8.98	
	DI	0.635	5.122**	2.166	0.399		−0.85
Third	CO2		0.304	1.306	2.51	0.5223	−0.92**
	TFP	9.04**		10.11***	4.292	1.831	
	GDP	1.668	3.365		0.282	0.5458	−0.17
	FDI	8.655	10.216	2.476		8.423**	−0.39
	DI	0.561	4.044	0.984	6.704***		−0.53**

Note *** and ** represent significant at 5 and 10% levels respectively

thereby emphasizing EKC effects; and changes in foreign direct investment cause domestic investment and TFP in the short run. However, there is no evidence of bidirectional or feedback effect between these variables. The statistically significant coefficient of the error correction term confirms the results of the bound tests.

In the case of first period, the results show that there is unidirectional causal nexus in the short run: from TFP to CO_2, GDP to CO_2, domestic investment to TFP and domestic investment to FDI. The significance of these findings implies that changes in TFP and GDP cause changes in CO_2 and change in domestic investment causes FDI but no evidence of feedback effect is reported in this period. Also, the long run causality is supported by the coefficient of the lagged error correction term which reported negative and statistically significant between the variables in this period. In the case of the second period, there is also evidence of unidirectional causal flow: from TFP to CO_2, FDI to CO_2 and TFP to domestic investment. This finding implies that changes in both TFP and FDI cause changes in CO_2. However, there is no evidence of long run causality between the variables due to insignificance of the error correction term and this findings also confirm the bound test result that show no long run relationships. In the third period, there is evidence of unidirectional causal nexus running from CO_2 to TFP, GDP to TFP, domestic investment to FDI and FDI to domestic investment. These findings also show that there is evidence of bidirectional causality relationships between foreign direct investment and domestic investment. Also, the statistically significant negative coefficient of the error correction terms shows that there is a long run relationship between the variables and these results is in consonance with the bound test results. In summary, the results show that Granger causality running from other variables to CO_2 emission is common in all the time horizons and there is only evidence of bidirectional causality effect between foreign direct investment and domestic investment in third period only.

5 Concluding Implications and Future Research Directions

The study has examined the differential capacities of FDI and domestic investments to drive green growth in Nigeria. The results show that domestic investment increases CO_2 emissions in the short run while foreign investment decreases CO_2 emissions in the long run. When the dataset is decomposed into three sub-samples in the light of cycles of investments within the trend analysis, findings of the third sub-sample (i.e. 2001–2017) reveal that both types of investments decrease CO_2 emissions in the long run while only domestic investment has a negative effect on CO_2 emissions in the short run. Therefore, the result showed that FDI inflows has not hampered green growth over the study period, thereby causing a rejection of the pollution haven hypothesis in Nigeria as far as CO_2 emissions are concerned. But via positive insignificant effects on pollution, domestic investments and GDP has challenged green growth in Nigeria; thus indicating EKC effects.

Also, the study was able to articulate that FDI caused (stimulates) domestic investment over the study period; however, there was no reverse causality in this regard.

Finally, this study was able to ascertain the causality between FDI and the different growth indicators. While FDI was seen to cause productivity growth in Nigeria, the causality estimates revealed neutrality effects for economic growth and green growth. However, the uni-directional causality result was consistent with the ARDL estimates of a flow from domestic investments to the green growth.

In all, given the outcome of the most recent period, this study indicates that positive spillovers for CO_2 emissions abound via FDI for Nigeria and even domestic investments which appears to dominate total investments in Nigeria. This study therefore concludes that as short-run distortions even out in the long-run, FDI and domestic investments has prospects for sustainable development in Nigeria through green growth. However, exploring some other components of green growth like land use, oil exploration may affirm or refute this conclusion.

Appendix

Graphical Representation of CUSUM and CUSUMQ

Whole Period (1970–2017)

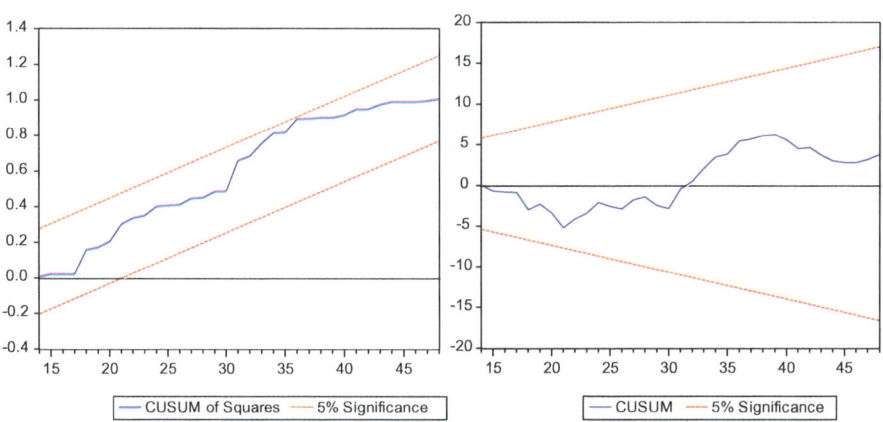

First Period (1970–1985)

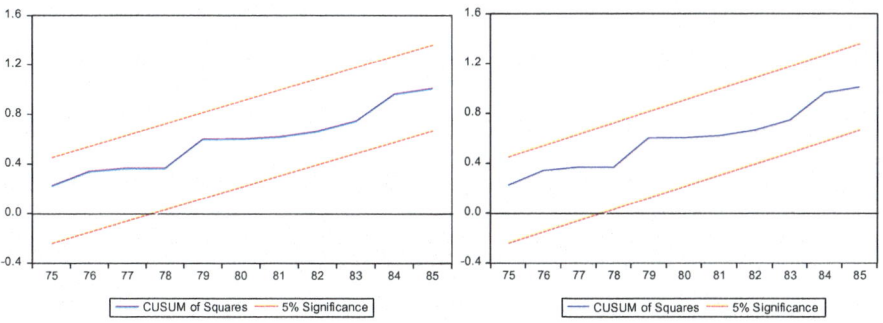

Second Period (1986–2000)

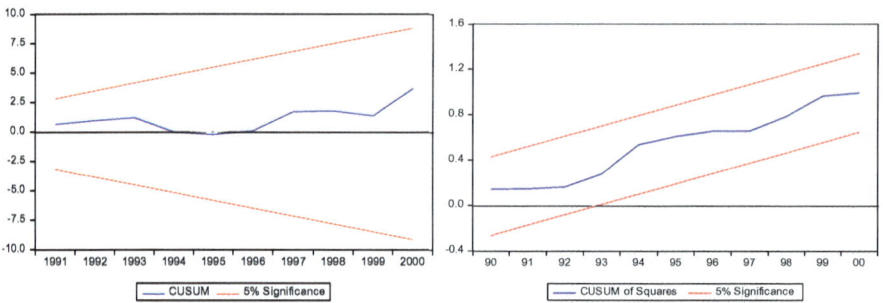

Third Period (2001–2017)

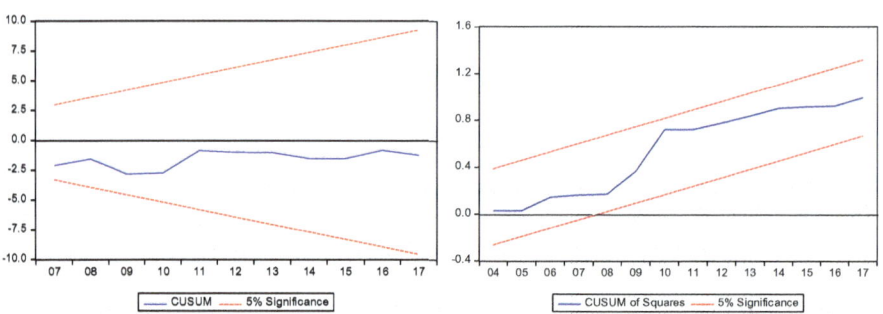

References

Adams S (2009) Foreign direct investment, domestic investment, and economic growth in Sub-
    Saharan Africa. J Policy Model 31(6):939–949
Agosin MR, Machado R (2005) Foreign investment in developing countries: does it crowd in
    domestic investment? Oxf Dev Stud 33(2):149–162
Aitken BJ, Harrison AE (1999) Do domestic firms benefit from direct foreign investment? Evidence
    from Venezuela. Am Econ Re 89(3):605–618, 75:131–150
Alege PO, Ogundipe AA (2015) Environmental quality and economic growth in Nigeria: a fractional
    cointegration analysis. Int J Dev Sustain 2(2)
Al-Mulali U, Tang CF (2013) Investigating the validity of pollution haven hypothesis in the gulf
    cooperation council (GCC) countries. Energy Policy 60:813–819
Aliyu MA (2005, June) Foreign direct investment and the environment: pollution haven hypothesis
    revisited. In: Eight annual conference on global economic analysis. Lübeck, Germany (pp. 9–11)
Asghari M (2013) Does FDI promote MENA Region's environmental quality? Pollution halo or
    pollution have hypothesis. Int J Sci Res Soc Sci (6):92–100
Asongu SA (2014a) Correcting inflation with financial dynamic fundamentals: which adjustments
    matter in Africa? J Afr Bus 15(1):64–73
Asongu SA (2014b) Does money matter in Africa? New empirics on long- and short-run effects of
    monetary policy on output and prices. Indian Growth Dev Rev 7(2):142–180
Asongu SA (2014c) Linkages between investment flows and financial development: causality evi-
    dence from selected African countries. Afr J Econ Manag Stud 5(3):269–299

Asongu SA, El-Montasser G, Toumi H (2016) Testing the relationships between energy consumption, CO_2 emissions and economic growth in 24 African countries: a panel ARDL approach. Environ Sci Pollut Res 23(7):6563–6573

Asongu SA (2018) CO2 emission thresholds for inclusive human development in Sub-Saharan Africa. Environ Sci Pollut Res. https://doi.org/10.1007/s11356-018-2626-6

Bahmani-Oskooee M, Chomsisengphet S (2002) Stability of M2 money demand function in industrial countries. Appl Econ 34(16):2075–2083

Balasubramanyam VN, Salisu M, Sapsford D (1996) Foreign direct investment and growth in EP and IS countries. Econ J, 92–105

Barrel R, Pain N (1999) The growth of foreign direct investment in Europe. In: Barrel R, Pain N (eds) Innovation, investment and diffusion of technology in Europe: German direct investment and economic growth in Postwar Europe. Cambridge University Press, Cambridge, pp 19–43

Beak J, Koo WW (2009) A dynamic approach to FDI-environment nexus: the case of China and India. J Int Econ Stud 13(2):1598–2769

Bende-Nabende A (2017) Globalisation, FDI, regional integration and sustainable development: theory, evidence and policy. Routledge

Blomström M, Kokko A (1998) Multinational corporations and spillovers. J Econ Surv 12(2):1–31

Borensztein E, De Gregorio J, Lee JW (1998) How does foreign direct investment affect economic growth? 1. J Int Econ 45(1):115–135

Boyd JH, Smith BD (1992) Intermediation and equilibrium allocation of investment capital: implication for economic development. J Monetary Econ, 409–432

Brown RL, Durbin J, Evans JM (1975) Techniques for testing the constancy of regression relations over time. J Roy Stat Soc B 37:149–163

Choe JI (2003) Do foreign direct investment and gross domestic investment promote economic growth? Rev Dev Econ 7(1):44–57

Chow G (1960) Test of equality between sets of coefficients in two linear regressions. Econometrica 28:591–605

Cole MA (2004) Trade, the pollution haven hypothesis and the environmental Kuznets curve: examining the linkages. Ecol Econ 48(1):71–81

Cole MA, Elliott RJ, Zhang J (2011) Growth, foreign direct investment, and the environment: evidence from Chinese cities. J Reg Sci 51(1):121–138

Copeland BR, Taylor MS (2004) Trade, growth, and the environment. J Econ Lit 42(1):7–71

Crespo N, Fontoura MP (2007) Determinant factors of FDI spillovers—what do we really know? World Dev 35(3):410–425

Damijan JP, Knell M, Majcen B, Rojec M (2003) The role of FDI, R&D accumulation and trade in transferring technology to transition countries: evidence from firm panel data for eight transition countries. Econ Syst 27(2):189–204

De Mello Jr LR (1997) Foreign direct investment in developing countries and growth: a selective survey. J Dev Stud 34(1):1–34

Dickey DA, Fuller WA (1981) Likelihood ratio statistics for autoregressive time series with a unit root. Econometrica: J Econometric Soc 1057–1072

Dollar D (1992) Outward oriented developing economies really do grow more rapidly: evidence from 95 LDCs, 1976–85. Econ Dev Cult Change 40:523–544

Engle R, Granger C (1987) Cointegration and error correction representation: estimation and testing. Econometrica 55:251–276

Eskeland GS, Harrison AE (2003) Moving to greener pastures? Multinationals and the Pollution Haven Hypothesis. J Dev Econ 70:1–23

Feder G (1983) On exports and economic growth. J Dev Econ 12(1–2):59–73

Girma S, Greenaway D, Wakelin K (2001) Who benefits from foreign direct investment in the UK? Scott J Polit Econ 48(2):119–133

Grimes P, Kento J (2003) Exporting the greenhouse: foreign capital penetration and CO_2 emissions 1980–1996. J World-Syst Res (2):261–275

Grossman G, Krueger A (1991) Environmental impacts of a North American free trade agreement. National Bureau of Economics Research Working Paper, No. 3194, NBER, Cambridge

Grossman GM, Krueger AB (1995) Economic growth and the environment. Quart J Econ 110:353–377

Görg H, Greenaway D (2004) Much ado about nothing? Do domestic firms really benefit from foreign direct investment? World Bank Res Observer 19(2):171–197

Halicioglu F (2009) An econometric study of CO_2 emissions, energy consumption, income and foreign trade in Turkey. Energy Policy 37(3):1156–1164

Hansen BE (1992) Tests for parameter instability in regressions with I(1) processes. J Bus Econ Stat 10:321–335

Hansen H, Johansen S (1993) Recursive estimation in cointegrated VAR-models, vol 1. Institute of Mathematical Statistics, University of Copenhagen, Copenhagen

He J (2006) Pollution haven hypothesis and environmental impacts of foreign direct investment: the case of industrial emission of sulfur dioxide (SO_2) in Chinese province. Ecol Econ 60:228–245

Herzer D, Klasen S, Nowak-Lehmann FD (2008) In search of FDI-led growth in developing countries: The way forward. Econ Model 25:793–810

Hoffmann R, Lee CG, Ramasamy B, Yeung M (2005) FDI and pollution: a granger causality test using panel data. J Int Dev: J Dev Stud Assoc 17(3):311–317

Hsiao C, Shen Y (2003) Foreign direct investment and economic growth: the importance of institutions and urbanization. Econ Dev Cult Change 51(4):883–896

Johansen S (1988) Statistical analysis of cointegrating vectors. J Econ Dyn Control 12:231–254

Johansen S (1996) Likelihood-based inference in cointegrated vector auto regressive models. Oxford University Press, Oxford, Seconded

Johansen S, Juselius K (1990) Maximum likelihood estimation and inference on cointegration—with application to the demand for money. Oxford Bull Econ Stat 52:169–210

Keho Y (2016) Trade openness and the impact of foreign direct investment on CO_2 emissions: econometric evidence from ECOWAS countries. J Econ Sustain Dev 17(18). ISSN (Paper)2222-1700; ISSN (Online)2222-2855

Kivyiro P, Arminen H (2014) Carbon dioxide emissions, energy consumption, economic growth, and foreign direct investment: causality analysis for Sub-Saharan Africa. Energy 74:595–606

Lall S (1980) Vertical inter-firm linkages in LDCs: an empirical study. Oxford Bull Econ Stat 42(3):203–226

Lee GC (2010) Foreign direct investment pollution and economic growth: evidence from Malaysia. Appl Econ 41:1709–1716

Lee JW (2013) The contribution of foreign direct investment to clean energy use, carbon emissions and economic growth. Energy Policy 55:483–489

Lin P, Saggi K (2004) Multinational firms and backward linkages: a survey and a simple model. Unpublished manuscript, Lingnan University and Southern Methodist University

Lipsey RE (2003) Home and host country effects of FDI. National Bureau of Economic Research Cambridge, MA) Working PaperNo. 9293, October 2002

Makki SS, Somwaru A (2004) Impact of foreign direct investment and trade on economic growth: evidence from developing countries. Am J Agr Econ 86(3):795–801

Maku O, Adegboyega ES, Oyelade AO (2018) The impact of foreign direct investment on CO_2 emission in Nigeria (1980–2014). Int J Sci Eng Res 9(2):2061. ISSN 2229-5518

Markusen J, Venables A (1999) Foreign direct investment as a catalyst for industrial development. Eur Econ Rev 43(2):335–356

OECD (2010) Organisation for economic co-operation and development. Interim Report of the green growth strategy: implementing our commitment for a sustainable future. OECD Publishing, 2010

Omri A, Kahouli B (2014) The nexus between foreign investment, domestic capital and economic growth: empirical evidence from the MENA region. Res Econ 68:257–263

Omri A, Nguyen DK, Rault C (2014) Causal interactions between CO_2 emissions, FDI, and economic growth: Evidence from dynamic simultaneous equation models. Econ Model 42:382–389

Pesaran MH, Shin Y, Smith RJ (2001) Bounds testing approaches to the analysis of level relationships. J Appl Econometrics 16(3):289–326

Phillip PC, Hansen BE (1990) Statistical inference in instrumental variables regression with I (1) processes. Rev Econ Stud 57(1):99–125

Phillip PCB, Perron P (1988) Testing for a unit root in time series regression. Biometrika 75(2):335–346

Porter M (1991) America's green strategy. Sci Am 264(4):168

Porter M, van der Linde C (1995) Toward a new conception of the environment competitiveness relationship. J Econ Perspect 9:97–118

Palmer K et al (1995) Tightening environmental standards: the benefits-cost or the no-cost paradigm? J Econ Perspect 9(4):119–132

Panayotou T (1993) Empirical tests and policy analysis of environmental degradation at different stages of economic development. WEP 2-22 Working Paper No. 238, International Labour Office, Technology and Employment Programme, Geneva

Pao HT, Tsai CM (2011) Modeling and forecasting the CO_2 emissions, energy consumption, and economic growth in Brazil. Energy 36(5):2450–2458

Ram R (1985) Exports and economic growth: some additional evidence. Econ Dev Cult Change 33(2):415–425

Rodriguez-Clare A (1996) Multinationals, linkages, and economic development. Am Econ Rev 86(4):852–873

Rodriguez F, Rodrik D (1999) Trade policy and economic growth: a skeptic's guide to the cross-country evidence. Working Paper, National Bureau of Economic Research, Cambridge, MA

Salvatore D, Hatcher T (1991) Inward oriented and outward oriented trade strategies. J Dev Stud 27(3):7–25

Sachs J, Werner A (1995) Economic reform and the process of global integration. Brookings Papers Econ Act 1:1–118

Shahbaz M, Nasreen S, Abbas F, Anis O (2015) Does foreign direct investment impede environmental quality in high-, middle-, and low-income countries? Energy Econ 51:275–287

Smarzynska BK, Wei SJ (2001) Pollution havens and foreign direct investment: dirty sector or popular myth? NBER Working Paper No. 8465

Smarzynska Javorcik B (2004) Does foreign direct investment increase the productivity of domestic firms? In search of spillovers through backward linkages. Am Econ Rev 94(3):605–627. https://doi.org/10.1257/0002828041464605

Solow RM (1957) Technical change and the aggregate production function. Rev Econ Stat 39(3):312–320

Tamazian A, Bhaskara Rao B (2010) Do economic, financial and institutional developments matter for environmental degradation? Evidence from transitional economies. Energy Econ 32(1):137–145

UNCTAD (2018) World investment Report—investment and new industrial policies. UNCTAD, Geneva

Wang JY, Blomström M (1992) Foreign investment and technology transfer: a simple model. Eur Econ Rev 36(1):137–155

Xing Y, Kolstad C (2002) Do lax environmental regulations attract foreign investment? Environ Resour Econ 21:1–22

Zhang YJ (2011) The impact of financial development on carbon emissions: an empirical analysis in China. Energy Policy 39:2197–2203

Openness and Greenness: Pay-Offs or Trade-Offs for the Nigerian Economy

Oluwabunmi O. Adejumo

Abstract Overtime, income level and level of industrialization are utilized in classifying economies as far as development is concerned. However, in these times where
sustainable development is concerned, the level of industrialization may no longer
suffice in categorizing countries. Indeed the level of greenness and the extent to which
countries exacerbate or preserve the environment vis-à-vis economic activities should
be the new order of economic classification. Nigeria, in its post-independence era,
pursued the structuralist ideals of an import substitution industrialization policy. But,
by 1986, during the economic reforms, the neoliberal policies of an open economy
where imports and exports flowed freely were embraced. In view of these divergent
policies, this chapter focused on the extent to which these policies have been consistent with the promoting or digressing from the ideals of eco-sustainability. Using
causal analysis and interactive regressions within the Autoregressive Distributed Lag
(ARDL) model, an assessment of the payoffs or otherwise of these openness policies
given the period 1970–1985 when Import Substitution Industrialization (ISI) policies was in operation as against the 1986–2015 when openness policies came into
operation. This is with a view to assessing the prospects for sustainable development
as Nigeria keeps its economy open.

Keywords Openness · Greenness · Sustainable development · Nigeria

1 Introduction

In the quest for development, most developing economies have historically embraced
different ideologies as regards economic systems ranging from communism, social-
ism, mixed systems and the laissez faire. This is equally typical of the quest for industrial advancement where developing economies had to battle contesting strategies that
are available in industrial development ideology. The structuralist theory modelled

O. O. Adejumo (✉)
Institute for Entrepreneurship and Development Studies, Obafemi Awolowo University, Ile-Ife,
Nigeria
e-mail: adejumobum@oauife.edu.ng; jumobum@gmail.com

© Springer Nature Switzerland AG 2020
W. Leal Filho et al. (eds.), *International Business, Trade and Institutional
Sustainability*, World Sustainability Series,
https://doi.org/10.1007/978-3-030-26759-9_51

its proposition in favour of import substitution industrialisation (ISI) with an aversive view for openness; while the neoliberals proposed export oriented industrialisation with an irrepressible direction for openness (Colclough and Manor 1993; Hague and Loader 1999). During the reforms tagged 'structural adjustment programme (SAP)' of 1986, the neoliberal development propositions fast outmatched structuralism in most developing economies; and tilted global development efforts in the direction of openness and liberalization which was largely rooted in competition and laissez faire. The import of this is that openness to external participation became a *sine qua non* to creating an environment of industrial dynamism, supplement resource-gap and increase commodity choices available to consumers.

The initial ISI policies engineered hostilities from most developing economies towards multi-lateral corporations (MNCs). But, the liberal shift resulted in massive external flow of resources in the amounts and the pattern of flows from industrial countries to developing economies in the 1980s and the 1990s when compared to earlier decades as seen in Fig. 1. For instance, from Fig. 1, long-term debt used to be a major source of financing in the 1970s as it accounted for almost 45% of external resources. However, owing to the debt crisis of the 1980s, debt became a less attractive source of income and Foreign Direct Investment (FDI) and government grants became more dominant in the 1990s when the neo-liberal policies became fully operational in developing economies. As a result, Multinational Enterprises (MNEs) had to play a long major role in resource mobilization between developed and developing economies. Meanwhile, given the increase demand for resources among these economies, developing economies have begun to compete for these

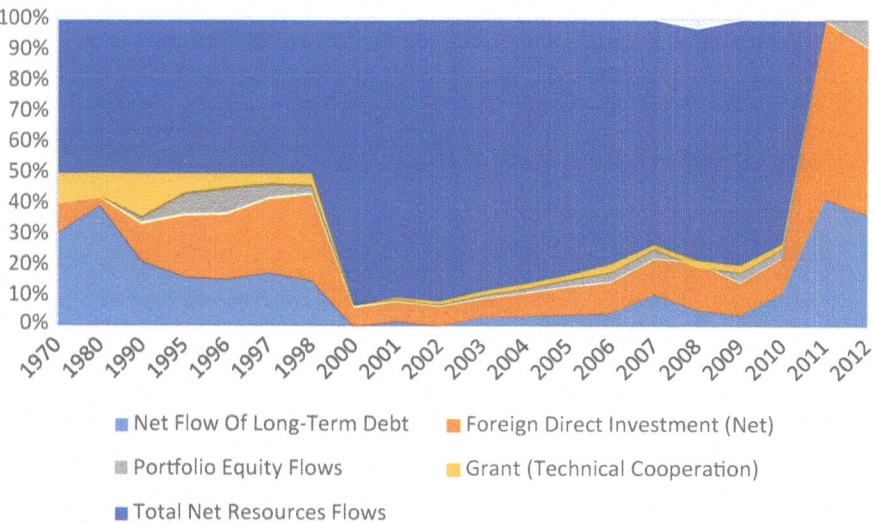

Fig. 1 Percentage of net capital flows to developing countries, 1970–2012. *Source* World Bank Debtor Reporting System, IMF, Bank for International Settlements, Global Development Finance, and OECD, various years

resources as well as position themselves to attract inflows which have tilted in a dwindling pattern of capital flows in the mid-2000s. But, since 2009, there has been a gradual upsurge in capital flow to developing economies. Despite this upsurge, inflows to sub-Saharan Africa remain low compared to other regions of the world. For instance, Fig. 2 shows that there has been a low trend in the volume of inflows in the past two decades compared to other regions, with a decrease more recently for the region. Generally, compared to other regions of the World, countries in sub-Saharan Africa (SSA) (for instance) see less foreign investment flow, which is about 13 times lower than those flows to East Asia and Pacific (EAP) in 2015 alone; thus depicting the intense competition via openness.

In the sub-Saharan region, major recipients of Capital inflows include South Africa, the Republic of Congo, Mozambique, Egypt and Nigeria. In 2013 and 2014, statistics revealed that the Nigerian economy is one of the leading beneficiaries of capital inflows in the region as she ranked number six and five respectively in each year; which equally depicts the extent of openness in Nigeria (UNCTAD 2016).

Since the reform era in 1986, given these quantum of inflows, there is no doubt that sub-Saharan economies have remained inclined to being open and even compete for or depend on trade, aid, capital and even remittances as alternative sources of income. However, the direction of pay-offs or trade-off of openness for greenness remains contestable. For instance, Zerbo (2015) argued that trade openness was not sufficient to improve environmental quality in Kenya, but it was complementary in South Africa. Thus, given the rising global emissions of greenhouse gases (GHG) and climate change challenges (Meinshausen et al. 2009); the extent to which openness policy is consistent with the principles of green growth and environmental sustainability is the crux of this study. Therefore, the rest of this chapter is structured as follows. The overview of the Nigerian policy-shift history is discussed in Sect. 2

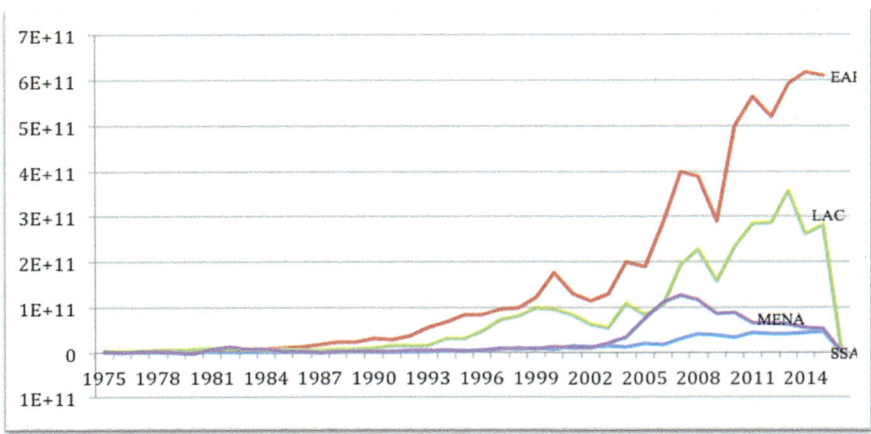

Fig. 2 Trend in foreign direct investment (Net inflows) across regions. *Source* UNCTAD (2018). *Notes* SSA—Sub Sahara Africa; MENA—Middle East and North Africa; LAC—Latin America and Caribbean; EAP—East Asia and Pacific

while Sect. 3 gives a theoretical-empirical overview of stylized literature on the openness-greenness nexus; while the conceptual framework is presented in Sect. 4, the methodology is shown in Sect. 5. The empirical results are presented in Sect. 6 and Sect. 7 concludes the chapter.

2 An Overview of Nigerian Policy Shift History

After the Nigerian independence in 1960, the most radical policy that came to the fore was the "Indigenization decree" of 1972 which is also referred to as the Nigerian Enterprises Promotion (NEP) decree. Prior to 1970, the National Economic Council (NEC) and the Joint Planning Committee (JPC) were the two intergovernmental organs responsible for the formulation of industrial development plan. These were replaced in 1970 by the Supreme Military Council and the Joint Planning Board, respectively. The industrial policy was premised upon the need to put an end to the imperialistic tendencies then evident in the Nigerian economy. In this period, the government adopted nationalization as a strategy to support local control of the economy and to achieve the level of industrial development. This philosophy was rooted in the structuralist' development thought[1] which earmarked a greater role for government in bringing the national economy to desirable levels of industrialization. Specifically, policy-making in early 1970s failed to support the small- and medium-sized industries which created a vacuum in Nigeria's entrepreneurship and industrial development path. Such policies were inappropriate for a country seeking out foreign direct investment, as policy makers failed to understand that foreign investors rarely wish to invest in a country where they have no voice or control over their capital.

Since, the NEP policy could not achieve its aim of a nationalised or partially open economy as most industrialists were faced by increasing foreign dominance. Though NEP policies were revised in 1977 in favour of infant industries and increased government participation as reflected in the establishment of state-owned, capital intensive industries; and to safeguard the growth of the new state-owned industries, the government introduced different types of import restrictions. Although this held sway for a while but the crash of oil prices which was used to finance government enterprises was challenged. Therefore, the indigenization policy eventually gave way to the adoption of the Structural Adjustment Programme (SAP) in 1986 which was driven by the ideals of the neo-liberals[2] (Ikpeze 1991).

In 1988, an independent National Planning Commission (NPC) was established as part of the civil service reform (Dauda 1993). The privatization and commercialization decree No. 25 of 1988 was promulgated with an autonomous NPC established to oversee the implementation of the SAP-induced privatization programme. The

---

[1]The structuralist emphasizes broad government themes through planning and policies. *See* Larner (2000).

[2]The neoliberals emphasize less government participation and is rooted in the competitive-market principles

SAP created a new economic and industrial development policy direction based on market-oriented principles and private sector participation. The fundamental aim was to boost confidence of domestic and foreign entrepreneurs in the economy. The measures taken included shrinking government involvement in production through privatization of state-owned enterprises. According to the World Bank Report (1994), SAP involves freeing market forces so that competition can help improve the allocation of resources, getting price signals right and creating an environment that allows businesses to respond to those signals in ways that increase the returns to investment.

On the economic front, since the introduction of SAP, Ikpeze (1991) and Aremu (2003) noted the reform dimensions as follows:

- The Second Tier Foreign Exchange Market was created in order to enhance foreign currency as well as facilitate imports and exports.
- The tariff structure was simplified, privatization of public enterprises, and promotion of small scale enterprises as well as rationalization of institutions for industrial development were also implemented.
- Foreigners were no longer restricted to the proportion of shareholdings in their investments.
- Companies and Allied Matters Act (CAMA) was promulgated in 1990 and it became the principal law governing the incorporation of businesses in Nigeria.
- The Exchange Control Act of 1984 was repealed and was replaced by foreign exchange Decree No. 17 of 1995. An Autonomous Foreign Exchange Market (AFEM) was created where the modalities for capital importation for foreign investors were deregulated and simplified; thus, making made it possible for entrepreneurs to invest in any Nigerian enterprise or securities.
- In 1995, Nigerian Investment Promotion Commission (NIPC) was also established. The NIPC Act No. 16 of 1995 was designed to promote, coordinate and monitor all investments in Nigeria.
- In 1999, the Act establishing the Investment and Securities was signed into law with the aim of deregulating the Nigerian Capital Market in order to facilitate the inflow of FDI into the economy.
- The transition to democracy in Nigeria in 1999 gave rise to a new economic policy framework tagged National Economic Empowerment and Development Strategy (NEEDS) which was still largely driven by a market-competitive economic system. The policy targets involved wealth creation, poverty reduction, employment generation, and value re-orientation.

Indeed, since the 1986 reforms, the Nigerian economy has witnessed trade flows that have caused it to be described as an import-dependent economy. For instance, from Fig. 3, imports accounted for more than 50% of trade flow from the early 1980s till early 2000s.

While pondering on these economic imbalances via trade, a more pertinent issue is the extent of environmental balance created via openness, and the implication for the Nigerian economy. The question that comes to mind could be: in a bid to expand resource physically, materially and financially, has Nigeria been cautious of preserving its environment? or otherwise to what extent has income growth fostered from

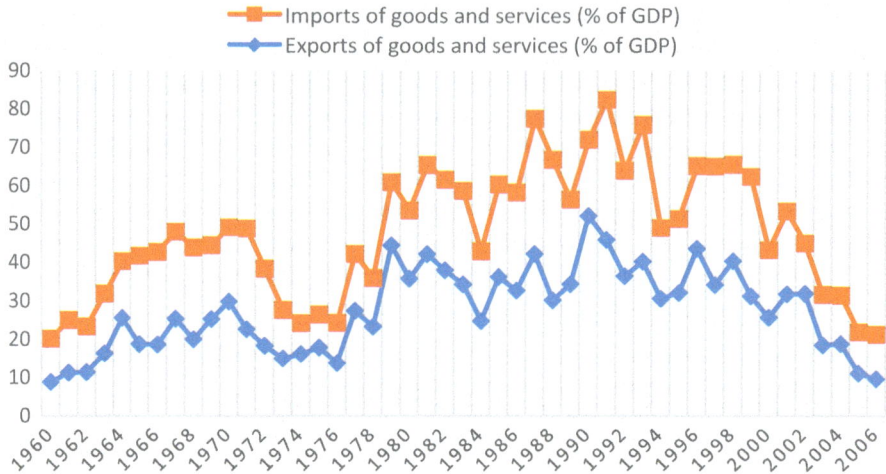

Fig. 3 Import and export of goods and services for Nigeria. *Source* WDI (2018)

openness-promoted green growth? Therefore, as the Nigerian economy grows and integrates globally, the attention given to environmental quality becomes a subject of policy discourse.

3 Literature Review

In view of the recent clamour for sustainable patterns environmentally, the issue of utmost concern is if policy shifts for liberalism have paid-off or indeed been a trade-off of economic freedom and advancement for posterity distress. In this decade of climate change, charting a global course towards combating climate irregularities via adaptations and mitigation strategies has put the discourse of openness and greenness on the front burner. This stems from the popular Environmental Kuznets Curve (EKC) hypothesis of Kuznets (1955) which was popularized by the works of Grossman and Krueger (1991). The birth of the openness-greenness nexus by Grossman and Krueger, which centered on the influence of the environment on the North American Free Trade Agreement (NAFTA) has since shaped discussions in this direction. Arguing from the Kuznets EKC hypothesis of 1955, Choi et al. (2010) explained environmental goods and their quality as normal good; such that as remunerations accumulate from free trade there would be an increase in the demand for greater environmental quality. This implies that amongst determinants of economic growth, activities that drive liberalization which include flow of information, technology, capital and trade should be consistent with maintaining environmental quality. Although, at initial developmental phases, irrespective of increase on income, little attention is given to maintaining environmental quality in terms of re-allocation from financial upsurge; nonetheless, the EKC expects that as per capita income rises beyond certain threshold, environmental degradation should be well-checked.

In view of the foregoing, several studies have examined the environmental consequences of liberalization vis-à-vis growth (Ang 2008; Sari and Soytas 2009; Choi et al. 2010; Saboori et al. 2012; Omri 2014; Omri et al. 2015); trade (Choi et al. 2010), Information and communication Technologies (ICT) (Asongu 2018; Park et al. 2018) and even capital flows (Jamel and Maktouf 2017). Some studies observed that openness advances the course of greenness; while some others submitted otherwise and even some observed mixed results. For instance, Sannassee and Seetanah (2016) investigated the link between trade openness and the environment in the case of Mauritius and showed that trade has served to increase the level of carbon dioxide (CO_2) emissions in the country in both the long run and short run. In this regard, the results demonstrate that a 1% increase in trade openness is accompanied by an upshot of 0.60% in CO_2 emission. The results were similar to with the findings of Onoja et al. (2014), Khalil and Inam (2006), Chebbi et al. (2011) and Asongu (2018) for developing country cases.

Zhang et al. (2017) support the existence of hypothetical EKC in ten newly industrialized countries; and while real GDP and energy have increased emissions, they indicated that trade openness promoted greenness. This is unlike Choi et al. (2010) who examined the environmental consequences according to openness and economic growth using instances from a newly industrialized economy (Korea), emerging economy (China) and developed economy (Japan) and found mixed results. While the EKC component in China revealed an inverted U-shaped curve, Japan had a U-shaped curve. Also, the relation for CO_2 emissions and openness showed the case of Korea and Japan as an inverted U-shaped curve, while China shows a U-shaped curve which implies that as openness increases, CO_2 increases as well. In like manner, using 23 selected economies in the European Union, Park et al. (2018) examined the impact of, economic growth, and trade openness among other variables on carbon dioxide (CO_2) emissions. Their findings suggested that while the ICT and electricity consumption increased CO_2 emissions significantly, economic growth and financial development have a diminishing negative impact on CO_2 emissions in the EU. Meanwhile trade openness was seen to be responsible for increasing CO_2 emissions only in one country, which is Bulgaria. But other countries such as Croatia, Czech Republic, Finland, France, Hungary, Italy, Netherland, Portugal, Romania, and Slovenia were not challenged from openness as far as CO_2 emissions are concerned. Therefore, Park et al. (2018) argued that most countries within the EU have efficient methods of trade which results in the reduced negative impact on CO_2 emissions and the environment. This is also because the relationship between trade and CO_2 emissions were deemed to be insignificant for the rest of Austria, Belgium, Cyprus, Denmark, Germany, Greece, Ireland, Luxembourg, Poland, Spain, Sweden, and the UK. They further argued that the heterogeneous effect of trade differences on CO_2 emissions within the EU may be that several countries achieve the level where trade can lessen CO_2 emissions. Meanwhile, according to the analysis by Park et al. (2018), for the remaining countries within the EU with insignificant results, it just depicts that they are moving in a right direction that might achieve the level that trade may help in reduction of CO_2 emissions.

Similarly, using 40 countries in the EU, Jamel and Maktouf (2017) empirically investigated the causal relations between economic growth (GDP), CO_2 emissions (environmental degradation), financial development, and trade openness. The findings suggest a bidirectional Granger causal linkages among GDP and pollution, GDP and financial sector development, GDP and trade openness, financial sector development and trade openness, and trade openness and pollution in the case of European economies. They validated the existence/confirm the validity of the environmental Kuznets curve hypothesis for the sampled EU economies. Also, they confirmed the feedback directional causality among trade openness and financial sector development. Meanwhile, the neutrality hypothesis linked carbon emissions and financial sector development inflows; in addition, a bidirectional nexus is seen between GDP and financial sector development and among GDP and trade openness in the European economies. The panel causality equally verified that bidirectional causal connection is found between economic growth, environmental degradation (CO_2), financial development, and trade openness.

Ertugrul et al. (2016) analyzed the relationship between carbon dioxide (CO_2) emissions, trade openness, real income and energy consumption in the top ten CO_2 emitters among 10 economies to include China, India, South Korea, Brazil, Mexico, Indonesia, South Africa, Turkey, Thailand and Malaysia. They validated the EKC hypothesis for Turkey, India, China and Korea. A long-run relationship existed among the variables of interest for Thailand, Turkey, India, Brazil, China, Indonesia and Korea; and real income, energy consumption and trade openness are the main determinants of carbon emissions in the long run, They found that energy consumption stimulates environmental pollution in most of the analyzed countries, and trade openness increases CO_2 emissions in Turkey, India, China and Indonesia while it has no effect on the environment in Thailand, Brazil and Korea. Hence, pollution-haven hypothesis is supported for Turkey, India, China and Indonesia while the net effect of trade based on scale, composition and technique effects was insignificant in Thailand, Brazil and Korea. Destek et al. (2016) investigated the relationship between CO_2 emission, real GDP, energy consumption, urbanization and trade openness for 10 for selected Central and Eastern European Countries (CEECs), including, Albania, Bulgaria Croatia, Czech Republic, Macedonia, Hungary, Poland, Romania, Slovak Republic and Slovenia for the period of 1991–2011. The results show that the environmental Kuznets curve (EKC) hypothesis holds for these countries. Meanwhile, the results reveal that increases in energy consumption leads to increases in CO_2 emissions. And the causality method show that there is bidirectional causal relationship between CO_2 emissions and realGDP and energy consumption.

Furthermore, country specific studies of Mustascu (2018) which investigated the interaction between openness and CO_2 and gas emissions for different sub-periods of time and frequencies in France. The disintegration into time-periods reflected mixed results. First, there was no co-movement at high frequency between trade openness and gas emissions, thus confirming the 'neutral hypothesis' on short term. Meanwhile, CO_2 emissions positively drove trade openness at medium frequency, thereby depicting the inexistence of strong environmental rules that stimulate international trade, especially the exports obtained based on 'pollutant capacities'. Also, trade openness positively ran gas emissions at low frequency which caused Mustascu (2018) to conclude that on long term, the interaction between trade and CO_2 emissions is driven by the business cycle.

Zhang et al. (2017) suggested that policymakers should encourage and expand the trade openness in these countries, not only to restrain CO_2 emissions but also to boost growth. Meanwhile, Asongu (2017) suggested ICT as a measure for checking environmental degradation that may result from openness; although Park et al. (2018) argued differently that internet usage lowers environmental quality as evidence from the EU region suggest unidirectional causality running from Internet use to CO_2 emissions; which implied that the European Union countries did not achieve the level of green information and telecommunication (ICTs) consumption. Therefore Park et al. (2018) argued that ICT cannot just be employed for green-growth; there is a level of greenness that ICT attain before it can be relevant in mitigating challenges that may arise from economic openness. Although, in addition to openness are other suggested components for reducing CO_2 such as population stabilization policy, development of alternative energy sources such as renewable and to use green and clean technologies (Ohlan 2015).

In all, by applying models such as EKC model and cyclical models, most of these studies have provided a better understanding of the environmental consequences of a liberal economy through international trade and opined that economic growth could be employed for environmental improvements and sustainability. Even in Nigeria, some studies like have examined the openness-greenness nexus (Essien 2011; Nnaji et al. 2013; Akin 2014; Alege and Ogundipe 2015; Ali et al. 2015, 2016; Rafindadi 2016); however, most previous studies in Nigeria have not taken into account the differential impacts of the time-variants given the history of pre-reforms and post-reforms era.

4 Conceptual Framework

In order to ascertain the trade-offs or pay-offs openness for greenness in Nigeria, this study synthesizes the economic application of the ideals of the structuralist and the neo-liberals to ascertain if the EKC position is satisfied ad conceptualized in Fig. 4. This hypothesis, suggests that as an economy grows, environmental sustainability is threatened to a certain threshold. However, an increase in economic growth is expected to influence income growth per capita to a point beyond a certain threshold where environmental sustainability will be assured- that is an inverted U-shaped relationship between economic growth and green growth which implies a non-linear relationship that is applicable to many areas.

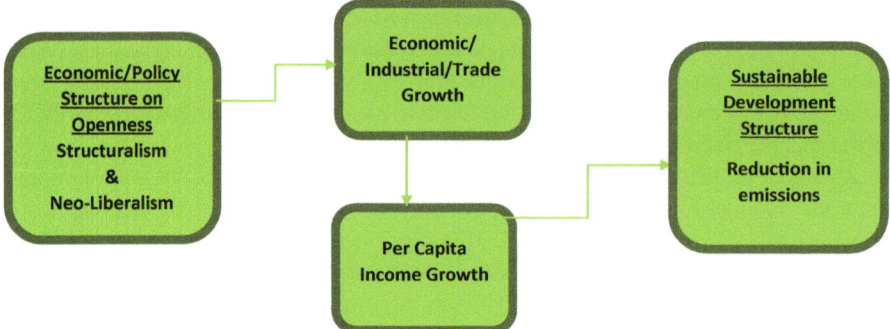

Fig. 4 Economic openness structure and sustainable development. *Source* Author

5 Methodology

The EKC framework[3] of an inverted U-framework with per capita income growth is compliant with explaining local economy pollutions. According to the EKC concept, pollutants are expected to have a positive relationship with the level of income or trade liberalization before the EKC threshold and then a negative relationship beyond the threshold. Given the climate change phenomenon, CO_2 has been largely used to proxy consequences of the global energy system and use. This is because Carbon dioxide (CO_2) accounts for the largest portion of greenhouse gas emissions and its interaction with other GHG can result in environmental problems. Thus, it is meaningful to examine the causal relationships between environmental pollution, trade liberalization, and economic growth. Following the concept of the EKC, a negative relationship is expected between CO_2 emissions and openness, then GHG emissions are likely to decrease as the country becomes more exposed to open markets (Choi et al. 2010). However, if a positive relationship is seen between CO_2 emissions and openness, it can be deduced that the Nigerian economy has not experienced optimum via trade liberalization.

Following the work of Ali et al. (2016), whose method was based on Hossain (2012) and Farhani et al. (2013), we equally incorporated income per capita growth (Y), trade openness (TO), foreign direct investment (FDI), population growth ($POPU$) in urban areas into the regression that explained CO_2 emissions.[4] Therefore, our baseline model is:

$$CO_2 = A\alpha_0, Y\alpha_1, TO\alpha_2, FDI\alpha_3, POPU\alpha_4 \tag{1}$$

Taking the natural logarithm of Eq. (1) and Stating the Eq. (1) explicitly with interactive regressions of the and PCI growth-trade openness ($PCIXTO$), and quadratic income growth Y^2, we have:

---

[3]For more on the decision for EKC *see* Adejumo (2018).

[4]See Appendix 1 for variable sources and measurements

$$\ln CO_2 = \alpha_0 + \alpha_1 \ln Y + \alpha_2 \ln Y^2 + \alpha_1 \ln TO + \alpha_4 \ln(Y \ X \ TO)$$
$$+ \alpha_5 \ln(Y \ X \ TO)^2 + FDI + \alpha_6 POPU_3 \qquad (2)$$

From Eq. (2), we formulated our autoregressive distributed lags (ARDL) that is estimated in order to find the links among the variables under investigation as shown in Eq. (3) below;

$$\Delta \ln CO_{2t} = \alpha_0 + \alpha_1 \ln Y_{t-j} + \alpha_2 \ln Y_{t-j}^2 + \alpha_3 \ln TO_{t-j} + \alpha_4 \ln(Y \ X \ TO)_{t-j}$$
$$+ \alpha_5 \ln(Y \ X \ TO)_{t-j}^2 + \alpha_6 \ln FDI_{t-j} + \alpha_7 \ln POPU_{t-j}$$
$$+ \sum_{J=0}^{P} \partial_j \Delta \ln CO_{2t-j} + \sum_{j=0}^{q} \beta_j \Delta \ln Y_{t-j} + \sum_{j=0}^{r} \emptyset_j \Delta \ln Y_{t-j}^2$$
$$+ \sum_{J=0}^{s} \Psi_j \Delta \ln TO_{t-j} + \sum_{j=0}^{t} \omega_j \Delta \ln(Y \ X \ TO)_{t-j}$$
$$+ \sum_{j=0}^{u} \varphi_j \Delta \ln(Y \ X \ TO)_{t-j}^2 + \sum_{j=0}^{v} \pi_j \Delta \ln FDI_{(t-j)}$$
$$+ \sum_{j=0}^{w} \Omega_j \Delta \ln POPU_{(t-j)} + \mu_t. \qquad (3)$$

Subscript t was used to the denote time period; while $\alpha, \partial, \beta, \emptyset, \Psi, \omega, \varphi, \pi,$ and Ω were coefficients of the selected variables; and μ represent the error term accommodating other explanatory variables of CO_2. To determine the existence of a long-run relationship among the variables, the ARDL bounds test is employed. Specifically, the null hypothesis of no cointegration ($H0 = \alpha1 = \alpha2 = \alpha3 = \alpha4 = \alpha5 = 0$) was tested against the alternate hypothesis of cointegration ($Ha \neq \alpha1 \neq \alpha2 \neq \alpha3 \neq \alpha4 \neq \alpha5 \neq 0$). The decision rule as suggested by Pesaran et al. (2001) is that if the estimated F-test value is higher than the upper bound critical value, null hypothesis will be rejected which signifies the presence of long-run relationship. However, if the estimated value is less than the lower critical value, null hypothesis cannot be rejected which means no element of cointegration exist among the variables, while if the value falls within the lower and upper bound critical bounds the result is inconclusive (Pesaran and Pesaran 1997). Meanwhile, Eq. (3) was also used to estimate the short-run and long-run coefficients of ARDL for CO_2.[5]

Meanwhile, the openness-greenness causal relations were ascertained via the bivariate Granger causality. Test for causality was used to ascertain the predictive ability of variables of interest. In this case, TO_t and CO_{2t} is contained solely in the time series data on these variables. The test involves estimating the regression of the form:

---

[5]See Appendix 4 for stability test using CUSUM residuals and CUSUM sum of squares residuals.

$$TO_t = \sum_{i=1}^{n} \alpha_i CO_{2t-i} \sum_{j=1}^{n} \beta_j TO_{t-j} + u_{it} \qquad (5)$$

$$CO_{2t} = \sum_{i=1}^{n} \theta_i TO_{t-i} + \sum_{j=1}^{m} \delta_j CO_{2t-j} + u_{2t} \qquad (6)$$

The disturbances from u_{1t} and u_{2t} were assumed to be uncorrelated. Equation (5) implied that current value of TO is related to past values of TO and CO_2; and Eq. (6) states same for CO_2. From the causal estimates, four cases are expected to include *Unidirectional causality* from CO_2 to TO or *Unidirectional causality* from TO to CO_2 is indicated if the estimated coefficients on the lagged CO_2 in Eq. (5) or TO in Eq. (6)are statistically different from zero as a group (i.e., $\sum \alpha_i \neq 0$; $\sum \alpha_i \theta_i \neq 0$) and the set of estimated coefficient on the lagged TO in Eq. (6) and CO_2 in Eq. (5) is not statistically different from zero (i.e., $\sum \theta_j = 0$); $\sum \alpha_j = 0$). While, a *Feedback, bilateral causality,* is seen to occur when the sets of CO_2 and TO coefficients are statistically different from zero in both regressions; while *independence* or *inconclusive* is suggested when the sets of CO_2 and TO coefficients are not statistically significant in both the regressions(Gujarati 2009). Since the past is used to predict the future, if variable CO_2 causes variable TO, then changes in CO_2 should *precede* changes in TO and Vice versa.

In appraising the Nigerian policy-shifts, we considered the degree of openness vis-à-vis her concerns for green growth. The appraisal is divided into two broad regime shifts in policy formulation. Accordingly, the appraisal is conducted beginning with the pre-reform period of 1970–1985, which corresponded to structuralist inward-looking import substitution industrialization (ISI) strategy. It then proceeded to cover both the reform and post-reform periods, 1986–2015, which fell within the neo-liberal policy regime of liberalization and outward (export) oriented industrialization strategy (EOI). The period in focused span from 1970–2015. Specifically, for the purpose of analysis, the period is therefore disintegrated into three main periods to cover the pre-reform period (1970–1985); immediate post-reform-era (1986–2000); and the post-reform era (2001–2015).

6 Results

The descriptive statistics of the three period[6] showed the skewness and kurtosis on the symmetry of the probability distribution of various data series as well as the thickness of the tails of these distributions respectively. The data showed that the distribution of the data are normal, given the skewness and kurtosis values. Except for Y in the third model, the skewness of the variable all the variables lie within 1.0

[6]Seen Appendix 2 for the descriptive statistics of the models.

Table 1 ARDL bounds test (test for a long-run relationship)

Period	Test statistic	Value
Period I	F-F-statistic	4.0944*
Period II	F-statistic	5.7245*
Period III	F-statistic	1.2656
Critical value bounds	I(0) bound	I(1) bound
Significance		
10%	2.26	3.35
5%	2.62	3.79
2.5%	2.96	4.18
1%	3.41	4.68

Source Author's Computation
Note *depicts significance at 5%

and -1.0; and can be said that all the distribution of the variables are symmetrical. Also a Gaussian distribution is expected to have kurtosis of 3.0 (Wooldridge 2013); thus, since all the variables lie within the range of 3, the distribution is normal. The normality assumption is further buttressed by the nearness of the mean and median values for these series. The closer the mean and median values of a data series, the greater the probability that such series will be normally distributed. Hence, the variables selected are relevant within the study.

Having performed the test for stationarity,[7] the bounds test according to Peasaran et al. (2001), is accompanied to depict the existence or co-movement of variables in the long-run in Table 1. Given the computed critical values, of I(0) and I(1) bounds, the null hypothesis is stated that *no long-run relationship exist*. Since the computed F-statistics exceeds the critical value bounds at all level of significance, the null hypothesis is rejected for period I and II showed that the variables move together in the long-run; but in period III, the co-movement diminishes. The implication of this for the openness-greenness nexus is made clearer via the ARDL long-run coefficients shown in Table 2.

From the ARDL estimate in Table 2, period 1 revealed that neither of the individual economic growth (Y) nor trade openness (TO) caused CO_2 intensity to increase, and these negative relation was insignificant which is consistent with the findings of Ali et al. (2016). But the economic growth that modulates openness appears to increase the intensity of CO_2; although these increases were insignificant in the linear augment, it became more prominent in the quadratic augment. The period I (1970 and 1985) in Nigeria history was the oil boom era; and where economic planning was done based on oil proceeds. Therefore, despite that the Nigerian economy embraced structuralism which was evident in its planning system, the opportunity for investment in the activities that stimulated growth from MNEs oil investment was still permitted. Apart from agricultural output, oil wealth became a prime source of growth whose proceeds were subsequently used for economic, social and industrial

---

[7]*See* Appendix 3.

Table 2 ARDL estimates (dependent variable: CO_2)

Period	Variable	Long-run coefficients	Short-run coefficients
Period I	$CO_2(-1)$		0.1959
	Y	−0.0139	−0.0112
	Y^2	−0.0028	−0.0022
	TO	−0.0049	−0.0039
	(Y X TO)	0.0008	−0.0007
	$(Y X TO)^2$	2E − 0.6**	1.65E − 0.6*
	FDI	0.0557	0.0470
	POPU	0.0209	0.0168
	C	0.7395	0.5946
	ECT(-1)		−0.8040*
Period II	$CO_2(-1)$		1.6923*
	Y	0.1733	0.1199*
	Y^2	−0.0068	0.0047
	TO	0.0106	−0.0073
	(Y X TO)	−0.0043**	0.0030*
	$(Y X TO)^2$	5E − 0.6	−3.24E − 0.6
	FDI	0.1283*	−0.0889*
	POPU	−1.4439*	0.9997*
	C	23.8037*	−16.48
	ECT(−1)		0.6**
Period III	CO2(−1)		0.6878*
	Y	−0.0698	−0.0218
	Y^2	0.0057	0.0017
	TO	−0.0041	−0.0013
	(Y X TO)	0.0009	0.0003
	$(Y X TO)^2$	−2E − 0.6	−6.90E − 0.7
	FDI	−0.1124	−0.0351*
	POPU	0.5208	0.1626*
	C	−6.9963	−2.1844
	ECT(−1)		−0.3122**

Test statistic			
Period (I)		Period (II)	Period (III)
R^2 0.79		0.88	0.95
Adjusted R^2 0.51		2.70	0.88
F-Statistics 3		4.7913	12.965
Durbin Watson 1.9		2.4	2

Source Author's computation
Note * (**) depicts significance at 5% (10%)

development in Nigeria during the period 1970–1985. Incidentally, the production crude oil have associated challenges particularly for the environment. Thus, from this period I, given the more rigorous analysis via the augmenting components, it can be argued that the EKC effects held partially. This is because growth-stimulating activities affected the environment both in its linear and quadratic augments; and as such a trade-off process can be submitted for the openness-greenness nexus

In Period II (1986–2000), which is the immediate post-reform era where neo-liberal ideals were just been embraced, it is seen that economic growth (Y) and openness caused CO_2 intensity to rise; while Y^2 caused CO_2. intensity to decrease; thus indicating EKC inverted U-effects. However, the growth-openness ($YXTO$) nexus indicated a U-shaped EKC effect; given the linear augment reducing CO_2 intensity and quadratic augment increasing CO_2 intensity. The complexity of this period of a change from structuralist ideals to neo-liberal ideals explains the conflicting results found in period II. Thus, the findings of the openness-greenness nexus can be submitted a inconclusive of neither pay-offs nor trade-offs since the effects of the reforms are yet to fully materialize.

In period III, economic growth (Y) and trade openness (TO) caused reduction in CO_2 intensity but Y^2 had an increasing effect on in CO_2 intensity which U-shaped EKC effects. While the growth-openness ($YXTO$) nexus indicated inverted U-shaped EKC effect; given the linear augment increasing CO_2 intensity and quadratic augment reducing CO_2 intensity. The implication of this result indicates a reverse of and an improvement over period I. Although, there are indications that improved economic growth (Y^2) challenge greenness which is also manifested in the linear growth-openness augments ($YXTO$); but the quadratic augment ($YXTO$)^2, indicate prospects for sustainable development through the growth-openness augments. This position is further buttressed by the result of $FDI - CO_2$ relation in period III which unlike period I and II, is also negative. Therefore, from this result, we can argue that policy-shifts for openness has large prospects for improving the growth-openness-greenness relations; thus, indicating relative pay-offs (Table 3).

The Causal estimates indicate neutrality in period I and II, a unidirectional relationship is from CO_2 to TO. This implies that TO does not cause CO_2, but CO_2 causes TO; which indicates that although there are challenges for the environment from openness, it is still within acceptable limits where continues openness is not threatened.

Table 3 Causal estimates for openness and greenness

Period	Test statistic	Value	Direction
Period I	Granger (F-statistic)	Neutrality	TO $\rightarrow$ CO_2
Period II	Granger (F-statistic)	Unidirectional	TO $\rightarrow$ CO_2
Period III	Granger (F-statistic)	Neutrality	TO $\rightarrow$ CO_2

Source Author's computation

7 Conclusion

This chapter has been able to articulate the trade-openness nexus via the EKC growth-environment framework. The conclusion of the findings is based augmented effects of growth-trade openness nexus in promoting green growth. The findings revealed that the period where the structuralist ISI policies in Nigeria's history, which spanned between 1970–1985, was characterized more by trade-offs in promoting economic growth through openness; but at the expense of environment. But, the era of liberalism, 1986–2000 and 2001–2015, has been characterized by different results. Indeed, unlike period III, period II depicts in its linear augment pay-offs between openness and greenness which implies positive gains from reform; but based on the EKC hypothesis, it does not appear sustainable as shown by the positive relation of the non-linear augment. But given the non-linear augment in period III, an indication of pay-offs between trade openness and greenness is verified. This implies that the longer the fallouts from the period of reform, the more the prospect for green-growth in Nigeria; or better still, as Nigeria keeps its economy open, the higher the prospect for sustainable development. This finding is similar to the proposition of Ali et al. (2016) who argued for initiating more open economy policies in the Nigerian economy as the openness leads to the reduction of pollutants from the environment particularly CO_2 emissions which is the major gas that deteriorates physical environment.

Some studies have argued as to some of the unsustainable practices in Nigeria especially through energy production and consumption. For instance, as the 10th largest producer of oil in the world, Nigeria possesses abundance of natural resources especially as regards hydrocarbons. Most of the oil produced in Nigeria comes from the Niger-Delta region and are exported through terminals and floating production vessels. With over 600 oil fields and 176 trillion million (Tcf) cubic natural gas reserves, the proceeds from oil accounts for more than 90% of the foreign earnings of the Nigerian economy.[8] Since the discovery and mining of oil in commercial quantity, efforts to submerge associated environmental problems such as slow water poisoning, destruction of vegetation and farmlands from oil spills and gas flaring from mining activities have been a green challenge in Nigeria. More worrisome, is the evasive and little concerns for this environmental hazards both by the Nigerian government and MNEs involved. The question then is how is the openness expected to achieve green-growth?

A clue to the question above can be got from the argument of Asongu (2017) who argued in favor of the efforts of governance through globalization geared towards green-growth; as well as the utilization of improved technology adoption and diffusion, as well as ICTs to check unsustainable practices. Figure 5 which centers on CO_2 intensity (kg per kg of energy used) gives a clue to this argument. Between 1970s and mid-1980s, Nigeria recorded the highest CO_2 emissions of an average of 1%. But since the reform era, CO_2 intensity has been kept below 1%. This suggest that as the Nigerian economy keeps growing, prospects for greenness through openness can be

---

[8]*See* https://infoguidenigeria.com/problems-nigeria-oil-and-gas-industry/.

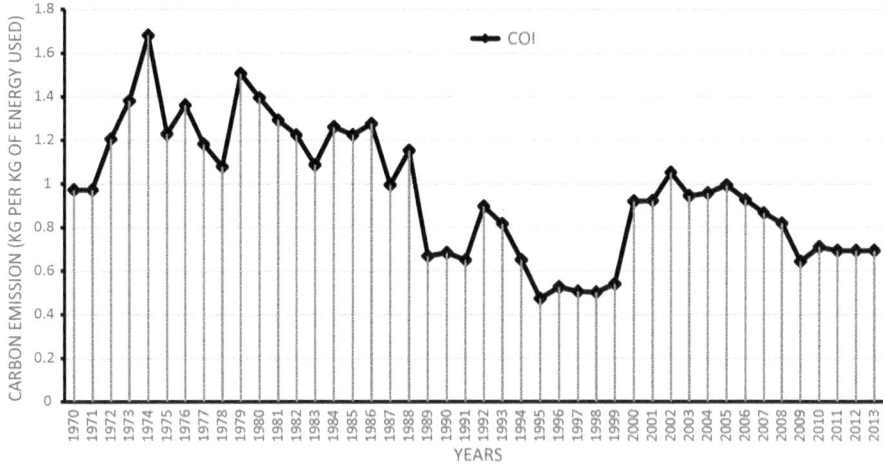

Fig. 5 CO_2 intensity (kg per kg of energy used). *Source* WDI (2014)

assured if policies for openness are carefully managed in this direction. Meanwhile, the neutrality causality found in period I and III suggest opportunity for openness-greenness policies in favour of sustainable development in Nigeria. Already reforms were implemented in period I in favour of openness, therefore, in this era of sustainability, policy direction can be geared towards achieving green growth by annexing the global opportunities given the existing open economic policy.

Appendix 1: Variable Source and Measurement

Variable	Source	Measurement
Carbondioxide (CO_2)	World Development Indicators (2018)	Carbon intensity (Kg per Kg of Energy used)
Economic growth (Y)	World Development Indicators (2018)	GDP per capita
Trade openness (TO)	World Development Indicators (2018)	Trade as a percentage of GDP
Foreign direct investment (FDI)	World Development Indicators (2018)	Foreign direct investment inflows
Urban population growth (POPU)	World Development Indicators (2018)	Urban population growth (POPU)

Appendix 2: Descriptive Statistics

Note: *(**) denotes the acceptance of the null hypothesis that the variables are normally distributed at 5% and (10%) significant level

Period I:1970–1985

	CO_2	Y	TO	FDI	POPU
Mean	0.785702	1.12203	35.43243	1.234835	15.33607
Median	0.82771	2.061785	38.74775	1.031162	15.4426
Maximum	1.007021	22.18323	48.57131	3.114868	16.27505
Minimum	0.383739	−15.4548	19.6206	−1.15086	14.16165
Std. dev.	0.154767	9.078312	10.21695	0.979935	0.726052
Skewness	−1.02373	0.370255	−0.17472	−0.27119	−0.34704
Kurtosis	4.099779*	3.246329*	1.474108	3.867146*	1.754796**
Jarque-Bera	3.601071	0.406021	1.633637	0.697417	1.35486
Probability	0.16521	0.81627	0.441835	0.705599	0.507921
Sum	12.57124	17.95248	566.9188	19.75736	245.377
Sum sq. dev.	0.359294	1236.236	1565.792	14.40409	7.907268

Period II: 1986–2000

	CO_2	Y	TO	FDI	POPU
Mean	0.51762	−0.81039	55.20951	4.064811	16.93079
Median	0.461303	−0.4332	58.10985	3.060115	17.05337
Maximum	0.853538	9.894914	76.85999	10.83256	17.2341
Minimum	0.32204	−13.0645	23.71676	0.932437	16.36697
Std. Dev.	0.168234	5.731556	14.09973	2.809191	0.285711
Skewness	0.60631	−0.54247	−0.70412	1.267692	−0.8543
Kurtosis	2.168025*	3.444972*	2.934896*	3.509345*	2.24265*
Jarque-Bera	1.351642	0.85943	1.242098	4.17975	2.183071
Probability	0.508739	0.650695	0.537381	0.123703	0.335701
Sum	7.764305	−12.1558	828.1427	60.97216	253.9619
Sum sq. dev.	0.396238	459.9103	2783.232	110.4818	1.14283

Period III: 2001–2015

	CO_2	Y	TO	FDI	POPU
Mean	0.620194	4.908238	53.22293	2.614536	15.86006
Median	0.613328	3.492157	52.7941	2.697522	15.81663
Maximum	0.759211	30.35658	81.81285	5.04766	16.84566
Minimum	0.462536	−0.02224	21.12435	0.650345	15.00262
Std. dev.	0.090199	7.315691	17.13841	1.336438	0.591606
Skewness	0.040167	3.050055	−0.24832	0.155742	0.172692
Kurtosis	1.849219**	11.24435*	2.254516*	2.0116*	1.794079**
Jarque-Bera	0.83172	65.73788	0.501501	0.671223	0.983461
Probability	0.659773	0	0.778217	0.714901	0.611567
Sum	9.302912	73.62357	798.344	39.21804	237.901
Sum sq. dev.	0.113903	749.2707	4112.153	25.00493	4.899962

Appendix 3: Unit Root Test

Variables	ADF		d	PP		d	KPSS		d
	Intercept & no trend	Intercept & trend		Intercept & no trend	Intercept & trend		Intercept & no trend	Intercept & trend	
CO_2	−1.92355	−3.01153		−1.85599	−3.033445		0.523153	0.113968	I(0)
ΔCO_2	−8.13129	−8.06662	I(1)	−8.168986	−8.094495	I(1)	0.114166	0.103664	I(1)
GDPC	−0.15251	−0.04779		−0.262600	−0.239471		0.212609	0.196220	
ΔGDPC	−5.77036	−6.40731	I(1)	−5.847613	−6.404826	I(1)	0.349977	0.138411	I(1)
TO	−1.43542	−2.06032		−1.352630	−2.074307		0.497362	0.119459	I(0)
ΔTO	−7.43229	−7.34153	I(1)	−7.455886	−7.363946	I(1)	0.088057	0.084058	
FDI	−1.56043	−2.86881	I(1)	−2.868805	−5.866385	I(0)	0.7183	0.1002	I(1)
ΔFDI	−11.42159	−11.2928		−32.94902	−39.88021	I(1)	0.2434	0.2434	
POPU	−2.65034	−1.17987	I(0)	−3.22163	−3.199678	I(0)	0.346864	0.218938	
ΔPOPU	−1.66282	−4.19234	I(1)	−6.62565	−6.395505	I(1)	0.708312	0.100955	I(1)

Critical regions for unit root test

	Mackinnon critical values				Asymptotic critical values			Asymptotic critical values
Level								
1%	−3.61045	−4.21186		−3.610453	−4.211868	0.739000		0.216000
5%	−2.93898	−3.52975		−2.938987	−3.529758	0.463000		0.146000
10%	−2.60906	−3.19831		−2.607933	−3.196411	0.347000		0.119000
1st difference								
1%	−3.61558	−4.21912		−3.615588	−4.219126			
5%	−2.94114	−3.53308		−2.941145	−3.533083			
10%	−2.60906	−3.19831		−2.609066	−3.198312			

Appendix 4: Stability Test

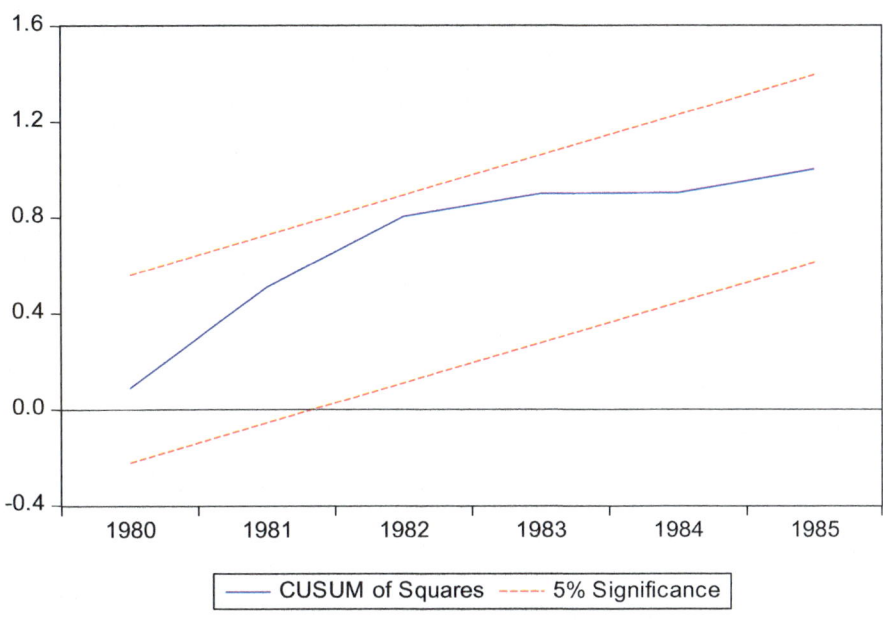

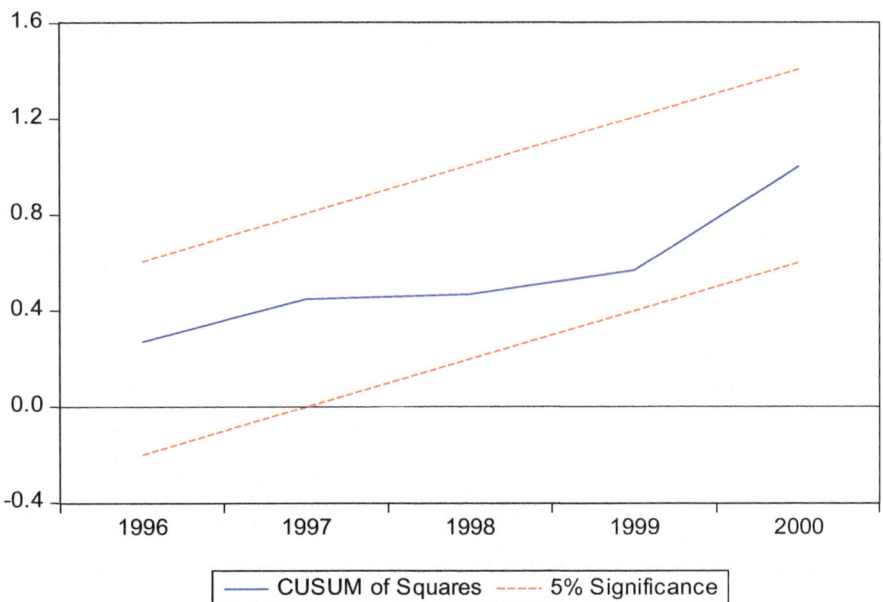

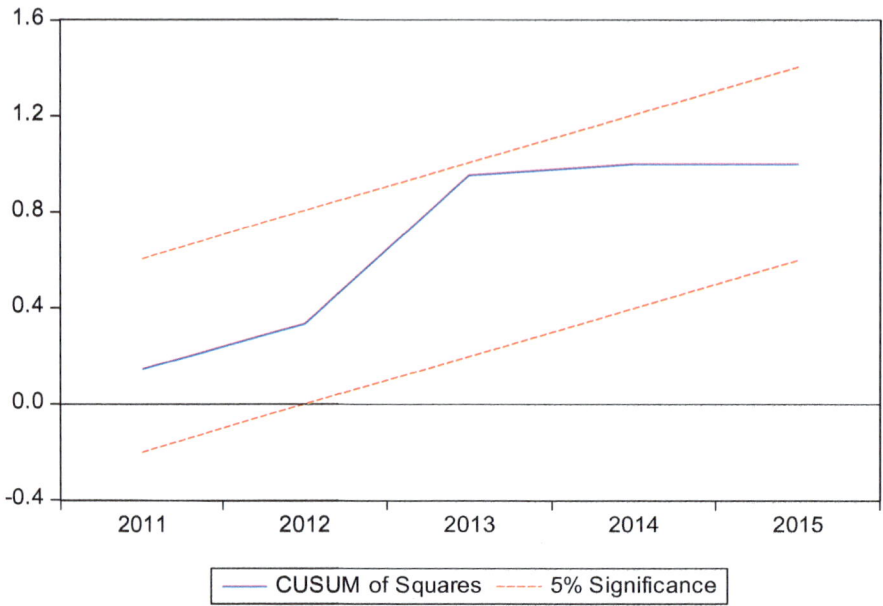

References

Adejumo OO (2018) Issues in sustainable development: the environment–income relationship. In Financing sustainable development in Africa. Palgrave Macmillan, Cham, pp 317–338

Akin CS (2014) The impact of foreign trade, energy consumption and income on CO_2 emissions. Int J Energy Econ Policy 4(3):465–475

Alege PO, Ogundipe AA (2015) Environmental quality and economic growth in Nigeria: a fractional cointegration analysis. Int J Dev Sustain 2(2)

Ali HS, Yusop ZB, Hook LS (2015) Financial development and energy consumption nexus in Nigeria: An application of ARDL bound testing approach. Int J Energy Econ Policy 5(3):816–821

Ali HS, Law SH, Zannah TI (2016) Dynamic impact of urbanization, economic growth, energy consumption, and trade openness on CO_2 emissions in Nigeria. Environ Sci Pollut Res 23(12):12435–12443

Ang JB (2008) Economic development, pollutant emissions and energy consumption in Malaysia. J Policy Model 30(2):271–278

Aremu, J. A. (2003). Foreign direct investment and performance. A Paper delivered at a workshop on foreign investment policy and practice organized by the Nigerian Institute of Advanced Legal Studies, Lagos

Asongu SA (2017) ICT, openness and CO_2 emissions in Africa. Environ Sci Pollut Res, 1–9

Asongu SA (2018) CO_2 emission thresholds for inclusive human development in sub-Saharan Africa. Environ Sci Pollut Res 25(26):26005–26019

Chebbi HE, Olarreaga M, Zitouna H (2011) Trade openness and CO_2 emissions in Tunisia. Middle East Dev J 3(01):29–53

Choi E, Heshmati A, Cho Y (2010) An empirical study of the relationships between CO_2 emissions, economic growth and openness

Colclough C, Manor J (eds) (1993) States or markets?: Neo-liberalism and the development policy debate. Oxford University Press

Dauda B (1993) Industrial policy and the Nigeria Bureaucracy, 1900–1988. In: Eldredge E, Falola T, Lovejoy P, Maier D, O'Hear A (eds) African economic history

Destek MA, Balli E, Manga M (2016) The relationship between CO_2 emission, energy consumption, urbanization and trade openness for selected CEECs. Res World Econ 7(1):52

Essien AV (2011) The relationship between economic growth and CO_2 emissions and the effects of energy consumption on CO_2 emission patterns in Nigerian economy

Ertugrul HM, Cetin M, Seker F, Dogan E (2016) The impact of trade openness on global carbon dioxide emissions: evidence from the top ten emitters among developing countries. Ecol Ind 67:543–555

Farhani S, Shahbaz M, Arouri MEH (2013) Panel analysis of CO_2 emissions, GDP, energy consumption, trade openness and urbanization for MENA countries

Grossman GM, Krueger AB (1991) Environmental impacts of a North American free trade agreement (No. w3914). National Bureau of Economic Research

Gujarati DN (2009) Basic econometrics. Tata McGraw-Hill Education

Hague BN, Loader B (eds) (1999) Digital democracy: discourse and decision making in the information age. Psychology Press

Hossain S (2012) An econometric analysis for CO_2 emissions, energy consumption, economic growth, foreign trade and urbanization of Japan. Low Carbon Econ 3(3)

Ikpeze NI (1991) New industrial policies and perspectives for manufacturing in Nigeria. In: Wohlmuth K, Oesterdiekhoff P, Kappel R, Hansohm D, Worch B, Bass HH, Grawert G, Zdunnek G, Franz J, Conrad M, Kleine K (eds) Industrialisation based on agricultural development, African development perspectives yearbook, vol 2. Muster/Hamburg, Lit, pp 585–608

Jamel L, Maktouf S (2017) The nexus between economic growth, financial development, trade openness, and CO_2 emissions in European countries. Cogent Econ Finance 5(1):1341456

Khalil S, Inam Z (2006) Is trade good for environment? a unit root cointegration analysis. Pak Dev Rev 1187–1196

Kuznets S (1955) Economic growth and income inequality. Am Econ Rev 45(1):1–28

Larner W (2000) Neo-liberalismi policy, ideology, governmentality. Stud Polit Econ 63(1):5–25

Meinshausen M, Meinshausen N, Hare W, Raper SC, Frieler K, Knutti R, Frame DJ, Allen MR (2009). Greenhouse-gas emission targets for limiting global warming to 2 °C. Nature 458(7242):1158

Mutascu M (2018) A time-frequency analysis of trade openness and CO_2 emissions in France. Energy Policy 115:443–455

Nnaji CE, Chukwu JO, Uzoma CC (2013) An econometric study of CO_2 emissions, energy consumption, foreign trade and economic growth in Nigeria. In: Conference proceedings of the 5th Annual NAEE/IAEE International Conference organized by Nigerian Association for Energy Economics, held at Abuja, Nigeria, pp 489–508

Ohlan R (2015) The impact of population density, energy consumption, economic growth and trade openness on CO_2 emissions in India. Nat Hazards 79(2):1409–1428

Omri A (2014) An international literature survey on energy-economic growth nexus: evidence from country-specific studies. Renew Sustain Energ Rev 38:951–959

Omri A, Daly S, Rault C, Chaibi A (2015) Financial development, environmental quality, trade and economic growth: what causes what in MENA countries. Energ Econ 48:242–252

Onoja AO, Ajie EN, Achike AI (2014) Econometric analysis of influences of trade openness, economic growth and urbanization on greenhouse gas emission in Africa (1960–2010). J Econ Sustain Dev 5(10):85–93

Park Y, Meng F, Baloch MA (2018) The effect of ICT, financial development, growth, and trade openness on CO_2 emissions: an empirical analysis. Environ Sci Pollut Res 25(30):30708–30719

Pesaran M, Pesaran B (1997) Working with Microfit 4.0: interactive economic analysis. Oxford University Press, Oxford

Pesaran MH, Shin Y, Smith RJ (2001) Bounds testing approaches to the analysis of level relationships. J Appl Econ 16:289–326

Rafindadi AA (2016) Does the need for economic growth influence energy consumption and CO_2 emissions in Nigeria? Evidence from the innovation accounting test. Renew Sustain Energy Rev 62:1209–1225

Saboori B, Sulaiman J, Mohd S (2012) Economic growth and CO_2 emissions in Malaysia: a cointegration analysis of the environmental Kuznets curve. Energ Policy 51:184–191

Sannassee RV, Seetanah B (2016) Trade Openness and CO_2 emission: evidence from a SIDS. In: Handbook of environmental and sustainable finance, pp 165–177

Sari R, Soytas U (2009) Are global warming and economic growth compatible? Evidence from five OPEC countries? Appl Energ 86(10):1887–1893

UNCTAD (2016) Structural transformation and industrial policy. UNCTAD, Geneva

UNCTAD (2018) World investment report—Investment and new industrial policies. UNCTAD, Geneva

Wooldridge JM (2013) Introductory econometrics, A modern approach. South-Western Cengage Learning. ISBN- 13: 978-1-111-53439-4

World Development Indicator (WDI) (2014) The World Bank, Washington, D. C., U.S.A.

World Development Indicator (WDI) (2018) The World Bank, Washington, D. C., U.S.A.

Zerbo E (2015) CO_2 emissions, growth, energy consumption and foreign trade in Sub-Sahara African countries

Zhang S, Liu X, Bae J (2017) Does trade openness affect CO_2 emissions: evidence from ten newly industrialized countries? Environ Sci Pollut Res 24(21):17616–17625

Ecotechnology as Mechanism of Development in Disadvantaged Regions of Mexico

Lorena del Carmen Álvarez-Castañón and Daniel Tagle-Zamora

Abstract The aim of this paper is to analyse the pertinence of ecotechnologies in vulnerable dwellings in five socially disadvantaged municipalities in Guanajuato (Mexico). The methodological design is integrated in two phases, one quantitative and other qualitative, to show the contributions of ecotechnologies in three categories: social, environmental and economic. The main finding shows its pertinence in the axes of: family relationships, health, environment, income. It is provided evidence on the potential contribution of ecotechnologies to the improvement of the conditions of the dwelling, to the environmental care and to the families' economy.

Keywords Ecotechnology · Social adoption · Technology transfer · Environmental education · Sustainable development

1 Introduction

Guanajuato is a Mexican federal entity formed by 46 municipalities. It is a territory located in the centre of the country where dynamism and innovation coexist with strong socioeconomic lags. Guanajuato has developed a strong industrial vocation. Its institutional system has taken advantage of its privileged geographic location to motivate an ecosystem based on the promotion of innovation to strengthen social development and building capacity in the territory (Álvarez and Estrada 2017; IPLANEG 2018). However, among its most obvious limitations and its territorial disparities are the level of marginalization, the low level of schooling, the high degree of disturbance in its ecosystems, the overexploitation of aquifers (SEDESOL 2017; SEMARNAT 2017). The absence of technology in the productive capability of rural areas has led to intensive use of water, soil degradation, deforestation, loss of biodiversity and worsening air quality (CONAGUA 2016; Mahdi et al. 2018).

L. C. Álvarez-Castañón (✉) · D. Tagle-Zamora
University of Guanajuato, Campus Leon, Guanajuato, Mexico
e-mail: lc.alvarez@ugto.mx

D. Tagle-Zamora
e-mail: da.tagle@ugto.mx

© Springer Nature Switzerland AG 2020
W. Leal Filho et al. (eds.), *International Business, Trade and Institutional Sustainability*, World Sustainability Series,
https://doi.org/10.1007/978-3-030-26759-9_52

This socio-productive and environmental complexity highlights the lag of the dwelling in Guanajuato, mainly in its disadvantaged localities. In the state, 9.8% of its population reports deficiencies in quality and spaces of their dwelling and 14.9% lack of access to basic services (CONEVAL 2010). In Guanajuato, ecotechnologies have been instrumented to mitigate the multidimensional problem of dwelling unable to satisfy a basic line of well-being (SEDESHU 2015). Social lag, climate change, absence of basic services, non-availability of water, among others, are challenges that have generated a space of action for ecotechnologies (Ortiz et al. 2014; Seema 2012). For decades several NGOs have promoted ecotechnologies as instruments that mitigate socio-environmental challenges in Guanajuato; the state government has implemented a series of programs to improve and equip households with solar heaters, photovoltaic panels, rainwater harvesters, ecological stoves, dry toilets with bio-digester, among others (Álvarez et al. 2018).

This paper shows the experience of transfer and social adoption of ecotechnologies in five socially disadvantaged Mexican municipalities; these ecotechnologies were transferred from 2012 to 2017 in Apaseo el Alto, Comonfort, Pénjamo, San Felipe and Tierra Blanca; five municipalities with a high degree of social marginalization (CONEVAL 2010). The main productive vocation of these municipalities is agriculture and one of their main problems is the detriment of their dwelling. In Apaseo el Alto and Comonfort, six out of ten people are in poverty and it is estimated that around 10% are in extreme poverty; in Pénjamo, San Felipe and Tierra Blanca, seven out of ten are in poverty and it is estimated that 23% are in extreme poverty (SEDESOL 2017).

Ecotechnologies are ecological technologies designed to mitigate environmental impact, to diminish pollution and to optimize the use of resources; however, the lack of a technology transfer process or the low level of social adoption of these ecotechnologies generates complex socio-environmental externalities (Álvarez et al. 2018). Ecotechnologies are rational artefacts designed to promote a social relationship consistent with environmental impact; however, these have been used as strategies in the design of social public policies focused on the improvement of the quality of housing (Tagle et al. 2017). Previous studies on the ecological history of Guanajuato show that socio-institutional efforts have been made to try to respond to a social demand for better dwelling; however, the weak social adoption of technology generates socio-environmental externalities greater than those that were presented before its transfer (Tagle et al. 2017).

A research, made by the University of Guanajuato and funded by the SEDESHU, on the social adoption of ecotechnologies in the underprivileged municipalities of the state reported that there are endogenous and exogenous factors that facilitate or inhibit their social adoption (Álvarez et al. 2018; Tagle et al. 2017). The exogenous factors demonstrate the total absence of environmental education during the process of transfer of the ecotechnology (TET) and the scarce articulation between the different governmental institutions involved in the TET; the endogenous factors show the community's lack of interest for the environmental care, its economic insolvency and the overvaluation for governmental support (Álvarez et al. 2018; Tagle et al. 2017).

The aim of this paper is to analyse the social, environmental and economic pertinence of ecotechnologies in housing in five municipalities of Guanajuato (Mexico) in social disadvantage from 2012 to 2017. The document is structured in three sections. The first section presents the methodological aspects of the research in two phases, quantitative and qualitative, due to the nature of the ecotechnology. The second section shows the development of the research: first, the characterization of study-object in the state of Guanajuato through a quantitative analysis of these ecotechnologies in their environmental and economic aspects; second, the social part based on the tangible and intangible benefits perceived by the beneficiaries of ecotechnologies. Finally, the third section presents the conclusions.

2 Materials and Methods

The research described herein is explanatory and transversal because it was essential to study social trends, to emphasize the meaning and to understand social action around ecotechnologies. The methodological design was based on a qualitative comparative approach, since it allows combining quantitative and qualitative tools whenever units of analysis are comparable and the categories to be studied are carefully operationalized (Rihoux 2006; Ragin 2014). First, the literature was thoroughly reviewed to build the state of art of the research, as well as to know the technical details about ecotechnologies. The purpose of this research is to study the TET in disadvantaged areas in five municipalities of Guanajuato: Apaseo el alto, Comonfort, Penjamo, San Felipe and Tierra Blanca.

The methodological process was integrated in two phases. The first phase consisted on the characterization of the five municipalities in the three areas of development—social, environmental and economic. The second phase consisted on the analysis of the TET process; therefore, the dense description of the interactions between the different actors involved (Cliford 2003) in the TET was collected through interviews and one focus group in each municipality (Table 1); this description allowed to capture the meaning of ecotechnologies through habits and practices. Both collector tools had three axes: perceptions, uses and benefits. The process was followed by the systematization of the data collected.

The main limitation of this paper is the coverage of the study, since only five municipalities were studied. Although the qualitative strategy limits the generalization of results, the qualitative comparative approach allows to assume that these disadvantaged municipalities could achieve sustainable development. This is relevant due to the explanatory nature of the research. Although it is not possible to infer about other municipalities, the results allow to generate a concrete idea of how ecotechnologies transfer processes are executed and what are the social, environmental and economic impacts that ecotechnologies generate.

Table 1 Data collection instruments

Semi-structured interviews with an average duration of one hour	Between 30 and 40 beneficiaries of the ecotechnology programs in each municipality
	Two suppliers of ecotechnologies
In-depth interviews with an average duration of one hour and a half per dwelling	Between five and eight dwellings by municipality, which had two ecotechnologies transferred The interview was made to family members in the house during the visit, since they are relevant actors in the TET process (they detailed about their daily life and their conviviality spaces with ecotechnologies)
Focus groups by municipality with sessions of minimum two hours	Beneficiaries of the ecotechnologies programs of the Social Development Directorates in each municipality

Source The authors

3   Results and Analysis

3.1   Characterization of Study-Object

The main productive vocation in the five municipalities was agriculture; in Comonfort and Apaseo el Alto, it follows the secondary sector, while in San Felipe, it follows the commerce. As previously mentioned, the lack of quality, spaces and basic services in housing are indisputable issues. According to official statistics, the quality and services in housing has improved in Guanajuato in the last twenty-five years; it was reported that there is a stationary decrease in shortcomings of housing (SEDESOL 2017). According to Secretariat of Social Development (SEDESOL 2017), during the period of 2010 to 2015, there was an evident improvement in the electricity access services in housing as this gap decreased 57.8% because the proportion of the population without this service was 1.6% in 2010 and 0.7% in 2015; furthermore, the access to piped water in homes increased because the proportion of the population without this service was 9.0% in 2010 and 4.1% in 2015 as the gap decreased 54.0%. However, the issues of access to water and drainage are still a pending debt (SEDESOL 2017), despite the fact that homes have been equipped with ecotechnologies to improve their conditions (SEDESHU 2015).

Ecotechnologies are practical social efforts to try to improve the capacities of exploitation of the natural resources used in the relation housing-environment; these have multidimensional benefits, since they directly and indirectly favour social, environmental and economic aspects (Tagle et al. 2017). According to INEGI (2015), the presence of ecotechnologies in Guanajuato has grown significantly as 7.7% of dwelling in Guanajuato have a solar ecotechnology installed. This is equivalent to more than one hundred thousand homes equipped with solar water heater and nine thousand homes with photovoltaic panels; therefore, there were more than four hun-

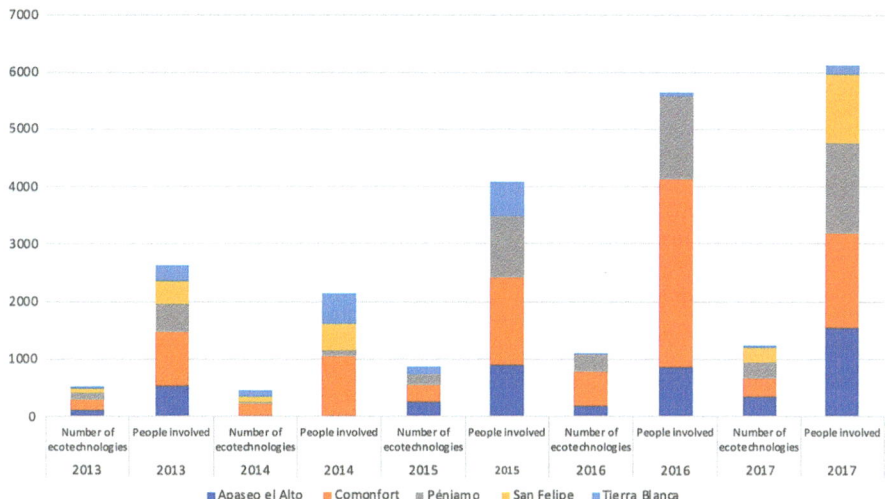

Fig. 1 The ecotechnologies installed in disadvantaged regions by social programs. *Source* The authors based on the data of SEDESHU (2018)

dred thousand people involved in the techno-environmental management. About the environmental aspect in Guanajuato, the dwellings with solar heater could eliminate one million tons of CO_{2eq} in 10 years based on two assumptions: there will be no increasement in the number of homes with this ecotechnology; the avoided emissions of greenhouse gases from a solar heater with a functional life of 10 years are 10.5 tons of CO_{2eq} (INEGI 2015; SEMARNAT 2017).

From 2012 to 2018, the state government through its social programs has installed more than eight thousand four hundred ecotechnologies; therefore, this is equivalent to approximately 38 thousand six hundred beneficiaries (SEDESHU 2018). These ecotechnologies were distributed as follows: 39.5% solar water heaters, 7.0% photovoltaic panels, 43.0% bathrooms with bio-digester and 10.5% ecological stoves; with a budget of 17.9 million dollars (SEDESHU 2018). In the last decade, SICES (2018) has financed 19 ecotechnology R&D projects, the research products have been experimentally installed in localities with a high degree of social marginalization with a budget around 1.6 million dollars. The ecotechnologies installed in these municipalities by social programs are shown in Fig. 1.

3.2 Impacts of the Transfer and Social Adoption of Ecotechnologies

In general, municipalities had exposed lack of information, training and monitoring of the technologies installed to date. Although traces of the training attempt on ecotechnologies were observed, there is no evidence of environmental education or presence of a participatory process in the TET. In the word cloud of the dense description, 179 elements were identified; 71 of these accumulated 80% of the mentions, which were stratified into eight codes as shown in the co-occurrence table (Table 2).

In the five municipalities studied, there were shown the multiple benefits of ecotechnologies. The ecotechnologies related to energy—solar heater and photovoltaic modules—showed the least resistance; the ecotechnologies that showed the most resistance and problems were the bathrooms with bio-digester, nevertheless, some dwelling have transformed the bathroom with bio-digester into a septic tank or they have connected it to the effluent network (it means a modification in the environmental function, however, ecotechnology maintains the social function); water-harvesters are widely used in the higher parts of the basins.

San Felipe and Tierra Blanca were atypical municipalities; in San Felipe there was a high level of interest in ecotechnologies and in Tierra Blanca it was common to find houses using two or more of them. In Penjamo, the TET has attenuated the negative perception of people towards the government's oblivion. In Apaseo and Comonfort, there was the greatest resistance to ecotechnologies. In the five municipalities, most of the time, women were the receiver of ecotechnologies during the TET process. The uses of ecotechnologies in these municipalities are shown in Table 3.

Frequently, the population manifested the attributes and benefits of the daily use of ecotechnologies. They said that these improve their family relationships when they use an ecological stove because there is less smoke in the kitchen space and they can spend more time there; when they harvest water because pregnant women or elderly people do not have to haul it from faraway places, therefore, conflicts between them are avoided. Others said that when they used the solar collectors, they got sick less frequently because they bathed in hot water; hence, they reduced their medical expenses and they had less absences to their jobs due to health issues. Others achieved a significant reduction in the cost of LP gas or firewood, or their time allocated to the collection of firewood was reduced.

The Table 4 shows the groups of benefits recognized by the people in each municipality. The impacts are achieved in four areas: family relationships, health, environment and income. As previously mentioned, with the harvest of water they manage to have less stress due to its access and they reduce conflicts by defining the member assigned to its search; consumption expenditure is reduced and women can devote their time to productive activities; visits to the river are reduced and time idle is

Table 2 Co-occurrence code (percentages)

Technical context	12.8	Sociocultural context	10.60
Technical constraint	13.54	Sociocultural constraint	23.22
Territorial constraint	3.89	Institutional constraint	11.20
Socio-environmental governance	14.05	Environmental education with five categories (social, cultural, gender perspective, ecological and technological)	8.97

Source The authors based on fieldwork

Table 3 Co-occurrence codes of ecotechnologies usage (percentages)

Usage	Kitchen-garden (orchard)	Ecological stove	Rainwater harvester	Solar heater	Photovoltaic panel	Dry toilet (bio-digester)
Feeding (cooking food, plucking chickens, heating drinks)	0.07	0.07	0.07	0.04	–	0.01 (biogas)
Exchange with neighbours (barter or commercial transaction)	0.06	–	0.03	–	–	–
Maintenance (orchard, plants, animals)	–	–	0.06	–	–	0.03
Home clean (wash kitchen-tools, do the laundry, discharge the bathrooms, take a shower, increase hygiene)	–	0.04	0.07	0.07	–	–
Comfortable homes (lighting, cooling or heating temperature)	–	0.03	–	–	0.07	–
Cellar or warehouse (water storage provided by the municipality or download grey water)	–	–	0.03	–	–	0.07
Unused	0.03	–	0.03	–	–	0.07

Source The authors based on fieldwork

Table 4 Co-occurrence codes of benefits (percentages)

	Kitchen-garden (orchard)	Ecological stove	Rainwater harvester	Solar heater	Photovoltaic panel	Dry toilet (bio-digester)
Family relationships	0.03	0.03	0.05	0.03	0.03	0.02
Family relationships (time)	0.02	0.03	0.02	0.03	0.02	0.03
Health (get sick less frequently)	0.03	0.02	–	0.02	–	0.02
Health (hygiene)	–	–	0.03	0.02	–	0.02
Health (food)	0.03	–	–	–	0.02	0.02
Environment (polluting gases)	–	0.02	–	0.02	0.02	0.02
Environment (deforestation)	–	0.02	–	0.02	0.02	–
Environment (rational use of natural resources)	0.03	–	0.05	–	–	0.02
Income (reduction in job absences)	–	–	–	0.02	–	–
Income (less expenses)	0.03	0.05	0.02	0.06	0.03	0.03
Income (barter or commercial transaction)	0.03	–	0.02	–	–	0.02

Source The authors based on fieldwork

avoided when waiting for the common use hose; better use of natural resources, reduces the consumption of river water and the extraction of wells. In the health axis there is evidence of less respiratory diseases; ecological stoves reduce consumption of firewood; solar heaters reduce gas consumption, and family members are bathed in hot water.

The actors involved in the TET do not know the word ecotechnology, although they identify positive impacts from their daily use (Fig. 2). The tangible and intangible benefits of the ecotechnologies identified in these municipalities are grouped into:

1. Social: better family coexistence is identified, and family relationships are favoured. Likewise, there is a positive impact on health because diseases in the respiratory tract are reduced (showers with hot water) and hygiene is improved.
2. Environmental: a significant reduction of greenhouse gases emissions and less impact on deforestation is identified, since the use of ecotechnologies decreases the consumption of firewood and gas.
3. Economic: expenses are reduced because the consumption of LP gas and the purchase of firewood decreased. Also, the time allocated to the transportation of water and the collection of firewood to heat the water is eliminated. Additionally, medical expenses and loss of income are reduced since job-absences due to health issues decrease.

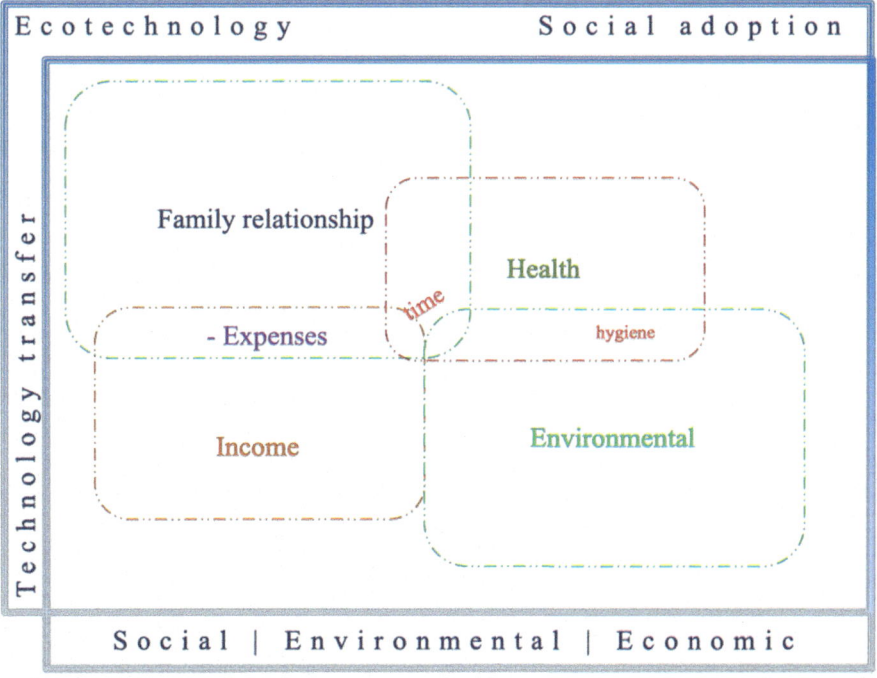

Fig. 2 Ecotechnology as mechanism of development. *Source* The authors based on fieldwork

It is possible to show that the TET is viable in the disadvantaged dwellings of the marginalized localities in Guanajuato; its viability depends on the type of ecotechnology and the socio-technical characteristics of the territory where this transfer occurs, since the TET is correlated with social, cultural and ecological diversity of the municipalities. The evidence on the ecotechnologies' potential is shown in the improvement of the dwelling, the environmental care and the families' economy. Their social adoption is strengthened through participatory processes with a gender perspective and environmental education based on the mixture of technical and vernacular knowledge of the territory. Thus, it is shown that ecotechnologies generate a positive social, environmental and economic development.

4   Conclusions

The results show positive impacts and the contribution of ecotechnologies is achieved in the axes of: family relationships, health, environment and income. The leadership of the territory has been fundamental. The leading role of women in the TET is evident, as they are the principal contact and main receiver of the technology. The dialogue of technical and local knowledge is strategic to maintain cooperative and trustworthy environments. The environmental education directed to the whole community is fundamental in the TET to accumulate a technological memory that ensures the continuous and standardized use of the technology.

Evidence was found to show that ecotechnologies generate positive socioeconomic impacts on fragile households to where they have been transferred, if and only if these are socially adopted. The technology management model is heterogeneous, and it has sought to improve the quality of housing in socially disadvantaged communities. In each locality, ecotechnologies have generated externalities in their use and appropriation; in some, there are committees that manage collective water harvesters, in others, the ecotechnology has been socially appropriated and the houses use two or more.

The social adoption of the technology is the main decisive factor to achieve an impact from ecotechnologies on the development of disadvantaged localities in these five municipalities. The TET depends on endogenous and exogenous factors that facilitate or inhibit this social adoption. Another fundamental factor is the provision of public resources necessary to incorporate ecotechnologies in dwelling that are not able to satisfy a basic line of welfare. Finally, the construction of local technological capabilities is the key factor to achieve the development of these localities.

References

Álvarez L, Estrada S (2017) El sistema estatal de innovación en una región en desarrollo. Las capacidades de innovación en Guanajuato. In: Montiel O, Rodríguez C (eds) Emprendimiento hoy. Multidimensionalidad, cambio e innovación. UACJ – Western New Mexico University, México, pp 255–282

Álvarez L, Tagle D, Romero M (2018) Transference of ecotechnology in disadvantaged regions of Mexico, towards sustainable development. In: Leal-Filho W, Noyola-Medellín P, Vargas V (eds) Sustainable development research and practice in Mexico and selected Latin American countries. Springer World Sustainability Series. Springer, Cham, pp 139–152. https://doi.org/10.1007/978-3-319-70560-6_9

Cliford G (2003) Interpretación de las culturas. GEDISA Editorial, España

CONAGUA (National Water Commission) (2016) Statistical. Retrieved from: https://www.gob.mx/conagua/documentos/estadisticas-agricolas-de-los-distritos-de-riego. Last accessed 31 May 2018

CONEVAL (National Council for the Evaluation of Social Development Policy) (2010) Annual report on the situation of poverty and social backwardness. Retrieved from: www.coneval.org.mx. Last accessed 31 May 2018

INEGI (National Institute of Statistic and Geography) (2015) Mexico in Figures: households, housing and urbanization. Retrieved from: http://www.beta.inegi.org.mx/app/areasgeograficas/?ag=11#. Last accessed 30 April 2018

IPLANEG (Institute of Planning, Statistics and Geography of the State of Guanajuato) (2018) Statistics. Retrieved from: http://geoinfo.iplaneg.net/maps/80/view. Last accessed 31 March 2018

Mahdi R, Ramazani S, Delshad A (2018) Sustainable development of agriculture: a system dynamics model. Kybernetes 47(1):142–162

Ortiz J, Masera O, Fuentes A (2014) La ecotecnología en México. Unidad de Ecotecnologías UNAM-CIECO, México

Ragin C (2014) The comparative method: moving beyond qualitative and quantitative strategies. Moving beyond qualitative and quantitative strategic. University of California Press, USA

Rihoux B (2006) Qualitative comparative analysis (QCA) and Related SYSTEMATIC comparative methods. Int Sociol 21(5):679–706. https://doi.org/10.1177/0268580906067836

SEDESHU (Secretariat of Social and Human Development of the state of Guanajuato) (2018) Database of ecotechnologies. Guanajuato, México. Retrieved from: https://portalsocial.guanajuato.gob.mx/cedulas-estadisticas. Last accessed 31 March 2018

SEDESHU (Secretariat of Social and Human Development of the state of Guanajuato) (2015) Citizen handbook. Guanajuato, México. Retrieved from: http://www.desarrollosocial.guanajuato.gob.mx/pdf/manual-ciudadano-2015.pdf. Last accessed 31 March 2018

SEDESOL (Secretariat of Social Development) (2017) Annual report on the situation of poverty and social backwardness 2017. Retrieved from: www.sedesol.gob.mx. Last accessed 30 April 2018

Seema J (2012) Role of science and technology for agricultural revival in India. World J Sci Technol Sustain Dev 9(2):108–119. https://doi.org/10.1108/20425941211244261

SEMARNAT (Secretariat of Social Development and Natural Resources) (2017) Diagnostic by entity 2017. Retrieved from: https://www.gob.mx/semarnat#369. Last accessed 31 March 2018

SICES (Innovation, Science and Higher Education) (2018) Mix fund of Guanajuato, México. Retrieved from: http://sices.guanajuato.gob.mx/fondoMixto. Last accessed 31 May 2018

Tagle D, Ramírez R, Caldera A (2017) Retos sociales y ambientales en la implementación gubernamental de ecotecnias en Guanajuato, México. Administración y Organizaciones 19(37):163–184

Lorena del Carmen Álvarez-Castañón is professor (titular A) at the University of Guanajuato, campus Leon, Mexico. Member of the National System of Researchers (2013–2020). Awards:

National Research Award for Scholars by the National Association of Colleges and Schools of Accounting and Administration (2012); Merit for academic excellence in the doctoral program; Merit for academic excellence by the National Association of Colleges and Schools of Engineering. She is member of the research group Water, Energy and Climate Change.

Daniel Tagle-Zamora is professor at the University of Guanajuato, campus Leon, Mexico and invited professor in the multidisciplinary program Environment and Resources Management (ENREM) at the University of Applied Sciences of Cologne (Germany) and the Autonomous University of San Luis Potosí (Mexico). Member of the National System of Researchers (2013–2019). He has participated in numerous research projects; he is leader of the research group Water, Energy and Climate Change.

Mapping the Industrial Water Demand from Metropolitan Region of Curitiba (Brazil) for Supporting the Effluent Reuse from Wastewater Treatment Plants

Carlos Henrique Machado, Patrícia Bilotta and Karen Juliana do Amaral

Abstract Water scarcity has become a major problem in the urban environment. However, a potential means of increasing the volume of water available for human necessities is the consumption by the industrial sector of effluent from wastewater treatment plants (WWTPs), provided the quality of this effluent is ensured. This work presents a map of the water demand of various industrial sectors of the Metropolitan Region of Curitiba (Brazil), to be used in support of an initiative to reuse the effluent from wastewater treatment plants. The method used to produce the map consisted of the following steps: (a) collection of data (number of employees and type of industrial production activity) provided by the Federation of Industries in Paraná State (FIEP); (b) application of the coefficients proposed by the Brazilian Water Agency (ANA); (c) construction of a geographic database that included the locations of the industries; (d) selection of WWTPs capable of providing the industries with treated effluent. Data were gathered from the municipalities of Campo Largo, São José dos Pinhais, and Curitiba (districts of Boqueirão, Cidade Industrial, and Xaxim only). The results identified 935 companies, which were classified according to their respective industrial sectors. Of these companies, 44 exhibit a high estimated water demand (a total of 88.73 L s^{-1}), and 36 are located within 5 km of the closest WWTP (a total of 68.91 L s^{-1}). The results also revealed that 9 WWTPs have the capability to meet the industrial demand for treated effluent. Assuming a effluent reuse potential of 15%, this represents a total average savings in drinking water in the order of 10.3 L s^{-1} for the municipalities studied; however, this percentage may be higher

C. H. Machado
Graduate Program in Civil Engineering, Positivo University, Rua Prof. Pedro Viriato Parigot de Souza, no. 5300, Curitiba, Paraná, Brazil
e-mail: carloshm27@gmail.com

P. Bilotta (✉)
Master's and Doctorate Program in Environmental Management, Positivo University, Rua Prof. Pedro Viriato Parigot de Souza, no. 5300, Curitiba, Paraná, Brazil
e-mail: pb.bilotta@gmail.com

K. J. do Amaral
Master's Program in Urban and Industrial Environment, Universität Stuttgart / UFPR, Av. Coronel Francisco Heráclioto dos Santos, no. 100, Curitiba, Paraná, Brazil
e-mail: karenjamaral@gmail.com

© Springer Nature Switzerland AG 2020
W. Leal Filho et al. (eds.), *International Business, Trade and Institutional Sustainability*, World Sustainability Series,
https://doi.org/10.1007/978-3-030-26759-9_53

(30–40%) for industries related to the manufacture of cellulose and paper, drinks, and food processing from animal sources.

Keywords Spatial analysis · Geographic database · CNAE category

1 Introduction

Studies indicate that global water demand is expected to rise from the current 4.5 billion m^3 per day to 6.9 billion by 2030 if the economic growth is maintained and no optimization in the industrial processes occurs regarding the use of water. This means that water consumption is about 40% greater than global water availability. In some regions of the planet, the situation may be even more serious, since 1/3 of the world population may live in watersheds having a water deficit greater than 50%. Another factor that aggravates the current scenario is that the amount of water actually available for consumption is less than the total amount of water present in the environment, due to requirements such as accessibility, reliability (quality) and the existence of an environmentally sustainable supply (2030 Water Research Group 2009).

One of several potential solutions to this crisis is the reuse of water, which directly benefits the environment by increasing the supply of available water and reducing the negative effects of effluent on the quality of the water in streams by redirecting it for reuse in various processes. Moreover, this practice allows for less dependence on greater water resource centralization and the need for expensive engineering works to remedy problems related to water supply. Furthermore, the conservation of water resources by means of reuse systems for public supply or other more restrictive uses can contribute to the universalization of sanitation, improving the cost-effectiveness of sewage treatment (Mierzwa and Hespanhol 2005).

The reuse of water is an old practice, originating in the use of land for disposal of treated effluents. With the installation of sewage collection systems in the 19th century, effluents were destined to locations called "sewage farms"; by the year 1900, there were several farms of this nature in Europe and United States. In these farms, effluents were used for irrigating plantations or other purposes that brought benefits for humans (Metcalf and Eddy 2004).

In the present, water reuse is becoming increasingly common, in conjunction with a growing awareness in the cities of the local and regional hydrological environment. Thus, one may consider the emergent forms of water reuse as an evolution of the classic linear anthropogenic hydrological cycle. The closure of this cycle brings new dimensions of safety and control of water quality and public health, and partially allows cities to disconnect from environmental restrictions (Wilcox et al. 2016; Dunhu et al. 2013).

In urbanized areas such as big cities and metropolitan areas, a large concentration of both population and economic and industrial activities leads to a growing demand for water. In the case of the industrial sector, in addition to the direct reuse both treated and untreated industrial effluents generated by the company itself, it is also possible to use effluents from domestic sewage treatment plants (WWTPs) for use in production processes (Kossar et al. 2013). This strategy uses the effluent to replace drinking water, thus increasing the offer of water available for human consumption in these areas.

In Brazil, a national example of the use of WWTP effluent for use in industrial processes is the AQUAPOLO PROJECT, in which about $650 \, L \, s^{-1}$ of treated effluent (from the tertiary level) is transported over a 17 km stretch to supply the petrochemical pole of the ABC Paulista district in the city of São Caetano and Santo André (Osorio 2013).

From an economic perspective, the treatment of effluents aims to achieve a standard of quality that is adequate for a specific type of reuse, thus increasing the potential for cost recovery. Effluent reuse becomes even more competitive in contexts where freshwater prices reflect the opportunity cost of their use, and where rates for the discharge of pollutants into water bodies reflect the cost of removing pollutants from wastewater streams. The strategy of using treated or partially treated effluents can increase the efficiency of water resources and provide environmental and social benefits such as a reduction in freshwater abstraction, the recycling and reuse of nutrients, an increase in the water available for fishing activities, and the maintenance of aquatic ecosystems, as a result of a reduction in water pollution and a renewal of existing aquifers (United Nations Educational, Scientific and Cultural Organization World Water Assessment Program 2017). Therefore, effluent reuse is strongly related to the Sustainable Development Goals (SDG), which were born during the United Nations Conference in 2012 (UNDP 2019), mainly to the SDG 6, 7, 9, 11, 13, and 14.

In view of the above benefits, the current work presents the results of a research project carried out in order to identify the highest industrial water consumers in the Metropolitan Region of Curitiba (MRC), located in the state of Paraná, Brazil. The data contained in these results was collected in order provide a basis for making future decisions and to support further initiatives involving the replacement of clean water by WWTP effluent in projects that display technical and economic feasibility. It is therefore a tool for water resource management in urban areas. This project was inserted into a macroproject financed by the Araucária Foundation and started in December 2016, under the coordination of Prof. Dr. Patrícia Bilotta from Graduate Program in Environmental Management (Master and Ph.D.)—Positivo University (PGAMB-UP), and in partnership with Prof. Dr. Karen Juliana do Amaral from the Graduate Program in Urban and Industrial Environment—University of Stuttgart (Germany).

2 Method

2.1 Strategy for Preparing the Inventory

The methodology proposed in the original project (Fig. 1), as proposed by Osório (2013), included conducting a survey of companies (industries) in the metropolitan area of Curitiba (MRC) and within a 10 km radius of WWTPs, in order to identify those that could potentially receive treated effluent for use in their operations.

Information regarding the WWTP coordinates and the available effluent flow for each plant were provided by the sanitation company. The coordinates were originally given in a decimal system and later converted to a DMS system (degrees, minutes, seconds).

After using the Google Maps tools to pinpoint the locations of the WWTPs within the MRC, industries located within a 10 km radius of these plants were determined visually. Subsequently, a geographic database containing the following headings was prepared: (a) WWTP identification; (b) WWTP municipality; (c) WWTP latitude and (d) longitude; (e) industry name; (f) industry address; (g) industry website; (h) industry activity sector (according to its website); and (i) radial distance between the industries and the closest WWTP.

This process, however, was severely limited, as became evident from the sparse amount of data collected for each of the companies; for example, it was impossible to obtain information regarding the scale of company operations or even the number of employees. In some cases, no company website was found. In order to correct these flaws in the original methodology, a new procedure was proposed, as described below.

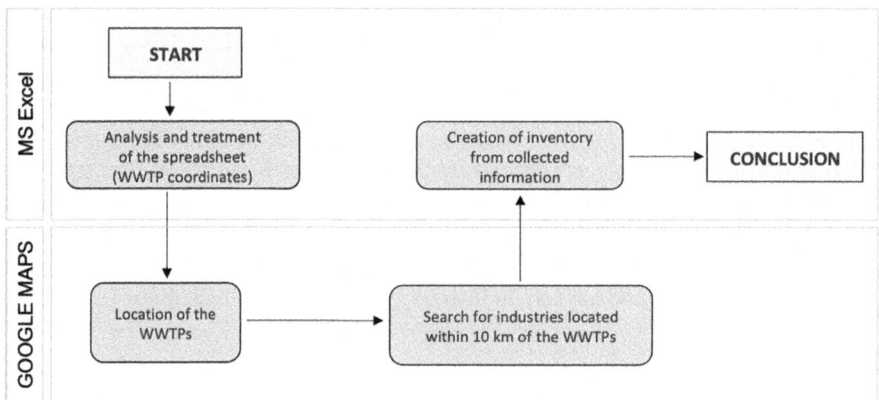

Fig. 1 Diagram of the methodology used for the inventory design

2.2 New Strategy for Inventory

A flowchart (Fig. 2) was drawn up for a new methodological approach to designing the geo-referenced database. The proposed method formed the basis of an article published in the Third Symposium on the Urban and Industrial Environment (MAUI), organized by the Federal University of Paraná (Machado et al. 2018).

Step I—Obtaining and organizing the data

The first step of the proposed method (Fig. 2) consisted of collecting and compiling the information as presented in Tables 1, 2, 3, 4, 5 and 6.

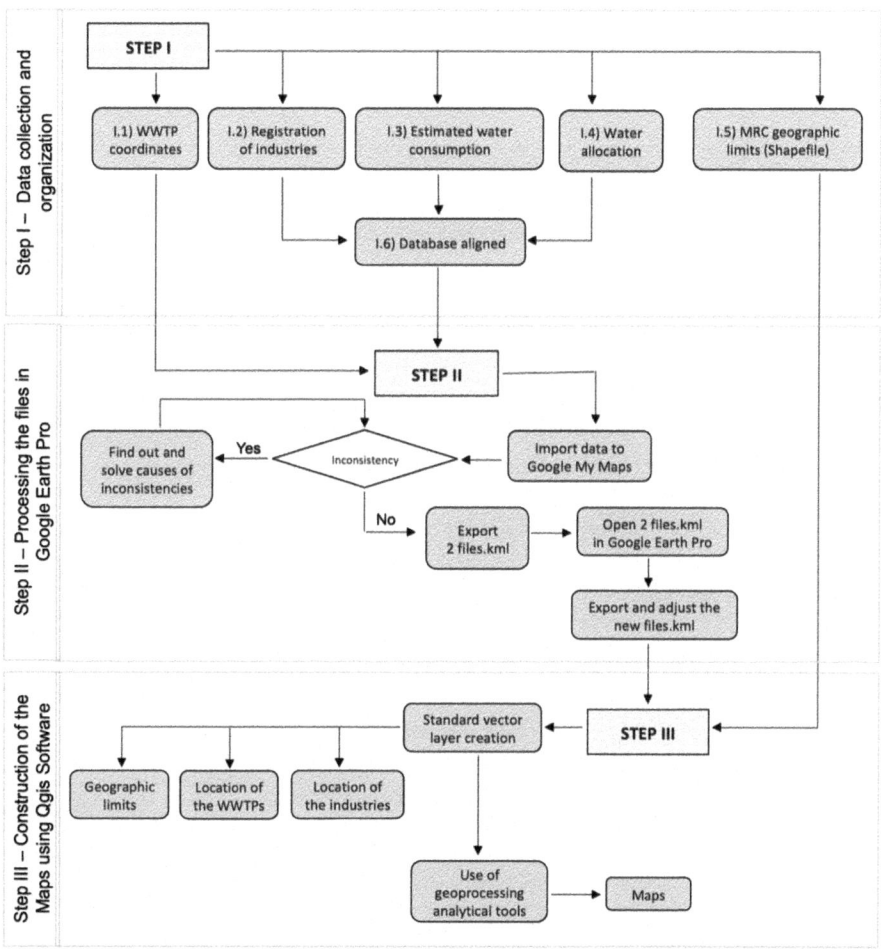

Fig. 2 Application of the proposed method for preparing an inventory of industrial water demand

Table 1 WWTP coordinates

Information	WWTP coordinates and the flow currently available
Data source	The sanitation company of the respective municipality
Action	Data systematized into columns on a spreadsheet. Column layout: WWTP code, WWTP identification, available flow (in L s^{-1}), latitude and longitude (in the GMS system, using a period as decimal separator instead of a comma)

Table 2 Industry registration data

Information	Data from each identified industry in the municipalities of Campo Largo, São José dos Pinhais, and Curitiba (districts: Boqueirão, Cidade Industrial, and Xaxim)—MRC
Data source	Federation of Industries in the State of Paraná (FIEP 2017), based on data from National Register of Industries
Action	Data systematized into columns on a spreadsheet. Column layout: name of the entity, address, CNPJ (National Registration Number for Legal Entities), number of employees, National Classification of Economic Activities (CNAE) and productive sector

Table 3 Industry water consumption

Information	Coefficient of water consumption (kc), according to CNAE
Data source	National Water Agency (ANA), obtained from the study "Water in industry: use and technical coefficients" (ANA 2017)
Action	Data systematized into columns on a spreadsheet. Column layout: CNAE, CNAE description, withdrawal coefficient and consumption coefficient

Table 4 Authorization for water abstraction

Information	List of companies that have a permission and exemptions for water abstraction and disposal of effluents
Data source	Paraná Water Institute (IAP 2017a)
Action	Data systematized into columns on a spreadsheet. Column layout: company name, CNPJ, contractual status and effective date of the contract

Table 5 Shapefile of the MRC geographical limits

Information	File containing a map with geographical limits for the area under consideration.
Data source	Brazilian Institute of Geography and Statistics (IBGE 2016)
Action	Map with the MRC area (file reserved for further use in Step III)

Table 6 Adjusted database

Information	File listing companies that do not have authorization for water abstraction and are not exempt from this agreement
Data source	File obtained by crossing worksheets I.2, I.3 and I.4, taking the CNPJ as reference in the Microsoft Excel vertical search mechanism (PROCV). Only those companies that were present in the industry registration but did not simultaneously appear in the Authorization database with current contract or exemption were considered
Action	Data systematized into columns on a spreadsheet. Column layout: industry name, address, CNPJ, number of employees, CNAE, productive sector, estimated water consumption (in $L\ s^{-1}$)

The estimated water flow consumed by the companies ($Q_{ind\_consumption}$) was obtained from the coefficient of water consumption (k_c), which is based on the number of employees in the company (NE), according to the model proposed by ANA (2017). This information was provided by the Ministry of Labor in its Annual Social Information Report (RAIS), which takes into account the National Classification of Economic-Fiscal Activities (CNAE) established by the Brazilian Institute of Geography and Statistics (IBGE 2016). Equation (1) presents the calculation model proposed by ANA (2017).

$$Q_{ind\_consumption} = \frac{NE * k_c}{86,400} \tag{1}$$

In addition to RAIS, the number of employees of a company registered in the Federation of Industries in Paraná State database may also be used (FIEP 2017); this was the source chosen for the research conducted for the current work.

The CNAE codes proved to be compatible with the classification of industrial activities adopted by the National Water Agency (ANA 2017), although this is only true of the codes that relate to the processing industries. Thus, it was possible to calculate the average required water flow for the industries located in the MRC and registered at FIEP and use this information to construct the geographic database.

Due to the large number of companies registered in the FIEP database, it was decided to include only those located in municipalities of São José dos Pinhais, Campo Largo, and Curitiba (of the Boqueirão, Cidade Industrial and Xaxim districts). Once the proposed method has proved to be robust and adequate for its intended purpose, data from companies located in other municipalities or in other districts of Curitiba may be incorporated into the inventory, even after the conclusion of the current research project.

Step II—File processing in Google Earth Pro

(a) *Phase I*

The contents of Table 1 were uploaded to the Google My Maps tool (Google 2018a), and the latitude and longitude columns were considered as search criteria. Next, a file containing the points generated in the Keyhole Markup Language extension (kml) was exported. The content of Table 6 was also transferred to Google My Maps; however, in this case, the company's name and address were adopted as search criteria (Google 2018a). To eliminate inconsistencies in rows that were not imported, corrections were made to remove typographical errors, missing accent marks, and incorrectly named districts. Subsequently, another file having a kml extension was generated. At this stage, although the files were natively supported by QGIS software, used in the spatial analysis of data, loading and pre-processing of information in Google Earth Pro, tests were performed to verify whether the points (company and WWTP location) were visible in the correct position on the map using the SIG application (Geographic Information System) (Google 2018b).

(b) *Phase II*

Two new files having a kml extension were saved using Google Earth Pro and opened in notepad, after which the following tags were deleted from the files: <MultiGeometry>, </MultiGeometry>, <LinearRing>, </LinearRing> (Google 2018b). This procedure was necessary in order to create the vector layers correctly within the QGIS application.

Step III—Creating the map in the QGIS

From the files obtained in the previous step, the following three vector layers were created in the QGIS software:

i.   Layer Shapefile (.shp), referring to the content of Table 5, containing the spatial limits of the MRC. The SIRGAS 2000 reference system was used.
ii. Layer Keyhole Markup Language (.kml), containing the coordinates of the WWTPs. The WGS 84 reference system was used.
iii. Layer Keyhole Markup Language (.kml), containing the coordinates of the industries. The WGS 84 reference system was used.

3 Results and Discussion

The inventory compiled in this paper includes a total of 1215 industries located in the municipalities of São José dos Pinhais, Campo Largo and Curitiba (districts of Boqueirão, Cidade Industrial, and Xaxim only). Table 7 compares the data obtained for each municipality, presenting the number of companies registered in the FIEP database (1st column), the adjusted number after excluding those that receive funding or are exempt from formalities involving the Paraná Water Institute (IAP 2017b) (2nd column), and the number of companies whose CNAEs are not covered by ANA (2017) (3rd column). In the end, 77.0% of the companies included in the FIEP database were admitted (935 industries).

According to the ANA method (2017), only processing industries (covered under CNAE codes 10 to 33) are eligible to have their water demand estimated. For this

Table 7 Analysis of the inventory results

Municipality	FIEP (industries)	After filter (IAP)	After filter (ANA)	Efficiency[a] (%)	$Q \geq 0.5\,\mathrm{L\,s^{-1}}$ (no. of industries)
Campo Largo	165	155	126	81.3	4
Curitiba	589	566	467	82.5	23
São José dos Pinhais	461	410	342	83.4	17
Total	1215	1131	935	82.7	44

[a]The percentage of industries selected by the proposed method was calculated after narrowing the results using the IAP and ANA filters

reason, the proposed method achieved an average efficiency of 82.7% in the selection of the companies composing the inventory.

Figure 3 shows the distribution of estimated water demand according to the sector of industrial activity. The highest consumption was related to the chemical, food, cellulose and paper, and auto industries that correspond to 65.3 L s^{-1} (73.4% of the total estimated water demand).

Figures 4, 5 and 6 show the spatial location of the 44 companies selected in the municipalities of São José dos Pinhais, Campo Largo, and part of Curitiba, that were classified as having a high water demand (greater than 0.5 L s^{-1}). It also indicates the closest corresponding WWTPs (9 in total) that may meet the industrial demand for treated effluent for each industry within a 10 km radius.

Table 8 shows the estimated water consumption and the corresponding CNAE codes. The total accumulated amount flow required by the 44 industries was 88.73 L s^{-1}. When a 10 km radius between each company and the closest WWTP

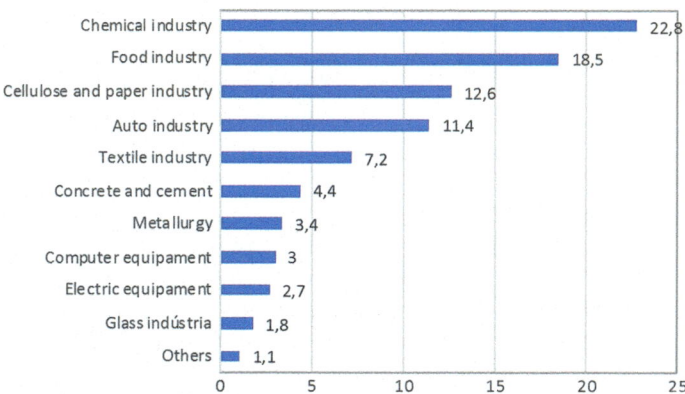

Fig. 3 Distribution of the estimated water demand for the 44 companies selected in the municipalities of Campo Largo, Curitiba (three districts), and São José dos Pinhais

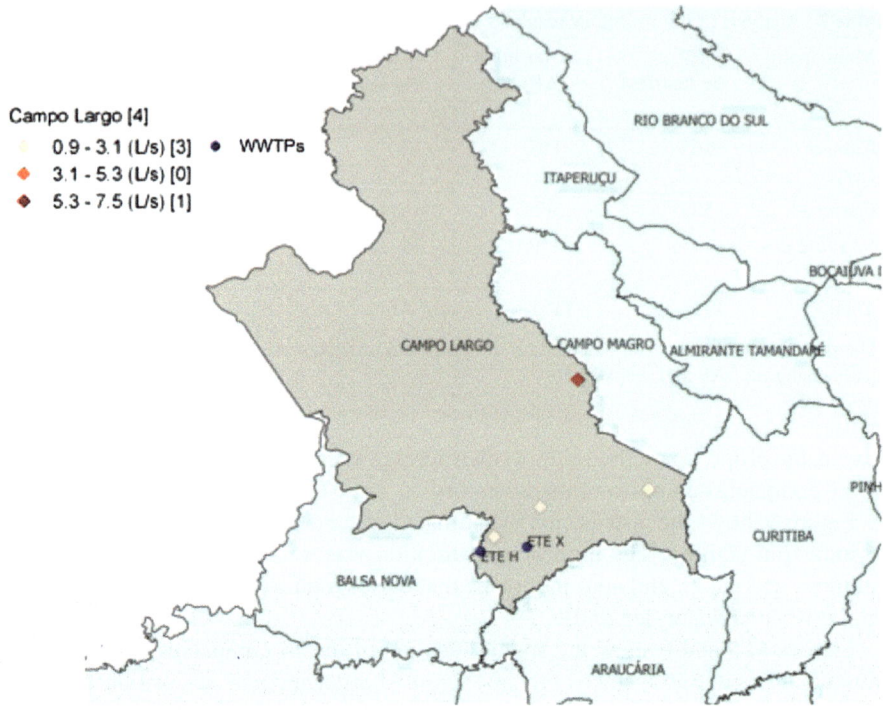

Fig. 4 Location of industries with high water demand and WWTPs in Campo Largo

was adopted, according to the method proposed by Osório (2013), data overlaps occurred, meaning that some companies were located within two or more radii. It was therefore decided to reduce the radius to 5 km; in cases where such overlaps still occurred, the company was assigned to the WWTP that presented the highest effluent flow available for serving the industrial sector.

The 5 km radius around each WWTP was traced using the buffers feature, provided by the MMQGIS snap-in. The intersection functionality was applied in order to link the industries to the WWTPs, The results obtained are presented in Table 9, which indicates the number of industries corresponding to each WWTP, in descending order of estimated water demand.

Assuming a potential 15% effluent reuse rate by the 36 industries selected in this study, and in agreement with the results achieved by several authors, it was concluded that 9 WWTPs possess sufficient flow to meet the industrial water demand. This represents a total average savings in drinking water in the order of 10.3 L s^{-1} for the municipalities studied; however, this percentage may be higher (30–40%) for industries related to the manufacture of cellulose and paper, drinks, and food processing from animal sources (Osório 2013; Kossar et al. 2013; Thoren et al. 2012; Viaro 2007).

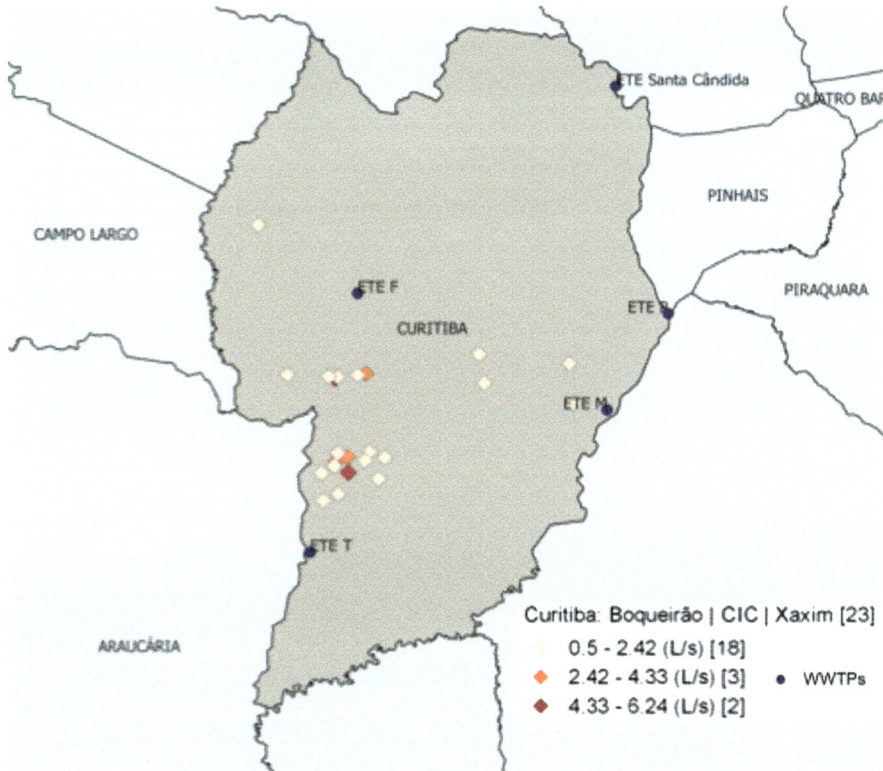

Fig. 5 Location of industries with high water demand and WWTPs in Curitiba (districts of Boqueirão, Cidade Industrial and Xaxim)

4 Conclusion

The research carried out for this work resulted in a structured inventory of 935 transformation companies registered in the FIEP database (Federation of Industries of Paraná State) and located in the municipalities of São José dos Pinhais, Campo Largo and Curitiba (districts of Boqueirão, Cidade Industrial and Xaxim only) in the Metropolitan Region of Curitiba. From this total, 44 companies may exhibit water demands in excess of 0.5 L s^{-1} or more, according to the water consumption estimation method developed by the Brazilian Water Agency (ANA) for industrial processes, these companies being considered the high-volume consumers for the purposes of this study.

The authors recommend the following for future investigation:

- an in-depth survey of methodologies for estimating the potential of replacing water with WWTP effluent for various industrial sectors, in order to refine the results obtained in this work.

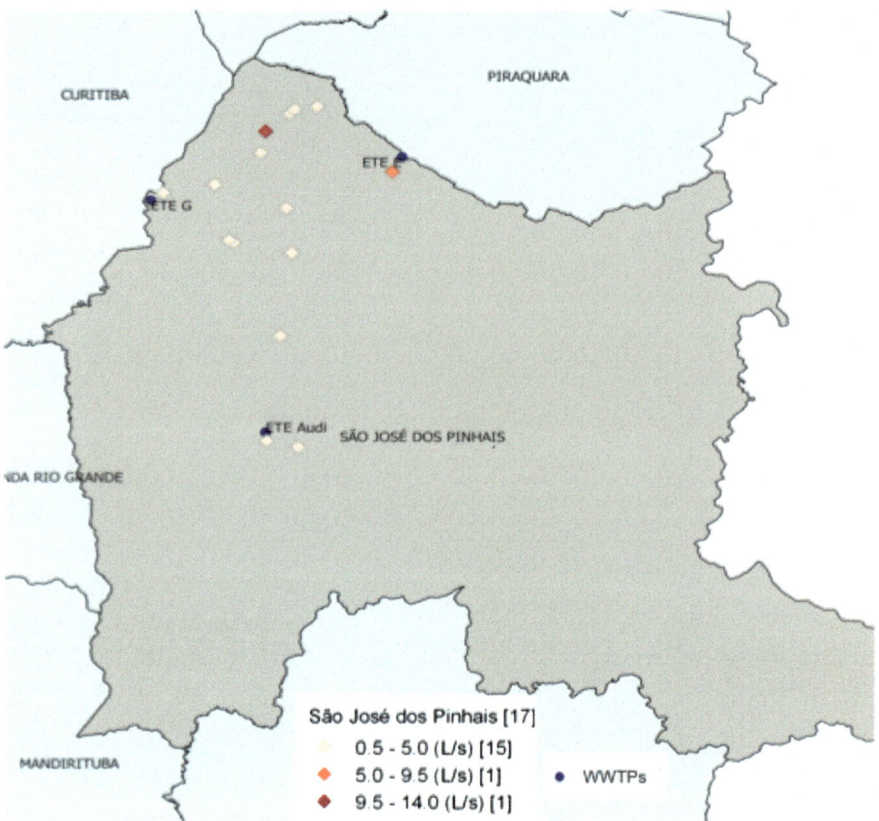

Fig. 6 Location of industries with high water demand and WWTPs in São José dos Pinhais

- a survey or the elaboration of techniques to estimate the water demand in those companies that may not be classified as processing industries but are included in the FIEP database, such as civil construction.
- a follow-up investigation in order to search for criteria to determine the minimum flow, since it would help in discerning which values are representative.
- a verification of the data available in the FIEP register, by accessing the company website or contacting it directly, in order to ensure a greater degree of reliability for the inventory.

Financial support This study was financially supported by the Araucaria Research Foundation of Parana State (Fundação Araucária de Apoio ao Desenvolvimento Científico e Tecnológico do Paraná), Brazil.

Table 8 Data for companies presenting a high estimated water demand in the three municipalities

CNAE codes	No. of industries	$Q_{ind\_consumption}$ (L s^{-1})
1011-2/05—Slaughterhouse (excluding the slaughter of pigs)	1	1.12
1069-4/00—Grinding and manufacture of vegetable products	2	2.90
1099-6/99—Manufacture of other food products	3	4.41
1111-9/02—Manufacture of other spirits and distilled beverages	1	0.61
1112-7/00—Manufacture of wine	1	1.33
1121-6/00—Manufacture of bottled water	1	7.47
1122-4/01—Manufacture of soft drinks	1	0.61
1323-5/00—Weaving of artificial and synthetic fibers	1	0.84
1340-5/01—Stamping and texturing (yarns, fabrics, textile articles, garments)	1	0.91
1354-5/00—Manufacture of special fabrics, including artifacts	1	4.09
1359-6/00—Manufacture of other textiles	1	1.34
1721-4/00—Manufacture of paper	1	2.31
1731-1/00—Manufacture of paper packaging	1	1.74
1732-0/00—Manufacture of paper and paper products	1	1.51
1742-7/01—Manufacture of disposable diapers	1	6.24
1742-7/99—Manufacture of paper products for household and sanitary purposes	1	0.76
2029-1/00—Manufacture of organic chemicals	1	3.03
2033-9/00—Manufacture of elastomers	1	5.12
2063-1/00—Manufacture of cosmetics, perfumery and toilet preparations	2	14.63
2311-7/00—Manufacture of flat glass and security glass	1	1.05
2319-2/00—Manufacture of glassware	1	0.78
2330-3/01—Manufacture of prefabricated reinforced concrete structures	1	0.63
2330-3/02—Manufacture of other fabricated metal products	1	0.63
2330-3/99—Manufacture of concrete articles, cement, asbestos-cement, gypsum	1	3.10
2449-1/99—Metallurgy of non-ferrous metals and their alloys	2	1.01

(continued)

Table 8 (continued)

CNAE codes	No. of industries	$Q_{ind\_consumption}$ (L s^{-1})
2599-3/99—Manufacture of fabricated metal products	3	2.40
2621-3/00—Manufacture of computer equipment	1	2.53
2622-1/00—Manufacture of computer peripherals	1	0.51
2731-7/00—Manufacture of electricity distribution, control apparatus	1	0.91
2751-1/00—Manufacture of equipment for domestic use (parts and accessories)	1	1.16
2790-2/99—Manufacture of electrical equipment, not specified elsewhere	1	0.58
2910-7/01—Manufacture of cars, vans and SUV's	1	5.22
2941-7/00—Manufacture of parts and accessories for motor vehicles	1	2.99
2949-2/99—Manufacture of other parts and accessories for motor vehicles not	3	3.18
3250-7/01—Manufacture of instruments for medical, dental and laboratory use	1	1.09
Total	44	88.73

Table 9 Number of industries and corresponding flow associated with each WWTP and estimated water demand

WWTP	No. of industries[a]	Flow (L s^{-1})	$Q_{ind\_consumption}$ (L s^{-1})
M	10	1195.3	25.54
T	9	458.9	17.38
F	6	410.3	11.23
E	1	8.0	5.2
G	2	38.0	4.89
P	3	1150.8	4.24
X	2	41.0	2.24
Audi	2	Unknown	1.40
H	1	5.9	0.91
Total	36	3308.2	68.91

[a]Located within 5 km of the WWTPs

References

ANA, Brazilian Water Agency (2017) Água na indústria: uso e coeficientes técnicos. ANA, Brasília. Retrieved from http://arquivos.ana.gov.br/imprensa/noticias/20170920_AguanaIndustria-UsoeCoeficientesTecnicos-VersaoFINAL.pdf

Dunhu C, Zhong M, Xin W (2013) Framework of wastewater reclamation and reuse policies (WRRPs) in China: comparative analysis across levels and areas. Environ Sci Policy 33:41–52

FIEP, Parana State Industry Federation (2017) Cadastro das Indústrias 2017. EBGE, Curitiba, 2017. 1 CD-ROM

Google (2018a) Google My Maps. www.google.com/intl/pt-BR/maps/about/mymaps/

Google (2018b) Google Earth Pro for computers. Retrieved from www.google.com/intl/pt-BR/earth/desktop/

IAP, Parana Water Institute (2017a) Outorgas de Captações. IAP, Curitiba. Retrieved from www.aguasparana.pr.gov.br/arquivos/File/OUTORGAS_DOWNLOAD/Out_Captacao_20171003.zip

IAP, Parana Water Institute (2017b) Cadastro de Usuários de Captações dispensados de outorga. IAP, Curitiba. www.aguasparana.pr.gov.br/arquivos/File/OUTORGAS_DOWNLOAD/DUI_Captacao_20171212.zip

IBGE, Brazilian Institute of Statistics and Geography (2016) Malha municipal do Estado do Paraná. Brasília, DF. Retrieved from http://servicodados.ibge.gov.br/Download/Download.ashx?u=geoftp.ibge.gov.br/organizacao_do_territorio/malhas_territoriais/malhas_municipais/municipio_2016/UFs/PR/pr_municipios.zip

Kossar MJ, Amaral J, Erbe MCL (2013) Proposal for water reuse in the Kraft pulp and paper industry. Water Pract Technol 8(3–4)

Machado CH, Bilotta P, Amaral KJ (2018) Concepção de inventário dos setores industriais com potencial de demanda de reúso de esgoto tratado. In: III Symposium of urban and industrial environment, Federal University Federal of Paraná, Curitiba

Metcalf and Eddy, Inc. Reviewed by Tchobanoglous G, Franklin L, Burton H, Stensel D (2004) Wastewater engineering: treatment and reuse, 4 edn. McGraw-Hill, New York

Mierzwa JC, Hespanhol I (2005) Água na indústria – Uso racional e reúso. Oficina de Textos, São Paulo

Osório RCF (2013) Reúso não potável de efluentes domésticos: proposta para Pólo Industrial da região sul do Brasil. Master Thesis, Graduate Program in Industrial and Urban Environment. Federal University of Paraná, Brazil

Thoren R, Atwater J, Berube P (2012) A model for analyzing water reuse and resource potential in urban areas. University of British Columbia, Canada

UNDP, United Nation Development Programme (2019) Background on the goals. Retrieved from www.undp.org/content/undp/en/home/sustainable-development-goals/background.html

United Nations Educational, Scientific and Cultural Organization—World Water Assessment Programme (2017) Wastewater the untapped resource. The United Nations World Water Development Report. Retrieved from http://unesdoc.unesco.org/images/0024/002475/247552e.pdf

Viaro VL (2007) Critérios para reúso de água de indústrias: aproveitamento do efluente da estação de tratamento de esgoto do Piçarrão, Campinas – SP. Doctorate Thesis, State University of Campinas, Brazil

Water Research Group (2009) Charting our water future: economic frameworks to inform decision-making. Water Research Group. Retrieved from https://goo.gl/G0Urr1

Wilcox J, Nasiri F, Bell S, Rahaman S (2016) Urban water reuse: A triple bottom line assessment framework and review. Sustain Cities Soc 27:448–456

Carlos Henrique Machado Graduate in Civil Engineering from Positivo University.

Patrícia Bilotta Master's and Ph.D. in Hydraulics and Sanitation from the University of São Paulo (EESC-USP), specialist in sustainable projects and climate change mitigation from the Federal University of Paraná (UFPR). Full Professor in Master and Doctorate Program in Environmental Management (PGAMB), Positivo University (UP), Curitiba (Brazil). Full member of the Paraná (State) and Curitiba (Municipality) Climate Change Forum. Topics of interest: sustainable effluent treatment plants (energy recovery and effluent reuse), low-carbon economy (energy efficiency and GHG emission management), and circular economy.

Karen Juliana do Amaral Master's and Ph.D. in Water Resources from the Federal University of Rio de Janeiro (COPPE/UFRJ). Researcher in the Institute of Sanitary Engineering, Water and Waste Quality at the University of Stuttgart (Germany), and professor in Master's Program in Urban and Industrial Environment (MAUI) at the Federal University of Paraná, SENAI and the University of Stuttgart. Topics of interest: technologies for effluent treatment, industrial water reuse, water resources legislation, computational modeling for streams.

Selection of Best Practices for Climate Change Adaptation with Focus on Rainwater Management

Jessica Andrade Michel, Giovana Reginatto, Janaina Mazutti, Luciana Londero Brandli and Rosa Maria Locatelli Kalil

Abstract The adaptation of cities, whether large, medium or small, in relation to extreme weather events, is a great challenge. The society development culminated in urbanization, often without planning and with great concentration of the population in the urban environment. Across the world, this process has increased tensions between urban and natural environments, with serious consequences for human well-being and for the efficiency of infrastructure systems. Hydrological events are one of the main concerns of cities in many parts of the world. Heavy rains, with large daily or monthly accumulations, usually results in floods at the urban environment. These situations, in addition to generating economic losses, endanger the life of the population. In this scenario, this paper aims to identify structuring and non-structural measures used by cities around the world, to sustainable development, through strategies that make them less vulnerable to climatic events, which can be incorporated into urban governance, in order to contribute to adapt to climate change and to build cities that are more resilient. The main practices identified were Low Impact Development, Water Sensitive Urban Design, environmental sanitation legislation and events focused on awareness of the importance of urban water management.

Keywords Rainwater management · Best practices · Climate change

J. A. Michel (✉) · G. Reginatto · J. Mazutti · L. L. Brandli · R. M. L. Kalil
Graduate Program in Civil and Environment Engineering, University of Passo Fundo, Campus I, BR 285, Passo Fundo, RS 99052-900, Brazil
e-mail: arq.jessicamichel@gmail.com

G. Reginatto
e-mail: gioreginato@gmail.com

J. Mazutti
e-mail: janainamazutti@gmail.com

L. L. Brandli
e-mail: brandli@upf.br

R. M. L. Kalil
e-mail: kalil@upf.br

© Springer Nature Switzerland AG 2020
W. Leal Filho et al. (eds.), *International Business, Trade and Institutional Sustainability*, World Sustainability Series,
https://doi.org/10.1007/978-3-030-26759-9_54

915

1   Introduction

The process of developing society and the urbanization of cities without adequate and sustainable planning throughout the world has increased tensions and environmental imbalance, generating consequences that interfere with human well-being and the sustainability of global communities. Associated with the unbridled urbanization process is the population growth in considered risk areas. The increase in urban population has been corroborated by the expansion of suburban areas and resulting in segregated, degraded urban environments with serious environmental problems.

Moura et al. (2009) state that throughout the history of humankind there has always been a connection between cities and watercourses, which are crucial for their very existence. The proximity of urban agglomerations and rivers was an important development factor, due the provided water for supply, possibilities for waste disposal, ensure of transportation, energy and military protection, among other factors (Miguez et al. 2014).

However, human presence has caused many disturbances in these ecosystems. The process of urbanization of these areas tends to alter the urban hydrological cycle and river systems tend to flood built environments that were previously much more permeable to rainwater (Miguez and Veról 2017).

The risk of periodic flooding was relatively well accepted until the mid-nineteenth century, and it was considered a "price to pay" for the availability of water near the city (Baptista et al. 2005). The increase of urban agglomerations from the nineteenth century caused the first disturbances resulting from the precarious infrastructure of rainwater and wastewater. The lack of rain infrastructure and unhealthy conditions led to the emergence of the great epidemics of cholera and typhus that plagued Europe at this time. In view of this, the search for hygiene began, which advocated the evacuation, as far as possible, of water of any nature present in urban areas (Moura et al. 2009).

Thus, water and its relations with urban spaces have been gaining attention from public management regarding the use and occupation of the soil. Yet the threats and challenges that extreme events have to do with water, concern and mobilize local and global actions.

As an aggravating factor of the hydrological events in the urban environment, the climatic change stands out. Despite the efforts of several global climate conventions, annual global emissions of carbon dioxide from fuel combustion grew by about 38% between 1990 and 2009, with the fastest growth rate after 2000 than in the 1990s. Therefore, it still cannot be said that the world is actively taking action to combat climate change (United Nations 2012).

In the same direction follow the public policies for adapting cities in relation to natural disasters, especially in developing countries, the effective implementation of water resources management instruments should promote a more sustainable reality in cities, which is not currently observed (Peixoto et al. 2016).

Given the global relevance of the climate change discussion, the United Nations Sustainable Development Goals (SDG) include combating climate change through its SDG 13. The SDG 13 has as its first goal "Strengthen resilience and adaptive capacity to climate-related hazards and natural disasters in all countries" and one of its indicators is the number of countries with strategies of disaster risk reduction (United Nations 2018).

Within this context, this paper aims to highlight the successful strategies applied to cities for better adaptation to hydrological events and climate change.

2   Climate Change and Cities

In the case of environmental disasters, the World Bank, in the document "Cities and Floods", states that flooding is the most frequent of all natural disasters (Jha et al. 2012).

The IPCC report (2012) indicates that there will be a likely frequency of heavy precipitation or the proportion of total precipitation of heavy rains will increase throughout the 21st century in many areas of the globe. Heavy rains associated with tropical cyclones are likely to increase with continued warming.

The impacts of climate change range from economic losses to health risk to population. Events with the most significant impacts are already perceived and are classified as direct effects of climate change, including hurricanes, storms and floods—especially in urban areas—hurricanes, landslides and heat waves. The indirect ones are: the loss of agricultural production, causing nutritional impact, and a decrease in personal and environmental hygiene standards (PBMC 2016).

With regard to hydrological impacts, floods cause the greatest disturbances in the urban environment, affecting buildings, urban infrastructure, depreciating the housing market in flood areas, disrupting public services and transport, spreading waterborne diseases. In addition, it can cause injuries and deaths, among other aspects (Miguez and Magalhães 2010).

Disorganized urbanization without planning and without supervision makes cities more sensitive to the obsolescence of their drainage systems, intensifying the occurrence of urban floods. Thus, traditional drainage systems are not the most adequate solution for these cities, and it is necessary to combine them with more sustainable drainage alternatives capable of accompanying the city's development process, maintaining its functionality (Miguez et al. 2015).

3   Methodology

The methodology of this study consists of the classification of good practices and subsequent identification of cities where they have been applied to serve as case studies.

The World Bank in partnership with the Global Fund for Disaster Reduction and Recovery says that an integrated approach to urban flood risk management is a combination of measures that can, on the whole, successfully reduce the risk of urban flooding. As a result, these institutions propose a division of flood management measures in the document Cities and Floods (2012), and the result is the classification of measures into structural and non-structural measures.

Structural measures aim to reduce the risk of flooding by controlling the flow of water both outside and inside urban settlements. These measures range from heavy engineering works and structures such as flood defenses and drainage channels to the most natural and sustainable complementary or alternative measures such as wetlands and natural buffers.

Non-structural measures, however, are linked to people's capacity building in dealing with hydrological events in their environments. Non-structural measures can be categorized into four basic objectives (Jha et al. 2012):

(a) Emergency planning and management including alert and evacuation;
(b) Increased preparedness through awareness-raising campaigns, which include urban flood risk reduction management procedures, such as maintenance of clean sewage piping through improved waste management;
(c) Conditions to prevent flooding by land use planning as observed in the legislation;
(d) Acceleration of the recovery and use of post-flood to increase resilience through the improvement of construction projects and the construction itself.

Alternative urban drainage practices have various nomenclatures that are used around the world, depending on the region and country where they are employed. The best known are the Low Impact Development (LID), Rainwater Measurement (SCM) and Best Management Practices (BMP) practices in the USA, the Water Sensitive Urban Project (WSUD) in Australia and Urban Drainage Systems (SUDS) in Europe (Eckart et al. 2017; Macedo et al. 2019).

Structural and non-structural measures do not oppose each other, and more successful strategies must combine both types. For this reason, the practices were divided into two groups and subdivided, as shown in Table 1, in order to facilitate the selection of best practices in each area.

Table 1 Best practices

Type	Practices
Structural measures	Low impact development practices—LID
	Water-sensitive urban design—WSUD
Non-structural measures	Legislation on environmental sanitation or specifically on urban drainage
	Events focused on awareness of the importance of urban water management

3.1 Low Impact Development—LID

Low impact development is a method of rainwater management based on the simulation of natural hydrological conditions, not causing changes in urban hydrological characteristics. This practice uses ecological measures, source control and distributed control measures to manage and use rainwater (Seo et al. 2017).

The approach focuses on maintaining or restoring a site's natural hydrological processes, providing opportunities for the occurrence of natural processes.

The principles of low-impact development include preservation of natural site resources, integrated small-scale rainwater management controls, minimizing and disconnecting impervious areas, controlling rainwater as close to its source as possible, flow times and create multifunctional landscapes.

The best LID management practices are techniques that rely on natural processes to manage the quantity and quality of water, including: absorption; infiltration; evaporation; evapotranspiration; filtration through plant material and soil layers; potential collection of pollutants by selected vegetation; biodegradation of pollutants by microbial soil communities.

The properties of natural materials such as soil, gravel, vegetation and mulch reduce the volume and peak flow rates that reach the streams and increase the quality of the rainwater entering the water bodies.

There are three main rainwater management objectives that typically target low-impact development applications. These are: rainwater volume control; peak flow control of rainwater; and improving the quality of rainwater.

3.2 Water Sensitive Urban Design—WSUD

The urban green drainage infrastructure, known as WSUD—Water Sensitive Urban Design, is increasingly being implemented in cities around the world to combat the effects of climate change and urbanization (Kuller et al. 2018).

This practice, created in Australia, aims to integrate urban planning with the management, protection and conservation of the urban water cycle, to protect and restore natural waterways, to reduce the risk and severity of floods, and to diversify sources of water supply (Dietz 2007; Wongand Brown 2009).

The use of WSUD provides solutions for urban water supply, wastewater and rainwater in a more economical way by introducing "green" technologies distributed in the urban landscape, causing less damage to the environment (Kuller et al. 2018). Among its main objectives are:

- Reducing demands for drinking water through more efficient uses of water, in addition to a more efficient approach to alternative water sources;
- Reduction of the generation of effluents, and treatment of these before being sent to receiving water bodies;

- Treatment of rainwater to meet quality standards when reused or dumped into receiving water bodies;
- Restore or preserve the natural hydrological regime of the river basins;
- Improve the health of water bodies;
- Improve the aesthetics and connection between water resources and the residents of the areas in which the systems are adopted.

3.3 Legislation on Environmental Sanitation or Specifically on Urban Drainage

This practice aims to ensure the benefits of environmental health to the entire population, through articulated action between the Union, the States, and the Municipalities.

Environmental sanitation is the set of socioeconomic actions that aim to achieve increasing levels of environmental health, through the supply of drinking water, collection and sanitary disposal of liquid, solid and gaseous wastes, promotion of sanitary discipline of land use and occupation, urban drainage, and vector and communicable disease control to protect and improve living conditions in urban centers, rural communities and poorer rural properties.

The Municipalities should promote the organization, planning and execution of the public functions of environmental sanitation, characterized as of local interest.

Legislation should be drawn up on the basis of a diagnosis that demonstrates the reality of local systems of urban drainage, collection of waste, water supply, among others. From this point on, improvements are proposed in these areas, and laws guarantee the creation of tools for these improvements to be implemented.

3.4 Events Focused on Awareness of the Importance of Urban Water Management

The events focused on raising awareness of the importance of urban water management aim to discuss solutions to urban drainage problems through: understanding the challenges, analyzing the necessary solutions and identifying investment opportunities in infrastructure.

The events also seek to create capacities so that the population can deal with hydrological disasters and act preventively so that there is a minimization of the negative effects of these occurrences.

The four subgroups bring practices with similar goals and impacts. For each practice, case studies are presented that help to understand the implementation of these initiatives.

The selection of good practices occurred through research on websites such as Sustainable Cities, WRI Cities, New Water Ways, among others. Besides these sources, the search was also made in scientific articles on the subject, which were selected searching for the keywords: good practices in rainwater management, urban design or water sensitive design and low impact development. The search encompassed a worldwide context, only restricting the period of time, where the selected good practices were implemented from the year 2000.

4 Results

Throughout this study, it was possible to observe that the LID and WSUD practices make use of the same tools of control, vegetative swales, bioretention cell, and permeable paving, green roof, among others. However, there are small differences between the two practices. The WSUD involves more initiatives aiming to reduce pollutants, water supply and reuse of rainwater. The LID practice also aims to reduce the pollutants that infiltrate the groundwater, but this practice uses more techniques to improve urban drainage and reduce floods, with the reduction of pollutants as a positive additional impact.

4.1 Good Practices for Storm Water Management

Table 2 presents good practices that contribute to urban drainage in cities, as well as their impacts and the places where they have already been implemented. The practices were selected and classified according to methodology proposed by Jha et al. (2012) as structural and non-structural measures. The application of these measures is observed through the case studies.

In order to better visualize, Fig. 1 presented the location of cities where good practices were implemented around the world.

4.2 Low Impact Development Applications—LID

Researchers have already demonstrated the efficiency of the LID system in their studies. Dietz (2007) performed an experimental comparison using LID based devices such as swales, biorretention cell and permeable paving. The results showed that the LID devices can decrease drainage area flow, and that the green roof can decrease 63% of the direct flow. Alfredo et al. (2009) also carried out experiments for the green

Table 2 Good practices for storm water management

Type	Practice	Impact	Case study
Structural measures	Low impact development—LID	Assistance in the flow of rainwater in a natural and sustainable way, reducing the occurrence of hydrological events	• Kuala Lumpur, Malaysia • Pune, India • Los Angeles, USA • Long Island, USA • Edmonton, Canada
	Water sensitive urban design—WSUD	Rainwater harvesting reduces the transport of pesticides, fertilizers, oils, metals, salt, bacteria, rubbish, and other materials into the water table; also helps to reduce the use of drinking water that has no consumption purpose	• Berlim, Germany • Rio de Janeiro, Brazil • Melbourne, Australia • Pune, India • Perth, Australia • Edmonton, Canada
Non-structural measures	Legislation on environmental sanitation or specifically on urban drainage	The measures implemented by the environmental sanitation legislation contribute to the reduction of floods and floods, and also has an impact on the public health of a local population	• Porto alegre, Brazil • Guarulhos, Brazil • Curitiba, Brazil • Indianapolis, USA • USA • France
	Events focused on awareness of the importance of urban water management	Knowledge about natural disasters helps the population to make decisions in situations of risk, reducing the number of homeless, wounded and dead	• Global • Porto Alegre, Brazil • Saijo, Japan • Doha, Qatar • Bonn, Germany

Source Environment Canada (2014), Ahiablame et al. (2013), Dietz (2007), Wong and Brown (2009)

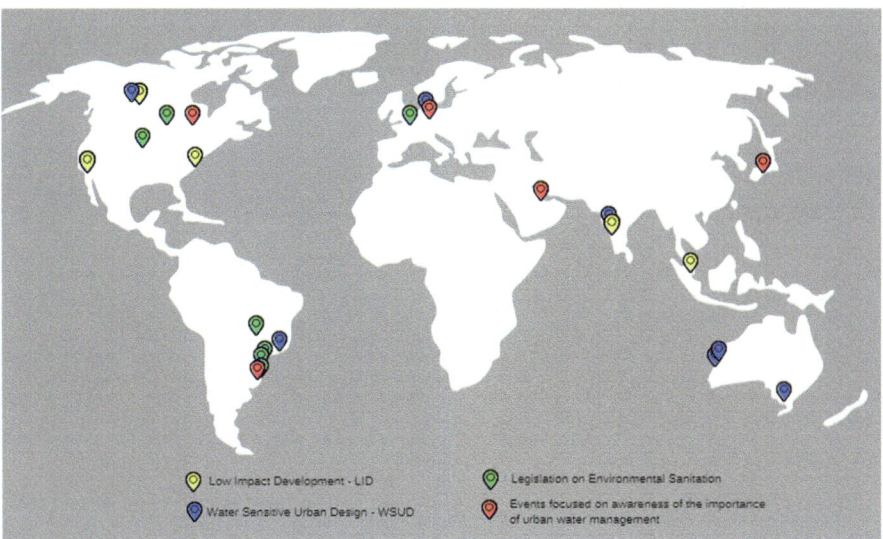

Fig. 1 Distribution of case studies of each best practice

roof, obtaining satisfactory results, which show that the magnitude of the temporal arrest and the flood of the peak flow can be promoted between 22 and 70%.

Liu et al. (2014) evaluated and quantified the peak and flood flows reduced by the Green Infrastructure (GI). The results showed that the flood reduction capability of a single device was very limited. However, through the layout of the general devices, the total rainfall flow of different return periods was reduced by 85–100%, and peak flows also decreased by 92–100%.

Different management objectives need to be stipulated for different levels of development (Huang et al. 2018). According to the results obtained by Fresno et al. (2005), urban drainage should be performed through a series of engineering solutions rather than a single device.

Below, there are the practices found for low impact development (Table 3).

4.3 Water Sensitive Urban Design—WSUD

The benefits of WSUD in sustainable water management and drinking water economy have been proven by several studies.

In the city of Rockingham, Australia Evermore Heights is a blending of 374 urban plots with urban water-sensitive design. In this allotment the residents have access to a double system of water supply. One incentive package in particular included a 3000 L rainwater tank that channels non-potable water to the laundry room and toilets. Project guidelines include the requirement for permeable paving, and the use

Table 3 Best low impact development practices LID

Year/Place	Description	Impact
2007/Kuala Lumpur, Malaysia	Road tunnel and storm management, with the purpose of draining large volumes of flood water to a reservoir (UNISDR 2012)	Reducing the risk of floods and urban floods, consequently reducing the number of displaced persons and public expenses in assisting the homeless of the floods
2010/Pune, India	Municipal action plan for the restoration of natural drainage, aiming to apply methodologies for natural infiltration of water in the soil such as streams expansion and vegetation conservation (UNISDR 2012)	Reducing the risk of floods and urban floods, which consequently reduces the number of displaced persons and public expenditure on assistance to the homeless
2011/Los Angeles, USA	Implantation of rainwater cisterns and barriers; drains with metal grids; permeable floors. Objective, control floods and reduce pollution (United States 2018a)	Reduction of water use costs due to possible reuse. In addition, there is reduced risk of flooding and reduced pollution
2013/Long Island, USA	Hunter's Point South linear park. Green infrastructure for retention and absorption of rainwater, searching for spaces for water infiltration without using conventional drainage systems (Strengari et al. 2015)	Reducing the risk of floods and urban floods, which consequently reduces the number of displaced persons and public expenditure on assistance to the homeless
2014/Edmonton, Canada	Gardens to infiltration of rainwater, drainage ditches with vegetation cover, green roof and permeable pavements. Provide better infiltration of rainwater (Environment Canada 2014)	Reducing the risk of floods and urban floods, which consequently reduces the number of displaced persons and public expenditure on assistance to the homeless. Saving water through possible reuse

of plants that do not require much water in landscaping. The result obtained by these initiatives was positive, generating a saving of 68% of drinking water (New Water Ways 2012).

Gilroy and McCuen (2009) also analyzed the effects of location and quantity of bioremediation cisterns and wells. A model was developed to determine the effects of these practices on single-family, residential, and commercial lots. The results show that tanks alone are able to control the runoff of small storms.

Table 4 lists more cities that use WSUD for rainwater management.

4.4 Implementation of Legislation for Environmental Sanitation

Table 5 describes legislation aimed at environmental sanitation in the cities where they are in force.

4.5 Events Focused on Awareness of the Importance of Urban Water Management

Immediate loss of lives due to flooding is increasing more slowly or even decreasing over time, reflecting the positive impact of implementing disaster risk management actions (Jha et al. 2012).

Despite this, fatalities are still significant in developing countries, where flooding has a greater impact on the low-income population, particularly women and children.

Listed below are initiatives that discuss the importance of stormwater management and the need to create capacity for the population to be able to act in hazardous situations (Table 6).

5 Conclusions

Alternative drainage systems are sustainable measures that can bring benefits to small and long-term if implemented in cities. Many measures do not represent such significant expenditure and significantly improve the drainage of rainwater. However, it is of great importance that there be a combination of structural and non-structural measures. So, that the population understands the changes that will occur in the dynamics of cities. The change in the behavior of the population in relation to the hydrological events is of great relevance, since knowing how to act in situations of risk can represent the difference between life and death.

Table 4 Good practices in water-sensitive urban design—WSUD

Year/Place	Description	Impact
2004/Berlim, Germany	Roof with rainwater capture installed at the Max Planck Institute of Physics (Rola et al. 2016)	Reduction of floods. Exploitation of rainwater
2006/Rio de Janeiro, Brazil	Water catchment in the "Samba City". Rainwater harvesting and storage in reservoirs in the internal and external cleaning depots and for the fire-fighting system (Rola et al. 2016)	Reduction of expenses with the use of drinking water due to possible reuse; reducing the risk of hydrological events
2009/Melburne, Australia	Rainwater tanks. Rainwater tanks collect stormwater runoff from impermeable surfaces such as roofs, reducing the amount entering the waterways. They are equipped with an overflow mechanism, which means that when a tank is full, excess water is redirected to the rainwater drainage system (Dietz 2007; Wong and Brown 2009)	Minimize the use of water when used in the bathroom, laundry or garden; reduce tension in the rainwater drainage system; retain water near the source; reduce site runoff and flood spikes
2010/Pune, India	Tax incentives to properties. Objective motivate owners to store rainwater for domestic use (UNISDR 2012)	Access to water improves the HDI; preservation of water resources; collaborates in minimizing water demand; and provides underground recharge and groundwater maintenance
2011/Perth, Australia	Refill of Cottesloe Aquifers. Drained stormwater in Perth was launched into the sea by ten underground pipes, these pipes were closed and rainwater now supplies the water table. In case of extreme rain event the pipes can be opened to prevent the occurrence of flooding. For this reason it was necessary (New Water Ways 2012)	Sustainable management of rainwater, avoiding the loss of 180 thousand liters of rainwater that is annually released into the sea and now supplies the water table. Reduction of floods
2014/Edmonton, Canada	Rainwater harvesting with cistern at the University of Alberta (Environment Canada 2014)	Reduce the volume and flow rates of rainwater while storing water for irrigation

Table 5 Good practices focused on legislation

Year/Place	Description	Impact
1999/Porto Alegre, Brazil	Master Plan that introduces articles on urban drainage. Specifies the need to reduce the outflow of rainwater in critical areas due to the degree of urbanization. To this end, it provides for the holding of this water (Tucci and Silveira 2001; Brazil 2010)	It defines the risk areas for occurrence of hydrological events, and assists in the non-occupation of these areas, reducing the number of people affected, such as homeless, dead and injured
2000/Guarulhos, Brazil	Code of Works. Mandatory (for residences with at least 1 hectare) rainwater detention system for reuse (Souza 2012)	Decrease of rainwater in catchment areas, reducing the number of floods, floods or floods
2003/Curitiba, Brazil	Program for the Conservation and Rational Use of Water in Buildings (PURAE) seeks to encourage the rational use and conservation of water, as well as the use of alternative sources of rainwater harvesting (Brazil 2016)	Reduces runoff through water storage, avoiding urban drainage problems, and promotes the conscious use of potable water
2013/IndianapolisUSA	Stormwater Credit Manual (Manual de Crédito de Águas Pluviais). Oferece créditos para quem utiliza a água da chuva, diminuindo gastos com o consumo da água potável (United States 2016)	Encourage environmental education; reduce water consumption
2014/USA (National)	Best Management Practices—BMPs, best practices for urban drainage, the United States Environmental Protection Agency (EPA), requires US municipalities to be adept at practices. These practices address critical criteria for the management of urban stormwater such as the volume of rainwater entering the sewer system and the maximum flow rate for the combined system (USEPA 2014)	It promotes the natural movement of water rather than allowing it to be carried in storm drains, reduces the amount of pollutants entering the water collection system, increases water quality and improves the water treatment process at reduce the risk of flooding
2015/France (National)	France creates environmental legislation where it decrees that roofs of new commercial buildings be covered with solar panels and plants. This practice aims to retain rainwater, and clean energy generation (UNFCCC 2015)	The impact of this practice is related to reduced flow, avoiding urban drainage problems and also helps to save energy, either by the solar panels or the thermal inertia that the green roof provides

Table 6 Events that discuss the importance of urban water management

Year/Place	Description	Impact
2001–2018 Global	International Day of Disaster Reduction. It aims to build a culture of risk reduction (UNESCO 2011)	Promote education to increase the resilience of communities as a whole. Education to reduce disaster risks is important to create new ways of thinking and new practices that allow societies to respond and adapt to the pressures of climate change
2017/Prague, Czech Republic	14th International Conference on Urban Drainege (ICUD) It aimed to present the latest advances and innovative approaches in fundamental and applied research on urban drainage, taking into account meteorological, hydrological, hydraulic, water quality and socioeconomic aspects worldwide. (ICUD 2017)	Discuss and share innovative alternatives to urban drainage. At the end of the event is generated a document compiling the best solutions for urban drainage, so that these can be applied in cities
2012/Saijo, Japan	Children and Communities Study Mountains and Urban Risks. Soon in kindergarten, Japanese schools embed children in education on how to identify and react to disasters, performing simulated routines and 'watching disasters' (UNISDR 2012)	This long-term investment has undoubtedly saved many lives, as the aging of the city's population presents a particular challenge. Young people, who have more physical strength, are very important for community systems of mutual aid and emergency preparedness
2016/Doha, Qatar	3rd International Conference on Sustainable Urban Drainage Systems—Middle East. This conference provides a platform for stakeholders to gain valuable insight into current and emerging urban water infrastructure projects in Qatar and throughout the Middle East region (ICSUDS 2016)	Discusses solutions for the management of rainwater in urban areas, providing infrastructure project sharing as well as improvement. The implementation of these projects reduces floods, floods and urban floods
2010–2018/Bonn, Germany	Forum of Adaptation of Mayors. Each year, mayors meet to discuss issues such as urban risk management, urban logistics for resilience, funding for resilient cities, food security in cities, intelligent infrastructure for resilience, and others (RCC 2016)	Opportunity to share experiences on urban risk management in order to build more resilient cities

References

Ahiablame LM, Engel BA, Chaubey I (2013) Effectiveness of low impact development practices in two urbanized watersheds: Retrofitting with rain barrel/cistern and porous pavement. J Environ Manag 119:151–161. https://www.sciencedirect.com/science/article/pii/S0301479713000571. Last accessed 12 July 2018

Alfredo K, Montalto F, Goldstein A (2009) Observed and modeled performances of prototype green roof test plots subjected to simulated low-and high-intensity precipitations in a laboratory experiment. J Hydrol Eng 15(6):444–457. https://ascelibrary.org/doi/pdf/10.1061/%28ASCE%29HE.1943-5584.0000135. Last accessed 12 July 2018

Baptista MB, Nascimento NO, Barraud S (2005) Técnicas Compensatórias em Drenagem Urbana. ABRH (Associação Brasileira de Recursos Hídricos), Porto Alegre, 266 p

Dietz ME (2007) Low impact development practices: a review of current research and recommendations for future directions. Water, Air, and Soil Pollution 186(1–4):351–363. https://doi.org/10.1007/s11270-007-9484-z. Last accessed 12 July 2018

Brazil (2010) Master plan of urban ENVIRONMENTAL development of Porto Alegre city. http://lproweb.procempa.com.br/pmpa/prefpoa/spm/usu_doc/planodiretortexto.pdf. Last accessed 12 July 2018

Brazil (2016) Programa de Conservação e Uso Racional da Água nas Edificações - PURAE. Curitiba, PR

Eckart K, Mcphee Z, Bolisetti T (2017) Performance and implementation of low impact development—a review. Science Total Environ 607–608:413–432. https://www.sciencedirect.com/science/article/pii/S0048969717316819. Last accessed 12 July 2018

Environment Canada (EC) (2014) Low impact development —best management practices—design guide. Report from City of Edmonton —Edmonton, Alberta. https://www.edmonton.ca/city_government/documents/PDF/LIDGuide.pdf. Last accessed 12 July 2018

Fresno DC, Bayón JR, Hernández JR, Muñoz FB (2005) Sistemas urbanos de drenaje sostenible (SUDS). Interciencia 30(5):255–260

Gilroy KL, McCuen RH (2009) Spatio-temporal effects of low impact development practices. J Hydrol 367(3–4):228–236. https://doi.org/10.1016/j.jhydrol.2009.01.008. Last accessed 12 July 2018

Huang CL, Hsu NS, Liu HJ, Huang YH (2018) Optimization of low impact development layout designs for megacity flood mitigation. J Hydrol 564:542–558. https://doi.org/10.1016/j.jhydrol.2018.07.044. Last accessed 12 July 2018)

ICSUDS (2016) 3rd international conference on sustainable urban drainage systems—Middle East. https://drainageandsewerageme.iqpc.ae/mediapartners. Last accessed 6 Nov 2018

ICUD (2017) 14th international conference on urban drainage. September 10 to 15, 2017, Prague, Czech Republic. http://www.icud2017.org/. Last accessed 12 July 2018

IPCC (Intergovernmental Panel on Climate Change) (2012) Managing the risks of extreme events and disasters to advance climate change adaptation. A special report of working groups I and II of the intergovernmental panel on climate change [Field CB, Barros V, Stocker TF, Qin Q, Dokken DJ, Ebi KL, Mastrandrea MD, Mach KJ, Plattner GK, Allen SK, Tignor M, Midgley PM (eds)]. Cambridge University Press, Cambridge, UK, and New York, NY, USA, 582 p

Jha AK, Bloch R, Lamond J (2012) Cities and flooding: a guide to integrated urban flood risk management for the 21st century. The World Bank. http://www.zaragoza.es/contenidos/medioambiente/onu/805-eng.pdf. Last accessed 12 July 2018

Kuller M, Bach PM, Ramirez-Lovering D, Deletic A (2018) What drives the location choice for water sensitive infrastructure in Melbourne, Australia? Landscape and Urban Plan 175:92–101. https://doi.org/10.1016/j.landurbplan.2018.03.018. Last accessed 12 July 2018

Liu W, Chen W, Peng C (2014) Assessing the effectiveness of green infrastructures on urban flooding reduction: a community scale study. Ecol Model 291:6–14. https://doi.org/10.1016/j.ecolmodel.2014.07.012. Last accessed 12 July 2018

Macedo MB, Lago CAF, Mendiondo EM (2019) Stormwater volume reduction and water quality improvement by bioretention: potentials and challenges for water security in a subtropical catchment. Science Total Environ 647:923–931. https://doi.org/10.1016/j.scitotenv.2018.08.002. Last accessed 12 July 2018

Miguez MG, Magalhães LPC (2010) Urban flood control simulation and management-an integrated approach. In: Pina Filho AC, de Pina AC (eds) Methods and techniques in urban engineering. InTech, Vukovar, Croatia, pp 131–160

Miguez MG, Veról AP (2017) A catchment scale Integrated Flood Resilience Index to support decision making in urban flood control design. Environ Plan B: Urban Analytics City Sci 44(5):925–946

Miguez MG, Rezende OM, Veról AP (2014) City growth and urban drainage alternatives: Sustainability challenge. J Urban Plan Dev 141(3):1–10. https://ascelibrary.org/doi/pdf/10.1061/%28ASCE%29UP.1943-5444.0000219. Last accessed 12 June 2018

Miguez MG, Veról AP, Sousa MM, Rezende OM (2015) Urban floods in lowlands—levee systems, unplanned urban growth and river restoration alternative: a case study in Brazil. Sustainability 7(8):11068–11097. https://www.mdpi.com/2071-1050/7/8/11068. Last accessed 12 July 2018

Moura PM, Baptista MB, Barraud S (2009) Avaliação multicritério de sistemas de drenagem urbana. Rega - Revista de Gestão de Água da América Latina 6:31–42

New Water Ways (2012) Australia government of western, Australia department of planning. https://www.newwaterways.org.au/projects/evermore-heights. Last accessed 12 July 2018

PBMC (Painel Brasileiro de Mudanças Climáticas) (2016) Mudanças Climáticas e Cidades. Relatório Especial do Painel Brasileiro de Mudanças Climáticas [Ribeiro SK, Santos AS (eds)]. PBMC, COPPE – UFRJ. Rio de Janeiro, Brasil. 116 p. ISBN: 978–85-285-0344-9

Peixoto F, Studart T, Campos J (2016) Gestão das águas urbanas: questões e integração entre legislações pertinentes. Revista de Gestão de Água da América Latina 13(2):160–174

RCC (2016) Resilient cities congress. Mayors adaptation forum at resilient cities. http://resilient-cities.iclei.org/index.php?id=833. Last accessed 12 July 2018

Rola SM, Silva NF, Vazquez EG (2016) Águas Pluviais e Resiliência Urbana ou os Impactos da Vulnerabilidade Hídrica em Áreas Rurais e Urbanas no Brasil. Cadernos de Pós-Graduação em Arquitetura e Urbanismo 15(1):127–154. http://editorarevistas.mackenzie.br/index.php/cpgau/article/view/2015.1.Rola/5563. Last accessed 12 July 2018

Seo M, Jaber F, Srinivasan R, Jeong J (2017) Evaluating the impact of Low Impact Development (LID) practices on water quantity and quality under different development designs using SWAT. Water 9(3):193. https://www.mdpi.com/2073-4441/9/3/193. Last accessed 12 July 2018

Souza SA (2012) Água juridicamente sustentável: um estudo da educação ambiental como instrumento de efetividade do programa de conservação e uso racional da água nas edificações de curitiba, PR. Ius Gentium 5(9):110–119. www.uninter.com/iusgentium/index.php/iusgentium/article/viewFile/51/pdf. Last accessed 12 July 2018

Strengari LAB, Kacuta LG, Dalessandro NT, Bianchi RH, Margato V (2015) Cidades Resilientes a inundações. Polytechnic School of the University of São Paulo. Department of Hydraulic and Environmental Engineering—PHA. Water in Urban Environments. http://www.pha.poli.usp.br/LeArq.aspx?id_arq=14206. Last accessed 12 July 2018

Tucci CE, Silveira A (2001) Gerenciamento da drenagem urbana. Revista Brasileira de Recursos Hídricos 7(1):5–27. http://rhama.com.br/blog/wp-content/uploads/2017/01/GEREN02.pdf. Last accessed 12 July 2018

UNESCO (United Nations Educational, Scientific and Cultural Organization) (2011) Dia Internacional para a Redução de Desastres Naturais. www.unesco.org/new/pt/brasilia/about-this-office/single-view/news/international_day_for_disaster_reduction_2011_13_october/. Last accessed 12 July 2018

UNFCCC (United Nations Framework Convention on Climate Change) (2015) France mandates green roofs. https://unfccc.int/news/france-mandates-green-roofs. Last accessed 12 July 2018

UNISDR (United Nations International Strategy for Disaster Reduction) (2012) Como Construir Cidades Mais Resilientes: Um Guia Para Gestores Públicos Locais (2005–2015). Genebra. https://www.unisdr.org/files/26462_guiagestorespublicosweb.pdf. Last accessed 12 July 2018

United Nations (2012) Resilient people, resilient planet: a future worth choosing. Secretary-General's High-level Panel on Global Sustainability. New York: United Nations

United Nations (2018) Sustainable development knowledge platform—SDG 13: targets & indicators. https://sustainabledevelopment.un.org/sdg13. Last accessed 12 Aug 2018

United States (2016) City of Indianapolis, Department of Public Works. Indianapolis Storm Water Credit Manual. http://www.indy.gov/eGov/City/DPW/StormWaterProgram/Documents/Stormwater%20Credit%20Manual%206.28.16.pdf. Last accessed 12 July 2018

United States (2018a) City of Los Angeles Stormwater Program. Low Impact Development: Stormwater Mitigation Plans SUSMP. https://www.lastormwater.org/green-la/low-impact-development/. Last accessed 12 July 2018

USEPA (United States Environmental Protection Agency) (2014) Best management practices (BMPs). https://www.epa.gov/water-research/best-management-practices-bmps-siting-tool. Last accessed 12 July 2018

Wong THF, Brown RR (2009) The water sensitive city: principles for practice. Water Sci Technol 60(3):673–682. https://pdfs.semanticscholar.org/fe3b/ede163a66c9eccb71491437459b5d8262cec.pdf. Last accessed 12 July 2018

Jessica Andrade Michel is graduated in Architecture and Urbanism (2011) and has a Specialization in Interior Architecture (2014). She is currently a master's student in Civil and Environmental Engineering at Passo Fundo University, in South of Brazil.

She is Professor in the Integrated Regional University of Alto Uruguai and the Missions, south of Brazil, working in at the courses of Architecture and Urbanism and Civil Engineering. Her interests are related to adaptation to hydrological events, urbanization and climate change.

Giovana Reginatto is a student of the graduation course in Civil Engineering at Passo Fundo University, in South of Brazil. She is also a scholarship student Pibic/CNPq. Her interests are related to sustainable management projects, sustainable development goals and energy efficiency.

Janaina Mazutti is graduated in Environmental Engineering (2018) and she is currently a master student in Civil and Environmental Engineering at Passo Fundo University, in South of Brazil. Her interests are related to sustainable management projects, sustainable development goals and sustainable and smart cities.

Professor Luciana Londero Brandli is graduated in Civil Engineering (1995), master's degree in Civil Engineering (1998) and Ph.D. in Production Engineering (2004). Pos Doctorial Research at Hamburg University of Applied Sciences (2014). She is currently Associate Professor in the University of Passo Fundo, south of Brazil, working in the Master Program in Engineering and Environment. Her current research interests include sustainability in high education and green campus, environment management, management of urban infrastructure, sustainable cities and green buildings.

Professor Rosa Maria Locatelli Kalil is graduated in Architecture and Urbanism (1978), and in Economic Sciences (1993), master's degree in Engineering (1983) and a Ph.D. in Architecture and Urbanism (2001). She is currently a Ph.D. Professor at the University of Passo Fundo where she works as a researcher and extensionist in the courses of Architecture and Urbanism, Specialization in Urban Management and Municipal Development, Specialization in Socio-Environmental Education and Post-Graduation Program in Civil and Environmental Engineering at UPF.

Her current research interests include urban periphery, social housing, urban planning and design, regional planning and development, urban sociology, post-occupation evaluation and sustainable development.

Implementing the SDG 2, 6 and 7 Nexus in Kenya—A Case Study of Solar Powered Water Pumping for Human Consumption and Irrigation

Izael Da Silva, Geoffrey Ronoh, Ignatius Maranga, Mathew Odhiambo and Raymond Kiyegga

Abstract The paper deals with the importance of the 17 SDGs as a force to align initiatives at a global level towards a green circular economy. The specific scope of this work is the nexus formed by the SDGs 2, 6 and 7 which deal with agribusiness, water management and renewable, reliable and affordable energy respectively. These three goals merge perfectly in the case of solar PV powered water pumps which can work for human consumption and irrigation. In a country such as Kenya where 75% plus of the population still live in rural areas and are deprived of both electricity and water, the lack of such basic resources drives people to an economy of subsistence with little or no surplus to generate income. In order to develop a sector there is a fundamental need to build capacity. That is why Strathmore Energy Research Centre (SERC) has, with the help of industry, developed a syllabus for theoretical and practical training on how to design, install and maintain such systems in different sizes. This initiative attracted NGO (Non-Governmental Organizations) such as Oxfam, Norwegian Refugee Council (NRC) and the International Organization for Migration (IOM) to partner and fund the training course. SERC has built a hands-on kit such that once gone through the course, participants are fully prepared to implement such units all over the geography of Kenya and other countries around. This nonetheless does not tell the whole story as aspects related to awareness amongst the rural population, access to line of credit and suitable regulations are also essential aspects required for adequate penetration of this specific technology. This paper thus analyses the holistic setting up of an ecosystem which has a potential to transform rural economy and eventually reduce urban migration.

Keywords SDGs · Capacity-building · Irrigation · Safe water · Solar PV · Agribusiness

I. Da Silva (✉) · G. Ronoh · I. Maranga · M. Odhiambo · R. Kiyegga
Strathmore University, P.O. Box 59857-00200, Ole Sangale Road, Nairobi, Kenya
e-mail: idasilva@strathmore.edu

© Springer Nature Switzerland AG 2020
W. Leal Filho et al. (eds.), *International Business, Trade and Institutional Sustainability*, World Sustainability Series,
https://doi.org/10.1007/978-3-030-26759-9_55

1 Introduction

1.1 World Situation

Globally, 780 million people do not have access to clean water (Centers for Disease Control and Prevention n.d.). Children are disproportionately affected by this global crisis since they are more prone to getting diseases as a result of ingesting dirty water (World Vision n.d.). Educational outcomes, especially of girls, are also negatively impacted by the burden of fetching water, which consumes an estimated 200 million hours each day (United Nations World Water Assessment Programme 2018).

Moreover, the world demand for water increases at a rate of 1% every year. The demand will continue to increase and the most affected countries will be developing countries. Further to this, the global water cycle is becoming more intense due to climate change; the wetter regions are getting wetter and drier ones drier. Currently, an estimated 3.6 billion people live in regions with scarcity of water at least one month per year. This population can increase to a projected 4.8–5.7 billion by 2050 (United Nations World Water Assessment Programme 2018).

1.2 Sub-Saharan Africa Situation

The water demand of the agricultural sector in Sub-Saharan Africa is immense. There is a growing number of people living in refugee camps and internally displaced camps which suffer from water shortage for human consumption and for agriculture of subsistence. Cost effective, reliable and sustainable water irrigation systems are needed in the long term to ensure the essential water supply.

The rural areas also struggles to produce high quality products for the national and international market as well as survival for the basic need in poor families in rural areas whose livelihood is entirely dependent on subsistence agriculture.

1.3 Kenya Situation

Part of the country's agenda is to construct over 380 solar water pumps in the rural areas of Kenya where water is a scarcity. With a population of 46 million, 41% of Kenyans still rely on unimproved water sources, such as ponds, shallow wells and rivers, while 59% of Kenyans use unimproved sanitation solutions. In view of that the 380 systems look very inadequate to address the problem. These challenges are especially evident in the rural areas and the urban slums.

As more than 95% of people in rural areas have no access to solar water pumping, they rely only on rain water. When, for whichever reason, the rain delays coming it is a crisis and hunger ensues. Moreover, there are very few dams in the country to collect water during rainy season. This calls for more sensitization and more innovative solutions to ameliorate the need.

Kenya Climate Innovation Centre (KCIC), a research based incubator for SMEs dealing with climate change mitigation and adaptation, is keen on supporting innovative green technology solutions in the water sector and these are businesses that are concerned with water efficiency and conservation. However, few of these look at a more sustainable solution.

There is also a growing number of refugees and asylum seekers who are in dire need of water support; an estimated number of 489,239 people half of which are children in Kenya, a large portion from Somalia and South Sudan (World Vision 2018). In 2018, there were an estimated 20,000 new arrivals from South Sudan. Strathmore Energy Research Centre (SERC) whose expertise is to train people in renewable energy and energy efficiency, has developed a syllabus to train technicians on design, construction and maintenance of water systems.

2 The Problem

There is a need to improve water availability, enhance water quality and disaster risk reduction in the more than half a million people living in refugee camps in Kenya. The growing numbers of people in need of sustainable water supply for basic needs and commercial are growing exponentially. This is a silent crisis of water shortage that is denying many a better livelihood. Although the country depends on natural water resources, a big part of it is dry and not suitable for crop cultivation (World Bank n.d.).

Agriculture is the largest contributor of Kenya's GDP, 2 years ago it shot up to 30% and agriculture alone contributes to 80% of freshwater withdrawals (World Bank 2019). Water shortage causes a ripple effect to the entire economy and therefore must be addressed. The problem is bigger than what is perceived at the moment. In 2000, the country's per capita water availability was 647 m^3. This dropped to 502 m^3 in 2012, and will be 235 m^3 by 2025 (Mati 2018).

In Nairobi alone, the water supply is 525,000 m^3/day and is expected to reach 1 million m^3/day by 2030. Water supply methods are not sufficient to respond to the growing population e.g. rope and washer pumps, motorized petrol and diesel pumps, submersible electric pumps. High fuel costs, lack of expert knowledge for repairs and lack of spare parts are some of the causes of failed water projects in the region (Südafrika AHK 2017). Thus a need to improve them with more sustainable solutions.

3 Methodology

3.1 Training at SERC

Solar water pumping is still an untapped area in the field of renewable energy, at least in Africa. Although there are a number of innovative ideas to spread this to the affected areas, there were no set curricula nor training centres within the region that would offer practical trainings on how to design, install and maintain solar water pumps.

Practically all companies offering water systems solutions do not engage in training. Therefore SERC filled this gap by providing the first solar water pumping training course in East Africa with curriculum approved by the National Industrial Training Authority (NITA). It is hoped that this will build capacity to strengthen the water sector and thus set pace to the deployment of more solutions to the water crisis.

Two trainings already took place and the participants (Government officials, development organisations officers, farmers, professionals—technical and nontechnical, etc.) will play a significant role in achieving the sustainable development goals in question in a very specific manner.

3.2 Partnership with Civil Society, Government and Private Sector

SERC, in partnership with EU Civil Protection and Humanitarian Aid (ECHO), Norwegian Refugee Council (NRC), International Organization for Migration (IOM) and Oxfam, in a collaborative manner, developed the course content, the hands-on training equipment and even dug a borehole at SERC to make the training as practical as possible.

The collaboration of Government of Kenya (GoK) will be to create regulations to make this training obligatory for all counties, especially the ones where there is water scarcity. Energy Regulation Commission (ERC), the regulatory body, will also ensure quality of such systems installation by requiring technicians to sit for practical exams repeated annually. The existence of such relevant policies will see the successful integration of solar PV technology into the water sector. The role GoK plays in this area is very important. For instance, the current Water Bill foresees a substantial increase in the number of irrigation projects.

Through Kenya Water Institute (KEWI), one of SERC's partners in capacity building, it is hoped to extend the training to thousands of people, especially the youth taking advantage that KEWI offers training at technician level which has been conducting specialized training in the water and sanitation sector since 1960.

The role of private sector is of essence because unless some companies have revenue through the sales and installation of solar water pumping, all the above mentioned activities will eventually come to nothing. With this in mind, SERC has approached a number of private sector players to participate in this venture. Our main partner is Lorentz, a German company which is a global player in this sector. They supplied surface and submersible pumps plus the electronic control array to give trainees the perfect hands-on experience. KCIC has been quite strategic for SERC as their clients (mostly SMEs) become captive market for this SERC training. Some of them are: SwissQuest Water Suppliers Company Limited who install and manage prepaid smart water metering solutions.; AfricAqua, a company which offers water treatment, packaging and distribution solution through its micro distribution centres. MajiMilele, a company which offers safe water at specific prepaid water points (KCIC).

Under the above mentioned collaborations, SERC is fully capacitated to offer training in various counties within the country. Since its inception in 2012, SERC has been carrying out training in the solar sector and boasts of training over 2500 technicians and engineers and has equipped them with skills which enhances the adoption of decentralized renewable energy technologies throughout Kenya and the neighbouring countries such as Somalia and South Sudan.

Finally, this SERC initiative is very much aligned with agribusiness which is one of the "Big Four Agenda" the Government of Kenya (GoK) plans to support in view of having Kenya reaching the goals of the Vision 2030. There is the need to expose the training participants on the solar powered water pumping to all the above information such that no matter which role and for which stakeholder he/she works that person will try aligning his/her plans with the laws and regulations already stipulated by the Government, be it Ministry of Water and Irrigation, Ministry of Energy or the Energy Regulation Commission.

3.3 The Solar Water Pumping Training and Its Benefits

The syllabus was developed by SERC in partnership with Oxfam and the International Organization of Migration (IOM) as part of the objectives of the Global Solar and Water Initiative. The syllabus incorporated both theoretic and practical aspect of the training.

These were developed from the needs assessment obtained from past experiences in solar water pumping installations and from an assessment of the knowledge one may require to have before installation of a solar water pumping system. The theoretic part of the training includes six main topics; Solar PV Basics, Design and Installation,

Fluid Dynamics, Pump Mechanics, Hydrology, Irrigation and the Use of common design software from two manufacturers—Grundfos and Lorentz. The practical part of the training includes an actual installation of a ground mounted pump and a submersible one.

Two different types of water pumping systems were installed as training kits— a ground mounted solar water pump and a submersible solar water pump. These pumps draw water from a 6 m deep well and discharge water back to the well for demonstration purposes.

This system consists of a three phase ground mounted 2.2 kW Grundfos pump, one string of 15,250 Wp solar PV modules connected in series, and a Grundfos pump controller. This system can be powered by AC from the grid and DC from the solar PV modules. The pump controller also serves as an inverter which converts DC from the solar PV modules to AC (Figs. 1, 2, 3 and 4).

This system consists of a three phase submersible 0.7 kW Lorentz pump, two strings of 3190 Wp solar PV modules, AC/DC Converter and a Lorentz pump controller. This system can be powered by AC from the grid and DC from the solar PV modules. The pump controller also serves as an inverter which converts DC from the solar PV modules to AC.

Since SERC is registered with NITA as a training institution, it can submit the syllabus of the solar water pumping system training for approval. This approval brings with it two main advantages. The first is that companies sending technicians for the training can be refunded for costs by the government. The second is that the training can also be valid for points towards continuous professional development. Yet another

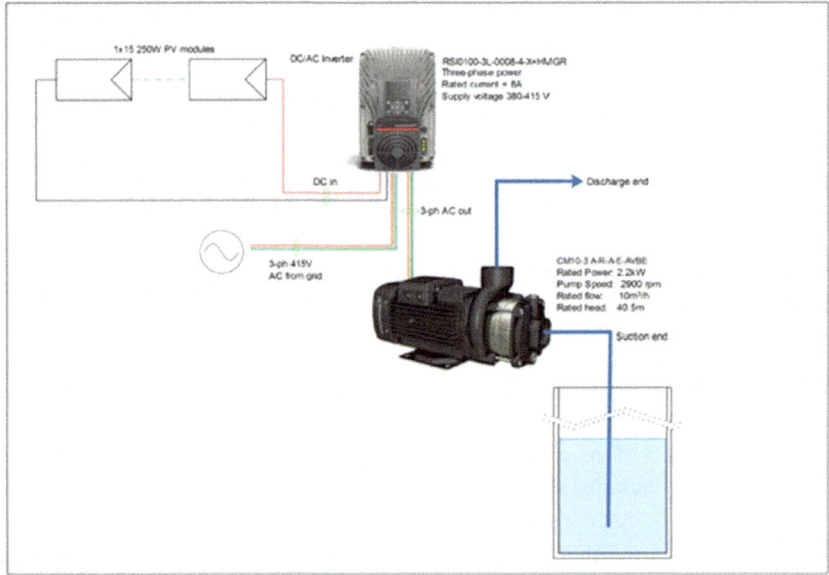

Fig. 1 Ground mounted solar water pumping system. *Source* SERC

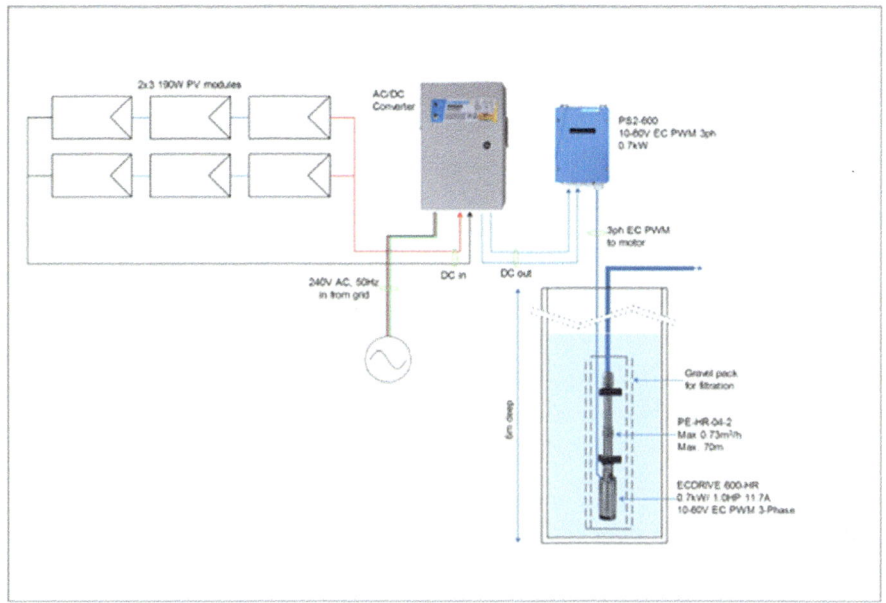

Fig. 2 Solar panels that power the pump at Strathmore University. *Source* SERC

Fig. 3 The solar water pump controllers at Strathmore University. *Source* SERC

Fig. 4 Submersible solar water pumping system. *Source* SERC

initiative to get the training mainstreamed would be to request the regulatory body (Energy Regulatory Commission) to approve the training and make it part of the licensing process for engineers and technicians.

Solar has the advantage of minimum recurrent costs and unlike other solar PV installations, solar water pumping systems do not need energy storage. Solar water pumping kits cost US$350–1100 on market with running costs of zero and no pollutants compared to the diesel and petrol pumping generators.

The Energy Sector Management Assistance Program (ESMAP) of the World Bank, projects the return on investment of a solar powered system to be 1–4 years for solar panels with a lifespan of an estimation of 25 years a reduction in cost over life of the systems of −40 to −90% in comparison to diesel generators (World Bank 2017).

4   Results and Discussion

In order to consolidate the progress on this area so far, a possible solution is to use the Train of Trainer system to avail this capacity building program to all the geographical areas of the country. In the past, SERC has, with the support of USAID and twenty-five Technical Training Institutes (TTIs) trained close to 2000 technicians on general solar PV systems design and installation at three levels, namely T1, T2 and T3. These levels go from DC micro systems to utility-size, grid-tie systems. This same strategy should be implemented for the training being considered.

Another point is to pursue the implementation of the Water Bill, 2014 in partnership with the Ministry of Water and Irrigation. For instance, it contemplates an audit of the water situation of Daadab refugee camp, near Garissa. The same can be said of Kakuma refugee camp in the north of Kenya.

As mentioned above, the support of NITA is of essence to achieve a large number of trained people. For that the approval of the syllabus must be fast-tracked. On this same note, it will be strategic to insert the solar PV water pumping training into the regular syllabus of the courses offered by KEWI as they have a wide reach of students in their five training centres located outside of Nairobi.

The financial aspect of such venture is a positive one as solar powered water pumping installations are, in the long run, more affordable than diesel generator based ones. The only issue is that either the Government or the Civil Society needs to meet the initial capital investment with the hope of recovering it through the sale of water. Refugee camp dwellers are already used to paying for water.

Kenya is divided into 47 counties and it is a common knowledge that the authorities in charge of water in those local government are quite ignorant of such technological innovation. This means that SERC can avail training not only to engineers to work on the installations but also to educate the decision makers and other stakeholders on the financial and social suitability of such ventures.

Finally, there is the need to carry out a national inventory on this area to find out which are the priority counties where the solar water pumping systems are more needed and could have a greater social impact.

5   Conclusions and Recommendations

Triple Helix is a concept created by Prof Michael Potter, a Harvard economist in the late seventies. It basically states that in order to have a wider and long lasting impact any venture has to work on a concerted manner with Government/Civil Society, Private Sector and Academia as each other these have a very specific contribution to the solution of the problem in case.

Therefore, the question of addressing the water challenge in Kenya would be more sustainable if triple helix based solutions were applied. That is because currently each stakeholder is working within a "silos mentality" which blinds it to the unique intervention the other two can contribute. Having them working together almost never happens without a purposeful effort to implement the triple helix.

Another recommendation is to avail training participants with a case study where practical situations beyond the hands-on aspect will arise. Most importantly the syllabus has to be revised to something better as mentioned in the case study. The training will be designed with a more practical approach to the local problems.

One that is more challenge based towards achieving a more efficient and effective framework for learning while solving real local challenges.

To a attain a more collaborative interaction between the industry and the participants in order to ask thoughtful questions, identify bigger ideas, investigate and solve challenges in solar water pumping, etc.

Bibliography

Barasa M et al (2018) Evaluation of the sustainability of solar powered water supply systems in Kenya. EED Advisory Limited. Retrieved from https://energypedia.info/images/5/51/Evaluation_of_the_Sustainability_of_SPWSS_in_Kenya.pdf

Centers for Disease Control and Prevention. (n.d.). Global WASH fast facts. Retrieved from https://www.cdc.gov/healthywater/global/wash_statistics.html

Jamal A (2000) Minimum standards and essential needs in a protracted refugee situation: a review of the UNHCR programme in Kakuma, Kenya. United Nations High Commissioner for Refugees Evaluation and Policy Analysis Unit. Retrieved from https://www.unhcr.org/research/evalreports/3ae6bd4c0/minimum-standards-essential-needs-protracted-refugee-situation-review-unhcr.html

Kenya Climate Innovation Centre. (n.d.). The water sector in Kenya. Retrieved from https://kenyacic.org/sites/default/files/publications/THE%20WATER%20SECTOR%20IN%20KENYA.pdf

Mati BM (2018) Innovations in solar water pumping systems: taking Kenya's development to the next level. Retrieved from http://www.jkuat.ac.ke/departments/warrec/wp-content/uploads/2018/09/Innovations-in-Solar-Water-Pumping-in-Kenya-Prof.-Mati-3.pdf

Nasrullah A (2018) UNHCR operational update: Dadaab, Kenya. United Nations High Commissioner for Refugees. Retrieved from https://www.unhcr.org/ke/wp-content/uploads/sites/2/2018/05/15-April-Dadaab-Bi-weekly-Operational-Update.pdf

Südafrika AHK (2017) German solar training week on solar water pumping, irrigation and treatment. Retrieved from http://suedafrika.ahk.de/en/events/current-events/event-details/events/german-solar-training-week-on-solar-water-pumping-irrigation-and-treatment/?cHash=d1f511d514af5cbe9fab9df7413c7c2

United Nations World Water Assessment Programme (2018) The United Nations World Water Development Report 2018: nature-based solutions for water. Paris. Retrieved from http://www.unesco.org/new/en/natural-sciences/environment/water/wwap/wwdr/2018-nature-based-solutions/

World Bank (2017) Energy sector management assistance program annual report 2017 (English). Washington, D.C. Retrieved from http://documents.worldbank.org/curated/en/171191523257054613/Energy-sector-management-assistance-program-annual-report-2017

World Bank. (n.d.). Kenya. Retrieved from https://data.worldbank.org/country/kenya

World Vision (2018) Kenya - November 2018 situation report. Retrieved from https://www.wvi.org/kenya/publication/kenya-november-2018-situation-report

World Vision. (n.d.). The dangers of dirty water, the power of a safe, clean well. Retrieved from https://www.worldvision.com.au/global-issues/work-we-do/climate-change/dangers-of-dirty-water-power-of-a-safe-well

Sustainability Reporting

Sustainability Reporting in Australian Universities: Case Study of Campus Sustainability Employing Institutional Analysis

Gavin Melles

Abstract In an era of climate change and greater demands for organisational commitment, the global university sector has taken up sustainability reporting in a modest fashion. Recent studies of the sector show extensive variation in coverage and quality of reporting even by comparison to the corporate sector. From a materiality point of view, sustainability reporting is an aspect of environmental management that aims via efficiency actions to reduce GHG emissions. Recent studies and reports criticise the Australian University sector for its lack of transparency and modest accomplishments in campus sustainability compared to the corporate sector and other regions. This study examines these claims and introduces organisational theory of institutions as a justification for the strategies and tactics adopted by the sector. Existing sector studies and reports have rarely employed organisational theory and preferred unrepresentative samples and definitions of reporting and scope, albeit admittedly due to the poor public availability of environmental and other data. This short case study combines desk analysis of publicly available Australian University sustainability reports, strategy plans, compliance reports of energy and water, built environment initiatives, and qualitative interviews with five (The interview component, which is on-going, has currently recruited and interviewed twenty participants. This study incorporates analysis of the first five interviews completed at the time of writing.) past and current campus sustainability managers and officers from the Australian University sector to (1) show the current state of reporting and commitment and (2) demonstrate the value of the organisational theory approach. The study finds that organisational theory is the missing theoretical lens to help explain why compliance rather than commitment and legitimacy claims in a context of weak national and sector expectations explains the varied leadership and commitment in this region. Reporting on the initial analysis of sources, this study identifies the concept of transparency and accountability across the academic and professional divide as key to understanding a sector compliance approach.

G. Melles (✉)
School of Design, Swinburne University, Melbourne, Australia
e-mail: gmelles@swin.edu.au

© Springer Nature Switzerland AG 2020
W. Leal Filho et al. (eds.), *International Business, Trade and Institutional Sustainability*, World Sustainability Series,
https://doi.org/10.1007/978-3-030-26759-9_56

945

Keywords Sustainability reporting · Australian universities · Institutional theory · Case study

1 Introduction: Sustainability in the University Sector—A Material Focus

There are multiple normative, socio-cultural and coercive forces driving organisations to adopt or avoid sustainability reporting. Moreover, the relationship between such reports and the sustainability credentials and practices of the organisation is complex and multi-faceted. From a materiality point of view, sustainability reporting is an aspect of environmental management in organisations, which should therefore report on a reduction in GHG (CO_2e) emissions. Beyond environmental aspects, reporting should also include a broader narrative about the overall triple bottom line (TBL) engagement with sustainability in the organisation (Gray 2011).[1] However, narratives about student engagement, teaching and research are complementary TBL aspects rather than a substitute for measured outcomes.

This study defines reporting in broad terms to include compliance (environmental) reporting of energy, water etc. omissions, and voluntary reporting of other activities, e.g. installation of renewable energy sources, building efficiency measures, and the publication of strategies and plans to achieve carbon neutrality or reductions. The inclusion of strategies and plans is important as these are publicly available claims of targets, and plans constitute a monitoring and evaluation (M&E) framework for annual reports themselves. It can be argued that whoever has no measured future vision, e.g. strategies, is reporting idiosyncratically.

In the current climate of climate change mitigation and sustainable development, the question this study addresses is to what extent university organisations 'walk the talk', and to what extent the university sector is performing well in relation to other sectors. For example, by comparison with the nearly 50% of S&P companies globally publish such reports, using the Global Reporting Initiative (GRI) format,[2] universities in general are not leading the way. Despite media and academic reports of a more positive nature[3] recent reports in the popular academic media channel—the Conversation (Feb 8, 2017)[4] suggest few Australian universities have made any specific carbon reduction targets, compulsory reporting on energy use showed increases over time, and planned building and development budgets rarely elicit any reference to significant sustainability initiatives. Indeed, a number of mitigating and hidden

---

[1] Note that Gray suggests that labelling triple bottom line reporting or TBL as equivalent to sustainability reporting is inaccurate.

[2] According to an IFC World Bank Presentation, available at https://www.povertyactionlab.org/sites/default/files/documents/Day2_Sustainability_Reporting.pdf.

[3] See http://100percentrenewables.com.au/universities-demonstrating-sustainable-energy-leadership/.

[4] https://theconversation.com/australias-universities-are-not-walking-the-talk-on-going-low-carbon-72411.

factors, such as the deep ties to corporates and industry of Australian universities raises questions about the rhetoric and reality of sustainability claims (Baer and Gallois 2018). In this chapter I question some of these claims while others must be nuanced.

A recent study examined five Australian university GRI based reports (in a sector of 43 organisations) from 2013 for content and disclosure, finding modest outcomes in the quality and quantity of reporting (Gamage and Sciulli 2017). This study focuses on a broader definition of reporting and examined a much wider number of organisations, while particularly focussing on measured environmental reporting or targets rather than just content analysis of a small sample. This study sees institutional theory of organisations as the missing lens to address current limitations and bring a theoretical lens to current discussions these gaps and explain the state of reporting in the sector.

2 Sustainability Reporting in the Global University Sector

Before considering the specific nature and scope of sustainability reporting, it is important to understand and critique the claims to leadership of the sector. The evidence for the role of universities as leaders in sustainability and sustainable development education is mixed (Lozano et al. 2013). Several studies have pointed to the limited engagement of universities with measuring the materiality of their environmental (e.g. Chang and Deegan 2009) and sustainability impacts. Thus, Ferrero-Ferrero et al. (2018) note the low expectation of sustainability transparency and materiality by internal stakeholders, e.g. faculty and students, in the university sector. Signing Declarations, e.g. Talloires,[5] seems to be a popular and potentially indicative strategy of commitment (Lozano et al. 2015) but also of limited or no significance to materiality in the relevant sense (e.g. see Ramirez 2015), since like many sustainability declarations they are aspirational and non-binding. Some have argued that unless declarations are converted into accountable strategy targets they are merely another form of greenwash (Bekessy et al. 2007).

Recent studies also document patchy integration in curriculum, such that education for sustainability (EfS) 'is not yet widely practiced in university classrooms across disciplines' (and see Sherren 2006; Christie et al. 2015b, 405). Although this varies with organisations, there appears to be very modest commitments to sustainability as a key graduate attribute (Lee et al. 2013), and change agents in specific organisations are challenged to promote curriculum change (Higgins and Thomas 2016). These mixed results are all the more exacerbated by the ignorance or opposition of some academics to sustainability (Christie et al. 2015a). Combined with similar knowledge gaps and mixed attitudes of students both before and after learning (Vermeulen et al. 2014) this means that in general the overall institutional sector

---

[5]See Talloires Network https://talloiresnetwork.tufts.edu/.

environment is not conducive to change. Thus, there are serious doubts about an environment of sustainability leadership.

This lack of actual leadership is also reflected in the absence of publicly available transparent reports on energy, water and other processes for the Australian university sector. Data collection and analysis for this study was very compromised by the lack of such data, and the absence of national sector frameworks for comparison, such as STARS in the USA, to judge progress and failure. Such a status quo can be seen as evidence of weak expectations in the institutional environment for transparency and accountability—this situation gives also further evidence for the need for theoretical rather than only common sense explanations.

3 Sustainability Reporting

While corporate and other organisations are required to report on energy emissions, pollution, and so forth, sustainability reporting as discussed here is voluntary. Current sustainability reporting practices of public and private organisations remain globally of modest quality and transparency (Guthrie and Farneti 2008; Perez-Batres et al. 2011; Hahn et al. 2015), including in Australia (Higgins et al. 2015; Liu et al. 2017; The Reporting Exchange 2018). The 2006 Parliamentary Report on Sustainability reporting in the Australian corporate sector noted a very low response from Australian companies relative to others, e.g. UK, Japan, USA, while continuing to recommend voluntary rather than mandatory publication.[6]

Organisational uptake is affected by a number of factors, including perceived and real internal and external stakeholder relationships and pressures (Fernandez-Feijoo et al. 2012, 2014). Others argue that the inherent commitment to growth of companies and the need for full transparency are in conflict and hence affect the low uptake and poor quality of reporting (Gray 2011). As this study shows, isolated arguments about the rationality or strategy behind acquiescence or refusal to report can be given a more coherent answer once organisational theory of organisations is invoked, especially where stakeholder appeals and legitimacy concepts are included under the banner of institutional analysis.[7]

The dominant corporate reporting standard remains the Global Reporting Initiative (GRI) model. This divides reporting into nine aspects and suggests indicators for each (Gray 2011).[8] As Livesey and Kearnis (2002) note in their analysis of the Body

---

[6]Parliamentary report available at https://www.aph.gov.au/About_Parliament/Parliamentary_ Departments/Parliamentary_Library/Browse_by_Topic/ClimateChangeold/responses/economic/ sustainability.

[7]There is a tendency in the literature to favour either legitimacy theory or stakeholder theory to explain sustainability reporting but in fact, institutional theory of organisations already alludes to both as factors in the tactics and strategic aims of organisations.

[8]The nine aspects are: materials, energy, water, biodiversity, emissions and waste, products and services, compliance, transport and overall. And see https://www.globalreporting.org/Pages/default. aspx for details and standards.

Shop and Shell versions of early sustainability reporting, the establishment of GRI standards aimed to create less idiosyncratic interpretations than hitherto. Brown et al. (2009b) observe that the original intent of the GRI to act as a prompt to discussion of the responsibility of business actors towards society has been replaced by a tool for the management of sustainability, reputation and brand management (and see Gray 2011). They conclude that the history of the GRI 'illustrates how an institution emerges in a dynamic fashion as a result of interactions among many actors, how interests and logics evolve, and how the trajectory of the institution can deviate from the intentions of the entrepreneurs' (Brown et al. 2009b, 579). GRI standards still play a role in the university sector but can not be described by any means as dominant.

4 Universities and Sustainability Reporting

Recent overview studies of this sector have found fragmented research approaches to sustainability reporting in the sector, limited evidence of third-party auditing or influence on organisational change (Ceulemans et al. 2015). Recent overview studies in Canada (Fonseca et al. 2011; and see Vaughter et al. 2015; Sassen and Azizi 2017), Germany (Azizi et al. 2018) and Australia (Gamage and Sciulli 2017) show poor sector uptake, varied standards, and limited transparency and materiality. Partial transparency with sharing data is also a characteristic of the university sector (Larrán Jorge et al. 2018). Approaches demanding strong materiality and engagement, e.g. ecological footprint analysis, are rare (e.g. Flint 2001). Recent studies also show that despite the potential for energy data to be an element of social and stakeholder dialogue universities implement but rarely divulge data openly (Disterheft et al. 2015).

Adams (2013) notes that despite the leadership potential and obligation of universities (as publicly funded organisations) in sustainability a minority in the sector report, strategic plans pay lip service, and beyond compliance there is limited deep engagement. She further notes studies showing limited and diverse incorporation of principles in curriculum, and contends that a lack of 'whole-of-institution approach' is everywhere in evidence (Adams 2013, 385). Adams further notes that materiality and measuring commitments is poor in the sector, and international initiatives such as the UI Green Metric[9] have unclear criteria, allowing for example courses to be counted with no measure of coverage (Adams 2013, 387).

In Australia no reporting framework or standards enjoys universal or even majority acceptance. GRI, the Living in Future Environments (LiFe) Index[10] and more recently SDGs have been relevant standards for university reporting; these different standards demand degrees of material transparency. In Australia, there are several important sector networks who engage in reporting on and rewarding sector efforts.

---

[9]A recent (Indonesian) initiative with two participants from Australia, see http://greenmetric.ui.ac.id/.

[10]Promoted by ACTS and adapted from the UK—materiality is one of four dimensions.

The Australian Universities Towards Sustainability (ACTS)[11] focus on promoting and rewarding campus sustainability and reporting, the Asia Pacific SDSN,[12] which is focused on promoting an SDG focus, and the Australian Student Environment Network (ASEN) are three key networks, which speak of a lively engagement with sustainability and development issues. However, closer analysis shows that an integrated multi-stakeholder 'whole-of institution approach', engaging professional, academic, student and community participants on common projects and change is rare. Poor student engagement, the practical failure of declarations and inconsistent reporting, inter alia, puts such activism in perspective.

5   Institutional Theory Analysis of Organisations and Sustainability

It is institutional theory of organisations that offers a framework for understanding the nature and quality of higher education's engagement with sustainability and reporting (and see Jensen and Berg 2012). In their review of sustainability reporting, Hahn and Kühnen (2013) observe that the majority of sustainability reporting studies refer to no theory at all or if they do mention stakeholder theory. They argue for the urgent need to include legitimacy, stakeholder and institutional theory in such analyses, which would imply 'a shift from the dominance of content analysis of published documents towards more exploratory and confirmatory methodological approaches such as interviews, surveys, and experimental studies' (Hahn and Kühnen 2013). In fact, institutional theory analysis, especially in recent interactive formulations, include reference to legitimacy and stakeholder values.[13]

Institutions are regulative, normative and cognitive 'beliefs, rules, roles, and symbolic elements capable of affecting organizational forms independent of resource flows and technical requirements' (Berthod 2016). Organisations, e.g. banks, schools, local governments, are the 'players' operating in institutional environments through specific arrangements and activities. As Iarossi et al. (2013a) note 'Institutional theory states that the business environment in which a firm operates exerts pressure on the firm. Pressures from these systems elicit different responses as firms seek legitimacy in order to "survive and thrive" in their environment … Institutional theory recognizes, however, that organizations are not passive actors and can respond to institutional demands in a variety of ways from conformance to reshaping those pressures' (Iarossi et al. 2013a, 78).

---

[11] See https://www.acts.asn.au/ the association recruits in NZ and Australia.

[12] See http://ap-unsdsn.org/. The organisation aims to promote SDG implementation and funds projects, e.g. decarbonisation pathways modelling. Targets and membership are voluntary. It recruits academic rather than professional staff and projects.

[13] The business management preference for stakeholder theory and values is also less relevant to the strategic approach of organisations as 'players' in an institutional environment relative to sustainability demands.

Much of the current focus of institutional theories of organisations is explaining stability ('isomorphism') and change in organisations due to coercive, normative and socio-cognitive drivers. Such an approach questions the rationality or efficiency arguments about organisational behaviour, i.e. that organisations are driven by rational or efficiency goals (Chatelain-Ponroy and Morin-Delerm 2016, 890). Thus, texts on organisational research assume a still prevailing rationalist approach to organisational behaviour, and one looks in vain for references to institutional theory (e.g. Symon and Cassell 2012; Jones 2014). Institutional theories predict strategically and ideologically driven isomorphism among organisations albeit acknowledging how local cultural and political specifics affect this convergence (Glover et al. 2014). A central role in organisations is played by habituation and habit (Hodgson 2007, 108).

As Scott (2008) observes early thinking about the determinate influence of environmental factors has given way to more flexible accounts of organisational responses to such pressures, 'providing impetus for field conflict and change' (2008, 140). Interactive approaches see potential for agency and differentiation, which may emerge through the influence of institutional entrepreneurship and change makers in organisations (Arroyo 2012). Thus, differentiation may also be to sub—sector market differentiation (e.g. Bebbington et al. 2009), weak obligations, entrepreneurial 'leadership', and internal conflict. While some describe the initial impetus for GRI as an example of sustainability entrepreneurship (Brown et al. 2009a) as such it has failed to promote societal dialogue due to the dominance of corporate social performance or 'a fixation on the organisation itself'(Buhr et al. 2014, 51) over societal dialogue (Brown et al. 2009b).

A key paper in the theory on interactive approaches institutional theory was that of Olivier (1991) who suggested that organisations respond with different strategies and tactics to institutional drivers. Note that such strategies and tactics are legitimacy moves with respect to perceived stakeholders (Table 1).

Without such theoretical contextualisation, content analysis of reports or single case studies of implementation remain under-theorised, strategically unmotivated, and particularistic. Thus, responses of universities to sustainability requirements often show multiple tactics employed. While it is currently difficult for the sector to openly *defy* the need to respond to climate change, individual organisations appear to be employing all other strategies and tactics in current approaches. Thus, some (a minority) in the sector have *acquiesced* to carbon neutrality as the new 'normal' standard, in part as a compromise response to internal and external stakeholders. The majority in the sector continue to *avoid*, e.g. no targets, or partially compromise or manipulate requirements by declaring other commitments, e.g. student or staff initiatives or minor solar investments, which in fact make a limited contribution to CO_2e reductions.

According to insitutional theory, Imitative, normative and coercive pressures drive organisations towards isomorphic responses to specific external and internal pressures, e.g. sustainability declarations, initiatives and reporting. Thus, many universities have subscribed to the UN Global Compact (UNGC) to promote the SDGs,

Table 1 Oliver (1991), p. 152. Strategic responses to institutional processes

Strategies	Tactics	Examples
Acquiescence	Habit	Following invisible taken for granted norms
	Imitate	Mimicking institutional models
	Comply	Obeying rules and accepted norms
Compromise	Balance	Balancing the expectations of multiple constituents
	Pacify	Placating and accommodating institutional elements
	Bargain	Negotiating with institutional stakeholders
Avoid	Conceal	Disguising nonconformity
	Buffer	Loosing institutional attachments
	Escape	Changing goals, activities or domains
Defy	Dismiss	Ignoring explicit norms and values
	Challenge	Contesting rules and requirements
	Attack	Assaulting the sources of institutional pressure
Manipulate	Co-opt	Importing influential constituents
	Influence	Shaping values and criteria
	Control	Dominating institutional constituents and processes

and have signed a range of aspirational[14] declarations, e.g. Talloires, s a potential compromise with social obligations (but which require no material commitments). In an environment where sustainability is strongly institutionalised, successful organisations in a specific sector will be imitated or aspired to by others (imitation), social and stakeholder expectations will be addressed (normative), and regulations, e.g. emissions reporting (coercive) will constrain variation (Zucker 1987).

The low degree of conformity to reporting, and the lack of consensus on standards suggests that the sector is still in a process of moving from broad habituation, i.e. having sustainability as a norm, to objectification (Suddaby and Lefsrud 2010). Prima facie, this current variation in Australia suggests a low degree of institutionalisation and isomorphism of sustainability in the sector. Institutional theory tells us that we should expect weak responses and isomorphism among sector organisations if the external drivers and expectations are similarly weak. One example in this context to is how certain universities in certain particular configurations of people and environment produced one off sustainability reports, conforming to specific standards but subsequently resourcing and reporting disappear, e.g. see Swinburne University (SWI) and University of the Sunshine Coast (USC) in table below.

Cho et al. (2015) note how current studies of sustainability reporting from a broad institutional theory approach have analysed reporting as legitimacy 'window-dressing where, 'sustainability reports serve not only to promote the interests of individual corporations, but also collectively to present current structural arrangements within society as able and willing to act on escalating sustainability challenges'

---

[14]The term aspirational here is used to define commitments that are not measured.

(Cho et al. 2015, 80). The authors discuss organized hypocrisy and facades as the results of corporations attempting to respond to conflicting demands. In their study, they identify a rational, innovation and reputations facades in corporate reporting, variously emphasizing different narratives about their commitments and practices.

6 A Broad National Context of Disinterest and Diluted Measures

Above I have alluded to low sector expectations and mixed messages about leadership. Meyer (2017) notes cultural and social forces influence such expectations and the adoption of institutional models. At the broadest socio-political level, there are questions about Australia's engagement at all with climate change and sustainable development (Stefanova et al. 2016), and recent national surveys bear this out. Thus, the CSIRO five year analysis (Leviston et al. 2014) shows less than half (46%) of Australians believe in human-induced climate change, and this disbelief has remained stable over five years (2010–2014). A more recent study of attitudes between 2001–2017 (World Wildlife Foundation 2018) show a decrease in belief in the need to act now to solve environmental problems (89 > 81%), and increased belief in the exaggeration of threats to the environment (26%). Thus, at the broadest level we can see the modest response of the university sector as a result of social and political apathy to sustainability in Australia.

In the contexts of weak commitments and proliferation of reporting frameworks, new trends away from discrete reporting to global guidelines, e.g. GRI, are also evident. One trend in the sector (and Australia) is so-called integrated reporting, in which sustainability issues and measures are included in annual reports and strategy rather than reported separately. Recent studies suggest it has had limited transformative impact, involves a diluted focus, and still constitutes a marginal phenomenon (Stubbs and Higgins 2014; Brusca et al. 2018). Another trend, evident also in this study is a move to use the SDGs as a sustainability reporting framework, which has come about inter alia through the coercive force of the Times Higher Education rankings including a dimension of achievement of SDGs. However this move does little to increase material transparency, although is perhaps a better social contract, i.e. engaging in a global concern (Chatelain-Ponroy and Morin-Delerm 2016) between the sector and society.

7 Theory and Methodology

Two approaches seem to dominate in studies of sustainability reporting in higher education: single institution case studies and content analysis of reports, (e.g. Buhr et al. 2014, 51). Thus, Corocran et al. (2004) note the proliferation of sustainability case studies of single institutions which 'rarely included any information on the theoretical approach to the methodology or on the methods used to gather the data. Instead, stories of successes were reported and the data supporting these successes are not readily available for public critique (Corcoran et al. 2004, 14). In response to this call, this study employs case study methodology explicitly informed by institutional theory of organisations.

For analysis purposes a more inclusive framework was required to overcome the unrepresentative analysis of reporting in this sector (e.g. n = 5, Gamage and Sciulli 2017). This study identifies five levels of sustainability reporting as part of an organisational culture of reporting (see diagram below). First, is the existence of a campus environmental or sustainability policy—this is an aspirational statement about desired behaviours with respect to university operation, e.g. sustainable procurement. Most universities have these it appears. Secondly, is the existence of sustainability plans or strategies with targets, baselines and timelines; this study identified 15/43 (36%) with draft, partial or complete plans. These may be broken down into separate aspects, e.g. transport, energy, etc., and do include targets, e.g. carbon neutrality or percentage emissions reductions. Third is the level of reporting whether or not to a specific standard, e.g. Learning in Future Environments (LiFe),[15] GRI, Scorecards, and whether or not reported in printable (e.g. PDF) format or not. Fourth, there is compliance reporting—and this refers to compulsory reporting on energy, waste, water and pollutants. This contains no narrative and may or may not be publicly available. Finally, there is voluntary reporting to other frameworks, e.g. in Australia Tertiary Education Facilities Management (TEFMA),[16] which may or may not (not in Australia) be available for public dissemination.

8 Prior Research Approaches

The case study specifically employs content analysis of sustainability reports, strategies and plans, and five qualitative interviews with sustainability reporting experts from the sector. NVivo 12 Mixed Methods Software was used to develop the analysis. CAQDAS software can help manage the 'messy' process of analysis (Sinkovics and Alfoldi 2012). Case study research employs multiple methods to focus on analysing events, change processes, organisations per se, and other phenomenon (Symon and Cassell 2012). Document analysis and interviews are common methods for researching organisations, including for in-depth fieldwork and broader case study approaches (Jones 2014). In cases where analysis is theory driven, progressive focussing of the topic through constant comparison of concepts and themes is a hallmark of analysis. Document content analysis is a key method in organisational research and is complemented by other methods in case study design (Lee 2012). Interview with key informants meanwhile is a subcategory of interview research, and selection of

---

[15] See https://life.acts.asn.au/.

[16] See https://www.tefma.com/—restricted password access to benchmarking reports.

participants is important, as well as taking account of the constructed 'active' nature of interview interaction (Holstein and Gubrium 1995; Alvesson and Ashcraft 2012). Given the breadth of information and data, the focus on demonstrating the value of institutional theory in this context, and the limits of this chapter, the study includes five interviews (from a cohort of twenty) as analysed and integrated with other sources.

9 Methodology: Case Study Employing Grounded Theory Approach

The existing literature on sustainability reporting in university and other sectors, as well as institutional theory studies allowed for the identification of a priori constructs, which themselves became inputs into the interview and document analysis (Eisenhardt 1989). The desktop survey of all universities allowed for early identification of report formats or absences—and strategies—as well as any potential links to data sources. This report first addresses document content analysis followed by interview thematic analysis and discussion focused on integrated identification of case insight concepts.

10 Transparency and Evidence on Energy Emissions CO_2e

It is impossible to get accurate sector emissions data since this is not publicly available nor reported to any voluntary sector framework at any scope, such as the Sustainability, Tracking Assessment and Rating System (STARS), USA.[17] One independent source, which ten universities (23%) have (ever) reported to on scope 1 and scope 2 energy emissions, is the so-called national NGER (Annual Corporate GHG emissions) database. As several universities admit, reporting of scope 3 emissions—the most indirect but including travel (e.g. Glover et al. 2017), would inflate final totals and such totals do significantly under-report CO_2e emissions. Such full Scope 1–3 totals are required for carbon neutral certification, e.g. University of Tasmania (UTAS) and Charles Sturt University (UTAS), or may be voluntarily reported, e.g. Monash University.[18]

To give some measure of progress towards emissions reductions extracted and report five years for ten institutions on the database. This tends to show with few exceptions, e.g. UQ, increased total (Scope 1 and Scope 2) emissions for the period and therefore little clear evidence of progress except in relative terms, e.g. energy intensity. Where universities aim to explain such totals in annual reports they usually

---

[17] See https://stars.aashe.org/.

[18] See   http://www.cleanenergyregulator.gov.au/NGER/About-the-National-Greenhouse-and-Energy-Reporting-scheme/Greenhouse-gases-and-energy for details.

allude to lower emissions per EFTSL or m2 of infrastructure. Of course, only absolute rather than relative emission reductions will lead to achieving climate change goals.

	2018	2017	2016	2015	2014	2013
ANU	597,635	621,322	595,101	601,250	611,336	587,937
DKU	337,351	344,349	338,009	336,700	327,547	297,635
GRF	265,960				261,188	253,077
LTB	496,622	528,521	464,287	488,988	483,732	608,054
MON	742,948	753,613	744,267	700,657	701,842	707,893
RMIT	428,623	360,837	354,434	371,725	360,358	365,057
UQ[a]		555,859	586,229		593,281	584,775
UOM	736,825	745,872	738,821	722,200	683,942	678,759
UNSW	449,427	419,904	411,692	369,098	395,032	366,880
USY	544,534	523,852	537,218	509,600	485,088	455,272

[a]In fact in the UQ Annual report for 2016–2017, p. 54–55 they appear to report Scope 1–3 emissions totals for the last six years showing a 20% decrease (110,133 2016–2017) since 2011 (133,963 MJ GHG). http://www.uq.edu.au/about/docs/annualreport/annual-report-17/10_ManagementAndResources.pdf

It was possible to approximate and extract figures from some other university sites as reported in sustainability websites or reports. Thus University of Western Australia data for energy use (gas and electricity) 2008–2016 and per GJ/m2 (the latter evaluated against 2020 target)[19] shows with a small reduction in 2014–2015 increasing energy use and energy use per m2/GFA (2020 target 0.55) is off target and has generally been for some years (2016 0.75). Total carbon emissions for the same period are not on track to meet their 2020 target although the target 0.9 GFA was achieved in 2015 (but has been exceeded again). Mains water use for 2013–2016 does show some downward trend. Other sources such as reported energy consumption and production in annual reports publicly available also show similar lack of clear progress.

11 Manipulating and Acquiescing to Carbon Neutrality?

Tightly bound up with reducing energy emissions are campus environment efficiencies, increased inclusion of renewable energy supplies, e.g. solar PV, including off and on-campus arrays, battery storage options, and Power Purchase Agreements (PPA) arrangements to purchase.[20] Large-scale renewable energy commitments—whether on or off-site—are the exception rather than the norm, and in the sector mostly make

[19]https://www.cm.uwa.edu.au/_data/assets/image/0012/3077769/CM-sustainability-graph-energy-2016.jpg.

[20]See recently https://www.pv-magazine-australia.com/2018/11/10/long-read-from-rooftops-to-innovative-ppa-structures-australias-universities-go-solar/.

a minimum energy contribution; divestment from fossil fuels may or may not be part of such strategies. The emerging benchmark has become declarations and certifications of carbon neutrality[21] through a range of offsetting, campus efficiencies and tactics. Other targets among subsector networks, such as the five members of the Australian Technology Network (ATN) sector, e.g. RMIT University announcing a 25% reduction in GHG emissions by 2020 to a 2007 baseline have also been made (Riedy and Daly 2010). This study identified ten organisations (~20%) making an explicit commitment to carbon neutrality at some point. Some universities make no declaration regarding this or suitably vague commitments (no targets) to reduce emissions either on websites or plans, e.g. Murdoch University.[22]

In the reputation competition towards carbon neutrality a tactic of some (especially G08[23]) universities is to question the achievement of others, University of Tasmania (UTAS),[24] or lower ranked institutions, e.g. Charles Sturt University (CSU),[25] in achieving (certified) carbon neutrality.[26] Carbon neutrality certification requires declaration of scope 3 emissions—and is therefore can be seen as a leadership commitment.[27] Monash University Zero emission strategy, which develops 'the most ambitious of its kind' 100% micro-grid facility to deliver carbon neutrality by 2030.[28] The University of Melbourne, which has committed to carbon neutrality before 2030 (in electricity by 2021),[29] has likewise questioned achievements of others in the sector. In response to pressure from students, staff, and other stakeholders, Newcastle University has recently signed a 100% renewable energy purchase agreement—seven-year contract to deliver three campuses[30] by 2020. RMIT University, which aims for carbon neutrality by 2030 aims to achieve this with 50% sourced from on-site efficiencies and 50% from off-site, including through its participation in the

---

[21] The government website and conditions can be seen here https://www.environment.gov.au/climate-change/government/carbon-neutral.

[22] As in Murdoch University Environmental Sustainability Charter and Strategy 2014–2017.

[23] The G08 is a sub-sector grouping of eight 'research intensive' universities in Australia, https://go8.edu.au/.

[24] UTAS appears to have achieved carbon neutrality largely through off-sets abroad and domestically but also through campus efficiencies. http://www.environment.gov.au/climate-change/government/carbon-neutral/certified-businesses/university-tasmania.

[25] CSU offset projects in Australia, India and China can be seen here https://www.csu.edu.au/csugreen/about-us/carbon-neutral-university/offset-projects.

[26] See https://www.environment.gov.au/climate-change/government/carbon-neutral/certification.

[27] It si notable that UTAS in its carbon neutral disclosure for 2017 shows that scope 3 emissions for 2015–2017 are approximately double its scope 1 and 2 emissions. This is also the case for UQ in their annual report, which lists scope 1–3 emissions, see http://www.uq.edu.au/about/docs/annualreport/annual-report-17/10_ManagementAndResources.pdf. Hence current compulsory sector reporting of scope 1 and 2 could be doubled to get an approximate fuller account.

[28] See https://www.monash.edu/news/articles/monash-to-become-australias-first-100-per-cent-renewable-energy-powered-university.

[29] See https://ourcampus.unimelb.edu.au/sustainability.

[30] See https://www.newcastle.edu.au/newsroom/featured-news/first-australian-university-to-sign-100-renewable-electricity-contract.

Melbourne Renewable Energy Project (MERP)[31] involving a PPA for off-site wind generated energy.

12 Sustainability Plans and Targets (Including Carbon Neutrality and Divestment)

In addition to universities with clear plans and targets, some universities, e.g. University of New England (UNE), mention such plans but they are not publicly available. Other universities mention a figure on facilities websites, e.g. Australian Catholic University (ACU) (7% reduction of energy and water by 2020—bench-line 2014) without specifics about actions. ACU also advises on its CO_2e emissions for particular years (2017 25,600) including broader (scope 3) emissions, e.g. travel. They include (broken) links to buildings deemed to follow sustainable design principles albeit not measured by any standard, e.g. GBCA. Others have websites announcing initiatives, e.g. Victoria University (Melbourne), but without revealing or disclosing publicly any trends or benchmarking. Others again, e.g. Flinders University (FLIN), have (available) draft plans, which announce some targets. Thus, from a transparency and materiality context, there is sector leadership among the minority who have made clear commitments to carbon neutrality and even more ambitious leadership by those looking at renewable generation rather than offsets to achieve this in the foreseeable future, e.g. UQ, UNSW, Monash (MON), Melbourne (UOM).

13 Sustainability Reporting: Voluntary Reporting Minimal and Haphazard

Multiple frameworks for reporting, where relevant, now typify the sector. Although GRI remains a force for some or an orientation framework for others, new guidelines such as the sector specific Learning in Future Environments (LiFe) framework[32] and most recently the SDGs have become popular. The GRI global database[33] lists (inaccurately) nine universities who have previously reported although not typically using GRI standards; the drive to inclusivity irrespective of standards has influenced this. Thus, universities can deposit reports that meet the various standards level, allude to them or do not address GRI standards. Sometimes related (accountability)

---

[31]See   https://www.melbourne.vic.gov.au/business/sustainable-business/mrep/Pages/melbourne-renewable-energy-project.aspx.

[32]See https://life.acts.asn.au/about-life/. It appears Charles Sturt University (CSU), James Cook University (JCU) and Macquarie University use this. LiFe appears to be somewhat popular in Europe, see https://www.eauc.org.uk/the_life_index_whole_institution_engagement.

[33]Available at https://www.globalreporting.org/Pages/default.aspx—we exclude from the list the sustainable campus group entry and only include individual organisations.

standards, e.g. AA1000 are mentioned alongside GRI.[34] Scope of reports of whatever nature varies from 2 to 3 page scorecard or overview documents to full independent reports (the minority). Also there is a trend to report directly to websites rather than formulate downloadable documents.

More universities have developed discrete reports, which are not lodged in the GRI, e.g. RMIT reports to GRI but no reports are listed. As noted above both LiFe (MacQuarie, James Cook University (JCU), University of Tasmania (UTAS) and Charles Sturt University (CSU) representing 9% of sector) and SDGs may be the new frameworks alluded to in non-GRI (GRIN below). In the table below, Citing GRI (GRIC) means alluding to the framework. RMIT and the University of Melbourne (UOM) currently continue to report using GRI standards (GRIS). Some universities, e.g. UTAS and more recently LaTrobe (LTR),[35] while they have no discrete reports have websites where multiple dimensions over the years are reported. Others, e.g. Griffith University (GRF)[36] and Curtin University (CUR), Newcastle University (NEW) have developed websites on the different aspects of reporting but provide little or no data on performance over the years often referring to their upcoming or current initiatives.

14 Infrastructure and Buildings

A key performance measure for sustainability is the campus built environment as both an environment for emissions and a source through buildings of emissions. One recent study sows that reputation rather than environment per se are key motivations for green buildings on campus (Li et al. 2013). Universities come with degrees of built environment legacy, multi-campus sites, and ownership issues, which make reporting and measuring up challenging. Most CO_2e emissions reported to NGERS or voluntary are linked to the built environment and in addition to other strategies, e.g. development of renewable energy production and efficiencies, new 'green' buildings are the trend to overall emission increases on campus suggest otherwise. Universities typically advertise their commitments to new building and energy initiatives, e.g. renewable solar installations, as evidence of leadership.

In Australia, the key benchmark for sustainability building is the Green Building Council Association (GBCA) Green Star rating system.[37] This assess buildings against sustainability criteria and awards in the range of four to six stars; it is comparable to other schemes, e.g. LEEDS (USA) and indicates 'potential' carbon reductions (Chen et al. 2013). These awards are for 'as built' potential and not a measure

---

[34] See https://www.accountability.org/standards/.

[35] Curitn University also seems to have followed this path but their website provides no data.

[36] Gritth University refer to a carbon management plan but it is not publicly available.

[37] Project directory accessed 20/02/19 https://new.gbca.org.au/green-star-projects/. The rating system and details are not discussed here in detail. Scoring for 1–3 is not awarded and the stars have scores: 4 (45–59); 5 (60–74) and 6 (76+).

of actual performance; the link between GBCA ratings and actual operation is an open research question (Zuo and Zhao 2014).. Broadly speaking four star is standard practice; five star 'best practice' and six star purportedly 'world's best practice'.[38] A performance rating system, NABERS[39] does provide an 'occupancy' rating and few if any universities report these ratings. As Mitchell (2010) notes there is an urgent need to integrate both reporting frameworks.

I used the GBCA register as a proxy for commitment to sustainable buildings. This study used university projects listed on the GBCA register and their ratings as a proxy for material commitment to a green built environment (n = 26, 60% of sector). Based on this fourty per cent of the sector have made no commitment and a number have made a minor commitment of one building. The table and results are not an indication of overall built environment performance but proxy indication of commitment, and a simple way of quantifying actual commitment and getting beyond rhetorical claims.

UNV	TOT	AVG	4*	5*	6*	REG	*TOT
ACU	225	75		1	2	1	17
ADL	159	79.5			2		12
ANU	222	74		1	2		13
BON	83	83			1		06
Charles Darwin University (CDU)	47	47	1				04
Canberra University (CAN)	68	68		1			05
Charles Sturt University (CSU)	135	67.5		1	1		11
Curtain University (CUR)	202	67.3		2	1[a]	3	16
Griffith University (GRF)	84	84			1	1	06
Latrobe University (LTR)	316	67.2		5			25
Macquarie University (MAQ)	124	62		2		2	10
Monash University (MON)	627	69.7		7	2	4	47
Newcastle University (NEW)	115	67.5		1		1	05
Swinburne University (SWI)	63	63		1			05
Royal Melbourne Institute of Technology (RMIT)	423	60.4		6			30
Queensland University of Technology (QUT)	200	66.7		3		2	15
University of New South Wales (UNSW)	77	77			1		06
University of Queensland (UOQ)	249	83		1	2	1	17
University of Melbourne (UOM)	479	68.4		6	2	4	42
University of Southern Australia (USA)	320	64		5			25

(continued)

[38] There are discussions at present to match GBCA rating system with the National Construction Code (NCC) energy requirements.

[39] https://www.nabers.gov.au/ for NABERS rating processes etc.

(continued)

UNV	TOT	AVG	4*	5*	6*	REG	*TOT
University of Southern Queensland (USQ)	70	70		1			05
University of Tasmania (UTA)	203	67.7		2	1	2	16
University of Technology Sydney (UTS)	344	68.8		3	2	1	27
Western Sydney (UWS)	371	61.8		5		4	25
Victoria University (VIC)	139	69.5			1	1	11
Wollongong University (WOL)	131	65.5	1	1			09

[a]The one six star building registered to Curtin is a Catholic Boys School.

The total column (TOT) indicates points awarded for all registered projects, and the average column (AVG) the awarded points per building; REG refers to registered projects yet to be scored. The *TOT column multiples number of buildings by star rating to give a proxy total. (Universities elsewhere analysed in this study who do not appear at all may be involved in built environment efficiencies but not using this standard). Monash, Melbourne and then RMIT lead clearly in this respect with then a group clustered some way back at 25–30 (n = 5); a third group at 15–17 and then a fourth group 9–14; followed by a few around 5–6 and then the absent 40% making no commitment on this standard.

15 Future Emissions and Energy Commitments; Degrees of Carbon Neutrality

A comprehensive search of organisational websites, reports and media reports show four levels of commitment to Co_2e reductions and (renewable) energy production. These include: (1) declarations about future reductions, e.g. UTS 30% reduction relative to 2007 by 2021; (2) declarations about partial reductions, including specific campuses, e.g. Griffith University MBA program, or energy sources, e.g. electricity neutrality at Flinders (2020); (3) specific targets and timelines for carbon neutrality,[40] which include offsets elsewhere; and (4) targets for zero emissions without carbon neutrality through on-site and off-site renewable production and energy efficiencies, e.g. Monash, UNSW, University of Melbourne. Thus Monash University Net Zero strategy aiming for carbon neutrality in 2030 includes divestment from fossil fuel energy sources and a combination of on-site and off-site energy production, as well as efficiencies, to achieve carbon neutrality. The status of universities globally with respect to carbon neutrality is at least judging by an internet search is modest and therefore Australia seems representative.

For some universities public reporting is minimal, e.g. UWA with one page dedicated to energy, water and recycling use. Most universities included in this study

---

[40]Guidelines for Carbon Neutrality Certification are at http://www.environment.gov.au/climate-change/government/carbon-neutral/certification.

show increases over time in energy consumption and water use, and variable results in other issues, e.g. recycling or renewable power generation. Many report missing target levels, achieving positive results in some areas, and may convert absolute figures into improvement per m^2 or per EFTSL (Table 2).

Some reports indicate periods of activity associated with particular individuals, which then change. For example, La Trobe (LTB) was a leader at one point in this area during a time when they also had appointed a PVC for Sustainability. With the end of that appointment discrete standards-based reporting was dropped although GRI aligned web-based reporting has continued (GRICW). The same is true of Swinburne University (SWIN), which conducted two reports, one uploaded here, and that engagement was also linked to the commitment of a particular individual who later left.

16   Interview Recruitment and Analysis

The Australian Campuses Towards Sustainability (ACTS) network[41] was used to identify sector members with a commitment to reporting. From the ACTS list of members, university websites were searched for evidence of sustainability reporting, strategy and officers. Email and phone details provided for individuals on office sites were used to contact potential participants with officer or manager status. All interviewees were provided with consent, information, and interview prompt information. Subsequent to consent, all interviews with one exception were conducted via phone and lasted on average thirty minutes. In several cases, it was necessary to obtain internal approval to speak with individuals or interviews were delegated to individuals (typically by managers) within the office. On this basis, twenty individuals were recruited to the study between October to December 2018. Given the limitations of this chapter, analysis of five interviews were combined with the content analysis to produce themes below.

1. Interviews were transcribed and analysed using NVIVO v.12 software with themes developed in relation to the existing literature and the trends and characteristics identified above from plans and strategies. The analysis process is staged and involved:
2. Preliminary themes identified from the literature, e.g. academic and professional divide, organisational tactics, e.g. carbon neutrality claims and counterclaims
3. Further themes developed from desktop analysis of organisational websites, e.g. range of styles and approaches and limited data transparency
4. Detailed thematic analysis of interview transcripts and theme development and clarification. This typically involves substantive coding of interviews and association with themes above

---

[41] See https://www.acts.asn.au/.

Table 2 Formal reporting, plans, policy and targets by Australian Universities

	GH or CN	Plan(s)	2008	2009	2010	2011	2012	2013	2014	2015	2016	2017	2018
ADL	CN 2050[a]												
ANU	35%GH 2020	Yes	GRIN	GRIN	GRIN	GRIN	GRIN	GRIN	GRIN	GRIN	GRIN		
CSU	CN 2016		GRIN	GRIN	GRIN	GRIN	GRIN	GRIN	GRIN	GRIN	GRIN	GRIN	GRIN
DKN	CN 2030								GRI4	GRI4	GRI4	GRIN	GRIN
FDU		Yes			GRIN	GRIN	GRIN	GRIN	GRIN	GRIN	GRIN		
LTB	25% 2022				GRI3	GRI3	GRI3	GRI3	GRI3	GRICW	GRICW	GRICW	GRICW
MCQ	40%GH 2020	Yes	GRIC	GRIC	GRIC	GRIC	GRIC	GRIC	GRIC				
MON	CNR 2030					GRIC	GRIC	GRIC	GRIN	GRIN	GRIN	GRIN	GRIN
RMIT	CN 2030[b]									GRIC	GRIC		GRIC
SCU									GRI3				
SWIN								GRIN[c]		GRIN	GRIN		GRIN
UOM	CN 2030[d]	Yes									GRI4	GRIS	GRIS
UNSW	CN 2020							GRIC	GRIC	GRIC	GRIC	GRIC	
UOQ			GRIN	GRIN	GRIN	GRIN	GRIN			SC	SC		
USC	CN 2025							GRIN					

(continued)

Table 2 (continued)

	GH or CN	Plan(s)	2008	2009	2010	2011	2012	2013	2014	2015	2016	2017	2018
USQ	(CN 2020)[e]		GRI3							GRIN	GRIN	GRIN	
UTAS[f]	CN 2017	Yes				GRINW	GRINW	GRINW	GRINW	GRINW	GRINW	GRINW	GRINW
UTS[g]	30% 2021				GRIN	GRIN	GRIN	GRIN	GRIN	GRIN	GRIN	GRIN	
WSU	12% 2020									GRIN	GRIN	GRIN	

[a] The new sustainability plan talks of net zero emissions but it is difficult to know if these means carbon neutral or zero emissions

[b] RMIT claims CN by 2030 applies to building

[c] I am very grateful to Gitanjali Bedi of Monash University Sustainability Institute for bringing this document to my attention

[d] And zero net for electricity by 2021

[e] Reported on one website but then removed

[f] University of Tasmania, similar to LaTrobe, provides annual data for a number of dimensions on its website while also having developed a number of discrete sustainability plans and targets

[g] UTS reports between 2010–2015 are short 2–3 page documents from the annual report. In 2016 and 2017 fuller reporting begins

The overall case study analysis process followed a constructivist grounded theory approach in integrating various sources and analytical codes into memo insights (see Timmermans and Tavory 2012). Driven by expectations from organisational theory of organisations, analysis sought to generate theory consistent with such expectations Throughout the process, coded themes are entered with brief descriptions of their meaning. Memos are subsequently written which integrate literature and findings into insights on specific issues. As others have noted (Birks et al. 2008), memoing in this sense is a valuable analytical tool at all stages of analysis, leading to identification of insights across data sources. These memos may be attached to multiple codes and other items, e.g. whole interviews. This is an on-going and iterative process consistent with thematic analysis.

From the interviews and interview protocol were developed a set of interrelated substantive categories, whose content is summarised as follows:

- EMS and reporting: links and purported relationships between EMS data and reports, including quality of data—note that data as such represent a small part of the overall narrative and that some aspects, e.g. water, more compromised than others
- Institutional theory claims: responses of participants to the idea that organisations respond in the strategic ways identified by theory—general acquiescence to the idea of influence of stakeholder (external and internal) apathy and enthusiasm on reports
- Intersectoral comparison: allusions to organisation sense of comparison with other organisations in the sector—clear perceptions about leaders and also scepticism about some green wash narrative by competitors
- Organisational challenges: clarification on the different challenges that represent the process of reporting, including organisational attitudes and culture—resourcing, executive comprehension and commitment being key issues
- Report and sustainability commitment: relationship between reporting (however done) and overall campus sustainability commitments—recognition of ideal (report reflects commitment) but tensions and missed opportunities
- Approaches and frameworks: opinions on the different frameworks employed in the sector—general consensus on the lack of common ground for comparison and a culture of individual choice
- Sector progress in the next five years: views on where the sector was heading—doubts about major changes but also beliefs about greater integration of technology and real time data in the future

Note that the categories above help organise responses into content focus of responses and are not of themselves theoretically inspired themes—these are addressed below. The move from substantive categories to theory-inspired themes, e.g. institutional theory of organisations, is a complex move (Nowell et al. 2017).

17   Five Themes

Intersecting these themes were insights developed in memos that enabled theory formulation consistent with the grounded approach adopted. I note that qualitative memos integrate and are associated with literature, document analysis, interviews and any other relevant data, e.g. NGERS or similar EMS. This enabled (at the time of writing) development of five overarching themes linked to institutional theory expectations and suggested by prior literature and current analysis.

18   Transparency and Accountability

Running throughout the interviews and the literature and from a theoretical point of view is the notion of transparency and accountability as strategic responses and tactics. Internal and external accountability varies and the perceived value of this to stakeholders, e.g. students, public and others also varies. A compliance mentality and fear of reputation loss through reporting facts is a negative sign. The degree of obscurity or quality in data, as well as executive desire to not report anything that may damage reputation drives approaches also. While competitors do not report scope 3 emissions or otherwise offer full transparency this is used as a reason for minimal compliance reporting. Given the (perceived) low expectation from the public and university community for real commitment, there is also no drive for transparency.

19   Academic and Professional Divide

Sustainability reporting potentially crosses boundaries between material and non-material aspects of sustainability commitments. Divided commitments and engagements to reporting at the intra-institutional level signals some of the internal tensions between rhetoric and reality. This implies in the future closer integration and dialogue between academics and professionals. The former may be distributed across the university in discipline silos or potentially co-located in sustainability institutes; the latter are located typically in facilities and management. Holistic living lab approaches would see both groups working more closely together. In an ideal scenario there would be a real alignment and collaboration between property and academic and this would be reflected in reporting and ultimately in whole of campus approaches. It's apparent in the interviews that properties and facilities and their work is only partly integrated with this other work, meaning whole of campus living lab type scenarios are rare. And this separation is exacerbated by executive or university taking a compliance attitude to reporting and goal setting.

20 Clear Awareness of Inter-sectoral Comparison and Competition

The question of comparisons between organisations comes with benchmarking, e.g. TEFMA, and with publicly available data, e.g. NGERS, but also in the discourses that universities develop for themselves, e.g. carbon neutral competition. The topic is explicit in some interviews and implicit in others. Comparison also emerges in relation to other sectors, e.g., Corporate CSR, as well as in relation to global comparisons, e.g. Green Impact. And clearly a lack of common frameworks makes comparisons impossible. According to some competition among organisations is limited while for others the opposite is true. In the interviews and other discussions universities are treated as public institutions dispensing a public good—education. However, there are clear indications this is only partially true in the market competition for students. Even interviewees acknowledge the influence of competition and differentiation on driving quality of reporting. Ultimately therefore is the question of whether isomorphism or difference rules.

21 Lack of Common Frameworks

The lack of common frameworks for benchmarking alongside a preference for individual choice seemed an unresolvable tension as well as an evident weakness in the desktop survey and content analysis. As noted above, it leads to difficulties in assessing sector progress at all, and probably evidences the low coercive and normative pressure on organisations to respond. One interviewee commented on moves away from GRI as a move away from measurable global standards. The Macquarie University (MAQ) meanwhile talks about the LiFe framework—which has four current adherents in Australia—as being more responsive to sector needs. However, participation is minimal, and measured commitments form only one part of the LiFE framework. In addition, the whole discussion depends on the existence of consensus about the need for reporting and from an institutional point of view, there is no requirement as no socio-political or legislative or other pressure exists. GRI for example has never exerted a strong pressure in this sector.

The fact that GRI does not report learning and teaching etc. and focuses on measurable outcomes rather than the narrative is viewed by some as a barrier to uptake. However, it is precisely the lack of measurement and vagueness of initiatives that disguise inaction as action. This is reinforced in a compliance non-transparency culture, which gets support from a socio-political apathy in Australia. So in fact, the small minority reporting to GRI and or revealing publicly energy statistics (as TEFNA and NGER are closed) are actually making the stronger social commitment.

22 Desire for Holistic Living Lab Integration of All Stakeholders

Among the desiderata for the sector is a more integrative approach bringing together all internal stakeholders (Mcmillin and Dyball 2009; Hancock and Nuttman 2014). While popular evidence is limited of such whole-of—campus approaches (Adams 2013). Such a notion implies greater transparency and real leadership. In this respect, the notion of living labs is popular. The idea of holistic sustainability transformation on campus invests the term with another meaning, i.e. the campus as a living lab in its approach to sustainability. Providing opportunity for staff and students to promote things.

For mainly professional staff the key current initiative is the Green Impact Program.[42] Bt contrast although many universities encourage student engagement with community sustainability initiatives this is not necessarily mirrored by access to campus data and issues for resolution. For direct engagement of students and faculty with facilities environmental challenges, UTAS has a Sustainability Program for Integration of Students (SIPS) program[43] that relates to this, Adelaide University Ecoversity Internships[44] also offer opportunities for direct engagement of students with facilities. For James Cook University (JCU) this is Tropeco[45]—the program, which includes roles for internships. The level of transparency and access to campus data will probably vary here. The five interviewees all mentioned the desirability of such approaches.

23 Conclusion

This study reports analysis of data from an on-going study informed by organisational theory of university commitment to sustainability reporting. The study finds a sector making slow and mixed progress towards campus sustainability with a few organisations showing some leadership in transparency and action. Holistic cross-campus living lab type projects exist but are a minority and there scope needs expansion. The study chose to interview and analyse what falls normally under environmental management of organisations rather than focus on the research, e.g. grants, and teaching, or declarative commitments of universities, e.g. declarations signed. These declarative or aspirational claims are by their very nature non-material, and there is some evidence in the literature already of the limits of their impact; and the fact that teaching is haphazard anyway. Current criticism of universities hinges more on the inconsistency between a rhetoric and reality on campus.

---

[42] See https://greenimpact.acts.asn.au/.

[43] See http://www.utas.edu.au/infrastructure-services-development/sustainability/SIPS.

[44] See https://www.adelaide.edu.au/ecoversity/action/internships/.

[45] See https://www.jcu.edu.au/tropeco-sustainability-in-action.

It is the potential contradiction between claims to be sustainability leaders or champions etc. and material processes on campus the key issue not just whether a compelling reporting narrative by leaders and marketing can be developed. From this perspective it was important to interview sustainability officers and managers in specific institutions rather than academics with no particular 'skin in the game' so to speak. The second aspect was taking a broad view of sustainability reporting not limited to formal reports following specific guidelines, e.g. GRI, as these guidelines and that kind of reporting is marginal in the sector, and hence would mis-represent sector commitments. The third innovative element with respect to the above was employing institutional theory of organisations rather than 'common sense' to explain the state of reporting in this sector. While there has been some initial work in this area it remains marginal to the literature on the topic.

Gray (2011) notes on current approaches to sustainability reporting that 'For an organization to be contributing to sustainability it needs to be, somewhat simplistically perhaps, reducing its ecological footprint and increasing equality of access to environmental resources. Neither of these conditions is very likely for an organization with commercial imperatives and the first will be unlikely for an organization that is growing' (Gray 2011, 441). This study finds with some exceptions, the university sector contribution to fit this pattern. In their study of corporate reporting from an institutional theory perspective, Iarossi et al. (2013b) note the strategic responses of companies to acquiesce to external and internal demands. From this perspective, local higher education arrangements, e.g. regarding sustainability activities, are dependent on the broader institutional environment, and reflect cultural scripts and organizational rules (Meyer et al. 2006). Such a view is particularly important for theorising sustainability commitments and reporting rather than allowing such commitments to be analysed using the rhetoric of the university itself.

Meyer et al. (2006) have pointed to the highly isomorphic nature of the university sector globally in structure and practices. Sustainability reporting and commitments of universities is no exception, as demonstrated in this study. This study has employed institutional theory of organisations to sustainability reporting in Australian universities, and in so doing implicitly highlighted the corporate nature of the sector. Although it is frequent for universities to claim a moral cum educational role as public institutions, recent commentators point to how universities operate as firms in a market delivering education as 'human capital formation' (Connell 2013, 104). Thus, similar to firms universities compete to market educational goods and services to student customers and clients (Olssen and Peters 2005).

Beynagi et al. (2016) note that in the post UN Decade for Education in Sustainable Development (UNDESD 2005–2014), universities are faced with different social, environmental or economic scenarios as a response to implementing the SDGs in an era of climate change. Insitutional theory of organisations allows us to understand the current and future reponses of organisations to sustainability, and potentially promote change. While there is clearly potential and a mandate for universities to provide leadership, current recommendations for how this can be implemented (e.g. Ralph and Stubbs 2014), repeat aspirational advice from previous years and fail to address institutional barriers.

References

Adams CA (2013) Sustainability reporting and performance management in universities: challenges and benefits. Sustain Acc Manage Policy J 4:384–392. https://doi.org/10.1108/SAMPJ-12-2012-0044

Alvesson M, Ashcraft KL (2012) Interviews. In: Qualitative organizational research: core methods and current challenges. SAGE Publications, Inc., London, EC1Y 1SP, pp 239–257. https://doi.org/10.4135/9781526435620.n14

Arroyo P (2012) Management accounting change and sustainability: an institutional approach. J Acc Org Change 8:286–309. https://doi.org/10.1108/18325911211258317

Azizi L, Bien C, Sassen R (2018) Recent trends in sustainability reporting by German universities. NachhaltigkeitsManagementForum|Sustainability Management Forum. https://doi.org/10.1007/s00550-018-0469-8

Baer HA, Gallois A (2018) How committed are Australian universities to environmental sustainability? A perspective on and from the University of Melbourne. Crit Sociol 44:357–373. https://doi.org/10.1177/0896920516680857

Bebbington J, Higgins C, Frame B (2009) Initiating sustainable development reporting: evidence from New Zealand. Acc Audit Acc J 22. https://doi.org/10.1108/09513570910955452

Bekessy SA, Samson K, Clarkson RE (2007) The failure of non-binding declarations to achieve university sustainability. Int J Sustain Higher Educ 8:301–316. https://doi.org/10.1108/14676370710817165

Berthod O (2016) Institutional theory of organizations. In: Farazmand A (ed) Global encyclopedia of public administration, public policy, and governance. Springer International Publishing AG, Cham, pp 1–5. https://doi.org/10.1007/978-3-319-31816-5_63-1

Beynaghi A, Trencher G, Moztarzadeh F, Mozafari M, Maknoon R, Leal W (2016) Future sustainability scenarios for universities : moving beyond the United Nations decade of education for sustainable development. J Clean Prod 112:3464–3478. (Elsevier Ltd). https://doi.org/10.1016/j.jclepro.2015.10.117

Birks M, Chapman Y, Francis K (2008) Memoing in qualitative research: probing data and processes. J Res Nurs 13:68–75. https://doi.org/10.1177/1744987107081254

Brown, HS, de Jong M, Levy DL (2009b) Building institutions based on information disclosure: lessons from GRI's sustainability reporting. J Clean Prod 17:571–580 (Elsevier Ltd). https://doi.org/10.1016/j.jclepro.2008.12.009

Brown HS, de Jong M, Lessidrenska T (2009) The rise of the global reporting initiative: a case of institutional entrepreneurship. Environ Polit 18:182–200. https://doi.org/10.1080/09644010802682551

Brusca I, Labrador M, Larran M (2018) The challenge of sustainability and integrated reporting at universities: a case study. J Clean Prod 188:347–354 (Elsevier Ltd.). https://doi.org/10.1016/j.jclepro.2018.03.292

Buhr N, Gray R, Milne, MJ (2014) Histories, rationales, voluntary standards and future prospects for sustainability reporting: CSR, GRI, IIRC and beyond. In: Gibassier D, Unerman J (eds) Sustainability accounting and acocuntability, 2nd ed. Routledge, London, pp 51–71. https://doi.org/10.1023/a:1026216110583

Ceulemans K, Molderez I, Van Liedekerke L (2015) Sustainability reporting in higher education: a comprehensive review of the recent literature and paths for further research. J Clean Prod 106:127–143 (Elsevier Ltd.). https://doi.org/10.1016/j.jclepro.2014.09.052

Chang HC, Deegan C (2009) Environmental management accounting and environmental accountability within universities: current practice and future potential. In: Environmental management accounting for cleaner production, vol 24. Springer Science & Business Media B.V., pp 301–320. https://doi.org/10.1007/978-1-4020-8913-8_16

Chatelain-Ponroy S, Morin-Delerm S (2016) Adoption of sustainable development reporting by universities: an analysis of French first-time reporters. Acc Audit Acc J 29:887–918. https://doi.org/10.1108/AAAJ-06-2014-1720

Chen Q, Zuo J, Xia Bo, Skitmore M, Pullen S (2013) Green star points obtained by australian building projects. J Arch Eng 19:302–308. https://doi.org/10.1061/(asce)ae.1943-5568.0000121

Cho CH, Laine M, Roberts RW, Rodrigue M (2015) Organized hypocrisy, organizational façades, and sustainability reporting. Accounting, organizations and society 40. Elsevier Ltd., pp 78–94. https://doi.org/10.1016/j.aos.2014.12.003

Christie BA, Miller KK, Cooke R, White JG (2015a). Environmental sustainability in higher education: what do academics think? In: Environmental Education Research, vol 21. Routledge, pp 655–686. https://doi.org/10.1080/13504622.2013.879697

Christie BA, Miller KK, Cooke R, White JG (2015b) Environmental sustainability in higher education: what do academics think? Environ Educ Res 21:655–686. https://doi.org/10.1080/13504622.2013.879697

Connell R (2013) The neoliberal cascade and education: an essay on the market agenda and its consequences. Critic Stud Educ 54:99–112. https://doi.org/10.1080/17508487.2013.776990

Corcoran PB, Walker KE, Wals AEJ, Walker KE (2004) Case studies, make your case studies, and case stories: a critique of case study methodology in sustainability in higher education. Environ Educ Res 10:7–21. https://doi.org/10.1080/1350462032000173670

Disterheft A, Caeiro S, Azeiteiro UM, Filho WL (2015) Sustainable universities—a study of critical success factors for participatory approaches. J Clean Prod 106:11–21 (Elsevier Ltd.). https://doi.org/10.1016/j.jclepro.2014.01.030

Eisenhardt KM (1989) Building theories from case study research. Acad Manage Rev 14:532. https://doi.org/10.2307/258557

Fernandez-Feijoo B, Romero S, Ruiz S (2012) Transparent information system for sustainability. Sect Anal Procedia Technol 5:31–39. https://doi.org/10.1016/j.protcy.2012.09.004

Fernandez-Feijoo B, Romero S, Ruiz S (2014) Effect of stakeholders' pressure on transparency of sustainability reports within the GRI framework. J Bus Eth 122:53–63. https://doi.org/10.1007/s10551-013-1748-5

Ferrero-Ferrero I, Fernández-Izquierdo MA, Muñoz-Torres MJ, Bellés-Colomer L (2018) Stakeholder engagement in sustainability reporting in higher education: an analysis of key internal stakeholders' expectations. Int J Sustain Higher Educ 19:313–336. https://doi.org/10.1108/IJSHE-06-2016-0116

Flint K (2001) Institutional ecological footprint analysis—a case study of the University of Newcastle, Australia. Int J Sustain Higher Educ 2:48–62. https://doi.org/10.1108/1467630110380299

Fonseca A, Macdonald A, Dandy E, Valenti P (2011) The state of sustainability reporting at Canadian universities. Int J Sustain Higher Educ 12:22–40. https://doi.org/10.1108/14676371111098285

Gamage P, Sciulli N (2017) Sustainability reporting by Australian universities. Aust J Public Admin 76:187–203. https://doi.org/10.1111/1467-8500.12215

Glover JL, Champion D, Daniels KJ, Dainty AJD (2014) An institutional theory perspective on sustainable practices across the dairy supply chain. Int J Prod Econ 152:102–111 (Elsevier). https://doi.org/10.1016/j.ijpe.2013.12.027

Glover A, Strengers Y, Lewis T (2017) The unsustainability of academic aeromobility in Australian universities. Sustain Sci Pract Policy 13:1–12 (Informa UK Limited, trading as Taylor & Francis Group). https://doi.org/10.1080/15487733.2017.1388620

Gray R (2011) Reporting and accounting. In: Brady J, Ebbage A, Lunn R (eds) Environmental management in organizations: the IEMA handbook, 2nd ed. Earthscan, Abingdon

Guthrie J, Farneti F (2008) Public money and management GRI sustainability reporting by Australian public sector organizations. Public Money Manage 28:361–366. https://doi.org/10.1111/j.1467-9302.2008.00670.x

Hahn R, Kühnen M (2013) Determinants of sustainability reporting: a review of results, trends, theory, and opportunities in an expanding field of research. J Clean Prod 59:5–21 (Elsevier Ltd.). https://doi.org/10.1016/j.jclepro.2013.07.005

Hahn R, Reimsbach D, Schiemann F (2015) Organizations, climate change, and transparency: reviewing the literature on carbon disclosure. Org Environ 28:80–102. https://doi.org/10.1177/1086026615575542

Hancock L, Nuttman S (2014) Engaging higher education institutions in the challenge of sustainability: sustainable transport as a catalyst for action. J Clean Prod 62:62–71 (Elsevier Ltd.). https://doi.org/10.1016/j.jclepro.2013.07.062

Higgins B, Thomas I (2016) Education for sustainability in universities: challenges and opportunities for change. Aust J Environ Educ 32:91–108. https://doi.org/10.1017/aee.2015.56

Higgins C, Milne MJ, van Gramberg B (2015) The uptake of sustainability reporting in Australia. J Bus Eth 129:445–468 (Springer Netherlands). https://doi.org/10.1007/s10551-014-2171-2

Hodgson GM (2007) Institutions and individuals: interaction and evolution. Org Stud 28:95–116. https://doi.org/10.1177/0170840607067832

Holstein J, Gubrium J (1995) The active interview. SAGE Publications, Inc., Thousand Oaks, California, USA. https://doi.org/10.4135/9781412986120

Iarossi J, Miller JK, O'Connor J, Keil M (2013a) Addressing the sustainability challenge: insights from institutional theory and organizational learning. J Leadersh Acc Eth 10:76–91. https://doi.org/10.2139/ssrn.1839802

Iarossi J, Miller JK, O'Connor J, Keil M (2013b) Addressing the sustainability challenge: insights from institutional theory and organizational learning. J Leadersh Acc Eth 10:76–91. https://doi.org/10.2139/ssrn.1839802

Jensen JC, Berg N (2012) Determinants of traditional sustainability reporting versus integrated reporting. an institutionalist approach. Bus Strateg Environ 21:299–316. https://doi.org/10.1002/bse.740

Jones M (2014) Researching organizations: the practice of organisational fieldwork. SAGE Publications, London, Thousand Oaks, New Delhi

Larrán Jorge M, Andrades Peña FJ, Madueño JH (2018) An analysis of university sustainability reports from the GRI database: an examination of influential variables. J Environ Plan Manage 0568:1–26 (Taylor & Francis). https://doi.org/10.1080/09640568.2018.1457952

Lee B (2012) Using documents in organizational research. In: Qualitative organizational research: core methods and current challenges. SAGE Publications, Inc., London EC1Y 1SP, pp 389–407 https://doi.org/10.4135/9781526435620.n22

Lee KH, Barker M, Mouasher A (2013) Is it even espoused? An exploratory study of commitment to sustainability as evidenced in vision, mission, and graduate attribute statements in Australian universities. J Clean Prod 48:20–28 (Elsevier Ltd.). https://doi.org/10.1016/j.jclepro.2013.01.007

Leviston Z, Greenhill M, Walker I (2014) Australian attitudes to climate change and adaptation: 2010–2014. Canberra, ACT. https://doi.org/10.4225/08/584af21158fe9

Li X, Strezov V, Amati M (2013) A qualitative study of motivation and influences for academic green building developments in Australian universities. J Green Build 8:166–183. https://doi.org/10.3992/jgb.8.3.166

Liu Z, Abhayawansa S, Jubb C, Perera L (2017) Regulatory impact on voluntary climate change–related reporting by Australian government-owned corporations. Financ Acc Manage 33:264–283. https://doi.org/10.1111/faam.12124

Livesey SM, Kearins K (2002) Transparent and caring corporations? Org Environ 15:233–258. https://doi.org/10.1177/1086026602153001

Lozano R, Lozano FJ, Mulder K, Huisingh D, Waas T (2013) Advancing higher education for sustainable development: international insights and critical reflections. J Clean Prod 48:3–9. https://doi.org/10.1016/j.jclepro.2013.03.034

Lozano R, Ceulemans K, Alonso-Almeida M, Huisingh D, Lozano FJ, Waas T, Lambrechts W, Lukman R, Hugé J (2015) A review of commitment and implementation of sustainable development in higher education: results from a worldwide survey. J Clean Prod 108:1–18 (Elsevier Ltd.). https://doi.org/10.1016/j.jclepro.2014.09.048

Mcmillin J, Dyball R (2009) Developing a whole-of-university approach to educating for sustainability. J Educ Sustain Dev 3:55–64. https://doi.org/10.1177/097340820900300113

Meyer JW (2017) Reflections on institutional theories of organizations. In Greenwood R, Oliver C, Lawrence TB, Meyer RE (eds) The sage handbook of organizational institutionalism, 2nd ed. SAGE Publications Ltd, London, New York, pp 788–809

Meyer JW, Ramirez F, Frank DJ, Schofer E (2006) Higher education as an institution. In: Gumport Patricia J (ed) Sociology of higher education: contributions and their contexts. John Hopkins University Press, Baltimore, pp 187–221

Mitchell LM (2010) Energy efficient cities. In: Bose RK (ed) Energy efficient cities: assessment tools and benchmarking practices. The World Bank, Washington DC. https://doi.org/10.1596/978-0-8213-8104-5

Nowell LS, Norris JM, White DE, Moules NJ (2017) Thematic analysis: striving to meet the trustworthiness criteria. Int J Qual Methods 16:1–13. https://doi.org/10.1177/1609406917733847

Oliver C (1991) Strategic responses to institutional processes. Acad Manage Rev 16:145–179. https://doi.org/10.5465/amr.1991.4279002

Olssen M, Peters MA (2005) Neoliberalism, higher education and the knowledge economy: from the free market to knowledge capitalism. J Educ Policy 20:313–345. https://doi.org/10.1080/02680930500108718

Perez-Batres LA, Miller VV, Pisani MJ (2011) Institutionalizing sustainability: an empirical study of corporate registration and commitment to the United Nations global compact guidelines. J Clean Prod 19:843–851 (Elsevier Ltd.). https://doi.org/10.1016/j.jclepro.2010.06.003

Ralph M, Stubbs W (2014) Integrating environmental sustainability into universities. Higher Educ 67:71–90. https://doi.org/10.1007/s10734-013-9641-9

Ramirez M (2015) Commitments of university leaders to the talloires declaration: are they evidenced in industrial design teaching and learning? In: Filho WL, Brandli L, Kuznetsova O, Paço A (eds) Integrative approaches to sustainable development at University Leve. Springer International Publishing, Cham, pp 225–244. https://doi.org/10.1007/978-3-319-10690-8_16

Riedy C, Daly J (2010) Targeting a low-carbon university: a greenhouse gas reduction target for the australian technology network of universities. In: Filho WL (ed) Universities and climate change. Springer, Cham, pp 151–162. https://doi.org/10.1007/978-3-642-10751-1_12

Sassen R, Azizi L (2017) Voluntary disclosure of sustainability reports by Canadian universities. J Bus Econ 88:97–137 (Springer Berlin Heidelberg). https://doi.org/10.1007/s11573-017-0869-1

Scott WR (2008) Approaching adulthood: the maturing of institutional theory. Theory Soc 37:427–442. https://doi.org/10.1007/s11186-008-9067-z

Sherren K (2006) Core issues: reflections on sustainability in Australian University coursework programs. Int J Sustain Higher Educ 7:400–413. https://doi.org/10.1108/14676370610702208

Sinkovics RR, Alfoldi EA (2012) Facilitating the interaction between theory and data in qualitative research using CAQDAS. In Qualitative organizational research: core methods and current challenges. SAGE Publications, Inc.,, London EC1Y 1SP, pp 109–131. https://doi.org/10.4135/9781526435620.n7

Stefanova K, Menzies L, Connor J (2016) Climate of the Nation 2016: Australian attitudes on climate change. Sydney, NSW. https://doi.org/10.1017/cbo9781107415324.004

Stubbs W, Higgins C (2014) Integrated reporting and internal mechanisms of change. Acc Audit Acc J 27:1068–1089. https://doi.org/10.1108/AAAJ-03-2013-1279

Suddaby R, Lefsrud L (2010) Institutional theory, old and new. In Mills A, Durepos G, Wiebe E, (ed) Encyclopedia of case study research. SAGE Publications, Inc., California, United States, pp 465–473. https://doi.org/10.4135/9781412957397

Symon G, Cassell C (2012) Qualitative organizational research: core methods and current challenges. SAGE Publications, Inc., London EC1Y 1SP. https://doi.org/10.4135/9781526435620

The Reporting Exchange (2018) Sustainability reporting in Australia: jumping into the mainstream

Timmermans S, Tavory I (2012) Theory construction in qualitative research. Sociol Theory 30:167–186. https://doi.org/10.1177/0735275112457914

Vaughter P, Wright T, Herbert Y (2015) 50 shades of green: an examination of sustainability policy on Canadian campuses. Can J Higher Educ 45:81–100 (Manuscript in press)

Vermeulen WJV, Bootsma MC, Tijm M (2014) Higher education level teaching of (master's) programmes in sustainable development: analysis of views on prerequisites and practices based on a worldwide survey. Int J Sustain Dev World Ecol 21:430–448 (Taylor & Francis). https://doi.org/10.1080/13504509.2014.944956

World Wildlife Foundation (2018) Australian attitudes to nature 2017

Zucker LG (1987) Institutional theories of organization. Ann Rev Sociol 13

Zuo J, Zhao ZY (2014) Green building research-current status and future agenda: a review. Renew Sustain Energy Rev 30:271–281 (Elsevier). https://doi.org/10.1016/j.rser.2013.10.021

Sustainability Reporting in Higher Education Institutions: What, Why, and How

Naif Alghamdi

Abstract The main purpose of this research is to address three important questions concerning sustainability reporting in universities: what to report, why reporting is necessary, and how reporting should be carried out. A desk study method was used to systematically review key peer-reviewed scientific articles as well as sustainability reports of higher education institutions. The paper discusses theoretical and empirical research of sustainability reporting in the higher education sector. The paper shows that there is a focus on the widely used criteria of sustainability, which are in line with the literature. The study also illustrates how universities can report their sustainability advancement through highlighting the type of data collected, process of preparing the report, the parties involved, and channels where reports can be submitted to. Ultimately, this paper seeks to justify the significance of such reporting and why all higher education institutions should develop and regularly review the reporting of their sustainability performance. The added value of this research is that higher education institutions, particularly those at the early stages of assessing, documenting, and reporting sustainability practices, can be greatly assisted.

Keywords Sustainability · Higher education · Sustainability reporting · Universities

1 Introduction

Sustainability issues have become progressively prominent (Thomashow 2014). As the president of Cornell University, David Skorton, said 'Sustainability is no longer an elective. It is a prerequisite' (Sharp and Shea 2012, 79). Although there are many higher education institutions worldwide that take sustainability issues into account, these institutions represent a small proportion of the total number of institutions

N. Alghamdi (✉)
Department of Architecture and Building Science (DABS), College of Architecture and Planning (CAP), King Saud University (KSU), King Abdullah Road, P.O. Box 230670, Riyadh 11321, Saudi Arabia
e-mail: naag@ksu.edu.sa

© Springer Nature Switzerland AG 2020
W. Leal Filho et al. (eds.), *International Business, Trade and Institutional Sustainability*, World Sustainability Series,
https://doi.org/10.1007/978-3-030-26759-9_57

in the world (Lozano et al. 2013). Failure to take sustainability seriously is largely due to the lack of awareness, lack of support, over-crowded curricula, resistance to change, lack of accountability, and lack of resources (Davis et al. 2003; Velazques et al. 2006; Chau 2007; Bekessy et al. 2007; Alghamdi et al. 2017).

Higher education institutions occupy a central role in supporting the advancement of sustainable development. Several of these institutions offer solutions to overcome societal issues as well as conduct essential research and studies on sustainability (Sassen et al. 2014). Colleges and universities are where future decision makers are being trained and educated and hence institutions can influence today's students for a better and greener future (Adams 2013). Universities around the world deal with the impacts of their practices and operations, which can be addressed through sustainable measures (Alonso-Almeida et al. 2015). Against all these opportunities and constraints, higher education institutions can make a huge impact on sustainability aspects, both on and off campus (Waas et al. 2012).

One of the challenges of sustainability in higher education is reporting. On the one hand, there has not been a great deal of research on sustainability reporting by universities (Ceulemans et al. 2015). On the other hand, sustainability reporting in the higher education sector 'is still in a very early stage' (Huber and Bassen 2018, 218). Sassen and Azizi (2018, 1159) confirm this and point out that "sustainability reporting by universities is still in its infancy compared with that of firms." This is not only because a low percentage of universities do actually report their sustainability progress (Fonseca et al. 2011; Lopatta and Jaeschke 2014; Alonso-Almeida et al. 2015; Sassen et al. 2015), but also because of the concerns in both the depth of reporting as well as the content of what is reported (Lopatta and Jaeschke 2014). An additional reason is the lack of accepted sustainability reporting standard for universities (Adams 2013), unlike the Global Reporting Initiative (GRI) guidelines which are the most common standards used by firms (KPMG 2013). To address the challenges of sustainability reporting in the higher education sector, this paper focuses on the three key questions: what to report, why reporting is necessary, and how reporting should be carried out. These questions assist colleges and universities in understanding the purpose of reporting sustainability (the 'why' question) and to simplify the process of reporting (the 'what' and 'how' questions).

The key terms to be clearly defined in this study are sustainability, sustainable university, reporting, and sustainability reporting. The Sustainable Development (SD) was defined as a development that "meets the needs of the present without compromising the ability of future generations to meet their own needs"—World Commission on Environment and Development (United Nations 1987). Scoullos (2010, 06) succinctly defined sustainable university as "a university which contributes to Sustainable Development (SD), a university which is able to deliver the message of integration and progress in all aspects of SD, to promote socially just, economically prosperous, and environmentally benign development, through the concepts, principles, and methods of Education for Sustainable Development (ESD) and where its staff, governance, operations, and infrastructures reflect commitment to SD". As for the word 'report', it means "An account given of a particular matter, especially in the form of an official document, after thorough investigation or consideration by an appointed person or body" (Oxford Dictionary 2019). The word reporting is defined as 'Give a spoken or written account of something that one has observed, heard, done, or investigated' (Ibid). Regarding the definition of sustainability reporting, it is defined as:

> communication of information regarding the sustainability performance of an organization to its stakeholders by any media. This may take the form of a print media standalone report, an online presentation of sustainability information, or the inclusion of sustainability information as a clearly defined subsection of another report to stakeholders (for example as part of an annual report). (Richardson and Kachler 2016, 6)

This paper has been divided into four parts. The first part introduces the methodology used in this research. The second part deals with the issue of what to report, through which universities can be informed about what should be reported. The third part explains the reasons why higher education institutions need to report their sustainability performance. The fourth part presents how sustainability practices and operations in universities can be reported. The last part highlights some final remarks and recommendations for colleges and universities to advance their sustainability reporting.

2 Research Methodology

The methodological approach taken in this paper is a desk study. Desk study was used to review earlier research to gain a broad understanding of the issue (Travis 2016). A desk study method incorporated reviewing research papers, sustainability frameworks, and reporting systems' websites. In this research, the method was used to systematically review key peer-reviewed scientific articles as well as sustainability reports of higher education institutions. The paper discusses theoretical and empirical research of sustainability reporting in the higher education sector. Examples of key references which extensively reviewed sustainability reporting issues in higher education sector include for example Huber and Bassen (2018), Sassen and Azizi (2018), Yanez et al. (2018), Zorio-Grima et al. (2018), Richardson and Kachler (2016), Thijssens et al. (2016), Ceulemans et al. (2015), Lozano (2013), and Lozano (2011). Such comprehensive research was undertaken to assist us in answering the three main questions raised in this present study. As for the sustainability reporting systems, three main sustainability systems have been reviewed regarding their reporting systems. These systems were the Global Reporting Initiative (GRI), UI Green Metric, and Sustainability Tracking, Assessment and Rating System (STARS). Content analysis was used for a sample of 4 online reports to gain a detailed understanding of the sustainability criteria and indicators merely explain what to report about and how. The randomly selected sustainability reports were of Stanford University (United States of America), National University of Ireland (Republic of Ireland), University of Waterloo (Canada), and American University of Sharjah (United Arab Emirates). Some key lessons were drawn from these real sustainability reports submitted to STARS.

3 What to Report

Sustainability reports reveal important information on an institution's 'triple bottom line' (Guidry and Patten 2010). The 'triple bottom line' places an emphasis on people, planet, and profit, which represent the social, environmental, and economic aspects of any organisation (Flint 2010). Similarly, scholars, in general, see sustainability through the three E's: ecology, economy, and equity. These reports represent essential data, which consists of descriptions, facts, and figures about the sustainability performance of higher education institutions. Sustainability reporting is a cycle that includes "a regular program of data collection, communication, and responses" (GRI 2019).

Reporting sustainability is a 'voluntary activity' which has two universal aims: first, is to 'assess the current state of an organisation's economic, environmental and social dimensions', and second, is to "communicate a company's efforts and sustainability progress to their stakeholders" (Dalal-Clayton and Bass 2002; GRI 2007; Hamann 2003; Lozano 2013). Richardson and Kachler (2016, 6) point out that the purpose of a sustainability report is to "provide stakeholders with sufficient information to hold the organization accountable for its sustainability performance". In order to manage sustainability reporting, a number of guidelines and standards have been developed (Lozano and Huisingh 2011; Cole 2003; Perrini and Tencati 2006). Currently, the most widely used guidelines are: the ISO 14000 family (ISO 2019) and the GRI Sustainability Guidelines (GRI 2019).

The sustainability reports are described as balanced but often compartmentalised (Lozano 2013). Lozano (Ibid, 64) explains this by highlighting 'the lack of inclusion' of sustainability reporting, and hence encourages 'a more systemic approach' to reporting through actively looking for "the interlinking issues and dimensions, in order to gain new insights with a view to reducing, or even avoiding, conflicts between/among issues."

What can help in addressing sustainability reporting is understanding what it takes to deliver a successful report. Higher education institutions should be informed about the possible factors that might influence sustainability reporting. In order to facilitate such understanding, Thijssens et al. 2016 have raised a number of key questions for companies regarding their sustainability reporting. These questions are equally applicable in the higher education sector and can also be answered by colleges and universities. The questions are as follows (Ibid, 91):

- How is sustainability reporting represented in the company? Do affiliated employees devote 100% of time? How many FTE? What is the estimated annual budget for sustainability reporting? How many management layers are involved in sustainability reporting? Who is responsible for what? Is there final approval? Who decides on the budget? Who coordinates the department or reporting process?

- Is there a formal sustainability reporting strategy? Is it written down? Has this document been published to the employees/public? Who formulates the strategy? How many years ahead is it formulated? Who are the target stakeholder groups? Are they all equally important?

- Do you have a content management system which contains all sustainability reporting information? What are the 3 most important systems generating the sustainability information? Do you have any certifications relating to sustainability reporting? Is the sustainability report verified by an external party? If yes: auditor or consultant? Do you have an external review committee?

- How does the communication within the sustainability department take place? What is the frequency of meetings?

- What is the professional background of the members of the sustainability team? Any members with NGO background? What kinds of training opportunities exist for group members? Which of your company skills would you call superior with regard to sustainability reporting? What is the most important document regarding sustainability reporting? When was it first published?

- How many employees are currently working for the sustainability reporting department? What are the job description and responsibilities of each team member? In case of a matrix structure: which departments do members come from? Are there any groupings within the team? On what are they based?

- How do you create awareness about sustainability reporting? E.g.: how are sustainability reporting award winnings communicated to employees? How do you encourage your employees to contribute to sustainability reporting? What is the most important reason for your company to engage in sustainability reporting?

The above-mentioned questions help us to understand what is needed for effective sustainability reporting. These questions can also yield interesting information on many essential elements of sustainability reporting. These elements include 'structure, strategy, systems, style, skills, staff, and shared values' (Waterman et al. 1980). The origin of these elements is the 7-S model, which is a framework that "systematically takes into account the breadth of the potential factors that may influence sustainability reporting" (Thijssens et al. 2016, 89).

Another important step towards overcoming challenges in sustainability reporting was the development of tools and systems, which were fundamental to the reporting of sustainability issues in higher education institutions. These tools and systems consist of a number of criteria, sub-criteria, and indicators. Scholars have reviewed the development of these sustainability tools and systems (Shriberg 2002; Cole 2003; Alshuwaikhat and Abubakar 2008; Kamal and Asmuss 2013; Gómez et al. 2015). These criteria have been developed to assess sustainability in higher education institutions. Also, they have been used to report the advancement of sustainability in colleges and universities. Examples of these tools and systems include, but not limited to: Sustainability Assessment Questionnaire—SAQ (ULSF 2019), Graphical Assessment of Sustainability in University—GASU (Lozano 2006a), University Environmental Management System—UEMS (Alshuwaikhat and Abubakar 2008), Assessment Instrument for Sustainability in Higher Education—AISHE (Roorda 2002), Benchmarking Indicators Questions—Alternative University Appraisal BIQ-AUA (AUA 2019), Sustainable Campus Assessment System—SCAS (PSPE 2019), Sustainability Tracking, Assessment, and Rating System—STARS (STARS 2019), Sustainability Code for Higher Education (Huber and Bassen 2018), and Green Metric—GM (GM 2019). The common criteria mentioned in the majority of these tools and systems were grouped into five aspects: management, academia, environment, engagement, and innovation (Alghamdi 2018).

Therefore, it can be said that these are the main criteria that higher education institutions use for reporting their advancement in sustainability aspects. Reporting on the management aspect may include the institution's sustainability vision, mission, commitment, statement, strategies, policies, planning, investment, governance. Reporting on the academic aspects might include the institution's formal and informal sustainability education, curriculum, sustainability research, scholarships. Reporting on the environmental aspect may include the institution's greening the infrastructure, renewable energy, conserving water, reducing waste, recycling, materials, optimal use of facilities and spaces of the institutions. Reporting on the engagement aspect may include the institution's social responsibilities, outreach programs, campus community engagement, public engagement. Reporting on the innovation aspect might include the institution's innovative and creative solutions to the sustainability challenges, which show leadership in sustainability.

4 Why Reporting

There are many reasons for organisations, including higher education institutions, to report their sustainability performance. Reporting benefits the institutions, both internally and externally. Benefits of reporting have been summed up by the Global Report Initiative (GRI 2019) as follows:

Internal benefits for companies and organizations can include:

- Increased understanding of risks and opportunities
- Emphasizing the link between financial and non-financial performance
- Influencing long term management strategy and policy, and business plans
- Streamlining processes, reducing costs and improving efficiency
- Benchmarking and assessing sustainability performance with respect to laws, norms, codes, performance standards, and voluntary initiatives
- Avoiding being implicated in publicized environmental, social and governance failures
- Comparing performance internally, and between organizations and sectors.

While external benefits of sustainability reporting can include:

- Mitigating – or reversing – negative environmental, social and governance impacts
- Improving reputation and brand loyalty
- Enabling external stakeholders to understand the organization's true value, and tangible and intangible assets
- Demonstrating how the organization influences, and is influenced by, expectations about sustainable development.

The Association for the Advancement of Sustainability in Higher Education (AASHE 2019) summarises the advantages which higher education institutions can gain from using their tool (STARS) to assess their sustainability performance as follows:

- Gain international recognition for your sustainability efforts.
- Generate new ideas.
- Engage your community.
- Create a baseline for continuous improvement.
- Inform strategic planning and budgeting.
- Integrate sustainability into the curriculum.
- Make real progress towards sustainability.
- Be part of a global community.

The University of Indonesia (Universitas Indonesia) (UI), which initiated a world university ranking in 2010 known as UI Green Metric World University Ranking (GM 2019), indicates that colleges and universities participate in the reporting and ranking are benefited through:

- Internationalisation and recognition.
- Increasing awareness of sustainability issues.
- Social change and action.
- Networking.

In summary, higher education institutions are encouraged to report their sustainability performance given the internal and external benefits they gain. Furthermore, colleges and universities should lead the way in sustainable practices and operations; leading by example. Cortese (2003, 19) raises an interesting question saying that "If higher education does not lead the sustainability effort in society, who will?" Similarly, Thomashow (2014, 3) believes that "university leadership is our last best hope for addressing the global climate challenge, and campus sustainability initiatives are the foundation of that leadership."

5 How to Report

Sustainability reports contain both qualitative and quantitative data on sustainability aspects in one or more of the following approaches: accounts, narrative, or indicator-based (Lozano 2006b). The indicator-based approach in reporting sustainability performance is the most recommended of all. Dalal-Clayton and Bass (2002, 135) point out that:

> Indicators enable assessments to be comprehensive yet selective: because they can be selective, they are better equipped than accounts to cover the wide array of issues necessary for an adequate portrayal of human and environmental conditions.

Lozano (2006b, 971) indicates that "indicator-based assessments offer higher levels of transparency, consistency, and usefulness for decision-making." Dalal-Clayton and Bass (2002, 135) add that "An indicator is fully representative if it covers the most important parts of the component concerned, and it shows trends over time and differences between places and groups of people." Therefore, most sustainability reporting systems use indicator-based approaches, given the advantages they have.

Colleges and universities can prepare, review, and publish their sustainability reports so that the public is aware of the institution's effort to sustainability advancement. Higher education institutions are also encouraged to also submit their reports regularly to one or more of the existing sustainability reporting systems. Examples of these sustainability systems include Global Report Initiative (GRI 2019), UI Green Metric (GM 2019), and Sustainability Tracking, Assessment, and Rating System (STARS 2019), to name but a few. Universities can use one of these reporting standards when preparing their sustainability reports. However, they can also generate a custom framework suitable for the institution in order to report their sustainability performance.

STARS, which was initially launched in 2010, is a well-known system and it is one of the most comprehensive and advantageous systems (Kamal and Asmuss 2013; Alghamdi et al. 2017). Some scholars argue that STARS has recently become one of the most popular frameworks (Gómez et al. 2015; Saadatian and Salleh 2011). Therefore, this study focuses on STARS reporting system to explain how higher education institutions can report.

As of April 2019, almost one thousand universities from across the globe have submitted their sustainability reports to STARS reporting system. Figures 1 and 2 show a sample of 4 universities which represent the 4 rating system:

- Platinum (Stanford University, United States of America).
- Gold (University College Cork—National University of Ireland, Republic of Ireland).
- Silver (University of Waterloo, Canada).
- Bronze (American University of Sharjah, United Arab Emirates).

The figures also demonstrate the aspects that these institutions reported on. It can be seen that there is some basic information that describes the characteristics of the institution. This includes institutional type and boundary, operational characteristics, academics and demographics. What follows is reporting on the common aspects of sustainability in the higher education sector. These aspects are:

- Academics: In this aspect, there are two main domains to be reported on. First is the Curriculum, which includes Academic Courses, Learning Outcomes, Undergraduate Program, Graduate Program, Sustainability Literacy Assessment, Incentives for Developing Courses, and demonstrating the Campus as a Living Laboratory. Second is the Research, which includes Sustainability Research and Scholarship, Support for Sustainability Research, and Open Access to Research.
- Engagement: In this aspect, there are two main domains to be reported on. First is the Campus Engagement, which includes Student Educators Program, Student Orientation, Student Life, Outreach Materials and Publications, Outreach Campaign, Assessing Sustainability Culture, Employee Educators Program, Employee Orientation, and Staff Professional Development. Second is the Public Engagement,

Stanford University

Stanford, CA, US

Rating	Score	Liaison	Submission Date	Executive Letter
Platinum	88.00	Moira Hafer	Feb. 22, 2019	Download

Institutional Characteristics

Institutional Characteristics

Academics

Curriculum	38.20 / 40.00
Research	17.00 / 18.00

Engagement

Campus Engagement	20.50 / 21.00
Public Engagement	18.85 / 20.00

Operations

Air & Climate	9.02 / 11.00
Buildings	3.73 / 8.00
Energy	7.37 / 10.00
Food & Dining	3.67 / 8.00
Grounds	3.00 / 4.00
Purchasing	4.71 / 6.00
Transportation	5.21 / 7.00
Waste	5.75 / 10.00
Water	8.00 / 8.00

Planning & Administration

Coordination & Planning	7.25 / 8.00
Diversity & Affordability	9.87 / 10.00
Investment & Finance	2.33 / 7.00
Wellbeing & Work	6.07 / 7.00

Innovation & Leadership

Exemplary Practice	2.50
Innovation	2.00

University College Cork - National University of Ireland, Cork

Cork, Co. Cork, IE

Rating	Score	Liaison	Submission Date	Executive Letter
Gold	66.39	Maria Kirrane	July 20, 2018	Download

Institutional Characteristics

Institutional Characteristics

Academics

Curriculum	19.59 / 40.00
Research	12.96 / 18.00

Engagement

Campus Engagement	19.00 / 21.00
Public Engagement	14.58 / 20.00

Operations

Air & Climate	5.28 / 11.00
Buildings	0.86 / 8.00
Energy	2.76 / 10.00
Food & Dining	5.59 / 8.00
Grounds	2.13 / 4.00
Purchasing	3.70 / 6.00
Transportation	4.83 / 7.00
Waste	7.66 / 10.00
Water	2.46 / 6.00

Planning & Administration

Coordination & Planning	7.75 / 8.00
Diversity & Affordability	6.44 / 10.00
Investment & Finance	6.00 / 7.00
Wellbeing & Work	3.81 / 7.00

Innovation & Leadership

Exemplary Practice	0.50
Innovation	4.00

Fig. 1 Examples of Platinum and Gold rating universities

which includes Community Partnerships, Inter-Campus Collaboration, Continuing Education, Community Service, Participation in Public Policy, and Trademark Licensing.

• Operations: In this aspect, there are nine main domains to be reported on. First is the Air and Climate, which include Greenhouse Gas Emissions, and Outdoor Air Quality. Second is the Buildings, which include Building Operations and Maintenance, and Building Design and Construction. Third is the Energy, which includes Building Energy Consumption, and Clean and Renewable Energy. Forth is Food and Dining, which include Food and Beverage Purchasing, and Sustainable Dining. Fifth is the Grounds, which include Landscape Management, and Biodiver-

University of Waterloo
Waterloo, ON, CA

Rating	Score	Liaison	Submission Date	Executive Letter
Silver	45.51	Mat Thijssen	Nov. 6, 2018	Download

Institutional Characteristics

Institutional Characteristics	

Academics

Curriculum	18.59 / 40.00
Research	12.88 / 18.00

Engagement

Campus Engagement	10.28 / 21.00
Public Engagement	6.50 / 20.00

Operations

Air & Climate	5.57 / 11.00
Buildings	0.00 / 8.00
Energy	2.86 / 10.00
Food & Dining	1.73 / 8.00
Grounds	1.00 / 3.00
Purchasing	1.46 / 6.00
Transportation	2.30 / 7.00
Waste	3.95 / 10.00
Water	3.17 / 6.00

Planning & Administration

Coordination & Planning	6.00 / 8.00
Diversity & Affordability	3.06 / 10.00
Investment & Finance	0.00 / 7.00
Wellbeing & Work	3.26 / 7.00

Innovation & Leadership

Exemplary Practice	0.00
Innovation	4.00

American University of Sharjah
Sharjah, Ash Shariqah, AE

Rating	Score	Liaison	Submission Date	Executive Letter
Bronze	34.02	Rose Armour	Sept. 16, 2018	Download

Institutional Characteristics

Institutional Characteristics	

Academics

Curriculum	19.77 / 37.00
Research	12.20 / 18.00

Engagement

Campus Engagement	7.75 / 21.00
Public Engagement	7.00 / 20.00

Operations

Air & Climate	1.67 / 11.00
Buildings	0.00 / 8.00
Energy	2.52 / 10.00
Food & Dining	0.25 / 8.00
Grounds	0.00 / 2.00
Purchasing	1.00 / 6.00
Transportation	3.74 / 7.00
Waste	1.97 / 10.00
Water	0.00 / 8.00

Planning & Administration

Coordination & Planning	3.75 / 8.00
Diversity & Affordability	0.75 / 10.00
Investment & Finance	0.00 / 7.00
Wellbeing & Work	3.00 / 7.00

Innovation & Leadership

Exemplary Practice	0.00
Innovation	1.00

Fig. 2 Examples of Silver and Bronze rating universities

sity. Sixth is the Purchasing, which includes Sustainable Procurement, Electronics Purchasing, Cleaning and Janitorial Purchasing, and Office Paper Purchasing. Seventh is the Transportation, which includes Campus Fleet, Student Commute Modal Split, Employee Commute Modal Split, and Support for Sustainable Transportation. Eighth is the Waste, which includes Waste Minimization and Diversion, Construction and Demolition Waste Diversion, and Hazardous Waste Management. Ninth is the water, which includes Water Use, and Rainwater Management.

- Planning and Administration: In this aspect, there are four main domains to be reported on. First are the Coordination and Planning, which include Sustainability Coordination, Sustainability Planning, and Participatory Governance. Second are the Diversity and Affordability, which include Diversity and Equity Coordination, Assessing Diversity and Equity, Support for Underrepresented Groups, and Affordability and Access. The Third are the Investment and Finance, which include the Committee on Investor Responsibility, Sustainable Investment, and Investment Disclosure. Forth are the Wellbeing and Work, which include Employee Compensation, Assessing Employee Satisfaction, Wellness Program, and Workplace Health and Safety.

• Innovation and Leadership: In this aspect, there are two main domains to be reported on; Exemplary Practice, and Innovation. Both domains include original and imaginative solutions to sustainability challenges. This means providing 'out-of-the-box thinking' in approaching sustainability. Bicycle Friendly University, Green Athletics, Green Laboratories, Green Event Certification are some cases in point.

It should be highlighted that the data provided in the reporting system of STARS is self-reported. Similarly, the UI Green Metric is a self-reporting system, where colleges and universities submit their reports using a framework. This means that each higher education institution is fully responsible about their information, since "AASHE staff review portions of all STARS reports" and hence "the data in STARS reports are not verified by AASHE" (STARS 2019). STARS reporting system offers a correction arrangement whereby institutions can review and amend the credit criteria. This can be carried out by completing the 'Data Inquiry Form' (Ibid). Additionally, registration is free of charge both in the STARS and the UI Green Metric.

This leads us to an important issue, which is how quality can be ensured when reporting the progress made towards sustainability goals in an organisation such as a university. This issue is extensively examined by Richardson and Kachler (2016). They define the quality of a sustainability report as:

> the extent to which the report provides valid and reliable data to meet stakeholder information needs. This definition requires that we specify the decision-model that stakeholders use to evaluate the organization and that the data reported is a reliable and valid indicator of the dimensions of sustainability performance used in that model. (Ibid, 12)

Richardson and Kachler (2016, 15) argue that in order to ensure the quality in sustainability reports, higher education institutions have to use an independent audit. They underline that "the broader the scope of an audit, the higher the quality of the report." They add that "the greater the independence and competence of the auditor, the more credibility that stakeholders are likely to attribute to the sustainability report." Another quality assurance is by early involvement of the stakeholders. "The use of stakeholder engagement processes signals the higher quality of reporting" (Ibid). Stakeholders in universities can be classified into two groups; internally (students, staff, faculty, and administrators) and externally (alumni, governments, business, and professional associations and society) (Ahmad and Crowther 2013, 162).

Given that reporting of sustainability advancement is self-reporting, this needs not only a committed team, but also coordination between various university departments and agencies. This is where an Office of Sustainability (common name in most

American countries) or Green Office (common name in most European countries) can be of great importance. Such an office can act as a leader for sustainability initiatives in any higher education institution. It serves as a catalyst and an advocate for sustainability in colleges and universities. It tracks progress made and report the advancement in sustainability practices and operations on campus. Reports can also be prepared by one or more university representatives such as sustainability committee, sustainability coordinator, or student initiatives (Sassen and Azizi 2018).

Another issue to highlight is that when using any of the sustainability reporting system specifically geared to higher education institutions, it may happen that institutions intentionally omit entering some information for several reasons. These reasons are summarised in the G4 Sustainability Reporting Guidelines by The Global Report Initiative (GRI Guidelines 2019, 13) as follows:

- A Standard Disclosure, part of a Standard Disclosure, or an Indicator is not applicable
- The information is subject to specific confidentiality constraints
- The existence of specific legal prohibitions
- The information is currently unavailable.

6 Conclusions

This study set out to answer three important questions concerning sustainability reporting in universities: what to report, why reporting is necessary, and how reporting should be carried out.

This study has identified the main common criteria and indicators which higher education institutions can use to report their sustainability performance. These criteria are management, academia, environment, engagement, and innovation. The paper shows that these widely used criteria of sustainability are in line with the literature.

This study has shown that why higher education institutions should report their sustainability practices and operations. They are encouraged to report their sustainability performance given the internal and external benefits they gain. Universities have a duty to lead the way in sustainable practices and operations; leading by example. Not only should they be charting the course, but also walking the talk. Who else has the expertise, resources and responsibility to take sustainability forward? Higher education institutions are different in many ways: student body, size and number of campuses, capacity, financial resources, and conditions. This has not prevented some colleges and universities from exercising their leadership role in advancing sustainability aspects in their institutions and beyond. Reporting sustainability practices and operations means that the institution is committed to making a noticeable change for the good of itself, its community, and the world.

The present study has also shown how universities can report their sustainability advancement. It has highlighted the type of data collected, process of preparing the report, the parties involved, and channels where reports can be submitted to (e.g. GRI, UI GM, and STARS).

The findings of this research will be of interest to higher education institutions, particularly those at the early stages of assessing, documenting, evaluating, reviewing and reporting sustainability practices and operations. The added value of this research is that it shows how quality can be insured when universities preparing and reviewing the sustainability reports of.

The scope of this study was limited in terms of the number of sustainability reports it analysed. This was because of the fact that the idea was to simply explain what higher education institutions need to report and how. That is why only 4 reports were presented and evaluated. Another limitation of this study is that it did only explain the contents of the sustainability reports submitted to STARS, while it did not in other reporting systems such as GRI and UI Green Metric.

Further research is required to show how reports can be prepared for submission to reporting systems such as GRI and UI Green Metric. Further studies need to be carried out in order to illustrate how quality can be insured in sustainability reports.

References

AASHE (2019) The association for the advancement of sustainability in higher education: why participate. Available at: https://stars.aashe.org/about-stars/why-participate/. Accessed 19 Jan 2019

Adams CA (2013) Sustainability reporting and performance management in universities. Chall Benef Sustain Acc Manage Policy J 4(3):384–392

Ahmed J, Crowther D (2013) Education and corporate social responsibility: international perspectives. Emerald Group Publishing Limited, Bingley, UK

Alghamdi N (2018) University campuses in Saudi Arabia: sustainability challenges and potential solutions. Doctoral dissertation, Delft University of Technology, Delft, The Netherlands

Alghamdi N, Den Heijer A, De Jonge H (2017) Assessment tools' indicators for sustainability in universities: an analytical overview. Int J Sustain Higher Educ 18(1):84–115

Alonso-Almeida MDM, Marimon F, Casani F, Rodriguez-Pomeda J (2015) Diffusion of sustainability reporting in universities: current situation and future perspectives. J Clean Prod 106:144–154

Alshuwaikhat HM, Abubakar I (2008) An integrated approach to achieving campus sustainability: assessment of the current campus environmental management practices. J Clean Prod 16(16):1777–1785

AUA (2019) Alternative university appraisal model for ESD in higher education institutions. Available at http://sustain.oia.hokudai.ac.jp/aua/. Accessed 19 Jan 2019

Bekessy SA, Samson K, Clarkson RE (2007) The failure of non-binding declarations to achieve university sustainability: a need for accountability. Int J Sustain Higher Educ 8(3):301–316

Ceulemans K, Molderez I, Van Liedekerke L (2015) Sustainability reporting in higher education: a comprehensive review of the recent literature and paths for further research. J Clean Prod 106(Special Issue):127–143

Chau KW (2007) Incorporation of sustainability concepts into a civil engineering curriculum. J Prof Issues Eng Educ Pract 133(3):188–191

Cole L (2003) Assessing sustainability on Canadian university campuses: development of a campus sustainability assessment framework. M. A. Environment and Management, Royal Roads University, Victoria, Canada

Cortese A (2003) The critical role of higher education in creating a sustainable future. Plan Higher Educ 31(3):15–22

Dalal-Clayton B, Bass S (2002) Sustainable development strategies, 1st edn. Earthscan Publications, London

Davis SA, Edmister JH, Sullivan K, West CK (2003) Educating sustainable societies for the twenty-first century. Int J Sustain Higher Educ 4(2):169–179

Flint RW (2010) Symbolism of sustainability: means of operationalizing the concept. Synesis J Sci Technol Eth Policy 01(01):25–37

Fonseca A, Macdonald A, Dandy E, Valenti P (2011) The state of sustainability reporting at Canadian universities. Int J Sustain Higher Educ 12(1):22–40

GM (2019) UI's Green Metric University sustainability ranking. Available at http://greenmetric.ui.ac.id/. Accessed 20 Jan 2019

Gómez F, Sáez-Navarrete C, Lioi S, Marzuca V (2015) Adaptable model for assessing sustainability in higher education. J Clean Prod 107:475–485

GRI (2007) Reports database. Available at: http://www.globalreporting.org/ReportsDatabase/. Accessed 11 Jan 2019

GRI (2019) Global report initiative: benefits of reporting. Available at https://www.globalreporting.org/information/sustainability-reporting/Pages/reporting-benefits.aspx. Accessed 23 Jan 2019

GRI Guidelines (2019) G4 sustainability reporting guidelines. Available at https://www.globalreporting.org/resourcelibrary/grig4-part1-reporting-principles-and-standard-disclosures.pdf. Accessed 23 Feb 2019

Guidry R, Patten D (2010) Market reactions to the first-time issuance of corporate sustainability reports: evidence that quality matters. Sustain Acc Manage Policy J 1(1):33–50

Hamann R (2003) Mining companies' role in sustainable development: the 'why' and 'how' of corporate social responsibility from a business perspective. Dev South Afr 20(2):234–254

Huber S, Bassen A (2018) Towards a sustainability reporting guideline in higher education. Int J Sustain Higher Educ 19(2):218–232

ISO (2019) ISO 14000 family—environmental management. Available at https://www.iso.org/iso-14001-environmental-management.html. Accessed 26 Jan 2019

Kamal A, Asmuss M (2013) Benchmarking tools for assessing and tracking sustainability in higher education institutions: identifying an effective tool for University of Saskatchewan. Int J Sustain Higher Educ 14(4):449–465

KPMG (2013) International survey of corporate responsibility reporting 2013. Available at www.kpmg.com/Global/en/IssuesAndInsights/ArticlesPublications/corporate-responsibility/Documents/corporateresponsibility-reporting-survey-2013-exec-summary.pdf. Accessed 10 Mar 2019

Lopatta K, Jaeschke R (2014) Sustainability reporting at German and Austrian universities. Int J Educ Econ Dev 5(1):66–90

Lozano R (2006a) A tool for a graphical assessment of sustainability in universities (GASU). J Clean Prod 14(9/11):963–972

Lozano R (2006b) Incorporation and institutionalization of SD into universities: breaking through barriers to change. J Clean Prod 14:787–796

Lozano R (2011) The state of sustainability reporting in universities. Int J Sustain High Educ 12(1):67–78

Lozano R (2013) Sustainability inter-linkages in reporting vindicated: a study of European companies. J Clean Prod 51:57–65

Lozano R, Huisingh D (2011) Inter-linking issues and dimensions in sustainability reporting. J Clean Prod 19(4):99–107

Lozano R, Lukman R, Lozano F, Huisingh D, Lambrechts W (2013) Declarations for sustainability in higher education: becoming better leaders, through addressing the university system. J Clean Prod 48:10–19

Oxford Dictionary (2019) Report. Available at https://en.oxforddictionaries.com/definition/report. Accessed 1 Jan 2019

Perrini F, Tencati A (2006) Sustainability and stakeholder management: the need for new corporate performance evaluation and reporting systems. Bus Strateg Environ 15:296–308

PSPE (2019) The platform for sustainability performance in education: sustainable campus assessment system. Available at https://www.osc.hokudai.ac.jp/en/action/assc. Accessed 20 Feb 2019

Richardson AJ, Kachler MD (2016) University sustainability reporting: a review of the literature and development of a model. In: Handbook of sustainability in management education. Available at: http://scholar.uwindsor.ca/odettepub/101. Accessed 11 Jan 2019

Roorda N (2002) Assessment and policy development of sustainability in higher education with AISHE. In: Fillo WL (ed) Teaching sustainability at universities: towards curriculum greening. Peter Lang, New York, NY

Saadatian O, Salleh E (2011) Identifying strength and weakness of sustainable higher educational assessment approaches. Int J Bus Soc Sci 2(3):137–146

Sassen R, Azizi L (2018) Assessing sustainability reports of US universities. Int J Sustain Higher Educ 19(7):1158–1184

Sassen R, Dienes D, Beth C (2014) Nachhaltigkeitsberichterstattung deutscher Hochschulen (Sustainability reporting of German higher education institutions). Zeitschrift Für Umweltpolitik & Umweltrecht (J Environ Policy Environ Law) 37(3):258–277

Sassen R, Dienes D, Wedemeier J (2015) Sustainability reporting of British higher education institutions. Working paper, University of Hamburg

Scoullos M (2010) What makes a university sustainable? Available at http://mio-ecsde.org/project/residential-training-workshop-on-universities-education-for-sustainable-development-amfissa-greece-23-28-may-2010-sustainable-mediterranean-issue-no-63-64-03-042010/. Accessed 11 Jan 2019

Sharp L, Shea C (2012) Institutionalising sustainability: achieving transformation from inside. In: Martin J, Samels J (eds) The sustainable university: green goals and new challenges for higher education leaders. The Johns Hopkins University Press, Baltimore, Maryland, pp 63–82

Shriberg M (2002) Institutional assessment tools for sustainability in higher education: strengths, weaknesses, and implications for practice and theory. Int J Sustain Higher Educ 3(3):254–270

STARS (2019) Sustainability tracking, assessment, and rating system. Available at https://stars.aashe.org/. Accessed 26 Jan 2019

Thijssens T, Bollen L, Hassink H (2016) Managing sustainability reporting: many ways to publish exemplary reports. J Clean Prod 136:86–101

Thomashow M (2014) The nine elements of a sustainable campus. MIT Press, Cambridge, Massachusetts

Travis D (2016) Desk research: the what, why and how. Available at https://www.userfocus.co.uk/articles/desk-research-the-what-why-and-how.html. Accessed 4 Mar 2019

ULSF (2019) University leaders for a sustainable future—sustainability assessment questionnaire for colleges and universities (SAQ). Available at http://ulsf.org/sustainability-assessment-questionnaire/. Accessed 11 Jan 2019

United Nations (1987) Towards sustainable development. From A/42/427—our common future: report of the World Commission on Environment and Development. Available at http://www.un-documents.net/ocf-02.htm. Accessed 1 Jan 2019

Velazques L, Munguia N, Platt A, Taddei J (2006) Sustainable university: what can be the matter? J Clear Prod 14(9/11):810–819

Waas T, Hugé J, Ceulemans K, Lambrechts W, Vandenabeele J, Lozano R, Wright T (2012) Sustainable higher education: understanding and moving forward. Flemish Government—Environment, Nature and Energy Department, Brussels

Waterman RH, Peters TJ, Philips JR (1980) Structure is not organisation. Bus Horiz 23:14–26

Yanez S, Uruburu A, Moreno A, Lumbreras J (2018) The sustainability report as an essential tool for the holistic and strategic vision of higher education institutions. J Clean Prod 207:57–66

Zorio-Grima A, Sierra-García L, Garcia-Benau M (2018) Sustainability reporting experience by universities: a causal configuration approach. Int J Sustain Higher Educ 19(2):337–352

"Reaching for the STARS": A Collaborative Approach to Transparent Sustainability Reporting in Higher Education, the Experience of a European University in Achieving STARS Gold

Maria J. Kirrane, Chris Pelton, Pat Mehigan, Mark Poland, Ger Mullally and John O'Halloran

Abstract In 2010, University College Cork (UCC) was the first institution in the world to be awarded a "Green Flag" from the Foundation for Environmental Education. Since 2011, UCC has been ranked highly in the UI Green Metric World University Rankings and, in 2018, the university participated in the Sustainability Tracking, Assessment & Rating System (STARS). STARS—a programme of the Association for the Advancement of Sustainability in Higher Education (AASHE)—is a transparent, self-reporting framework for colleges and universities to comprehensively measure their sustainability performance. In 2018, UCC became the first university outside of the US and Canada to achieve a STARS Gold Rating. This paper documents the experience of UCC in undertaking the STARS assessment for the first time and explores how the process is helping to shape the sustainability agenda at UCC. The STARS submission process involves a collaborative review stage, and the use of this collaboration to shape further development of the STARS programme will be also be examined. Transparent reporting tools, such as STARS, present opportunities for further engagement with campus stakeholders. Recommendations for campuses embarking on the programme, and for the assessment of sustainability in higher education will be presented.

Keywords STARS · Sustainability assessment · Stakeholder engagement · Higher education · Green campus

M. J. Kirrane (✉) · P. Mehigan · M. Poland · G. Mullally · J. O'Halloran
University College Cork, Cork, Ireland
e-mail: m.kirrane@ucc.ie

C. Pelton
Association for the Advancement of Sustainability in Higher Education (AASHE), Philadelphia, PA, USA
e-mail: chris.pelton@aashe.org

© Springer Nature Switzerland AG 2020
W. Leal Filho et al. (eds.), *International Business, Trade and Institutional Sustainability*, World Sustainability Series,
https://doi.org/10.1007/978-3-030-26759-9_58

1 Introduction

In 2010, University College Cork (UCC) became the first university in the world to be awarded a Green Flag from the international, Denmark-based, Foundation for Environmental Education (FEE) following the implementation of a pilot programme facilitated by FEE's national operator in Ireland, An Taisce (Reidy et al. 2015). The Green Campus programme is an environmental education and awards system, that takes a themed approach across the operational aspects of the campus. The development of the Green Campus programme in Ireland is documented in Ryan-Fogarty et al. (2016), while the implementation of the pilot programme at UCC is documented in Reidy et al. (2015). Of significance is the fact that the programme provided a recognizable framework for the university and its community to collaborate on environmental sustainability projects. The seven-step Green Flag programme requires all key stakeholders to play an active role in its implementation. The assessment is both qualitative and quantitative, and flexible to the individual campus; campuses must demonstrate that they have made improvements from their own baseline. UCC successfully renewed its Green Flag in 2013, 2016 and 2019, and in 2018 was the only university outside of North America to be awarded a Gold STARS award from the Association for the Advancement of Sustainability in Higher Education (AASHE). This paper explores how, following over a decade of implementing the Green Flag approach, UCC embarked on the STARS programme.

In Ireland, the Green Campus programme has been entirely voluntary. At a sector level, the biggest influence has been the National Energy Efficiency Action plan, which places a responsibility on public sector organisations to lead the way in reaching the national targets for energy efficiency (Department of Communications, Energy Efficiency and Natural Resources 2009). The National Strategy on Education for Sustainable Development has encouraged the incorporation of Education for Sustainable Development (ESD) into higher education (HE), and the extension of the Green Campus programme to all Higher Education Institutions (HEIs). The international ranking system UI Green Metric has been adopted as an assessment by a number of Irish institutions. In 2018, four of the seven Irish universities took part in the ranking, with University College Cork ranked 9th, DCU ranked 12th, the University of Limerick in 20th place and Maynooth University in 130th out of 719 global universities.

University College Cork, established in 1845, is a comprehensive and research-intensive university ranked in the top 2% of universities worldwide (University College Cork 2017). The university is one of the largest employers in the southern region of Ireland and, as of December 2017, serves 21,000 students including 3,300 international students from 100 countries worldwide. The university undertakes outreach programmes in 40 locations nationwide, while its centres, such as the Centre for Global Development, engage in international development and sustainability issues globally. UCC's vision is to be a leading university of independent thinkers and its mission is "Creating, understanding and sharing knowledge and applying it for the good of all". In 2015, the Sustainable Development Goals (SDGs) were agreed by the United Nations and, in 2016, UCC published its Sustainability Strategy (UCC 2016), reflecting the broad focus of the SDGs but also their interrelatedness and local application. In 2018, UCC's Academic Strategy adopted a "connected curriculum" setting out sustainability as a key attribute in the university's curriculum.

1.1   The STARS Programme

The Sustainability Tracking, Assessment & Rating System (STARS) is a transparent, self-reporting framework for HEIs to measure their sustainability performance (AASHE 2019a). STARS was developed by AASHE in response to a call from The Higher Education Associations Sustainability Consortium (HEASC) for a "voluntary assessment and rating system for campus sustainability that would become the 'standard' for the higher education community" (HEASC 2006). The first version of STARS (1.0) was launched by AASHE in 2010 after a four-year stakeholder engagement process and a pilot project involving nearly 70 institutions.

STARS was developed, in part, as an alternative to green rankings published by organisations outside the higher education sector. The programme was designed to:

- Provide a framework for understanding sustainability in all sectors of higher education.
- Enable meaningful comparisons over time and across institutions using a common set of measurements developed with broad participation from the international campus sustainability community.
- Create incentives for continual improvement toward sustainability.
- Facilitate information sharing about higher education sustainability practices and performance.
- Build a stronger, more diverse campus sustainability community.

STARS serves as both a free online reporting and transparency platform (the STARS Reporting Tool) and a fee-based rating system. There are four ratings available to paid participants that elect to publish a scored report: Bronze, Silver, Gold and Platinum. An institution may instead opt to publish an unscored report at no cost, earning the STARS Reporter designation. STARS only provides positive recognition for participants and all of the information reported is publicly accessible on the STARS website.

One of the other distinguishing characteristics of STARS is the level of support AASHE provides for programme participants. In addition to hosting the Reporting Tool and public STARS reports, AASHE:

- Administers formal governance bodies and stakeholder engagement mechanisms to help ensure that the programme reflects the latest developments in the field and is responsive to the campus sustainability community.
- Provides ongoing staff support for participants, including responding to questions by email and telephone, maintaining an online help center, and making available data collection templates and tools.

- Supports mechanisms to enhance the quality of public reports and data, protect the credibility of the system, and provide a fair and transparent means for resolving questions about the accuracy of reported information. These mechanisms include a collaborative review and revision process through which AASHE staff reviewers work with each institution to ensure that its reports are completed in adherence with the criteria published in the STARS Technical Manual.
- Publishes the Sustainable Campus Index (SCI) annually to highlight best practices and top performers in STARS by impact area and institution type.

As of March 2019, more than 900 institutions worldwide were using the STARS Reporting Tool in some capacity. Participants include research universities, Master's and Baccalaureate-level institutions, community colleges and other short-cycle institutions, college preparatory schools and adult education centres. STARS reports are valid for three years and about a third of participants hold a valid Bronze, Silver, Gold or Platinum Rating at any point in time.

Although initially developed for institutions in the US and Canada, STARS currently has participants in 39 different countries. AASHE has helped facilitate and support this growth in a number of ways. In 2011, AASHE launched a two-year International Pilot programme to explore the possibilities and potential challenges of making STARS available globally. As a result, the STARS rating process was opened to institutions outside North America in 2013.

Among the conclusions that AASHE has drawn from the pilot and subsequent international participation is that an institution is typically more likely to complete the reporting process and find it valuable when its peers are also participating and/or there is alignment with other sustainability-related objectives, for example, benchmarking, conducting gap analyses or reporting on the institution's contributions towards the SDGs. Building on these conclusions, AASHE has recently:

- Partnered with Australasian Campuses Toward Sustainability (ACTS) to support a cohort of institutions in that region with their STARS reporting (AASHE 2018),
- Released an online benchmarking tool to allow institutions to compare their performance over time and in comparison to other STARS participants, and
- Included information mapping STARS to the SDGs in version 2.2 of the STARS Technical Manual (AASHE 2019b).

1.2   Embarking on the STARS Programme

Numerous attempts have been made at developing robust and reliable methods of assessing sustainability in higher education. There is an extensive literature on the subject, reviewed by Findler et al. (2018). From an operational perspective a university which chooses to report transparently on its sustainability activities is generally doing so voluntarily, reflecting not only a desire for good governance but a "quality mark" for the institution at a time when SDGs are on the agenda. As pointed out by Findler et al. (2018), an important consideration when deciding on what tool to use for a university is balancing the quality of assessment with the effort required to gather the data. At UCC, the decision was also based on the potential for the data gathering exercise to further align activities and structures going forward. In addition, the comprehensive nature of the assessment enables it to be used as a gap analysis to further inform action plans in future years. UCC's decision to embark on STARS was also informed by a review undertaken by the Environmental Association of Universities and Colleges (EAUC), of which UCC is a member, that found STARS to be "one of the top whole institution approach tools" (EAUC 2017).

2 Methodology

This paper takes a case study approach of a university outside of the US system, but which has a strong sustainability agenda, in achieving STARS Gold. An analysis of the University's STARS rating and how it was informed by the existing programmes across the university will be undertaken. UCC undertook both the STARS assessment and UI Green Metric Assessment in 2018. Therefore, in addition, by combining STARS categories to reflect the broader categories of the UI Green Metric Assessment, a comparison of how the university performed in both metrics will be undertaken. The STARS methodology awards points under 18 "subcategories" (Fig. 1). Within those subcategories are 63 standard credits, 4 open-ended Innovation credits, and a catalogue of Exemplary Practice credits (Annex 1).

UCC's procedure for submitting to the STARS programme can be categorised into three phases.

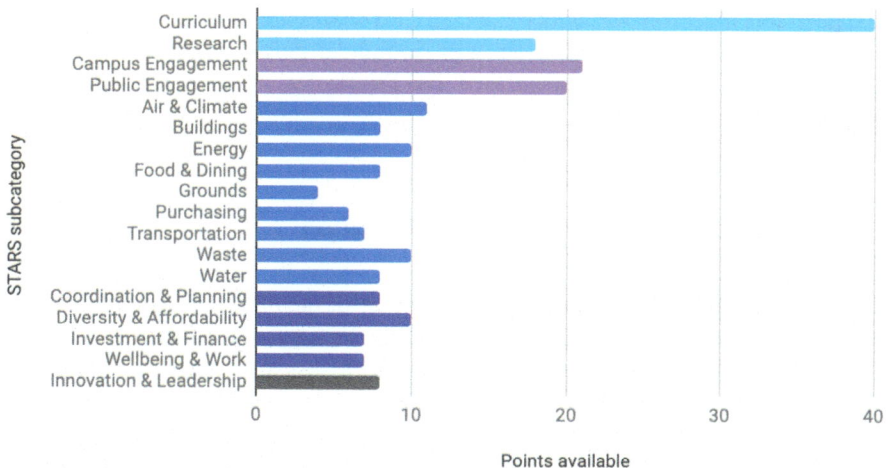

Fig. 1 STARS subcategories by points available (version 2.1)

1. Over the period from January to April 2018, UCC undertook a review of all data available within the university corresponding to each of the STARS criteria and indicators.
2. From April to July In July 2018, an extensive data gathering exercise took place across the university, coordinated by the Green Forum chairs and the university's Sustainability Officer. This required the completion of a number of inventories including greenhouse gas emissions and sustainability inclusion in the curriculum. In July 2018 an initial submission was made.
3. From July to October 2018, the collaborative review process with the STARS programme manager and STARS reviewers was undertaken and UCC was awarded a Gold Rating in October 2018.

The first phase, of reviewing available data, enabled the university to make decisions regarding what inventories could be completed immediately, using readily available data, and those that required a greater time investment. A decision was then made on what information to seek in this submission and what information should be included within the action plan for the university's next submission, after three years. For example, as the university is currently in the process of upgrading its academic registry undertaking a review of sustainability inclusion in module descriptions would give both the Green Forum, and the systems administration office, a better idea of reporting requirements going forward.

2.1 Carbon Emissions Calculation

The university had previously calculated its carbon emissions as part of its Green Flag assessment, using the Greenhouse Gas Protocol Corporate Standard and so an update of this inventory was undertaken. STARS scores are awarded based on Scope 1 and Scope 2 emissions reduction, with additional points for inclusion of some scope three emissions. This mirrors the GHG Protocol where Scope 3 emissions are elective.

2.2 Sustainability Inclusion in Curriculum

This was arguably the most time consuming part of the data collection process but was undertaken due to the reasons outlined above. A keyword search, using the keywords developed by Kingston University to assess sustainability inclusion in curricula, was used (Hands and Anderson 2016). Courses in STARS equate to what the Irish HE system would refer to as modules, i.e. 5, 10, or 15 credit Bologna aligned units. All courses that returned three search terms or more were manually assessed (reading of course description and learning outcomes). Those that were identified to be "sustainability courses" were marked with the most relevant SDG. If the course contained a significant element of sustainability or directly addressed the theme of sustainable development then it was designated as sustainability focused. Courses that directly addressed one or more SDG, but didn't specifically mention sustainability or sustainable development were designated as courses that included sustainability. For foundation courses, those that specified the inclusion of real world examples that addressed environmental or sustainability themes were included. This method likely underestimates the number of sustainability relevant courses in UCC.

2.3 Purchasing Inventories

STARS requires purchasing inventories under a number of procurement categories. A number of these were available from on campus suppliers, while those that required interrogation of the universities purchasing system were assessed as being too laborious to collect in the initial submission.

In all, UCC submitted under 59 of the 67 credits that were applicable to the institution. STARS takes a "whole institution" approach, awarding points in categories across operations; academics; engagement; planning and administration; and innovation and leadership. The case study analysis will examine UCC's assessment across these broad categories.

3 Results and Analysis

UCC's performance in STARS reflects the strengths of the Green Flag programme and ISO 50001 management tool. The university scored high across engagement and coordination categories reflecting the fact that the basis of the Green Flag programme is a committee made up of all campus stakeholders. The ISO 50001 approach of establishing teams in Significant Energy Using buildings also contributed to this score.

3.1 Operations

The "operations" element of the STARS assessment saw the greatest variation in scores for the University (Fig. 2). Not surprisingly, as the original impetus for the establishment of the pilot Green Flag programme, "Waste" achieved the highest score in this category. While the university has made significant improvements in waste management on campus (reducing total waste tonnage by 25% since 2012), it remains a focal point for engagement across campus. Waste is potentially the most visual and "social" element of campus sustainability; as a normative behaviour (Barr 2007), recycling rates will be influenced by the culture of the institution. Waste is

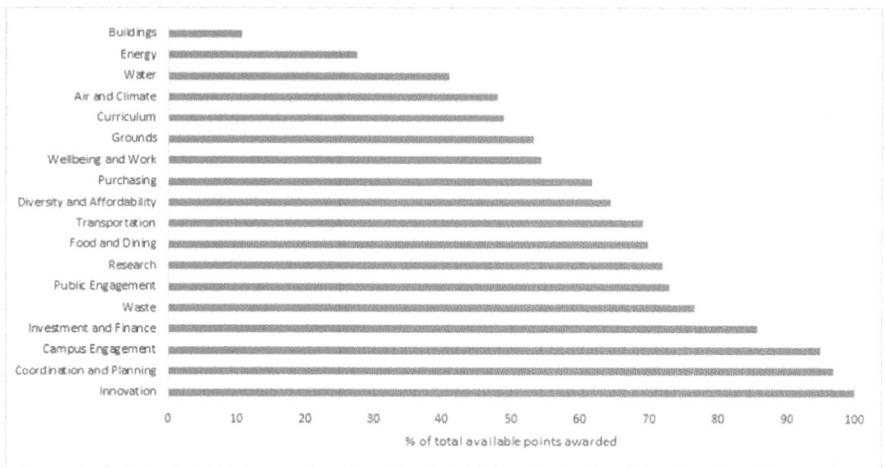

Fig. 2 UCC's % of points awarded across STARS criteria

an aspect that both students and staff can control. Chaplin and Wyton (2014), in a study of student engagement in sustainability, reported a feeling of hopelessness in the face of the enormous challenges posed. Through targeted campaigns that both create a physical environment where pro-environmental behaviour is enabled and a cultural environment where it becomes the norm, it is argued that these individuals may be empowered (Stern 2000). In many ways, the university has also acted as a microcosm for broader society, reflecting the global drive to reduce plastic pollution that has emerged over the last number of years. Another area that performed strongly was "Food and Dining" reflecting the university's "Farm to Fork" programme and the local sourcing of food items by the campus catering contractor KSG. On the other end of the scale "Buildings" and "Energy" scored low. The latter may seem surprising given the university's ISO 50001 certification. This is down to a number of reasons. The calculation of energy intensive floor area differs in Ireland's Public Sector Energy Efficiency Reporting and the STARS methodology and the STARS "Energy" category includes points for on-site renewable energy generation, of which UCC has relatively low amounts. As part of the university's Sustainability Action Plan over the next number of years, PPA to decarbonise the energy system are being investigated.

Within the operations category, the impact of regulatory structures in the two different jurisdictions (Ireland and US) did become apparent. For example, in Ireland food waste produced by catering companies is much more highly regulated than in the US. On the other hand, in the US, until recently, carbon emissions from HEIs gained much more attention than in Ireland. This reflects the benefit in comparing institutions across a wide geographical spread using a multi-criteria rating such as STARS; the overall benchmark becomes an amalgamation of best practice across both jurisdictions. It is vital in the long term however, that these benchmarks continue to lead institutions in achieving significant improvements.

3.2 Academics

The STARS programme breaks the "Academics" category into "Research" and "Curriculum" components. In turn, UCC's Strategic Plan (2017–2022) states, as a key strategic goal, that the curriculum be research-led and commits to "position interdisciplinarity as a core academic mission of the university". As part of the university's submission to the UI Green Metric ranking, a broad assessment of courses that address the environmental dimension of sustainability had been undertaken. However, the level of detailed inventory required by STARS had only been undertaken once in the university, in an assessment of research-led teaching (O'Mahony et al. 2015). Both the process of the inventory, and the outcome, were valuable sources of institutional learning and will be key in the development of UCC's academic administration systems and the implementation of UCC's new Academic Strategy, where sustainability and inter-/trans-disciplinarity are two components of the university's new "Connected Curriculum" (UCC 2018a, b). Instilling students with sustainability "competencies" as opposed to pure knowledge is recognised as one of the most difficult elements of incorporating sustainability into higher education (Molderez and Fonseca 2018). It is generally accepted that encouraging transdisciplinarity in the curriculum is most likely to achieve this, as well as the recognition of how knowledge that comes from outside the walls of the university can complement that which is created within (Byrne et al. 2017). New forms of learning including multi-stakeholder interaction, value-based learning, and applied problem solving are recognised as key in incorporating sustainability into higher education (Wals and Corcoran 2012). Despite the fact that research universities are thought to lag behind in ESD (Wals and Corcoran 2012), UCC has worked to instill sustainability competencies in its students through, for example, a university wide module on Sustainability, event-based pedagogies, non-degree awards, and community engagement in teaching and learning.

The university wide module is open to all students and staff of UCC, as well as the general public. This inclusivity reflects the principles of the Green Campus programme; staff from 15 different disciplines in the university, as well as the office of Buildings and Estates deliver sessions during the module. Reflecting the local focus of the Sustainable Development Goals, the module title is "Putting sustainability in its place: putting yourself in the picture". In addition to the University Wide Module, which is open to members of the public, UCC has brought learning directly to the community through the Learning Neighbourhoods project. This project, which is part of Cork's designation as a UNESCO Learning City, brings life-long and life-wide learning to areas that experience educational disadvantages in Cork city. Whilst still in its early stages, the inclusive methodologies being utilised have great potential to open the university up to more marginalised communities (O'Sullivan et al. 2018). Through dissolving the traditional boundary between "town and gown", the co-creation of knowledge to develop solutions for a more sustainable society can be achieved (Mero 2011; König 2015). In September 2017, Cork city hosted the UNESCO Learning Cities Festival, in a partnership between the city council,

UCC, Cork Institute of Technology and the Cork Education and Training Board. The outcome of the festival was a "Cork Call to Action for Learning Cities", which included amongst other things, a commitment to "achieving sustainable development in all its dimensions, recognizing the links between all of its social, environmental and economic aspects in order to secure a sustainable future for all" (UNESCO 2017).

The University's "Research" score reflects its strengths in hosting the Environmental Research Institute (ERI). The ERI is UCC's flagship institute for interdisciplinary research that furthers environmental sustainability. It is not only a physical space, but also a focal point for collaboration within the university. The ERI hosts the United Nations GEMS Water Capacity Development Centre, the implementing partner for SDG Indicator 6.3.2 focusing on Water Quality. The Institute also holds observer status on the United Nations Framework Convention on Climate Change (UNFCCC). In 2011, UCC's Research Office identified 5 thematic areas with associated research priorities areas, which were interdisciplinary in nature and focused on solving global challenges in topics that UCC research excelled in. The designation of an "Environmental Citizenship" Research Priority Area brought together researchers across the university in meetings, workshops, conferences and symposia which eventually led to the publication of the book "Transdisciplinary Perspectives on Transitions to Sustainability" (Byrne et al. 2017). This book provides a useful insight into the "ongoing dialogue" within UCC in transdisciplinary approaches to sustainability research. In turn, numerous successful funding bids have emerged from this initiative, which have further strengthened these inter and trans-disciplinary relationships. Byrne (2017) describes these collaborations within UCC as "spaces which both legitimise and promote disciplinary openness in a quest for emergent knowledge and 'greater than the sum of the parts' understandings". These collaborations have seen the SDGs explored through disciplines as diverse as creative practice, English literature, law and marketing. Many of these are hosted within this ERI, for example an Irish Environmental Protection Agency funded project on Climate Change, Behaviour and Community Response, which sees the co-supervision of doctoral candidates by a sociologist and an engineer. These inter- and trans-disciplinary dialogues are reflected in the fact that over half of the university's "research-producing departments" are engaged in sustainability research (STARS AC-9).

3.3 Engagement

The first step of the Green Campus programme is to ensure a committee is in place with involvement of all key stakeholders. UCC also implements the ISO 50001 approach to energy management, which engages across the institution to improve energy efficiency. University College Cork is obliged under the Irish Government's Public Sector Energy Efficiency Action Plan (PSEEAP) to improve energy efficiency by 33% by 2020 (Department of Communications, Energy Efficiency and Natural Resources 2009). In order to reach the PSEEAP targets, the decision was taken to focus on individual buildings that were significant energy users (SEUs). A review of energy usage across campus revealed that 13 of UCC's buildings were responsible for 87% of the university's overall energy consumption. Eight of the 13 buildings house significant laboratory space. This is a common feature of universities; a study at the University of Copenhagen revealed, for example, that two thirds of all energy consumed at the university was related to laboratory use (Faghihi et al. 2015). Other significant energy users included the library, which has one of the largest footfalls within the university, and the main quadrangle which is made up of buildings that date back to the establishment of the university in the 1800's. It therefore became apparent that a one-size fits all approach would not work and it was decided that a pilot programme, working with the building users to monitor energy usage and incentivise efficiency improvements would be rolled out. At the same time, the energy budget for the building was decentralised to the department. "Green Revolving Funds" have been used to significant success in higher education institutions in a number of countries, with at least 79 in operation in North America (Maiorano and Savan 2015). In UCC, a "Saver Saves" scheme was launched, whereby any monetary savings accrued due to energy improvements would be retained by the department, to be spent on further environmental improvements. The ISO 50001 approach is to establish teams in Significant Energy Using buildings (SEUs). While the Green Campus committee has membership from both staff and students, it is in the main "student-led". The ISO 50001 approach also targets both groups, but is more focused on staff. Therefore the combination of both approaches resulted in the University scoring extremely high across engagement categories.

UCC takes the approach that campaigns and events that engage the student body in environmental issues are part of the "informal curriculum". Experience-based learning, which includes hands-on exploration and the investigation of real-world local issues, was identified by Ballantyne and Packer (2009) as being most effective in facilitating learning for sustainability. In 2014, UCC Green Campus students developed the idea of running an intervarsity BioBlitz on campuses across Ireland. A BioBlitz is an organised event where members of the public work alongside experts to identify as many species of flora and fauna as possible within a set time period. Competitive BioBlitzes have been run by the National Biodiversity Data Centre (NBDC) in Ireland for a number of years. The Green Campus Bioblitz was developed with students from UCC, in collaboration with An Taisce, and the National Biodiversity Data Centre (Ryan-Fogarty et al. submitted). The event took place in 2014, 2015 and 2016 with students, staff and members of local wildlife organisations taking part. Students participants ran outreach events to the local hospital children's ward and local wildlife experts visited the campus to impart their knowledge in species identification. A number of students were trained in database management by the NBDC. The 24-hour event takes students from various backgrounds and immerses them in a learning experience in nature where they can see, in real time, the diversity of species on their own campus grounds. In October 2017, UCC Green Campus, together with the Environmental Research Institute and Blackstone LaunchPad at UCC, ran the Cork Climathon, as part of the wider global Climate-KIC initiative. The Cork event challenged participants to develop more sustainable modes of transport for UCC campus and its surrounds. As an event that focuses on developing business solutions, the

Climathon attracted a wide range of disciplines and highlighted the potential for universities to transcend disciplines in order to develop solutions to the world's most complex problems.

3.4   Coordination and Planning

UCC's scoring in this category is a clear reflection of university's commitment to sustainability across all levels. According to the review by Parnell (2016), the common thread amongst tested solutions to sustainability challenges at HE was "top-down commitment, combined with bottom-up participation in the development and implementation of new efforts in sustainability education and operations". In UCC, this is epitomised by the Green Campus committee, which is grassroots in nature and the Green Forum, which represents the higher-level leadership element. Both are connected by key members of staff and representatives from the Student Union. Molderez and Fonseca (2018) argue that within sustainability initiatives in higher education, the inter-linkages between operations, teaching, and research remain modest due to the discipline focused nature of HE. Within UCC however, both the Green Campus committee and the Green Forum have representation from operations, teaching, research and the student body. Alignment is now also evident in the various university strategies; The university's Sustainability Strategy covers all of these aspects while sustainability is also highlighted in the university's new Strategic Plan, Academic Strategy, and Civic Engagement Plan. The approach taken in developing the Green Campus Pilot and then further developing the programme to incorporate research and learning more broadly, can be seen as a Living Laboratory, in itself, for the incorporation of sustainability into higher education.

As a "framework" for sustainability in higher education, the Green Flag programme focuses heavily on including all levels and units within the university. The first step of the programme is to assemble a representative committee and evidence senior level commitment. What began as a "student-led" campaign therefore developed into a programme that saw both the bottom-up and top-down structures combine. As the programme progressed, and the university embarked on its ISO 50001 certification, the establishment of "green energy teams" mainly comprised of local managers and technical staff saw a flourishing of action in the middle - the "middle-out" phenomenon (Parag and Janda 2014) (Fig. 3). This was mirrored in other parts of the university through, for example, an Environmental Citizenship Research Priority Area, which brought together researchers and academics to work on transdisciplinary projects (Byrne et al. 2017). Therefore through a combination of frameworks the university has organically progressed a "whole-institution" approach, which is both recognized in the literature as being best-practice, and reflected within the structure of the more comprehensive assessments such as STARS.

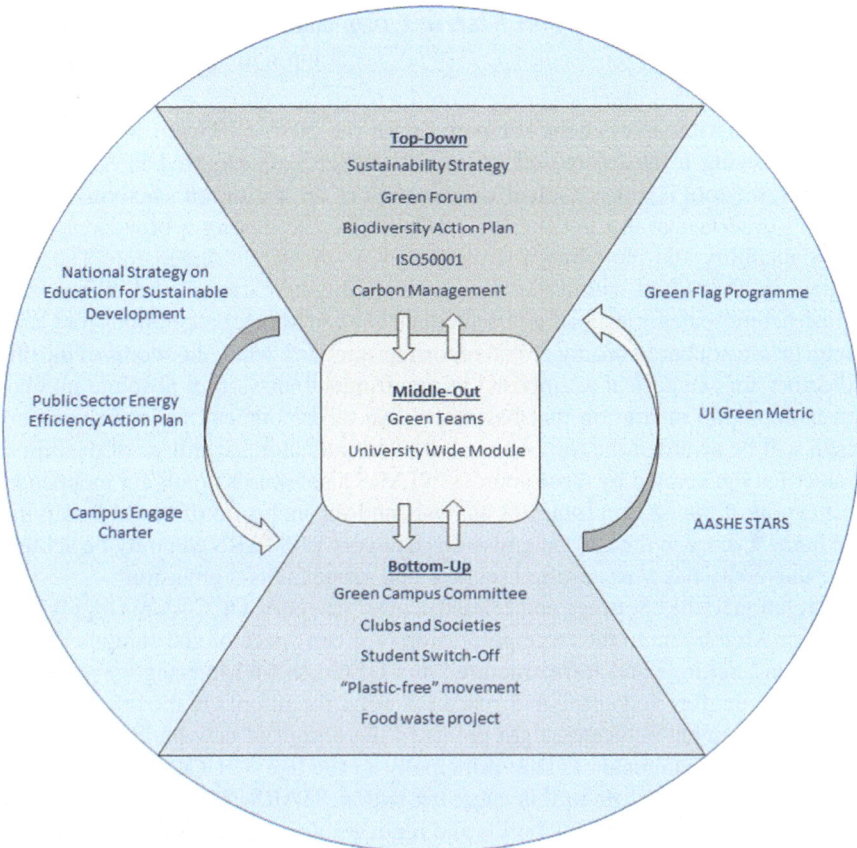

Fig. 3 Diagram illustrating UCC's Sustainability "system"

3.5 Innovation

While in the main the STARS criteria are quite rigid, there is an opportunity to showcase achievements unique to individual campuses in the "Innovation & Leadership" category. Again the broad reach of UCC's sustainability activities is evident here. Collaborations with the Glucksman Gallery around educational projects and exhibitions with local schools stemmed from that building's engagement with the University's ISO 50001 programme. This category also opens opportunities to engage with other activities on campus that would not traditionally be associated with the "green"/sustainability agenda, for example UCC's designation as a University of Sanctuary.

3.6 STARS and UI Green Metric Comparison

Across all categories, the University scored a lower percentage of total points available in its STARS assessment compared with the 2018 UI Green Metric ranking (Fig. 4). Owing to the increased granularity of the data required in STARS submissions, the tool is a more sensitive reflection of an institution's activities. It also includes criteria that are not present in Green Metric, taking a broader approach to sustainability and including a strong focus on the social dimension. The greatest deviation occurred within the "Energy and Climate Change" and "Settings and Infrastructure" categories. In the former, the UI Green Metric calculations for Carbon Footprint are not based on any global reporting standard, while the scores of the other indicators are categorical as opposed to requiring submission of absolute numbers. For example, an institution that has more than three sources of renewable energy onsite will be awarded the top points under that indicator, regardless of the amount of electricity generated by those sources. STARS also awards points for independent verification of the carbon footprint analysis and for inclusion of scope 3 activities. The highest score in the carbon emissions category of STARS can only be achieved if the university has zero adjusted scope 1 and scope 2 GHG emissions.

In relation to the "Settings and Infrastructure" category, UCC scores highly in the UI Green Metric due to the large proportion of green space on the campus. Despite the title of "Settings and Infrastructure", the UI Green Metric category refers only to open, green, forested space and space for water retention. On the other hand, the STARS assessment dedicates eight points to the design of new buildings and their operations and maintenance. Due principally to the age of UCC's building stock, the University scores low in this category within STARS. The adoption of Green Building standards across new builds and renovations is being looked at by UCC as part of its Action Plan going forward. Again this demonstrates the value of STARS in providing benchmarks that the University can aspire to. There is nevertheless value

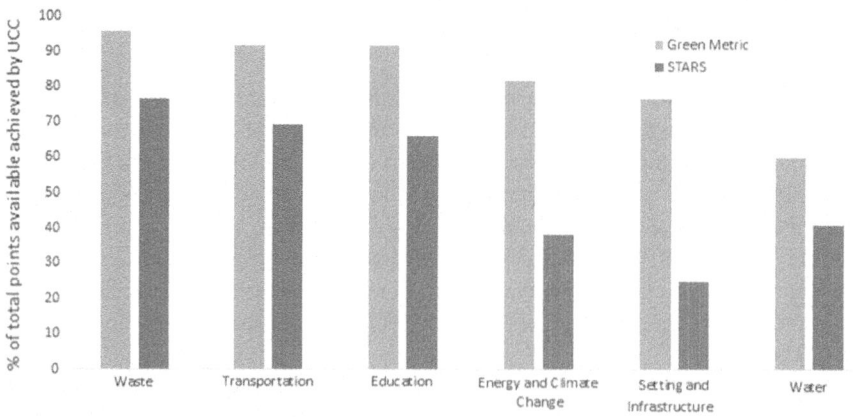

Fig. 4 UCC's performance in STARS versus UI Green Metric broad categories for 2018

in taking part in the UI Green Metric assessment as it provides a global ranking of universities at various stages on their journey to becoming more sustainable. This reflects the original aim of the Green Metric ranking that it be "an entry-level tool" … "of interest and accessible to universities in developing countries as well as to those in developed countries" (Lauder et al. 2015).

4 Conclusions

The STARS reporting process and outcome reflect a university's commitment to embedding sustainability across all aspects of its activities. The cultural norms and legislative measures within a jurisdiction can impact the outcome, but there is enough flexibility to enable an institution to adequately highlight, and gain recognition for, its own individual innovations. Engagement across the whole campus community and adequate resources in terms of time and system capabilities are essential before embarking on the STARS assessment process. For UCC the process is as important as the result, in providing a guide for future action plans. Care needs to be taken to ensure that the STARS programme continues to reflect best practice and innovative approaches to achieving genuine sustainability on university campuses.

Annex 1

STARS credits (version 2.1)

Academics (AC)	Curriculum 40 points available	AC 1: Academic courses
		AC 2: Learning outcomes
		AC 3: Undergraduate program
		AC 4: Graduate program
		AC 5: Immersive experience
		AC 6: Sustainability literacy assessment
		AC 7: Incentives for developing courses
		AC 8: Campus as a living laboratory
	Research 18 points available	AC 9: Research and scholarship
		AC 10: Support for sustainability research
		AC 11: Open access to research

(continued)

(continued)

Engagement (EN)	Campus engagement 21 points available	EN 1: Student educators program EN 2: Student orientation EN 3: Student life EN 4: Outreach materials and publications EN 5: Outreach campaign EN 6: Assessing sustainability culture EN 7: Employee educators program EN 8: Employee orientation EN 9: Staff professional development
	Public engagement 20 points available	EN 10: Community partnerships EN 11: Inter-campus collaboration EN 12: Continuing education EN 13: Community service EN 14: Participation in public policy EN 15: Trademark licensing
Operations (OP)	Air and climate 11 points available	OP 1: Greenhouse gas emissions OP 2: Outdoor air quality
	Buildings 8 points available	OP 3: Building operations and maintenance OP 4: Building design and construction
	Energy 10 points available	OP 5: Building energy consumption OP 6: Clean and renewable energy
	Food and dining 8 points available	OP 7: Food and beverage purchasing OP 8: Sustainable dining
	Grounds 4 points available	OP 9: Landscape management OP 10: Biodiversity
	Purchasing 6 points available	OP 11: Sustainable procurement OP 12: Electronics purchasing OP 13: Cleaning and janitorial purchasing OP 14: Office paper purchasing
	Transportation 7 points available	OP 15: Campus fleet OP 16: Student commute modal split OP 17: Employee commute modal split OP 18: Support for sustainable transportation
	Waste 10 points available	OP 19: Waste minimization and diversion OP 20: Construction and demolition waste diversion OP 21: Hazardous waste management
	Water 8 points available	OP 22: Water use OP 23: Rainwater management
Planning and administration (PA)	Coordination and planning 8 points available	PA 1: Sustainability coordination PA 2: Sustainability planning PA 3: Participatory governance
	Diversity and affordability 10 points available	PA 4: Diversity and equity coordination PA 5: Assessing diversity and equity PA 6: Support for underrepresented groups PA 7: Affordability and access

(continued)

(continued)

	Investment and finance 7 points available	PA 8: Committee on investor responsibility PA 9: Sustainable investment PA 10: Investment disclosure
	Wellbeing and work 7 points available	PA 11: Employee compensation PA 12: Assessing employee satisfaction PA 13: Wellness program PA 14: Workplace health and safety
Innovation and leadership	4 bonus points available	A catalog of exemplary practice and open-ended innovation credits available for selection

References

Association for the Advancement of Sustainability in Higher Education (2018) New AASHE & ACTS partnership brings STARS to Australasia. https://www.aashe.org/news/aashe-acts-partnership/. Last accessed 28 Mar 2019

Association for the Advancement of Sustainability in Higher Education (2019a) About STARS. https://stars.aashe.org/about-stars/. Last accessed 28 Mar 2019

Association for the Advancement of Sustainability in Higher Education (2019b) STARS Technical Manual. https://stars.aashe.org/resources-support/technical-manual/. Last accessed 28 Mar 2019

Ballantyne R, Packer J (2009) Introducing a fifth pedagogy: experience-based strategies for facilitating learning in natural environments. Env Ed Res 15(2):243–262

Barr S (2007) Factors influencing environmental attitudes and behaviors, a U.K. case study of household waste management. Environ Behav 39(4):435–473

Byrne E (2017) Paradigmatic transformation across the disciplines; snapshots of an emerging complexity informed approach to progress, evolution and sustainability. In: Byrne E, Mullally G, Sage C (eds) Transdisciplinary perspectives on transitions to sustainability. Routledge, Abingdon, pp 65–82

Byrne E, Mullally G, Sage C (2017) Transdisciplinary perspectives on transitions to sustainability. Routledge, Abingdon

Chaplin G, Wyton P (2014) Student engagement with sustainability: understanding the value–action gap. Int J Sust Higher Ed 15(4):404–417

Department of Communications, Energy Efficiency and Natural Resources (2009) Maximising Ireland's energy efficiency the national energy efficiency action plan 2009–2020. Department of Communications, Energy Efficiency and Natural Resources, Dublin

Environmental Association of Universities and Colleges (2017) Mapping sustainability assessment and reporting in the UK Tertiary education. https://www.eauc.org.uk/eauc_launches_new_research_on_sustainability_re. Last accessed 28 Mar 2019

Faghihi V, Hessami AR, Ford DN (2015) Sustainable campus improvement program design using energy efficiency and conservation. J Clean Prod 107:400–409

Findler F, Schönherr R, Lozano R, Stacherl R (2018) Assessing the impacts of higher education institutions on sustainable development—an analysis of tools and indicators. Sust 11:59–78

Hands V, Anderson R (2016) Benchmarking sustainability research: a methodology for reviewing sustainable development research in universities. In: Leal Filho W (ed) Sustainable development research at universities in the United Kingdom. Springer, Heidelberg, pp 27–43

Higher Education Associations Sustainability Consortium (2006) Call for a system for assessing & comparing progress in campus sustainability. https://stars.aashe.org/wp-content/uploads/2019/03/HEASC-call.pdf. Last accessed 28 Mar 2019

König A (2015) Changing requisites to universities in the 21st century: organizing for transformative sustainability science for systemic change. Curr Opin Env Sust 16:105–111

Lauder A, Sari RF, Suwartha N, Tjahjono G (2015) Critical review of a global campus sustainability ranking: Green Metric. J Clean Prod 108:852–863

Maiorano J, Savan B (2015) Barriers to energy efficiency and the uptake of green revolving funds in Canadian universities. Int J Sust Higher Ed 16(2):200–216

Mero T (2011) Town and gown unite. Sustain J Record 4(4):169–173

Molderez I, Fonseca L (2018) The efficacy of real-world experiences and service learning for fostering competences for sustainable development in higher education. J Clean Prod 172:4397–4410

O'Mahony C, Higgs B, Alexander D, Kilcommins S, Ryan AC, Blackshields D, McCarthy M, O'Sullivan K, Cronin JGR (2015) A review of scholarly work and educator understanding of Pedagogies of transition: implications for impact. National Forum for the Enhancement of Teaching and Learning, Dublin

O'Sullivan S, O'Tuama S, Kenny L (2018) Universities as key responders to education inequality. Global Discourse 7(4):527–538

Parag Y, Janda KB (2014) More than filler: Middle actors and socio-technical change in the energy system from the "middle-out". Energy Res Soc Sci 3:102–112

Parnell R (2016) Grassroots participation integrated with strong administration commitment is essential to address challenges of sustainability leadership: tools for successfully meeting in the middle. J Env Stud Sci 6(2):399–404

Reidy D, Kirrane M, Curley B, Brosnan D, Koch S, Bolger P, Dunphy N, McCarthy M, Poland M, Ryan-Fogarty Y, O'Halloran J (2015) A journey in sustainable development in an urban campus. In: Leal Filho W, Brandli L, Kuznetsova O, Paço AMF (eds) Integrative approaches to sustainable development at university level. World Sustainability Series. Springer International Publishing, pp. 599

Ryan-Fogarty Y, O'Regan B, Moles R (2016) Greening healthcare: systematic implementation of environmental programmes in a university teaching hospital. J Clean Prod 126:248–259

Ryan-Fogarty Y, Carlin C, Reidy DT, Kirrane MJ (submitted) Intervarsity BioBlitz as a sustainability reporting tool for the higher education sector: a case study of Irish Universities

Stern PC (2000) Toward a coherent theory of environmentally significant behavior. J Soc Iss 56(3):407–424

UNESCO (2017) Cork call to action for learning cities. https://uil.unesco.org/lifelong-learning/learning-cities/cork-call-action-learning-cities. Accessed 30 Jan 2019

University College Cork (2016) UCC sustainability strategy. https://www.ucc.ie/en/media/support/greencampus/UCCSustainabilityStrategy_interactive.pdf. Accessed 30 Jan 2019

University College Cork (2017) UCC civic engagement plan: together With and for the community. https://www.ucc.ie/en/media/centralmedia/UCC_Civic_Engage_2017a.pdf. Accessed 30 Jan 2019

University College Cork (2018a) UCC ranks in world's top 10 'Green' universities. Available at: https://www.ucc.ie/en/news/ucc-ranks-in-worlds-top-10-green-universities.html. Accessed 30 Jan 2018

University College Cork (2018b) The connected university: academic strategy (2018–2022). https://www.ucc.ie/en/media/support/regsa/dpr/academicstrategy/AcademicStrategy2018-2022.pdf. Last accessed 28 Mar 2019

Wals AEJ, Corcoran PB (2012) Learning for sustainability in times of accelerating change. Wageningen Academic Publishers

Comfortable Environment: The Formation of Students-Architects' Professional Consciousness in the Paradigm of Sustainable Development

Olga Melnikova

Abstract The concept of Sustainable Development attracts more and more specialists from various scientific fields. For a long time for architecture there was an understanding of sustainability as a "green", "ecological" or "energy-efficient" architecture. Environmental psychology offered a new interpretation of the concept of "environment" giving the architectural space not only objective, but also psychological components. Preservation of historical and cultural heritage and the emergence of new architecture through the orientation of future architects' professional consciousness to meet the needs of various architectural environment's users is considered in this article as the main aspect on the way to sustainability in architecture. This work is devoted to the study of professionally valuable psychological qualities of the man who has chosen the profession of an architect in the paradigm of Sustainable Development. This article demonstrates the research results of creative imagination features of students-architects; the special attention is given to the change of values-based orientations in the process of professionalization. At present, there is a growing interest in the interaction of an architectural space with users which, in turn, requires increased attention to the professional architect's formation at all stages of developing. This paper will be useful for teachers who work directly with students-architects, as well as for experts who form educational programs for them.

Keywords Sustainable development · Sustainable architecture · Creative imagination · Architectural space · Environmental psychology · Professional consciousness · Environment's users

O. Melnikova (✉)
Faculty of Architecture, Saint Petersburg State University of Architecture and Civil Engineering (SPbGASU), 2-Ya Krasnoarmeiskaya St., 4, St. Petersburg, Russia190005
e-mail: melnikova.gasu@yahoo.com

© Springer Nature Switzerland AG 2020
W. Leal Filho et al. (eds.), *International Business, Trade and Institutional Sustainability*, World Sustainability Series,
https://doi.org/10.1007/978-3-030-26759-9_59

1 Introduction

The modern stage of society development puts forward a number of fundamentally new requirements for graduates in the system of higher education in developing countries, including Russia. On the one hand, these requirements are due to the intensive development of the economy and technologies that require the appearance of a competitive specialist in the labor market. On the other hand—the principles and provisions of the Sustainable development concept, which were formulated and adopted by the General Assembly as the outcome document of the United Nations summit on 18 September 2015. The document was entitled "Transforming our world: the 2030 Agenda for sustainable development" and it was noted in its preamble that "this Agenda is a plan of action for people, the planet and prosperity" (UN General Assembly Resolution 2015). Along with global challenges, such as the eradication of poverty worldwide, the Agenda includes such challenges as the realization of human potential, as well as the right to exist in a healthy environment.

Humanity is continuously developing, and progress has not stopped for a moment, improving the economic and social conditions of human existence which leads to an increase in the importance of the factors that ensure physical and psychological health. That is why the creation of a healthy, comfortable, eco-friendly architectural environment is a top priority for the modern architect (Crabbé et al. 2017). However, this task is complex and it requires active implementation of an integrated and interdisciplinary approach, closer cooperation of specialists from various fields, including psychologists, power engineers, economists, builders, environmentalists and, of course, appropriately trained architects.

Modern urbanists face many questions: is it possible in a modern metropolis with the current system of city management to reduce the time that a person spends on commuting to work? Is it possible in one area to connect the whole range of social processes that occur around the human home, and the functions of public life which usually can be found in the city center? What should be the scale and the capacity of a sustainable city? How and by what means it is possible to connect the main aspects of the sustainable city: a service-oriented economy, the use of eco-friendly and new materials, renewable energy, regeneration of water resources, separate collection and recycling of waste, etc. In such environment great attention must be paid to the landscaping, building open parks and greenhouses on rooftops, and, perhaps, special farms with artificial lighting inside.

Solving problems of this kind requires a non-standard integrated approach, and, as a result, the emergence of new environmental specialists, urbanists, architects—carriers of a bright creative idea, consistent in its implementation and able to inspire colleagues, stimulating community development and progress. This idea is now consistent with the principles of sustainable architecture in the concept of Sustainable Development. Of exceptional importance for the sustainable architecture development in our time is the analysis of the ongoing planetary processes and the search for new concepts of creating a spatial environment from the standpoint of the universal values approval. The mindset of an architect, an urban planner, or a builder is an important component of the future architectural and spatial environment formation (Esaulov 2014).

It is obvious that for the emergence of a professional architect, capable of creating a healthy sustainable architectural environment and fully meeting modern requirements, it is necessary to make the transition from the traditional educational paradigm which is based on the reproduction of a high level of student's theoretical knowledge to creative professional education, focused on the development of student's creative potential as a future graduate of a new generation. In the future, such an architect will be able to successfully operate in rapidly changing socio-economic conditions and instantly focus on moral criteria in the environment transformation, as well as to seek and find non-standard and effective solutions to various problems. These assumptions are also reflected in the project "Russian education 2020: an education model for a knowledge-based economy". The project defines the development of the individual creative potential not only as a resource for economic and social development, but also as the ability to adapt quickly to the needs and requirements of a rapidly changing world.

It should be noted that the emergence of a professional architect of the new formation is associated with overcoming many objective obstacles and contradictions, and one of the main contradictions is the degree of matching between the personality and the profession. This problem is solved in the process of professionalization—one of the personality development directions. Professionalization is the process of becoming a professional which includes:

- a person's choice of profession according to his/her own abilities;
- mastering the rules and regulations of the profession;
- self-awareness as a professional;
- personal development by means of profession.

In general, professionalization is one of the aspects of socialization, just as the professional formation is one of the aspects of personal development.

Today there is an increasing interest in psychological support of professional activity which, in turn, requires increased attention to the professional formation at all stages of becoming a specialist.

Some researchers determine that "the opportunity for the individual to become a person and the disproportion between personal and individual development of professional activity are due to the fact that this development occurs in various structures: professional specialization is currently carried out mainly in the formalized structures of the profession, the architect as a person is formed in the unformalized structures of socio-cultural and social activities" (Stepanov et al. 1993). Others note that tertiary education, workplace education and lifelong learning, not to mention communal learning, are not coordinated with the plan for formal schooling (Jucker 2011).

At the present time, when the basics of human professional activity are sufficiently studied, the most important things are professional training for future activity and its modeling, ideas about future activity, especially the formation of images related to professional activities. Carrying out the training of specialists in the system of vocational education, it is necessary to have an idea of the initial, final and intermediate result of educational activities, which is impossible without the formed image of the student's profession. E. A. Klimov considers the image of the profession as the presence of an adequate system of subjectively interpreted value representations of a specialist on the professional knowledge basis, as a person's idea of his profession and its components.

The initial stage of professional self-determination—the stage of professional intentions and professional choice made by secondary school graduates is described in special literature. The subsequent stages of the personality's professional development should continue to form its attitude to itself as to the subject of professional activity. Today the second stage of professional development is of particular interest, which is, first of all, studying in a higher education institute. The main psychological new formations at this stage are professional orientation, professional and ethical value orientations, spiritual maturity, origin and development of professionally important qualities, readiness for professional activity.

Several groups of "images" take part in the formation of the architect's professional consciousness. The idea of architects about their future activity is formed into the "image of the profession". This image is largely prognostic and consists of several components. Until now, there were three components: who builds (an architect), what builds (a building) and how builds (styles of activity). At the present stage which is focused on sustainable architecture they added another component—for whom an architect builds (a user). These components eventually turn into images—regulators which are an integral part of the architect's professional activity at all stages of professionalization. But it is the lack of understanding of the ultimate user that leads to impoverishment and rejection of the architectural environment.

2 Architectural Space and Architectural Environment

Space organization is the main object of activity for an architect in all periods of history; "space has always been the essence of architecture of all times both in buildings and in cities" (Dzevi 1972). Throughout the history of mankind, ideas about the surrounding space have developed in interaction with the way of human life, type of his activity, the current worldview and in close dependence on the form of consciousness (Stepanov et al. 1993).

There are several main categories of visually perceived subject-spatial environment. These categories include the space itself as well as color, light, texture, shape, proportions, image. Each of these categories is filled with philosophical meanings and has its own clear definitions. With regard to the artificial environment, in general, and architectural one, in particular, the concepts of artistic or architectural space are the most commonly used. It should be noted that these concepts are different, and have different definitions. Thus, the artistic space is a space of the work of art, including such categories as unity and completeness, designed primarily to unite people with the help of aesthetics and other visual means. The artistic space formed for people should naturally have an accessible language and include a system of generally understood signs. Due to the fact that "we are talking about the attitude of a man to something, the artistic space should have a system of specific values that determine these attitudes of a man to the real world" (Maslova 2007).

Architectural space is an artistic space, but it is supplemented by a set of specific qualities. First of all, it is an idea, the bearer of which is society, and the spokesman—the architect. And this idea, which is undergoing transformation under the influence of the architect's personality, can and should shape people's lives in a certain way. "In architecture, space clearly acts as a carrier of the event" (Maslova 2007).

The concepts of artistic and architectural images are the key concepts of artistic and architectural space; and they differ respectively. Such factors as trends, style and fashion inherent in current historical period have a significant impact on the artistic image formation, whereas the architectural image is formed under the influence of a large number of "non-artistic" factors. The characteristics of any structure include such concepts as benefit, strength, capacity, ergonomics, etc.; today the characteristic of "sustainability" is also added. The term "sustainable architecture" is widespread but not unambiguous. Such concepts as "green architecture", "sustainable construction", "ecological architecture", "low-cost architecture", "high-tech architecture", "bioclimatic architecture", "energy efficient and smart construction" are often used along with the term "sustainable architecture" (Salmina and Bystrova 2015). Many definitions have an economic, social or technological bias, but architecture is not confined to any of these areas. All presented concepts relate primarily to the economy and technology of construction, but do not reflect the aesthetic, symbolic, artistic and psychological aspects of architectural activity. In architecture there is artistry, history, culture, conceptual design; it has its own expressive means (Salmina and Bystrova 2015) which cannot be reduced to a set of only technical or economically sound indicators.

In the middle of XX century, with the advent of environmental psychology, the architectural environment was seen as a set of phenomena that surround a man and involve him in a certain interaction. Now the architectural environment is increasingly becoming the object of study for various groups of specialists. This topic attracts designers, architects, artists, psychologists and sociologists. When designing the environment where people live or work, many factors should be taken into account. People are exposed to positive or negative impacts to the different extent by physical objects they use; and the impact depends on the correspondence to the characteristics of the human body, sensory and motor systems. Any type of architectural object imposes restrictions on its inhabitants, and the habitat developed according to the relevant projects can facilitate or complicate the performance of vital functions. Therefore, architects-designers of comfortable living and social environment should be interested in a deeper study of architectural "subject–spatial environment" problem.

3 Methods

The modern stage of society development is marked with a massive change of working conditions, automation and computerization of production, introduction of new technologies and changing monoprofessionalism into poliprofessionalism. This leads to the fact that the modern world is in dire need of such professionals who are able to quickly and effectively solve urgent problems, fully realize their potential and contribute to sustainable and versatile individual development in the rapidly changing socio-economic conditions.

The most important of the cognitive abilities that influence the formation of the architect professional consciousness is creative imagination. In a certain quality or amount the creative component is found in many professions, but architecture is one of the few ones where imagination plays a leading role. The abilities to creative imagination among architecture institution applicants are revealed at a stage of entrance examinations, and in the course of further training the student should solve more and more complex problems with each course. The educational process in architectural design in Russia is mainly based on the fact that the teacher gives a student a design assignment, which indicates the purpose of the building, the planned number of consumers, the master plan, geographical area, etc. A student on the basis of this task, designs a given object in several steps using professional, regulatory and special sources. In the design process (both student and professional), as a rule, the consumer characteristics—gender, age, professional, information about the family, interests, and many other characteristics are not included or some of them are taken into account insufficiently. All this proves the need to include the elements of modern creative professional education into traditional education.

In the course of the study, which was conducted in St. Petersburg (Russia) among students of the University of architecture and construction of SPbGASU in 2015, an attempt was made to identify the features of creative imagination. Professional productive creative imagination is the core of professional consciousness of the architect, and a necessary component of his professional activity. Research in this area is particularly relevant at the present time, which requires a new look at both the comfortable architectural environment and the specialists that form it. The undoubted novelty of our study was an attempt to shift the emphasis of designers with a pronounced stylistic and aesthetic functions of the architectural environment objects and focus on the identity of the consumer (Melnikova and Soloveva 2015).

In *the first part* of the task we provided students-architects with photos of several buildings of different styles. In this part, students had to, using only their own imagination, describe the user of these buildings, come up with the image of the man: his age, his profession, his hobby, character, marital status.

In the second part of the experiment, we offered to finish, i.e. to reconstruct a certain faceless urban space in accordance with their ideas about the values, inherent in the modern comfortable environment. Students were asked to choose, at their request, the object, among which was the area, the street, the intersection and the yard (Lynch 1982).

In total, the experiment involved three groups of students-architects of the 1st, 3rd and 5th year, a total of 166 people.

4 Results and Discussion

Table 1 describes the users of the three buildings with the lowest, highest and neutral ratings.

When processing the results, we noticed that the students-architects had difficulty describing the user of the building. They easily described the objects themselves, styles, authors of buildings, but noted that it was extremely difficult for them to switch to the image of a person. This is due to the fact that as a result of professional training and accumulated professional experience the imagination of the architect becomes inextricably linked with the object of activity and entirely complies with the tasks of forming the building appearance (Melnikova 2014). As noted above,

Table 1 Descriptions of users

S. No.	Objects	Users category
1.	With a lot of positive reviews 49%	Respondents described the resident as a modern architect or a designer who designed the house for himself and his family, keeping up with the times and holding a good position, which indicates his successful career. This is the way of how most young architects-respondents imagine their ideal house
2.	With a lot of negative reviews 80%	Respondents described the user of the building as an insecure person with an unstable psyche, who has no sense of proportion and taste, does not know what he needs and maybe he wants to stand out from the crowd due to such a defiant house. None of the respondents would like to live in this house
3.	With a lot of neutral reviews 31%	Respondents believe that the residents of this block of flats received apartments in this house under state programs or by inheritance, and are not able to change their place of residence. Residents are described as ordinary people, the average family with children and parents, and even the bright colors on the front elevation will not be able to diversify their lives

Table 2 Objects for reconstruction

		1st year	3rd year	5th year
1	Square	50%	25%	22%
2	Street	29%	25%	22%
3	Intersection	3%	20%	0%
4	Yard	18%	30%	56%

there are various tasks: aesthetic, stylistic, artistic; but at the moment the interest in the person as a user of the subject-spatial environment is extremely small. It should be noted that the object number 3 took the last places in all groups. The description of this house's inhabitant, most of all, included such characteristics as "ordinary people", "ordinary family". It would seem that in this case, the imagination of respondents, which was not bound up with any conventions, should have given a rise to a mass of different images, but this did not happen. The conclusion is that there is a connection between "ordinary man" and "ordinary, faceless" architecture, and no one would like to live in this ordinary house, no one wished to associate themselves with an ordinary resident of the given house.

The second part of our experiment was devoted to the reconstruction of the existing urban environment. Table 2 shows the percentages of the objects selected for reconstruction.

As it can be seen from Table 2, the greatest interest among the 1st year students was caused by the reconstruction of the square, and the smallest interest—the yard. This is due to the fact that the one, who came to this profession, does not yet perceive it adequately, believes that self-expression is the main thing in it. Public open space is ideal for it, and a chamber, intimate place (yard) does not play a significant role in the structure of the entire urban space.

On the contrary, the yard has caused the greatest interest among the 5th year students. By the end of the training students come to understand that their activities have a significant impact on the citizen's life. They begin to realize that all aspects of human life, social, and psychological, public and intimate are important for any person.

We were extremely interested to know what categories future architects will use to describe the modern urban environment and what methods of reconstruction they will choose. Table 3 shows the most typical responses in percentage terms.

As it can be seen from Table 3, the categories chosen for the reconstruction of the space are quite diverse. Among the ideas there are banal ones, absolutely relevant, and there are those that do not fit into the ecological ideas about the modern environment. The most controversial, in our opinion, are the ideas to cut down trees in the square and in the street. It stands to mention the lack of attention of almost all groups of respondents to such objects as garbage containers, pedestrian crossings and equipped parking space. But the modern comfortable urban environment is unthinkable without bins, benches, modern garbage containers, without public toilets and everything that makes life easier for the user and makes his existence comfortable and safe.

Table 3 Categories of the comfortable urban environment

	Object	Publicity, the space for walking	Public amenities	Place of interest	Front elevation decoration	Greenspace extension	Pedestrian crossing	Lights, street lamps	Cycle path	Playgrounds	Garbage containers	Parking space	Comfort, beauty, cleanliness	Cutting down trees
1 year	Square	28	28	7	21	–	–	14	–	–	–	–	–	7
	Street	12	60	12	36	48	–	36	–	–	–	12	36	–
	Intersection	–	–	–	–	–	100	–	–	–	–	–	–	–
	Yard	20	80	–	60	60	–	20	–	80	–	20	40	–
3 year	Square	40	100	40	40	40	40	60	40	–	–	–	–	–
	Street	–	100	–	80	40	–	20	60	–	–	–	–	20
	Intersection	50	–	75	–	50	50	25	75	25	25	–	–	–
	Yard	–	68	–	85	51	–	34	–	68	17	51	17	–
5 year	Square	66	100	100	66	66	–	33	–	–	–	–	–	–
	Street	33	100	33	100	33	–	–	66	–	–	–	–	–
	Intersection	–	–	–	–	–	–	–	–	–	–	–	–	–
	Yard	26	91	39	65	52	–	26	13	52	–	39	–	–

5 Conclusions

In recent years, the relationships between a man and the environment are noted to become more aggressive and tough due to the changes in the economic and socio-political life of various countries. This leaves an imprint on both the natural and artificial architectural subject-spatial and social environment (Melnikova 2011). The city becomes a platform where the interests of various participants and conflicts collide. One-sided economic justification of development projects leads to the destruction of green areas, architectural monuments, economically unprofitable recreational areas are desolated, and, as a result, become unfavorable and often criminal places that is not correlated with the main aspects of a sustainable city. New processes taking place in the life of an individual and society as a whole give rise to the new psychological problems. In this case, it can be assumed that the use of the latest knowledge, obtained from psychology, sociology, cultural studies and other sciences, by the architects will help to bring the architectural science to a new level.

Our research has proved that on the one hand, with the time of studying architects come to understand the importance of human personality in the design of a comfortable modern environment. And this, in turn, gives hope that over time we will be able to have the architect of the new formation, who will use the principles of Sustainable Development as the reference point in his work. The reports show that sustainable development in higher education institutions requires not only making statements and preparing documents but also taking concrete actions (del Mar Alonso-Almeida et al. 2015). The inclusion of the latest knowledge from various sciences in the educational programs, the modern architects' awareness of the special importance of the ongoing planetary processes and the human values prominence will make it possible to do a lot for the implementation and development of sustainable architecture. Over time, the contradiction between the professional specialization of the architect, carried out in the formal structures of the profession, and the formation of the personality of the architect, occurring in the informal social activity structures can be smoothed. The resolution of this contradiction is a complex task that requires thoughtful and careful study on the basis of the new non-standard scientific researches.

Modern architects who have chosen the principles of humanism, safety and environmental friendliness as the goal of their activity can and should form such an architectural environment that modern residents expect from them. Changing the mindset of the architect or the builder, in this case, seems to be an important component in the formation of comfortable architectural-spatial environment.

Society is increasingly sensitive about issues related to sustainable development (Lozano and Young 2013). The architect, professing the idea of sustainable development, in the future will be able to successfully operate in a rapidly changing socio-economic conditions and instantly focus on moral criteria in the transformation of reality, as well as to seek and find innovative and effective solutions to various problems. These assumptions are reflected in the project "Russian education 2020: a model of education for a knowledge-based economy" (Volkov et al. 2008) which defines the development of creative potential of the individual not only as a resource for economic and social development, but also as the ability to adapt quickly to the needs and requirements of a rapidly changing world.

Our experiment has also showed that, on the one hand, architects are aware that their activities are aimed at meeting human needs, but, on the other hand, they are not ready to imagine the specific identity of the user. Architects are able to influence users of the subject-spatial environment through their professional activities, to meet a variety of human needs, including the realization of human potential, as well as his right to exist in a healthy environment. In this case, a more complete picture of the specific person, for whom they work and create, will help to form new levels of professional consciousness of the architect, and to reach new levels of professional architectural activity which is fully consistent with the Sustainable Development principles.

References

Crabbé A, Bergmans A, Craps M (2017) Participation in spatial planning for sustainable cities: the importance of a learning-by-doing approach. lifelong learning and education in healthy and sustainable cities, pp 69–85

del Mar Alonso-Almeida M, Marimon F, Casani F, Rodriguez-Pomeda J (2015) Diffusion of sustainability reporting in universities: current situation and future perspectives. J Clean Prod 106:144–154

Dzevi B (1972) Masters of architecture about architecture: the chosen fragments from letters, articles, performances and treatises. M.: Art. 487

Esaulov G (2014) Sustainable architecture—from principles to development strategy. Vestnik TGASU No 6(9)

Jucker R (2011) ESD between systemic change and Bureaucratic obfuscation some reflections on environmental education and education for sustainable development in Switzerland. J Educ Sustain Dev 5(1):39–60

Lozano R, Young W (2013) Assessing sustainability in university curricula: exploring the influence of student numbers and course credits. J Clean Prod 49:134–141

Maslova LA (2007) Plener for architects. Ukhta: UGTU, 92

Melnikova O (2011) Concepts of subject-spatial environment in psychology and architecture. In: Actual problems of modern construction: 64th international scientific and technical conference of young scientists/Spbgasu. 3 h. h. III. SPb, p 264

Melnikova O (2014) Complex analysis of "city image" formation by students on the basis of semantic differential method. Bull of Civil Eng 3(44):302–307

Melnikova O, Soloveva E (2015) Respondent's subjective perceptions about types of users architectural objects. In: Materials of the all-russian scientific and practical conference « The city in the mirror of sciences—2015 » . Spbgasu, pp 179–184

Lynch K (1982) Image of the city. Translated from English V. L. Glazycheva; Comp. A. V. Ikonnikov. Under the editorship of A. V. Ikonnikov. M.: Stroyizdat, p 328

Salmina O, Bystrova T (2015) The principles of sustainable architecture creation. In: Academic Bulletin of the Ural Research institute project RAACH, vol 4

Stepanov AV, Ivanova GI, Nechayev NN (1993) Architecture and psychology: manual for higher education institutions. M.: Stroyizdat, p 295, 24

UN General Assembly Resolution A/RES/70/1. (2015). Transforming our world: The 2030 agenda for sustainable development. September 18, 2015. New York: United Nations Assembly Resolution. http://www.un.org/ga/search/view_doc.asp?symbol=A/RES/70/1&Lang=E

Volkov A, Remorenko I, Kuzminov Y, Frumin I, Jacobson L, Andrushchak G, Yudkevich M (2008) Russian education—2020: education model for the economy based on knowledge: to the IX International scientific conference "Modernization of Economy and Globalization". M.: State University HSE publishing house

Students and University Teachers Facing the Curricular Change for Sustainability. Reporting in Sustainability Literacy and Teaching Methodologies at UNED

A. Coronado-Marín, M. J. Bautista-Cerro and M. A. Murga-Menoyo

Abstract This paper presents the results of a research framed in the strategy of *Conferencia de Rectores de Universidades Españolas* (CRUE. Conference of Chancellors of Spanish Universities) towards the Education for Sustainable Development, Target 4.7 from UN 2030 Agenda. It focuses on the Master's degree in Compulsory Secondary Education and Baccalaureate, Vocational Training and Language Education Teacher Training taught at National University of Distance Education (UNED); with two main objectives: (a) to know students' literacy level in matters of sustainability (Sustainability literacy): (b) to know how university teachers perceive their teaching practice in substantial issues of the approach to education for sustainable development. In this empiric study the data have been collected through two questionnaires, one applied to the students, one applied to the university teachers. The results show, on one side, the insufficient literacy of students about the challenges of the current socio-ecological crisis; and on the other side, the teacher's low awareness of their role as agents of educational change for sustainability, although they declare to use usual teaching methodologies of this type of education. It is confirmed the need for institutional strategies to strengthen the training of both groups, students and teachers, in the competencies that sustainable development demands.

Keywords Education for sustainability · Sustainability literacy · Teaching practice · Higher education · Curricular sustainability · Teaching methodologies

A. Coronado-Marín (✉)
International Doctoral School, National University of Distance Education, UNED, Bravo Murillo St. N°14, 28040 Madrid, Spain
e-mail: acoronado@edu.uned.es

M. J. Bautista-Cerro · M. A. Murga-Menoyo
UNESCO Chair in Environmental Education and Sustainable Development, National University of Distance Education, UNED, Juan Del Rosal St. No 14, 28040 Madrid, Spain
e-mail: mjbautistac@edu.uned.es

M. A. Murga-Menoyo
e-mail: mmurga@edu.uned.es

1   Introduction

Sustainable development, a model adopted by United Nations, requires a social change that involves multiple strategies. Education is among these strategies, which is urged to a deep reconsideration, not only of its purpose, but also of its essence (UNESCO 2015). This idea aspires to change the educational system in every level, but perhaps, higher education may be particularly affected due to its complexity.

Nowadays, traditional higher education system is a controversial topic. They are required to address processes of an innovative and adaptive nature, guiding their management towards sustainability and increasing the personal and collective resilience of all their members. They are obliged not only to improve the quality of education, but also to reduce their own institutional vulnerability facing the environmental crisis, taking advantage of the many synergies that could result from the contributions of social movements (Murga-Menoyo et al. 2017).

Therefore, higher education deals with a multifaceted change process that requires a systemic approach of institutions and centres; a transformation certainly not without obstacles and difficulties, as highlighted by Leal Filho et al. (2017). From this perspective, sustainability should be a transversal axis of the curriculum and university life.

Among the lines of action, a main one is placed in the field of curricular innovation. On the occasion of the UN Decade of Education for Sustainable Development (DESD), the responsible agency for its coordination, the UNESCO, highlighted the need for: "questioning, rethinking, and revising education from pre-school through university to include more principles, knowledge, skills, perspectives and values related to sustainability in each of the three realms—environment, society, and economy—is important to our current and future societies" (UNESCO 2005, p. 29).

This change requires that the principles and values of sustainable development inspire the teaching projects of the subjects, and these include among their training objectives the acquisition of competencies in sustainability (UNESCO 2014, 2017) by students—the teachers of the future. Because, following Jucker and Mathar (2015), the challenge lies not so much in transmitting knowledge about the pressing socio-ecological problems but, above all, in activating the behaviours that allow to counteract them.

This process, called '*curricular sustainability*' in the paper, focuses on disciplinary contents and training methodologies, and, therefore, on the role played by university teachers as agents of the training process. The decisive role that teachers play as agents of education for sustainable development is contained as one of the explicit goals of the UN 2030 Agenda, Target 4.7 (UNGA 2015); but the knowledge that the students have about the socioecological problems is also a prerequisite of their possible commitment to tackle them (Leal Filho et al. 2018).

In the classrooms, the central agents of the *curricular sustainability* process are: faculty, as curriculum designers and implementers of the projects, and students, as main characters of the formative processes. Therefore, knowing their affinities with the approach of sustainability is an unavoidable starting point for curricular innovation. It is necessary to establish, on the one hand, if the students are sensitized and have, at least, an elementary level of "literacy" in the current most significant problems of sustainable development. But it is also required to know what teachers think about the environmental crisis, if they perceive their role as an agent of change in this crisis, and finally, if their teaching methodologies facilitate the goals of sustainable development.

The issue presented by the *curricular sustainability* is, therefore, a relevant research subject of broad trajectory and a necessary field of teaching innovation. The following empirical study contributes to this line of work, focused on the Master's degree on Compulsory Secondary Education and Baccalaureate, Vocational Training and Language Education Teacher Training taught at National University of Distance Education (UNED), based in Madrid (Spain).

2 Context of the Empirical Study

In Spain, as in other countries with a similar sociocultural environment, universities are involved in institutional processes of transition towards sustainability. For more than a decade, the *Conferencia de Rectores de Universidades Españolas* (CRUE: Conference of Chancellors of Spanish Universities) has a working group, composed by more than a dozen universities, whose specific purpose is to promote the *curricular sustainability* of university degrees, process that consists of including the acquisition by students of competencies in sustainability as a first level training objective. The group carries out an intense activity, using diverse strategies (CRUE 2005; CRUE-CADEP 2009, 2015), among which is included an offer of teacher training courses.

There are also research groups with extensive experience and specialized bibliographic production. Among the main research groups are the following:

(a) ACES network (*Red de Ambientalización Curricular de los Estudios Superiores.* Curricular Greening of Higher Education) which, made up of 5 European universities and 6 from Latin America, was constituted in 2000 within the Alpha Program of the European Union (Geli de Ciurana and Leal Filho 2006; Junyent and Geli de Ciurana 2008; among others).

(b) *Red de investigación en Educación para la Sostenibilidad* (EDUSOST. Research Network on Education for Sustainability) of Catalonia (Cebrián and Junyent 2015; Medir et al. 2016; Cebrián 2017; Garcia et al. 2017; Viciana et al. 2017; among others).

(c) *EDUCAMDES* group, composed of members of the UNESCO Chair of Environmental Education and Sustainable Development, housed in the UNED, whose interdisciplinary strategy for the training of professionals about sustainability's challenges has been consolidated over three decades (Bautista-Cerro and Díaz González 2017; Novo et al. 2010; Murga-Menoyo 2014; Novo and Murga-Menoyo 2014; Cutanda and Murga-Menoyo 2014; among others).

(d) *Ambientalización curricular en la universidad* group (ACUVEG. Curricular Greening at the university), of the *Universitat de València* (Aznar Minguet et al. 2016; Mascarell and Vilches 2016; Vilches and Gil-Pérez 2016; among others).
(e) The research group of the *Universidad Complutense de Madrid* (Saban et al. 2017; Saenz-Rico et al. 2015; Sánchez et al. 2017; Gonzalo et al. 2017; among others).

All of this research groups follow the legislative framework of higher education in Spain, the legal regulation of the training of university students that allude in their text to the need to meet the demands of sustainable development. This is the case of the Organic Law 4/2007 of Universities, which states explicitly: "the University will develop quality research and effective management of the transfer of knowledge and technology, with the aim of contributing to (…) a responsible, equitable and sustainable development". And it adds later: "it will promote the realization of activities and initiatives that contribute to the promotion of the culture of peace, sustainable development and respect for the environment, as essential elements for solidarity progress".

Likewise, Royal Decree 1393/2007, of October 29, which establishes the organization of official university education, states: "it must be taken into account that training in any professional activity must contribute to knowledge and development of Human Rights, democratic principles, the principles of equality between women and men, solidarity, environmental protection, universal accessibility and design for all, and the promotion of a culture of peace".

Taking into consideration these points, it can be said that the national context offers a solid framework for the goals of the research presented in this article. The scenario is located in the National University of Distance Education (UNED).

The UNED is the largest and oldest distance learning university in Spain. Its foundation dates back to 1972, and is the first of those that have been emerging around the world following the model of the Open University of the United Kingdom. Today, it is among the so-called mega-universities, a type of institutions whose emergence, as Sir John Daniel claimed (1999, 37), can be considered: "the most important development in higher education from the 20th century that will influence the evolution of the idea of the university in the 21st century".

In the 2016–17 academic year, the UNED welcomed 206,629 students, 1363 professors, 6044 teacher tutors and 1232 members of administrative and service staff. It offered 27 Degree titles, 74 official master's degrees, 523 Lifelong Learning courses, 14 language courses, 135 Summer Courses, and 502 University Extension activities. In the research field the university has 109 research groups with a consolidated track record, and 145 doctoral theses were finalized last year (UNED 2017).

The postgraduate training offer includes, among others, the Master's Degree in Teacher Training of Compulsory Secondary Education and Baccalaureate, Vocational Training and Language Teaching, which has been analysed in this empirical study. The students are graduates from different degrees and wish to practice the teaching profession in the secondary stage of the school system (compulsory and post-compulsory), prior to higher education. Students are required by the Spanish law to do this official postgraduate degree.

3 Research Objectives

The study aims at two specific objectives:

(a) To know the level of literacy of students in matters of sustainability.
(b) To know the perception that university teachers have about the socio-environmental crisis and their role in it, as well as the affinity of their teaching practice with the model of education for sustainable development.

4 Methodology

The study is descriptive and has been carried out using a quantitative methodology. The data has been collected through a knowledge test applied to students and an opinion questionnaire to teachers. To achieve the first objective, the data have been collected via a knowledge test applied to students; to reach the second objective, an opinion questionnaire has been chosen for teachers.

5 Population and Sample

The population under study is composed of two groups of subjects: (a) students of the Master's Degree in Secondary Education Teacher Training of the UNED enrolled in the academic year 2016–17; (b) professors with teaching in this Degree. Being the population a total of 629 students and 299 teachers, respectively. All of them were invited to participate voluntarily in the research. Teachers were invited through a letter endorsed by the academic authorities.

6 Instruments for Data Collection

6.1 Sulitest

The instrument used to collect data from students has been the 'Sustainability Literacy Test' (Sulitest) (The Sulitest Organization 2018). Its function is to measure the Sustainability Literacy. That "is the knowledge, skills and mind-sets that allow individuals to become deeply committed to building a sustainable future and help them to make informed and effective decisions to this end" (The Sulitest Organization 2016).

It is a powerful tool that was developed within the framework of the Higher Education Sustainability Initiative (HESI 2019), in order to offer a response to the commitment of the United Nations General Assembly to education for sustainable development (UNGE 2012). They are founding members of HESI, the following entities, agencies and international organizations: United Nations Department of Economic and Social Affairs (UN-DESA), UNESCO, United Nations Environment Programme (UNEP), Principles for Responsible Management Education (PRME), United Nations University (UNU) and the United Nations Human Settlements Programme (UN-Habitat).

The Sulitest consists of 30 questions; each of them allows you to obtain: 4 points for the correct answer, 1 point for 'I am not sure' and 0 for a not correct answer. It offers four response options (only one of them valid) and one more for those who do not wish to express their opinion.

The questions refer to basic knowledge about sustainability. They cover 15 different issues which have been articulated in four themes: (a) Sustainable humanity and ecosystems; (b) Global and local human-constructed systems; (c) Transition towards sustainability; (d) Role to play, individual and systemic change.

The test has been designed with a systemic approach, so that the questions are interrelated reciprocally through a mesh of 44 tags, all of them significant terms extracted from the text of the Earth Charter and the Sustainable Development Goals. Each question has been associated with between 1 and 3 tags and has been stored in an extensive database, from which 30 different questions have been extracted in each evaluation session, ensuring an appropriate balance between the relevant topics.

The Sulitest has proven its value in the international context and it has allowed a first report on the knowledge gaps that students of higher education have regarding the problems of sustainable development (HESI 2017). It is, therefore, a useful instrument to detect educational needs within the framework of the 2030 Agenda (UNGA 2015).

In the UNED case, this tool has been chosen because of its reliability for the purposes of this investigation. Additionally, it has been considered that, as a collateral effect, the completion of the test itself could contribute to the students' awareness of the problems of sustainability; and motivate them to ask questions and find information to answer them. Once each test question has been completed, students receive a feedback that shows them the correct answer, the selected answer, and some additional comments.

The data for this research was collected in 2017, using the version of the questionnaire available at that time.

6.2 "Sustainability and Teaching Practice" Questionnaire

This questionnaire was elaborated ad hoc (Murga-Menoyo et al. 2016a, b) and includes a total of 26 questions referring to two main thematic areas:

1. The perception that teachers have about:

   (a) the social and ecological crisis (4 items);
   (b) the teaching practice itself as an expression of commitment to sustainability (5 items);
   (c) the strengths of their discipline to contribute to education for sustainable development (1 item).

2. The self-reflection of teachers on their own teaching practice, focused on the following aspects:

   (a) competencies in sustainability that are proposed to be trained in (6 items);
   (b) usual methodologies and strategies (6 items and one open question);
   (c) aspects of the teaching project that explicitly contribute to training for sustainability (3 items).

Items adopt a Likert scale, with five possible levels of agreement from (1) "totally agree" to (5) "strongly disagree".

The questionnaire was developed taking into consideration other instruments with a similar purpose (Aznar Minguet et al. 2011, 2014; Murga-Menoyo et al. 2016a, b; García Esteban and Murga-Menoyo 2012, 2015). It is based on the education model for sustainable development promoted by UNESCO (2014, 2015, 2017) and other institutions with similar approaches (UNENCE 2009, 2013; QAA/HEA 2014; HEFCE 2014; Higher Education Academy 2014).

The questionnaire was submitted to expert judgment, and after appropriate improvements it was implemented, inviting all faculty of the Master's degree object of study to complete it.

7 Results

7.1 Students Facing the Curricular Change for Sustainability

The Sulitest was answered by 314 students, 49.92% of the population. The global data and subscale data are shown in Fig. 1. UNED students have responded correctly to 52% of the questions. They are two points above the national average (50%) and two points below the world average (54%), differences that are not remarkable.

The results in the subtests maintain a similar trend. UNED students reach a middle position, where the bigger difference is regarding the knowledge about *Transition towards sustainability*, up to five points below their international peers.

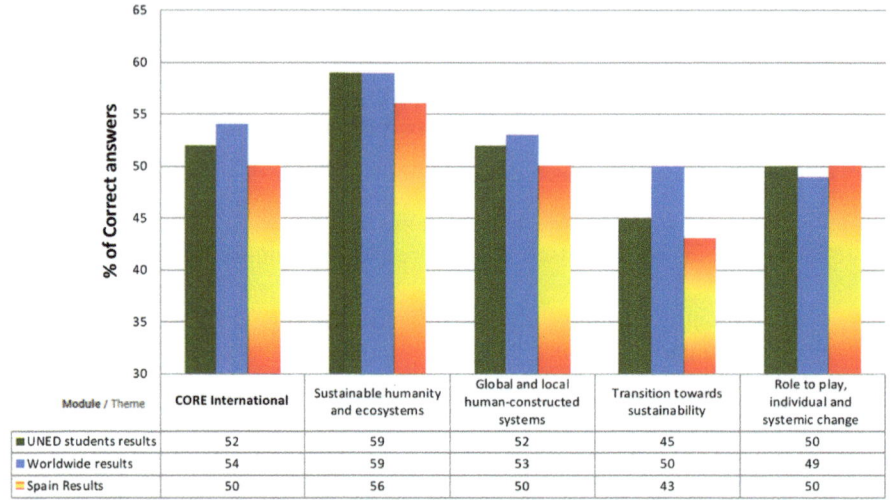

Fig. 1 Global results. Results of the SULITEST from UNED students in the national and international context

Regarding the more specific contents assessed, grouped according to the categories proposed in Sulitest, the best results were obtained in the questions referring to *Formal education and life-long learning*, with a 87% correct answers, and *Trade (local, international, fair, etc.)*, with 74% (Fig. 2).

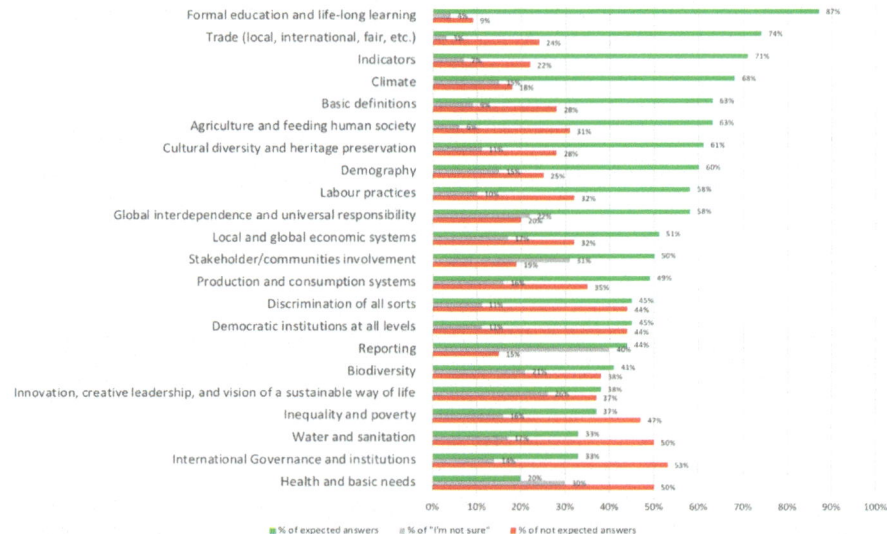

Fig. 2 Results per tag. Percentage of responses depending on the tag associated with the question, ordered from highest to lowest percentage of correct answers

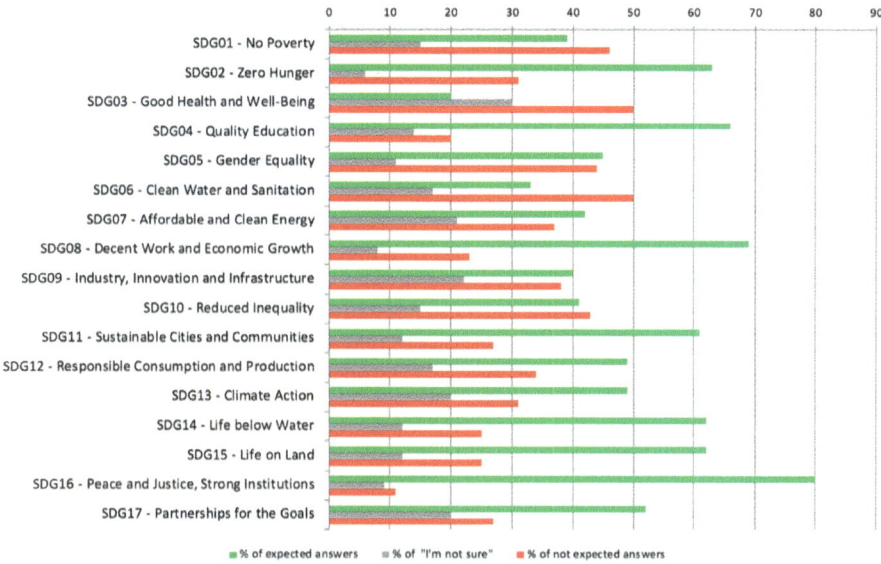

Fig. 3 Results per SDG. Knowledge of the UNED students on the sustainable development goals

50% of correct answers would allow students to be considered at a sufficient level of knowledge; but students did not exceed this level in 9 of the 22 tags present in the questionnaire. Among them, the following were the lowest scores: International *Governance and institutions* (33%), *Health and basic needs* (33%) and *Water and sanitation* (20%).

An interesting strength of the Sulitest is that allows to analyse the outcomes from the perspective of the Sustainable Development Goals (SDGs). Questions have been related to these objectives, a relationship that can be observed also in the tags assigned to them.

Figure 3 shows that eight SDGs have a score higher than 50%; and in two of them even a higher percentage: *SDG No. 16, Peace and Justice, Strong Institutions*, with 80% correct answers, and *SDG No. 8, Decent Work and Economic Growth*, with 69% of correct answers.

On the other hand, the results are inadequate in SDG No. 3 *Good Health and Well-Being*, and SDG No. 6 *Clean Water and Sanitation*, in both cases with only 20% and 33% valid answers respectively.

7.2 Teachers Facing the Curricular Change for Sustainability

In regards to how the data was collected, there are some considerations that need to be taken into account. 79 teachers from the Master's degree accepted to take part in the study, which only represents the 26.42% of the total. The data, then, should be interpreted with caution given the sample size. Additionally, since the participants volunteered for the study, the results may not be representative of the different specialties of the Master's degree. The analysis of the data available, however, allow a first approximation to the teachers' awareness of the social and ecological crisis and also to the coherence of their professional practice with education for sustainable development, as detailed below.

– *Teachers facing the socio-ecological crisis: professional commitment*

A high percentage of teachers (71%) recognise the anthropic nature of the socio-environmental crisis; 42% of the participants claim that it is the result of human activity; 71% of them consider this crisis an urgent challenge for humankind, in this case with a position of 52% of the subjects. This position does not contradict the fact that 20% of teachers consider that, 'always' or 'many times', this threat can be used for political purposes (Table 1).

According to the data, a preliminary conclusion would address the need to increase the number of university teachers aware of the environmental challenge (social and ecological), since there is a large group of them that recognises its existence already. Therefore, the expectations for this group to have a professional commitment associated with that matter contradicts the data.

As shown in Table 2, the issues related to teachers awareness is lower than 50% (if the number of teachers that answered 'agree' or 'totally agree' are combined): only 35% recognise an active professional commitment to sustainability; 45% seek to train students in the principles and values of sustainability; and 42% educate their students in the necessary skills and competencies to contribute to sustainable societies. The data shows a low level of sensitisation by the teachers about the need of social change towards sustainable development. However, at the same time, 57% of teachers declare that they do train students in the defence and respect of human rights, which informs the previous interpretation. Given these results, it has become apparent that teachers fail to see a relationship between sustainability and human rights, an aspect of significant importance in matters of sustainability (Gil Pérez and Vilches 2017).

In one hand, 42% of teachers seem to give to their subject an irrelevant role in the achievement of sustainability, considering that they only have to encourage the acquisition of content in their respective subject matter. This response can be interpreted as a lack of motivation among teachers to get involved in innovation processes oriented towards curricular sustainability. On the other hand, a similar percentage (41%) is involved in this kind of process , as we will see below.

Table 1 How teachers perceive the socio-ecological issues

How do you agree with the following statements about the current socio-ecological crisis?	Strongly disagree [5]	Disagree [4]	Neutral [3]	Agree [2]	Totally agree [1]	Average	Mode	Standard deviation
Result of human beings activities	6% (5)	11% (9)	11% (9)	29% (23)	42% (33)	2114	1	1243
An urgent challenge for humankind	11% (9)	5% (4)	13% (10)	19% (15)	52% (41)	2051	1	1368
A threat that can be manipulated for political purposes	38% (30)	24% (19)	18% (14)	14% (11)	6% (5)	3734	5	127
A cyclical phenomenon	18% (14)	25% (20)	34% (27)	16% (13)	6% (5)	3316	3	1131

In parentheses the number of responses in each case

Table 2 How teachers perceive their teaching practice facing the challenge of sustainability

To what extent does the exercise of your profession contribute to the achievement of the following goals?	Strongly disagree [5]	Disagree [4]	Neutral [3]	Agree [2]	Totally agree [1]	Average	Mode	Standard deviation
Make explicit teacher's commitment to sustainable development	9% (7)	14% (11)	42% (33)	20% (16)	15% (12)	2.810	3	1.126
Training students in the principles and values of sustainability	9% (7)	13% (10)	34% (27)	27% (21)	18% (14)	2684	3	1164
Training students in key competencies for sustainable development	11% (9)	13% (10)	34% (27)	28% (22)	14% (11)	2797	3	1173
Training students in the defence and respect for human rights	9% (7)	10% (8)	24% (19)	28% (22)	29% (23)	2418	1	1249
Exclusively a training on the contents of your subject	14% (11)	27% (21)	18% (14)	27% (21)	15% (12)	2975	4	1302

In parentheses, the number of responses in each case

It is reasonable to conclude that there is a need to create strategies to raise teachers' awareness of the implications of their practice. It is well known the positive or negative impact that classroom training has on the environmental crisis.

– *Teachers analysing their own practice: reflective self-evaluation*

The data shown in Table 3 reflect that around 30% of teachers explicitly include content, activities and competencies related to sustainability in their teaching project, and also foresee how to evaluate achievements. The three areas are of compulsory attention for a teaching practice actively committed to sustainability.

Delving the central core of the processes of the *curricular sustainability*, the training of competencies in sustainability, the data collected are shown in Table 4. Most of the teachers affirm that they include, always (totally agree) or many times (agree), critical analysis (75%) and systemic reflection (71%) among their learning objectives; and 50% promote a sense of responsibility towards present and future generations (50%). There is, however, a minority that drives collaborative decision making (37%). UNESCO (2014) named these four competencies as key competencies for sustainability.

Additionally, the group of teachers who claim to enhance creativity (59%) and interdisciplinary work (61%), both necessary abilities for the performance of the competencies named above, represents the majority.

The methodologies submitted in the questionnaire have been widely recognized in education for sustainable development (UNESCO 2017). However as shown in the results in Table 5, the responses indicate that its use is not yet generalized. It is worth mentioning the percentage of teachers who declare to use 'never' or 'very little' (few times) conceptual maps (43%), peer evaluation (57%), collaborative learning (41%), case studies (26%) and ethical-environmental dilemmas (49%). This situation indicates that the dissemination of these methodologies needs a boost, since their beneficial implications for the improvement in the quality of training processes in the framework of sustainability has been proven (Tilbury and Ryan 2012). Around 25% of teachers have used these methodologies, a clue that points to the possible success. Therefore, we expect that they must have an elementary knowledge of them.

It should be noted that the 'case study' (mode = 2) is the strategy most widely used by teachers ('often' or 'always', in 52% of the subjects).

Finally, in response to the open question included in the questionnaire, 14 people claimed to use other strategies such as: project learning, problem-based learning, service learning and ICT practices.

Table 3 Curricular sustainability of the teaching project

To what extent does your teaching project contribute to education for Sustainable Development?	Strongly disagree [5]	Disagree [4]	Neutral [3]	Agree [2]	Totally agree [1]	Average	Mode	Standard deviation
The program includes contents explicitly related to Sustainable Development	24% (19)	28% (22)	22% (17)	15% (12)	11% (9)	3468	4	1,31
Activities to train competencies in sustainability are proposed	9% (7)	24% (19)	28% (22)	27% (21)	13% (10)	2899	3	1,16
The evaluation of the results includes the acquisition of competencies in sustainability	32% (25)	20% (16)	23% (18)	14% (11)	11% (9)	3468	5	1,36

In parentheses the number of responses in each case

Table 4 Competencies in Sustainability included among the learning objectives

To what extent does your usual teaching practice seek to empower students with the following competencies?	Strongly disagree [5]	Disagree [4]	Neutral [3]	Agree [2]	Totally agree [1]	Average	Mode	Standard deviation
Critical analysis	8% (6)	8% (6)	8% (6)	27% (21)	48% (38)	2051	1	1330
Systemic reflection	8% (6)	13% (10)	9% (7)	38% (30)	33% (26)	2241	2	1245
Collaborative decision making	4% (3)	20% (16)	39% (31)	29% (23)	8% (6)	2835	3	0960
Responsibility towards present and future generations	11% (9)	16% (13)	23% (18)	27% (21)	23% (18)	2671	2	1300
Creativity	4% (3)	8% (6)	29% (23)	35% (28)	24% (19)	2316	2	1038
Interdisciplinarity	6% (5)	15% (12)	18% (14)	23% (18)	38% (30)	2.291	1	1284

In parentheses the number of responses in each case

Table 5 Methodologies and strategies of regular use

Do you use the following methodological tools in your teaching practice?	Nothing/never [5]	Few times [4]	Sometimes [3]	Often [2]	Always [1]	Average	Mode	Standard deviation
Conceptual maps	18% (14)	25% (20)	25% (20)	24% (19)	8% (6)	3215	3	1208
Peer evaluation	30 (24)	27% (21)	20% (16)	16% (13)	6% (5)	3582	1	1249
Collaborative learning	22% (17)	19% (15)	28% (22)	27% (21)	5% (4)	3253	3	1206
Case study	10% (8)	16% (13)	22% (17)	33% (26)	19% (15)	2.658	2	1242
Ethical-environmental dilemmas	30% (24)	19% (15)	22% (17)	19% (15)	10% (8)	3.405	1	1355

In parentheses the number of responses in each case

8 Conclusions

As a conclusion, it can be said that the *'curricular sustainability'* of the Masters' degree, aim of the research, is in a very early stage and it is necessary to counteract the obstacles that hold it back. Two main ones have been detected. On the one hand, the students have a moderate level of literacy in matters of sustainability. And, on the other hand, university teachers show insufficient awareness of the responsibilities of their professional role as agents of change for sustainable development.

Regarding the students, although they are in a similar situation to their international and national counterparts, there is a wide margin of improvement that requires attention through transversal courses and training programs, aimed at providing them with knowledge and key competencies in sustainability. In the context of distance education universities, such as the UNED, these processes of generalized literacy are facilitated by the scale impact of this type of higher education institution.

Concerning teachers, they cannot be considered agents actively committed to sustainability change; neither by their awareness of the urgency that the challenge raises, nor above all, by the conception merely disciplinary of their subjects. From their answers, it can be interpreted that they have not yet come to realise that both the contents of their subjects and the classroom methodologies can and should contribute to the training of competencies in sustainability.

There is still a long way to go. This study contributes with its results to make it possible, despite the limits derived from the minor participation in it of the teaching staff. This is a fact that must be mentioned since it endorses the urge to increase the means and resources for dissemination in this collective of needs, principles and values of sustainability.

Acknowledgements The authors would like to thank the financial support from the Spanish Ministry of Economy, Industry and Competitiveness. This work is part of the Project entitled: *"(Re) orienting teaching practice towards Sustainability: virtual and face-to-face environments for training Secondary Ed. Teachers"* (EDU2015-66591-R).

References

Aznar Minguet P, Martínez Agut MP, Palacios B, Piñero A, Ull Solís MA (2011) Introducing sustainability into university curricula: an indicator and baseline survey of the views of university teachers at the University of Valencia. Environ Educ Res 17(2):145–166

Aznar Minguet P, Ull Solís MA, Piñero A, Martínez Agut MP (2014) Autodiagnóstico de la inclusión de la sostenibilidad en las actividades docentes del profesorado universitario. Universidad de Valencia. Licence creative commons by.nc-nd. OTRI (UV): 126721

Aznar Minguet P, Martínez-Agut MP, Piñero A, Ull MA (2016) Competencies for sustainability in the curricula of all new degrees from the university of Valencia (Spain). In: Barth M, Michelsen G, Rieckmann M, Thomas I (eds) Handbook of higher education for sustainable development. Routledge Publishers, London, pp 434–444

Bautista-Cerro MJ, Díaz González MJ (2017) La sostenibilidad en los grados universitarios: presencia y coherencia. Rev Teoría de la educación. Rev Interuniversitaria 29(1):161–187. http://dx.doi.org/10.14201/teoredu291

Cebrián G (2017) A collaborative action research project towards embedding ESD within the higher education curriculum. Int J Sustain High Educ 18(6):857–876. https://doi.org/10.1108/IJSHE-02-2016-0038

Cebrián G, Junyent M (2015) Competencies in education for sustainable development: exploring the student teachers' views. Sustainability 7(3):2768–2786. https://doi.org/10.3390/su7032768

CRUE (2005) Guidelines for the inclusion of sustainability in the curriculum. Document approved by the CRUE General Assembly on 27 October. Available at http://www.crue.org/Documentos%20compartidos/Declaraciones/Directrices_Ingles_Sostenibilidad_Crue2012.pdf. Accessed 16 Apr 2018

CRUE-CADEP (2009) Barreras clave identificadas para sostenibilizar los curricula universitarios. In: Conclusiones del encuentro del grupo de trabajo de la CRUE para la calidad ambiental, desarrollo sostenible y la prevención de riesgos "Inclusión de aspectos ambientales y de prevención de riesgos en los planes de estudio", Granada, España, 21 Apr 2009. Available at http://www.crue.org/Documentos%20compartidos/Res%C3%BAmenes%20y%20Conclusiones/9._INCLUSION_ASPECTOS_AMBIENTALES.pdf. Accessed 16 Apr 2018

CRUE-CADEP (2015) Institucionalización del Aprendizaje-Servicio como estrategia docente dentro del marco de la Responsabilidad Social Universitaria para la promoción de la Sostenibilidad en la Universidad. Working paper, Comité Ejecutivo y el Plenario de la Comisión de Sostenibilidad (Grupo CADEP), León (Spain), 29 May 2015. Available at http://www.crue.org/Documentos%20compartidos/Recomendaciones%20y%20criterios%20tecnicos/2.%20APROBADA%20INSTITUCIONALIZACION%20ApS.pdf. Accessed 16 Apr 2018

CRUE-CADEP (n.d.) Curso de formación al profesorado: Introducción de la sostenibilidad en la docencia universitaria. Available at http://www.crue.org/Documentos%20compartidos/Formaci%C3%B3n/CURSO_FORMACION_CADEP%202014.pdf. Accessed 16 Apr 2018

Cutanda GA, Murga-Menoyo MªÁ (2014) Analysis of mythical-metaphorical narratives as a resource for education in the principles and values of sustainability. J Teacher Educ Sustain 16(2):18–38. https://doi.org/10.2478/jtes-2014-0009

Daniel J (1999) The mega-universities. Some issues. In: Veinticinco años de la UNED, 25–38. UNED, Madrid

García Esteban FE, Murga-Menoyo MA (2012) Cuestionario de auto percepción de la propia práctica docente. Un instrumento para la evaluación de la calidad docente a la luz del modelo UNESCO de educación para el desarrollo sostenible. en: Gaviria JL, Palmero MC, Alonso P (eds) Entre Generaciones: educación, herencia y promesas. Proceedings of the XV National Congress and V Iberoamerican Pedagogy, Spanish Society of Pedagogy, Burgos, pp 852–863

García Esteban FE, Murga-Menoyo MA (2015) El profesorado de educación infantil ante el desarrollo sostenible. Necesidades formativas. Enseñ Teach 33(1):121–142. http://dx.doi.org/10.14201/et2015331121142

Garcia MR, Junyent M, Fonolleda M (2017) How to assess professional competencies in Education for Sustainability? An approach from a perspective of complexity. Int J Sustain High Educ 18(5):772–797. https://doi.org/10.1108/IJSHE-03-2016-0055

Geli de Ciurana AM, Leal Filho W (2006) Education for sustainability in university studies: Experiences from a project involving European and Latin American universities. Int J Sustain High Educ 7(1):81–93. https://doi.org/10.1108/14676370610639263

Gil Pérez D, Vilches A (2017) Educación para la Sostenibilidad y Educación en Derechos Humanos: dos campos que deben vincularse. Teoría de la Educación Revista Interuniversitaria 29(1):79–100. http://dx.doi.org/10.14201/teoredu291

Gonzalo V, Sobrino MR, Benítez L, Coronado-Marín A (2017) Revisión sistemática sobre competencias en desarrollo sostenible en educación superior. Rev Iberoamericana de Educación 73:85–108. Available at https://rieoei.org/RIE/article/view/289. Accessed 16 Apr 2018

HEFCE (2014) Sustainable development in higher education: consultation outcomes. Higher Education Funding Council for England. Available at http://www.hefce.ac.uk/media/hefce/content/pubs/2014/201430/HEFCE2014_30a.pdf. Accessed 16 Apr 2018

HESI (2017) Mapping awareness of the global goals. Report from the Sulitest, Tangible implementation of the HESI & contributor to the review of the 2030 Agenda. Available at https://www.sulitest.org/hlpf2017report.pdf. Accessed 16 Apr 2018

HESI (2019) Higher education sustainability initiative. Available at https://sustainabledevelopment.un.org/sdinaction/hesi. Accessed 14 Marv 2019

Higher Education Academy (2014) Education for sustainable development: guidance for UK higher education providers. Available at http://www.qaa.ac.uk/en/Publications/Documents/Education-sustainable-development-Guidance-June-14.pdf. Accessed 16 Apr 2018

Jucker R, Mathar R (2015) Schooling for Sustainable development in Europe. Springer, London

Junyent M, Geli de Ciurana AM (2008) Education for sustainability in university studies: a model for reorienting the curriculum. Br Edu Res J 34(6):763–782. https://doi.org/10.1080/01411920802041343

Leal Filho W, Jim Wu Y-C, Londero Brandli L, Veiga Avila L, Miranda Azeiteiro U, Caeiro S, da Rosa Rejane, Gama Madruga L (2017) Identifying and overcoming obstacles to the implementation of sustainable development at universities. J Integr Environ Sci 14(1):93–108. https://doi.org/10.1080/1943815X.2017.1362007

Leal Filho W, Raatt S, Lazzarini B, Vargas VR, de Souza L, Anholon R, Quelhas OLG, Haddad R, Klavins M, Orlovic VL (2018) The role of transformation in learning and education for sustainability. J Cleaner Prod (199):286–295. https://doi.org/10.1016/j.jclepro.2018.07.017

Mascarell L, Vilches A (2016) Química para la Sostenibilidad en la formación del profesorado. Indagatio Didáctica 8(1):16–29

Medir RM, Heras R, Magin C (2016) Una propuesta evaluativa para actividades de educación ambiental para la sostenibilidad. Educación XX1 19(1):331–355. https://doi.org/10.5944/educXX1.14226

Murga-Menoyo MÁ (2014) learning for a sustainable economy: teaching of green competencies in the university. Sustainability 6:2974–2992

Murga-Menoyo MA, Bautista-Cerro MJ, Borderías Uribeondo P, Galán Gonzalez MA (2016) Cuestionario "Sostenibilidad y práctica docente". Territorial Registry of Intellectual Property, ref.:03/283438.9/17

Murga-Menoyo MA, Correia F, Espinosa A (2016) AACSB criteria and sustainability in the school's curricula. School's "AACSB Assurance of Learning (AoL) Review Workshop, 18 May 2016, Hull University Cusiness School, Hull, UK

Murga-Menoyo MA, Espinosa A, Novo M (2017) What do we imagine the campuses of tomorrow will be like? Universities' transition toward sustainability in the light of the transition initiatives. In: Leal Filho W, Azeiteiro UM, Alves F, Molthan-Hill P (eds) Handbook of theory and practice of sustainable development in higher education, vol 4. Springer, Berlin, pp 193–232

Novo M, Murga-Menoyo MA (2014) The processes of integrating sustainability in higher education curricula: a theoretical-practical experience regarding key competences and their cross-curricular incorporation into degree courses. In Leal Filho W (ed) Transformative approaches to sustainable development at universities. Springer, pp 119–136

Novo M, Murga-Menoyo MÁ, Bautista-Cerro Ruiz MJ (2010) Educational advances and trends for sustainable development: a research project on educational innovation. J Baltic Sci Educ 9(4):302–314

QAA/HEA (2014) Education for sustainable development: guidance for UK higher education providers. Quality Assurance Agency for Higher Education/Higher Education Academy. Available at http://www.qaa.ac.uk/en/Publications/Documents/Education-sustainable-development-Guidance-June-14.pdf. Accessed 16 Apr 2018

Saban C, Gonzalo V, Gómez-Jarabo I, Ruiz B, Coronado-Marín A, Pérez S, Sáenz-Rico B (2017) Ruta ODS. Material didáctico sobre los Objetivos de Desarrollo Sostenible. Madrid. ISBN 978-84-697-3510-7

Sáenz-Rico B, Benítez L, Neira JM, Sobrino MR, D'Angelo E (2015) Perfiles profesionales de futuros maestros para el desarrollo sostenible desde un modelo formativo centrado en el diseño de ambientes de aprendizaje. Foro de Educación 13(19):141–163. http://dx.doi.org/10.14516/fde. 2015.013.019.007

Sánchez B, Gomez-Jarabo I, Saban MC, Sáenz-Rico, B (2017) Sostenibilización del perfil profesional del educador social. Necesidades y demandas compartidas. Rev Iberoamericana de Educación 73:109–130

The Sulitest Organization (2016) Vision & Mission. [Online] sulitest.org. Available at https://www.sulitest.org/en/vision-mission.html. Accessed 16 Apr 2018

The Sulitest Organization (2018) The sustainability literacy test. [Online] Sulitest.org. Available at https://www.sulitest.org/en/the-sustainability-literacy-test.html. Accessed 16 Apr 2018

Tilbury D, Ryan A (2012) Guide to quality and education for sustainability in higher education. From project: leading curriculum change for sustainability: strategic approaches to quality enhancement. Available at http://efsandquality.glos.ac.uk/user_quide_to_this_resource.htm. Accessed 16 Apr 2018

UNECE (2009) Learning from each other. The UNECE Strategy for Education for Sustainable Development. United Nation: New York and Geneva. Available at https://sustainabledevelopment.un.org/content/documents/798ece5.pdf. Accessed 16 April 2018

UNECE (2013) Empowering educators for a sustainable future. Tools for policy and practice workshops on competencies in education for sustainable development. Geneva, United Nations, EC/CEP/165

UNED (2017) Summary of the Report of the Course 2016–17. Available at http://portal.uned.es/pls/portal/docs/PAGE/UNED_MAIN/LAUNIVERSIDAD/VICERRECTORADOS/SECRETARIA/MEMORIAS/RESUMEN_MEMORIA_2016_2017.PDF. Accessed 16 Apr 2018

UNESCO (2005) United Nations Decade of Education for Sustainable Development (2005–2014): International implementation scheme. ED/DESD/2005/PI/01. Available at http://unesdoc.unesco.org/images/0014/001486/148654e.pdf. Accessed 16 Apr 2018

UNESCO (2014) Roadmap for Implementing the Global Action Programme on Education for Sustainable Development. Available at http://unesdoc.unesco.org/images/0023/002305/230514e.pdf. Accessed 16 Apr 2018

UNESCO (2015) Rethinking education: towards a global common good? Available at http://unesdoc.unesco.org/images/0023/002325/232555e.pdf. Accessed 16 Apr 2018

UNESCO (2017) Education for sustainable development goals. Learn Objectives. Available at http://unesdoc.unesco.org/images/0024/002474/247444e.pdf. Accessed 16 Apr 2018

UNGA (2015) Transforming our world: the 2030 Agenda for sustainable development. Resolution adopted by the General Assembly on 25 September A/RES/70/1. Available at http://www.un.org/ga/search/view_doc.asp?symbol=A/RES/70/1&Lang=E_. Accessed 16 April 2018

UNGE (2012) The future we want. Resolution adopted by the General Assembly on 27 July 2012. A/RES/66/288 Available at http://www.un.org/ga/search/view_doc.asp?symbol=A/RES/66/288&Lang=E. Accessed 16 Apr 2018

Viciana S, Junyent M, Cebrian G (2017) Análisis de un modelo formativo para avanzar en la ambientalización curricular: transferencia en diversidad de contextos. Enseñanza de las ciencias 3137–3142. https://ddd.uab.cat/record/184006

Vilches A, Gil-Pérez D (2016) Química Verde y Sostenibilidad en la Educación en Ciencias en Secundaria. Enseñanza de las Cienc 34(2):25–42

Alfonso Coronado-Marín Master's degree in Advanced Studies in Pedagogy at the Complutense University of Madrid. Also has two degrees, in Pedagogy and in Teaching: Special Education. He has experience in research groups such as 'Studies on Communication and Languages for Inclusion and Educational Equity (ECOLE)' and the 'Permanent Seminar to strengthen the professional profile of the social educator as a promoter of inclusive societies for sustainable human development (Project SOCIDES—UCM)'. He is currently carrying out doctoral studies in the framework of the UNESCO Chair of Environmental Education and Sustainable Development of the UNED, funded by a fellowship for Training of Research Staff.

María José Bautista Cerro Ph.D. in Pedagogy (Universidad Nacional de Educación a Distancia, 2009). Degree in Political Science (Universidad Complutense de Madrid, 1995). Lecturer and researcher at the Department of Educational Theory and Social Pedagogy. Deputy Vice Rector of Methodological Design and Innovation (UNED, Spain). She is member of the UNESCO Chair in Environmental Education and Sustainable Development and Director of Post-graduate Open Course in Sustainability and Education (UNED). She is president of working group "Curricular Sustainability" at *Conferencia de Rectores de Universidades Españolas* (CRUE. Conference of Chancellors of Spanish Universities). Her research focuses on higher distance education, curricular sustainability and education for sustainable development. She is part of national and international research teams.

María Ángeles Murga-Menoyo Ph.D. in Philosophy and Education Sciences (Universidad Nacional de Educación a Distancia, 1983). Is member of the UNESCO Chair of Environmental Education and Sustainable Development at the National University of Distance Education (UNED), Spain. Senior lecturer and researcher at the Department of Educational Theory and Social Pedagogy. Her research activities, integrated in EDUCAMDES group, are focused in two main working lines: environmental education for sustainable development and teaching innovation in distance learning educational contexts. She is currently leading a project funded by the Spanish Ministry of Economy, Industry and Competitiveness for the curricular sustainability of the Master's degree in Compulsory Secondary Education and Baccalaureate, Vocational Training and Language Education Teacher Training. She is the Coordinator of the Doctoral degree in Education Program of the International Doctoral School of the UNED (Spain).

Unfolding the Complexities of the Sustainability Reporting Process in Higher Education: A Case Study in The University of British Columbia

Kim Ceulemans, Carol Scarff Seatter, Ingrid Molderez, Luc Van Liedekerke and Rodrigo Lozano

Abstract Sustainability reporting is an important tool for the assessment and communication of sustainability performance in the corporate world; however, it is still in its early stages in higher education. Only a limited number of universities around the world have published sustainability reports, and there is still a lack of empirical evidence of how the sustainability reporting process is organized and how it contributes to sustainability integration in higher education. The aim of this paper is to provide in-depth insights into the complexities of the sustainability reporting process. A case study is presented on sustainability reporting in The University of British Columbia (UBC), Canada, covering the period of 2008–2014. The case consists of document analysis and interviews with the internal actors of the sustainability reporting process at UBC. In the study, it was found that even in an institution with a deliberate

K. Ceulemans (✉)
Toulouse Business School, 1 Place Alphonse Jourdain, CS 66810, 31068 Toulouse Cedex 7, France
e-mail: k.ceulemans@tbs-education.fr

C. Scarff Seatter
Faculty of Education, University of British Columbia, Kelowna, BC, Canada
e-mail: carol.scarff@ubc.ca

I. Molderez
Centre for Economics and Corporate Sustainability, KU Leuven—University of Leuven, Warmoesberg 26, 1000 Brussels, Belgium
e-mail: ingrid.molderez@kuleuven.be

L. Van Liedekerke
BNP Chair Ethics, Finance and Sustainability, University of Antwerp, Prinsstraat 13, 2000 Antwerp, Belgium
e-mail: luc.vanliedekerke@uantwerpen.be

Center for Economics and Ethics, KU Leuven, Naamsestraat 69, 3000 Leuven, Belgium

R. Lozano
Department of Engineering and Sustainable Development, University of Gävle, Kungsbäcksvägen 47, 80176 Gävle, Sweden
e-mail: rodrigo.lozano@hig.se

Organisational Sustainability, Ltd., 40 Machen Place, Cardiff CF11 6EQ, UK

© Springer Nature Switzerland AG 2020
W. Leal Filho et al. (eds.), *International Business, Trade and Institutional Sustainability*, World Sustainability Series,
https://doi.org/10.1007/978-3-030-26759-9_61

1043

sustainability strategy and integration process, sustainability reporting is a complex undertaking. Some of the key elements adding complexity to the reporting process are the search for suitable lead authors and relevant content for the reports in consecutive reporting cycles; the different aims that reporting tries to achieve simultaneously for the sustainability integration process; and suitably addressing the different internal stakeholders that are involved in the reporting process in higher education. In order to reduce complexity in universities, assigning an internal hybrid actor to facilitate the reporting process could be helpful. Unfolding the key complexities arising when undertaking consecutive reporting cycles can help internal actors of other institutions to better design and implement the process of sustainability reporting and integration in their universities.

Keywords Sustainability reporting · Higher education · Sustainability performance · Sustainability integration strategies

1 Introduction

Higher education institutions (HEIs) around the world have been actively integrating principles of sustainable development (SD) into their organisations, in response to calls from the international community to contribute to a sustainable society (Buckler and Creech 2014; Sedlacek 2013). HEIs are important actors for SD integration, because of their reach and potential to facilitate, promote and encourage a societal response to a diverse array of SD challenges facing communities around the world (Stephens et al. 2008). Their engagement for SD is seen as crucial because our future leaders, decision-makers and managers are formed and shaped within the world's HEIs (Barth and Timm 2011; de Lange 2013).

Traditionally, SD initiatives can be implemented in the four core functions of HEIs; i.e. education, research, operations and outreach (Cortese 2003; Velazquez et al. 2006). Based on the key areas of SD integration addressed in declarations, charters and initiatives on SD integration in HEIs, Lozano et al. (2013b) identified extra dimensions in which SD integration can be sought: the institutional framework, university collaboration, on-campus experiences, assessment and reporting, and educate-the-educator programmes, which together comprise the HEI system. SD initiatives can range from being '*ad hoc*', to '*flagship*' activities, and to fully '*integrated*' or completely institutionalised into the HEI's system (Desha and Hargroves 2014). They should also ideally be interlinked to achieve their full potential (Cortese 2003; Lozano et al. 2013b).

Because SD integration in HEIs remains challenging, Barth (2013) stressed the necessity to study drivers and barriers of sustainability integration processes in HEIs, and pointed to the lack of research on how they influence each other, how they are linked, and how they may change during the process. Examples of barriers and challenges that should be overcome when integrating SD in HEIs are the lack of policies to promote sustainability and the vagueness of the concept of sustainability (for a full overview, see Blanco-Portela et al. 2017; Lozano 2006a). Some of the main barriers to change for SD in HEIs are related to human factors, such as resistance, communication, empowerment, and lack of involvement (Verhulst and Lambrechts 2015). Some barriers are related to the stakeholders involved in HEIs, such as the divide between different academic and operational departments, and the disciplinary structure of faculties (Bero et al. 2012; Krizek et al. 2012). In the Higher Education for Sustainable Development (HESD) literature, this has been linked to Weick's (1976) concept of 'loose coupling' (or loosely coupled systems), explaining the low degree of coordination and collaboration between HEIs' departments and thus internal stakeholders (see Albrecht et al. 2007; Ceulemans et al. 2014).

As a means to demonstrate their commitment and implementation of SD to a large range of stakeholders, some HEIs have engaged in sustainability reporting (Fonseca et al. 2011; Lozano 2011). Sustainability reporting is a voluntary activity aimed at communication and accountability on SD impacts towards internal and external stakeholders, and at the assessment and improvement of SD performance in an organisation (Adams and Frost 2008; Dalal-Clayton and Bass 2002; Joseph 2012). Organisations, including companies and HEIs, have published sustainability reports for reasons other than communication or assessment, such as benchmarking against other organisations, cost savings identification, or planning changes towards SD in the organisation (Adams and McNicholas 2007; Kolk 2010; Lozano et al. 2016; Maubane et al. 2014). Herzig and Schaltegger (2011) identified two perspectives that drive sustainability reporting: the 'inside-out' perspective, with SD strategy and internal performance measurement as the main elements driving SD, and the 'outside-in' perspective, where external stakeholder requests are the main drivers for an organisation's sustainability reporting. Lozano et al. (2016) found that a combined approach based on internal and external motivations could be identified for consecutive sustainability reporting cycles in companies.

Some of the main reasons for HEIs' engagement in the practice of sustainability reporting include facilitating transparency of SD activities, assessing SD efforts, engaging with stakeholders, improving SD performance, improving SD reputation, and facilitating change towards SD in higher education (Ceulemans et al. 2015b; Kamal and Asmuss 2013). In general, sustainability reporting in HEIs requires management support (Adams 2013), yet sustainability reports have also been developed in cases where management support was low (Moudrak and Clarke 2012).

The publication of sustainability reports has evolved alongside the development of some higher education-specific SD assessment tools to measure and monitor the SD performance of HEIs, such as the Sustainability Tracking, Assessment and Rating System (STARS) (AASHE 2013), the Sustainability Tool for Assessing University's Curricula Holistically (STAUNCH (RTM)) (Lozano and Peattie 2011), the tool for a Graphical Assessment of Sustainability in Universities (GASU) (Lozano 2006b), and the Auditing Instrument for Sustainability in Higher Education (AISHE) (Roorda 2002).[1]

Research on sustainability reporting in the HESD literature has focused mainly on the development or improvement of SD assessment tools for HEIs, and on measuring the state of sustainability reporting in HEIs at a certain point in time (often through content analysis of sustainability reports) (Ceulemans et al. 2015b). Far less has been said about the *process* of sustainability reporting in HEIs. As Adams (2013: 388) stated, *"Whilst there can be no 'one size fits all' approach to managing sustainability in universities, knowledge of best practice sustainability management in other sectors and knowledge of the university sector leads to the identification of important features of a framework for managing sustainability"*. Or, in other words, there is a lack of in-depth guidance on how to organize the sustainability reporting process in organizations as complex and distinctly organized as HEIs (Lozano et al. 2013a; Ceulemans et al. 2018).

The aim of this chapter is to provide in-depth insights in a real-world context on some of the complexities of the sustainability reporting process and how it contributes to the sustainability integration process. In order to address this gap, a case study was conducted on sustainability reporting at The University of British Columbia (UBC), Canada (covering the period of 2008–2014). UBC, a leader in sustainability initiatives and innovation, offers a unique case of sustainability reporting in higher education because of its long track record of strategic sustainability integration and consecutive reporting. The study is focused on the internal actors of sustainability reporting, unpacks some of the main complexities of the reporting process in higher education, and offers recommendations on sustainability reporting for the higher education sector.

In the following sections, the methods of this chapter will be presented (Sect. 2), followed by the findings of the UBC case study (Sect. 3), the discussion uncovering the complexities of the sustainability reporting process (Sect. 4), and the conclusion and recommendations for scholars and practitioners (Sect. 5).

2 Methods

The case study method was used to empirically study the complexities of the sustainability reporting process in higher education. This was done to gain rich insights on the dynamics and driving forces of the topic in a real-world context (as suggested by Yin 2009; Eisenhardt and Graebner 2007). Because of the nature of the topic, i.e. an underdeveloped topic such as sustainability reporting in HEIs with a low number of HEIs carrying out this practice, a single case study has been used for this research.

The University of British Columbia (UBC) was chosen for this case study through theoretical sampling (Eisenhardt et al. 2016) because of:

---

[1]For an overview of SD assessment tools for HEIs, see Findler et al. (2018), Kamal and Asmuss (2013), or Yarime and Tanaka (2012).

(1) Its broadly recognised and well-documented leadership in SD integration and innovation, with strong management support around SD integration and sustainability reporting (see Beaudoin 2014; Gudz 2004; Moore 2005);

(2) Its inclusion in Lozano's (2011) seminal study on sustainability reporting worldwide, covering 12 early-adopter HEIs;

(3) Its continuing reporting cycles for over six consecutive years (a rare feature in the higher education sector, see Alonso-Almeida et al. 2015); and

(4) The inclusion of verifiable performance indicators in its sustainability reports.

Because of these different elements, the case can provide insights of the complexities of the sustainability reporting process in HEIs and offers recommendations for researchers and policy makers.

2.1 Data Collection and Analysis

The data collection for the case study took place in June–October 2014 at UBC Vancouver and Okanagan Campuses and was primarily focused on UBC's sustainability reporting activities between 2008 and 2014. It was carried out via retrospective document analysis and semi-structured interviewing, allowing for methodological triangulation to take place (Yin 2009). The main focus of this study has been on the Vancouver Campus because of its size and its role as a frontrunner in SD integration, yet documents and participants from the Okanagan Campus have been included as well.

The retrospective document analysis covered the period of 1997–2015, including publications such as the Annual Sustainability Reports, the STARS Report, and others (see Table 1). Semi-structured interviews were conducted with the main actors of sustainability reporting and SD integration at UBC. In total, 16 interviews (of approx. 40–60 min each) were executed with 14 UBC staff, faculty, and students. Each of these persons were involved in the preparation of the sustainability reports at UBC. Table 2 presents the interviewees of the case study. The interviews with each of the participants were recorded and transcribed, and sent to the participants for approval and further explanations.

Content analysis (Jupp 2012; Schreier 2012) was used to analyse the documents of this case study (see Table 1 for an overview of the studied documents) and describe the broader context of sustainability integration and reporting at UBC. The data analysis of the interviews took place through the use of constant comparative analysis of Grounded Theory (see Glaser and Strauss 1999; Corbin and Strauss 2008) and was used to get a view on the complexities of the sustainability reporting process from the viewpoint of its internal actors.

Table 1 Type of documents studied in UBC case study

Annual Reports
Annual Sustainability Reports (2008–2014)
Carbon Neutral Action Reports and Climate Action Reports (2009–2013)
ISCN Reports (2013–2014)
College Sustainability Report Card (2007–2011)
Policy Documents
Policy #5
Place and Promise (Plan and Reports)
Climate Action Plan
Sustainability Academic Strategy
Inspirations and Aspirations (Strategy and Final Report)
20-year Sustainability Strategy
Scientific Articles
Gudz (2004) on 'the implementation of UBC's SD policy'
Moore et al. (2005) on 'UBC's engagement with SD integration in education'
Moore (2005) on 'UBC's engagement with SD integration in policies and plans'
Other Reports, Publications and Communications
STARS Report (2011)
UBC Webpages (e.g. sustain.ubc.ca and strategicplan.ubc.ca)
UBC 'Case Study' Publications [Institutionalising Sustainability; Green Building; Sustainability + Residence Coordinators; Energy + Climate Management; Social Ecological Economic Development Studies (SEEDS)]
Sustainability Benchmarking Document, MacEwan University (Beaudoin 2014)

2.2 Limitations of the Method

The reliability of the study may be affected by the data collection in this case study. Bias may be present related to the participants and to the researchers. The participants of a study can provide socially desirable answers (refer to Brinkmann and Kvale 2015; Saunders et al. 2012), and hence misrepresent the sustainability reporting process. However, the case presented in this paper is not a simple 'story of success' (see Corcoran et al. 2004). A broad set of participants was interviewed, including students, in order to avoid this. The participants raised critical voices about the process of sustainability reporting at UBC, in an attempt to learn from their own SD integration experiences and to provide a learning experience for other HEIs.

The findings and discussion of the study may be influenced by the researchers involved in the data analysis process, which affects the internal validity. This was dealt with through the involvement of multiple researchers in the study, and through addressing rival explanations (see Yin 2009) during the data analysis. The findings and discussion of this study are also influenced by the time during which the case

Table 2 Interviewed case study participants

Vancouver Campus: Campus + Community Planning (C+CP)
Director Operational Sustainability
Sustainability Engagement and Reporting Coordinator (2 interviews)
Manager Strategic Planning
Manager of Sustainability Engagement
Manager UBC SEEDS Program
Vancouver Campus: UBC Sustainability Initiative (USI)
Associate Provost Sustainability
Associate Director of the Teaching, Learning and Research Office
Director of Communications and Community Engagement
Engagement Specialist (2 interviews)
Manager of Advising and Student Involvement
Vancouver Campus: Students
Graduate Student and Member of Student Sustainability Council
Undergraduate Student and Member of Student Sustainability Council
Okanagan Campus
Associate Director Sustainability Operations, Sustainability Office
Director of the Okanagan Sustainability Institute

study was undertaken. Repeating this study at another time in the same context would probably accomplish different results, because the case reflects a reality at a certain moment. Because of the presence of a case study protocol, the conditions under which the case study took place can be reproduced by other researchers.

The context in which the HEI is situated also affects the findings and discussion of this study. UBC is a large, research-intensive, North American university with strong SD leadership, and thus the findings may have been different in other contexts. While sustainability reporting in higher education needs a contextualised approach, depending on, for example, the regional context and the stakeholders involved in the HEI (Madeira et al. 2011; Ceulemans et al. 2015b), lessons can still be learned between UBC and other HEIs around the world since (1) the approach to sustainability reporting is adaptable to different situations, and (2) only theoretical generalisations are made in this case study.

2.3   Presentation of the University Under Study: The University of British Columbia

The University of British Columbia (UBC) is a global centre for research and teaching, and a public institution with a $2 billion CAD annual operating budget (UBC 2014b). The university was founded in 1915, and is situated in the Province of British Columbia, Canada, on the Vancouver Campus in Point Grey, Vancouver, and since 2005 also in the Okanagan Valley on the Okanagan Campus in Kelowna. In 2014, UBC had over 58,000 students, and over 15,000 staff and faculty (UBC 2014a). UBC's Vancouver Campus has a large residential community of over 20,000 people (students, faculty, staff and others), while the Okanagan Campus houses around 1700 students (UBC 2014a).

UBC is a university with a long track record in SD integration and institutionalisation (UBC 2013; Lozano 2011; Gudz 2004), and is a recognised leader and model HEI for SD integration amongst its peers (Beaudoin 2014). The university has been involved in systematic integration of SD on the Vancouver Campus since 1997, when UBC became Canada's first university to adopt a sustainability policy, named '*Policy #5*' (UBC 2010a, 2015a). The Okanagan Campus has been engaged in SD integration since its opening in 2005 (UBC 2014b), and the SD initiatives on both campuses have been streamlined over the last years.

There has been a presence of strong leadership and management support for SD integration in UBC over the last twenty years, driven by UBC's '*Policy #5*', i.e. UBC's first Sustainability Policy (established in 1997); '*Inspirations and Aspirations*', i.e. UBC's sustainability strategy for 2006–2010; '*Place and Promise*', i.e. UBC's strategic plan for 2010 and beyond in which SD is one of nine strategic priorities; and the '*20-year Sustainability Strategy*' for UBC's Vancouver Campus, developed in 2014 (UBC 2015c).

Since its focus on SD integration in 1997, UBC has developed a number of departments around its expanding SD activities. UBC became the first Canadian university with a 'Campus Sustainability Office' in 1998, situated within the operational department of 'Land and Building Services' (UBC 2010a, 2014a). The Campus Sustainability Office's mission was to "*create a culture of sustainability at UBC*" (UBC 2010a). At the time of the case study, operational SD was the responsibility of the Vice-President of 'Finance, Resources and Operations', under the 'Campus + Community Planning' (C+CP) Department on the Vancouver Campus. Within C+CP, different staff members with main responsibilities on SD were located in the 'Sustainability and Engineering Department', and the 'Campus Programs and Animations' Department. The Okanagan Campus had a separate 'Sustainability Office' responsible for operational SD integration since 2010, and the 'Okanagan Sustainability Institute' (OSI) was its academic counterpart.

On the Vancouver Campus, the creation of the *UBC Sustainability Initiative* (USI) in 2010 made the academic pillar of SD integration in UBC more prominent. The USI's mission was to integrate UBC's academic and operational SD efforts, through working around two cross-cutting themes, i.e. '*Campus as a Living Laboratory*' and *'the University as an Agent of Change'*. The USI was led by the Associate Provost Sustainability, a UBC Professor and member of the Centre for Interactive Research on Sustainability (CIRS). The USI employed around 15 full-time staff members in 2014, and consisted of a 'Central Office', a 'Teaching, Learning and Research Office', a 'Communications and Community Engagement Office', and a 'Management Team' (UBC 2015e).

In total, around 40 full-time staff equivalents were working on various SD activities in UBC in 2014 (Beaudoin 2014). Important SD achievements in UBC included, amongst others (UBC 2010b, c, d, 2014a):

– the Social Ecological Economic Development Studies (SEEDS) Program (aimed at linking students, staff, and faculty through real-life SD projects on campus);
– the Sustainability Coordinator Program (focused on changing staff's individual environmental behaviour);
– the Sustainability Ambassador Program (i.e. a voluntary sustainability education outreach program for students);
– the Sustainability Learning Pathways (aimed at providing students a holistic view on SD through a systematic inclusion in the curriculum); and
– the construction of Green Buildings (linked to UBC's Climate Action Targets).

While each of these separate programs and activities would be interesting for further research, the focus of this case study is on UBC's sustainability reporting efforts.

UBC has published and distributed Annual Sustainability Reports to the general public since 2008, resulting in a total of six consecutive reports by 2014, which have been presented annually to UBC's Board of Governors. The university had an annual reporting requirement to the Provincial Government of British Columbia on greenhouse gas (GHG) emissions because of the *'Greenhouse Gas Reduction Targets Act, Bill 44 – 2007'* (Province of British Columbia, Canada). This led to the publication of Carbon Neutral Action Reports (CNAR) for the Vancouver and Okanagan Campuses since 2009 (UBC 2015b). UBC also utilises SD assessment tools, i.e. the College Sustainability Report Card from 2007–2010 (UBC 2015d), and the Sustainability Tracking, Assessment and Rating System (STARS) since 2009 (UBC 2009). It has also been reporting on the implementation of the International Sustainable Campus Network (ISCN)-Gulf Sustainable Campus Charter since 2013 (UBC 2013).

3 Findings: The Sustainability Integration and Reporting Process at UBC

In this section, the sustainability integration process at UBC will be analysed, followed by an overview of the sustainability reporting activities that are part of this integration process. Afterwards, the different functions of the sustainability reporting process at UBC will be discussed, as well as the role of a hybrid internal actor in the process: the *UBC Sustainability Initiative* (USI).

3.1 Analysis of the Sustainability Integration Process at UBC (1990–2014)

UBC has been involved in SD integration and institutionalisation for over 20 years. A representation of UBC's SD integration process between 1990 and 2014 is shown in Fig. 1 (based on Desha and Hargroves 2014 and own findings). It shows that the university was in an ad hoc stage of SD integration in the early 1990s. In this stage, UBC's approach towards SD integration was still disconnected and informal, resulting in a lack of documentation of SD integration efforts. An example of SD milestones in this stage were the membership of the University Leaders for a Sustainable Future (ULSF), the signing of the Talloires Declaration in 1990, and the construction of the '*C.K. Choi Building*', UBC's first green building, in 1996 (Gudz 2004; UBC 2010b, 2014a).

The HEI moved to a *flagship* stage (see Fig. 1) for SD integration around 1997, with the creation of the SD policy, i.e. '*Policy #5*'. This stage featured the presence of individual champions, staff support for SD integration, and a formal recognition of the SD activities, which became more connected than in the previous stage. UBC's SD integration is well documented at this stage because of the scientific case studies on the implementation of UBC's SD policy (see Gudz 2004; Moore 2005). The studies showed that SD remained unaddressed in many areas of the university at the time and the SD integration efforts were criticized for their operational focus with limited academic ties. Yet, despite low staff numbers (i.e. four full-time equivalents)

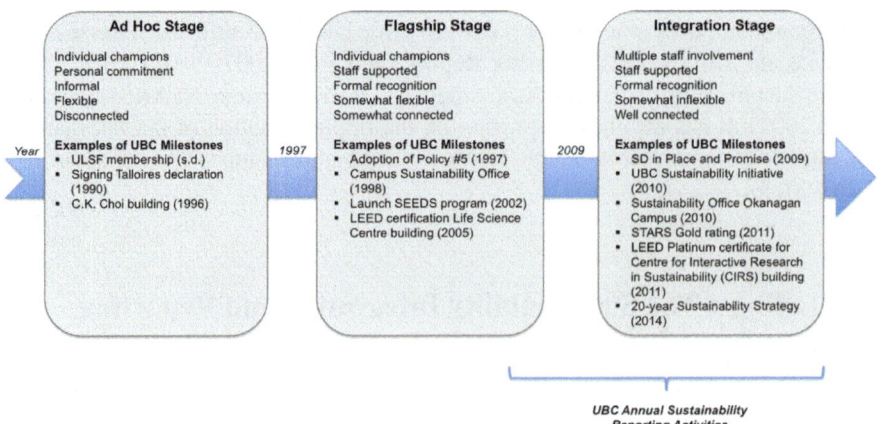

Fig. 1 UBC's stages of SD integration (1990–2014). *Source* Based on Desha and Hargroves (2014) and own findings, 2014

many SD initiatives were implemented in UBC, and cost savings generated from water and energy efficiency initiatives were reinvested back into SD activities (Gudz 2004). SD milestones of this period included, amongst others, the establishment of a Campus Sustainability Office, and the Leadership in Energy and Environmental Design (LEED) Gold Certification for UBC's green building, '*Life Sciences Centre*'.

UBC progressed to the current stage of *integration* (see Fig. 1) around 2009, with the incorporation of SD as a core pillar in UBC's highest level strategic plan (i.e. Place and Promise, the strategic vision of UBC's 12th President, Stephen J. Toope). This stage was characterized by increased staff support and involvement in SD integration and a formal recognition of well-connected SD activities. The creation of the USI on the Vancouver Campus in 2010, the STARS Gold rating in 2011, and the development of a 20-Year Sustainability Strategy for the Vancouver Campus in 2014 have been some SD milestones for this period.

UBC's approach towards SD can be described as strategic since the *flagship* stage in 1997, because of the strong institutionalisation and systematisation of SD efforts. UBC's previous president, Stephen J. Toope (2006–2014), his appointed Vice-Presidents, and the Associate Provost of Sustainability have been identified by the UBC interviewees of this case study as strong leaders and agents of change for SD in the institution. They can be seen as key persons for UBC's strategic approach towards SD and have led the university to the *integration* stage.

3.2 Overview of UBC's Sustainability Reporting Efforts (2008–2014)

UBC's sustainability reporting initiatives started in the *flagship* period (see Fig. 1), around a decade after the establishment of the 'Campus Sustainability Office'. UBC documents (i.e. UBC 2009, 2010a) mentioned that the university started reporting in 2007 (covering the period 2006–2007), but the first publicly distributed sustainability report is the 2008/2009 Annual Sustainability Report.

'*Policy #5*' required UBC's Director of Sustainability to "*coordinate reporting on all related University efforts, including progress (and lack of progress), and plans for long-term development*" (UBC 2010a). Consequently, UBC's first (and unpublished) sustainability reports were created as a way to directly track the university's progress in relation to '*Policy #5*'. In 2009, when SD was elevated to the President's Office through inclusion in '*Place and Promise*', the first published Sustainability Reports materialised (UBC 2010a).

An overview of UBC's published *Annual Sustainability Reports* can be found in Table 3. Annual sustainability reporting was seen by UBC as a way to measure their progress, and showcase their main achievements, and leadership for SD to a broader audience:

As UBC has a long history of institutionalizing sustainability and developing ambitious sustainability plans and targets, compiling and publishing sustainability reports has been

Table 3 Overview of UBC's published annual sustainability reports (2009–2014)

Year of publication	Title of sustainability report	Campus included in report	Lead author(s) of report	Content included in report	Number of pages	Inclusion of indicators
2009	2008/09 UBC Sustainability—Annual Report on Implementation of Sustainability Initiatives	Vancouver Campus	UBC Sustainability Office	Introduction; Contextual Overview; Highlight Achievements (including Policy and Governance, Campus Engagement, Academic Integration, Model Campus, Communications and Outreach, External Leadership and Influence); Future Priorities; Final Words	30	–
2010	Campus Sustainability Office—Annual Report to Board 2009/10	Vancouver Campus	Campus Sustainability Office	Executive Summary; Highlight Achievements (including Policy and Governance, Social Programs, Ecological, Economic, Communications and Outreach); Future Priorities 2010/2011; Measures of Sustainability (2009/2010)	19	✓

(continued)

Table 3 (continued)

Year of publication	Title of sustainability report	Campus included in report	Lead author(s) of report	Content included in report	Number of pages	Inclusion of indicators
2011	UBC Annual Operational Sustainability Report— Vancouver Campus 2010/2011	Vancouver Campus	Campus + Community Planning, Campus Sustainability	Executive Summary; Highlight Achieve- ments (including Policy and Governance, Social Programs, Ecological Sustainabil- ity, UTOWN@UBC Community Develop- ment, Economic Sustainabil- ity); Inspirations and Aspirations: Wrap Up Report; Future Priorities 2011/2012; Measures of Sustainability (2010/11); Acknowl- edgements	30	✓
2012	UBC Annual Operational Sustainability Report— Vancouver Campus 2011/2012	Vancouver Campus	Campus + Community Planning, Campus Sustainability	Executive Summary; Highlight Achieve- ments (including Engagement, Climate and Energy, Green Buildings, Water and Waste, Trans- portation, UTOWN@UBC); Future Priorities 2012/2013; Measures of Sustainability (2011/12)	15	✓

(continued)

Table 3 (continued)

Year of publication	Title of sustainability report	Campus included in report	Lead author(s) of report	Content included in report	Number of pages	Inclusion of indicators
2013	UBC Sustainability Annual Report 2013— Vancouver Campus— 2012/2013 Fiscal Year	Vancouver Campus	Campus + Community Planning	Letter from the President; Introduction; Part A: Summary Report (including Achievement Highlights); Part B: Performance Report (including Academic Sustainability, Operational Sustainability, Communications and Engagement, Integration, Future Priorities); Appendix A: Detailed Operational Achievements; Appendix B: ISCN Reporting Requirements	58	✓
2014	Annual Sustainability Report 2013–2014	Vancouver Campus and Okanagan Campus	UBC Sustainability Initiative	Message from the Associate Provost; Introduction; Teaching, Learning and Research; Campus as a Living Lab; Operations and Infrastructure; Community	44	✓

an integral part of measuring our progress in achieving our sustainability goals, as well as celebrating and communicating our sustainability achievements. (Sustainability Engagement and Reporting Coordinator, C+CP)

Table 3 shows that the content of UBC's Sustainability Reports have ranged from highlighting overall SD initiatives in 2009 and 2010, to Sustainability Reports focusing on operational SD achievements in 2011 and 2012, to a more in-depth coverage of academic and operational SD initiatives on campus in 2013 and 2014. There has also been a shift in the lead department authoring the Sustainability Reports (see Table 3).

From 2009 until 2013, UBC's Annual Sustainability Reports focused on the largest campus, i.e. the Vancouver Campus, while the 2013–2014 report was the first joint report for the Vancouver Campus and the Okanagan Campus (see Table 3). UBC's different campuses collaborate and identify synergies in SD integration and sustainability reporting. The joint 2013–2014 Annual Sustainability Report was one of the visible consequences of the streamlining of SD activities and on-going collaborations between the campuses.

3.2.1 Focusing on UBC's Various Sustainability Reports in Fiscal Year 2014

The Annual Sustainability Reports presented in Sect. 3.2 developed from UBC's internal motivations, and were directed at internal and external audiences (e.g. the Board of Governors or peer institutions). However, because of the large range of initiatives UBC was involved in, Annual Sustainability Reports alone would not have captured all the SD activities that were happening on campus on the strategic and the operational level.

UBC was also engaged in a large number of other reporting practices to disclose its SD activities to different stakeholders and networks. Table 4 shows a summary of UBC's sustainability reporting efforts during the fiscal year 2014. This includes information on the outcome (i.e. type of report), the directive (or reason for reporting), the intended audience, and the departments leading and supporting the preparation of the report. The table shows that UBC reports on its SD efforts via a variety of channels, including the Annual Sustainability Report, Place and Promise, the Carbon Neutral Action Report, and STARS.

The *Annual Sustainability Reports* mainly focused on communicating past achievements and setting up future priorities. They were mostly directed at demonstrating the highest-level achievements and comparing them to the outlined SD plans. The reporting process was initially coordinated by the 'Campus Engagement and Reporting Coordinator' (C+CP), but during the data collection of the case study, the 'Engagement Specialist' (USI) became the coordinator.

The report for '*Place and Promise*' was also an internally motivated type of report that was required by the President's Office. It was aimed at reporting on the achieved goals set out in the President's strategic plan, and used fragments of information

Table 4 Overview of UBC's reports on SD activities (Fiscal Year 2014)

Report	Directive	Audience	Lead authors (Support authors)
Place and Promise	Required by President's Office	Internal and external	USI (C+CP) and Sustainability Office Okanagan Campus
Annual Sustainability Report	Required by Board of Governors (BoG), Policy #5, ISCN	BoG, ISCN, internal and external stakeholders, peer institutions	USI and C+CP
Board Report	Required by BoG, Policy #5	BoG	USI
International Sustainable Campus Network (ISCN) Report	Required as a signatory to the ISCN charter	International peers	USI and C+CP
Operational Sustainability Strategy	Policy #5 (each department to have a sustainability plan)	Internal stakeholders	C+CP
Carbon Neutral Action Report (CNAR)	Required by Provincial legislation (Bill 44); also serves as UBC's internal Climate Action Plan report	Province of British Columbia, other public sector organisations, internal and external audiences interested in GHG reporting	C+CP and Sustainability Office Okanagan Campus
Sustainability Tracking, Assessment and Rating System (STARS)	Voluntary external benchmarking report	Internal and external stakeholders, peer institutions	USI and C+CP

Source UBC (2014c)

available in the Annual Sustainability Reports. This process was led by the USI, and supported by C+CP.

UBC's main operational departments also reported on their achievements against the *Operational Sustainability Strategy*. This was another internally motivated report, where the operational departments, in close cooperation with C+CP, annually chose relevant metrics and actions to follow up and report on. This process was led by the 'Manager Strategic Planning' and the 'Campus Engagement and Reporting Coordinator' (see Table 4).

The university used *STARS reporting* for external benchmarking purposes. According to UBC interviewees, STARS' detailed and clearly outlined credits allowed in-depth comparison to the university's peer institutions in North America. Case study participants such as the 'Manager of Sustainability Engagement', the 'Manager UBC SEEDS Program', and the 'Manager of Advising and Student Involvement' (see Table 4) provided information for the STARS reports.

UBC's membership of the *ISCN network* offered the university the possibility to share and exchange learning experiences and best practices on SD in campus operations, research, and teaching with other leading HEIs around the world. The involvement in the network required annual updates on achievements on SD, coordinated by USI and C+CP.

UBC's *Carbon Neutral Action Reports* (CNARs) were mandatory reports, required by the Provincial government of British Columbia, and led by the 'Campus Engagement and Reporting Coordinator' (Vancouver Campus), and the 'Associate Director Sustainability Operations' (Okanagan Campus). The interviewees mentioned that this type of mandatory reporting created a dynamic of very ambitious goal-setting for GHG emission reduction in UBC, leading to decreasing emissions despite increasing building floor space and student enrolment.

3.3 The Functions of Sustainability Reporting at UBC

The interviews with the internal actors of the sustainability reporting process at UBC showed that there were various reasons for the university to engage in consecutive SD reporting cycles. In this section we will discuss each of these functions: the monitoring of sustainability performance, communication on sustainability activities and engaging with stakeholders.

3.3.1 Monitoring Sustainability Performance

In UBC, sustainability reporting was largely driven by the need for assessment of SD performance in the institution. The overall goal of the assessment and reporting efforts were to have a clear view on what has been done on SD, which areas still needed to be addressed, and how the university was achieving its strategic goals over time. This was linked to the elevation of SD to the Presidential level: as sustainability became a key strategic pillar, the Board of Governors needed to be annually updated on the progress in this field.

a. *Monitoring Operational Performance*

A managerial mind set of 'what gets measured, gets done' was visible in the responses of most of the interviewed staff members. According to the respondents, SD assessment through sustainability reporting was a way to monitor the implementation of the SD integration process in the organisation, by measuring the outcomes of SD actions and being accountable for their results.

> The reporting process is crucial, because you've got to document what you're doing. You want to have targets, but you have to show that you're achieving them, right? So the reporting is the 'ground truthing' of the whole thing. (...) So the reporting is, I think, holding your feet

to the fire, in a way, and saying 'are you really doing it?' (Associate Provost Sustainability, USI)

Mainly in the operational units at UBC, clear SD targets and indicators were set up, and annual monitoring and adjustments of SD targets and plans were made. For example, in 2013–2014, UBC achieved a 14% reduction in absolute greenhouse gas (GHG) emissions compared to 2007 levels for its Vancouver Campus (compared to a target of 33% by 2015) (UBC 2014a). For materials and waste, UBC's Vancouver Campus had an overall waste diversion rate of 61% in 2013–2014 (compared to a target of 70% by 2016) (UBC 2014a).

Reporting on such targets, and disclosing the progress made, helped UBC to achieve incremental progress in the targeted areas of the campus operations. While the targets were not always reached, UBC valued transparent communication on achieved (or lacking) progress. For example, the waste diversion rate was not achieved in 2010, after which a Waste Audit was conducted and a new Waste Action Plan was set up after stakeholder consultations (UBC 2011).

b. *Monitoring Academic Performance*

On the academic side, reporting on the progress towards SD was seen as a more complex process in UBC, and its contribution to the SD integration process was identified in a different way. According to the members of the Teaching, Learning and Research Office (TLRO), UBC reported on the implementation of SD activities in its core activities of education and research through STARS credits, of which some were included in the Annual Sustainability Reports.

The TLRO collected, organised, and made information on these core activities accessible, through course inventories, updated SD program lists, lists of faculty members with SD research interests etc. For example, in 2013–2014, UBC's Vancouver Campus had over 500 SD-related courses,[2] over 40 SD-related academic programs, and over 50 SD-related Ph.D. theses (UBC 2014a). Nevertheless, the interviewees were not convinced of use of the academic SD indicators for improving SD performance: according to the TLRO, counting the number of SD-related courses at UBC did not motivate faculty members to include SD in their course. Yet this served another purpose:

> Creating an inventory into which a student or a faculty member could go and search around and find colleagues and find collaborators, I think that the richness of the information can support change and support the creation of new networks and new opportunities. (Associate Director of the Teaching, Learning and Research Office, USI)

Reporting on the HEI's core activities was viewed in terms of its benefits of connecting people who are interested in collaborating on SD at UBC, rather than as a means to improve sustainability performance.

---

[2] As defined within AASHE's STARS 2.0 framework, credit AC1 comprising of 'sustainability courses' and 'courses that include sustainability' (see AASHE 2015).

3.3.2   Communicating Sustainability Achievements in UBC

UBC used sustainability reporting to communicate directly and indirectly on its SD efforts and achievements to internal and external stakeholders. This was elaborated on by different participants, for example:

> I think communicating about what we're doing and communicating the successes is really, really important. Important to show the value of the work that's being done. So that you continue to have that mandate. And so that someone can't come along and say 'do we actually really need this department? Or do we really need this role in the university?' (…) It's a way of justifying the added value. (Director Operational Sustainability, C+CP)

> It's important to demonstrate how sustainability performance is being achieved on campus. Internally, it enables us to measure progress against our goals and celebrate our achievements. Externally, it helps us demonstrate our actions in response to key issues such as climate change. In this way, sustainability reporting can inspire others to take action. (Associate Director Sustainability Operations, Sustainability Office, Okanagan Campus)

Hence, sustainability reporting was a way for UBC to celebrate its success around SD (internally and externally), to keep the SD integration process at UBC going, and to inspire other internal and external UBC stakeholders.

3.3.3   Engaging Internal Stakeholders for Sustainability in UBC

Sustainability reporting was also used to engage stakeholders in the SD integration process at UBC. Mostly internal stakeholders were engaged in the sustainability reporting process (i.e. staff, faculty, and students), yet their engagement was often limited to the sustainability reporting data collection process. For example, through internships at the C+CP, some students were responsible for the collection of STARS data within the reporting process. According to the Director of Operational Sustainability, sustainability reporting created a valuable type of engagement between the C+CP and the various units around the Vancouver campus.

Certain operational departments were more strongly involved in sustainability reporting and engaged in SD priority setting than others, e.g. through the use of the 'Departmental Level Sustainability Frameworks'. The involved participants noticed changes in the way these departments perceived SD and cooperated on SD initiatives because of their involvement in the reporting process. According to the TLRO, faculty members (in comparison to staff members) were generally more difficult to reach and engage in the sustainability reporting process, and their input was solely used to report on STARS data.

There was a difference between the stakeholder engagement processes on the two main campuses. During the interviews, the smaller scale of the Okanagan Campus in comparison with the Vancouver Campus, and the consequences for the sustainability reporting process were discussed, e.g.:

> It's an opportunity and a challenge. A smaller campus enables us to directly engage key stakeholder groups during the reporting process. We use this process as an engagement

tool to understand challenges, brainstorm options, and integrate sustainability at the unit level, for greater impact. (Associate Director Sustainability Operations, Sustainability Office, Okanagan Campus)

The Okanagan Campus had more direct possibilities for engaging internal and external stakeholders because of its smaller scale. The Vancouver Campus needed a more formal way of engaging its stakeholders, because of the large number of persons, entities and groups working and living on the campus.

3.4 The Role of the UBC Sustainability Initiative in the Sustainability Reporting Process

During the data collection process of the 2013–2014 Annual Sustainability Report, there was a change in the lead authorship of the sustainability report. There was a mutual consent between the C+CP and USI departments that the coordination of the sustainability reporting process should be a responsibility of the USI, because of its focus on linking academic and operational SD. This created a new dynamic in the organisation, and due to this change in responsibilities, the USI started rethinking and gathering ideas for improving the role of UBC's sustainability reports.

Having the USI, a department responsible for combining academic and operational SD integration at UBC, as a leader and facilitator of the sustainability reporting process, could be important for the further development of sustainability reporting at UBC, and for its potential to foster stakeholder engagement and help further the SD integration process.

We [i.e. the USI] are intended to be a catalyst. We are linked in a networked way to our colleagues at the farm, the botanical gardens, at various student associations and groups. We are helping them along their journey of talking about sustainability, engaging in sustainability programs and undertakings. So we have an interesting role to play, as a catalyst group, an enabling group. (Director of Communications and Community Engagement, USI)

The USI had an enabling function for SD within the institution: the department created opportunities for faculty and staff to collaborate and reach mutual SD achievements on campus. Reporting was one of these opportunities where collaboration with different internal stakeholders was sought, with the aim of furthering the SD integration process.

4 Discussion: Unpacking the Complexities of the Reporting Process

In this section, the findings of the UBC case are analysed and discussed by focusing on four elements: strategic sustainability integration, the various reporting efforts of the university, the different aims of the sustainability reporting process, and the potential of an internal actor taking responsibility over the reporting process.

4.1 Strategic Sustainability Integration and Sustainability Reporting

The sustainability reporting process at UBC was mainly driven by a strong focus of the university on performance management. Hence, these were internal motivations, analogously to what has been found in most reporting HEIs (see Ceulemans et al. 2015a). This corresponds to Herzig and Schaltegger's (2011) 'inside-out' perspective of sustainability reporting, where the implementation of sustainability strategy leads to sustainability reporting activities. In UBC, the participants stressed the importance of internal SD strategies and policies for sustainability reporting, and of the presence of management support and leadership for sustainability reporting. Additionally, the mandatory Carbon Neutral Action Reporting (CNAR), as required by the BC Government, also provided an external push for sustainability reporting (see Herzig and Schaltegger's (2011) 'outside-in' perspective). All these elements indicate that internal leadership (i.e., top management support), as well as external parties (e.g., governments) can play a significant role in the adoption and continuation of sustainability reporting for an HEI that is engaged in strategic sustainability integration. This concurs with the combined approach based on internal and external motivations, found for consecutive sustainability reporting in companies by Lozano et al. (2016).

4.2 Analyzing UBCs Various Reporting Efforts

Tables 3 and 4 provided an overview of the various reporting efforts UBC is engaged in. Table 3 showed the complex process of an entity that is reporting during consecutive reporting cycles in a sector for which sustainability reporting was still in its early stages. There was an internal search and shift over the years in terms of the content to be included in the reports (e.g., covering only operational content, academic topics or both), the lead authors of the reports (e.g., an operational department or an academic department), and the entities to include (e.g., only the Vancouver Campus or also the Okanagan Campus). This concurs with the learning process identified by Ceulemans et al. (2015a) for sustainability reporting in universities, where HEIs are using consecutive reporting cycles to optimise sustainability data collection and management. Additionally, this learning process may lead to shifts in responsibilities in the reporting process in the search for providing impactful reports.

Table 4 showed the various reporting requirements UBC was engaged in during one fiscal year (2014). As a leader in sustainability integration, UBC was part of a large variety of networks/frameworks, resulting in reporting requirements for each of these frameworks. While some reports are very distinct (e.g., the voluntary STARS reporting versus the mandatory CNAR reports), it would reduce the complexity of the reporting process if more synergies could be found between some of these frameworks and the reporting requirements imposed by the involved actors (see also Ceulemans et al. 2015b). As Findler et al. (2018) also stressed, it might be necessary for HEIs to take a more holistic perspective and come up with new approaches that complement or replace some of the currently existing and fragmented reporting frameworks. Moreover, having many different lead authors for each of these frameworks, as was the case at UBC, makes the process more complex since responsibilities are dispersed over the university and reduces the chance of finding synergies between the different reporting requirements.

4.3 Analysing the Functions of Sustainability Reporting

The analysis of the case shows that universities are engaged in sustainability reporting to achieve multiple aims simultaneously. Reporting can highlight past SD achievements of an HEI, can help the university monitor and improve its operational sustainability performance, and can help engage internal actors for sustainability integration. This concurs with the main aims of reporting found in the corporate sector (see Dalal-Clayton and Bass 2002; Joseph 2012; Lozano et al. 2016).

In UBC's case, there was a clear focus on sustainability assessment through reporting, in order to achieve improvements in sustainability reporting for the institution. Thus, the monitoring and assessment approach was dominant at UBC. However, in terms of its effects on the sustainability performance of the university, there was a difference visible between the monitoring of academic versus operational performance. The operational performance monitoring led to direct incremental improvements in the operational sustainability performance of the university (which concurs with Ceulemans et al. 2015a). However, according to the interviewees, the academic performance monitoring was not directly improving the academic sustainability performance of the university. The internal actors of the sustainability reporting process questioned the potential of assessing academic indicators for increasing sustainability performance (e.g., on sustainability course content or SD research activities), but identified an engagement potential in the collected academic sustainability information.

The analysis of the UBC case shows that sustainability reporting can contribute to the sustainability integration process in a university through its monitoring and assessment function, when a strong inclination towards performance management is present in the university, linked to deliberate strategic goal setting for sustainability. However, sustainability reporting may not be as effective as an assessment tool for sustainability integration to measure academic performance, as it is to measure operational performance. Or, this could mean that the internal actors have not found the ideal tools yet to effectively assess their academic impacts in relation to sustainability. This supports Findler et al.'s (2018) call for new approaches to measure the sustainability impacts of universities.

UBC also used the sustainability reporting process as a means to communicate their SD integration achievements to internal and external stakeholders. In terms of two-way communication and engagement, the reporting process mostly engaged internal stakeholders in the SD integration process (as opposed to external stakeholders). This corresponds to the stakeholder engagement processes observed in most HEIs who publish sustainability reports (see Ceulemans et al. 2015a; Disterheft et al. 2015). For example, in UBC, sustainability reporting contributed to SD integration through the interaction between the C+CP and various operational departments, or through the interaction of the USI with faculty for reporting on academic SD activities. Thus, one of the assets of sustainability reporting may be that it interconnects internal stakeholders who in other circumstances do not interact frequently. This concurs with Albrecht et al.'s (2007) research indicating that sustainability reporting has the ability to mobilise internal stakeholders in HEIs for learning and change towards SD.

In this chapter's case study, a stronger ease of communication or interaction with staff was identified, in comparison to the ease of interaction with faculty members on sustainability reporting. Academic (faculty) and operational (staff) members belong to different internal groups of the university, and therefore may react differently to sustainability reporting. This implies that in order to enhance sustainability integration through using the engagement function of sustainability reporting, a party should be involved that can interact effectively with both of these groups and/or a diversified approach for these two groups may be desirable.

While a university may have a stronger focus on the sustainability monitoring and assessment function of sustainability reporting than on its use for communication purposes, an HEI may still use the communication function of reporting to help sustainability integration in the institution. In the case of UBC, the larger Vancouver Campus had a stronger focus on assessment, while the smaller Okanagan Campus used the engagement function of sustainability reporting more intensively. This might also be the case for different HEIs around the world: depending on their distinct campus culture and context (see Madeira et al. 2011), HEIs may require a focus on a certain function of sustainability reporting to achieve a fit with its SD activities, and thereby contribute to SD integration in the HEI.

4.4 The Role of a Hybrid Internal Actor to Lead the Reporting Process

An important element highlighted in this chapter is the role of a mixed internal stakeholder group facilitating the sustainability reporting process, and thus contributing to the SD integration process. The *UBC Sustainability Initiative* (USI) was a hybrid internal stakeholder group working beyond the boundaries of the traditional groups of a university, with the aim of fostering SD integration in UBC. The USI addressed the divide between academic and operational departments, an issue that is often present in HEIs (see Bero et al. 2012; Krizek et al. 2012). This type of internal stakeholder may be an important actor for sustainability integration and learning in higher education (see also Barth 2013). More specifically for the process of sustainability reporting in HEIs, a hybrid internal actor can help reduce some of the identified complexities of the sustainability reporting process. An actor facilitating the communication between internal stakeholder groups is necessary for the reporting process, since these disconnected subgroups will have to collaborate and communicate intensively to create a report that includes academic and operational information on the HEI, which can both be considered material aspects for HEIs (see Ceulemans et al. 2015b). Hybrid internal stakeholder groups are particularly valuable in HEIs to tackle the highly decentralised and loosely-coupled structures of HEIs (see Weick 1976).

5 Conclusion

While sustainability reporting is currently not widely spread in higher education, the practice of sustainability reporting has been identified as one of the tools to foster sustainability integration into HEIs. This research is one of the first case studies aimed at unpacking the complexities of the process of sustainability reporting in HEIs. Through the use of The University of British Columbia (UBC) as a case study, the sustainability reporting process was analysed, the main complexities of the process were discussed (with a focus on the HEIs internal actors), and some recommendations specific to the HEI sector were proposed. The study was based on in-depth interviews with the main actors involved in the sustainability reporting process at UBC, and used document analysis to outline its context and triangulate the findings from the interviews.

In the study, it was found that even in an institution with a deliberate sustainability strategy and integration process, sustainability reporting is a complex undertaking. Some of the main elements adding complexity to the reporting process are: the search for suitable lead authors and relevant content for the reports in consecutive reporting cycles; the different aims that reporting tries to achieve simultaneously for the SD integration process; and suitably addressing the different internal stakeholders that are involved in the reporting process in HEIs. In order to reduce complexity in the HEI, assigning an internal hybrid actor to facilitate the reporting process may be helpful, as was the case with the *UBC Sustainability Initiative* (USI) at UBC.

While sustainability reporting is a complex process, universities may want to engage in it because of its clear contributions to the sustainability integration process. Reporting can highlight past SD achievements of an HEI, can help the university monitor and improve its operational sustainability performance, and can help engage internal actors for sustainability integration. The context in which the HEI is situated will also affect the way the reporting process will play out and this will be different for other HEIs around the world (hence requiring a contextual approach). However, unfolding the key complexities arising when undertaking consecutive reporting cycles can help internal actors of other institutions to better design and implement the process of sustainability reporting and integration in their universities.

Further research should study the topic of sustainability reporting more closely, for example in cases where management support is lacking or where the reporting process started as a response to external pressure from stakeholders. Future studies could also focus more in detail on the influence of sustainability leadership on the sustainability reporting process and its effect on the SD integration process in HEIs, or on the potential of sustainability reporting for SD integration in relation to a university's material academic impacts.

Acknowledgements The authors of this paper would like to acknowledge the participants of this case study at The University of British Columbia, the FWO (Research Foundation—Flanders, Belgium), Prof. Valérie Cappuyns, Prof. Tomás Ramos, and Prof. Didier Van Caillie for their support and valuable input provided for this study.

References

AASHE (2013) STARS, a program of AASHE. Association for the Advancement of Sustainability in Higher Education. https://stars.aashe.org. Last accessed 28 Mar 2019

AASHE (2015) AC1 academic courses. Association for the Advancement of Sustainability in Higher Education. https://drive.google.com/file/d/0BzY7o-k46NLgTThoQU41c05NN3M/view. Last accessed 28 Mar 2019

Adams C (2013) Sustainability reporting and performance management in universities: challenges and benefits. Sustain Account Manage Policy J 4(3):384–392

Adams C, Frost GR (2008) Integrating sustainability reporting into management practices. Account Forum 32:288–302

Adams CA, McNicholas P (2007) Making a difference: sustainability reporting, accountability and organisational change. Account Auditing Accountability J 20(3):382–402

Albrecht P, Burandt S, Schaltegger S (2007) Do sustainability projects stimulate organizational learning in universities? Int J Sustain High Educ 8(4):403–415

Alonso-Almeida MDM, Marimon F, Casani F, Rodriguez-Pomeda J (2015) Diffusion of sustainability reporting in universities: current situation and future perspectives. J Clean Prod 106:144–154

Barth M (2013) Many roads lead to sustainability: a process-oriented analysis of change in higher education. Int J Sustain High Educ 14(2):160–175

Barth M, Timm J-M (2011) Higher education for sustainable development: students' perspectives on an innovative approach to educational change. J Soc Sci 7(1):13–23

Beaudoin C (2014) Sustainability benchmarking document. MacEwan University, Office of Sustainability, Edmonton, Canada

Bero BN, Doerry E, Middleton R, Meinhardt C (2012) Challenges in the development of environmental management systems on the modern university campus. Int J Sustain High Educ 13(2):133–149

Blanco-Portela N, Benayas J, Pertierra LR, Lozano R (2017) Towards the integration of sustainability in higher education institutions: a review of drivers of and barriers to organisational change and their comparison against those found of companies. J Clean Prod 166:563–578

Brinkmann S, Kvale S (2015) InterViews: learning the craft of qualitative research interviewing, 3rd edn. SAGE Publications Inc, Thousand Oaks, CA

Buckler C, Creech H (2014) Shaping the future we want. UN decade of education for sustainable development (2005–2014). Final report. DESD monitoring and evaluation. UNESCO, Paris, France

Ceulemans K, Van Caillie D, Molderez I, Van Liedekerke L (2014) A management control perspective of sustainability reporting in higher education: in search of a holistic view. ACRN J Entrepreneurship Perspect 3(1):1–17

Ceulemans K, Lozano R, Alonso-Almeida M (2015a) Sustainability reporting in higher education: interconnecting the reporting process and organisational change management for sustainability. Sustainability 7:8881–8903

Ceulemans K, Molderez I, Van Liedekerke L (2015b) Sustainability reporting in higher education: a comprehensive review of the recent literature and paths for further research. J Clean Prod 106:127–143

Ceulemans K, Stough T, Lambrechts W (2018) Pioneering in sustainability reporting in higher education: experiences of a belgian business faculty. In: Leal Filho W (ed) Handbook of sustainability science and research. Springer International Publishing, Switzerland, pp 211–223

Corbin J, Strauss AL (2008) Basics of qualitative research. Techniques and procedures for developing grounded theory, 3rd edn. SAGE Publications, California, Thousand Oaks

Corcoran PB, Walker KE, Wals AEJ (2004) Case studies, make your case studies, and case stories: a critique of case study methodology in sustainability in higher education. Environ Educ Res 10(1):7–21

Cortese AD (2003) The critical role of higher education in creating a sustainable future. Plan High Educ 31(3):15–22

Dalal-Clayton B, Bass S (2002) Sustainable development strategies. Earthscan Publications Ltd, London

de Lange DE (2013) How do universities make progress? Stakeholder-related mechanisms affecting adoption of sustainability in university curricula. J Bus Ethics 118(1):103–116

Desha C, Hargroves K (2014) Higher education and sustainable development. A model for curriculum renewal. Routledge, Oxon, UK

Disterheft A, Caeiro S, Azeiteiro UM, Leal Filho W (2015) Sustainable universities—a study of critical success factors for participatory approaches. J Clean Prod 106:11–21

Eisenhardt KM, Graebner ME (2007) Theory building from cases: opportunities and challenges. Acad Manag J 50(1):25–32

Eisenhardt KM, Graebner ME, Sonenshein S (2016) From the editors: grand challenges and inductive methods: rigor without rigor mortis. Acad Manage J 59(4):1113–1123

Findler F, Schönherr N, Lozano R, Stacherl B (2018) Assessing the impacts of higher education institutions on sustainable development—an analysis of tools and indicators. Sustainability 11(59)

Fonseca A, Macdonald A, Dandy E, Valenti P (2011) The state of sustainability reporting at Canadian universities. Int J Sustain High Educ 12(1):22–40

Glaser BG, Strauss AL (1999) The discovery of grounded theory: strategies for qualitative research. Aldine de Gruyter, New York

Greenhouse Gas Reduction Targets Act, Bill 44 - 2007 (Province of British Columbia, Canada). Legislative Assembly of British Columbia. http://www.leg.bc.ca/38th3rd/1st_read/gov44-1.htm. Last accessed 28 Mar 2019

Gudz NA (2004) Implementing the sustainable development policy at the university of British Columbia: an analysis of the implications for organisational learning. Int J Sustain High Educ 5(2):156–168

Herzig C, Schaltegger S (2011) Chapter 14: corporate sustainability reporting. In: Godemann J, Michelsen G (eds) Sustainability communication: interdisciplinary perspectives and theoretical foundations. Springer, Dordrecht, pp 151–169

Joseph G (2012) Ambiguous but tethered: an accounting basis for sustainability reporting. Crit Perspect Account 23(2):93–106

Jupp V (ed) (2012) The SAGE dictionary of social research methods. Sage Publications Ltd, London, UK

Kamal ASM, Asmuss M (2013) Benchmarking tools for assessing and tracking sustainability in higher educational institutions: identifying an effective tool for the University of Saskatchewan. Int J Sustain High Educ 14(4):449–465

Kolk A (2010) Trajectories of sustainability reporting by MNCs. J World Bus 45(4):367–374

Krizek KJ, Newport D, White J, Townsend AR (2012) Higher education's sustainability imperative: how to practically respond? Int J Sustain High Educ 13(1):19–33

Lozano R (2006a) Incorporation and institutionalization of SD into universities: breaking through barriers to change. J Clean Prod 14(9–11):787–796

Lozano R (2006b) A tool for a graphical assessment of sustainability in universities (GASU). J Clean Prod 14(9–11):963–972

Lozano R (2011) The state of sustainability reporting in universities. Int J Sustain High Educ 12(1):67–78

Lozano R, Peattie K (2011) Assessing Cardiff University's curricula contribution to sustainable development using the STAUNCH [RTM] system. J Educ Sustain Dev 5(1):115–128

Lozano R, Llobet J, Tideswell G (2013a) The process of assessing and reporting sustainability at universities: preparing the report of the University of Leeds. Revista Internacional de Sostenibilidad, Tecnología y Humanismo 8:85–113

Lozano R, Lukman R, Lozano FJ, Huisingh D, Lambrechts W (2013b) Declarations for sustainability in higher education: becoming better leaders, through addressing the university system. J Clean Prod 48:10–19

Lozano R, Nummert B, Ceulemans K (2016) Elucidating the relationship between sustainability reporting and organisational change management for sustainability. J Clean Prod 125:168–188

Madeira AC, Carravilla MA, Oliveira JF, CaV Costa (2011) A methodology for sustainability evaluation and reporting in higher education institutions. High Educ Policy 24(4):459–479

Maubane P, Prinsloo A, Van Rooyen N (2014) Sustainability reporting patterns of companies listed on the Johannesburg securities exchange. Public Relat Rev 40:153–160

Moore J (2005) Policy, priorities and action: a case study of the University of British Columbia's engagement with sustainability. High Educ Policy 18(2):179–197

Moore J, Pagani F, Quayle M, Robinson J, Swada B, Spiegelman G, Van Wynsberghe R (2005) Recreating the university from within: collaborative reflections on the University of British Columbia's engagement with sustainability. Int J Sustain High Educ 6(1):65–80

Moudrak N, Clarke A (2012) Developing a sustainability report for a higher education institution. In: Leal Filho W (ed) Sustainable development at universities: new horizons. Peter Lang Scientific Publishers, Frankfurt

Roorda N (2002) Assessment and policy development of sustainability in higher education with AISHE. In: Filho WL (ed) Teaching sustainability at universities: towards curriculum greening, environmental education, communication and sustainability. Peter Lang, Frankfurt

Saunders M, Lewis P, Thornhill A (2012) Research methods for business students, 6th edn. Pearson Education Ltd, Essex, UK

Schreier M (2012) Qualitative content analysis in practice. SAGE Publications Ltd, London, UK

Sedlacek S (2013) The role of universities in fostering sustainable development at the regional level. J Clean Prod 48:74–84

Stephens JC, Hernandez ME, Román M, Graham AC, Scholz RW (2008) Higher education as a change agent for sustainability in different cultures and contexts. Int J Sustain High Educ 9(3):317–338

UBC (2009) 2008/09 UBC sustainability: annual report in implementation of sustainability initiatives. UBC Sustainability Office, Vancouver, CA

UBC (2010a) Case study: institutionalizing sustainability. UBC Campus Sustainability Office, Vancouver, CA

UBC (2010b) Case study: green building. UBC Campus Sustainability Office, Vancouver, CA

UBC (2010c) Case study: social ecological economic development studies. UBC Campus Sustainability Office, Vancouver, CA

UBC (2010d) Case study: sustainability + residence coordinators. UBC Campus Sustainability Office, Vancouver, CA

UBC (2011) Inspirations and aspirations: UBC sustainability strategy 2006–2010. Final report. UBC Campus + Community Planning, Vancouver, CA

UBC (2013) 2013 ISCN-GULF sustainable campus charter report. University of British Columbia, Vancouver Campus. 2012/2013 Fiscal Year. UBC, Vancouver, CA

UBC (2014a) Annual sustainability report 2013–2014. UBC, Vancouver, CA

UBC (2014b) Addendum to the 2013–2014 annual sustainability report. ISCN-Gulf sustainable campus charter report. UBC, Vancouver, CA

UBC (2014c) 14/15 UBC sustainability reporting summary. Internal Document Campus + Community Planning. UBC Vancouver Campus, Vancouver, CA

UBC (2015a) UBC sustainability: who we are. UBC. http://sustain.ubc.ca/our-commitment/our-story. Last accessed 28 Mar 2019

UBC (2015b) Carbon neutral action report. UBC. http://sustain.ok.ubc.ca/reports/cnar.html. Last accessed 28 Mar 2019

UBC (2015c) Strategic plans, policies and reports. UBC http://sustain.ubc.ca/our-commitment/strategic-plans-policies-and-reports. Last accessed 28 Mar 2019

UBC (2015d) UBC sustainability: external benchmarks. http://sustain.ubc.ca/our-commitment/strategic-plans-policies-reports/external-benchmarks. Last accessed 15 Feb 2015

UBC (2015e) UBC sustainability initiative. http://www.johnrobinson.ires.ubc.ca/ubc-sustainability-initiative/. Last accessed 28 Mar 2019

Velazquez L, Munguia N, Platt A, Taddei J (2006) Sustainable university: what can be the matter? J Clean Prod 14(9–11):810–819

Verhulst E, Lambrechts W (2015) Fostering the incorporation of sustainable development in higher education. Lessons learned from a change management perspective. J Clean Prod 106:189–204

Weick KE (1976) Educational organizations as loosely coupled systems. Adm Sci Q 21(1):1–19

Yarime M, Tanaka Y (2012) The issues and methodologies in sustainability assessment tools for higher education institutions. A review of recent trends and future challenges. J Educ Sustain Dev 6(1):63–77

Yin RK (2009) Case study research: design and methods. In: Applied social research methods series, vol 5, 4th edn. Sage, Thousand Oaks, p 219

Measuring the Use of Sustainable Modes of Transport at a University

Jan Silberer, Thomas Bäumer, Patrick Müller, Payam Dehdari and Stephanie Huber

Abstract Measuring the use of sustainable modes of transport at a university can confer many benefits. Investigating how university members arrive on-site can create a sense of urgency for sustainable mobility, and showing how the use of these modes of transport affects the university's total ecological footprint can be an effective way of doing so. Furthermore, scenarios about the effects of methods for increasing transport sustainability can be calculated, although they should be regarded with caution. It can also be crucial to examine *why* university members use a specific mode of transport. This enables sustainability officers to adjust sustainability measures to needs and environmental conditions that can be critical for their success. Therefore, this paper introduces a four-stage approach to assessing the ecological footprint of specific modes of transportation based on university members' attitudes and motives, the situation and environment, their actual behaviour and the resulting environmental impact. In doing so, the construct of mobility can be analysed in a way that does justice to its complexity and multifaceted nature. This is demonstrated with an analysis of two studies at the Hochschule für Technik (University of Applied Sciences) Stuttgart in which all of these aspects were used to define target groups and adjust and prioritise corresponding measures targeting their mobility behaviour.

Keywords Sustainable mobility · Technology acceptance

J. Silberer (✉) · T. Bäumer · P. Müller · P. Dehdari · S. Huber
Schellingstr. 24, 70174 Stuttgart, Germany
e-mail: jan.silberer@hft-stuttgart.de

T. Bäumer
e-mail: thomas.baeumer@hft-stuttgart.de

P. Müller
e-mail: patrick.mueller@hft-stuttgart.de

P. Dehdari
e-mail: payam.dehdari@hft-stuttgart.de

S. Huber
e-mail: stephanie.huber@hft-stuttgart.de

© Springer Nature Switzerland AG 2020
W. Leal Filho et al. (eds.), *International Business, Trade and Institutional Sustainability*, World Sustainability Series,
https://doi.org/10.1007/978-3-030-26759-9_62

1   Introduction

Climate change was long considered a phenomenon that was only happening in distant countries outside Europe. However, its effects have now arrived in countries like Germany, with far-reaching consequences (Brasseur et al. 2017). An international research consortium has impressively demonstrated that climate change is anthropogenic (Le Quéré et al. 2018). In order to avert catastrophic effects, the global community set itself the goal in 2016 of limiting global warming to below 2 °C by 2050, if possible to 1.5 °C (Bundesministerium für Umwelt et al. 2016). A special report by the Intergovernmental Panel on Climate Change taking into account more than 6000 studies, as well as other research, shows that this target is currently on track to being missed and that warming of 3.2° is more likely by 2050 (Figueres et al. 2018; Masson-Delmotte 2018; United Nations Environment Programme 2019).

The commitment of the world's nations is essential if this is to be changed. Zhang et al. (2017) demonstrate this by calculating the negative effects of the United States' withdrawal from the Paris Agreement on budget cuts to climate change programs. However, the EU is not on course to meet its targets either (Le Quéré et al. 2018). As the largest economy in the EU, Germany plays a key role here. The transport sector should be particularly targeted to change negative developments with respect to global warming. Approximately 14% of total CO_2 emissions come from the transport sector (Pachauri and Mayer 2015).

Universities have a special responsibility to assist society in effectively tackling this problem. Thus, it is crucial for them to not only to teach students sustainable and responsible behaviour, but also to set an example for society. Before this can be achieved, it is necessary to elucidate the status quo and identify areas for improvement. This represents a challenge because mobility is a complex construct. Mobility is influenced by numerous different aspects that need to be taken into account if changes in people's mobility behaviour are to be successfully achieved.

2   Stages of Mobility: A Framework

As previously mentioned, mobility is an extremely complex construct that can be viewed from a variety of perspectives. This complexity means that a broad range of factors must be taken into account when considering the impact of mobility on the environment as a whole as well as when planning effective measures to change mobility behaviour. Therefore, previous research on the environmental influences of mobility has considered a number of factors (Hunecke et al. 2007), albeit largely individually or in narrow groups. For example, psychological research has attempted to gain a better understanding of people's attitudes towards certain types of mobility (Sato et al. 2014). However, the concrete effects of behaviour on the environment have rarely been considered. This left it unclear to what extent interventions aimed at modifying these factors can have a substantial effect on the environment. Other studies concentrated on the effects of individual behaviours on the environment, e.g. François et al. (2017). Here, however, less attention was paid to general conditions and their influence on the psychological processes that precede such behaviours (for an exception, see Hunecke et al. 2007). Without taking psychological acceptance into account, it is difficult to assess whether the possible effects of behaviour modification in the real world are realistic at all.

For these reasons, a holistic view of the mobility phenomenon that simultaneously considers and systematically organises numerous individual factors appears highly relevant. Therefore, the present paper proposes a four-stage model providing a holistic view of the impact of mobility on the environment. This framework is the basis of the studies reported in Sect. 3 (mobility surveys) of this paper. As can be seen from Fig. 1, the first step when considering mobility should be to record the *attitudes and motives* of the persons involved. This elucidates the intra-psychological processes that underlie their decisions to use certain means of transport and also offers insights for the development of interventions. However, environmental psychology research has long realized that individuals' attitudes and motives are not the only relevant factors affecting their behaviour (Huijts et al. 2012). They stand in a complex interplay with environmental factors that make it easier, more difficult or even impossible for people to live out their personal attitudes and motives. These factors should also be thoroughly assessed in a second stage considering the *situation and environment*. Subsequently, in a third stage, the *behaviour* resulting from the interplay of *attitudes and motives* and the *situation and environment* should be measured.

Finally, these three stages serve as a basis for demonstrating *environmental impact* in the fourth stage. A life cycle assessment can be drawn up modelling the current or past behaviour of members of an organization such as a university. In addition, potential savings that would be possible through interventions to alter individual behaviour indirectly (by changing *attitudes and motives* and/or the *situation and environment*) can be identified and their size estimated.

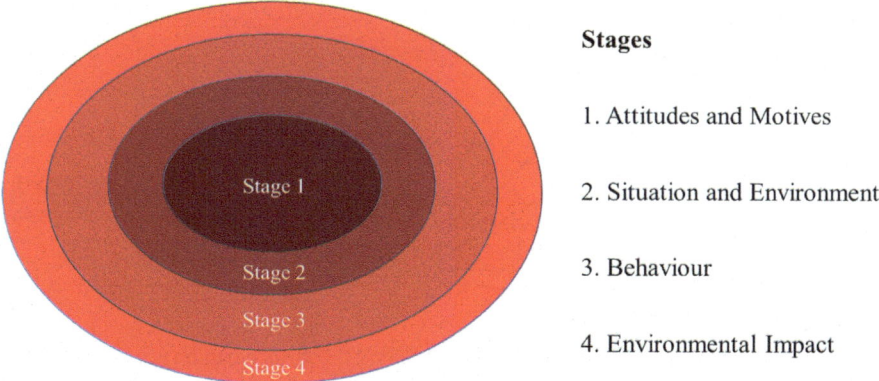

Stages

1. Attitudes and Motives

2. Situation and Environment

3. Behaviour

4. Environmental Impact

Fig. 1 Stages of mobility

As described in the previous section, there are different factors that influence mobility behaviour. In the following paragraphs, the most important factors will be presented on the basis of a short literature review. Many of the models dealing with *attitudes and motives*, the psychological parameters that influence decisions about means of transport (cf. Level 1), are based on the theory of the planned behaviour by Ajzen (1985). A central aspect of this theory are individuals' *beliefs* about the effects that behaviour will have on them. These expectations influence their *attitudes* towards different behaviours (evaluation), which in turn has an impact on actual behaviour through the intention to behave in a certain way (Ajzen 1985). Therefore, if a person expects public transport to be good for the environment, and evaluates this aspect positively, then his/her intention to use public transport will increase.

Factors that contribute to the formation of attitudes in the context of environmentally friendly technologies are discussed in the model by Huijts et al. (2012) (see Fig. 2), which was first introduced as a way of understanding the acceptance of new environmentally friendly technologies. This context can be regarded as particularly important for university members' education concerning mobility behaviour, since many new means of transport are new (often environmentally friendly) technologies that people have no experience with. This can, for example, be seen in the use of sharing systems (Dialego n.d.).

In this model, *trust in the functionality of a technology* is considered an important component for the formation of attitudes. Trust has an impact on *affective reactions* as well as *cognitive expectations about costs and benefits* of using a technology

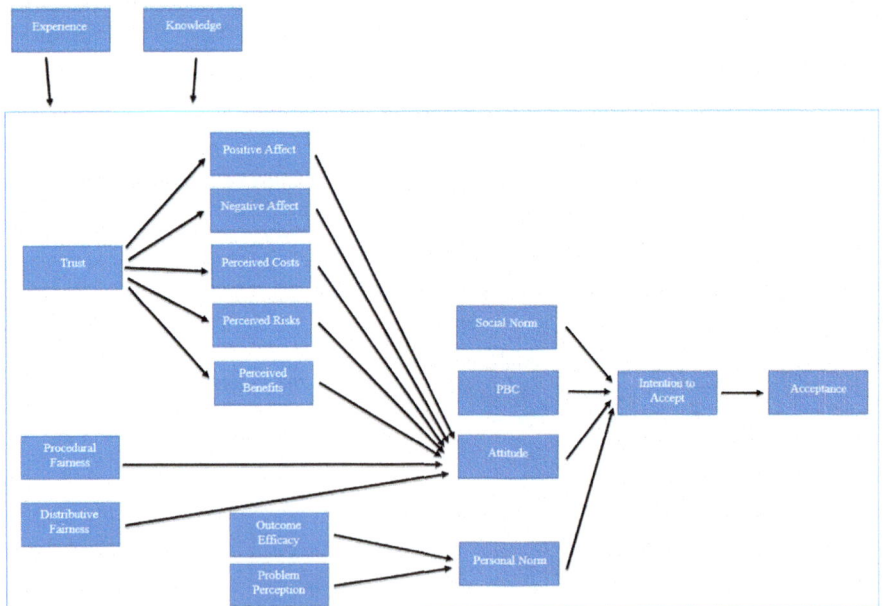

Fig. 2 Acceptance model for environmentally friendly technologies based on Huijts et al. (2012)

(Midden and Huijts 2009; Montijn-Dorgelo 2009; Siegrist et al. 2007). For example, the reason for the low use of sharing systems for travel to the university might be concerns that the loan process will not function properly, causing delays. In principle, a distinction can be made between *trust in the technology itself* and *in its operators*. Especially when people know little about a (new) technology, the authority or provider responsible for it often serves as the basis for trust (Midden and Huijts 2009; Siegrist 2010). Being aware of this and adjusting measures accordingly can especially be beneficial during the introduction of new technologies.

The use of sustainable transport can also be seen as prosocial behaviour, as others benefit from the positive effects of an individual's behaviour, e.g. through cleaner air. According to Schwartz's (1968) norm activation model, *personal norms* are another important factor influencing such behaviour. Huijts et al. (2012) include them as a further component influencing the acceptance of environmentally friendly technologies. A *personal norm* refers to a sense of moral obligation to refrain from or carry out certain actions. This feeling is activated when people understand the negative consequences of not exhibiting the prosocial behaviour (*problem perception*) and believe that they can act effectively to stop these negative consequences, e.g. by choosing to use a different means of transport (*outcome efficacy*). The difference to *attitudes* is that *personal norms* foreground the unconscious, unplanned component of predicting behaviour. For example, students may have a negative attitude towards cycling because they expect a loss of time and comfort compared to taking a car. At the same time, they could feel morally obliged to cycle to work due to their knowledge of the negative environmental impact of using the car. Measures to activate a *personal norm* can help to reduce CO_2. Graham et al. (2011) show that feedback on CO_2 savings from use of a bicycle compared to a car can induce people to use the latter less often.

In addition to the factors described so far, i.e. *attitudes and motives* (cf. Stage 1 in Fig. 1), Huijts et al. (2012) also integrate two other aspects from the theory of planned behaviour into their model: *social norms* and *perceived behavioural control*. These factors are related to the *situation and environment* (cf. Stage 2 in Fig. 1).

A *social norm* refers to a person's assessment of whether important people in his or her environment would support his or her behaviour. The effect of a social norm on the intention to behave in a certain way also depends on the individual's motivation to align their behaviour with that of the people in their environment (Ajzen 1985). At a university, for example, the behaviour of professors could have an influence on other university members, as they have a certain role model function for both students and staff.

Perceived behavioural control refers to a person's perceived and/or actual ability to exhibit certain behaviour. Rölle (2010) shows that an information package about public transport and a one-day test ticket can help promote environmentally friendly mobility behaviour among new organizational members. For such measures to be successful, however, it is important to analyse in advance whether a certain target group is even able to change their behaviour. Hence, which means of transport (e.g., bicycle, car) are generally available to the target group should be assessed. It is also important to identify how well the target group's homes are connected to public transport and whether there are sufficient Park and Ride facilities.

Finally, *experience* with an (environmentally friendly) technology has a general impact on all of the factors described above by influencing its acceptance (Huijts et al. 2012). Therefore, Stage 3 in our model *(prior) behaviour* (cf. Fig. 1), should be assessed particularly thoroughly. Which means of transport students are accustomed to using and their frequency of using a certain means of transport (for a certain part of their total journey) have a strong influence on future behaviour (Schneider 2013).

The model by Huijts et al. (2012) described here was used as a basis for two surveys conducted at the Hochschule für Technik (University of Applied Sciences) Stuttgart.

3   Mobility Surveys

Two surveys were conducted to investigate the mobility behaviour of students at the university. The first survey took place in summer 2016 and focused on actual mobility behaviour. The aim was to determine the students' CO_2 footprint in terms of mobility. The second study took place in winter 2018 and focused on the influencing factors behind students' mobility behaviour and the situational framework conditions in order to enable a more realistic assessment of the potential CO_2 savings. Both studies took the form of student projects. The target group for both surveys was the university's students, as this stakeholder group, encompassing approximately 4,000 persons, has the greatest influence on mobility-related CO_2 emissions. Therefore, the greatest savings potential for the entire university was seen here.

3.1   Survey 1: Mobility Behaviour

As described above, the first survey covered which means of transport students use to get to (and return from) university.

3.1.1   Method

The survey was distributed in the form of a self-assessment questionnaire to students as they entered the university grounds. Potential participants were approached by a student researcher and were able to complete the survey on a tablet computer. The researchers stood at the main entrances to the four (out of eight) university buildings in which most lecture halls are located. The survey took place over one week during Summer Semester 2016. Participants were asked which means of transport they had used to get to university that day and what distances they covered with different means of transport during their journey. We chose to ask students about their most recent journey in order to minimize possible memory distortions. To support route estimation, a link to *Google Maps* was included to help participants estimate the route's distance more reliably.

A total of $N = 321$ students took part in the survey, but 4 students were excluded because they stated 0 km or >100 km as daily distances. Of the remaining $N = 317$ students, 33% were female and 67% male. The participants were between 18 and 37 years old ($M = 23.1$, $SD = 3.17$). Completing the entire questionnaire took less than 5 min on average ($Md = 3.82$).

3.1.2 Results and Discussion

On average, the students travelled a distance of $M = 17.83$ km ($Md = 10$; $SD = 19.64$) to get from their home to the university (with a range of less than 1 km to a maximum of 100 km). The sample can be divided into three groups: 36% travel up to 5 km, 34% 6–20 km and 30% more than 20 km. The total distance travelled was then divided into sub-sections covered with different means of transport (see Table 1). Overall, one can see that the longest distance is travelled by rail, especially among students who live more than 5 km away from the university. Cars are also important for students who live far away (>20 km) from the university. One can assume that public transport coverage is weaker in these regions or that *Park & Ride* facilities may not be sufficiently available. Although bicycles are more important as a means of transport for students who live close to the university (up to 5 km), public transport services are also used to a large extent by this group.

Table 1 Average distances (in km) travelled with different means of transport

	Distance travelled from home to university			
	0–5 km (n = 113)	6–20 km (n = 109)	>20 km (n = 95)	All respondents (N = 317)
	M	M	M	M
Train (regional/long-distance)	0.00	0.30	19.86[a,b]	6.06
Tram/subway	1.19	10.86[a]	15.49[a,b]	8.80
Bus (local)	0.46	0.37	1.22[b]	0.65
Car (as driver)	0.03	0.22	3.69[a,b]	1.19
Car (as passenger)	0.03	0.18	0.76	0.30
Motorbike/moped	0.17	0.07	0.25	0.16
Bicycle	0.31[c]	0.15	0.00	0.16
On foot	0.28[b,c]	0.07	0.11	0.16

[a] significantly higher than group "0–5 km". [b] significantly higher than group "6–20 km". [c] significantly higher than group "> 20 km" (95% significance level; with Bonferroni correction)

The results provide an indication that measures to increase the sustainability of mobility should differentially target different groups. Students living further away from the university (>20 km) could use more sustainable means of transport (e.g. public transport) instead of cars. However, it is important to find out whether, for example, sufficient *Park & Ride* facilities are available for this group. Additional potential seems to lie in persuading students who live close to the university (0–5 km) to use bicycles as opposed to public transport. However, both approaches necessitate an examination of the students' personal attitudes and situational conditions, which was done in the second survey.

3.2   Survey 2: Choice of Transport Mode

The second survey was conducted in order to get a better understanding of the reasons why students choose a given means of transport. On the one hand, criteria for choosing a means of transport were assessed here. On the other hand, the structural conditions effecting the choice of a means of transport were examined more closely: Which means of transport are generally available to students, are there *Park & Ride* facilities, and how do students evaluate the cycling infrastructure? Evaluating these aspects should enable a more precise estimation of the potential for CO_2 savings.

3.2.1   Method

The survey was carried out online. An invitation link was distributed via e-mail distribution lists at the university. Posters on campus were also used to draw attention to the survey. Various prizes (e.g. reusable coffee cups) were raffled off among the participants as incentives. The survey took place over a period of 2.5 weeks during Winter Semester 2018. The survey addressed, on the one hand, the relevance of various factors for the choice of a means of transport and, on the other hand, which means of transport are available and used by students for their journey to university. Car users were also asked whether *Park & Ride* facilities were available and in use. Students living in the vicinity of the university who cycled to class were asked to evaluate the cycling infrastructure. In addition, interest in improving the cycling infrastructure was assessed.

A total of $N = 1,314$ students took part in the survey. Of these, 53% were female and 47% male. The participants were between 17 and 55 years old ($M = 23.4$, $SD = 3.59$). Completing the entire questionnaire took between 5 and 10 min on average ($Md = 7.49$).

3.2.2 Results and Discussion

In principle, almost all students have access to public transport: 92% have a semester ticket for public transport and can use it to travel to university. For 39%, regional transport services are an option. In addition, almost half of the students have a car[1] (47%) or bicycle (43%) at their disposal which can be used to travel to university. 47% of the students can walk at least part of the way to the campus. The availability of different means of transport differed among the students living close (*up to 5* km; 23%) to the university, at a medium distance (6–20 km; 43%) and far away (>20 km; 34%). 67% of the students living near the university have access to a bicycle, but only 25% have access to a car. For students living far away, 65% have a car at their disposal, but only 30% have access to a bicycle. Accordingly, the use of each means of transport also depends on the distance to be covered.

The decisive factors for the choice of transport from the students' point of view are reliability (important for 64% asked on a scale of 1= *unimportant* to 5= *important*), travel time (60%) and costs (56%). In addition, the following factors were of medium importance: flexibility (39%), security (35%) and weather independence (34%). Sustainability was also moderately important (24%). Students for whom this topic is important use bicycles more frequently (21% vs. 10%) and cars less frequently (11% vs. 18%).

The first survey identified two target groups whose mobility behaviour could be more sustainable: First, it might be possible for students who live further away from the university and use a car (for at least part of the journey) to make more use of public transport services. The second group are students who live close to the university and do not cycle. This raises the question of whether the use of bicycles in this target group could be increased.

65% of the students living far away (and 44% living at a middle distance) from the university have a car at their disposal. 32% (12%) use a car for at least part of their trip to university: 5% (2%) for most of the distance, and 2.5% (1.6%) for the whole distance. Students who use a car for at least part of the journey (n = 215) are more interested in the following aspects than others: reliability (70%), flexibility (45%), comfort (17%) and driving pleasure (11%). In contrast, costs (49%) and environmental aspects (16%) are less important for students using a car in comparison to the whole sample of students. Against this background, it is questionable whether public transport services can be made attractive for this group. Nevertheless, 44% of this group use existing *Park & Ride* facilities, while 18% do not. For the remaining 38%, no such facilities seem to be available yet or the students do not know about them. It can therefore be deduced that better infrastructure would be the primary means of inducing car drivers to switch to more sustainable means of transport. The majority of car drivers would use *Park & Ride* facilities if they were available.

67% of the students living close to the university have a bicycle at their disposal. 26% also use their bicycles to travel to university, 13% for most or all of the distance. Students in this group who at least partially cycle to university (n = 81) place

---

[1] with combustion engine.

more importance than others on the following aspects: flexibility (56%), sustainability (43%), sports/health (31%) and driving pleasure (11%). Weather independence (19%), on the other hand, is comparatively less important to them.

This group of cyclists expressed limited satisfaction with the condition of the cycling infrastructure on the way to university with respect to the following aspects: quality of cycle paths (70% [rather] dissatisfied | surveyed on a scale of 1 = *unsatisfied* to 5 = *satisfied*), availability of cycle paths (70%), directions to university (33%) and bicycle parking facilities (22%). Among students living close to the university who do not yet cycle to university (n = 216), 83% would at least occasionally cycle to university if there was a better cycle path for the route. 47% could even imagine completely forgoing the semester ticket for public transport during the summer semester in this case. The majority of this group are also interested in the development of a coherent network of cycle paths throughout the city (78% [major] interest surveyed on a scale from 1 = *no interest* to 5 = *major interest*) and the establishment of a rapid cycle route network (74%). Hence, providing the appropriate infrastructure would also induce this group to switch to commuting by bicycle. In summary, we see potential to change mobility behaviour (cf. Level 3 in Fig. 1) by increasing the availability of *Park & Ride* facilities for students who live far away from the university. For students who live close to the university, increasing the availability of cycle paths and improving their quality could have an effect. However, the question remains as to what influence such measures would have on CO_2 savings.

4   Calculation of Environmental Impacts

The environmental impact (cf. Level 4 in Fig. 1) of the aforementioned measures was assessed in terms of their influence on driving time and carbon dioxide equivalent (CO_2e) emissions (François-Marie Bréon 2013).

4.1   Method

It is assumed that arrival time at the university is 7.50 a.m. on a Tuesday. Distances are calculated using *BRouter*. BRouter's data is based on *OpenStreetMap* and is usually updated once a week (BRouter).

The Google Maps function "Arrival at" was used to forecast the average journey time per car given the selected arrival time (Google Maps 2019). Google Maps takes into account the expected traffic load for the corresponding route at the desired time in its forecast. Predicted travel time with local public transit was made using the travel information provided by the German national railway company (Deutsche Bahn 2019b). According to their own analysis, Deutsche Bahn achieved a punctuality rate of 94.6% in local public transport in January 2019 (Deutsche Bahn 2019a). Punctual arrival at the train station means less than six minutes delay. In the lower valuation, a conservative approach was chosen. Thus, an average of 20% more travel time was expected.

There are four different approaches for calculating CO_2e. With decreasing accuracy, but increasing practicality, either

- the direct output can be measured,
- a conclusion can be drawn about fuel consumption,
- both the load weight and the transport distance can be used
- or only the transport distance (Jofred 2011).

The most common calculation tools use emission factors multiplied by the transport distance. For increased accuracy, different factors are used depending on the vehicle type or average consumption. The data sources used in the specific calculation tool play a significant role in the bandwidth of the result (Schächtele 2007).

The German Federal Environment Agency's CO_2 calculator uses emission factors that take IPCC specifications for the national greenhouse gas inventory into account (Schächtele 2007; Umweltbundesamt 2017). The Ifeu institute provides the basis for updating the calculations and emission factors. To ensure consistency between the emissions target presented in the introduction and the emissions savings calculations, we use the Federal Environment Agency's CO_2 calculator to compute the latter.

4.2 Results and Discussion

4.2.1 >20 Km Group and Comparison of Car to Public Transport

The average distance travelled by the group living >20 km away from the university (with n = 95; c.f. Table 1) is 41.38 km. We conducted the calculations for a student living in Holzhausen. This is a village southeast of Stuttgart and without a direct railway connection. The nearest train station is Uhingen, which is 1.5 km away. The distance by individual motorized traffic from Holzhausen to the Hochschule für Technik (University of Applied Sciences) Stuttgart is 39.1 km. Google Maps forecasts a travel time of 45–80 min for the selected arrival time. 62.5 min on average was assumed for the comparison as well as a further 5 min for parking one's vehicle in a nearby multi-storey car park and walking to the university. This sums up to a total of 67.5 min (Google Maps 2019). With 200 single trips per year, a small car (petrol) with an average consumption of 6.7 L per 100 km in a mix of highway and city traffic would generate 1.45 t CO_2e (Umweltbundesamt 2017).

This is compared with intermodal transport. The route from Holzhausen to the train station in Uhingen is completed by bicycle. The student then takes the Deutsche Bahn regional train to Stuttgart Main Station. The last mile of the route is walked. The nearest train station is 1.5 km by bicycle from the student's place of residence. Assuming an average speed of 15 km/h, 6 min would be required. Another estimated 5 min will be needed for locking the bike and for waiting time at the station. The regional train takes 30 + 6 min as the assumed average delay (Deutsche Bahn 2019b).

Car target group

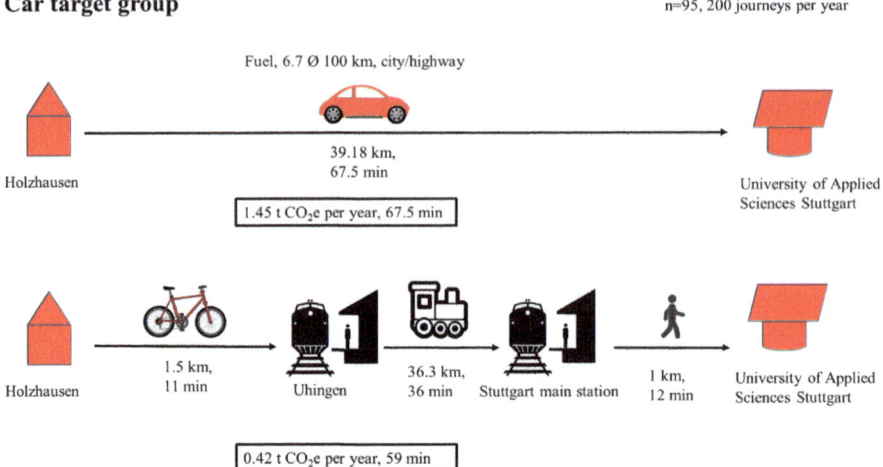

Fig. 3 Effects of a change in mobility behaviour for the car target group

The distance travelled by railway is 36.3 km (BRouter). The last 1 km of the route will take 12 min with an average walking speed of 5 km/h. On average, intermodal transport takes 59 min in total. This generates 0.42 t CO_2e for one person taking 200 journeys per year (Umweltbundesamt 2017). See Fig. 3 for an overview of these results.

4.2.2 Group 0–5 Km and Comparison of Public Transport to Bicycle Use

The average distance travelling in the group living close to campus (0–5 km distance; with n = 113, c.f. Table 1) is 2.47 km. We conducted the calculations for a student living at Alexanderstrasse 152 in the south of Stuttgart, i.e. 2.5 km from away from the Hochschule für Technik (University of Applied Sciences) Stuttgart by bicycle (BRouter). Assuming an average speed of 15 km/h, 10 min would be required. Another 2 min are estimated for locking the bike. In total, we assume 12 min of travel time. No CO_2e is emitted.

This is compared with intermodal transport. The student walks for one minute to the bus stop on Falbenhennenstraße and waits on average two minutes for the bus. The bus travels from Falbenhennenstraße to the bus stop on Berliner Platz in 12 + 3 min average delay. The distance covered is 3.1 km. From there it is 550 m and thus a 6 min walk to the University of Applied Sciences Stuttgart (BRouter). Therefore, intermodal transport takes 24 min in total. This generates 0.04 t CO_2e in 200 journeys per year (Umweltbundesamt 2017). The results of this analysis are shown in Fig. 4.

Public transport target group

n=113, 200 journeys per year

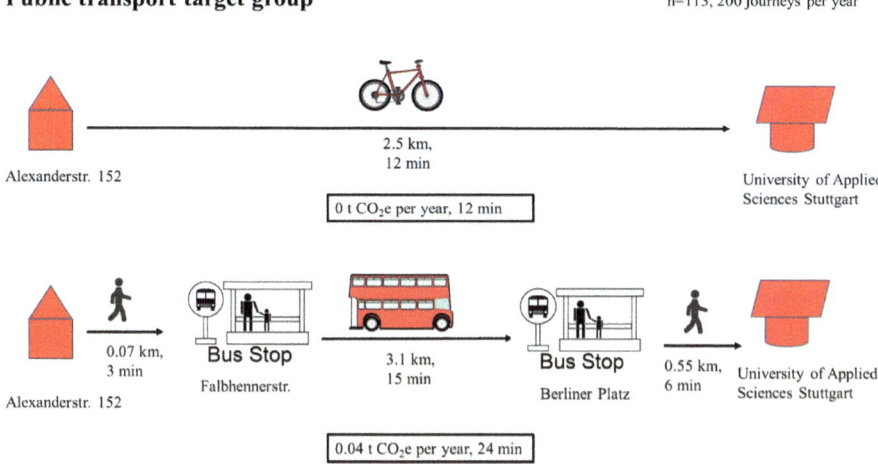

Fig. 4 Effects of a change in mobility behaviour for the public transport target group

5 Discussion

The studies and analyses carried out in this paper show that integrating work in several disciplines helps do justice to the complex construct of mobility. Different perspectives facilitate the assessment of how realistic it is to expect changes in behaviour in certain target groups and what effects would have on the environment. This allows potential mitigation measures to be assessed, prioritised and evaluated in a more comprehensive way. Nevertheless, the results should be regarded with caution. The sample sizes of the target groups should be increased in future studies on the status quo of mobility behaviour in order to increase validity. Social norms also ought to be included in follow-up studies as they could be another important factor influencing mobility behaviour. Furthermore, a more systematic approach of selecting locations from which to calculate CO_2e emissions should be applied in upcoming studies in order to make the calculations more precise.

The results of the present study seem to indicate that measures such as building more bike paths and an app providing user ratings of the quality of existing bike paths could be a way to motivate *the public transport target group* (*living close to the university*). For the *car target group*(*living far away from the university*), increasing train service and a database of *Park & Ride* facilities could support a behaviour change. Evaluating the measures shows that initiating behaviour change in the bicycle target group is more realistic, as they have a high rate of bicycle access but low rates of actual use. In addition, changing their mobility behaviour would decrease their travel time by half. However, the environmental impact would be comparably low. In the *car target group*, the travel time would decrease rather minimally and it could be hard for public transport to meet their expectations of high reliability and flexibility, but the ecological footprint effects would be large.

Measures aimed at the *public transport target group* can be expected to result in small quick wins. Large but hard to realise successes are anticipated for the *car target group*. Thus, by integrating multiple disciplines, it can be seen that a mix of measures targeting both groups could be the optimal solution.

Acknowledgements Parts of the surveys were funded by the Ministry of Science, Research and the Arts of Baden-Württemberg as part of the Strategy Dialogue Automotive Industry.
We would like to thank the following persons who were significantly involved in the planning and execution of the surveys: Lara Dobisch, Lucia Fuchs, Jannik Grünebaum, Rebecca Heckmann, Vera Knappe, Teresa Konrad, Sarah Lorenzen, Benedikt Mentzel, Jessica Preuss, Anna Sauter, Patrick, Schwille, Karen Unverfärth, Stefan Zimmermann.

References

Ajzen I (1985) From intentions to actions: a theory of planned behavior. In: Kuhl J, Beckmann J (eds) Springer series in social psychology. Action control. From cognition to behavior, 1st edn. Springer, Berlin, pp 11–39

Brasseur GP, Jacob D, Schuck-Zöller S (2017) Klimawandel in Deutschland. Springer, Berlin, Heidelberg

BRouter. Online-service der BRouter routing engine. Retrieved from https://brouter.damsy.net/latest/#map=6/50.990/9.860/OpenStreetMap. Accessed on 10 Mar 2019

Bühler R, Kunert U (2010) Trends des Verkehrsverhaltens in den USA und in Deutschland. Internationales Verkehrswesen

Bundesministerium für Umwelt, Naturschutz und nukleare Sicherheit (2016) Deutschland startet Ratifikation des Pariser Klimaschutzabkommens. From www.bmu.de/PM6578

Deutsche Bahn (2019a) Erläuterung Pünktlichkeitswerte Januar 2019. Retrieved from https://www.deutschebahn.com/de/konzern/konzernprofil/zahlen_fakten/puenktlichkeitswerte-1187696. Accessed on 10 Mar 2019

Deutsche Bahn (2019b) Reiseauskunft. Retrieved from https://www.deutschebahn.com/de. Accessed on 10 March 2019

Dialego (n.d.) Aus welchen der folgenden Branchen haben Sie bereits Sharing-Dienste genutzt? From https://de.statista.com/statistik/daten/studie/943745/umfrage/umfrage-zur-nutzung-von-sharing-diensten-in-deutschland-nach-branche/

Figueres C, Le Quéré C, Mahindra A, Bäte O, Whiteman G, Peters G, Guan D (2018) Emissions are still rising: ramp up the cuts. Nature 564(7734):27–30

François C, Gondran N, Nicolas J-P, Parsons D (2017) Environmental assessment of urban mobility: combining life cycle assessment with land-use and transport interaction modelling—application to Lyon (France). Ecol Ind 72:597–604

François-Marie Bréon WC (2013) Working group I contribution to the IPCC fifth assessment report (AR5), climate change 2013: the physical science basis: Chapter 8: Anthropogenic and Natural Radiative Forcing-Final Draft Underlying Scientific Technical Assessment

Google Maps (2019) Fahrt uhingen->HfT. Retrieved from https://www.google.de/maps/dir/Hochschule + für + Technik + Stuttgart +-+HFT + Stuttgart, + Schellingstraße, + Stuttgart/Holzhausen, + 73066 + Uhingen/@48.7438888,9.2342534,11z/data =!3m1!4b1!4m14!4m13!1m5!1m1!1s0x4799db371328f901:0x6946c5ec40a9542!2m2!1d9.1726453!2d48.7803251!1m5!1m1!1s0x4799a3565643d665:0xe540835cb9759cca!2m2!1d9.5873981!2d48.7200358!3e0. Accessed on 10 Mar 2019

Graham J, Koo M, Wilson TD (2011) Conserving energy by inducing people to drive less. J Appl Soc Psychol 41(1):106–118

Huijts NMA, Molin EJE, Steg L (2012) Psychological factors influencing sustainable energy technology acceptance: a review-based comprehensive framework. Renew Sustain Energy Rev 16(1):525–531

Hunecke M, Haustein S, Grischkat S, Böhler S (2007) Psychological, sociodemographic, and infrastructural factors as determinants of ecological impact caused by mobility behavior. J Environ Psychol 27(4):277–292

Jofred PÖ (2011) CO_2 emissions from freight transport and the impact of supply chain management: a case study at Atlas Copco Industrial Technique

Katharina Schächtele HH (2007) Die CO2 Bilanz des Bürgers: Recherche für ein internetbasiertes Tool zur Erstellung persönlicher CO2 Bilanzen

Le Quéré C, Andrew RM, Friedlingstein P, Sitch S, Hauck J, Pongratz J et al (2018) Global carbon budget 2018. Earth Syst Sci Data 10(4):2141–2194

Masson-Delmotte V (2018) Global warming of 1.5 °C: an IPCC special report on the impacts of global warming of 1.5 °C above pre-industrial levels and related global greenhouse gas emission pathways, in the context of strengthening the global response to the threat of climate change, sustainable development, and efforts to eradicate poverty, from IPCC https://www.ipcc.ch/sr15/

Midden CJH, Huijts NMA (2009) The role of trust in the affective evaluation of novel risks: the case of CO2 storage. Risk Anal Off Publ Soc Risk Anal 29(5):743–751

Montijn-Dorgelo FNH (2009) On the acceptance of sustainable energy systems: explicit and implicit effects in perceived value

Pachauri RK, Mayer L (eds) (2015) Climate change 2014: synthesis report. Intergovernmental Panel on Climate Change, Geneva, Switzerland

Rölle D (2010) Beim nächsten ort wird alles anders!? analysen des mobilitätsverhaltens von zuzüglern auf der basis der ipsativen handlungstheorie. Umweltpsychologie

Sato K, Yuki M, Norasakkunkit V (2014) A socio-ecological approach to cross-cultural differences in the sensitivity to social rejection. J Cross Cult Psychol 45(10):1549–1560

Schneider RJ (2013) Theory of routine mode choice decisions: an operational framework to increase sustainable transportation. Transp Policy 25:128–137

Schwartz SH (1968) Words, deeds and the perception of consequences and responsibility in action situations. J Pers Soc Psychol 10(3):232–242

Siegrist M (2010) Trust and confidence: the difficulties in distinguishing the two concepts in research. Risk Anal Off Publ Soc Risk Anal 30(7):1022–1024

Siegrist M, Cousin M-E, Kastenholz H, Wiek A (2007) Public acceptance of nanotechnology foods and food packaging: the influence of affect and trust. Appetite 49(2):459–466

Umweltbundesamt (2017) CO_2-Rechner. From http://www.uba.co2-rechner.de/de_DE

United Nations Environment Programme (2019) Emissions gap report 2018. [S.l.]: UNEP

Zhang Y-X, Chao Q-C, Zheng Q-H, Huang L (2017) The withdrawal of the U.S. from the Paris Agreement and its impact on global climate change governance. Adv Clim Change Res 8(4):213–219

Sustainability and Institutions: Achieving Synergies

Walter Leal Filho

Abstract This final paper summarises some of the lessons learned from the book and outlines some aspects related to sustainability at organisations, which may advance it further.

Keywords Sustainable development · Corporations · Businesses · Integration · Synergies

1 Introduction

Sustainable development as a whole, and sustainability in particular, are matters of great relevance to organisations. Indeed, social responsibility and sustainability go hand in hand (Leal Filho 2018). Figure 1 describes some of the elements which illustrate the importance of sustainability to organisations.

In view of a global economy, where processes, interactions and interdependences are constantly changing, the incorporation of sustainability components in the operations of organisations is becoming an increasing trend. Apart from being helpful in assisting an organisation's attempts to become more efficient from an environmental perspective, sustainability efforts can greatly assist an organization in their attempts to gain the trust and support from their communities.

But in order to succeed, it is important that organisations establish interactive relationships with their internal and external stakeholders. By doing so, they can achieve long-term results. Ashley et al. (2000) point out that institutional (or corporate) social responsibility has been consolidated as a concept linked to a "systems approach, focused on the relationships between stakeholders directly or indirectly related to the business". Therefore it can be stated that socially responsible organisations are the ones that adopt modern management strategies, aimed at not only optimising the environmental and socio-economic aspects of their operations,

W. Leal Filho (✉)
European School of Sustainability Science and Research, Hamburg University of Applied Sciences, Ulmenliet 20, 21033 Hamburg, Germany
e-mail: walter.leal2@haw-hamburg.de

© Springer Nature Switzerland AG 2020
W. Leal Filho et al. (eds.), *International Business, Trade and Institutional Sustainability*, World Sustainability Series,
https://doi.org/10.1007/978-3-030-26759-9_63

Fig. 1 The relevance of sustainability to organisations

but which also foster transformational relationships with (and among) their stakeholders.

The international perspectives presented by Leal Filho and Idowu (2009) show that companies often need to rethink the basis on which they establish relationships with their stakeholders, especially with neighboring communities and their facilities. This is so because business which prioritise values such as citizenship, ethics, autonomy, empowerment, equity, transparency and sustainability, are more likely to:

(a) achieve higher levels of productivity (Louche et al. 2010)
(b) foster employees' satisfaction
(c) be more widely regarded by customers
(d) have greater levels of competitiveness.

It should be taken into account that the paradigm of sustainability is at least partly based on the integration and interaction of the various factors which are either present, or which influence the institutional environment. From this premise, this short chapter intends to describe how the inclusion of sustainability components may assist organisations in becoming more efficient.

2 Fostering Better Relations Towards Greater Efficiency

There are various strategic factors which may help to promote more sustainable relationships between institutions (including companies) and communities. Some of them are:

(a) an outline of the management aspects which should be enhanced or main-
    streamed so as to cater for a feeling of sustainability;
(b) a description of the status of the relationship between companies and commu-
    nities with a view to identifying where improvements are needed;
(c) the identification of the social and environmental aspects which may account
    for possible conflicts;

In socially responsible management and sustainability-oriented management, two
terms with completely different meanings but which are mutually complementary,
three main factors ought to be considered: the implications of the organization and
of its operations to the planet (environmental concerns), the implications of the
organization and its operations to people (social concerns) and the implications of
the organization and its operations towards its profitability (economic concerns). By
means of a combination of these dimensions in a balanced way, an institution shall
be able to open the way for the creation of a good reputation thus increasing the
trust and loyalty amidst its stakeholders, be them employees, partners, customers
and suppliers.

In respect of fostering better relations, it should be considered that the interrela-
tionships between environmental and social systems often lead to conflicts, be it in
response to the use of manpower, in the overexploitation of resources, or in respect
of little emphasis to cooperation and joint efforts.

Since the first UN Conference on Environment and Development held in Rio de
Janeiro in 1992, to the latest one held also in the same city in 2012, the concept
of sustainable development has been the subject of much debate, and has evolved
to the point that the current emphasis is on the implementation of the "Sustainable
Development Goals" which are:

- GOAL 1: No Poverty
- GOAL 2: Zero Hunger
- GOAL 3: Good Health and Well-being
- GOAL 4: Quality Education
- GOAL 5: Gender Equality
- GOAL 6: Clean Water and Sanitation
- GOAL 7: Affordable and Clean Energy
- GOAL 8: Decent Work and Economic Growth
- GOAL 9: Industry, Innovation and Infrastructure
- GOAL 10: Reduced Inequality
- GOAL 11: Sustainable Cities and Communities
- GOAL 12: Responsible Consumption and Production
- GOAL 13: Climate Action
- GOAL 14: Life Below Water
- GOAL 15: Life on Land
- GOAL 16: Peace and Justice Strong Institutions
- GOAL 17: Partnerships to Achieve the Goal.

As it can be expected, a sustainability element is behind each SDG. The idea is that, due to the complexity involved in their implementation, which is partly due to their scope which goes beyond the traditional view of economic growth and economic development against social-economic-environmental ecosystems, realizing the SDGs requires significant changes in the ways of thinking and the *modus operandi* of organisations. This is because the strategic orientation of each company towards sustainability is directly associated with the principles and values of its organizational culture and legal and cultural environment, and from the context of this business is conducted.

Knowledge of the SDGs may in any case help institutions to move beyond conservative viewpoints, and assist them in pursuing more progressive and critical goals, deploying concepts such as democracy, participation, equity, efficiency, citizenship, autonomy, decentralization, and belonging.

3 Conclusions

As this overview paper has outlined, the concept of sustainability embedded at the institutional level-going beyond Corporate Social Responsibility-brings many advantages to an institution. The theory and practice behind this concept has been progressively consolidated over the past three decades, and has received a new momentum with the launching of the SDGs, whose objectives are expected to be achieved by 2030.

But it needs to be stated that the perceptions and the implementation of sustainability at institutional level are currently found at different levels of evolution. This is due to the many differences in the ways sustainability is perceived and practised, and to the fact that in many cases, its inherently interdisciplinary and multidimensional dimensions are not fully understood.

Yet, in the face of a globalized economy, organisations, be them public or private, are asked to play a new role and adopt a different attitude, especially in respect of ethics and transparency, and in terms of their organizational behavior and the relationships they establish with their stakeholders.

In this context, sustainability factors may act as drivers, mobilising the various stakeholders and encouraging them to become active participants in the value creation process.

References

Ashley PA, Coutinho RBG, Tomei PA (2000) Corporate social responsibility and corporate citizenship: a comparative analysis. In: EnANPAD 2000: annual meeting, Florianópolis, Brazil
Louche C, Idowu SO, Leal Filho W (eds) (2010) Innovative CSR: from risk management to value creation. Greenleaf, Sheffield
Leal Filho W (ed) (2018) Social responsibility and sustainability. Springer, Berlin
Leal Filho W, Idowu S (eds) (2009) Global practices of corporate social responsibility. Springer, Berlin

Printed by Printforce, the Netherlands